# Mechanism in Protein Chemistry

*Dedicated to my teachers*

Richard Ramette
Charles Carlin
John Edsall
Gustav Lienhard
Klaus Weber
Guido Guidotti
Jon Singer
Russell Doolittle
Charles Perrin

# Mechanism in Protein Chemistry

**Jack Kyte**

Professor of Chemistry
University of California at San Diego

**Garland Publishing, Inc.**
**New York & London**
**1995**

*Erst die Theorie entscheidet darüber, was man beobachten kann.*

Attributed to Einstein by Heisenberg



**Library of Congress Cataloging-in-Publication Data**

Kyte, Jack.
Mechanism in protein chemistry/ Jack Kyte.
p. cm.
Includes bibliographical references (p. – ) and index.
ISBN 0-8153-1700-X
1. Proteins. 2. Chemical reactions. I. Title.
QD431.K97 1995 1995
547.7'5804594—dc20 95-18614
CIP

Artwork for the cover was provided by Brian W. Matthews,
Howard Hughes Medical Institute Research Laboratories,
University of Oregon



Manufactured in the United States of America

# Contents

# Preface

Over the past forty years, our understanding of the strategies that are used by enzymes to catalyze their respective reactions has expanded so dramatically that it seems to be complete. At the moment, it seems unlikely that some new, as yet undescribed strategy will be discovered, except perhaps for the few enzymes that employ homolytic mechanisms. It is the purpose of this text to present a comprehensive description of these strategies. It is not, however, intended to be a comprehensive, encyclopedic catalogue of the enzymes, systematically presenting the mechanisms of each in its appropriate category, although this is sometimes done.

It is also the intention of this book to describe the many techniques that are used to study enzymatic mechanisms. Each of these methods provides information about a particular strategy and can be discussed as the respective strategy is discussed. The most informative observation about the mechanism of an enzyme is a crystallographic molecular model of that enzyme with its active site occupied by an analogue of a high energy intermediate in the normal reaction. Unfortunately, the number of these crystallographic molecular models that are presently available is probably fewer than ten. The majority of our understanding of the mechanisms of enzymatic reactions comes from chemical studies of these mechanisms. In fact, even the design of an analogue of a high energy intermediate usually requires that a significant amount of chemical information has been gathered. Furthermore, even when a crystallographic molecular model of an analogue of a high energy intermediate within an active site is available, its interpretation usually relies heavily on the available chemical information. At the moment, a comprehensive understanding of the methods that are used to study enzymatic mechanisms, as well as the limitations of these methods, is necessary, and for this reason these methods are discussed critically in this text.

Because an understanding of enzymatic mechanisms requires a firm knowledge of bonding and mechanistic organic chemistry and because an understanding of the techniques used to study these mechanisms requires a firm knowledge of kinetics, thermodynamics, synthetic organic chemistry, and analytical chemistry, the text places severe demands on the student. The concepts of bonding and chemical mechanism are incorporated into structural drawings of individual molecules and chemical reactions. The kinetics and thermodynamics are represented in mathematical equations. The results of the particular techniques used to study enzymes are presented as graphs and tables from the original literature. The crystallographic molecular models are observed in stereo images. The student must have the background necessary to read each of these languages.

The course created by this text is intended to bridge the gap between introductory chemistry and biochemistry classes and the research literature. There are suggested readings in which the concepts discussed in each section are applied in an experimental setting, and there are also citations within the text that direct the student to the research literature. At the completion of this course, the student should be equipped to take charge of her own education by reading critically the experimental literature. To do this she must be able to understand the experiments performed and be able to reach the same conclusions as did the authors or to realize that the authors were mistaken. It is my intention to develop in the student the ability to draw her own conclusions from only the experimental results. To this end, the problems at the end of each section are usually based on actual experimental results which are to be evaluated by the student.

Over the last few months I have had the onerous task of reading the proofs of this book. I have been impressed by the fact that each time I have read through a proof, I have found errors I missed before. I am certain that many still exist and would be grateful to any reader who draws them to my attention.

It is a pleasure to thank everyone who has helped me in the preparation of this book. My wife, Francey, beyond the call of duty, typed the first draft of the manuscript. My secretary, Brenda Leake, transferred the manuscript onto the computer, a major task because of the mathematical equations. My daughter Sara assisted Brenda in this job. All of the chemical drawings of molecular structures and reactions were created by Scott Ralph. This required all of Scott's remarkable knowledge of both bonding and mechanism. Jack Kirsch read the entire manuscript with remarkable care, understanding of the chemistry, and attention to detail. The copyeditor, Heather Whirlow Cammarn, turned the manuscript into scientific text. Individual sections of the manuscript were reviewed critically by Charles Perrin, Tom Bruice, Bill Parson, Paul MacPherson, Ted Traylor, David Dolphin, Wallace Cleland, Jonathan Cohen, Howard Schachman, Paul Bartlett, David Matthews, Jeremy Knowles, Judy Klinman, Craig Townsend, Fred Hartman, Paul Saltman, Don Tilley, Bill Trogler, Marion O'Leary, Paul Boyer, and Bruce Kemp. Each of them provided many helpful comments.

Jack Kyte
April 1995
La Jolla, California

# Stereo Drawings

There are a number of stereo drawings included in the text. It is now standard practice to publish drawings of crystallographic molecular models in this format, sometimes in color. To appreciate the results of crystallographic studies, one must be able to view these images. Although a few individuals can view them effortlessly by crossing their eyes, the rest of us require a stereo viewer. The best I have found is the stereo viewer distributed by Luminos Photo Corporation, P.O. Box 158, Yonkers, New York 10705 (800-586-4667). It has been my experience that a novice will usually complain that although everyone else can learn to use one of these viewers, he cannot. It has also been my experience that everyone learns to use one. It is essential that anyone interested in the structures of proteins learn to view crystallographic molecular models in stereo. The drawings have been placed vertically rather than in their usual horizontal orientation and each has been placed on the outside edge of a page. This has been done to allow each image to be spread as flat as possible for the best viewing.

## Chapter ONE

# Reaction Mechanisms

An enzyme is a protein that is responsible for catalyzing a specific chemical reaction. The reactions that occur on the active sites of enzymes proceed by pathways that often resemble very closely the normal chemical mechanisms of similar organic reactions, and an enzyme usually accelerates a particular reaction by controlling the steps that would normally occur in its absence. It follows that the first step in any examination of the mechanism of a particular enzyme is to write out in exhausting detail one or more realistic chemical mechanisms for the reaction of interest and focus upon the specific features of those mechanisms that are vulnerable to manipulation by the enzyme. Therefore, a logical place to begin this discussion is to examine the chemical mechanisms of reactions similar to those that are catalyzed by enzymes.

## Proton Transfer

It is usually assumed that the mechanism of a particular type of chemical reaction most relevant to the mechanism that proceeds within the active site of an enzyme is the one that takes place in aqueous solution in the absence of the enzyme. Unless otherwise noted, all of the mechanisms discussed in this chapter will be for reactions occurring in water. Water is a peculiar solvent because it is both an acid and a base.

When it is fulfilling the role of an acid, water can provide protons in two ways. The **hydronium cation**, unavoidably present in aqueous solution, is composed of one proton and either four molecules of water[1]

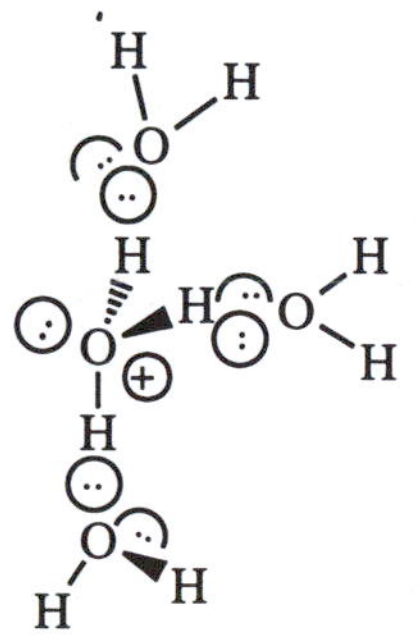

1–1

or 21 molecules of water.[2,3] At the periphery of the $H_9O_4^+$ cation or the $H_{43}O_{21}^+$ cation are six or 10 hydrogen atoms, respectively. Each of these hydrogens is capable of being transferred to a base in the solution or capable of becoming the proton of another hydronium cation simply by being transferred within a hydrogen bond to a neighboring molecule of water. This arrangement makes each isolated proton in an aqueous solution multivalent. It also enhances the rate of diffusion of the proton, which appears to be limited not by transfer among the waters but by rearrangement of the hydration shells around the cation as it steps through the solvent. For these reasons, in addition to its low $pK_a$ (–1.7), the hydrated proton, usually abbreviated $H^+$, is a very efficient acid, a fact that more than makes up for its low concentration. Water itself is an acid as well, albeit a weak one ($pK_a$ = 15.7), but its concentration (55 M) can sometimes help to overcome this deficit. Every basic lone pair of electrons on any molecule in an aqueous solution is adjacent to the hydrogen on a neighboring water molecule, usually in a hydrogen bond, and can immediately accept a proton from that molecule of water.

When it is fulfilling the role of a base, water can provide lone pairs of electrons in two ways. The **hydroxide ion**, also unavoidably present in aqueous solution, can be considered as a proton hole. This hole is shared among the central oxygen and three or perhaps more of its nearest neighbors

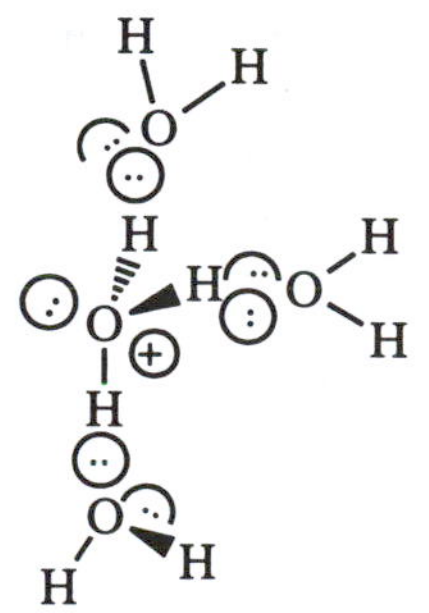

1–2

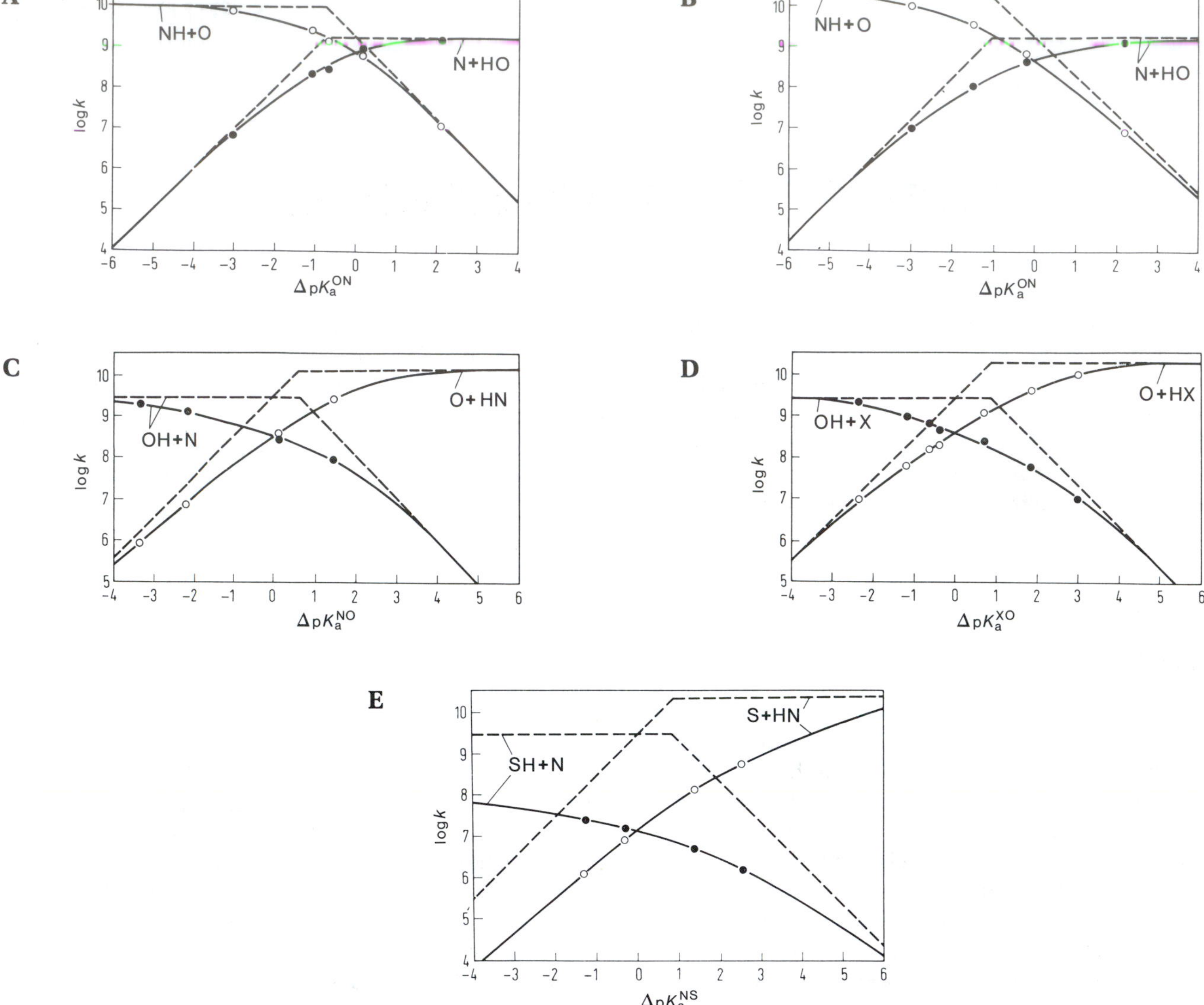


**Figure 1–1:** Bimolecular rate constants for transfer of a proton between an acid and a base dissolved in aqueous solution.[1] In each panel, the logarithm of the composite rate constant for proton transfer from an acid to a base, $k_F$, is presented as well as the composite rate constant, $k_R$, for proton transfer from the respective conjugate base to the conjugate acid. The designations of acid and its conjugate base and of base and its conjugate acid for each pair are arbitrary. The logarithms of the rate constants are plotted as a function of the difference between the p$K_a$ of the acid and the p$K_a$ of the conjugate acid of the base, as defined by Equation 1–8. (A) Base (N), ammonia; conjugate acid (NH), ammonium cation; acids (OH), glucose ($\Delta pK_a = -3$), bicarbonate ($\Delta pK_a = -1.1$), phenol ($\Delta pK_a = -0.7$), boric acid ($\Delta pK_a = +0.1$), and *p*-nitrophenol ($\Delta pK_a = +2.1$); filled circles, transfer from acid (OH) to base (N); open circles, transfer from conjugate acid (NH) to conjugate base (O). (B) Base (N), imidazole; conjugate acid (NH), imidazolium cation; acids (OH), phenol ($\Delta pK_a = -3$), hydrogen pyrophosphate trianion ($\Delta pK_a = -0.7$), *p*-nitrophenol ($\Delta pK_a = -0.1$), and acetic acid ($\Delta pK_a = +2.3$); filled circles, transfer from acid (OH) to base (N); open circles, transfer from conjugate acid (NH) to conjugate base (O). (C) Base (O), acetate anion; conjugate acid (OH), acetic acid; acids (NH), hydrazinium cation ($\Delta pK_a = -3.3$), imidazolium cation ($\Delta pK_a = -2.3$), anilinium cation ($\Delta pK_a = +0.1$), and *m*-chloroanilinium cation ($\Delta pK_a = +1.4$); open circles, transfer from acid (NH) to base (O); filled circles, transfer from conjugate acid (OH) to conjugate base (N). (D) Base (O), phenolate anion; conjugate acid (OH), phenol; acids (XH), hydrogen phosphate dianion ($\Delta pK_a = -2.4$), piperidinium cation ($\Delta pK_a = -1.3$), propylammonium cation ($\Delta pK_a = -0.6$), bicarbonate anion ($\Delta pK_a = -0.4$), ammonium cation ($\Delta pK_a = +0.7$), hydrazinium cation ($\Delta pK_a = +1.9$), and imidazolium cation ($\Delta pK_a = +2.9$); open circles, transfer from acid (XH) to base (O); closed circles, transfer from conjugate acid to conjugate base. (E) Base (S), thiolate anion of thioglycol; conjugate acid (SH), thioglycol; acids (NH), dimethylammonium cation ($\Delta pK_a = -1.2$), trimethylammonium cation ($\Delta pK_a = +0.3$), hydrazinium cation ($\Delta pK_a = +1.4$), and imidazolium cation ($\Delta pK_a = +2.4$); open circles, transfer from acid (NH) to base (S); filled circles, transfer from conjugate acid (SH) to conjugate base (N). The horizontal dashed lines are asymptotes based on the assumption that a sufficiently exothermic proton transfer should have a diffusion-controlled bimolecular rate constant. The dashed lines with slope +1 or slope −1 are the endothermic asymptotes calculated by Equation 1–9. Adapted with permission from ref 1. Copyright 1964 VCH.

As with the hydronium cation, this arrangement makes the hydroxide ion a base that can diffuse more rapidly through the solution than any other base and extends its reach simultaneously to at least six lone pairs of electrons. The hydroxide ion is also a strong base ($pK_a = 15.7$).

Any time a stronger base than hydroxide ion is added to water at neutral pH, it will be converted into its conjugate acid while hydroxide ion is produced in equimolar amount

$$B^{\ominus} + H_2O \rightleftharpoons BH + OH^- \tag{1–1}$$

At equilibrium, the relationship between the concentration of the unprotonated strong base and the concentration of hydroxide ion is

$$[B] = \frac{K_a^{BH}}{K_w}[OH^-]^2 \tag{1–2}$$

where $K_a^{BH}$ is the acid dissociation constant of the conjugate acid of the base and $K_w$ is the water ionization constant ($10^{-14}$ $M^2$). Because $[OH^-] < [B]_{TOT}$, where $[B]_{TOT}$ is the total molar concentration of the base in all of its forms, if $K_a^{BH} << K_w$ and $[B]_{TOT} < 1$ M, then $[B] << [OH^-]$ and most of the added base will have been converted into hydroxide ion. This is called the **leveling effect** of water. Because some of the free base is always present, however, it cannot be said that hydroxide ion is the strongest base that can exist in water.

Water itself is a base, weak ($pK_a = -1.7$) but concentrated. In aqueous solution, every acidic proton is adjacent to one of the lone pairs of a molecule of water, usually in a hydrogen bond, and can be immediately transferred to it.

What these facts mean is that the mechanism of a reaction occurring in water is rarely hindered for want of protons that can be captured or lone pairs of electrons onto which protons can be dumped, and these are taken for granted when the mechanism is written. The decision that the appropriate mechanism to use as a model for the one taking place within the active site of an enzyme is the mechanism occurring in aqueous solution implies that an additional decision has been made about the nature of the active site of an enzyme; namely, that the active site is at least as efficient a source of protons and lone pairs of electrons as an aqueous solution.

The transfer of a proton between two lone pairs of electrons proceeds with the intermediate formation of a hydrogen bond between acid and base.[3a] The **kinetic mechanism** of this reaction* can be represented schematically[1]

$$AH + {\ominus}B^{(-)} \underset{k_{-1}}{\overset{k_1}{\rightleftharpoons}} AH{\ominus}B^{(-)} \underset{k_{-2}}{\overset{k_2}{\rightleftharpoons}} {}^{(-)}A{\ominus}HB \underset{k_{-3}}{\overset{k_3}{\rightleftharpoons}} {}^{(-)}A{\ominus} + HB \tag{1–3}$$

This mechanism is an expanded version of the overall acid–base reaction

$$AH + {\ominus}B^{(-)} \underset{k_R}{\overset{k_F}{\rightleftharpoons}} {}^{(-)}A{\ominus} + HB \tag{1–4}$$

where $k_F$ and $k_R$ are composite rate constants. It can be shown[1] that

$$k_F = \frac{k_1 k_2 k_3}{(k_{-1} + k_2)k_3 + k_{-1}k_{-2}} \tag{1–5}$$

and

$$k_R = \frac{k_{-1} k_{-2} k_{-3}}{(k_{-1} + k_2)k_3 + k_{-1}k_{-2}} \tag{1–6}$$

It follows that

$$K_{eq}^{AB} = \frac{K_a^{HA}}{K_a^{HB}} = \frac{[A]_{eq}[HB]_{eq}}{[AH]_{eq}[B]_{eq}} = \frac{k_F}{k_R} = \frac{k_1 k_2 k_3}{k_{-1} k_{-2} k_{-3}} \tag{1–7}$$

It is usually assumed that acid–base reactions are so rapid as to be always in equilibrium, but their actual rates (Figure 1–1) are of importance when reactions proceed as rapidly as those within the active site of an enzyme.

The difference in the $pK_a$ between an acid, AH, and the conjugate acid, HB, of a base, B, can be defined by

$$\Delta pK_a^{AB} = pK_a^{HB} - pK_a^{HA} = \log\frac{K_a^{HA}}{K_a^{HB}} \tag{1–8}$$

Usually, when $\Delta pK_a^{AB}$ is positive and greater than 4, so that the proton transfer is significantly **exothermic**, the composite rate constant $k_F$ of the proton transfer from the acid to the base falls between $10^9$ and $10^{11}$ $M^{-1}$ $s^{-1}$ (right asymptotes in Figure 1–1). The particular value in a given situation is very close to that calculated theoretically for the rate at which collisions occur between molecules of HA and B.[1] This is referred to as the diffusion-controlled bimolecular rate constant.

The **diffusion-controlled bimolecular rate constant** is the rate constant for a reaction in which every collision between the two reactants results in product. It is referred to as diffusion-controlled because the only factor that causes the reaction to be less than instantaneous is the requirement that molecules of the two reactants diffuse through the solution until they collide with their opposite members. If the two respective molecules that must collide in solution are both neutral and of the normal size of small molecules, the diffusion-controlled bimolecular rate constant for the reaction

*Because this is a general kinetic mechanism for all acid–base reactions, the charge on the species A and B may vary. The negative charges in parentheses indicate that when a proton is transferred from HA to $B^{(-)}$, A gains one negative charge and B loses one negative charge.

should be between $10^9$ and $10^{10}$ $M^{-1}$ $s^{-1}$ at 25 °C. The particular value of the diffusion-controlled rate constant is greater by a factor of about 6 if a positively charged acid reacts with a negatively charged base than if either a neutral acid reacts with a neutral or charged base or a neutral base reacts with a neutral or charged acid, because electrostatic attraction increases the rates of collision on the former situation.[4]

Because the rates of exothermic proton transfers are diffusion-controlled, it necessarily follows that each time the acid and the base collide in solution, a hydrogen bond forms and the proton transfer within that hydrogen bond occurs faster than the rate at which the two molecules can separate. One major exception to this rule occurs when the proton is held in the acid within a strong intramolecular hydrogen bond.[4a] In this case, the difficulty lies in forming the initial hydrogen bond, but once it is formed the reaction proceeds normally.

In a situation where no collision is required to occur, for example, when the base is the solvent and is already adjacent to the acid, the proton transfer becomes unimolecular. If the proton transfer is also exothermic ($\Delta pK_a^{AB} > 4$), the first-order rate constant for proton transfer[5] at 25 °C usually falls between $10^{10}$ and $10^{11}$ $s^{-1}$, which is the value calculated theoretically for a diffusion-controlled dissociation of two pre-associated molecules. This dissociation is governed by the unimolecular rate constant $k_3$ in Mechanism 1–3. This again demonstrates that in exothermic proton transfers the reaction is not limited by the transfer of the proton within the hydrogen-bonded complex. Therefore, this transfer, governed by the unimolecular rate constant $k_2$ in Mechanism 1–3, must be faster than $10^{10}$ $s^{-1}$. The value for the unimolecular rate constant of proton transfer, based on absolute rate theory, if no energy barrier existed, would be $k_B Th^{-1}$, or $6 \times 10^{12}$ $s^{-1}$ at 25 °C, where $k_B$ is Boltzmann's constant and $h$ is Planck's constant. This calculated value is consistent with an observed value of $10^{-13}$ s for the lifetime of the $H_3O^+$ ion in ice, where the movement of this ion through the solid phase depends only on the transfer of a proton in a hydrogen bond.[1]

Within a particular class of proton transfers, for example, all of the transfers of a proton from a neutral oxygen to a neutral nitrogen to create a negatively charged oxygen and a positively charged nitrogen, the proton transfers from AH to B in the exothermic direction ($\Delta pK_a^{AB} > 4$) are always diffusion-controlled and have the same values for their second-order rate constants $k_F$. Therefore, the rates of the same proton transfers in the reverse direction, or the endothermic direction ($\Delta pK_a^{BA} < -4$) from BH to A, must be governed only by $\Delta pK_a^{AB}$ because, from Equation 1–7, it follows that

$$\log k_R = \log k_F - \Delta pK_a^{AB} \tag{1–9}$$

Because $k_F$ is always the same in the exothermic direction, the rate constant for proton transfer in the endothermic direction should decrease by a factor of 10 for every increase in $\Delta pK_a^{AB}$ of one logarithmic unit. This expectation has been demonstrated experimentally. For example, Equation 1–9 defines the diagonal asymptotes in Figure 1–1. What this suggests is that the probability the proton will leave the hydrogen-bonded complex on the weaker base is governed only by the difference in $pK_a$ between the conjugate acids of the two bases and reflects the distribution of the proton between the two lone pairs in the hydrogen-bonded complex.

The behavior of proton transfer when the reaction is neither strongly exothermic nor strongly endothermic ($-4 < \Delta pK_a^{AB} < 4$) deviates smoothly from these limits. An example is the transfer of a proton between a negative oxyanion as one base and a neutral amino nitrogen as the other base (Figure 1–1A)

$$RNH_3^+ + {}^-OR \rightleftharpoons RNH_2 + HOR \tag{1–10}$$

When the transfer is from oxygen to nitrogen and exothermic, the diffusion-controlled second-order rate constant is less than that for an exothermic transfer from nitrogen to oxygen because of the charge asymmetry (horizontal asymptotes in Figure 1–1A). In either direction, when the proton transfers have become strongly endothermic, the rate constants fall off as exponential functions of $\Delta pK_a^{ON}$. The intersection of the horizontal asymptote defined by the rate constant for the diffusion-controlled exothermic proton transfer from oxygen to nitrogen intersects the asymptote defined by the rate constant for endothermic transfer from nitrogen to oxygen when $\Delta pK_a^{ON} = 0$ as required by Equation 1–9. When $\Delta pK_a^{ON} = \Delta pK_a^{NO} = 0$, the actual rate constants for proton transfer in the two directions are equal as required by Equation 1–7.

For proton transfer in one direction, the distance that the smooth curve defining the actual second-order rate constant of a proton transfer lies below the intersection of the two asymptotes, horizontal and diagonal, governing its class of proton transfers (Figure 1–1) provides information about the rate of proton transfer in the hydrogen-bonded complex. If the rate constant $k_1$ in Mechanism 1–3 remains constant and diffusion-controlled as $\Delta pK_a^{AB}$ is varied and $k_2$ and $k_{-2}$ remain very fast, it can be shown that the composite rate constant for proton transfer $k_F$ should have half its asymptotic diffusion-controlled value (horizontal asymptotes in Figure 1–1) at the $\Delta pK_a^{AB}$ at which the horizontal and diagonal asymptotes intersect. If this is the case, the logarithm of the actual rate constant of proton transfer should lie a distance of $\log 2 = 0.30$ below the point of intersection. If, however, the rate constant $k_2$ is less than the rate constant $k_{-1}$ in Mechanism 1–3 because proton transfer in the hydrogen-bonded complex is slow compared to diffusion of the reactants apart, then, at the $\Delta pK_a^{AB}$ corresponding to the intersection of the two asymptotes, the distance the smooth curve defining the logarithm of the actual value for the composite rate constant of proton transfer lies below the intersection of the asymptotes will be approximately $\log (k_2/k_{-1})$, and it will be dictated by the rate constant $k_2$ of the proton transfer in the hydrogen-bonded complex.

For a particular acid–base on an amino acid in a protein, there is a scale of efficiency for proton transfer to or from another acid–base with the same $pK_a$. Such an isoergonic proton transfer to or from a primary amine in a hydrogen-bonded complex is the most rapid (Figure 1–1A). Even in this case, however, the overall reaction in solution is limited by the value of $k_2$ because the logarithm of the actual rate constant $k_F$ is more than 0.3 unit below the intersections of the asymptotes. The primary amine is followed in its efficiency by an imidazole, a carboxylic acid, and a phenol (Figure 1–1B–D). Nevertheless, even with carboxylic acids and phenols the distance below the intersection of the asymptotes is only about 1.2 logarithmic units, and $k_2/k_{-1} \cong 0.06$. If the rate constant for separation of the products, $k_{-1}$, remains diffusion-controlled at $10^{10}$ $s^{-1}$, then the rate constant for proton transfer $k_{-2}$ is still $>10^8$ $s^{-1}$, and the rate of proton transfer, once the hydrogen-bonded complex has formed, is still very rapid. This should be true for all proton transfers within a hydrogen bond between nitrogen and oxygen, nitrogen and nitrogen, or oxygen and oxygen in the active site of an enzyme as long as the proton transfer is isothermic or exothermic. If it is endothermic, the rate of proton transfer will be limited by the unfavorable distribution of the proton between the donor and the acceptor in the hydrogen bond, as determined by $\Delta pK_a^{AB}$.

Two other types of proton transfer in enzymatic reactions, however, are intrinsically much slower. First, the most inefficient heteroatomic acid in proton transfer to nitrogen and oxygen is a thiol (Figure 1–1E). This type of proton transfer is anomalously slow because a thiol is a poor donor, if a donor at all, in a hydrogen bond. Second, proton transfer to and from carbon is much slower than that between heteroatoms.

The rate constants for proton transfer to and from carbon exhibit large negative deviations from the asymptotic behavior exemplified by Figure 1–1 in the midrange of $\Delta pK_a^{AB}$. If the reaction is severely exothermic, such as the transfer of a proton from hydronium cation ($pK_a$ = −1.7) to the carbanion of acetone ($pK_a \cong 20$), it proceeds at diffusion-controlled rates,[1] but even the exothermic transfer of a proton from the acidic carbon on acetylacetone to hydroxide anion ($\Delta pK_a^{AB}$ = +7)

$$OH^- + H_3C\text{–}C(=O)\text{–}CH_2\text{–}C(=O)\text{–}CH_3 \xrightarrow{4 \times 10^4\ M^{-1}\ s^{-1}} H_2O + H_3C\text{–}C(=O)\text{–}CH{=}C(O^-)\text{–}CH_3 \quad (1\text{–}11)$$

or the exothermic transfer of a proton from hydronium cation to the carbanion of acetylacetone ($\Delta pK_a^{AB}$ = +11)

$$H_3O^+ + H_3C\text{–}C(=O)\text{–}CH{=}C(O^-)\text{–}CH_3 \xrightarrow{1 \times 10^7\ M^{-1}\ s^{-1}} H_2O + H_3C\text{–}C(=O)\text{–}CH_2\text{–}C(=O)\text{–}CH_3 \quad (1\text{–}12)$$

have second-order rate constants of $4 \times 10^4$ $M^{-1}$ $s^{-1}$ and $1 \times 10^7$ $M^{-1}$ $s^{-1}$, respectively, which are considerably less than diffusion-controlled rates. When the proton removal is endothermic, so that it should be less rapid than a diffusion-controlled reaction anyway, the removal of a proton from carbon is extraordinarily slow. For example, the rate constant for the removal of a proton from acetone by hydroxide ion ($\Delta pK_a^{AB} \cong -5$)

$$OH^- + H_3C\text{–}C(=O)\text{–}CH_3 \xrightarrow{2.7 \times 10^{-1}\ M^{-1}\ s^{-1}} H_2O + H_2C{=}C(O^-)\text{–}CH_3 \quad (1\text{–}13)$$

is much less ($2.7 \times 10^{-1}$ $M^{-1}$ $s^{-1}$)[1] than the rate constant of about $10^5$ $M^{-1}$ $s^{-1}$ expected for the rate of proton transfer between two atoms that differ this much in acid dissociation constants (Figure 1–1).

The reasons for the sluggish behavior of carbon are, in one direction, the failure of carbon–hydrogen bonds to participate in hydrogen bonding and, in the other, the low free electron density on carbon in enolates. The hydrogen bond formed between a normal base and a normal protic acid lowers the free energy of activation for proton transfer by bringing the reactants closer together than the van der Waals radii of the atoms would normally permit. Because carbon does not participate in hydrogen bonding, this assistance is not available and proton transfer to and from carbon, a reaction often occurring in enzymatic reactions, is very slow when it occurs in free solution between two independent solutes. To overcome these difficulties and increase the rate of proton removal, the base can be positioned next to the hydrogen on carbon by forces other than direct hydrogen bonding. In the reverse direction, the rate of proton transfer to carbon can be increased if the $\pi$ electron density on an enolate can be pushed onto carbon, causing it to be more basic by making the oxygen less electronegative.

## *Suggested Reading*

Eigen, M. (1964) Proton Transfer, Acid–Base Catalysis, and Enzymatic Hydrolysis, *Angew. Chem., Int. Ed. Engl.* 3, 1–19.

# Nucleophilic Substitution

The simplest reaction in organic chemistry, other than proton transfer, is a **bimolecular nucleophilic substitution**, or an $S_N2$ reaction. Although such reactions are rarely encountered in enzymatic mechanisms, they are often used to study proteins experimentally. An example would be the carboxamidomethylation of a cysteine residue with chloroacetamide

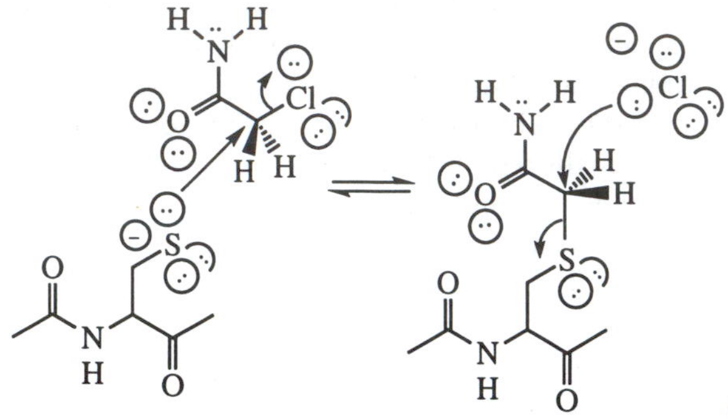

(1–14)

In this reaction, a bond between carbon and chlorine has been substituted with a bond between the same carbon and sulfur. The effect of pH on the concentration of the nucleophile in this reaction has already been discussed in detail.[5a]

The electrons participating in this reaction are confined initially in the atomic and molecular orbitals of the reactants; they end up confined in the atomic and molecular orbitals of the product; and, at the apogee of the reaction, they are confined in the **atomic and molecular orbital systems of the transition state**. For this to occur, the atomic and molecular orbitals of the reactants that will mix to form the atomic and molecular orbitals of the transition state must be able to approach each other closely enough and in the proper orientation to **overlap**, and the atomic and molecular orbitals of the product must be such that they can be derived from a permitted unmixing of the atomic and molecular orbitals of the transition state.

To be able to produce a mechanism consistent with these requirements, the atomic and molecular orbitals of the reactants that overlap during the reaction to produce the transition state and the electrons that occupy this mixture of orbitals should be identified. In the case of the nucleophilic substitution described in Reaction 1–14, one of the atomic orbitals on sulfur containing one of its lone pairs of electrons mixes with the molecular orbital system of the $\sigma$ bond between carbon and chlorine to produce the molecular orbital system of the transition state. This process can be viewed as a smooth transition in which the occupied atomic orbital on sulfur overlaps with the unoccupied antibonding molecular orbital of the $\sigma$ bond. This initial overlap leads to the molecular orbital system of the transition state, composed from the atomic orbital on sulfur, a $p$ atomic orbital from carbon, and an atomic orbital from chlorine (Figure 1–2). Therefore, the molecular orbital system of the transition state has three molecular orbitals, and the two of lowest energy are occupied by the four electrons that are being rearranged. These electrons are the two originally in the lone pair on sulfur and the two originally in the $\sigma$ bond between carbon and chlorine.

It is when the structure of the chloroacetamide is examined closely that the **energy barrier** encountered by the lone pair of electrons on sulfur is appreciated

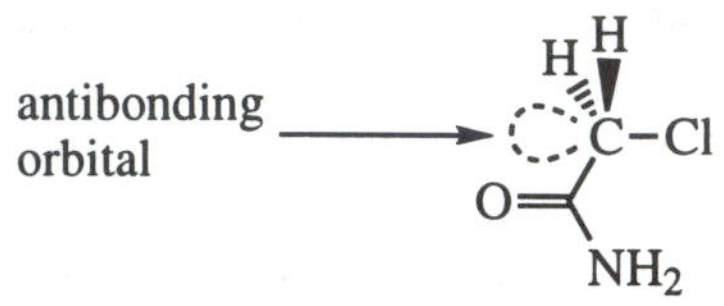

antibonding orbital

**1–3**

The vacant antibonding orbital is in a depression formed from three of the $\sigma$ bonds to the electrophilic carbon, and the six electrons in these three $\sigma$ bonds repel the two electrons in the lone pair on sulfur. This **electron repulsion** also results from orbital overlap, but overlap between two filled orbitals. If along the trajectory of the reaction the atomic orbital containing the lone pair of electrons on sulfur overlaps the occupied bonding molecular orbitals of any one of these three $\sigma$ bonds, electron repulsion occurs, and this is unfavorable. There is also both repulsion produced by the occupied molecular orbital of the carbon–chlorine bond and a barrier to the reaction resulting from the requirement to desolvate the thiolate anion. The repulsive barriers of the three $\sigma$ bonds, however, are more significant in dictating the steric course of the reaction, and it is they that confine the trajectory of the nucleophile. These three $\sigma$ bonds are pushed away by the lone pair of electrons on the nucleophile as the reaction progresses.

As they are pushed away, these three $\sigma$ bonds flatten into a plane normal to the $\sigma$ bond between the carbon and the chlorine

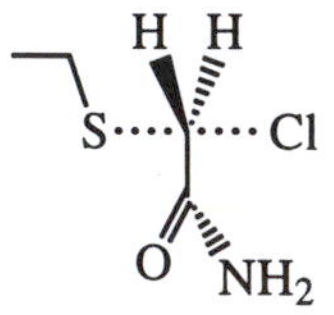

**1–4**

This reorientation causes the mutual repulsion of these three $\sigma$ bonds to increase. The transition state involves too much electron repulsion to be stable, and it breaks down to yield the products.

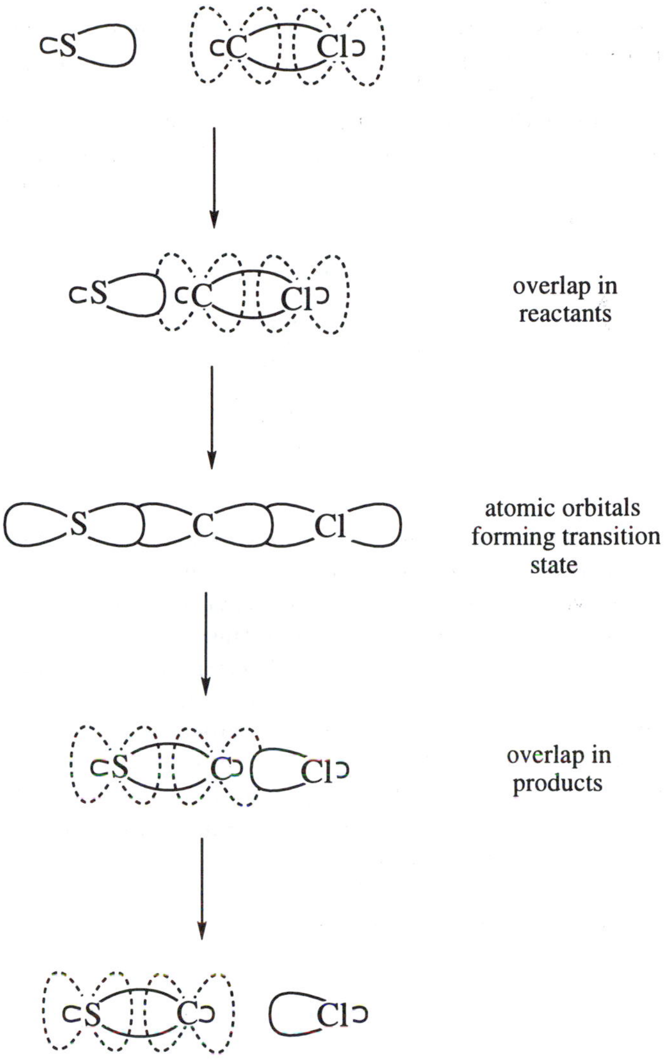

**Figure 1–2:** Progress of an $S_N2$ reaction. An atomic orbital on the nucleophile, occupied by a lone pair of electrons (solid line), overlaps the antibonding molecular orbital (dashed lines) of the carbon–chlorine bond and the overlap proceeds to the transition state (center) formed from a combination of three atomic orbitals (solid lines), one from each of the three atoms, sulfur, carbon, and chlorine. The transition state decomposes through an overlap between an atomic orbital on chlorine (solid line), which will contain a lone pair of electrons in the product, and the antibonding molecular orbital (dashed lines) of the carbon–sulfur bond in the other product.

overlap in reactants

atomic orbitals forming transition state

overlap in products

To effect the reaction, the lone pair of electrons on sulfur must approach a location on the carbon opposite the carbon–chlorine bond with sufficient kinetic energy or, just as effectively, be forced against the carbon with sufficient potential energy to overcome the electron repulsion inherent in both the reorientations of the $\sigma$ bonds and the crowding in the transition state. In solution, the kinetic energy of the reactants prior to the collision is all the energy that is available, and only collisions between two molecules with relative kinetic energy greater than this minimum amount of energy can be productive.

The **leaving group** in a nucleophilic substitution, for example, the chloride in Reaction 1–14, can be improved by rendering its conjugate acid more acidic. Because it leaves carbon with a lone pair of electrons, the departure of the leaving group from carbon resembles in its character the departure of the same constellation of atoms from a proton in an acid dissociation, for example

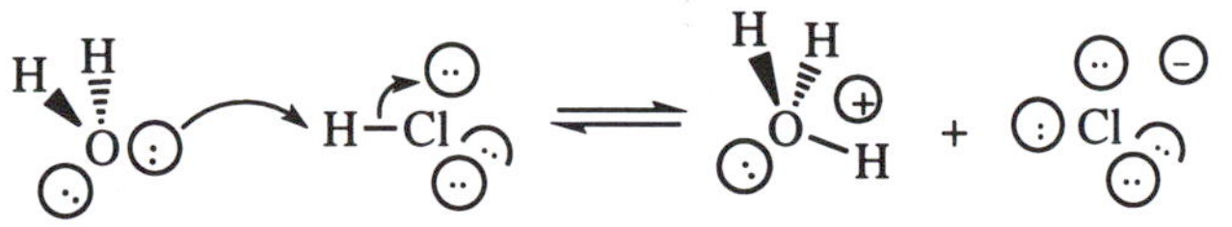

$$H_2O + H{-}Cl \rightleftharpoons H_3O^+ + Cl^- \qquad (1\text{–}15)$$

As a result, any alteration in the leaving group that increases the p$K_a$ of its conjugate acid improves its capacity for departure. For example, $I^-$ is a better leaving group than $Cl^-$ because HI is a stronger acid than HCl.

Other examples of such improvements usually involve the addition of electrophiles to the leaving group. When the oxygen of a hydroxyl group has been protonated, it is a better leaving group because the conjugate acid of water, $HOH_2^+$ (p$K_a$ = –1.7), is a stronger acid than the conjugate acid of hydroxide, HOH (p$K_a$ = 15.7). If a hydroxyl group is replaced with a phosphate, $-OPO_3H^-$, the bonded oxygen is a better leaving group because the conjugate acid of phosphate, $HOPO_3H^-$ (p$K_a$ = 7.2), is a stronger acid than the conjugate acid of hydroxide ion, HOH (p$K_a$ = 15.7). The protonation of that phosphate on another of its oxygens to form $-OPO_3H_2$ produces an even better leaving group because its conjugate acid, $HOPO_3H_2$ (p$K_a$ = 2.1), is even more acidic. When the oxygen of a hydroxyl group has been acylated with an electrophilic functional group such as an acetyl group, $-COCH_3$, it is a better leaving group because the conjugate acid of an acetate anion, $HOCOCH_3$ (p$K_a$ = 4.7), is a stronger acid than the conjugate acid of a hydroxyl ion, HOH (p$K_a$ = 15.7). The protonation of the imidazole nitrogen of the guanine in guanosine

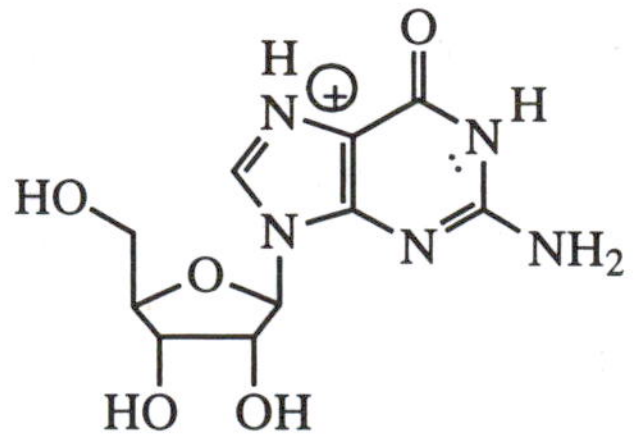

**1–5**

causes the guanine to become a better leaving group because the cationic conjugate acid of guanine ($pK_a = 3.3$) is a stronger acid than guanine itself ($pK_a = 9.2$).

If the leaving group in a nucleophilic substitution reaction at carbon is improved sufficiently that it dissociates from the carbon before the nucleophile collides, it leaves behind a **carbocation**

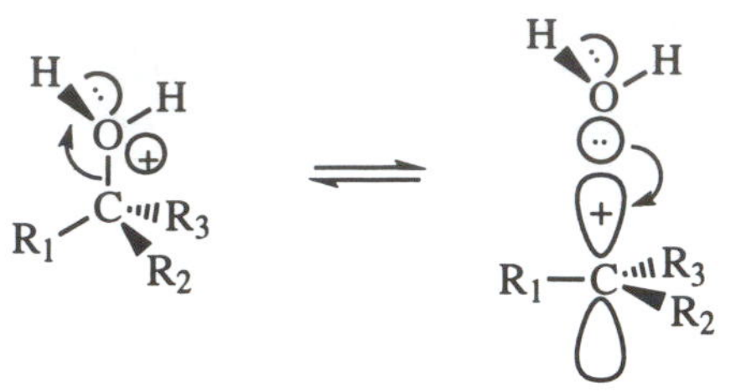

(1–16)

The carbocation, because it has finite existence, is an **intermediate** in the reaction rather than a transition state. If it reacts with a nucleophile, a **unimolecular nucleophilic substitution**, or an $S_N1$ reaction, occurs. This reaction is referred to as a unimolecular substitution because the rate-limiting step is the unimolecular cleavage of the bond between carbon and the leaving group to produce the carbocation.

In Reaction 1–16, the two molecular orbitals of the $\sigma$ bond, bonding (filled) and antibonding (vacant), have been smoothly turned into the filled atomic orbital containing the lone pair of electrons on the leaving group and the vacant *p* orbital on carbon, while the three substituents around carbon have entered the plane normal to the axis of the *p* orbital

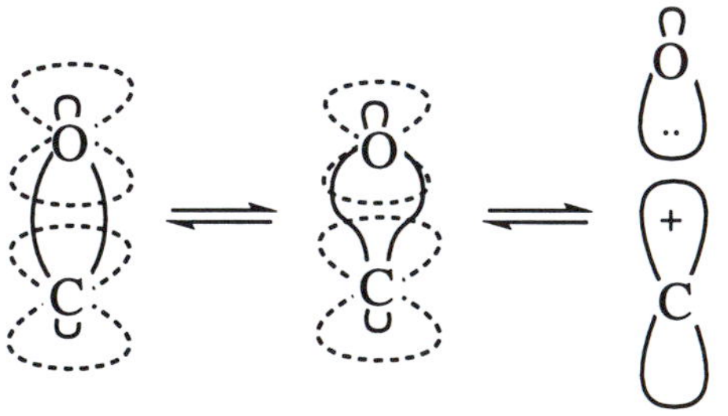

(1–17)

Carbocations are hybridized *p*, $sp^2$, $sp^2$, $sp^2$ because of electron repulsion among the three occupied molecular orbitals of the three $\sigma$ bonds to the substituents and the decrease in promotion energy produced by bringing the six electrons in these $\sigma$ bonds closer to the nucleus. The reaction is endothermic because a covalent bond—in the example cited, that between carbon and oxygen—is broken with no recompense other than changes in solvation. It is the heterolytic cleavage of this bond that creates the energy barrier to the reaction. In the reverse direction, however, no significant energy barrier exists and the combinations of nucleophiles with carbocations are diffusion-controlled reactions.

Aliphatic carbocations are rarely intermediates in biochemical reactions, presumably for the same reasons they are avoided in organic synthesis, which are the problems of elimination and rearrangement. One particular type of carbocation, however, is a common intermediate in biochemical reactions. This is the carbocation formed from the removal of a leaving group from an **acetal** ($R_2$ = H) or a **ketal** ($R_2$ and $R_3$ are both alkyl)

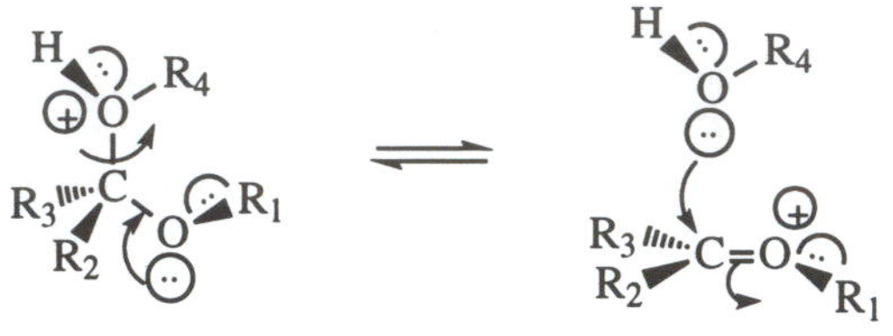

(1–18)

or from a **hemiaminal**

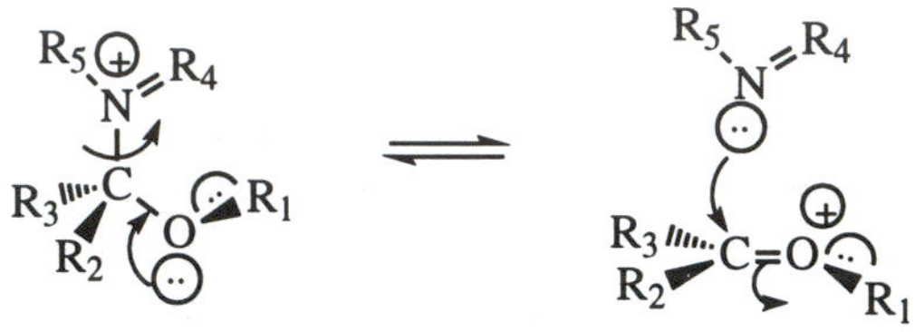

(1–19)

The resonance form drawn in Reactions 1–18 and 1–19 for the cationic product is the form in which all atoms have octets, and this resonance form indicates why the cation is more stable than a simple carbocation. This structure is referred to as an **oxocarbocation.**

A complete nucleophilic substitution involving such an oxocarbocation would be that catalyzed by uracil phosphoribosyl transferase

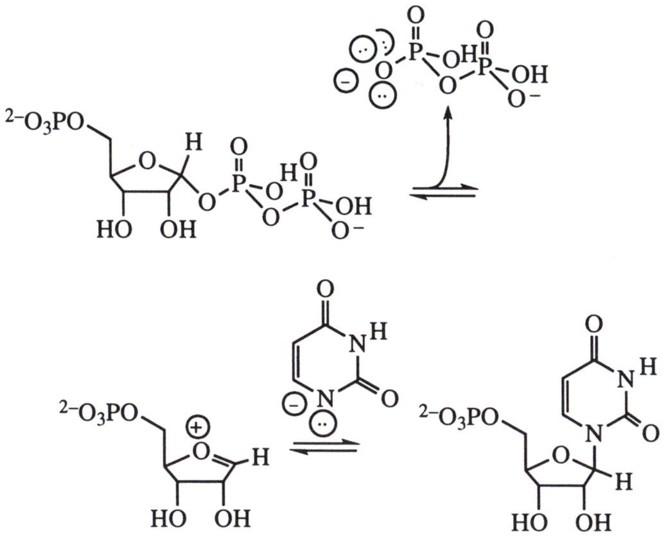

(1–20)

or the acid-catalyzed hydrolysis of adenosine[6]

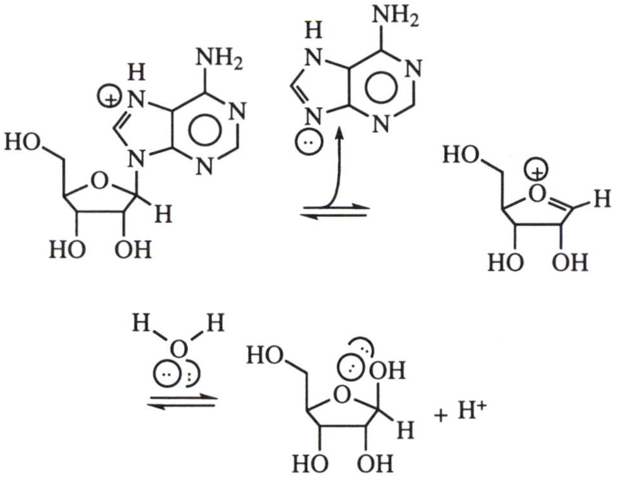

(1–21)

The oxocarbocation formed during the acid-catalyzed hydrolysis of acetophenone dimethyl ketal has been identified by trapping it with sulfite[7]

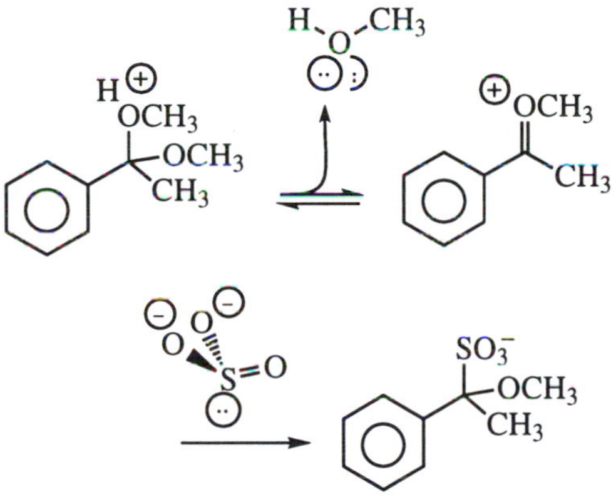

(1–22)

In most of these reactions, there is an absolute requirement for protonation of the leaving group, as with the oxygen in an acetal or a ketal,[8] but the dissociation of the $\sigma$ bond between carbon and the leaving group is still the rate-limiting step.[9]

Another reaction in which an oxocarbocation is an intermediate is the formation of vinyl ethers in a reaction formally equivalent to an $E_1$ elimination, which by definition is a reaction proceeding through a carbocation

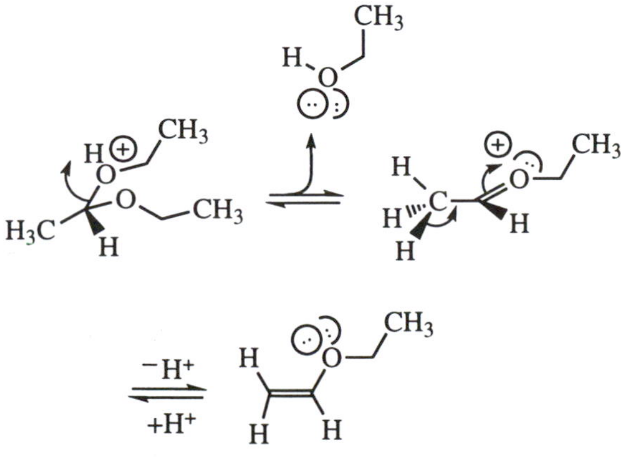

(1–23)

Because proton loss, the second step, is the rate-limiting step in this reaction,[9] it can be prevented from occurring by denying the proton a base and capturing the oxonium cation by a nucleophile before it can eliminate the proton.

During the opening of imidazolines

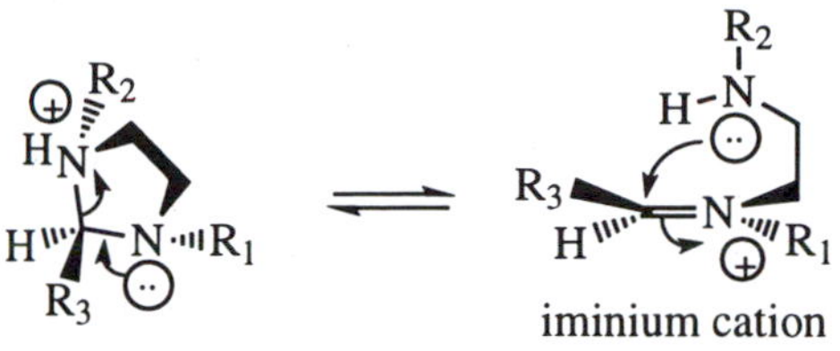

iminium cation (1–24)

the less basic nitrogen, which is determined by the identity of $R_1$ and $R_2$, is the leaving group, but only after it is protonated.[10] Reaction 1–24 serves as an example of a nitrogen rather than an oxygen providing the lone pair of electrons stabilizing the carbocation by forming an **iminium cation** rather than an oxocarbocation.

In the oxocarbocations or iminium cations that are formed in these reactions, the $\pi$ molecular orbital system between carbon and oxygen or carbon and nitrogen contains two electrons. This leaves the constellation with a net positive formal charge. This can be represented by resonance structures as

**1–6**

**1–7**

The orbitals of the reactant that mix to form the transition state that lies between the hemiaminal and the products in Reaction 1–19 are a $p$ atomic orbital on the adjacent oxygen, an atomic orbital on carbon, and an atomic orbital on the nitrogen (Figure 1–3). The molecular orbital system of the transition state created from these three atomic orbitals has three molecular orbitals, and the two with lowest energy are occupied by the four electrons participating in the reaction.

In the reverse direction, the addition of the nitrogen to the oxocarbocation can be viewed as being initiated by overlap between the atomic orbital on nitrogen occupied by the $\sigma$ lone pair of electrons and the vacant antibonding molecular orbital of the $\pi$ molecular orbital connecting the carbon and the adjacent oxygen in the oxocarbocation. This transition state in turn unravels into the two molecular orbitals of the $\sigma$ bond between carbon and nitrogen and the atomic orbital on oxygen containing the lone pair of electrons.

In the forward direction, the reaction can be considered to be the expulsion of the leaving group accelerated by the **push** of a lone pair on the adjacent oxygen.[11] This can be abbreviated

**Figure 1–3:** Dissociation of a hemiaminal to produce an oxocarbocation (Reaction 1–19). The atomic orbital on oxygen bearing the lone pair of electrons (solid line) overlaps the unoccupied antibonding molecular orbital (dashed lines) of the carbon–nitrogen $\sigma$ bond to produce the transition state (center) formed from the overlap of three atomic orbitals, one from each of the three atoms, nitrogen, carbon, and oxygen. The transition state decomposes through a form in which the unoccupied antibonding molecular orbital (dashed lines) of the $\pi$ molecular orbital of the oxocarbocation overlaps the atomic orbital on nitrogen (solid line) occupied by the lone pair of electrons present in the product.

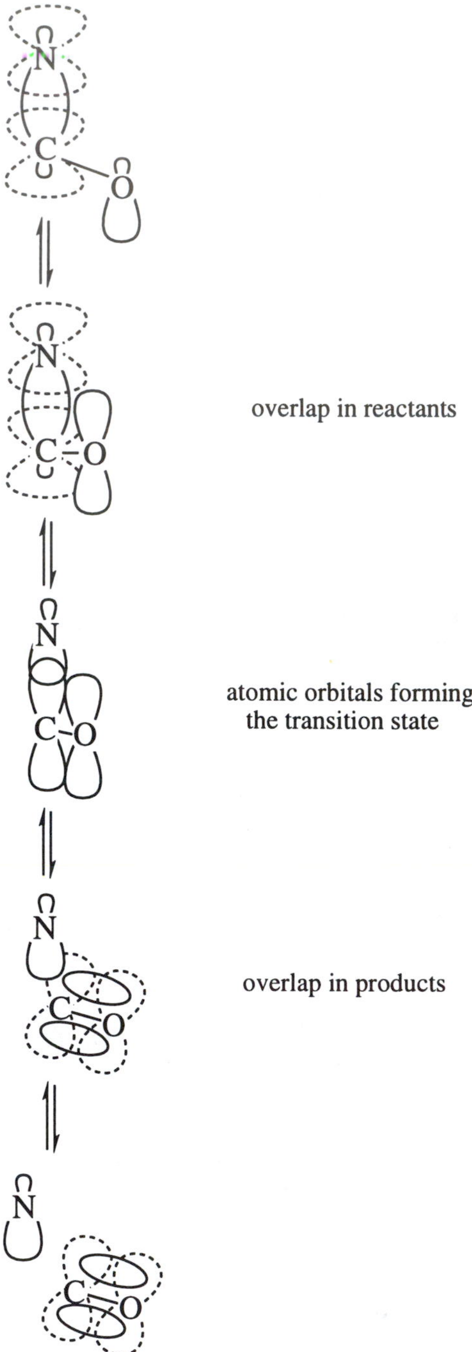


**1–8**

There is little doubt that a push is operating in displacements of this type because they are more rapid than the same reaction in which the oxygen or nitrogen is replaced by carbon. There is, however, an interesting issue surrounding the stereochemical requirements imposed upon the electrons providing the push.

The atomic orbital holding the lone pair of electrons on oxygen or nitrogen that will provide the push has been shown in Figure 1–3, Reaction 1–18, or Structure **1–8** as assuming an antiperiplanar orientation to the departing nitrogen. In an **antiperiplanar** arrangement, the lone pair of electrons providing the push is parallel to the $\sigma$ bond being broken and on the opposite side of the constellation of atoms

**1–9**

In Structure **1–9** the atom X, the $\sigma$ bond between X and the carbon, the carbon itself, the $\sigma$ bond between the carbon and Y, the atom Y, and the lone pair of electrons on Y are all in the plane of the page. This coplanarity is always necessary for the proper overlaps to occur. The lone pair of electrons on Y, however, is drawn on the opposite side of the $\sigma$ bond between carbon and Y from the pair of electrons in the $\sigma$ bond between carbon and X. There is a proposal, described as **stereoelectronic control**,[12,13] that such an orientation is a necessity of the reaction because the push must be provided, through the antibonding orbital (Figure 1–3), from the opposite side of the carbon from which the nitrogen is leaving. There is recent experimental evidence,[14] however, indicating that although the electrons providing the push must always be in an atomic orbital parallel to the breaking bond, there is no significant preference for an antiperiplanar orientation over a **synperiplanar** orientation. For example, the synperiplanar analogue of Reaction 1–18 would be

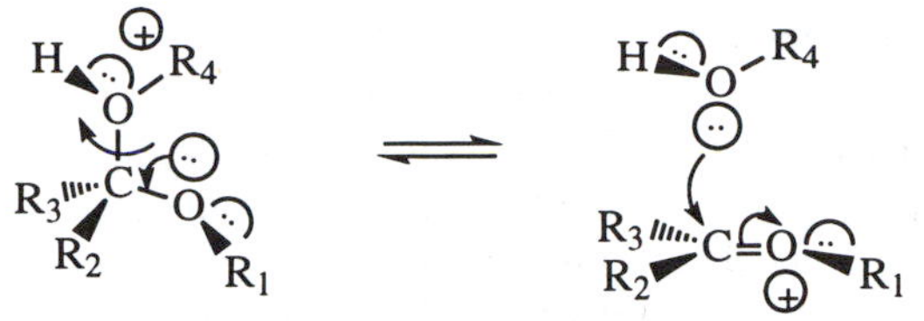

(1–25)

In the reverse direction, that of the addition of a lone pair of electrons on a nucleophile to a carbon–oxygen or carbon–nitrogen $\pi$ bond (Reaction 1–19 and Figure 1–3), a portion of the energy barrier is again due to electron repulsion

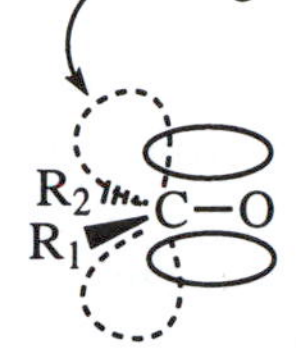

lobe of antibonding orbital

**1–10**

The lone pair of electrons of the nucleophile on a trajectory toward the unoccupied antibonding molecular orbital finds itself surrounded by the two pairs of electrons in the two $\sigma$ bonds between the carbon and its two other substituents and the pair of electrons in the occupied bonding $\pi$ molecular orbital between carbon and oxygen. The two electrons in the $\pi$ bond are delocalized and more concentrated over oxygen because of the coulomb effect, so the electron repulsion is not so great as that experienced by a nucleophile in a bimolecular nucleophilic substitution. Nevertheless, either the kinetic energy of the nucleophile as it collides or the potential energy of squeezing the nucleophile against the carbon is required to overcome the energy barrier.

In reactions involving the addition of a nucleophile to an oxocarbocation, the oxygen atom has an alkyl substituent attached to it ($R_1$ in Reaction 1–19 and Structure **1–6**). This substituent fixes a positive formal charge on the oxygen and thereby increases its coulomb effect and draws the two electrons in the $\pi$ bond more strongly over it (Structure **1–6**). Nitrogen, because it is a less electronegative element, exerts a weaker coulomb effect, even though it also bears a positive formal charge in an iminium cation (Reaction 1–24 and Structure **1–7**). If the alkyl substituent on the oxygen is replaced by a lone pair of electrons, the oxocarbocation becomes a **carbonyl** group

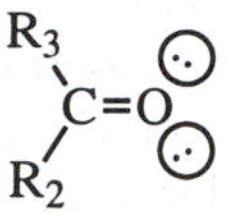

**1–11**

Although it is more stable than an oxocarbocation because oxygen is no longer required to bear a formal positive charge, the carbonyl also is susceptible to nucleophilic addition. The energy barrier experienced by the nucleophile, however, is greater in carbonyl addition because of the greater electron density over carbon in the bonding molecular orbital of the $\pi$ molecular orbital system.

### *Suggested Reading*

Young, P.R., & Jencks, W.P. (1977) Trapping of the Oxocarbonium Ion Intermediate in the Hydrolysis of Acetophenone Dimethyl Ketals, *J. Am. Chem. Soc. 99*, 8238–8247.

# Nucleophilic Addition to a Carbonyl

The reverse reaction to the expulsion of a leaving group from a carbon adjacent to an oxygen or a nitrogen (Reactions 1–18 and 1–19) is the addition of a nucleophile to a carbonyl carbon. The compounds that can be formed in this type of a reaction are hemiacetals, hemiketals, acetals, ketals, thiohemiacetals, thiohemiketals, carbinolamines, and imines (Table 1–1). A class of these reactions is the addition of a thiolate anion to an aldehyde[15] or a ketone[16] to produce a **hemithioacetal** or **hemithioketal**. An example of such a reaction would be

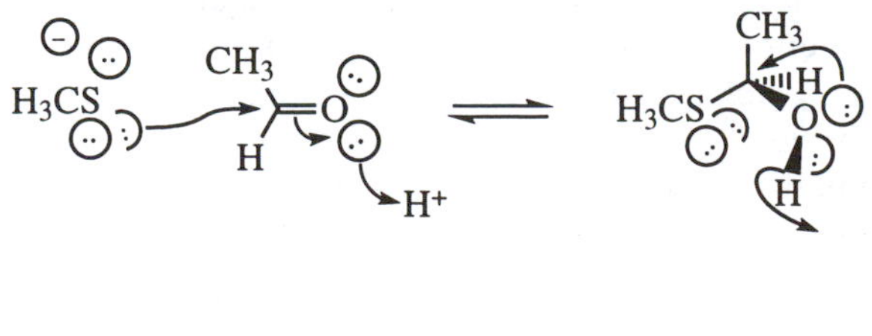

(1–26)

In this reaction, the initial contact between the reactants can be viewed as the overlap between the atomic orbital on sulfur occupied by the lone pair of electrons and the unoccupied antibonding orbital of the $\pi$ molecular orbital system of the carbonyl (Figure 1–3). In the process, electron density builds on oxygen at the expense of the nucleophile as the molecular orbital system of the transition state, produced from the three overlapping atomic orbitals, is formed. As the transition state decomposes to the product, the new $\sigma$ bond is formed by the mixing of an atomic orbital from sulfur and an atomic orbital from carbon as an atomic orbital bearing the lone pair withdraws onto the oxygen. In the first step, electron repulsion develops from overlap between the nucleophilic lone pair of electrons on sulfur and either the occupied, bonding molecular orbital of the $\pi$ molecular orbital system or one of the occupied $\sigma$ bonds to the carbon. Therefore, productive collisions occur only when the trajectory of the lone pair of electrons on sulfur is from above or below the plane of the carbonyl and from a direction distal to the oxygen

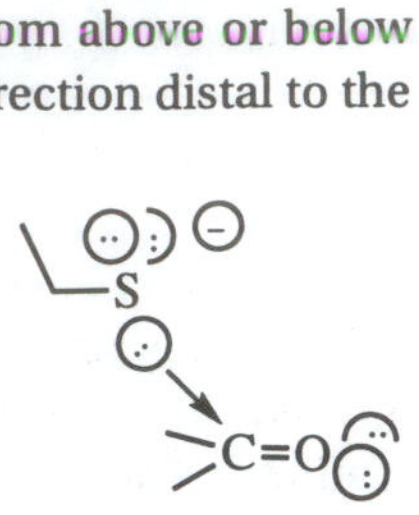

**1–12**

In the formation of a hemithioacetal or hemithioketal, two steps can be formally recognized: formation of the carbon–sulfur bond and protonation of the oxygen. At values of pH greater than 3, the behavior of the reaction as a function of pH is consistent with formation of the carbon–sulfur bond before the protonation of the oxygen[17]

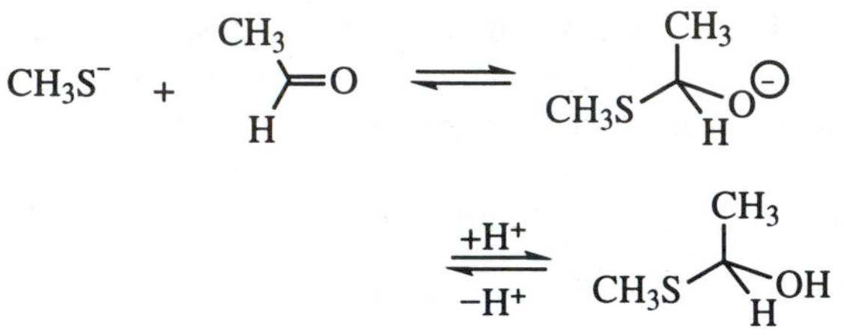

$$CH_3S^- + CH_3CHO \rightleftharpoons CH_3S\text{–}CH(CH_3)\text{–}O^- \underset{-H^+}{\overset{+H^+}{\rightleftharpoons}} CH_3S\text{–}CH(CH_3)\text{–}OH \qquad (1\text{–}27)$$

The intermediate anion is unstable and the rate of the reaction is determined by the rate at which the intermediate is trapped by proton addition from water or other acids in the solution.[18]

An analogous reaction can occur with alcohols to produce hemiacetals and hemiketals. Under **base catalysis** (Figure 1–4),[19] two different reactions, the formation of the oxygen–carbon bond and the removal of a proton from the attacking alcohol by a base that is hydrogen-bonded to it beforehand, occur simultaneously. The $\sigma$ lone pair of electrons on the attacking oxygen is mixing with the $\pi$ molecular orbital system of the carbonyl (Figure 1–3) while a proton is being transferred from a different lone pair on the same oxygen to the base. Under **acid catalysis** (Figure 1–4),[19] the reaction proceeds by nucleophilic addition of the alcoholic oxygen to the carbonyl carbon (Figure 1–3) at the same time that proton transfer is occurring between the preassociated acid and a lone pair of electrons on the carbonyl oxygen distinct from and orthogonal to the developing lone pair of electrons derived from the retraction of the $\pi$ molecular orbital system of the carbonyl onto the same oxygen. A biochemical example of such a reaction occurs in sugars

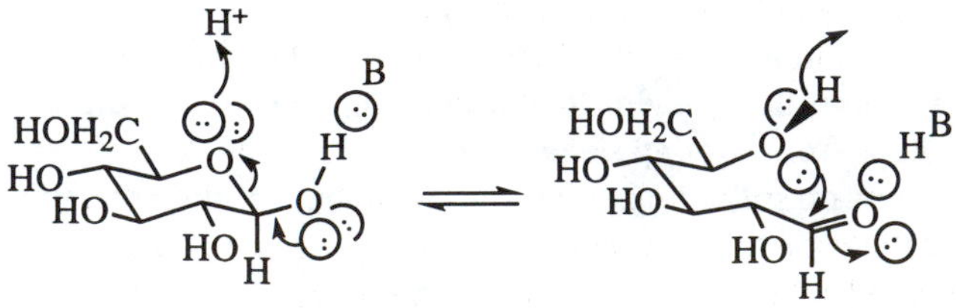

The hemithioacetals, hemithioketals, hemiacetals, or hemiketals formed in all of these reactions represent a class of products that result from addition to a carbonyl by a heteroatomic nucleophile. Many biochemical metabolites contain such products of carbonyl addition reactions. In each of these products the **carbonyl carbon** is recognized as the carbon bonded directly to two heteroatoms and to two other carbons or to one carbon and one hydrogen or to two hydrogens (Table 1–1). Another paradigm of this class is the **carbinolamine** (Structure **1–13** in Figure 1–5) formed by the reaction between a primary amine and a carbonyl compound. It is an unstable compound produced by the addition of an amine to a carbonyl carbon (Figure 1–3) to produce a zwitterion followed by a tautomerization.

The tautomerization of a carbinolamine (Figure 1–5) proceeds by transfer of a proton from the ammonium cation to a base in the solution and addition of a proton to the oxyanion from an acid in the solution. It cannot occur by intramolecular proton transfer because the acidic ammonium cation is only two atoms away from the basic oxyanion. Although there are examples of slow intramolecular proton transfers between an acid and a base in the same molecule that are three atoms apart,[23] intramolecular proton transfer between an acid and a base two atoms apart is sterically impossible. The *s* atomic orbital of the acidic hydrogen cannot approach closely enough the two $sp^3$ orbitals containing the two lone pairs, found on oxygen and nitrogen, respectively, to overlap simultaneously and produce the transition state required for proton transfer

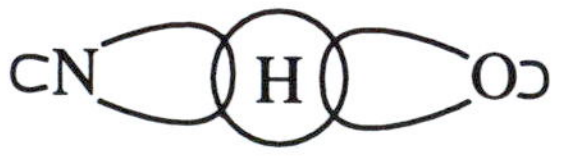

**1–14**

Although this transition state is not required to be linear as drawn,[23] overlap of the atomic orbitals creating it is an inescapable requirement. Such overlap in the arrangement

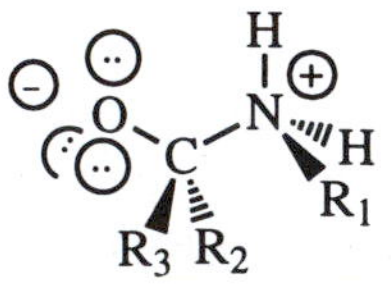

**1–15**

would require the formation of a four- or three-membered ring, depending on whether or not the hydrogen nucleus is counted as one of the vertices, and this would produce severe and unacceptable angular strain in the transition state for proton transfer.

Tautomerizations involving the removal of a proton from an ammonium cation and the addition of a proton to an

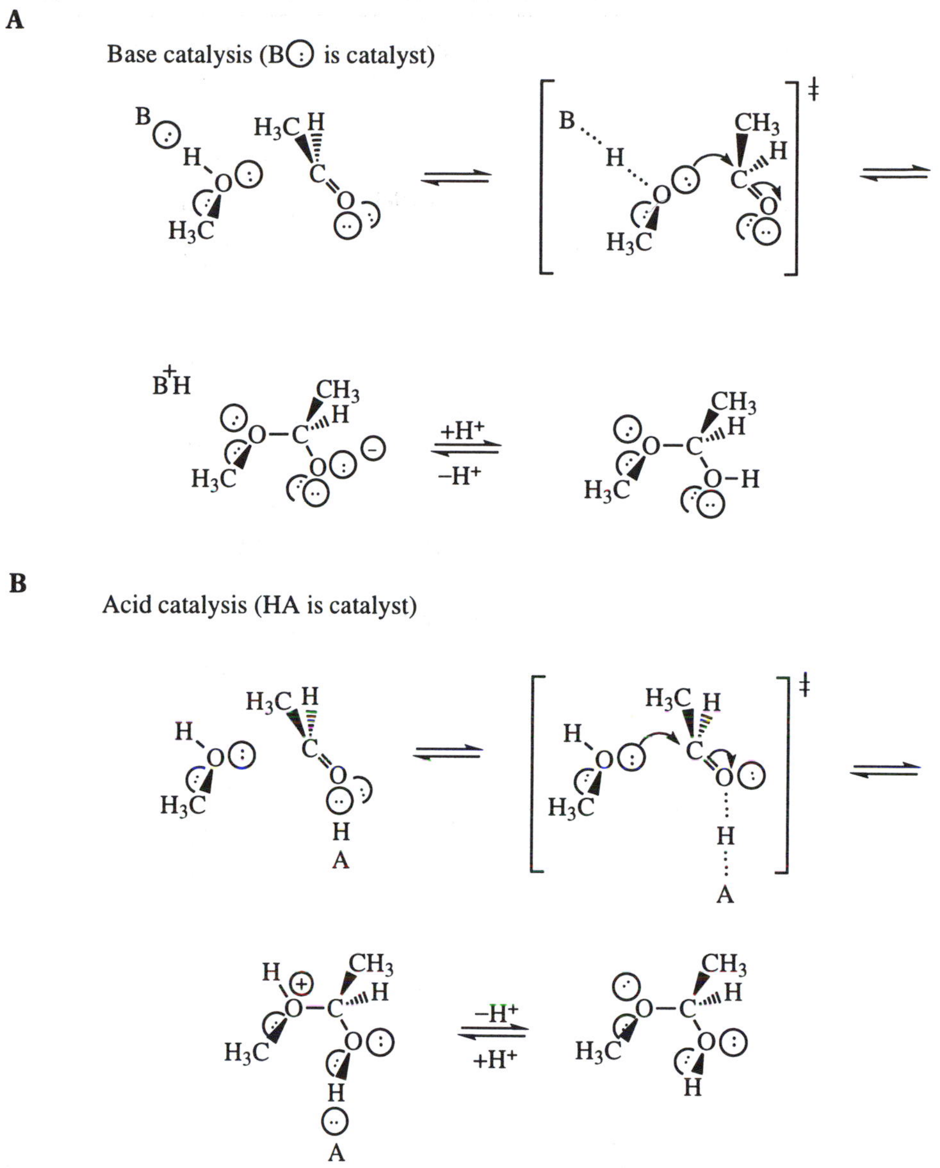

**Figure 1–4:** General base and general acid catalysis of the formation of a hemiacetal.[19] (A) In general base catalysis, the base B, hydrogen-bonded to the alcohol, removes the proton from the oxygen at the same time another lone pair of electrons on the same oxygen adds to the carbonyl carbon. The two orthogonal reactions are distinguished by using dotted lines for one and arrows for the other. The second step in the reaction is protonation of the oxyanion to form the hemiacetal. (B) In general acid catalysis, the acid HA, hydrogen-bonded to the carbonyl oxygen, adds a proton to this oxygen at the same time that the oxygen of the alcohol adds to the carbonyl carbon. The two orthogonal reactions are distinguished by using dotted lines for one and arrows for the other. The second step in the reaction is dissociation of the proton from the oxygen of the alcohol.

A

Base catalysis (B⊙ is catalyst)

B

Acid catalysis (HA is catalyst)

alkoxide two atoms away or the reverse are encountered frequently in biochemical mechanisms. The ability of bifunctional acid–bases such as phosphate and bicarbonate to catalyze reactions involving such tautomerizations 2–3 orders of magnitude more effectively than monofunctional acids or bases with similar values of p$K_a$ has been presented as evidence for the cyclic transfer of the two protons[24]

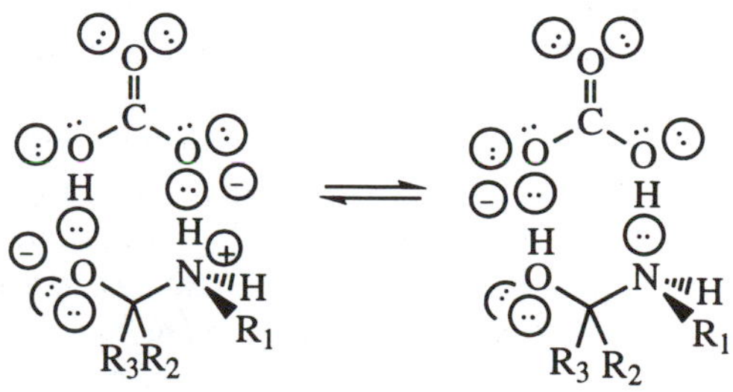

(1–29)

A similar cyclic intermediate has been invoked to explain the much less remarkable catalysis of such tautomerizations by primary amines[25,26]

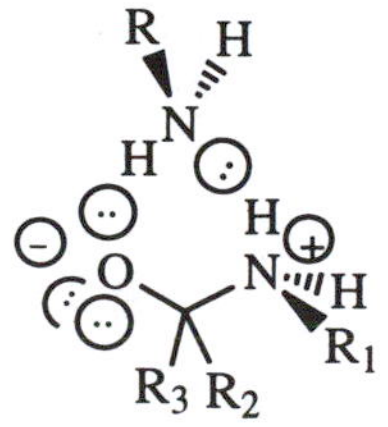

**1–16**

A carbinolamine (Structure **1–13** in Figure 1–5) is a symmetric intermediate. There are lone pairs on oxygen able to push away nitrogen to return to the ketone or aldehyde and

**Table 1–1: Derivatives of Acetaldehyde or Acetone as Examples of Carbonyl Compounds[a]**

| derivatives of acetaldehyde[b] | | derivatives of acetone[b] | |
|---|---|---|---|
| hemiacetal | $H_3C$–C(OH)(H)($OCH_3$) | hemiketal | $H_3C$–C(OH)($CH_3$)($OCH_3$) |
| acetal | $H_3C$–C($OCH_3$)(H)($OCH_3$) | ketal | $H_3C$–C($OCH_3$)($CH_3$)($OCH_3$) |
| thiohemiacetal | $H_3C$–C(OH)(H)($SCH_3$) | thiohemiketal | $H_3C$–C(OH)($CH_3$)($SCH_3$) |
| carbinolamine | $H_3C$–C(OH)(H)–N(H)–$CH_3$ | carbinolamine | $H_3C$–C(OH)($CH_3$)–N(H)–$CH_3$ |
| imine | $H_3C$–C(H)=NH | imine | $H_3C$–C($CH_3$)=NH |

[a] Derivatives formed from methanol, methanethiol, or methylamine are presented for the sake of simplicity. [b] Only one enantiomer of each derivative is presented.

there is a lone pair on nitrogen able to push away oxygen to produce an **imine**, the nitrogen analogue of the ketone or aldehyde. The expulsion of oxygen can occur under either acid catalysis or base catalysis. Under base catalysis, the breaking of the bond between carbon and oxygen, assisted by the push of the lone pair on nitrogen (Figure 1–5), occurs simultaneously[20] with the removal of a proton by the base from another lone pair on nitrogen. The removal of the proton from the other lone pair on nitrogen during the formation of the transition state increases the electron density on nitrogen and amplifies the push of the initially present lone pair by electron repulsion so that it is strong enough to push out the poor leaving group, hydroxide anion. Under acid catalysis (Figure 1–5) the breaking of the bond between carbon and oxygen, assisted by the push of the lone pair on nitrogen, occurs simultaneously[21] with the protonation[22] of one of the available lone pairs on the oxygen of the leaving group. This lone pair is distinct from the lone pair created during the dissociation of the carbon–oxygen bond. The addition of a proton to oxygen decreases its electron density and attenuates the carbon–oxygen $\sigma$ bond by pulling its electrons toward the oxygen so that it is weak enough to be broken by the relatively feeble push of a lone pair of electrons on a neutral nitrogen.

The susceptibility of a particular carbonyl carbon to addition can be evaluated by referring to the **standard free energy of hydration**, $\Delta G^\circ_{HYDRAT}$, where hydration is the reaction

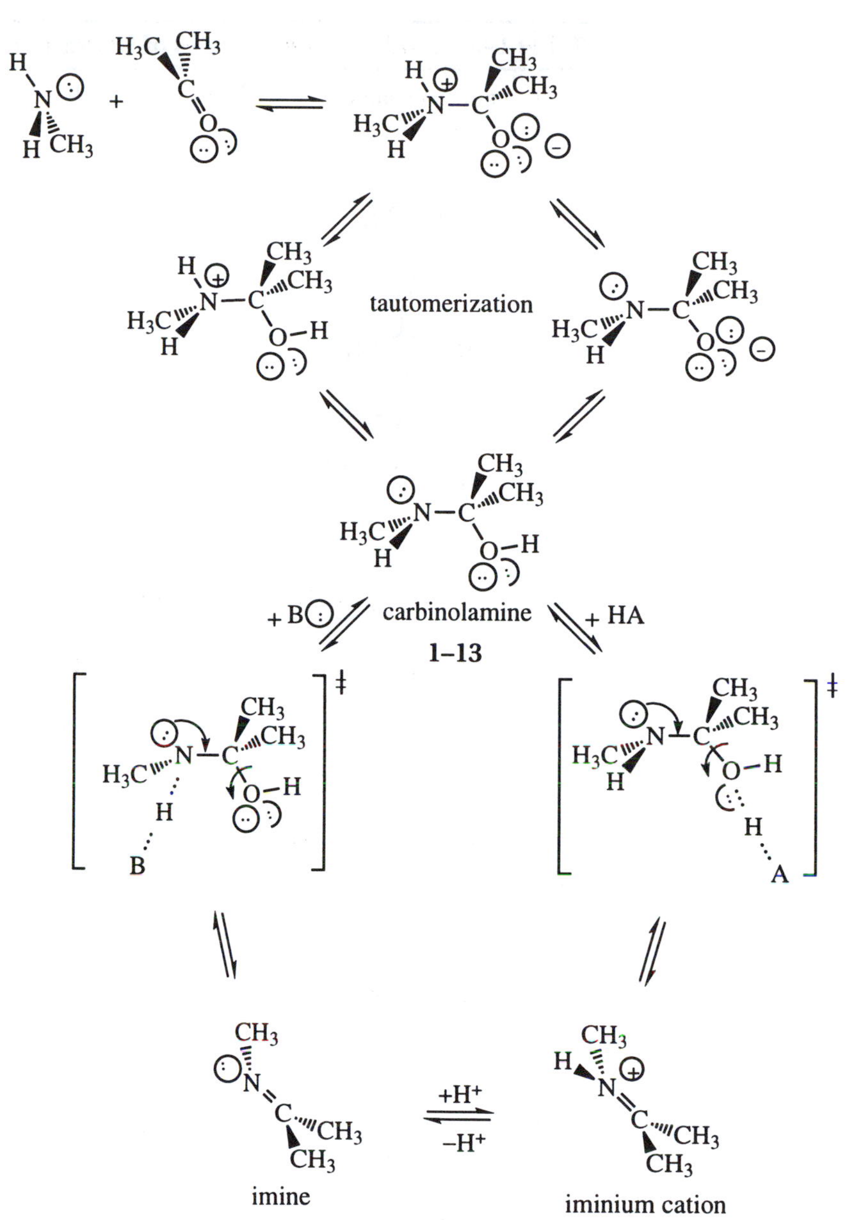

**Figure 1–5:** Addition of an amine to a carbonyl compound. For simplicity the addition of methylamine to acetone was chosen as an example. The free base of the amine adds to the carbonyl to produce the zwitterion. A base in the solvent removes a proton from the ammonium cation and an acid in the solvent adds a proton to the oxyanion to produce the carbinolamine. The carbinolamine is susceptible to elimination in a reaction catalyzed by a general base[20] or a general acid.[21,22] In the general-base-catalyzed reaction a proton is removed from the nitrogen at the same time that another lone pair of electrons on the nitrogen pushes out the leaving group, a hydroxide ion. These two reactions are distinguished by using dotted lines and arrows, respectively. In the general-acid-catalyzed reaction, a proton is added to the departing oxygen at the same time that the lone pair of electrons on nitrogen is pushing out the oxygen. These two reactions are distinguished by using dotted lines and arrows, respectively.

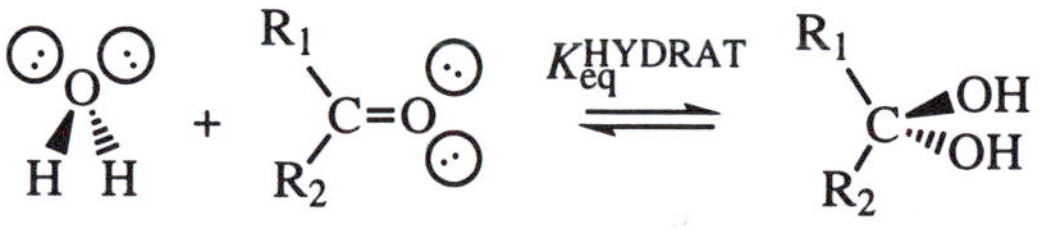

tautomerization

$+B$ carbinolamine **1–13** $+HA$

imine $\underset{-H^+}{\overset{+H^+}{\rightleftharpoons}}$ iminium cation

$$H_2O + R_1R_2C{=}O \underset{}{\overset{K_{eq}^{HYDRAT}}{\rightleftharpoons}} R_1R_2C(OH)_2 \quad (1\text{–}30)$$

The equilibrium constant for hydration is defined as

$$K_{eq}^{HYDRAT} = \frac{[R_1R_2C(OH)_2]}{X_{H_2O}[R_1R_2C{=}O]} \quad (1\text{–}31)$$

where the standard state is an infinitely dilute solution of carbonyl compound and its hydrate and $X_{H_2O}$ is the mole fraction of water, which can be less than 1.00 if mixed solvents are used. The standard free energy of hydration is defined as

$$\Delta G^\circ_{HYDRAT} = -RT \ln K_{eq}^{HYDRAT} \quad (1\text{–}32)$$

The more negative the value of $\Delta G^\circ_{HYDRAT}$, the more susceptible is the carbonyl carbon to addition. The use of water as a reference for addition to a carbonyl resembles the use of hydronium cation as the reference acid and water as the reference base in the definition of the acid dissociation constant. Values for the standard free energy of hydration of a large number of aldehydes and ketones are available (Table 1–2).[15,16,27] Electron-donating substituents, especially those capable of conjugation or hyperconjugation, decrease the susceptibility of the carbonyl to addition (compare formaldehyde, acetaldehyde, and acetone). Electron-withdrawing substituents increase its susceptibility (compare acetaldehyde to chloroacetaldehyde or acetone to chloroacetone). This is because of the ability of such substituents to stabilize or destabilize, respectively, the $\pi$ system of the unhydrated carbonyl group relative to the saturated *gem*-diol by either donating electron density to the $\pi$ system or withdrawing

**Table 1–2: Standard Free Energies of Hydration of Selected Aldehydes and Ketones at 25 °C[a]**

| carbonyl compound[b] | $\Delta G^{\circ}_{HYDRAT}$ (kJ mol$^{-1}$) | carbonyl compound[b] | $\Delta G^{\circ}_{HYDRAT}$ (kJ mol$^{-1}$) |
|---|---|---|---|
| $HC(=O)H$ | −19 | $HC(=O)CH_3$ | −1 |
| $H_3CC(=O)CH_3$ | +15 | $HC(=O)CH_2Cl$ | −9 |
| $H_3CC(=O)CH_2Cl$ | +1 | $ClCH_2C(=O)CH_2Cl$ | −6 |
| $H_3CC(=O)CH_2OCH_3$ | +10 | $HC(=O)CH_2OH$ | −5 |
| $H_3CC(=O)C(=O)OCH_3$ | −2 | $H_3CCH_2C(=O)C(=O)OCH_3$ | −1 |
| $H_3CCH_2C(=O)C(=O)OH$ | −1 | $H_3CCH(CH_3)C(=O)C(=O)OH$ | 0 |
| $H_3CC(=O)C(=O)O^-$ | +6 | $HC(=O)CH_2NH_3^+$ | −17 |
| $HC(=O)CH(OH)CH_2OH$ | −8 | $HC(=O)CH(OH)CH_2OPO_3H^-$ | −14 |
| $HC(=O)CH(OH)CH_2OPO_3^{2-}$ | −8 | $HC(=O)C_6H_5$ | +11 |

[a]Standard free energy of hydration[15,16,27] based on a standard state of the infinitely dilute solutions and an activity of pure water of 1.
[b]The table is arranged intentionally to permit various comparisons in both the horizontal and the vertical directions.

electron density from the $\pi$ system. An unconjugated hydroxyl group or methoxyl group is less electron-withdrawing than a conjugated carboxylic acid or ester (compare methoxyacetone with methyl pyruvate). Neighboring negative charge increases electron density in the carbon–oxygen double bond (compare pyruvate anion with the neutral pyruvate) and neighboring positive charge decreases electron density (compare ammonioacetaldehyde with acetaldehyde). A phosphate in its uncharged acidic form is strongly electron-withdrawing, but it becomes less so as it is titrated (compare glyceraldehyde with glyceraldehyde 3-phosphate monoanion and glyceraldehyde 3-phosphate dianion). Therefore, lowering the pH will increase the extent of hydration of a carbonyl near a phosphate in the same molecule because hydration itself neither consumes nor produces protons (Reaction 1–30) and cannot be affected by pH directly. The second methyl group on an $\alpha$-carbon both donates electrons and exerts a steric effect and these two opposite tendencies cancel each other (compare 2-oxobutyrate to 3-methyl-2-oxobutyrate).

In aqueous solution aldehydes and ketones are in equilibrium with their **hydrates**. Because the concentration of water is expressed in mole fraction in Equation 1–31, any carbonyl compound with a standard free energy of hydration less than 0 will be present mostly as the hydrate and any carbonyl compound with a standard free energy of hydration greater than 0 will be present mostly as the unhydrated carbonyl in aqueous solution. Recall that 5.7 kJ $mol^{-1}$ is equivalent to a factor of 10 in equilibrium constant. In all cases, this equilibrium has the effect of decreasing the concentration of unhydrated carbonyl, which is usually the reactive species.

The standard free energies for the addition of thiols to carbonyls are linearly correlated to the standard free energies of hydration.[15,16] This is not surprising since the formation of a thiohemiacetal (Reaction 1–26) is homologous to the formation of the hydrate (Reaction 1–30). Therefore, the relative susceptibility of different carbonyls to addition by a thiol can be estimated from the corresponding standard free energies of hydration.

### *Suggested Reading*

Hine, J., Cholod, M.D., & Chess, W.K. (1973) Kinetics of the Formation of Imines from Acetone and Primary Amines. Evidence for Internal Acid-Catalyzed Dehydration of Certain Intermediate Carbinolamines, *J. Am. Chem. Soc. 95*, 4270–4276.

# Acyl Exchange

An acyl exchange reaction resembles a reaction involving addition to a carbonyl. An example of such an acyl exchange reaction would be the aminolysis of a **thioester**

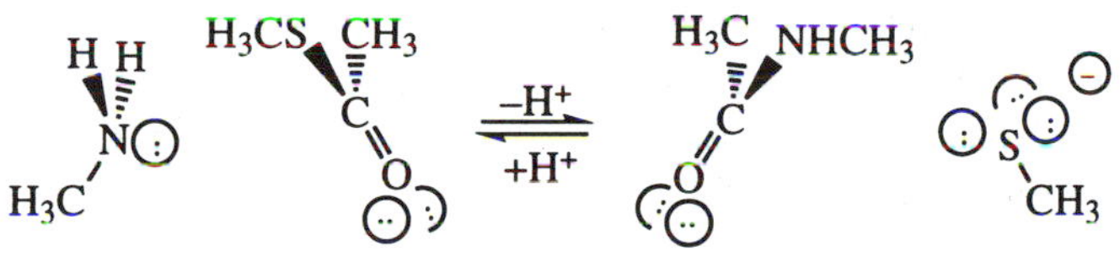

H H N H₃C   H₃CS CH₃ C O   $\underset{+H^+}{\overset{-H^+}{\rightleftharpoons}}$   H₃C NHCH₃ C O   S CH₃

(1–33)

in which the nitrogen exchanges with the sulfur to produce an **amide**.

The central intermediate in all acyl exchanges resembles the carbinolamine that lies between an aldehyde or a ketone and an imine (Figure 1–5). It can be referred to as a **tetravalent intermediate** and forms by the addition of the nucleophile to the initial acyl derivative, for example[28]

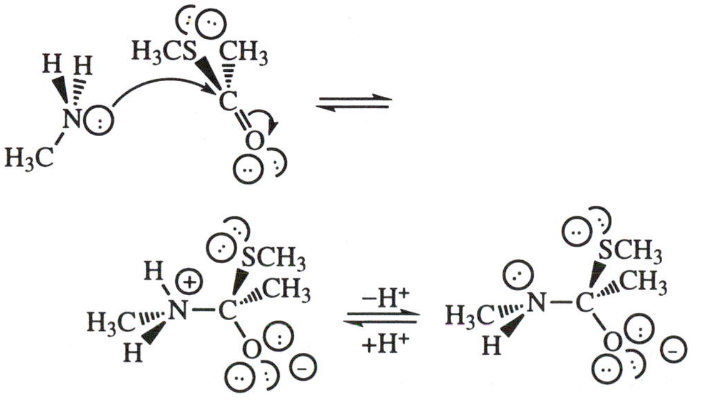

H H N H₃C   H₃CS CH₃ C O   $\rightleftharpoons$

H N⊕ H₃C H — C SCH₃ CH₃ O   $\underset{+H^+}{\overset{-H^+}{\rightleftharpoons}}$   H₃C N H — C SCH₃ CH₃ O

tetravalent intermediates

(1–34)

The addition of a nucleophile to an acyl derivative and the expulsion of a leaving group from a tetravalent intermediate (forward and reverse of Reaction 1–34, respectively) proceed through the overlap of the atomic orbital bearing the lone pair on the nucleophile with the antibonding $\pi$ molecular orbital of the carbon–oxygen double bond and the overlap of the atomic orbital providing the push from oxygen with the antibonding orbital of the $\sigma$ bond between carbon and the leaving group, respectively (Figure 1–3). All of the considerations of structure of the transition state, electron repulsion, stereochemistry of the attack, and energy barriers to the formation of the transition state pertinent to nucleophilic addition to carbonyls apply to acyl exchange reactions.

The difference between a tetravalent intermediate and a carbinolamine is that there is a third heteroatom attached to carbon in the former. The third heteroatom renders a tetravalent intermediate much less stable than the immediate product of an addition to a carbonyl group. This can be demonstrated by a comparison of equilibrium constants[15,29]

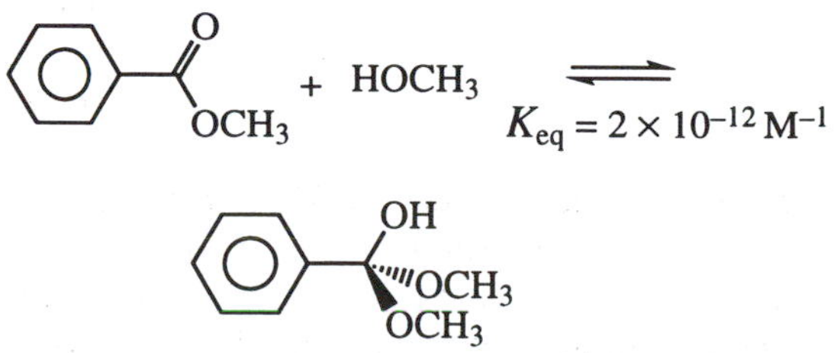

O OCH₃ + $HOCH_3$ $\rightleftharpoons$   $K_{eq} = 2 \times 10^{-12}\ M^{-1}$

OH OCH₃ OCH₃

(1–35)

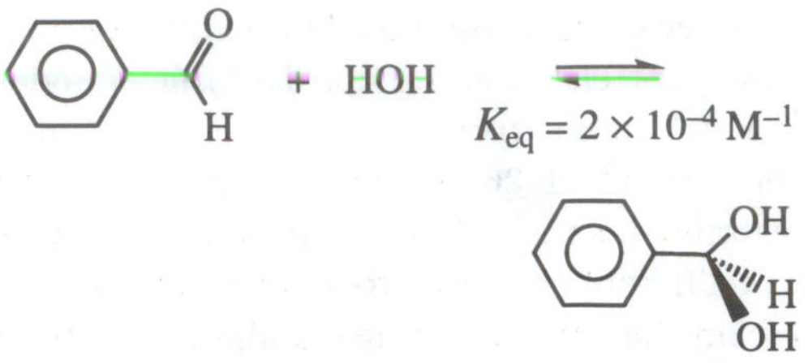

$$\text{PhCHO} + \text{HOH} \underset{K_{eq} = 2 \times 10^{-4}\,M^{-1}}{\rightleftharpoons} \text{PhCH(OH)}_2 \quad (1\text{–}36)$$

It is the instability of the tetravalent intermediate that causes its formation to be readily reversible (Reaction 1–34). This is advantageous because often the tetravalent intermediate must be formed and revert to reactants many times before the expulsion producing a product occurs. If the nucleophile that produces the tetravalent intermediate in the first place is the most likely leaving group, then the yield of product for every collision that produces a tetravalent intermediate will be low. Even with the rapid reversal, the reaction will proceed slowly. The additional heteroatom also creates a situation in which two lone pairs of electrons can be simultaneously periplanar to the leaving group and provide greater push

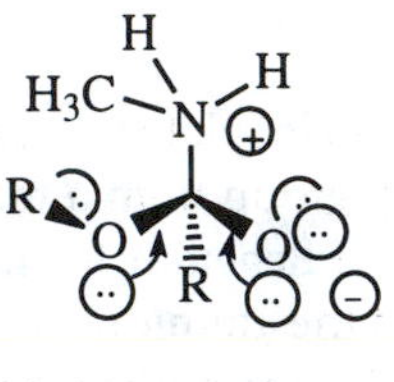

**1–17**

In every tetravalent intermediate this double push is possible.

Because there are three heteroatoms in a tetravalent intermediate and any one of the three can depart, there are always two possible products that can form in addition to the one formed by the departure regenerating the reactants. For example, when ethyl thionobenzoate is hydrolyzed, two products in addition to the reactant can form

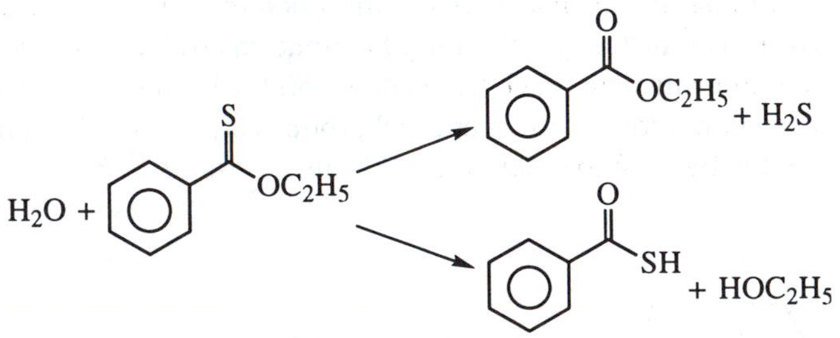

$$H_2O + \text{PhC(S)OC}_2H_5 \rightarrow \text{PhC(O)OC}_2H_5 + H_2S$$
$$\rightarrow \text{PhC(O)SH} + HOC_2H_5 \quad (1\text{–}37)$$

The ethyl ester predominates under weakly acidic conditions, but the fraction of thiolobenzoic acid increases significantly as the pH is lowered below 1.0.[30] That this is due to a shift in the preference for expulsion within the tetravalent intermediate

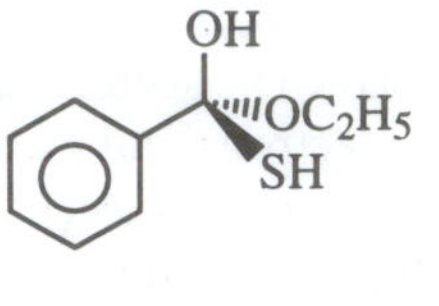

**1–18**

was demonstrated by generating an analogous tetravalent intermediate through a completely different route involving the decomposition of an ortho ester. The same increase in the yield of thiolobenzoate relative to benzoate was observed as the pH was lowered.[31] An explanation for the shift in the ratio of products would be the increased protonation of the alkoxy oxygen within tetravalent intermediate **1–18** at the lower pH, which would render it the better leaving group. This reaction provides an example of a protonation determining the identity of the leaving group.

Another example of the importance of protonation in determining the identity of the leaving group from a tetravalent intermediate occurs in the acyl exchange reaction between trifluoroethoxide anion and acetylimidazole (Figure 1–6). If the imidazole is unprotonated (right column), expulsion of trifluoroethoxide from the tetravalent intermediate predominates; if the imidazole is protonated (left column), its expulsion predominates.[32] This explains why either the protonated or the methylated[33] acetylimidazolium cation produces the trifluorethyl ester 200 times more rapidly than unprotonated acetylimidazole.[32]

The hydrolysis of ethyl *N*-methyl-*N*-phenylacetimidate[34] proceeds by the addition of hydroxide anion or water to the cationic conjugate acid of the imidate ester in a slow, rate-limiting step to produce the initial tetravalent intermediate **1–19**

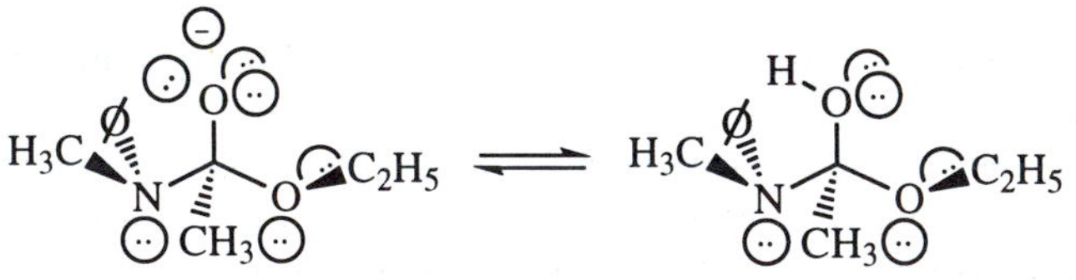

**1–21** ⇌ **1–19** 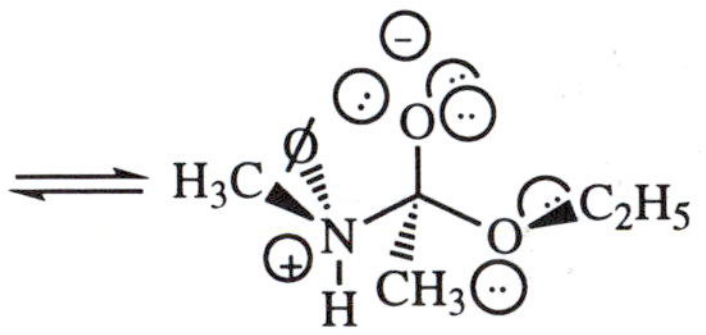 ⇌ **1–20**

This can tautomerize to tetravalent intermediate **1–20** and expel *N*-methylaniline or lose a proton to produce tetravalent intermediate **1–21** and expel the ethoxide. The partition between these two outcomes is influenced secondarily by acids and bases in the solution, but expulsion of the aniline always increases as the pH of the solution is lowered and protonation of the nitrogen is increased.

**Figure 1–6:** Effect of protonation of the leaving group on the partition in a tetravalent intermediate.[32] Addition of 2,2,2-trifluoroethoxide anion to *N*-acetylimidazole, in either its protonated or unprotonated form, produces a tetravalent intermediate from which either the trifluoroethoxide anion or the imidazole is a potential leaving group. When the imidazole is protonated, it is the preferred leaving group because its conjugate acid ($pK_a = 7.05$) is more acidic than the conjugate acid of the trifluoroethoxide anion ($pK_a = 12.4$). Consequently, imidazole departs rapidly from the tetravalent intermediate relative to the trifluoroethoxide anion (fast against moderate). When the imidazole is unprotonated and must leave as the imidazolate anion, the trifluoroethoxide anion is the preferred leaving group because its conjugate acid ($pK_a = 12.4$) is more acidic than the conjugate acid of the imidazolate anion ($pK_a = 14.5$). Because the major change in acidity has been to the imidazole, the major change in rate occurs in its expulsion. Upon deprotonation, the loss of imidazole becomes slower than the loss of trifluoroethoxide from the tetravalent intermediate (slow against moderate).

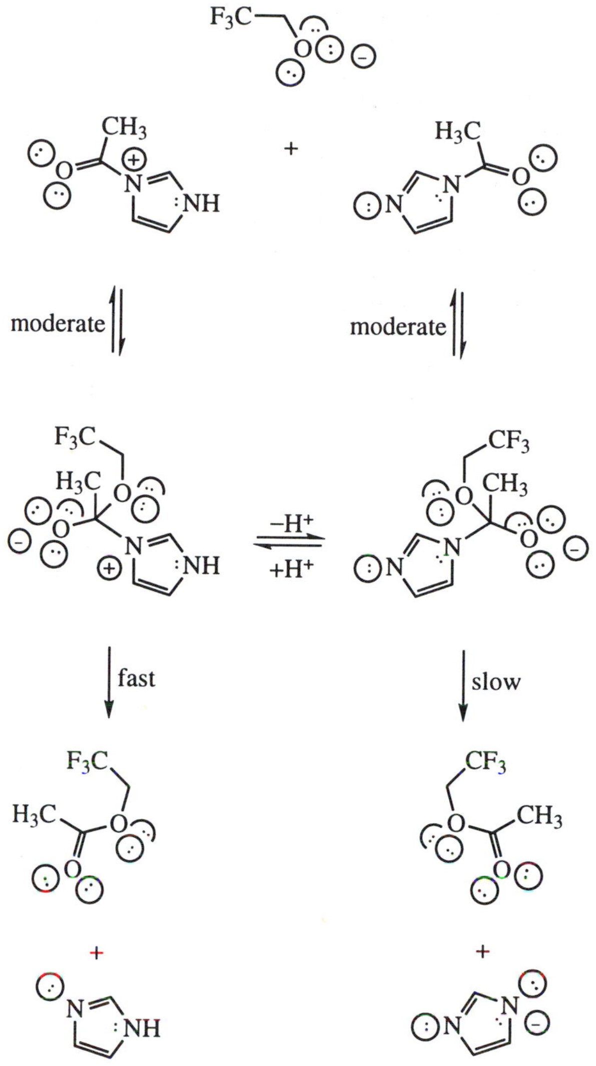

In addition to protonation, the magnitude of the push available in the tetravalent intermediate has significant influence on the choice of the leaving group. This can be appreciated by examining the product formed from the collapse of each of the following carbinolamines and tetravalent intermediates under conditions of acid catalysis[11]

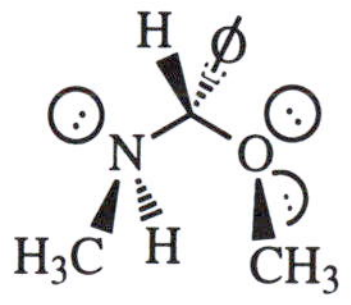

**1–22** 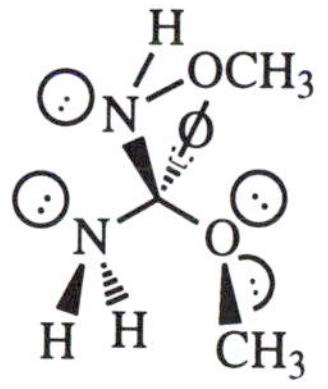 **1–23** 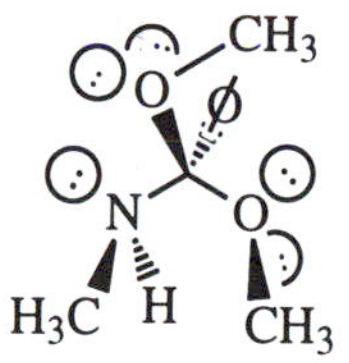 **1–24**

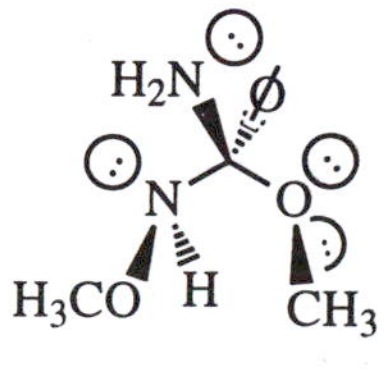

**1–25** 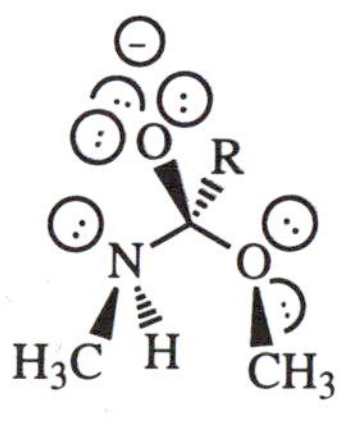 **1–26**

In each of these compounds, the two potential leaving groups have been placed to the right and left sides of the carbon, respectively. In each case the lone pair of electrons on the oxygen to the right is less basic than the lone pair of electrons on the nitrogen to the left, usually by about 10 units of $pK_a$. Under acid catalysis, the nitrogen will always be protonated more readily and should be the better leaving group.

The fact that the lone pair on nitrogen is the more basic, however, also means that it can provide more push. In the carbinolamine **1–22**, nothing other than one of the two potential leaving groups can provide push, so push becomes more important than protonation, and carbinolamines preferentially expel the alkoxy group under acid catalysis. In either the methoxyamino tetravalent intermediate **1–23** or the methoxy tetravalent intermediate **1–24**, the third heteroatom is able to provide only a weak push, but it is enough to enhance the expulsion of nitrogen sufficiently so that its rate of departure becomes equivalent to that of the oxygen under acid catalysis. In the amino tetravalent intermediate **1–25**, the push of the more basic lone pair of electrons on the amino nitrogen predominates over the lone pair on the leaving nitrogen and the departure of the methoxylamine, the more readily protonated leaving group, is several hundred times more rapid than that of methanol. Finally, in the alkoxy tetravalent intermediate **1–26**, push from the lone pair on

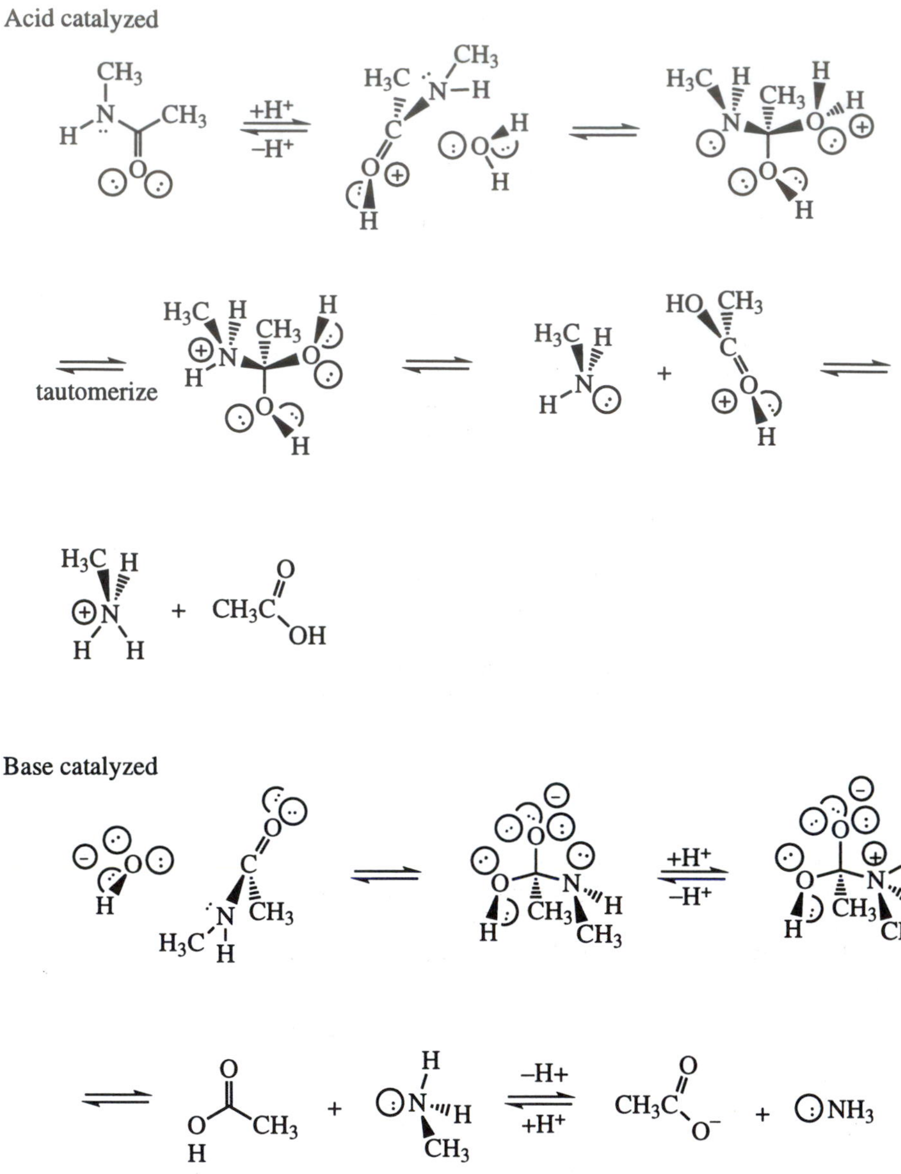

**Figure 1–7:** Mechanisms of hydrolysis of an amide under acid catalysis (top) or base catalysis (bottom). Under acid catalysis, a molecule of water adds to the amide, protonated on oxygen, to produce the tetravalent intermediate. Under base catalysis, a hydroxide anion adds to the neutral amide to produce the tetravalent intermediate. In each case, the nitrogen in the tetravalent intermediate must be protonated before it can leave.

Acid catalyzed

Base catalyzed

nitrogen becomes irrelevant in comparison to the push of the lone pairs of the alkoxide. In the presence of such an abundance, which is able to be directed at either leaving group promiscuously, protonation of the nitrogen, which is the more basic leaving group, dictates the preference, and alkoxy tetravalent intermediate **1–26** decomposes almost exclusively through expulsion of amine under acid catalysis.

Acyl exchange reactions can be catalyzed either by acids or by bases. A relevant example is the hydrolysis of amides, which takes place readily in the presence of strong acid or strong base (Figure 1–7). In strong acid, the acyl oxygen is protonated[35] in a regular acid–base reaction

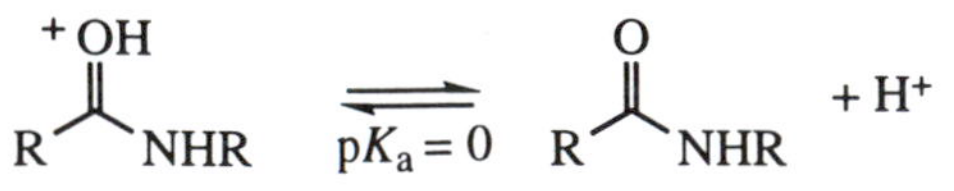

$$\mathrm{R{-}C({=}\overset{+}{O}H){-}NHR} \underset{pK_a = 0}{\rightleftharpoons} \mathrm{R{-}C({=}O){-}NHR} + \mathrm{H^+} \qquad (1\text{–}38)$$

This causes the acyl carbon to be electrophilic enough to react readily with water even though this remains the slowest step in the reaction. At values of pH where the acyl oxygen cannot be significantly protonated, the reaction of water as a nucleophile at the acyl carbon is extremely slow.[36] The initial tetravalent intermediate tautomerizes, and this tautomerization causes the nitrogen rather than the oxygen to be the leaving group. Before the tetravalent intermediate is formed, the lone pair on nitrogen is unavailable for protonation ($pK_a$ = –7.3),[35] but after the tetravalent intermediate has formed, the nitrogen has become the nitrogen of an amine rather than the nitrogen of an amide. It can be rapidly protonated. The tetravalent intermediate in strong acid resembles tetravalent intermediate **1–24**, and therefore, the expulsion of nitrogen and the expulsion of oxygen occur with about equal probability. Following its expulsion as the leaving group, the amine is protonated under the acidic conditions, and this protonation provides the free energy necessary to render the hydrolysis favorable.[36]

In strong base, hydroxide, as a strong nucleophile, readily adds to the acyl carbon in a rapid reaction.[36] This creates the alkoxide oxygen ready to push out the leaving group. The amino nitrogen created during this first step is only slowly protonated because the hydronium ion concentration is negligible, and the transfer of a proton from water is endothermic. Proton transfer to nitrogen is the slow step under basic conditions.[36] At neutral pH, proton transfer to an amino nitrogen occurs readily because of the much higher concentration of hydronium ion, but the concentration of hydroxide anion is too low and nucleophilic attack becomes the slow step in the reaction. Upon its protonation, the nitrogen leaves in preference to oxygen because the tetravalent intermediate resembles tetravalent intermediate **1–26**. The carboxylic acid produced loses its proton under the basic conditions, and it is this ionization that provides the free energy necessary to make the reaction favorable.[36] At neutral pH, both the amine and the carboxylic acid are ionized following the hydrolysis, and it is both of these ionizations together that provide sufficient favorable free energy to cause the hydrolysis to be isoergonic rather than endergonic.[36]

Acyl exchange reactions are often susceptible to **nucleophilic catalysis** as well as catalysis by acids and bases. For example, pyridine is a nucleophilic catalyst for acyl derivatives with good leaving groups. This acceleration of the reaction results from the fact that pyridine enters and leaves tetravalent intermediates readily, yet acylpyridinium cations are fairly unstable. Therefore, pyridine reacts rapidly with the initial acyl derivative to produce the acylpyridinium cation

$$R{-}C(=O){-}N^{+}C_5H_5$$

**1–27**

which then reacts rapidly with the nucleophile to produce the product and regenerate the catalyst.[37]

In an acyl exchange reaction there are always at least two steps: the combination of acyl derivative and nucleophile to create the tetravalent intermediate and the expulsion of the leaving group to create the products. Either of these steps can be the rate-limiting step. The **rate-limiting step** in the mechanism of a reaction is the last step in the sequence that exerts any influence on the overall rate.[38] By this definition, all of the steps that follow the rate-limiting step must be so fast that they occur immediately relative to the passage through the rate-limiting step. The rate-limiting step in a reaction can be distinguished from a rate-determining step. A **rate-determining step** in a reaction is any step the rate of which affects the rate of the overall reaction. In other words, if a step is rate-determining, an increase or decrease in its individual rate will cause a change in the overall rate of the reaction, but not necessarily of the same magnitude. As nucleophiles,[39] acyl derivatives, or catalysts[40] are varied within a given series of acyl exchange reactions, the major effect of each of these variations will usually be felt by only one step in the mechanism. It often happens that, at some point, the step that was rate-limiting becomes faster than either a step preceding it or a step following it and the new slower step then becomes the rate-limiting step. This is referred to as a **change in the rate-limiting step.** For example, an acyl exchange occurs between thiolates and *p*-nitrophenyl acetate[39]

$$RS^- + CH_3C(=O)OC_6H_4NO_2 \longrightarrow RSC(=O)CH_3 + HOC_6H_4NO_2 \qquad (1\text{–}39)$$

A kinetic mechanism, incorporating the tetravalent intermediate, can be written for this reaction

$$RS^- + CH_3C(=O)OC_6H_4NO_2 \underset{k_{-1}}{\overset{k_1}{\rightleftharpoons}} RSC(O^-)(CH_3)OC_6H_4NO_2 \xrightarrow{k_2} RSC(=O)CH_3 + {}^-OC_6H_4NO_2 \qquad (1\text{–}40)$$

When $k_{-1}$ is slow relative to $k_2$, every tetravalent intermediate formed proceeds to product. In this situation the step governed by $k_1$ is rate-limiting because the rate of the reaction depends entirely on the rate at which the tetravalent intermediate is formed. When $k_{-1}$ is fast relative to $k_2$, a preequilibrium governs the concentration of tetravalent intermediate and the rate of the reaction is limited by the rate at which tetravalent intermediate breaks down to produce product through the step governed by $k_2$. Under these circumstances, however, all three of the rates, those governed by $k_1$, $k_{-1}$, and $k_2$, are rate-determining because a change of any of these rate constants will affect the rate of the overall reaction even though only the step governed by $k_2$ is the rate-limiting step. When the thiolate used in the reaction described in Mechanism 1–40 becomes sufficiently acidic ($pK_a < 8$), the expulsion of the thiolate from the tetravalent intermediate becomes faster than the expulsion of nitrophenolate and the rate of the reaction changes from one in which the initial formation of the tetravalent intermediate is the rate-limiting step to one in which the expulsion of the nitrophenolate anion is the rate-limiting step.[39]

**Table 1–3: Standard Free Energies of Hydrolysis for Various Acyl Derivatives at 25 °C[a]**

| acyl derivative | $\Delta G^\circ_{HYDRO}$ (kJ mol$^{-1}$) | p$K_a$ of conjugate acid of the leaving group |
|---|---|---|
| $H_3C-C(=O)-O-C(=O)-CH_3$ | –66 | 4.7 |
| $H_3C-C(=O)-N$ (imidazole, N:) | –40 | 14.5 |
| $H_3C-C(=O)-OPO_3H^-$ | –34[b] | 7.2 |
| $H_3C-C(=O)-O-C_6H_5$ | –31 | 10.0 |
| $H_3C-C(=O)-OPO_3^{2-}$ | –28[c] | 12.4 |
| $H_3C-C(=O)-S-CH_2CH_3$ | –18 | 10.5 |
| $H_3C-C(=O)-OH$ | –10 | 15.7 |
| $H_3C-C(=O)-O-CH_2CH_3$ | –7 | 16 |
| $H_3C-C(=O)-NH-CH_2CH_3$ | +34 | 34 |

[a]Standard free energy of hydrolysis (Reaction 1–41) based on a standard state of 1 M concentrations of the uncharged reactants and products and an activity of pure water of 1.[42,43] [b]Based on a p$K_a$ of inorganic phosphate of 7.2 and of an acyl phosphate of 6.2. Value for the monoanions of acetyl phosphate and phosphate. [c]Value for the dianions of acetyl phosphate and phosphate.

It is often the case that the initial nucleophilic attack or the expulsion of the leaving group proceeds more rapidly when it occurs simultaneously with a particular proton transfer (Figures 1–4 and 1–5). It is usually assumed that proton transfers cannot be rate-limiting steps. At least one instance, however, of a simple proton transfer that is the rate-limiting step in an acyl exchange reaction has been reported.[41]

Biochemically relevant acyl derivatives can be arranged in order of their standard free energies of hydrolysis, $\Delta G^\circ_{HYDRO}$ (Table 1–3).[42] These are the changes in standard free energy for the reaction

$$H_3C-C(=O)-Y + H_2O \xrightleftharpoons{\Delta G^\circ_{HYDRO\,(Y)}} H_3C-C(=O)-OH + HY \qquad (1\text{–}41)$$

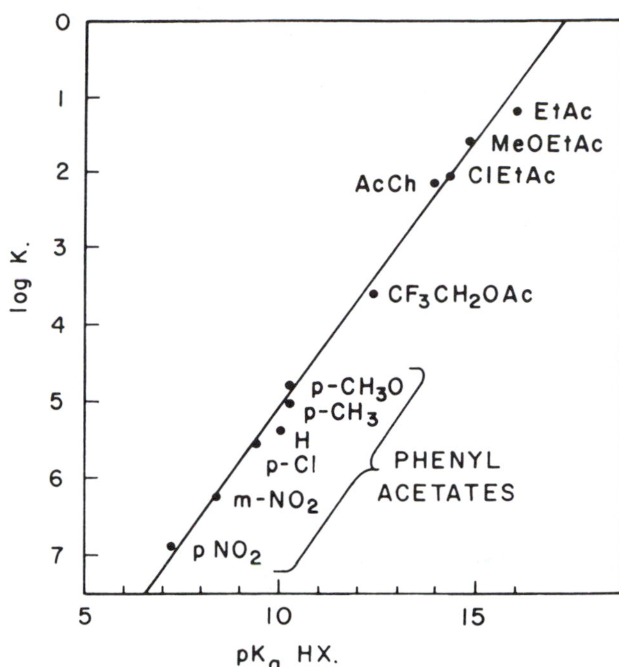


**Figure 1–8:** Comparison of the logarithms of the equilibrium constants, $K$, for the hydrolysis of the respective acetate esters of a series of alcohols and phenols with the p$K_a$ of the alcohol or phenol.[44] The values for the equilibrium constants were based on Reaction 1–41 for uncharged reactants, ester and water, and uncharged products, carboxylic acid and alcohol. The activity of water was taken to be 1.0. The line drawn through the points is defined by Equation 1–46, with a slope $\beta_{lg}$ of 0.70. Reprinted with permission from ref 44. Copyright 1964 American Chemical Society.

Water acts as a reference nucleophile, permitting the various acyl derivatives to be placed on the same scale. The change in standard free energy of any particular acyl exchange reaction

$$H_3C{-}C(=O)Y + HZ \xrightleftharpoons{\Delta G^\circ_{YZ}} H_3C{-}C(=O)Z + HY \quad (1\text{–}42)$$

can be estimated by taking differences between the two values for the two relevant standard free energies of hydrolysis

$$\Delta G^\circ_{YZ} = \Delta G^\circ_{HYDRO\,(Y)} - \Delta G^\circ_{HYDRO\,(Z)} \quad (1\text{–}43)$$

For example, the change in standard free energy for the aminolysis of an ester

$$H_3C{-}C(=O)OC_2H_5 + H_2NC_2H_5 \rightleftharpoons H_3C{-}C(=O)N(H)C_2H_5 + HOC_2H_5 \quad (1\text{–}44)$$

a reaction resembling the formation of a peptide bond during protein synthesis, should be –41 kJ mol$^{-1}$. When the acyl exchanges of imidates are being considered

$$H_3C{-}C(=NR)Y + HZ \rightleftharpoons H_3C{-}C(=NR)Z + HY \quad (1\text{–}45)$$

the same scale of standard free energies of hydrolysis can be used to estimate roughly the standard change in free energy for the reaction. Consistent with Equation 1–43, the acyl exchange reactions that are commonly observed are those that convert an acyl derivative with a more negative standard free energy of hydrolysis into one with a less negative standard free energy of hydrolysis.

As the leaving groups become weaker bases and their conjugate acids become stronger acids, the standard free energies of hydrolysis decrease (Table 1–3). The acidity of the parent acids, however, has much less of an effect, if any.[43] The departure of the leaving group from an acyl carbon resembles its departure from a proton. The reaction used as the reference (Reaction 1–41), however, formally involves the subsequent protonation of the leaving group. Therefore, the tabulated standard free energies of hydrolysis are actually correlated to the standard free energy for transfer of the leaving group from an acyl carbon to a proton. If a proton and an acyl carbon were electrophilically equivalent to each other, there would be no dependence of the standard free energy of hydrolysis on the p$K_a$ of the leaving group, and yet there is. For example, within a series of oxygen esters, the equilibrium constant for Reaction 1–41, $K_{eq}^{HYDRO}$, is correlated to the respective acid dissociation constants (Figure 1–8) by the equation

$$-\log K_{eq}^{HYDRO} = A + \beta_{lg}\, pK_a \quad (1\text{–}46)$$

where $\beta_{lg}$, the Brønsted coefficient for the leaving group, is 0.70.[44] For thioesters the Brønsted coefficient is 0.38,[39] and for amides it is 0.51.[43] These positive values indicate that the bond between an acyl carbon and the leaving group is more greatly affected by the basicity of the leaving group than the bond between a proton and the leaving group.

One of the significant differences between an acyl group and a proton is the $\pi$ molecular orbital system of the former. Consider an ester, in which the atom bound to the acyl carbon is an oxygen. As the leaving group becomes more basic, the electron density in the $\sigma$ bond between the acyl carbon and the acyl oxygen increases, just as the electron density in the oxygen–hydrogen bond of the alcohol increases, but so does the electron density in the lone pairs of electrons on the alkoxy oxygen

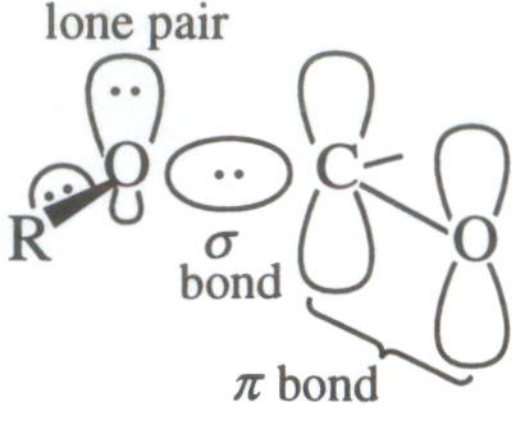


**1–28**

Either one of the lone pairs can overlap with the π molecular orbital system between the acyl carbon and an acyl oxygen, and this overlap increases the strength of the bond between the acyl carbon and the oxygen. This could explain why the free energies of hydrolysis for oxygen esters increase more rapidly than do the free energies for the acid dissociations of their leaving groups, and also why $\beta_{lg}$ for oxygen esters is more positive than $\beta_{lg}$ for thioesters.

### *Suggested Reading*

Gilbert, H.F., & Jencks, W.P. (1982) Mechanism of the Aminolysis of Alkyl Benzimidates, *J. Am. Chem. Soc. 104*, 6769–6779.

# Phosphoryl Exchange

Derivatives of phosphoric acid are susceptible to exchange of substituents on phosphorus just as derivatives of carboxylic acids are susceptible to exchange of substituents on carbon. An example of such a phosphoryl exchange would be

$H_3CO$–P(=O)(–O–$CH_2CH_2$–O–) + $H_2O$ ⇌ HO–P(=O)($OCH_3$)–O–$CH_2CH_2$–OH

(1–47)

in which the oxygen of water has exchanged with one of the two oxygens of ethylene glycol. The behavior of reactions of this type when the pH of the solution or the identity of the leaving groups is varied suggests that the reaction proceeds through at least one intermediate in at least two steps.[45] The intermediate normally invoked, in analogy to the tetravalent intermediate in acyl exchange reactions, is a **pentavalent intermediate.**[46]

Pentavalent phosphorus, unlike pentavalent carbon, can exist as a stable structure because phosphorus is an atom from the third row of the periodic table. This causes it to form longer bonds and also makes *d* atomic orbitals available to hybridize with its 3*s* atomic orbital and three 3*p* atomic orbitals. The most commonly encountered hybridization that produces five bonding atomic orbitals on phosphorus is an $sp^3d$ mixture that creates a **trigonal bipyramid** (Figure 1–9).[47] In a trigonal bipyramid, three of the substituents (designated

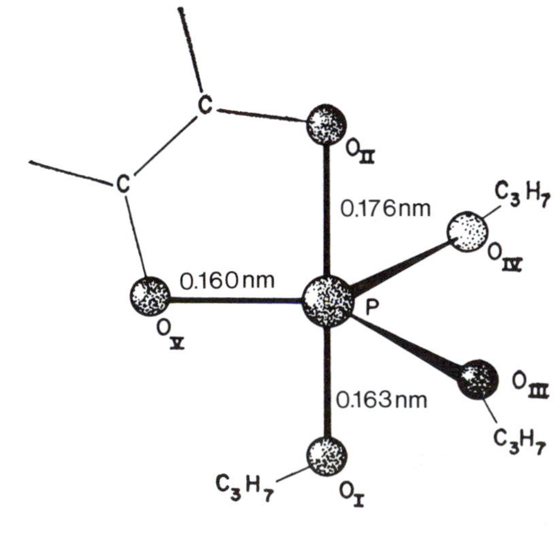


**Figure 1–9:** Arrangement of the five oxygen atoms around the pentavalent phosphorus in the crystallographic molecular model of the adduct between triisopropyl phosphite and 9,10-phenanthroquinone.[47] The individual oxygen atoms are designated as isopropyl oxygens ($C_3H_7$) or as vicinal oxygens from the 9,10-phenanthroquinone. The structure is a nearly perfect trigonal bipyramid with the oxygens of the five-membered ring occupying one equatorial and one apical position. Bond lengths are in nanometers. Reprinted with permission from ref 47. Copyright 1965 American Chemical Society.

**equatorial**) lie in a plane at 120° angles to each other, and two of the substituents (designated **apical**) are located above and below the plane, respectively, on a line passing through the phosphorus and normal to the plane.

In a phosphoryl exchange, the pentavalent intermediate is formed by attack of the nucleophile on the **electrophilic phosphorus** atom and decomposes by pushing out the leaving group

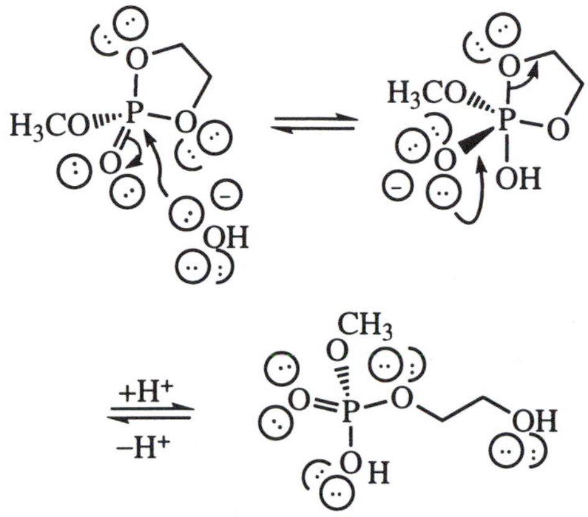

(1–48)

It is possible to study the stereochemistry of phosphoryl exchange reactions. For example, in a phosphoryl exchange reaction involving a simple monoester, the three oxygens can be labeled respectively with the three isotopes of oxygen, $^{16}O$, $^{17}O$, and $^{18}O$, and the absolute stereochemistry of reactant and product can be determined.[48,49] In all phosphoryl exchanges of simple monoesters

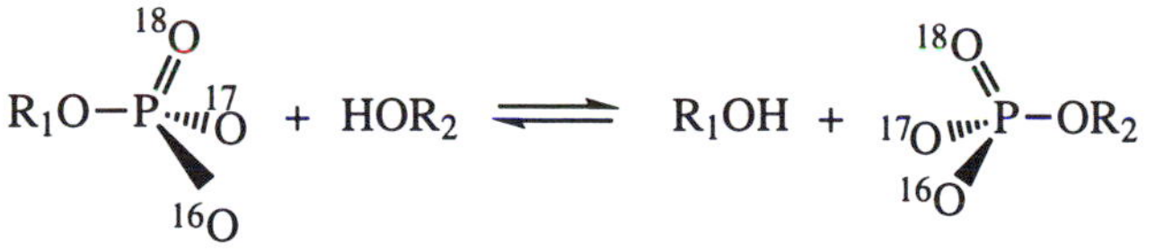

$$R_1O-P(^{18}O)(^{17}O)(^{16}O) + HOR_2 \rightleftharpoons R_1OH + (^{17}O)(^{16}O)(^{18}O)P-OR_2$$

(1–49)

the reaction proceeds with **inversion of configuration**.[48,50,51] This means that the nucleophile enters the pentavalent intermediate at an apical position and the leaving group leaves from the other apical position as shown in Reaction 1–48. If entry and exit were from equatorial positions (Figure 1–9), the reaction would have to proceed with **retention of configuration**. It is assumed that, in all phosphoryl exchange reactions involving a pentavalent intermediate, the rule of apical entry and apical exit applies. In order to accomplish apical entry, the lone pair of electrons of the nucleophile approaches the phosphorus atom in the cavity formed by three of its $\sigma$ bonds, much as the nucleophile approaches the electrophilic carbon in an $S_N2$ reaction (Reaction 1–14). For the same reason, namely, the electron repulsion of these three $\sigma$ bonds, the nucleophile has to approach on a trajectory bisecting the three $\sigma$ bond angles and in line with the fourth $\sigma$ bond to phosphorus.



The crystallographic results presented in Figure 1–9 are for the more relevant case of phosphorus surrounded by five oxygens, but the most commonly encountered pentavalent compounds of phosphorus are fluorides and chlorides. Phosphorus pentafluoride has the infrared spectrum of a trigonal bipyramid[52] but the fluorines all have an equivalent absorption in nuclear magnetic resonance. It has been proposed that the five fluorines remain apical and equatorial, respectively, on the time scale of infrared spectroscopy but equilibrate among the apical and equatorial positions of the trigonal bipyramid over the much longer time scale of nuclear magnetic resonance.[45]

The motion in a trigonal bipyramid that accomplishes this equilibration most efficiently is a **pseudorotation**.[45] One of the pseudorotations available to the trigonal bipyramid of Figure 1–9 would move $O_{III}$ back into the page to fall in line with $O_V$ and move $O_I$ and $O_{II}$ forward out of the page to form a trigonal plane with $O_{IV}$. In this pseudorotation, $O_I$ and $O_{II}$ would move from apical to equatorial positions and $O_{III}$ and $O_V$ from equatorial to apical positions.

The rate at which pseudorotation proceeds depends upon the identity of the five substituents around phosphorus. When all five are fluorine, it is rapid (>1000 $s^{-1}$). When three are fluorine and two are carbon, it is slower (<1000 $s^{-1}$), and the carbons assume two equatorial positions almost exclusively.[53] If, however, the two carbons and the phosphorus are in a five-membered ring, the pseudorotation becomes rapid (>1000 $s^{-1}$) when the other three substituents are fluorines. The reason for this is that the bond angle of the five-membered ring (108°) fits neither the angle of 90° between an apical and an equatorial position nor the angle of 120° between an equatorial and an equatorial position in the trigonal bipyramid but rather the transition state among the electronic isomers, and this strain accelerates the equilibration among them. When all of the atoms are chlorine, as in $PCl_5$, pseudorotation is very slow (<1 $min^{-1}$).[54] In the pentaoxyphosphorane of Figure 1–9, pseudorotation, at least within the crystal, must be completely inhibited because only one crystallographic structure was observed. The existence of this unique arrangement is probably due to the fact that the 9,10-phenanthrenediol, with its two trigonal carbons with bond angles of 120° and its two oxygens with bond angles of 109°, can occupy only an apical and an equatorial position, with a bond angle of 90°, to fit the 540° for the sum of the interior angles of a pentagon. Even so, the upper apical oxygen–phosphorus bond is quite long (Figure 1–9), a fact that might indicate some unrelieved strain. The rate of pseudorotation when all five substituents around phosphorus are oxygens, when none of them participates in a ring, and when the compound is in free solution would be the rate relevant to most phosphoryl exchanges in biochemistry. If pseudorotation under these circumstances is much slower than the lifetime of the pentavalent intermediate, then pseudorotation would be of no consequence.

In instances where pseudorotation cannot occur, the nucleophile would enter the pentavalent intermediate from an apical position, and the leaving group would be determined only by the direction from which the nucleophile attacked the phosphorus, because the leaving group is required to be the substituent that occupies the other apical position in the pentavalent intermediate. There would be two ways to control the identity of the leaving group. First, if the nucleophile were forced to enter at only one of the four available points of entry on phosphorus, the leaving group would always be the substituent opposite that point of entry. Second, if only one of the substituents surrounding the phosphorus in the reactant was a better leaving group than the nucleophile itself, then each pentavalent intermediate in which that substituent was not apical would decompose to regenerate reactants. The ability of a substituent to leave the pentavalent intermediate can be controlled, as in the situation of acyl exchange reactions (Reaction 1–37), by protonations or deprotonations.

The products from Reaction 1–48 can be both the one shown, which arises from cleavage of the ring, or the product of exocyclic cleavage

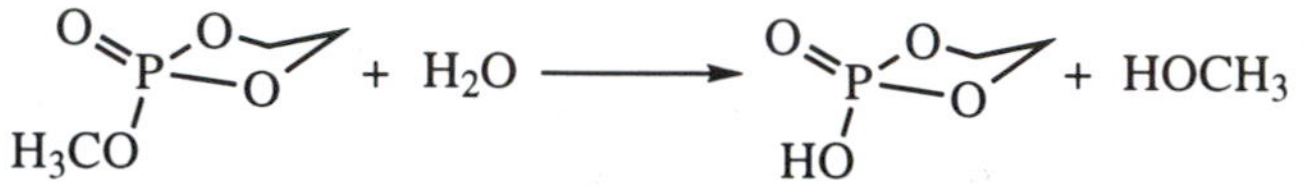

(1–50)

The ratio between these two products varies with pH, and it has been proposed that this variation can be explained if pseudorotation could occur in the pentavalent intermediate for this reaction.[55] Certainly, the saturated five-membered ring in the pentavalent intermediate of Reaction 1–48 could increase the rate of pseudorotation significantly. If pseudorotation is fast enough, this introduces a third step into the mechanism, namely, the pseudorotation itself. This additional step would provide an explanation for the rather complex changes with pH in the ratio between the two products, because as the pH was varied there could be changes in the rate-limiting step of the mechanism.

It has been shown that the direct transfer of phosphate between the two adjacent oxygens in a 1,2-diol

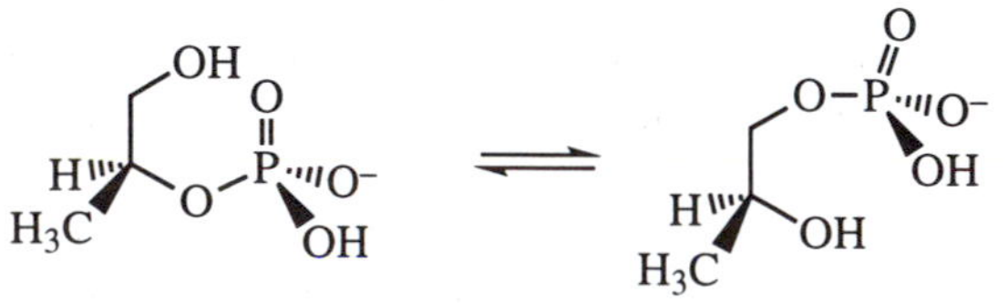

(1–51)

another reaction involving a five-membered ring in the pentavalent intermediate, proceeds with retention of configuration.[56] This has been explained by invoking pseudorotation, but it is hard to see how this reaction could proceed with inversion of configuration under any circumstances. More to the point, if entry into and exit from the pentavalent intermediate must always occur from apical positions and if Reaction 1–51 can occur without the intermediate formation of the cyclic phosphodiester

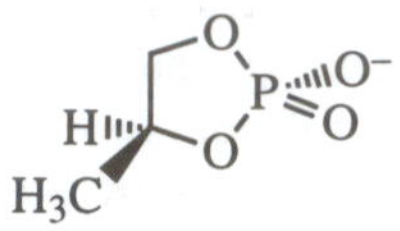

**1–29**

then it must involve a pseudorotation. This example illustrates the main advantage of pseudorotation, which is to permit one of the three equatorial substituents in the initial pentavalent intermediate to become the leaving group.

When the leaving groups in phosphoryl exchange reactions are significantly better than hydroxide or alkoxide, the character of the reaction changes. Examples of phosphoryl derivatives with good leaving groups are *N*-phosphonopyridine (**1–30**), *N*-phosphonoguanidine (**1–31**), and phenylphosphate (**1–32**)

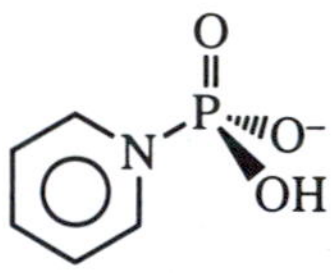

**1–30** 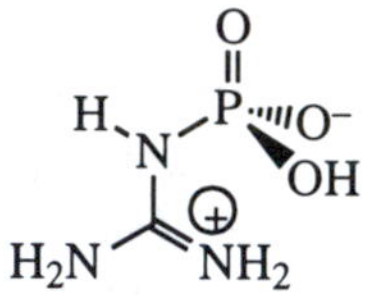 **1–31** 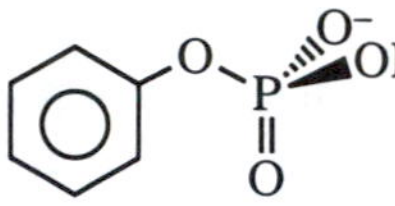 **1–32**

There are several features[57] of phosphoryl exchange reactions involving these types of leaving groups that were at one time explained[58] by postulating a unimolecular loss of the leaving group, in a reaction analogous to an $S_N1$ nucleophilic substitution. This dissociation would produce **monomeric metaphosphate**[59–61]

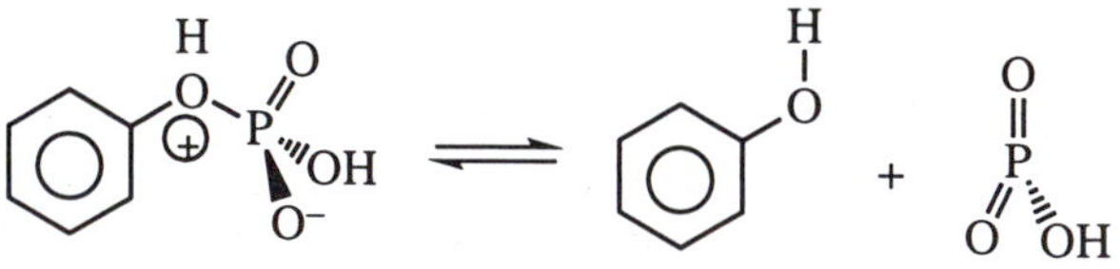

(1–52)

This compound is analogous to an oxocarbocation, and it would react immediately with nucleophiles in the solution. The unequivocal test for an $S_N1$ mechanism is that it proceeds with racemization at the central carbon, yet it has been demonstrated that phosphoryl exchanges involving derivatives with good leaving groups nevertheless proceed with inversion of configuration,[57] a fact that rules out the involvement of free monomeric metaphosphate.

Unlike phosphoryl exchanges involving leaving groups such as alkoxide and hydroxide ions, however, phosphoryl exchanges involving good leaving groups do not proceed through a pentavalent intermediate but seem to involve only one transition state.[62] In phosphoryl exchanges between pyridines (Structure **1–30**), the transition state seems to be one in which most of the bond (80%) between leaving group and phosphorus has been broken at the same time that only a small fraction (20%) of the bond between nucleophile and phosphorus has been formed.[62,63] As such, the transition state can be considered as a monomeric metaphosphate sandwiched between the lone pairs of electrons of the two pyridines

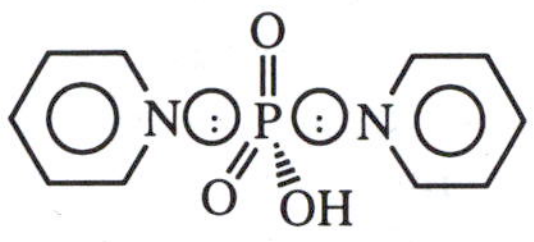

**1–33**

but not as free monomeric metaphosphate.

Many biochemical phosphoryl exchange reactions involve good leaving groups. The mechanisms of these reactions may also lack pentavalent intermediates and involve single transition states such as that in the case of phosphoryl exchange between pyridines. For example, the transfer of a phosphoryl group between ATP and acetate is a transfer between two excellent leaving groups, and the transition state in this reaction might be represented as

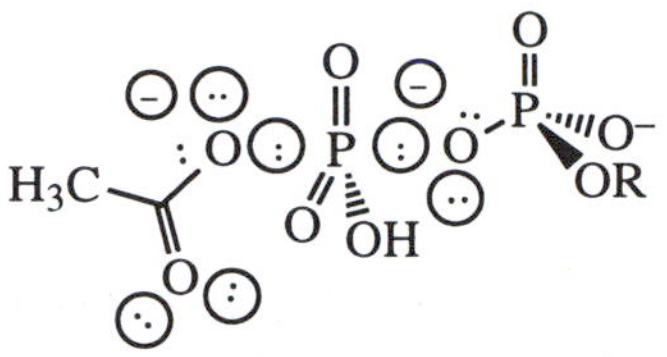

**1–34**

In a transition state of this type, only inversion of configuration at phosphorus can occur because there is no time for pseudorotation and the nucleophile must enter at a point opposite the location of the leaving group, if monomeric metaphosphate adequately represents the central structure in this transition state.

### *Suggested Reading*

Kluger, R., Covitz, F., Dennis, E., Williams, L.D., & Westheimer, F.H. (1969) pH-Product and pH-Rate Profiles for the Hydrolysis of Methyl Ethylene Phosphate. Rate-Limiting Pseudorotation, *J. Am. Chem. Soc. 91*, 6066–6072.

# Formal Hydride Transfer

Each of the functional groups in a metabolite serves as a representative of one of the five **oxidation levels of carbon**. When carbon is surrounded only by other carbons and hydrogens and not participating in any $\pi$ bonds, it is reduced and fully saturated. If one of the hydrogens on carbon is replaced by a hydroxyl group, an oxidation of that carbon has occurred

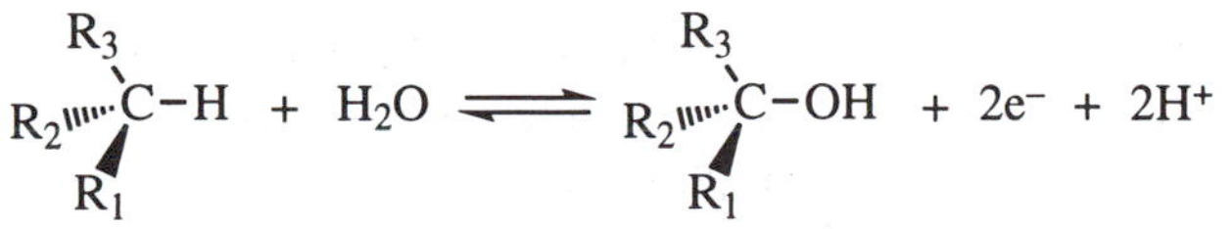

$$R_1R_2R_3C{-}H + H_2O \rightleftharpoons R_1R_2R_3C{-}OH + 2e^- + 2H^+ \quad (1\text{–}53)$$

and it is now at the **alcoholic level of oxidation**. Alcohols can be turned into ethers, thiols, thioethers, and amines without any net oxidation or reduction. If one of the carbons next to a hydroxyl group bears a hydrogen, an alcohol can also be converted into an alkene without any net oxidation or reduction

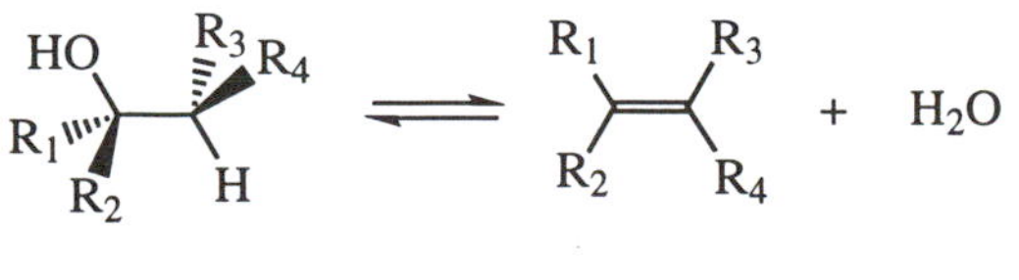

$$R_1R_2C(OH){-}CHR_3R_4 \rightleftharpoons R_1R_2C{=}CR_3R_4 + H_2O \quad (1\text{–}54)$$

An alcohol can be **oxidized** to a ketone or an aldehyde, or a ketone or an aldehyde can be **reduced** to an alcohol

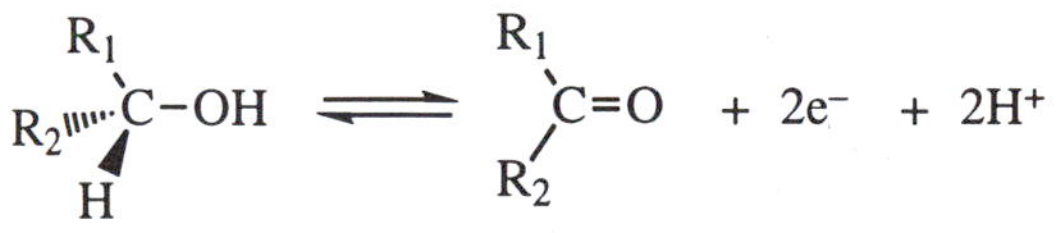

$$R_1R_2HC{-}OH \rightleftharpoons R_1R_2C{=}O + 2e^- + 2H^+ \quad (1\text{–}55)$$

The carbon in a ketone or an aldehyde is at the **carbonyl level of oxidation** and can be converted into a hemiacetal, a hemiketal, an acetal, a ketal, a thiohemiacetal, a thiohemiketal, a carbinolamine, or an imine (Table 1–1) without any net oxidation or reduction. A carbonyl carbon can be identified in a molecule by the fact that it has two bonds, either two $\sigma$ bonds or a $\sigma$ bond and a $\pi$ bond, to the heteroatom or heteroatoms oxygen, nitrogen, or sulfur and two other bonds to either carbon or hydrogen.

An aldehyde or the derivative of an aldehyde can be oxidized to an acyl derivative, or an acyl derivative can be reduced to an aldehyde or the derivative of an aldehyde

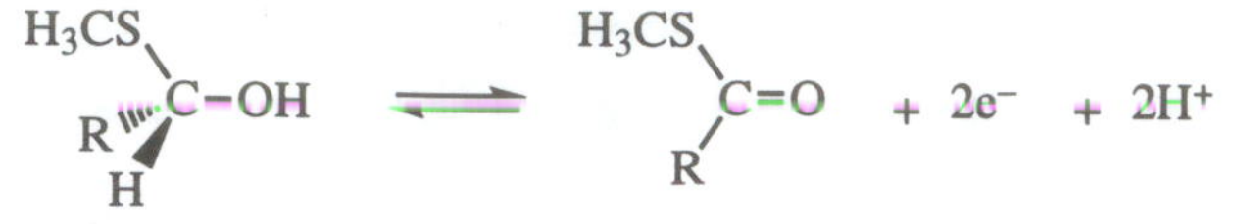

(1–56)

Acyl exchange reactions among acyl derivatives occur without any net oxidation or reduction. A carbon at the **acyl level of oxidation** can be identified in a molecule as a carbon that has three bonds, either $\sigma$ bonds or $\pi$ bonds, to heteroatoms and only one $\sigma$ bond to carbon or hydrogen. Extreme examples of acyl carbons are the carbon in a nitrile (**1–35**) or the carbon in an orthoester (**1–36**)

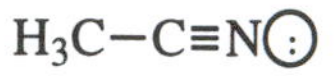

**1–35** 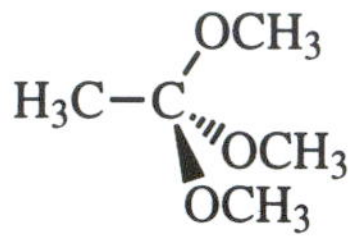 **1–36**

All acyl carbons can be converted into all of the other types of acyl carbons in reactions that do not involve net oxidation or reduction.

A formate can be oxidized to a carbonate, or a carbonate can be reduced to a formate; for example

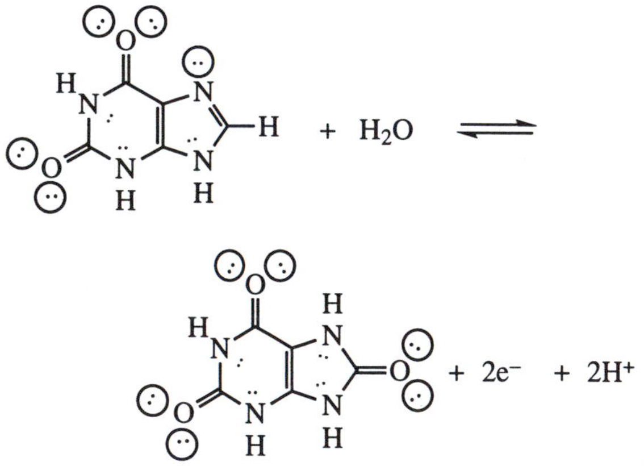

(1–57)

Acyl exchange reactions also occur among carbons at the **carbonate level of oxidation** without any net oxidation or reduction. A carbonate carbon can be identified in a molecule as a carbon that is entirely bonded to heteroatoms.

In enzymatic reactions, oxidations and reductions connecting the alcoholic level with the carbonyl level of oxidation and the carbonyl level with the acyl level of oxidation usually involve the formal transfer of a **hydride anion** to the nicotinamide ring in oxidized nicotinamide adenine dinucleotide ($NAD^+$) or the formal transfer of a hydride anion from reduced nicotinamide adenine dinucleotide (NADH), respectively

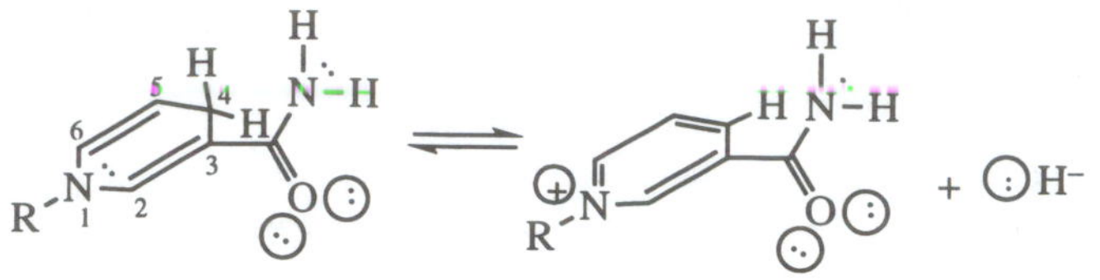

NADH NAD$^+$

(1–58)

Upon reduction of $NAD^+$ by the addition of the elements of a hydride ion to carbon 4, the aromatic $\pi$ molecular orbital system is broken because carbon 4 becomes saturated and can no longer support a $p$ orbital. The compensation for the loss of the aromaticity during the reaction is the formation of an extended $\pi$ molecular orbital system that includes the exocyclic amide, the two carbon–carbon double bonds, and the nitrogen within the ring

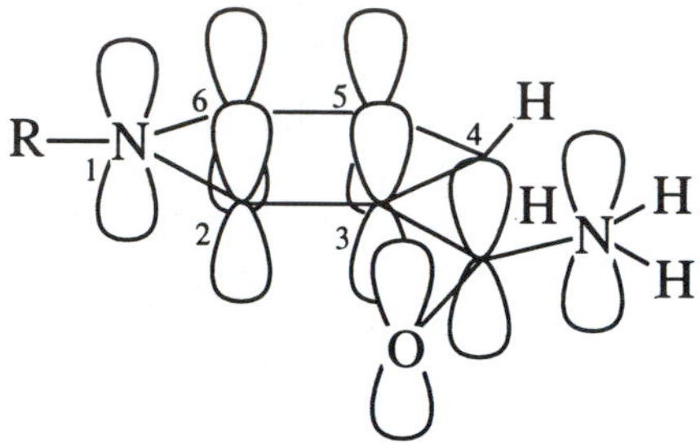

**1–37**

This provides an unbroken set of 8 $p$ orbitals that mix to form a $\pi$ molecular orbital system containing 10 $\pi$ electrons. This $\pi$ molecular orbital system explains the fact that the six atoms in the ring, the three atoms in the amide, and the carbon of the functional group attached to the nitrogen in the ring are in the same plane.[64–66] The bond lengths, however, reflect the uneven distribution of double-bond character over the 7 bonds in this molecular orbital system. If the lengths of the bonds in reduced nicotinamide are compared to the lengths of the same bonds in oxidized nicotinamide,[67] the bonds between carbons 2 and 3 and between carbons 5 and 6 each shorten by 7 pm to a length equal to the bond length of a normal olefin,[65] while the bonds between nitrogen 1 and carbon 2 and between nitrogen 1 and carbon 6 each lengthen by 7 pm.[65,66] These changes demonstrate that the carbon–carbon bonds become more olefinic while the carbon–nitrogen bonds lose double-bond character upon reduction of nicotinamide. This is consistent with the fact that resonance structures placing double bonds between carbon and nitrogen are unfavorable because of separation of charge and cross-conjugation

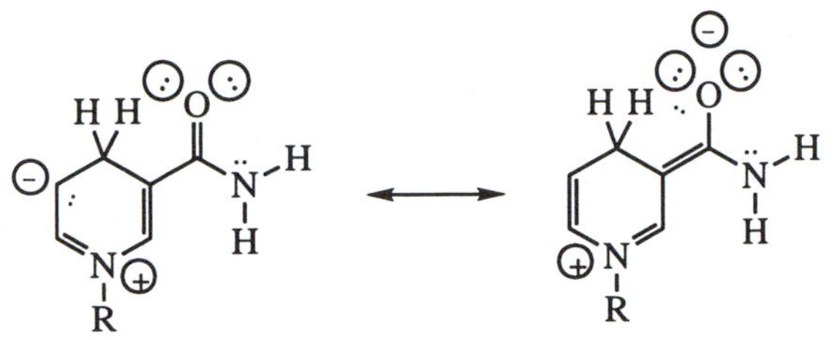

**1–38**

It is the extended π molecular orbital system that is responsible for the absorbance displayed by NADH or by its phosphorylated derivative, NADPH, at 340 nm.

There are several facts known about the enzymatically catalyzed oxidation–reduction reactions involving $NAD^+$ and NADH. The substrate must have at least one basic lone pair of electrons on an oxygen or a nitrogen attached to the carbon to be reduced. The lone pair of electrons provides push to the leaving hydride. These must be lone pairs that can be positioned parallel to the σ bond between carbon and hydrogen

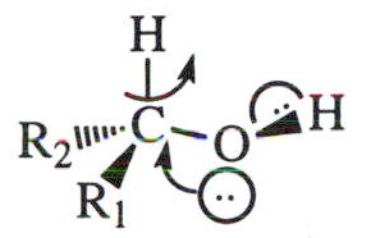

**1–39** 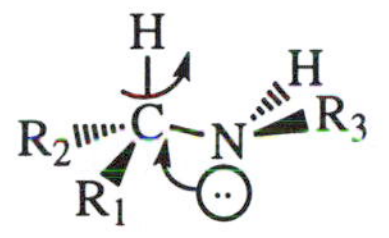 **1–40**

It is also important to have a proton on the heteroatom before the reduction takes place so that, as the hydride leaves, the proton can be simultaneously removed so as to increase the basicity of the lone pair that is providing the push. In aldehydes, a similar arrangement is achieved by forming hemiacetals, hemithioacetals, or carbinolamines

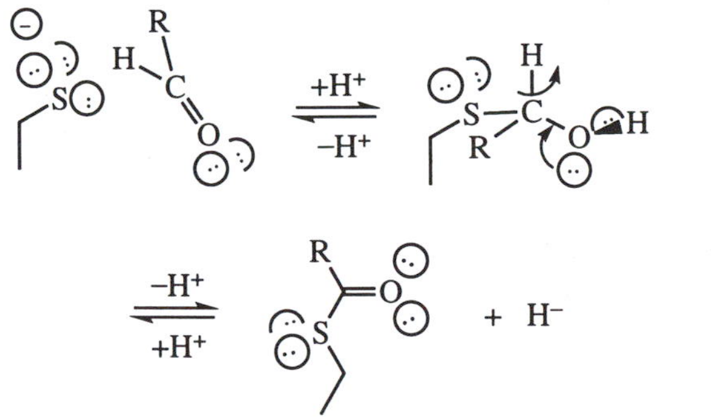

(1–59)

In these enzymatic reactions, the nicotinamide is positioned so that its carbon 4 is immediately adjacent to the location from which the hydrogen is removed or to which the hydrogen is to be added

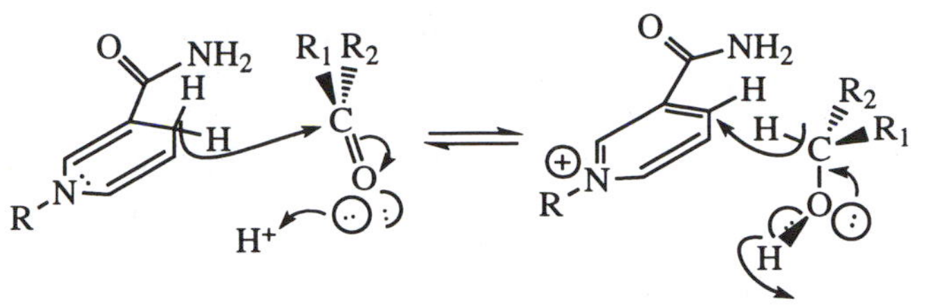

(1–60)

Perhaps the most graphic demonstration of this fact is that a derivative of $NAD^+$, (3*S*)-5-(3-carboxy-3-hydroxypropyl)-nicotinamide adenine dinucleotide

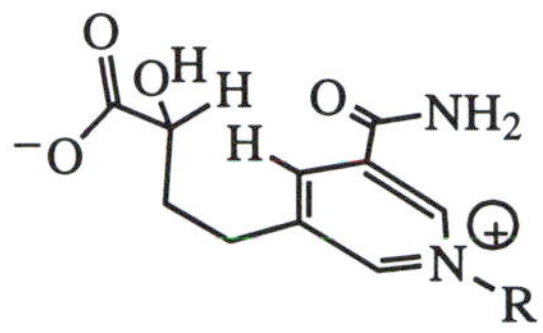

**1–41**

which is a substrate for the enzyme lactate dehydrogenase[68]

$$CH_3CHOHCOO^- + NAD^+ \rightleftharpoons CH_3COCOO^- + NADH + H^+ \quad (1\text{–}61)$$

fits snugly into the active site of the enzyme with the hydrogen that is removed during the enzymatic reaction immediately adjacent to carbon 4 of the nicotinamide ring.[69] This fact demonstrates that, in the normal reaction (Reaction 1–61), the hydrogen on the carbon oxidized by the nicotinamide is immediately adjacent to carbon 4 of the oxidant.

It is unknown whether the hydride is transferred in these intimate complexes in one step or several steps. In nonaqueous solvents, which presumably mimic the interior of a molecule of protein, the electrochemical reduction of 1-methylnicotinamide occurs in a stepped process[70]

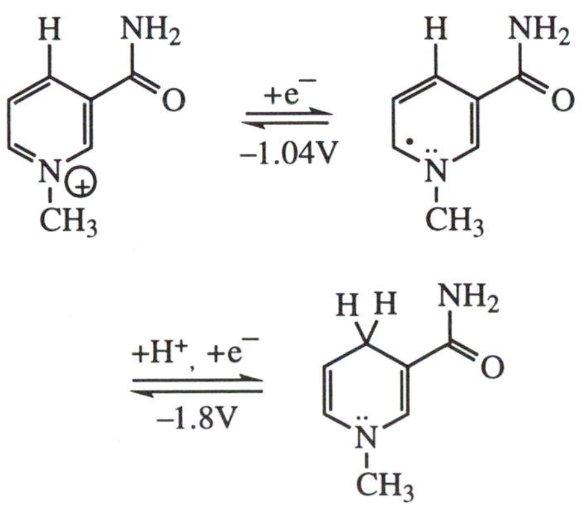

(1–62)

that involves initial electron transfer to cationic $NAD^+$ to form a neutral radical, followed by simultaneous transfer of the second electron and a proton to the radical. The reduction potentials for the two steps, however, are quite negative (–1.04 and

−1.8 V, respectively), and fairly strong one-electron reductants would be required. This seems to rule out an alcohol. In contrast, the reduction potential for $NAD^+$ in a two-electron process is +0.32 V at pH 7 in aqueous solution. In an analogous reaction, dihydropteridines can be oxidized by ferrocyanide in two consecutive one-electron processes in a 20% aqueous solution of acetonitrile.[71] In the first step, one electron is removed to produce the radical cation

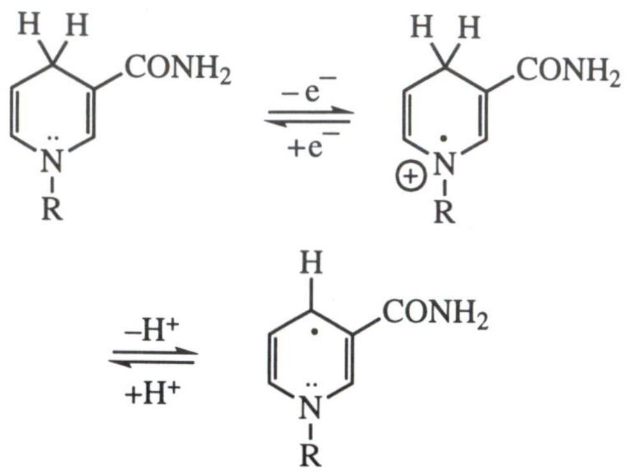

(1–63)

that is subsequently unprotonated at carbon in a second step to give the neutral radical. The initial electron removal by ferrocyanide is rapidly reversible if sufficient ferricyanide is present in the solution. The second step displays a deuterium kinetic isotope effect consistent with rate-limiting proton removal. The neutral radical is then further oxidized in a one-electron step that is rapid relative to the removal of the proton from carbon in the radical cation. Both of these reactions (Mechanisms 1–62 and 1–63) demonstrate the ability of 1-substituted 3-nicotinamides to participate in one-electron processes under appropriate circumstances. An electron–proton–electron mechanism such as that shown in Mechanism 1–62 or 1–63, however, has been ruled out in the nonenzymatic reduction of 4-cinnamoylpyridines by 4,4-dideuterio-1,4-dihydropyridines because this reaction involves transfer of the deuterium directly from the dihydropyridine to the substrate.[72] Likewise, the stable semiquinone radical of flavin does not react so readily (>$10^5$ times slower) in a one-electron reaction with NADH as does fully oxidized flavin in a two-electron, hydride transfer.[73] Flavin is the most logical of the naturally occurring participants in oxidation-reduction reactions with $NAD^+$–NADH to participate in one-electron oxidation-reduction. Therefore, evidence has been presented both for and against a stepped electron-proton-electron transfer and for hydride transfer in reactions involving nicotinamide adenine dinucleotides.

### Suggested Reading

Powell, M.F., Wu, J.C., & Bruice, T.C. (1984) Ferricyanide Oxidation of Dihydropyridines and Analogues, *J. Am. Chem. Soc. 106*, 3850–3856.

# One-Carbon Chemistry

Formaldehyde and formic acid are one-carbon compounds involved in metabolism. These reactive molecules are rarely found free in solution but are, like the hydride ion, associated with a larger carrier of single carbons, **tetrahydrofolate**

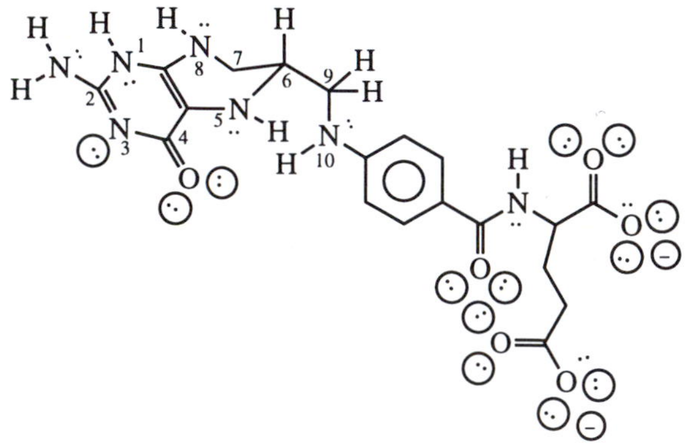

**1–42**

The planar pyrimidine moiety of tetrahydrofolate (the ring containing atoms 1–4) is analogous to the heterocyclic bases of the nucleic acids; and, as with the nucleic acid bases, the three nitrogens exocyclic to the pyrimidine, attached to carbon 2 and at positions 5 and 8, are planar nitrogens[74] with $\pi$ lone pairs of electrons. The nitrogen at position 10 is analogous to the nitrogen in aniline. The one-carbon fragments are carried by nitrogen 5 or nitrogen 10 or by both of them simultaneously. As both of these active nitrogens are adjacent to cyclic $\pi$ molecular orbital systems and can be considered to be nearly planar aniline nitrogens, the active portion of tetrahydrofolate can be abbreviated as

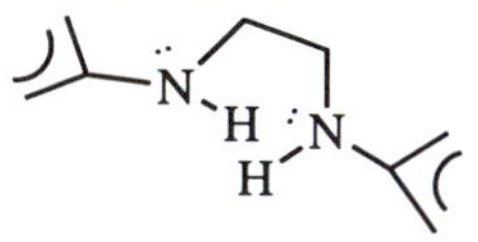

**1–43**

Formaldehyde can be carried as the iminium cation at one of the two aniline nitrogens

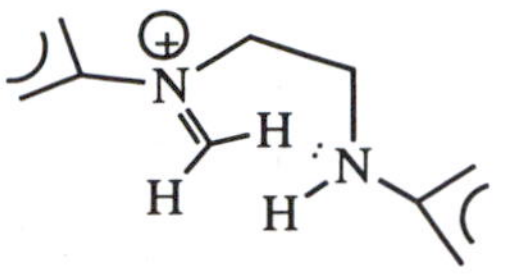

**1–44**

or, more stably,[75] as the *gem*-diamine in $N^5$,$N^{10}$-methylenetetrahydrofolate

carbon of the *gem*-diamine

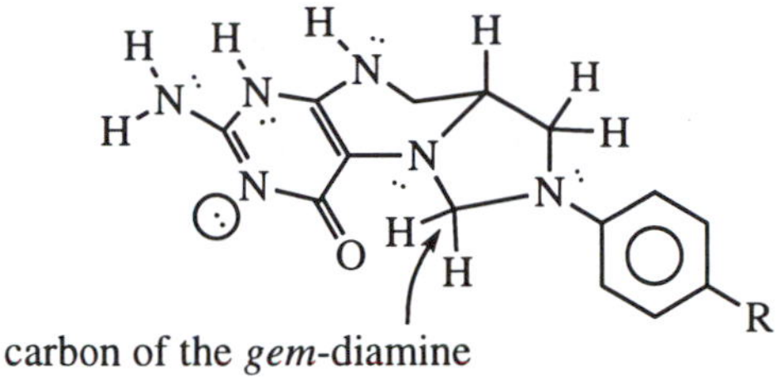

**1–45**

Formaldehyde itself rarely exists in a biochemical situation as the free molecule in solution. Presumably, this is because it reacts with proteins and nucleic acids to cross-link them among themselves in as yet poorly understood reactions. Nevertheless, its direct reaction with tetrahydrofolate to form $N^5$,$N^{10}$-methylenetetrahydrofolate may occur under certain circumstances. This process has been studied in a simpler system. In the reaction between formaldehyde and a model compound for tetrahydrofolate[75]

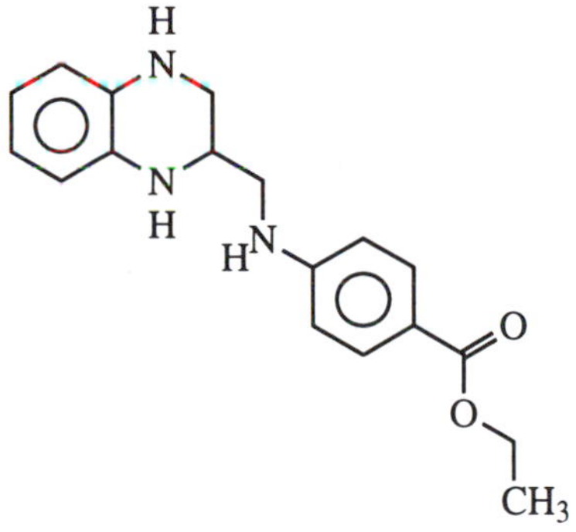

**1–46**

the iminium cation of one of the two nitrogens is formed in a rapid equilibrium by a reaction involving a carbinolamine that dehydrates under acid catalysis (Figure 1–5). The *gem*-diamine then forms from the iminium cation by direct nucleophilic addition

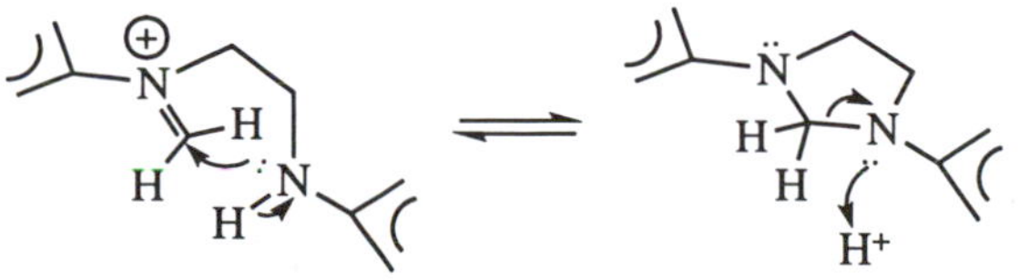

(1–64)

Because the nucleophile is an aniline nitrogen, this step can be catalyzed by bases, which simultaneously remove the proton from the nitrogen making it more nucleophilic.[75] In the *gem*-diamine, the two nitrogens remain conjugated to the respectively adjacent $\pi$ molecular orbital systems of the rings.

Formic acid is carried as the formanilide at either nitrogen

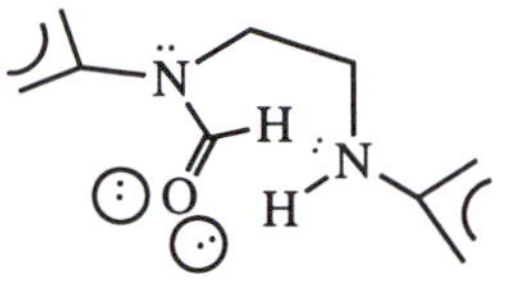

**1–47**

as in $N^5$-formyltetrahydrofolate or $N^{10}$-formyltetrahydrofolate, or as the formamidinium cation between the two nitrogens, as in $N^5$,$N^{10}$-methenyltetrahydrofolate

carbon of the formamidinium cation

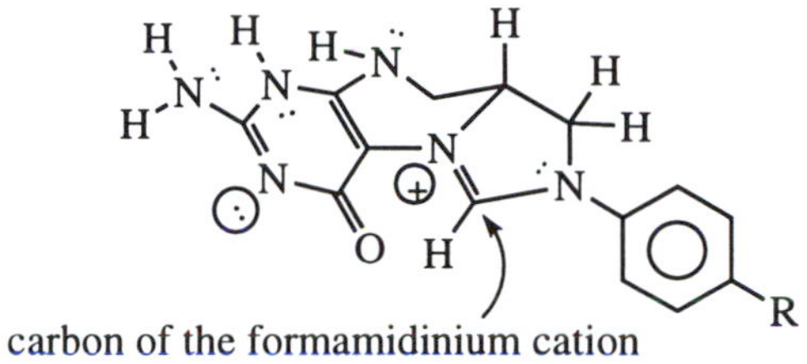

**1–48**

The formamidinium cation is formed from $N^5$-formyltetrahydrofolate by an acyl exchange reaction (Figure 1–10) in which a base simultaneously removes the proton from the attacking aniline nitrogen to render it a better nucleophile.[76] In the product, $N^5$,$N^{10}$-methenyltetrahydrofolate (Figure 1–11), both the *p*-benzoyl ring attached to nitrogen 10 and the pyrimidine ring attached to nitrogen 5 are rotated away from conjugation with the lone pairs on these nitrogens because of the strong conjugation within the planar amidine.[74]

Most of the reactions involving the adduct of tetrahydrofolate with either formaldehyde or formic acid have mechanisms that avoid the formation of free formaldehyde or free formic acid and that transfer these fragments with the minimum number of steps. In the case of $N^5$,$N^{10}$-methylenetetrahydrofolate, the reactions proceed through the iminium cation (reverse of Mechanism 1–64) that can transfer the single carbon to another nucleophile directly. For example, in the reaction catalyzed by serine transhydroxymethylase

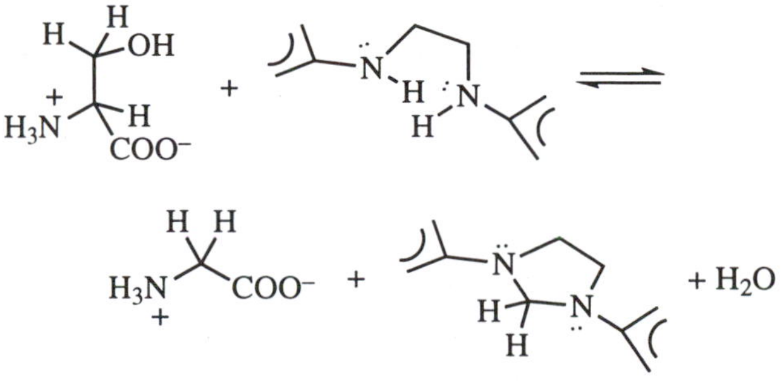

$$+ H_2O$$

(1–65)

**Figure 1–10:** Transformation of $N^5$-formyltetrahydrofolate to $N^5$,$N^{10}$-methenyltetrahydrofolate. Formation of the tetravalent intermediate occurs with catalysis by base and the oxygen of the tetravalent intermediate must be protonated before it can leave.

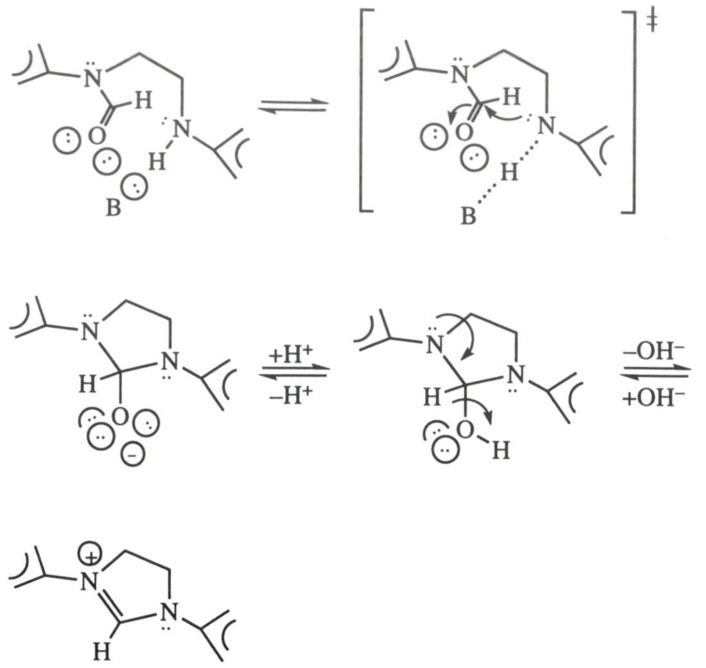

the single carbonyl carbon is transferred directly between the carbon of glycine and nitrogen 5 of the tetrahydrofolate.[77]

Formate is often transferred from $N^5$,$N^{10}$-methenyltetrahydrofolate (**1–48**) to yet a third amine, as in the reaction catalyzed by phosphoribosylglycinamide formyltransferase

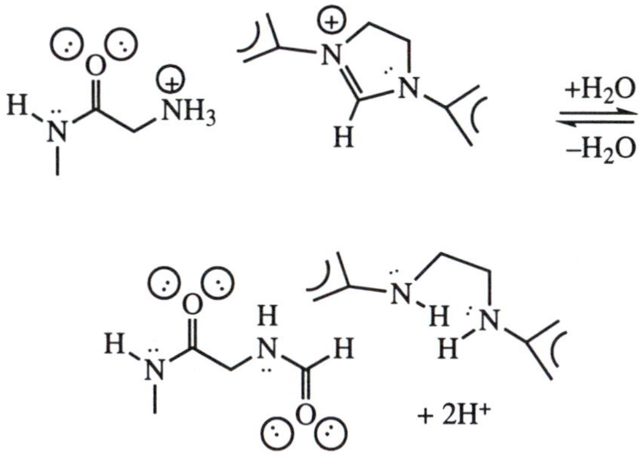

$$\xrightleftharpoons[-H_2O]{+H_2O} \quad + 2H^+ \qquad (1\text{–}66)$$

The first step in a reaction of this type is the attack of the amine of the acceptor upon the $N^5$,$N^{10}$-methenyltetrahydrofolate (Figure 1–12).[78,79] The first tetravalent intermediate has three possible leaving groups, each an amine. Although protonation of either of the $\pi$ lone pairs on the nitrogens of the tetrahydrofolate to make it a leaving group is more difficult than protonating the aliphatic amine that just attacked, no reaction occurs until one of them is protonated. The product of the first acyl transfer reaction is a mixed amidine between one of the nitrogens of tetrahydrofolate, usually[79] nitrogen 10, and the nitrogen of the amine. To complete the transfer of the formyl group, hydroxide ion attacks this mixed amidine to form a second tetravalent intermediate. It is at this point that a difficulty arises. In free solution the leaving group from this tetravalent intermediate is always the aliphatic amine rather than the aniline[78] because the aliphatic nitrogen is more basic, and hence more readily protonated, than the aniline nitrogen [p$K_a$(aniline) = 4.6; p$K_a$(glycinamide) = 8.0]. Presumably, in the enzymatic reactions some strategy is employed to change the preference for the leaving group so that efficient transfer of the formyl group to the aliphatic amine can occur.

When they are adducts of tetrahydrofolate, one-carbon compounds are interconverted by an oxidation–reduction reaction involving hydride transfer[80]

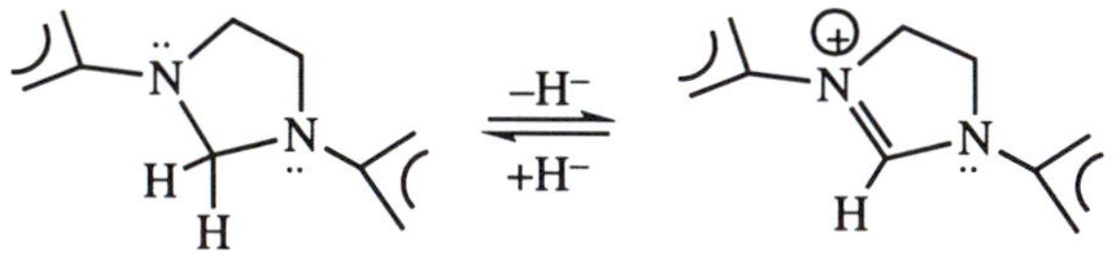

$$\xrightleftharpoons[+H^-]{-H^-} \qquad (1\text{–}67)$$

$$\xrightleftharpoons[-H^-]{+H^-} \qquad (1\text{–}68)$$

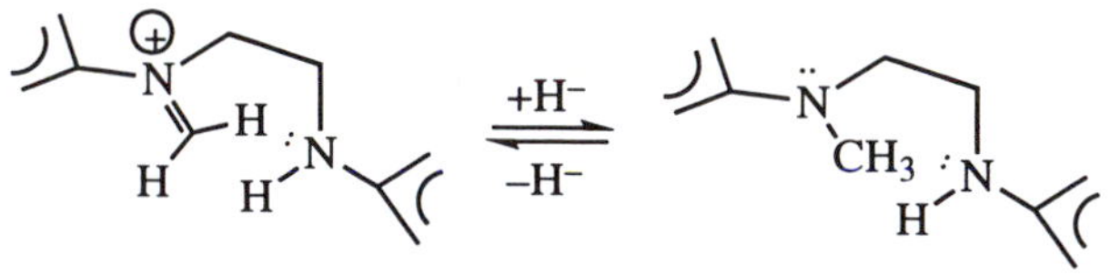

These hydride transfers are directly mediated by either nicotinamide adenine dinucleotide or nicotinamide adenine dinucleotide phosphate. The latter reaction (Reaction 1–68) produces $N^5$-methyltetrahydrofolate that can act as a simple

**Figure 1–11:** Crystallographic molecular model of $N^5$,$N^{10}$-methenyltetrahydrofolate.[74] The cation of $N^5$,$N^{10}$-methenyltetrahydrofolate was produced from $N^5$-formyltetrahydrofolate in hydrochloric acid and crystallized as the bromide salt from 50% hydrobromic acid. A map of electron density was derived from X-ray diffraction of these crystals. The asterisks denote the two chiral carbons in this, the naturally occurring diastereomer. The formamidinium cation is the planar complex of the one carbon between the two nitrogens. Reprinted with permission from ref 74. Copyright 1979 American Chemical Society.

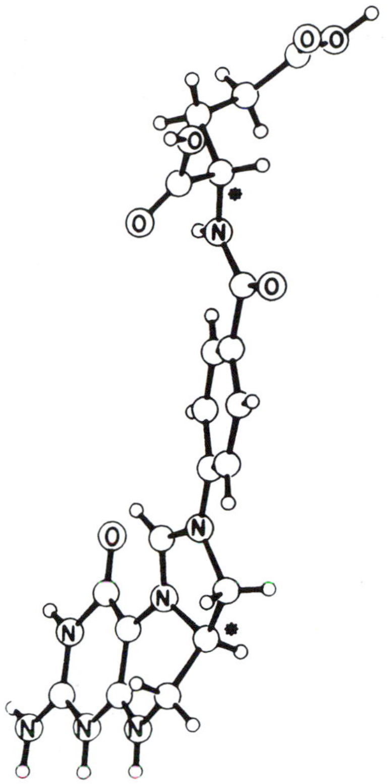

**Figure 1–12:** Transfer of the formyl group from $N^5$,$N^{10}$-methenyltetrahydrofolate to the primary amine on phosphoribosylglycinamide. The reaction involves two consecutive acyl exchange reactions: the first in which the amine from the glycinamide enters and a nitrogen of tetrahydrofolate leaves, and the second in which the oxygen of hydroxide enters and the other nitrogen of tetrahydrofolate leaves.

tetravalent intermediate

tetravalent intermediate

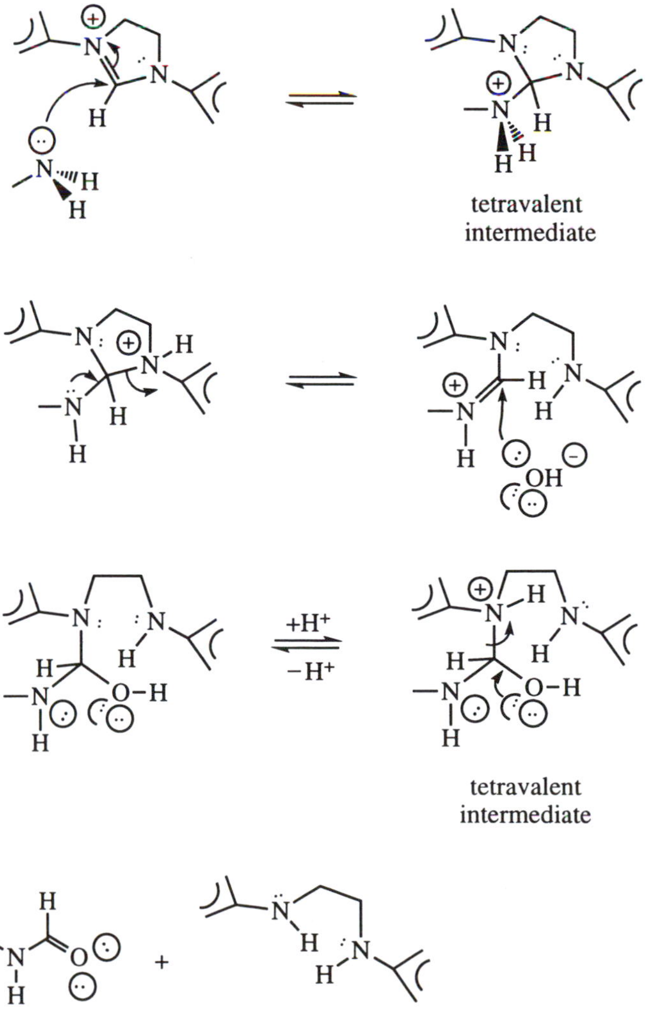

methylating agent in an $S_N2$ nucleophilic substitution with strong nucleophiles such as thiolate anion.[81]

Tetrahydrofolate itself is susceptible to oxidation–reduction by hydride transfer in a reaction catalyzed by dihydrofolate reductase

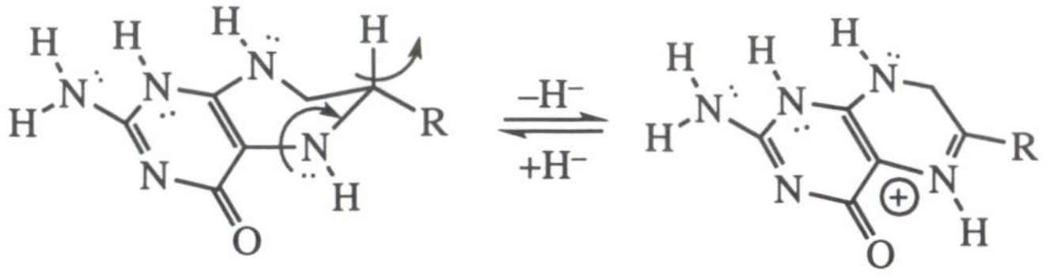

(1–69)

This reaction regenerates tetrahydrofolate from the 7,8-dihydrofolate produced during the synthesis of thymidine, and it also participates in the production of tetrahydrofolate from folate.

### Suggested Reading

Benkovic, S.J., Barrows, T.H., & Farina, P.R. (1973) Studies on Models for Tetrahydrofolic Acid. IV. Reactions of Amines with Formamidinium Tetrahydroquinozaline Analogs, *J. Am. Chem. Soc.* 95, 8414–8420.

# Enolates, Enols, and Enamines

The hydrogen on a carbon immediately adjacent to a carbonyl group or a neutral acyl derivative is acidic because of the ability of the heteroatom two atoms away from the acidic carbon to support the excess electron density of the conjugate base. The simplest example of this acidity is acetaldehyde ($pK_a = 17.6$)[82]

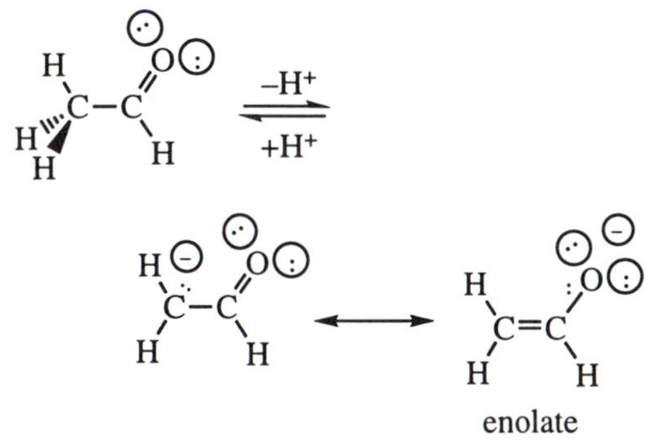

enolate (1–70)

The resonance structures demonstrate the assistance of the neighboring oxygen in the **enolate ion**. It is this conjugation stabilizing the enolate ion that causes the acid dissociation constant of acetaldehyde to be much less than that of propane ($pK_a \cong 40$). As hydroxide ion can remove a proton from acetaldehyde many orders of magnitude more rapidly than it can from propane, the neighboring $\pi$ molecular orbital system must be involved in the transition state for proton transfer (Figure 1–13). The two electrons in the occupied bonding molecular orbital of the original carbon–hydrogen $\sigma$ bond remain behind in the $\pi$ molecular orbital system of the enolate. As the transition state is approached, this occupied $\sigma$ molecular orbital of the carbon–hydrogen bond can be viewed as overlapping the unoccupied antibonding $\pi$ molecular orbital of the $\pi$ molecular orbital system between carbon and oxygen. Also during the approach to the transition state, the occupied atomic orbital of the lone pair of electrons on the base can be viewed as overlapping the unoccupied antibonding molecular orbital of the $\sigma$ bond between carbon and hydrogen to gain access to the proton. In the transition state, the atomic orbital of the base, the $1s$ orbital of the hydrogen, two atomic orbitals on the two carbons, and a $p$ orbital on the oxygen mix, and three of the five resulting molecular orbitals of the transition state are occupied by the six electrons participating in the reaction. The transition state can be viewed as decomposing into the protonated base and the $\pi$ molecular orbital system of the enolate through an overlap between the unoccupied antibonding molecular orbital of the $\sigma$ bond between the base and the proton and the highest occupied nonbonding $\pi$ molecular orbital of the enolate. It is this highest occupied nonbonding $\pi$ molecular orbital of the product enolate that is nucleophilic because it bears the electron excess of the enolate and it accounts for the resonance structures (Mechanism 1–70) that assign negative charge only to the two end atoms of the three atoms participating in the enolate. For the acid–base reaction to occur in this way, the proton removed from carbon must be in a $\sigma$ bond parallel to the $\pi$ molecular orbital system to permit the necessary overlaps and mixing in the transition state

acidic hydrogen

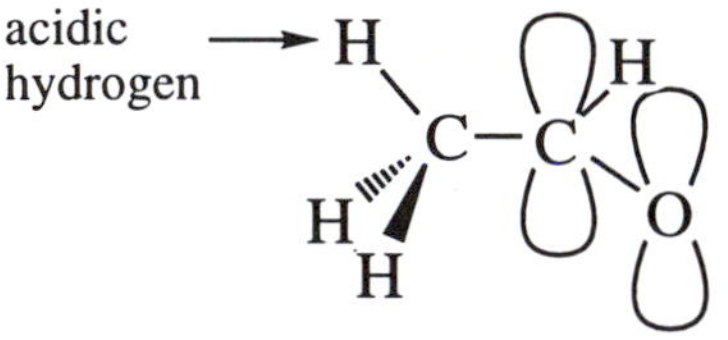

**1–49**

It has already been noted that the removal of a proton from the $\alpha$-carbon of a carbonyl compound or acyl derivative by a base, even when the reaction is exothermic, is a slow process.

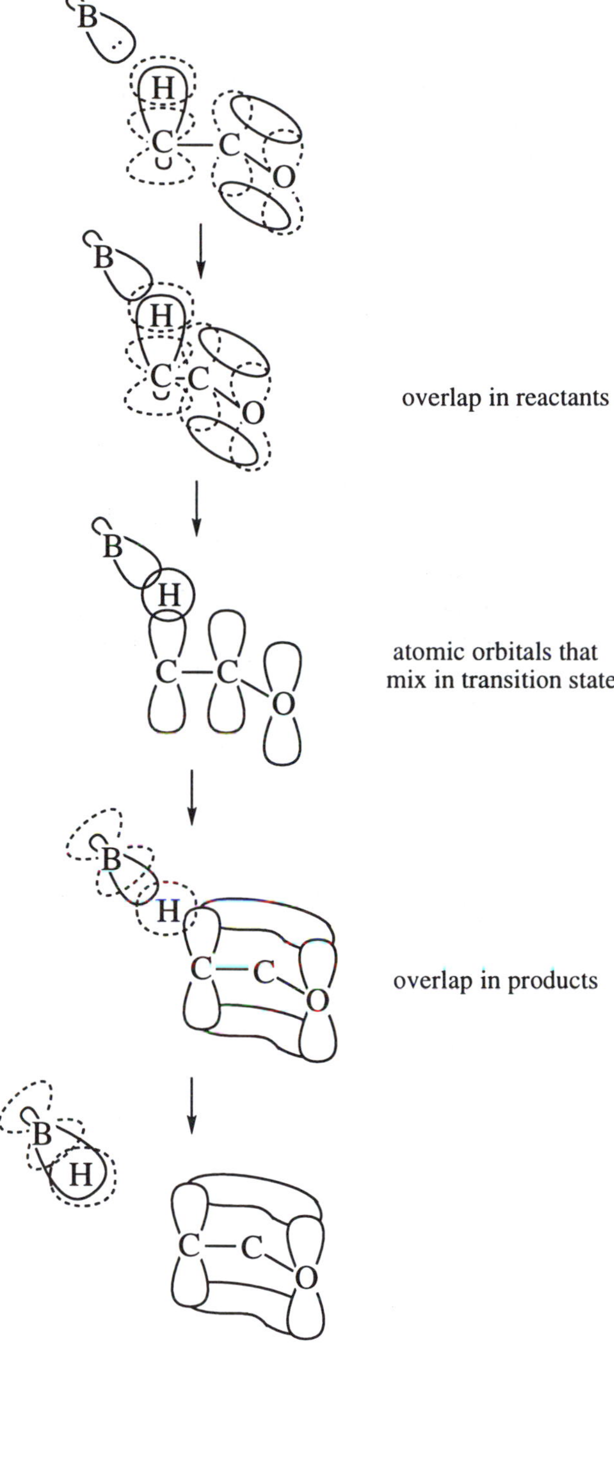


**Figure 1–13:** Removal of a proton by a base, B, from a carbon adjacent to a carbonyl group. As the atomic orbital occupied by the lone pair on the base (solid line) overlaps the antibonding molecular orbital of the carbon–hydrogen bond (dashed lines), the bonding molecular orbital of the carbon–hydrogen bond (solid line) overlaps the antibonding molecular orbital of the $\pi$ bond between carbon and oxygen (dashed lines). The five atomic orbitals that mix in the transition state are the atomic orbital of the base, the $s$ orbital of the hydrogen, and three $p$ orbitals on the two carbons and the oxygen. The transition state decomposes through a form in which the antibonding molecular orbital of the $\sigma$ bond between the base and the proton (dashed lines) overlaps the nonbonding $\pi$ molecular orbital (composed of the two peripheral $p$ orbitals) of the enolate.

The enolate ion (Mechanism 1–70) can be protonated on oxygen to form the **enol**

**1–50**

The enol of acetaldehyde is acidic at its hydroxyl group and is a much stronger acid ($pK_a$ = 11.1)[83] than acetaldehyde itself. Removal of the proton from the oxygen by a strong base such as hydroxide anion ($\Delta pK_a$ = 4.6) is a diffusion-controlled reaction.[1] This behavior is consistent with an uncomplicated removal of a proton from a lone pair on oxygen. The enol can be produced directly from acetaldehyde in an acid-catalyzed reaction in which either protonation at oxygen precedes deprotonation at carbon

(1–71)

or protonation at oxygen occurs simultaneously with deprotonation at carbon[82]

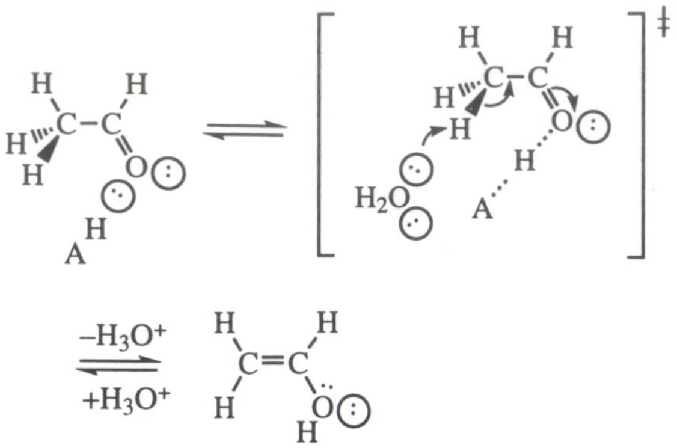

$-H_3O^+$ / $+H_3O^+$

(1–72)

In either case, the proton is added to a lone pair on oxygen orthogonal to the developing $\pi$ molecular orbital system to increase the electronegativity of oxygen and increase its ability to withdraw electron density from the carbon–hydrogen $\sigma$ bond.

It is much easier to remove a proton from a carbon $\alpha$ to a carbonyl if the oxygen of the carbonyl is protonated simultaneously by a general acid (Mechanism 1–72). In the active site of an enzyme there will probably always be a Brønsted acid hydrogen-bonded to one of the lone pairs on the oxygen of a carbonyl to provide the necessary proton or a metallic Lewis acid to accomplish the same purpose. The enol that results is also more stable than the respective enolate anion. For all of these reasons, enolates are unlikely intermediates in enzymatic reactions relative to the respective enols, and reactions involving enol–enolates from this point on will be written with the enols as intermediates.

An enol is less nucleophilic than its conjugate enolate. One way to see this is to examine the effect of protonation of the enolate oxygen on the distribution of electron density in the nonbonding $\pi$ molecular orbital of the enolate

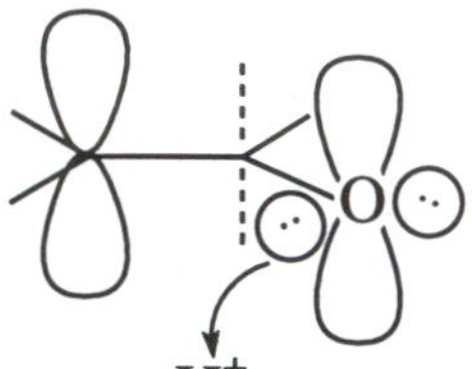

$H^+$

**1–51**

This molecular orbital is the nucleophile because it is the molecular orbital that places excess electron density on the $\alpha$-

**Table 1–4: Acid Dissociation Constants for Carbon Acids**[a]

| compound | p$K_a$ | compound | p$K_a$ |
|---|---|---|---|
| 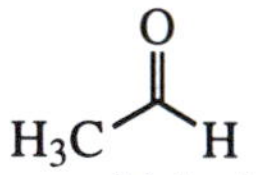 acetaldehyde | 17.6 | 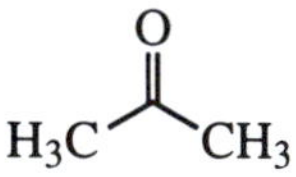 acetone | 20 |
| 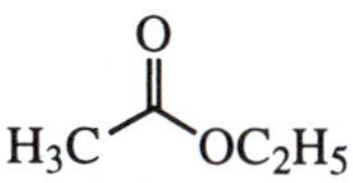 ethyl acetate | 24.5 | 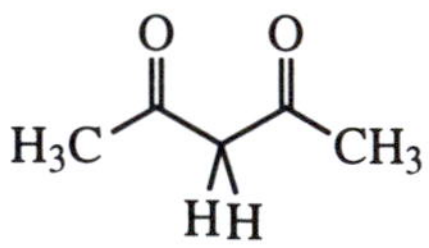 acetylacetone | 8.9 |
| 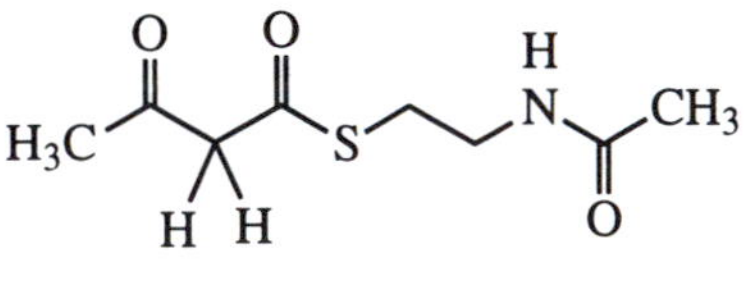 $H_3C$–CO–$CH_2$–CO–S–$CH_2CH_2$–NH–CO–$CH_3$ | 8.5 | 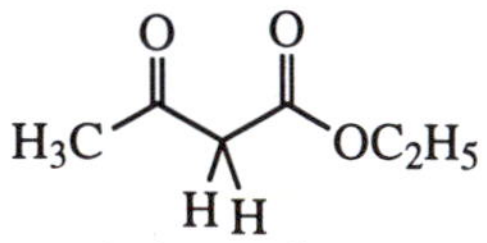 ethyl acetylacetate | 10.7 |
| | | 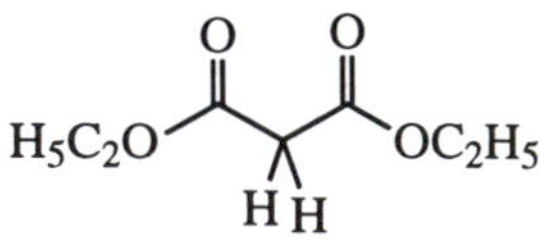 diethyl malonate | 13.3 |

[a] The acid dissociation constants are arranged in the table to facilitate comparisons horizontally, vertically, and diagonally.

carbon. In the enolate, the electron excess on the unprotonated oxygen increases the electron density on the $\alpha$-carbon relative to the oxygen. When one of the $\sigma$ lone pairs of electrons on the oxygen of the enolate is protonated, the oxygen loses the electron excess, and the resulting coulomb effect of its underlying electronegativity increases the $\pi$ electron density on oxygen at the expense of the electron density on carbon. Nevertheless, the carbon in an enol does retain excess electron density, as illustrated in the right-hand resonance form of Structure **1–50**. This, on paper, is an unlikely resonance form because of separation of charge, but it does indicate that the carbon shares the $\pi$ electron density in the highest occupied molecular orbital and remains sufficiently nucleophilic to participate in the reactions encountered biochemically.

The acidity of the proton on the carbon adjacent to a carbonyl varies as the substituents around the carbon itself or the adjacent carbon–oxygen double bond are changed (Table 1–4). When the hydrogen on the carbonyl carbon of acetaldehyde ($pK_a$ = 17.6) is changed to a methyl group, the $pK_a$ increases [$pK_a$(acetone) ≅ 20]. This reflects the overlap between the carbon–hydrogen $\sigma$ bonds of the methyl group and the $\pi$ system of the enolate anion. The pairs of electrons in these $\sigma$ bonds push into the $\pi$ system of the enolate and destabilize the anion relative to the conjugate acid. When the hydrogen on the carbonyl carbon of acetaldehyde is changed to an ethoxy group, the $pK_a$ increases even further [$pK_a$(ethyl acetate) ≅ 24.5] because the lone pairs of electrons on the additional oxygen push electron density even more strongly into the $\pi$ molecular orbital system than does a methyl group. When a second carbonyl group is substituted for one of the methyl hydrogens of acetone, the proton between the two carbonyl groups is much more acidic [$pK_a$(acetylacetone) ≅ 8.9]. This reflects the stabilization brought about by delocalization of the negative charge through a $\pi$ molecular orbital system spread over five atoms of the enolate anion

$$\left[ H_3C-C(O)-C-C(O)-CH_3 \right]^-$$

**1–52**

When one of the methyl groups of acetylacetone is replaced with an ethoxy group, the acidity decreases [$pK_a$(ethyl acetylacetate) = 10.7], and when the other is replaced by an ethoxy as well, there is a further decrease [$pK_a$(diethyl malonate) = 13.3]. These decreases reflect the stronger $\pi$ electron donation of the lone pairs of electrons on oxygen than the pairs of electrons in the carbon–hydrogen $\sigma$ bonds of the methyls. The reach of a carbonyl or an acyl group can be extended by a conjugated carbon–carbon double bond

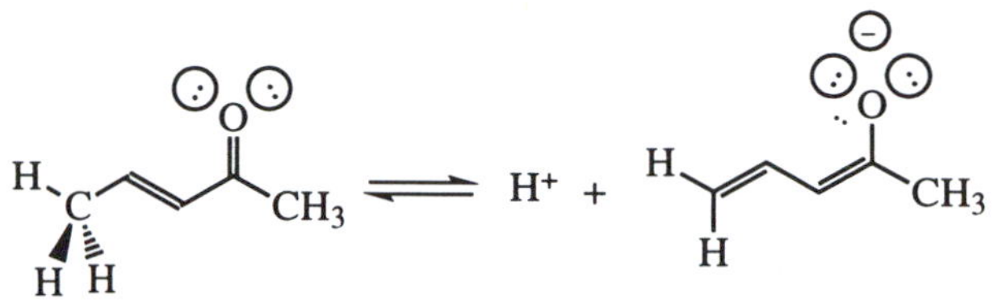

$$H_3C-CH=CH-C(=O)-CH_3 \rightleftharpoons H^+ + H_2C=CH-CH=C(O^-)-CH_3 \qquad (1\text{–}73)$$

The proton on a carbon next to an ester is less acidic than that next to a ketone because of overlap between the lone pair of electrons on the ester oxygen and the $\pi$ molecular orbital system of the enolate. The proton on a carbon adjacent to a thioester, however, is more acidic than that adjacent to an oxygen ester, and enolates of thioesters are more readily formed than enolates of oxygen esters. From the yields of products in reactions in which the conjugate bases of thioesters and oxyesters are competing against each other as enolates for active electrophiles or from the rates of reactions in which the conjugate bases of both thioesters and oxyesters participate as enolates, it has been estimated that the proton on a carbon adjacent to a thioester is 40–100 times more acidic than the proton adjacent to an oxyester.[84] The acidic proton on a thioester of acetylacetic acid has a $pK_a$ 2 logarithmic units lower than that of the respective oxyester (Table 1–4)[85] and is quite close to that of the analogous ketone, acetylacetone ($pK_a$ = 8.9).

The greater acidity of a thioester indicates that there is less overlap between the lone pairs of electrons on the sulfur and the $\pi$ molecular orbital system of the enolate than there is with the lone pairs of electrons on oxygen in an oxyester. Consistent with this conclusion is the fact that in the neutral thioester itself, unlike in an oxygen ester, there is no significant overlap between the lone pairs on sulfur and the $\pi$ molecular orbital system of the double bond between carbon and oxygen. This conclusion is reached from the fact that, unlike in the structure of an oxygen ester, in a thioester there is no shortening of the bond between the acyl carbon and the sulfur relative to the bond between a sulfur and an aryl or a vinyl carbon.[86] These observations suggest that the double bond between carbon and oxygen in a thioester resembles the isolated double bond in a ketone in its properties.

The failure of the lone pairs on sulfur to overlap with either the double bond between carbon and oxygen in a neutral thioester or the $\pi$ molecular orbital system in an enolate results from the fact that sulfur is a third-row element whose lone pairs in the two atomic orbitals overlap poorly with the *p* orbitals of a $\pi$ molecular orbital system between carbon and oxygen. Sulfur also has vacant 3*d* orbitals whose shape is such that they should overlap well with the $\pi$ molecular orbital system of an enolate, and it has been proposed that one of these might delocalize its excess electron density

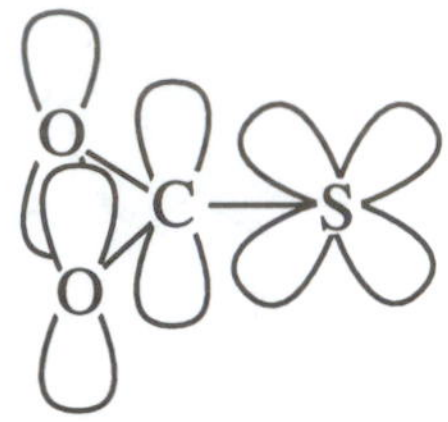

vacant *d* orbital

**1–53**

This, however, is unlikely because in the highest occupied molecular orbital of an enolate there is almost no contribution from the *p* orbital on the central carbon, so the overlap would be negligible. This inability of sulfur to overlap effectively is also illustrated by the resonance structure

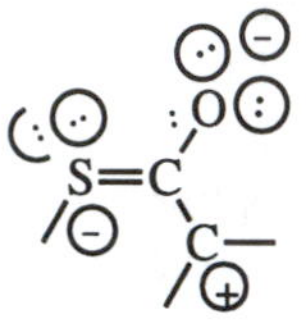

**1–54**

It follows that $d p\pi$ overlap is probably inconsequential and the acidity of a thioester results entirely from the fact that it resembles the more acidic ketone than the less acidic ester. In biochemical reactions, when a proton must be removed from the $\alpha$-carbon of an acyl derivative to produce the corresponding enol, that acyl derivative will usually be a thioester formed with coenzyme A, with a cysteine on a protein, or with a pantetheine attached to a protein as a posttranslational modification.

It is also possible to activate a proton on a carbon adjacent to a carbonyl by converting the carbonyl into an iminium cation[87]

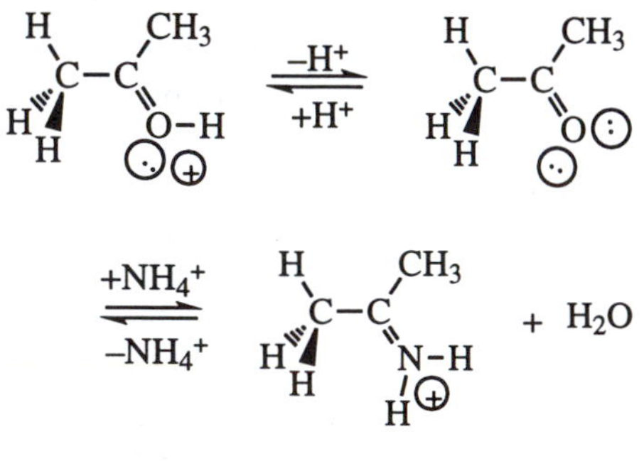

(1–74)

It has been estimated that the p$K_a$ of a proton on acetone is decreased by around 12 units (from p$K_a \cong 20$ to p$K_a \cong 8$) by this substitution.[87] It has also been observed that, depending on the strength of the base, the rate of proton removal from the $\alpha$-carbon increases by a factor of $10^5$–$10^8$ in the iminium cation compared to the parent carbonyl compound.[87,88] The rates of proton removal from the $\alpha$-carbon of an iminium cation are a factor of $10^3$ slower than the rates of proton removal from the respective protonated carbonyl compound (Reaction 1–74), but protonating an imine is much easier (p$K_a \cong 8$) than protonating a carbonyl oxygen (p$K_a \cong -6$). An **enamine** is the conjugate base resulting from the removal of a proton from the carbon atom of an iminium cation

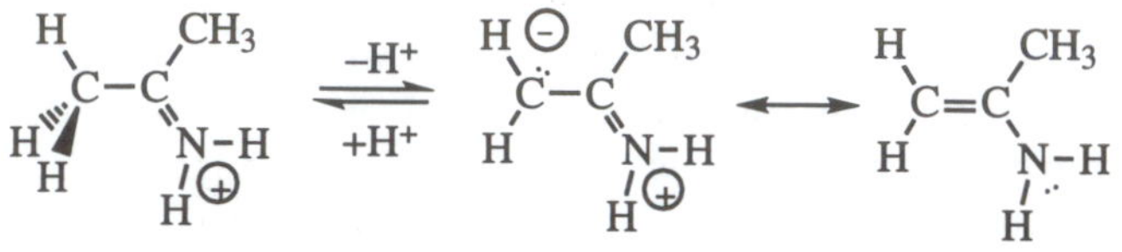

(1–75)

An enamine is more nucleophilic at the anionic carbon than the corresponding enol (**1–50**) because the coulomb effect of the protonated nitrogen is less than the coulomb effect of protonated oxygen. All of these considerations illustrate the advantages of an enamine at neutral pH in the absence of strong bases and acids.

Enols, enamines, and the enols of thioesters participate as nucleophiles in additions to carbonyls and in acyl exchange reactions. An enzymatic example of the addition of an enamine to a carbonyl is the reaction catalyzed by fructose-bisphosphate aldolase

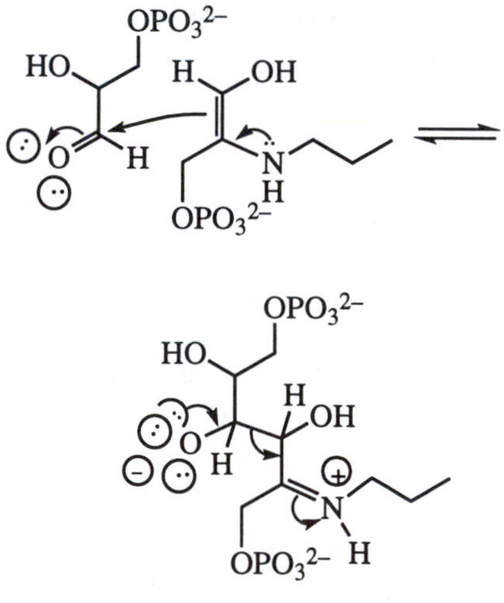

(1–76)

An enzymatic example of the participation of the enol of a thioester in an acyl exchange reaction is that catalyzed by fatty acid synthase

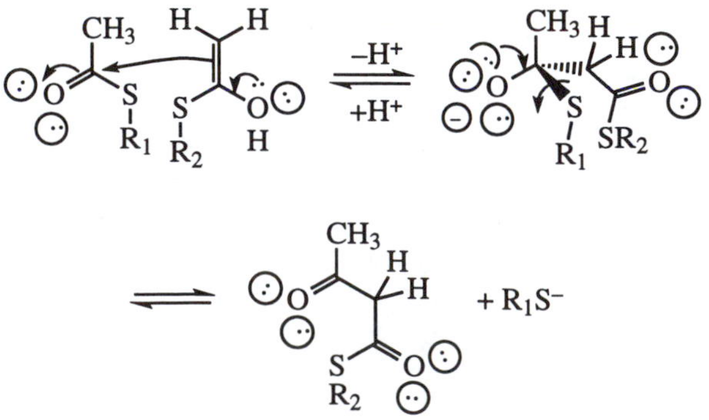

(1–77)

In the reverse reactions of the first steps in these two examples, one of the lone pairs on the incipient carbonyl oxygen or incipient acyl oxygen, respectively, participates in the breaking of the carbon–carbon bond by providing push. For example, in the reverse of Reaction 1–76

(1–78)

The orbitals that mix to form the transition state and that must be able to overlap as the transition state is approached are the atomic orbital bearing the lone pair of electrons on the oxygen, the $\sigma$ molecular orbital system of the $\sigma$ bond that is broken, and the $\pi$ molecular orbital system of the carbon–nitrogen double bond (Figure 1–14). In the reverse reaction, the orbitals that mix to form the transition state are the $\pi$ molecular orbital system of the enamine and the $\pi$ molecular orbital system of the carbonyl group. In order that the transition state be accessible, these molecular and atomic orbitals, or all of their composite atomic orbitals, must be parallel to each other, as indicated in Reaction 1–78 and Figure 1–14. By

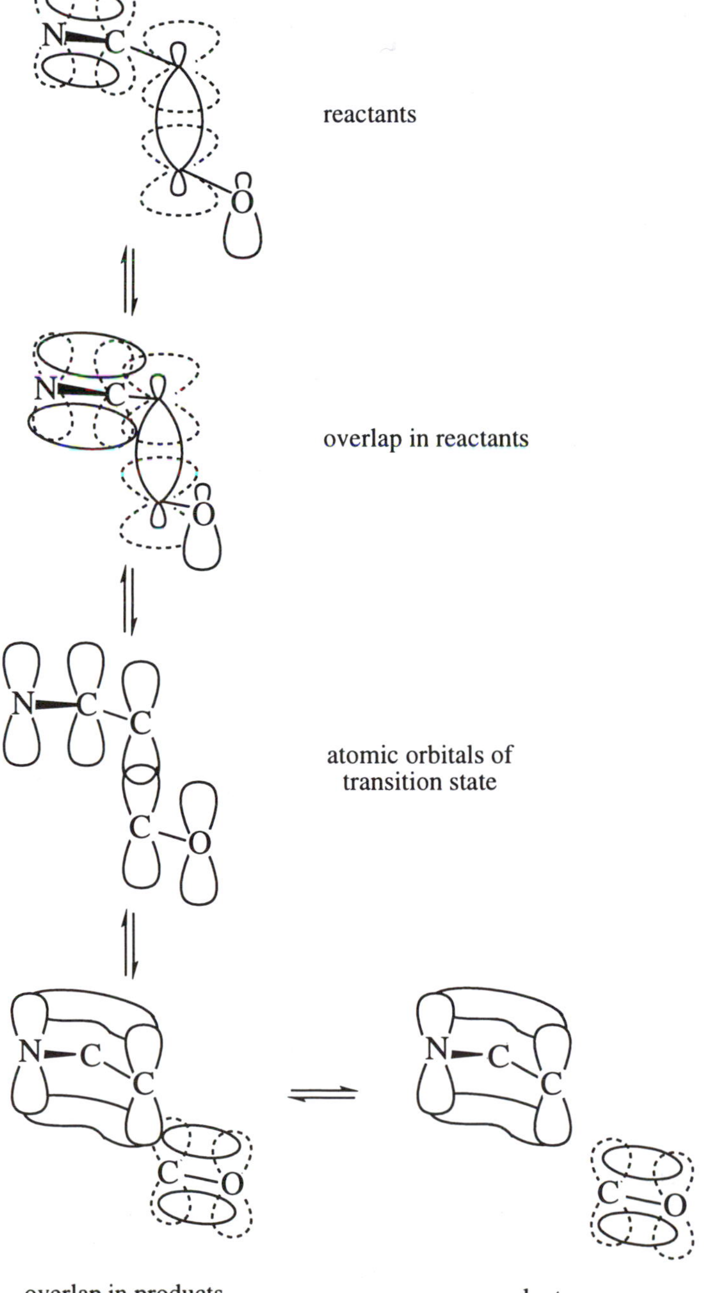


**Figure 1–14:** Breaking or forming of a carbon–carbon bond adjacent to an imine (Reaction 1–76). The reaction breaking the carbon–carbon bond is initiated by overlap of the atomic orbital occupied by the lone pair of electrons on oxygen (solid line) and the antibonding $\sigma$ molecular orbital of the carbon–carbon bond (dashed lines) and by overlap of the bonding $\sigma$ molecular orbital of the carbon–carbon bond (solid lines) and the antibonding $\pi$ molecular orbital of the carbon–nitrogen bond (dashed lines). In the transition state five *p* orbitals, one from each of the five atoms, overlap. The transition state decomposes through overlap of the occupied nonbonding $\pi$ molecular orbital of the enamine (the two peripheral *p* orbitals of the enamine) and the antibonding $\pi$ molecular orbital (dashed lines) of the carbon–oxygen double bond. Formation of the carbon–carbon bond is the reverse reaction.

microscopic reversibility, if the $sp^3$ atomic orbital on the oxygen providing the push enters the reaction parallel to the $\sigma$ bond being broken, it must emerge from the transition state in the reverse reaction parallel to the newly formed $\sigma$ bond.

Enolates, enols, enamines, or carbanions of thioesters are also intermediates in carboxylations and **decarboxylations.** Carbon dioxide can be considered to be a dicarbonyl compound

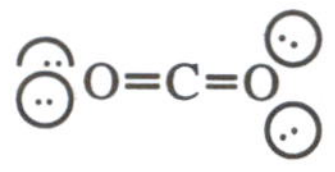

O=C=O

**1–55**

in which the central carbon is electrophilic and hence a Lewis acid, as is a proton. As a Lewis acid, the central carbon reacts with lone pairs of electrons. It is susceptible to carbonyl addition reactions, the simplest again being hydration

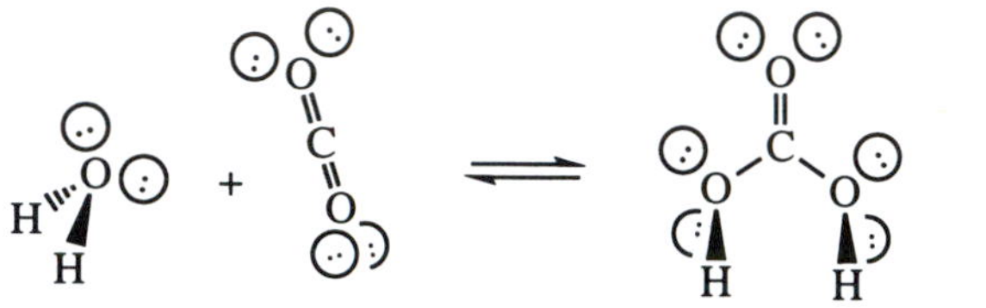

(1–79)

The mechanism of this reaction involves removal of a proton from water and addition of a proton to the carbonyl oxygen. As a carbonyl compound, carbon dioxide can participate in carbonyl addition reactions with enols

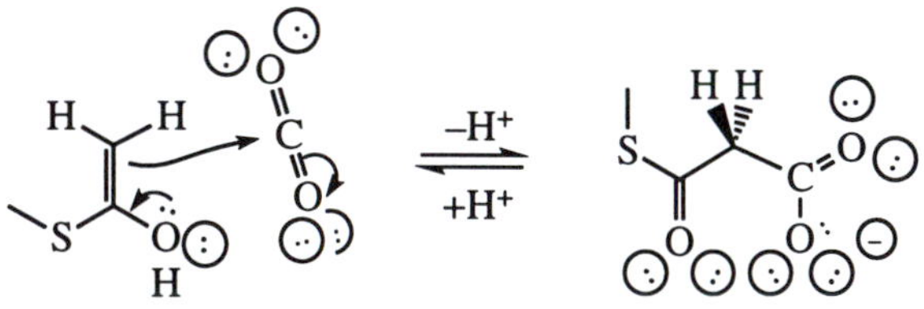

−H⁺ / +H⁺

(1–80)

Decarboxylations, because they are the reverse reactions of the addition of a simple carbonyl to a carbanion, are assisted by push from the lone pairs on the carboxylate

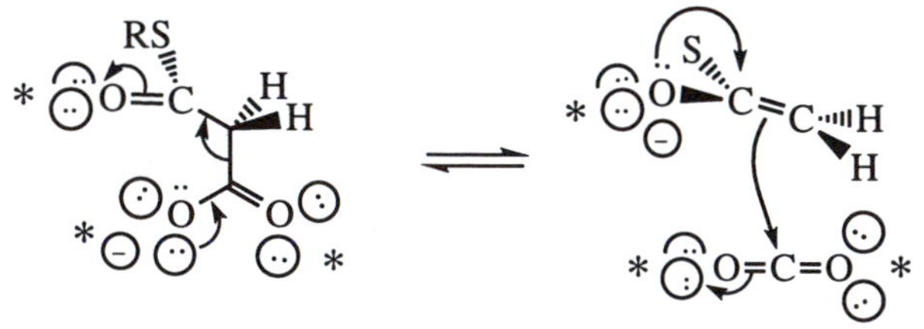

(1–81)

In this reaction the negative charge is moved among the three oxygens whose lone pairs of electrons are indicated by asterisks. In solution, decarboxylations of $\beta$-oxo acids occur more rapidly when the carboxylate is protonated than when it is anionic. This is usually explained as the result of an intramolecular transfer of the proton between the lone pairs of electrons on the oxygen of the carboxylic acid and the oxygen of the incipient enol, respectively

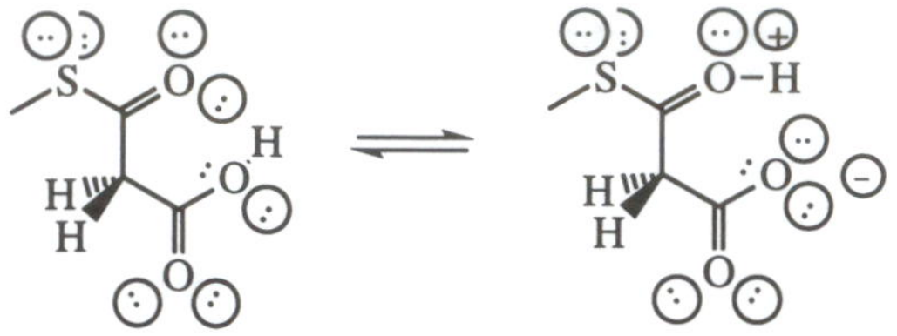

(1–82)

This proton transfer between two oxygen atoms that are four atoms apart from each other enhances the ability of the acyl group or carbonyl group to withdraw electrons from the carbon–carbon bond to be broken and increases the push of the lone pair of electrons on the carboxyl oxygen simultaneously. Because it is between two lone pairs both orthogonal to the orbitals mixing in the transition state for decarboxylation and sterically cannot occur while the atomic orbitals of the transition state are aligned, the proton transfer should be considered as a preceding and independent process from the decarboxylation itself

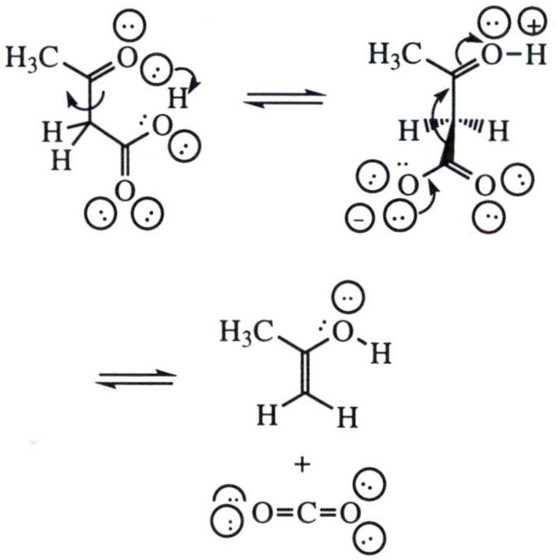

(1–83)

In the decarboxylation of $\beta$-oxo acids, the transfer of the proton from the carboxylic acid to the carbonyl oxygen permits the enol to be the immediate product of the decarboxylation (Reaction 1–83). Such decarboxylations are also enhanced by forming the imine at the carbonyl carbon. For example, the decarboxylation of the bicyclic $\beta$-oxo acid

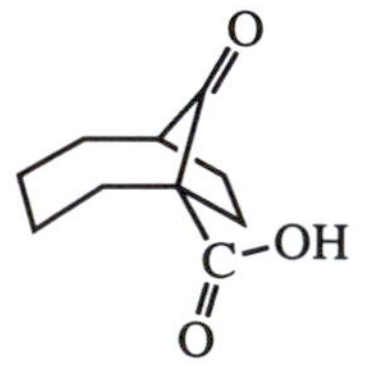

**1–56**

is enhanced $10^6$ times by the formation of the anilinoimine at the carbonyl group, and the behavior of the reaction is consistent with involvement of proton transfer between the carboxylic acid and a lone pair of electrons on the nitrogen of the imine in the transition state.[89] This rather peculiar $\beta$-oxo acid was chosen for this study because the decarboxylation of its imine is slow enough to measure conveniently, owing to the strain present in the enamine that is the direct product of the reaction.

Metal ions, such as cupric cation, can catalyze the decarboxylation of $\beta$-oxodicarboxylic acids, presumably by forming a complex involving the carbonyl oxygen

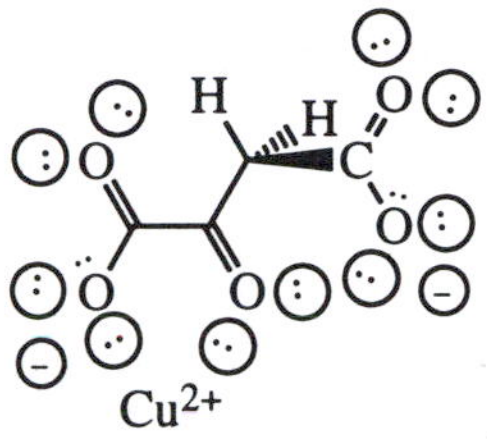

$Cu^{2+}$

**1–57**

and neutralizing the negative charge forming on that oxygen during the reaction.[90] A lone pair of electrons on the other carboxylate (Reaction 1-81) provides an additional point of attachment for the copper, the main function of which is to act as a Lewis acid at the oxygen of the incipient enolate and stabilize it by a coulomb effect as if it were a proton.

The decarboxylation of 1,3-dimethylorotic acid is unusual in that it proceeds through a carbanion that cannot be stabilized by an adjacent carbonyl or acyl functionality because it is orthogonal to any that are available

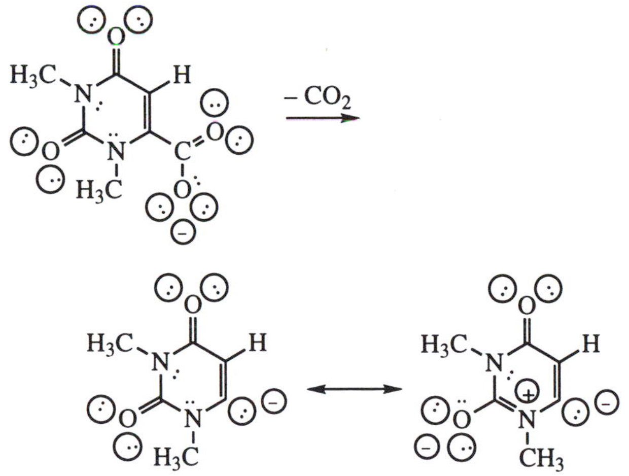

$- CO_2$

(1–84)

It is believed that the carbanion is stabilized as a partial ylide with the electron-deficient nitrogen immediately adjacent to it.[91] The decarboxylation is enhanced when the exocyclic oxygen is protonated, which would increase the positive charge on the nitrogen.

An enol is also the intermediate in the isomerization of an $\alpha$-hydroxyketone or $\alpha$-hydroxyaldehyde.[92] An example of such a reaction is that catalyzed by triosephosphate isomerase

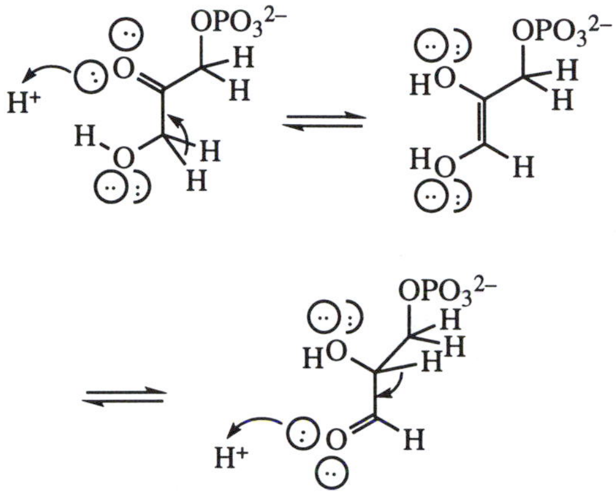

(1–85)

in which the carbonyl group moves between two adjacent carbons. A similar isomerization can interconvert a carbonyl group and an acyl group by the same mechanism[93,94] as in the reaction catalyzed by lactoylglutathione lyase

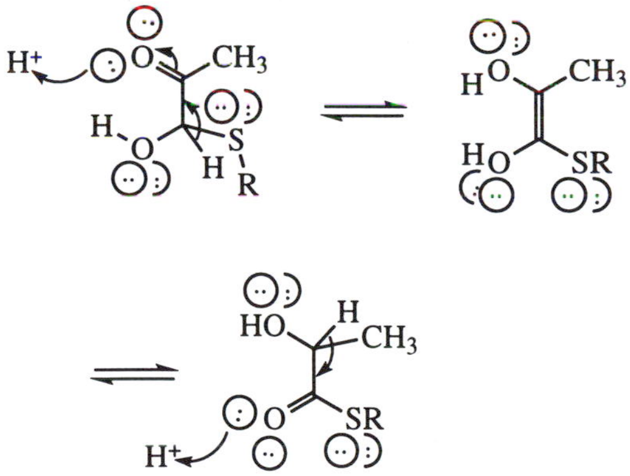

(1–86)

## *Suggested Reading*

Bender, M.L., & Williams, A. (1966) Ketimine Intermediates in Amine-Catalyzed Enolization of Acetone, *J. Am. Chem. Soc. 88*, 2502–2508.

# Dehydration–Hydration and Elimination–Addition

In an elimination–addition reaction, an acid and a base are removed from adjacent carbons to produce a carbon–carbon double bond or an acid and a base are added to an olefin to produce an adduct. For example, in the elimination[95]

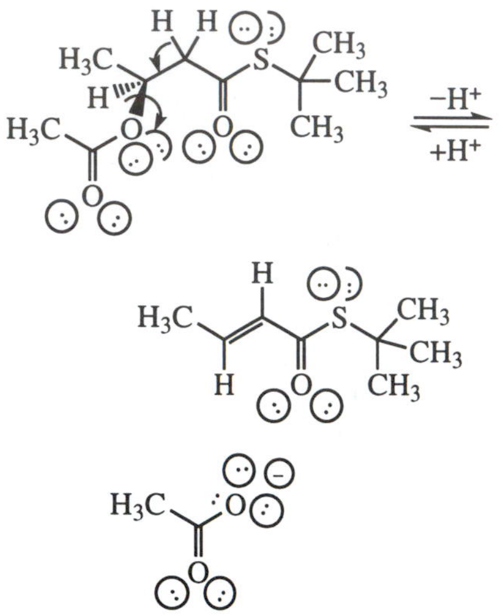

(1–87)

a proton and an acetate anion are removed from the two adjacent carbons to produce the olefin, and in the addition, which is the reverse reaction, a proton and an acetate anion are added to the olefin.

Elimination–addition reactions can proceed through either carbocations or enolates, or they can be concerted reactions. The hydration–dehydration of simple aliphatic olefins such as propene in solutions of strong, nonnucleophilic acids proceeds through a **carbocation**[96]

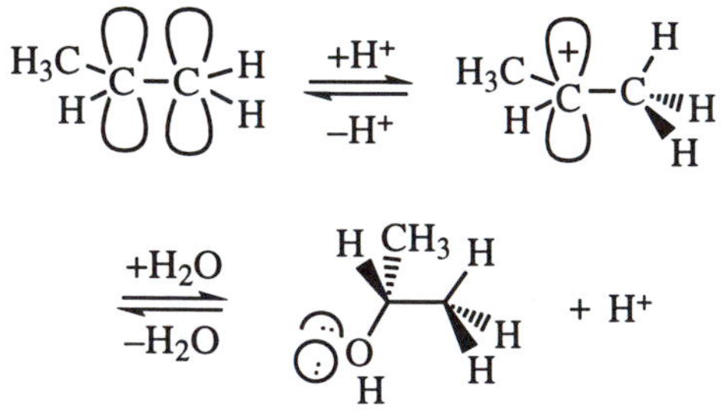

(1–88)

The $\beta$-elimination–addition of methanol from 4-methoxy-2-butanone under basic conditions to give the respective $\alpha,\beta$-unsaturated ketone proceeds through an **enolate anion**[97]

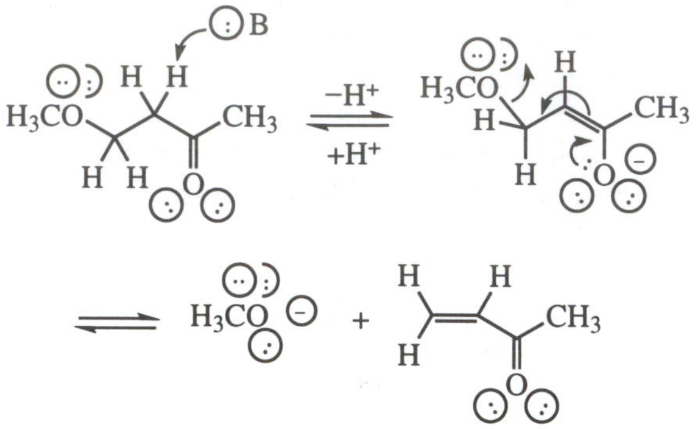

(1–89)

A similar $\beta$-elimination occurs when glyceraldehyde 3-phosphate is exposed to base[92]

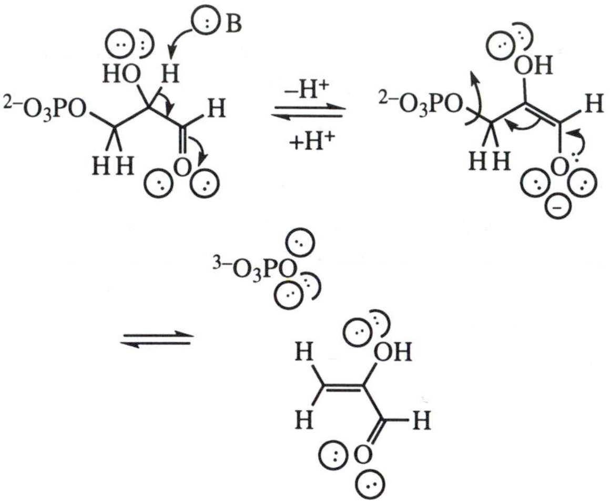

(1–90)

In both of these latter instances, the removal of an acidic hydrogen on a carbon adjacent to a carbonyl initiates the elimination. In the elimination effected by the push of the enolate in the second step of Reaction 1–89 or 1–90, the $\sigma$ bond between oxygen and carbon must be aligned parallel to the $\pi$ system of the enolate

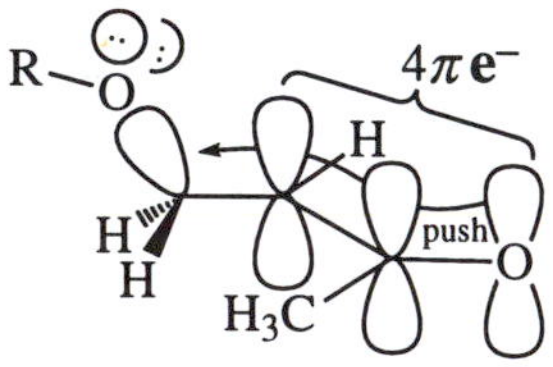

**1–58**

The $\pi$ molecular orbital system enters carbon 4 through the antibonding $\sigma$ molecular orbital of the carbon–oxygen $\sigma$ bond as the four $\pi$ electrons push away the two $\sigma$ electrons of the leaving group.

The elimination of toluenesulfonic acid from *trans*-2-phenylcyclopentyl tosylate in the presence of alkoxide in alcoholic solutions[98] is a **concerted elimination**

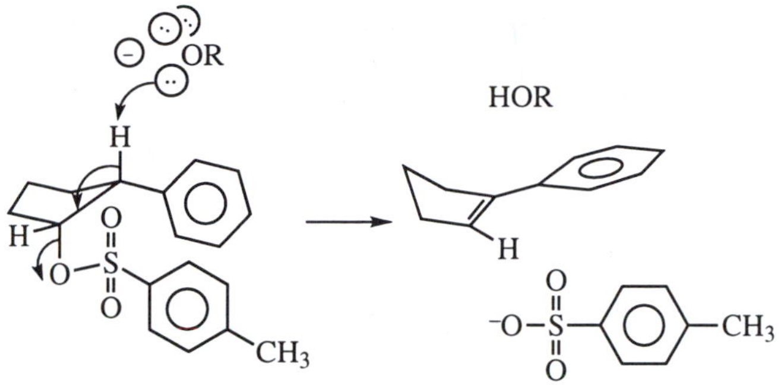

(1–91)

In such a concerted elimination, the pair of electrons in the carbon–hydrogen $\sigma$ bond, as they are stripped of the proton, provide the push to expel the leaving group. In Reaction 1–90, the atomic orbitals that mix in the transition state

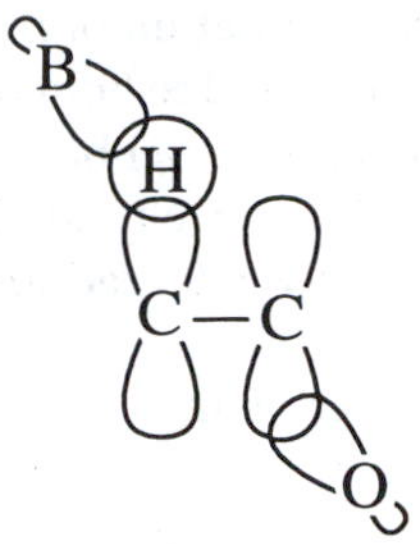

**1–59**

are the atomic orbital of the alkoxide containing the basic lone pair of electrons, the *s* atomic orbital of the hydrogen, the two respective atomic orbitals on the two adjacent carbons, and the atomic orbitals on the leaving oxygen. All must be parallel to each other.

Concerted eliminations are usually ***anti*** as shown in Reaction 1–91, with proton and leaving group departing on opposite sides of the carbon–carbon bond. In a ***syn*** **elimination**, the orbitals are still parallel, but they are so crowded as to be sterically disfavored. This crowding can be ameliorated. The relative yield of product from the *syn* isomer, when Reaction 1–91 is run with a diastereomeric mixture of reactants, is enhanced by the presence of alkali metal ions that are

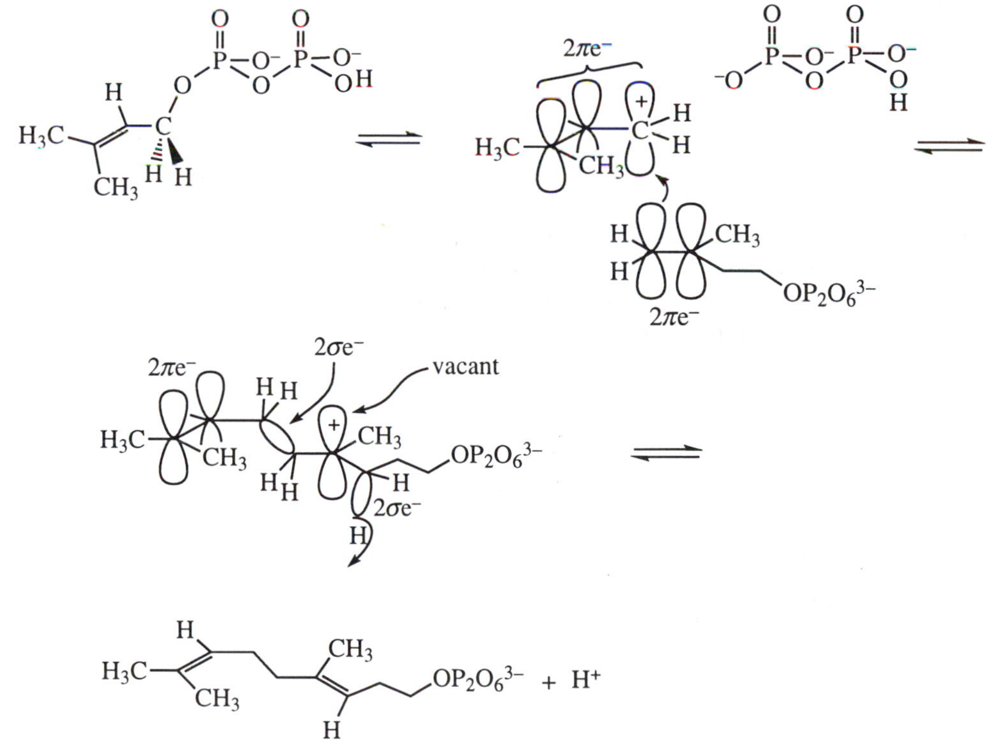

**Figure 1–15:** Addition of 3,3-dimethylallyl carbocation to isopentenyl diphosphate. 3,3-Dimethylallyl carbocation is formed from the elimination of pyrophosphate from 3,3-dimethylallyl diphosphate. This allylcarbocation adds to the carbon–carbon double bond of isopentenyl diphosphate to form a new carbon–carbon $\sigma$ bond in a carbocationic adduct. The carbocationic adduct loses a proton to produce geranyl diphosphate.

thought to promote the formation of a complex between the leaving group and the base and to hold them on the same side of the carbon–carbon bond. Under basic conditions, Reaction 1–87 proceeds with 15% *syn* elimination.[95] Therefore, *syn* eliminations are not unprecedented and often occur under enzymatic catalysis.

An instructive example of an addition reaction involving a Lewis acid other than a proton occurs in terpene biosynthesis in a reaction catalyzed by dimethylallyl *trans*transferase. The carbocation formed from 3,3-dimethylallyl diphosphate adds to isopentenyl diphosphate and the resulting carbocation eliminates a proton to yield the product, geranyl diphosphate (Figure 1–15). The first step in the reaction is the formation of an allyl carbocation consisting of three parallel *p* orbitals creating an allyl $\pi$ molecular orbital system with only 2 $\pi$ electrons. It is a carbocation, and hence a Lewis acid, and it adds as would a proton to the carbon–carbon double bond in isopentenyl diphosphate. This produces a tertiary carbocation that does not rearrange. It eliminates a proton from an adjacent carbon to form the olefin.

The **isomerization of olefins** can proceed through either an addition–elimination reaction or a tautomeric rearrangement. The isomerization of simple aliphatic olefins under acidic conditions proceeds through a carbocation

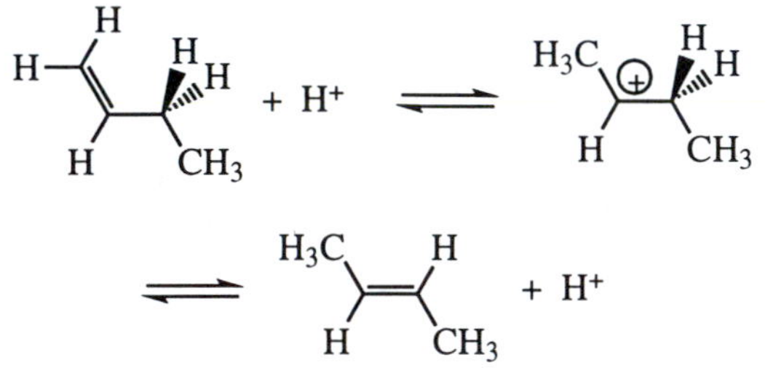

(1–92)

but isomerizations of olefins $\alpha$ to carbonyl groups or acyl derivatives can occur under basic conditions by proceeding through an enolate. For example, the isomerization of *tert*-butyl thiolbut-3-enoate to *tert*-butyl thiolbut-2-enoate[99]

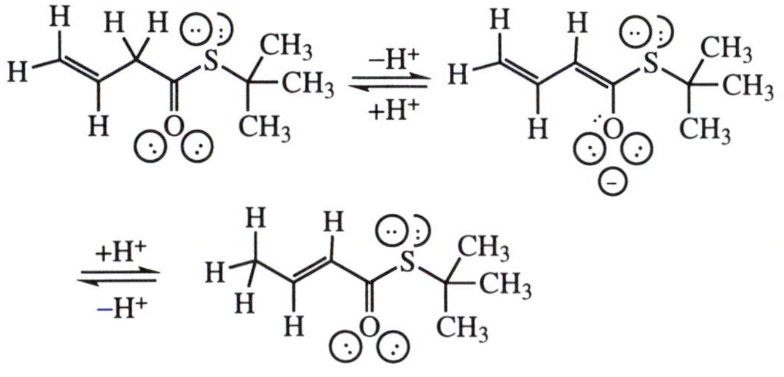

(1–93)

proceeds through the enolate. The excess negative charge is delocalized over five atoms, the four carbons and the oxygen.

The rate of an elimination reaction that proceeds through an intermediate enolate under catalysis by a base can be enhanced by the formation of an enamine[100]

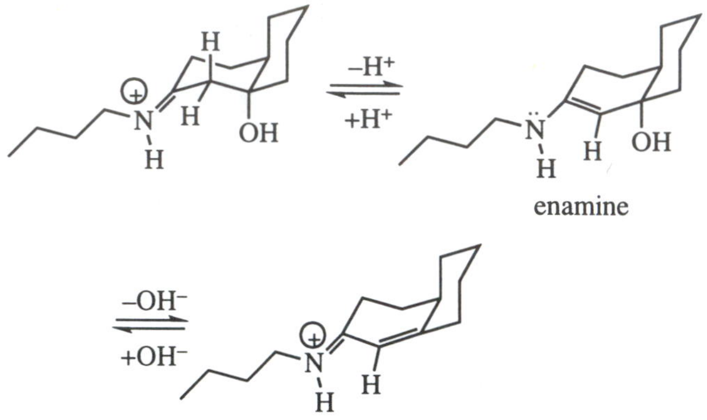

(1–94)

In this reaction the slow step in the forward direction is the second one. The $\pi$ lone pair of electrons on the enamine must provide the push to expel the hydroxide ion, a poor leaving group, by $\beta$-elimination

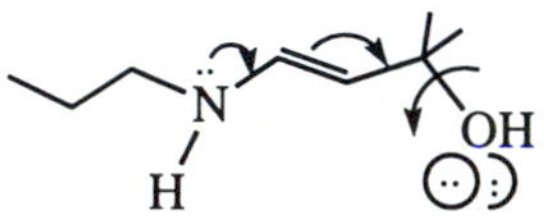

**1–60**

The highest occupied $\pi$ molecular orbital of the enamine must be parallel to the $\sigma$ bond between the carbon and the oxygen of the leaving group to exert the necessary push (see **1–58**).

## *Suggested Reading*

Hupe, D.J., Kendall, M.C.R., & Spencer, T.A. (1973) Amine Catalysis of $\beta$-Ketol Dehydration. II. Catalysis *via* Iminium Ion Formation. General Analysis of Nucleophilic Amine Catalysis, *J. Am. Chem. Soc.* 95, 2271–2278.

# Electrophilic Aromatic Substitution

Examples of electrophilic aromatic substitution occasionally are encountered in biochemical reactions. For example, the dimethylallyltransferase[101] in the biosynthetic pathway for glyceollin catalyzes a Friedel–Crafts alkylation

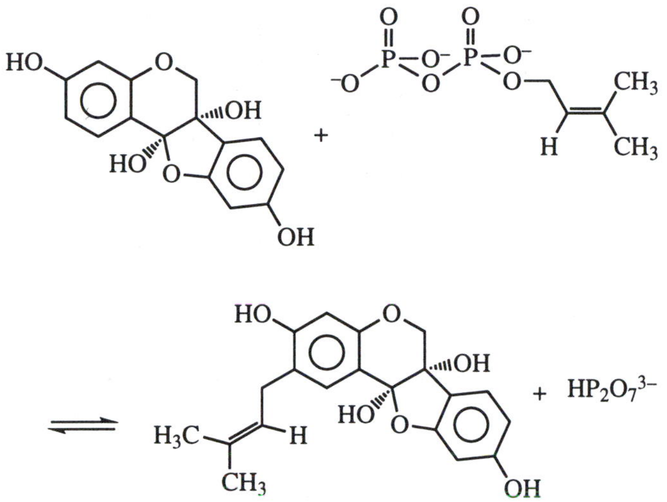

(1–95)

The electrophile is the allyl carbocation produced when the pyrophosphate leaves the dimethylallyl diphosphate

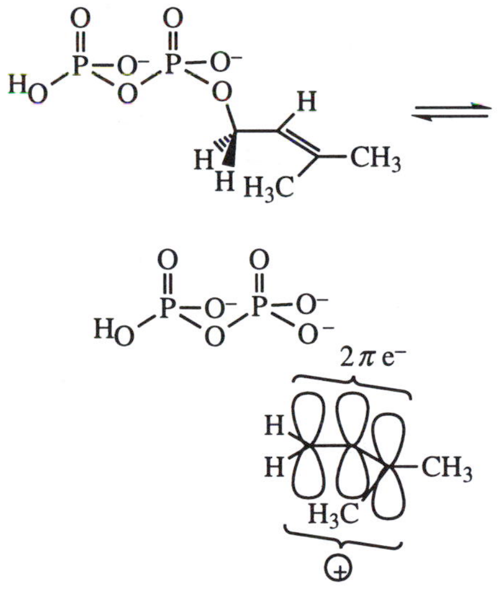

(1–96)

It adds to the nucleophilic $\pi$ molecular orbital system of the phenyl ring at a position *ortho* to the hydroxyl and *para* to the ether oxygen to produce a **pentadienyl cation** as an intermediate. The cation can be represented by resonance

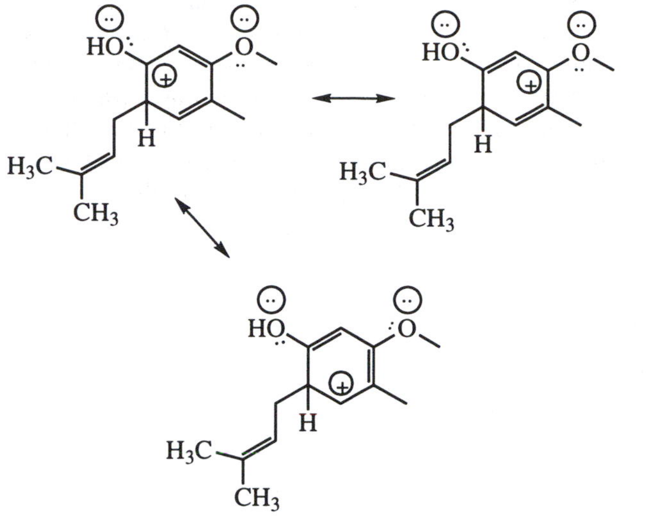

(1–97)

or by its lowest unoccupied $\pi$ molecular orbital

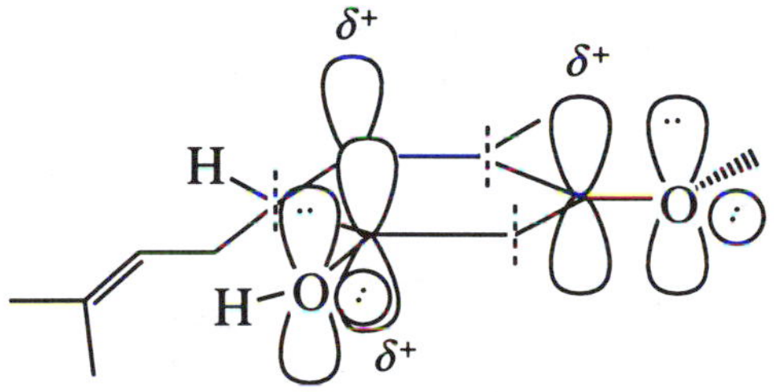

**1–61**

The pentadienyl cation has a $\pi$ molecular orbital system composed from five consecutive *p* orbitals, and the two bonding $\pi$ molecular orbitals are occupied by two pairs of electrons, respectively. The lobes of the unoccupied nonbonding $\pi$ molecular orbital of pentadienyl cation **1–61** define the electron-deficient positions of the $\pi$ molecular orbital system, and these positions are the same locations at which the carbocation is found in the resonance structures. The oxygens at two of these three locations stabilize the carbocation by $\pi$ electron donation, and this is responsible for the regioselectivity of the reaction. This electron donation can be represented by two resonance structures in addition to those in Reaction 1–97

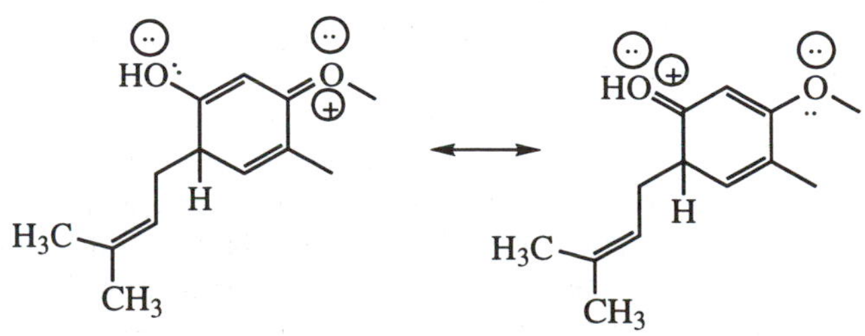

**1–62**

The addition of the dimethylallyl carbocation to the phenyl ring destroys the aromatic $\pi$ molecular orbital system by introducing a saturated carbon. The only way the aromatic $\pi$ system can be regained is by loss of either the original electrophile or the proton from the same saturated carbon. Loss of a proton leaves behind the two electrons formerly in the carbon–hydrogen $\sigma$ bond and also permits that carbon to rehybridize $sp^2$, $sp^2$, $sp^2$, $p$. The $p$ orbital of this carbon, the two electrons, and the pentadienyl cation together can form the aromatic $\pi$ molecular orbital system of the product

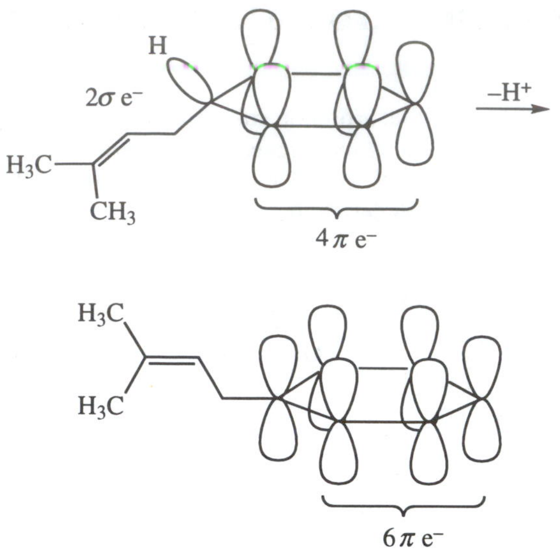

H $2\sigma$ e$^-$ $H_3C$ $CH_3$ $4\pi$ e$^-$ $-H^+$

$H_3C$ $H_3C$ $6\pi$ e$^-$

(1–98)

# Concerted Reactions

The most unequivocal concerted reaction in biochemistry is the reaction catalyzed by chorismate mutase

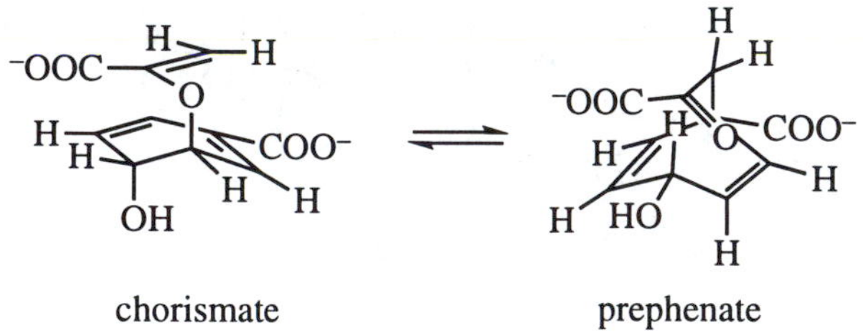

chorismate ⇌ prephenate

(1–99)

In this reaction, six atoms are involved in the rearrangement of the electrons. A $\sigma$ bond is broken, a new $\sigma$ bond is formed, and two double bonds migrate to adjacent positions. The reaction proceeds through an **aromatic transition state** in which all six of the atoms involved are hybridized $sp^2$, $sp^2$, $sp^2$, $p$; in which neither of the $\sigma$ bonds involved in the rearrangement is present; and in which the cyclic $\pi$ molecular orbital system formed from the six parallel $p$ orbitals is occupied by six $\pi$ electrons (Figure 1–16). The transition state is approached by two overlaps: one between the unoccupied antibonding molecular orbital of the $\sigma$ bond and the occupied bonding $\pi$ molecular orbital of one of the double bonds and the other between the occupied bonding molecular orbital of the $\sigma$ bond and the unoccupied antibonding $\pi$ molecular orbital of the other double bond. All six of the $p$ atomic orbitals mixing in the transition state can readily align parallel to each other and no serious electron repulsion occurs in the transition state. The carboxylate of the migrating enol ether sits above a hydrogen in the plane of the ring, and the uninvolved double bond in the ring provides an empty antibonding $\pi$ molecular orbital for overlap with the $\pi$ molecular orbital system of the carboxylate. The ring, because it bears only one saturated carbon in the transition state, can adopt a completely planar geometry (Figure 1–16), and this assists the overlaps. Although aromatic, the transition state is unstable with respect to reactant and product because it has one less $\sigma$ bond and only one more $\pi$ bond. It therefore readily decomposes to the reactant or product, either of which is more stable than the transition state.

Reactions that have been shown to proceed through concerted mechanisms always display this **economy of mechanism**. Simultaneous overlaps between occupied and unoccupied atomic or molecular orbitals, respectively, are stereochemically efficient; but, perhaps even more important, overlaps between two occupied molecular or atomic orbitals that would lead to electron repulsion are avoided. Many more concerted mechanisms have been proposed than have been demonstrated to occur. In many instances, a proposed concerted mechanism, when examined closely, has many problems associated with it.

Consider the following mechanism proposed for carboxyl transfer from biotin[102]

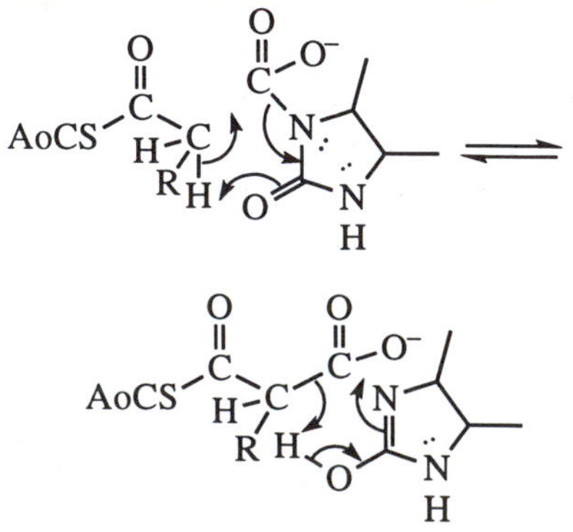

(1–100)

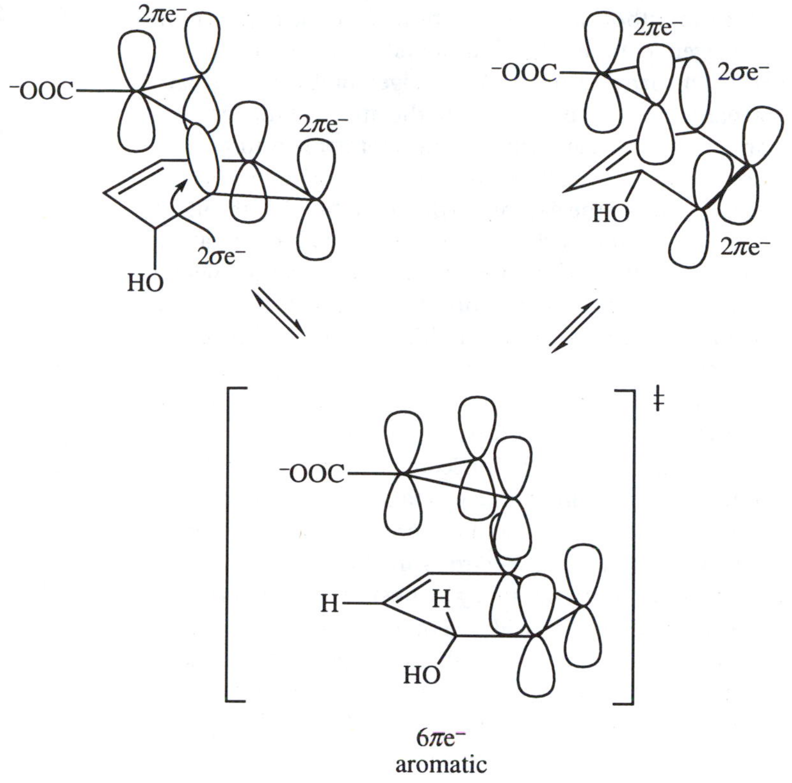

**Figure 1–16:** Reactant, transition state, and product of the reaction catalyzed by chorismate mutase. The reactant, chorismate (upper left), has a $\sigma$ bond between oxygen and carbon flanked by two carbon–carbon double bonds. In the transition state (bottom) the carbon–oxygen bond has broken to produce two additional *p* orbitals, and these new *p* orbitals and the four *p* orbitals in the reactant now form an unbroken ring of six *p* orbitals. The $\pi$ molecular orbital system is occupied with six electrons. This aromatic $\pi$ system holds together the two fragments that are no longer connected by a $\sigma$ bond. When a $\sigma$ bond forms between two *p* orbitals on the opposite side of the ring, the product prephenate (upper right) is produced. It also has a $\sigma$ bond flanked by two double bonds.

Superficially, it resembles the mechanism of the chorismate mutase reaction

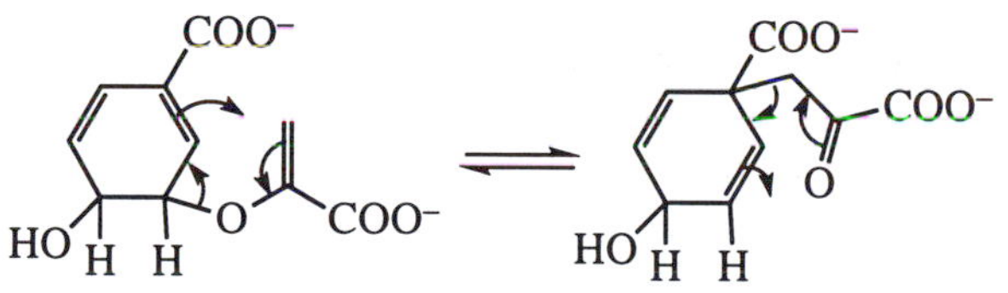

(1–101)

When the details of Mechanism 1–100 are examined, however, severe difficulties arise. It is more informative to try to build molecular models, but a drawing will indicate some of the problems

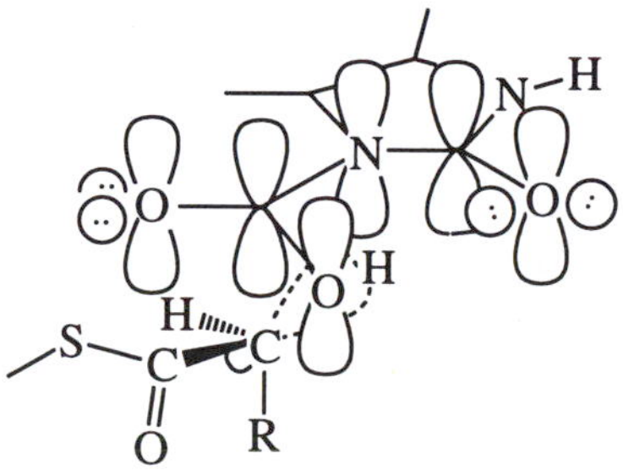

**1–63**

Mechanism 1–100 states that the pair of $\pi$ electrons between the carbon and the oxygen of the *N*-carboxyurea act as a general base to remove a proton from carbon. Although $\pi$ electrons of this type are not usually basic, Mechanism 1–100 states that they are coupled to the incipient carbanion on the carbon $\alpha$ to the carbonyl, and this enhances their basicity. The coupling proposed, however, has as one of its links an overlap between the $\sigma$ bond between the carbon of the carbamate and the nitrogen of the *N*-carboxyurea. This $\sigma$ bond is orthogonal (Structure **1–63**) to the $\pi$ bond between the carbon and the oxygen of the *N*-carboxyurea and cannot overlap efficiently. This problem, however, can be remedied, at least on paper, by proposing that the electrons in the carbon–nitrogen $\sigma$ bond are pushed onto the nitrogen simultaneously with the movement of the free $\pi$ lone pair on the nitrogen toward the acyl oxygen

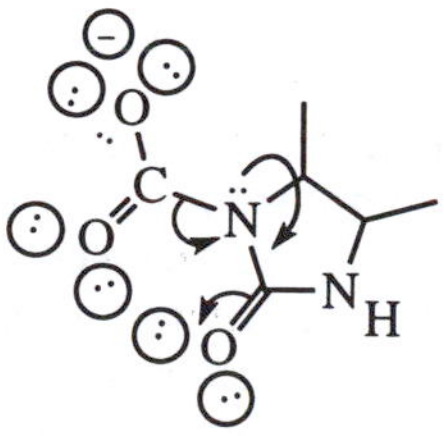

**1–64**

These two movements could not be electronically concerted because they are orthogonal to each other, but they save the mechanism from one of its problems. It seems unlikely, however, that the basicity of the incipient carbanion can be relayed through this link or that protonation of the oxygen of the urea even with a strong acid, let alone a carbon acid, could cause the $\pi$ bond between the carbon and the oxygen to enhance the withdrawal of electrons from an orthogonal carbon–nitrogen $\sigma$ bond significantly. Mechanism 1–100 also states that the leaving group, the nitrogen of the urea, departs the carbamate simultaneously with the attack of the incipient carbanion without the formation of a tetravalent intermediate, which is unprecedented. Mechanism 1–100 states that the pair of electrons in the carbon–hydrogen $\sigma$ bond of the incipient carbanion attacks the carbon of the carbamate at the same time that the $\pi$ electrons of the carbon–oxygen double bond are removing the proton. This implies that a carbon–hydrogen $\sigma$ bond is nucleophilic, which is highly unlikely. The mechanism would also involve a pentavalent carbon in the transition state as the proton is departing and the incipient carbon–carbon bond is forming, and during this process the atomic orbital on the carbon losing the proton would have to overlap with both an atomic orbital on the carbamate carbon and the *s* atomic orbital of the proton being transferred. The attack of the pair of electrons in the carbon–hydrogen $\sigma$ bond must occur from directly below the carbon of the carbamate to avoid overlap with occupied $\pi$ molecular orbitals (Structure **1–63**), and this places the hydrogen in this bond too far from the $\pi$ electrons between the carbon and the oxygen of the *N*-carboxyurea for efficient overlap. It is not surprising that it has been demonstrated that this reaction occurs in steps rather than by a concerted mechanism.[103,104]

## *Suggested Reading*

Stubbe, J., Fish, S., & Abeles, R.H. (1980) Are Carboxylations Involving Biotin Concerted or Nonconcerted? *J. Biol. Chem.* 255, 236–242.

## PROBLEM 1–1

The following reactions are simplified versions of actual reactions that occur in metabolism. In most of them the complicated functional groups of the actual metabolites have been replaced by methyl or ethyl groups to make it simpler to write the compounds. Therefore, these specific reactions do not necessarily occur anywhere. Nevertheless, as written, the chemistry is identical to the chemistry of the reactions between the more complicated biological molecules that are catalyzed by enzymes. The majority of the chemical reactions that occur in metabolism are represented here. The remainder of the reactions in metabolism are catalyzed by enzymes that use coenzymes, and these will be discussed in the next chapter. These reactions are variations or combinations of eight distinct reaction classes: reactions of acetals and ketals, nucleophilic addition to a carbonyl, acyl exchange, phosphoryl exchange, formal hydride transfer, reactions of enols, decarboxylation, and addition–elimination. Assume each of these reactions occurs in aqueous solution so that proton donors (acids) and proton acceptors (bases) are always available. Write complete chemical mechanisms for these reactions. Where hydrides are shown it is assumed that $NAD^+$ would be the carrier of them.

(A) $CH_3CH_2OH \rightleftharpoons CH_3CHO + H^- + H^+$

(B) $CH_3COCH_2OH \rightleftharpoons CH_3CH(OH)CHO$

(C) $CH_3CH(OH)CHO + HSCH_2CH_2OH \rightleftharpoons CH_3CH(OH)COSCH_2CH_2OH + H^- + H^+$

(D) $CH_3CH_2NH_3^+ + CH_3COCH_2OH \rightleftharpoons CH_3CH_2\overset{+}{N}HC(CH_3)CH_2OH$

(E) $CH_3CH_2\overset{+}{N}HC(CH_3)CH_2OH + CH_3CHO \rightleftharpoons CH_3CH_2\overset{+}{N}HC(CH_3)CH(OH)CH(OH)CH_3$

(F) $CH_3CH_2NH_3^+ + HCHO + H^- + H^+ \rightleftharpoons CH_3CH_2\overset{+}{N}H_2CH_3 + H_2O$

(G) $CH_3CH_2CONHCH_3 + H_2O \rightleftharpoons CH_3CH_2COO^- + H_3\overset{+}{N}CH_3$

(H) $CH_3COSCH_3 + CH_3CHO + CH_3CHO \rightleftharpoons CH_3CH(OH)CH_2COSCH_3$

(I) $CH_3COOCH_3 + CH_3\overset{+}{N}H_3 \rightleftharpoons CH_3CONHCH_3 + H^+ + HOCH_3$

(J) $CH_3COCH_2COO^- + H_3O^+ \rightleftharpoons 2CH_3COO^-$

(K) $2CH_3COSCH_3 \rightleftharpoons CH_3COCH_2COSCH_3 + CH_3SH$

(L) $CH_3CHCHCOCH_3 + CO_2 \rightleftharpoons {}^-OOCCH_2CHCHCOCH_3 + H^+$

(M) $CH_3COSCH_3 + CO_2 \rightleftharpoons {}^-OOCCH_2COSCH_3 + H^+$

(N) $CH_3COSCH_3 + HOPO_3^{2-} \rightleftharpoons CH_3COOPO_3^{2-} + HSCH_3$

(O) $CH_3COSCH_3 + CH_3OH \rightleftharpoons CH_3COOCH_3 + HSCH_3$

(P) $H_2NCOOPO_3^{2-} + CH_3\overset{+}{N}H_3 \rightleftharpoons H_2NCONHCH_3 + HOPO_3^{2-} + H^+$

(Q) $CH_3CONHCH_3 + C_2H_5\overset{+}{N}H_3 \rightleftharpoons CH_3(NC_2H_5)NHCH_3 + H_2O + H^+$

(R) $H_2NCH\overset{+}{N}H_2 + H_2O \rightleftharpoons H_2NCONH_2 + H^- + 2H^+$

(S) $H^+ + CH_3CONH_2 + CH_3COO^- \rightleftharpoons CH_3CONHCOCH_3 + H_2O$

(T) $CH_3COOPO_3^{2-} + H^- \rightleftharpoons CH_3CHO + HOPO_3^{2-}$

(U) $CH^3CHO + H_2O \rightleftharpoons CH_3COO^- + H^- + 2H^+$

(V) $CH_3C(OCH_2CH_3)_2CH_3 + H_2O \rightleftharpoons CH_3COCH_3 + 2C_2H_5OH$

(W) $CH_3C(OCH_2CH_3)_2CH_3 + HOPO_3^{2-} \rightleftharpoons$

$CH_3C(OCH_2CH_3)(OPO_3^{2-})CH_3 + CH_3CH_2OH$

(X) $CH_3CH(OCH_2CH_3)(OPO_3^{2-}) + CH_3\overset{+}{N}H_3 \rightleftharpoons$

$CH_3CH(OCH_2CH_3)(\overset{+}{N}H_2CH_3) + HOPO_3^{2-}$

(Y) $CH_3CH(OH)CH_2COSCH_3 \rightleftharpoons CH_3CHCHCOSCH_3 + H_2O$

(Z) *cis*-$CH_3CHCHCH_3 \rightleftharpoons$ *trans*-$CH_3CHCHCH_3$

(a) $CH_3CHCHCH_2OPO_3^{2-} \rightleftharpoons CH_2CHCHCH_2 + HOPO_3^{2-}$

(b) $CH_2CHCHCH_2 + H_2O \rightleftharpoons CH_3CH_2CH_2CHO$

(c) $CH_3CH(OH)CH_2COO^- + H^+ \rightleftharpoons CH_3CHCH_2 + CO_2 + H_2O$

(d) $CH_3CH(OH)CH_2OPO_3^{2-} + CH_3COCH_3 \rightleftharpoons$

$CH_3COCH_2C(CH_3)_2OH$

(e) HO–(cyclohexadiene)–$COO^-$ $\rightleftharpoons$

(benzene) $+ H_2O + CO_2$

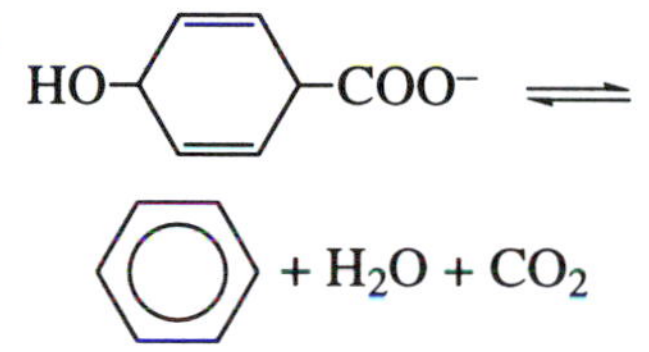

(f) $H_2C(OH)CH(\overset{+}{N}H_2CH_3)(OC_2H_5) \rightleftharpoons$

$HCOCH_2\overset{+}{N}H_2CH_3 + C_2H_5OH$

(g) $CH_3CH(OH)CH_2COO^- \rightleftharpoons CH_3CH_2CH(OH)COO^-$

(h) $CH_3COCH_2COO^- \rightleftharpoons CH_3COCH_3 + CO_2$

(i) $CH_2 = CHCH_2OPO_3^{2-} + CH_2CHCH_3 \rightleftharpoons$

$CH_2CHCH_2CH_2CHCH_2 + HOPO_3^{2-}$

(j) $CH_3CH_2OH + HOPO_3^{2-} \rightleftharpoons CH_3CH_2OPO_3^{2-} + H_2O$

(k) $CH_3COOPO_3^{2-} + CH_3OH \rightleftharpoons$

$CH_3COO^- + CH_3OPO_3^{2-} + H^+$

(l) $CH_3COO^- + (CH_3)_2\overset{+}{S}C_2H_5 \rightleftharpoons CH_3COOCH_3 + CH_3SC_2H_5$

**PROBLEM 1–2**

Write the mechanisms for the following reactions

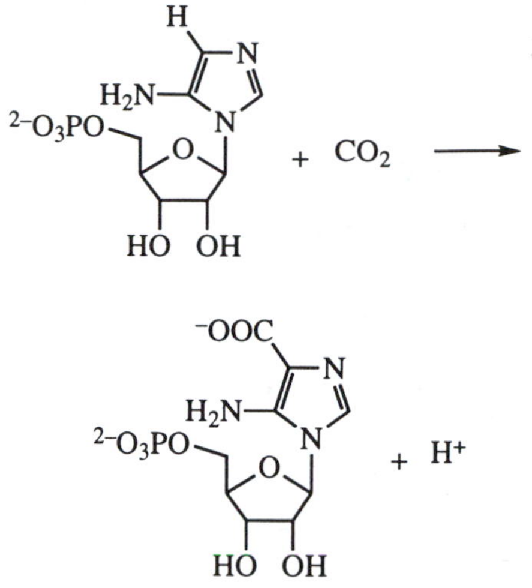

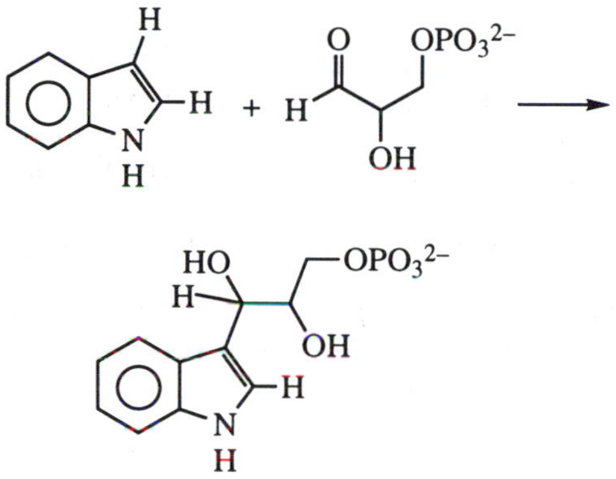

H, N, $H_2N$, $^{2-}O_3PO$, N, O, HO OH $+ CO_2 \longrightarrow$

$^-OOC$, N, $H_2N$, $^{2-}O_3PO$, N, O, HO OH $+ H^+$

H, H, N, H $+$ H, O, $OPO_3^{2-}$, OH $\longrightarrow$

HO, H, $OPO_3^{2-}$, OH, H, N, H

**PROBLEM 1–3**

In animal cells, D-glucosamine is made from the ammonium cation ($NH_4^+$) and D-fructose. Write the complete chemical mechanism for this reaction.

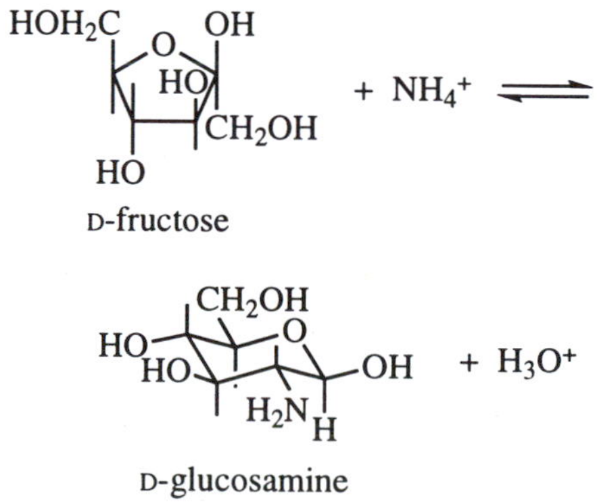

$HOH_2C$, O, OH, HO, $CH_2OH$, HO $+ NH_4^+ \rightleftharpoons$

D-fructose

$CH_2OH$, O, HO, HO, OH, $H_2N$, H $+ H_3O^+$

D-glucosamine

## PROBLEM 1–4

Write a mechanism for this reaction that involves only carbonyl chemistry rather than electrocyclic chemistry.

## PROBLEM 1–5

Criticize the following proposal for a concerted reaction:

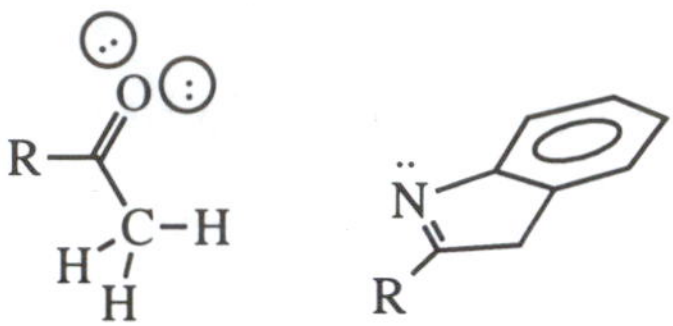

## References


1. Eigen, M. (1964) *Angew. Chem., Int. Ed. Engl. 3,* 1–19.
2. Yang, X., & Castleman, A.W. (1989) *J. Am. Chem. Soc. 11,* 6845–6846.
3. Wei, S., Shi, Z., & Castleman, A.W. (1991) *J. Chem. Phys. 94,* 3268–3278.

3a. Kyte, J. (1995) *Structure in Protein Chemistry,* p 159, Garland Publishing, New York.

4. Debye, P. (1942) *Trans. Electrochem. Soc. 82,* 265–272.

4a. Kyte, J. (1995) *Structure in Protein Chemistry,* p 177, Garland Publishing, New York.

5. Perrin, C.L. (1986) *J. Am. Chem. Soc. 108,* 6807–6808.

5a. Kyte, J. (1995) *Structure in Protein Chemistry,* pp 358–360, Garland Publishing, New York.

6. Romero, R., Stein, R., Bull, H.G., & Cordes, E.H. (1978) *J. Am. Chem. Soc. 100,* 7620–7624.
7. Young, P.R., & Jencks, W.P. (1977) *J. Am. Chem. Soc. 99,* 8238–8247.
8. Kresge, A.J., & Weeks, D.P. (1984) *J. Am. Chem. Soc. 106,* 7140–7143.
9. Kankaanpera, A., Salomaa, P., Juhala, P., Aaltonen, R., & Mattsen, M. (1973) *J. Am. Chem. Soc. 95,* 3618–3624.
10. Fife, T.H., & Pellino, A.M. (1980) *J. Am. Chem. Soc. 102,* 3062–3071.
11. Gilbert, H.F., & Jencks, W.P. (1982) *J. Am. Chem. Soc. 104,* 6769–6779.
12. Deslongchamps, P. (1975) *Tetrahedron 31,* 2463–2490.
13. Perrin, C.L., & Arrhenius, G.M. (1982) *J. Am. Chem. Soc. 104,* 2839–2842.
14. Perrin, C.L., & Nuñez, O. (1986) *J. Am. Chem. Soc. 108,* 5997–6003.
15. Kanchuger, M.S., & Byers, L.D. (1979) *J. Am. Chem. Soc. 101,* 3005–3010.
16. Burkey, T.J., & Fahey, R.C. (1983) *J. Am. Chem. Soc. 105,* 868–871.
17. Barnett, R.E., & Jencks, W.P. (1969) *J. Am. Chem. Soc. 91,* 6758–6765.
18. Gilbert, H.F., & Jencks, W.P. (1977) *J. Am. Chem. Soc. 99,* 7931–7947.
19. Funderburk, L.H., Aldwin, L., & Jencks, W.P. (1978) *J. Am. Chem. Soc. 100,* 5444–5459.
20. Sayer, J.M., & Jencks, W.P. (1977) *J. Am. Chem. Soc. 99,* 464–474.
21. Funderburk, L.H., & Jencks, W.P. (1978) *J. Am. Chem. Soc. 100,* 6708–6714.
22. Hine, J., Cholod, M.S., & Chess, W.K. (1973) *J. Am. Chem. Soc. 95,* 4270–4271.
23. Menger, F.M., Chow, J.F., Kaiserman, H., & Vasquez, P.C. (1983) *J. Am. Chem. Soc. 105,* 4996–5002.
24. Cox, M.M., & Jencks, W.P. (1981) *J. Am. Chem. Soc. 103,* 580–587.
25. Kirsch, J.F., & Kline, A. (1969) *J. Am. Chem. Soc. 91,* 1841–1847.
26. Bruice, T.C., Hegarty, A.F., Felton, S.M., Donzel, A., & Kundu, N.G. (1970) *J. Am. Chem. Soc. 92,* 1370–1376.
27. Bell, R.P. (1966) *Adv. Phys. Org. Chem. 4,* 1–29.
28. Satterthwait, A.C., & Jencks, W.P. (1974) *J. Am. Chem. Soc. 96,* 7018–7044.
29. McClelland, R.A., & Patel, G. (1981) *J. Am. Chem. Soc. 103,* 6912–6915.
30. Edward, J.T., & Wong, S.C. (1977) *J. Am. Chem. Soc. 99,* 7224–7228.
31. Santry, L.J., & McClelland, R.A. (1983) *J. Am. Chem. Soc. 105,* 3167–3172.
32. Oakenfull, D.G., & Jencks, W.P. (1971) *J. Am. Chem. Soc. 93,* 178–188.
33. Oakenfull, D.G., Salvesen, K., & Jencks, W.P. (1971) *J. Am. Chem. Soc. 93,* 188–194.
34. Lee, Y.N., & Schmir, G. (1978) *J. Am. Chem. Soc. 100,* 6700–6707.
35. Williams, A. (1976) *J. Am. Chem. Soc. 98,* 5645–5651.
36. Fersht, A.R. (1971) *J. Am. Chem. Soc. 93,* 3504–3515.
37. Fersht, A.R., & Jencks, W.P. (1970) *J. Am. Chem. Soc. 92,* 5432–5442.
38. Rocek, J., Westheimer, F.H., Eschenmoser, L., Moldovnyi, L., & Schreiber, J. (1962) *Helv. Chim. Acta 45,* 2554–2567.
39. Hupe, D.J., & Jencks, W.P. (1977) *J. Am. Chem. Soc. 99,* 451–465.
40. Cox, M.M., & Jencks, W.P. (1981) *J. Am. Chem. Soc. 103,* 572–580.
41. Barnett, R.E., & Jencks, W.P. (1969 *J. Am. Chem. Soc. 91,* 2358–2369.
42. Jencks, W.P. (1970) in *Handbook of Biochemistry* (Sober, H.A., Ed.) pp J-181–J-186, CRC Press, Cleveland, OH.
43. Fersht, A.R., & Requena, Y. (1971) *J. Am. Chem. Soc. 93,* 3499–3504.
44. Gerstein, J., & Jencks, W.P. (1964) *J. Am. Chem. Soc. 86,* 4655–4663.
45. Berry, R.S. (1960) *J. Chem. Phys. 32,* 933–938.
46. Dennis, E.A., & Westheimer, F.H. (1966) *J. Am. Chem. Soc. 88,* 3432–3433.

47. Hamilton, W.C., LaPlaca, S.J., & Ramirez, F. (1965) *J. Am. Chem. Soc. 87*, 127–128.
48. Abbott, S.J., Jones, S.R., Weinman, S.A., Bockhoff, F.M., McLafferty, F.W., & Knowles, J.R. (1979) *J. Am. Chem. Soc. 101*, 4323–4332.
49. Buchwald, S.L., & Knowles, J.R. (1980) *J. Am. Chem. Soc. 102*, 6601–6602.
50. Buchwald, S.L., & Knowles, J.R. (1982) *J. Am. Chem. Soc. 104*, 1438–1440.
51. Usher, D.A., Richardson, D.I., & Eckstein, F. (1970) *Nature 228*, 663–665.
52. Gutowsky, H.S., & Liehr, A.D. (1952) *J. Chem. Phys. 20*, 1652–1653.
53. Schmutzler, R. (1965) *Angew. Chem., Int. Ed. Engl. 4*, 496–508.
54. Downs, J.J., & Johnson, R.E. (1955) *J. Am. Chem. Soc. 77*, 2098–2102.
55. Kluger, R., Covitz, F., Dennis, E., Williams, L.D., & Westheimer, F.H. (1969) *J. Am. Chem. Soc. 91*, 6066–6072.
56. Buchwald, S.L., Pliura, D.H., & Knowles, J.R. (1984) *J. Am. Chem. Soc. 106*, 4916–4922.
57. Buchwald, S.L., Friedman, J.M., & Knowles, J.R. (1984) *J. Am. Chem. Soc. 106*, 4911–4916.
58. Kirby, A.J., & Varvoglis, A.G. (1967) *J. Am. Chem. Soc. 89*, 415–423.
59. Butcher, W.W., & Westheimer, F.H. (1955) *J. Am. Chem. Soc. 77*, 2420–2424.
60. Barnard, P.W.C., Bunton, C.A., Llwellyn, D.R., Oldham, K.G., Silver, B.L., & Vernon, C.A. (1955) *Chem. Ind. (London)* 760–763.
61. Clapp, C.H., & Westheimer, F.H. (1974) *J. Am. Chem. Soc. 96*, 6710–6714.
62. Bourne, N., & Williams, A. (1984) *J. Am. Chem. Soc. 106*, 7591–7596.
63. Skoog, M.T., & Jencks, W.P. (1984) *J. Am. Chem. Soc. 106*, 7597–7606.
64. Filman, D.J., Bolin, J.T., Matthews, D.A., & Kraut, J. (1982) *J. Biol. Chem. 257*, 13663–13672.
65. Karle, I.L. (1961) *Acta Crystallogr. 14*, 497–502.
66. Koyama, H. (1963) *Z. Kristall. 118*, 51–68.
67. Wright, W.B., & King, G.S.D. (1954) *Acta Crystallogr. 7*, 283–288.
68. Grau, U., Kapmeyer, H., & Trommer, W.E. (1978) *Biochemistry 17*, 4621–4626.
69. Grau, U.M., Trommer, W.E., & Rossmann, M.G. (1981) *J. Mol. Biol. 151*, 289–307.
70. Santhanam, K.S.V., & Elving, P.J. (1973) *J. Am. Chem. Soc. 95*, 5482–5490.
71. Powell, M.F., Wu, J., & Bruice, T.C. (1984) *J. Am. Chem. Soc. 106*, 3850–3856.
72. Gase, R.A., & Pandit, U.K. (1979) *J. Am. Chem. Soc. 101*, 7059–7064.
73. Powell, M.F., Wong, W.H., & Bruice, T.C. (1982) *Proc. Natl. Acad. Sci. U.S.A. 79*, 4604–4608.
74. Fontecilla-Camps, J.C., Bugg, C.E., Temple, C., Rose, J.D., Montgomery, J.A., & Kisliuk, R.L. (1979) *J. Am. Chem. Soc. 101*, 6114–6115.
75. Benkovic, S.J., Benkovic, P.A., & Comfort, D.R. (1969) *J. Am. Chem. Soc. 91*, 5270–5279.
76. Benkovic, S.J., Bullard, W.P., & Benkovic, P.A. (1972) *J. Am. Chem. Soc. 94*, 7542–7549.
77. Jordon, P.M., & Akhtar, M. (1970) *Biochem. J. 116*, 277–286.
78. Benkovic, S.J., Barrows, T.H., & Farina, P.R. (1973) *J. Am. Chem. Soc. 95*, 8414–8420.
79. Burdick, B.A., Benkovic, P.A., & Benkovic, S.J. (1977) *J. Am. Chem. Soc. 99*, 5716–5725.
80. Tan, L.U., Drury, E.J., & MacKenzie, R.E. (1977) *J. Biol. Chem. 252*, 1117–1122.
81. Whitfield, C.D., Steers, E.J., & Weissbach, H. (1970) *J. Biol. Chem. 245*, 390–401.
82. Capon, B., & Zucco, C. (1982) *J. Am. Chem. Soc. 104*, 7567–7572.
83. Guthrie, J.P., & Cullimore, P.A. (1979) *Can. J. Chem. 57*, 240–248.
84. Bruice, T.C., & Benkovic, S.J. (1966) *Bioorganic Mechanisms*, Vol. I, pp 294–297, Benjamin, New York.
85. Lynen, F. (1953) *Fed. Proc. 12*, 683–691.
86. Zacharias, D.E., Murray-Rust, P., Preston, R.M., & Glusker, J.P. (1983) *Arch. Biochem. Biophys. 222*, 22–34.
87. Bender, M.L., & Williams, A. (1966) *J. Am. Chem. Soc. 88*, 2502–2508.
88. Roberts, R.D., Ferran, H.E., Gula, M.J., & Spencer, T.A. (1980) *J. Am. Chem. Soc. 102*, 7054–7058.
89. Taguchi, K., & Westheimer, F.H. (1973) *J. Am. Chem. Soc. 95*, 7413–7418.
90. Raghaven, N.V., & Leussing, D.L. (1976) *J. Am. Chem. Soc. 98*, 723–730.
91. Beak, P., & Siegel, B. (1976) *J. Am. Chem. Soc. 98*, 3601–3605.
92. Richard, J.P. (1984) *J. Am. Chem. Soc. 106*, 4926–4936.
93. Hall, S.S., Doweyko, A.M., & Jordon, F. (1978) *J. Am. Chem. Soc. 100*, 5934–5939.
94. Okuyama, T., Komoguchi, S., & Fueno, T. (1982) *J. Am. Chem. Soc. 104*, 2582–2587.
95. Mohrig, J.R., Schultz, S.C., & Morin, G. (1983) *J. Am. Chem. Soc. 105*, 5150–5151.
96. Chwang, W.K., Nowlan, V.J., & Tidwell, T.T. (1977) *J. Am. Chem. Soc. 99*, 7233–7238.
97. Fedor, L.R. (1969) *J. Am. Chem. Soc. 91*, 908–913.
98. Bartsch, R.A., Mintz, E.A., & Parlman, R.M. (1974) *J. Am. Chem. Soc. 96*, 4249–4252.
99. Fedor, L., & Gray, P.H. (1976) *J. Am. Chem. Soc. 98*, 783–787.
100. Hupe, D.J., Kendall, M.C.R., & Spencer, T.A. (1973) *J. Am. Chem. Soc. 95*, 2271–2278.
101. Hagmann, M.L., Heller, W., & Grisebach, H. (1984) *Eur. J. Biochem. 142*, 127–131.
102. Rétey, J., & Lynen, F. (1965) *Biochem. Z. 342*, 256–271.
103. O'Keefe, S.J., & Knowles, J.R. (1986) *Biochemistry 25*, 6077–6084.
104. Stubbe, J., Fish, S., & Abeles, R.H. (1980) *J. Biol. Chem. 255*, 236–242.

## Chapter TWO

# Coenzymes

In addition to the 20 amino acids and the peptide bonds, which are provided unavoidably by the protein as catalysts for the chemical transformation performed by an enzyme, there are sometimes more complicated catalytic functional groups associated with the enzyme, referred to as coenzymes. A **coenzyme** is any functional group, other than the amino acids of the polypeptide, that is incorporated into an enzyme to assist in its catalysis. Coenzymes are often attached covalently to the protein as **posttranslational modifications.** In this instance, the coenzyme is similar to an amino acid in that it is a covalent participant in the structure of the protein. Often, however, a coenzyme is merely tightly but **noncovalently bound** to the protein. Although a tightly bound coenzyme does not leave the enzyme readily and may remain associated with it throughout its lifetime, it is possible for it to dissociate slowly from the protein even when the protein is in its native state. A noncovalently bound coenzyme always dissociates when the protein is unfolded. As far as the reaction catalyzed by the enyzme is concerned, however, a noncovalently bound coenzyme is a permanent feature of the molecular structure of the enzyme; and, in this sense, it is also indistinguishable from any of its amino acids.

An example of a covalently bound coenzyme would be the 4′-phosphopantetheine attached covalently to a serine residue as a posttranslational modification of animal fatty acid synthase

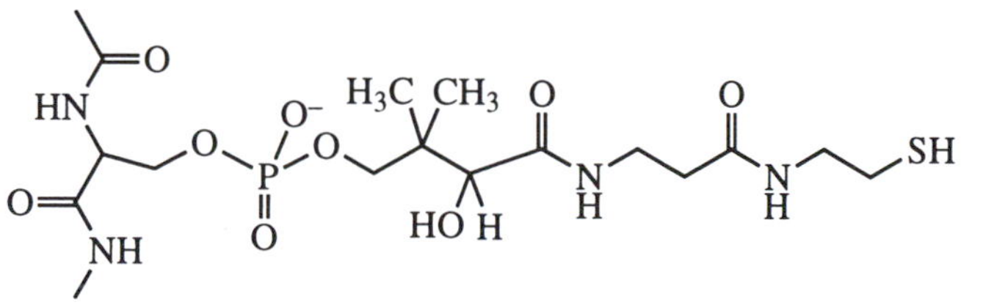

2–1

The thiol at the distal end of the coenzyme is acylated during enzymatic catalysis, and the thioester thus formed participates as an enol in the enzymatic reaction. The role of the 4′-phosphopantetheine as a covalently attached coenzyme in these reactions should be distinguished from its role as a substrate when it is incorporated into coenzyme A. A substrate, as distinguished from a coenzyme, enters and leaves the active site of the enzyme during every turnover. Ironically, coenzyme A is a substrate rather than a coenzyme.

An example of a tightly but noncovalently bound coenzyme would be the oxidized nicotinamide adenine dinucleotide that is tightly bound to UDP-galactose:UDP-glucose epimerase. In the reaction catalyzed by this epimerase

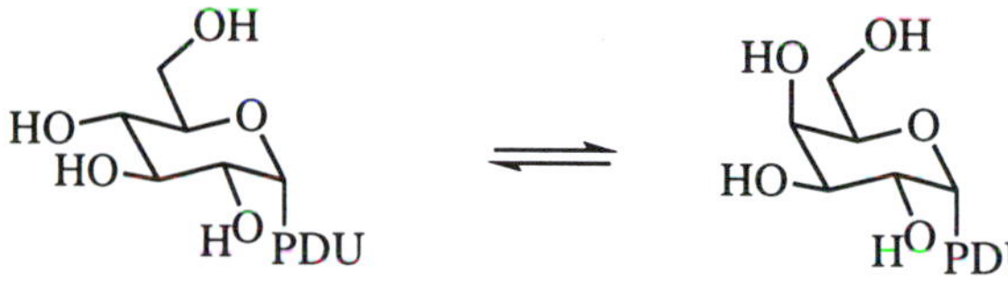

UDP-glucose ⇌ UDP-galactose (2–1)

a hydride ion is removed from the substrate by the $NAD^+$ and then added back with the epimeric stereochemistry to produce the product.[1] Because the $NAD^+$ is regenerated after each catalytic cycle, there is no need for it to enter and leave the active site of the enzyme, and it remains continuously bound as a coenzyme. The role of the $NAD^+$ as a coenzyme in these reactions should be distinguished from its role in almost every other situation where it is a simple substrate that enters and leaves the active site during each turnover.

Coenzymes exist because they effect chemical transformations beyond those that can be accomplished by the side chains of the amino acids. Therefore, a discussion of the mechanisms of coenzymes should precede an analysis of the mechanisms of enzymatic reactions in general.

## Pyridoxal Phosphate

Pyridoxal phosphate has the structure

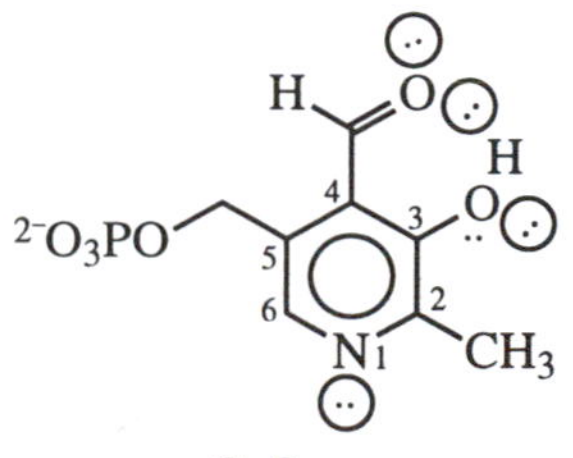

2–2

Because it is a benzylic aldehyde, it is not prone to the many unfortunate side reactions in which aliphatic aldehydes engage with the amino acids of proteins.[1a] Pyridoxal phosphate, when acting as a coenzyme, is usually bound covalently to the resting enzyme as the **pyridoximine** of a lysine[2–4]

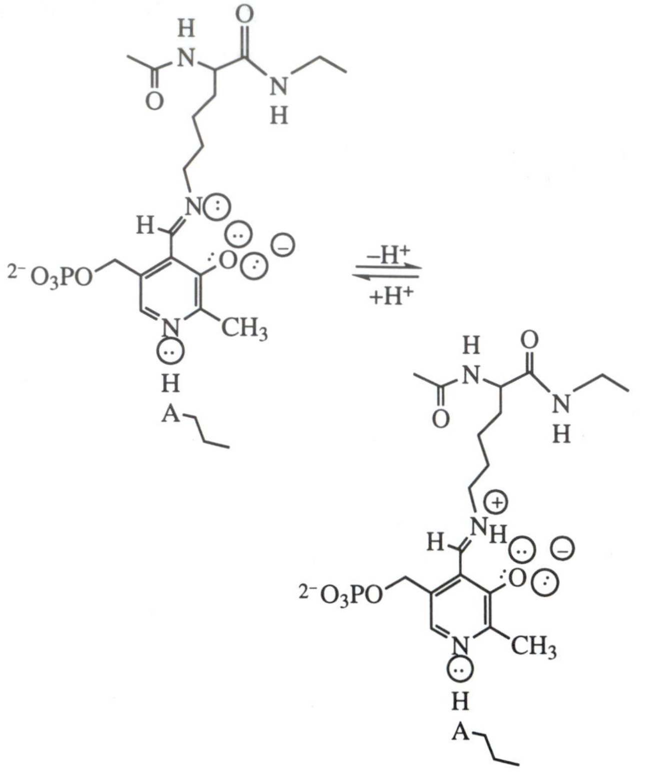

(2–2)

The unprotonated phenolate form of the pyridoximine has its maximum of absorbance at 365 nm,[2,5] and upon protonation of the iminium nitrogen[6] the maximum of absorbance shifts to 415 nm (Figure 2–1).[2,5] This absorption at longer wavelengths arises from the *N*-protonated imine, which is a **vinylogous amide,** a fact most clearly demonstrated by writing the resonance structure[6]

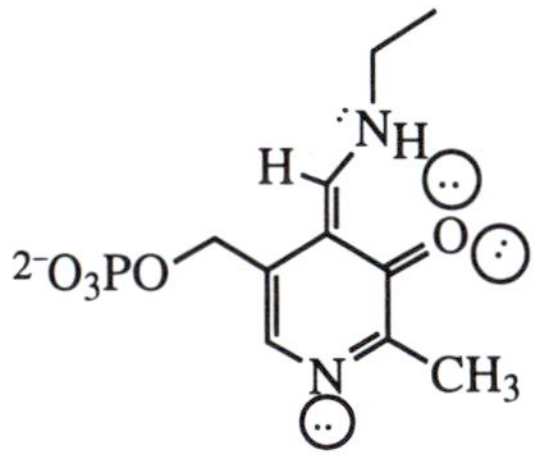

**2–3**

This resonance structure states that a pyridoximine can be viewed as a $\pi$ molecular orbital system encompassing nine atoms

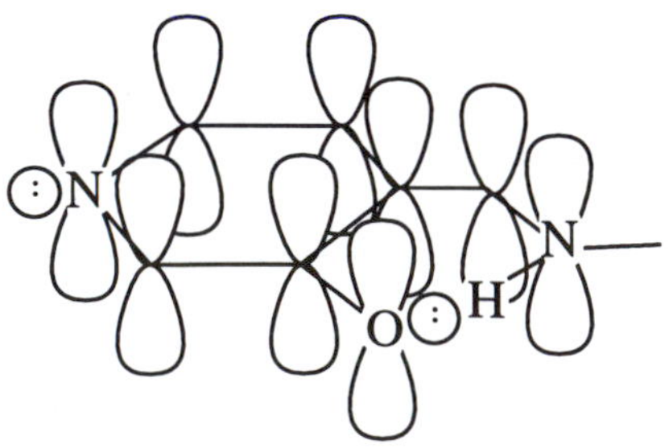

**2–4**

whose nine molecular orbitals are occupied by five pairs of electrons.

The p$K_a$ for the acid–base reaction[2] of Equation 2–2 is near 6 for enzyme-bound pyridoximine. Because the values for the p$K_a$ of pyridoximines formed from amines other than enzymes are usually around 10, the low values for the p$K_a$ of enzyme-bound pyridoximine may reflect either the formation of a strong hydrogen bond between the enzyme and the pyridinium nitrogen of the pyridoximine or the protonation of the pyridoximine at the pyridinium nitrogen by a strong acid on the enzyme. Protonations of the pyridinium nitrogens cause no significant changes in the spectra of the imines of pyridoxal phosphate.[5] Another factor that affects the p$K_a$ of pyridoxal phosphate when it is in imine linkage to an enzyme is the binding of anions.[7] A bound anion, presumably located adjacent to the pyridoximine, can raise the value of this p$K_a$ electrostatically.

The formation of an imine between pyridoxal phosphate and the primary amine of a lysine on the enzyme is a simple addition to a carbonyl and proceeds through the carbinolamine. The slow step in the formation of an imine in aqueous solution is usually the loss of water from the carbinolamine, a step susceptible to both base and acid catalysis (Figure 1–5). In the case of pyridoxal phosphate, the phenolic oxygen catalyzes this dehydration intramolecularly.[8] This catalysis is thought to be due to its ability to act as a base

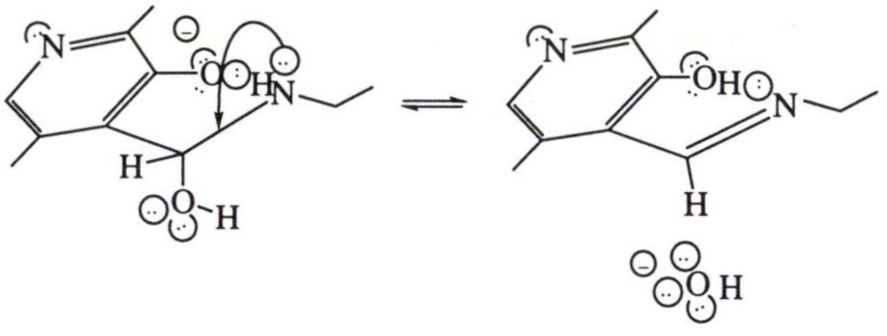

(2–3)

and remove the proton from nitrogen temporarily or simultaneously to enhance the push of the lone pair periplanar to the leaving group. Such intramolecular assistance could be very important within the active site of an enyzme that is shielded from the acids and bases in the surrounding solution.

All of the substrates upon which pyridoxal phosphate acts coenzymatically are primary amines. The first step in every one of these reactions is to form the imine between the nitrogen of the primary amine of the substrate and the carbonyl carbon of the pyridoxal phosphate. If the pyridoxal phosphate is in the form of an imine with a lysine of the enzyme, **transimination**[9,10] must ensue (Figure 2–2). On an enzyme at neutral pH, the reaction would begin with the cationic conjugate acid of the amine and the phenolate anion of the pyridoximine.[11] When there is no phenolic oxygen in an ortho position of a benzylic imine, a transimination involving this imine displays a requirement for both acid and base catalysis. This is consistent with the need to pull protons from the amine that is entering and the coincident need to add protons to the amine that is leaving.[10] When the transimination involves imines of pyridoxal phosphate, however, no requirement for acid or base

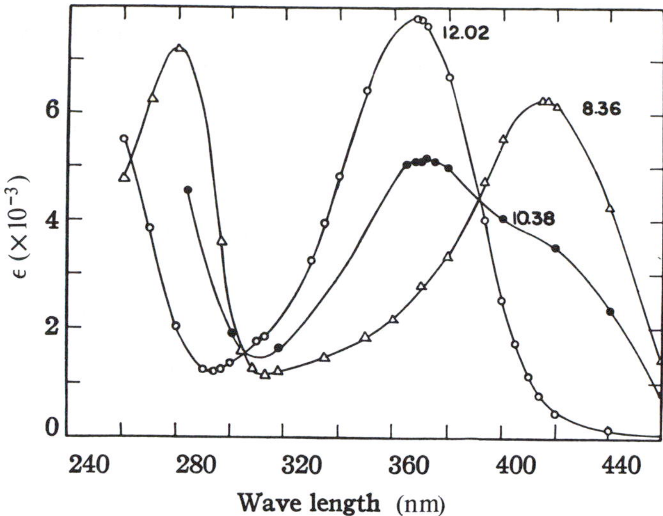


**Figure 2–1:** Spectral shifts associated with protonation of the imine nitrogen of a pyridoximine.[5] A solution was prepared that was 0.2 mM pyridoxal phosphate and 0.5 M valine. At these concentrations the imine between the $\alpha$-amino group of valine and the aldehyde of pyridoxal phosphate forms quantitatively. The absorption spectra as a function of pH were measured. Molar extinction coefficient ($\varepsilon \times 10^{-3}$) is plotted as a function of wavelength (nanometers) for three values of pH (numbers beside the curves). The spectra display two isosbestic points, demonstrating that a clean titration between a base and its conjugate acid is occurring. The p$K_a$ for this acid–base, calculated from a series of such curves, is 9.5. Adapted with permission from ref 5. Copyright 1957 American Chemical Society.

**Figure 2–2:** Transimination catalyzed by the phenolic oxygen of pyridoxal phosphate. The pyridoxal phosphate is initially the imine with lysine (right side) and the reactant in the transformation is the conjugate acid of a primary amine (left side). As the reaction progresses, protons are transferred from the nitrogen of the reactant to the nitrogen of the lysine. The phenolate oxygen could shuttle as many as three protons from the entering ammonium cation to the departing lysinium cation, but other acid–bases may be involved as well. The phenolate would perform this transfer by swinging back and forth between the two nitrogens in this symmetrical reaction.

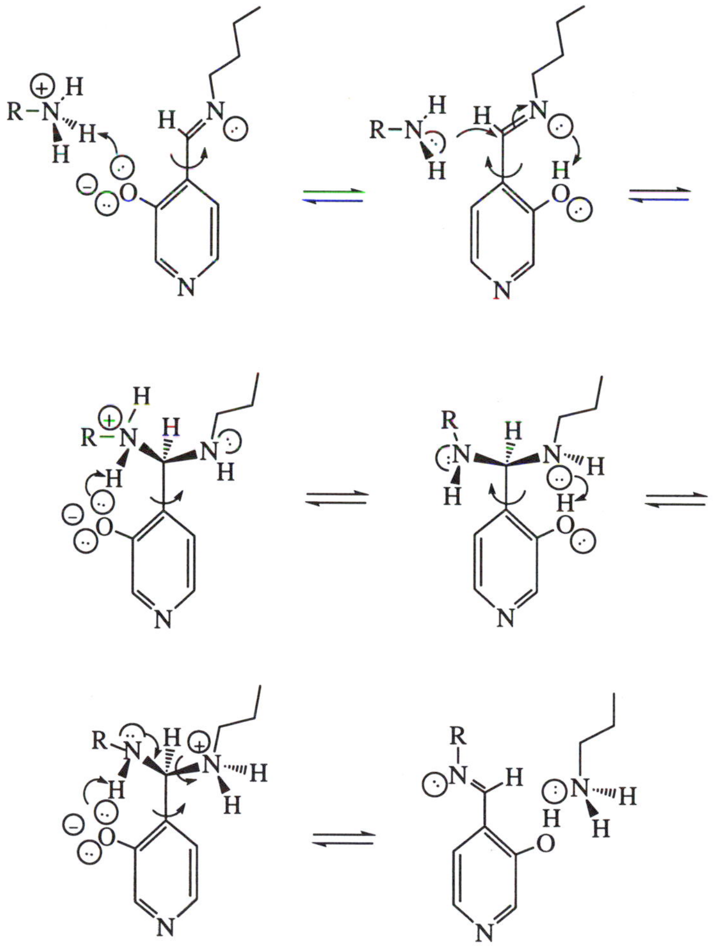

catalysis is observed.[9] This lack of catalytic requirement can be readily explained by the fact that the phenolic oxygen is four atoms away from both of the nitrogens and can easily shuttle protons between them (Figure 2–2). As the protons are shuttled, the aromatic ring would act as if it were a door swinging between the two nitrogens. This intramolecular proton transfer would be of even greater importance on the active site of an enzyme shielded from the solution.

There had been some doubt[12] as to whether or not the intermediates in the transimination of pyridoxal phosphate would be stable enough that the reaction could proceed through the *gem*-diamine as shown in Figure 2–2 rather than passing through the aldehyde. It was subsequently shown,[9] however, that such a direct transimination in aqueous solution proceeds about twice as fast as the exchange through the indirect route in which the aldehyde, pyridoxal phosphate itself, is the intermediate. In an active site, shielded from water, the indirect route which requires water necessarily, to provide the oxygen of the aldehyde, would be even slower.

When an amine is coupled as an imine to pyridoxal phosphate, and the pyridinium nitrogen of the pyridoximine is protonated, the three $\sigma$ bonds on the carbon to which the nitrogen of that amine is attached become electron-deficient and susceptible to heterolytic cleavage.[6] This results from the coulomb effect exerted by this pyridinium nitrogen withdrawing electron density from the distal end of the $\pi$ molecular orbital system. This, in turn, withdraws electron density from whichever of the three $\sigma$ bonds happens to be aligned parallel to the $\pi$ molecular orbital system

$\pi$ electron withdrawal

$\sigma$ electron withdrawal

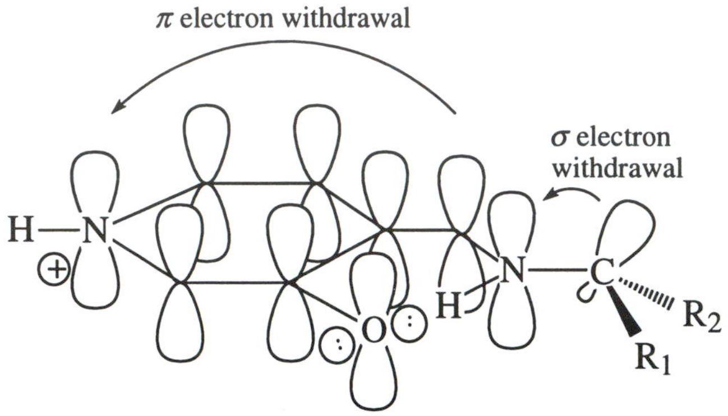

**2–5**

This electron withdrawal can be symbolized by writing the resonance structure

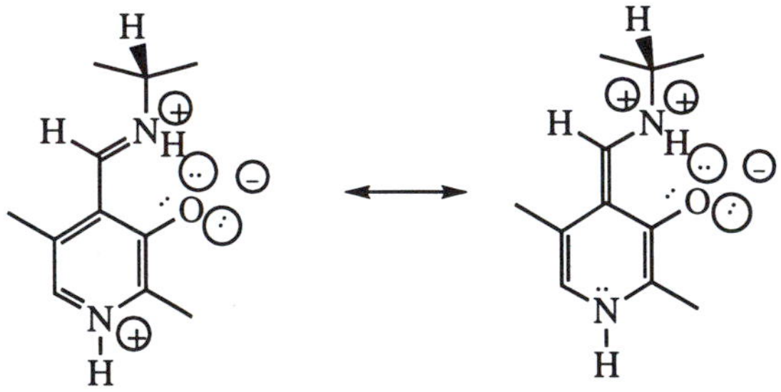

**2–6**

This resonance structure is not so satisfying as Structure **2–5** because it involves a dication missing two valence electrons on the iminium nitrogen. This resonance structure, however, illustrates that an imine of pyridoxal phosphate, with its pyridinium nitrogen protonated, is a vinylogous iminium cation

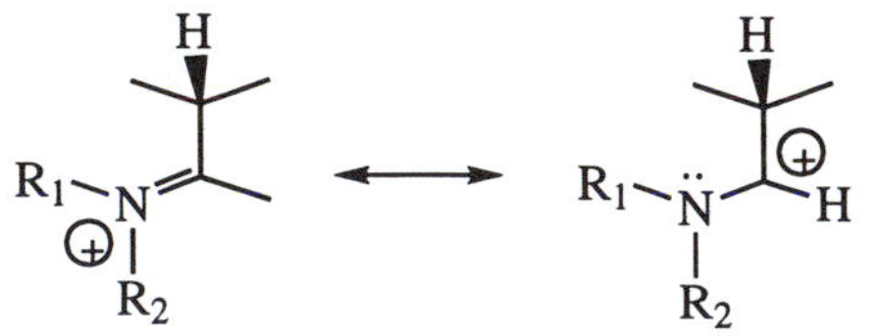

**2–7**

and has the same effect of weakening a parallel $\sigma$ bond on the adjacent carbon.

If the carbon $\alpha$ to a pyridoximine is bonded to an incipient electrophile, that electrophile can depart. For example, if the $\alpha$-carbon of a pyridoximine contains a **hydrogen,** that hydrogen can depart as a proton. This occurs in the reaction catalyzed by 4-aminobutyrate aminotransferase, the first step of which is the loss of a proton from the $\alpha$-carbon[13] of the imine formed between pyridoxal phosphate and 4-aminobutyrate

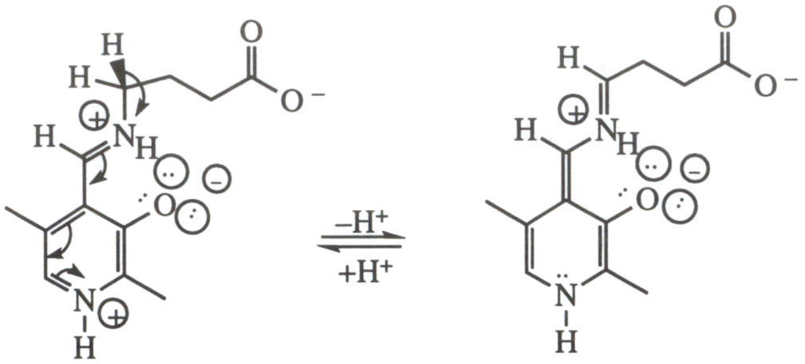

$-H^+$ / $+H^+$

(2–4)

If the $\alpha$-carbon of a pyridoximine contains a **carboxylate,** that carboxylate can depart as a molecule of the dicarbonyl compound, carbon dioxide. An example of this is the first step in the forward reaction catalyzed by 5-aminolevulinate synthase,[14] which proceeds with the loss of carbon dioxide from the imine formed between glycine and pyridoxal phosphate

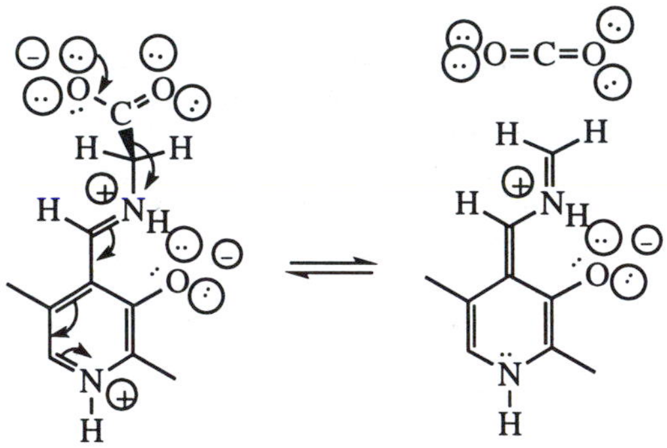

(2–5)

If the $\alpha$-carbon of a pyridoximine contains an **incipient carbonyl group** or **imine** in the form of a $\beta$-carbon to which a hydroxyl group or an amino group is attached, respectively, that carbon can depart as a carbonyl group. For example, one of the steps in the reaction catalyzed by glycine hydroxymethyltransferase is the departure of an iminum cation formed between formaldehyde and one of the nitrogens on tetrahydrofolate from the imine formed between pyridoxal phosphate and a derivative of serine in which the nitrogen from tetrahydrofolate has replaced the hydroxyl of serine[15]

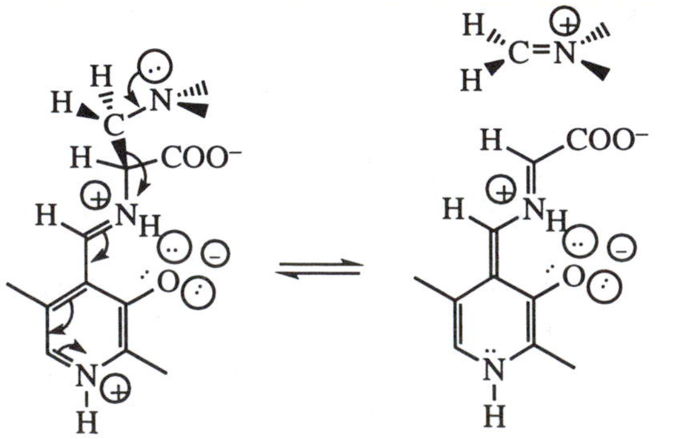

(2–6)

A similar reaction, uncomplicated by tetrahydrofolate, is catalyzed by threonine aldolase from *Candida humicola*[16]

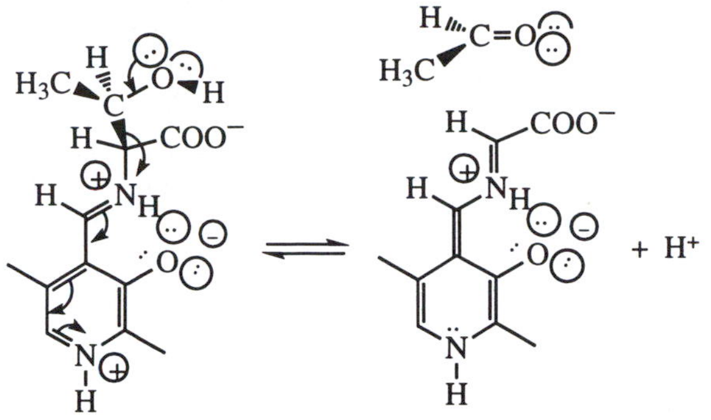

(2–7)

If the $\alpha$-carbon of a pyridoximine contains an **incipient acyl compound** in the form of a $\beta$-keto group, that $\beta$-carbon can depart as an acyl derivative. For example, in the reverse reaction catalyzed by 5-aminolevulinate synthase,[14] the second step is loss of succinyl-SCoA

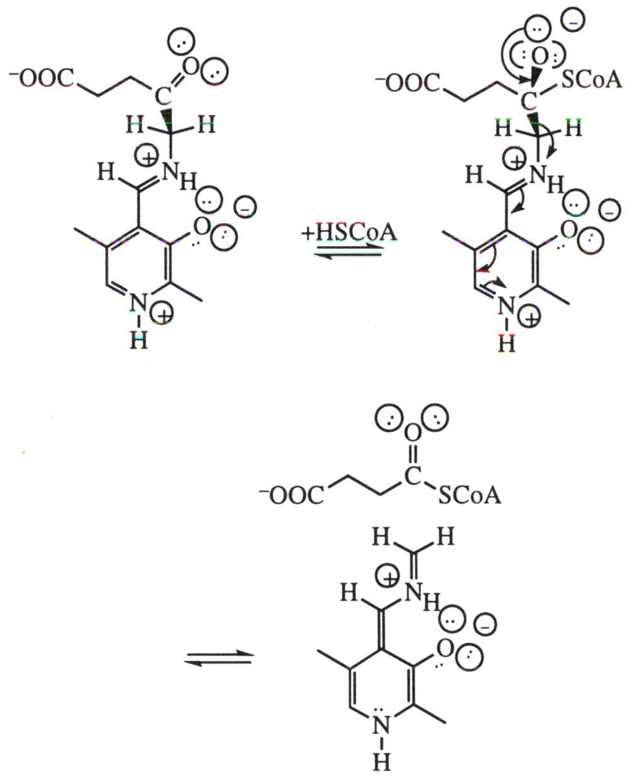

(2–8)

As indicated by Structure **2–5** and Reactions 2–4 through 2–8, the $\sigma$ bond that is weakened by the adjacent pyridoxal must be aligned parallel to the $\pi$ molecular orbital system for the necessary overlap to occur.[17] In enzymatically catalyzed reactions, the necessary alignment is made as the substrates are bound in the active site. This serves as one of the ways that the enzyme can dictate the outcome of the reaction. In free solution, pyridoxal phosphate, often supplemented with metal cations[18] such as $Cu^{II}$, $Fe^{III}$, or $Al^{III}$, can catalyze some of these departures of electrophiles from the carbon adjacent to an amine with which it has formed an imine. Because free rotation around the carbon–nitrogen bond can occur in solution, however, all three of the substituents on the $\alpha$-carbon of the amine are vulnerable. If the $\alpha$-carbon contains a hydrogen, the usual reaction is loss of a proton leading to transimination.[19] If, however, there is a carboxylate on the $\alpha$-carbon and no hydrogens, as in 2-aminoisobutyric acid or $\alpha$-methylserine, or two carboxylates, as in aminomalonic acid,[20] decarboxylation can occur. The addition of metal cations to these nonenzymatic reactions suppresses decarboxylation and promotes the loss of alternate electrophiles for the $\alpha$-carbon. Metal cations, however, are not involved in enzymatically catalyzed reactions.

The net effect of any one of the departures of these electrophiles from the respective $\alpha$-carbons to form these **quinonoid intermediates** is the expansion of the $\pi$ molecular orbital system of the pyridoximine by one carbon and by the pair of electrons that was in the $\sigma$ bond and that has neutralized the positive formal charge on the pyridinium nitrogen. For example, the product of Reaction 2–4 can be represented as

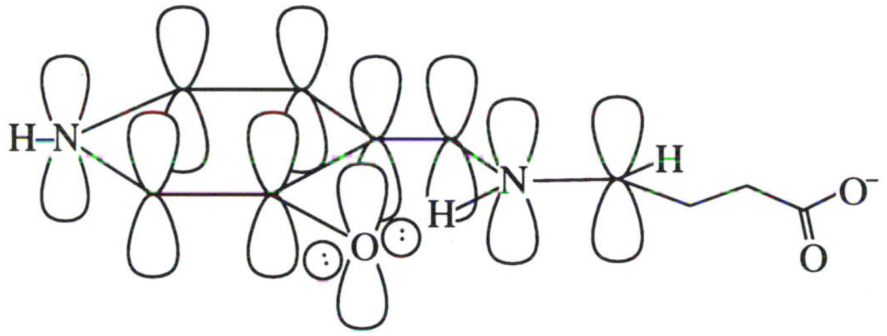

**2–8**

In all of these quinonoid intermediates, because 10 $p$ orbitals are combined, the new $\pi$ molecular orbital system has 10 $\pi$ molecular orbitals, and the six of lowest energy are each occupied by a pair of electrons. This excess electron density renders the quinonoid intermediate electron-rich and nucleophilic. It cannot absorb another $\sigma$ bond, but it can either push or be protonated. In these two roles the quinonoid intermediate is acting as a vinylogous enamine

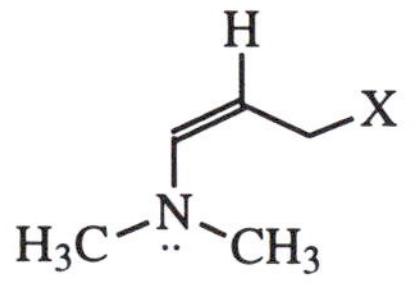

enamine **2–9** vinylogous enamine **2–10**

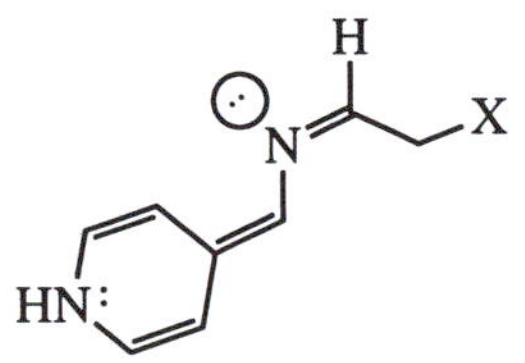

An enamine is electron-rich and basic. It can push out a leaving group (X in Structure **2–9**), or it can be protonated on carbon to form the imine.

When the $\pi$ molecular orbital system of the quinonoid intermediate is used to provide push, a ***β*-elimination** occurs.[21] An example of such a $\beta$-elimination is encountered in one of the steps in the reaction catalyzed by tryptophanase

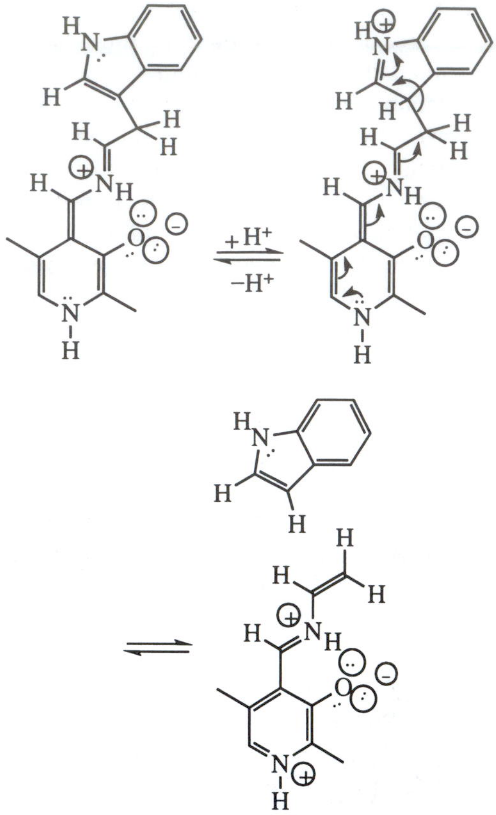

(2–9)

The indole, when it has been protonated on carbon, is an excellent leaving group because the two electrons in the σ bond connecting it to the quinonoid intermediate can enter its π molecular orbital system. The bond is pushed away by the pyridoxal phosphate and pulled into the indole

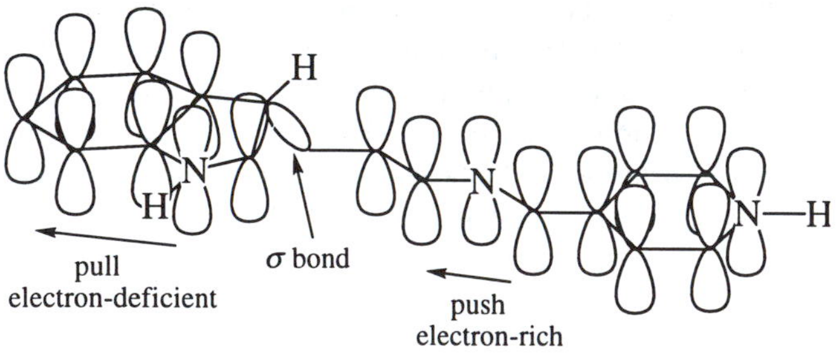

pull electron-deficient σ bond push electron-rich

**2–11**

The pyridoximine product of a β-elimination (Reaction 2-9) can add a nucleophile at the distal carbon, just as it has pushed away a nucleophile. If this occurs, then the reaction can proceed in reverse until the amine is liberated from the pyridoxal phosphate with the nucleophile that added to the β-carbon substituting for the nucleophile that left it. The pyridoximine formed by the expulsion of the leaving group can also be released directly by transimination

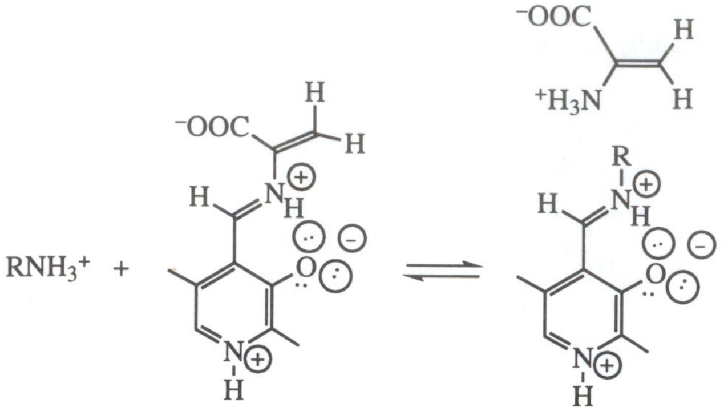

$RNH_3^+$ +

(2–10)

In this example, the product is an enamine that would rapidly hydrolyze to pyruvate and ammonia

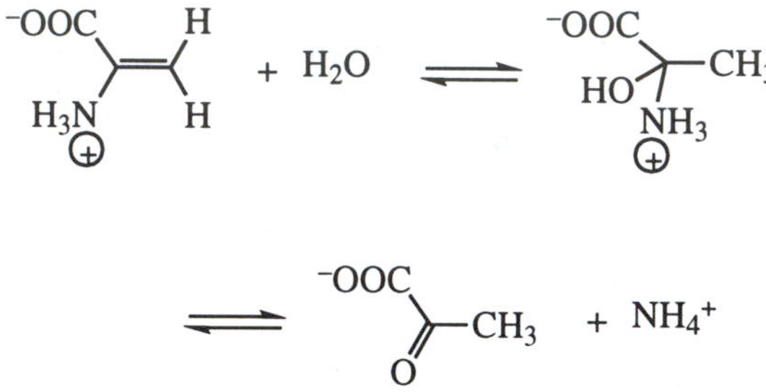

(2–11)

Protonation of the π molecular orbital system of a quinonoid intermediate is the second step in a **transamination** reaction. In a transamination, the proton is added to the quinonoid intermediate at the former carbonyl carbon of the pyridoxal phosphate

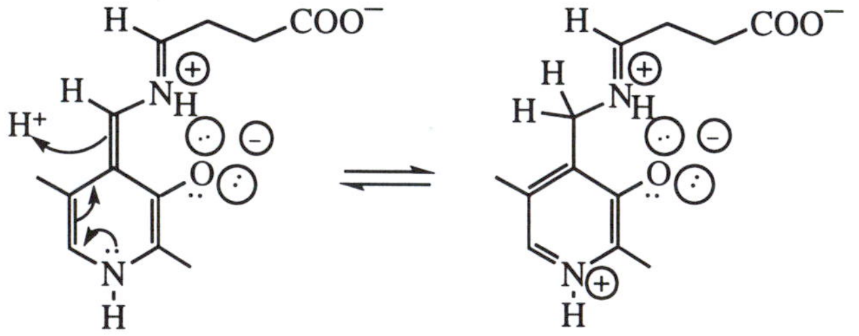

(2–12)

Two of the 12 π electrons of the π molecular orbital system of the reactant become the two electrons in the new σ bond of the product. The remaining *p* orbitals and π electrons are divided into two unequal sets by the protonation. Eight π electrons remain in the π molecular orbital system of the 3-hydroxypyridine. This π molecular orbital system is composed of seven *p* orbitals. Two π electrons remain in the double bond of the peripheral imine. This peripheral imine can be hydrolyzed readily to the corresponding aldehyde or ketone and **pyridoxamine phosphate**

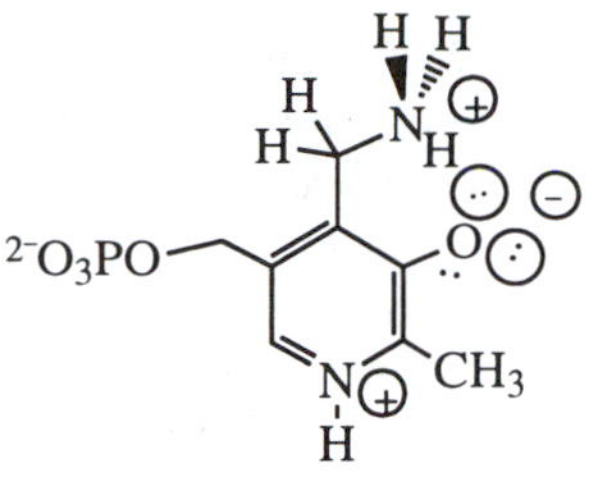

**2–12**

The overall reaction has been the conversion of the carbon that originally bore the amine into a ketone or aldehyde coincident with the conversion of the carbonyl of the pyridoxal phosphate into an amine. The second half of a transamination is to convert a ketone or aldehyde that is the other reactant into an amine by the reverse reaction, which simultaneously regenerates the carbonyl of the pyridoxal phosphate so that it can start again. For example, the reaction catalyzed by 4-aminobutyrate aminotransferase is

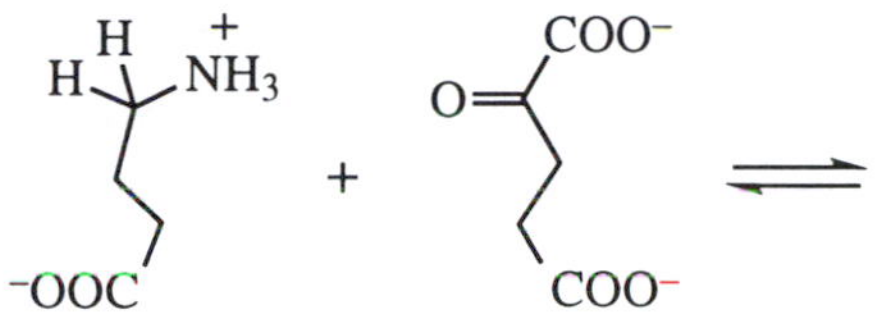

4-aminobutyrate 2-oxoglutarate

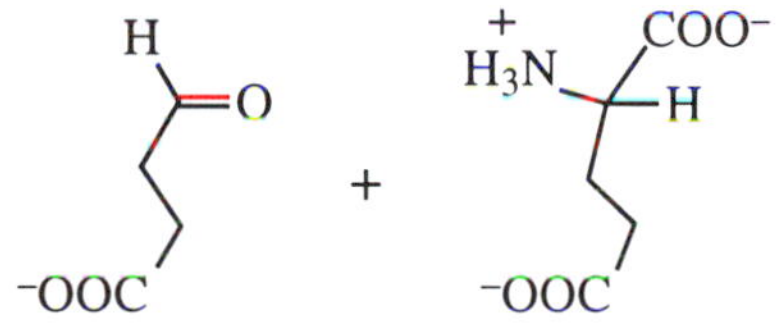

4-oxobutyrate L-glutamate (2–13)

In this reaction the pyridoximine of 4-aminobutyrate is formed, the $\alpha$-proton is removed, the carbonyl carbon of the quinonoid intermediate is protonated, the peripheral imine is hydrolyzed to produce 4-oxobutyrate and pyridoxamine, the 2-oxoglutarate replaces the 4-oxobutyrate, and the reaction proceeds in reverse. Transaminations are also catalyzed nonenzymatically by pyridoxal phosphate in the presence of metal cations. For example, alanine and 2-ketoglutarate are transformed into glutamate and pyruvate by pyridoxal phosphate in the presence of $CuSO_4$.[22]

The central step in a transamination reaction involving pyridoxal phosphate can be considered as the tautomerization of an iminium cation (the sum of Reactions 2–4 and 2–12)

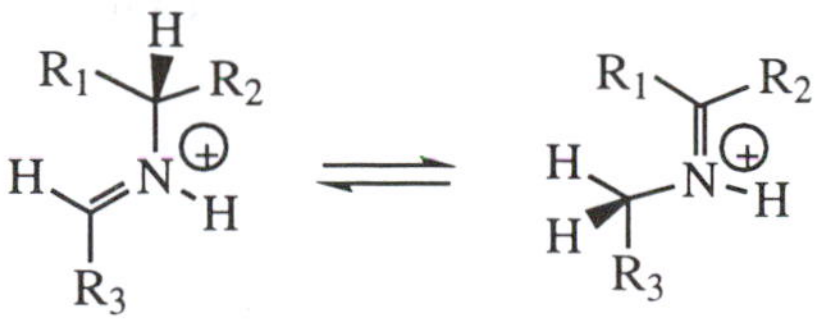

(2–14)

In nonenzymatic situations, simple tautomerizations of this type are catalyzed by bases,[23] which remove the proton from carbon during the slow step of the reaction.[24] Both a weak acid and a weak base together can catalyze the tautomerization of a pyridoximine, presumably by catalyzing the removal of the proton from the one carbon and the addition of the proton to the other carbon.[8] An analogue of pyridoxal phosphate in which an imidazole has been appropriately incorporated as an intramolecular catalyst can undergo this tautomerization

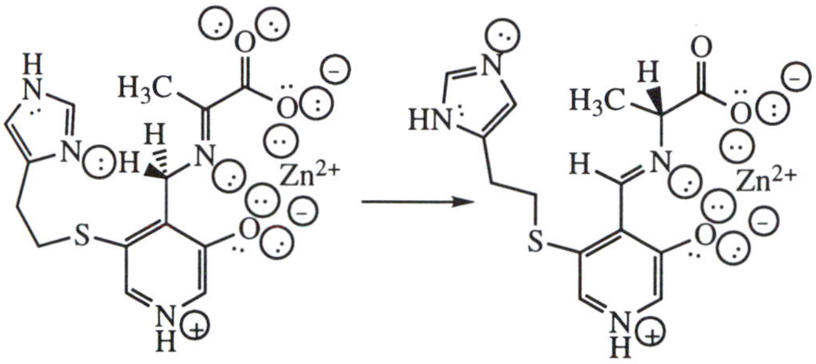

(2–15)

80 times more rapidly than another analogue identical to the first with the exception that the imidazole has been replaced by a methyl group.[25] In this case $Zn^{2+}$ was used as a Lewis acid to promote the reaction. Even with the assistance of the pyridoxal phosphate in weakening these $\sigma$ bonds between carbon and hydrogen, the removal of a proton from carbon remains a more difficult reaction than the carbonyl additions necessary to form the pyridoximine in the first place.

Following such tautomerizations (Reactions 2–4 and 2–12, or Reaction 2–14), an iminium cation is formed that can withdraw electrons from a $\sigma$ bond on the $\beta$-carbon of the substrate and promote the **expulsion of an electrophile** from that $\beta$-carbon. For example, in one of the steps catalyzed by *O*-succinylhomoserine (thiol)-lyase, the hydrogen on the $\beta$-carbon of *O*-succinylhomoserine is lost as a proton[26] from such an imine

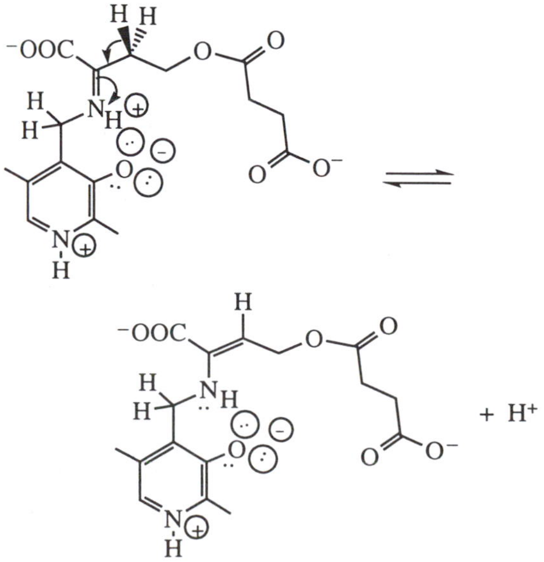

(2–16)

This is a simple manifestation of the acidity of a proton on the carbon adjacent to an iminium cation. The acid–base reaction produces an enamine. In the next step of the reaction catalyzed by this enzyme, the electron excess in the enamine pushes out the leaving group to extend the reach of the catalyst to the $\gamma$-carbon of the amine

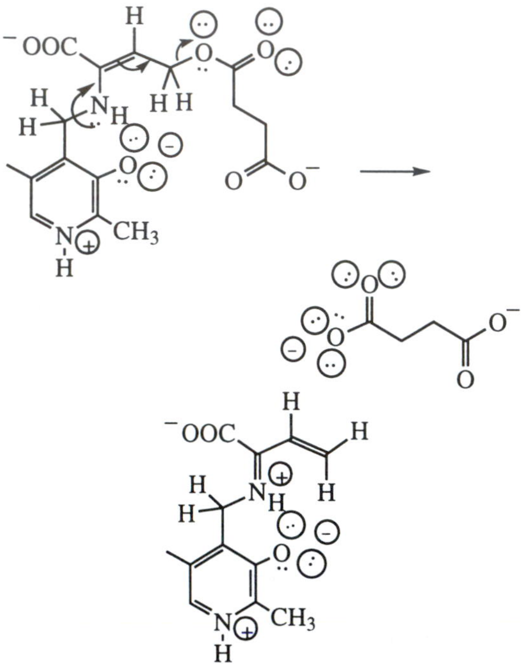

(2–17)

It is at this point (Reaction 2–17) that the design of pyridoxal phosphate is displayed at its extreme. The pyridinium nitrogen at the base of the ring functions to withdraw electrons from $\sigma$ bonds around the $\alpha$-carbon of the amine and, upon removal of an electrophile from this carbon, to push away leaving groups from the $\beta$-carbon of the amine. The nitrogen of the amine itself functions to withdraw electrons from $\sigma$ bonds around the $\beta$-carbon of the amine and, upon removal of an electrophile from this carbon, to push away leaving groups from the $\gamma$-carbon of the amine. The basis for the ability of the nitrogens to promote opposite reactions on the same carbon and the same reaction at alternate carbons is that they are an odd number of atoms apart from each other in the initial pyridoximine. The influence of these two nitrogens could be propagated even beyond the $\gamma$-carbon of the amine if incipient electrophiles and leaving groups were properly distributed so that the expanding $\pi$ molecular orbital systems were not required to become excessively electron-rich or excessively electron-poor.

At each step in the extension of the $\pi$ systems over the substrate, the electrophile that has just departed can be replaced by a different electrophile or the leaving group that has just departed can be replaced by another nucleophile. In the reaction catalyzed by glycine hydroxymethyltransferase (Reaction 2–6), the iminium cation of formaldehyde is replaced by a proton. In the reaction catalyzed by $O$-succinylhomoserine (thiol)-lyase (Reaction 2–17), the oxygen of the succinate is replaced by the sulfur of a cysteine. After these replacements are made the reaction can proceed in reverse, passing through the same steps that led to the loss of the electrophile or expulsion of the leaving group to regenerate the amine in which only the one substitution has occurred. This is what happens with glycine hydroxymethyltransferase

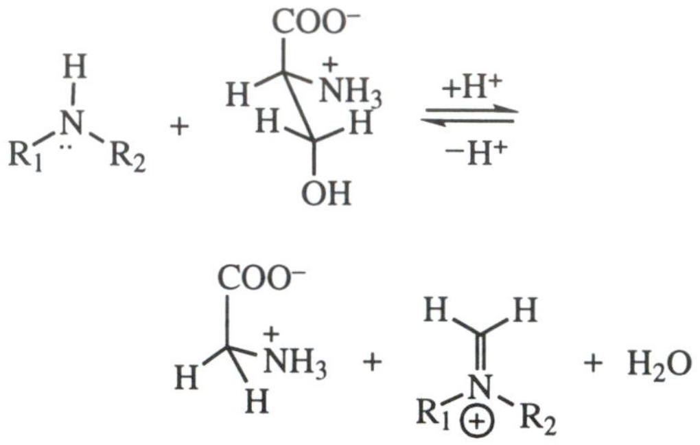

(2–18)

and $O$-succinylhomoserine (thiol)-lyase

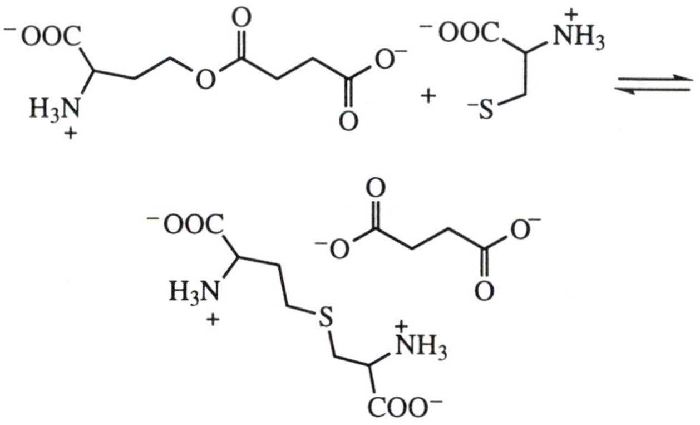

(2–19)

The most remarkable aspect of the reaction catalyzed by $O$-succinylhomoserine (thiol)-lyase is that the reaction involves the replacement of a good leaving group, a carboxylate anion, by a strong nucleophile, a thiol, at a primary carbon. Rather than proceed through the simplest possible mechanism, an $S_N2$ reaction, the enzyme patiently uses pyridoxal phosphate to strip protons away and replace the carboxylate by a $\beta$-elimination–addition involving four steps in each direction from the initial pyridoximine.

At each step in the reactions catalyzed by pyridoxal phosphate, on the way back or on the way forward, the nitrogen of the original amine is formally an imine that can be hydrolyzed to regenerate the pyridoxal phosphate and a significantly altered derivative of the original amine (Reaction 2–10) or pyridoxamine phosphate and the significantly altered or the unaltered ketone or aldehyde of the amine (Reaction 2–12).

## Suggested Reading

Posner, B.I., & Flavin, M. (1972) Cystathione γ-Synthase: Studies of Hydrogen Exchange Reactions, *J. Biol. Chem. 247*, 6402–6411.

**PROBLEM 2–1**
Write a complete mechanism for Reaction 2–13 under catalysis by pyridoxal phosphate.

**PROBLEM 2–2**
Write a mechanism involving a $\beta$-elimination catalyzed by pyridoxal phosphate for replacing the hydroxyl of serine with the amino group of tetrahydrofolate that will form the iminium cation of the product in the reaction catalyzed by glycine hydroxymethyltransferase (Reaction 2–18; see also Reaction 2–6).

**PROBLEM 2–3**
Write mechanisms for the following reactions that are mediated by pyridoxal phosphate

(A) + HSCoA ⇌ + $H_3C$–CO–SCoA

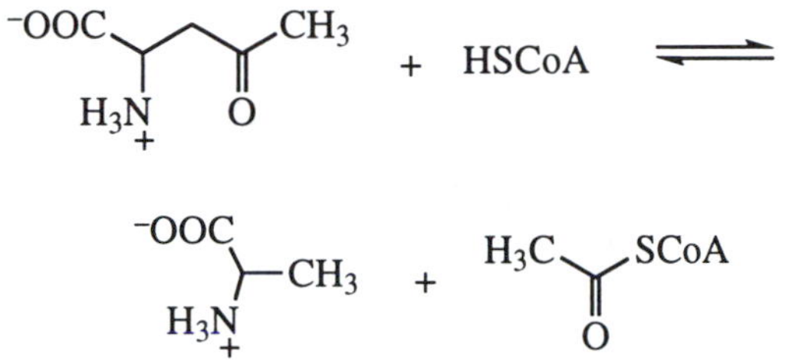

(B) ⇌ +

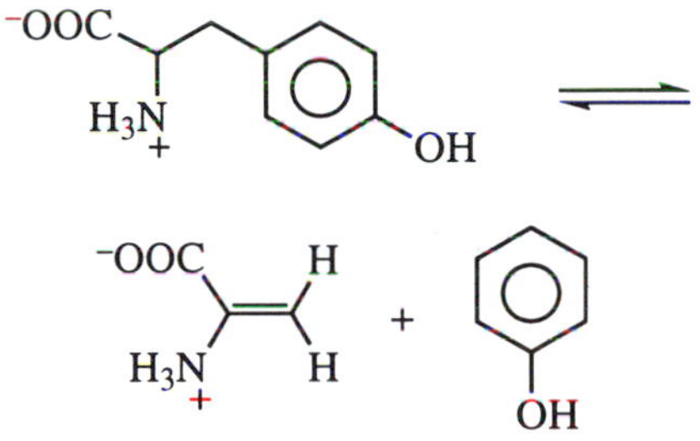

(C) ⇌ $^{-}OOC$–CO–$CH_3$ + $NH_4^+$ + HCl

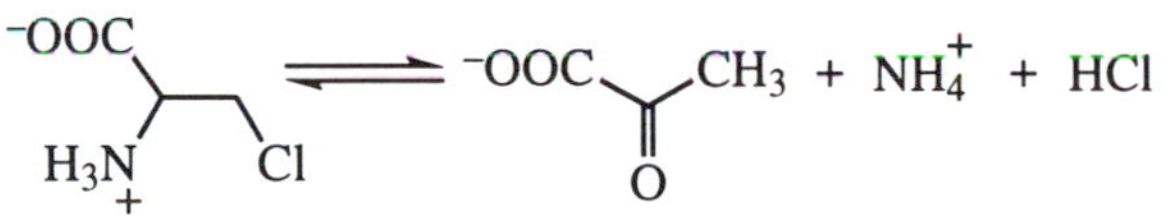

(D) + ⇌ + $CH_3SH$

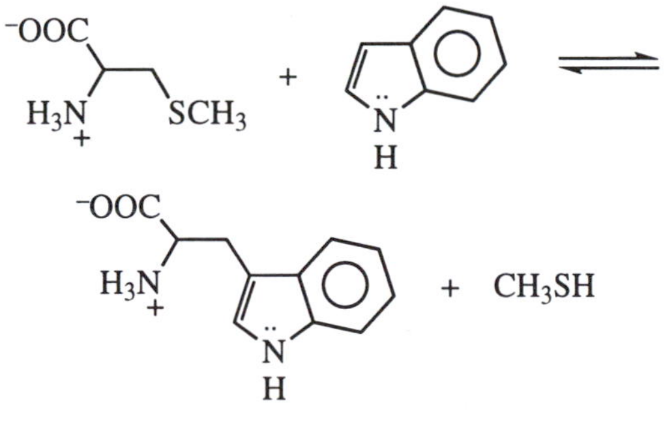

(E) + ⇌ + HCl

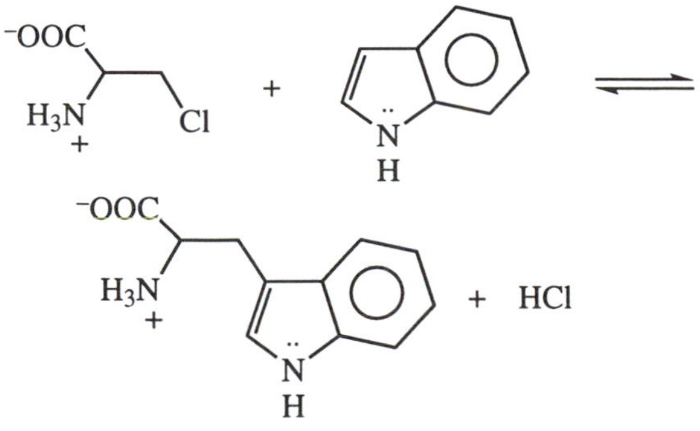

(F) + + $H_2O$ ⇌ +

(G) + ⇌ + $CO_2$ + HSCoA

(H) + $H_2O$ ⇌ + $NH_4^+$

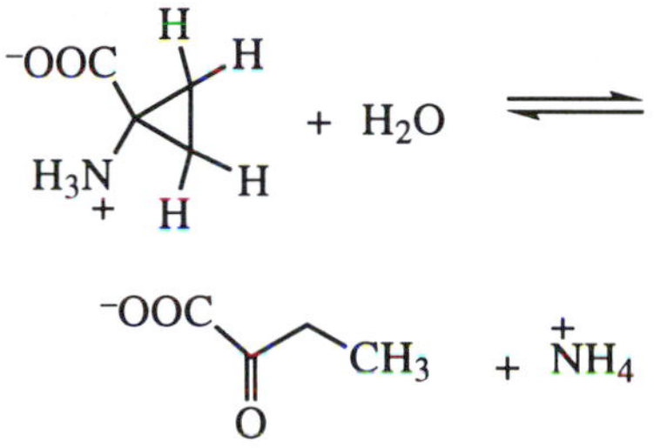

(I) ⇌ + Se (elemental)

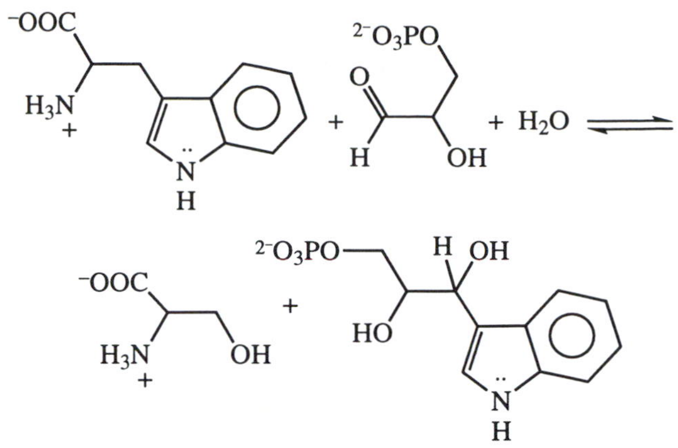

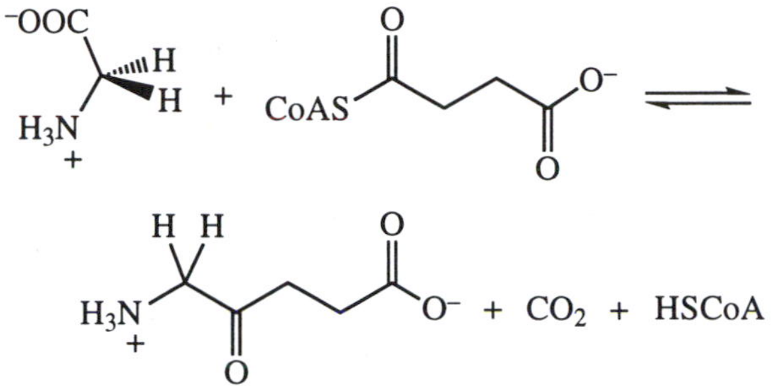

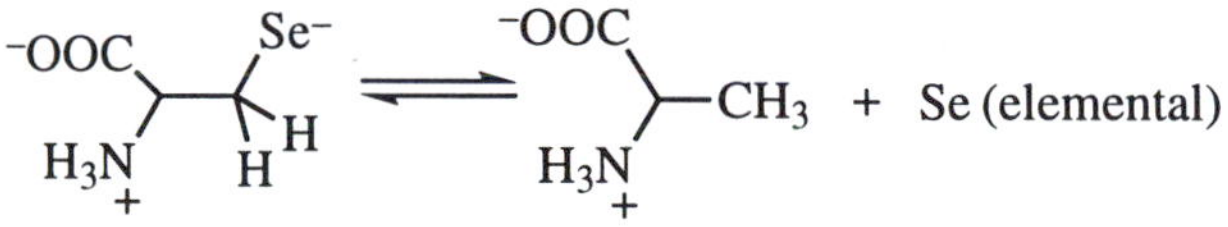

(J) + ⇌ +

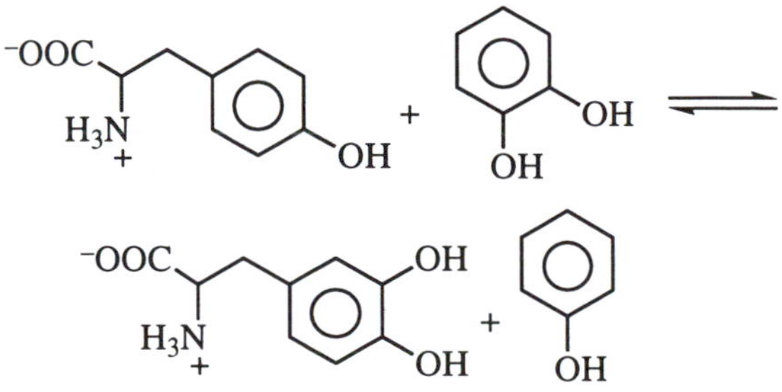

(K) $\xrightarrow{+H^+}$ + $CO_2$

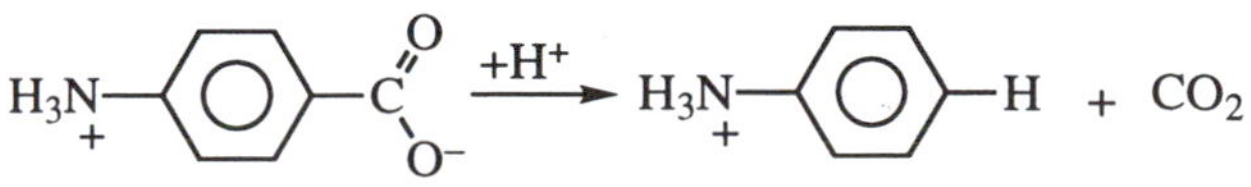

**PROBLEM 2–4**
The compound gabaculine

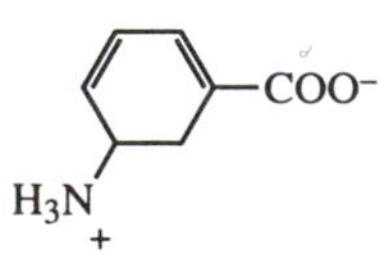

inhibits 4-aminobutyrate aminotransferase by forming an aromatic, anilinium adduct with the pyridoxal phosphate that is the coenzyme for this enzyme. Write the mechanism for the formation of this adduct.

### PROBLEM 2–5

Write the mechanism of the D-serine dehydrase reaction

$$\text{D-serine} \rightleftharpoons \text{pyruvate} + NH_4^+$$

as it is catalyzed by the pyridoxal phosphate bound to the enzyme. Show clearly the role of the pyridoxal in each step of the reaction. In each step label every $sp^2$ carbon conjugated to the aromatic ring. Watch closely the expansion and then contraction of the $\pi$ system as atoms are turned from $sp^3$ into $sp^2$ and back into $sp^3$. What step in the reaction requires the pyridoxal the most?

### PROBLEM 2–6

Histidine decarboxylase from *Lactobacillus* 30a catalyzes the reaction

$$\text{histidine} \rightleftharpoons \text{histamine} + CO_2$$

The enzyme has a pyruvyl group incorporated as a posttranslational modification of one of its amino termini.[26a] The enzyme was mixed with [$^{14}$C]histidine, and $NaBH_4$ was added. Sodium borohydride reduces carbonyl carbons to alcoholic carbons

$$>C{=}X + BH_4^- + H_2O \longrightarrow H{-}C{-}X{-}H + HOBH_3^-$$

Following this treatment the enzyme had covalently incorporated $^{14}$carbon. It was submitted to acid hydrolysis and two radioactive products were isolated

Write a mechanism for the reaction that explains the role of the pyruvate acting as a coenzyme.

Phosphopantothenoylcysteine decarboxylase catalyzes the reaction

This enzyme also has a carbonyl as a coenzyme. Write a mechanism for this decarboxylation using a carbonyl as the coenzyme.

# Thiamin Pyrophosphate

Thiamin pyrophosphate

**2–13**

(2–20)

is a coenzyme that is tightly, but noncovalently, bound to the active sites of the enzymes in which it has a functional role. The molecule is complicated enough that several locations in its structure were sequentially proposed to be the functionally important site[27] before it was demonstrated that it was located at carbon 2, sandwiched between the quaternary nitrogen and the sulfur in the **thiazolium ring.**[28]

This carbon is acidic and readily exchanges with a deuterium in solution[28] by forming the anion

The apparent rate constant for the exchange of the proton on carbon 2 of thiamin,[29] at 30 °C and pH 3.5, is $1.8 \times 10^{-4}\ s^{-1}$, and this apparent rate constant is directly proportional to $[OH^-]$.[30] The p$K_a$ for the dissociation of the proton from carbon 2 of thiamin (p$K_a$ = 12.7) has been measured directly using a rapid kinetic titration[31] to avoid the rapid decomposition of the thiazolium ring that occurs at high concentrations of hydroxide anion.[32] There are several reasons why thiamin is such a strong carbon acid.

First, because it is within an aromatic heterocycle, the acidic carbon is hybridized $sp^2$, $sp^2$, $sp^2$, $p$. The $\sigma$ bond to the hydrogen, because it is formed from one of the $sp^2$ atomic orbitals of the carbon atom, holds the pair of bonding electrons closer to the nucleus and the $sp^2$ atomic orbital holds the lone pair of electrons formed upon dissociation of the proton closer to the nucleus than if the carbon were hybridized $sp^3$, $sp^3$, $sp^3$, $sp^3$. These properties render the carbon–hydrogen bond more acidic and the carbanion less basic[27] than in carbon acids in which carbon is hybridized $sp^3$, $sp^3$, $sp^3$, $sp^3$.

Second, the carbanion of thiamin is an **ylide.** An ylide is an arrangement in which a carbon with an unshared pair of $\sigma$ electrons and a fixed negative charge is adjacent to another atom bearing a fixed positive charge. Because the valence shell of the adjacent atom bearing the positive charge is full, it cannot accept electrons from the adjacent carbanion. Nevertheless, the negative charge on carbon is stabilized by the effect of the electrostatic field of the adjacent positive charge on nitrogen. This stabilization can be seen in semiempirical calculations of the electronic structure of the carbanion of thiamin,[33] and in the ability of 1,3-dimethylimidazoliums to display carbon acidity similar to that of 3-methylthiazoliums.[32] The adjacent, positively charged nitrogen also weakens the $\sigma$ bond between the hydrogen and carbon 2 in thiamin. This effect is manifested in the deshielding of carbon 2 observed in the nuclear magnetic resonance spectrum of thiamin.[34]

Third, the neighboring sulfur stabilizes the negative charge in the $\sigma$ lone pair on carbon 2 of the carbanion of thiamin pyrophosphate by $dp\pi$ overlap. This effect can be observed by comparing the rate of hydrogen exchange of 3,4-dimethylthiazolium cation and 1,3,4-trimethylimidazolium cation.[30] Although the respective carbon–hydrogen bonds in these two cations are very similar, the hydrogen on carbon 2 of the thiazolium exchanges 3000 times faster than does the hydrogen on carbon 2 of the imidazolium.

Finally, even the identity of the heterocyclic substituent on nitrogen 3 of thiamin enhances the acidity of the coenzyme. 3,4-Dimethylthiazolium cation exchanges the hydrogen on carbon 2 about 10-fold less readily than thiamin,[34] and this presumably results from the fact that the 4-amino-2-methylpyrimidine attached through a methylene at nitrogen 3 of the thiazole in thiamin pyrophosphate (Structure **2–13**) withdraws electrons inductively from that nitrogen. This inductive electron withdrawal is also reflected in the greater susceptibility[35] of thiamin than 3,4-dimethyl-5-(2-hydroxyethyl)thiazolium cation to addition of hydroxide anion at carbon 2

(2–21)

This electron withdrawal must be inductive and take place through $\sigma$ bonds between nitrogen 3 and the pyrimidines because there is a saturated benzylic carbon between the two rings. In the crystallographic structure of a derivative of thiamin, the two rings are almost perpendicular to each other (Figure 2–3)[36] and cannot interact through an overlap of their $\pi$ molecular orbital systems.

**Figure 2–3:** Drawing of a crystallographic molecular model of 2-(1-hydroxyethyl)thiamin.[36] The adduct between thiamin and acetaldehyde was formed at alkaline pH and then crystallized as the dichloride of the dication formed by protonation of the pyrimidine. The thiamin is the free alcohol rather than the pyrophosphate, and the normal 2-hydroxyethyl substituent at carbon 5 is in the upper right corner. The 1-hydroxyethyl substituent formed upon addition of acetaldehyde to carbon 2 is at the bottom; the ethyl backbone is viewed end-on in the staggered conformation. The pyrimidine ring is almost perpendicular to the page; the thiazole ring, almost parallel to the page. One of the chloride ions is shown at the bottom associated with the hydroxyl group of the 1-hydroxyethyl substituent. Adapted with permission from ref 36. Copyright 1974 American Chemical Society.

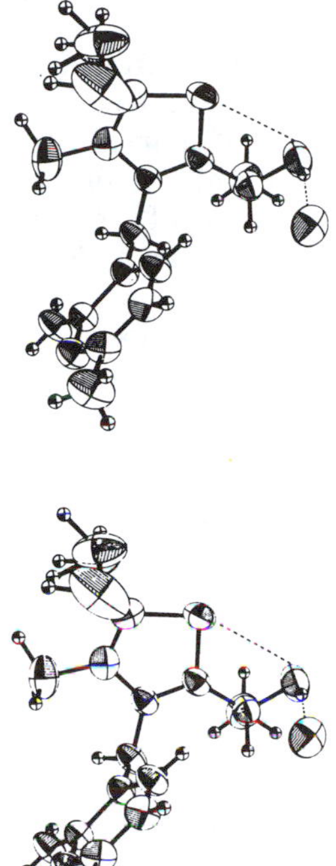

Once the ylide has been formed by removal of the proton, it can react with an aldehyde or a ketone to form a product of carbonyl addition

**2–14**

(2–22)

similar to the cyanohydrin that forms between an aldehyde or a ketone and cyanide

(2–23)

It is this product of addition of the ylide of thiamin to a ketone or an aldehyde (Structure **2–14**) that explains the ability of thiamin to catalyze the reactions for which it is responsible.

There is no doubt that the thiazole ring is aromatic because every bond within the ring displays the shortening expected for a conjugated heterocycle.[36] The 3*p* atomic orbitals of sulfur, however, do not overlap the 2*p* atomic orbitals of its two neighbors very effectively. This causes the nitrogen within the ring to be less like the nitrogen in an alkylpyridinium cation and more like the nitrogen of an iminium cation. This also explains the facility with which a hydroxide anion adds to carbon 2 of thiamin itself to form the carbinolamine (Reaction 2–21)[32] even though this necessarily destroys the aromaticity of the ring. Because it resembles the nitrogen in an iminium cation, the nitrogen in the thiazole ring of thiamin withdraws electron density from the $\sigma$ bonds around the former carbonyl carbon within the adduct (Structure **2–14**) and promotes the departure of an incipient electrophile.

In the product of an addition between thiamin and an aldehyde, one of the substituents bonded to the former carbonyl carbon is a **hydrogen**, and this hydrogen can depart as a proton. For example, thiamin, like cyanide, can catalyze the formation of acyloins,[32] as in the condensation of benzaldehyde and acetaldehyde[37]

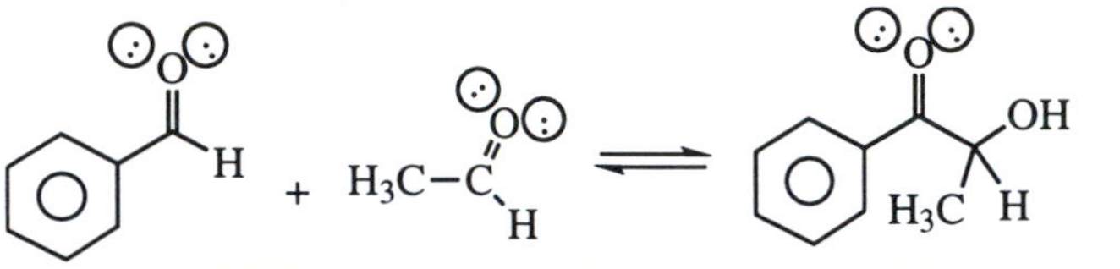

(2–24)

In this reaction the acetaldehyde, as an electrophile, has replaced the hydrogen originally on the carbonyl carbon of the benzaldehyde. In the adduct that forms between the benzaldehyde and the thiamin, the adjacent thiazole ring causes the hydrogen on the aldehyde carbon of benzaldehyde to become acidic because the iminium cation is positioned properly to withdraw electron density from the carbon–hydrogen bond

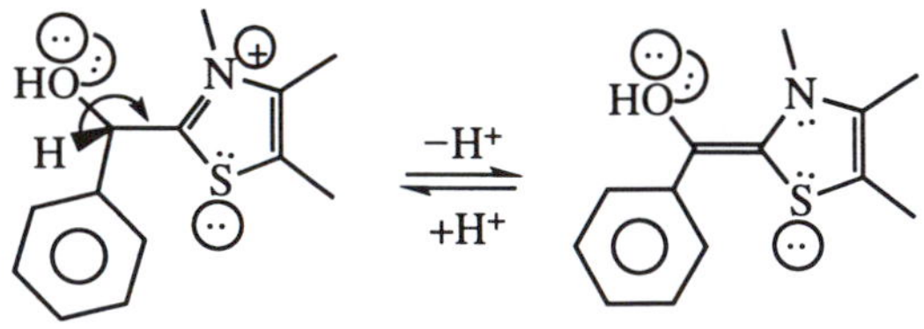

(2–25)

The proton that has left is replaced by the acetaldehyde

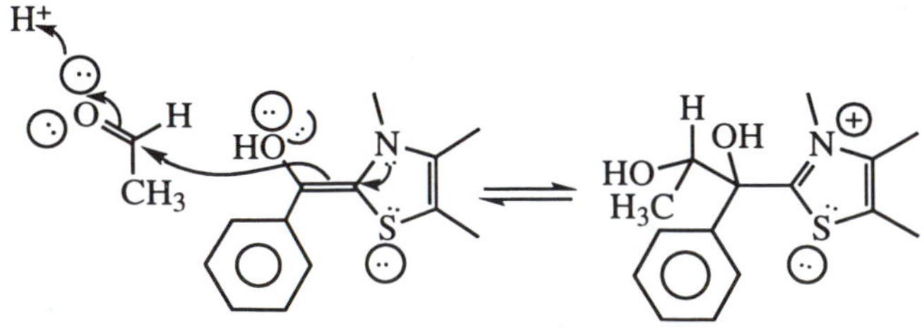

(2–26)

through the addition of the enamine to the carbonyl carbon. This is accompanied by restoration of the aromaticity in the thiazole ring.

The products of addition between aldehydes and thiamin (Figure 2–3) are readily synthesized stable compounds, and hydrogen exchange at the former carbonyl carbon can be monitored.[37] This observation demonstrates the acidity of this location.

In the product of addition between thiamin and an $\alpha$-keto acid, one of the substituents bonded to the former carbonyl carbon is a **carboxylate**, and it can depart as carbon dioxide. For example, in the reaction catalyzed by the enzyme pyruvate decarboxylase

$$\text{pyruvate} \rightleftharpoons \text{acetaldehyde} + CO_2 \qquad (2\text{–}27)$$

the immediate adduct between thiamin pyrophosphate and pyruvate is decarboxylated

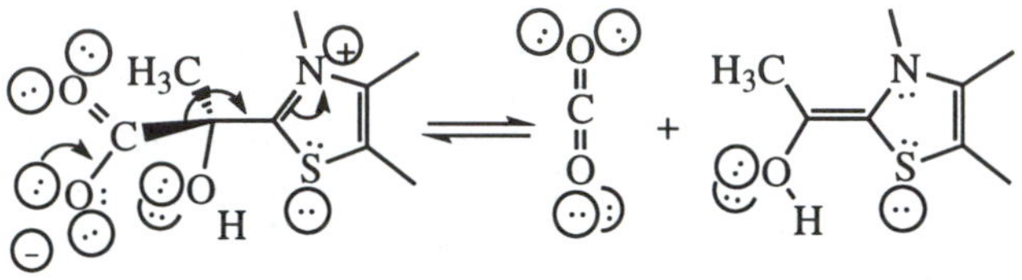

(2–28)

The analogous product of the addition of 3,4-dimethylthiazole and pyruvate has been synthesized and shown to decarboxylate readily[38] at values of pH greater than the p$K_a$ of the carboxy group where it would be unprotonated as required. The product of addition between thiamin itself and pyruvate has also been synthesized, and the rate constants and equilibrium constants for various steps in the mechanism of its decarboxylation (Figure 2–4) have been measured.[39]

The product of the reaction in which the carboxylate of pyruvate has been replaced by a hydrogen is in turn the product of addition of thiamin to acetaldehyde.[40] In addition to catalyzing the decarboxylation of pyruvate, **pyruvate decarboxylase** also catalyzes the deprotonation of acetaldehyde, in a completely analogous reaction, during its catalysis of the production of acetoin[41]

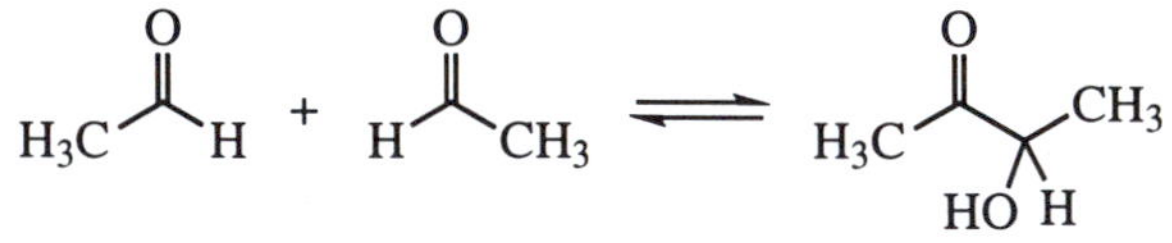

(2–29)

The slowest step in this reaction as it occurs on the enzyme is the removal of the proton from the former aldehyde carbon in the product of addition between the thiamin pyrophosphate and acetaldehyde. This conclusion follows from the fact that 1-deuterioacetaldehyde reacts 4.5 times more slowly than acetaldehyde in the acetoin reaction catalyzed by the enzyme.

An interesting observation made during the study of the decarboxylation of the product of addition between 3,4-dimethylthiazole (Reaction 2–28) and pyruvate was that, as expected from a reaction in which formal charge is neutral-

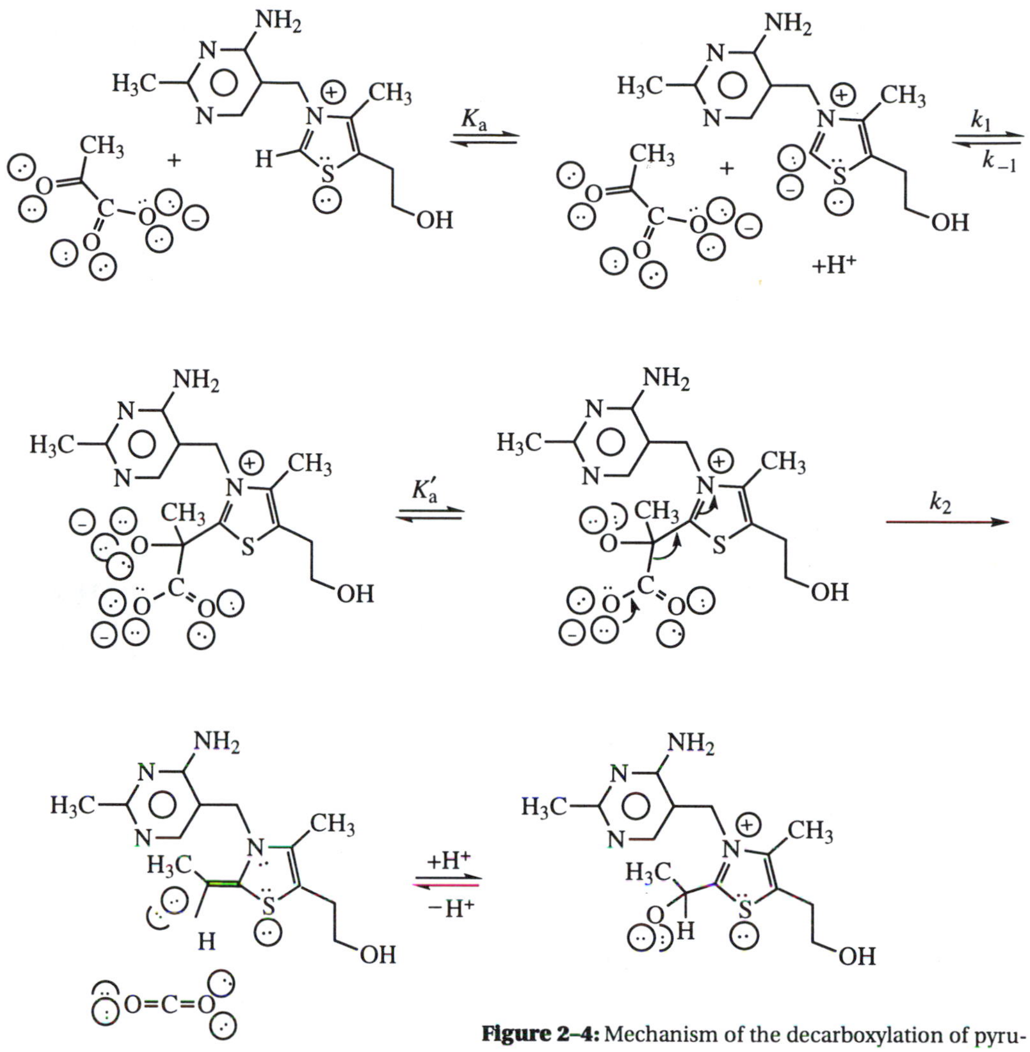

**Figure 2–4:** Mechanism of the decarboxylation of pyruvate catalyzed by thiamin (25 °C).[39] The ylide is formed by removing the proton from carbon 2 (p$K_a$ = 12.7). The adduct between thiamin and pyruvate forms in a bimolecular reaction ($k_1 = 6.5 \times 10^{-2}$ $M^{-1}$ $s^{-1}$; $k_{-1} = 0.5$ $s^{-1}$). The oxyanion protonates to form the conjugate acid (p$K'_a$ = 13.6), referred to as $\alpha$-lactylthiamin. The $\alpha$-lactylthiamin then decarboxylates ($k_2 = 4 \times 10^{-5}$ $s^{-1}$) to form the enamine, which protonates to give the final product. The final product is in turn the adduct of thiamin and acetaldehyde.

ized, the decarboxylation is accelerated by performing it in nonpolar solvents rather than polar solvents. In water at 46 °C, the rate of the decarboxylation[38] was $0.48 \times 10^{-3}$ $s^{-1}$; but, in ethanol at 26 °C, the rate was $220 \times 10^{-3}$ $s^{-1}$. The suggestion has been made that pyruvate decarboxylase may catalyze this reaction in part by providing such a nonpolar environment.[38] An acceleration of $10^4$ in the rate of the decarboxylation of the product of addition between pyruvate and thiamin itself also occurs when the solvent is changed from water to ethanol.[39]

Likewise, the removal of the acidic proton from carbon 2 of thiamin, which is the first step in all of these reactions, involves the formation of an ylide from a pair of separated, oppositely charged molecules, the hydroxide ion and the thiazolium cation (Reaction 2–20). When the deprotonation of 3,4-dimethylthiazolium cation is carried out in ethanol with ethoxide anion, it is 500 times faster ($2.7 \times 10^{11}$ $M^{-1}$ $s^{-1}$ at 25 °C) than it is when it is carried out in water with hydroxide anion.[42] Because the former reaction is a diffusion-controlled removal of a proton from carbon, thiamin must be a much stronger acid than ethanol under these circumstances. Again, a nonpolar active site might favor removal of the proton from carbon 2 of thiamin in the first step of these reactions.

In the product of the addition between thiamin and an $\alpha$-hydroxyketone or thiamin and an $\alpha$-hydroxyaldehyde, one of the substituents bonded to the former carbonyl carbon will be an **incipient carbonyl group**, and it can depart as a carbonyl compound. For example, in one of the reactions catalyzed by transketolase

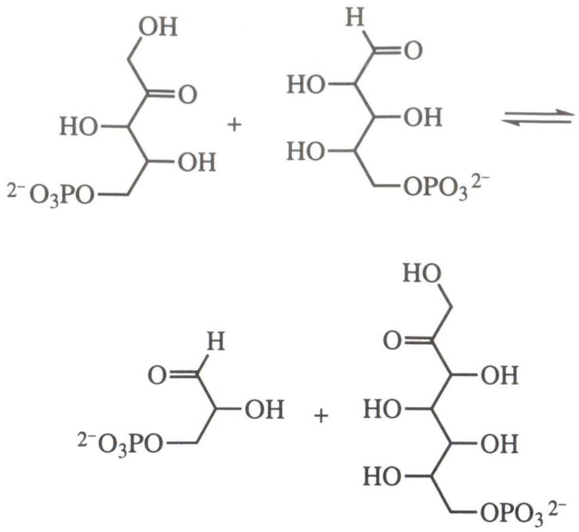

(2–30)

in which a two-carbon fragment is passed between two aldehydes, the product of the addition between either of the ketones and thiamin pyrophosphate loses the respective aldehyde as a product

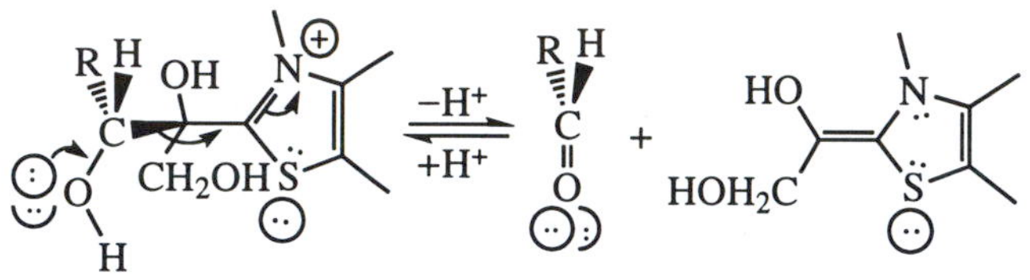

(2–31)

In all of these reactions, the $\sigma$ bond being broken has been drawn as parallel to the $\pi$ molecular orbital system of the double bond between carbon 2 and nitrogen 3 of the thiamin and, by extension, the whole aromatic heterocycle. This permits the overlap necessary for the coulomb effect of the cationic nitrogen to withdraw electron density from the $\sigma$ bond between the former carbonyl carbon and the departing electrophile

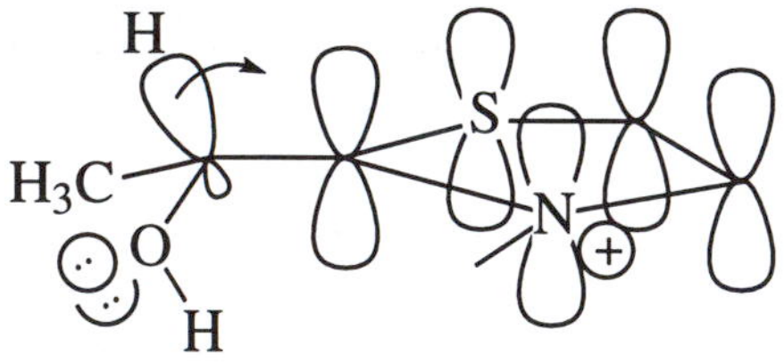

**2–15**

In the crystallographic molecular model (Figure 2–3) of the addition compound between acetaldehyde and thiamin, the dihedral angle along the bond between carbon 2 of the thiamin and the former aldehyde carbon assumes a particular orientation due to both crystal packing forces and electrostatic interactions between the hydroxyl on this carbon and both an adjacent chloride ion and the adjacent sulfur atom of the thiazole ring. In this orientation the $\sigma$ bond to the methyl group is placed parallel to the $\pi$ molecular orbital system of the thiazole ring, which would be a nonproductive alignment. This example illustrates the fact that the hydroxyl group on all of these products of addition, both because of its ability to form hydrogen bonds and because it is not a candidate for departure, can act as a handle upon the former carbonyl carbon that can be used to fix the dihedral angle in the appropriate orientation to weaken the proper $\sigma$ bond.

In all of these reactions the immediate product from loss of the electrophile is a **2-hydroxyenamine**

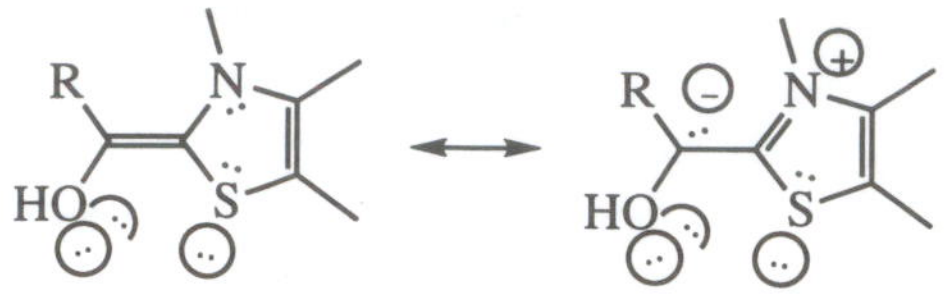

**2–16**

The $\pi$ molecular orbital system of this compound is formed from six atoms

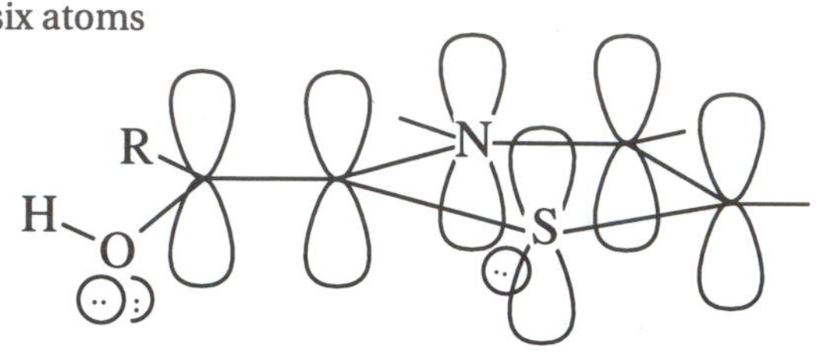

**2–17**

and it contains four pairs of $\pi$ electrons, making it electron-rich and nucleophilic at the exocyclic carbon. If it adds to a proton, the product of addition between thiamin and an aldehyde is formed; if it adds to carbon dioxide, the product of addition between thiamin and an $\alpha$-keto acid is formed. If it adds to an aldehyde or ketone, the product of addition between thiamin and an $\alpha$-hydroxyketone is formed (Reactions 2–24, 2–29, and 2–30). Each of these reactions can be considered to be the addition of an electrophile to the synthon

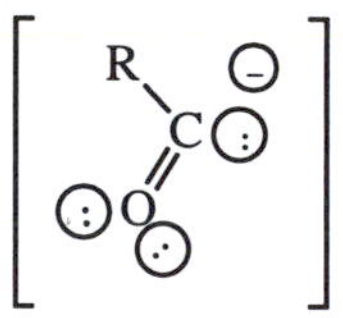

**2–18**

that was produced by the departure of an electrophile from a carbonyl carbon. Considered in this way, all of the enzymatic reactions in which thiamin pyrophosphate is used as a coenzyme are reactions in which one electrophile departs from this synthon and a different electrophile adds. One electrophile is exchanged for another, carbon dioxide for a proton, one aldehyde for another aldehyde, or carbon dioxide for a disulfide.

The final step in a reaction catalyzed by thiamin pyrophosphate is the regeneration of the carbonyl carbon by ejecting the thiazole. This is assisted by the fact that the hydroxyl on the incipient carbonyl carbon is fairly acidic (p$K_a$ = 12),[38] and the removal of its proton by base[38,39] initiates the regeneration

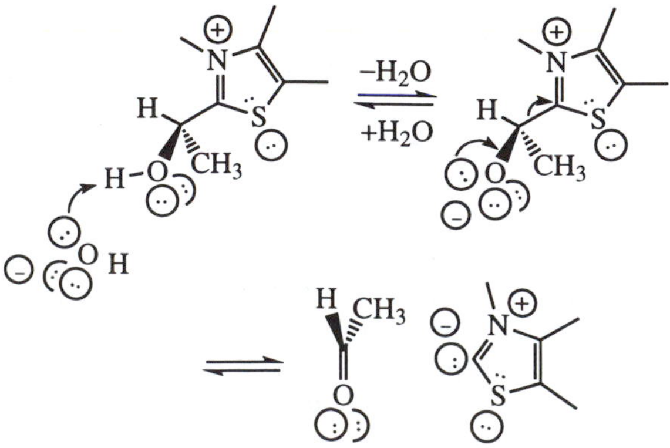

(2–32)

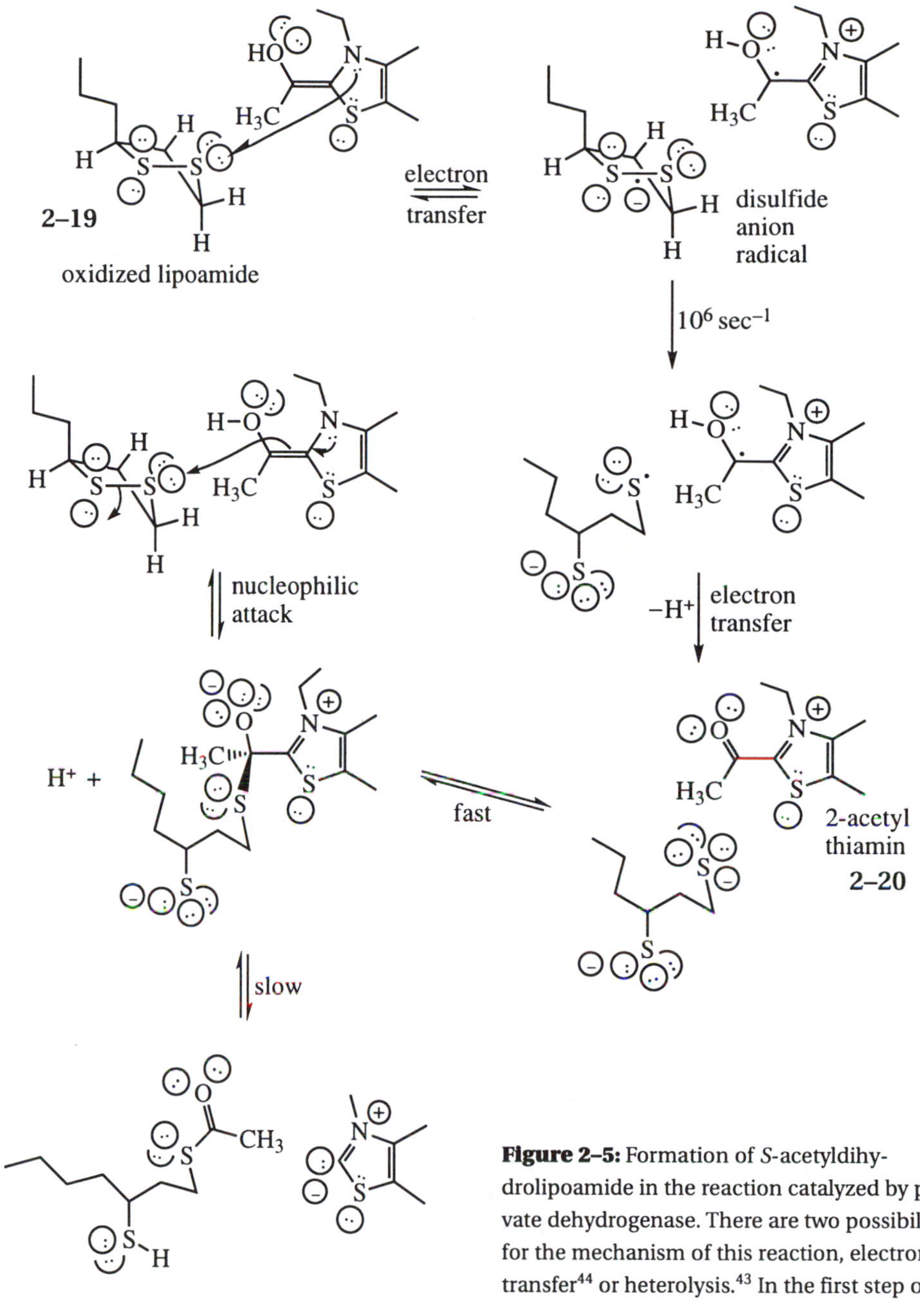

**Figure 2–5:** Formation of *S*-acetyldihydrolipoamide in the reaction catalyzed by pyruvate dehydrogenase. There are two possibilities for the mechanism of this reaction, electron transfer[44] or heterolysis.[43] In the first step of electron transfer (top line), the electron-rich enamine loses an electron to the disulfide to from the disulfide radical anion and thiamin radical cation. The disulfide radical anion fragments to thiolate anion and thiol radical and the second electron is transferred to the radical, to give the bis-thiolate and 2-acetylthiamin. In the first step of the heterolytic reaction (middle left), the enamine as a nucleophile displaces a thiolate anion from the disulfide and the resulting tetravalent intermediate collapses to give the bis-thiolate and 2-acetylthiamin. At this point the two mechanisms converge and the acetyl group is transferred through the tetravalent intermediate to one of the thiolates of dihydrolipoamide.

In the case of the reaction catalyzed by pyruvate dehydrogenase, however, the steps following the decarboxylation of pyruvate do not produce acetaldehyde as indicated in Reaction 2–32. Enamine **2–16** formed immediately upon the decarboxylation of the pyruvate (Reaction 2–28) is not protonated but reacts with a different electrophile, the oxidized form of the coenzyme, **lipoamide** (Structure **2–19**, Figure 2–5), to produce the acetyl thioester of the reduced form of the same coenzyme, referred to as **dihydrolipoamide**. There are two possible mechanisms by which this reaction could take place (Figure 2–5). This enamine of thiamin is nucleophilic enough to attack one of the two sulfurs in an acyclic disulfide,[43] and within the active site of pyruvate decarboxylase, it

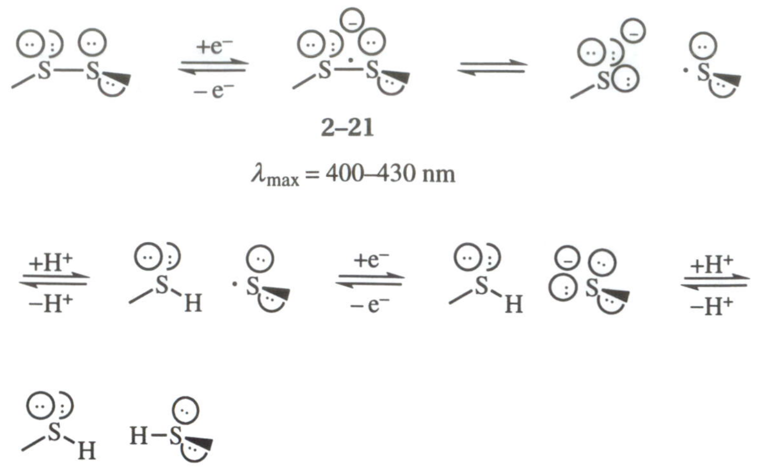

2–21

$\lambda_{max}$ = 400–430 nm

**Figure 2–6:** One-electron reduction of a disulfide.[45] In the first step, an electron is provided by a reductant to produce the disulfide radical anion **2–21** with a characteristic electronic absorption at 400–430 nm. This fragments rapidly ($10^6\ s^{-1}$) to form a thiolate and a thiol radical. The thiolate can be protonated, and the thiol radical can accept a second electron to produce a second thiolate anion.

should be able to react as a nucleophile with oxidized lipoamide acting as the electrophile. The **nucleophilic attack** of this electron-rich enamine resembles the nucleophilic attack of a thiolate anion during a disulfide exchange,[43a] and it liberates one of the sulfurs as a thiolate anion while the other sulfur becomes bonded to the former carbonyl carbon. This creates a tetravalent intermediate with the thiazolium as a potential leaving group. Upon its departure, initiated by removal of the proton from the hydroxyl group, the thioester would be the product.

An alternative mechanism, first proposed on the basis of the ability of this enamine (Reaction 2–28) of thiamin to reduce ferricyanide,[44] would involve two one-electron transfers between the enamine and the disulfide of the oxidized lipoic acid. A **disulfide radical anion**[45] would be a transient intermediate in this electron transfer. Disulfides can be reduced in two one-electron steps in a reaction involving such a disulfide radical anion as an intermediate (Figure 2–6). The rate at which the disulfide radical anion spontaneously fragments to a thiolate anion and a thiol radical is $10^6\ s^{-1}$, more than fast enough for it to act as an intermediate in the reaction. The thiol radical would then accept the second electron either directly or by forming a bond to the exocyclic carbon by radical collapse.

It has been shown that 2-acetylthiamin (Structure **2–20**, Figure 2–5), when it is formed on the appropriate active site of pyruvate dehydrogenase, produces acetyllipoamide as rapidly as does the enamine of thiamin.[46] This has been cited as evidence for the mechanism involving electron transfer. If the tetravalent intermediate proposed for the nucleophilic mechanism is examined closely, however, it is clear that the thiolate is the better leaving group. It necessarily follows that, in either mechanism, 2-acetylthiamin would be formed and unformed many times before the tetravalent intermediate would decompose to yield acetyllipoamide.

The acetyl functional group on *S*-acetyldihydrolipoamide is transferred to the substrate, coenzyme A, in a simple acyl exchange reaction liberating dihydrolipoamide

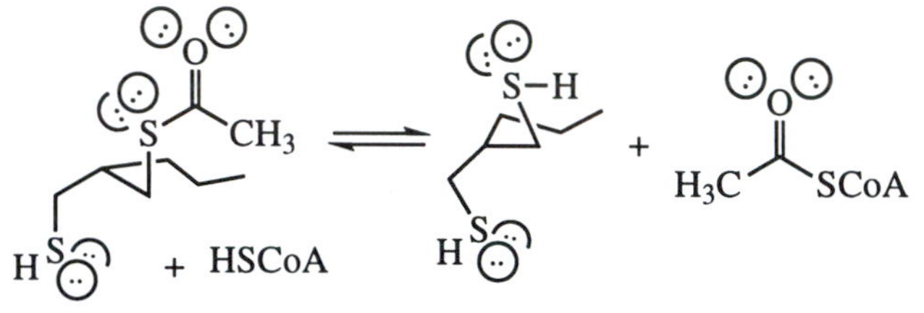

(2–33)

Dihydrolipoamide is covalently bound through an **amide linkage to a lysine** in pyruvate dehydrogenase

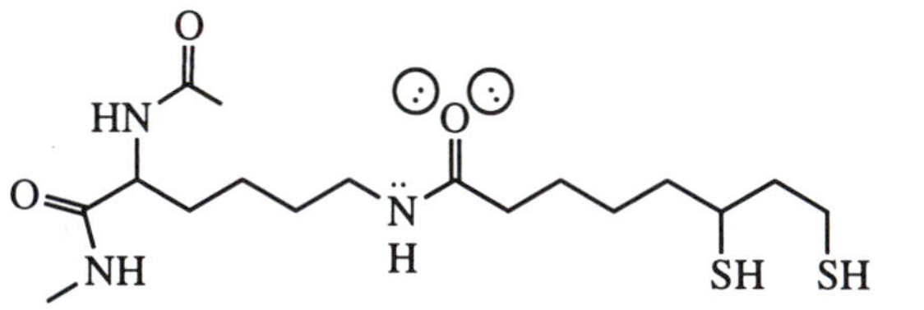

2–22

It can be independently transformed between the reduced and the oxidized form

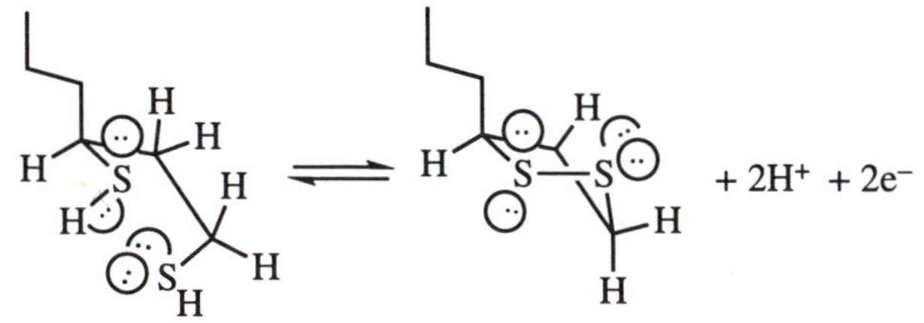

(2–34)

and this oxidation–reduction is catalyzed by the coenzyme, flavin.

## *Suggested Reading*

Kluger, R., Chin, J., & Smyth, T. (1981) Thiamin-Catalyzed Decarboxylation of Pyruvate. Synthesis and Reactivity Analysis of the Central, Elusive Intermediate, α-Lactylthiamin, *J. Am. Chem. Soc. 103*, 884–888.

**PROBLEM 2–7**
Write a mechanism for the transketolase reaction (Reaction 2–30) catalyzed by thiamin pyrophosphate.

**PROBLEM 2–8**
There are two enzymes that can catalyze, respectively, the two reactions

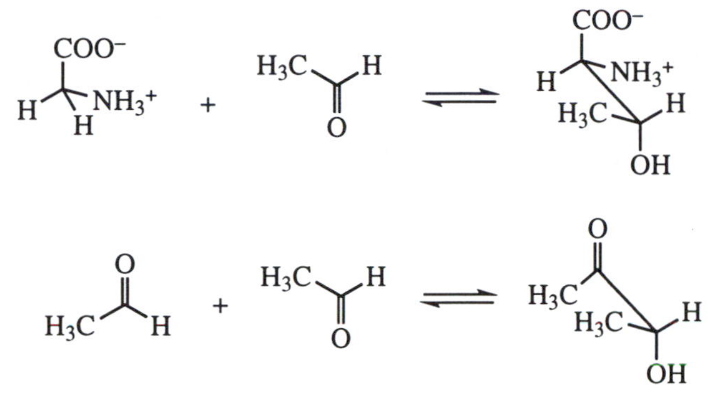

In both of these reactions a proton is replaced by acetaldehyde. The two enzymes, however, use different coenzymes. Write a mechanism for each reaction using the proper coenzyme in each case.

**PROBLEM 2–9**
4-(*p*-Chlorophenyl)-2-ketobut-3-enoic acid is an irreversible inhibitor of pyruvate decarboxylase (Reaction 2–27).

(A) Write the mechanism of the reaction between this inhibitor and the thiamin on the enzyme.
(B) Show by drawing resonance structures why this compound forms an irreversible, covalent adduct with the thiamin within the enzyme.

# Biotin

The coenzyme biotin is a catalyst for **carboxylation** reactions, for example, the reaction catalyzed by propionyl-CoA carboxylase

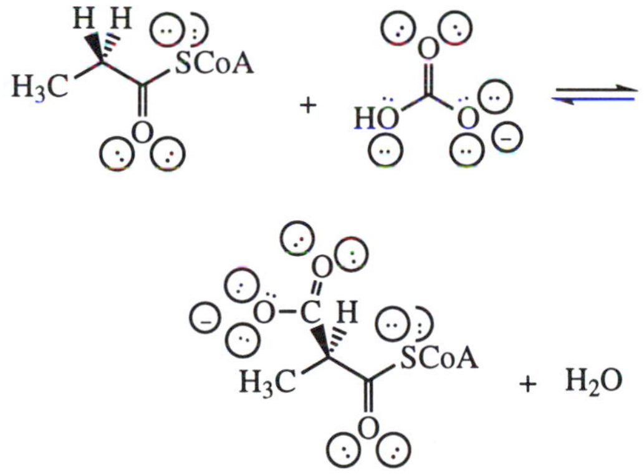

(2–35)

In such reactions, an acidic hydrogen on carbon is removed as a proton and is replaced by the electrophilic acyl carbon of bicarbonate[47] in an acyl exchange reaction in which water is the formal leaving group.

Decarboxylations, the reverse of reactions such as Reaction 2–35, are always significantly exergonic; and, if the carboxylation is not to be endergonic, it must be coupled to a reaction that yields energy. In the case of carboxylation under catalysis by biotin, this energy-yielding reaction is the hydrolysis of MgATP to MgADP and inorganic phosphate. Each time a molecule of carboxylated product is produced, a molecule of MgATP is hydrolyzed. For example, the decarboxylation of oxaloacetate is significantly exergonic[48]

$$\text{oxaloacetate}^{2-} \rightleftharpoons \text{pyruvate}^{-} + HCO_3^{-} \qquad (2\text{–}36)$$

$$\Delta G^\circ = -26\ \text{kJ mol}^{-1}$$

Although this reaction is easy to catalyze, it involves a large waste of free energy, and it is almost never catalyzed in living organisms. In one rare exception, the oxaloacetate decarboxylase of *Klebsiella aerogenes* stores the free energy of this reaction in the form of an electrochemical gradient of sodium cation.[49] In a much more common reaction, however, pyruvate is carboxylated by pyruvate carboxylase

$$\text{MgATP} + \text{pyruvate}^{-} + HCO_3^{-} \rightleftharpoons \text{oxaloacetate}^{2-} + \text{MgADP} + HOPO_3^{2-} \qquad (2\text{–}37)$$

$$\Delta G^\circ(\text{pH } 7) = -11\ \text{kJ mol}^{-1}$$

under catalysis by biotin. It is the biotin that couples the hydrolysis of MgATP to the carboxylation.

Biotin is covalently attached to the active site of an enzyme by an amide between the carboxylate of the coenzyme and a lysine from the protein[50]

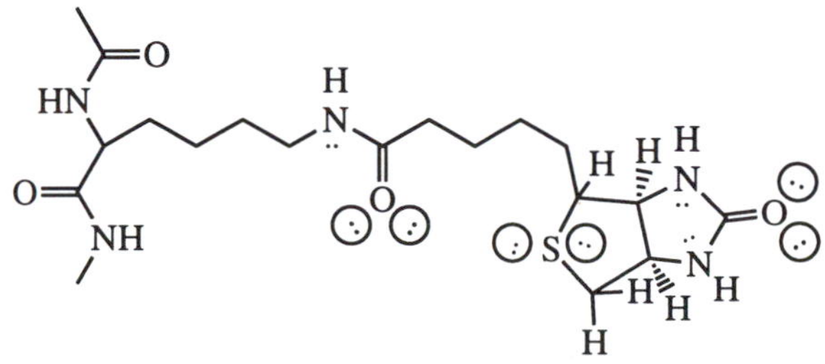

**2–23**

The catalytically active portion of the coenzyme is bicyclic. At one end of the two fused rings is an ***N,N′*-dialkylurea.** The bond lengths between the carbonate carbon of the urea and the acyl oxygen and the carbonate carbon and the two nitrogens, respectively, are very close to those in urea itself[51]

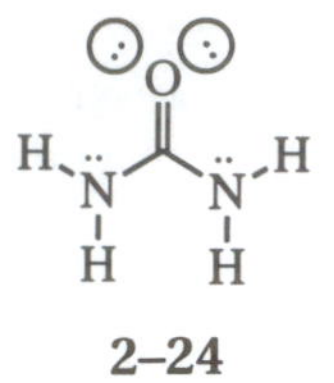

**2–24**

and this suggests that the incorporation of this functional group into the one ring is electronically inconsequential. The ring containing the urea is absolutely flat (Figure 2–7),[51] consistent with the planar structure enforced by the $\pi$ molecular orbitals. Because the two bridgehead carbons are tetrahedral, the two rings are tilted with respect to each other at an angle of 122° as if they were the wings of a butterfly. The cyclic thioether is fully saturated, and the sulfur occupies the puckered position of the five-membered ring. It is puckered toward the carbonate carbon of the urea rather than away. This allows the alkyl group of the side chain of biotin to occupy an equatorial position, and presumably this steric effect explains the stereochemistry. This stereochemistry moves the sulfur closer to the urea, perhaps to assist in reactions at this location by a transannular effect,[52] but the sulfur is too far from the carbon of the urea for significant overlap to occur.[51] The sulfur, however, may in part be responsible for the planarity of the ureido ring because in dethiobiotin this ring is puckered.

**Figure 2–7:** Drawing of the crystallographic molecular model of D-(+)-biotin. Crystals were obtained from an aqueous solution of the free acid of D-(+)-biotin by cooling a saturated solution from 95 °C. The fused rings are inclined at 122° with the *N,N′*-dialkylurea at the top.[51] Reprinted with permission from ref 51. Copyright 1976 American Chemical Society.

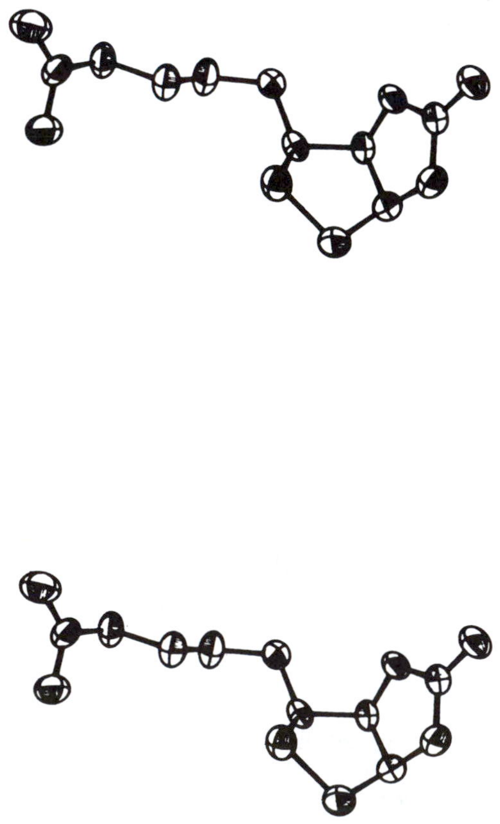

The urea moiety is the catalytic center of biotin. The central intermediate in all reactions catalyzed by biotin is **1′-(*N*-carboxy)biotin**[53,54]

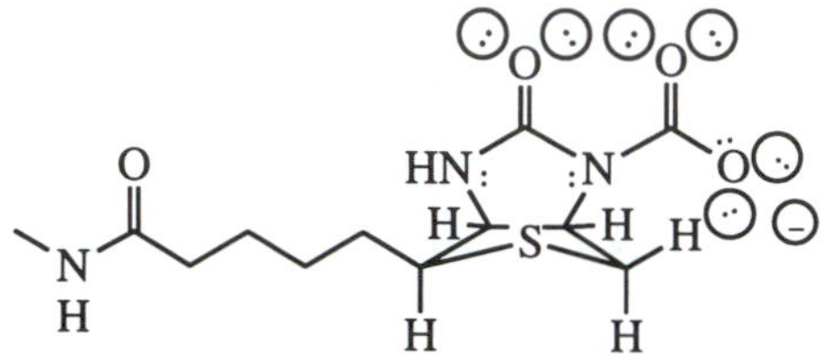

**2–25**

in which one of the oxygens on the bicarbonate has been replaced by one of the nitrogens of the urea. It is during the carboxylation of the biotin that the coupling to the hydrolysis of ATP occurs.

During the carboxylation of biotin coincident with the hydrolysis of MgATP, one of the three oxygens of bicarbonate is incorporated into the inorganic phosphate formed during the reaction and two of the oxygens end up in the new carboxylate formed during the reaction.[55] This transfer of $^{18}O$ from [$^{18}O$]bicarbonate to the fourth position of the phosphate produced in the enzymatic reaction is also observed[56] in the slow bicarbonate-dependent ATPase reaction

$$\text{MgATP} + \text{HC}^{18}\text{O}_3^- + \text{H}_2\text{O} \rightleftharpoons \text{MgADP} + \text{H}^{18}\text{OPO}_3^{2-} + \text{HOC}^{18}\text{O}_2^- \qquad (2\text{–}38)$$

catalyzed by biotin carboxylase in the absence of biotin. This result demonstrates that the transfer of the oxygen does not require any intermediate involving the biotin itself. The enzymatically catalyzed carboxylation of biotin also proceeds with inversion of configuration at the terminal phosphate of ATP.[57] All of these results are consistent with a mechanism involving the intermediate formation of carboxyphosphate (Structure **2–26,** Figure 2–8).

The nucleophile attacking the carboxyphosphate would be either the anionic conjugate base of the urea (Figure 2–8) or the $\pi$ lone pair on the nitrogen of the urea coincident with the removal of the proton on that same nitrogen by a base participating in the reaction.[58] The model reaction for the first possibility is the rapid (>12 $s^{-1}$ at pH 7.4), base-catalyzed exchange of the protons on the nitrogens of the urea in biotin,

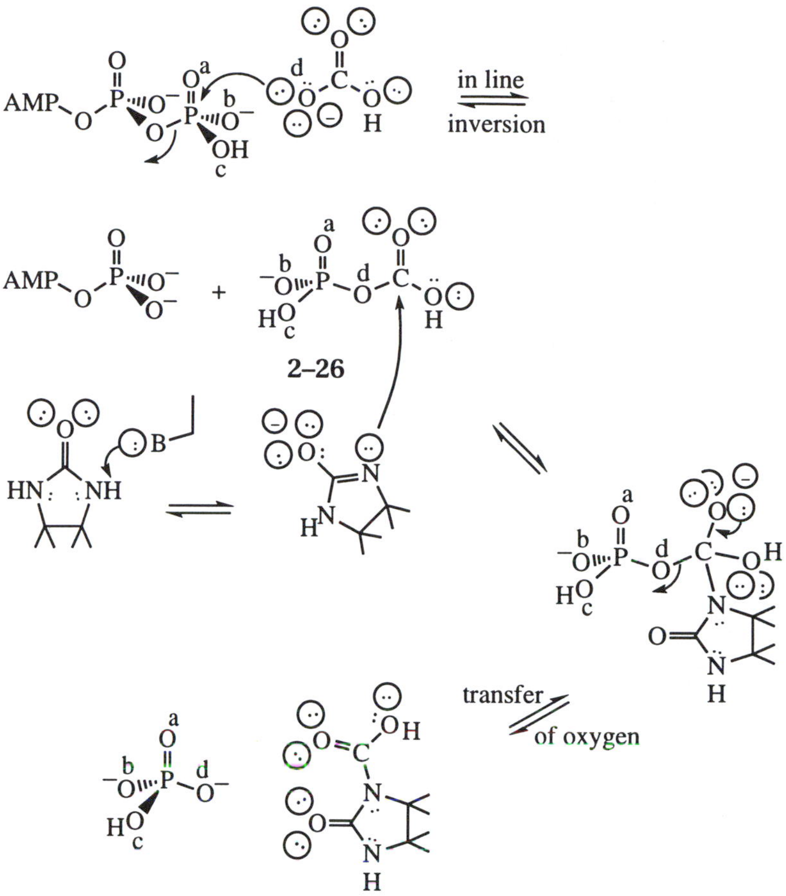

**Figure 2–8:** Formation of 1′-(*N*-carboxy)-biotin simultaneously with the hydrolysis of MgATP. Bicarbonate accepts phosphate in an in-line transfer with inversion of configuration at phosphorus. The intermediate carboxyphosphate **2–26** is then attacked by the 1′ amide nitrogen on biotin to form a tetravalent intermediate that decomposes with loss of inorganic phosphate to produce 1′-(*N*-carboxy)biotin. In the reaction, one of the oxygens of bicarbonate (marked d) ends up in the inorganic phosphate, as required by the observation that phosphate produced during carboxylation of biotin with [$^{18}O$]$HCO_3^-$ bears an equivalent of $^{18}O$.

in line
inversion

**2–26**

transfer of oxygen

which indicates that these protons can be readily removed to form the conjugate base.[51] The model reaction for the second possibility is the acid-catalyzed proton exchange of biotin[58] that proceeds by the protonation of the $\pi$ lone pair of one of the ureido nitrogens,[59] which is somewhat more basic than the nitrogen of an amide because of cross conjugation, to produce the ureido cation

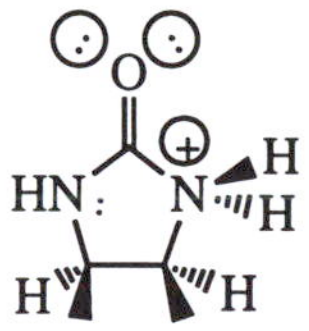

**2–27**

In this latter acid–base reaction, the $\pi$ lone pair of the nitrogen has participated as a nucleophile whose reaction with an electrophile, the proton, has preceded the removal of the other proton.

Regardless of the exact identity of the nucleophile, the reaction is a simple **acyl exchange** at the carbonate carbon of carboxyphosphate with the nitrogen of the biotin as the nucleophile and the inorganic phosphate as the leaving group. Phosphorylation of the oxygen of the bicarbonate by ATP has turned a poor leaving group ($HO^-$) into a good leaving group ($HO_3PO^{2-}$).

1′-(*N*-Carboxy)biotin (**2–25**) is the derivative of carbonic acid that transfers the carboxylate to an enol in an acyl exchange reaction

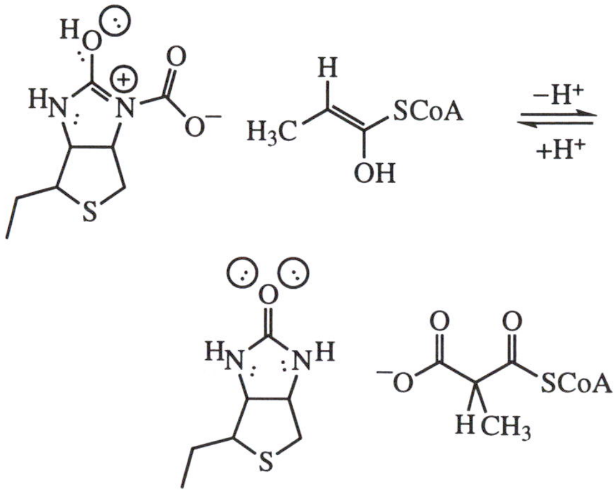

$-H^+$ / $+H^+$

(2–39)

Protonation of the biotin on the exocyclic oxygen is probably a necessary step in designating it as the leaving group. A very similar, intramolecular carboxyl transfer occurs in the model reaction[60]

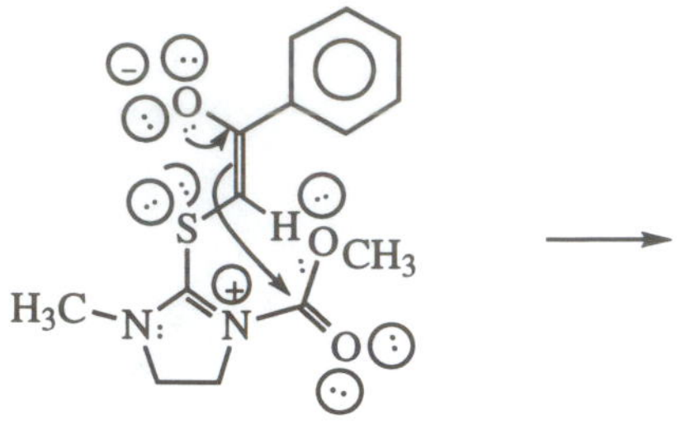

2–28

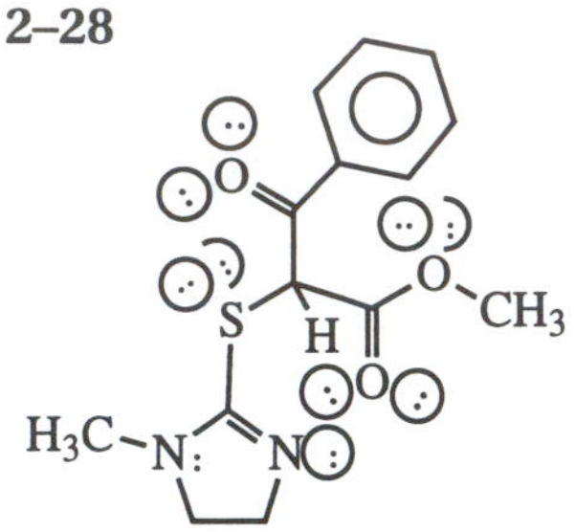

(2–40)

but only if the other nitrogen is methylated. The *N*-methyl-*S*-alkyl model compound **2–28** is equivalent to the *N*-protonated *O*-protonated biotinyl cation in Reaction 2–39. The biotin is a reasonably good leaving group, as illustrated by the fact that the free energy of hydrolysis of 1′-(*N*-carboxy)biotin is –20. kJ $mol^{-1}$ at pH 7 and 0 °C.[48]

### *Suggested Reading*

Kohn, H. (1976) Model Studies on the Mechanism of Biotin-Dependent Carboxylations, *J. Am. Chem. Soc. 98*, 3690–3694.

### PROBLEM 2–10

Write a mechanism for Reaction 2–35.

# Flavin

The various flavins have, as their catalytically active portion, a **7,8-dimethylisoalloxazine** substituted at nitrogen 10

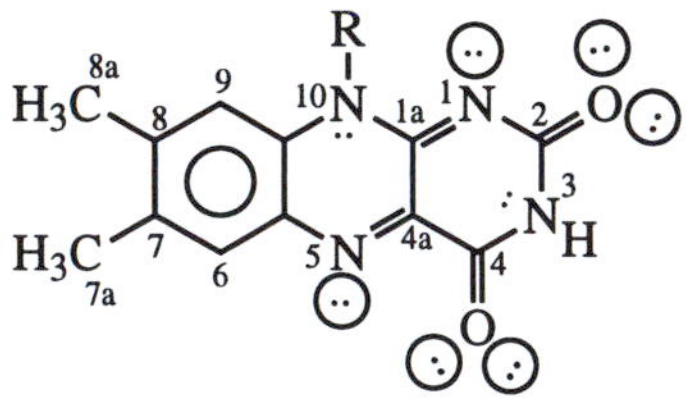

2–29

Flavins are incorporated into enzymes either covalently or noncovalently. In every case, the flavin is never a substrate but always a true coenzyme, remaining bound tightly to the protein during the entry of substrates and the exit of products. When it is covalently attached to a protein, the covalent bond is usually between the carbon of the 8a methyl group and a nucleophilic amino acid such as histidine[61] or cysteine.[62] This presumably results from a nucleophilic attack on the vinylogous $\alpha,\beta$-unsaturated imine present in tautomer **2–30** of the 8-methylisoalloxazine of oxidized flavin

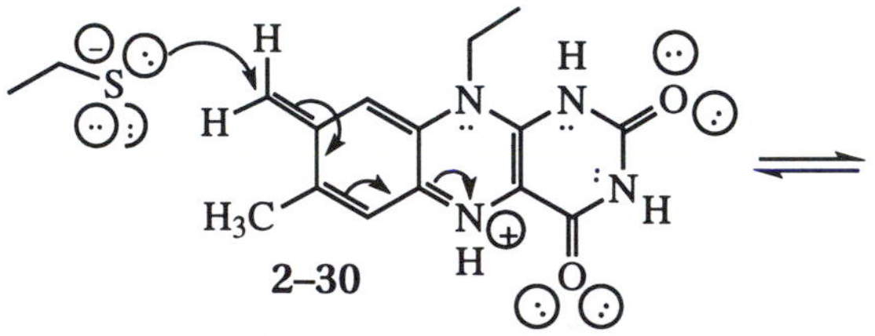

2–30

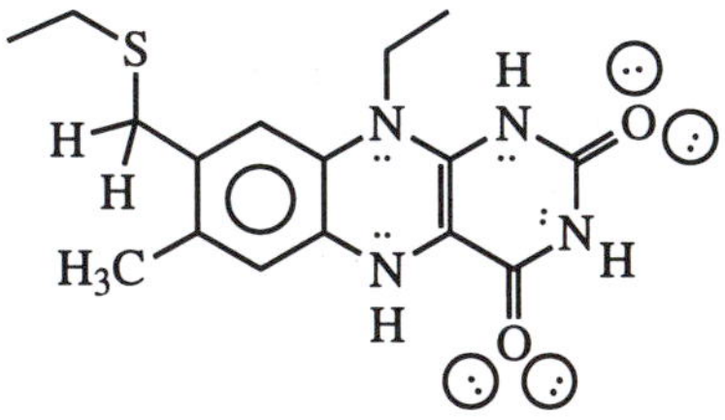

(2–41)

and the addition is catalyzed by the active site of the enzyme itself. Because flavin is either covalently bound or very tightly bound to a protein and also has a bright yellow color, a protein with bound flavin can be purified by monitoring this color. The characteristic absorption spectrum of such a protein can be used to demonstrate that it has the coenzyme bound to it, and this spectrum designates the protein as a **flavoprotein** or **flavoenzyme.** These terms are sometimes used to designate a protein before its specific function has been determined and a descriptive name has been assigned to it.

There are three stable forms of flavin that differ sequentially by single electrons in their oxidation states (Figure 2–9). Neutral, **oxidized flavin** ($HFl_{ox}$, Figure 2–9) is the most electron-deficient form occurring naturally. It contains 18 $\pi$ electrons in a $\pi$ molecular orbital system constructed from 16 *p* orbitals. When it is completely unprotonated as the anion ($Fl_{ox}^-$, Figure 2–9) it has seven $\sigma$ lone pairs of electrons available for protonation, which provide ample opportunity for tautomers. In the abbreviations that follow, it will be assumed that the completely deprotonated flavin is this anion, and the number of protons in each abbreviation will be based on this designation. The addition of an electron to the neutral form of oxidized flavin ($HFl_{ox}$, Figure 2–9) forms the radical anion ($HFl^{\cdot -}$, Figure 2–9). At neutral pH, the radical anion takes up a proton to become the uncharged species.

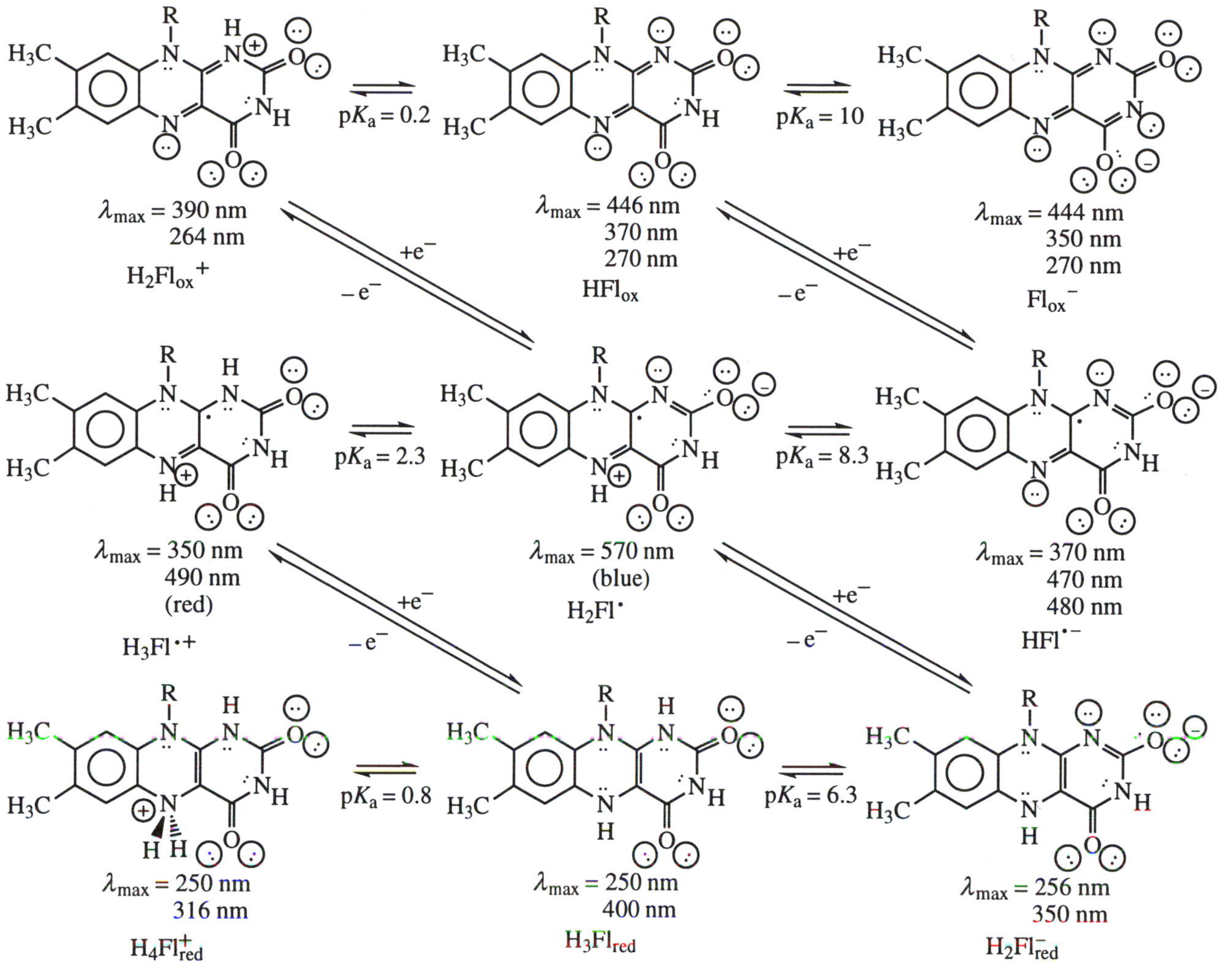

The neutral **flavin radical** ($H_2Fl^{\cdot}$, Figure 2–9) absorbs light of long wavelength ($\lambda_{max}$ = 570 nm) because of its zwitterionic character, but it is bleached when either a proton is added to one of the $\sigma$ lone pairs to generate the radical cation ($H_3Fl^{\cdot +}$, Figure 2–9) or a proton is removed from one of the $\sigma$ lone pairs to generate the radical anion ($HFl^{\cdot -}$, Figure 2–9). In all forms of the flavin radical, however, there are formally 19 $\pi$ electrons in a $\pi$ molecular orbital system constructed from 16 $p$ orbitals, and in its neutral form it has seven $\sigma$ lone pairs on heteroatoms available for protonation and two of these are occupied by two protons. In the flavin radical, the resonance form presented (Figure 2–9) emphasizes the ability of an adjacent lone pair of electrons to stabilize an unpaired electron

**Figure 2–9:** The three oxidation states of flavin and the normally accessible conjugate acids and bases at each oxidation level.[63–65] Three conjugate acid–bases of fully oxidized flavin are presented on the top row with the characteristic absorptions and their abbreviations. Addition of one electron produces the radical (middle row). The most unusual of the three conjugate acid–bases of the radical is the blue neutral species ($\lambda_{max}$ = 570). In forming this radical ($H_2Fl^{\cdot}$) directly from the cationic oxidized flavin ($H_2Fl_{ox}^{+}$), the two nitrogens switch basicity and the neutral radical is the opposite tautomer.[65] Addition of a second electron produces fully reduced flavin (bottom row), three conjugate acid–bases of which are presented. The addition of each electron, as expected, increases the basicity of the acid–base to which the electron was added.

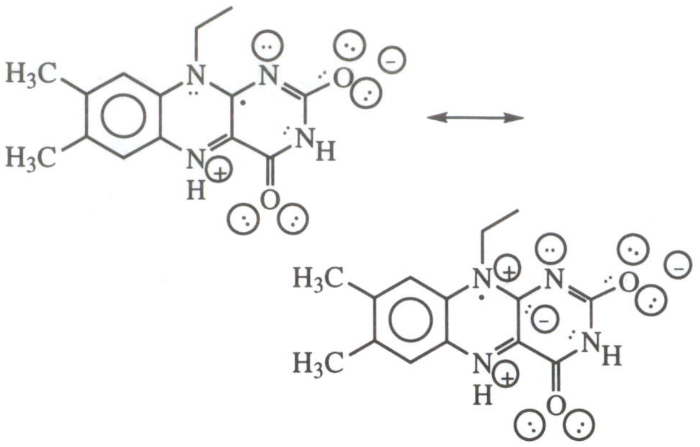

**2–31**

The addition of an electron to the neutral flavin radical produces the reduced flavin anion, $H_2Fl^-_{red}$. Its conjugate acid is reduced flavin ($H_3Fl_{red}$).

The dihydrodimethylisoalloxazine that constitutes neutral **reduced flavin** ($H_3Fl_{red}$, Figure 2–9)

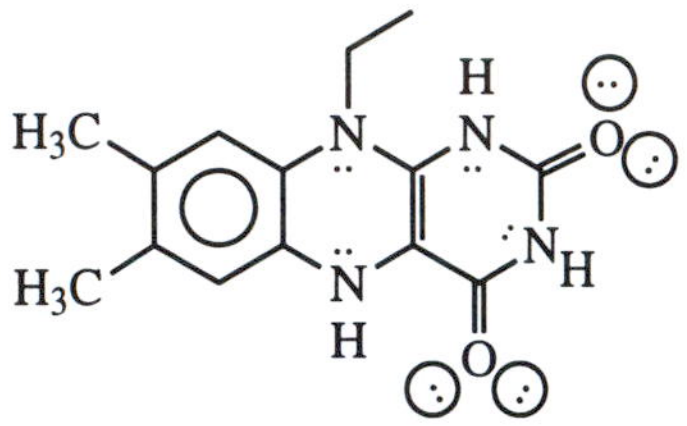

**2–32**

is conveniently divided in the middle along the axis of nitrogen 5 and nitrogen 10 into 1,2-diamino-4,5-dimethylbenzene

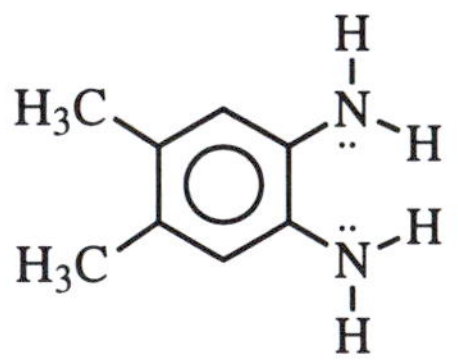

**2–33**

and 5,6-diaminouracil

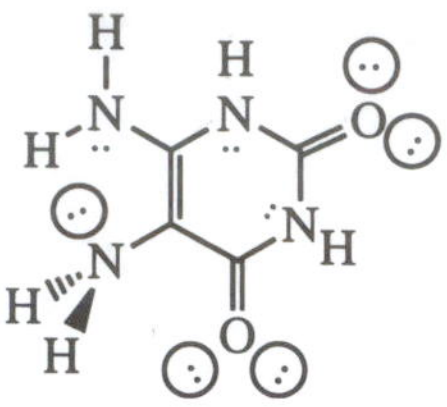

**2–34**

In this dissection, the isoalloxazine is described as a hybrid of a dianiline and the base of a nucleic acid. The diaminouracil, however, has a normal exocyclic nitrogen at position 6 whose lone pair of electrons is conjugated to the heterocycle.

Neutral reduced flavin ($H_3Fl_{red}$), if it is fully conjugated, has 20 $\pi$ electrons in a $\pi$ molecular orbital system constructed from 16 $p$ orbitals. This is an unusually electron-rich $\pi$ molecular orbital system, and it may be so electron-rich that it cracks in half along its central axis. One way to relieve the pressure of the electron density would be to bend the molecule along an axis between nitrogen 5 and nitrogen 10. This would cause the hybridization on these two nitrogens to be [$sp^3$, $sp^3$, $sp^3$, $sp^3$] and allow them each to confine a pair of those 20 electrons as a $\sigma$ lone pair. This would split the $\pi$ molecular orbital system into two halves—the diaminodimethylbenzene and the diaminouracil. For this to happen the planes of these two rings must meet at an angle.

The analogue of reduced flavin that displays the largest angle between the plane of the diaminodimethylbenzene and the diaminouracil is 5-acetyl-9-bromo-1,3,7,8,10-pentamethyl-1,5-dihydroisoalloxazine (Figure 2–10). In this compound the angle between the two rings is 144.5°. It has been implied that this deviation from planarity is electronic in origin and reflects the hybridization on the two nitrogens. There is, however, convincing evidence that the deviation is steric[66] in origin. In 5-acetyl-9-bromo-1,3,7,8,10-pentamethyl-1,5-dihydroisoalloxazine, the methyl group ($C_{20}$ in Figure 2–10) on nitrogen 10 is forced below the plane by the unfavorable steric repulsion of the bromine on carbon 9 and the methyl on nitrogen 1, and the acetyl group on nitrogen 5

**Figure 2–10:** Drawing of a crystallographic molecular model of 5-acetyl-9-bromo-1,3,7,8,10-pentamethyl-1,5-dihydroisoalloxazine.[68] The dimethylbenzene ring is to the left and the pyrimidine ring is to the right. The view is parallel to the planes of the two rings down the axis defined by nitrogen 5 and nitrogen 10. Reprinted with permission from ref 68. Copyright 1970 Munksgaard.

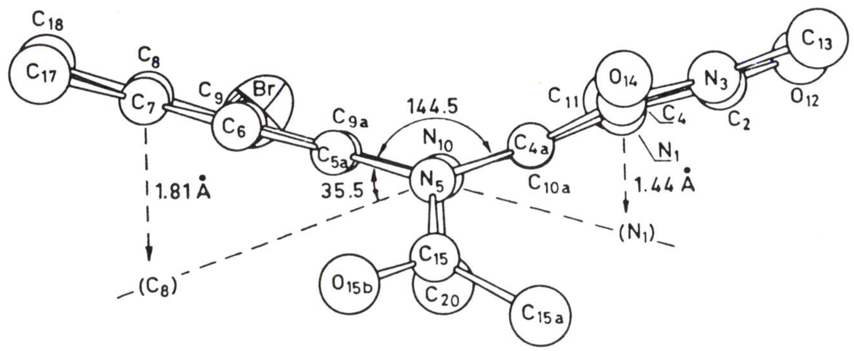

is forced below the plane by the unfavorable steric repulsion of the oxygen ($O_{14}$ in Figure 2–10) on carbon 4 and the hydrogen on carbon 6. As reduced flavin itself has three hydrogens respectively replacing the bromine on carbon 9, the methyl on nitrogen 1, and the acetyl on nitrogen 5, these steric effects would be greatly diminished in the biochemically active compound. In fact, when the acetyl group on nitrogen 5 in the analogue is replaced by a hydrogen, the angle between the rings increases to 159° even though the clearly observed steric repulsion among the bromine, the methyl, and the methyl at the back side of the ring remains. In 4a-isopropyl-3-methyl-4a,5-dihydrolumiflavin

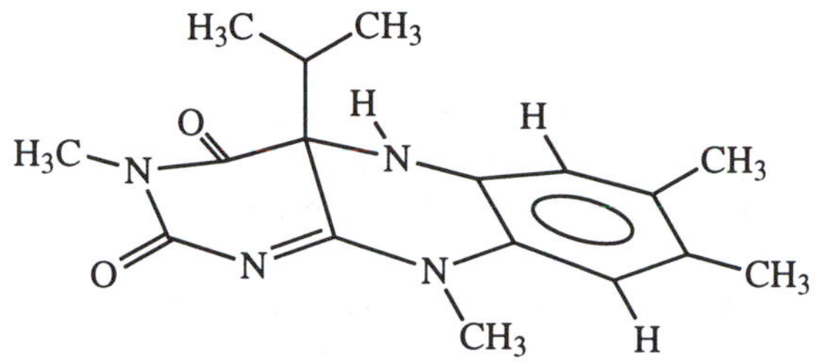

**2–35**

in which the bromine on carbon 9 and the methyl on nitrogen 1, as well as the methyl on nitrogen 5, are all replaced with hydrogens, the *N*-acyl-*N*-methylurea of the pyrimidine is coplanar with the *o*-phenylenediamine.[67]

All of these considerations based on crystallographic molecular models suggest that the isoalloxazine ring in reduced flavin would be planar. Nuclear magnetic resonance spectra of reduced flavins in both apolar and polar solvents are also consistent with the conclusion that these molecules are planar.[69] This conclusion would also be consistent with the acid–base behavior of the two nitrogens on 1,2-diaminobenzene itself ($pK_{a1}$ = 1.3 and $pK_{a2}$ = 4.5), which are considerably less basic than those in 1,2-diaminoethane ($pK_{a1}$ = 7.5 and $pK_{a2}$ = 10.0). This fact indicates that these two nitrogens have $\pi$ lone pairs of electrons and are themselves planar. Furthermore, the exocyclic nitrogens of nucleic acid bases, such as 6-aminouracil, are usually planar. These considerations would explain the particularly low $pK_a$ ($pK_a$ = 0.8) for nitrogen 5 of reduced flavin (Figure 2–9), which cannot be mistaken for the $pK_a$ of a tetrahedral nitrogen ($pK_a$ = 10) and more closely resembles the values of the $pK_a$ for the exocyclic nitrogens in nucleic acid bases. This would also be consistent with the observation that the reduced flavin within the protein, flavodoxin, is planar.[70]

Oxidized flavin ($HFl_{ox}$) is generally assumed to be completely planar owing to the conjugation that connects every atom in the three rings. This conclusion is usually based on the crystallographic molecular models of several 10-alkylisoalloxazines, which are perfectly planar (Figure 2–11).[71] Most of these analogues, however, have been cocrystallized with naphthalene-2,3-diol,[71–73] which sandwiches the isoalloxazines in the crystal and could thereby force them to be planar even if they were not already planar. Therefore, the distinction that is usually made between planar, rigid oxidized flavin and flexible, bent reduced flavin may not be so clear-cut as has been assumed, and it is probably not the case that constraining the flavin to one geometry can affect its oxidation–reduction potential. The nature of the acid–bases around the flavin is probably far more important.[74]

A reduced molecule and an oxidized molecule that can be interconverted by adding or subtracting electrons, respectively, are an example of a **redox pair.** The **reduction potential,** ***E,*** of a redox pair under a given set of conditions is the voltage it can provide to move electrons when they are transferred from

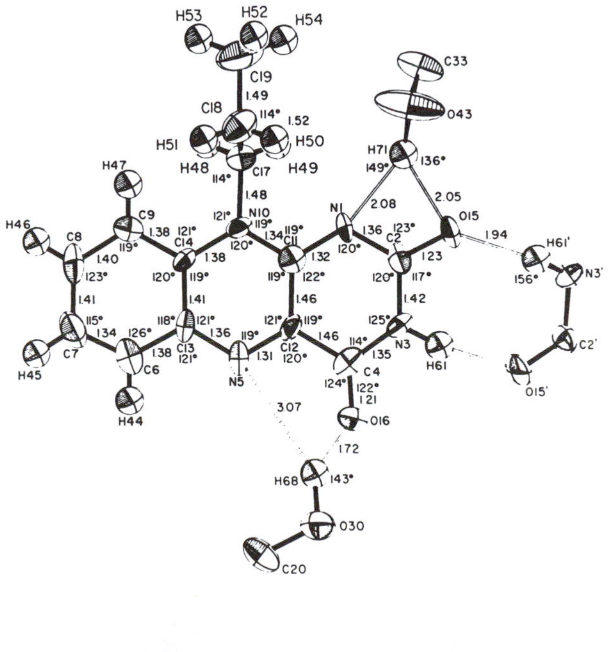

**Figure 2–11:** Drawing of the crystallographic molecular model of 10-propylisoalloxazine.[71] The oxidized flavin was cocrystallized with bis(naphthalene-2,3-diol) by cooling an aqueous solution saturated with both compounds. The propyl group on nitrogen 10 of the flavin is at the top of the central molecule. The three rings of the oxidized flavin are all in the same plane (±0.005 nm). Hydrogen bonds with several neighbors are indicated. A pair of hydrogen bonds link the pyrimidine ring of one isoalloxazine to the pyrimidine ring of a neighboring isoalloxazine (right side). Only carbon 2′, nitrogen 3′, and oxygen 15′ of this rotationally symmetric neighbor are displayed. Hydrogen bonds between the isoalloxazine and two neighboring naphthalene diols (top right and bottom) are also shown, to nitrogen 1 and oxygen 16, respectively. Only the carbon, oxygen, and hydrogen of the alcohol are displayed in each case. Reprinted with permission from ref 71. Copyright 1974 Munksgaard.

a standard hydrogen electrode to the oxidized molecules of the redox pair to form the reduced molecules. The standard hydrogen electrode

$$H_2 \rightleftharpoons 2H^+ + 2e^- \qquad (2\text{–}42)$$

is diatomic hydrogen at an activity of 1 atm in equilibrium with two protons at pH = 0 in aqueous solution. A simple example of a redox pair would be ferricyanide and ferrocyanide

$$Fe^{III}(CN)_6 + e^- \rightleftharpoons Fe^{II}(CN)_6 \qquad (2\text{–}43)$$

The voltage that this redox pair can produce to move electrons from the molecules of $H_2$ gas at a standard hydrogen electrode to ferricyanide to produce ferrocyanide, when the conditions are such that the activities of the ferricyanide and ferrocyanide are equal, is +0.36 V. In such a simple unimolecular oxidation–reduction, the reduction potential provided by the redox pair to move electrons from the standard hydrogen electrode to a mixture of the two members of the redox pair at equal activity is the **standard reduction potential, $E°$.** The more positive the reduction potential, the easier it will be to reduce the oxidized molecules of the redox pair and the harder it will be to oxidize the reduced molecules. Electrons move from redox pairs of more negative standard reduction potential to redox pairs of more positive standard reduction potential. For example, the standard reduction potential for the redox pair of oxidized flavin and flavin radical anion

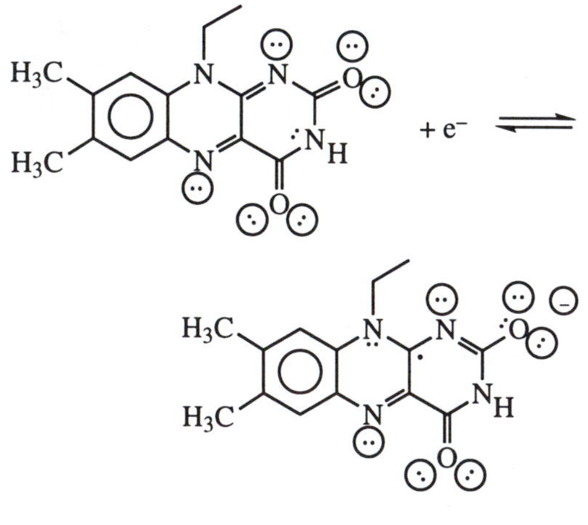

(oxidized flavin) + e⁻ ⇌ (flavin radical anion) (2–44)

is −0.33 V.[75] Therefore, in the complete chemical reaction when reactants and products are at equal activity

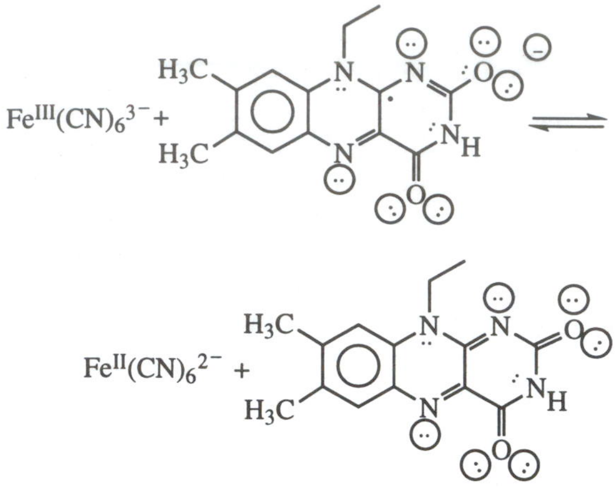

$Fe^{III}(CN)_6^{3-}$ + (flavin radical anion) ⇌ $Fe^{II}(CN)_6^{2-}$ + (oxidized flavin) (2–45)

the electron is moving from a redox pair with a standard reduction potential of −0.33 V to a redox pair with a standard reduction potential of +0.36 V, and this reaction is favorable in the direction written.

The **free energy change** of an oxidation–reduction reaction under a particular set of conditions, $\Delta G$, is related to the reduction potentials of the two redox pairs under those same conditions by the equation

$$\Delta G = -n\mathcal{J}\Delta E \qquad (2\text{–}46)$$

where $n$ is the number of electrons transferred from reductant to oxidant for the stoichiometric reaction as written, $\mathcal{J}$ is Faraday's constant (96,485 J $V^{-1}$ $mol^{-1}$) and

$$\Delta E = E\,(\text{oxidized reactant/reduced product}) - E\,(\text{oxidized product/reduced reactant}) \qquad (2\text{–}47)$$

where the values of $E$ are the reduction potentials of the two redox pairs under the particular conditions of the reaction. The more positive the difference in the reduction potentials, the more free energy is released in the oxidation–reduction.

The reduction potential, $E$, varies, as does the free energy, with the respective concentrations of the two molecules of the redox pair. As the concentration of oxidized species is increased, the pair becomes easier to reduce, and contrariwise. Quantitatively

$$E = E° - \frac{RT}{n\mathcal{J}} \ln \frac{\gamma_{red}[\text{reduced}]}{\gamma_{ox}[\text{oxidized}]} \qquad (2\text{–}48)$$

where $\gamma_{red}$ and $\gamma_{ox}$ are the activity coefficients of the reduced and oxidized species, respectively, and [reduced] and [oxidized] are their respective molar concentrations. It follows from Equation 2–46 that

$$\Delta G° = -n\mathcal{J}\Delta E° \qquad (2\text{–}49)$$

where $\Delta G°$ is the change in standard free energy.

In most situations, the addition of one or more electrons to the oxidized molecule of a redox pair dramatically changes the $pK_a$ of its lone pairs of electrons and causes the reduced molecule to be more basic. For example, the addition of an electron to the neutral form of oxidized flavin ($HFl_{ox}$) to form the radical anion ($HFl^{\cdot-}$) raises the p$K_a$ of the conjugate acid from 0.2 to 8.3 (Figure 2–9) because, with the additional electron, the flavin becomes more basic. The addition of the electron and the addition of a proton are linked

$$\begin{array}{ccc} e^- + H_2Fl_{ox}^+ & \overset{K_a^{ox}}{\rightleftharpoons} & HFl_{ox} + H^+ + e^- \\ \Big\Updownarrow E^\circ_{H_2} & & \Big\Updownarrow E^\circ_{H} \\ H_2Fl^{\cdot} & \underset{K_a^{rad}}{\rightleftharpoons} & HFl^{\cdot-} + H^+ \end{array} \tag{2–50}$$

where $K_a^{ox}$ and $K_a^{rad}$ are acid dissociation constants and $E^\circ_{H_2}$ and $E^\circ_{H}$ are standard reduction potentials. The linkage is through the standard free energies

$$E^\circ_{H_2} - E^\circ_{H} = \frac{RT}{n\mathcal{F}} \ln \frac{K_a^{ox}}{K_a^{rad}} \tag{2–51}$$

Therefore, the protonation of neutral oxidized flavin to form the cation ($H_2Fl_{ox}^+$) makes it easier to reduce in a one-electron reaction by 0.48 V. This illustrates the fact that the nature of the acid–bases that provide hydrogen-bond donors to the $\sigma$ lone pairs around the flavin when it is in the active site of an enzyme have a significant effect on the affinity of the redox pair of oxidized flavin and flavin radical for an electron.

Usually when one or more electrons are added to a molecule, the resulting increase in basicity causes it to take up one or more protons immediately. How many protons are taken up depends on the pH of the solution. The number of protons taken up affects the apparent reduction potential. Under a particular set of circumstances, the **apparent standard reduction potential,** $E^\circ_{app}$, is defined as the potential measured when the total concentration of reduced reactant, $[\text{reduced}]_{TOT}$, in all of its forms, is equal to the total concentration of oxidized reactant $[\text{oxidized}]_{TOT}$, in all of its forms, and the conditions are such that the activity coefficients cancel, usually because the concentrations of the reactants are small. By this definition

$$E = E^\circ_{app} - \frac{RT}{n\mathcal{F}} \ln \frac{[\text{reduced}]_{TOT}}{[\text{oxidized}]_{TOT}} \tag{2–52}$$

A simple example of the effect of pH on the apparent reduction potential is encountered in the reduction of the two-electron redox pair of anthroquinone-2-sulfonate

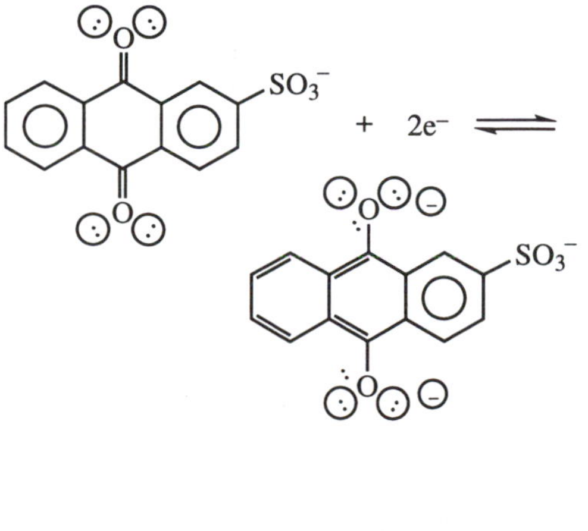

(2–53)

The addition of two electrons to the quinone raises the values of the p$K_a$ for the $\sigma$ lone pairs of electrons on the two oxygens from below 0 to p$K_{a1}$ = 8.0 and p$K_{a2}$ = 11.3 for the first and second ionizations, respectively[76]

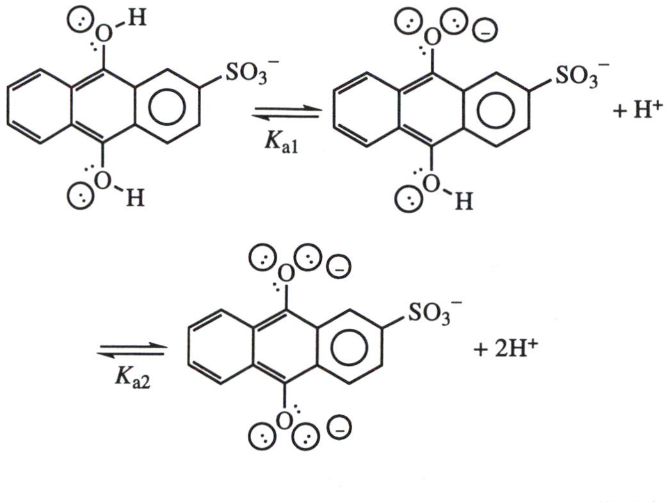

(2–54)

It is easily shown that

$$[Q^{3-}_{red}] = [Q_{red}]_{TOT} \frac{K_{a1}K_{a2}}{K_{a1}K_{a2} + K_{a1}[H^+] + [H^+]^2} \tag{2–55}$$

where $[Q^{3-}_{red}]$ is the concentration of the trianion of the quinol and $[Q_{red}]_{TOT}$ is the total concentration of quinol in all forms. The reduction of the monoanionic quinone to the trianion of the quinol (Reaction 2–53) has a standard reduction potential, $E^\circ$, of –0.38 V.[76] Therefore the observed reduction potential, $E$, will be equal to

$$E = -0.38\text{V} - \frac{RT}{2\mathcal{F}} \ln \frac{[Q^{3-}_{red}]}{[Q_{ox}]} \tag{2–56}$$

if $\gamma_{red}/\gamma_{ox}$ is assumed to be 1. As $[Q_{ox}] = [Q_{ox}]_{TOT}$

$$E = -0.38\text{ V} - \frac{RT}{2\mathcal{F}} \ln \frac{[Q_{red}]_{TOT}}{[Q_{ox}]_{TOT}} - \frac{RT}{2\mathcal{F}} \ln \frac{K_{a1}K_{a2}}{K_{a1}K_{a2} + K_{a1}[H^+] + [H^+]^2} \tag{2–57}$$

By definition (Equation 2–52), when $[Q_{red}]_{TOT}/[Q_{ox}]_{TOT} = 1$, $E = E^\circ_{app}$ Therefore,

$$E^\circ_{app} = -0.38\text{ V} - \frac{RT}{2\mathcal{F}} \ln \frac{K_{a1}K_{a2}}{K_{a1}K_{a2} + K_{a1}[H^+] + [H^+]^2} \tag{2–58}$$

Equation 2–58 describes the behavior, as a function of the pH, of the apparent standard reduction potential of anthroquinone-1-sulfonate (Figure 2–12).[76] When pH > p$K_{a2}$ and $[H^+] < K_{a2}$

$$E^\circ_{app} = -0.38\text{ V} = E^\circ \tag{2–59}$$

When p$K_{a2}$ > pH > p$K_{a1}$ and $K_{a2} < [H^+] < K_{a1}$

$$E^\circ_{app} = -0.38\text{ V} - \frac{2.303RT}{2\mathcal{F}}(\text{pH} - \text{p}K_{a2}) \tag{2–60}$$

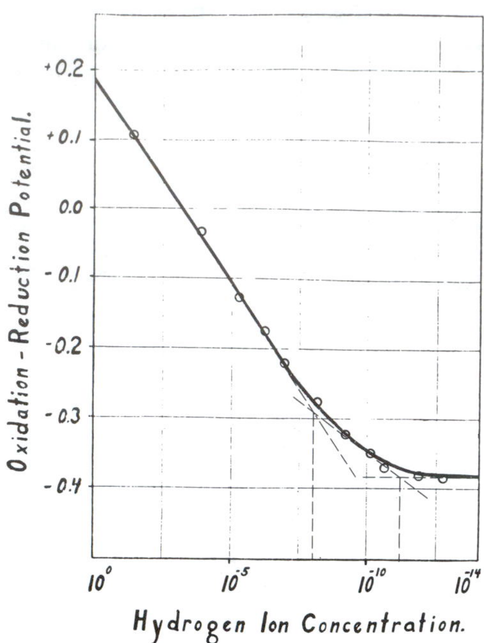


**Figure 2–12:** Apparent standard reduction potential of anthroquinone-2-sulfonate as a function of pH.[76] A 3 mM solution of the quinone at the desired pH was titrated with the strong reducing agent sodium hydrosulfite. The reduction potential of the solution was monitored directly with a platinum electrode after each addition. The solution was connected by a salt bridge to a calomel electrode to complete the circuit. The measured potential at the point at which half of the quinone had been reduced, determined by the direct titration, was taken as the apparent standard reduction potential at that pH. This is the potential displayed in the figure as a function of pH. The solution also contained a hydrogen electrode that was used to calibrate the readings. The three dashed lines of increasing negative slope from right to left are the lines generated by Equations 2–59, 2–60, and 2–61, respectively, and the two values for the p$K_a$ are at the intersections of these lines, as indicated by the two vertical dashed lines. Adapted with permission from ref 76. Copyright 1922 American Chemical Society.

Because $-2.303\ RT/2\mathcal{F} = -0.029$ V at 25 °C, the slope of the line in this region is $-0.029$ V (unit of pH)$^{-1}$. When p$K_{a1}$ > pH and $K_{a1} < [H^+]$

$$E^\circ_{app} = -0.38\ V - \frac{2.303RT}{2\mathcal{F}}(2pH - pK_{a2} - pK_{a1}) \qquad (2\text{–}61)$$

and the slope of the line in this region is $-0.059$ V (unit of pH)$^{-1}$. By solving for the intersection of the functions described by Equations 2–59 and 2–60 and the intersection of the functions described by Equations 2–60 and 2–61, it can be shown that these two intersections occur when pH = p$K_{a2}$ and pH = p$K_{a1}$, respectively.

At values of pH greater than p$K_{a2}$, the apparent standard reduction potential of the reaction is invariant with pH because it involves no proton uptake (Equation 2–53). At values of pH less than p$K_{a1}$, the reaction involves the stoichiometric uptake of two protons for every two electrons

$SO_3^-$ + $2e^-$ + $2H^+$ $\rightleftharpoons$ $SO_3^-$ (2–62)

and the slope of the apparent standard reduction potential reflects this stoichiometry. At values of pH between the two p$K_a$ values, the reaction involves the uptake of only one proton for every two electrons and the slope is half the maximum value.

Unfortunately, the standard reduction potentials for most redox pairs are tabulated for the sum of the electron uptake and the proton uptake as well as any subsequent chemical transformations. This causes the reduction potentials to be much more complicated and harder to understand. For example, the reaction tabulated for the reduction of sulfite catalyzed by sulfite reductase

$$H_2SO_3 + 6H^+ + 6e^- \rightarrow H_2S + 3H_2O \qquad E^\circ = +0.347\ V \qquad (2\text{–}63)$$

is the reaction as written at standard state, which is equimolar sulfurous acid and hydrogen sulfide at an activity of protons of unity (pH = 0). Corrections for both the actual pH of the reaction and for the ionization of the reactant and product are required to obtain the apparent standard reduction potential for this reaction at a particular pH.

The apparent standard reduction potentials for flavin (Figure 2–13)[77] reflect its various **acid dissociation constants** (Figure 2–9). In the redox pair of oxidized flavin and reduced flavin, involving the transfer of two electrons, the apparent standard reduction potential ($E_m$ in Figure 2–13) shows two inflections, one at the p$K_a$ of the oxidized flavin (p$K_a$ = 10) and one at the p$K_a$ of the reduced flavin (p$K_a$ = 6.3). Below pH 6.3 and above pH 10 two protons are taken up for each two electrons, and the slope of the function is 0.059 V (unit of pH)$^{-1}$. When the pH is between the p$K_a$ of the reduced flavin and the p$K_a$ of the oxidized flavin, only one proton is taken up for

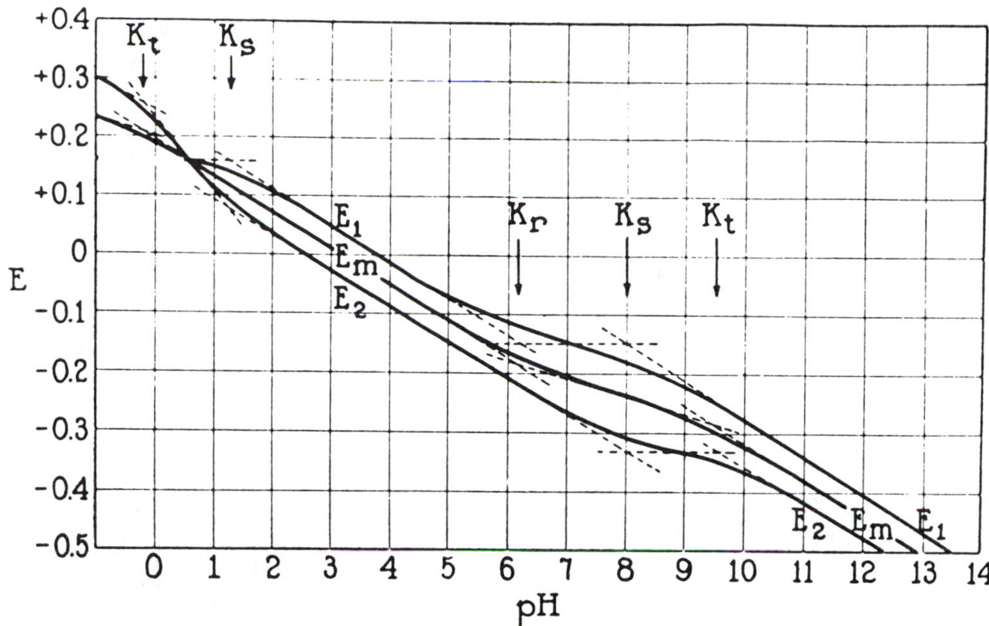


**Figure 2–13:** Apparent reduction potentials (E) of riboflavin as a function of pH.[77] A dilute solution (0.3 mM) of oxidized riboflavin buffered at the desired pH was flushed with nitrogen in a vessel containing a platinum electrode with which the reduction potential of the solution could be monitored. Measured volumes of a solution of the strong reducing agent sodium hydrosulfite were added to first reduce all of the remaining oxygen and then systematically reduce the riboflavin. The potential of the solution was followed as a function of the volume of reductant added to produce a titration curve of the reduction at the chosen pH. The complete titration curve was that of a two-electron reduction and the final product was reduced flavin. The apparent reduction potential at the half-reduced point ($E_m$) is the apparent standard reduction potential for the complete two-electron reduction of riboflavin. Both the individual titration curves of the reduction and the spectra of the solutions at intermediate reduction indicated that significant concentrations of the riboflavin radical were present at intermediate stages of reduction. The exact concentrations present were determined by fitting the direct titration curves of the reductions with a function derived on the assumption that the two-electron reduction of the riboflavin occurred in two one-electron steps. From the concentrations of riboflavin radical in equilibrium with oxidized and reduced riboflavin, the apparent standard reduction potential for the reduction of oxidized riboflavin to the radical ($E_2$) and the apparent standard reduction potential for the reduction of riboflavin radical to reduced riboflavin ($E_1$) could be calculated. The values of the p$K_a$ established by the intersections of the dotted line in the figure were later shown to be slightly in error. The presently used p$K_a$ values[64] are listed in Figure 2–9. Adapted with permission from ref 77. Copyright 1936 *Journal of Biological Chemistry.*

every two electrons, and the slope of the function is 0.029 V (unit of pH)$^{-1}$. In the redox pair of oxidized flavin and flavin radical, involving the transfer of one electron, the apparent standard reduction potential ($E_2$ in Figure 2–13) shows two inflections, one at the p$K_a$ of the oxidized flavin (p$K_a$ = 10) and one at the p$K_a$ of the flavin radical (p$K_a$ = 8.3). Below pH 8.3 and above pH 10, one proton is taken up for each electron, and the slope of the function is 0.059 V (unit of pH)$^{-1}$. When the pH is between the two respective values for the p$K_a$, no protons would be taken up for each electron, but the respective values for the p$K_a$ are too close together to witness the invariance of the function with pH. In the redox pair of flavin radical and reduced flavin, involving the transfer of one electron, the apparent standard reduction potential ($E_1$ in Figure 2–13) shows two inflections, one at the p$K_a$ of flavin radical (p$K_a$ = 8.3) and one at the p$K_a$ of reduced flavin (p$K_a$ = 6.3).

In contradistinction to all of these explicable variations in oxidation–reduction potential when flavin is free in solution, when it is bound as a coenzyme to an enzyme, its oxidation–reduction potential is less affected and unpredictably affected by the pH because it is surrounded by the bases and acids of the protein that form hydrogen bonds with its $\sigma$ lone pairs of electrons and acidic protons. These bases and acids remove and add, respectively, the protons transferred from and to the $\sigma$ lone pairs of the flavin during the oxidation–reduction. They are fixed in position next to the acidic protons and $\sigma$ lone pairs of electrons on which they operate.

If all locations around the flavin were occupied by bases and acids of the protein, which in turn were isolated from the solution, the reduction potential of the flavin would be unaffected by the pH of the solution and determined only by the intrinsic p$K_a$ values for these bases and acids. Consider the simplest situation in which the flavin is surrounded by the protein and the protein provides only one acidic amino acid, AH, and it is adjacent to the most basic of the lone pairs of electrons of the neutral flavin radical, the lone pair of nitrogen 1. The addition of an electron to the neutral flavin radical, $H_2Fl^{\cdot}$, to produce reduced flavin anion, $H_2Fl^{-}_{red}$, and the subsequent transfer of the proton to produce the neutral reduced flavin, $H_3Fl_{red}$, can be described by the following mechanism

$$\begin{array}{ccc} AH{\odot}H_2Fl^{\cdot} + e^- & \rightleftharpoons & AH{\odot}H_2Fl^{-}_{red} \\ \updownarrow & & \updownarrow \\ A{\odot}H_3Fl^{\cdot+} + e^- & \rightleftharpoons & A{\odot}H_3Fl_{red} \end{array} \quad (2\text{–}64)$$

where the four complexes denote hydrogen bonds between the acid, AH, or its conjugate base, A, and the bound flavin. If these are the only forms present

$$[Fl_{red}]_{TOT} = [AH{\odot}H_2Fl^{-}_{red}] + [A{\odot}H_3Fl_{red}] \quad (2\text{–}65)$$

and

$$[Fl^{\cdot}]_{TOT} = [AH{\odot}H_2Fl^{\cdot}] + [A{\odot}H_3Fl^{\cdot+}] \quad (2\text{–}66)$$

Because the acid–base reactions are intramolecular proton transfers

$$\frac{K_a^{AH}}{K_a^{H_3Fl_{red}}} = \frac{[A{\odot}H_3Fl_{red}]}{[AH{\odot}H_2Fl^{-}_{red}]} \quad (2\text{–}67)$$

where the $K_a{}^i$ are the respective microscopic acid dissociation constants of the respective acids when they are in the complex. The reduction potential, $E$, for the flavoprotein is

$$E = E^{\circ}_{H_2Fl} - \frac{RT}{\mathcal{F}} \ln \frac{[H_2Fl^{-}_{red}]}{[H_2Fl^{\cdot}]} \quad (2\text{–}68)$$

where $E^{\circ}_{H_2Fl}$ is the standard reduction potential for the deprotonated flavin. When Equations 2–65 through 2–68 are combined

$$E = E^{\circ}_{H_2Fl} - \frac{RT}{\mathcal{F}} \ln \left[ \frac{(K_a^{H_3Fl_{red}})(K_a^{H_3Fl^{\cdot+}}) + (K_a^{H_3Fl_{red}})(K_a^{AH})}{(K_a^{H_3Fl^{\cdot+}})(K_a^{H_3Fl_{red}}) + (K_a^{H_3Fl^{\cdot+}})(K_a^{AH})} \right] - \frac{RT}{\mathcal{F}} \ln \frac{[Fl_{red}]_{TOT}}{[Fl^{\cdot}]_{TOT}} \quad (2\text{–}69)$$

and the apparent reduction potential, $E^{\circ}_{app}$, is

$$E^{\circ}_{app} = E^{\circ}_{H_2Fl} - \frac{RT}{\mathcal{F}} \ln \left[ \frac{(K_a^{H_3Fl_{red}})(K_a^{H_3Fl^{\cdot+}}) + (K_a^{H_3Fl_{red}})(K_a^{AH})}{(K_a^{H_3Fl^{\cdot+}})(K_a^{H_3Fl_{red}}) + (K_a^{H_3Fl^{\cdot+}})(K_a^{AH})} \right] \quad (2\text{–}70)$$

This is the expression that would have resulted for the dependence of $E^{\circ}_{app}$ on pH in free solution if $K_a^{AH}$ were replaced by $[H^+]$. Therefore, if the intrinsic values of the p$K_a$ for flavin were to remain the same when it was within the proteins, the plot of $E^{\circ}_{app}$ against pH for the one-electron reaction ($E_1$ in Figure 2–13) would be equivalent to the plot of $E^{\circ}_{app}$ against the p$K_a$ of the acid in the protein that provides the proton to nitrogen 1 during the one-electron reduction. These equations show how the intrinsic acidity of the acid hydrogen-bonded to nitrogen 1 can control the reduction potential of the flavin when it is within a flavoprotein and, by extension, how the acid–bases surrounding the flavin control its reduction potential.

At all values of pH greater than 1 (Figure 2–13) the reduction potential for the reaction between oxidized flavin and flavin radical ($E_2$) is more negative than that for the reaction between flavin radical and reduced flavin ($E_1$). This means that, in solution, flavin radical participates reversibly in a disproportionation

$$H_2Fl^{\cdot} + H_2Fl^{\cdot} \rightleftharpoons HFl_{ox} + H_3Fl_{red} \quad (2\text{–}71)$$

with a negative standard free energy change as written ($K_{eq}$ = 9.1 at pH 4.0 and $K_{eq}$ = 50 at pH 11.5).[75] Below pH 0, however, the flavin radical is the more stable species ($K_{eq} < 1$).[77] As Reaction 2–71 requires collision of the two flavin radicals, the flavin radical buried in an enzyme is not so prone to disproportionation as it is in solution. There is, however, an enzyme, electron transfer factor–coenzyme Q oxidoreductase, that can catalyze the disproportionation of flavin when it is bound tightly to the flavoprotein known as electron transfer factor[78]

$$ETF{\cdot}H_2Fl^{-}_{red} + ETF{\cdot}HFl_{ox} \rightleftharpoons 2ETF{\cdot}HFl^{\cdot-} + H^+ \quad (2\text{–}72)$$

In this reaction, the oxidoreductase removes an electron from a molecule of reduced electron transfer factor and transfers it to a molecule of oxidized electron transfer factor to produce the two molecules of electron transfer factor radical.

At neutral pH, the flavin on electron transfer factor behaves as the anion ($H_2Fl^{-}_{red}$ in Figure 2–9) when it is fully reduced and as the radical anion ($HFl^{\cdot-}$ in Figure 2–9) when it is the radical, respectively.[78] These observations can be explained if the protein provides an acid at nitrogen 4 with a p$K_a$ greater than 8.3 but acidic enough to protonate the dianion of reduced flavin and if the protein provides a hydrogen-bond donor at nitrogen 1 that has a p$K_a$ significantly greater than 6.3. Such an arrangement would enforce the retention of the anion of flavin in both oxidation states (Figure 2–9).

In its role in biochemical oxidation–reduction reactions, the ability of flavin to participate both in the one-electron oxidation–reductions, involving the stable enzyme-bound radical, and in two-electron oxidation–reductions, not involving flavin radical, is essential. Flavin sits at the junction between two-electron reductants, such as NADH and butyryl coenzyme A, which provide the formal equivalent of a hydride ion, and one-electron reductants, such as iron–sulfur centers and ubiquinone, which receive single electrons from the reduced flavin and the flavin radical. It is the latter ability of flavin to be reduced by formal hydride donors that introduces the heterolytic chemistry of the flavin ring. These classes of reactions, involving the transfer of two electrons and one proton between the donor and flavin, have been proposed to proceed by direct hydride transfer, by passing through radical intermediates, by the formation of enols as intermediates, or by electron transfer.

The evidence for **direct hydride transfer** in two-electron oxidation–reductions of flavin comes from experiments with 5-carba-5-deazaflavins, in which nitrogen 5 of the isoalloxazine has been substituted by a carbon. These analogues of flavin retain the hydrogen nucleus that has been transferred in a carbon–hydrogen bond rather than losing the proton through acid–base exchange at a nitrogen–hydrogen bond. In free aqueous solution, NADH is able to reduce 5-carba-5-deaza-3,10-dimethylisoalloxazine

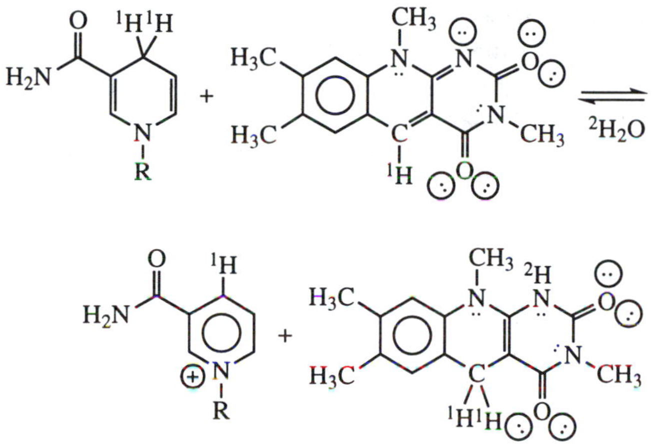

(2–73)

by the direct transfer of its hydrogen to carbon 5 of the deazaflavin with no interchange with protons in the solvent.[79] This observation is consistent with direct hydride transfer and rules out electron transfer followed by proton transfer because it is unlikely that a proton lost from carbon 4 of NADH would be transferred directly to carbon 5 of the deazaflavin in free solution. 5-Carba-5-deaza-1,5-dihydro-3,10-dimethylisoalloxazine, a reduced deazaflavin, is also able to transfer the hydrogen on carbon 5 directly to the carbonyl carbon of pyridoxal phosphate,[80] and this ability demonstrates the ability of reduced flavin to transfer a hydride to a substrate. The situation is much more ambiguous on the active site of an enzyme, however, where the two reactants are held adjacent to each other and hydride transfer cannot be distinguished from a directed proton transfer between the reactants. Nevertheless, in several enzymatically catalyzed reactions, the hydride donor, the hydrogen of which is labeled as either tritium or deuterium, has been shown to transfer the labeled hydrogen directly to carbon 5 of a 5-carba-5-deazaflavin, ostensibly as a hydride ion

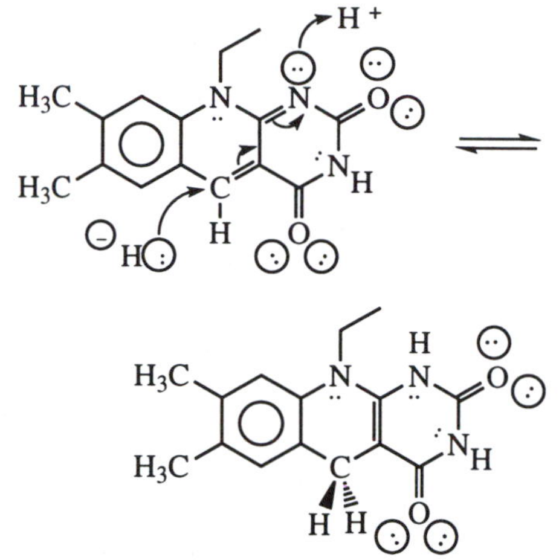

(2–74)

The hydride donors in the respective enzymatically catalyzed reactions have been [$^3$H]NADH itself[81]

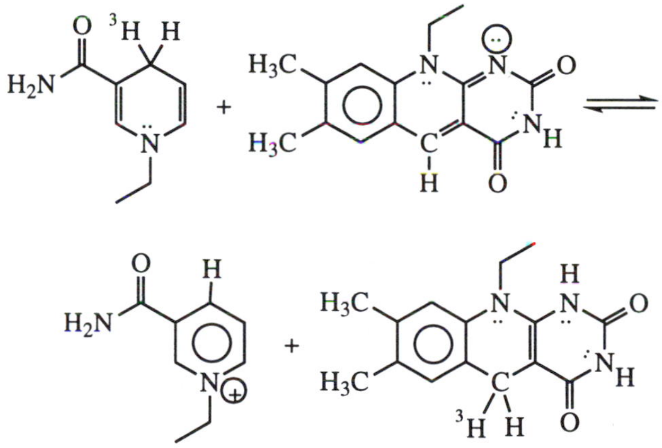

(2–75)

D-[$\alpha$-$^3$H]alanine[82]

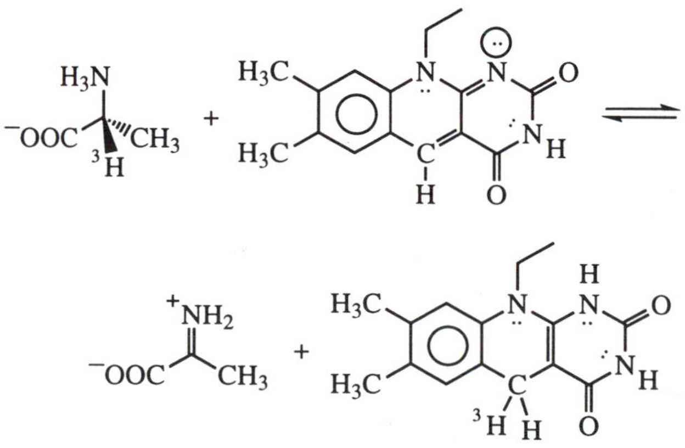

(2–76)

L-[$\alpha$-$^3$H]glutamate[83]

(2–77)

or the enol of [3-$^2$H]butyryl-SCoA[84]

(2–78)

In each case, the deazaflavin ends up with the nucleus of the hydrogen that was originally on the donor.

When the hydride transfer from reduced 5-carba-5-deazaflavin to the $\beta$-carbon of crotonyl-SCoA (reverse of Reaction 2–78) was carried out by the enzyme butyryl-CoA:(acceptor) oxidoreductase, the rate was equal to the rate of hydride transfer from reduced flavin to the $\beta$-carbon of crotonyl-SCoA catalyzed by the same enzyme.[84] This suggests that the reaction of the deaza analogue reproduces the normal reaction, if the rate-limiting step in the enzymatic reaction actually is hydride transfer. There is, however, evidence that, when the coenzyme is flavin, rather than 5-carba-5-deazaflavin, transfer of the equivalent of a hydride ion from the enol of butyryl-SCoA to the flavin takes place in steps involving electron transfers and a proton transfer,[85] and the reasonableness in general of one-electron mechanisms involving flavin has been established by calculating standard free energies of formation for **radical intermediates.**[86] These suggestions are reminiscent of the disagreements surrounding hydride transfer in general.

The reaction between flavin and D-alanine catalyzed by D-amino-acid oxidase (Reaction 2–76) is formally equivalent in the opposite direction to the reduction of a carbonyl compound by reduced flavin. The results with 5-carba-5-deazaflavin suggest that this is a straight hydride transfer. Reduced 3,5,7,8,10-pentamethylisoalloxazine, an analogue of reduced flavin, however, will react with various electron-deficient carbonyl compounds to yield the flavin radical, recognized by its blue color, and presumably the carbonyl radical anion.[87] This has been proposed to be the first step in the ultimate reduction of the carbonyl to the alcohol. These results also raise the possibility that a stepped mechanism involving radical intermediates could be involved in what is formally a hydride transfer, but discrete intermediates in such processes have been difficult, if not impossible, to observe.

In many reactions in which flavin oxidizes a reactant with an acidic carbon, in either an enzymatically catalyzed reaction or a nonenzymatic reaction,[88] the enolate or **enol** of this carbon acid is involved as an intermediate. The evidence for these mechanisms in enzymatic reactions is the ability of various flavoenzymes to catalyze the dehydrohalogenation of halogenated analogues of their substrates. In several of the enzymatic reactions in which the acidic hydrogen on the reactant is transferred to 5-carba-5-deazaflavin directly, there is also evidence that it is the enol of the reactant that is oxidized by flavin. These two observations seem to be self-contradictory. For example, D-amino-acid oxidase, the same flavoenzyme in which hydrogen is transferred between D-[$\alpha$-$^3$H]alanine and 5-carba-5-deazaflavin (Reaction 2–76),[82] readily dehydrohalogenates $\beta$-chloralanine,[89] presumably through the $\beta$-elimination from the enol

(2–79)

Lactate oxidase, a flavoenzyme responsible for converting lactate to acetate and carbon dioxide, has as its first step the oxidation of lactate to pyruvate. This enzyme dehydrohalogenates 3-chlorolactate to

(2–80)

and this reaction is consistent with the involvement of the enol of lactate as an intermediate in the normal enzymatic reaction. Flavocytochrome $b_2$, an enzyme also oxidizing lactate to pyruvate, can catalyze the dehalogenation of bromopyruvate in a reaction that simultaneously oxidizes reduced flavin. In this reaction, reduced flavin produces the enol,

which then dehalogenates[91]

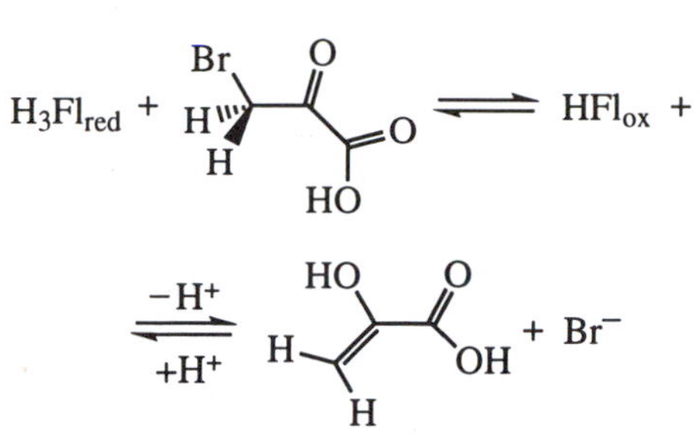

(2–81)

All of these observations have been interpreted to mean that the enol of the respective substrate is first formed, and then two of the electrons in the electron-rich $\pi$ molecular orbital system are transferred to the oxidized flavin to oxidize the substrate and reduce the flavin.

There are, however, several problems reconciling the involvement of an enol with the observation of direct transfer of hydrogen from the acidic carbon to position 5 of the flavin. If an enol is involved, the proton must be removed from carbon by a base and yet it must eventually end up on nitrogen 5 of the flavin. Nitrogen 5 of oxidized flavin is not a base ($pK_a \leq 0.2$, Figure 2–9) but it becomes a strong base after two electrons are transferred to the flavin to produce the dianion of reduced flavin, the conjugate acid of which has a $pK_a > 16$.[64] Therefore, if a base on the enzyme removed the proton from the carbon of the substrate to form the enol, the same proton could be removed from the same base by nitrogen 5 of the flavin after the electrons had been transferred. The problem with this mechanism is that the proton in neither D-alanine nor L-lactate is very acidic, and a strong base would be required to remove either. Once the proton has been removed, however, the oxidation of the enol by the flavin would be facile; before the proton is removed, the initial reactant is very difficult to oxidize by anything like a flavin. This is why a mechanism involving direct electron transfer from the conjugate carbon acid to $HFl_{ox}$ followed by proton transfer to the $HFl^{\cdot -}$ anion (Figure 2–9) and electron transfer to the resulting $H_2Fl^{\cdot}$ radical, although it could explain both the transfer of a hydrogen between a reactant and a 5-carba-5-deazaflavin and the dehydrohalogenation catalyzed by these enzymes, seems unrealistic.

In reactions in which an enol or enolate is formed as an intermediate, the question arises of how the two electrons pass from the enol to the flavin during the oxidation. There is evidence that the two electrons can be passed in two one-electron steps. This conclusion follows from the observations that scavengers of free radicals can lower the yields of oxidations by flavin that pass through enolates.[92] It is also possible, however, that the enol or enolate is oxidized through an intermediate produced by nucleophilic addition to the isoalloxazine. Hidden within oxidized flavin is a **1,4-diazabutadiene**

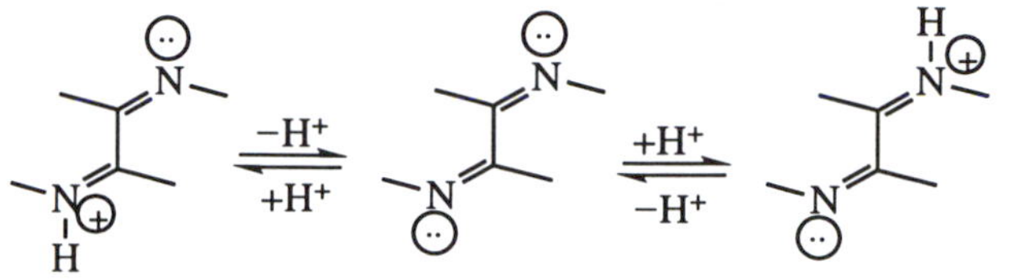

(2–82)

If flavin participates in either an adduct with an enol or direct hydride transfer, the protonated 1,4-diazabutadiene could be acting as an electrophile with nitrogen 5 resembling the electrophilic $\beta$-carbon in an $\alpha,\beta$-unsaturated iminium cation

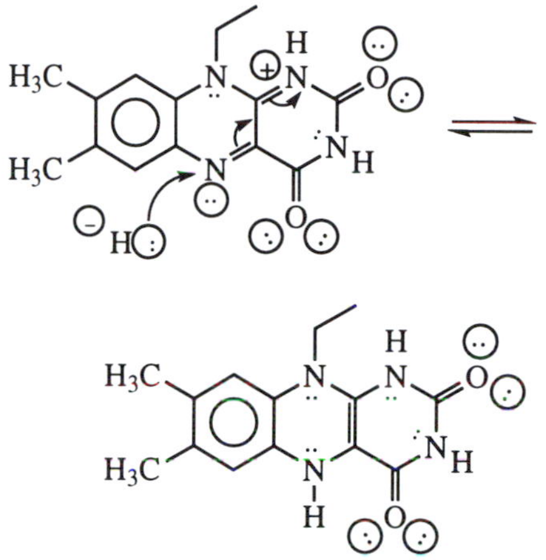

(2–83)

That nitrogen 5 in flavin is electrophilic has been demonstrated by using sulfite as a nucleophile.[93] When 7,8-dimethyl-1,10-ethenoisoalloxazinium cation is modified with sulfite, a crystalline product is obtained that results from addition of the nucleophile to nitrogen 5

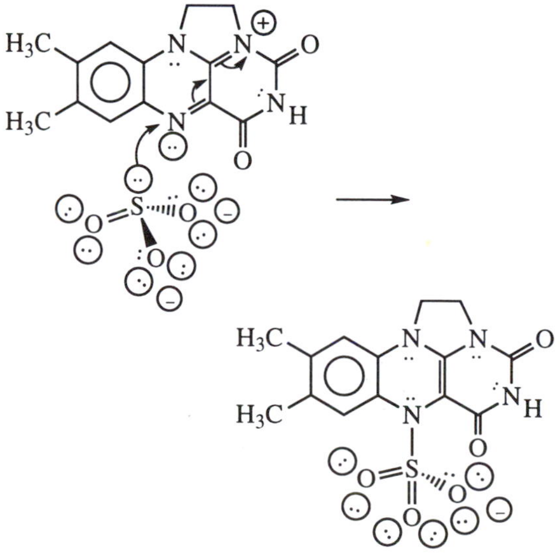

(2–84)

This reaction illustrates that when one nitrogen in a diazobutadiene is cationic, either because it is alkylated or because it is protonated, the other nitrogen is electrophilic. On the basis of the electrophilicity of nitrogen 5, a mechanism has been proposed that involves a covalent intermediate in the transfer of a pair of electrons from a nucleophile to oxidized flavin to produce an electrophile

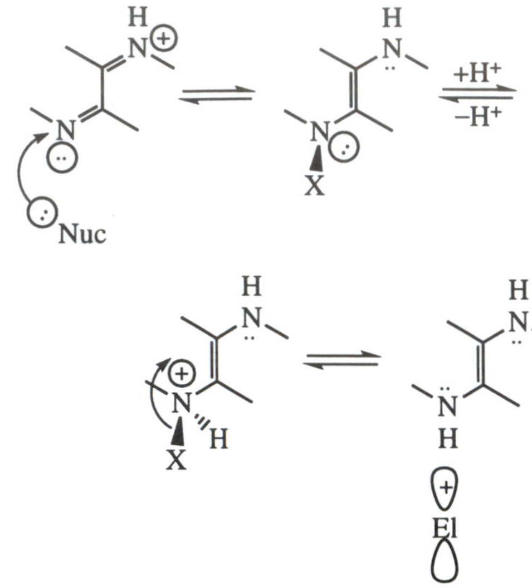

(2–85)

This mechanism has been proposed for the steps in the mechanism of lactate oxidase[94] that convert lactate to pyruvate

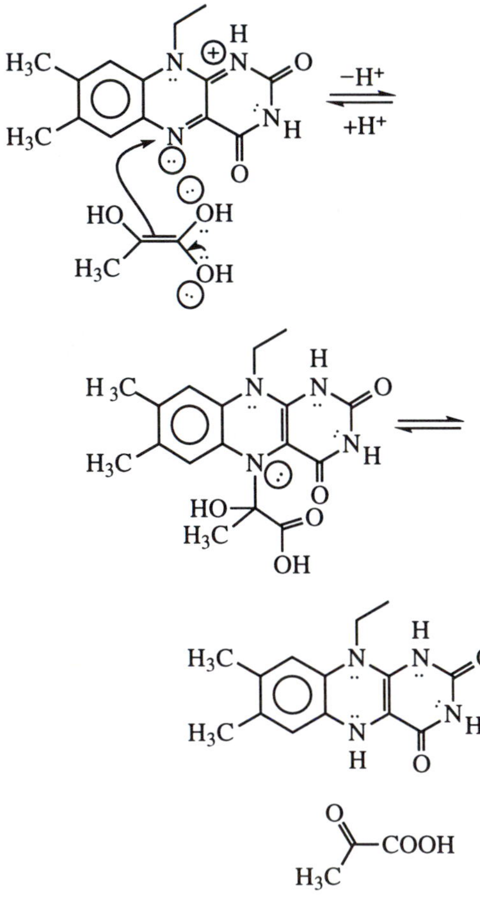

(2–86)

When ethyl pyruvate, however, was reduced in a nonenzymatic reaction by 3-(carboxymethyl)lumiflavin, even though the carbinolamine (central intermediate in Reaction 2–86) was formed, it participated mainly in the formation of the imine at nitrogen 5 of the flavin. This side reaction was off the main path of the reduction, which proceeded through an independent route that did not seem to involve the carbinolamine.[86] If Reaction 2–86 does, nevertheless, occur on the enzyme, as written, two of the $\pi$ electrons of the enol add as a nucleophile to nitrogen 5 of the flavin and are left behind in the reduced flavin while the electrophile, a ketone, departs. In this mechanism, the leaving group from the carbinolamine is a protonated nitrogen and this step would be driven by the return of the pair of electrons to the $\pi$ molecular orbital system of the reduced flavin. In the reverse direction, however, the addition to the electrophile would be by nitrogen 5 of the reduced flavin and this nitrogen is not nucleophilic (p$K_a \leq 0.8$, Figure 2–9). This difficulty is not encountered if an adduct at carbon 4a is proposed rather than the adduct at nitrogen 5.

On paper, a diazobutadiene (Equation 2–82) is also electrophilic at either of its two iminium carbons, which would be carbon 1a and carbon 4a of flavin. Several products have been isolated[95] that seem to arise from the nucleophilic attack of hydroxide anion, ammonia, or hydride anion at carbon 1a of 1,3,7,8,10-pentamethylisoalloxazine, but the addition of these heteroatomic nucleophiles leads unavoidably to the decomposition of the pyrimidine ring, an outcome that would not be salutary for a flavoenzyme

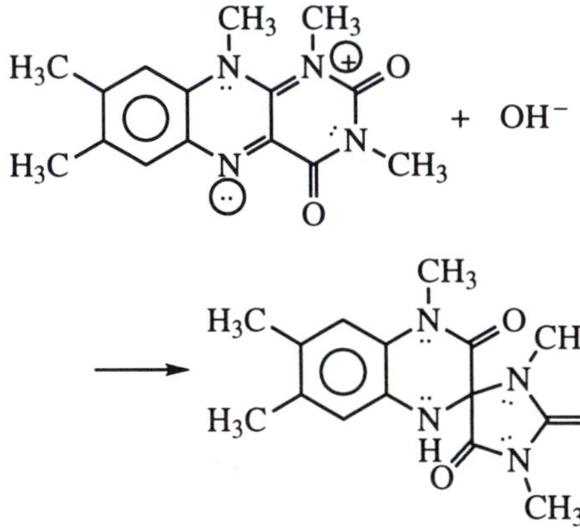

(2–87)

Sulfite adds to 10-(2,6-dimethylphenyl)-6,8-disulfo-3-methylisoalloxazine[96] to produce a stable derivative at carbon 4a

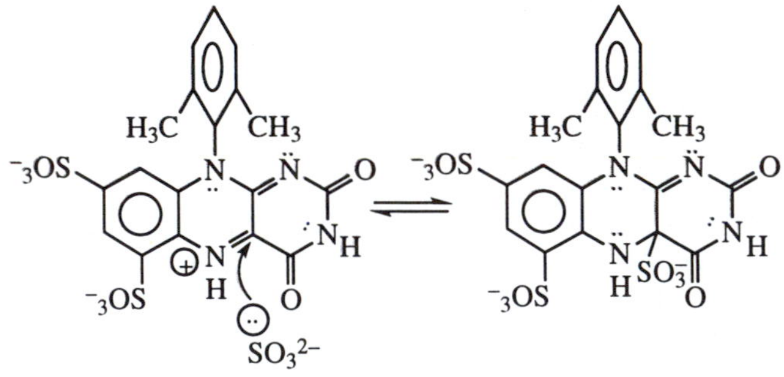

(2–88)

Two drawbacks of this observation are that the sulfonates on the 6 position and the 8 position, respectively, mildly withdraw electron density ($\sigma_p = 0.1$) from carbon 4a and that the sulfonate on carbon 6 sterically blocks nucleophilic attack at nitrogen 5. Even under these favorable circumstances, an equilibrium mixture of adducts at both carbon 4a and nitrogen 5 is obtained.[97] When the sulfonates are not present, sulfite adds exclusively[93] to nitrogen 5.

This would suggest that, in normal flavins, nitrogen 5 is much more electrophilic than carbon 4a. This impression may arise form the fact that the addition of a nucleophile to nitrogen 5 can proceed spontaneously without the need for protonation of nitrogen 1

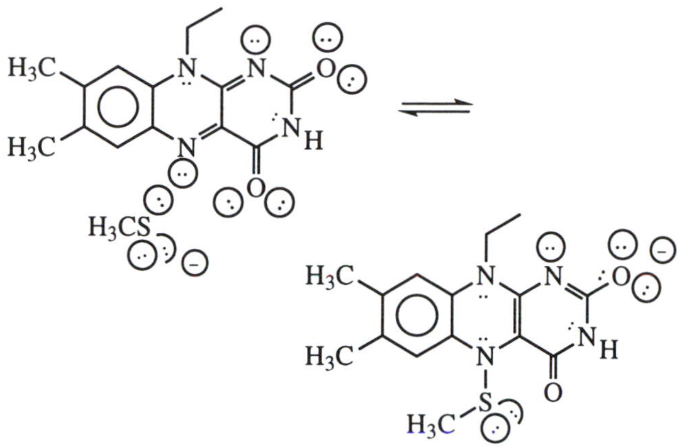

(2–89)

while addition at carbon 4a requires acid catalysis

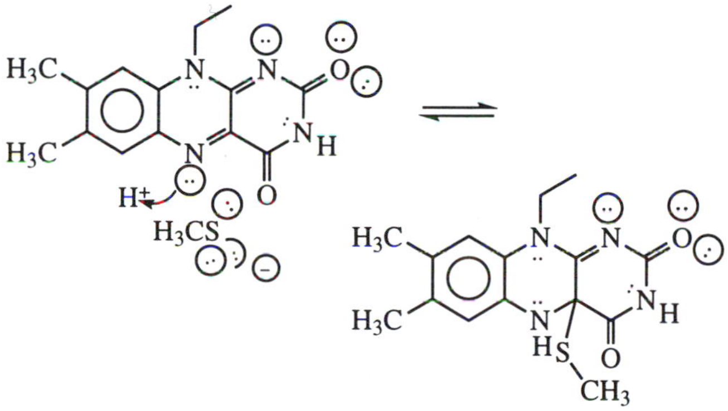

(2–90)

These conclusions result from a consideration of the values of the p$K_a$ for these nitrogens (Figure 2-9). Stable nucleophilic adducts at carbon 4a have been observed with oxidized 5-ethyl-3,7,8,10-tetramethylisoalloxazine. This flavin can add reversibly either hydroxide anion[98]

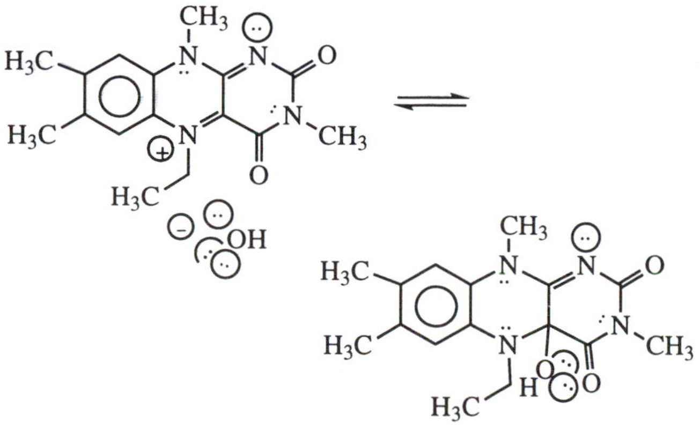

(2–91)

or nitromethane anion ($^-$:$CH_2NO_2$)[99] to carbon 4a. In this reaction, the ethyl group accomplishes the same purpose as protonation of nitrogen 5, only permanently, and this result suggests that, under acid catalysis at nitrogen 5, carbon 4a is electrophilic. The behavior of the rate constant for the decomposition of 4a-(indolylmethyl)-3-methyllumiflavin as a function of pH can be explained[100] as the result of a rapid equilibrium between the adduct at carbon 4a and the adduct at nitrogen 5. If this were the correct explanation, then these two locations would have almost equivalent electrophilicities.

The desire for adducts at carbon 4a in oxidized flavin to be transient intermediates in the reactions catalyzed by flavin arises from the appeal of the following mechanism for electron withdrawal from a $\sigma$ bond

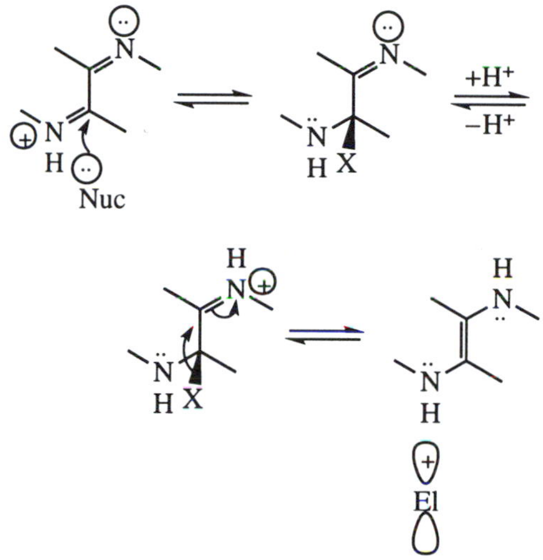

(2–92)

The nucleophile of the reduced reactant enters the electrophilic position at carbon 4a with a lone pair of electrons that forms the new $\sigma$ bond. That new $\sigma$ bond is now parallel to the $\pi$ molecular orbital of an adjacent iminium cation that withdraws its electrons to release an electron-deficient, oxidized product and produce reduced flavin. The two electrons entering with the nucleophile remain in the flavin when the electrophile is released. In the reverse direction, the enamine of the reduced flavin would attack the electrophile to form the $\sigma$ bond to carbon 4a and the electrons in this $\sigma$ bond would leave with the nucleophile. It is the $\sigma$ bond to carbon 4a that is the transfer point for the two electrons that reduce the flavin or the electrophile, respectively. In the crystallographic molecular model of an isoalloxazine tetrahedral at carbon 4a, the $\sigma$ bond to the exocyclic substituent at carbon 4a and the $\pi$ bond of the adjacent imine at carbon 1a and nitrogen 1 are parallel (Structure **2–35**),[67] as required in Reaction 2-92.

This mechanism for transfer of two electrons to flavin through an adduct at carbon 4a has been proposed for several of the reactions catalyzed by the coenzyme. For example, a mechanism for steps in the D-amino-acid oxidase reaction involving an adduct at carbon 4a rather than nitrogen 5 (as in Equation 2-86) can be written

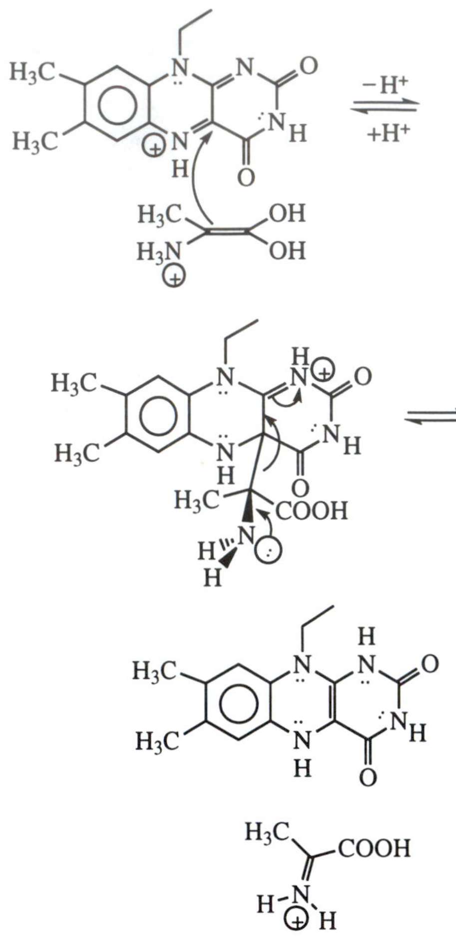

(2–93)

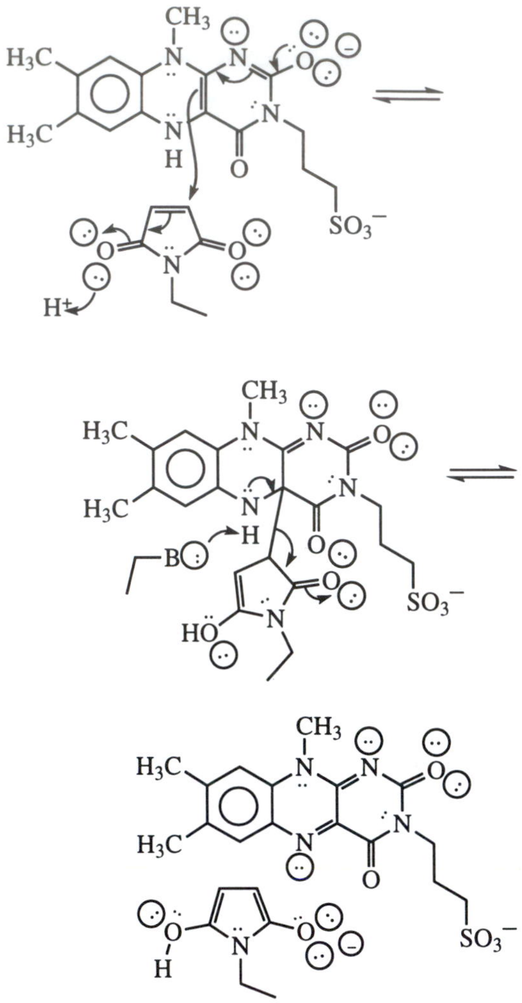

Two of the $\pi$ electrons of the enol form the $\sigma$ bond at carbon 4a and the two electrons in this $\sigma$ bond are left behind on the flavin as the iminium cation departs. During the reduction of maleimides by reduced 3-(3-sulfopropyl)-7,8,10-trimethyl-isoalloxazine, an adduct forms between the substrate and the flavin as an intermediate in the reaction.[101] The absorption spectrum of this intermediate identifies it as an adduct at carbon 4a

(2–94)

that would result from the nucleophilic addition of the anion of reduced flavin to the $\alpha,\beta$-unsaturated imide. The adduct is formed in a reaction catalyzed by acid and decomposes in a reaction catalyzed by base.[101]

Glutathione reductase is a flavoenzyme that catalyzes the reaction

$$H^+ + GSSG + NADPH \rightleftharpoons GSH + HSG + NADP^+ \quad (2\text{–}95)$$

where GSSG is oxidized glutathione, which contains an intermolecular cysteine formed from two molecules of reduced glutathione, GSH

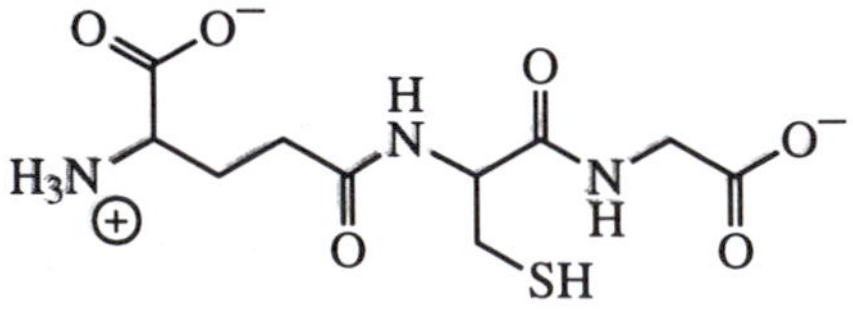

reduced glutathione

**2–36**

The flavoenzyme, lipoamide dehydrogenase, catalyzes the reaction

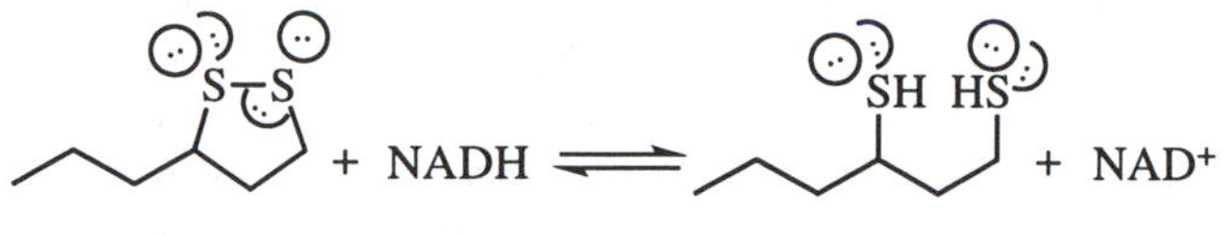

$$\text{lipoamide (S–S)} + NADH \rightleftharpoons \text{dihydrolipoamide (SH HS)} + NAD^+ \quad (2\text{–}96)$$

which is equivalent to that catalyzed by glutathione reductase. Both enzymes use flavin as a coenzyme. The two enzymes have homologous sequences of almost the same length[102] and must have superposable tertiary structures and very similar if not identical mechanisms. Both enzymes reduce the disulfide by first reducing a cystine in their native structure to two cysteines and then using the two cysteines to reduce the substrate by disulfide interchange.[43a]

Lipoamide dehydrogenase can be reduced, and one of the two resulting cysteines in the protein can be alkylated with iodoacetamide. When $NAD^+$ is added to the alkylated enzyme, a spectral change occurs that was tentatively assigned as arising from an adduct between the remaining thiolate anion and carbon 4a of oxidized flavin.[103] This has been used as evidence for the formation of an adduct at carbon 4a during the normal oxidation–reduction of the cysteines on the unalkylated, native enzyme

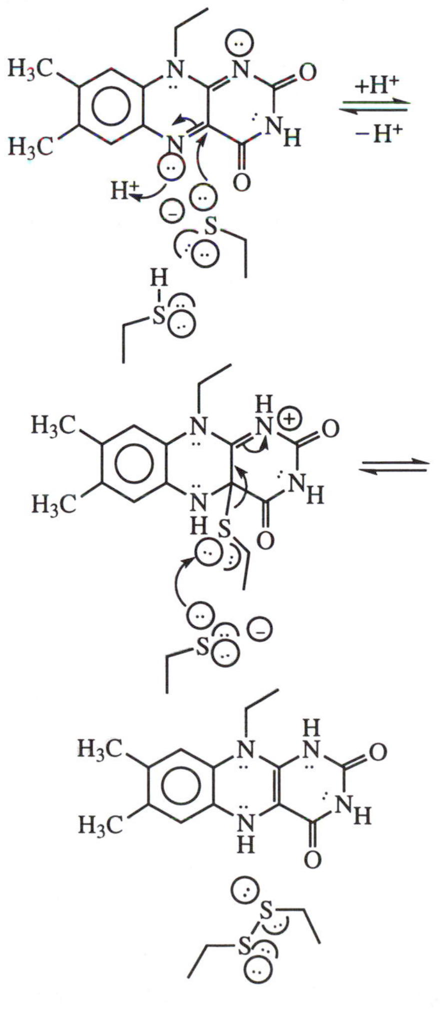

$$\xrightleftharpoons[-H^+]{+H^+} \quad \rightleftharpoons \quad (2\text{–}97)$$

This is the mechanism by which thiols reduce flavins in solution.[104] In the reverse direction, the reaction of the reduced flavin is the attack of an enamine on a disulfide and is equivalent to the attack of the enamine of thiamin on oxidized lipoic acid (Figure 2–5). In the forward direction, a lone pair on one of the cysteines becomes a $\sigma$ bond that is withdrawn into the flavin in the second step.

Glutathione reductase can also be reduced and alkylated on only one of the two catalytic cysteines.[105] A crystallographic molecular model of this reduced and alkylated enzyme, in a complex with $NADP^+$, is available.[106] Although the unalkylated cysteine is the one closest to the flavin and although the closest atom in the flavin to the sulfur is carbon 4a, there is no bond between the sulfur of the cysteine and carbon 4a of the oxidized flavin. Therefore, the postulated adduct (Reaction 2–97) cannot form, at least in glutathione reductase.

In the crystallographic molecular model of glutathione reductase with NADPH bound at the active site, the amino group of a lysine is blocking the approach of carbon 4′ of the nicotinamide ring to the flavin even though the closest atom in the flavin to hydrogen 4′ of the nicotinamide is nitrogen 5 (Figure 2–14)[106] and the arrangement of the flavin and the nicotinamide is such that in-line hydride transfer would be impossible. From an examination of the crystallographic molecular model, it was concluded that a hydride could not be transferred from carbon 4′ of the nicotinamide to nitrogen 5 of the flavin.[106] Instead, it was proposed that during the reduction, two electrons would flow from the nicotinamide to the cysteine through the flavin and the hydrogen of NADPH would be removed as a proton by the nitrogen of that lysine. The stacking of the nicotinamide on the flavin would promote the movement of the electrons. Rather than relying on hydride transfer and covalent intermediates, the oxidation–reduction reaction would use the flavin as a wire to connect ultimate oxidant and ultimate reductant. This process would be a pure **electron transfer,** accompanied by local acid–base reactions.

A disulfide radical anion[45] (Figure 2–6) would be a transient intermediate in this electron transfer. It would spontaneously fragment to a thiolate anion and a thiol radical. The thiolate radical would then accept the second electron. As nicotinamide can also give up electrons one at a time (Reaction 1–62), the electrons could pass individually through the flavin.

The sequence of events that would accomplish the reaction most efficiently can be described by referring to Figure 2–14B. One electron would be transferred from NADPH to flavin to produce the respective radical cation and radical anion. A proton would be removed from carbon 4 of the NADPH radical cation by the lysine. A proton would be added to oxygen 4 of the flavin radical anion from the lysine cation. A second electron would be transferred from NADP radical to the flavin radical to produce $NADP^+$ and reduced flavin anion. An electron would be transferred from reduced flavin anion to the cystine to produce the disulfide radical anion and the flavin radical. The disulfide radical anion would be protonated on its distal sulfur by the adjacent histidine. The protonated disulfide radical would cleave to produce the protonated thiol and the thiol radical (Figure 2–6). The proton

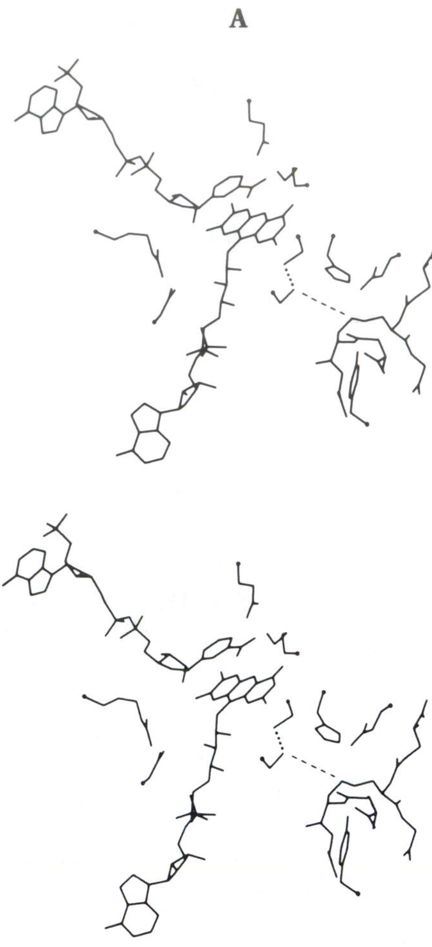


**Figure 2–14:** Crystallographic molecular model[106] of the active site of oxidized glutathione reductase from human erythrocytes with bound NADPH and oxidized glutathione (A) and a diagrammatic representation of the arrangement of NADPH, flavin, and the internal disulfide (B). Crystals of glutathione reductase were transferred to solutions containing either 10 mM NADPH or 10 mM $NADP^+$ and 10 mM oxidized glutathione. Data sets to Bragg spacings of 0.3 nm were collected from crystals that had been soaked under each set of conditions. These data sets, the data set from the unliganded enzyme, and the refined phases from the previous crystallographic study of the unliganded enzyme were used to produce maps of difference electron density into which either the NADPH or the $NADP^+$ and the oxidized glutathione were inserted. The figure is a composite of the two resulting crystallographic molecular models. From bottom left to top right the molecular models of oxidized glutathione, the internal disulfide, the flavin adenine dinucleotide, and the NADPH are presented. A few amino acid side chains, in particular Lysine 66 and Histidine 467, are also included. The disulfide between Cysteine 58 and Cysteine 63 is indicated by a dotted line. The distance between Cysteine 58 and the closest sulfur on oxidized glutathione is represented by a dashed line. (B) A diagram of the crystallographic molecular model in panel A. For the sake of clarity only the central ring and the pyrimidine of flavin have been drawn. The crystallographers note that direct contact between nitrogen 5 of the flavin and the hydride of nicotinamide is blocked sterically by the side chain of Lysine 66 (bottom of the figure). This would prevent direct hydride transfer to nitrogen 5 of the flavin. Although the distal sulfur of the internal disulfide (behind the flavin) is adjacent to carbon 4a of the flavin, in the complex between alkylated enzyme and $NADP^+$, the thiolate does not form an adduct with carbon 4a of oxidized flavin. The histidine displayed is Histidine 267, adjacent to the disulfide between Cysteine 63 and Cysteine 58 (panel A). Reprinted with permission from ref 106. Copyright 1983 *Journal of Biological Chemistry.*

B

would be removed from oxygen 4 of the flavin radical. An electron would be transferred from the flavin radical anion to the adjacent thiol radical to form the thiolate anion and regenerate the oxidized flavin. With the exception of the decomposition of the disulfide radical anion, no covalent bonds other than those between protons and bases need to be made or broken. If this mechanism is correct, the strong absorbance of the reduced and alkylated lipoamide dehydrogenase would result from a charge transfer complex between the proximal thiolate anion and the oxidized flavin rather than a covalent adduct. This charge transfer complex would reflect the ability of an electron to equilibrate between oxidized flavin and the thiolate anion. An observation that might contradict this interpretation is that $NADP^+$ is able to oxidize the flavin in glutathione reductase reconstituted with reduced 5,5-dihydro-[5-$^3$H]-8-demethyl-8-hydroxy-5-deaza-5-carbaisoalloxazine.[107] Unfortunately, the identity of the tritiated product of this reaction was not established. In any case, the mechanism just described serves as a convenient way to introduce electron transfers involving flavin.

What has been discussed so far can be summarized. In most of the oxidation–reductions in which flavin participates, half of the oxidation–reduction can be written as a formal hydride transfer. The reduced substrate has a $\sigma$ bond between hydrogen and another atom such as carbon or sulfur. The two electrons in that $\sigma$ bond are transferred to oxidized flavin during the oxidation of the substrate, or they are transferred from reduced flavin to the substrate during its reduction. Whether or not the nucleus of the hydrogen atom is simultaneously and directly transferred is unclear. At the moment there are several proposals about how this process might occur. First, it might be a simple hydride transfer between substrate and flavin, as has been proposed for nicotinamide. Second, an electron-rich enol may be formed initially on the reactant by removal of the nucleus of the hydrogen atom as a proton by a base, and two electrons may be transferred from this $\pi$ molecular orbital system to the neighboring flavin. Third, a covalent adduct between this enol as a nucleophile and the flavin as an electrophile may form at either nitrogen 5 or carbon 4a, and the $\sigma$ bond resulting from that reaction may be withdrawn into the flavin as the oxidized substrate departs. Fourth, the reduction may involve an electron transfer followed by proton removal followed by an electron transfer and thereby involve radical intermediates. It is unknown if there is a common mechanism among all the flavoenzymes for these formal hydride transfers or if there are different mechanisms for different classes of flavoenzymes. It may well be that flavins are such versatile coenzymes for oxidation–reduction because they can participate in several different types of mechanisms.

Following the formal removal of a hydride ion from the substrate, regardless of the steps involved in this process, the flavin is left in the reduced form and the substrate in the oxidized form produced by the formal hydride transfer. Usually the oxidized form of the substrate is the final product. The flavin, however, because it is only a catalyst, must be returned to its initial oxidized form.

There are several ways the flavin is reoxidized. The reduced flavin can be oxidized by nicotinamide

$$H_3Fl_{red} + NAD^+ \rightleftharpoons HFl_{ox} + NADH + H^+ \qquad (2\text{–}98)$$

as in the reaction catalyzed by lysine dehydrogenase

$$H_3\overset{+}{N}\text{–}(CH_2)_4\text{–}CH(\overset{+}{N}H_3)\text{–}COO^- + NAD^+ \rightleftharpoons$$

$$\text{(cyclic imine: }\overset{\ominus}{N}\text{=C–}COO^-\text{)} + NH_4^+ + NADH + 2H^+ \qquad (2\text{–}99)$$

The reduced flavin can react directly with molecular oxygen

$$H_3Fl_{red} + O_2 \rightleftharpoons HFl_{ox} + H_2O_2 \qquad (2\text{–}100)$$

as in the reaction catalyzed by D-amino-acid oxidase

$$H_2O + \text{D-alanine} + O_2 \rightleftharpoons \text{pyruvate} + NH_4^+ + H_2O_2 \qquad (2\text{–}101)$$

The reduced flavin can use molecular oxygen to oxidize the product further, coincident with its own oxidation

$$H_3Fl_{red} + H_3C\text{–}C(=O)\text{–}X + O_2 \rightleftharpoons$$

$$HFl_{ox} + H_3C\text{–}C(=O)\text{–}OH + H_2O + X \qquad (2\text{–}102)$$

For example, in the reaction catalyzed by lactate oxidase

$$\text{lactate} + O_2 \rightleftharpoons \text{acetate} + CO_2 + H_2O \qquad (2\text{–}103)$$

the flavin is reoxidized while an oxidative decarboxylation of pyruvate is being performed

$$H_3Fl_{red} + H_3C\text{–}C(=O)\text{–}C(=O)\text{–}O^- + O_2 \rightleftharpoons$$

$$HFl_{ox} + H_3C\text{–}C(=O)\text{–}OH + CO_2 + H_2O \qquad (2\text{–}104)$$

The reduced flavin can also pass on its electrons to some other oxidant, usually in one-electron transfers

$$H_3Fl_{red} + \text{acceptor}_{ox} \rightleftharpoons HFl_{ox} + H_2\text{acceptor}_{red} \qquad (2\text{–}105)$$

as in the reaction catalyzed by succinate dehydrogenase

$$\text{succinate} + \text{ubiquinone} \rightleftharpoons \text{fumarate} + \text{ubiquinol} \qquad (2\text{–}106)$$

which proceeds with two one-electron transfers to ubiquinone. Each of these latter three reactions distinguishes flavin from nicotinamide, which can neither react directly with oxygen nor participate in one-electron transfer. These types of reactions will now be considered.

## *Suggested Reading*

Venkataran, U.V., & Bruice, T.C. (1984) On the Mechanisms of Flavin-Catalyzed Dehydrogenation $\alpha,\beta$ to an Acyl Function. The Mechanism of 1,5-Dihydroflavin Reduction of Maleimides, *J. Am. Chem. Soc. 106*, 5703–5709.

### PROBLEM 2–11

Sarcosine oxidase is a flavoenzyme that catalyzes the reaction[108]

$$H_2O + \text{sarcosine} + HFl_{ox} \rightleftharpoons \text{glycine} + \text{formaldehyde} + H_3Fl_{red}$$

(A) Write a mechanism for the reaction as a direct hydride transfer. What provides the push?

(B) Write a mechanism for the reaction involving an adduct at nitrogen 5 of the flavin.

(C) Write a mechanism for the reaction involving an adduct at carbon 4a of the flavin.

(D) Write a mechanism involving electron transfer between sarcosine and flavin. Have the hydrogen nucleus removed as a proton by a base.

### PROBLEM 2–12

The following reaction is catalyzed by flavin and thiamin pyrophosphate acting together. Write a mechanism involving both coenzymes.

$$H_3C\text{–}C(=O)\text{–}COO^- + 2\text{ ferriheme } b_1 + H_2O \rightleftharpoons$$

$$2H^+ + H_3C\text{–}C(=O)\text{–}O^- + 2\text{ ferroheme } b_1 + CO_2$$

**PROBLEM 2–13**
Show how the reaction catalyzed by the flavoenzyme dihydroorotate oxidase[109]

$$\text{dihydroorotate} + \text{HFl}_{ox} \rightleftharpoons \text{orotate} + \text{H}_3\text{Fl}_{red}$$

can be written as a hydride transfer to flavin and how it resembles the reaction catalyzed by D-amino-acid oxidase.

**PROBLEM 2–14**
2-Hydroxy-3-butynoate is capable of inactivating the flavoenzyme lactate oxidase in the following reaction

$$\text{2-hydroxy-3-butynoate} + \text{HFl}_{ox} \rightleftharpoons$$

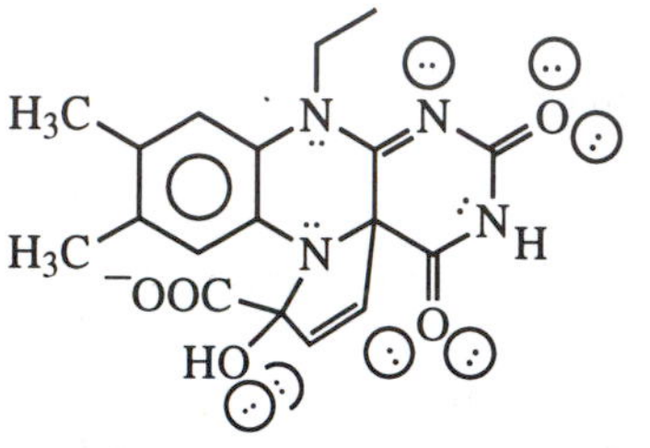

Write a mechanism for this reaction involving proton removal from the 2-hydroxy-3-butynoate at the carbon analogous to the carbon from which a hydrogen is removed in the regular reaction of lactate with the enzyme.

**PROBLEM 2–15**
Each of the following reactions is catalyzed by either pyridoxal phosphate, thiamin pyrophosphate, or flavin. Decide which coenzyme is required and write a mechanism for the reaction involving that coenzyme. If there are several possible mechanisms, choose only one. There is no exclusively correct mechanism, so the choice is one of personal preference, but the chemistry must be correct.

**A.**

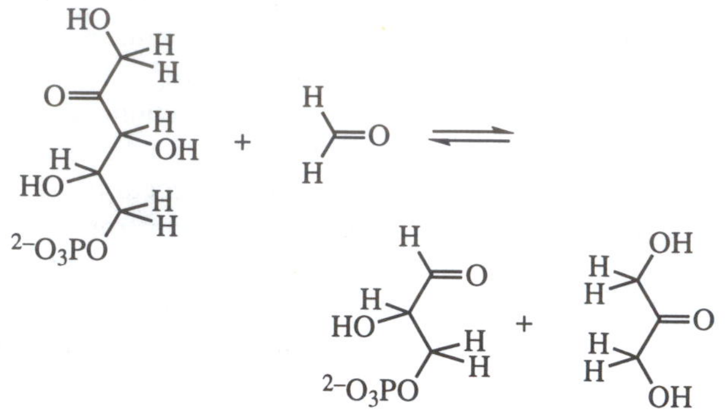

**B.** + NAD⁺ ⇌ + NADH

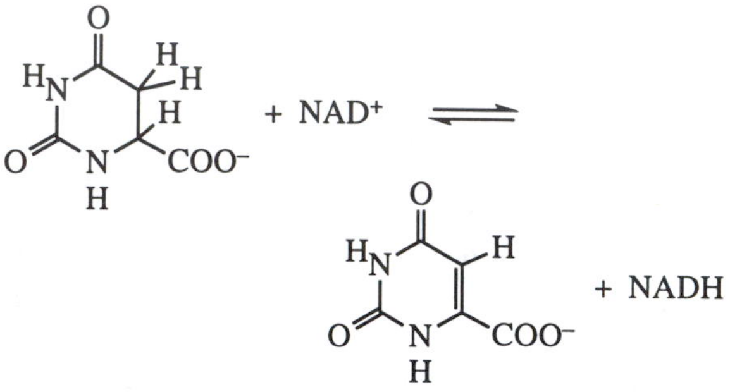

**C.** H⁺ + ⇌ + $CO_2$

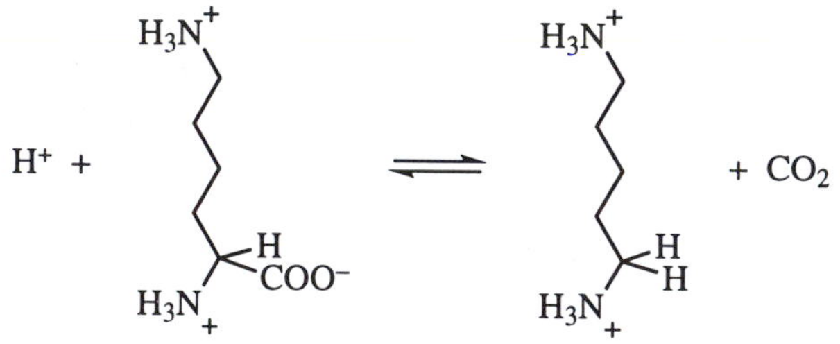

## One-Electron Transfer

A coenzyme capable of one-electron transfer is a chemical substance that is capable of undergoing an oxidation–reduction reaction involving only one electron. A simple chemical example of such a substance, which is not a coenzyme, is ferricyanide (Reaction 2–43). The iron in ferricyanide readily accepts or gives back one electron and passes reversibly between $Fe^{III}$ and $Fe^{II}$ because the electron in question enters or then leaves a 3*d* atomic orital already occupied by one electron. The entry into that half-filled 3*d* atomic orbital is favorable because it fills an orbital to create a pair of electrons. The exit from that filled *d* orbital is favorable because the departure of the electron creates a half-filled 3*d* subshell in which all five *d* orbitals have only one electron. A one-electron coenzyme resembles the iron in ferrocyanide because the two states of oxidation–reduction separated by the one electron are both stable structures.

There are one-electron coenzymes that do not contain transition metals such as iron but only some combination of carbon, nitrogen, oxygen, sulfur, or hydrogen. Flavin is an example of such an organic, one-electron coenzyme. Again, the ability of flavin to participate in one-electron oxidation–reduction is due to the individual stabilities of the three forms of flavin: oxidized, radical, and reduced. As a result, both members of either redox pair differing by one electron are stable compounds. The reason that organic one-electron carriers are significantly different from inorganic one-electron carriers such as ferricyanide is that organic one-electron carriers do not have half-filled *d* atomic orbitals. Small organic molecules are constructed on the octet rule whereby every molecular orbital is filled with a pair of electrons. Removing one electron from a $\sigma$ bond, a $\sigma$ lone pair of electrons, or a small $\pi$ molecular orbital system creates a radical that is both too unstable, so that it is inaccessible under normal circumstances, and too reactive, so that it would abstract a hydrogen atom from other molecules in its surroundings. The reason the flavin radical is stable and unreactive is that the unpaired electron is delocalized over a large $\pi$ molecular orbital system. Therefore, coenzymes of one-electron transfer that have no transition metals are based on a different electronic strategy than those that do.

There are many one-electron oxidation–reductions that occur in biological situations. They can be isolated events in single enzymes, such as the two one-electron transfers between peroxide and two molecules of cytochrome *c* catalyzed by cytochrome *c* peroxidase, or a series of coenzymes of one-electron transfer can be linked in a long sequence, as in the electron transport chain of mitochondria.

The **electron transport chain of mitochondria**, because it incorporates several of the universally used coenzymes and because it is familiar, serves as a convenient introduction to the coenzymes of one-electron transfer. In the electron transport chain of mitochondria, substrates that can participate only in two-electron oxidation–reduction and that have negative enough reduction potentials to reduce oxidized flavin, within the active site of the appropriate enzyme, are the initial sources of the electrons. Such substrates are succinate ($E^{\circ}_{\text{app,pH7}}$ = +0.03 V), butanoyl-SCoA ($E^{\circ}_{\text{app,pH7}}$ = −0.02 V), and NADH itself ($E^{\circ}_{\text{app,pH7}}$ = −0.31 V). The ultimate receptacle of the electrons is molecular oxygen ($E^{\circ}_{\text{app,pH7}}$ = +0.82 V). The electrons are passed by flavin from the reducing substrates to ubiquinone, then from ubiquinone to cytochrome *c*, and finally from cytochrome *c* to oxygen

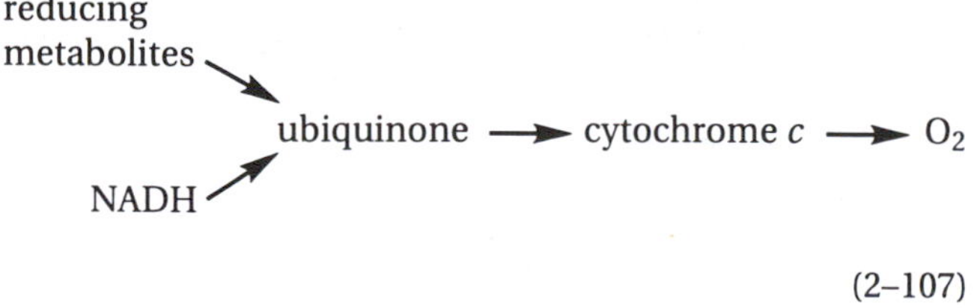


(2–107)

where reducing metabolites are succinate, butyryl-SCoA, and electron-transferring flavoprotein among others. Each of the arrows in Scheme 2–107 represents a different enzyme. An example of a flavoenzyme that oxidizes a reducing metabolite would be succinate dehydrogenase (ubiquinone), and the flavoenzyme that oxidizes NADH is NADH dehydrogenase (ubiquinone). The enzyme oxidizing the ubiquinol produced in these two reactions is ubiquinol–cytochrome *c* reductase, and the enzyme oxidizing cytochrome *c* is cytochrome *c* oxidase. In this electron transport chain, flavin plays a key role because of its ability to participate in either one-electron or two-electron oxidation–reductions.

Flavin accepts electrons from substrates that are required to give up two electrons in the same step of the reaction. Unless these two electrons are transferred directly to oxygen when flavin is acting as the coenzyme in an oxidase, they are usually transferred one at a time from flavin to an iron–sulfur cluster that participates in one-electron transfer. Once the electrons have been divided by flavin and the accompanying iron–sulfur cluster, they then pass individually among sets of coenzymes capable of engaging in one-electron transfer until they are dumped onto oxygen, to which they are transferred by one or more atoms of iron or copper. Between the entry at flavin and the exit at oxygen, there are a varied assortment of coenzymes of one-electron transfer responsible for shuttling the electrons. These are iron–sulfur clusters, hemes, and ubiquinone. Associated with each of these coenzymes of one-electron transfer is an apparent standard reduction potential at pH 7 (Table 2–1).

**Table 2–1: Apparent Standard Reduction Potentials for Coenzymes of One-Electron Transfer[a]**

| redox pair | $E^{\circ}_{\text{app,pH7}}$ (V) |
|---|---|
| ferredoxin($Fe^{III}Fe^{III}$)[b] + $e^-$ ⇌ ferredoxin ($Fe^{II}Fe^{III}$) | −0.43 |
| ferredoxin($Fe_2{}^{II}Fe_2{}^{III}$)[c] + $e^-$ ⇌ ferredoxin ($Fe_3{}^{II}Fe^{III}$) | −0.40 |
| $NAD^+ + H^+ + 2e^-$ ⇌ NADH | −0.31 |
| $HFl_{ox} + e^- + H^+ \rightleftharpoons H_2Fl^{\cdot}$ | −0.27 |
| $HFl_{ox} + 2e^- + H^+ \rightleftharpoons H_2Fl_{red}{}^-$ | −0.21 |
| $H_2Fl^{\cdot} + e^- \rightleftharpoons H_2Fl_{red}{}^-$ | −0.15 |
| cytochrome $b_5$($Fe^{III}$) + $e^-$ ⇌ cytochrome $b_5$($Fe^{II}$)[d] | −0.10 |
| rubredoxin($Fe^{III}$) + $e^-$ ⇌ rubredoxin($Fe^{II}$) | −0.06 |
| crotonyl-SCoA + $2e^-$ + $2H^+$ ⇌ butanoyl-SCoA | −0.02 |
| fumarate + $2e^-$ + $2H^+$ ⇌ succinate | +0.03 |
| ubiquinone[e] + $2e^-$ + $2H^+$ ⇌ ubiquinol | +0.08 |
| cytochrome $c_1$($Fe^{III}$)[f] + $e^-$ ⇌ cytochrome $c_1$($Fe^{II}$) | +0.22 |
| cytochrome *c*($Fe^{III}$) + $e^-$ ⇌ cytochrome *c*($Fe^{II}$) | +0.25 |
| cytochrome *a*($Fe^{III}$)[f] + $e^-$ ⇌ cytochrome *a*($Fe^{II}$) | +0.29 |
| HIPIP($Fe^{II}Fe_3{}^{III}$)[g] + $e^-$ ⇌ HIPIP ($Fe_2{}^{II}Fe_2{}^{III}$) | +0.35 |
| $Fe^{III}(CN)_6{}^{3-} + e^- \rightleftharpoons Fe^{II}(CN)_6{}^{4-}$ | +0.36 |
| ferredoxin($Fe_3{}^{III}$)[h] + $e^-$ ⇌ ferredoxin ($Fe^{II}Fe_2{}^{III}$) | +0.42 |
| $Fe^{III}(OH_2)_6{}^{3+} + e^- \rightleftharpoons Fe^{II}(OH_2)_6{}^{2+}$ | +0.77 |
| $O_2 + 4e^- + 4H^+ \rightleftharpoons 2H_2O$ | +0.82 |

[a]The apparent standard reduction potentials presented in the text have been gathered in this table. The references, unless otherwise noted, are those for the values as they are presented in the text. A more complete table can be found in refs 110 and 111. [b]Spinach. [c]*Peptococcus aerogenes.* [d]Liver.[111] [e]In a bilayer of phospholipids. [f]Heart muscle.[111] [g]*Chromatium vinosum.* [h]*Azotobacter vinelandii.*

**Iron–sulfur clusters** are arrays of iron atoms and sulfur atoms. Each of these arrays contains one or more iron atoms. Each iron atom is surrounded by four sulfur atoms, either as free sulfide, $S^{2-}$, which associates with two atoms of iron simultaneously, or as thiolate, $RS^-$, which can associate with only one atom of iron. The free sulfide incorporated into the iron–sulfur clusters is generated during their assembly by the enzymes thiosulfate sulfurtransferase[112,113] or 3-mercaptopyruvate sulfurtransferase,[114] and the thiolates are cysteines from the protein carrying the iron–sulfur cluster as a coenzyme.

**Mononuclear** iron–sulfur clusters, containing one atom of iron, necessarily have all of the four positions around the iron occupied by the thiolates of cysteines in the protein. In the mononuclear iron–sulfur cluster of rubredoxin,[115] the four sulfurs of the four cysteines surround the iron in a slightly distorted tetrahedron (bond angles [S–Fe–S] range from 101° to 115°) and with fairly constant iron–sulfur bond distances (0.21–0.23 nm).

**2–37**

Rubredoxin has an apparent standard reduction potential at pH 7 ($E^{\circ}_{\mathrm{app,pH7}}$) of –0.06 V[116] for the reaction

$$\mathrm{Fe^{III}(SR)_4^{-} + e^{-} \rightleftharpoons Fe^{II}(SR)_4^{2-}} \qquad (2\text{–}108)$$

**Binuclear** iron–sulfur clusters, containing two atoms of iron, have two of the positions around each iron occupied by a sulfide and two of the positions around each iron occupied by the thiolates of cysteines. This arrangement involves two irons, two sulfides, and four thiolates. A small compound that contains such a cluster has been synthesized, and a crystallographic molecular model has been obtained (Figure 2–15A).[117] In the central rhombus the angles at the sulfurs (75°) are more acute than the angles at the irons (105°). The compound crystallizes in its oxidized state, $[\mathrm{Fe_2^{III}S_2(RS_2)_2}]^{2-}$, and this state has been shown to display electronic, Mössbauer, and nuclear magnetic resonance spectra similar to those of the oxidized form of a binuclear iron–sulfur cluster in a protein.[117] If the oxidation level of the iron in the oxidized state of the model compound is the same as the oxidation level of the iron in the oxidized state of the protein, then the apparent standard reduction potential for a protein such as spinach ferredoxin ($E^{\circ}_{\mathrm{app,pH7}}$ = –0.43 V) is for the reaction

$$\mathrm{Fe_2^{III}\,S_2\,(RS)_4^{2-} + e^{-} \rightleftharpoons Fe^{II}Fe^{III}\,S_2\,(RS)_4^{3-}} \qquad (2\text{–}109)$$

Because the two irons are symmetrically coupled by the bridging sulfides, the odd electron in the reduced state is shared between them equally by occupying a molecular orbital spread evenly among the irons and the sulfurs.[119] Because the reduced form of the coenzyme has an odd number of electrons, however, it gives an electron paramagnetic spectrum. The oxidized form has an even number of electrons and no unpaired electron spins because the two ferric irons, each formally a species with an odd number of electrons, share the two odd electrons as a pair in a molecular orbital spread over the rhombus.

**Tetranuclear** iron–sulfur clusters have three positions occupied by sulfide around each iron (Figure 2–15B).[118] Each iron is bonded also to the thiolate of a cysteine from the protein. The model compound in which the four positions that would be occupied by cysteines in a protein are occupied by benzyl thiolates (Figure 2–15B)[118] is indistinguishable in its crystallographic molecular model from the tetranuclear iron–sulfur clusters found in proteins,[120] and it can be assumed that only small displacements of the positions of the sulfur atoms of the cysteines occur[120] as a result of the native structure of the protein. In a tetranuclear iron–sulfur cluster, the iron atoms occupy vertices with wider angles (104°) than the vertices occupied by the sulfurs (74°). These are the same angles as those of a binuclear iron–sulfur cluster (Figure 2–15A). Each bond length within the box, however, is almost the same (0.23 nm) because they are electronically equivalent. These geometric constraints give the structure a distorted but rotationally symmetric shape, with three orthogonal 2-fold rotational axes of symmetry.

Because there are so many iron atoms in a tetranuclear iron–sulfur cluster, its successive, one-electron reduction potentials are fairly close to each other.[118,120] The two biochemically accessible oxidation–reductions are

$$\mathrm{Fe^{II}Fe_3^{III}\,S_4\,(RS)_4^{-} + e^{-} \rightleftharpoons Fe_2^{II}Fe_2^{III}\,S_4\,(RS)_4^{2-}} \qquad (2\text{–}110)$$

$$\mathrm{Fe_2^{II}Fe_2^{III}\,S_4\,(RS)_4^{2-} + e^{-} \rightleftharpoons Fe_3^{II}Fe^{III}\,S_4\,(RS)_4^{3-}} \qquad (2\text{–}111)$$

The intermediate species between these two reactions $\mathrm{Fe_2^{II}Fe_2^{III}S_4(RS)_4^{2-}}$, has an even number of electrons and all are paired. This species lacks an electron spin resonance absorption. All of the irons are coupled, so several electron pairs are probably shared among them by occupying molecular orbitals spread over the cluster. If one electron is removed from one of these extended molecular orbitals in the reverse of Reaction 2–110 or one electron is added to a vacant molecular orbital in Reaction 2–111, each of the two species formed has an unpaired electron and each has an electron spin resonance absorption.

Iron–sulfur clusters in proteins that normally participate in Reaction 2–110 are referred to as **high-potential**, tetranuclear iron–sulfur clusters. The iron–sulfur cluster in the high-potential iron protein (HIPIP) from *Chromatium vinosum* has an apparent standard reduction potential for Reaction 2–110 of +0.35 V at pH 7. Tetranuclear iron–sulfur clusters in proteins that normally participate in Reaction 2–111 are **low-potential**, tetranuclear iron–sulfur clusters because they have more negative reduction potentials. For example, the ferredoxin from *Peptococcus aerogenes* has an apparent standard reduction potential of –0.40 V at pH 7 for Reaction 2–111.

**Trinuclear** iron–sulfur clusters, containing three atoms of iron, are tetranuclear iron–sulfur clusters in which one of the four irons has dissociated from the cluster because the cysteine in the protein that would normally form a strong bond with this iron is unable to.[121] Usually, a trinuclear iron–sulfur cluster can be easily converted into a tetranuclear cluster by adding ferrous ion.[122] In the crystallographic molecular model, at present levels of resolution, a trinuclear iron–sulfur cluster is indistinguishable from a tetranuclear iron–sulfur cluster (Figure 2–15B) except for the missing iron atom.

In its most oxidized state, a trinuclear iron–sulfur cluster has an unpaired electron because it contains three $\mathrm{Fe^{III}}$ atoms, each with an odd number of electrons, but pairs of electrons from every other substituent. Upon reduction ($E^{\circ}_{\mathrm{app,pH7}}$ = +0.42 V),[123] the trinuclear iron–sulfur cluster of *Azotobacter vinelandii* has no unpaired electrons and does not give an

**A**

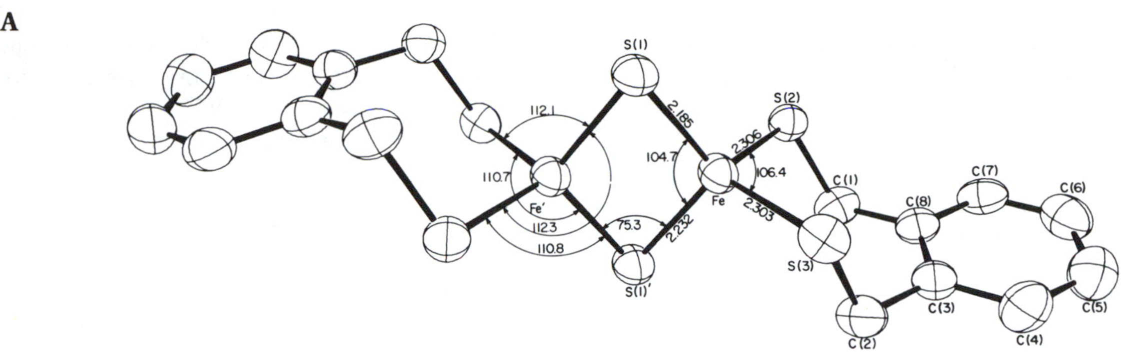


**B**

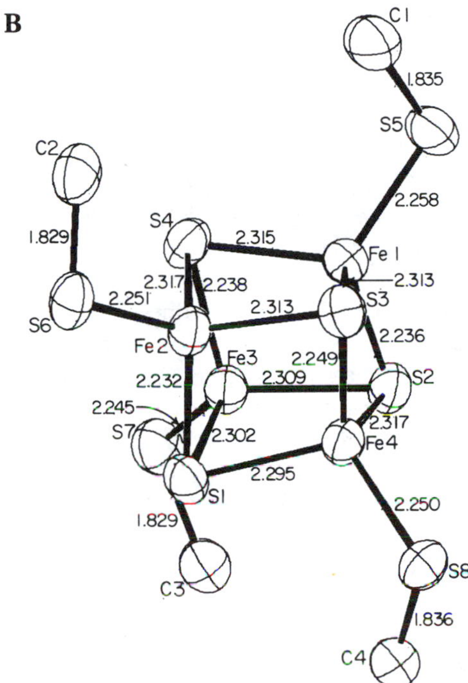


**Figure 2–15:** Drawings from crystallographic molecular models of (A) a binuclear iron–sulfur cluster[117] and (B) a tetranuclear iron–sulfur cluster.[118] (A) The anion $\{Fe_2S_2[(SCH_2)_2C_6H_4]_2\}^{2-}$ was produced by reaction of $FeCl_3$ with *o*-bis(mercaptomethyl)benzene and then $Na_2S$. It was crystallized as the tetraethylammonium salt from *N,N*-dimethylformamide and methanol. Only the molecular model of the anion is presented. The model is perfectly centrosymmetric; bond angles are noted to the left and bond lengths (in ångstroms) to the right. The planes defined respectively by FeS(1)S(1)′ and FeS(2)S(3) are perfectly perpendicular to each other and the central rhombus is perfectly planar. Panel A reprinted with permission from ref 117. Copyright 1973 National Academy of Sciences. (B) The anion $[Fe_4S_4(SCH_2C_6H_4)_4]^{2-}$ was prepared by reaction of $FeCl_3$, $NaOCH_3$, $Na_2S$, and (mercaptomethyl)benzene in methanol and crystallized as the tetraethylammonium salt from acetonitrile. Only the central portion of the crystallographic molecular model is presented; each of the four peripheral carbons is bonded to a phenyl ring that is not shown. Bond lengths are in ångstroms. Panel B reprinted with permission from ref 118. Copyright 1972 National Academy of Sciences.

absorption in electron spin resonance. Therefore, the formal reaction in which this iron–sulfur cluster participates is

$$Fe_3^{III}\,S_4\,(RS)_4^{3-} + e^- \rightleftharpoons Fe^{II}Fe_2^{III}\,S_4\,(RS)_4^{4-} \qquad (2\text{–}112)$$

There is spectral evidence that the pair of electrons formed upon addition of the one electron is distributed over only two of the three iron atoms.[124]

All of the iron–sulfur clusters use the ability of iron to exist in either the **ferrous** or **ferric** state, $Fe^{II}$ or $Fe^{III}$, respectively, as stable species in aqueous solution. When iron is surrounded by six water molecules, the standard reduction potential for the reaction

$$Fe^{III}\,(OH_2)_6^{3+} + e^- \rightleftharpoons Fe^{II}\,(OH_2)_6^{2+} \qquad (2\text{–}113)$$

is +0.77 V. When it is surrounded by six cyanide anions, the standard reduction potential for the reaction

$$Fe^{III}(CN_2)_6^{3-} + e^- \rightleftharpoons Fe^{II}\,(CN)_6^{4-} \qquad (2\text{–}114)$$

is +0.36 V. When it is surrounded by the four sulfurs of rubredoxin, the apparent reduction potential at pH 7 is −0.06 V. In this series, the iron successively becomes less willing to accept an electron and more willing to give it back.

In the **cytochromes**, iron is surrounded by six heteroatoms in **octahedral coordination**. Four of the six heteroatoms are the nitrogens of a **porphyrin** (Figure 2–16A).[125] The several naturally occurring porphyrins

$R_1$ $R_2$ $R_8$ $R_3$ :NH HN: $R_7$ $R_4$ $R_6$ $R_5$

**2–38**

differ in the substitutents around the periphery. For example, one of the most common is protoporphyrin IX ($R_1 = R_3 = R_5 = R_8 = -CH_3$; $R_2 = R_4 = -CH{=}CH_2$; $R_6 = R_7 = -CH_2CH_2COO^-$). The formal reaction for inserting iron into a porphyrin is*

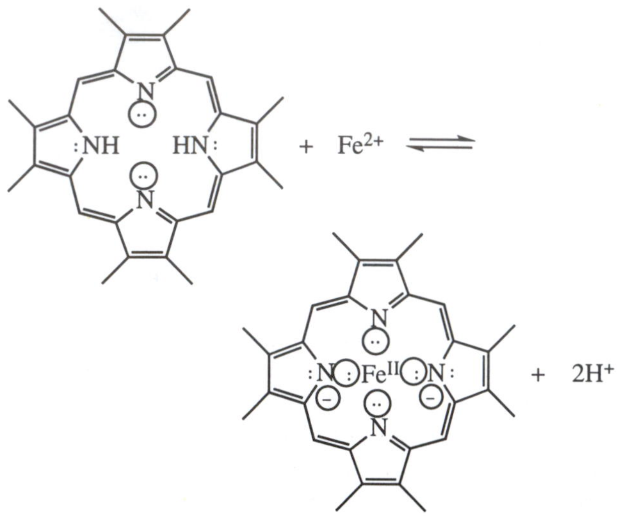

(2–115)

In this reaction, the iron, as a Lewis acid, has formally replaced two protons, also Lewis acids, and the product formed is a **heme**.

In all cytochromes except those that react directly with oxygen, such as cytochrome $a_3$ and cytochrome P450, the other two vertices of the octahedron around the iron, one on either side of the plane of the heme and both on an axis normal to it (Figure 2–16A) are referred to as the axial positions. They are occupied by the lone pairs of nucleophilic amino acids from the folded polypeptide of the cytochrome. Cytochrome *c* and cytochrome $c_3$ are two of the many cytochromes for which crystallographic molecular models[125a] are available. In cytochrome *c*, the two axial positions are occupied by the lone pair of electrons of the imidazole of a histidine and the lone pair of electrons of a methionine, respectively

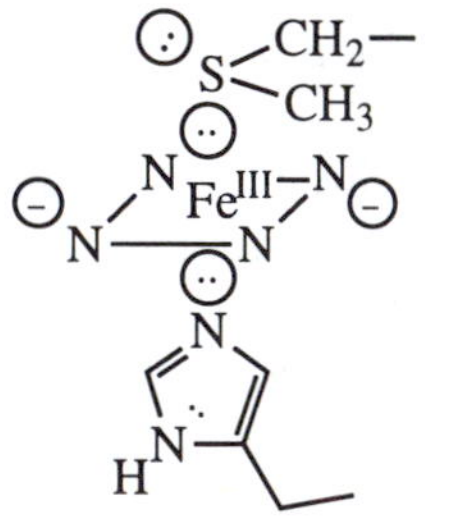

**2–39**

The structure of the heme in cytochrome *c* is almost indistinguishable[127] from the structure of a model compound (Figure 2–16B)[126] built to incorporate the central features of the heme in cytochrome *c*, but the protein does move its axial ligands slightly so that they are no longer directly above and directly below the iron.[127]

The apparent reduction potential for the heme in cytochrome *c*

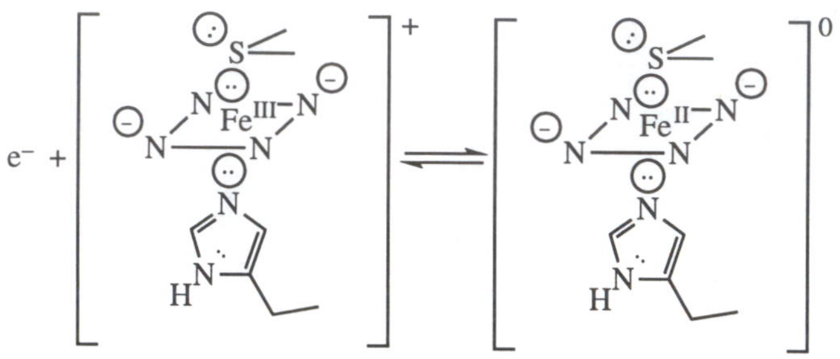

(2–116)

at pH 7 is +0.25 V.[128] Upon addition of the electron, the central core of the heme loses its positive charge and several amino acids near the sulfur rearrange their orientations in response to this change in charge. Between pH 8 and 3 the reduction of cytochrome *c* proceeds without proton uptake,[128] so these rearrangements must reflect responses of the structure of the protein only to the change in charge. Above pH 8, however, a proton is taken up with the electron and no change in charge on the protein occurs during the reaction.

Depending on the type of porphyrin used, the identity of the two axial ligands, and the perturbations engendered by the structure of the protein surrounding the heme, the reduction potential for the reaction

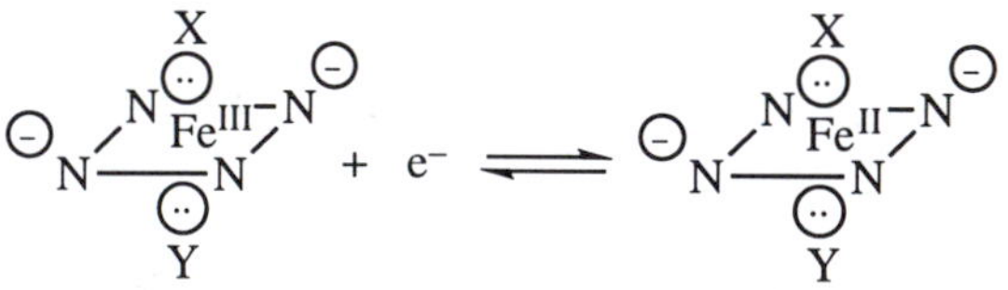

(2–117)

can vary between –0.15 and +0.35 V at pH 7. As such, the hemes in cytochromes lie between low-potential binuclear and tetranuclear iron–sulfur clusters and oxygen in reduction potential, and electrons are usually transferred from iron–sulfur clusters to hemes.

**Ubiquinones** have the structure

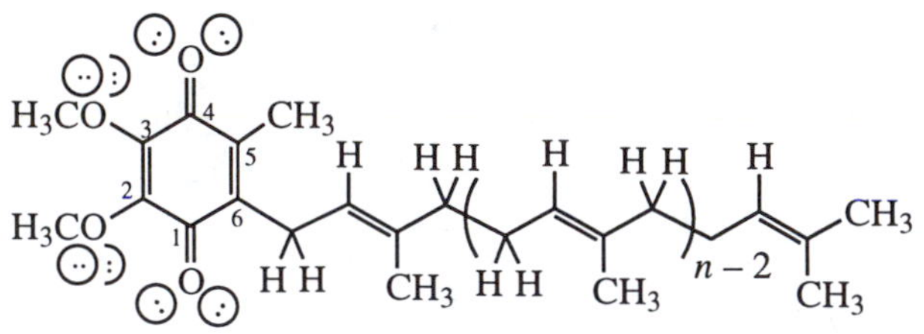

**2–40**

The different ubiquinones differ only in the number, *n*, of isoprenoid units that are attached to carbon 6 of the quinone. A ubiquinone is an **amphipathic compound**, with the quinone ring providing the hydrophilic end and the polyisoprenoid tail the hydrophobic end. When not tightly bound by a mem-

*The complexes between transition metals and Lewis bases will be treated as complexes between the metallic cations and the lone pairs of electrons on the ligands. The formal charge on each of the heteroatoms of the bases will be the formal charge that each would bear if it were uncomplexed to the cation.

**Figure 2–16:** Drawings of the crystallographic molecular models of two different porphyrins occupied by metallic cations. (A) Molecular model of the dichloro complex of tetraphenylporphyrin occupied by tin(IV).[125] Because of its symmetry, this molecule presents the simplest picture of a flat, planar, aromatic porphyrin with the metal ion in the exact center of the porphyrin and with two equivalent axial ligands respectively occupying the fifth and sixth coordination sites on the tin. The porphyrin is substituted only at the four methine positions with four equivalent phenyl groups each orthogonal to the plane of the porphyrin. Panel A reprinted with permission from ref 125. Copyright 1971 American Association for the Advancement of Science. (B) Molecular model of a heme[126] that incorporates the ligation found in cytochrome *c*. The metallic cation is iron(II) and the porphyrin is tetraphenylporphyrin as in panel A. To one of the phenyl groups an *o*-amino group has been added, and this was used to attach *N*-(4-carboxybutyl)imidazole in amide linkage. This provided an imidazole as the fifth ligand to the iron(II). The sixth ligand is tetrahydrothiophene and this provides a thioether similar to the methionine in cytochrome *c*. The imidazole is above the porphyrin in the drawing, the tetrahydrothiophene is below the porphyrin, and the iron(II) is in the center of the porphyrin. Panel B reprinted with permission from ref 126. Copyright 1979 American Chemical Society.

A

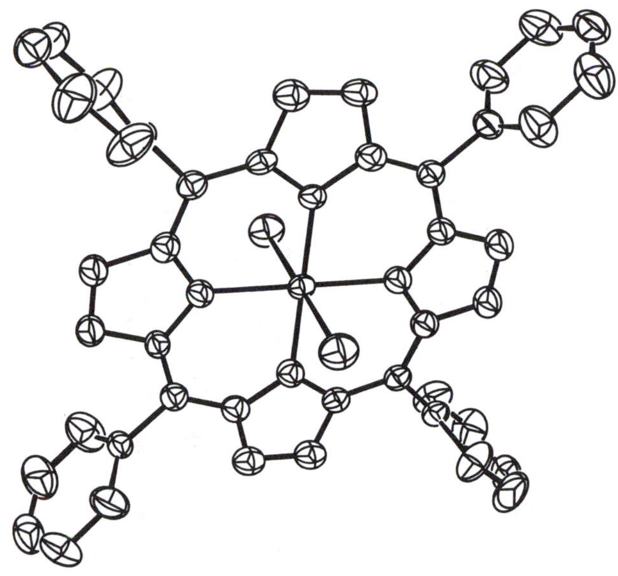

B

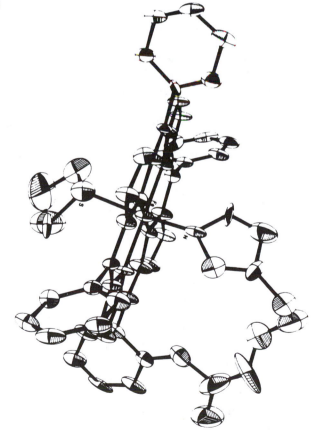

brane-bound protein as coenzymes, ubiquinones are solutes dissolved in the phospholipid bilayer of the plasma membrane, the mitochondrion, the chromatophore, or the chloroplast with which they are associated. As an amphipathic solute, a ubiquinone has the diffusional properties of a phospholipid, and most of the time it translates in two dimensions through the bilayer with the quinone at the interface between the alkane and the water. The hydrophilic portion of a ubiquinone, however, is much less hydrophilic than that of a phospholipid and ubiquinones can pass rapidly back and forth across the bilayer between the interfaces. Nevertheless, the physical and chemical properties of the unbound coenzyme when it is examined in homogeneous solutions of organic solvents can be misleading; rather, these properties should be studied while the quinone is in vesicles of phospholipid.

When it is a free solute in a phospholipid bilayer, ubiquinone ($Q_{ox}$) participates in the two-electron oxidation–reduction that produces **ubiquinol** ($H_2Q_{red}$)

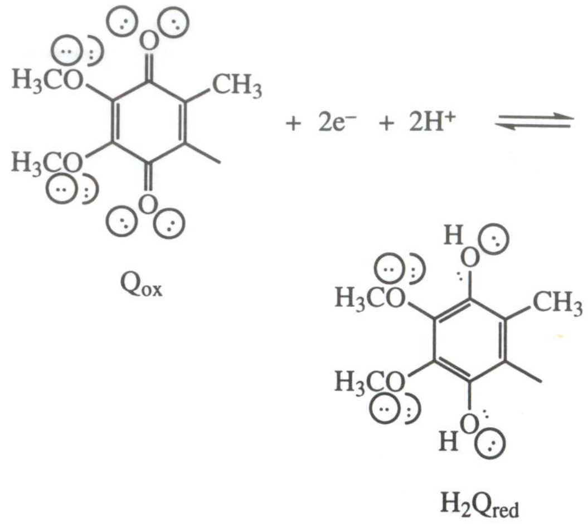

$$Q_{ox} + 2e^- + 2H^+ \rightleftharpoons H_2Q_{red} \qquad (2\text{–}118)$$

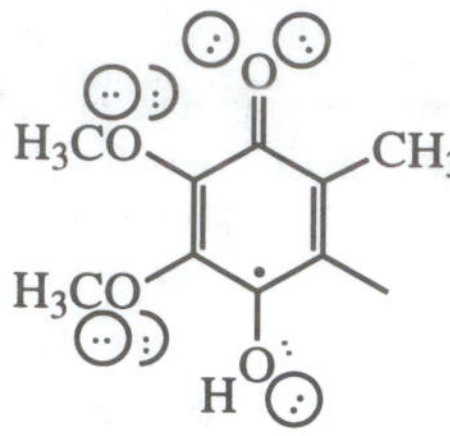

2–41

Because the two sequential values for the p$K_a$ of the quinol are greater than or equal to 13, respectively,[129] the transfer of the two electrons to the $\pi$ molecular orbital system of the quinone occurs with the uptake of two protons at its $\sigma$ lone pairs (Reaction 2–54). This causes $E^{\circ}_{app}$ to vary with pH (as in Figure 2–12) with a slope[130,131] of −0.060 V (unit of pH)$^{-1}$. When it is a free solute in the membrane of a biological organelle, the apparent reduction potential of ubiquinone at pH 7 is +0.08 V.[130,131]

The biological significance of ubiquinone is its ability to transfer protons across membranes. Both the neutral quinone, $Q_{ox}$, and the neutral quinol, $H_2Q_{red}$, because they are very hydrophobic molecules, readily pass back and forth across a bilayer of phospholipids. When ubiquinone is reduced to ubiquinol, two protons are taken up to make the neutral quinol (Equation 2–118). Those two protons must come from some aqueous solution. It seems to be the case that when ubiquinone is reduced to ubiquinol, the active sites of the enzymes in which that reduction occurs face the interior of the thylakoid, the mitochondrion, or the bacterium. This causes the protons to be taken up at that surface from the interior solution. When the ubiquinol produced is reoxidized in the next step of the respective electron transport chain, the active sites of the enzymes at which it is oxidized face the exterior of the thylakoid, the mitochondrion, or the bacterium, and the protons are released into the external solution. Consequently, two protons are taken up from the interior of the organelle during the reduction of ubiquinone by two electrons, the ubiquinol diffuses across the membrane, and two protons are released to the exterior during its reoxidation. Because ubiquinone is reduced by a coenzyme with a more negative reduction potential than the coenzyme by which it is oxidized, the movement of protons up their electrochemical gradient is driven by the movement of electrons down a gradient of reduction potential. The system works because ubiquinone is a very weak base, ubiquinolate anion is a very strong base, and neither the cationic conjugate acid of ubiquinone nor the anionic conjugate base of ubiquinol can pass across the membrane.

In addition to the quinone and the quinol (Equation 2–118), ubiquinone can exist as the ubiquinone radical ($HQ^{\cdot}$)

In methanol,[132] the p$K_a$ for the ubiquinone radical is 6.5, and when it is formed by electron transfer to the quinone (p$K_a < 0$) the ubiquinone radical will pick up a proton if the pH is less than its p$K_a$ under the particular circumstances

$$e^- + Q_{ox} \rightleftharpoons Q^{\cdot -} \underset{-H^+}{\overset{+H^+}{\rightleftharpoons}} HQ^{\cdot} \qquad (2\text{–}119)$$

In a solution, such as the solution of quinones dissolved in a biological membrane, the ubiquinone radical is in equilibrium with the quinol and the quinone by a one-electron transfer

$$Q_{ox} + H_2Q_{red} \rightleftharpoons 2HQ^{\cdot} \qquad (2\text{–}120)$$

in analogy to the similar disproportionation in which flavin participates (Equation 2–71). The equilibrium constant for Reaction 2–120 is the formation constant for the ubiquinone radical

$$K_f^{HQ^{\cdot}} = \frac{[HQ^{\cdot}]^2}{[Q_{ox}][H_2Q_{red}]} \qquad (2\text{–}121)$$

When ubiquinone is in free solution in biological membranes, no ubiquinone radical can be detected regardless of the ratio between the concentrations of quinone and quinol.[130] When, however, ubiquinone is bound tightly to particular proteins involved in electron transfer, significant quantities of the ubiquinone radical are observed as the concentrations of quinone and quinol are varied reciprocally.[133,134] This probably reflects the fact that in the proteins the quinones are fixed in orientations that preclude the conversion of two quinone radicals into a quinone and a quinol. In solution, such a disproportionation is facile because all of the species can collide with each other. In any case, the ability of ubiquinone to form a stable radical is essential because ubiquinone is required to accept electrons one at a time from iron–sulfur clusters or other coenzymes.

The various coenzymes involved in one-electron transfer, with the exception of ubiquinone when it is a free solute in a membrane, are always tightly bound to particular proteins. Iron–sulfur clusters are bound through the cysteines. The hemes of cytochromes can also be covalently bound through peripheral locations on the heme as is the case with cytochrome *c*, in which two of the substituents on two of the pyrroles of the heme participate in thioethers with cysteines of the protein. The hemes in most cytochromes, however, are noncovalently bound by the protein that embraces them. Neither iron–sulfur clusters nor hemes ever leave the protein in which they are enclosed during its lifetime.

The proteins to which these coenzymes are bound vary considerably in their size, location within the cell, and complexity. The largest of these proteins are the electron transfer complexes that catalyze the reactions in Scheme 2–107. These enzymes are membrane-bound, and they are large enough to dwarf the phospholipid bilayers in which they are found. The paradigm of this class of **large membrane-bound electron transport enzymes** is cytochrome oxidase.[134a] These enzymes are combined together to constitute the various electron transport chains that pass electrons from the hydride ions removed from various substrates to the ultimate oxidants such as molecular oxygen (Scheme 2–107). As the electrons pass through these proteins, the free energy inherent in their passage from reductants of more negative reduction potential to oxidants of more positive reduction potential is captured and used to produce a currency of chemical energy, usually MgATP.

Operating as lighters among these argosies are small proteins that transfer electrons by diffusing rapidly through the cytoplasm or some other cellular space or by diffusing over a biological membrane. **Ferredoxins** are small, water-soluble proteins with iron–sulfur clusters. Although they are more common in photosynthetic organisms, ferredoxins are also present in animal cells.[135] There are also **small, water-soluble flavoproteins** involved in shuttling electrons. An example would be the electron-transferring flavoprotein, which is a small $\alpha_2$ dimer ($n_{aa}$ = 240 for each polypeptide) containing one flavin in each subunit.[136] It is responsible for transferring electrons from cytoplasmic flavoenzymes, involved in hydride transfer, to flavoproteins of the mitochondrial electron transport chain, embedded in the mitochondrial membrane. There are **small, water-soluble cytochromes**, such as cytochrome *c*, and there are also small anchored, **membrane-bound cytochromes**, such as cytochrome $b_5$. Cytochrome $b_5$ can diffuse rapidly over the membrane of the endoplasmic reticulum carrying an electron. Nicotinamide adenine dinucleotide itself is a small, water-soluble hydride carrier among flavoenzymes, and ubiquinone is a small membrane-bound electron carrier among the larger complexes of the electron transport chains. The large membrane-bound electron transfer complexes receive electrons from one of these mobile carriers and pass them on to another mobile carrier of more positive reduction potential.

Reduced nicotinamide adenine dinucleotide dehydrogenase (ubiquinone), the first enzyme in the electron transport chain (Scheme 2–107) accepts electrons from NADH and transfers them to ubiquinone. The enzyme from mitochondria is a large membrane-bound protein constructed from at least nine polypeptides ranging in length from 80 to 660 amino acids.[137] The complete complex contains one flavin and several iron–sulfur clusters for each set of polypeptides.[137] The iron–sulfur clusters are thought to be of two types, binuclear and tetranuclear,[138,139] but the number of moles of each for every mole of protein has not been unambiguously ascertained. All of the iron–sulfur clusters have the midpoint potentials[140] and electron spin resonance spectra[141] of low-potential iron–sulfur clusters (Reactions 2–109 and 2–111), in which the reduced forms have unpaired spins. There appear to be four distinct iron–sulfur clusters that can be distinguished in the electron paramagnetic resonance spectrum of the enzyme.[142] All can be reduced by the addition of excess NADH. Additional absorptions in electron spin resonance spectra have also been detected,[141] but these may result from artifactual heterogeneity. The transfer of electrons through the coenzymes in NADH dehydrogenase (ubiquinone) begins with formal hydride transfer from NADH to oxidized flavin. The electrons are then passed singly from flavin to the iron–sulfur clusters and from the iron–sulfur clusters to ubiquinone.

A membrane-bound protein involved in NADH dehydrogenase (ubiquinone) activity has been purified from the bacterium *Escherichia coli*. It is composed of one polypeptide ($n_{aa}$ = 430) and contains one flavin but seems to lack iron–sulfur clusters.[143,144] It is able to transfer electrons from NADH to ubiquinone, but this reaction may employ the substoichiometric levels of iron–sulfur clusters still contaminating the preparation. This observation, however, raises the possibility that, in certain circumstances, electrons could be transferred directly between flavin and ubiquinone, both of which function readily as two-electron carriers.

Succinate dehydrogenase (ubiquinone) receives electrons from succinate and transfers them to ubiquinone (Scheme 2–107). The enzyme from mitochondria is a large membrane-spanning[145] protein composed of at least two polypeptides ($n_{aa}$ = 620 and 240).[146] Two additional short polypeptides ($n_{aa}$ = 120 and 60) may also be components.[147,148] The enzyme contains one flavin for each set of polypeptides[146] and three iron–sulfur clusters.[149,150] One of the iron–sulfur clusters is binuclear, one is trinuclear, and one is tetranuclear.[151] The tetranuclear cluster has the reduction potential and electron spin resonance behavior of the high-potential class of tetranuclear clusters ($E^{\circ}_{app,pH7}$ = +0.06 V). It has been reported that a cytochrome *b* also is associated with succinate dehydrogenase (ubiquinone),[148] but the molar stoichiometries of this cytochrome in the several preparations are variable and may represent contaminants.

The transfer of electrons through the coenzymes of succinate dehydrogenase (ubiquinone) begins with formal hydride transfer to flavin from succinate. The electrons are then passed singly from flavin to ubiquinone. Because the high-potential, tetranuclear iron–sulfur cluster is absolutely essential for transfer of the electrons to ubiquinone[150] but not essential for transfer of electrons from flavin to the other iron–sulfur clusters[146,149] and because the high-potential, tetranuclear cluster has the most positive reduction potential ($E^{\circ}_{app,pH7}$ = +0.06 V), it is assumed that it passes the electrons, one at a time, to ubiquinone ($E^{\circ}_{app,pH7}$ = +0.06 V).

Electron-transferring flavoprotein–ubiquinone oxidoreductase is a mitochondrial enzyme that transfers electrons from the small water-soluble electron-transferring flavoprotein to ubiquinone.[152] The enzyme is membrane-bound and composed of one polypeptide ($n_{aa}$ = 580). It contains one flavin and one tetranuclear iron–sulfur cluster for each polypeptide. The initial electron transfer is from the flavin of the electron-transferring flavoprotein to the flavin of the oxidoreductase. The oxidoreductase rapidly catalyzes the disproportionation of the electron-transferring flavoprotein (Reaction 2–72), and this requires that its flavin participate in one-electron transfers.[78] The electrons are then passed one at a time through the iron–sulfur cluster to ubiquinone.

In mitochondria, the three reactions catalyzed by NADH dehydrogenase (ubiquinone), succinate dehydrogenase (ubiquinone), and electron-transferring flavoprotein–ubiquinone oxidoreductase receive electrons from diverse sources, but each transfers the electrons to ubiquinone to form ubiquinol (Scheme 2–107). It is this common product that passes the electrons derived from each of these enzymes to the next membrane-bound protein in the electron transport chain.

The overall reaction catalyzed by ubiquinol–cytochrome *c* reductase is the transfer of electrons from ubiquinol to cytochrome *c* (Scheme 2–107). The enzyme from mitochondria is a large, membrane-spanning protein composed of nine polypeptides that range in length from 70 to 500 amino acids.[153] For each complete set of polypeptides, the protein contains two hemes of cytochrome *b*, one heme of cytochrome $c_1$, and one binuclear iron–sulfur cluster. The polypeptides of the cytochrome $c_1$ and the iron–sulfur protein have been identified by isolating them as separate properly folded polypeptides.[154,155] The apparent reduction potential of the binuclear iron–sulfur center at pH 7 is between +0.2 and +0.3 V.[156] Because this particular iron–sulfur center gives an electron spin resonance spectrum upon reduction,[155] it must participate in the normal oxidation–reduction reaction of a binuclear iron–sulfur cluster (Table 2–1) but with a much higher reduction potential. This peculiar type of binuclear iron–sulfur cluster has two histidines as ligands to the irons,[157,158] and this may explain its peculiar electronic and electrochemical properties.

The single polypeptide enclosing the hemes *b* of ubiquinol–cytochrome *c* reductase, and thereby forming the cytochrome *b* subunit of the complete protein, has also been identified by purifying it from the intact enzyme.[159] The polypeptide ($n_{aa}$ = 380) forming this subunit[160] enfolds two hemes *b*,[161] and this observation explains the fact that chemical analyses of either mitochondria or purified ubiquinol–cytochrome *c* reductase have always detected approximately 2 mol of heme *b* for every mol of heme $c_1$,[156,162] even though the purest preparations of the enzyme seem to have an equimolar ratio between the two polypeptides responsible for forming cytochrome *b* and cytochrome $c_1$.[153,163]

The complete, intact ubiquinol–cytochrome *c* reductase is an $\alpha_2$ dimer within the membrane with two complete sets of folded polypeptides arrayed around a 2-fold rotational axis of symmetry normal to the plane of the membrane[153] as in the structure of cytochrome oxidase.[134a] In fact, in both size and shape the protein resembles cytochrome oxidase quite closely.

Part or all of the transfer of electrons through the coenzymes of ubiquinol–cytochrome *c* reductase begins with the transfer of electrons from ubiquinol to the peculiar binuclear iron–sulfur cluster. When ubiquinol is tightly bound to ubiquinol–cytochrome *c* reductase, ubiquinol radical is formed at 10–20% of the total bound ubiquinol.[133,134] This presumably reflects the requirement that a single electron be transferred to the iron–sulfur cluster before being passed on. The unusually positive reduction potential of the iron–sulfur cluster may be necessary to form the levels of ubiquinone radical observed when the ubiquinol is associated with the enzyme. The sequence of the steps in the path of electron transfer through ubiquinol–cytochrome *c* reductase have yet to be established.

It has been proposed[164] that one of the purposes of the ubiquinol dissolved in the mitochondrial membrane is to shuttle pairs of electrons from the large membrane-bound flavoenzymes to ubiquinol–cytochrome *c* reductase. As the concentration of ubiquinone and ubiquinol is varied, the kinetics of electron transfer between either NADH dehydrogenase (ubiquinone) or succinate dehydrogenase (ubiquinone), acting as the donor, and ubiquinol–cytochrome *c* reductase, acting as the acceptor, are consistent with the following sequence of events. The ubiquinone bound to the donor receives the electrons. The resulting ubiquinol dissociates from the donor, diffuses at random over the surface of the mitochondrial membrane, collides with the acceptor, associates with the acceptor, and transfers the electrons to it.[164,165] When ubiquinone is added back to mitochondria that have been depleted of endogenous ubiquinone, the behavior of the increase in electron transfer among these enzymes is also consistent with the postulate that each of them has access to the same mobile pool of ubiquinone and ubiquinol.[166]

It has also been observed, however, that a specific complex can form between NADH dehydrogenase (ubiquinone) and ubiquinol–cytochrome *c* reductase when the enzymes are dissolved in solutions of detergent and that electron transfer in such a solution occurs only within this complex rather than by a shuttling performed by ubiquinone as it plies between the isolated proteins.[167,168] Whether a fixed complex between these large membrane-bound proteins of the electron transport chain can form and is absolutely required for electron transfer within the intact mitochondrion or is simply an artifactual requirement for electron transfer between these two proteins when they have been dissolved from the membrane of the mitochondrion is unknown.

A similar dispute arises at the next step in the electron transport chain in mitochondria, the transfer of the electrons between ubiquinol–cytochrome *c* reductase and cytochrome *c* oxidase. This transfer is mediated by cytochrome *c*, which accepts an electron from the reductase and donates that electron to the oxidase. Cytochrome *c* is a small, water-soluble protein that could perform this transfer by accepting an electron from ubiquinol–cytochrome *c* reductase, dissociating from it, translationally diffusing until it collides with cytochrome *c* oxidase, associating with it, and donating the electron. It could also form a permanent bridge between molecules of ubiquinol–cytochrome *c* reductase and cytochrome *c* oxidase through which the electrons would pass.

At physiological ionic strength, cytochrome *c* is tightly bound to the cytoplasmic surface of the inner mitochondrial membrane, presumably to ubiquinol–cytochrome *c* reductase or cytochrome *c* oxidase or to both of them simultaneously. At low ionic strength, cytochrome *c* binds specifically and with high affinity to cytochrome $c_1$ purified from ubiquinol–cytochrome *c* reductase.[169] When cytochrome *c* is bound to the mitochondrial membrane and then cross-linked to it, the protein to which it becomes covalently attached is cytochrome *c* oxidase.[170] A similar result is obtained when mixtures of cytochrome *c* and purified cytochrome *c* oxidase are cross-linked.[171] These results raise a third possibility, which is that cytochrome *c* is continuously bound to cytochrome *c* oxidase and is reduced by collision of the complex of cytochrome *c* and cytochrome *c* oxidase with ubiquinol–cytochrome *c* reductase.

If cytochrome *c* were to bridge the reductase and the oxidase, there would have to be unique and exclusive sites on its surface for interacting with each protein simultaneously. Cytochrome *c* has been modified with numerous electrophilic reagents and the individual products resulting from the modification of particular amino acids have been isolated. The derivatives of cytochrome *c* with Monoiodotyrosine 74, Formyltryptophan 59, Monoiodotyrosine 67, or (Carboxymethyl)methionine 80 cannot be reduced by ubiquinol–cytochrome *c* reductase but are readily oxidized by cytochrome *c* oxidase, while the bis(phenyl)glyoxal derivative of Arginine 13 can be readily reduced by ubiquinol–cytochrome *c* reductase but cannot be oxidized by cytochrome *c* oxidase.[172] A polyclonal set of immunoglobulins raised against cytochrome *c* have been isolated that can also block the oxidation of cytochrome *c* by cytochrome *c* oxidase but not its reduction by ubiquinol–cytochrome *c* reductase.[173] These observations imply that the sites to which ubiquinol–cytochrome *c* reductase and cytochrome *c* oxidase bind on the surface of cytochrome *c* are distinct.

The experiment, however, has been performed in reverse. When a complex is formed between cytochrome *c* and cytochrome *c* oxidase, Lysines 8, 13, 72, 73, 86, and 87 in the amino acid sequence of cytochrome *c* become shielded from acetylation by acetic anhydride.[174] When cytochrome *c* is in a complex with ubiquinol–cytochrome *c* reductase, the same lysines and no others are shielded. These results indicate that the binding sites on cytochrome *c* for the reductase and the oxidase overlap extensively. If this is the case, however, cytochrome *c* cannot bridge these two proteins. Furthermore, it must dissociate entirely from the ubiquinol–cytochrome *c* reductase before associating with cytochrome *c* oxidase, and it necessarily operates as a shuttle diffusing between the two. The fact that the transfer of electrons between purified cytochrome $c_1$ and cytochrome *c* is a purely bimolecular reaction at normal ionic strengths[175] is also inconsistent with the existence of a strong, long-lived complex between these two proteins in the mitochondrial membrane.

Cytochrome *c* oxidase contains two hemes *a* for each set of its polypeptides. The two hemes can be readily distinguished spectroscopically and have been designated cytochrome *a* and cytochrome $a_3$. There are also two copper ions for every set of polypeptides. The transfer of electrons from cytochrome *c* to cytochrome *c* oxidase begins with the transfer to cytochrome *a*,[176] then to the coppers and cytochrome $a_3$, and

From what is known about the electron transport chains in mitochondria, bacterial plasma membranes, the chromatophores of photosynthetic bacteria, and the chloroplasts of plants, it seems clear that the pattern of large membrane-spanning proteins composed from several polypeptides and several coenzymes connected by small electron shuttles capable of diffusing rapidly between their respective reductases and oxidases is a general strategy. For example, under different growth conditions *E. coli* uses one or the other of at least two electron transport proteins to transfer electrons to oxygen in the last step of its electron transport chain. Two of these proteins, the cytochrome *d* terminal oxidase[177] and the cytochrome *o* terminal oxidase,[178] have been purified. Both are large membrane-bound proteins, composed of two polypeptides ($n_{aa}$ = 510 and 380) or four polypeptides ($n_{aa}$ = 590, 310, 200, and 150), respectively. The cytochrome *d* terminal oxidase contains three hemes for each set of polypeptides, and the cytochrome *o* terminal oxidase, two hemes. Both of these terminal oxidases seem to accept electrons from ubiquinol and transfer them to oxygen, but they lack iron–sulfur clusters.[177] Chloroplasts contain a cytochrome *b* the sequence of which is homologous to that of the cytochrome *b* of ubiquinol–cytochrome *c* reductase from mitochondria.[160] It is contained in a membrane-bound protein, closely related to ubiquinol–cytochrome *c* reductase, composed from several polypeptides, and incorporating two moles of heme *b* (cytochrome $b_6$), one mole of heme *c* (cytochrome *f*), and one binuclear iron–sulfur cluster.[179] Another binuclear iron–sulfur cluster similar to the high-potential binuclear iron–sulfur cluster in ubiquinol–cytochrome *c* reductase is present in chloroplasts and may be associated with this large membrane-bound protein containing cytochrome $b_6$ and cytochrome *f*.[180]

In addition to the large, membrane-spanning electron transport proteins of the electron transport chains, there are a number of large **water-soluble electron-transfer proteins** that carry out one-electron transfer among tightly bound coenzymes. Sulfite reductase from *E. coli* catalyzes the reaction[181]

$$7H^+ + 6e^- + HSO_3 \rightleftharpoons HSO_3^- + 3H_2O \qquad (2\text{–}122)$$

The six electrons are derived from NADPH. The enzyme is a large complex formed from two polypeptides with the stoichiometry $\alpha_8\beta_4$.[182] The $\alpha$ subunit ($n_{aa}$ = 550) contains tightly bound flavin. It exists as an $\alpha_8$ octomer in the absence of the $\beta$ subunit and forms the core of the complex. The $\beta$ subunit contains[181] one tetranuclear iron–sulfur center and one siroheme (in Structure **2–38**, $R_1 = R_3 = R_5 = R_8 = -CH_2COO^-$; $R_2 = R_4 = R_6 = R_7 = -CH_2CH_2COO^-$; olefins between $R_1$ and $R_2$ and $R_3$ and $R_4$ are missing $R_1' = R_3' = -CH_3$). It exists as a $\beta$ monomer ($n_{aa}$ = 490) in the absence of the $\alpha$ subunit and is peripherally attached to the core in the complex.[182] The siroheme on the $\beta$ subunit binds the $HSO_3^-$ at its iron atom and dumps the electrons into it.[181] The tetranuclear iron–sulfur cluster is of the low-potential class ($E^{\circ}_{app,pH7}$ = −0.4 V) and has an unpaired electron in its reduced form (Reaction 2–111). Electrons arrive at the flavin on the $\alpha$ subunit from NADPH. They are transferred to the flavin as a formal hydride anion, passed through the iron–sulfur cluster to the siroheme, and donated to the $SO_3^{2-}$.[181,183] Because the protein is not a membrane-spanning complex, none of the free energy of the oxidation ($\Delta E^{\circ}_{app,pH7}$ = +0.44 V) is conserved.

A very similar reaction

$$8H^+ + 6e^- + NO_2^- \rightleftharpoons NH_4^+ + 2H_2O \qquad (2\text{–}123)$$

is carried out by the nitrite reductase from spinach.[184] This enzyme is a much smaller, water-soluble protein composed of a single polypeptide ($n_{aa}$ = 560) as a monomer. Because the electrons used in this reaction are derived from reduced ferredoxin, a one-electron shuttle, flavin is not necessary to split a pair of electrons and a flavoprotein subunit is not present in the enzyme. Each monomeric protein does have one binuclear iron–sulfur center to which each electron is passed from reduced ferredoxin and one siroheme where the electrons are passed, one at a time, onto bound nitrite. These properties resemble the structure and function of the monomeric $\beta$ subunit of sulfite reductase.[181]

Flavocytochrome $b_2$, the L-lactate dehydrogenase (cytochrome) from yeast, catalyzes the oxidation[185]

$$\text{L-lactate} + 2\,\text{ferricytochrome } c \rightleftharpoons \text{pyruvate} + 2\,\text{ferrocytochrome } c \qquad (2\text{–}124)$$

In the first step, the electrons from the lactate, in the form of a formal hydride ion, are passed directly to flavin. In the final step, the electrons are transferred to the heme $c$ of the cytochrome $c$ from the heme $b$ of the cytochrome $b_2$ bound to the enzyme.[186] Because the enzyme contains no iron–sulfur clusters, the electrons are passed one at a time from flavin directly to the heme $b$. One of the intermediate steps in this reaction is[186]

$$\text{HFl}^-_{red} + \text{heme } b_{ox} \rightleftharpoons \text{HFl}^{\cdot} + \text{heme } b_{red} \qquad (2\text{–}125)$$

in which the stability of flavin radical is used to permit one-electron transfer. This enzyme provides an informative exception to the usual situation in which the transfer of single electrons from flavin proceeds through iron–sulfur clusters.

Xanthine oxidase catalyzes the reaction

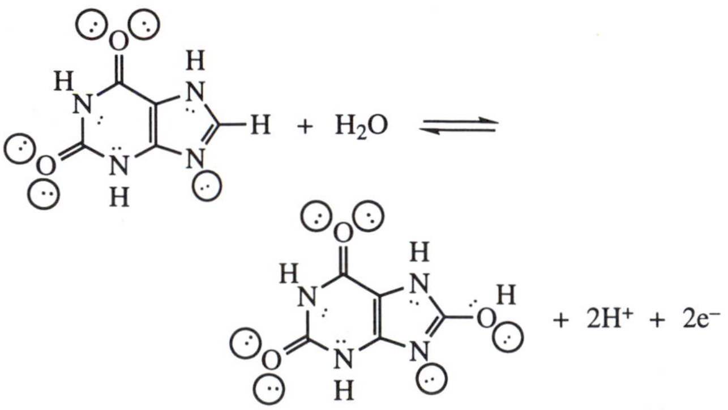

$$\text{xanthine} + H_2O \rightleftharpoons \text{uric acid} + 2H^+ + 2e^-$$

(2–126)

The hydroxyl added to the xanthine is derived from water[187] in an oxidative hydration catalyzed by a coenzyme containing **molybdenum**. On the basis of indirect chemical evidence, a structure for this coenzyme, when it is bound to the protein, has been proposed[188]

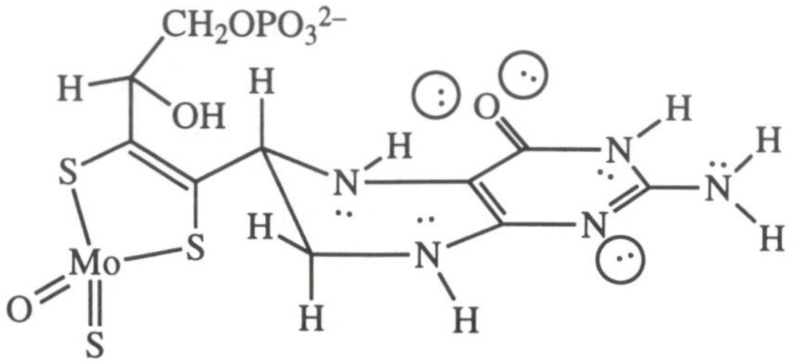

**2–42**

The heterocycle associated with the enedithiol is a pterin, which is the same heterocycle found in tetrahydrofolate **1–42**. It has been proposed[187] that the hydrate of xanthine is the species oxidized

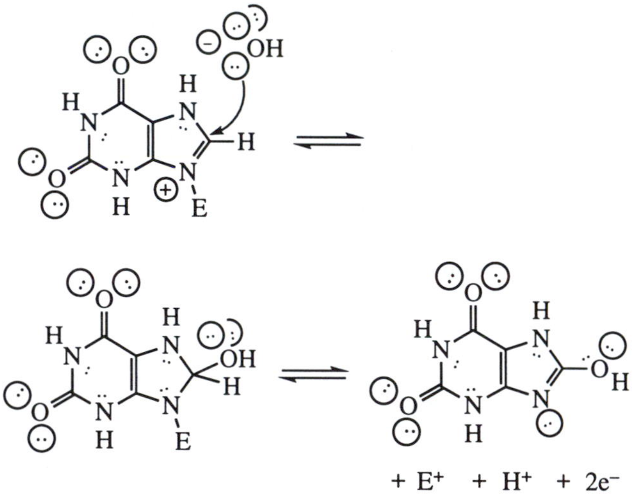

+ E⁺ + H⁺ + 2e⁻

(2–127)

where E is an electrophilic group on the coenzyme that combines with the $\sigma$ lone pair on nitrogen 9 of xanthine. Whether the electrons leave as the formal hydride from the carbinolamine on carbon 8 of the hydrate or are withdrawn with the electrophile as it departs is unknown. Nevertheless, the two electrons enter the electron transfer chain of the protein on molybdenum. The molybdenum of the protein can exist in three states, each differing by one electron[189]

$$Mo^{VI} + e^- \rightleftharpoons Mo^{V} \qquad E^{\circ}_{app,pH7.7} = -0.38\ \text{V} \qquad (2\text{–}128)$$

$$Mo^{V} + e^- \rightleftharpoons Mo^{IV} \qquad E^{\circ}_{app,pH7.7} = -0.41\ \text{V} \qquad (2\text{–}129)$$

and the oxidation of the xanthine transfers two electrons, transforming $Mo^{VI}$ to $Mo^{IV}$.

Xanthine oxidase is a water-soluble $\alpha_2$ dimer of two identical polypeptides ($n_{aa}$ = 1230) and contains one pterin, one molybdenum, two binuclear iron–sulfur clusters, and one flavin for each polypeptide.[189] The electrons are passed from molybdenum through the iron–sulfur clusters to flavin.[190] From the flavin they are added to oxygen.

The large membrane-spanning proteins of an electron transport chain such as the one in mitochondria differ from these water-soluble electron transfer proteins because they are able to conserve the free energy of electron transfer. The donor of electrons to each of the membrane-bound proteins in the mitochondrial electron transport chain has a more negative reduction potential than the oxidant that ultimately carries away those electrons. For example (Scheme 2–107), the apparent reduction potential of NADH at pH 7 is –0.32 V, that of ubiquinone dissolved as a solute in a phospholipid bilayer at pH 7 is +0.08 V, that of cytochrome *c* at pH 7 is +0.25 V, and that of oxygen is +0.81 V (Table 2–1). The differences in apparent standard reduction potential among these steps (0.17–0.56 V) translate into differences of free energy of –16 to –54 kJ for every mole of electrons passing through each protein. This is the energy that is conserved and used for a number of purposes such as the synthesis of MgATP or the transport of metabolites against their concentration gradient. The energy is stored in the form of an electrochemical gradient of protons created by coupling electron transport to the transport of protons across the membranes in which these enzymes and coenzymes are located. This process is understood in molecular detail in only one instance, that of the photosynthetic reaction center.

A **photosynthetic reaction center**[190a] is a membrane-spanning protein that can reverse electron transport. It forces electrons to move from donors of more positive reduction potential to acceptors of more negative reduction potential by consuming the energy in visible light. A photon of visible light is absorbed to form the excited state of a coenzyme. This excited state has a much lower reduction potential than its ground state. As a result, the excited electron can pass spontaneously to other coenzymes that have more positive reduction potentials than the excited state, even though those other coenzymes have more negative reduction potentials than the ground state of the coenzyme that absorbs the light. A photosynthetic reaction center can be viewed as a cage that encloses and positions an array of coenzymes of one-electron transfer. This array of coenzymes (Figure 2–17)[191] describes the path taken by the excited electron. A reaction center also serves as an introduction to two of these coenzymes.

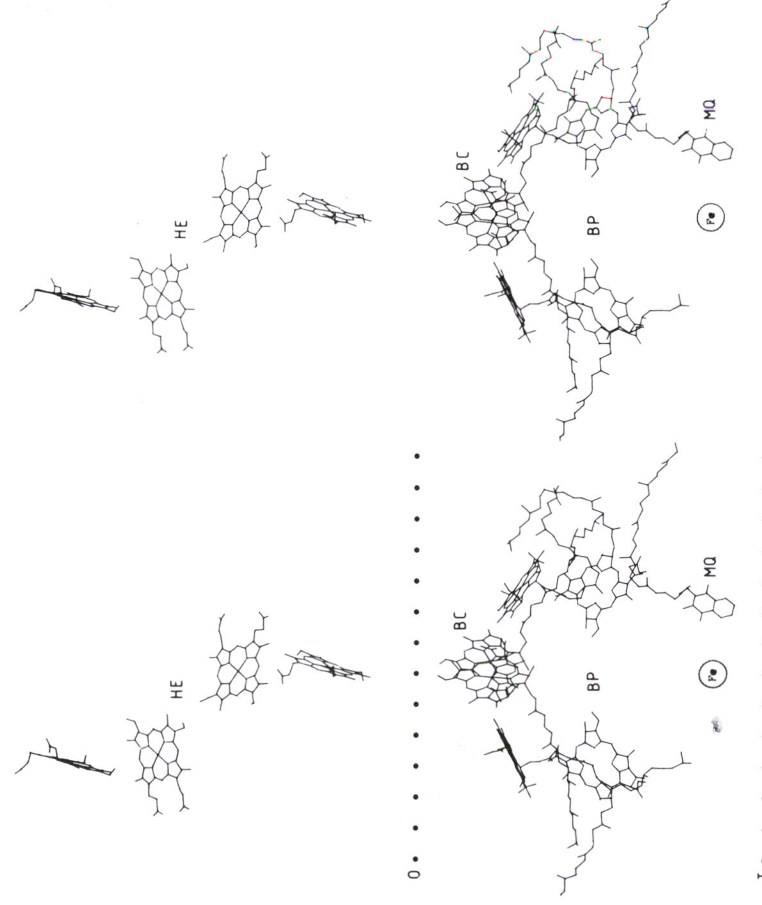


**Figure 2–17:** Drawing from the crystallographic molecular model of the reaction center of *Rhodopseudomonas viridis*.[191] The drawing presents only the coenzymes in their proper orientations. The path of the electron is from the top of the figure downward. It begins in the four hemes (HE) of the cytochrome and passes to the special pair, which is a rotationally symmetric dimer of bacteriochlorophyll *b* (BC). On each path, right and left, a bacteriochlorophyll *b* bridges the space between the special pair and each of the symmetrically arrayed bacteriopheophytins (BP). Below the bacteriopheophytin on the right path is a menaquinone (MQ), which is immediately adjacent to an iron cation (Fe). The membrane lies between the two horizontal dotted lines marked I and O to indicate the inside (cytoplasmic) and outside (extracytoplasmic) surfaces of the plasma membrane of the bacterium. The local 2-fold rotational axis of pseudosymmetry within the reaction center itself runs vertically through the iron cation and the center of the special pair. The site for the loosely bound ubiquinone is on the left side, symmetrically located relative to the menaquinone. The electron is thought to pass through the right path in the orientation presented. Reprinted with permission from ref 191. Copyright 1984 Academic Press.

The two, as yet undescribed, coenzymes found in a reaction center are **chlorophyll** and **pheophytin**. Both chlorophylls and pheophytins are constructed from an aromatic heterocycle

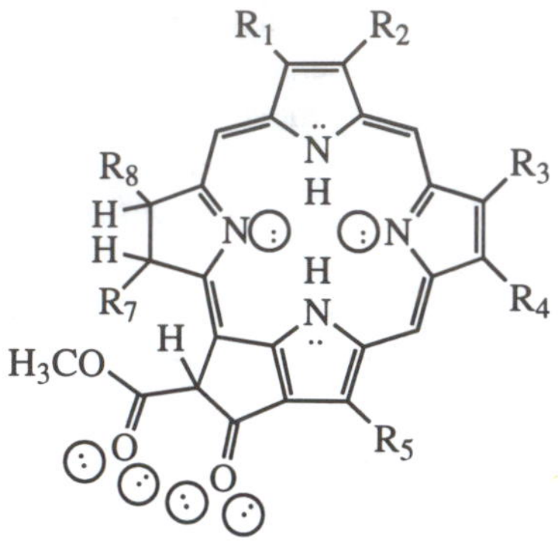

**2–43**

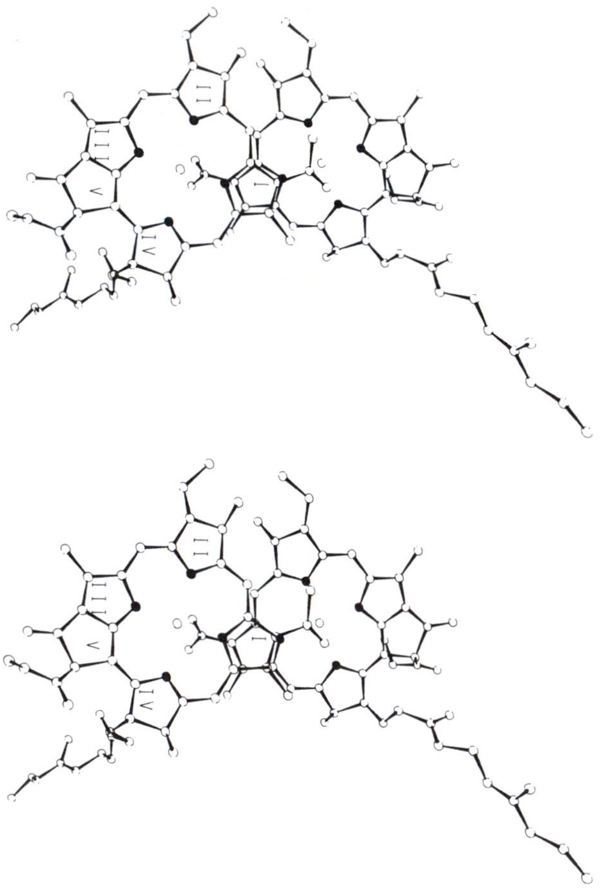

**Figure 2–18:** Drawing of the special pair in the crystallographic molecular model of the photosynthetic reaction center from *Rps. viridis.*[191] The two bacteriochlorophylls *b* of the special pair are held in orientations displayed symmetrically about the 2-fold rotational axis of pseudosymmetry of the protein. The rotational axis runs horizontally in the plane of the page between the two bacteriochlorophylls *b* and causes them to overlap at pyrrole I. The polyisoprene tails were not drawn along their entire length. Reprinted with permission from ref 191. Copyright 1984 Academic Press.

that differs somewhat from the porphyrins (**2–38**) used to construct the hemes of cytochromes. One of the peripheral olefins uninvolved in the aromaticity (between $R_7$ and $R_8$) has been reduced, and an extra oxacyclopentane has been incorporated at the periphery (bottom left of **2–43**). As with the hemes and porphyrins, the several chlorophylls and pheophytins differ in the substituents found at positions $R_1$–$R_8$. These substituents are hydrogens, methyls, ethyls, ethenyls, formyls, or acetyls except at carbon 7. The substituent at carbon 7 ($R_7$ in **2–43**) is either an acrylyl (–CH=CH–COO–) or a propionyl ($-CH_2-CH_2-COO-$) which is esterified with a long polyisoprenyl alcohol, such as 3,7,11,15-tetramethylhexadec-2-en-1-ol (Figure 2–17). Pheophytins are the aromatic heterocycles alone, while the corresponding chlorophylls have a magnesium ($Mg^{2+}$) ion within the respective pheophytins. The magnesium is inserted in place of the two protons, as is the iron in the formation of a heme (Reaction 2–115). The particular pheophytin in the reaction center from *Rhodopseudomonas viridis* (Figure 2–17) is bacteriopheophytin *b* (olefin between $R_3$ and $R_4$ has been exocyclically isomerized into $R_4$; $R_1 = R_5 = R_8 = -CH_3$; $R_2 = -COCH_3$; $R_3 = -CHO$; $R_4 = {=}C{=}CH_3$), and the chlorophyll is bacteriochlorophyll *b*, which is identical to the pheophytin except for the magnesium.

Chlorophylls and pheophytins can participate in one-electron oxidation–reductions because they are extended aromatic heterocycles ($4n + 2 = 18$). An electron can be removed from the highest occupied $\pi$ molecular orbital of the aromatic heterocycle to yield the radical cation

$$\text{chlorophyll} \rightleftharpoons \text{chlorophyll}^{\cdot+} + e^- \qquad (2\text{–}130)$$

In the radical cation, the highest occupied $\pi$ molecular orbital contains an unpaired electron that coincidentally can reside next to uninvolved lone pairs of electrons on the nitrogens. The situation can also be viewed from a reciprocal perspective. Because the highest occupied $\pi$ molecular orbital of the radical cation now lacks the one electron necessary to satisfy the aromatic requirement, it can be defined as a $\pi$ molecular orbital containing a **hole** for an electron. The lowest unoccupied $\pi$ molecular orbital of the neutral, aromatic heterocycle can also accept an electron to yield the radical anion

$$\text{pheophytin} + e^- \rightleftharpoons \text{pheophytin}^{\cdot-} \qquad (2\text{–}131)$$

In the radical anion the previously unoccupied $\pi$ molecular orbital of lowest energy has become the highest occupied molecular orbital, and it contains the unpaired electron.

Quinones also are associated with a reaction center. They can be ubiquinones, **plastoquinones**

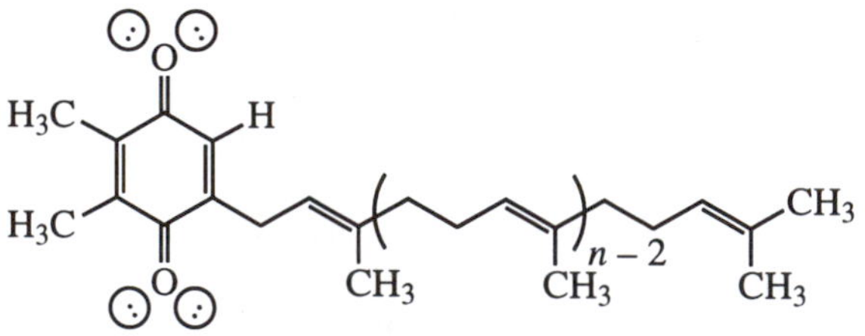

**2–44**

or **menaquinones**

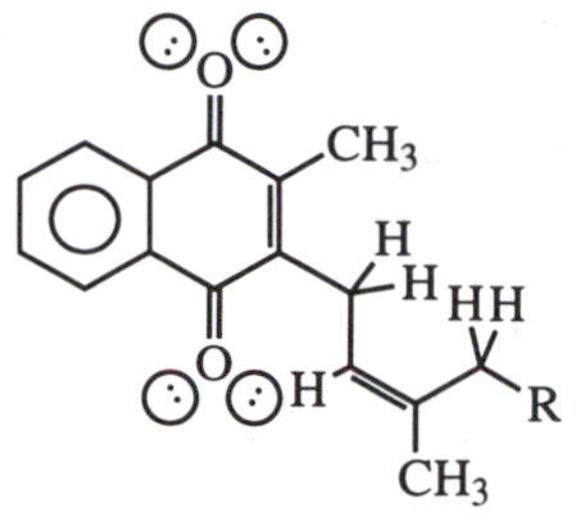

**2–45**

where R is any one of a number of polyisoprenes or reduced polyisoprenes. In the photosynthetic reaction center from *Rps. viridis*, the tightly bound quinone is a menaquinone with a nine-isoprene hydrocarbon [R = $(-CH_2CH{=}CCH_3CH_2-)_8H$].[192] Because plastoquinones and menaquinones usually have attached polyisoprenoid hydrocarbons, as free solutes, they, like the ubiquinones, are dissolved within the membranes of the organelles with which they are associated.

The central coenzyme in a bacterial photosynthetic reaction center is the **special pair** formed from two[193] bacteriochlorophyll *b* molecules (Figure 2–18).[191] The two chlorophylls of the special pair are arrayed about a 2-fold rotational axis of pseudosymmetry. The overlap of the aromatic systems permitted by this arrangement shifts the absorption of the two bacteriochlorophylls to longer wavelengths. This permits the special pair to be a much more efficient sink for the energy that is transferred to it from the coenzymes in the light-harvesting antenna proteins, which absorb light of shorter wavelengths. The 2-fold rotational axis of pseudosymmetry that produces this overlap exists because two of the three subunits[194] of the reaction center, subunits L and M, are homologous in sequence (30% identity with 2 gaps for every 100 amino acids),[195] superposable in their native structure,[196] span the membrane, and necessarily are arrayed around a 2-fold rotational axis of pseudosymmetry normal to the plane of the membrane. It is this rotational axis of pseudosymmetry within the protein that relates the two bacteriochlorophylls of the special pair. Because these two subunits in the protein arose from the same common ancestor, the first oligomer must have been an $\alpha_2$ dimer of identical membrane-spanning subunits that brought the two bacteriochlorophylls together in the proper orientation to display the unique properties of the special pair.

Because crystallographic molecular models are available only for the photosynthetic reaction center from the bacterium *Rps. viridis* (Figure 2–17), and for the photosynthetic reaction center from the bacterium *Rhodobacter sphaeroides*,[197,198] which closely resembles the reaction center from *Rps. viridis*, it is only in these instances that the individual steps in the one-electron transfers can be assigned to specific coenzymes. At the moment, these two very similar examples serve as models for photosynthetic reaction centers both in other photosynthetic bacteria and in plants. The special pair in the photosynthetic reaction center from *Rb. sphaeroides* absorbs light of wavelength 865 nm, where photons have energies of 1.43 eV, while the special pair in the reaction center from *Rps. viridis* absorbs light of wavelength 960 nm (1.29 eV).[199] The absorption of light by the special pair, or the absorption of an equivalent amount of energy transferred from the antenna proteins, produces an excited state in which one electron has been transferred from the highest occupied $\pi$ molecular orbital of the special pair to its lowest unoccupied $\pi$ molecular orbital. From here on, unless otherwise noted, the fate of this electron in the excited state of the reaction center of *Rb. sphaeroides* will be followed.

Normally, the electron in the excited state of the special pair would return within several nanoseconds to the $\pi$ molecular orbital of lower energy, which now contains only one electron, and this process would emit a quantum of light of longer wavelength (900 nm) as fluorescence. In fact, a short time after the absorption, such fluorescence emission can be observed from a population of excited reaction centers, but it decays rapidly ($t_{1/2}$ = 3 ps).[200] This decay in emission is referred to as **quenching**.

The quenching of the fluorescence is due to the transfer of the excited electron from the special pair to the lowest unoccupied $\pi$ molecular orbital on a bacteriopheophytin near the special pair.[200] This transfer is favorable because the excited special pair has become a better reductant than it was when it was in the ground state. Its reduction potential, after the relaxation of the excited state, has experienced a decrease equivalent to the energy in a quantum of light of a wavelength equal to its emission wavelength (1.37 eV). The transfer of the electron to the nearby bacteriopheophytin creates a radical anion on the bacteriopheophytin and leaves a radical cation behind on the special pair.[201] Because another bacteriochlorophyll, not involved in the special pair, lies between it and the nearest bacteriopheophytin (Figure 2–17), it is assumed that transfer occurs through this bacteriochlorophyll.[200]

If the excited electron could be delocalized to the bacteriopheophytin in the first place, by microscopic reversibility, it must be able to return to the $\pi$ molecular orbital it occupied in the excited state of the special pair from whence it came. If its transfer beyond the bacteriopheophytin is blocked by reducing the ultimate acceptor of the electron,[201] the trapped electron eventually returns to the $\pi$ molecular orbital of lower energy on the special pair, and the energy of the excited state is dissipated as heat and fluorescence. The rate of decay of the excited state under these circumstances ($t_{1/2}$ = 10 ns in *Rps. viridis*)[201] is longer than it would have been if all of the excited electron density had remained in the $\pi$ molecular orbital of higher energy on the special pair rather than spreading between the two locations. This is due to the fact that the reduction potential of the electron on the bacteriopheophytin is about 0.2 V more positive than in the excited state of the special pair,[199] and this difference in reduction potential causes most of the electron density to reside on the bacteriopheophytin. Under normal circumstances, however, the electron transfers ($t_{1/2}$ = 140 ps) from the bacteriopheophytin to the quinone adjacent to the bacteriopheophytin (the menaquinone in Figure 2–17) to form a quinone radical anion. In the reaction center from *Rps. viridis* this quinone is a menaquinone, while in the reaction center from *Rb. sphaeroides*, it is a ubiquinone.

At this point, the excited electron located in the quinone radical anion is a good distance (Figure 2–17) from the hole in the $\pi$ molecular orbital of lower energy in the special pair where it originated in the ground state. A small fraction of its density is, however, still associated with the $\pi$ molecular orbital of higher energy in the special pair that it entered upon excitation. The excited electron necessarily distributes among all of the electron carriers between and including the molecular orbital of higher energy on the special pair and the $\pi$ molecular orbital of the quinone; but, because the reduction potential of the quinone radical anion is 0.6 V more positive than the radical anion of the bacteriopheophytin and 0.8 V more positive than the excited state of the special pair (Figure 2–19), the electron is located almost exclusively on the quinone.

In exchange for the expenditure of 60 kJ mol$^{-1}$ of free energy associated with the increase in reduction potential between the bacteriopheophytin and the quinone, the electron has gained time. The situation can be described kinetically (Figure 2–19) as

$$P^{\cdot+}BPhQ_A^{\cdot-} \underset{k_2}{\overset{k_{-2}}{\rightleftharpoons}} P^{\cdot+}BPh^{\cdot-}Q_A \xrightarrow{k_{-1}} PBPhQ_A \tag{2–132}$$

where P is the special pair, BPh is the bacteriopheophytin, $Q_A$ is the quinone, and the plus and minus refer to radical cations and radical anions, respectively. The reaction as written has an overall negative change in free energy (Figure 2–19).

If the system is at steady state and it begins as $P^{\cdot+}BPh^{\cdot-}Q_A$, then

$$\frac{d[PBPhQ_A]}{dt} = \left(\frac{k_{-2}k_{-1}}{k_2 + k_{-1}}\right)\left[P^{\cdot+}BPhQ_A^{\cdot-}\right] \tag{2–133}$$

Because the state $P^{\cdot+}BPhQ_A^{\cdot-}$ should be about 60 kJ mol$^{-1}$ lower in free energy than the state $P^{\cdot+}BPh^{\cdot-}Q_A$

$$-RT\ln\left(\frac{k_{-2}}{k_2}\right) = -RT\ln K_{eq} = 60 \text{ kJ mol}^{-1} \tag{2–134}$$

Because $k_2$ is $5 \times 10^9$ s$^{-1}$, $k_{-2}$ should be 0.2 s$^{-1}$. The rate constant for the decay of the quinone radical anion to produce the ground state by passing back through the excited state of the special pair, a consequence that would dissipate the absorbed energy entirely as fluorescence, should be $3 \times 10^{-3}$ s$^{-1}$ (Equation 2–133) and this would give a half-time of 250 s.

The observed rate constant,[202] however, for the reaction

$$P^{\cdot+}BPhQ_A^{\cdot-} \rightleftharpoons PBPhQ_A \tag{2–135}$$

is 10 s$^{-1}$ ($t_{1/2}$ = 70 ms). This value is significantly larger than the estimate based on the difference in free energy between the quinone and the bacteriopheophytin. For this and other reasons, it is presumed that there is a short circuit in the system that bypasses the forward pathway. The rate constant for return of the electron to the special pair through this short circuit is still too large to provide sufficient time for the energy to be trapped efficiently. To overcome this dilemma, a second strategy is employed.

Long after the electron has been transferred from the bacteriopheophytin to the quinone, the hole in the highest occupied $\pi$ molecular orbital of the special pair, left behind by the electron, is filled ($t_{1/2}$ = 200 ns)[201] by an electron from among those present in the four hemes of the cytochrome attached to the reaction center (Figure 2–17).[190a] Because this is a reaction that proceeds with a negative free energy change ($\Delta E°$ = 0.2 V; $\Delta G°$ = –19 kJ mol$^{-1}$), the concentration of the species $P^{\cdot+}BPhQ^{\cdot-}$ is decreased by the amount of $PBPhQ^{\cdot-}$ that is consequently formed

$$\frac{[PBPhQ_A^{\cdot-}]}{[P^{\cdot+}BPhQ_A^{\cdot-}]} = 4 \times 10^4 \tag{2–136}$$

Because so little $P^{\cdot+}BPhQA^{\cdot-}$ is present at any instant, the rate constant of Reaction 2–135 is decreased significantly.

The transfer of this other electron to the hole in the special pair makes the overall reaction catalyzed by a photosynthetic reaction center

$$\text{cytochrome}_{red} + h\nu + \text{quinone}_A \rightleftharpoons \text{cytochrome}_{ox} + \text{quinone}_A \text{ radical anion} \tag{2–137}$$

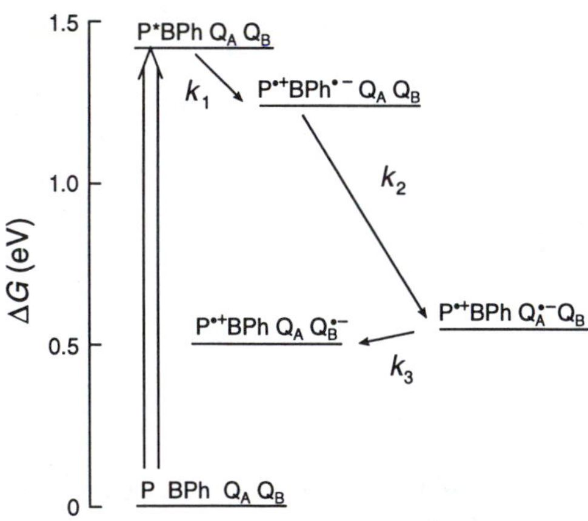


**Figure 2–19:** Energy diagram for the path of the electron in bacterial photosynthesis.[199] The four successive coenzymes on the path of the electron are the special pair (P), the bacteriopheophytin *b* (BPh), the tightly bound quinone ($Q_A$), and the loosely bound quinone ($Q_B$). The excited state of the special pair is indicated by P*, the radical cation, by $P^{\cdot+}$, and the successive radical anions, by $BPh^{\cdot-}$, $Q_A^{\cdot-}$, and $Q_A^{\cdot-}$. The energy levels of the various steps relative to the ground state are presented in electron volts (1 eV = 96.5 kJ mol$^{-1}$). The rate constants for transfer of the electron between the various steps are $k_1 = 2 \times 10^{11}$ s$^{-1}$ ($t_{1/2}$ = 3 ps); $k_2 = 5 \times 10^9$ s$^{-1}$ ($t_{1/2}$ = 140 ps); and $k_3 = 5 \times 10^{3}$s$^{-1}$ ($t_{1/2}$ = 100 $\mu$s). Before the electron has time to move from quinone A to quinone B, an electron from the cytochrome would normally fill the hole in the special pair ($k = 3.7 \times 10^6$ s$^{-1}$; $t_{1/2}$ = 200 ns). Adapted with permission from ref 199. Copyright 1987 Kluwer Academic Publishers.

The electron is moved from a donor of more positive reduction potential, the cytochrome, to an acceptor of more negative reduction potential, the quinone, and the energy of the photon is used to do this. At this point, the system is still at a higher energy than the ground state, and it is capable of returning, albeit slowly, to the ground state. This return is prevented and the energy is conserved because the electron on the quinone radical anion departs entirely from the reaction center before the energy can be dissipated.

In a reaction center under normal circumstances, in addition to the quinone that receives the electron directly from bacteriopheophytin (quinone A or $Q_A$), there is a second quinone,[203] usually a ubiquinone (quinone B or $Q_B$). This ubiquinone is loosely bound to the reaction center and it is assumed that under normal circumstances it is freely exchangeable with the pool of ubiquinone in the membrane, while quinone A is permanently bound to the reaction center. The site at which the loosely bound ubiquinone is located has been defined in the crystallographic molecular model of the reaction center from *Rb. sphaeroides*.[197,198] Its location is related by the 2-fold rotational axis of pseudosymmetry of the LM dimer to the site at which the tightly bound ubiquinone is found in this crystallographic molecular model, which is the same site at which the single menaquinone is found in the crystallographic molecular model of the reaction center from *Rps. viridis* displayed in Figure 2–17. The two quinones in a given reaction center are symmetrically displayed around the ferrous cation, which sits on the 2-fold rotational axis of pseudosymmetry. The electron is transferred from quinone A to quinone B at a rate[204] of $5 \times 10^3\ s^{-1}$. The role of the iron in this electron transfer is unclear. Although reaction centers depleted of iron display slower transfer rates, if the iron is replaced by other divalent cations the rates of transfer are essentially the same as those of a native reaction center.[205] Quinone B has a more positive reduction potential so that the equilibrium constant for the reaction

$$Q_A^{\cdot-}Q_B \rightleftharpoons Q_AQ_B^{\cdot-} \qquad (2\text{–}138)$$

is about 10. Under usual conditions of illumination the first electron ends up on quinone B as a radical anion, which remains tightly bound to the reaction center until the arrival of a second electron.

The second electron arrives, as did the first, from an excited state of the same special pair, which was regenerated upon the transfer of an electron from the cytochrome. After it was regenerated with the electron from the cytochrome, the special pair could absorb a second photon to generate a new excited state that passes its electron as before but to the quinone radical anion of quinone B rather than to the fully oxidized quinone B. If a solution of reaction centers is exposed to a saturating flash of light, the absorption spectrum of the ubiquinone radical anion in site B appears, which remains stable in amplitude over 5 s. When a second saturating flash is administered, the absorption of this ubiquinone radical anion disappears because the second electron has arrived at site B from the excited special pair and has produced the ubiquinolate dianion. If a third saturating flash is administered, however, the absorption of the ubiquinone radical anion at site B reappears.[206,207] This last observation means that the ubiquinolate dianion formed in the second flash has been replaced by a ubiquinone or the third electron could not have been received. Because the ubiquinone radical anion was bound at the instant of the second flash, the replacement must occur only after the ubiquinolate dianion is formed. Either the ubiquinolate dianion or one of its protonated forms must depart from the reaction center into the membrane and be replaced by a free ubiquinone from the membrane, or the electrons must depart the ubiquinolate dianion by transfer to a free ubiquinone in the membrane. It has been shown that the site-directed mutation of a glutamate in the reaction center decreases the rate at which the ubiquinolate dianion is protonated without affecting the rates of electron transfer, and this suggests that the dianion is protonated and then leaves the protein. In any case, the net result of the reaction is

$$\text{cytochrome}_{red} + 2h\nu + \text{free quinone} + 2H^+ \rightleftharpoons \\ 2\ \text{heme}_{ox} + \text{free quinol} \qquad (2\text{–}139)$$

The fact that ubiquinolate dianion is the immediate product of the two-electron transfer permits some of the energy in the photon to be conserved in an **electrochemical gradient of protons** across the membrane. The ubiquinolate dianion is formed at a site in the reaction center on the cytoplasmic side of the membrane (Figure 2–17). After the ubiquinolate dianion has been formed at site B on the reaction center, the lone pairs of electrons on the two quinolate oxygens, which were not previously basic ($pK_a < 0$), become extremely basic ($pK_a \geq 13$).[129] These lone pairs must pick up protons, and as is clear from the crystallographic molecular model,[194] the protons they pick up come from the cytoplasm of the bacterial cell.[208] In the normal cyclic process that occurs in bacterial photosynthesis, the ubiquinol produced at the cytoplasmic surface diffuses across the membrane and is oxidized at the extracytoplasmic surface by the ubiquinol–cytochrome reductase present in these membranes. This enzyme in turn passes the electrons back to the cytochrome associated with the extracytoplasmic surface of the reaction center. When the ubiquinol is oxidized by ubiquinol–cytochrome reductase, it releases two protons to the extracytoplasmic side of the membrane, opposite from the side on which they were absorbed during the neutralization of the ubiquinolate dianion (Figure 2–20). The net result is that two protons appear in the solution on the extracytoplasmic side of the membrane and two protons disappear from the cytoplasm for every two photons absorbed.[209]

Because the reaction center was formed during evolution from two identical subunits arrayed around a 2-fold rotational axis of symmetry and because each subunit contained two bacteriochlorophyll molecules, one molecule of bacteriopheophytin, and one molecule of quinone, each reaction center originally contained two identical pathways for electron transfer. When the gene duplicated and the two subunits began to evolve independently, one of the two pathways was shut down. It is believed that electron flow from the special pair to the tightly bound ubiquinone is confined to the path-

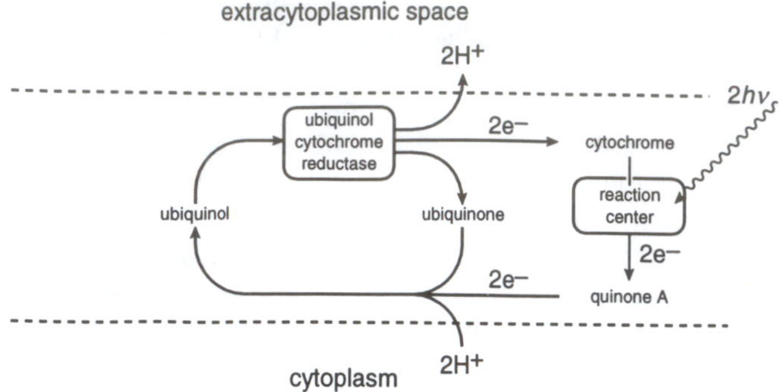


**Figure 2–20:** A Diagram of cyclic photosynthesis in the chromatophores of photosynthetic bacteria. In the process, two protons pass from the cytoplasm to the extracytoplasmic space driven by two photons.

way on the right side of the reaction center as presented in Figure 2–17.[210] The site for binding quinones in the left-hand branch (site B), however, is used as the site for binding the loosely bound quinone.

Each type of reaction center donates the excited electron to the particular electron acceptor that it is designed to use, and the hole created in the highest occupied $\pi$ molecular orbital of the special pair consumes an electron residing on the particular electron donor that it is designed to use. The reaction center of *Rb. sphaeroides* donates the excited electron to oxidized ubiquinone and the hole created in the special pair accepts an electron from one of the hemes in the attached cytochrome. The reaction center of photosystem II from the chloroplasts of spinach donates the excited electron to oxidized plastoquinone, and the hole created in the special pair is a strong enough oxidant to remove an electron from one of the four manganese ions[211] in an associated cluster of manganese cations.[212,213] This manganocluster in its oxidized state is in turn a strong enough oxidant to remove electrons from water to produce molecular oxygen.

The problem that has been solved by a photosynthetic reaction center is the disconnection of the excited electron from access to the hole it vacated in the molecular orbital of lower energy fast enough so that its return to the hole by the normal process of fluorescent decay can be prevented. To accomplish this disconnection, the electron is shuttled at an extremely rapid rate ($k = 5 \times 10^9\ s^{-1}$) to an electron carrier, quinone A, distant enough from the hole and at a low enough oxidation–reduction potential that the return of the electron to the hole becomes so slow it can be held long enough to be used as a reductant. This speed and this span can be accomplished because a continuous array of coenzymes connects the special pair with the loosely bound ubiquinone (Figure 2–17). This arrangement of the coenzymes in a reaction center is a solution to the problem of moving the net of one electron rapidly over a significant distance (about 4 nm). In this instance, the requirement that the electron travel over such a great distance is dictated by the fact that the greater its dilution, the slower its return to the ground state will be and the fact that the electron must be received by the hole at one side of the membrane and added to the departing quinolate dianion at the other side of the membrane to create a gradient of protons.

It is believed that, during its transfer from flavin to oxygen through the large, membrane-spanning electron transport proteins of the mitochondrion, an electron also must cover considerable distance. This intuition is partly based on the size of the complexes, the diameters of which are 5–8 nm,[134a] but it is also based on a consideration of the ability of these proteins to produce gradients of protons across the membranes in which they are situated. These proteins contain far fewer moles of the coenzymes of one-electron transfer for each gram of protein than does a reaction center. They are not, however, required to transfer electrons so rapidly because the movement is always exergonic. These two considerations have raised the possibility that electron transfer in these proteins occurs not along fixed arrays of coenzymes but through the amino acids and the polypeptide backbone of the protein itself.

During a one-electron oxidation–reduction reaction, an unaccompanied electron can transfer through unadulterated protein over a considerable distance (1.5–2.5 nm).[214,215] An experiment was designed to demonstrate this fact by following the transfer of an electron through the protein lying between two of the hemes in hemoglobin. By forming a species at one of the hemes with a negative reduction potential relative to the reduction potential of a species at another heme in the same molecule of hemoglobin, the rate of one-electron transfer from the heme of lower reduction potential to the heme of higher reduction potential could be followed. Hemoglobin is a protein that does not normally participate in oxidation–reduction reactions; and, therefore, it is a protein that has not evolved to be able to support the transfer of electrons through itself. The protein normally contains four hemes, one on each of its subunits. It is an $(\alpha\beta)_2$ heterotetramer, but $\alpha$ and $\beta$ polypeptides are homologous in sequence and the native structures of $\alpha$ and $\beta$ subunits are superposable. It is possible to separate $\alpha$ and $\beta$ subunits and reconstitute the protein from these separated subunits. Therefore, a hemoglobin in which the $\alpha$ subunits contained $Fe^{III}$·porphyrin and the $\beta$ subunits contained $Zn^{II}$·porphyrin could be constructed.

When a $Zn^{II}$·porphyrin is excited by absorbing a photon of light, the excited electron is particularly prone to a spin inversion that produces a triplet excited state, $^3Zn^{II}$·porphyrin, in good quantum yield. The lifetime of this phosphorescent triplet excited state is much longer (greater than 10 ms) than that of the fluorescent singlet excited state from which it was derived.[216] The triplet excited state is a strong enough reductant ($E^\circ_{app,pH7} = -1.1$ V)

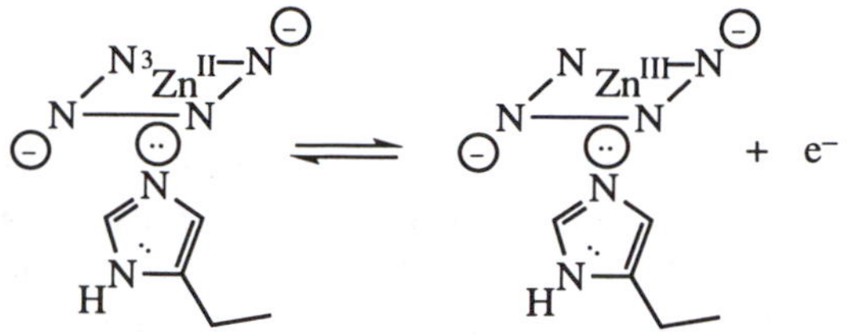

(2–140)

to transfer an electron to the iron of Fe$^{III}$·porphyrin ($E^\circ_{app,pH7}$ = +0.15 V)

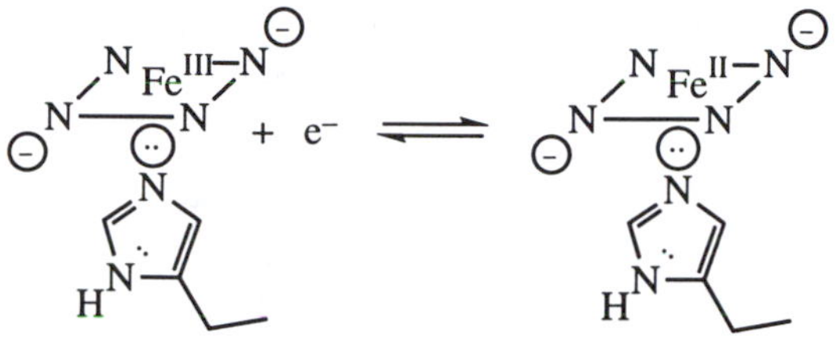

(2–141)

The free energy change of the overall reaction is quite favorable ($\Delta E^\circ_{app,pH7}$ = −1.25 V).

When hemoglobin containing a Zn$^{II}$·porphyrin in the $\beta$ subunits and an Fe$^{III}$·porphyrin in the $\alpha$ subunits is photolyzed, the decay of the triplet state of the Zn$^{II}$·porphyrin is more rapid than it is after the Fe$^{III}$·porphyrin is reduced to Fe$^{II}$·porphyrin prior to photolysis. If it is assumed that this increase in the rate of decay is due to the transfer of the electron from the $^3$Zn$^{II}$·porphyrin to the Fe$^{III}$·porphyrin, the first-order rate constant of that transfer is 100 s$^{-1}$ at 25 °C.[217] In the crystallographic molecular model of a molecule of hemoglobin, the shortest distance between a heme on a $\beta$ subunit and a heme on an $\alpha$ subunit is 2.5 nm.

A similar experiment has been performed with cytochrome *c*.[214] In this case, the transfer of an electron between a Ru$^{II}$ bound to Histidine 33[125a] and the Fe$^{III}$ in the heme of cytochrome *c* (a distance of 1.5 nm) had a rate constant of 20 s$^{-1}$ at 25 °C. In this case, it could have been argued that the electron transfer occurred because cytochrome *c* has evolved to participate in electron transfer; nevertheless, the rate constant was similar to that for electron transfer through hemoglobin, a protein that has not evolved to perform this role.

The rates at which electrons are transferred through proteins between these oxidants and reductants are far too slow to support photosynthetic electron transfer. The decay of the excited state, even when the electron can reach the bacteriopheophytin, is 10$^8$ s$^{-1}$. The rates of transfer of electrons through unadulterated protein, however, are rapid enough to account for the reactions that occur within the proteins of the electron transport chain of mitochondria. For example, the ubiquinol–cytochrome *c* reductase of mitochondria can be reconstituted with reaction centers from *Rb. sphaeroides*. In the presence of ubiquinone and cytochrome *c*, the reaction center, upon excitation with light, can add an electron to a heme *b* of the ubiquinol–cytochrome *c* reductase by producing ubiquinol and remove an electron from a heme $c_1$ of the ubiquinol–cytochrome *c* reductase by producing oxidized cytochrome *c*. The electron on the reduced heme *b* then transfers to the oxidized heme $c_1$ within the ubiquinol–cytochrome c reductase at a rate ($k_{obs}$ = 70 s$^{-1}$)[218] very similar to the rate at which an electron can be passed between the hemes in hemoglobin. The transfer of an electron from cytochrome *b* to cytochrome $c_1$ in ubiquinol–cytochrome *c* reductase, however, may not be a consequential reaction because current schemes for the transfer of electrons in this enzyme have cytochrome *b* and cytochrome $c_1$ on separate paths of electron transfer.[219]

## *Suggested Reading*


Siegel, L.M., Davis, P.S., & Kamin, H. (1974) Reduced Nicotinamide Adenine Dinucleotide Phosphate–Sulfite Reductase of Enterobacteria. The *Escherichia coli* Hemoflavoprotein: Catalytic Parameters and the Sequence of Electron Flow, *J. Biol. Chem. 249*, 1572–1586.

Siegel, L.M., Rueger, D.C., Barber, M.J., Krueger, R.J., Orme-Johnson, N.R., & Orme-Johnson, W.H. (1982) *Escherichia coli* Sulfite Reductase Hemoprotein Subunit: Prosthetic Groups, Catalytic Parameters, and Ligand Complexes, *J. Biol. Chem. 257*, 6343–6350.

Karlsson, B., Hovmöller, S., Weiss, H., & Leonard, K. (1983) Structural Studies of Cytochrome Reductase: Subunit Topography Determined by Electron Microscopy of Membrane Crystals of a Subcomplex, *J. Mol. Biol. 165*, 287–302.

Deisenhofer, J., Epp, O., Miki, K., Huber, R., & Michel, H. (1984) X-ray Structure Analysis of a Membrane Protein Complex: Electron Density Map at 3-Å Resolution and a Model of the Chromophores of the Photosynthetic Reaction Center from *Rhodopseudomonas viridis*, *J. Mol. Biol. 180*, 385–398.


### PROBLEM 2–16

The apparent reduction potential of cytochrome *c* is invariant with pH below pH 7.8 and decreases with a slope of 0.06 V (unit of pH)$^{-1}$ above pH 7.8. Explain these observations in terms of the p$K_a$ of an amino acid in oxidized cytochrome *c* and its p$K_a$ in reduced cytochrome *c*.

# Transfer of Energy

The photon responsible for the excitation of an electron in a photosynthetic reaction center is not absorbed directly from the ambient light by the special pair but is absorbed by an accessory **light-harvesting coenzyme**. Its energy is transferred to the special pair through a network of coenzymes. Transfers of energy occur within arrays of coenzymes, any one of which is able either to absorb an ambient photon or to act as a conductor of excitation between two of its neighbors. Once absorbed by one of these coenzymes, the energy of the photon is shuttled among them. Eventually the energy reaches the special pair and electron transfer takes over. Each step in the transfer of energy occurs by **radiationless transfer** that is identical to the radiationless transfer between fluorescent donor and acceptor.[220,220a]

The individual coenzymes catalyzing the transfer of energy are chlorophylls, carotenoids, and bilipigments such as phycocyanobilin. **Phycocyanobilins** are related to porphyrins and are covalently attached to the proteins in which they are found through thioethers to cysteine residues[221]

**2–46**

Running through a phycocyanobilin is an extended $\pi$ molecular orbital system that permits it to absorb visible light. Carotenoids also contain extended, but linear, $\pi$ molecular orbital systems. $\beta$-Carotene is a commonly encountered carotenoid

**2–47**

It is a hexaisoprenoid that contains a linear $\pi$ molecular orbital system of 22 carbon atoms. The molecule is *trans* and rigid throughout its length. Most, if not all, of the carotenoids in a chloroplast or chromatophore are tightly bound to the proteins. All carotenoids are related to $\beta$-carotene, and most differ from it only at the two ends, where the cyclohexene rings are substituted, further unsaturated, or opened. The coenzymes responsible for transferring energy are distributed in arrays enclosed within the proteins responsible for absorbing photons and transferring the energy of the resulting excited states.

An example of such a protein would be the bacteriochlorophyll *a* protein from *Prosthecochloris aestuarii*.[222,223] The protein as it is isolated is a trimer of identical subunits. Each subunit (Figure 2–21) resembles a cage in which are confined

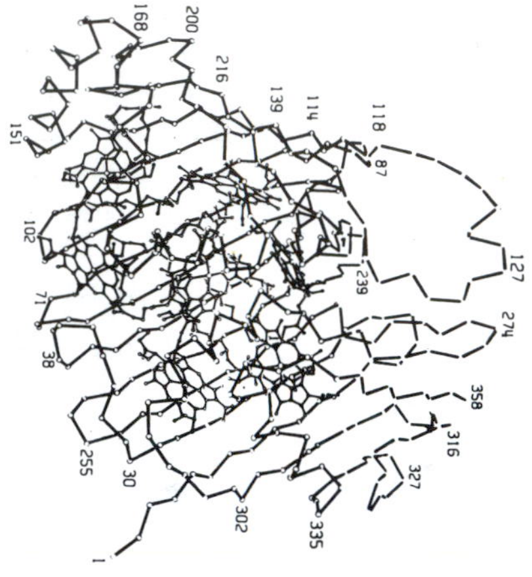

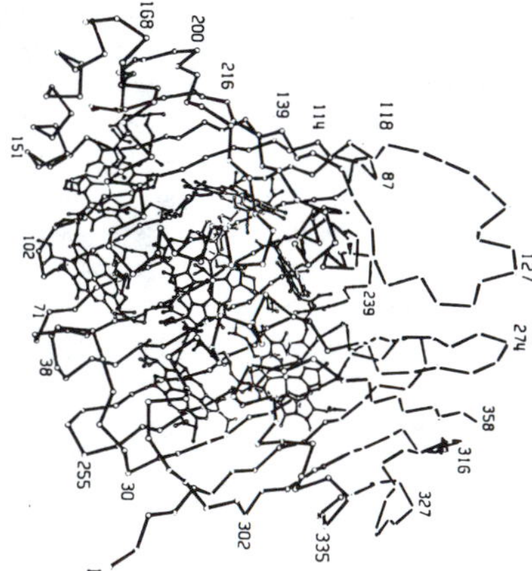

**Figure 2–21:** Drawing from the crystallographic molecular model of the bacteriochlorophyll *a* protein from the photosynthetic bacterium *P. aestuarii*.[222] In order to emphasize the bacteriochlorophylls *a*, each $\alpha$-carbon in the polypeptide has been located in space with a dot and the dots have been connected by line segments to indicate the polypeptide backbone. Only one of the three subunits of the trimer is presented. The seven molecules of bacteriochlorophyll *a* are clustered in the center of the protein. Each takes up a unique location and is fixed in that particular location by interactions with the protein. Reprinted with permission from ref 222. Copyright 1979 Academic Press.

seven molecules of bacteriochlorophyll *a* (olefin reduced between $R_3$ and $R_4$; $R_1 = R_3 = R_5 = R_8 = -CH_3$; $R_2 = -COCH_3$; $R_4 = -C_2H_5$; $R_7 = -CH_2CH_2COOC_{20}H_{39}$; **2–43**). The seven bacteriochlorophylls are bundled together in the very center of the subunit with the phytyl chains wound among the central rings.[222] The arrangement is such that the central rings of the bacteriochlorophylls end up evenly distributed over the volume enclosed by the shell of protein that forms the trimer (Figure 2–22). The distances between adjacent molecules of bacteriochlorophyll *a* are short enough that energy from an absorbed photon can be passed among them by radiationless transfer,[222] a process facilitated by evenly spaced donors and acceptors in proper orientation.

Another protein that performs the role of harvesting photons and transferring their energy is the C-phycocyanin from *Agmenellum quadruplicatum*. The coenzyme in this protein is phycocyanobilin (**2–46**). This protein is an $(\alpha\beta)_6$ hexamer but the $\alpha$ polypeptides are homologous to the $\beta$ polypeptides in sequence and the $\alpha$ subunits in the crystallographic molecular model are superposable on the $\beta$ subunits. Each $\alpha$ subunit has one covalently attached phycocyanobilin, and each $\beta$ subunit has two covalently attached phycocyanobilins, one in a location homologous to that in an $\alpha$ subunit. In the hexamer, the 18 coenzymes are distributed evenly over the volume of the protein (Figure 2–23).[224] The coenzymes are close enough together for efficient radiationless transfer to occur. Because the rate and hence the efficiency of radiationless energy transfer depends inversely on $r^6$, where $r$ is the distance between the phycocyanobilins,[220a] the even spacing is essential to maximize conduction among the coenzymes. It has been estimated that $R_0$, the distance at which the efficiency of energy transfer between two phycocyanobilins in a C-phycocyanin should be 50%, is 5.0 nm,[225] and many of the pairs of phycocyanobilins are less than 3.0 nm apart.[224]

The small individual proteins of light harvesting and energy transfer are assembled in the organism into larger oligomers, each focused on a reaction center. For example, the intact oligomeric unit containing both the light-harvesting proteins and the reaction center can be isolated from chromatophores of *Rps. viridis*.[226] The complete structure is a large hexameric wheel with a 6-fold rotational axis of symmetry. The tire of the wheel is constructed from the individual light-harvesting proteins, and the hub of the wheel is a reaction center.[227] The hexameric light-harvesting oligomer funnels the energy to the reaction center. In contrast to this arrangement, the phycocyanins are gathered into rods that are formed from $(\alpha\beta)_6$ hexameric rings stacked one on top of another.[224] These rods are then attached at one of their ends to other energy-transferring complexes that are coupled to reaction centers.[221] Energy is passed down the rods in the direction of the reaction centers, where it is absorbed and excites an electron in the special pair. The alignment of the coenzymes in the rod can be deduced from the stacking of the subunits of the C-phycocyanin in the crystallographic molecular model (Figure 2–23).[224]

**Figure 2–22:** Drawing of the arrangement of the 21 bacteriochlorophylls *a* in the crystallographic molecular model of a trimer of the bacteriochlorophyll *a* protein from *P. aestuarii*.[222] Only the skeletons of the 21 central rings of the bacteriochlorophylls *a* are arranged in space as they are in the crystallographic molecular model. The numbers refer to the seven bacteriochlorophylls *a* in each of the three subunits. The precise crystallographic 3-fold rotational axis of symmetry relating the three subunits is indicated by a dashed horizontal line. Reprinted with permission from ref 222. Copyright 1979 Academic Press.

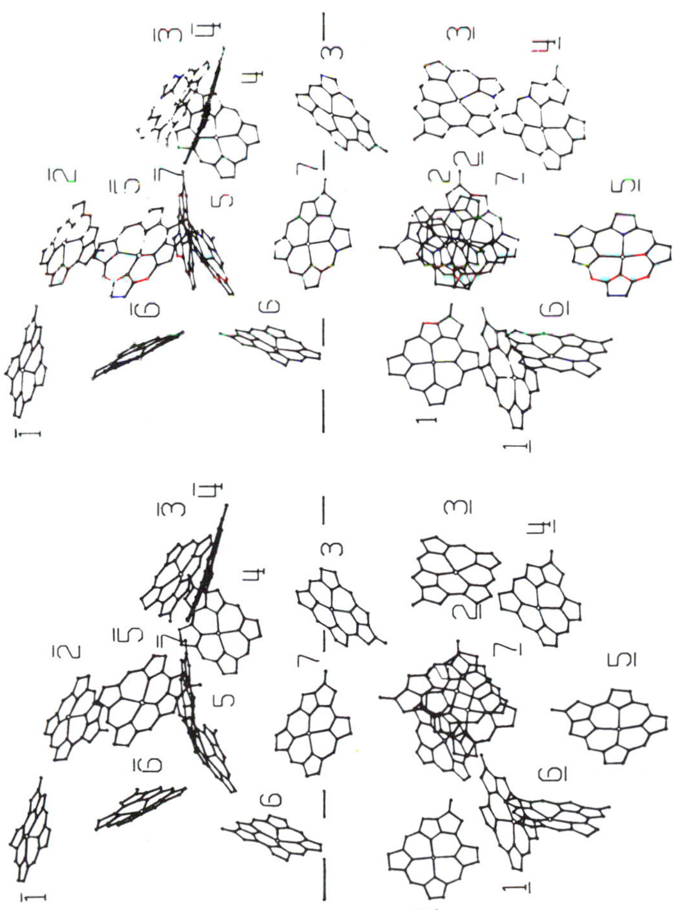

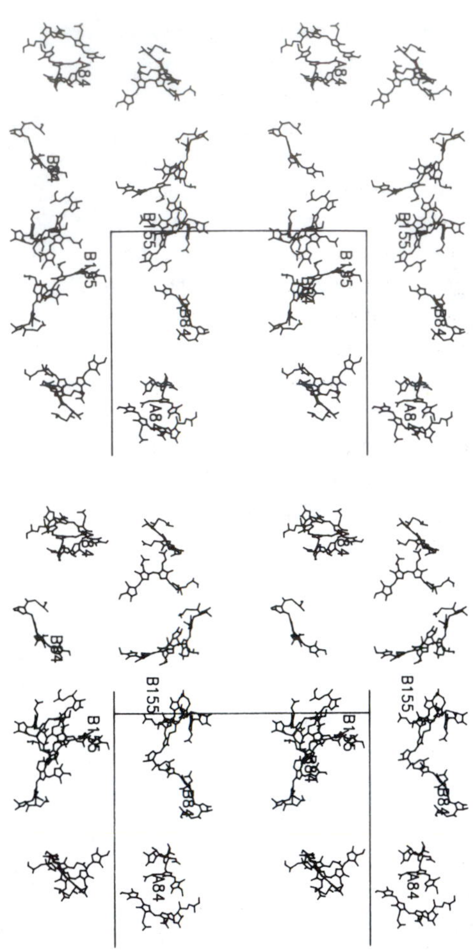


**Figure 2–23:** Drawing of the distribution of the phycocyanobilins within the crystallographic molecular model of C-phycocyanin from *A. quadruplicatum*.[224] In the bacterium, hexamers stack on top of each other to form cylinders and this arrangement is preserved in the crystal, which is formed from cylindrical columns of hexamers. In the figure, a segment, two hexamers in height, out of one of these columns is presented. A crystallographic 3-fold axis of symmetry runs through the center of each column and coincides with the vertical line in the center of the figure. The arrangement of coenzymes in the crystal is repeated indefinitely above and below the figure. Reprinted with permission from ref 224. Copyright 1986 Academic Press.

### *Suggested Reading*

Shirmer, T., Huber, R., Schneider, M., Rode, W., Miller, M., & Hackert, M.L. (1986) Crystal Structure Analysis and Refinement at 2.5 Å of Hexameric C-Phycocyanin from the Cyanobacterium *Agmenellum quadruplicatum:* The Molecular Model and Its Implications for Light Harvesting, *J. Mol. Biol. 188,* 651–676.

# Coenzymes of the Metabolism of Molecular Oxygen

**Molecular oxygen** is diatomic and has 12 valence electrons. The molecular orbital system for molecular oxygen is the same as the molecular orbital systems of the small diatomic molecules $N_2$, $F_2$, CO, and NO and the ion $CN^-$. Depending on the formalism preferred by the narrator, the electronic structure of molecular oxygen has been variously depicted.[228–230]

The various views seem to agree on several salient points (Figure 2–24).[231] Along the axis between the two oxygen nuclei, there is a $\sigma$ bond connecting the two oxygen atoms and occupied by the two valence electrons of lowest energy. Three other $\sigma$ molecular orbitals, of which the two of lowest energy are occupied by two pairs of electrons, are also cylindrically symmetric with this axis. The three filled orbitals comprise the $\sigma$ structure of the molecule. This $\sigma$ structure is formed by mixing the two 2*s* orbitals and the two $2p_z$ orbitals of the oxygen atoms and is occupied by six electrons. Parallel to the axis of the molecule and perpendicular to each other are two $\pi$ bonds, each containing two electrons. They are formed from the two $2p_x$ and the two $2p_y$ atomic orbitals, respectively, that overlap in phase to form two orthogonal, bonding $\pi$ molecular orbitals. The remaining two valence electrons are then found in the two separate antibonding $\pi$ molecular orbitals, formed from overlap out of phase of the two $2p_x$ and two $2p_y$ atomic orbitals, respectively

σ

π* ↑    ↑ π*

π ↑↓    ↑↓    ↑↓ π

↑↓

↑↓

**2–48**

This arrangement produces a symmetric distribution of electrons, each of which is fully delocalized over both atoms.

**Figure 2–24:** One of several ways to depict the molecular orbitals of the $O_2$ molecule.[231] The four $\sigma$ molecular orbitals in the center of the diagram are formed from the overlap of two *sp* atomic orbitals from each oxygen atom. The two pairs of $\pi$ molecular orbitals, which are orthogonal and degenerate to each other, are formed from *p* orbitals of the two oxygen atoms. Each energy noted is the ionization potential (in electron volts) for removing an electron from each of these orbitals in the ground state. The orbitals in the ground state have electronic configuration **2–48**. Reprinted with permission from ref 231. Copyright 1979 Annual Reviews, Inc.

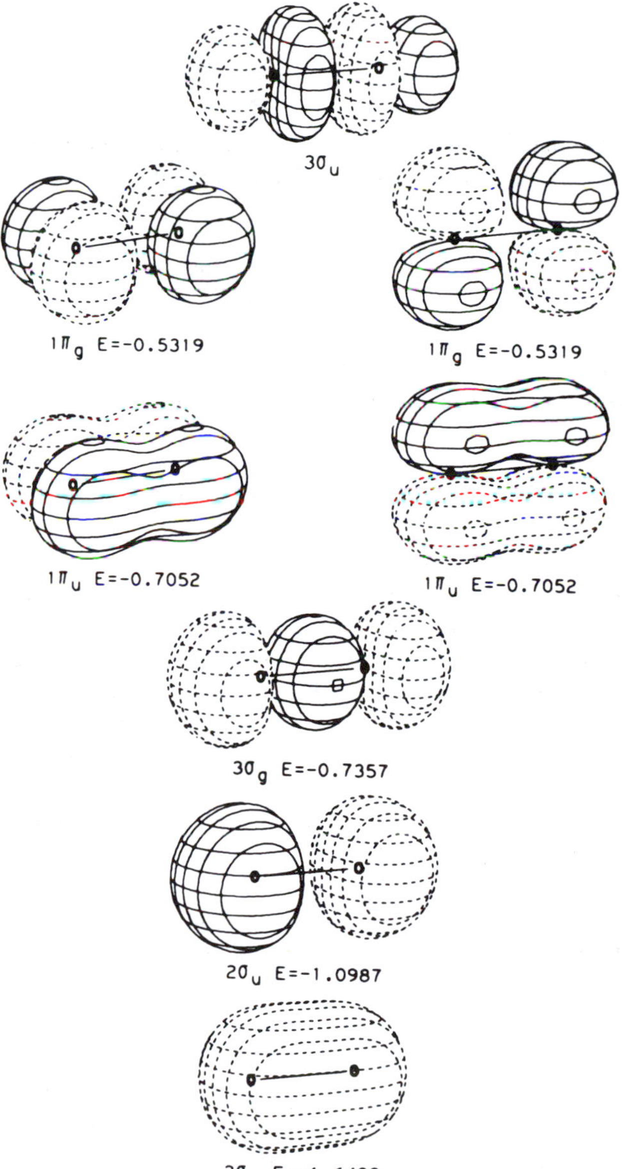


The two orthogonal, antibonding $\pi$ molecular orbitals, $\pi_x{}^*$ and $\pi_y{}^*$, are degenerate; and, by Hund's rule, each is occupied by an unpaired electron. This causes molecular oxygen in its ground state to be a diradical and paramagnetic. The two unpaired electrons have spins of the same sign in the ground state or triplet state. Upon excitation in the proper fashion, one of the two electrons can have its spin inverted to create a singlet state in which the two unpaired electrons are of opposite spin. The triplet ground state of molecular oxygen engages in radical chemistry and single electron transfer. The singlet excited state of molecular oxygen is a strong electrophile that can engage in heterolytic and electrocyclic chemistry, behavior demonstrating that the electrons in the singlet state are all paired.

The two $\pi$ molecular orbitals, each occupied by an unpaired electron, are antibonding but only half-filled. When one electron is added to one of these two antibonding orbitals, the **superoxide radical anion** is produced

$$O_2 + e^- \rightleftharpoons O_2^{\cdot-} \qquad (2\text{–}142)$$

This reaction, even though it fills an orbital, is not a favorable one [$E° = -0.16$ V (based on a standard state of 1 M dissolved $O_2$ and 1 M superoxide anion)][229] because the electron is added to an antibonding orbital. Unlike the other intermediates in the reduction of molecular oxygen, the superoxide anion is a mild reducing agent. The addition of a second electron to the remaining unfilled, antibonding $\pi$ molecular orbital renders the dioxygen diamagnetic because all electrons are now paired. It also produces an electron-rich molecule that is highly basic and picks up two protons at pH 7 to form two oxygen–hydrogen $\sigma$ bonds, one to each oxygen. This reduction and protonation forms **hydrogen peroxide**[232]

$$O_2^{\cdot-} + 2H^+ + e^- \rightleftharpoons H_2O_2 \qquad E^\circ_{\text{app,pH7}} = +0.94\ \text{V}^{211} \qquad (2\text{–}143)$$

The acid dissociation constants of the conjugate acid of superoxide anion and of hydrogen peroxide define the character of the respective lone pairs of electrons in these two molecules. Molecular oxygen is not basic because the two pairs of nonbonding $\sigma$ electrons it contains ($2\sigma_u$ and $3\sigma_g$ in Figure 2–24) are, at best, located in two *sp* atomic orbitals and, at worst, delocalized into molecular orbitals cylindrically symmetric to the axis of the molecule, and the two pairs of electrons in the two $\pi$ bonds it contains are, as always, poor bases. After an electron is added to produce the superoxide anion, the molecule becomes considerably more basic[233]

$$HO_2^{\cdot} \rightleftharpoons H^+ + O_2^{\cdot-} \qquad pK_a = 4.9 \qquad (2\text{–}144)$$

This suggests that the $\sigma$ electrons in superoxide anion are not all delocalized as they are in molecular oxygen. This acid dissociation constant is high enough to suggest that there is considerable *p* character in the $\sigma$ lone pair of electrons accepting the proton. This conclusion is reinforced by the observation that the p$K_a$ of dihydrogen superoxide cation[232]

$$H_2O_2^{\cdot +} \rightleftharpoons H^+ + HO_2^{\cdot} \qquad (2\text{–}145)$$

is 1.2, a value considerably higher than the p$K_a$ for the oxygen of an aldehyde

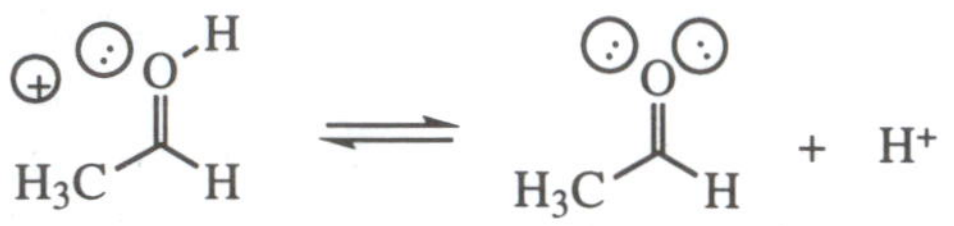

(2–146)

From these values for the two acid dissociation constants of superoxide, it is reasonable to assume that the two oxygens in the molecule are effective acceptors of hydrogen bonds. When it is dissolved in water, the superoxide anion should be hybridized so that in one plane each oxygen has two $sp^2$ lone pairs of electrons accepting hydrogen bonds; and, in the orthogonal plane, a $\pi$ molecular orbital system occupied by three electrons

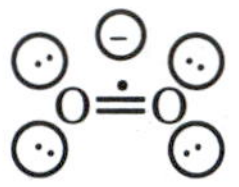

**2–49**

This electronic structure, at each oxygen, resembles the electronic structure at the oxygen of acetaldehyde (Equation 2–146), and this arrangement is promoted by the weakening of the $\pi$ system to the benefit of the $\sigma$ system owing to the addition of the extra electron to an antibonding $\pi$ molecular orbital on molecular oxygen. The breakdown of the $\pi$ system is also reflected in the lengthening of the oxygen–oxygen bond from 0.121 nm in molecular oxygen to 0.133 nm in superoxide anion.

After a second electron has been added to molecular oxygen, there should be as many antibonding $\pi$ electrons (four) as bonding $\pi$ electrons (four) were the molecular orbital structure of molecular oxygen (**2–48**) to be retained. For this reason, the $\pi$ molecular orbitals become even weaker; and, in a hydrogen-bonding solvent like water, they decompose into localized lone pairs of electrons on each oxygen. This is reflected in the acid dissociation constant of hydrogen peroxide

$$H_2O_2 \rightleftharpoons H^+ + HO_2^- \qquad pK_a = 11.7 \qquad (2\text{–}147)$$

which is that expected of a simple hydroxyl with a strong electron-withdrawing group attached to it [p$K_a$($CF_3CH_2OH$) = 12.5]. That only a single $\sigma$ bond remains is also reflected in the long bond length (0.148 nm) of hydrogen peroxide. The dihedral angle between the two hydrogens (92°) in hydrogen peroxide can also be explained by the consideration that, were the value of this angle 90°, each of the two $sp^3$ lone pairs on one oxygen would be as far as possible from being parallel to either of the two $sp^3$ lone pairs on the other oxygen.

There are two more one-electron steps in the reduction of oxygen to two molecules of water. When one electron is added to hydrogen peroxide, the $\sigma$ bond is cleaved and hydroxide anion and **hydroxyl radical** are produced[234]

$$H_2O_2 + H^+ + e^- \rightleftharpoons HO^{\cdot} + H_2O \qquad E^{\circ}_{app,pH7} = +1.35 \text{ V} \qquad (2\text{–}148)$$

The proton is picked up because the hydroxide anion produced upon heterolysis is a strong base. Hydroxyl radical is a strong oxidant that can be reduced in a one-electron reaction to hydroxide anion.[234]

$$HO^{\cdot} + e^- + H^+ \rightleftharpoons H_2O \qquad E^{\circ}_{app,pH7} = +2.33 \text{ V} \qquad (2\text{–}149)$$

In biochemical situations, these two steps (Reactions 2–148 and 2–149) are usually forced to be concerted[234]

$$H_2O_2 + 2H^+ + 2e^- \rightleftharpoons 2H_2O \qquad E^{\circ}_{app,pH7} = +1.35 \text{ V} \qquad (2\text{–}150)$$

to avoid the production of hydroxyl radical, which is able to abstract a hydrogen atom from almost anything

$$H\text{-}OH \rightleftharpoons H^{\cdot} + {}^{\cdot}OH \qquad \Delta H^{\circ} = +500 \text{ kJ mol}^{-1} \qquad (2\text{–}151)$$

There are several chemical features of molecular oxygen that must be taken into account by the coenzymes that react with it. Its $\sigma$ lone pairs of electrons are neither basic nor nucleophilic. It is a diradical with two unfilled $\pi$ molecular orbitals. These molecular orbitals are antibonding and occupation of them rearranges the electronic structure of the molecule.

**Hemes** are one class of coenzymes that react with either molecular oxygen or hydrogen peroxide. Unlike in the cytochromes, in hemes that bind oxygen or hydrogen peroxide, the sixth ligand site on the iron is not occupied by a basic amino acid from the protein. This sixth site is the location at which a $\sigma$ lone pair from $O_2$ (**2–50**), $H_2O_2$ (**2–51**), or $H_2O$ (**2–52**) is found, depending on the circumstances.

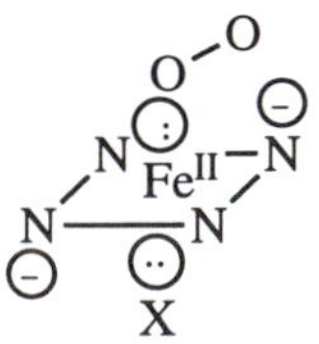

**2–50** 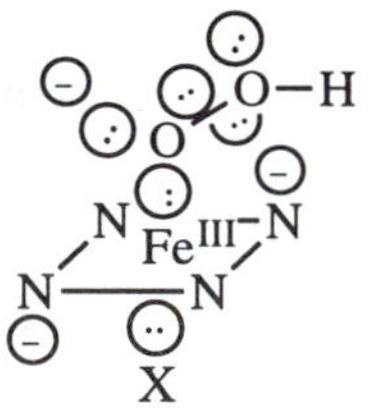 **2–51** 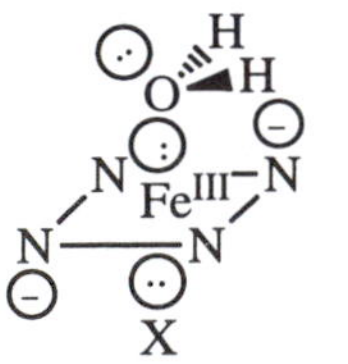 **2–52**

The hemoproteins that disproportionate hydrogen peroxide are referred to as **catalases**. In these hemoproteins the fifth ligand to the iron in the heme is a tyrosine (Tyrosine 357 in the beef liver enzyme).[235] The reaction catalyzed by a catalase is

$$2H_2O_2 \rightleftharpoons 2H_2O + O_2 \qquad (2\text{–}152)$$

The first step in the reaction of catalase is the replacement of a water on the ferric heme by a hydrogen peroxide.

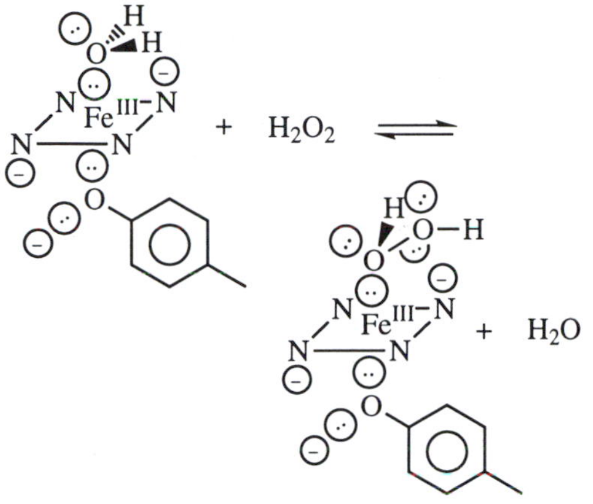

(2–153)

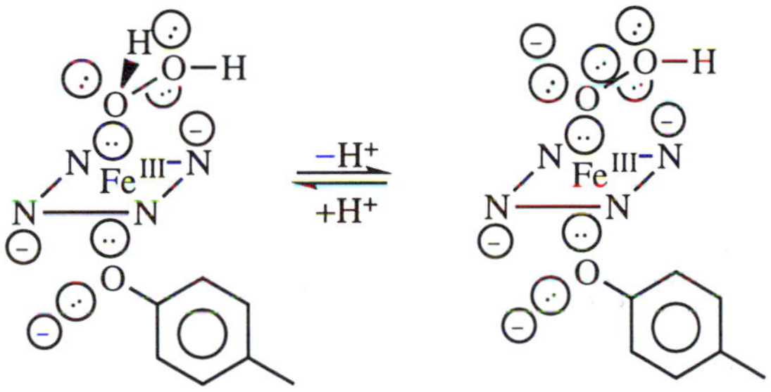

2–53

(2–154)

The bond formed between oxygen and iron is between a $\sigma$ lone pair of electrons from the conjugate base of hydrogen peroxide and the empty, axial $d_{z^2}$ atomic orbital on the iron.[228]

The bound hydrogen peroxide is then split heterolytically to produce an oxene intermediate[236] referred to as **compound I (2–54)**[*]

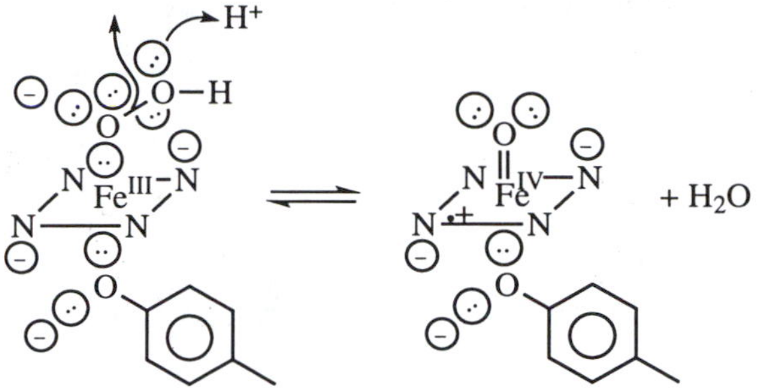

2–54

(2–155)

This reaction can be studied by using a simple model compound resembling the arrangement of the substituents in a hemoprotein[237]

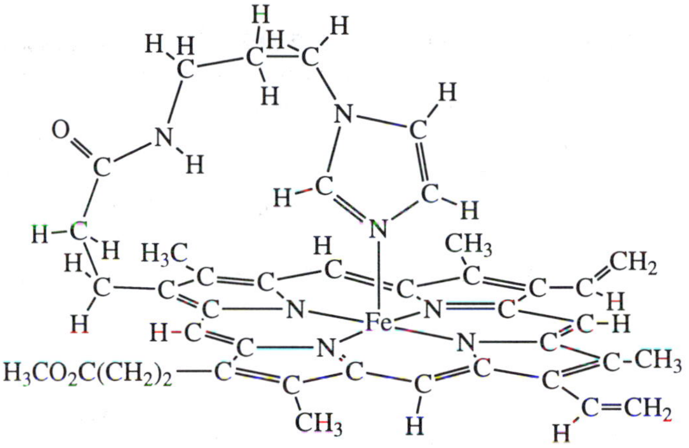

2–55

The formation of compound I in this model compound is catalyzed by the addition of a proton to one of the lone pairs of the leaving group (Reaction 2–155).[238] At the proper location relative to the heme in the crystallographic molecular model of catalase, there is a histidine, Histidine 57, the conjugate acid of which has been proposed to be the acid responsible for this protonation.[235]

Upon departure of the water to form compound I, two electrons are required to complete the octet around the oxygen. One is contributed by the iron and one by the $\pi$ molecular orbital system of the porphyrin itself. Formally, compound I is an oxide dianion complexed with an iron(IV)–heme radical cation. Compound I has been produced as a stable species in heme model compound (Figure 2–25).[236] Its properties are those expected of a compound containing **iron(IV)** and a **porphyrin cation radical** formed from removal of one electron from the highest occupied $\pi$ molecular orbital of the porphyrin.

In the next two steps of the reaction catalyzed by catalase, the second molecule of hydrogen peroxide is turned into oxygen. It has been demonstrated that compound I, formed

[*] The oxene is written as if there were a covalent double bond between oxygen and iron, and the formal negative charge of 2 that the oxide dianion would have if it were only a complexed base has been disregarded in this particular structure. The double bond is covalent in this instance because the iron(IV)–heme radical cation is so electron-deficient that the four electrons are probably shared equally between iron and oxygen.

in heme model compound **2–55**, can readily abstract a hydrogen atom from hydrogen peroxide[239] to produce hydrogen superoxide

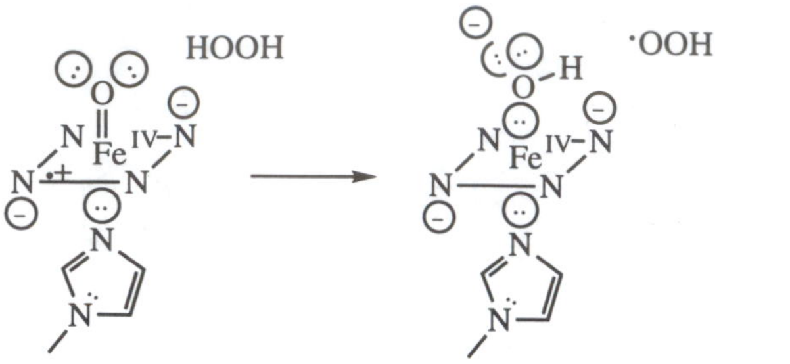

(2–156)

The bimolecular rate constant of this reaction at 25 °C is $10^7$ $M^{-1}$ $s^{-1}$ and this suggests that no catalysis is required at this step.

The product of this reaction (**2–56**) is formally equivalent to **compound II**, an intermediate formed when the compound I of catalase is reduced by a one-electron donor rather than a two-electron donor such as hydrogen peroxide. Compound II is formally equivalent to either a hydroxyl radical bound to $Fe^{III}$ or a hydroxide anion on $Fe^{IV}$. The last step in the enzymatic reaction of catalase could be electron transfer from hydrogen superoxide coupled to a proton rearrangement

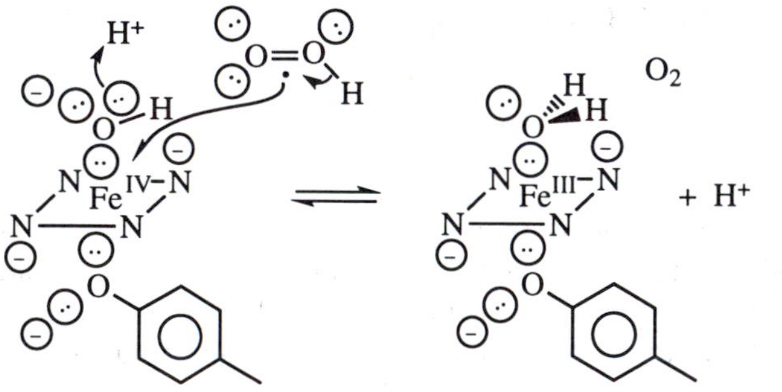

(2–157)

In this reaction compound II is acting as hydroxide anion bound to $Fe^{IV}$. There is only one acid–base, Histidine 74, in the vicinity of the heme in beef liver catalase. It is thought that it could remove the proton from the hydrogen superoxide, but it is too far away to add the proton to the hydroxide on the iron.[235] Therefore, an alternative possibility for the last step is the abstraction of a second hydrogen atom by compound II from the hydrogen superoxide that has turned around within the active site of the enzyme

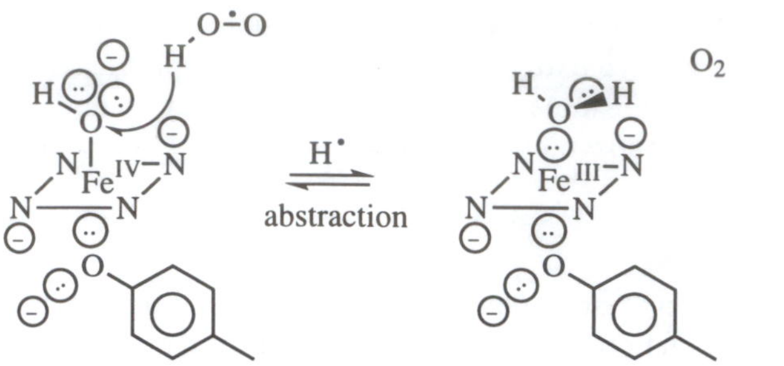

(2–158)

Such cage disproportionations are observed frequently in reactions of free radicals. In this reaction the oxygen on the heme would be functioning as a hydroxyl radical on $Fe^{III}$.

Compound I, when generated in heme model compound **2–55**, is capable of two reactions in addition to hydrogen abstraction from peroxides. It can epoxidize olefins,[236] as if it were a peroxyacid

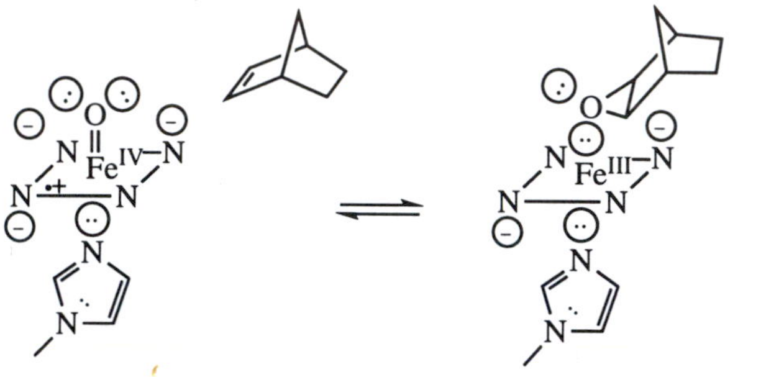

(2–159)

In this reaction the oxygen incorporated into the olefin is the oxygen of the oxene on iron. Compound I is also able to participate in **one-electron transfer**[239]

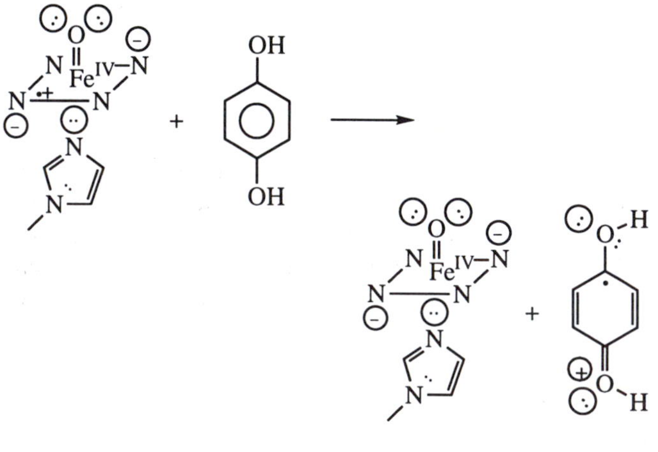

(2–160)

In the enzyme cytochrome *c* peroxidase from yeast, an intermediate analogous to compound I is formed from an organic peroxide (ROOH) in steps analogous to Reactions 2–153, 2–154, and 2–155 that have the alcohol (ROH) as their product. In this enzyme, the fifth ligand to the iron is an imidazole from a histidine residue in the protein (Histidine 174 in

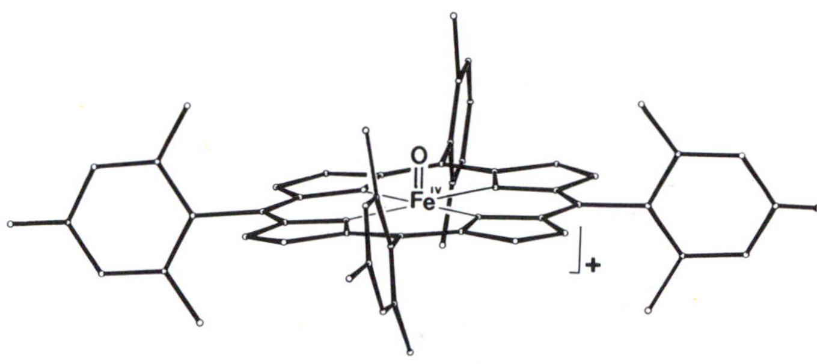

**Figure 2–25:** Structure of a synthetic model[236] for compound I of catalases and peroxidases. Iron(III) cation was inserted into 5,10,15,20-tetramesitylporphin to produce chloro-5,10,15,20-tetramesitylporphinatoiron(III). This compound was oxidized with 1.5 equiv of *m*-chloroperoxybenzoic acid in methylene chloride/methanol to produce the green model compound. Nuclear magnetic resonance, Mössbauer, visible, and electron paramagnetic resonance spectra of this compound indicate that it had the structure shown. Reprinted with permission from ref 236. Copyright 1981 American Chemical Society.

the enzyme from bakers' yeast).[240] Another feature of the oxene intermediate in cytochrome *c* peroxidase is that the cation radical is thought not to be in the porphyrin but somewhere in the protein ($P^{\cdot+}$) that surrounds the heme.[241] The compound I formed in cytochrome *c* peroxidase is reduced by two consecutive one-electron transfers from reduced cytochrome *c*

$$P^{\cdot+} \quad + 2e^- + 2H^+ \longrightarrow \quad P: \qquad (2\text{–}161)$$

**Peroxidases** are responsible for eliminating organic hydroperoxides from cells

$$ROOH + 2H^+ + 2e^- \rightarrow ROH + H_2O \qquad (2\text{–}162)$$

As with cytochrome *c* peroxidase, in the first half of the reaction, the production of the alcohol, the steps are identical to those for catalase (Reactions 2–153, 2–154, and 2–155). The problem faced by peroxidases is that the organic peroxides that are their reactants, unlike the hydrogen peroxide that is the reactant with a catalase, cannot be used to reduce compound I after it is formed. This is due to the fact that an organic peroxide has an alkyl group that cannot be removed as a proton or a hydrogen atom (Reaction 2–157). This problem is solved in one of several ways. Compound I can be reduced by successive one-electron transfers as in cytochrome *c* peroxidase. It can also be reduced by a two-electron donor such as ethanol, as in the peroxidatic reaction catalyzed by catalase.[235,242] In this case, compound I abstracts the hydrogen from the alcoholic carbon, and the compound II that is formed (Reaction 2–156) accepts a second electron by electron transfer from the resulting alkoxyl radical, while a proton is removed from the alcoholic carbon (Reaction 2–157).

Compound I of a peroxidase can also accept electrons one at a time from any one of a large array of small one-electron donors capable of either arriving or departing from the heme as stable radical species. In peroxidases that reduce compound I in this way, it is thought that the reductants react with the edge of the heme at a particular meso position.[243] A **meso position** on a heme is any one of the four carbons linking two pyrroles of the porphyrin. The conclusion that electrons enter the heme at the edge at a meso position is drawn from experiments with reductants, such as phenylhydrazine, that can yield reactive radical species during their oxidation by compound I or compound II. In the case of phenylhydrazine, one product of two successive one-electron oxidations is phenyldiazine

$$PhNHNH_2 \rightarrow 2e^- + 2H^+ + PhN{=}NH \qquad (2\text{–}163)$$

and the product of the one-electron oxidation of phenyldiazine is phenyl radical

$$PhN{=}NH \rightarrow e^- + H^+ + N_2 + Ph^{\cdot} \qquad (2\text{–}164)$$

If this later reaction (Reaction 2–164) occurs in the active site of the enzyme during the reduction of compound I, the phenyl radical can insert into the heme at a location immediately adjacent to where it was produced. The only observed product of this type is a heme phenylated at one of its meso positions, and this result suggests that the reduction of compound I by naturally occurring reductants proceeds by electron transfer while the reductant is adjacent to this one particular meso carbon.

Hemes also form reversible complexes with molecular oxygen itself. The hemoproteins that can form such complexes are the **globins**, such as hemoglobin and myoglobin. In these proteins the fifth ligand to the iron is the imidazole of a histidine in the protein and the sixth ligand position is the one occupied by the molecular oxygen. Iron(II) forms the complex with molecular oxygen. Because iron(II) is less acidic, it forms complexes with bases such as water or hydrogen peroxide very poorly. Consequently, when the heme is in aqueous solution in the absence of molecular oxygen, carbon monoxide, or similar ligands, the sixth site is vacant or fitfully occupied.

A

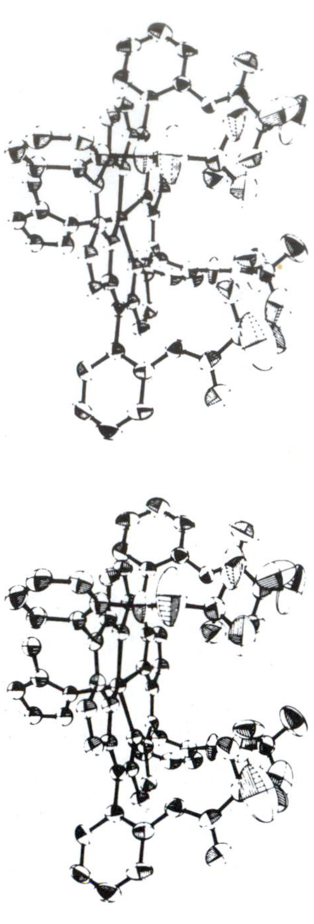

B

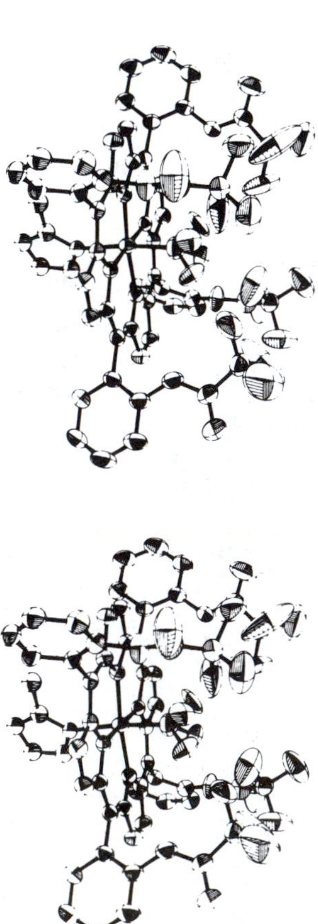

**Figure 2–26:** Drawings of the crystallographic molecular models of a model compound for the heme in an oxygen-binding hemoprotein in its unoccupied form (A) and its oxygenated form (B).[245] The porphyrin in this model compound is a tetra(*o*-aminophenyl)porphyrin to which four molecules of pivalic acid [$(CH_3)_3CCOOH$] have been attached in amide linkage. These four pivalamides form a dome over one side of the iron. The fifth ligand position on the iron is occupied by a 2-methylimidazole to give the heme properties similar to the heme in hemoglobin or myoglobin. In the unoccupied heme (A), the iron is pulled out of the plane of the porphyrin by the 2-methylimidazole. When oxygen binds (B), the iron returns to the plane of the heme. Because the entire molecule has a vertical 4-fold rotational axis of symmetry passing through the iron and because the molecular oxygen is tilted with respect to this axis, there are four equivalent orientations for the oxygen. Because the oxygen rapidly rotates to occupy successively each of these four locations, in the map of electron density each of these four orientations is occupied by a fourth of an oxygen molecule. This produces the strange branched structure seen on the iron in the drawing. Each of the four branches represents a fourth of an oxygen atom. Reprinted with permission from ref 245. Copyright 1980 American Chemical Society.

There are several facts that are known about the complex between oxygen and an $Fe^{II}$·porphyrin. It is **diamagnetic.**[244] This fact states that all electrons in the complex are paired. The **Fe–O–O angle** in the oxygen complex of one heme model compound (Figure 2–26)[245] is 129°. The Fe–O–O angles in myoglobin (120°)[246] and hemoglobin (160°)[247] are known with far less accuracy but nevertheless are similar, if not identical, to the angle in this model compound. The iron in the unliganded heme is **displaced from the plane** of the porphyrin toward the fifth ligand and moves into the plane of the porphyrin when the oxygen binds (Figure 2–26). The **oxygen–oxygen bond length** in the complex (0.122 nm)[245] is indistinguishable from that of molecular oxygen itself (0.121 nm) but is much shorter than that of superoxide anion (0.133 nm)[229] or hydrogen peroxide (0.148 nm). This fact states that most if not all of the $\pi$ bonding remains in the oxygen–oxygen bond.

Iron(II) has an even number of electrons (24); but, like oxygen, it is paramagnetic. This results from the fact that it has six electrons to distribute over five *d* orbitals of almost equal energy (Figure 2–27)[231] and four electrons end up unpaired. This produces a quintet state or high-spin state.[228] Upon the entry of the ferrous iron into the heme, the quintet state becomes less stable but is still more stable than the closest diamagnetic state in which all of the electrons would be

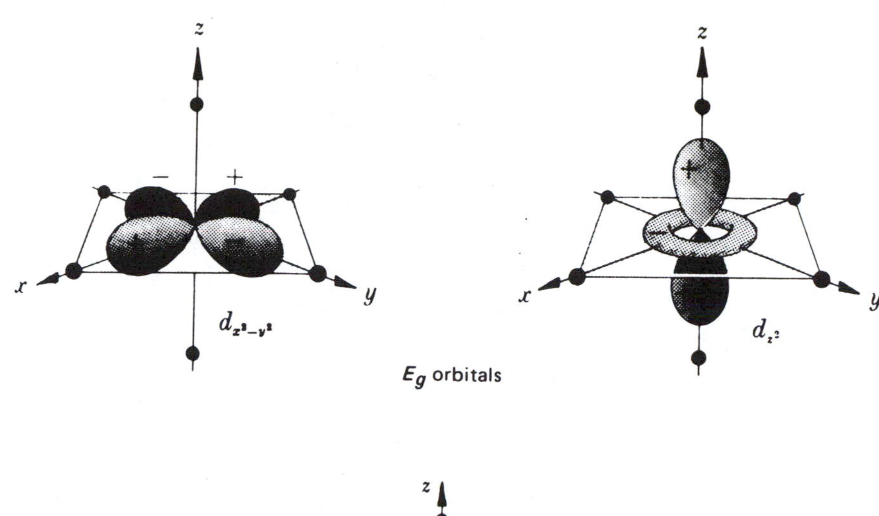


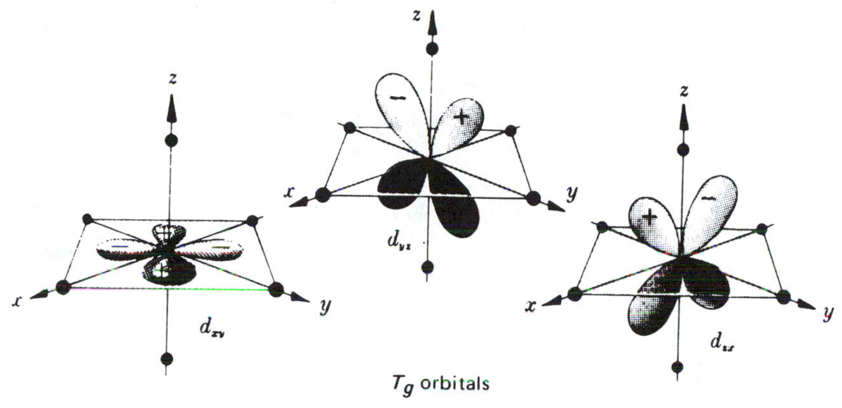


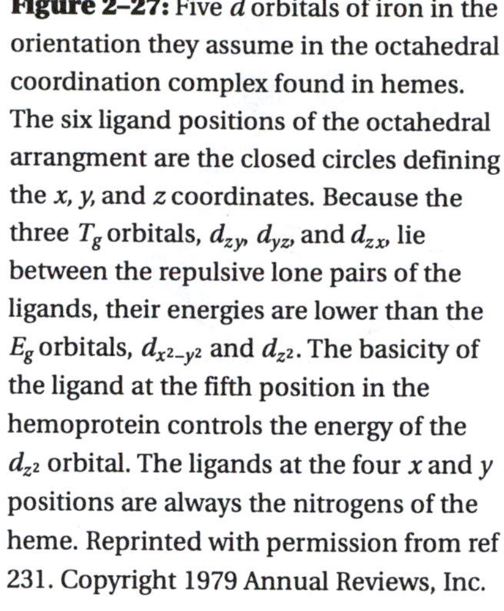

**Figure 2–27:** Five *d* orbitals of iron in the orientation they assume in the octahedral coordination complex found in hemes. The six ligand positions of the octahedral arrangment are the closed circles defining the *x*, *y*, and *z* coordinates. Because the three $T_g$ orbitals, $d_{zy}$, $d_{yz}$, and $d_{zx}$, lie between the repulsive lone pairs of the ligands, their energies are lower than the $E_g$ orbitals, $d_{x^2-y^2}$ and $d_{z^2}$. The basicity of the ligand at the fifth position in the hemoprotein controls the energy of the $d_{z^2}$ orbital. The ligands at the four *x* and *y* positions are always the nitrogens of the heme. Reprinted with permission from ref 231. Copyright 1979 Annual Reviews, Inc.

paired. Oxygen also has an even number of electrons but is paramagnetic. Upon formation of the complex, the six formerly unpaired electrons, four from iron and two from oxygen, must become paired, because the complex is diamagnetic.[244] In the process of pairing these electrons, one of the two oxygens becomes bonded to the iron.

Although there is no agreement on the details of the electronic structure, two features of the bonding are commonly emphasized. Overlap between one or more *d* orbitals on iron, either $d_{xz}$ or $d_{yz}$ or both, and two or more *p* orbitals on oxygen creates ***dpπ* molecular orbital systems** whose occupied, bonding *dpπ* molecular orbitals provide energy for the bond. This entry of *d* electrons from iron into antibonding π molecular orbitals on oxygen enhances the basicity of one of the σ lone pairs of electrons on oxygen so that it can form a σ bond with the $d_{z^2}$ orbital on iron.[228] This **σ bond** must have gained sufficient *p* character on oxygen to cause the oxygen from which it is derived to be hybridized in an $sp^2$ fashion because the Fe–O–O bond has an angle of 130°. If the σ orbitals on oxygen had not gained extra *p* character, they would have remained aligned with the axis of the two oxygen atoms and the Fe–O–O bond angle would have been 180°. Presumably, this extra *p* character results from the occupation of antibonding π molecular orbitals on oxygen, just as it does during the production of superoxide anion by the reduction of molecular oxygen.

As electron density, entering what were formerly antibonding orbitals, is gained from iron in the *dpπ* molecular orbitals, the net π bonding between the two oxygen atoms—the difference between occupation of bonding and antibonding orbitals—must not change because the oxygen–oxygen bond length remains the same. After these readjustments have taken place, there is excess electron density, creating a negative formal charge, on the distal oxygen.[248] Therefore, iron, in the electron-rich, ferrous state, has contributed more electron density to the *dpπ* molecular orbital system than oxygen, an electronegative element, has contributed to the σ bond. The unique feature of $Fe^{II}$ in this reaction is the ability of its *d* orbitals to overlap with the π molecular orbital system of oxygen and donate sufficient electron density through the *dpπ* molecular orbital system. This donation of π electron density is a common feature of the coenzymes that interact with molecular oxygen.

The reversible association between hemoproteins and oxygen can be duplicated quite closely with the heme model compound **2–55** and its homologues when they are dissolved in aqueous solutions of detergents.[237,249] The detergent provides micelles into which these hemes can dissolve. The duplication of the properties of the globins includes the changes in optical spectra upon ligation, the dissociation constants, and the on and off rates. These observations suggest that the protein of the monomeric globins such as myoglobin

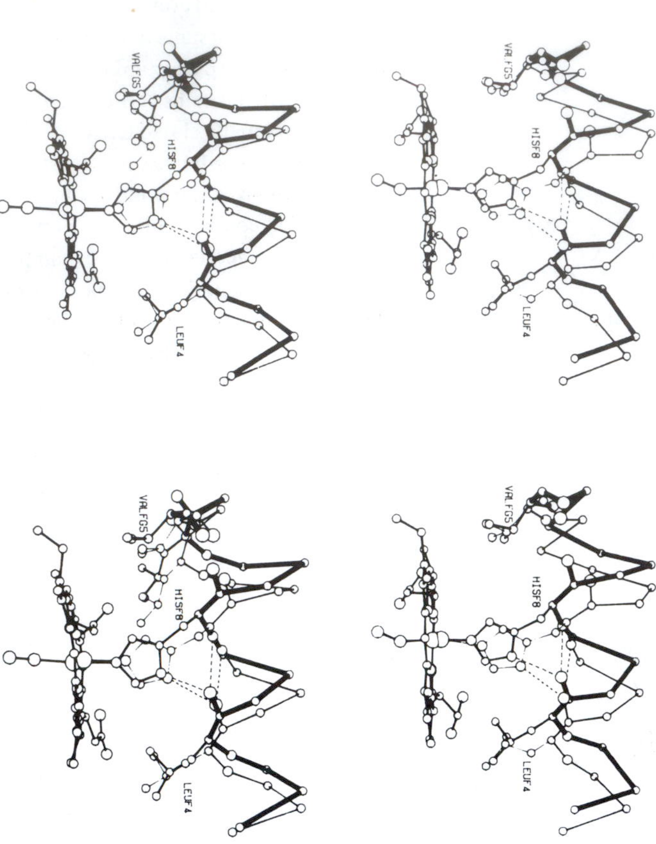

**Figure 2–28:** Drawings[252] of portions of the crystallographic molecular model of hemoglobin in the regions surrounding the histidine that is the fifth ligand to the iron in the heme from the $\alpha$ subunit (A) and the $\beta$ subunit (B). In each case, the $\alpha$ helix that contains the respective histidine (Histidine F8) is presented as well as the entire heme, viewed edge-on. The valine (Valine FG5) and leucine (Leucine F4) that form contacts with the hydrophobic surface of the heme are also drawn. Otherwise, only the peptide backbone of the $\alpha$ helix is presented. Each panel is a comparison of the structure of the unliganded protein (light lines) and the protein in which the heme is occupied by oxygen (heavy lines). In each case, the structure is fixed at the heme and the difference in the position of the $\alpha$ helix is presented before and after ligation. In the unliganded conformation, the position of the histidine is such that it pulls the iron out of the plane of the heme. Upon occupation by oxygen, the $\alpha$ helix shifts its position, the histidine comes closer to the center of the heme, and the iron drops into the plane of the porphyrin. Reprinted with permission from ref 252. 

simply provides a convenient way of simultaneously dissolving the otherwise insoluble heme in water, preventing the formation of heme dimers held together by peroxide dianion, and providing the fifth ligand.

The affinity for oxygen of the heme in oligomeric globins such as hemoglobin is further influenced by the structure of the protein. Depending on the conformation of the protein, the hemes have either a low or a high affinity for oxygen. Because the iron must move into the plane of the heme upon binding oxygen (Figure 2–26), any steric constraint that hinders this movement should decrease the affinity of the heme for oxygen. When alterations were made in the structure of heme model compound **2–55** that pulled the imidazole away from the plane of the porphyrin and consequently hindered the movement of the iron into the plane, the affinities for oxygen decreased.[250] It is also possible to compare crystallographic molecular models of hemoglobin in the conformation of high and low affinity.[251] In the conformation of low affinity, the $\alpha$ helix in which the histidine providing the fifth ligand is located has moved relative to its position in the conformation of high affinity. This movement pulls the imidazole farther away from the plane of the heme. Because the iron should respond to this tug, the movement of this $\alpha$ helix should be the explanation for the lower affinity for molecular oxygen (Figure 2–28).[252]

Although occasionally the iron–oxygen bond splits heterolytically to yield superoxide anion and $Fe^{III}$·porphyrin

$$O_2Fe^{II}\cdot\text{porphyrin} \rightleftharpoons O_2^{\cdot -} + Fe^{III}\cdot\text{porphyrin} \qquad (2\text{–}165)$$

the molecular oxygen bound reversibly to the hemes in hemoglobin or myoglobin is almost inert to reduction. The hemoprotein **cytochrome P450**, however, binds molecular oxygen to an $Fe^{II}$·porphyrin and then under normal circumstances participates in a reduction of the bound oxygen. In this enzyme, the fifth position on the heme is occupied by the thiolate anion of a cysteine from the protein.[253]

Molecular oxygen binds to the ferrous state of the heme in cytochrome P450,[254] as it would to hemoglobin, and is then reduced. In most animal cells, the flavoenzyme cytochrome P450 reductase[255] provides an electron to reduce the iron–oxygen complex. In the bacterium *Pseudomonas putida*, the same one-electron reduction is performed by a small protein, putidaredoxin, which contains an iron–sulfur cluster.[256] A similar iron–sulfur protein, adrenodoxin, provides the one-electron transfers for the cytochrome P450 of the adrenal glands, which is involved in the hydroxylations of steroids.[257] Each of these one-electron transfers reduces the respective $O_2{\cdot}Fe^{II}$·porphyrin and the addition of a proton produces the iron–hydrogen peroxide complex, $HO_2^{-}{\cdot}Fe^{III}$·porphyrin, which is the cysteinyl homologue of Structure **2–53**

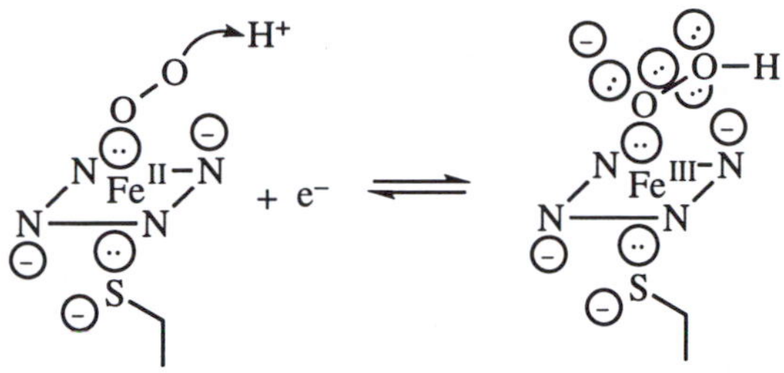

(2–166)

The addition of a second proton and the loss of water (Reaction 2–155) produces a compound I that is the cysteinyl homologue of Structure **2–54**.

The compound I formed in cytochrome P450 is capable of hydroxylating alkanes

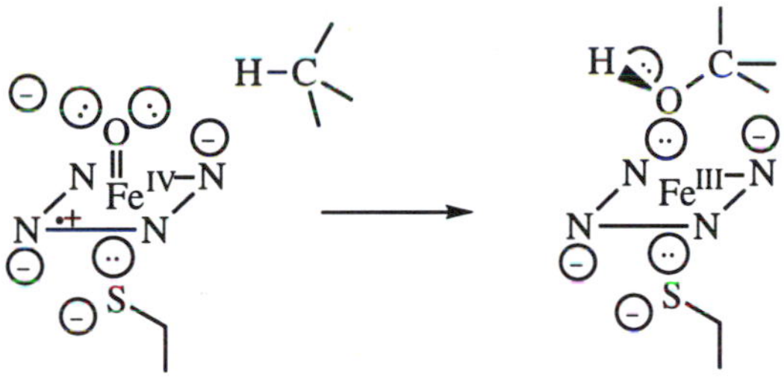

(2–167)

This reaction can be duplicated in a simple heme model compound in which a hydrocarbon chain is intramolecularly positioned over the iron and the compound I is formed from iodosobenzene[258]

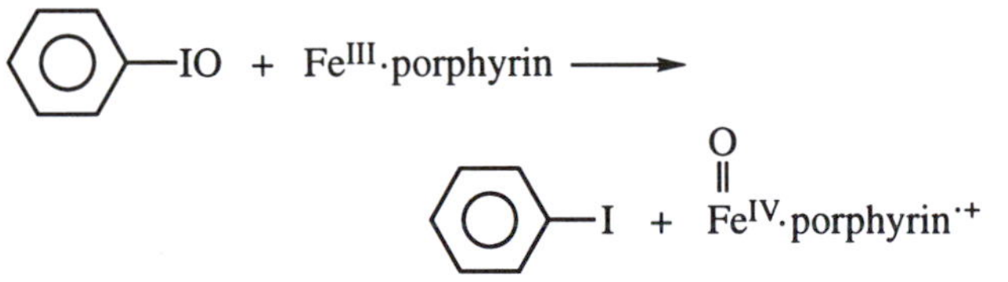

$$C_6H_5{-}IO + Fe^{III}{\cdot}\text{porphyrin} \longrightarrow C_6H_5{-}I + O{=}Fe^{IV}{\cdot}\text{porphyrin}^{\cdot+}$$

(2–168)

or other oxidants such as hydrogen peroxide. One of the methylenes in the hydrocarbon of the model compound is oxidized by the adjacent compound I to the secondary alcohol. Presumably, Reaction 2–167 proceeds by hydrogen abstraction (Reaction 2–156) followed by the collapse within the solvent cage of the carbon radical with the hydroxyl radical that was just formed on the iron. Consistent with this proposal is the fact that partial loss of stereochemistry is observed in model reactions and reactions of cytochrome P450 itself. The reduction of the $Fe^{III}$·porphyrin by a second electron, from the same source that provided the first electron, ejects the alcohol and regenerates the $Fe^{II}$·porphyrin

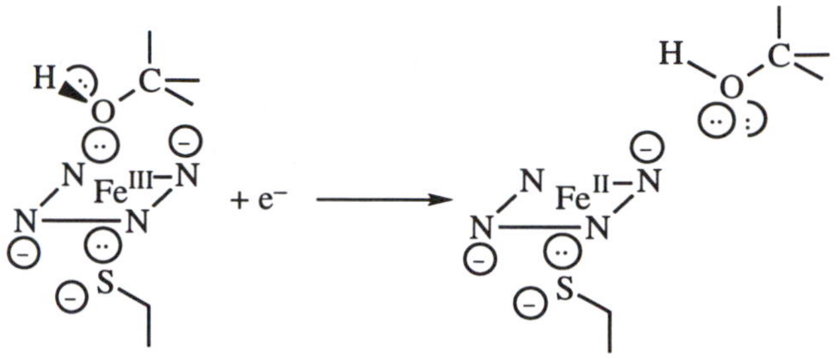

(2–169)

The hydroxylations of aromatic compounds performed by cytochrome P450 have as one of the possible intermediates[259] an arene oxide[260]

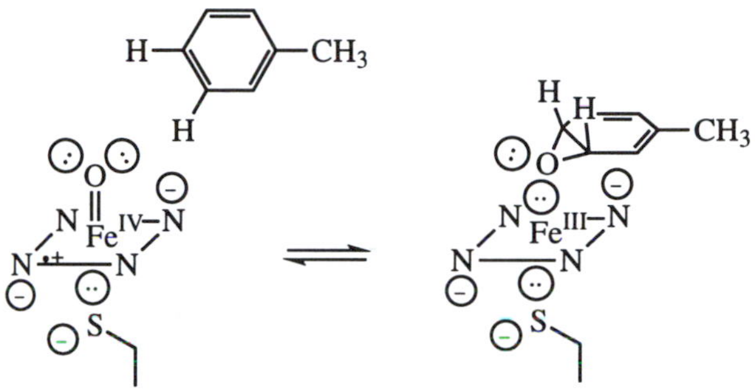

(2–170)

This reaction is analogous to the epoxidation of an olefin. It is the arene oxide that is responsible for the hydrogen migration seen in these reactions[260]

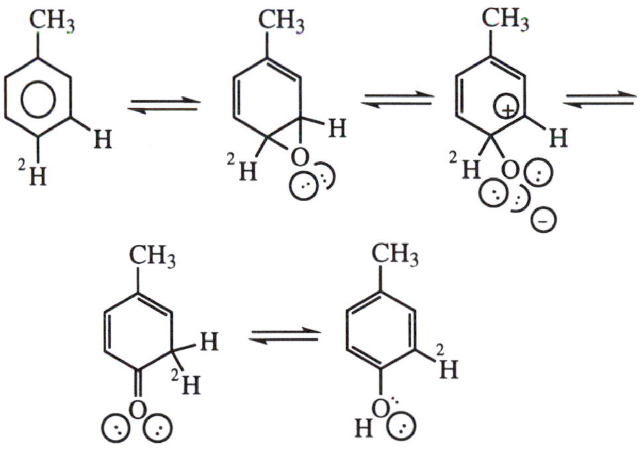

(2–171)

In the crystallographic molecular model of the cytochrome P450 from *P. putida*,[253] the region surrounding the sixth position on the iron is a large cavity whose walls are the hydrocarbon of valines, leucines, and phenylalanines. With the exception of Threonine 252, there are no acids present that could provide the protons necessary to form the compound I. Furthermore, the natural substrate, in this case camphor, is thought to occupy the active site before oxygen does. As the camphor directs only hydrocarbon toward the oxygen, its binding, by displacing neighboring molecules of water, renders the region around the sixth ligand even less acidic. Nevertheless, there may be a cluster of water molecules nearby capable of shuttling the necessary protons from the solvent. In this regard, the safest molecules, other than the substrate, with which to surround compound I within an active site would be molecules of water because molecules of water are the most difficult functional groups in the solution from which to abstract hydrogen atoms (the bond dissociation energy of $H_2O$ is 500 kJ $mol^{-1}$). For this reason, water would be a more logical general acid than any amino acid from the protein surrounding the active site.

There are also a group of hemoproteins that catalyze the incorporation of both atoms of molecular oxygen into their substrates. Examples are tryptophan-2,3-dioxygenase[261]

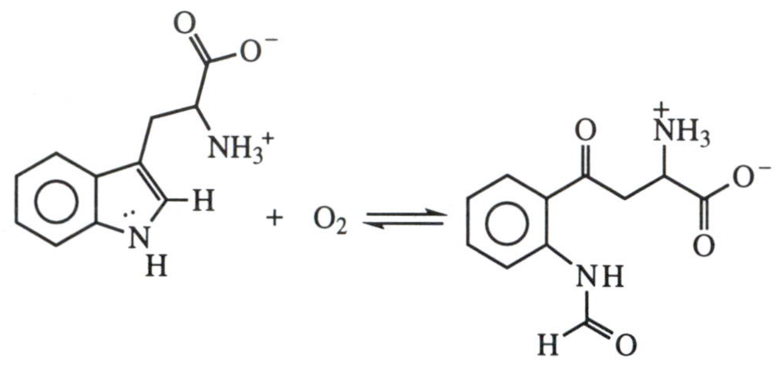

(2–172)

pyrocatechase[261]

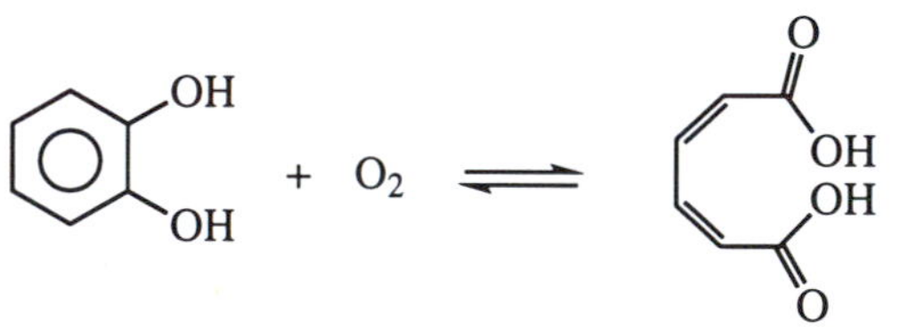

(2–173)

and a side-chain dioxygenase of tryptophan[262]

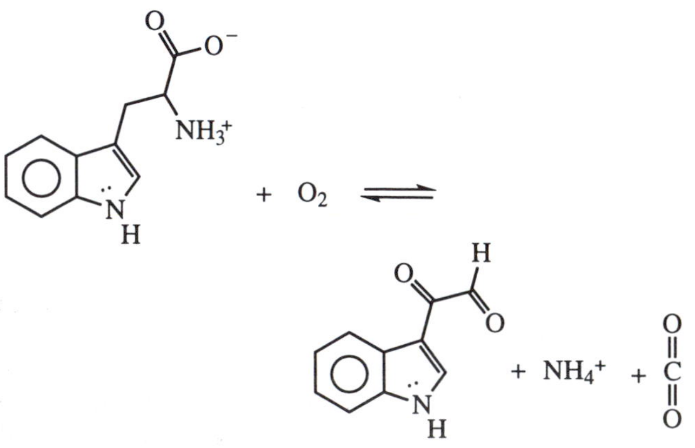

(2–174)

These enzymes are referred to as **dioxygenases,** while cytochrome P450 is referred to as a **monooxygenase.**

**Cytochrome *c* oxidase** is a hemoprotein that is able to reduce molecular oxygen completely

$$O_2 + 4e^- + 4H^+ \rightleftharpoons 2H_2O \tag{2–175}$$

Each molecule of the protein contains two easily distinguished hemes *a* and two easily distinguished copper ions. On the basis of measurements of the concentration of ferrous heme capable of binding carbon monoxide in fully reduced cytochrome *c* oxidase, it has been concluded that only one of the two hemes *a* can bind molecular oxygen directly.[263] In the fully reduced enzyme, one of the ferrous hemes is diamagnetic or low-spin and the other is paramagnetic or high-spin.[264] In analogy with hemoglobin, it is the high-spin heme in its ferrous state that binds oxygen. One of the coppers is strongly coupled antiferromagnetically to this high-spin iron,[265] and this indicates that it is very close to the iron atom. From a study of the rate of appearance of spectrally defined intermediates in the reduction of oxygen by reduced cytochrome *c* oxidase, it has been proposed that the first intermediate in the process is a normal complex between molecular oxygen and the ferrous heme. The next intermediate is thought to be peroxide dianion[266] that is held between the iron and the neighboring copper[265] and to which each metal has contributed one electron

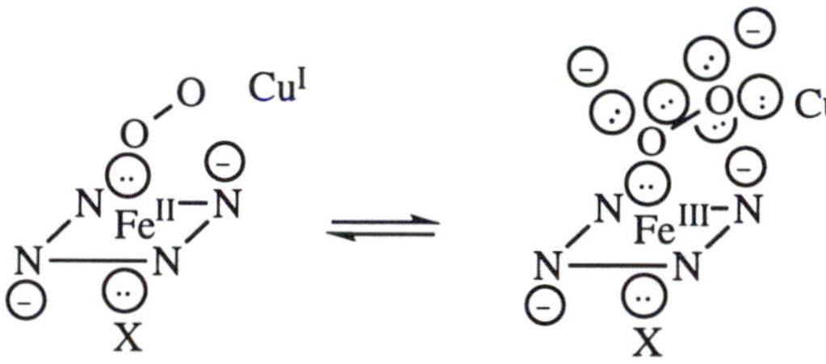

(2–176)

At this stage, two of the final four electrons have reached the molecular oxygen and the peroxide is tightly bound. This complex then receives one electron from the other heme *a* and one electron from the other copper, by electron transfer. Simultaneous with or subsequent to these electron transfers, protons are added to the σ lone pairs as they appear and increase in basicity until only water remains attached to the oxidized coenzymes

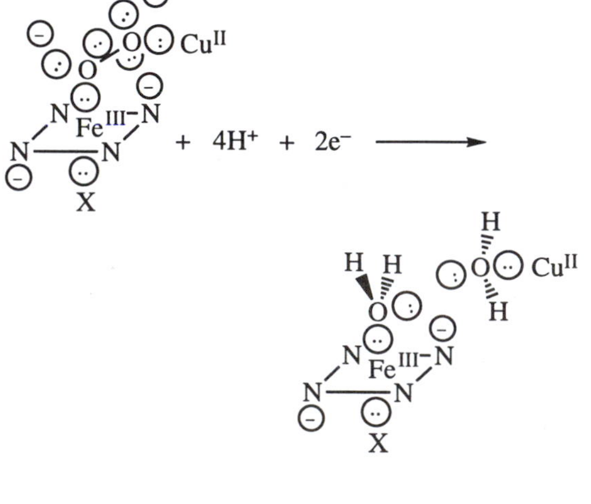

(2–177)

The heme *a* that provides one of these two electrons is spectrally identified as a simple hexacoordinated heme, as in the cytochromes, which is responsible for simple one-electron transfer.[264] The copper that provides the other electron is thought to be a tetracoordinate copper operating also as a one-electron coenzyme.[267] In this way the electrons entering the mitochondrial electron transport chain on a flavoprotein leave on molecules of water.

There is another class of metalloproteins that, like the hemoproteins, has members that can associate reversibly with oxygen or activate it for alkyl monooxygenase reactions. Members of this class of proteins include **hemocyanins, hemerythrins,** and **methane monooxygenase**.

The hemocyanins and hemerythrins are metalloproteins responsible for binding $O_2$ and transporting it to tissues in invertebrates. Although they share a common ancestor, the hemocyanins have two copper cations bound to each subunit and the hemerythrins have two iron cations bound to each subunit. In the hemocyanins in the deoxygenated state, the two copper cations are at the Cu(I) oxidation level, but when diatomic oxygen binds, they are both oxidized reversibly[268] to Cu(II) and the dioxygen is reduced to the peroxide dianion $O_2^{2-}$

$$\mathrm{Cu(I)Cu(I)} + O_2 \rightleftharpoons \mathrm{Cu(II)}O_2^{2-}\,\mathrm{Cu(II)} \qquad (2\text{–}178)$$

In the hemerythrins in the deoxygenated state, the two iron cations are at the Fe(II) oxidation level, but when diatomic oxygen binds they are both oxidized reversibly[269,270] to Fe(III) and the dioxygen is reduced to the hydroperoxide anion

$$H^+ + \mathrm{Fe(II)Fe(II)} + O_2 \rightleftharpoons \mathrm{Fe(III)}HO_2^-\,\mathrm{Fe(III)} \qquad (2\text{–}179)$$

Two high-resolution crystallographic molecular models of hemerythrin have been obtained.[271,272] Within the binuclear iron center, one iron cation is coordinated by three histidines from the protein and the other iron cation is coordinated by two histidines (Figure 2–29).[271] Two of the other ligands to each of the iron cations are provided by the respective oxygens of an aspartate and a glutamate. Finally, the two iron cations are bridged by a hydroxide anion. This constellation is referred to as a **binuclear $\mu$-oxo iron cluster**.

This arrangement of ligands causes the lower of the irons in the crystallographic molecular model presented in Figure 2–29 to be hexacoordinate and octahedral. The sixth position on the upper iron cation in the crystallographic molecular model is occupied by azide anion, $N_3^-$, which makes it also hexacoordinate and octahedral (Figure 2–29). The crystallographic molecular model is of the oxidized form of hemerythrin, methemerythrin, complexed with azide

$$\mathrm{Fe(III)Fe(III)} + N_3^- \rightleftharpoons \mathrm{Fe(III)}N_3^-\,\mathrm{Fe(III)} \qquad (2\text{–}180)$$

In the deoxygenated form of the competent ferrous state of the protein, this upper iron cation is believed to be pentacoordinate or hexacoordinate with $H_2O$ in the sixth position. It is at the vacant coordination site of the upper iron cation that the oxygen binds, and difference electron density maps[269]

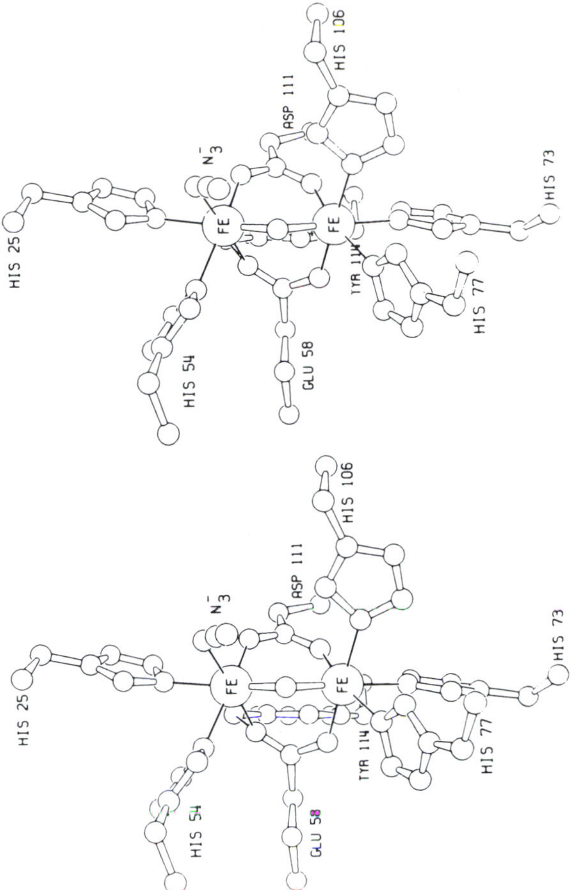


**Figure 2–29:** Crystallographic molecular model of the nine ligands to the two iron cations in sipunculan myohemerythrin.[271] The two irons of myohemerythrin purified from *Dendrostomum pyroides* were oxidized to the Fe(III) oxidation level, and a complex was formed with azide, $N_3^-$, an analogue of hydroperoxide, $HO_2^-$. Crystals of this derivative were grown in 55% $(NH_4)_2SO_4$ at pH 6.7. A data set was collected from these crystals to Bragg spacings of 0.15 nm. Difficulty with isomorphous replacements prevented phasing of the reflections to high resolution, but a map of electron density at low resolution from reflections to spacings of 0.6 nm could be calculated. This initial map allowed a crude initial crystallographic molecular model to be built from a molecular model of the polypeptide chain. Anomalous scattering from the two iron cations combined with information from this initial molecular model permitted phases of the reflections at smaller spacings to be estimated and a sufficiently detailed map of electron density was obtained to insert a model of the polypeptide unambiguously into it. This initial model was then refined against the complete data set. The upper $Fe^{3+}$ is coordinated by Histidines 25 and 54, and an azide anion ($N_3^-$). The lower $Fe^{3+}$ is coordinated by Histidines 73, 77, and 106. The octahedral constellations around each iron are completed by three ligands that are shared between them: the carboxylates of Glutamate 58 and Aspartate 111 and an hydroxide anion ($HO^-$) which occupies the front of the picture. Reprinted with permission from ref 271. Copyright 1987 Academic Press.

between the oxygenated ferrous form of the protein and the unliganded ferric form are consistent with the hydroperoxide assuming the location occupied by the azide anion in Figure 2–29. There is no indication for any interaction between the bound peroxide and the other fully coordinated iron atom, but the hydroperoxide in the crystallographic molecular model is in a plane that also contains the two irons and the oxygen that bridges the two irons

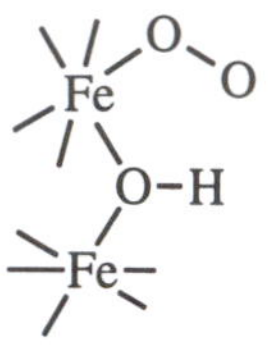

**2–57**

It is possible that a hydrogen bond exists between the peripheral oxygen of the peroxide and the oxygen bridging the irons and that it is this hydrogen bond that is responsible for the observed orientation of the peroxide.

The protein surrounding the binuclear copper site in the crystallographic molecular model of hemocyanin is superposable on the protein surrounding the binuclear iron site in the crystallographic molecular models of the hemerythrins,[273] and this observation demonstrates that these two proteins, hemocyanin and hemerythrin, share a common ancestor. In hemocyanin, histidines are found at positions both in the amino acid sequence and in the crystallographic molecular model equivalent to those of Histidines 25, 54, 73, 77, and 106 in hemerythrin, and they all provide ligands to the two copper cations, which themselves occupy positions equivalent to those of the two iron cations in the hemerythrin (Figure 2–29). In the hemocyanins, however, Glutamate 58 is replaced by a threonine and Aspartate 11 is replaced by an asparagine, neither of which is a ligand to either copper cation.[274] This change removes two negative charges from the constellation around the metal cations, consistent with the difference in charge between two $Fe^{2+}$ and two $Cu^{+}$. It also removes two ligands from each of the cations and is consistent with a decrease in coordination from hexacoordinate iron to tetracoordinate copper. There is, however, one additional ligand in the hemocyanins to compensate partially for the loss of the four carboxylate oxygens. Phenylalanine 29 in the amino acid sequence of the hemerythrins is replaced by a histidine in the hemocyanins that provides a third ligand to the one copper cation. This causes the two copper cations to be equivalent in their coordination as each has three histidines surrounding it. This may permit the diatomic oxygen to bind simultaneously to both coppers

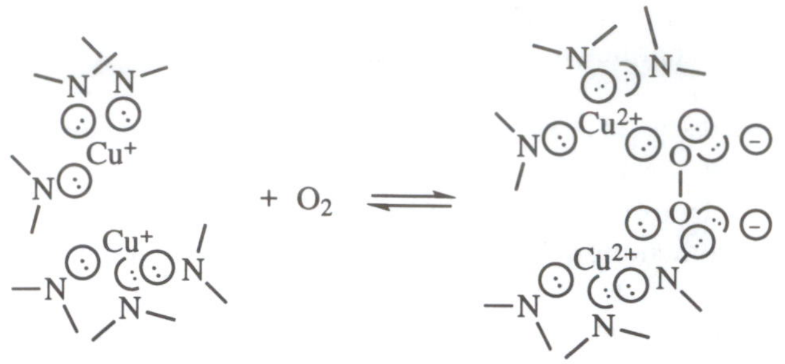

$$+ O_2 \rightleftharpoons \qquad (2\text{–}181)$$

and there is spectroscopic evidence in favor of this possibility.[275] The map of electron density is of such low resolution that it is not known whether or not the hemocyanins have a hydroxide bridging the two coppers, but no other Lewis base in the protein is close enough to act as a ligand to either copper cation.

Methane monooxygenase[276] is also a member of this family of binuclear metalloproteins. This enzyme catalyzes a reaction

$$2H^+ + CH_4 + O_2 + 2e^- \rightleftharpoons CH_3OH + H_2O \qquad (2\text{–}182)$$

formally identical to that catalyzed by a cytochrome P450

$$2H^+ + R_3CH + O_2 + 2e^- \rightleftharpoons R_3COH + H_2O \qquad (2\text{–}183)$$

The coenzyme in the active site of methane monooxygenase is a binuclear $\mu$-oxo iron clustser (Figure 2–29) closely resembling that in a hemerythrin.[277] The ligands, however, are somewhat different; Histidine 106 is replaced by a glutamate, Histidine 25 and the azide anion are replaced by the two oxygens of an aspartate, and Aspartate 111 is missing, which leaves vacant two sites for ligands on the iron.

Even though its coenzyme is a $\mu$-oxo iron cluster, the enzymatic properties of methane monooxygenase resemble those of a cytochrome P450. As with most cytochromes P450, methane monooxygenase can hydroxylate a large array of hydrocarbons.[278] As with cytochrome P450, the ferric form of methane monooxygenase can use hydrogen peroxide as a reactant

$$CH_4 + H_2O_2 \rightleftharpoons CH_3OH + H_2O \qquad (2\text{–}184)$$

to hydroxylate methane directly.[279]

These similarities, as well as the fact that hemerythrins, as do hemoglobins, bind oxygen reversibly, have prompted the conclusion[279] that the mechanism of methane monooxygenase is formally equivalent to that of a cytochrome P450. The ferrous form of the coenzyme would bind the molecular oxygen as a complex between hydroperoxide anion and the diferric cluster (Reaction 2–179). Protonation of the hydroperoxide followed by heterolytic cleavage of the oxygen–oxygen bond would lead to a compound I at one of the irons (Reaction 2–155). In this compound I, the second iron(III), by becoming oxidized to iron(IV), would provide the additional electron rather than the porphyrin. The compound I would hydroxylate the hydrocarbon to produce the alcohol bound to the diferric form of the cluster. This diferric form of methane monooxygenase is known to be reduced to the diferrous form by a separate reductase. This reductase is a flavoprotein containing a binuclear iron–sulfur cluster and passes two electrons from NADH to the two irons.[276]

So far, the coenzymatic reactions that have been considered are those that use molecular oxygen as a reactant. There is a coenzyme that creates molecular oxygen.

The coenzyme that catalyzes the reverse reaction of that catalyzed by cytochrome *c* oxidase (Reaction 2–175) is a manganese–oxygen cluster on photosystem II of green plants. This coenzyme provides four electrons, one at a time, in a sequence of clearly defined steps,[280] to the successive holes that develop during the absorption of four successive quanta of light by the associated reaction center. The four electrons provided to the reaction center by the manganese–oxygen cluster are removed by it from two molecules of water during the production of oxygen

$$2H_2O \rightarrow O_2 + 4e^- + 4H^+ \qquad (2\text{–}185)$$

Although there is some disagreement about the details, and no crystallographic molecular model is available, it is thought[281] that the manganese–oxygen cluster is similar to a tetranuclear iron–sulfur cluster in that the four manganese atoms are bridged by four oxygen atoms

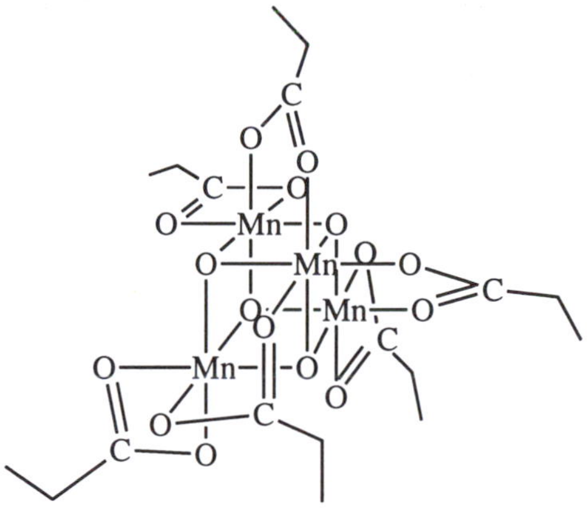

**2–58**

but that, unlike the iron atoms in an iron–sulfur cluster, the coordination of the manganese in the most reduced state remains octahedral owing to the fact that several carboxylate side chains of glutamates or aspartates are bidentate ligands to the metals in the cluster within the protein. The other difference is that most mechanisms for oxygen production involve the bridging oxygen atoms, formally oxide dianions derived originally from water, as the precursors of the oxygen. The manganese atoms in turn would be oxidized by the successive holes in the reaction center and in turn oxidize bridging oxides.

Although the formation of the oxygen–oxygen bond to produce the peroxide anion complexed at its two oxygens to two respective manganese atoms has yet to be duplicated, a binuclear manganese–peroxide complex

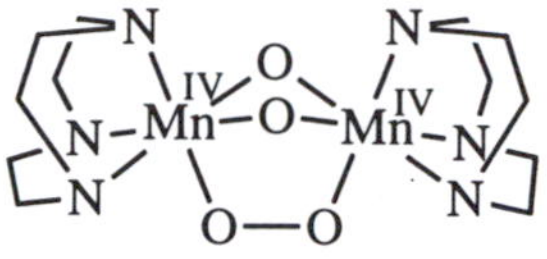

**2–59**

has been synthesized that can duplicate the last step in the reaction.[282] In this complex, two manganese(IV) are complexed with a bridging peroxide dianion. Upon acidification (pH 3) in 0.5 M NaCl, the peroxide is released as molecular oxygen and the manganese ends up as the chloride of manganese(III). In this reaction each manganese has oxidized each atom of oxygen by one electron.

Another coenzyme that reacts directly with molecular oxygen is **reduced flavin**. These reactions have one of two outcomes. In the simplest instance, the molecular oxygen is used to regenerate oxidized flavin from the reduced flavin formed during a prior hydride transfer (Reactions 2–100 and 2–101). In this case hydrogen peroxide is the product. There are also monooxygenases that use flavin as a coenzyme. Examples are melilotate-3-monooxygenase[283]

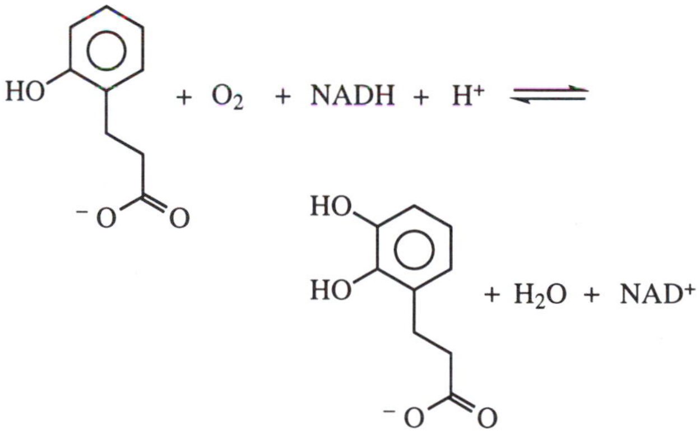

$$\text{(melilotate)} + O_2 + NADH + H^+ \rightleftharpoons \text{(3-hydroxymelilotate)} + H_2O + NAD^+ \qquad (2\text{–}186)$$

and cyclohexanone monooxygenase[284]

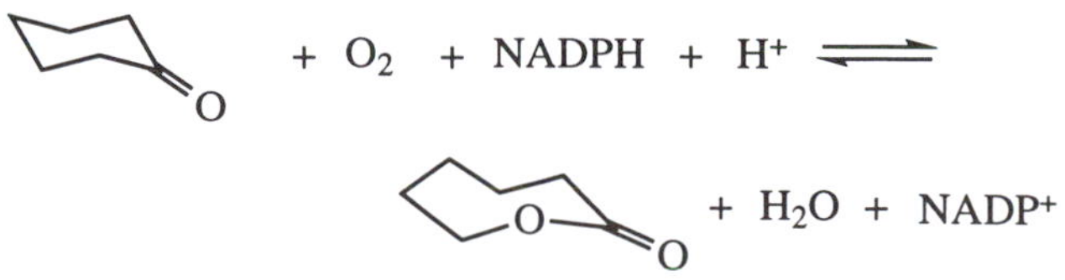

$$\text{(cyclohexanone)} + O_2 + NADPH + H^+ \rightleftharpoons \text{(caprolactone)} + H_2O + NADP^+ \qquad (2\text{–}187)$$

In these reactions one of the two oxygen atoms of molecular oxygen is incorporated into the product and the other into water.

When molecular oxygen is used as the final receptacle for electrons from a flavin that catalyzes formal hydride transfer, there is no reason to invoke any type of covalent intermediate between flavin and oxygen. As long as the conjugate acid of the base on the enzyme that removes the proton from reduced flavin upon the first one-electron step

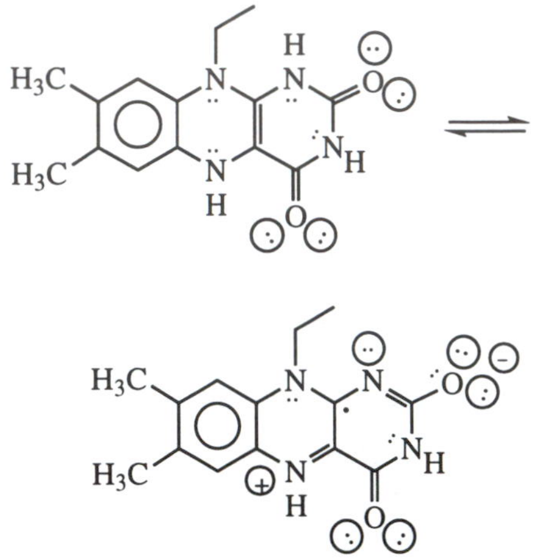

(2–188)

has a p$K_a$ greater than 7, the reduction potential of the flavin will be less than –0.15 V (Figure 2–13 and Equation 2–70). In this situation, a one-electron transfer to molecular oxygen ($E°$= –0.16 V) is favorable, and the arrival of molecular oxygen in the vicinity of the flavin would permit electron transfer to form superoxide[285]

$$O_2 + H_3Fl_{red} \rightleftharpoons O_2 \cdot H_3Fl_{red} \rightleftharpoons O_2^{\cdot -} \cdot H_3Fl^{\cdot +} \quad (2\text{–}189)$$

The superoxide formed transiently[286] would then accept the second electron from the flavin radical to form hydrogen peroxide, provided two protons were available from acids in the enzyme or from the water surrounding the enzyme

$$O_2^{\cdot -} \cdot H_3Fl^{\cdot +} \rightleftharpoons H_2O_2 + HFl_{ox} \quad (2\text{–}190)$$

Reduced flavins in solution, at early times in the reaction before other side reactions become important, react with molecular oxygen to produce hydrogen peroxide at significant rates.[64,286] The superoxide formed as an intermediate in these reactions during the first electron transfer does not exist long enough to depart from the vicinity of the flavin before it is reduced to hydrogen peroxide.[64] On an enzyme, the failure of the superoxide to depart before further reduction occurs should be even more complete. Therefore, the reduction of molecular oxygen to hydrogen peroxide performed by reduced flavin, as the terminal step in the reactions catalyzed by various flavoenzymes, can be thought of as simple two-electron transfer, within the encounter complex, to the $\pi$ molecular orbital system of the molecular oxygen concerted with proton transfer to the $\sigma$ lone pairs of the developing peroxydianion and hydroperoxyanion.

It is claimed, however, that this simple picture is inconsistent with the kinetic behavior of the reactions of reduced flavins in free solution with molecular oxygen to produce hydrogen peroxide.[64] It is possible to avoid the stated inconsistencies by postulating that a covalent intermediate is formed between superoxide anion and flavin radical in an irreversible step before any proton transfers occur. It has been shown that superoxide can react in dimethylformamide with the neutral radical of 5-ethyl-3,7,8,10-tetramethylisoalloxazine[287] to produce the **4a-peroxyflavin anion**

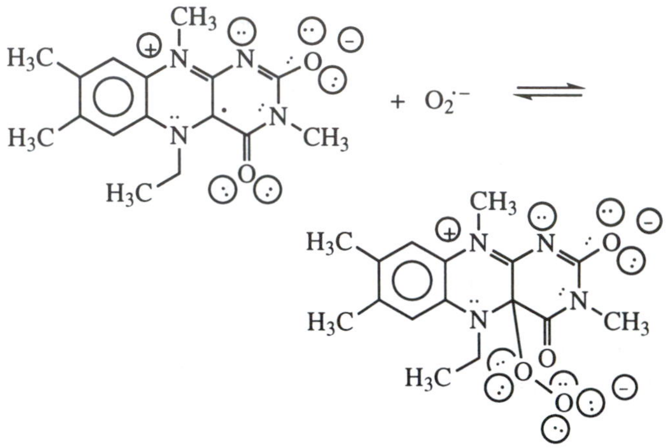

(2–191)

If this reaction occurs following the initial production of superoxide from molecular oxygen and a reduced flavin, the resulting 4a-peroxyflavin anion would be the irreversibly formed intermediate needed to explain the kinetic inconsistencies. The neutral form of this 4a-hydroperoxide has also been produced from the reaction between hydrogen peroxide and an oxidized flavin in nonpolar solvents[288]

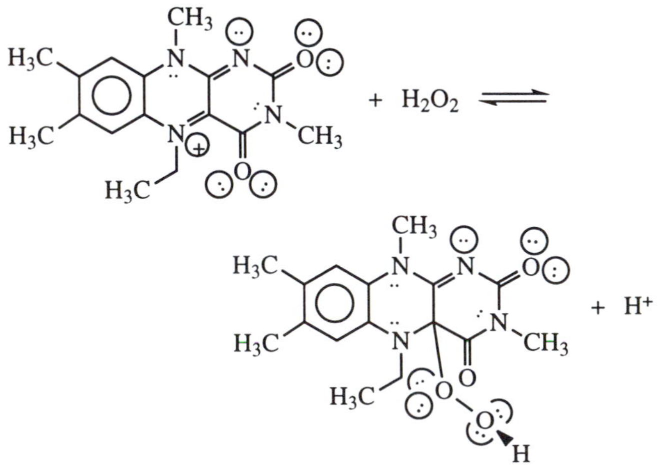

**2–60**

(2–192)

The neutral 4a-hydroperoxyflavin **2–60** reverts readily to reactants because the reaction is a simple addition of the conjugate base of hydrogen peroxide, $HO_2^-$ as a nucleophile, to the iminium cation of the oxidized 5-ethylflavin. Both its formation from superoxide anion and its reversion to hydrogen peroxide are consistent with the intermediacy of a 4a-hydroperoxyflavin in the reactions of oxygen with flavin that produce hydrogen peroxide.

The involvement of both superoxide and hydrogen peroxide in the reoxidation of reduced flavin by molecular oxygen is illustrated by the reoxidation of reduced xanthine oxidase.[289] Fully reduced xanthine oxidase has six more electrons than the fully oxidized enzyme: two on the manganese ion, one in each of the two iron–sulfur clusters, and two in the reduced flavin. The reoxidation of the reduced complex by molecular oxygen occurs at the flavin. As the first four electrons are removed by two molecules of oxygen during reoxidation, they leave rapidly enough that two molecules of hydrogen peroxide can be produced at the flavin. The last two electrons, however, are of positive enough reduction potential that they cannot be transferred efficiently through the flavin to oxygen, and two molecules of superoxide anion are made from two molecules of oxygen; each superoxide anion departs from the flavin with only one of the two electrons. As a result, when xanthine oxidase is functioning normally, it produces a variable mixture of hydrogen peroxide and superoxide.

There is a remaining, as yet undiscussed, and often unmentioned difficulty in either the two-electron transfer or the involvement of a covalent intermediate in the reaction

$$^{3}O_2 + H_3Fl_{red} \rightleftharpoons H_2O_2 + HFl_{ox} \tag{2–193}$$

Molecular oxygen is a diradical whose two unpaired electrons have parallel spins in the triplet ground state, $^{3}O_2$. Reduced flavin, hydrogen peroxide, and oxidized flavin are all singlet molecules whose electrons are present in pairs whose partners have opposite spins. Therefore during the overall reaction, one electron must invert its spin. Such a **spin inversion** is normally a slow process, but it can be catalyzed. For example, the combination of triplet oxygen with a heme to form oxyheme, which is diamagnetic,[244] necessarily involves a spin inversion. As this spin inversion is instantaneous, it can be concluded that it is catalyzed by the iron. It is in this way that the spin inversions required for the reactions catalyzed by cytochromes P450 occur upon the initial ligations of the oxygen and the iron(II). Spin inversions can also be accelerated by electron transfer. For example, after the electron transfer from reduced flavin to molecular oxygen to form $O_2^{\cdot -}$ and $H_3Fl^{\cdot +}$ the two unpaired electrons of the triplet state are now located, respectively, on flavin and superoxide rather than both being on molecular oxygen. Because they are unpaired electrons on different atoms, carbon and oxygen, in different molecules, flavin and superoxide, spin inversion becomes very rapid. All that is required to catalyze spin inversion by this mechanism is to have an electron donor in which the remaining unpaired electron will not reside on oxygen and which has a low enough reduction potential to transfer an electron to triplet molecular oxygen.

Transition metal ions other than iron can also be used as the one-electron reductants in the catalysis of spin inversion. Galactose oxidase catalyzes the reaction

$$\text{D-galactose} + O_2 \rightleftharpoons \text{D-}\textit{galacto}\text{-hexodialdose} + H_2O_2 \tag{2–194}$$

Formally, this is a simple dehydrogenase, but instead of the two electrons being transferred to singlet NAD$^+$, they are transferred to triplet molecular oxygen. During the reaction, spin inversion must take place because both products are singlet molecules. The redox pair in the active site of galactose oxidase consists of a complex between a copper cation and an *o*-cysteinyltyrosine.[290,290a] The reduced state of the redox pair consists of copper I complexed with the *o*-cysteinyltyrosine.[291] This complex is the immediate electron donor to triplet oxygen. Either the direct complexation of singlet oxygen with the copper[292] or an electron transfer from copper to triplet oxygen to split the triplet between oxygen and copper catalyzes the necessary spin inversion.

In the enzymatic reactions catalyzed by monooxygenases that use flavin as a coenzyme (Reactions 2–186 and 2–187), the species that reacts with molecular oxygen is reduced flavin.[283,293] It is formed during the reduction of the enzyme-bound flavin by the NADH, the NADPH, or the other reductants used by these enzymes. The reduced flavin reacts with molecular oxygen to produce an intermediate that either yields hydrogen peroxide and oxidized flavin if the substrate to be hydroxylated is absent or hydroxylates the substrate if it is present.[293] The spectral characteristics of these initial intermediates formed in the various flavin monooxygenases[284,294,295] are indistinguishable from those of 4a-hydroperoxyflavins (**2–60**, Reaction 2–192).

4a-Hydroperoxides are as active as **peroxy acids** in the monooxygenations of olefins, tertiary amines, alkyl sulfides, and iodide anion.[296] The oxygenated flavin formed in the enzymatic reaction of cyclohexane oxidase has also been shown to participate in a series of monooxygenations characteristic of a peroxy acid[284] in addition to the usual oxygenation (Reaction 2–187), which has all of the characteristics of a Baeyer–Villiger reaction.[297] These results suggest[284] that the hydroxylating species in all of these reactions is the 4a-hydroperoxyflavin itself (Figure 2–30) acting as if it were a peroxy acid. In the last step, the second oxygen atom of molecular oxygen has in turn hydroxylated the flavin but at a carbonyl carbon. The resulting carbinolamine will lose hydroxide anion in a reaction that is the reverse of the addition of hydroxide anion to oxidized flavin (Reaction 2–91).

The accepted mechanisms for the reactions of peroxy acids (Figure 2–31) rely on the ability of the acyl oxygen to be protonated and the resulting incipient carboxylic acid to function as the leaving group. Because the 4a-hydroperoxide of flavin resides on a fully saturated carbon, it is unclear how the flavin moiety can participate in a similar way to explain the reactivity of the 4a-hydroperoxyflavin. There is kinetic evidence for yet another intermediate in these reactions,[293] in addition to the 4a-hydroperoxyflavin, and various products of intramolecular reactions within the 4a-hydroperoxyflavin have been proposed as possible oxygenating species.[294,298]

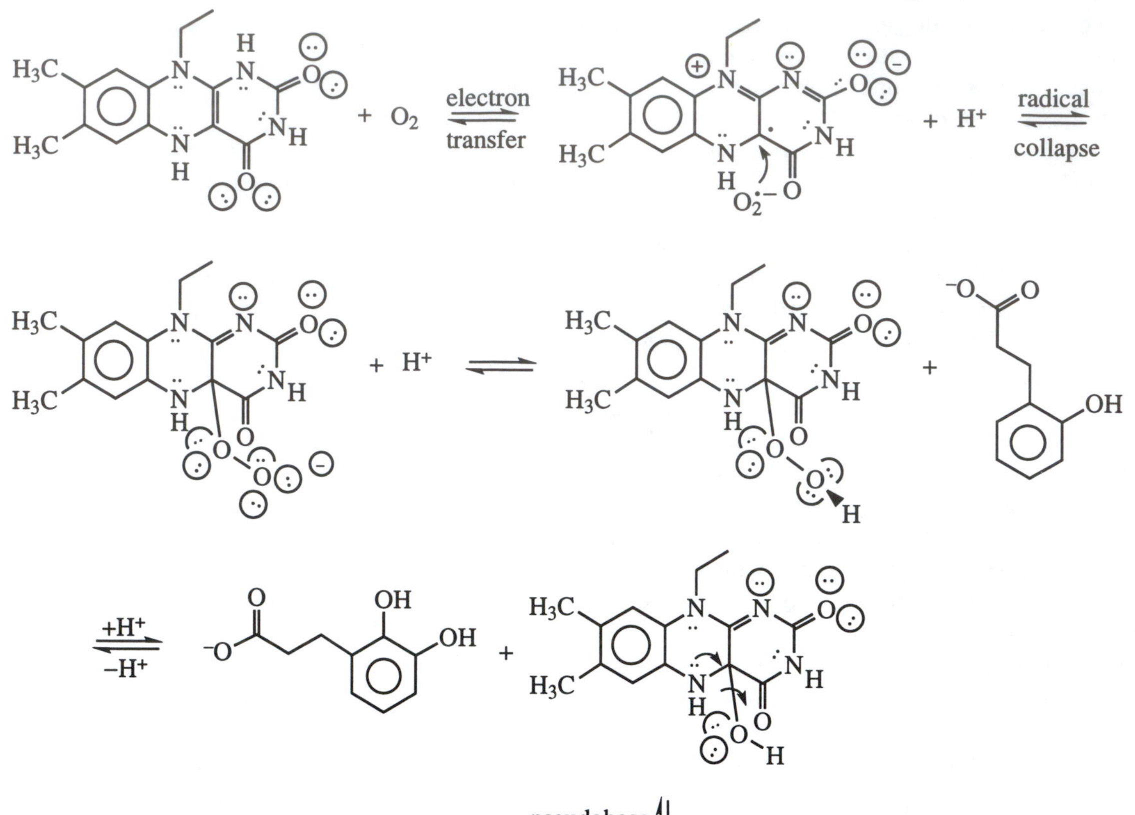

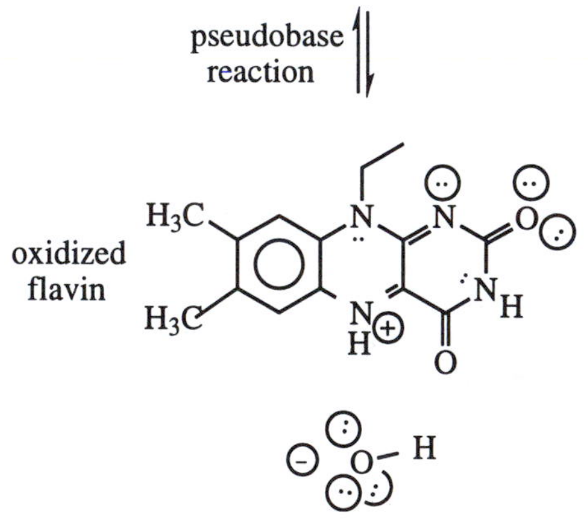

**Figure 2–30:** Mechanism for the formation of the oxygenating intermediate in the reaction catalyzed by the flavoenzyme melilotate monooxygenase (Reaction 2–185). Electron transfer produces superoxide anion and flavin radical. Radical collapse produces the 4a-hydroperoxyflavin. This behaves as a peroxy acid and hydroxylates the melilotate with the 4a-hydroxyflavin as the other product. This dissociates in the reverse of the pseudobase reaction to produce oxidized flavin.

Monooxygenations in which **pterins** are used as substrates rather than as coenzymes are entirely analogous to the monooxygenations using flavin as a coenzyme. Pterins, such as tetrahydrobiopterin, $H_4Pt_{red}$

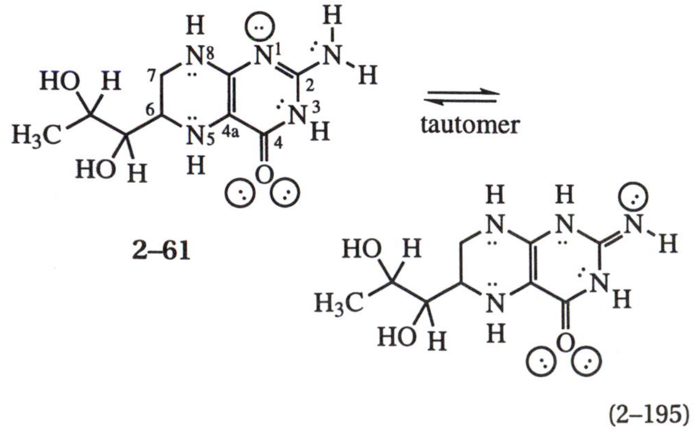

(2–195)

have almost the same bicyclic system as is present in flavin (**2–32**) with the exception that the 1,4-diazacyclohexene is saturated between carbon 6 and carbon 7

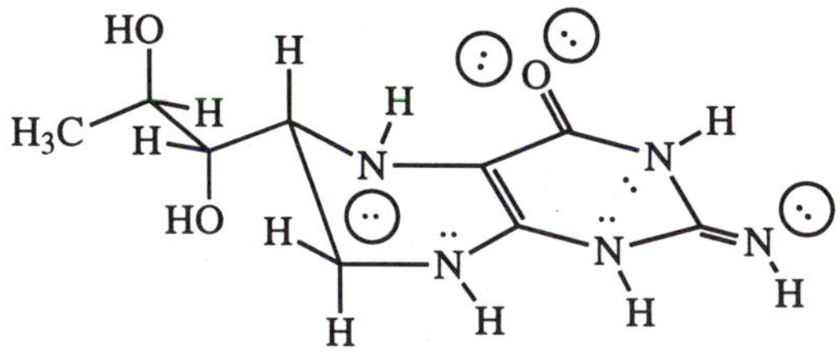

It is the 6*R* enantiomer (**2–62**) that is used biologically.[299] As with the flavin monooxygenases, the biopterin reacts with oxygen to produce an oxygenated intermediate[300] capable of monooxygenating the substrate. For example, phenylalanine

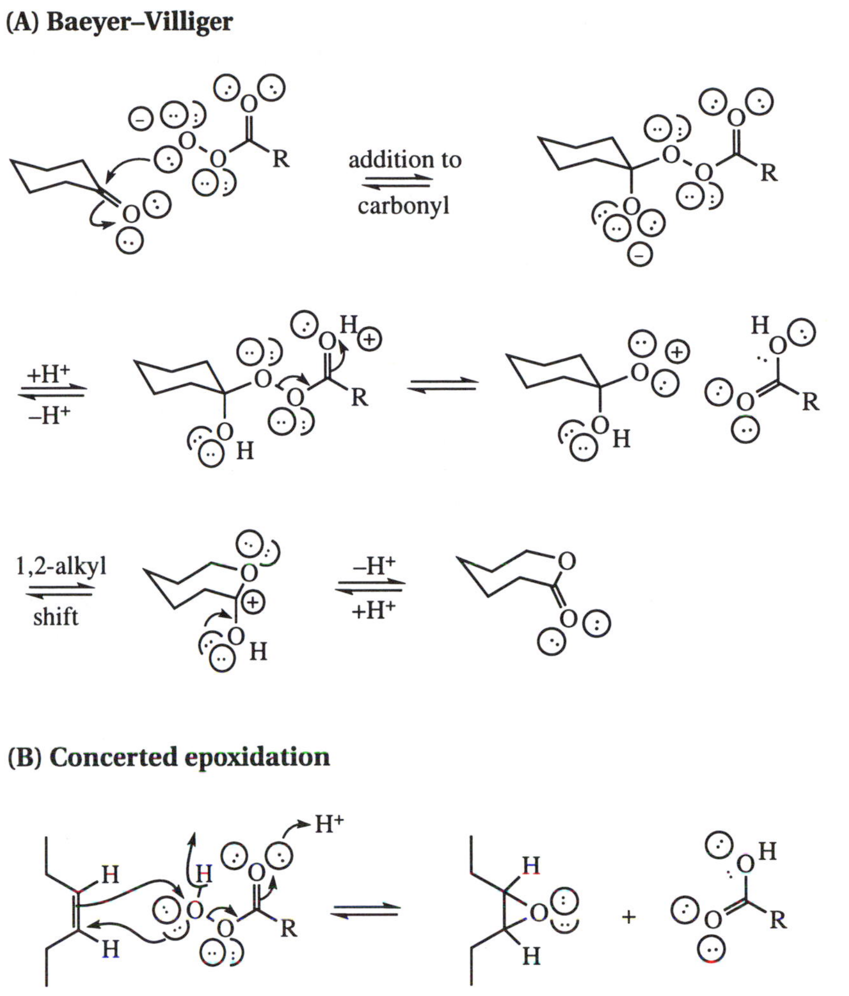

**Figure 2–31:** Mechanisms for two types of reactions in which peroxy acids oxygenate substrates. In the Baeyer–Villiger monooxygenation the oxygen atom is inserted into a carbon–carbon $\sigma$ bond. In a concerted epoxidation, the oxygen atom is inserted into a carbon–carbon $\pi$ bond.

**(A) Baeyer–Villiger**

addition to carbonyl

$+H^+$ / $-H^+$

1,2-alkyl shift

$-H^+$ / $+H^+$

**(B) Concerted epoxidation**

hydroxylase catalyzes the monooxygenation of phenylalanine

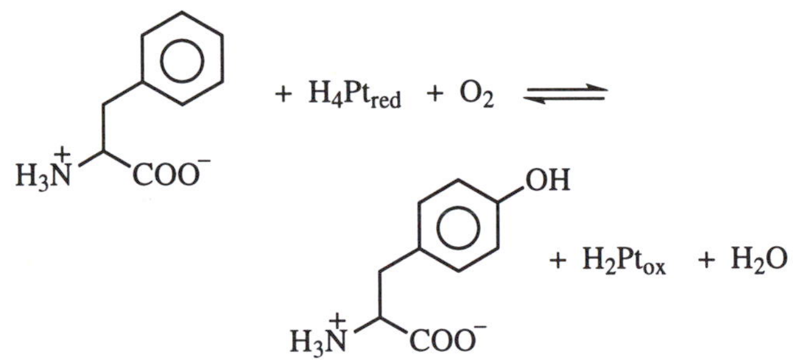

$$\text{Phe} + H_4Pt_{red} + O_2 \rightleftharpoons \text{Tyr} + H_2Pt_{ox} + H_2O \qquad (2\text{–}196)$$

where $H_2Pt_{ox}$ is the oxidized form of biopterin

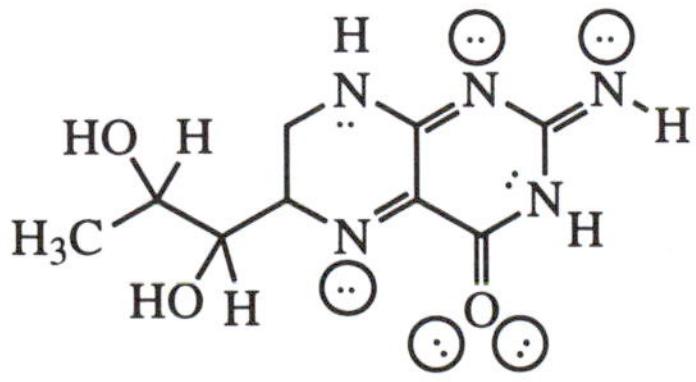

**2–63**

Because the biopterin is a substrate rather than a coenzyme, it is reduced by another enzyme using NADH or NADPH as the other substrate.

In the case of pterin monooxygenases, the immediate product of the reaction is the pseudobase or **4a-carbinol-amine**, which dehydrates spontaneously to the dihydro-pterin[299,301]

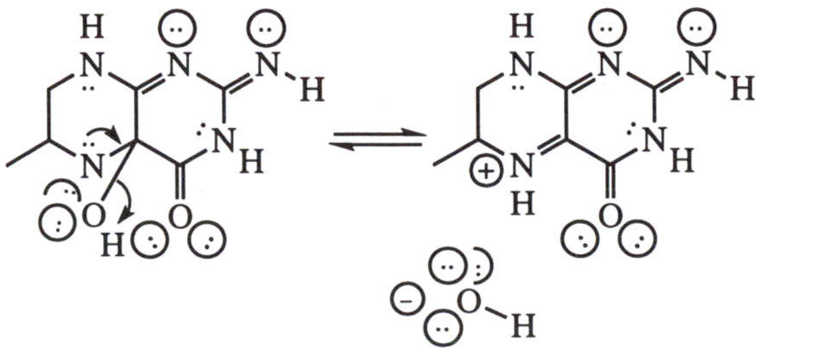

(2–197)

The involvement of this 4a-carbinolamine serves as a model for the last step believed to occur in flavin monooxygenases (Figure 2–30). Isotopic studies[302] of phenylalanine hydroxylase are consistent with the formation of a 4a-peroxypterin, analogous to the 4a-peroxyflavin in flavohydroxylases (Figure 2–30), as the hydroxylating species

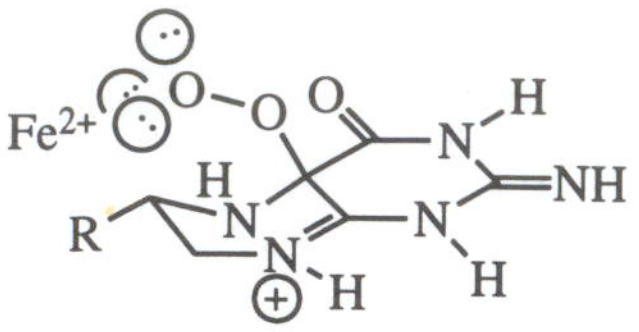

**2–64**

in which both the iminium cation of the pterin and the iron would make the peripheral oxygen electrophilic enough to form the arene oxide (Equation 2–170) thought to be the next intermediate in the reaction catalyzed by this enzyme[303]

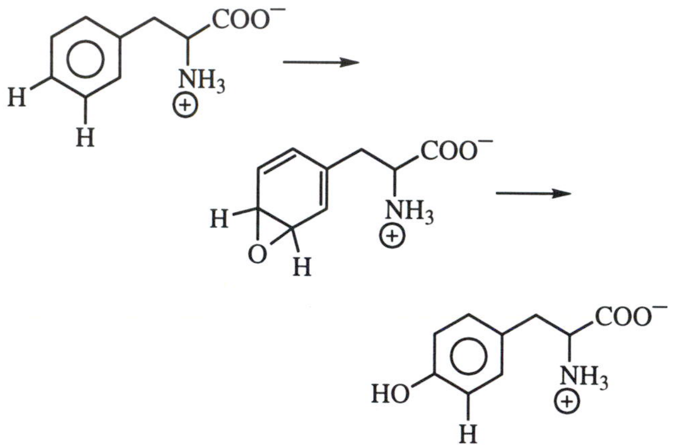

(2–198)

4a-Hydroperoxybiotin (**2–64**), however, has not been directly identified as the oxygenated intermediate in pterin monooxygenases. This may be due to the fact that the oxygenated intermediate in these reactions is rapidly discharged, in the absence of the substrate that is normally hydroxylated, by oxidizing another molecule of tetrahydrobiopterin.[300] It is also possible that the oxygenated intermediate may not be the 4a-hydroperoxybiopterin.

In phenylalanine hydroxylase, there is also a bound **non-heme iron atom** essential for enzymatic activity.[304,305] In enzyme that has been exposed to molecular oxygen in the absence of reduced pterin, the iron is in the ferric state, and the enzyme is inactive. Reduced pterin reactivates the enzyme by reducing the ferric iron to ferrous iron. Therefore, ferrous iron is required for enzymatic activity. This enzyme combines two of the molecules that can react with molecular oxygen, a reduced, flavinlike heterocycle and ferrous iron.

Another class of monooxygenases resemble phenylalanine hydroxylase in requiring, as a substrate, a carbonyl compound, as is the 4a-position in a pterin, and, as a coenzyme, ferrous iron. These are the oxygenases that use **2-oxoglutarate** as a substrate. An example of such an oxygenase is procollagen–proline, 2-oxoglutarate dioxygenase

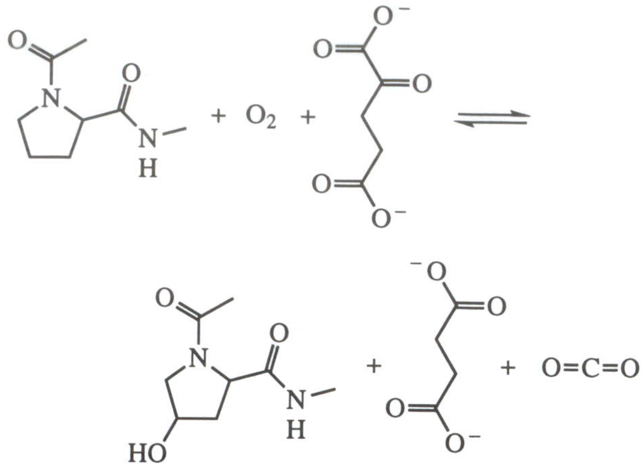

(2–199)

The substrates for the enzyme are prolines within the protein procollagen. The enzyme is a dioxygenase because one atom of the molecular oxygen ends up in proline and the other in the succinate.[306] Likewise, in a flavin monooxygenase or pterin monooxygenase one atom of molecular oxygen ends up in the flavin or the pterin, respectively, but is then lost immediately upon dehydration (Figure 2–30). Therefore, the different names do not necessarily signify different mechanisms.

In the absence of procollagen, 2-oxoglutarate is slowly decarboxylated by procollagen–proline, 2-oxoglutarate dioxygenase,[307] and this has been offered as evidence for the suggestion that a peracid is an intermediate in the reactions catalyzed by these enzymes

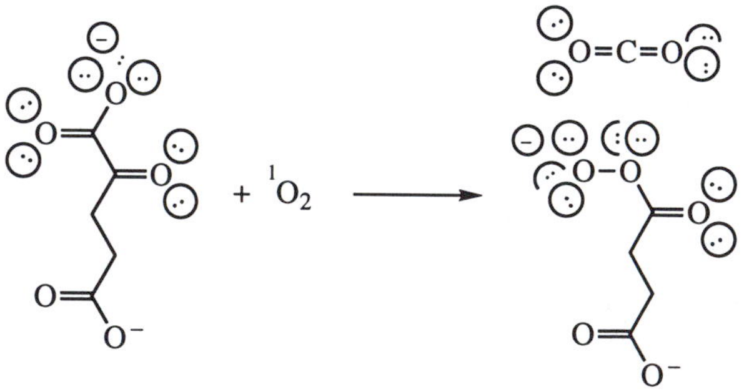

(2–200)

The difficulty with this reaction is that molecular oxygen in its triplet ground state cannot react with singlet 2-oxoglutarate to produce singlet $CO_2$ and singlet peroxy acid because electron spin would not be conserved in the reaction. Molecular oxygen, however, has an excited singlet state and one of the ways to produce singlet molecular oxygen is through binding to a transition metal such as iron. It has been shown that singlet molecular oxygen, $^1O_2$, can participate in Reaction 2–200 when it is generated photochemically.[308] This would explain the requirement for both iron and a 2-keto acid in these reactions. If a peroxy acid is an intermediate, it must be further activated to be able to abstract the hydrogen from an aliphatic position on the substrate (Reaction 2–199). This may involve the iron further in the mechanism of the reaction.

With many of the enzymes that activate molecular oxygen by using ferrous iron, in particular the 2-oxoglutarate dioxygenases, a serious flaw arises in practice. When molecular oxygen is bound to ferrous iron, there is always a significant probability that the complex will dissociate to produce superoxide anion and ferric iron

$$Fe^{II} + O_2 \rightleftharpoons Fe{\cdot}O_2 \rightleftharpoons Fe^{III} + O_2^{\cdot-} \qquad (2\text{–}201)$$

As superoxide anion is rapidly disproportionated in the cytoplasm, it does not accumulate sufficiently to reduce the ferric iron back to ferrous iron, and the enzyme is inactivated by its own reaction. In most oxygenases, especially those using nonheme iron, the usual mechanism for reducing the iron back to the ferrous form and reactivating the enzyme is to use the one-electron reductant **ascorbic acid** (Figure 2–32). When these oxygenases, in particular procollagen–proline, 2-oxoglutarate dioxygenase, are operating, ascorbate must be present continuously. Even at its maximum velocity, a molecule of procollagen–proline, 2-oxoglutarate dioxygenase can oxygenate only about 50–200 molecules of substrate in the absence of ascorbate before the iron becomes adventitiously oxidized. It can be rereduced only at the expense of one electron from ascorbate.[310,311] In other oxygenases, reductants other than ascorbic acid, and even superoxide ion itself, can substitute in this role.[312]

Perhaps the most peculiar substrates that can activate molecular oxygen are the **menaquinones**, or **vitamins K** (**2–45**). Usually menaquinones have a polyisoprenoid of significant length. This causes them to be incorporated exclusively as solutes into phospholipid bilayers, as is ubiquinone. A menaquinol can combine with molecular oxygen to produce a base strong enough to remove a proton from carbon 4 of glutamates in proteins so that the resulting enolate can be carboxylated[313]

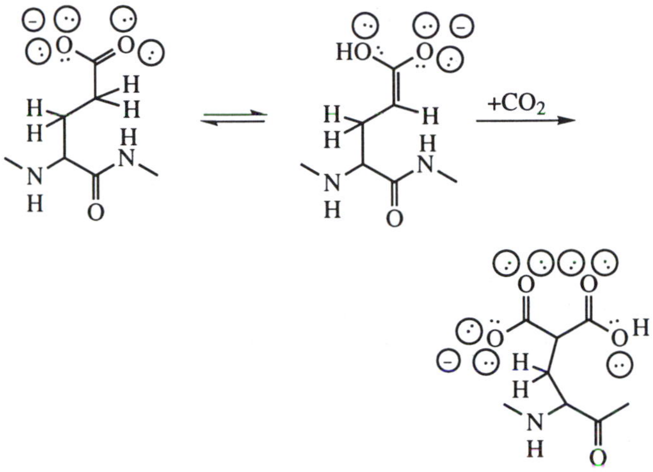

$+CO_2$

(2–202)

in a posttranslational modification. As the carboxylation occurs, the menaquinol is monooxygenated by one of the atoms of molecular oxygen to form the menaquinone epoxide[314]

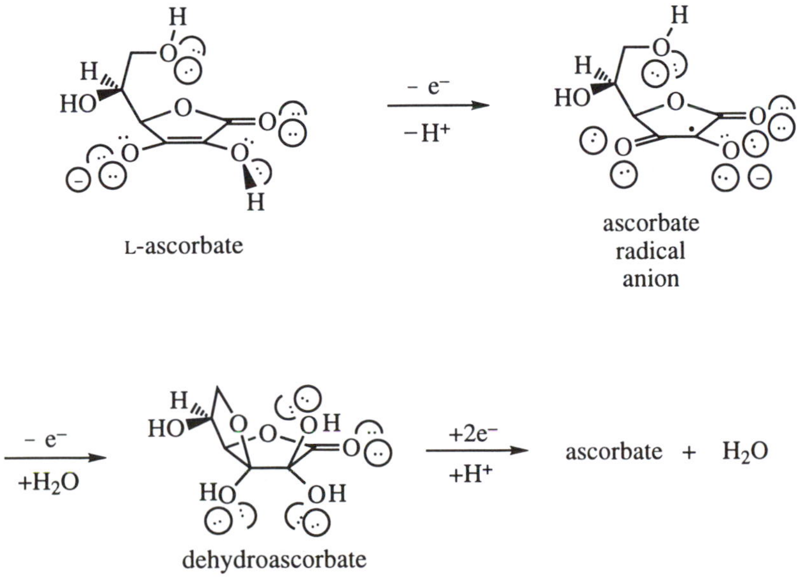

L-ascorbate

$-\ e^-$ / $-H^+$

ascorbate radical anion

$-\ e^-$ / $+H_2O$

dehydroascorbate

$+2e^-$ / $+H^+$ ascorbate + $H_2O$

**Figure 2–32:** Ascorbate as a reductant. L-Ascorbate is a reductant that can provide one electron in each of two steps because the radical anion of ascorbate[309] is a stable species. In this radical anion, the unpaired spin is delocalized over the three carbons and the three oxygens, mostly on the carbons but with significant spin density on the three oxygens. Dehydroascorbate is readily reduced to ascorbate by thiols such as glutathione.

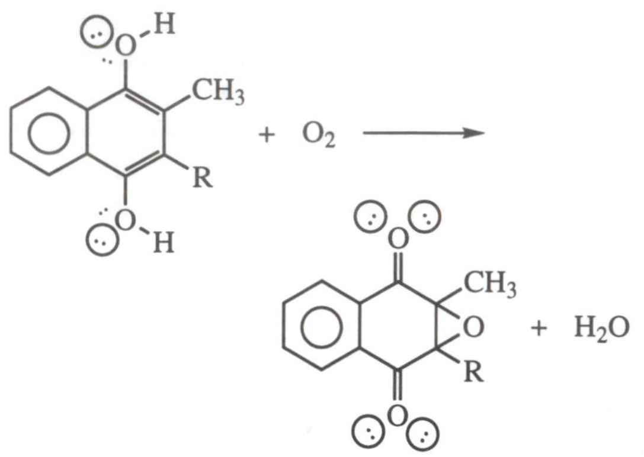

(2–203)

When there is sufficient carbon dioxide present, one mole of the menaquinone epoxide is produced for every mole of glutamic acid carboxylated.[315] When there is insufficient carbon dioxide present, the enolate formed picks up a proton from the solution to regenerate the glutamate.[313] Both the **carboxylation** (Reaction 2–202) and the **epoxidation** (Reaction 2–203) are performed by the same enzyme.[316] The menaquinone epoxide is readily reduced back to the dihydroquinone to be used again.

It is during the epoxidation of the vitamin K that the necessary base is formed.[313] One of the two oxygen atoms on a molecular oxygen ends up as water. As that oxygen atom is reduced during the epoxidation, its $\sigma$ lone pairs become stronger and stronger bases. If these $\sigma$ lone pairs on this one oxygen atom are positioned in such a way that the only proton in their vicinity is the hydrogen on carbon 4 of the glutamate, one of these lone pairs will be forced to remove that proton at some point in the reduction (Figure 2–33). In the extreme, oxide dianion could be formed if the reduction were sufficiently exergonic.

## *Suggested Reading*

Groves, J.T., Haushalter, R.C., Nakamura, M., Nemo, T.E., & Evans, B.J. (1981) High-Valent Iron–Porphyrin Complexes Related to Peroxidase and Cytochrome P-450, *J. Am. Chem. Soc. 103*, 2884–2886.

Branchaud, B.P., & Walsh, C.T. (1985) Functional Group Diversity in Enzymatic Oxygenation Reactions Catalyzed by Bacterial Flavin-Containing Cyclohexanone Oxygenase, *J. Am. Chem. Soc. 107*, 2153–2161.

### PROBLEM 2–17

The enzyme 4-hydroxyphenylpyruvate dioxygenase catalyzes the reaction

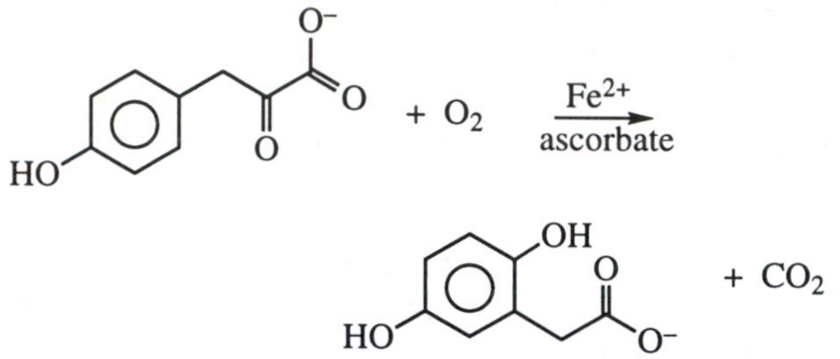

Assume that it is an intramolecular 2-oxoacid dioxygenase and write a mechanism to explain the reaction that involves a peroxy acid and an arene oxide.

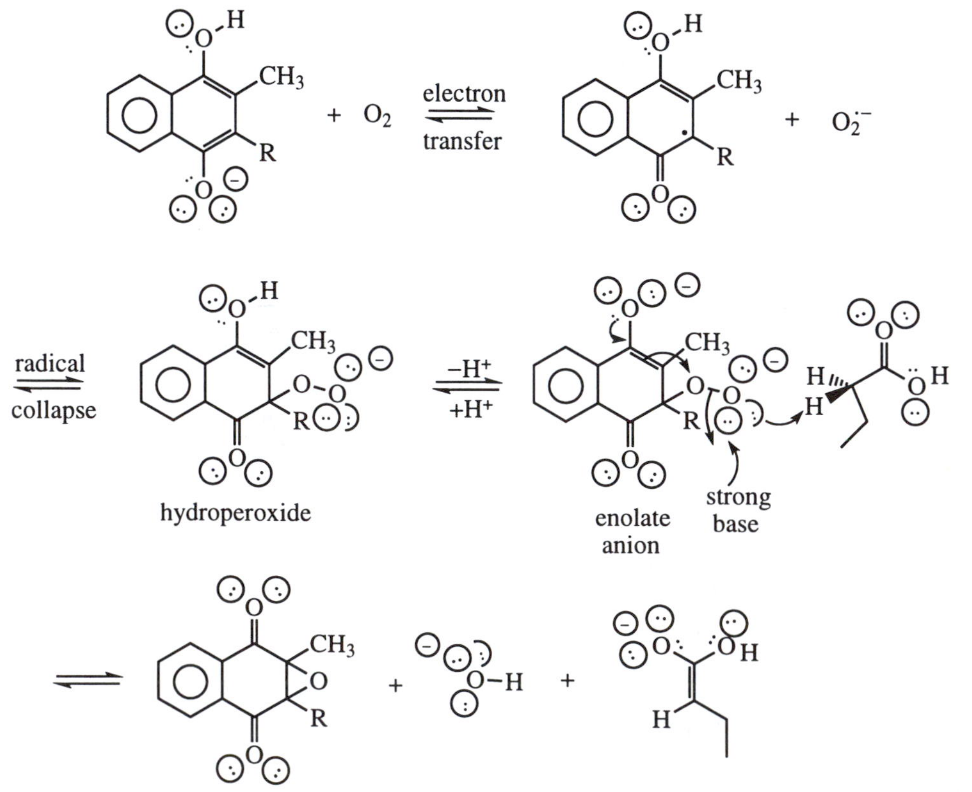

**Figure 2–33:** Mechanism for the formation of glutamyl enolate by oxygen and menaquinol. The menaquinol anion would reduce molecular oxygen to superoxide anion, which would form the peroxyquinone by spin inversion and radical collapse. An intramolecular reduction of the peripheral oxygen of the peroxy group to oxide dianion would produce a base strong enough to remove the $\alpha$-proton from the side chain of a glutamate to produce the necessary enolate. Menaquinone epoxide would be produced simultaneously.

# Selenocysteine

Selenium is the element in group 6 below sulfur on the periodic table. An atom of selenium functions as a coenzyme in several enzymes from diverse sources: glycine reductase, nicotinic acid hydroxylase, and xanthine dehydrogenase from clostridia; hydrogenase from *Methanococcus vanniellii*; formate dehydrogenase from obligate and facultative anaerobic bacteria; and glutathione peroxidase from mammals and birds.[317] All of these enzymes are engaged in either electron transfer or the processing of oxygen or peroxides. In the enzymes in which this has been determined, the selenium atom replaces the sulfur in one of the cysteines in the protein.[318–320] The selenocysteine is inserted into the amino acid sequence during translation rather than as a posttranslational modification.[317]

In the glycine reductase system from clostridia, the selenocysteine is present in a small protein that is thought to act as a shuttle for electrons in the reduction reaction.[318] This small protein also contains two ysteines, and it is possible that a mixed selenothiocystine is involved in its role in electron transfer

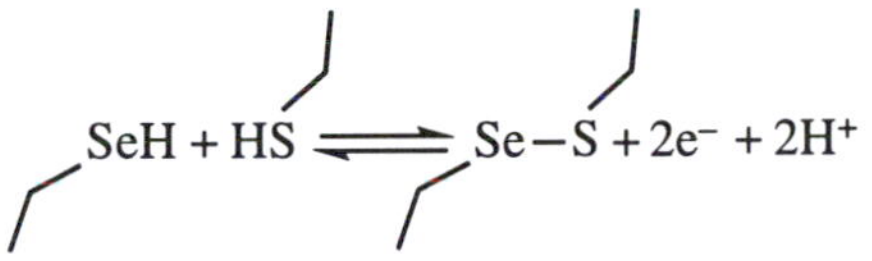

$$\text{SeH} + \text{HS} \rightleftharpoons \text{Se}-\text{S} + 2e^- + 2\text{H}^+ \qquad (2\text{–}204)$$

Selenocysteinyl radical is more stable than cysteinyl radical, and this capacity may permit the protein to participate in one-electron transfer with a radical intermediate. It is also possible that the selenothiocystine radical anion ($-Se^{\cdot}-S^{-}-$) has particularly advantageous properties.

Glutathione peroxidase serves as a paradigm for the interaction of selenocysteine with oxygen. The enzyme catalyzes the reaction

$$\text{ROOH} + 2\text{GSH} \rightleftharpoons \text{ROH} + \text{H}_2\text{O} + \text{GSSG} \qquad (2\text{–}205)$$

where GSH is glutathione. When the enzyme has been reduced by glutathione, the selenocysteine is in the selenol form (**2–65**)

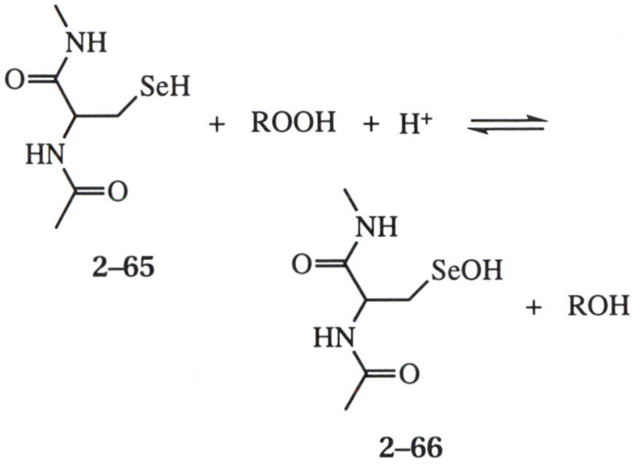

**2–65** + ROOH + H+ ⇌ **2–66** + ROH (2–206)

but when the enzyme is oxidized with a peroxide, the selenium is in the seleninyl form (**2–66**).[321] From these observations, a plausible mechanism for this reaction can be formulated (Figure 2–34), which is consistent with the kinetics of the enzymatic reaction.[322] In this mechanism, selenium is used because it is more nucleophilic than sulfur and the selenium–sulfur bond should be weaker than a sulfur–sulfur bond.

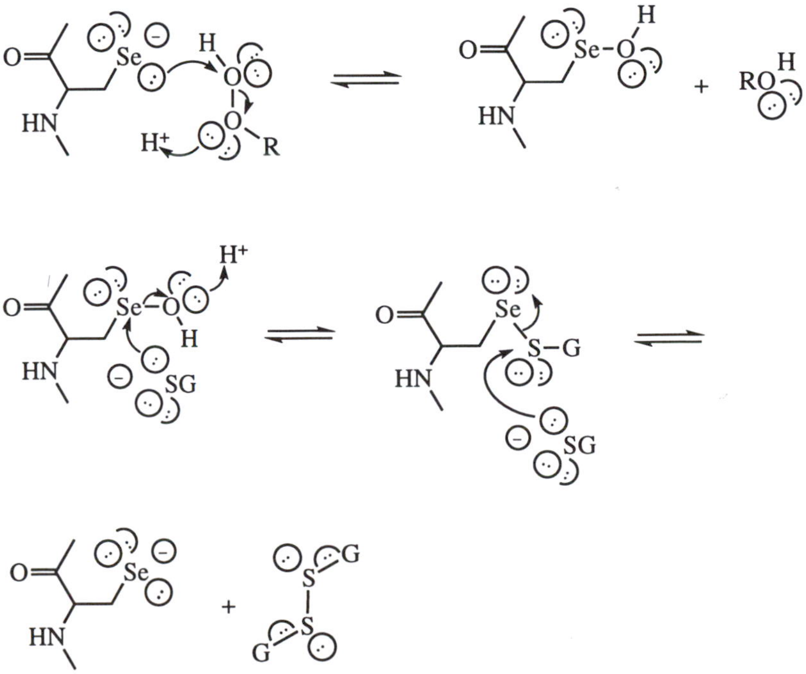

**Figure 2–34:** Mechanism for the reduction of peroxides catalyzed by glutathione peroxidase.[322] The mechanism is a series of reactions, each analogous to disulfide interchange.[43a] Selenium is used because it is a stronger nucleophile than sulfur and more able to cleave an oxygen–oxygen bond heterolytically.

# Coenzyme $B_{12}$

Coenzyme $B_{12}$ contains hexacoordinate cobalt in a macrocyclic structure vaguely resembling a porphyrin (Figure 2–35).[323,324] It is tightly but noncovalently bound to the enzymes in which it is a coenzyme. Cobalt is inserted into the central heterocycle in place of one proton to form a corrin

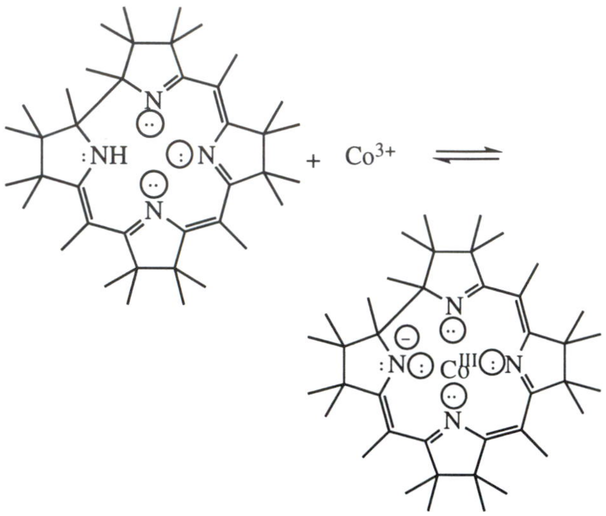

(2–207)

The corrin, unlike the hemes and chlorophylls, is not aromatic. The cobalt is coordinated at its fifth position by dimethylbenzimidazole

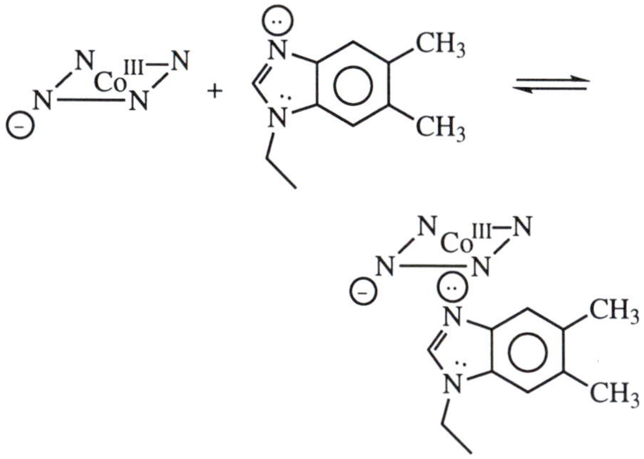

(2–208)

At its sixth position, the cobalt is formally occupied by the carbanion of 5′-deoxyadenosine*

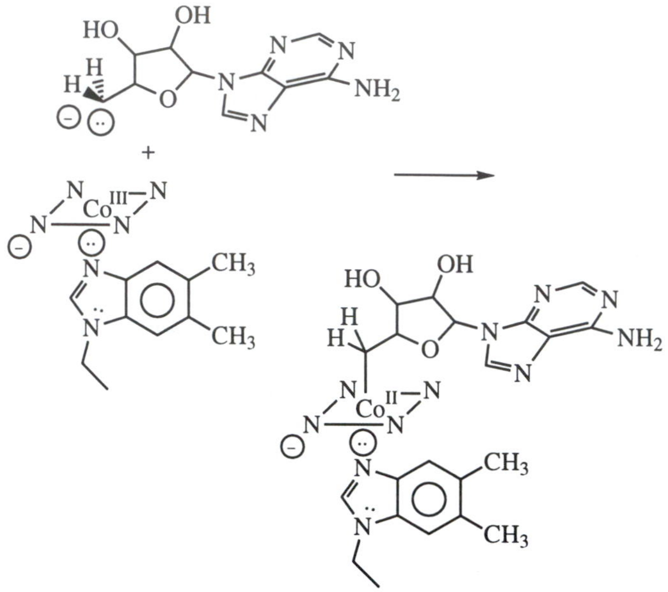

(2–209)

As such, coenzyme $B_{12}$ is an organocobalt compound. The carbon–cobalt bond has been identified as the active location in the coenzyme.

The majority of the reactions in which coenzyme $B_{12}$ participates as a coenzyme can be written as variations on a common theme. Formally, a hydrogen atom undergoes a 1,2 shift in concert with a reciprocal 1,2 shift by one of the substituents on the carbon to which the hydrogen moves

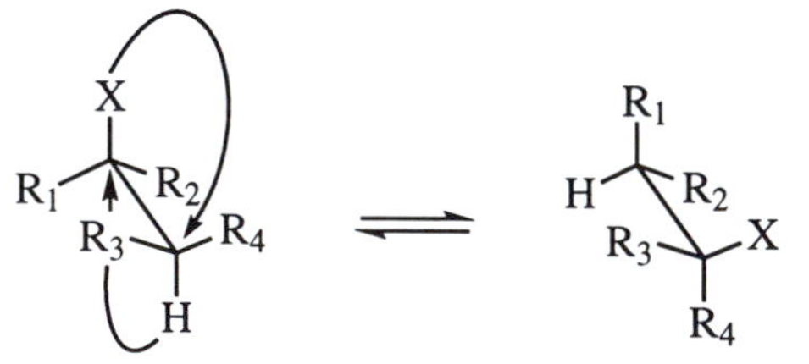

(2–210)

In Reaction 2–210 the transfers are shown with inversion of configuration, but either inversion of configuration or retention of configuration is observed, depending upon the particular enzyme,[325] and this requires that the mechanism of the catalysis performed by coenzyme $B_{12}$ be consistent with either stereochemical outcome. One of the two participants migrating is always a hydrogen but the other can be any one of a number of apparently unrelated functionalities (Table 2–2).

*The bond between carbon and cobalt is written as a covalent bond and the formal negative charge that the carbon would have if it were only a complexed base has been disregarded because of the covalency. The carbon–cobalt bond is covalent because the carbanion is so basic.

**Figure 2–35:** Drawing of a crystallographic molecular model of coenzyme $B_{12}$.[323] Crystals were produced by slowly diffusing acetone into a concentrated solution of coenzyme $B_{12}$ in water. The diagram[324] of the chemical structure of coenzyme $B_{12}$ (A) can be used as a guide to the stereo drawing (B) of the crystallographic molecular model[323] built from maps of electron density derived from a crystallographic analysis. Panel A reprinted with permission from ref 324; copyright 1976 American Chemical Society. Panel B reprinted with permission from ref 323; copyright 1968 Royal Society.

A

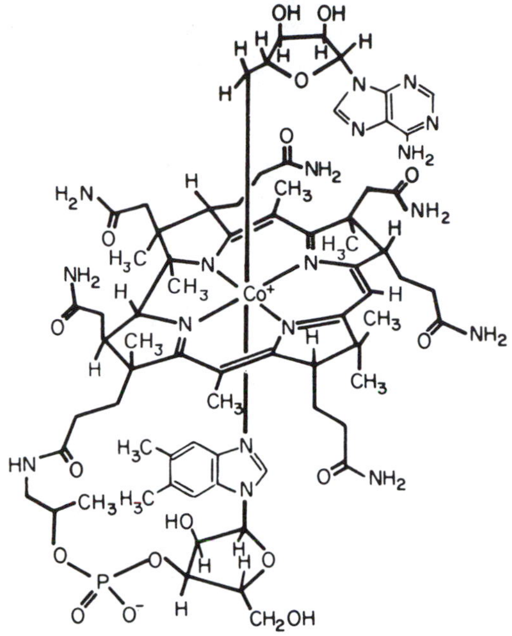

B

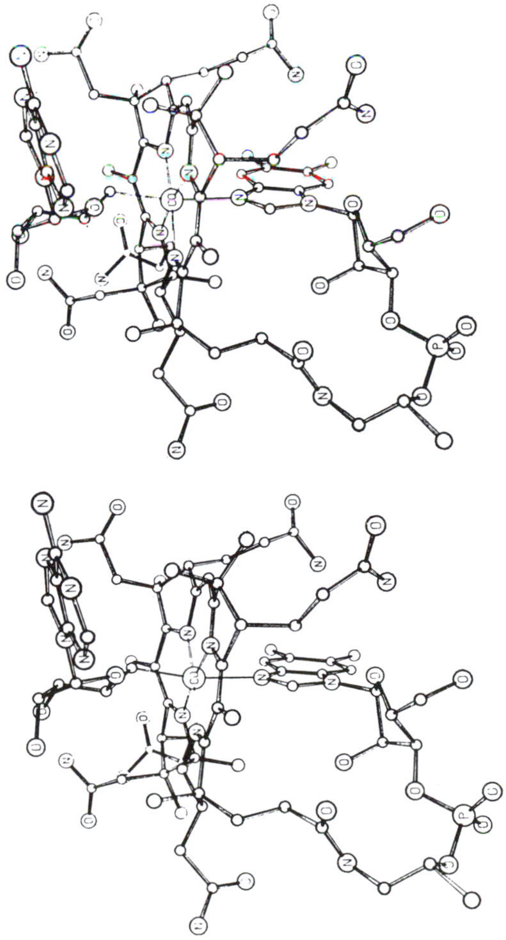

**Table 2–2: Substituents Migrating in Reactions Catalyzed by Coenzyme $B_{12}$**

| enzyme | X in Reaction 2–210 |
|---|---|
| ethanolamine ammonia–lyase[326] | $-NH_3^+$ |
| propanediol dehydratase[327] | –OH |
| β-lysine mutase | $-NH_3^+$ |
| methylaspartate mutase[328] | $-C(H)(COO^-)(NH_3^+)$ |
| methylmalonyl-CoA mutase[329] | $-C(=O)SCoA$ |
| 2-methyleneglutarate mutase[330] | $-C(=CH_2)COO^-$ |

An example of one of these reactions is that catalyzed by methylaspartate mutase. This reaction proceeds with inversion of configuration[325] when the 3-deuterosubstrate is used

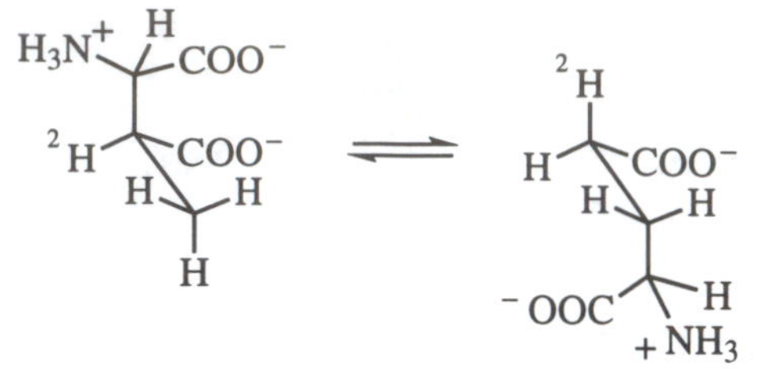

(2–211)

When the deoxyadenosine is removed from coenzyme $B_{12}$ and the sixth ligand site on cobalt is vacant, the resulting compound is referred to as **cobalamine**. The cobalt in cobalamine can exist in any one of three oxidation states.

As cobalt is element 27, cobalt(I )has an even number of electrons. When the cobalt in a cobalamine is cobalt(I), there is a nucleophilic lone pair of electrons on the cobalt at the site of the sixth ligand. **Cob(I)alamine** reacts nucleophilically with primary alkyl halides,[331] and with secondary alkyl halides under acidic conditions,[332] to yield alkylcobalamines

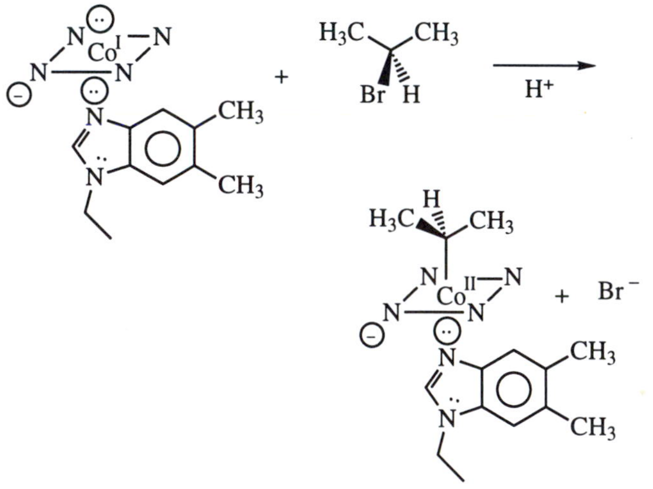

(2–212)

This nucleophilic form of the coenzyme is the one that reacts with methanol in the first step of methane production.[333] It has been proposed that this is an $S_N2$ reaction

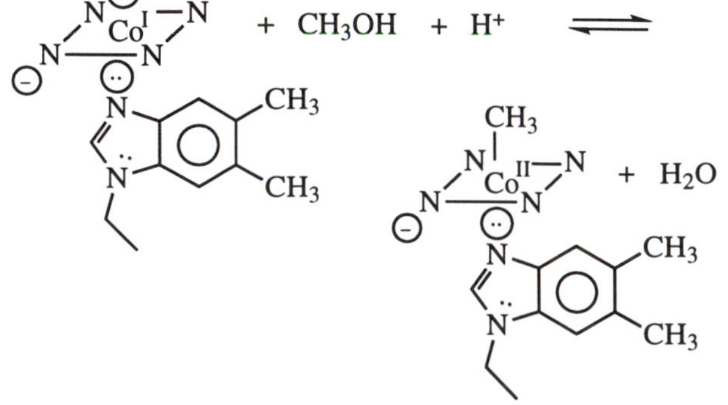

(2–213)

occurring at the active site of an enzyme under general acid catalysis to activate the methyl group.

When the carbon–cobalt bond of an alkylcobalamine breaks homolytically, the species generated is **cob(II)alamine**. In this state, the cobalt has one unpaired electron and is a radical. The bond dissociation energy for this carbon–cobalt bond[334] in an alkylcobalamine such as coenzyme $B_{12}$ is about 105 kJ $mol^{-1}$

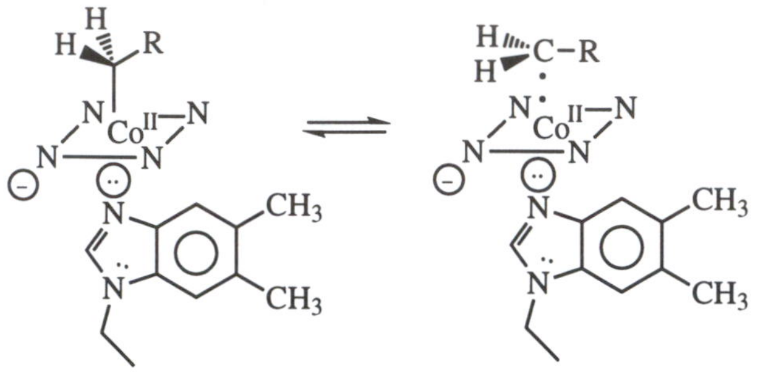

(2–214)

This carbon–cobalt bond also cleaves homolytically upon anaerobic photolysis to yield cob(II)alamine.[335]

The fully oxidized form of cobalamine, **cob(III)alamine**, is the form in which the coenzyme is prepared as a dietary supplement. Cob(III)alamine, like ferric heme, is electrophilic and reacts readily with anions such as cyanide to yield adducts in which the anion occupies the sixth ligand site. It is the cyanide adduct that is used commercially as a vitamin.

In a reaction catalyzed by coenzyme $B_{12}$, the hydrogen that migrates across the substrate (Scheme 2–210) does not exchange with protons in the solvent.[336] If, however, substrate labeled at this hydrogen is mixed with the appropriate enzyme, the hydrogens at the 5′ carbon of the deoxyadenosine of the coenzyme $B_{12}$ become labeled.[337] The reverse reaction also occurs

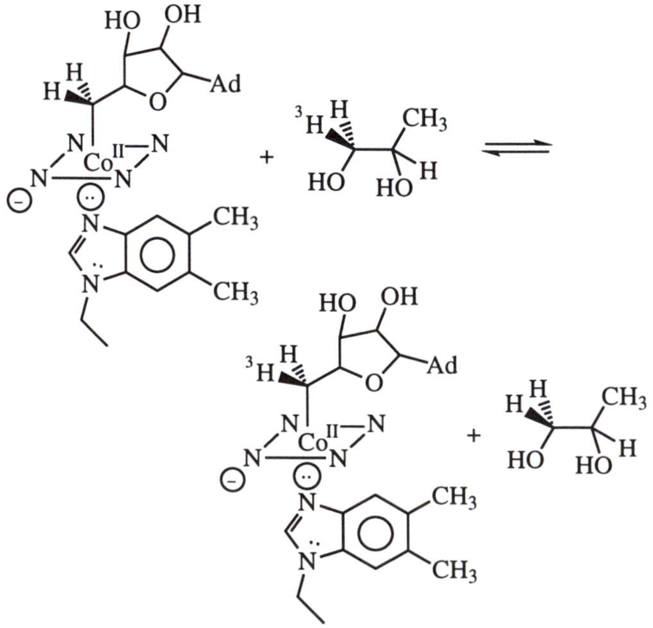

(2–215)

The hydrogens equilibrating with the deoxyadenosine of the coenzyme do so in such a way that they occupy one of three equivalent locations when they enter the 5′ carbon of the deoxyadenosine.[336] When 2-aminopropanol, rather than the physiological reactant, ethanolamine, is added to ethanolamine ammonia–lyase, the carbon–cobalt bond cleaves rapidly to produce 5′-deoxyadenosine and cob(II)alamine, and this enzyme, containing the cleaved coenzyme, is still fully active in the normal ethanolamine ammonia–lyase reaction.[338] All of these observations are consistent with, but do not prove, the existence of a step in the mechanism for the normal enzymatic reactions in which the adenosine–cobalt bond cleaves and is replaced by a bond between the substrate and cobalt

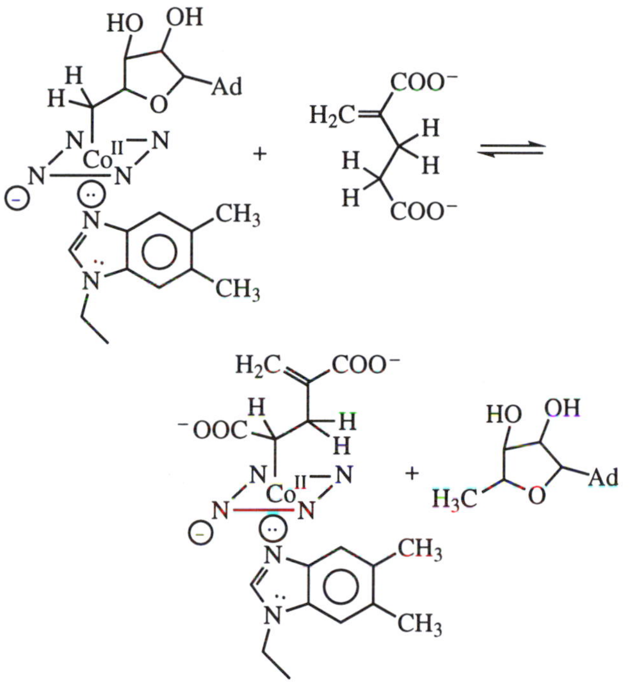

(2–216)

In this rearrangement, the hydrogen lost from the substrate is the hydrogen gained on the deoxyadenosine, and conversely.

Regardless of the mechanism proposed for Reaction 2–216, it is not possible to avoid the involvement of highly unstable and reactive species. The three possible immediate cobalt products, cob(I)alamine, cob(II)alamine, and cob(III)alamine, are all stable compounds. The three respective carbon fragments, a **deoxyadenosyl 5′-carbocation**, a **deoxyadenosyl 5′-radical**, or a **deoxyadenosyl 5′-carbanion**, are all unstable and highly reactive. If any one of these were used to remove the unactivated aliphatic hydrogen from the substrate, it should also be able to remove hydrogens from carbons of the amino acids in the surrounding protein. Various mechanisms for the transformation written in Reaction 2–216, usually involving cob(II)alamine radical and deoxyadenosyl radical,[279,286] have been proposed. It has also been pointed out, however, that Reaction 2–216 itself is unnecessary to explain the exchange of hydrogens among the substrates, products, and coenzymes,[339] and no direct evidence for a carbon–cobalt bond between coenzyme and substrate has been obtained.

Several organocobalt compounds, however, have been synthesized that can undergo isomerizations related to those involved in the reactions catalyzed by coenzyme $B_{12}$. In every instance, it is tacitly assumed that Reaction 2–216 in one of its particular forms is the first step in the respective overall reaction, and an alkylcobalamine or other alkylcobalt compound is synthesized in which the cobalt replaces the hydrogen on the substrate that would migrate in the enzymatically catalyzed reaction. Formally, the rearrangement that is studied, in analogy with Scheme 2–209, is

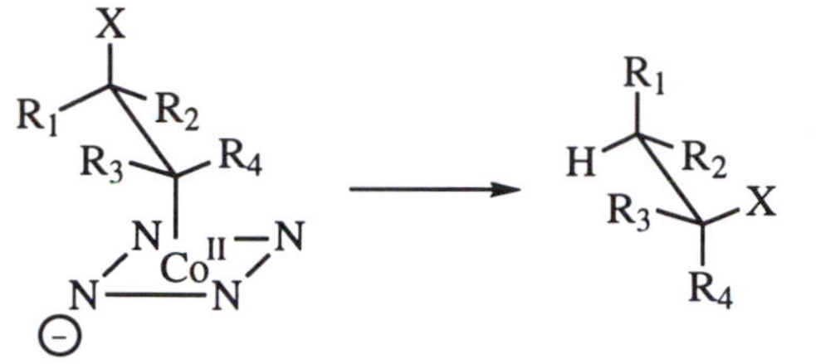

(2–217)

For example, it has been shown that the [(dimethoxy)carboxymethyl]cobalamine **2–67** produces succinic acid in 4% yield upon exposure to water at 25 °C[340]

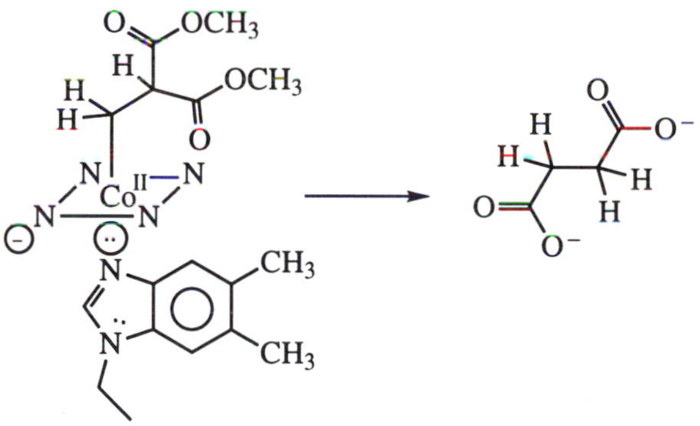

**2–67** (2–218)

and this is proposed as a model for the reaction catalyzed by methylmalonyl-CoA mutase (Table 2–2)

(2–219)

The state of the cobalt at the end of the reaction and the origin of the hydrogen that replaces the cobalamine provide insight into the mechanism of these model reactions. From the observations made so far, it is clear that no one mechanism is common to all of these reactions. The issues are the role of the cobalt and the status of the intermediates.

Electrophilic pyridine cob(III)aloxime

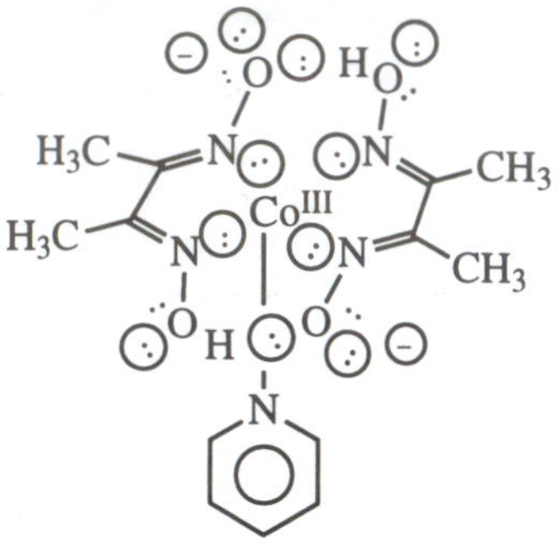

2–68

is a model compound for cob(III)alamine.[339] Pyridine cob(III)aloxime can react with an enol ether in the presence of ethanol to produce an alkylcobaloxime[341]

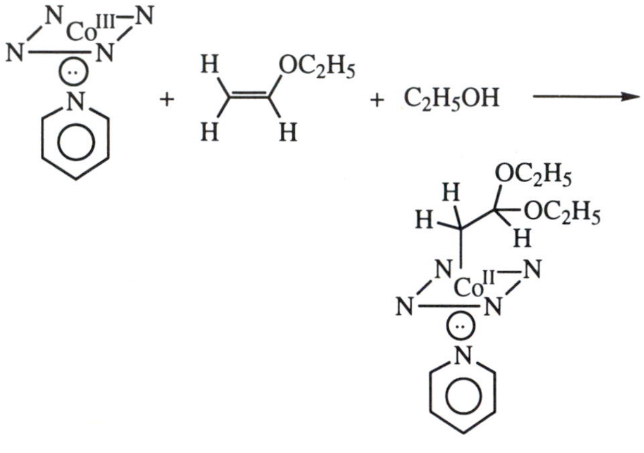

(2–220)

It has been proposed that an intermediate in this reaction is a $\pi$ complex between the electrophilic cobalt(III) and the electron-rich olefin of the enol ether

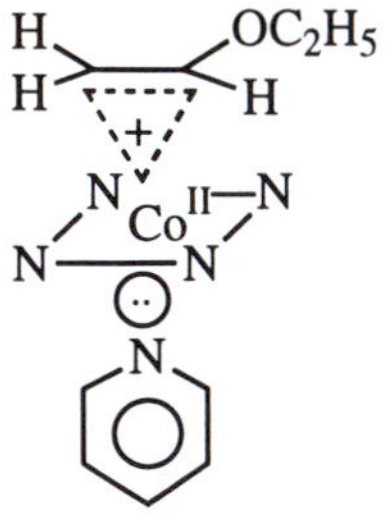

2–69

This adduct would be trapped by ethoxide anion to produce the product. Such an intermediate could also explain the isomerization occurring during the methanolysis of 2-[(methylcarbonato)-[2-$^{13}$C]ethyl]cobaloxime[324]

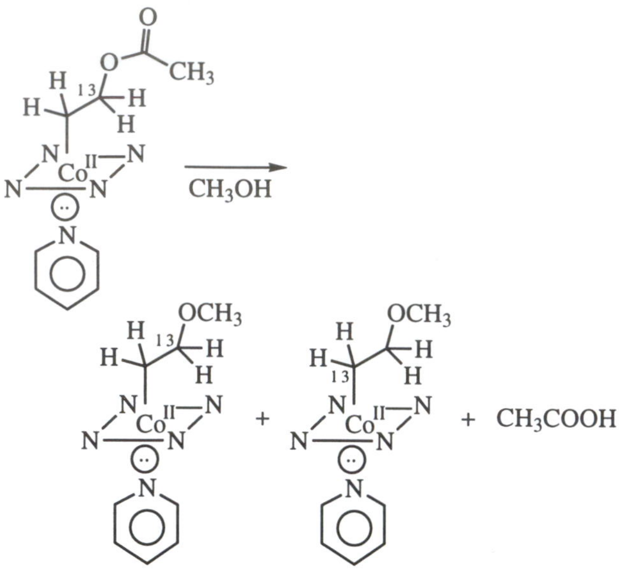

(2–221)

In this reaction the methylcarbonato functional group would be the leaving group in an $S_N1$ reaction, leading to the intermediate $\pi$ complex.

On the basis of these observations, the involvement of such a $\pi$ complex in the reaction catalyzed by propanediol dehydratase (Table 2–2)

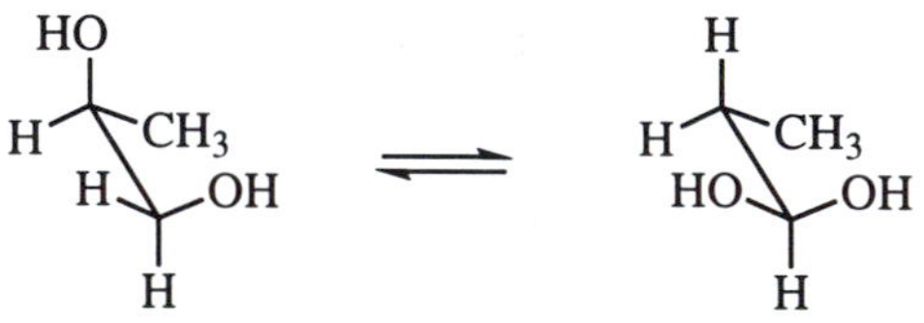

(2–222)

has been proposed. When, however, the postulated intermediate alkylcobaloxime **2–70** was prepared, it would not produce the expected 2-cobaloxime of acetaldehyde[342]

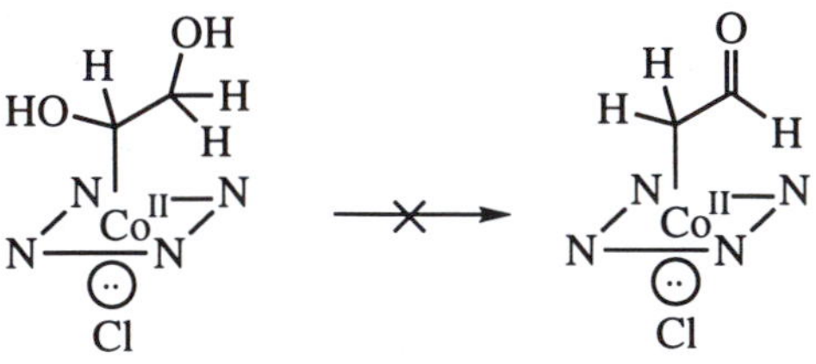

(2–223)

2–70

even though it would produce high yields of acetaldehyde.[343] The acetaldehyde produced was shown to result from homolytic cleavage of alkylcobaloxime **2–70** which produced a diol radical that rearranged and abstracted a hydrogen from the solvent.[342] The other product was the cob(II)aloxime

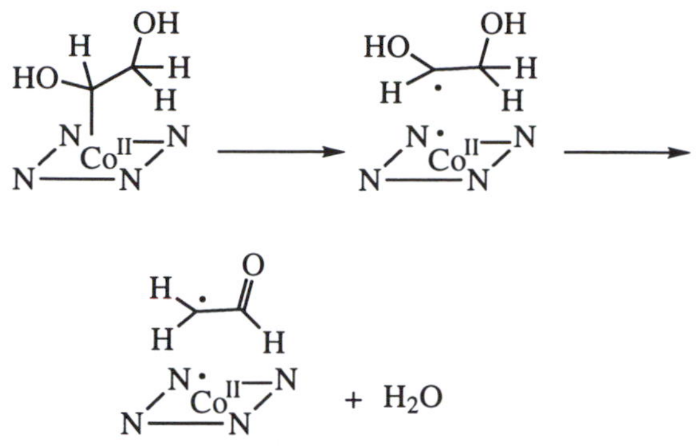

(2–224)

These results suggest that the involvement of an alkylcobalamine such as **2–67** in a reaction catalyzed by coenzyme $B_{12}$ may not be necessary; merely the formation of the radical might be sufficient.

Several other observations also suggest that there are rearrangements of the alkyl groups of alkylcobalamines and alkylcobaloximes which do not occur on cobalt but occur within the alkyl substituents themselves after they have been ejected from the cobalt. Alkylcobalamine **2–71**

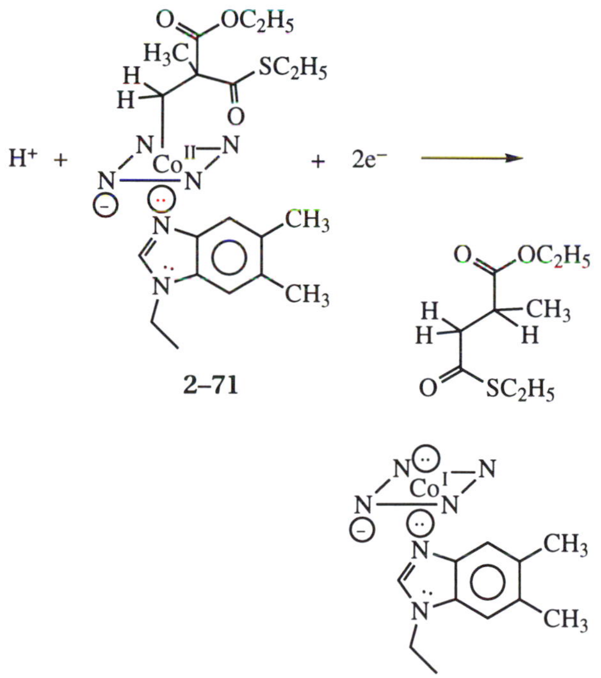

**2–71**

(2–225)

under reductive conditions, produces high yields of the product of the desired rearrangement, 1-ethyl-4-thioethyl-2-methylsuccinate.[335] This reaction serves as a model for the rearrangement catalyzed by methylmalonyl-CoA mutase (Table 2–2 and Reaction 2–219). A similar rearrangement has been observed with the alkylcobalamines of either methylitaconic acid[344] or bis(tetrahydropyranyl)methylitaconate,[330] which would be models for the reaction catalyzed by 2-methyleneglutarate mutase (Table 2–2)

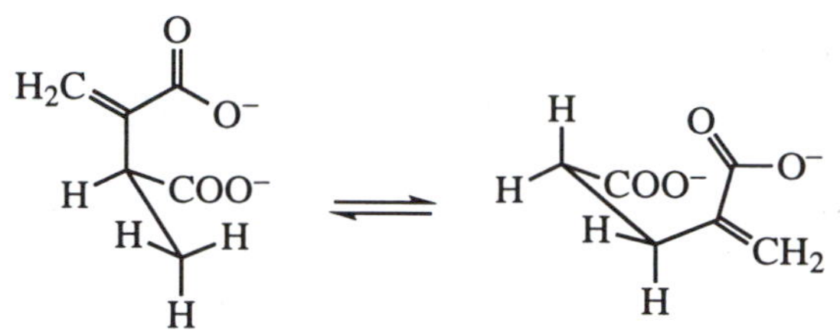

(2–226)

Unlike the model reaction for propanediol dehydratase (Reaction 2–224), however, homolytic cleavage of these alkylcobalamines produced no rearranged products. When the rearrangement of the bis(tetrahydropyranyl)-2-methylcobalaminylitaconate was performed in $^2H_2O$, a deuterium atom was found in the products at the locations which the hydrogens migrating during the enzymatic reaction would have occupied[330]

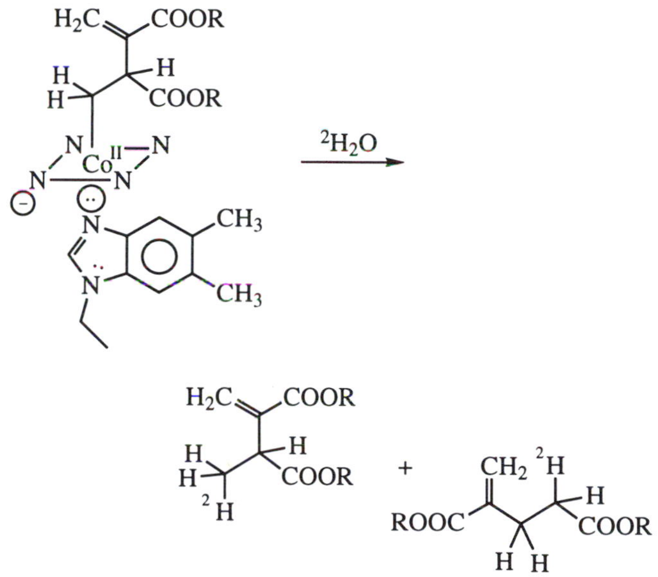

(2–227)

Although water gives up a proton readily, a hydrogen atom is abstracted from water only with great difficulty. This result, the requirement for coincident reduction of cobalt to yield rearranged products, and the lack of any rearranged products upon homolysis has led to the conclusion that these rearrangements occur within a carbanion released from the cobalt directly or upon reduction

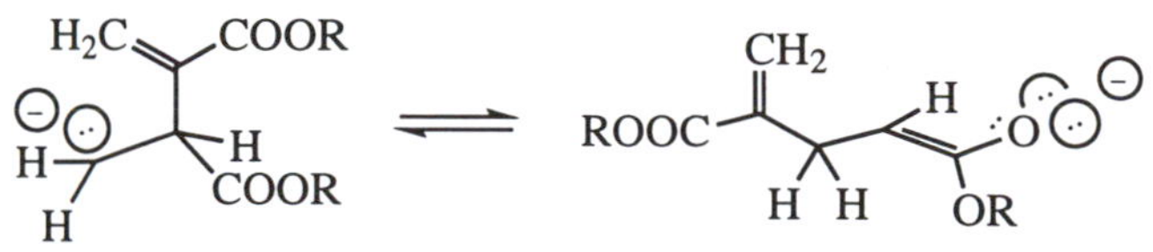

(2–228)

The carbanions released during these reactions would pick up a proton from the solvent following rearrangement.[330]

Therefore, model reactions involving rearrangements on cobalt (Reaction 2–221), rearrangements within radicals ejected from cobalt (Reaction 2–224), and rearrangements within carbanions ejected from cobalt (Reaction 2–228) have been presented as analogues of the rearrangements catalyzed within enzymes by coenzyme $B_{12}$. The ejection of the equivalent of a carbocation from alkylcobaloximes and cobalamines has also been documented.[324,339]

It is unclear how coenzyme $B_{12}$, when it is within an enzyme, catalyzes the reactions for which it is responsible or even if there is a mechanism common to these reactions.

## Suggested Reading

Grate, J.H., Grate, J.W., & Schrauzer, G.N. (1982) Studies on Vitamin $B_{12}$ and Related Compounds: Synthesis and Reactions of Organocobalamines Relevant to the Mechanism of the Methylmalonyl-CoA–Succinyl-CoA Mutase Enzyme, *J. Am. Chem. Soc. 104*, 1588–1594.

### PROBLEM 2–18

Each of the following reactions is catalyzed by an enzyme that has a coenzyme in its active site. What is the particular coenzyme used in each of these reactions? Do not write mechanisms.

**A.** $H^+$ + $^-$O(C=O)C(=O)SCoA ⇌ $CO_2$ + HC(=O)SCoA

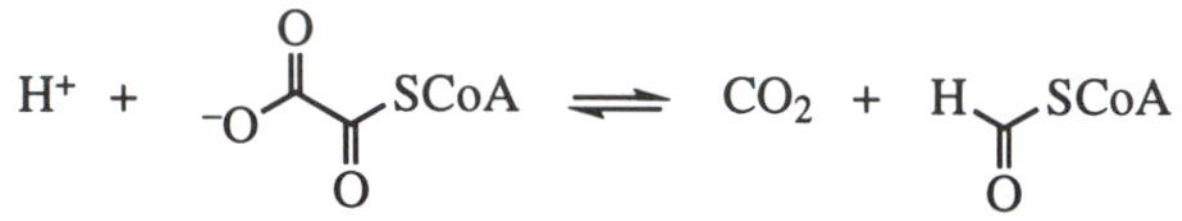

**B.** $2H^+ + 2e^- +$ [steroid: O=, $H_3C$, H, H, $H_3C$, O, $CH_2OH$, OH] $+ O_2$ ⇌ [steroid: O=, $H_3C$, HO, H, $H_3C$, O, $CH_2OH$, OH] $+ H_2O$

**C.** MgATP + $CO_2$ + $H_2N$C(=O)$NH_2$ + $H_2O$ ⇌ MgADP + $HOPO_3^{2-}$ + $2NH_3$ + $2CO_2$

**D.** [$H_3C$, $CH_3$, H, H, H, $COO^-$, $NH_3^+$] ⇌ [$H_3C$, $CH_3$, H, $H_3N_+$, H, $COO^-$, H]

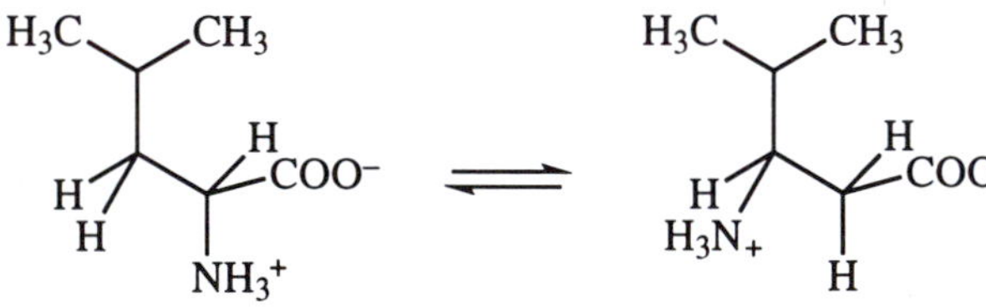

**E.** [S=O, H, $COO^-$, $NH_3^+$] ⇌ [S–OH] + [H, H, $COO^-$, $NH_3^+$]

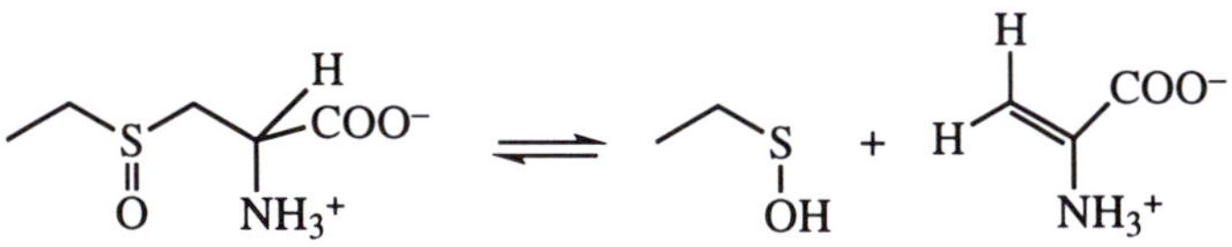

**F.** [$^{2-}O_3PO$, OH, HO, O, OH] + H(C=O)H ⇌ [$^{2-}O_3PO$, HO, O, H] + [HO, OH, O]

**G.** $H_2O$ + $^-O_3S$CH$_2$C(H)=O ⇌ $^{2-}O_3S$ + $CH_3$C(=O)OH + $H^+$

**H.** [long chain, OH, $H_3N_+$, $OPO_3^{2-}$] ⇌ [long chain, O, H] + $H_3N_+$CH$_2$CH$_2$$OPO_3^{2-}$

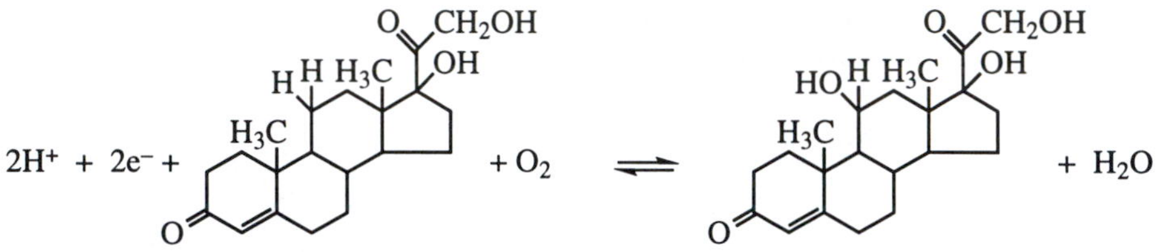

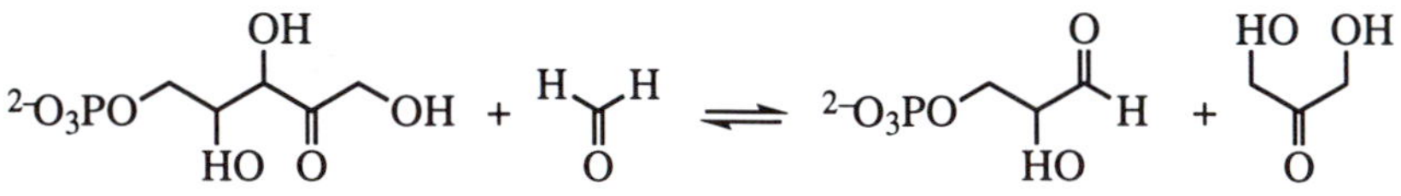

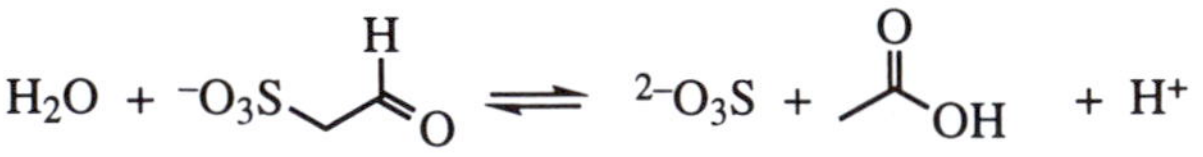

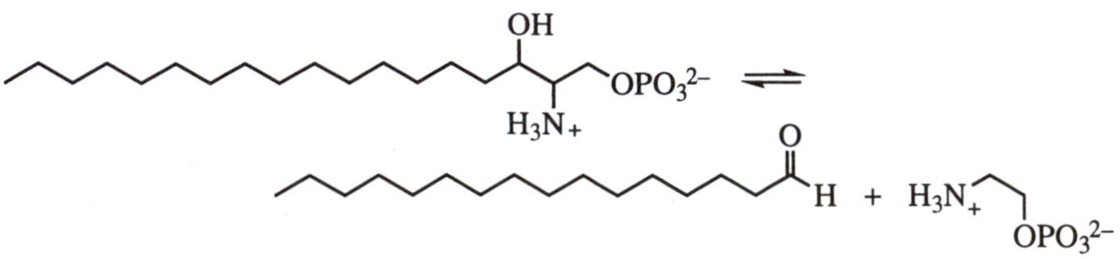

## References


1. Nelsestuen, G.L., & Kirkwood, S. (1971) *J. Biol. Chem. 246,* 7533–7543.

1a. Kyte, J. (1995) *Structure in Protein Chemistry,* p 99, Garland Publishing, New York.

2. Jenkins, W.T., & Sizer, I.W. (1957) *J. Am. Chem. Soc. 79,* 2655–2656.

3. Hughes, R.C., Jenkins, W.T., & Fischer, E.H. (1962) *Proc. Natl. Acad. Sci. U.S.A. 48,* 1615–1618.

4. Ford, G.C., Eichele, G., & Jansonius, J.N. (1980) *Proc. Natl. Acad. Sci. U.S.A. 77,* 2559–2563.

5. Metzler, D.E. (1957) *J. Am. Chem. Soc. 79,* 485–490.

6. Jencks, W.P. (1969) *Catalysis in Chemistry and Enzymology,* pp 133–146, McGraw-Hill, New York.

7. Bergami, M., Marin, G., & Scardi, V. (1968) *Biochem. J. 110,* 471–473.

8. Bruice, T.C., & Benkovic, S.J. (1966) *Bioorganic Mechanisms, Vol. II,* pp 226–300, Benjamin, New York.

9. Weng, S.H., & Leussing, D.L. (1983) *J. Am. Chem. Soc. 105,* 4082–4090.

10. Hogg, J.L., Jencks, D.A., & Jencks, W.P. (1977) *J. Am. Chem. Soc. 99,* 4772–4778.

11. Kiick, D.M., & Cook, P.F. (1983) *Biochemistry 22,* 375–382.

12. Abbott, E.H., & Martell, A.E. (1971) *J. Am. Chem. Soc. 93,* 5852–5856.

13. Jung, M.J., & Metacalf, B.W. (1975) *Biochem. Biophys. Res. Commun. 67,* 301–306.

14. Laver, W.G., Neuberger, A., & Udenfriend, S. (1958) *Biochem. J. 70,* 4–14.

15. Jordan, P.M., & Akhtar, M. (1970) *Biochem. J. 116,* 277–286.

16. Yamada, H., Kumagai, H., Nagate, T., & Yoshida, H. (1970) *Biochem. Biophys. Res. Commun. 39,* 53–58.

17. Dunathan, H.C. (1966) *Proc. Natl. Acad. Sci. U.S.A. 55,* 712–716.

18. Metzler, D.E., & Snell, E.E. (1952) *J. Am. Chem. Soc. 74,* 979–983.

19. Snell, E.E. (1945) *J. Am. Chem. Soc. 67,* 194–197.

20. Thanassi, J., & Fruton, J.S. (1962) *Biochemistry 1,* 975–982.

21. Kumagai, H., & Miles, E.W. (1971) *Biochem. Biophys. Res. Commun. 44,* 1271–1278.

22. Longenecker, J.B., & Snell, E.E. (1956) *Proc. Natl. Acad. Sci. U.S.A. 42,* 221–227.

23. Cram, D.J., & Guthrie, R.D. (1965) *J. Am. Chem. Soc. 87,* 397–398.

24. Blake, M.I., Siegel, F.P., Katz, J.J., & Kilpatrick, M. (1963) *J. Am. Chem. Soc. 85,* 294–297.

25. Zimmerman, S.C., Czarnik, A.W., & Breslow, R. (1983) *J. Am. Chem. Soc. 105,* 1694–1695.

26. Posner, B.I., & Flavin, M. (1972) *J. Biol. Chem. 247,* 6402–6411.

26a. Kyte, J. (1995) *Structure in Protein Chemistry,* p 94, Garland Publishing, New York.

27. Breslow, R. (1958) *J. Am. Chem. Soc. 80,* 3719–3726.

28. Breslow, R. (1957) *J. Am. Chem. Soc. 79,* 1762–1763.

29. Suchy, J., Mieyal, J.J., Bantle, G., & Sable, H.Z. (1972) *J. Biol. Chem. 247,* 5905–5912.

30. Haake, P., Bausher, L.P., & Miller, W.B. (1969) *J. Am. Chem. Soc. 91,* 1113–1119.

31. Hopmann, R.F.W., & Brugnoni, G.P. (1973) *Nature New Biol. 246,* 157–158.

32. Bruice, T.C., & Benkovic, S.J. (1966) *Bioorganic Mechanisms, Vol. II,* pp 204–226, Benjamin, New York.

33. Jordan, F. (1974) *J. Am. Chem. Soc. 96,* 3623–3630.

34. Gallo, A.A., & Sable, H.Z. (1974) *J. Biol. Chem. 249,* 1382–1389.

35. Yount, R.G., & Metzler, D.E. (1959) *J. Biol. Chem. 234,* 738–741.

36. Sax, M., Pulsinelli, P., & Pletcher, J. (1974) *J. Am. Chem. Soc. 96,* 155–165.

37. Mieyal, J.J., Bantle, G., Votaw, R.G., Rosner, I.A., & Sable, H.Z. (1971) *J. Biol. Chem. 246,* 5213–5219.

38. Crosby, J., Stone, R., & Lienhard, G.E. (1970) *J. Am. Chem. Soc. 92,* 2891–2900.

39. Kluger, R., Chin, J., & Smyth, T. (1981) *J. Am. Chem. Soc. 103,* 884–888.

40. Holzer, H., Kattermann, R., & Busch, D. (1962) *Biochem. Biophys. Res. Commun. 7,* 167–172.

41. Chen, G.C., & Jordan, F. (1984) *Biochemistry 23,* 3576–3582.

42. Crosby, J., & Lienhard, G.E. (1970) *J. Am. Chem. Soc. 92,* 5707–5716.

43. Rastetter, W.H., Adams, J., Frost, J.W., Nummy, L.J., Frommer, J.E., & Roberts, K.B. (1979) *J. Am. Chem. Soc. 101,* 2752–2753.

43a. Kyte, J. (1995) *Structure in Protein Chemistry,* p 100, Garland Publishing, New York.

44. Das, M., Koike, M., & Reed, L.J. (1961) *Proc. Natl. Acad. Sci. U.S.A. 47,* 753–759.

45. Asmus, K.D. (1983) in *Radioprotectors and Anticarcinogens* (Nygaard, O. F., & Simic, M. D., Eds.) pp 23–42, Academic Press, New York.

46. Flournoy, D.S., & Frey, P.A. (1986) *Biochemistry 25,* 6036–6043.

47. Cooper, T.G., & Wood, H.G. (1971) *J. Biol. Chem. 246,* 5488–5490.

48. Jencks, W.P. (1976) in *Handbook of Biochemistry and Molecular Biology, 3rd Edition, Physical and Chemical Data, Vol. I* (Fasman, G.D., Ed.) pp 296–304, CRC Press, Cleveland, OH.

49. Dimroth, P. (1982) *Eur. J. Biochem. 121,* 443–449.

50. Moss, J., & Lane, M.D. (1971) *Adv. Enzymol. Relat. Areas Mol. Biol. 35,* 321–442.

51. deTitta, G.T., Edmonds, J.W., Stallins, W., & Donohue, J. (1976) *J. Am. Chem. Soc. 98,* 1920–1926.

52. Fry, D.C., Fox, T.L., Lane, M.D., & Mildvan, A.S. (1985) *J. Am. Chem. Soc. 107,* 7659–7665.

53. Guchait, R.B., Polakis, S.E., Hollis, D., Fenselau, C., & Lane, M.D. (1974) *J. Biol. Chem. 249,* 6646–6656.

54. Knappe, J., Ringelmann, E., & Lynen, F. (1961) *Biochem. Z. 335,* 168–176.

55. Kaziro, Y., Hass, L.F., Boyer, P.D., & Ochoa, S. (1962) *J. Biol. Chem. 237,* 1460–1468.

56. Ogita, T., & Knowles, J.R. (1988) *Biochemistry 27,* 8028–8033.

57. Hansen, D.E., & Knowles, J.R. (1985) *J. Am. Chem. Soc. 107,* 8304–8305.

58. Perrin, C.L., & Dwyer, T.J. (1987) *J. Am. Chem. Soc. 109,* 5163–5167.

59. Martin, R.B. (1972) *J. Chem. Soc., Chem. Commun.,* 793–794.

60. Kohn, H. (1976) *J. Am. Chem. Soc. 98,* 3690–3694.

61. Walker, W.H., & Singer, T.P. (1970) *J. Biol. Chem. 245,* 4224–4225.

62. Walker, W.H., Kearney, E.B., Seng, R., & Singer, T.P. (1971) *Biochem. Biophys. Res. Commun. 44,* 287–292.

63. Bruice, T.C. (1976) *Prog. Bioorg. Chem. 4,* 1–87.

64. Eberlein, G., & Bruice, T.C. (1983) *J. Am. Chem. Soc. 105,* 6685–6697.

65. Kemal, C., & Bruice, T.C. (1976) *J. Am. Chem. Soc. 98,* 3955–3964.

66. Tauscher, L., Ghisla, S., & Hemmerich, P. (1973) *Helv. Chim. Acta 56,* 630–643.

67. Bolognesi, M., Ghisla, S., & Incoccia, L. (1978) *Acta Crystallogr. B34,* 821–828.

68. Werner, P.E., & Rönnquist, O. (1970) *Acta Chem. Scand. 24,* 997–109.

69. Moonen, C.T.W., Vervoort, J., & Müller, F. (1984) *Biochemistry 23*, 4859–4867.
70. Burnett, R.M., Darling, G.D., Kendall, D.S., leQuesne, M.E., Mayhew, S.G., Smith, W.W., & Ludwig, M.L. (1974) *J. Biol. Chem. 249*, 4383–4392.
71. Kuo, M.C., Dunn, J.B.R., & Fritchie, C.J. (1974) *Acta Crystallogr. B30*, 1766–1771.
72. Trus, B.L., Wells, J.L., Johnstone, R.M., Fritchie, C.J., & Marsh, R.E. (1971) *J. Chem. Soc., Chem. Commun.*, 751–73.
73. Fritchie, C.J., & Johnston, R.M. (1975) *Acta Crystallogr. B31*, 454–461.
74. Ludwig, M.L., Schopfer, L.M., Metzger, A.L., Pattridge, K.A., & Massey, V. (1990) *Biochemistry 29*, 10364–10375.
75. Draper, R.D., & Ingraham, L.L. (1968) *Arch. Biochem. Biophys. 125*, 802–808.
76. Conant, J.B., Kahn, H.M., Fieser, L.F., & Kurtz, S.S. (1922) *J. Am. Chem. Soc. 44*, 1382–1396.
77. Michaelis, L., Schubert, M.P., & Smythe, C.V. (1936) *J. Biol. Chem. 116*, 587–607.
78. Beckmann, J.D., & Frerman, F.E. (1985) *Biochemistry 24*, 3913–3921.
79. Brustlein, M., & Bruice, T.C. (1972) *J. Am. Chem. Soc. 94*, 6548–6549.
80. Shinkai, S., & Bruice, T.C. (1973) *J. Am. Chem. Soc. 95*, 7526–7528.
81. Fisher, J., & Walsh, C. (1974) *J. Am. Chem. Soc. 96*, 4346–4347.
82. Hersh, L.B., & Jorns, M.S. (1975) *J. Biol. Chem. 250*, 8728–8734.
83. Jorns, M.S., & Hersh, L.B. (1974) *J. Am. Chem. Soc. 96*, 4012–4014.
84. Ghisla, S., Thorpe, C., & Massey, V. (1984) *Biochemistry 23*, 3154–3161.
85. Schmidt, J., Reinsch, J., & McFarland, J.T. (1981) *J. Biol. Chem. 256*, 11667–11670.
86. Williams, R.F., & Bruice, T.C. (1976) *J. Am. Chem. Soc. 98*, 7752–7768.
87. Bruice, T.C., & Yano, Y. (1975) *J. Am. Chem. Soc. 97*, 5263–5270.
88. Bruice, T.C., (1980) in *Biomimetic Chemistry* (Dolphin, D., McKenna, C., Murakami, Y., & Tabushi, I., Eds.) pp 89–118, American Chemical Society, Washington, DC
89. Walsh, C.T., Schonbrunn, A., & Abeles, R.H. (1971) *J. Biol. Chem. 246*, 6855–6866.
90. Walsh, C., Lockridge, O., Massey, V., & Abeles, R. (1973) *J. Biol. Chem. 248*, 7049–7054.
91. Urban, P., & Lederer, F. (1984) *Eur. J. Biochem. 144*, 345–351.
92. Novak, M., & Bruice, T.C. (1980) *J. Chem. Soc., Chem. Commun.*, 372–374.
93. Müller, F., & Massey, V. (1969) *J. Biol. Chem. 244*, 4007–4016.
94. Ghisla, S., & Massey, V. (1980) *J. Biol. Chem. 255*, 5688–5696.
95. Dudley, K.H., & Hemmerich, P. (1967) *J. Org. Chem. 32*, 3049–3054.
96. Hevesi, L., & Bruice, T.C. (1973) *Biochemistry 12*, 290–297.
97. Bruice, T.C., Hevesi, L., & Shinkai, S. (1973) *Biochemistry 12*, 2083–2089.
98. Ghisla, S., Hartman, U., Hemmerich, P., & Müller, F. (1973) *Justus Liebigs Ann. Chem.*, 1388–1415.
99. Chan, T.W., & Bruice, T.C. (1978) *Biochemistry 17*, 4784–4793.
100. Clerin, D., & Bruice, T.C. (1974) *J. Am. Chem. Soc. 96*, 5571–5573.
101. Venkataram, U.V., & Bruice, T.C. (1984) *J. Am. Chem. Soc. 106*, 5703–5709.
102. Perham, R. N. (1987) *Biochem. Soc. Trans. 15*, 730-733.
103. Thorpe, C., & Williams, C. H. (1976) *J. Biol. Chem. 251*, 7726-7728.
104. Loechler, E. L., & Hollocher, T. C. (1980) *J. Am. Chem. Soc. 102*, 7312-7321.
105. Arscott, L.D., Thorpe, C., & Williams, C.H. (1981) *Biochemistry 20*, 1513–1520.
106. Pai, E.F., & Schulz, G.E. (1983) *J. Biol. Chem. 258*, 1752–1757.
107. Manstein, D.J., Pai, E.F., Schopfer, L.M., & Massey, V. (1986) *Biochemistry 25*, 6807–6816.
108. Jorns, M.S. (1985) *Biochemistry 24*, 3189–3194.
109. Pascal, R.A., Trang, N.L., Cerami, A., & Walsh, C. (1983) *Biochemistry 22*, 171–178.
110. Loach, P.A. (1976) in *Handbook of Biochemistry and Molecular Biology, 3rd Edition, Physical and Chemical Data, Vol. I* (Fasman, G.D., Ed.) pp 122–130, CRC Press, Cleveland, OH.
111. Henderson, R.W., & Morton, T.C. (1976) in *Handbook of Biochemistry and Molecular Biology, 3rd Edition, Physical and Chemical Data, Vol. I* (Fasman, G.D., Ed.) pp 131–150, CRC Press, Cleveland, OH.
112. Finazzi-Agro, A., Canella, C., Graziani, M.T., & Cavallini, D. (1971) *FEBS Lett. 16*, 172–174.
113. Horowitz, P.M., & Criscimagna, N.L., (1983) *Biochem. Biophys. Res. Commun. 111*, 595–601.
114. White, R.H. (1983) *Biochem. Biophys. Res. Commun. 112*, 66–72.
115. Watenpaugh, K.D., Sieker, L.C., Herriot, J.R., & Jensen, L.H. (1973) *Acta Crystallogr. B29*, 943–956.
116. Mahler, H.R., & Cordes, E.H. (1971) *Biological Chemistry*, 2nd ed., Harper & Row, New York.
117. Mayerle, J.J., Frankel, R.B., Holm, R.H., Ibers, J.A., Phillips, W.D., & Weiher, J.F. (1973) *Proc. Natl. Acad. Sci. U.S.A. 70*, 2429–2433.
118. Herskovitz, T., Averill, B.A., Holm, R.H., Ibers, J.A., Phillips, W.D., & Weiher, J.F. (1972) *Proc. Natl. Acad. Sci. U.S.A. 69*, 2437–2441.
119. Fee, J.A., & Palmer, G. (1971) *Biochim. Biophys. Acta 245*, 175–195.
120. Carter, C.W., Kraut, J., Freer, S.T., Alden, R.A., Sieker, L.C., Adman, E., & Jensen, L.H. (1972) *Proc. Natl. Acad. Sci. U.S.A. 69*, 3526–3529.
121. Stout, C.D. (1989) *J. Mol. Biol. 205*, 545–555.
122. Kent, T.A., Dreyer, J., Kennedy, M.C., Huynh, B.H., Emptag, M.H., Beinert, H., & Münck, E. (1982) *Proc. Natl. Acad. Sci. U.S.A. 79*, 1096–1100.
123. Sweeney, W.V., Rabinowitz, J.C., & Yoch, D.C. (1975) *J. Biol. Chem. 250*, 7842–7847.
124. Ghosh, D., O'Donnell, S., Furey, W., Robbins, A.H., & Stout, C.D. (1982) *J. Mol. Biol. 158*, 73–109.
125. Hoard, J.L. (1971) *Science 174*, 1295–1302.
125a. Kyte, J. (1995) *Structure in Protein Chemistry*, pp 139 and 258, Garland Publishing, New York.
126. Mashiko, T., Marchon, J., Musser, D.T., Reed, C.A., Kastner, M.E., & Scheidt, W.R. (1979) *J. Am. Chem. Soc. 101*, 3653–3655.
127. Takano, T., & Dickerson, R.E. (1981) *J. Mol. Biol. 153*, 79–94.
128. Rodkey, F.L., & Ball, E.G. (1950) *J. Biol. Chem. 182*, 17–28.
129. Morrison, L.E., Schelhorn, J.E., Cotton, T.M., Bering, C.L., & Loach, P.A. (1982) in *Function of Quinones in Energy Conserving Systems* (Trumpower, B.L., Ed.) pp 35–58, Academic Press, New York.
130. Takamiya, K.I., & Dutton, P.L. (1979) *Biochim. Biophys. Acta 546*, 1–16.
131. Urban, P.F., & Klingenberg, M. (1969) *Eur. J. Biochem. 9*, 519–525.
132. Land, E.J., & Swallow, A.J. (1970) *J. Biol. Chem. 245*, 1890–1894.
133. Yu, C.A., Nagaoka, S., Yu, L., & King, T.E. (1978) *Biochem. Biophys. Res. Commun. 82*, 1070–1078.
134. Tsai, A., & Palmer, G. (1983) *Biochim. Biophys. Acta 722*, 349–363.

134a. Kyte, J. (1995) *Structure in Protein Chemistry*, p 542, Garland Publishing, New York.
135. Yoon, P.S., & DeLuca, H.F. (1980) *Biochemistry 19*, 2165–2171.
136. Hall, C.L., & Kamin, H. (1975) *J. Biol. Chem. 250*, 3476–3486.
137. Ragan, I.C., Galante, Y.M., & Hatefi, Y. (1982) *Biochemistry 21*, 2518–2524.
138. Paech, C., Reynolds, J.G., Singer, T.P., & Holm, R.H. (1981) *J. Biol. Chem. 256*, 3167–3170.
139. Albracht, S.P.J., & Subramanian, J. (1977) *Biochim. Biophys. Acta 462*, 36–48.
140. Ingledew, W.J., & Ohnishi, T. (1980) *Biochem. J. 186*, 111–117.
141. Ohnishi, T. (1975) *Biochim. Biophys. Acta 387*, 475–490.
142. Orme-Johnson, N.R., Hansen, R.E., & Beinert, H. (1974) *J. Biol. Chem. 249*, 1922–1927.
143. Jaworski, A., Campbell, H.D., Poulis, M.I., & Young, I.G. (1981) *Biochemistry 20*, 2041–2047.
144. Jaworski, A., Mayo, G., Shaw, D.C., Campbell, H.D., & Young, I.G. (1981) *Biochemistry 20*, 3621–3628.
145. Merli, A., Capaldi, R., Ackrell, B.A.C., & Kearney, E.B. (1979) *Biochemistry 18*, 1393–1400.
146. Davis, K.A., & Hatefi, Y. (1971) *Biochemistry 10*, 2509–2516.
147. Capaldi, R., Sweetland, J., & Merli, A. (1977) *Biochemistry 16*, 5707–5710.
148. Hatefi, Y., & Galante, Y. (1980) *J. Biol. Chem. 255*, 5530–5537.
149. Ohnishi, T., Salerno, J.C., Winter, D.B., Lim, J., Yu, C.A., Yu, L., & King, T.E. (1976) *J. Biol. Chem. 251*, 2094–2104.
150. Ohnishi, T., Lim, J., Winter, D.B., & King, T.E. (1976) *J. Biol. Chem. 251*, 2105–2109.
151. Johnson, M.K., Morningstar, J.E., Bennett, D.E., Ackrell, B.A.C., & Kearney, E.B. (1985) *J. Biol. Chem. 260*, 7368–7378.
152. Ruzicka, F.J., & Beinert, H. (1977) *J. Biol. Chem. 252*, 8440–8445.
153. Karlsson, B., Hovmoller, S., Weiss, H., & Leonard, K. (1983) *J. Mol. Biol. 165*, 287–302.
154. Li, Y., Leonard, K., & Weiss, H. (1981) *Eur. J. Biochem. 116*, 199–205.
155. Trumpower, B.L., & Edwards, C.A. (1979) *J. Biol. Chem. 254*, 8697–8706.
156. Rieske, J.S. (1976) *Biochim. Biophys. Acta 456*, 195–247.
157. Gubriel, R.J., Batie, C.J., Sivaragja, M., True, A.E., Fee, J.A., Hoffman, B.M., & Ballou, D.P. (1989) *Biochemistry 28*, 4861–4871.
158. Britt, R.D., Sauer, K., Klein, M.P., Knaff, D.B., Kriauciunas, A., Yu, C.A., Yu, L., & Malkin, R. (1991) *Biochemistry 30*, 230–238.
159. Weiss, H. (1976) *Biochim. Biophys. Acta 456*, 291–313.
160. Widger, W.R., Cramer, W.A., Hermann, R.G., & Trebst, A. (1984) *Proc. Natl. Acad. Sci. U.S.A. 81*, 674–678.
161. T'sai, A., & Palmer, G. (1982) *Biochim. Biophys. Acta 681*, 484–495.
162. Erecinska, M., Wilson, D.F., & Miyata, Y. (1976) *Arch. Biochem. Biophys. 177*, 133–143.
163. Yu, L., Yu, C., & King, T.E. (1977) *Biochim. Biophys. Acta 495*, 232–247.
164. Kröger, A., & Klingenberg, M. (1973) *Eur. J. Biochem. 34*, 358–368.
165. Kröger, A., & Klingenberg, M. (1973) *Eur. J. Biochem. 39*, 313–323.
166. Norling, B., Glazek, E., Nelson, B.D., & Ernster, L. (1974) *Eur. J. Biochem. 47*, 475–482.
167. Ragan, C.J., & Heron, C. (1978) *Biochem. J. 174*, 783–790.
168. Heron, C., Ragan, C.I., & Trumpower, B.L. (1978) *Biochem. J. 174*, 791–800.
169. Chiang, Y., Kaminsky, L.S., & King, T.E. (1976) *J. Biol. Chem. 251*, 29–36.
170. Erecinska, M., Vanderkooi, J.M., & Wilson, D.F. (1975) *Arch. Biochem. Biophys. 171*, 108–116.
171. Birchmeier, W., Kohler, C.E., & Schatz, G. (1976) *Proc. Natl. Acad. Sci. U.S.A. 73*, 4334–4338.
172. Margoliash, E., Ferguson-Miller, S., Tulloss, J., Kang, C.H., Feinberg, B.A., Brautigan, D.L., & Morrison, M. (1973) *Proc. Natl. Acad. Sci. U.S.A. 70*, 3245–3249.
173. Smith, L., Davies, H.C., Reichlin, M., & Margoliash, E. (1973) *J. Biol. Chem. 248*, 237–243.
174. Rieder, R., & Bosshard, H.R. (1980) *J. Biol. Chem. 255*, 4732–4739.
175. König, B.W., Wilms, J., & VanGelder, B.F. (1981) *Biochim. Biophys. Acta 636*, 9–16.
176. Wharton, D.C., & Cusanovich, M.A. (1969) *Biochem. Biophys. Res. Commun. 37*, 111–115.
177. Miller, M.J., & Gennis, R.B. (1983) *J. Biol. Chem. 258*, 9159–9165.
178. Matsusjhita, K., Patel, L., & Kaback, H.R. (1984) *Biochemistry 23*, 4703–4714.
179. Hauska, G., Hurt, E., Gabellini, N., & Lockau, W. (1983) *Biochim. Biophys. Acta 726*, 97–133.
180. Malkin, R., & Bearden, A.J. (1978) *Biochim. Biophys. Acta 505*, 147–181.
181. Siegel, L.M., Rueger, D.C., Barber, M.J., Krueger, R.J., Orme-Johnson, N.R., & Orme-Johnson, W.H. (1982) *J. Biol. Chem. 257*, 6343–6350.
182. Siegel, L.M., & Davis, P.S. (1974) *J. Biol. Chem. 249*, 1587–1598.
183. Siegel, L.M., Davis, P.S., & Kamin, H. (1974) *J. Biol. Chem. 249*, 1572–1586.
184. Vega, J.M., & Kamin, H. (1977) *J. Biol. Chem. 252*, 896–909.
185. Labeyrie, F., Baudras, A., & Lederer, F. (1978) *Methods Enzymol. 53*, 238–256.
186. Capeillere-Blandin, C. (1982) *Eur. J. Biochem. 128*, 533–542.
187. Steifel, E.I. (1973) *Proc. Natl. Acad. Sci. U.S.A. 70*, 988–992.
188. Kramer, S.P., Johnson, J.L., Ribiero, A.A., Millington, D.S., & Rajagopalan, K.V. (1987) *J. Biol. Chem. 262*, 16357–16363.
189. Spence, J.T., Barber, M.J., & Siegel, L.M. (1982) *Biochemistry 21*, 1656–1661.
190. Barber, M.J., Salerno, J.C., & Siegel, L.M. (1982) *Biochemistry 21*, 1648–1656.
190a. Kyte, J. (1995) *Structure in Protein Chemistry*, p 524, Garland Publishing, New York.
191. Deisenhofer, J., Epp, O., Miki, K., Huber, R., & Michel, H. (1984) *J. Mol. Biol. 180*, 385–397.
192. Trüper, H.G., & Pfennig, N. (1978) in *The Photosynthetic Bacteria* (Clayton, R.K., & Sistrom, W.R., Eds.) pp 19–27, Plenum Press, New York.
193. Norris, J.R., Uphaus, R.A., Crespi, H.L., & Katz, J.J. (1971) *Proc. Natl. Acad. Sci. U.S.A. 68*, 625–628.
194. Okamura, M.Y., Steiner, L.A., & Feher, G. (1974) *Biochemistry 13*, 1394–1403.
195. Williams, J.C., Steiner, L.A., Feher, G., & Simon, M.I. (1984) *Proc. Natl. Acad. Sci. U.S.A. 81*, 7303–7307.
196. Deisenhofer, J., Epp, O., Miki, K., Huber, R., & Michel, H. (1985) *Nature 318*, 618–623.
197. Allen, J.P., Feher, G., Yeates, T.O., Komiya, H., & Rees, D.C. (1987) *Proc. Natl. Acad. Sci. U.S.A. 84*, 5730–5734.
198. Yeates, T.O., Komiya, H., Chirino, A., Rees, D.C., Allen, J.P., & Feher, G. (1988) *Proc. Natl. Acad. Sci. U.S.A. 85*, 7993–7997.
199. Kirmaier, C., & Holten, D. (1987) *Photosynth. Res. 13*, 225–260.
200. Woodbury, N.W., Becker, M., Middendorf, D., & Parson, W.W. (1985) *Biochemistry 24*, 7516–7521.
201. Holten, D., Windsor, M.W., Parson, W.W., & Thornberger, J.P. (1978) *Biochim. Biophys. Acta 501*, 112–126.
202. Kleinfeld, D., Okamura, M.Y., & Feher, G. (1984) *Biochim. Biophys. Acta 766*, 126–140.

203. Okamura, M.Y., Isaacson, R.A., & Feher, G. (1975) *Proc. Natl. Acad. Sci. U.S.A. 72*, 3491–3495.
204. Wraight, C.A. (1979) *Biochim. Biophys. Acta 548*, 309–327.
305. Debus, R.J., Feher, G., & Okamura, M.Y. (1986) *Biochemistry 25*, 2276–2287.
206. Vermeglio, A. (1977) *Biochim. Biophys. Acta 459*, 516–524.
207. Wraight, C.A. (1977) *Biochim. Biophys. Acta 459*, 525–531.
208. Paddock, M.L., McPherson, P.H., Feher, G., & Okamura, M.Y. (1990) *Proc. Natl. Acad. Sci. U.S.A. 87*, 6803–6807.
209. Junge, W., & Ausländer, W. (1973) *Biochim. Biophys. Acta 333*, 59–70.
210. Michel, H., & Deisenhofer, J. (1988) *Bull. Inst. Pasteur 86*, 37–45.
211. Yocum, C.F., Yerkes, C.T., Blankeship, R.E., Sharp, R.R., & Babcock, G.T. (1981) *Proc. Natl. Acad. Sci. U.S.A. 78*, 7507–7511.
212. Abramowicz, D.A., & Dismukes, G.C. (1984) *Biochim. Biophys. Acta 765*, 318–328.
213. Bowlby, N.R., & Frasch, W.D. (1986) *Biochemistry 25*, 1402–1407.
214. Winkler, J.R., Nocera, D.G., Yocum, K.M., Bordignon, E., & Gray, H.B. (1982) *J. Am. Chem. Soc. 104*, 5798–5800.
215. McGourty, J.L., Blough, N.V., & Hoffman, B.M. (1983) *J. Am. Chem. Soc. 105*, 4470–4472.
216. Zemel, H., & Hoffman, B.M. (1981) *J. Am. Chem. Soc. 103*, 1192–1201.
217. Peterson-Kennedy, S.E., McGourty, J.L., & Hoffman, B.M. (1984) *J. Am. Chem. Soc. 106*, 5010–5012.
218. Packham, N.K., Tiede, D.M., Mueller, P., & Dutton, P.L. (1980) *Proc. Natl. Acad. Sci. U.S.A. 77*, 6339–6343.
219. Rich, P.R. (1986) *J. Bioenerg. Biomembr. 18*, 145–156.
220. Knox, R.S. (1975) in *Bioenergetics of Photosynthesis* (Govindjee, Ed.) pp 183–221, Academic Press, New York.
220a. Kyte, J. (1995) *Structure in Protein Chemistry*, pp 413–418, Garland Publishing, New York.
221. Glazer, A. (1984) *Biochim. Biophys. Acta 768*, 29–51.
222. Matthews, B.W., Fenna, R. E., Bolognesi, M.C., Schmid, M.F., & Olson, J.M. (1979) *J. Mol. Biol. 131*, 259–285.
223. Tronrud, D.E., Schmid, M.F., & Matthews, B.W. (1986) *J. Mol. Biol. 188*, 443–454.
224. Schirmer, T., Huber, R., Schneider, M., Bode, W., Miller, M., & Hackert, M.L. (1986) *J. Mol. Biol. 188*, 651–676.
225. Grabowski, J., & Gantt, E. (1978) *Photochem. Photobiol. 28*, 39–45.
226. Jay, F., Lambillotte, M., Stark, W., & Mühlethaler, K. (1984) *EMBO J. 3*, 773–776.
227. Stark, W., Kuhlbrandt, W., Wildhaber, I., Wehrli, E., & Mühlethaler, K. (1984) *EMBO J. 3*, 777–783.
228. Goddard, W.A., & Olafson, B.D. (1979) in *Biochemical and Clinical Aspects of Oxygen* (Caughey, W., Ed.) pp 87–123, Academic Press, New York.
229. Ingraham, L.L., & Meyer, D.L. (1985) *Biochemistry of Dioxygen*, Plenum Press, New York.
230. Lapidot, A., & Irving, C.S. (1974) in *Molecular Oxygen in Biology* (Hayaishi, O., Ed.) pp 33–80, North Holland Publishing, Amsterdam.
231. Perutz, M.F. (1979) *Annu. Rev. Biochem. 48*, 327–386.
232. Wood, P.M. (1974) *FEBS Lett. 44*, 22–24.
233. Bielski, B.H.J. (1978) *Photochem. Photobiol. 28*, 645–649.
234. George, P. (1965) in *Oxidases and Related Redox Systems* (King, T.E., Mason, H.S., & Morrison, M., Eds.) pp 1–36, John Wiley, New York.
235. Fita, I., & Rossmann, M.G. (1985) *J. Mol. Biol. 185*, 21–37.
236. Groves, J.T., Haushalter, R.C., Nakamura, M., Nemo, T.E., & Evans, B.J. (1981) *J. Am. Chem. Soc. 103*, 2884–2886.
237. Chang, C.K., & Traylor, T.G. (1973) *Proc. Natl. Acad. Sci. U.S.A. 70*, 2647–2650.
238. Traylor, T.G., Lee, W.A., & Stynes, D.V. (1984) *J. Am. Chem. Soc. 106*, 755–764.
239. Traylor, T.G., & Xu, F. (1987) *J. Am. Chem. Soc. 109*, 6201–6202.
240. Poulos, T.L., & Kraut, J. (1980) *J. Biol. Chem. 255*, 8199–8205.
241. Edwards, S.L., Xuong, N., Hamlin, R.C., & Kraut, J. (1987) *Biochemistry 26*, 1503–1511.
242. Schonbaum, G.R., & Chance, B. (1976) in *The Enzymes* (Boyer, P., Ed.) *Vol. XIII, Part C*, pp 363–408, Academic Press, New York.
243. dePillis, G.D., & Ortiz de Montellano, P.R. (1989) *Biochemistry 28*, 7947–7952.
244. Pauling, L., & Coryell, C.D. (1936) *Proc. Natl. Acad. Sci. U.S.A. 22*, 210–216.
245. Jameson, G.B., Molinaro, F.S., Ibers, J.A., Collman, J.P., Rose, E., & Suslick, K.S. (1980) *J. Am. Chem. Soc. 102*, 3224–3237.
246. Phillips, S.E.V. (1978) *Nature 273*, 247–248.
247. Shaanan, B. (1982) *Nature 296*, 683–684.
248. Traylor, T.G., White, D.K., Campbell, D., & Berzinis, A.P. (1981) *J. Am. Chem. Soc. 103*, 4932–4936.
249. Traylor, T.G., Chang, C.K., Geibel, J., Berzinis, A., Mincey, T., & Cannon, J. (1979) *J. Am. Chem. Soc. 101*, 6716–6731.
250. Geibel, J., Cannon, J., Campbell, D., & Traylor, T.G. (1978) *J. Am. Chem. Soc. 100*, 3575–3585.
251. Baldwin, J., & Chothia, C. (1979) *J. Mol. Biol. 129*, 175–220.
252. Shaanan, B. (1983) *J. Mol. Biol. 171*, 31–59.
253. Poulos, T.L., Finzel, B.C., & Howard, A.J. (1987) *J. Mol. Biol. 195*, 687–700.
254. Macdonald, T.L., Burka, L.T., Wright, S.T., & Guengerich, F.P. (1982) *Biochem. Biophys. Res. Commun. 104*, 620–625.
255. Yasukochi, Y., & Masters, B.S.S. (1976) *J. Biol. Chem. 251*, 5337–5344.
256. Tyson, C.A., Lipscomb, J.D., & Gunsalus, I.C. (1972) *J. Biol. Chem. 247*, 5777–5784.
257. Baron, J., Taylor, W.E., & Masters, B.S.S. (1972) *Arch. Biochem. Biophys. 150*, 105–115.
258. Chang, C.K., & Kuo, M. (1979) *J. Am. Chem. Soc. 101*, 3413–3415.
259. Tomaszewski, J.E., Jerina, D.M., & Daly, J.W. (1975) *Biochemistry 14*, 2024–2031.
260. Daly, J.W., Jerina, D.M., & Witkop, B. (1972) *Experientia 28*, 1129–1149.
261. Hayaishi, O., & Nozaki, M. (1969) *Science 164*, 389–396.
262. Takai, K., Ushiro, H., Noda, Y., Narumiya, S., Tokuyama, T., & Hayaishi, O. (1977) *J. Biol. Chem. 252*, 2648–2656.
263. Wharton, D.C., & Gibson, Q.H. (1976) *J. Biol. Chem. 251*, 2861–2862.
264. Babcock, G.T., Vickery, L.E., & Palmer, G. (1976) *J. Biol. Chem. 251*, 7907–7919.
265. Petty, R.H., Welch, B.R., Wilson, L.J., Bottomley, L.A., & Kadish, K.M. (1980) *J. Am. Chem. Soc. 102*, 611–620.
266. Chance, B., Saronio, C., & Leigh, J.S. (1975) *J. Biol. Chem. 250*, 9226–9237.
267. Stevens, T.H., Martin, C.T., Wang, H., Brudvig, G.W., Scholes, C.P., & Chan, S.I. (1982) *J. Biol. Chem. 257*, 12106–12113.
268. Loehr, J.S., Freedman, T.B., & Loehr, T.M. (1974) *Biochem. Biophys. Res. Commun. 56*, 510–515.
269. Stenkamp, R.E., Sieker, L.C., Jensen, L.H., McCallum, J.D., & Sanders-Loehr, J. (1985) *Proc. Natl. Acad. Sci. U.S.A. 82*, 713–716.
270. Klotz, I.M., & Klotz, T.A. (1955) *Science 121*, 477–480.
271. Sheriff, S., Hendrickson, W.A., & Smith, J.L. (1987) *J. Mol. Biol. 197*, 273–296.
272. Stenkamp, R.E., Sieker, L.C., & Jensen, L.H. (1984) *J. Am. Chem. Soc. 106*, 618–622.

273. Volbeda, A., & Hol, W.G.J. (1989) *J. Mol. Biol. 206*, 531–546.
274. Gaykema, W.P.J., Volbeda, A., & Hol, W.G.J. (1986) *J. Mol. Biol. 187*, 255–275.
275. Solomon, E.I. (1988) in *Progress in Clinical and Biological Research, Vol. 274: Oxidases and Related Redox Systems* (King, T.E., Mason, H.S., & Morrison, M., Eds.) pp 309–329, Alan R. Liss, New York.
276. Fox, B.G., Froland, W.A., Dege, J.E., & Lipscomb, J.D. (1989) *J. Biol. Chem. 264*, 10023–10033.
277. Nordlund, P., Dalton, H., & Eklund, H. (1992) *FEBS Lett. 307*, 257–262.
278. Froland, W.A., Andersson, K.K., Lee, S., Liu, Y., & Lipscomb, J.D. (1992) *J. Biol. Chem. 267*, 17588–17597.
279. Andersson, K.K., Froland, W.A., Lee, S., & Lipscomb, J.D. (1991) *New J. Chem. 15*, 411–415.
280. Kok, B., Forbush, B., & McGloin, M. (1970) *Photochem. Photobiol. 11*, 457–475.
281. Wieghardt, K. (1989) *Angew. Chem., Int. Ed. Engl. 28*, 1153–1172.
282. Bossek, V., Weyhermüller, T., Wieghardt, K., Nuber, B., & Weiss, J. (1990) *J. Am. Chem. Soc. 112*, 6387–6388.
283. Strickland, S., & Massey, V. (1973) *J. Biol. Chem. 248*, 2953–2962.
284. Branchaud, B.P., & Walsh, C.T (1985) *J. Am. Chem. Soc. 107*, 2153–2161.
285. Favaudon, V. (1977) *Eur. J. Biochem. 78*, 293–307.
286. Kemal, C., Chan, T.W., & Bruice, T.C. (1977) *J. Am. Chem. Soc. 99*, 7272–7286.
287. Nanni, E.J., Sawyer, D.T., Ball, S.S., & Bruice, T.C. (1981) *J. Am. Chem. Soc. 103*, 2797–2802.
288. Kemal, C., & Bruice, T.C. (1976) *Proc. Natl. Acad. Sci. U.S.A. 73*, 995–999.
289. Parras, A.G., Olson, J.S., & Palmer, G. (1981) *J. Biol. Chem. 256*, 9096–9103.
290. Ito, N., Phillips, S.E.V., Stevens, C., Ogel, Z.B., McPherson, M.J., Keen, J.N., Yadav, K.D.S., & Knowles, P.F. (1991) *Nature 350*, 87–90.
290a. Kyte, J. (1995) *Structure in Protein Chemistry*, p 98, Garland Publishing, New York.
291. Whittaker, M.M., & Whittaker, J.W. (1990) *J. Biol. Chem. 265*, 9610–9613.
292. Whittaker, M.M., & Whittaker, J.W. (1988) *J. Biol. Chem. 263*, 6074–6080.
293. Entsch, B., Ballou, D.P., & Massey, V. (1976) *J. Biol. Chem. 251*, 2550–2563.
294. Wessiak, A., & Bruice, T.C. (1983) *J. Am. Chem. Soc. 105*, 4809–4825.
295. Ghisla, S., Hastings, J.W., Favaudon, V., & Lhoste, J. (1978) *Proc. Natl. Acad. Sci. U.S.A. 75*, 5860–5863.
296. Bruice, T.C., Noar, J.B., Ball, S.S., & Venkataram, U.V. (1983) *J. Am. Chem. Soc. 105*, 2452–2463.
297. Schwab, J.M., Li, W., & Thomas, L.P. (1983) *J. Am. Chem. Soc. 105*, 4800–4808.
298. Wagner, W.R., Spero, D.M., & Rastetter, W.H. (1984) *J. Am. Chem. Soc. 106*, 1476–1480.
299 Haavik, J., Døskeland, A.P., & Flatmark, T. (1986) *Eur. J. Biochem. 160*, 1–8.
300. Dix, T.A., & Benkovic, S.J. (1985) *Biochemistry 24*, 5839–5846.
301. Lazarus, R.A., deBrosse, C.W., & Benkovic, S.J. (1982) *J. Am. Chem. Soc. 104*, 6869–6871.
302. Dix, T.A., Bollag, G.E., Domanico, P.L., & Benkovic, S.J. (1985) *Biochemistry 24*, 2955–2958.
303. Guroff, G., Daly, J.W., Jerina, D.M., Renson, J., Witkop, B., & Udenfriend, S. (1967) *Science 157*, 1524–1530.
304. Fisher, D.B., Kirkwood, R., & Kaufman, S. (1972) *J. Biol. Chem. 247*. 5161–5167.
305. Marota, J.J.A., & Shiman, R. (1984) *Biochemistry 23*, 1303–1311.
306. Cardinale, G.J., Rhoads, R.E., & Udenfriend, S. (1971) *Biochem. Biophys. Res. Commun. 43*, 537–543.
307. Tuderman, L., Myllylä, R., & Kivirikko, K.I. (1977) *Eur. J. Biochem. 80*, 341–348.
308. Jefford, C.W., Boschung, A.F., Bolsman, T.A.B.M., Moriarity, R.M., & Melnick, B. (1976) *J. Am. Chem. Soc. 98*, 1017–1018.
309. Laroff, G.P., Fessenden, R.W., & Shuler, R.H. (1972) *J. Am. Chem. Soc. 94*, 9062–9073.
310. Myllylä, R., Kuuti-Savolainen, E., & Kivirikko, K.I. (1978) *Biochem. Biophys. Res. Commun. 83*, 441–448.
311. deJong, L., Albracht, S.P.J., & Kemp, A. (1982) *Biochim. Biophys. Acta 704*, 326–332.
312. Hayaishi, O. (1976) *J. Biochem. (Tokyo) 79*, 13p–21p.
313. Anton, D.L., & Friedman, P.A. (1983) *J. Biol. Chem. 258*, 14084–14087.
314. Tischler, M., Fieser, L.F., & Wendler, N.L. (1940) *J. Am. Chem. Soc. 62*, 2866–2871.
315. Larson, A.E., Friedman, P.A., & Suttie, J.W. (1981) *J. Biol. Chem. 256*, 11032–11035.
316. Wallin, R., & Suttie, J.W. (1982) *Arch. Biochem. Biophys. 214*, 155–163.
317. Zinoni, F., Birkmann, A., Stadtman, T.C., & Bock, A. (1986) *Proc. Natl. Acad. Sci. U.S.A. 83*, 4650–4654.
318. Cone, J.E., del Rio, R.M., & Stadtman, T.C. (1977) *J. Biol. Chem. 252*, 5337–5344.
319. Zakowski, J.J., Forstrom, J.W., Condell, R.A., & Tappel, A.L. (1978) *Biochem. Biophys. Res. Commun. 84*, 248–253.
320. Jones, J.B., Dilworth, G.L., & Stadtman, T.C. (1979) *Arch. Biochem. Biophys. 195*, 255–260.
321. Wendel, A., Pilz, W., Ladenstein, R., Sawatzki, G., & Weser, U. (1975) *Biochim. Biophys. Acta 377*, 211–215.
322. Ladenstein, R., & Wendel, A. (1976) *J. Mol. Biol. 104*, 877–882.
323. Lehnert, P.G. (1968) *Proc. R. Soc. London, A 303*, 45–84.
324. Silverman, R.B., & Dolphin, D. (1976) *J. Am. Chem. Soc. 98*, 4626–4633.
325. Sprecher, M., Clark, M.J., & Sprinson, D.B. (1964) *Biochem. Biophys. Res. Commun. 15*, 581–587.
326. Graves, S.W., Krouwer, J.S., & Babior, B.M. (1980) *J. Biol. Chem. 255*, 7444–7448.
327. Finlay, T.H., Valinsky, J., Mildvan, A.S., & Abeles, R.H. (1973) *J. Biol. Chem. 248*, 1285–1290.
328. Munch-Peterson, A., & Barker, H.A. (1958) *J. Biol. Chem. 230*, 649–653.
329. Eggerer, H., Stadtmann, E.R., Overath, P., & Lynen, F. (1960) *Biochem. Z. 333*, 1–9.
330. Dowd, P., Trivedi, B.K., Shapiro, M., & Marawaha, L.K. (1976) *J. Am. Chem. Soc. 98*, 7875–7877.
331. Johnson, A.W., Mervyn, L., Shaw, N., & Smith, E.L. (1963) *J. Chem. Soc.*, 4146–4156.
332. Grate, J.H., & Schrauzer, G.N. (1979) *J. Am. Chem. Soc. 101*, 4601–4611.
333. Wood, J.M., Moura, I., Moura, J.J.G., Santos, M.H., Xavier, A.V., LeGall, J., & Scandellari, M. (1982) *Science 216*, 303–305.
334. Halpern, J., Ng, F.T.T., & Rempel, G.L. (1979) *J. Am. Chem. Soc. 101*, 7124–7126.
335. Grate, J.H., Grate, J.W., & Schrauzer, G.N. (1982) *J. Am. Chem. Soc. 104*, 1588–1594.
336. Eagar, R.G., Baltimore, B.G., Herbst, M.M., Barker, H.A., & Richards, J.H. (1972) *Biochemistry 11*, 253–264.
337. Frey, P.A., & Abeles, R.H. (1966) *J. Biol. Chem. 241*, 2732–2733.
338. Babior, B.M., Carty, T.J., & Abeles, R.H. (1974) *J. Biol. Chem. 249*, 1689–1695.

339. Schrauzer, G.N., & Silbert, J.W. (1970) *J. Am. Chem. Soc. 92,* 1022–1030.
340. Dowd, P., & Shapiro, M. (1976) *J. Am. Chem. Soc. 98,* 3724–3725.
341. Silverman, R.B., & Dolphin, D. (1973) *J. Am. Chem. Soc. 95,* 1686–1687.
342. Finke, R.C., & Shiraldi, D.A. (1983) *J. Am. Chem. Soc. 105,* 7605–7617.
343. Finke, R.G., McKenna, W.P., Shiraldi, D.A., Smith, B.L., & Pierpont, C. (1983) *J. Am. Chem. Soc. 105,* 7592–7604.
344. Grate, J.W., & Schrauzer, G.N. (1984) *Z. Naturforsch. B: Anorg. Chem. Org. Chem. 39B,* 821–823.

## Chapter THREE

# Binding of Substrates to the Active Site

**Metabolism** is the coordinated sequences of chemical reactions that occur within a living cell to provide energy and to accomplish biosynthesis. Each step in any one of these sequences of reactions is catalyzed by a unique enzyme. An **enzyme** is a molecule of protein that catalyzes a particular chemical reaction. For example, fumarate hydratase is a molecule of protein that catalyzes the interconversion of fumarate and malate

$$\text{fumarate } ({}^{-}OOC\text{–}CH{=}CH\text{–}COO^{-}) + H_2O \rightleftharpoons \text{malate } ({}^{-}OOC\text{–}CH(OH)\text{–}CH_2\text{–}COO^{-}) \qquad (3\text{–}1)$$

a reaction that is found in the citric acid cycle. In most instances, the only role that a particular protein plays in a particular cell is to catalyze its unique chemical reaction.

The sequences of chemical reactions that constitute the metabolic pathways can be written out and submitted to memory systematically and in an organized structure. Each reaction uses as its reactants the products of the reaction or reactions that precede it, and the products of each reaction are the reactants for the enzyme or enzymes that follow it on the chart. The diagrams of metabolic pathways leave the impression that metabolism is a highly organized and neatly ordered system. This organization, however, does not extend to the proteins themselves, which, as the products of evolution by natural selection, are a motley collection. Within a particular metabolic sequence, each of the consecutive enzymes is usually a protein with a polypeptide whose amino acid sequence is unrelated to the amino acid sequences of the polypeptides comprising the other enzymes in the same sequence of reactions; each is a protein with a stoichiometry of protomers unrelated to the stoichiometries of the protomers of the other enzymes in the same sequence of reactions; and each is a protein with a native molecular structure unrelated to the native molecular structure of the other enzymes in the same sequence of reactions.

Consequently, when enzymes are discussed, they are not arranged in the order in which they occur in metabolic pathways. They can be presented as members of the various families of proteins that recapitulate their evolutionary relationships. This approach emphasizes structural homologies. Unfortunately, evolutionally related enzymes often catalyze chemically unrelated reactions. Enzymes can also be presented in categories reflecting the types of chemical reactions that they catalyze. Unfortunately, different enzymes will often catalyze homologous chemical reactions by entirely different chemical mechanisms. It is better to discuss strategies by which enzymes catalyze reactions and use particular enzymes as examples of those strategies.

As a **catalyst**, an enzyme displays particular abilities and is constrained by particular rules. In an enzymatically catalyzed reaction, the molecules interconverted by the enzyme are referred to as **substrates**. When an enzyme is added to a solution containing only the substrates appearing on one side of the chemical equation for the particular reaction for which it is responsible—for example, an aqueous solution of fumarate—all of the substrates appearing on the other side of the particular reaction—for example, malate—are formed at a **greater rate** than they would have been in the absence of the enzyme. This increase in the rate of a particular reaction is directly proportional to the molar concentration of the enzyme. The enzymatically catalyzed reaction is also **more specific** than the uncatalyzed reaction. For example, the enzyme fumarate hydratase produces only *S*-malate (Reaction 3–1) while the uncatalyzed reaction would produce both *S*-malate and *R*-malate, as well as other side products such as maleate. The very **low yield of undesired side products**, which would normally appear in significant yield in an uncatalyzed reaction, is an essential capability for each enzyme if the sequences of reactions comprising metabolism are to operate efficiently.

Under normal circumstances, tens to thousands of substrate molecules are interconverted during each second by one molecule of enzyme. Therefore, very little enzyme, in terms of its molar concentration relative to the molar concentrations of its substrates, must be added to produce a significant increase in the rate of the reaction. For this small amount of enzyme to be able to process these much larger amounts of substrates, each molecule of enzyme, ready to process the next molecule of reactant, must be **regenerated** in exactly the same state at the end of each turnover. The enzyme, because it is present in the solution in very small molar concentration and because it must be regenerated after each transformation, cannot be a stoichiometric participant in the chemical reaction. Therefore, an enzyme, under these normal circumstances, **cannot affect the equilibrium constant** and cannot affect the change in standard free energy of the reaction. There are exceptions to this last rule. These exceptions are the frequently encountered situations in which the molar concentration of an enzyme is artificially adjusted by an investigator in a particular experiment to within an order of magnitude of the molar concentrations of the substrates and the seldom encountered situations in which the concentration of an enzyme in the cell is greater than that of its substrates.[1]

Because an enzyme cannot change the equilibrium constant of a reaction as long as it is not itself a reactant, it necessarily follows that if the enzyme catalyzes the reaction in one direction, it must also catalyze the reaction in the other direction. The principle of **microscopic reversibility**, when applied to an enzymatically catalyzed reaction, is that the enzymatically catalyzed reaction must pass through the same transition states but in opposite order in the two respective directions. The enzyme accelerates the reaction by lowering the relative free energy of one or more of the transition states in this sequence. Therefore, it must accelerate the reaction in both directions. For this reason, enzymatically catalyzed reactions, as are all chemical reactions, are **approaches to equilibrium**. It just happens to be the case that most biological reactions have changes in standard free energy between 40 and –40 kJ $mol^{-1}$, so the equilibria are much more obvious. For example, at pH 7 and 25 °C, the ratio between the concentration of fumarate and the concentration of *S*-malate is 0.22 at equilibrium.[2]

## The Reaction or Reactions Catalyzed

Before the mechanism of a particular enzyme can be studied, stoichiometric and complete chemical equations must be established for the chemical reaction that it catalyzes. Although the **stoichiometric equation** seems clear in the cases of interconversions such as that between fumarate and malate, there are many instances in which it is obscured.

In many enzymatically catalyzed reactions, there are substrates that are not explicit in the reaction originally identified. Fatty acids can be converted into the thioesters of coenzyme A in an acyl exchange reaction

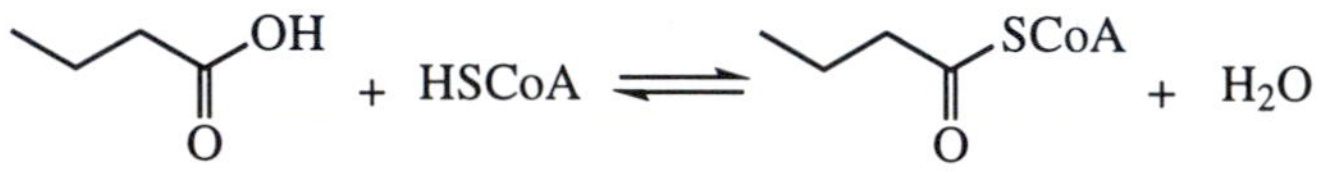

$$\text{RC(=O)OH} + \text{HSCoA} \rightleftharpoons \text{RC(=O)SCoA} + H_2O \qquad (3\text{–}2)$$

but the reaction as written is endergonic. Therefore, the reaction catalyzed by long-chain fatty acid-CoA ligase involves the hydrolysis of a phosphodiester of MgATP as well[3]

$$\text{MgATP} + \text{fatty acid} + \text{HSCoA} \rightleftharpoons \text{AMP} + HOP_2O_6^{3-} + Mg^{2+} + \text{fatty acyl-SCoA} \qquad (3\text{–}3)$$

Procollagen-lysine, 2-oxoglutarate 5-dioxygenase hydroxylates lysine in procollagen but only when stoichiometric amounts of 2-oxoglutarate are present to consume the other half of the molecular oxygen[4]

$$\text{procollagen-lysine} + \text{2-oxoglutarate} + O_2 \rightleftharpoons \text{procollagen-5-hydroxylysine} + \text{succinate} + CO_2 \qquad (3\text{–}4)$$

Nitrogenase is an enzyme that catalyzes the reduction of nitrogen to ammonia by transferring electrons

$$N_2 + 6e^- + 8H^+ \rightleftharpoons 2NH_4^+ \qquad (3\text{–}5)$$

The ultimate source of the electrons is reduced ferredoxin, but even so the reaction is endergonic. To render it exergonic, MgATP is hydrolyzed; 2 moles of MgATP are hydrolyzed for each mole of electrons transferred[5]

$$12\ MgATP^{4-} + N_2 + 6e^- + 8H^+ \rightleftharpoons 2NH_4^+ + 12\ MgADP^{2-} + 12HOPO_3^{2-} \qquad (3\text{–}6)$$

In many instances, a substrate in an enzymatic reaction exists in solution in several different forms in rapid equilibrium with each other, but the enzyme will recognize only one of these several forms. Although only 3% of the glyceraldehyde 3-phosphate in solution is the unhydrated aldehyde, that is the only form of this substrate recognized by glyceraldehyde-3-phosphate dehydrogenase.[6] In a solution of fructose 1,6-diphosphate, there is an equilibrium mixture containing 20% of the *α* anomer (**3–1**) and 80% of the *β* anomer (**3–2**) of the furanose[7]

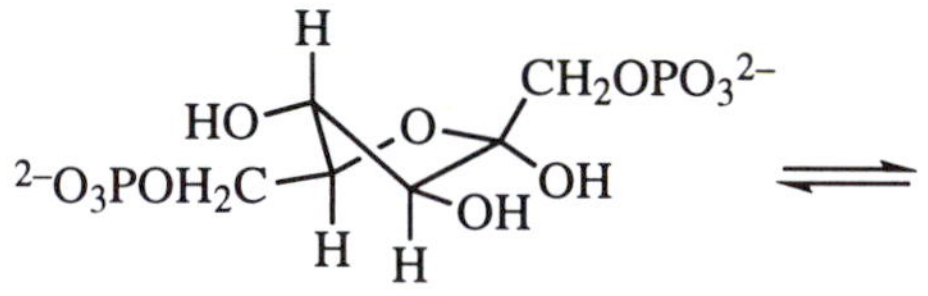

H, HO, $^{2-}O_3POH_2C$, O, $CH_2OPO_3^{2-}$, OH, OH, H, H ⇌

*α* anomer
**3–1**

H, HO, $^{2-}O_3POH_2C$, O, OH, $CH_2OPO_3^{2-}$, OH, H, H

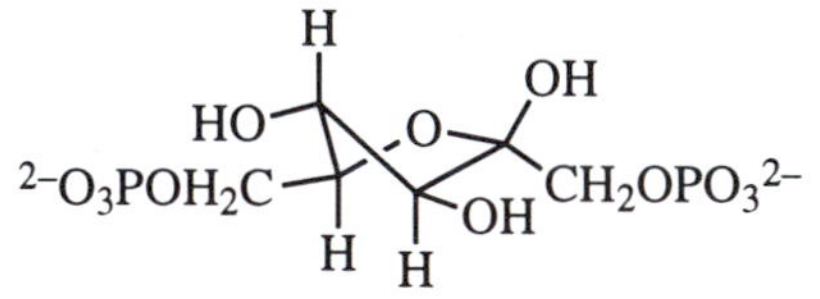

*β* anomer
**3–2**

(3–7)

Although these are interconverted through the open ketone, less than 1% of either the open ketone or its hydrate are present in solution. The enzyme fructose bisphosphatase acts preferentially, but not exclusively, on the *α* anomer.[8] Carbon dioxide and carbonic acid are in equilibrium with each other, and it is difficult to distinguish which is the substrate in a given enzymatic reaction. By following the uptake of protons, it could be demonstrated that the immediate product of the reaction catalyzed by tartronate-semialdehyde synthase from *Escherichia coli*, an enzyme containing thiamin pyrophosphate, is carbon dioxide[9]

$$H^+ + 2\ \text{glyoxylate} \rightleftharpoons \text{tartronate semialdehyde} + CO_2 \qquad (3\text{–}8)$$

Were carbonic acid the product, protons would have been released rather than being consumed.

Often the product actually released by the enzyme rapidly decomposes nonenzymatically to the ultimate product. As far as the mechanism of the enzyme is concerned, however, it is the immediate product that should be shown in the chemical equation written for the reaction. Amino-acid oxidase releases, as its immediate product, the imine between ammonia and the keto acid that is its ultimate product,[10] and the equation written

$$^{-}OOC{-}CH(R){-}NH_3^{+} + O_2 \rightleftharpoons {}^{-}OOC{-}C(R){=}NH_2^{+} + H_2O_2 \quad (3\text{–}9)$$

must reflect this fact. Glutamate-5-semialdehyde dehydrogenase converts glutamate 5-semialdehyde and $NADP^+$ into 5-oxopyrrolidine-2-carboxylate and NADPH, but only in the presence of inorganic phosphate.[11] There are several facts, however, which lead to the conclusion that the reaction actually catalyzed by the enzyme is the production of glutamate 5-phosphate

$$\text{glutamate 5-semialdehyde} + HOPO_3^{2-} + NADP^+ \rightleftharpoons \text{glutamate 5-phosphate} + NADPH \quad (3\text{–}10)$$

and the 5-oxopyrrolidine-2-carboxylate arises from a subsequent, nonenzymatic reaction

$$^{-}HO_3PO{-}C(=O){-}CH_2{-}CH_2{-}CH(NH_3^{+}){-}COO^{-} \rightleftharpoons HOPO_3^{2-} + \text{5-oxopyrrolidine-2-carboxylate} + 2H^+ \quad (3\text{–}11)$$

In many enzymatic reactions, a molecule of another protein serves as a substrate. The reaction catalyzed by acetyl-CoA carboxylase from avian liver is

$$\text{acetyl-SCoA} + HCO_3^{2-} + MgATP^{3-} \rightleftharpoons \text{malonyl-SCoA} + MgADP^{2-} + HOPO_3^{2-} \quad (3\text{–}12)$$

but the enzyme from *E. coli*[12] utilizes the carbamate of a biotin attached to a small protein ($\alpha_2$; $n_{aa} = 2 \times 210$), referred to as biotin carboxyl carrier protein (BCCPNH)[13]

$$\text{acetyl-SCoA} + \text{BCCPN-}CO_2^{-} \rightleftharpoons \text{malonyl-SCoA} + \text{BCCPNH} \quad (3\text{–}13)$$

The biotin carboxyl carrier protein is carboxylated in a separate reaction requiring the hydrolysis of MgATP.[14]

In a few enzymatic reactions, the molar concentration of the active sites of the enzyme in the cytoplasm is equal to or greater than the concentrations of the substrates. In such reactions, the enzyme must be considered as an explicit participant in the reaction. The enzyme glyceraldehyde-3-phosphate dehydrogenase at low concentrations catalyzes the reaction

$$\text{HC(=O)}{-}\text{CH(OH)}{-}CH_2OPO_3^{2-} + HOPO_3^{2-} + NAD^+ \rightleftharpoons H^{-}O_3PO{-}C(=O){-}CH(OH){-}CH_2OPO_3^{2-} + NADH \quad (3\text{–}14)$$

The D-glyceraldehyde 3-phosphate is a reactive aldehyde that would be either bound as an imine to other proteins in the cell or present in solution as the hydrate,[6] and neither of these forms would be available to the enzyme, which can process only the unhydrated aldehyde.[6] The 1,3-diphosphoglycerate is an acyl phosphate that is readily hydrolyzed. For both of these reasons, the strategy that seems to have evolved is to maintain the concentrations of these two substrates at very low levels in the cytoplasm but to have a high concentration of 3-phosphoglycerate bound at the active site of the enzyme as a thioester with a unique cysteine in the folded polypeptide.[1] For this reason, the molar concentration of the enzyme in the cytoplasm (1 mM) is greater than the molar concentrations of its substrates (<0.10 mM). As a result, two of the reactions in glycolysis have glyceraldehyde-3-phosphate dehydrogenase and phosphoglyceryl glyceraldehyde-3-phosphate dehydrogenase as reactants as well as catalysts

$$\text{glyceraldehyde-3-P} + \text{enzyme} + NAD^+ \rightleftharpoons \text{phosphoglyceryl enzyme} + NADH + H^+ \quad (3\text{–}15)$$

$$\text{phosphoglyceryl enzyme} + HOPO_3^{2-} \rightleftharpoons \text{enzyme} + \text{1,3-diphosphoglycerate} \quad (3\text{–}16)$$

and the equilibria explicitly incorporate the concentrations of enzyme and phosphoglyceryl enzyme. The equilibrium constant for the overall reaction, however, involving only free substrates and no forms of the enzyme, cannot be altered by any concentration of glyceraldehyde 3-phosphate dehydrogenase, because enzyme would not be a reactant in this reaction as it is written.

Some enzymes are able to process more than one reactant and produce more than one product. Transketolase can use either xylulose 5-phosphate or fructose 6-phosphate as the ketone in the reaction and, by using thiamin pyrophosphate, transfers the hydroxyacetyl group derived from either to ribose 5-phosphate[15]

xylulose 5-phosphate + ribose 5-phosphate ⇌
sedoheptulose 7-phosphate + glyceraldehyde 3-phosphate (3–17)

fructose 6-phosphate + ribose 5-phosphate ⇌
sedoheptulose 7-phosphate + erythrose 4-phosphate (3–18)

Fatty acid synthase produces a mixture of saturated fatty acids of variable chain length. The composition of the mixture depends on the tissue and the species, and it can be varied by the presence of distinct thioesterases in a particular tissue.[16] The enzyme also can produce significant quantities of butyryl-SCoA in addition to long-chain free fatty acids.[17]

The same enzyme often catalyzes transformations among a group of closely related substrates, and this ability is an important clue as to its mechanism. Bisphosphoglycerate mutase from erythrocytes can catalyze three reactions[18]

2-phosphoglycerate ⇌ 3-phosphoglycerate (3–19)

2,3-bisphosphoglycerate ⇌ 3-phosphoglyceroyl phosphate (3–20)

2,3-bisphosphoglycerate + $H_2O$ ⇌
3-phosphoglycerate + $H_2OPO_3^-$ (3–21)

The first two reactions are interrelated by the fact that the same phosphorylated enzyme is an intermediate in both. 2,3-Bisphosphoglycerate, a substrate in Reaction 3–20, is an intermediate in Reaction 3–19, and 3-phosphoglycerate, a substrate in Reaction 3–19, is an intermediate in Reaction 3–20. Reaction 3–21 is a result of the spontaneous hydrolysis of the phosphorylated enzyme. Reaction 3–19 can be expanded to include the phosphorylated enzyme (E-$OPO_3^{2-}$)

2-phosphoglycerate + E-$OPO_3^{2-}$ ⇌
2,3-bisphosphoglycerate + E ⇌
3-phosphoglycerate + E-$OPO_3^{2-}$ (3–22)

and Reaction 3–20 can be expanded to include the phosphorylated enzyme

2,3-bisphosphoglycerate + E ⇌
3-phosphoglycerate + E-$OPO_3^{2-}$ ⇌
3-phosphoglyceroyl phosphate + E (3–23)

Just as the different reactions among natural metabolites that can be catalyzed by an enzyme provide insight into the mechanism of that enzyme, so too do the transformations of unnatural substrates catalyzed by the enzyme. Fumarate hydratase, in addition to its normal reaction (Reaction 3–1), can also catalyze the stereospecific hydration of *trans*-2,3-epoxysuccinate[19]

$^-OOC$–CH(–O–)CH–$COO^-$ + $H_2O$ ⇌ $^-OOC$–CH(OH)–CH(OH)–$COO^-$ (3–24)

and any mechanism proposed for the natural reaction must explain this unnatural reaction. Because glyceraldehyde-3-phosphate dehydrogenase can catalyze the production of acetyl phosphate from acetaldehyde[20]

$H_3C$–C(=O)–H + $HOPO_3^{2-}$ + $NAD^+$ ⇌ $H_3C$–C(=O)–$OPO_3^{2-}$ + NADH (3–25)

it is clear that there is no portion of the normal substrate, glyceraldehyde 3-phosphate, other than the aldehyde, that is essential for the normal reaction. Glyoxylate is not a natural substrate for L-lactate dehydrogenase, the enzyme that catalyzes the reaction

$H_3C$–C(=O)–$COO^-$ + NADH ⇌ $H_3C$–C(HO)(H)–$COO^-$ + $NAD^+$ (3–26)

but the enzyme nevertheless reduces glyoxylate to glycolate

H–C(=O)–$COO^-$ + NADH ⇌ H–C(HO)(H)–$COO^-$ + $NAD^+$ (3–27)

and reduces oxalate to glyoxylate.

HO–C(=O)–$COO^-$ + NADH ⇌ HO–C(HO)(H)–$COO^-$ + $NAD^+$ (3–28)

This result suggests that either a hydrogen or a hydroxyl, respectively, can be mistaken by the enzyme for a methyl group. Both of these functional groups are smaller than a methyl group and would not be sterically excluded from substituting for it.

Once the chemical reaction or reactions catalyzed by the enzyme have been unambiguously defined, the steps in the enzymatic reaction can be defined. The usual first step in this definition is to examine the steady-state kinetics of the reaction.

*Suggested Reading*

Rhoads, R.E., & Udenfriend, S. (1968) Decarboxylation of α-Ketoglutarate Coupled to Collagen Proline Hydroxylase, *Proc. Natl. Acad. Sci. U.S.A. 60,* 1473–1478.

**PROBLEM 3–1**
Show how all of these conversions can be catalyzed by the same enzyme.

D-erythrose 4-phosphate ⇌ D-threose 4-phosphate
⇖ ⇗
D-erythrulose 4-phosphate

D-ribulose 5-phosphate ⇌ D-xylulose 5-phosphate

Why does the same enzyme not produce (*R,S*)-D-1,2,4,5-tetrahydroxy-3-ketopentane 5-phosphate?

# Steady-State Kinetics

If a purified enzyme is added to a solution containing a mixture of all of the substrates that appear on one side of the chemical equation for the reaction catalyzed by that enzyme but missing one or more of the substrates that appear on the other side, the concentrations of the reactants will begin to decrease and the concentrations of the products will begin to increase. The **reactants** are the complete set of substrates from the one side of the chemical equation. The **products** are the incomplete or entirely absent set of substrates from the other side of the chemical equation. In this artificial situation, what is a reactant and what is a product are defined by the investigator. The reaction will also proceed in one direction if all of the products are present but purposely added at concentrations below their equilibrium levels. In this case, however, the initial rate of the reaction does not represent a unidirectional process. It will turn out that unidirectional processes are easier to study.

The progress of the reaction can be followed by monitoring the concentration of one of the reactants or one of the products as a function of time because all changes in concentration are stoichiometrically related by the chemical equation for the reaction catalyzed. In the case of the reaction catalyzed by fumarate hydratase (Reaction 3–1), if the purified enzyme is added to a solution containing fumarate (F) and malate (M) is absent, the concentration of fumarate will decrease as the concentration of malate increases until the equilibrium concentrations are reached. At this point, the reaction cannot proceed further (Figure 3–1). Although no net change can occur at equilibrium, unidirectional changes are still taking place, albeit at rates somewhat less than the initial rates; it is just that the unidirectional rate of the reaction in the forward direction equals the unidirectional rate of the reaction in the reverse direction.

The curves describing the concentrations of fumarate and malate (Figure 3–1)[2] describe the **complete kinetic course of the approach to equilibrium**. It is possible to simulate these curves using equations whose form and parameters can be derived from simple kinetic mechanisms.[2] The strategy of fitting complete kinetic courses, however, is rarely applied because the form of the integrated rate equations are complicated, too many variables usually are changing at once, the enzyme may slowly inactivate with time, and the most characteristic changes occur late in the reaction, when the observations are most uncertain. Instead, initial velocity studies are performed, and kinetic mechanisms consistent with these observations are formulated.

The initial velocity ($v_0$ in Figure 3–1) of an enzymatically catalyzed reaction is the rate of decrease in the concentration of reactant, A, or the rate of increase in the concentration of product, P, measured over a period of time so short that the rate remains constant rather than slowly decreasing with time. Over this short period of time, only small changes in the concentrations of substrates occur. If necessary, an extrapolation of the curve can be performed by numerical analysis to obtain the rate of the reaction at zero time. If the initial velocity measured in either of these two ways is, in fact, one of the limits

$$v_0 = \lim_{t \to 0} \frac{d[P]}{dt} = -\lim_{t \to 0} \frac{d[A]}{dt} \tag{3–29}$$

where A and P have the same stoichiometry, then the concentration of any reactant governing the initial velocity must be the concentration initially present, or the **initial concentration**, $[A]_0$. The data collected in an experiment of this kind are the initial velocities of the reaction at a series of respective initial concentrations of reactants.

If the initial concentrations of all other reactants are kept constant and the initial concentration of only one reactant is varied, the behavior of the initial velocity, $v_0$, observed in most enzymatically catalyzed reactions is that defined by a **rectangular hyperbola** with a horizontal asymptote (Figure 3–1B). An analytical equation that defines a rectangular hyperbola with a horizontal asymptote at $y = a$ is

$$y = \frac{ax}{x+b} \tag{3–30}$$

In the case of the initial velocity of an enzymatic reaction, the variable $y$ is the initial velocity, $v_0$, the variable $x$ is the initial concentration of the varied reactant, $[A]_0$, the parameter $a$ is referred to as the **maximum velocity**, $V_{max}$, and the parameter $b$ is referred to as the **Michaelis constant**, $K_m$. The equation

A

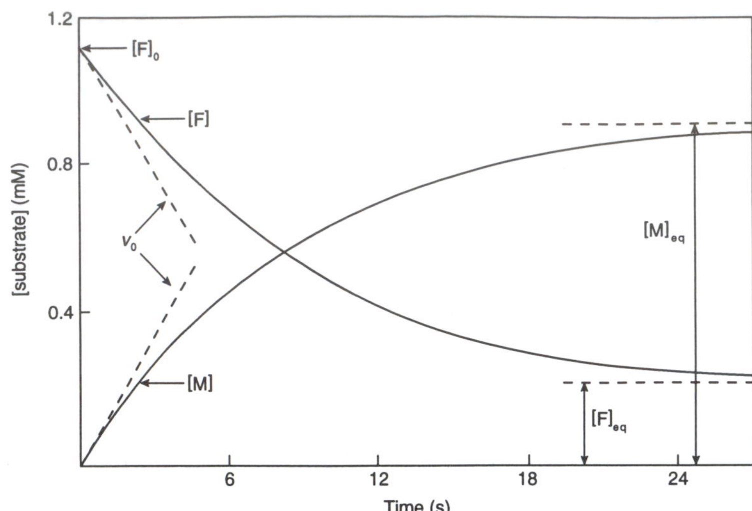

B

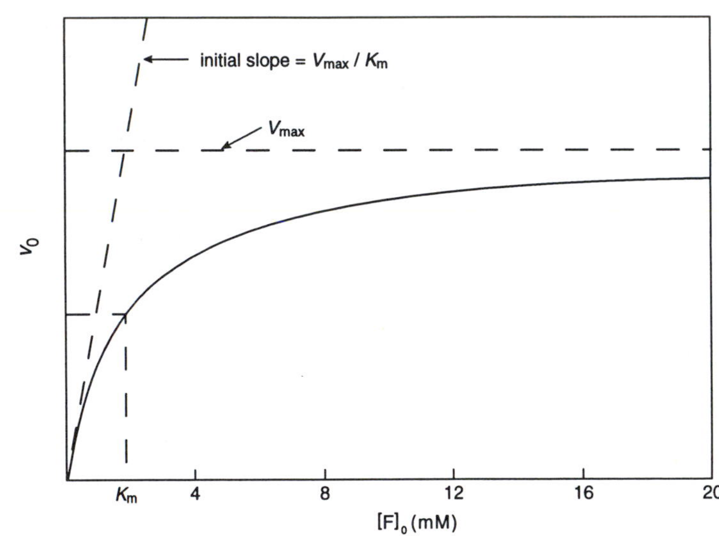


**Figure 3–1:** (A) Progress of the conversion of fumarate to malate catalyzed by the enzyme fumarate hydratase.[2] A solution of fumarate at pH 7 and 25 °C was added to the cuvette of a spectrophotometer. The reaction was initiated by adding fumarate hydratase to a final concentration of 15 $\mu$g mL$^{-1}$, and the initial concentration of fumarate was 1.13 mM. The decrease in the molar concentration of fumarate ([F]) was followed as a function of time by its absorbance at 270 nm. The increase in the concentration of malate ([M]) was estimated by difference. At times greater than 1 min, the concentration of fumarate no longer decreased because it had reached its equilibrium level ($[F]_{eq}$). The initial slope of the decrease of fumarate as a function of time is the initial velocity ($v_0$) in units of millimolar second$^{-1}$. (B) Behavior of the initial velocity of the reaction catalyzed by fumarate hydratase as a function of the initial concentration of fumarate.[21] Equivalent amounts of fumarate hydratase were added to a series of reaction mixtures at pH 6.4 and 23 °C which had various initial concentrations of fumarate. The decrease in absorbance at 295 nm was followed as a function of time for each mixture and the initial slope of each progress curve was measured and converted to an initial velocity ($v_0$). Initial velocity is plotted as a function of the initial concentration of fumarate ($[F]_0$). The parameter $V_{max}$ is the horizontal asymptote of the hyperbola, the parameter $K_m$ (1.8 mM) is the value of $[F]_0$ when $v_0$ is $^1/_2\ V_{max}$, and $(K_m)^{-1}$ is the initial slope of the hyperbola.

for the rectangular hyperbola is rewritten as

$$v_0 = \frac{V_{max}[A]_0}{[A]_0 + K_m} \tag{3–31}$$

As the concentration of the varied reactant, reactant A, is raised until $[A]_0 >> K_m$, the initial rate approaches the horizontal asymptote of the rectangular hyperbola, and this asymptote has the numerical value of the maximum velocity, $V_{max}$

$$\lim_{[A]_0 \to \infty} v_0 = V_{max} \tag{3–32}$$

When $x = b$ in Equation 3–30 or when $[A]_0 = K_m$ in Equation 3–31, $y = {}^1/_2\, a$ or $v_0 = {}^1/_2 V_{max}$, respectively (Figure 3–1B). This is one definition of $K_m$.

There is another form of the equation for a rectangular hyperbola

$$y = \frac{x}{c + dx} \tag{3–33}$$

In this form, the asymptote as $x$ approaches infinity is $d^{-1}$ and the initial slope of the curve as $x$ approaches zero is $c^{-1}$. If Equation 3–31 is recast in this form

$$v_0 = \frac{[A]_0}{\frac{K_m}{V_{max}} + \frac{[A]_0}{V_{max}}} \tag{3–34}$$

At low initial concentrations of reactant A, where $[A]_0 << K_m$, the initial velocity of the reaction increases in direct proportion to the initial concentration of the varied reactant

$$\lim_{[A]_0 \to 0} v_0 = \frac{V_{max}}{K_m}[A]_0 \tag{3–35}$$

This consideration and a consideration of Equation 3–32 demonstrate that the initial slope of the hyperbola, $V_{max}/K_m$, and its asymptote, $V_{max}$, are as capable of defining it as $K_m$ and $V_{max}$. It is only a historical accident that $K_m$ and $V_{max}$ were chosen as the defining parameters.

The most accurate values of the parameters $K_m$ and $V_{max}$ are obtained from experimental values of $v_0$ as a function of $[A]_0$ by numerical analysis. The data are a set of values of the initial velocity corresponding to a set of initial molar concentrations of the varied reactant, product, or inhibitor. At the present time, a rectangular hyperbola is fitted directly to the data by weighted, nonlinear least-squares analysis.[22] In the past, the data were first converted to a linear form before the fitting was performed, but that practice, although it is still done occasionally, has been supplanted by the nonlinear numerical analysis made possible by digital computers. Numerical values for $K_m$ and $V_{max}$ and standard deviations of those numerical values are the parameters obtained by the fitting procedure. These numerical values are the **experimental observations**. Although $K_m$ and $V_{max}$ are the parameters usually tabulated, $V_{max}/K_m$ and $V_{max}$ are the two parameters that have been found to vary independently of each other when pH, temperature, and concentrations of reactants, products, and inhibitors are varied. For this reason, tabulations of values of $V_{max}/K_m$ and $V_{max}$, which alone are also sufficient to define a unique rectangular hyperbola (Equation 3–33), are probably more useful.

While the values of the parameters $V_{max}/K_m$ and $V_{max}$ are obtained by direct numerical analysis, the data themselves are usually displayed graphically in one or the other of several linear forms. The most widely used linear form, the Michaelis–Menten equation, is that based on the inverse of Equation 3–31

$$\frac{1}{v_0} = \frac{1}{V_{max}} + \frac{K_m}{V_{max}[A]_0} \tag{3–36}$$

If the values of $v_0^{-1}$ are plotted against the corresponding values of $[A]_0^{-1}$, the points should define a straight line whose **intercept** at the ordinate is $V_{max}^{-1}$, whose **slope** is $K_m/V_{max}$, and whose **intercept** at the abscissa is $-K_m^{-1}$. Because this linear form, referred to as the **double-reciprocal plot**, separates all of the important parameters, $V_{max}$, $K_m$, and $V_{max}/K_m$, it is perhaps the most useful. For example, if water is disregarded, chymotrypsin can catalyze a reaction involving a single reactant, namely, any one in a series of phenyl hippurates

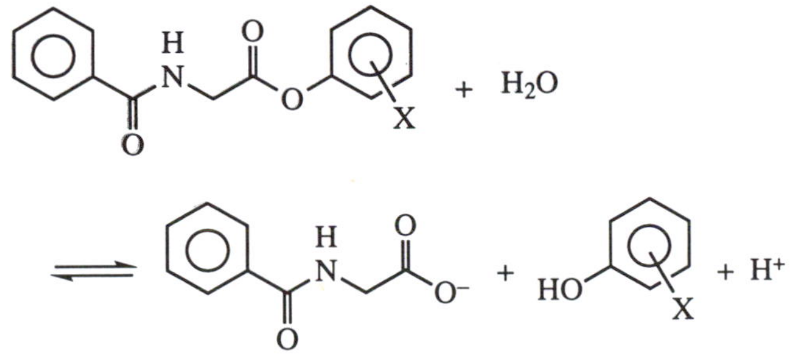

(3–37)

In this equation, X stands for a series of substituents on the phenyl group. The behavior of $v_0^{-1}$ as a function of $[\text{hippurate}]_0^{-1}$ is linear with positive slope for the whole series of phenyl hippurates that were studied (Figure 3–2).[23]

Another linear form can be obtained by a different rearrangement of Equation 3–31

$$v_0 = V_{max} - K_m \frac{v_0}{[A]_0} \tag{3–38}$$

If the values of $v_0$ are plotted against the corresponding values of $v_0/[A]_0$, the data should define a straight line whose intercept at the ordinate is $V_{max}$ and whose slope is $-K_m$. Because $v_0/[A]_0$ is the product of two independent variables, each with its own inaccuracy, the scatter of the data from the line is greater and more alarming.

By themselves, numerical values of $V_{max}$ and $K_m$, although often measured, mean very little. To provide an explanation of the observed behavior and to give meaning to these two parameters, a **kinetic mechanism** is proposed, and it is demonstrated mathematically that the kinetic mechanism predicts that the enzymatic reaction will display the observed behavior. Consider the simplest enzymatically catalyzed reaction in which reactant A is turned into product P. A kinetic mechanism that can explain the hyperbolic behavior of the initial velocity of this reaction as a function of $[A]_0$ is

$$\text{E} + \text{A} \underset{k_{-1}}{\overset{k_1}{\rightleftharpoons}} \text{E}\cdot\text{A} \overset{k_2}{\rightarrow} \text{E} + \text{P} \tag{3–39}$$

The complex, E·A, is formed reversibly between reactant A and the enzyme but does not involve the chemical transformation of reactant A. In the next step, the bound reactant A is then converted to product P and product P dissociates from the enzyme. This step is written as an irreversible process because there is no product P in the solution under the initial conditions. It should be emphasized that this mechanism was written by intuition rather than by deduction.

A rate equation is now derived from the proposed mechanism. A **rate equation** is a mathematical equation, derived from the written mechanism, that relates the initial velocity, $v_0$, of the reaction to the initial concentration of the reactant, $[A]_0$, and the total concentration of the enzyme $[E]_{TOT}$, both of which are known quantities

$$v_0 = f([E]_{TOT}, [A]_0) \tag{3–40}$$

In this rate equation the rate constants $k_1$, $k_{-1}$, and $k_2$ appear as parameters.

Because the only way in which product can be formed is through the step governed by $k_2$, it follows that

$$v_0 = \frac{d[P]}{dt} = k_2[\text{E}\cdot\text{A}] \tag{3–41}$$

But [E·A] is not a known concentration as are $[E]_{TOT}$ and $[A]_0$. The concentration of the complex between enzyme and reactant, [E·A], and the concentration of free enzyme, [E], are unknowns, and an expression for [E·A] only in terms of

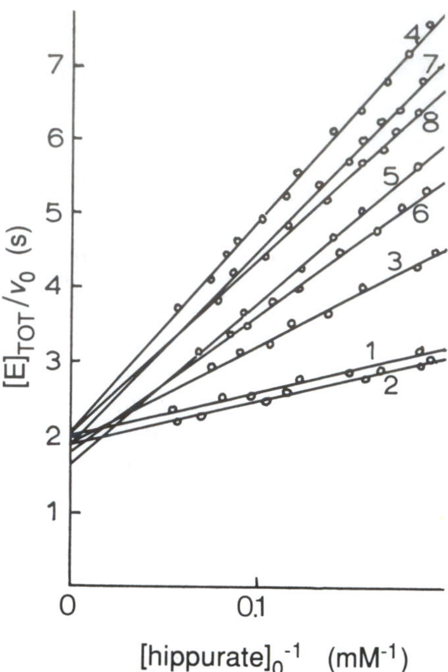


**Figure 3–2:** Double-reciprocal plot for the hydrolysis of a series of phenyl hippurates by chymotrypsin.[23] Equivalent amounts of chymotrypsin were mixed with a series of solutions containing various concentrations of a particular phenyl hippurate at an ionic strength of 0.1 M, at pH 6.91, and at 25 °C. The initial velocity of the release of the respective phenol was followed by its characteristic absorbance in the near ultraviolet. This provided a set of initial velocities, $v_0$, corresponding to the set of concentrations of the respective phenyl hippurate. The reciprocal of each initial velocity ($v_0^{-1}$ in millimolar$^{-1}$ second) was multiplied by the molar concentration of chymotrypsin ($[E]_{TOT}$) and this variable ($[E]_{TOT}/v_0$ in seconds) was plotted as a function of the reciprocal of the initial concentration corresponding to that initial velocity ($[\text{hippurate}]_0^{-1}$ in millimolar). The lines connect the data for the hydrolysis by chymotrypsin of the phenyl hippurate formed from (1) *p*-nitrophenol, (2) *p*-acetoxyphenol, (3) *m*-nitrophenol, (4) phenol, (5) *p*-cresol, (6) *p*-methoxyphenol, (7) *p*-fluorophenol, and (8) *p*-chlorophenol. Adapted with permission from ref 23. Copyright 1970 American Chemical Society.

defined quantities and rate constants must be obtained. To eliminate the unknown variable [E] and solve for the unknown variable [E·A], a set of **independent simultaneous equations** must be formulated such that there are as many independent simultaneous equations as there are unknown variables. The first simultaneous equation is always **conservation of enzyme**. The total amount of enzyme present, which is a known and constant quantity, must be equal to the sum of the concentrations of all forms of the enzyme. In the simple case of Mechanism 3–39

$$[E]_{TOT} = [E] + [E \cdot A] \tag{3–42}$$

The next group of simultaneous equations are the **steady-state relationships**. The assumption that is always made in derivations of this kind is that because the concentration of reactant is so much greater than the concentration of enzyme, after the first few molecules of the reactants have been turned into products by the enzyme, the concentration of each species involving the enzyme has reached a constant, **steady state**. This follows in the simple case of Mechanism 3–39 from the consideration that if [E·A] were increasing, [E] would be decreasing and the rate of formation of E·A would be decreasing while its rate of decay would be increasing. At some point, the rate of formation and the rate of decay of E·A would become equal, and its concentration would remain constant. A similar argument can be made for the situation in which [E·A] is decreasing. As long as $[A]_0$ remains unchanged, which is the definition of initial velocity (Equation 3–29), the individual concentrations of all forms of the enzyme will remain constant once the steady state has been reached. Direct measurements of the rate of attainment of steady state have shown that it is usually achieved within less than 0.1 s with most enzymes.

For Mechanism 3–39, the steady-state assumption is

$$\frac{d[E]}{dt} = -\frac{d[E \cdot A]}{dt} = 0 \tag{3–43}$$

Although this seems to define two equations, they turn out to be the same equation. The number of independent steady-state equations is always one less than the number of kinetically unique forms of the enzyme, and the shortfall is made up by the equation stating the conservation of enzyme (Equation 3–42). The single steady-state equation for Mechanism 3–39 is

$$\frac{d[E \cdot A]}{dt} = k_1[E][A]_0 - (k_{-1} + k_2)[E \cdot A] = 0 \tag{3–44}$$

Equation 3–44 states that the difference between the rate at which E·A is formed and the rate at which E·A is lost must be zero at steady state.

When the two independent simultaneous equations, Equations 3–42 and 3–44, are used to solve algebraically for [E·A] in terms of the known quantities, it is found that

$$[E \cdot A] = \frac{[E]_{TOT}[A]_0}{[A]_0 + \dfrac{k_2 + k_{-1}}{k_1}} \tag{3–45}$$

and that

$$v_0 = \frac{k_2[E]_{TOT}[A]_0}{[A]_0 + \dfrac{k_2 + k_{-1}}{k_1}} \tag{3–46}$$

A more general way to justify the assumption of Equation 3–43 is to write an explicit relationship for one of its terms, for example, Equation 3–44. Then, suppose Equation 3–43 is not valid because a steady state has not been established. If the terms in Equation 3–44 are not equal to zero, then Equation 3–46 would become

$$\frac{k_2}{k_{-1}+k_2+k_1[\mathrm{A}]_0}\left(\frac{\mathrm{d}[\mathrm{E\cdot A}]}{\mathrm{d}t}\right)+\frac{\mathrm{d}[\mathrm{P}]}{\mathrm{d}t}=\frac{k_2[\mathrm{E}]_{\mathrm{TOT}}[\mathrm{A}]_0}{[\mathrm{A}]_0+\dfrac{k_{-1}+k_2}{k_1}} \tag{3–47}$$

The term $k_2(k_1 + k_2 + k_1[\mathrm{A}]_0)^{-1}$ is less than 1 and

$$\frac{\mathrm{d}[\mathrm{E\cdot A}]}{\mathrm{d}t}<<\frac{\mathrm{d}[\mathrm{P}]}{\mathrm{d}t} \tag{3–48}$$

because [E] << [A]. Therefore, the change in [E·A] with time, although not zero because a steady state has not been established, nevertheless is negligible.

Equation 3–46 is of the form of Equation 3–31 and, therefore, provides an explanation of the observed behavior. It states that the measured value of $V_{\max}$ is $k_2[\mathrm{E}]_{\mathrm{TOT}}$ and the measured value of $K_{\mathrm{m}}$ is $(k_2 + k_{-1})(k_1)^{-1}$, if Mechanism 3–39 is the actual kinetic mechanism applicable to the enzyme in question. The point of the exercise is that the kinetic mechanism, Mechanism 3–39, which was formulated only by intuition and which did not follow by deduction from the observations, is nevertheless **consistent with** the behavior of the reaction as it is experimentally observed.

In Mechanism 3–39, the rate constant $k_2$ can have one of two quite different meanings, depending on the intentions of the author. If $k_2$ was intended to be a microscopic rate constant, then the step immediately following the formation of the complex E·A has been designated the rate-limiting step in the reaction. As the mechanism is written, however, there must be at least two steps between the complex E·A and the dissociated enzyme E and product P. These steps are the transformation of reactant A to product P on the enzyme and the dissociation of product P from the enzyme. If the author intended the rate constant $k_2$ to include the rate constants for these steps, then $k_2$ is a composite rate constant.

A rate constant that refers to a step involving only one transition state is a **microscopic rate constant**. On the basis of this formalism, a step involving only one transition state is a **microscopic step**. The ultimate ideal of a kinetic investigation is to define explicitly the complete sequence of microscopic steps in the mechanism of a reaction and to assign a numerical value to each of the microscopic rate constants. In simple nonenzymatic reactions, the chemical mechanism usually involves only a few steps, and the individual events, such as bond-making, bond-breaking, protonation, electron transfer, or noncovalent association, can be written as explicit steps, each of which seems to have only one transition state. Rate constants are then assigned systematically to these steps. In such situations, the measured rate constants are assumed to be microscopic rate constants associated with the explicit chemical events.

If the author of Mechanism 3–39 intended $k_2$ to be a microscopic rate constant, then $k_2$ cannot include the rate constant for the release of product because the release of product must involve a different transition state from the transition state that immediately follows the formation of the complex between enzyme and reactant. If the rate constant $k_2$ is intended to be the microscopic rate constant only for the step following the formation of the complex between enzyme and reactant, then the author of the mechanism has implicitly designated the step governed by the rate constant $k_2$ as the rate-limiting step of the overall reaction. This follows from the fact that the rate constants for all of the later steps are ignored.

Most students have been exposed to simple chemical kinetics before they encounter enzymatic kinetics and are trained to treat an unadorned rate constant as a microscopic rate constant. This is always a simplification because even straightforward chemical reactions probably involve more transition states, and hence more microscopic steps, than are presented. The reason this is experimentally inconsequential is that, even if there are several microscopic steps between two points in a mechanism, there is a rate at which the reaction proceeds between those two points and a rate constant that can be assigned to the flux between those two points. Because that rate constant is a function of the several microscopic rate constants for the several microscopic steps between the two points, it is a composite rate constant. The rate constant for a transformation that proceeds through more than one transition state and involves more than one microscopic step is a **composite rate constant.**

Suppose that $k_2$ is not a rate-limiting microscopic rate constant in Mechanism 3–39 and that it is a composite rate constant for two steps: the transformation of reactant to product and the dissociation of product

$$\mathrm{E\cdot A}\underset{k_{-6}}{\overset{k_6}{\rightleftharpoons}}\mathrm{E\cdot P}\overset{k_7}{\rightarrow}\mathrm{E+P} \tag{3–49}$$

In this expansion of the second step in Mechanism 3–39, the dissociation of product, governed by the rate constant $k_7$, is written as an irreversible process, as was the second step in Mechanism 3–39, because there is no product in the solution under the initial conditions. The rate at which product is produced

$$\frac{\mathrm{d}[\mathrm{P}]}{\mathrm{d}t}=k_7[\mathrm{E\cdot P}] \tag{3–50}$$

If it is assumed that E·P is in steady state, then

$$\frac{\mathrm{d}[\mathrm{E\cdot P}]}{\mathrm{d}t}=0=k_6[\mathrm{E\cdot A}]-(k_{-6}+k_7)[\mathrm{E\cdot P}] \tag{3–51}$$

and

$$\frac{\mathrm{d}[\mathrm{P}]}{\mathrm{d}t}=\left(\frac{k_7k_6}{k_{-6}+k_7}\right)[\mathrm{E\cdot A}] \tag{3–52}$$

In Mechanism 3–39

$$\frac{\mathrm{d}[\mathrm{P}]}{\mathrm{d}t}=k_2[\mathrm{E\cdot A}] \tag{3–53}$$

Therefore, if $k_2$ is a composite rate constant for the two steps in the expanded mechanism (Mechanism 3–49), then

$$k_2 = \frac{k_7 k_6}{k_{-6} + k_7} \tag{3–54}$$

If, however, $k_6$ is the rate-limiting step, then $k_7 >> k_{-6}$, $k_2$ equals $k_6$, and $k_2$ is the rate-limiting step.

Formally, the rate constant $k_1$ in Mechanism 3–39 is the second-order rate constant for the formation of the complex E·A and the rate constant $k_{-1}$ is the first-order rate constant for its dissociation. Although there is no necessity to assume it, the rate constants $k_1$ and $k_{-1}$ could also be composite rate constants. Suppose, for example, that there were three steps that preceded the conversion of reactant into product, one for association of the reactant with the enzyme and two involving isomerizations of enzyme and reactant, respectively

$$\mathrm{E+A} \underset{k_{-3}}{\overset{k_3}{\rightleftharpoons}} \mathrm{E \cdot A} \underset{k_{-4}}{\overset{k_4}{\rightleftharpoons}} \mathrm{E' \cdot A} \underset{k_{-5}}{\overset{k_5}{\rightleftharpoons}} \mathrm{E' \cdot A'} \tag{3–55}$$

The composite rate constant $k_1$ would then be the unidirectional rate constant for the conversion of E + A into E′·A′

$$\mathrm{E+A} \underset{k_{-3}}{\overset{k_3}{\rightleftharpoons}} \mathrm{E \cdot A} \underset{k_{-4}}{\overset{k_4}{\rightleftharpoons}} \mathrm{E' \cdot A} \overset{k_5}{\rightarrow} \mathrm{E' \cdot A'} \tag{3–56}$$

and

$$k_5[\mathrm{E' \cdot A}] = k_1[\mathrm{E}][\mathrm{A}]_0 \tag{3–57}$$

If it is assumed that [E·A] and [E′·A] are at steady state, then

$$k_3[\mathrm{E}][\mathrm{A}]_0 + k_{-4}[\mathrm{E' \cdot A}] = (k_{-3} + k_4)[\mathrm{E \cdot A}] \tag{3–58}$$

and

$$k_4[\mathrm{E \cdot A}] = (k_{-4} + k_5)[\mathrm{E' \cdot A}] \tag{3–59}$$

respectively. Solving for [E′·A]

$$[\mathrm{E' \cdot A}] = \frac{k_3 k_4 [\mathrm{E}][\mathrm{A}]_0}{k_{-3}k_{-4} + k_{-3}k_5 + k_4 k_5} \tag{3–60}$$

and

$$k_1 = \frac{k_3 k_4 k_5}{k_{-3}k_{-4} + k_{-3}k_5 + k_4 k_5} \tag{3–61}$$

The composite rate constant $k_{-1}$ is the unidirectional rate constant for the conversion of E′·A′ into E + A

$$\mathrm{E' \cdot A'} \underset{k_5}{\overset{k_{-5}}{\rightleftharpoons}} \mathrm{E' \cdot A} \underset{k_4}{\overset{k_{-4}}{\rightleftharpoons}} \mathrm{E \cdot A} \overset{k_{-3}}{\rightarrow} \mathrm{E+A} \tag{3–62}$$

where

$$k_{-3}[\mathrm{E \cdot A}] = k_{-1}[\mathrm{E' \cdot A'}] \tag{3–63}$$

If it is assumed that [E′·A] and [E·A] are at steady state, then

$$k_{-5}[\mathrm{E' \cdot A'}] + k_4[\mathrm{E \cdot A}] = (k_5 + k_{-4})[\mathrm{E' \cdot A}] \tag{3–64}$$

and

$$k_{-4}[\mathrm{E' \cdot A}] = (k_4 + k_{-3})[\mathrm{E \cdot A}] \tag{3–65}$$

respectively. Solving for [E·A]

$$[\mathrm{E \cdot A}] = \frac{k_{-4}k_{-5}[\mathrm{E' \cdot A'}]}{k_5 k_{-3} + k_5 k_4 + k_{-4}k_{-3}} \tag{3–66}$$

and

$$k_{-1} = \frac{k_{-3}k_{-4}k_{-5}}{k_5 k_{-3} + k_5 k_4 + k_{-4}k_{-3}} \tag{3–67}$$

The expressions for $k_1$, $k_{-1}$, and $k_2$, when they are the composite rate constants for the steps in the now expanded mechanism

$$\mathrm{E+A} \underset{k_{-3}}{\overset{k_3}{\rightleftharpoons}} \mathrm{E \cdot A} \underset{k_{-4}}{\overset{k_4}{\rightleftharpoons}} \mathrm{E' \cdot A} \underset{k_{-5}}{\overset{k_5}{\rightleftharpoons}} \mathrm{E' \cdot A'} \underset{k_{-6}}{\overset{k_6}{\rightleftharpoons}} \mathrm{E \cdot P} \overset{k_7}{\rightarrow} \mathrm{E+P} \tag{3–68}$$

contain only rate constants. Therefore, the rate equation for Mechanism 3–68 is mathematically equivalent to the rate equation for Mechanism 3–39 (Equation 3–46), but $k_1$, $k_{-1}$, and $k_2$ are replaced by their equivalents (Equations 3–61, 3–67, and 3–54, respectively). It follows that both Mechanisms 3–39 and 3–68 are consistent with the experimental observation that the initial velocity displays saturation with respect to the initial concentration of the reactant. The explanations of the two experimental parameters $K_m$ and $V_{max}$, however, are different. Steady-state kinetics cannot decide which mechanism is the complete mechanism, or if either of them is, because it is clear that further expansion of the kinetic mechanism is possible by supplying additional steps, and it is also clear that each expanded mechanism will also be consistent with the observations.

It is often stated that kinetics cannot prove a mechanism and that a mechanism can only be consistent with kinetic observations. There are two levels of meaning to this statement. First, kinetics cannot prove a mechanism because the possibility always exists that there is a significantly different mechanism with a different sequence of steps that is, nevertheless, also consistent with all of the experimental observations. In this sense the kinetic method, because it proceeds inductively, must suffer from this problem, which is common to all inductive reasoning. It follows that the most powerful use of kinetics is to disprove a proposed mechanism. The mistake should never be made, however, of assuming that since all of the mechanisms you could imagine but one have been disproven, the one remaining has been proven.

The second level of meaning is that even if a particular mechanism can be shown by independent experiments to be the correct mechanism, it may, nevertheless, be incomplete. The complete mechanism for a chemical reaction would include explicitly every microscopic step in that reaction. A mechanism in which one or more of the rate constants are composite rate constants is an incomplete mechanism for the reaction even if it is not an incorrect mechanism. Because any of the rate constants in a mechanism can be composite rate constants and because the number of steps covered by a composite rate constant can be large, no mechanism can be proven to be complete.

Regardless of whether $k_2$ in Mechanism 3–39 is the rate-limiting microscopic rate constant for the overall reaction or a composite rate constant including several steps, one of which is dissociation of the product, the transformation to which it refers, the conversion of reactant to product on the enzyme and the dissociation of the product from the enzyme, is kinetically irreversible. This follows from the fact that the step governing the dissociation of the product must be irreversible because no product is present in the solution at the initial point in the reaction. The creation of **irreversibility** by omitting one or more of the products of a reaction is one purpose behind measuring initial velocities. In this way, the rate measured is a unidirectional rate and the equations are greatly simplified.

All present explanations of the catalytic activity of enzymes assume that the reactants bind to the molecule of the enzyme and are converted to products while they are bound, as indicated in Mechanism 3–39. This hypothesis and kinetic mechanisms of the type represented by Mechanism 3–39 provide an explanation for the universally observed fact that the initial velocity of an enzymatic reaction approaches a horizontal asymptote as the initial concentrations of the reactants are increased. This behavior is referred to as **saturation** of the enzyme.

There is unequivocal evidence from crystallographic observations that enzymes have on their surfaces specific locations, known as **active sites**, at which reactants bind and are transformed into products. Most enzymes are rotationally symmetric oligomeric proteins, and there is usually one active site for each protomer. This fact defines the molar concentration of active sites in a given solution as being equal to the molar concentration of protomers. Because the molar concentration of active sites is fixed, the rate of an enzymatic reaction cannot increase once all of the active sites are occupied. This is why enzymatic reactions display saturation.

## *Mechanisms for Two Reactants*

If an enzymatic reaction has two reactants, A and B, the kinetic behavior resembles in some aspects that of a reaction that has only one reactant, but it has added complications. If the initial concentration of reactant B is held or fixed at a constant value while the initial concentration of reactant A is varied, the observed relationship between the initial velocity of the reaction and the initial concentration of reactant A usually defines a rectangular hyperbola, the parameters of which, for a given fixed concentration of reactant B, can be defined as $V^{A}_{max,app}$ and $K^{A}_{m,app}$. On linear plots of $v_0^{-1}$ as a function of $[A]_0^{-1}$ (Figure 3–3A),[24] the intercept of a given line at the ordinate is $V^{A}_{max,app}$ and the slope of that line is the respective value of $K^{A}_{m,app}/V^{A}_{max,app}$. Because the decision to examine reactant A was arbitrary, the same behavior is also observed when the initial concentration of reactant A is fixed and the initial concentration of reactant B is varied (Figure 3–3B).

The maximum velocity, $V^{A}_{max,app}$, and the Michaelis constant, $K^{A}_{m,app}$, observed when the concentration of A is varied are themselves variables of the concentration of reactant B. Usually, the behavior seen (Figure 3–4)[25] is that

$$\frac{1}{V^{A}_{max,app}} = \frac{1}{V^{A}_{max}} + \frac{K_m{}^{B}}{V^{A}_{max}[B]_0} \tag{3–69}$$

where $V^{A}_{max}$ is the maximum velocity when both $[A]_0$ and $[B]_0$ are saturating and $K_m{}^{B}$ is the Michaelis constant for reactant B. When either $(V^{A}_{max,app})^{-1}$ is plotted as a function of $[B]_0^{-1}$ or $(V^{B}_{max,app})^{-1}$ is plotted as a function of $[A]_0^{-1}$, the same intercept $(V_{max})^{-1}$ should be obtained (Figure 3–4C). In other words $V^{A}_{max}$ should equal $V^{B}_{max}$ and can be defined simply as $V_{max}$. The slopes $(K^{A}_{m,app}/V^{A}_{max,app})$ of the individual lines on the primary plots of $v_0^{-1}$ as a function of $[A]_0^{-1}$ usually satisfy the relationship (Figure 3–4D)

$$\frac{K^{A}_{m,\,app}}{V^{A}_{max,\,app}} = \frac{K_m{}^{A}}{V_{max}} + \frac{K_m{}^{AB}}{V_{max}[B]_0} \tag{3–70}$$

where $K_m{}^{A}$ is the Michaelis constant for reactant A and $K_m{}^{AB}$ is the dual Michaelis constant for reactants A and B. From a set of experimental determinations of the initial velocities at various values of $[A]_0$ and $[B]_0$, experimental values of $V_{max}$, $K_m{}^{A}$, $K_m{}^{B}$, and $K_m{}^{AB}$ can be determined by numerical analysis. The overall analytical equation that defines the observed behavior and that is used in the nonlinear, least-squares fitting procedure is

$$v_0 = \frac{V_{max}[A]_0[B]_0}{[A]_0[B]_0 + K_m{}^{B}[A]_0 + K_m{}^{A}[B]_0 + K_m{}^{AB}} \tag{3–71}$$

This relationship justifies the definition of the Michaelis constants. When $[B]_0$ is saturating but $[A]_0$ is very low

$$v_0 = \frac{V_{max}}{K_m{}^{A}}[A]_0 \tag{3–72}$$

When $[A]_0$ is saturating but $[B]_0$ is very low

$$v_0 = \frac{V_{max}}{K_m{}^{B}}[B]_0 \tag{3–73}$$

When both $[A]_0$ and $[B]_0$ are very low

$$v_0 = \frac{V_{max}}{K_m{}^{AB}}[A]_0[B]_0 \tag{3–74}$$

Kinetic mechanisms consistent with this observed behavior can be formulated. Assume for the moment that reactant A binds first to the active site of the enzyme, followed by reactant B, and that the products P and Q are released from the active site in a formally irreversible step because neither P nor Q is present at initial times

$$E + A \underset{k_{-1}}{\overset{k_1}{\rightleftharpoons}} E\cdot A + B \underset{k_{-2}}{\overset{k_2}{\rightleftharpoons}} E\cdot A\cdot B \overset{k_3}{\rightarrow} E + P + Q \tag{3–75}$$

A

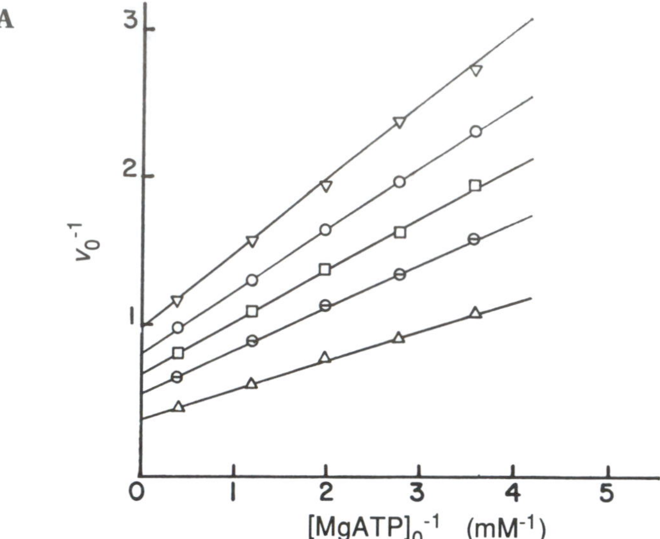


B

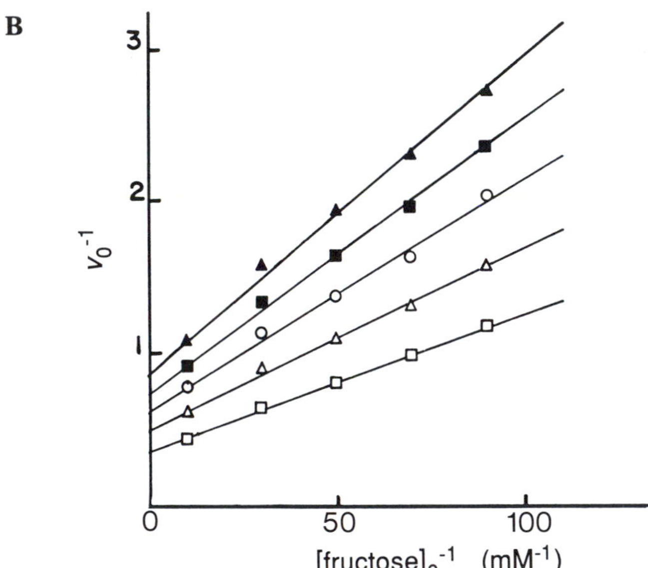


**Figure 3–3:** Double-reciprocal plots of the initial velocity of the reaction catalyzed by hexokinase from rat brain.[24] Solutions containing MgATP and fructose at various concentrations were prepared at pH 7.6 and 28 °C. The concentration of uncomplexed $Mg^{2+}$ cation was fixed at 1.0 mM throughout. The reaction was initiated by adding hexokinase, and the rate of production of fructose 6-phosphate was followed over the first few minutes to obtain initial velocities ($v_0$). The production of fructose 6-phosphate was followed in a coupled assay with glucose-6-phosphate isomerase, glucose-6-phosphate dehydrogenase, and $NAD^+$ by the increase in $A_{340}$ from the conversion of $NAD^+$ to NADH. Reciprocals of initial velocities ($v_0^{-1}$ in micromoles$^{-1}$ minute) are presented either (A) as a function of the reciprocal of the initial concentration of MgATP ([MgATP]$_0^{-1}$ in millimolar$^{-1}$) at 1.1 mM (▽), 1.43 mM (○), 2.00 mM (□), 3.33 mM (⊖), and 10.0 mM (△) fructose or (B) as a function of the reciprocal of the initial concentration of fructose ([fructose]$_0^{-1}$ in millimolar$^{-1}$) at 0.28 mM (▲), 0.36 mM (■), 0.50 mM (○), 0.83 mM (△), and 2.5 (□) MgATP. Adapted with permission from ref 24. Copyright 1971 *Journal of Biological Chemistry.*

In this mechanism as it is written, $k_3$ is either the microscopic rate constant for the rate-limiting step of the reaction or a composite rate constant incorporating the rate constant for the conversion of reactants to products and the individual rate constants for the formally irreversible dissociations of each of the products. The initial velocity of the reaction is

$$v_0 = k_3[\mathrm{E\cdot A\cdot B}] \tag{3–76}$$

The principle of conservation of enzyme states that

$$[\mathrm{E}]_{\mathrm{TOT}} = [\mathrm{E}]+[\mathrm{E\cdot A}]+[\mathrm{E\cdot A\cdot B}] \tag{3–77}$$

There are three unknowns; two other equations are required. They are the two steady-state equations

$$k_1[\mathrm{A}]_0[\mathrm{E}]+k_{-2}[\mathrm{E\cdot A\cdot B}]=(k_{-1}+k_2[\mathrm{B}]_0)[\mathrm{E\cdot A}] \tag{3–78}$$

and

$$k_2[\mathrm{B}]_0[\mathrm{E\cdot A}]=(k_{-2}+k_3)[\mathrm{E\cdot A\cdot B}] \tag{3–79}$$

Solving these three equations for [E·A·B] and entering that expression into Equation 3–76

$$v_0 = \frac{k_3[\mathrm{E}]_{\mathrm{TOT}}[\mathrm{A}]_0[\mathrm{B}]_0}{[\mathrm{A}]_0[\mathrm{B}]_0+\left(\dfrac{k_{-2}+k_3}{k_2}\right)[\mathrm{A}]_0+\left(\dfrac{k_3}{k_1}\right)[\mathrm{B}]_0+\left(\dfrac{k_{-1}k_{-2}+k_{-1}k_3}{k_1k_2}\right)} \tag{3–80}$$

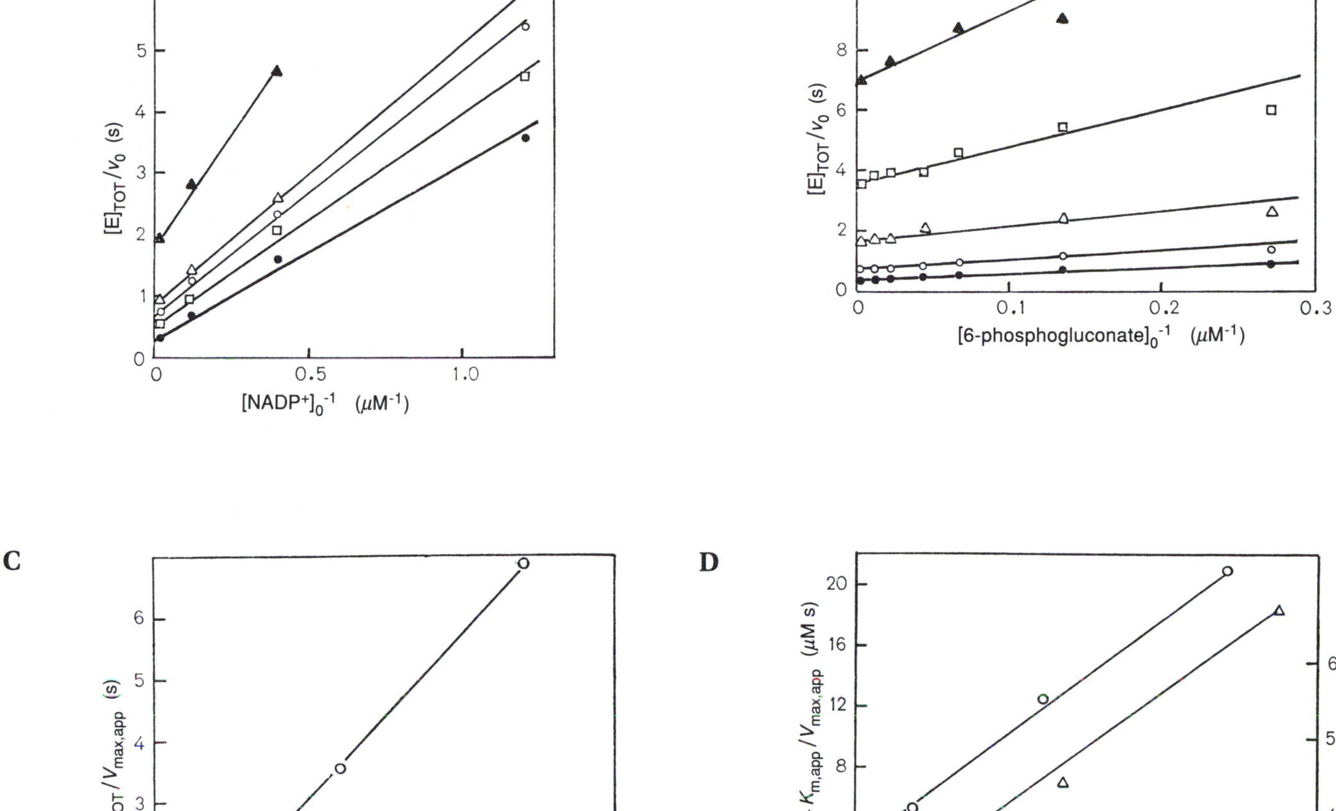


**Figure 3–4:** Double-reciprocal plots[25] of the initial velocity of the reaction catalyzed by phosphogluconate dehydrogenase (decarboxylating) from sheep liver and secondary plots of $(V_{max,app})^{-1}$ and $K_{m,app}/V_{max,app}$. Solutions containing $NADP^+$ and D-gluconate 6-phosphate at various concentrations were prepared at pH 7.0, an ionic strength of 0.13 M, and 25 °C. The reaction was initiated by the addition of enzyme, and the initial rate of the production of NADPH was followed by fluorescence. Reciprocals of the initial velocities $v_0^{-1}$ (in molar$^{-1}$ second) normalized with the molar concentration of enzyme ($[E]_{TOT}$ in molar) are presented either as a function of (A) the reciprocal of the initial concentration of nictotinamide adenine dinucleotide phosphate ($[NADP^+]_0^{-1}$ in micromolar$^{-1}$) at 1.8 $\mu$M (▲), 3.7 $\mu$M (△), 7.4 $\mu$M(○), 14.7 $\mu$M (□), or 36.8 $\mu$M (●) D-gluconate 6-phosphate or as a function of (B) the reciprocal of the initial concentration of D-gluconate 6-phosphate ($[6\text{-phosphogluconate}]_0^{-1}$ in micromolar$^{-1}$) at 0.4 $\mu$M (▲), 0.8 $\mu$M (□), 2.5 $\mu$M (△), 8.3 $\mu$M (○), or 82.5 $\mu$M (●) nicotinamide adenine dinucleotide phosphate. (C) The values of $[E]_{TOT}/V_{max,app}$ (seconds) for each fixed concentration of D-gluconate 6-phosphate in panel A are plotted (△) as a function of the reciprocal of the initial concentration of D-gluconate 6-phosphate ($[6\text{-phosphogluconate}]_0^{-1}$ in micromolar$^{-1}$), and the values of $[E]_{TOT}/V_{max,app}$ for each fixed concentration of nicotinamide adenine dinucleotide phosphate in panel B are plotted (○) as a function of the reciprocal of the initial concentration of $NADP^+$ ($[NADP^+]_0^{-1}$ in micromolar$^{-1}$). (D) The values of $[E]_{TOT}K_{m,app}/V_{max,app}$ (micromolar seconds) for each fixed concentration of D-gluconate 6-phosphate in panel A are plotted (△) as a function of the reciprocal of the initial concentration of D-gluconate 6-phosphate ($[6\text{-phosphogluconate}]_0^{-1}$ in micromolar$^{-1}$), and the values of $[E]_{TOT}K_{m,app}/V_{max,app}$ for each fixed concentration of nicotinamide adenine dinucleotide phosphate in panel B are plotted (○) as a function of the reciprocal of the initial concentration of $NADP^+$ ($[NADP^+]_0^{-1}$ in micromolar$^{-1}$). Adapted with permission from ref 25. 

Equation 3–80 is of the form of Equation 3–71, and it follows that Mechanism 3–75 is consistent with the behavior that is observed.

Again, there is an array of expanded kinetic mechanisms, each of which is also consistent with the observed behavior. As before, the introduction of steps after both A and B have bound to the active site produces expanded mechanisms consistent with the observations. The introduction of reversible steps between the binding of A to the active site and the binding of B to the active site also produces expanded mechanisms consistent with the observations for the same reason, namely, that composite rate constants are usually indistinguishable from microscopic rate constants by simple steady-state kinetics. There are, however, significantly different mechanisms that are also consistent with the observations. In particular, the order in which A and B add to the active site of the enzyme cannot be distinguished by initial velocity experiments in which only $[A]_0$ and $[B]_0$ are varied. This is due to the fact that $[A]_0$ and $[B]_0$ appear interchangeably in Equation 3–71. Because Equation 3–71 is the only relationship defined by the experimental observations, the experimental observations cannot distinguish between reactants A and B. It also happens that if the reactants A and B come to equilibrium rapidly with the active site and either can bind first, or in other words, if the mechanism is rapid-equilibrium-random, the kinetic equation derived has the form of Equation 3–71.

An interesting and informative special case of Equation 3–71, which has been occasionally encountered, is the case in which the $K_m$ for one of the reactants is zero. In this instance, unlike the more frequently encountered general case, the experiments do distinguish between reactants A and B; one of them has a $K_m$ of zero. Suppose that it is $K_m{}^A$ that is zero, so that the experimentally observed behavior of the initial velocity is

$$v_0 = \frac{V_{max}[A]_0[B]_0}{[A]_0[B]_0 + K_m{}^B[A]_0 + K_m{}^{AB}} \qquad (3\text{–}81)$$

This behavior is observed when glycerol kinase phosphorylates (*S*)-1-amino-2,3-propanediol

$$(S)\text{-1-amino-2,3-propanediol} + \text{MgATP} \rightleftharpoons (S)\text{-}N\text{-phosphono-1-amino-2,3-propanediol} + \text{MgADP} \qquad (3\text{–}81a)$$

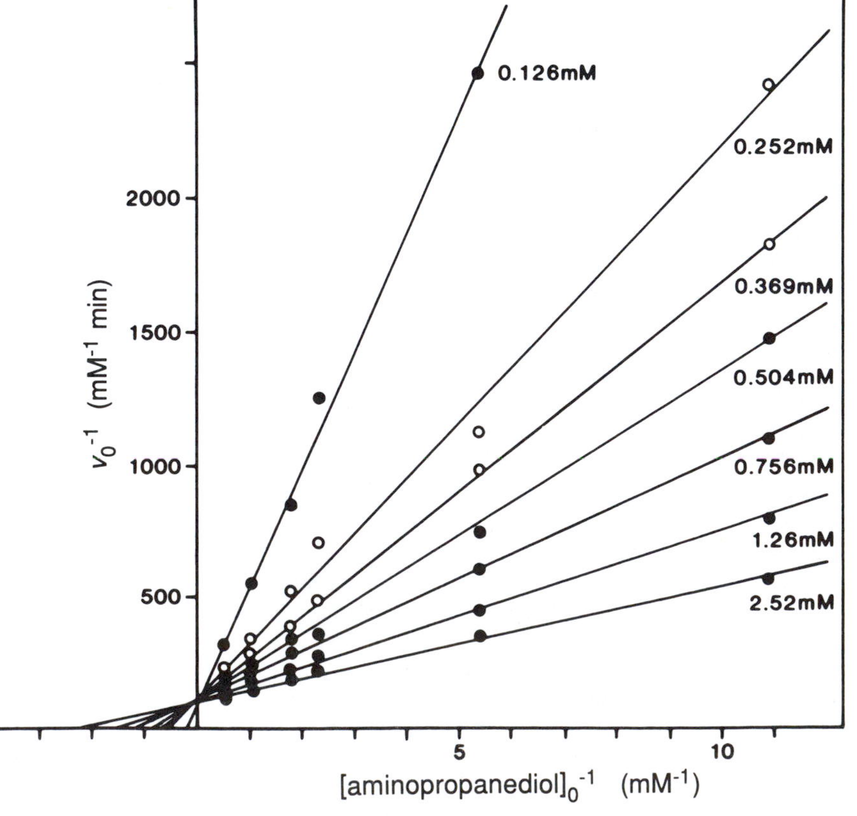


**Figure 3–5:** Double-reciprocal plots of the initial velocity of the phosphorylation of (*S*)-1-amino-2,3-propanediol by glycerol kinase from *Candida mycoderma*.[26] Solutions containing MgATP and aminopropanediol at various concentrations were prepared at pH 9.4 and 25 °C. The concentration of uncomplexed $Mg^{2+}$ cation was fixed at 7 mM throughout. The reaction was initiated by adding enzyme, and the rate of production of MgADP was followed over the first few minutes to obtain initial velocities ($v_0$). The production of MgADP was followed in a coupled assay with phospho*enol*pyruvate, pyruvate kinase, dihydro-nicotinamide adenine dinucleotide, and lactate dehydrogenase by the decrease in $A_{340}$ from the conversion of NADH into $NAD^+$. Reciprocals of the initial velocities ($v_0{}^{-1}$ in millimolar$^{-1}$ minute) are presented as a function of the reciprocal of the initial concentration of (*S*)-1-amino-2,3-propanediol ([*S*-aminopropanediol]$_0{}^{-1}$ in millimolar$^{-1}$) at the noted fixed concentrations of MgATP. Because $V_{max,app}^{aminopropanediol}$ is invariant with the concentration of MgATP, the Michaelis constant for MgATP, $K_m{}^{MgATP}$, must be zero (Equation 3–69). Adapted with permission from ref 26. Copyright 1989 American Chemical Society.

In Equation 3–81

$$V_{\max,\text{app}}^{\text{B}} = V_{\max} \quad (3\text{–}82)$$

which is invariant as the concentration of reactant A is varied. For glycerol kinase, this relation governs if reactant A is MgATP and reactant B is (*S*)-1-amino-2,3-propanediol (Figure 3–5).[26]

A simple mechanism consistent with this behavior is the equilibrium-ordered mechanism

$$\text{E}+\text{A}+\text{B} \overset{K_\text{d}^{\text{A}}}{\rightleftharpoons} \text{E}\cdot\text{A}+\text{B} \overset{K_\text{d}^{\text{BA}}}{\rightleftharpoons} \text{E}\cdot\text{A}\cdot\text{B} \overset{k_3}{\rightarrow} \text{E}+\text{P}+\text{Q} \quad (3\text{–}83)$$

where $K_\text{d}^{\text{A}}$ and $K_\text{d}^{\text{BA}}$ are dissociation constants. When the mechanism is written in this way, the author states that both the association of reactant A with the enzyme and the association of reactant B with the complex between enzyme and reactant A reach equilibrium so rapidly that the various complexes and free reactants are in equilibrium during the entire reaction. In such an equilibrium mechanism, the steady-state assumptions are unnecessary; the dissociation constants themselves provide the necessary relationships

$$K_\text{d}^{\text{A}} = \frac{[\text{E}][\text{A}]_0}{[\text{E}\cdot\text{A}]} \quad (3\text{–}84)$$

$$K_\text{d}^{\text{BA}} = \frac{[\text{E}\cdot\text{A}][\text{B}]_0}{[\text{E}\cdot\text{A}\cdot\text{B}]} \quad (3\text{–}85)$$

They can be substituted directly into Equation 3–77 to give the rate equation

$$v_0 = \frac{k_3[\text{E}]_{\text{TOT}}[\text{A}]_0[\text{B}]_0}{[\text{A}]_0[\text{B}]_0 + K_\text{d}^{\text{BA}}[\text{A}]_0 + K_\text{d}^{\text{A}} K_\text{d}^{\text{BA}}} \quad (3\text{–}86)$$

If glycerol kinase proceeds by Mechanism 3–83, MgATP must bind to the active site before (*S*)-1-amino-2,3-propanediol.

By the definition of the initial velocity (Equation 3–29), the initial concentrations of the reactants are constant over the period of time in which the value of the initial velocity pertains. Therefore, any terms in a kinetic derivation that have the form $k_i[\text{J}]_0$, where $k_i$ is the rate of step $i$ and J is reactant J, are formally pseudo-first-order rate constants. For example, Equation 3–80 can be rewritten as

$$v_0 = \frac{k_3[\text{E}]_{\text{TOT}} k_1[\text{A}]_0 k_2[\text{B}]_0}{k_1[\text{A}]_0 k_2[\text{B}]_0 + (k_{-2}+k_3)k_1[\text{A}]_0 + k_3 k_2[\text{B}]_0 + k_{-1}k_{-2} + k_{-1}k_3} \quad (3\text{–}87)$$

and Mechanism 3–75 can be rewritten as

$$\text{E} \underset{k_{-1}}{\overset{k_1[\text{A}]_0}{\rightleftharpoons}} \text{E}\cdot\text{A} \underset{k_{-2}}{\overset{k_2[\text{B}]_0}{\rightleftharpoons}} \text{E}\cdot\text{A}\cdot\text{B} \overset{k_3}{\rightarrow} \text{E}+\text{P}+\text{Q} \quad (3\text{–}88)$$

Thus recast, these equations illustrate the fact that any kinetic mechanism used to derive an equation for an initial velocity can be treated as a sequence of steps, the individual rate constants of whose forward and reverse reactions are both simple first-order rate constants, regardless of which steps involve the association of substrates with the active site of the enzyme or dissociation of substrates from the active site of the enzyme, respectively. For example, if Mechanism 3–88 is expanded by two steps to add an isomerization of the enzyme and to take account of the possibility that the dissociation of the first product is the rate-limiting step, rather than the conversion of reactants to products, then

$$\text{E} \underset{k_{-3}}{\overset{k_3[\text{A}]_0}{\rightleftharpoons}} \text{E}\cdot\text{A} \underset{k_{-4}}{\overset{k_4[\text{B}]_0}{\rightleftharpoons}} \text{E}\cdot\text{A}\cdot\text{B} \underset{k_{-5}}{\overset{k_5}{\rightleftharpoons}} \text{E}'\cdot\text{A}\cdot\text{B} \underset{k_{-6}}{\overset{k_6}{\rightleftharpoons}} \text{E}\cdot\text{P}\cdot\text{Q} \overset{k_7}{\rightarrow} \text{E}+\text{P}+\text{Q} \quad (3\text{–}89)$$

If Mechanism 3–68 is rewritten in the same way, then

$$\text{E} \underset{k_{-3}}{\overset{k_3[\text{A}]_0}{\rightleftharpoons}} \text{E}\cdot\text{A} \underset{k_{-4}}{\overset{k_4}{\rightleftharpoons}} \text{E}'\cdot\text{A} \underset{k_{-5}}{\overset{k_5}{\rightleftharpoons}} \text{E}'\cdot\text{A}' \underset{k_{-6}}{\overset{k_6}{\rightleftharpoons}} \text{E}\cdot\text{P} \overset{k_7}{\rightarrow} \text{E}+\text{P} \quad (3\text{–}90)$$

Because the concentrations of reactants remain constant at their initial values during the measurement of initial velocities, the terms in these two mechanisms involving concentrations, such as $k_2[\text{B}]_0$ and $k_1[\text{A}]_0$, are pseudo-first-order rate constants which are formally equivalent to the normal first-order rate constants. In this sense, these two mechanisms are mathematically equivalent. Therefore, both should give equations for $v_0$ as a function of $[\text{A}]_0$ of the form of Equation 3–46 where $k_1$, $k_{-1}$, and $k_2$ are defined by Equations 3–61, 3–67, and 3–54, respectively, with $k_4$ replaced by $k_4[\text{B}]_0$. The simplification that permits every kinetic mechanism for initial velocity to be written as a series of first-order steps has been exploited by King and Altman.[27,28] These investigators have formulated a series of rules by which an expression for $v_0$ can be obtained systematically from complicated kinetic mechanisms that have many steps, branching pathways, coalescing pathways, or all of these features. It is worth the time necessary to learn this system if one is required to derive rate equations for several long and complicated mechanisms and if the simplification provided by the use of composite rate constants is not sufficient.

Because of the fact that composite rate constants for a series of steps, none of which involves the entry or exit of a substrate to or from the active site of the enzyme, are indistinguishable from microscopic rate constants, steady-state kinetics based on initial velocities can provide information only about entry of substrates from the solution into the active site or exit of substrates to the solution from the active site. If a substrate must associate with the active site of the enzyme during the reaction, experimental variation of the concentration of that substrate will affect the observed initial velocity, usually in a rectangularly hyperbolic relationship (Figure 3–1B). Although the order in which substrates bind to the active site of an enzyme usually cannot be determined only by varying the initial concentrations of the reactants, it can be obtained from studies of product inhibition. To understand this strategy, the effects of inhibitors of enzymatic reactions on steady-state kinetics must first be explained.

## *Mechanisms Involving Inhibitors*

An **inhibitor** is any chemical substance that when added to the solution, decreases the rate of an enzymatically catalyzed reaction. **Reversible inhibitors** exert their effect by binding noncovalently to the enzyme and decreasing or eliminating its ability to catalyze the reaction. When a reversible inhibitor is removed from the solution, its effect on the enzymatic reaction disappears, or is reversed. A **competitive reversible inhibitor** with respect to reactant A is any reversible inhibitor the addition of which to the solution increases the observed values of $K^{A}_{m,app}/V^{A}_{max,app}$ while having no effect on the values of $V^{A}_{max,app}$. It must be emphasized that this is a classification attached to a particular experimental observation, not to an explanation of that observation. An inhibitor that is competitive with respect to reactant A is not necessarily competitive with respect to the other reactants.

One explanation for the observation of reversible competitive inhibition is that, for whatever reason, it is not possible for a molecule of the inhibitor and a molecule of the particular reactant to bind simultaneously to the enzyme. Although this explanation is often extended by the assumption that this exclusion results from the competition between the inhibitor and the reactant for the same location in an active site of the enzyme, there is no necessity for this extension. For example, if the binding of an inhibitor at a site distinct from the active site causes a conformational change in the protein that prevents reactants from binding at the active site and the binding of reactant at the active site causes a conformational change that prevents the inhibitor from binding to this site, one still observes competitive inhibition.

Consider the situation in which inhibitor I and reactant A cannot bind simultaneously to the enzyme. For the sake of argument, assume that reactant A binds to the active site before reactant B. One of the kinetic mechanisms consistent with these assumptions is

$$\mathrm{E\cdot I} \underset{k_{i}}{\overset{k_{-i}}{\rightleftharpoons}} \mathrm{I+E} \underset{k_{-1}}{\overset{k_1[\mathrm{A}]_0}{\rightleftharpoons}} \mathrm{E\cdot A} \underset{k_{-2}}{\overset{k_2[\mathrm{B}]_0}{\rightleftharpoons}} \mathrm{E\cdot A\cdot B} \xrightarrow{k_3} \mathrm{E+P+Q} \tag{3–91}$$

In Mechanism 3–91, inhibitor I binds to the enzyme; when it is bound, reactant A cannot bind, and no further reaction occurs until it is released. Such a situation is referred to as **dead-end inhibition.** If the steady-state assumption is applied to the form [E·I]

$$k_i[\mathrm{E}][\mathrm{I}] = k_{-i}[\mathrm{E\cdot I}] \tag{3–92}$$

and

$$\frac{[\mathrm{E}][\mathrm{I}]}{[\mathrm{E\cdot I}]} = \frac{k_{-i}}{k_i} \tag{3–93}$$

The ratio $k_{-i}(k_i)^{-1}$, therefore, is equal to the dissociation constant, $K_i$, for the complex between enzyme and inhibitor since

$$K_i = \frac{[\mathrm{E}][\mathrm{I}]}{[\mathrm{E\cdot I}]} \tag{3–94}$$

If the inhibition is of the dead-end type, kinetic measurements can be used to determine a numerical value for this dissociation constant.

The existence of the complex between enzyme and inhibitor expands the equation stating the conservation of enzyme (Equation 3–77)

$$[\mathrm{E}]_{TOT} = [\mathrm{E}] + [\mathrm{E\cdot I}] + [\mathrm{E\cdot A}] + [\mathrm{E\cdot A\cdot B}] \tag{3–95}$$

and the number of unknowns, but it also provides an additional simultaneous equation, that for the dissociation constant (Equation 3–94). It turns out that this is a trivial expansion because the unknown [E·I] is eliminated immediately

$$[\mathrm{E}]_{TOT} = \left(1 + \frac{[\mathrm{I}]}{K_i}\right)[\mathrm{E}] + [\mathrm{E\cdot A}] + [\mathrm{E\cdot A\cdot B}] \tag{3–96}$$

**Immediate eliminations** of this type account for the appearance in expressions for $v_0$ of terms such as $(1 + [\mathrm{I}]/K_i)$, $(1 + [\mathrm{H^+}]/K_a)$, or $(1 + [\mathrm{L}]/K_d^{\mathrm{L}})$, where $K_a$ is an acid dissociation constant and $K_d^{\mathrm{L}}$ is the dissociation constant for the complex of a ligand, L, and the enzyme. After combining Equation 3–96 with Equations 3–78 and 3–79 and solving for $v_0$

$$v_0 = \frac{k_3[\mathrm{E}]_{TOT}[\mathrm{A}]_0[\mathrm{B}]_0}{[\mathrm{A}]_0[\mathrm{B}]_0 + \left(\frac{k_{-2}+k_3}{k_2}\right)[\mathrm{A}]_0 + \left(1+\frac{[\mathrm{I}]}{K_i}\right)\left(\frac{k_3[\mathrm{B}]_0}{k_1} + \frac{k_{-1}k_{-2}+k_{-1}k_3}{k_1 k_2}\right)} \tag{3–97}$$

By examining Equations 3–97, 3–71, 3–69, and 3–70, it can be seen that $V^{A}_{max,app}$ at any fixed concentration of reactant B is unaffected by addition of the inhibitor and that

$$\frac{K^{A}_{m,app}}{V^{A}_{max,app}} = \left(\frac{K_m{}^{A}[\mathrm{B}]_0 + K_m{}^{AB}}{V_{max}[\mathrm{B}]_0}\right)\left(1+\frac{[\mathrm{I}]}{K_i}\right) \tag{3–98}$$

Because the maximum velocity of the reaction, $V^{A}_{max,app}$, is unchanged by addition of the inhibitor I but the values for $K^{A}_{m,app}/V_{max,app}$ observed when $[\mathrm{A}]_0$ is varied at any constant $[\mathrm{B}]_0$ increase by the factor $(1 + [\mathrm{I}]/K_i)$ upon addition of the inhibitor I, Mechanism 3–91 would be consistent with an observation that inhibitor I was a competitive inhibitor with respect to reactant A.

In Mechanism 3–91 both inhibitor I and reactant A compete for the same form, E, of the enzyme. If, however, reversible isomerizations of the enzyme occur between the release of I and the binding of A, their existence will have no effect on the algebraic form of the final equation for $v_0$. There is one type of reversible isomerization that can explain the ability of some enzymes to be inhibited competitively by inhibitors that nevertheless do not compete with any of the reactants for the same location in the active site. If one isomer of the enzyme binds only inhibitor I and not reactant A and the other isomer binds only reactant A and not inhibitor I, inhibitor I will be a competitive inhibitor with respect to reactant A regardless of where it binds to the enzyme.

A **noncompetitive reversible inhibitor** with respect to reactant A in an enzymatic reaction is a reversible inhibitor, the addition of which to the solution increases $K^{A}_{max,app}/V^{A}_{max,app}$ at the same time that it decreases $V^{A}_{max,app}$.

The effect of inhibitor I in Mechanism 3–91 on the initial velocity of the enzymatic reaction when reactant B is varied and reactant A is held constant is that of noncompetitive inhibition. From Equations 3–97 and 3–71

$$V^{B}_{max,app} = \frac{V_{max}[A]_0}{[A]_0 + \left(1+\frac{[I]}{K_i}\right)K_m{}^{A}} \qquad (3\text{–}99)$$

and

$$\frac{K^{B}_{m,app}}{V^{B}_{max,app}} = \frac{K_m{}^{B}[A]_0 + \left(1+\frac{[I]}{K_i}\right)K_m{}^{AB}}{V_{max}[A]_0} \qquad (3\text{–}100)$$

At fixed concentrations of reactant A below saturation the apparent maximum velocity $V^{B}_{max,app}$ decreases and the quantity $K^{B}_{max,app}/V^{B}_{max,app}$ increases as the concentration of inhibitor I is increased. Inhibitor I would have to be a noncompetitive inhibitor with respect to reactant B and a competitive inhibitor with respect to reactant A for Mechanism 3–91 to be consistent with the observations.

In theory, there is a special case of noncompetitive inhibition. If an inhibitor has an equal affinity for all forms of the enzyme and, whenever a molecule of inhibitor is bound, the enzyme is defunct, then only the values of $V_{max,app}$ should be affected by addition of the inhibitor and all of the values for $K_{m,app}$ should be unaffected.[29] In this case, values of $V_{max,app}$ decrease as the concentration of inhibitor is increased but all values of $K_{m,app}$ are invariant with changes in the concentration of inhibitor. It is, however, highly unlikely that an inhibitor, if it is capable of inhibiting the enzyme in the first place, would not have a different affinity for at least one of the forms of the enzyme. As a result, in most cases of reversible, noncompetitive inhibition where the values of $K_{m,app}$ appear to be invariant, the invariance is a fortuitous coincidence. If, however, the inhibitor binds stoichiometrically and so tightly as to be kinetically irreversible and if it prevents catalysis while it is bound, then for every mole of inhibitor added, a mole of active sites disappears. In this instance $[E]_{TOT}$ decreases by the appropriate factor, and this causes $V_{max}$ to decrease, but the values of $K_m$ are unaffected. This behavior will be considered in greater detail when covalent modification of the active site is discussed.

If an inhibitor decreases $V_{max,app}$ of an enzymatically catalyzed reaction without changing the value of $K_{m,app}/V_{max,app}$, it is referred to as an **uncompetitive inhibitor**. For the case of an enzyme with two reactants that proceeds by Mechanism 3–75, a kinetic mechanism consistent with uncompetitive inhibition would be

$$\mathrm{E} \underset{k_{-1}}{\overset{k_1[\mathrm{A}]_0}{\rightleftharpoons}} \mathrm{E{\cdot}A} \underset{k_{-2}}{\overset{k_2[\mathrm{B}]_0}{\rightleftharpoons}} \mathrm{E{\cdot}A{\cdot}B} \xrightarrow{k_3} \mathrm{E+P+Q}$$
$$\mathrm{E{\cdot}A} + \mathrm{I} \overset{K_i}{\rightleftharpoons} \mathrm{I{\cdot}E{\cdot}A} \qquad (3\text{–}101)$$

The equation for the conservation of enzyme is

$$[E]_{TOT} = [E]+[E{\cdot}A]+[I{\cdot}E{\cdot}A]+[E{\cdot}A{\cdot}B] \qquad (3\text{–}102)$$

After eliminating [I·E·A]

$$[E]_{TOT} = [E]+\left(1+\frac{[I]}{K_i}\right)[E{\cdot}A]+[E{\cdot}A{\cdot}B] \qquad (3\text{–}103)$$

Combining this equation with the necessary steady-state equations (Equations 3–78 and 3–79)

$$v_0 = \frac{k_3[E]_{TOT}[A]_0[B]_0}{[A]_0[B]_0 + \left(1+\frac{[I]}{K_i}\right)\left(\frac{k_{-2}+k_3}{k_2}\right)[A]_0 + \left(\frac{k_3}{k_1}\right)[B]_0 + \frac{k_{-1}k_{-2}+k_{-1}k_3}{k_1k_2}} \qquad (3\text{–}104)$$

From Equations 3–69 through 3–71, the apparent maximum velocity for reactant A at a fixed concentration of reactant B

$$V^{A}_{max,app} = \frac{k_3[E]_{TOT}[B]_0}{[B]_0 + \left(1+\frac{[I]}{K_i}\right)\left(\frac{k_{-2}+k_3}{k_2}\right)} \qquad (3\text{–}105)$$

and the ratio

$$\frac{K^{A}_{m,app}}{V^{A}_{max,app}} = \frac{1}{k_1[E]_{TOT}} + \frac{k_{-1}k_{-2}+k_{-1}k_3}{k_1k_2k_3[E]_{TOT}[B]_0} \qquad (3\text{–}106)$$

From these relationships, it can be seen that as long as $[B]_0$ is fixed below saturation, $V^{A}_{max,app}$ decreases in value as [I] is increased but $K^{A}_{m,app}/V^{A}_{max,app}$ is unaffected by changes in [I]. That I is an uncompetitive inhibitor with respect to reactant A would be consistent with Mechanism 3–101. In general, when an enzyme has two or more reactants, an inhibitor will display uncompetitive inhibition with respect to each reactant that must bind to the enzyme before the inhibitor itself can bind to the enzyme (see Mechanism 3–101).[30] Therefore, an uncompetitive inhibitor defines a point in the mechanism before which a particular set of reactants must have been bound.

The decision as to what type of inhibition is displayed by a particular inhibitor, I, with respect to a particular reactant, A, is usually made by examining graphical presentations of the experimental results gathered when all other reactants are at fixed concentrations. If $[A]_0$is varied at a fixed concentration of inhibitor I and the experimentally determined values for $v_0{}^{-1}$ are plotted against the respective values for $[A]_0{}^{-1}$, a linear relationship between the two variables is usually observed. That relationship is governed by $V^{A}_{max,app}$ and $K^{A}_{m,app}$. If another fixed concentration of inhibitor I is then chosen and the experiment is repeated, a second line defining the behavior of $v_0{}^{-1}$ as a function of $[A]_0{}^{-1}$ will be generated. The first and the second line will usually intersect at a point or be parallel to each other. If any number of fixed concentrations of inhibitor I are studied, including [I] = 0, each fixed concentration of inhibitor I will generate another line for $v_0{}^{-1}$ as a function of $[A]_0{}^{-1}$. All of these lines will usually pass through the same point or be parallel to each other. The location of the point at which all of the lines intersect on such reciprocal plots, or the fact that they are parallel, defines the type of inhibition.

Phosphoribosylformylglycinamidine synthase catalyzes the reaction

$$\text{MgATP} + 5'\text{-phosphoribosylformylglycinamide} + \text{L-glutamine} + H_2O \rightleftharpoons \text{MgADP} + HOPO_3^{2-} + \text{L-glutamate} + 5'\text{-phosphoribosylformylglycinamidine} \quad (3\text{–}107)$$

This enzymatic reaction is inhibited by 5′-adenylylmethylenediphosphonate

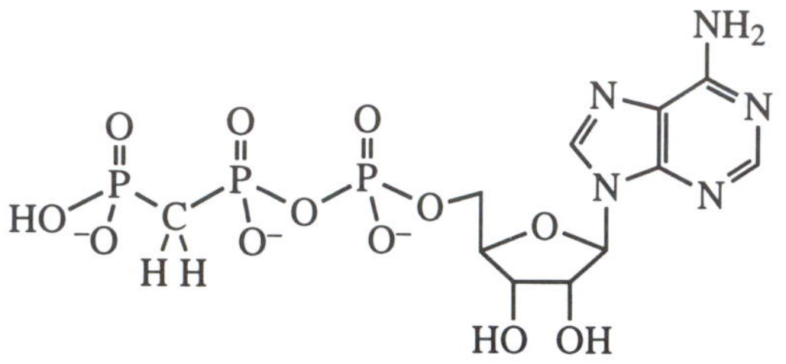

**3–3**

When MgATP (ATP) was the variable substrate at several fixed concentrations of the inhibitor, 5′-adenylylmethylenediphosphonate (AMPPCP), the lines describing the several sets of values for $v_0^{-1}$ and $[\text{MgATP}]_0^{-1}$ all intersected at the same point, $(V_{\text{max,app}}^{\text{A}})^{-1}$, on the ordinate (Figure 3–6A).[31] Because $V_{\text{max,app}}^{\text{A}}$ was invariant as the concentration of the inhibitor was varied, the inhibition is competitive with respect to MgATP. The slopes of these lines, governed by $K_{\text{m,app}}^{\text{A}}/V_{\text{max,app}}^{\text{A}}$ (Equation 3–36), vary directly (Equation 3–98) by the factor $1+[\text{AMPPCP}]/K_{\text{i}}^{\text{AMPPCP}}$. When these separate values of $K_{\text{m,app}}^{\text{ATP}}/V_{\text{max,app}}^{\text{ATP}}$ were plotted as a function of [AMPPCP], the intercept with the abscissa on this secondary plot was $-K_{\text{i}}^{\text{AMPPCP}}$. When 5′-phosphoribosylformylglycinamide (amide) was the variable reactant in the reaction catalyzed by phosphoribosylformylglycinamidine synthase at several fixed concentrations of 5′-adenylylmethylenediphosphonate, the inhibition was noncompetitive (Figure 3–6C).[31] When glutamine (Gln) was the variable reactant, all of the lines on the reciprocal plots were parallel to each other and did not intersect (Figure 3–6B),[31] and the inhibition was uncompetitive. Uncompetitive inhibition is a situation in which $(V_{\text{max,app}}^{\text{Gln}})^{-1}$, the intercept at the ordinate, increases systematically as [AMPPCP] is increased but $K_{\text{m,app}}^{\text{Gln}}/V_{\text{max,app}}^{\text{Gln}}$, the slope of each line, is invariant as [AMPPCP] is varied.

A note of caution should be sounded, however. In two instances, uncompetitive inhibition and a ping-pong bisubstrate mechanism, sets of parallel lines on such reciprocal plots are diagnostic. It is possible, however, if concentrations of inhibitors and substrates are high enough, to sample only a region of the field in which lines that are actually converging appear to be parallel. For example, in the case of phospho-2-dehydro-3-deoxygluconate aldolase, a set of apparently parallel lines generated by data at high concentrations of substrates clearly became a set of converging lines when much lower initial concentrations were employed to spread the data over a wider area.[32]

Inhibitors are often able to provide information about which of two or more reactants add first to the enzyme or if the reactants add at random.[33] The equation for $v_0$ derived from Mechanism 3–91 (Equation 3–97) states that inhibitor I should be a competitive inhibitor with respect to the first reactant that binds, reactant A, and a noncompetitive inhibitor with respect to the second reactant that binds, reactant B. This is due to the fact that both inhibitor I and reactant A compete for the free enzyme but reactant B cannot bind to the free enzyme. This particular diagnostic pattern of inhibition, defining the order in which the reactants bind to the active site of the enzyme, arises because inhibitor I can bind only to the free enzyme. For example, the inhibitor *N*-phosphonacetyl-L-aspartate (**3–4**)[34] binds to aspartate transcarbamylase only when its active site is unoccupied. This is due to the fact that *N*-phosphonacetyl-L-aspartate is a bisubstrate analogue composed of portions from both of the reactants, carbamyl phosphate (**3–5**) and L-aspartate (**3–6**)

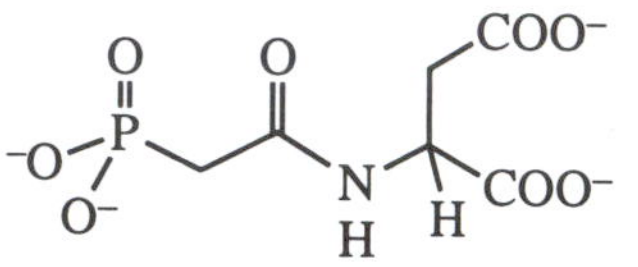

**3–4**

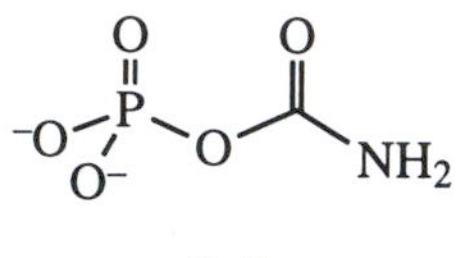

**3–5**

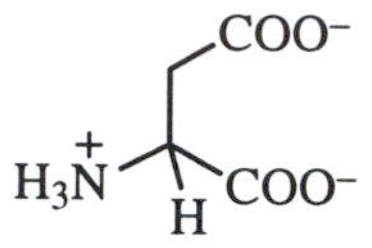

**3–6**

If carbamyl phosphate must bind to the active site of aspartate transcarbamylase before aspartate can bind, this would be consistent with the observation that *N*-phosphonacetyl-L-aspartate is a competitive inhibitor with respect to carbamyl phosphate but a noncompetitive inhibitor with respect to L-aspartate (Figure 3–7).[35] Had *N*-phosphonacetyl-L-aspartate been competitive with respect to both reactants, then either reactant could have been bound by the free enzyme and the order of their addition in the enzymatic reaction would have been random. If inhibitor I can bind only to the completely unoccupied enzyme, every reactant with respect to which it is a competitive inhibitor should also be able to bind to the free enzyme, and the order of addition to the active site among that set of reactants should be random.

Suppose, however, that two inhibitors are available, inhibitors I and J, that closely resemble reactants A and B, respectively, and that the binding of inhibitor I to the active site excludes reactant A but does not affect the binding of reactant B and the binding of inhibitor J to the active site excludes reactant B but does not affect the binding of reactant A. If reactants A and B bind at random to the active site, then inhibitor I will be competitive with respect to reactant A but noncompetitive with respect to reactant B, and inhibitor J will be competitive with respect to reactant B but noncompetitive with respect to reactant B. This was the observation made with skeletal muscle hexokinase when *N*-acetylglucosamine, which resembles glucose, and unchelated ATP,[36] which resembles the actual substrate $\text{MgATP}^{2-}$, were used as inhibitors (Figure 3–8).[33] The conclusion that the reactants glucose and $\text{MgATP}^{2-}$ add at random to the enzyme is consistent with this observation. The lines in the reciprocal plot, however, are almost parallel for the case in which $\text{ATP}^{4-}$ is the inhibitor with respect to glucose (Figure 3–8D). If they had been parallel, this

A

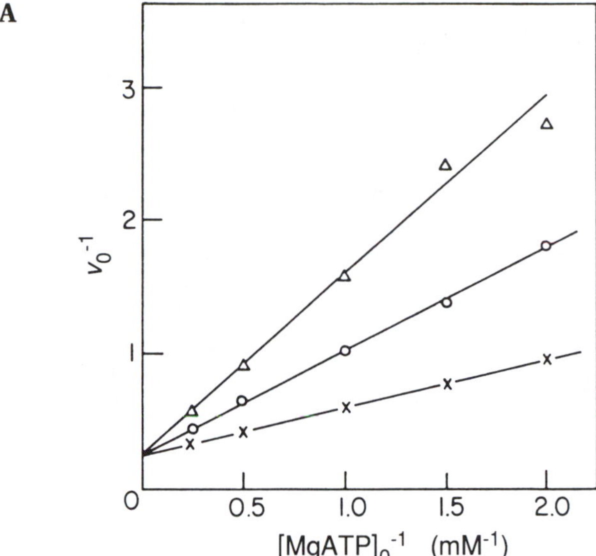


B

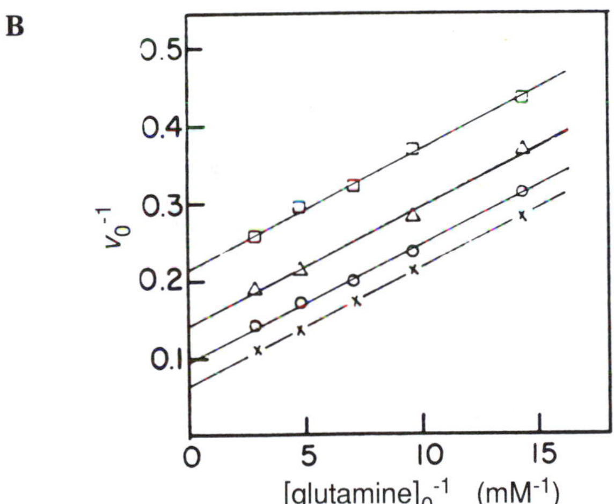


C

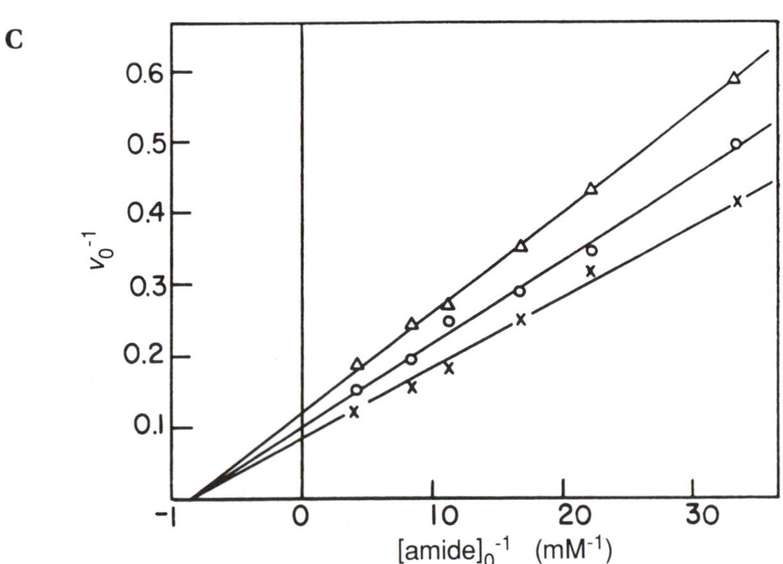


**Figure 3–6:** Double-reciprocal plots for the reversible inhibition of the reaction catalyzed by phosphoribosylformylglycinamide synthase from chicken liver.[31] Solutions containing MgATP, 5′-phosphoribosylglycinamide, and L-glutamine were prepared at pH 8.0 and 37 °C. The concentration of uncomplexed $Mg^{2+}$ cation was fixed at 1 mM throughout. The reaction was initiated by adding enzyme, and the rate of production of L-glutamine was followed over the first few minutes to obtain initial velocities ($v_0$). The production of L-glutamine was followed in a coupled assay with the 3-acetylpyridine analogue of $NAD^+$ and glutamate dehydrogenase by the increase in $A_{363}$ upon reduction of the analogue. (A) Reciprocals of the initial velocities ($v_0^{-1}$ in nanomoles$^{-1}$ minute) at 0.35 mM L-glutamine and 0.25 mM 5′-phosphoribosylformylglycinamide as a function of the reciprocals of the initial concentrations of MgATP ($[MgATP]_0^{-1}$ in millimolar$^{-1}$) at 0 mM (×), 0.5 mM (○), and 1.0 mM (△) 5′-adenylylmethylenediphosphonate. (B) Reciprocals of the initial velocities ($v_0^{-1}$ in nanomoles$^{-1}$ minute) at 0.25 mM 5′-phosphoribosylformylglycinamide and 2.0 mM MgATP as a function of the reciprocals of the initial concentrations of L-glutamine ($[glutamine]_0^{-1}$ in millimolar$^{-1}$) at 0.0 mM (×), 0.5 mM (○), 1.0 mM (△), and 1.5 mM (□) 5′-adenylylmethylenediphosphonate. (C) Reciprocals of the initial velocities ($v_0^{-1}$ in nanomoles$^{-1}$ minute) at 0.35 mM L-glutamine and 2.0 mM MgATP as a function of the reciprocals of the initial concentrations of 5′-phosphoribosylformylglycinamide ($[amide]_0^{-1}$ in millimolar$^{-1}$) at 0.0 mM (×), 0.5 mM (○), and 1.0 mM (△) 5-adenylylmethylenediphosphonate. Adapted with permission from ref 31. Copyright 1971 *Journal of Biological Chemistry.*

pattern of inhibition would have indicated that the reactants must enter the active site of hexokinase in order, glucose before MgATP.[37]

If reactant A must bind to the enzyme before reactant B can bind, inhibitor I will be competitive with respect to reactant A and noncompetitive with respect to reactant B, but inhibitor J will be competitive with respect to reactant B and uncompetitive with respect to reactant A (Mechanism 3–101). This is the observation with glycerol kinase when the chromium complex of ATP, CrATP, which closely resembles MgATP but cannot be substituted for it, is used as an inhibitor (Figure 3–9).[38] The conclusion that $MgATP^{2-}$ binds to the enzyme only after glycerol binds to the enzyme would be consistent with these observations. It has already been noted, however, that when glycerol kinase phosphorylates (*S*)-1-amino-2,3-propanediol, the observed kinetic behavior (Figure 3–5) is consistent with a mechanism in which MgATP binds to the enzyme before (*S*)-1-aminopropanol, an analogue of glycerol. Furthermore, glycerol kinase has a slow, but measurable, MgATPase activity, and this is another indication that MgATP can bind to the active site in the absence of glycerol.[26] Both of these observations seem to contradict the fact that CrATP is an uncompetitive inhibitor with respect to glycerol.

Both the almost uncompetitive inhibition of hexokinase by $ATP^{4-}$ with respect to glucose (Figure 3–8D) and the uncompetitive inhibition of glycerol kinase by CrATP with respect to glycerol (Figure 3–9) can be explained in the following way. If the binding of glucose to hexokinase or the binding of glycerol

A

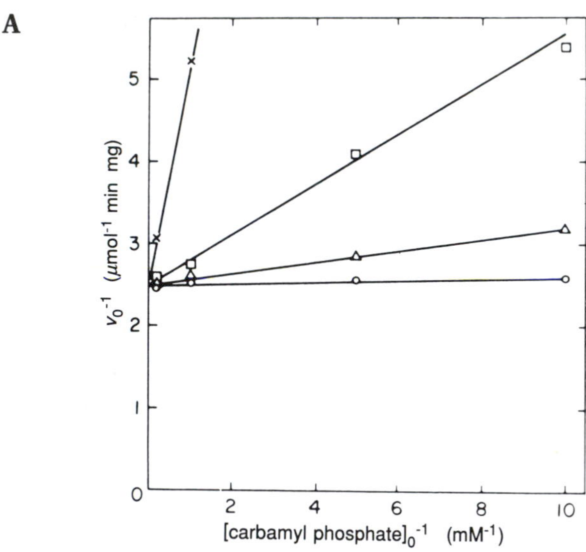


B

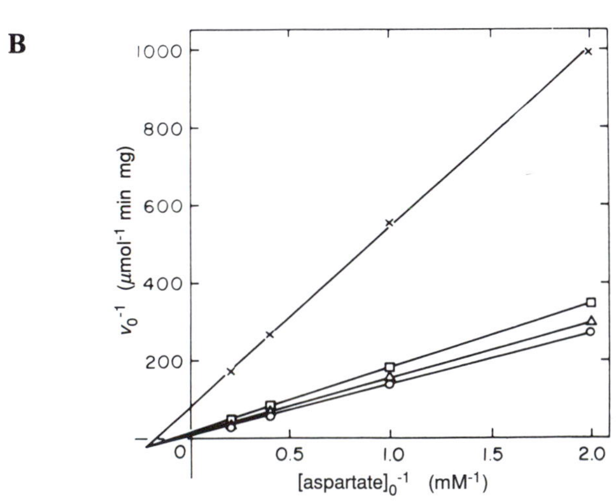


**Figure 3–7:** Double-reciprocal plots for the reversible inhibition of the reaction catalyzed by aspartate transcarbamylase from murine spleen.[35] Solutions containing L-aspartate and carbamyl phosphate were prepared at pH 7.4 and 37 °C. The reactions were initiated by adding enzyme. (A) The rate of production of *N*-carbamyl-[$^{14}$C]aspartate was followed over several minutes to obtain initial velocities ($v_0$). For this assay, a small amount of [$^{14}$C]aspartate was added to a mixture, samples were withdrawn at appropriate intervals after addition of enzyme and quenched with acetic acid, and the carbamyl-[$^{14}$C]aspartate was purified from each sample by cation-exchange chromatography. Reciprocals of initial velocities [$v_0^{-1}$ in micromoles$^{-1}$ minute (milligram of enzyme)] at 10 mM L-aspartate are presented as a function of the reciprocals of the initial concentrations of carbamyl phosphate (millimolar$^{-1}$) at 0 M (○), $10^{-7}$ M (△), $10^{-6}$ M (□), and $10^{-5}$ M (×) *N*-phosphonacetyl-L-aspartate. (B) The rate of production of *N*-[$^{14}$C]carbamylaspartate was followed over several minutes to obtained initial velocities ($v_0$). For this assay, a small amount of [$^{14}$C]carbamyl phosphate was added to a mixture, and samples were withdrawn at appropriate intervals after adding the enzyme and quenched with 10% perchloric acid. The strong acid hydrolyzed the [$^{14}$C]carbamyl phosphate, but not the [$^{14}$C]carbamylaspartate, to $^{14}CO_2$, and this was flushed from each sample with $^{12}CO_2$ to leave the enzymatically synthesized [$^{14}$C]carbamylaspartate. Reciprocals of initial velocities [$v_0^{-1}$ in micromoles$^{-1}$ minute (milligram of enzyme)] at 20 mM carbamyl phosphate are presented as a function of the reciprocals of the initial concentrations of L-aspartate (millimolar$^{-1}$) at $1 \times 10^{-8}$ M (○), $5 \times 10^{-8}$ M (△), $1 \times 10^{-7}$ M (□), and $1 \times 10^{-6}$ M (×) *N*-phosphonacetyl-L-aspartate. Adapted with permission from ref 35. Copyright 1974 Academic Press.

A

$v_0^{-1}$ ($\mu$M$^{-1}$ min)

[glucose]$_0^{-1}$ (mM$^{-1}$)

B

$v_0^{-1}$ ($\mu$M$^{-1}$ min)

[MgATP]$_0^{-1}$ (mM$^{-1}$)

C

$v_0^{-1}$ ($\mu$M$^{-1}$ min)

[MgATP]$_0^{-1}$ (mM$^{-1}$)

D

$v_0^{-1}$ ($\mu$M$^{-1}$ min)

[glucose]$_0^{-1}$ (mM$^{-1}$)

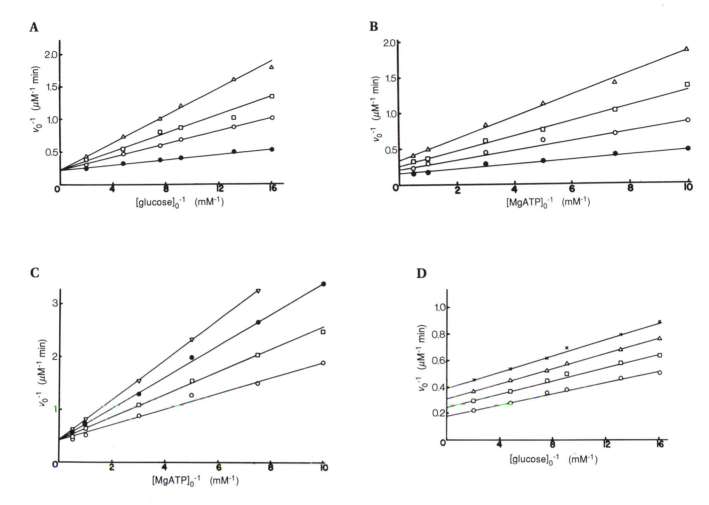

**Figure 3–8:** Double-reciprocal plots for the reversible inhibition of the reaction catalyzed by hexokinase from rat skeletal muscle.[33] Solutions containing MgATP and glucose were prepared at pH 7.7 and 28 °C. The initial velocities were followed by coupling the production of glucose 6-phosphate to the reduction of $NADP^+$ with glucose-6-phosphate dehydrogenase (Figure 3–3). Reciprocals of the initial velocities ($v_0^{-1}$ in micromolar$^{-1}$ minute) are presented as a function of either (B and C) the reciprocals of the initial concentrations of MgATP ([MgATP]$_0^{-1}$ in millimolar$^{-1}$) or (A and D) the reciprocals of the initial concentrations of glucose ([glucose]$_0^{-1}$ in millimolar$^{-1}$). (A and B) *N*-Acetylglucosamine was used as a reversible inhibitor of the enzymatic reaction at fixed concentrations of 0 mM (●), 0.1 mM (○), 0.2 mM (□), and 0.4 mM (△). (C and D) Uncomplexed ATP was used as a reversible inhibitor of the enzymatic reaction at fixed concentrations of 0 mM (○), 0.5 mM (□), 1.0 mM (●, △), and 1.5 mM (▽, ×). Adapted with permission from ref 33. Copyright 1974 *Journal of Biological Chemistry.*

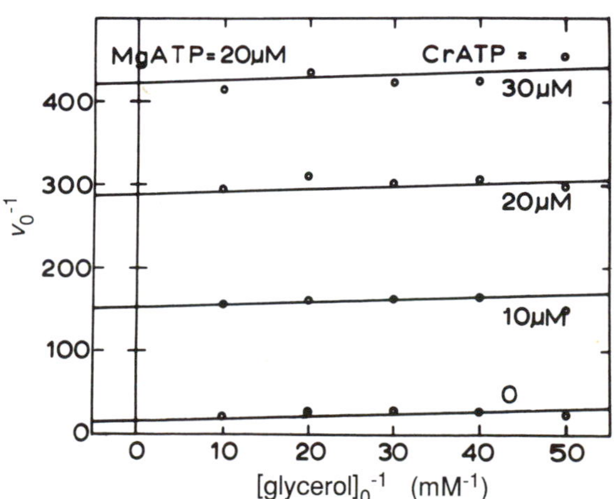


**Figure 3–9:** Double-reciprocal plot of the uncompetitive inhibition of glycerol kinase from *C. mycoderma* with respect to the reactant glycerol by the chromium complex of ATP.[38] Solutions containing glycerol and MgATP were prepared at pH 7.0 and 25 °C. The concentration of uncomplexed $Mg^{2+}$ cation was fixed at 3 mM throughout. The initial velocities were followed by coupling the production of MgADP to the oxidation of NADH (Figure 3–5). Reciprocals of the initial velocities ($v_0^{-1}$) are presented as a function of the reciprocals of the initial concentrations of glycerol ([glycerol]$_0^{-1}$ in millimolar$^{-1}$). The concentration of MgATP was fixed at 20 $\mu$M and the concentrations of the chromium complex of ATP ([CrATP]) are noted on each line. Adapted with permission from ref 38. Copyright 1974 *Journal of Biological Chemistry.*

to glycerol kinase simply increases the affinity of the respective enzyme for MgATP by a significant factor, rather than being obligatory for the binding of MgATP, then almost uncompetitive inhibition would be observed in both cases.[26] If this is the explanation, MgATP could bind to either empty active site, albeit weakly.[26] This weak binding of MgATP would allow it to precede (*S*)-1-aminopropanol onto the active site of glycerol kinase, even though it follows glycerol. This suggests that all mechanisms in which the reactants bind in a particular sequence are probably oversimplifications. Reactants appear to bind in a particular order because the binding of one reactant significantly, but not infinitely, increases the affinity of the active site for another.

## *Product Inhibition*

Often, the use of the products of an enzymatic reaction as inhibitors can further clarify the order in which reactants bind to the active site and products are released from the active site of an enzyme. Because an enzyme must catalyze its reaction in both directions and because most of the reactions catalyzed by enzymes have small changes in free energy associated with them, the products of an enzymatic reaction bind to the active site of the enzyme often as well as the reactants. By microscopic reversibility, the products must bind to the same active site as the reactants, so the binding of a product to the enzyme necessarily prevents the binding of a reactant that shares a portion of its structure with that product. For example, when malate as a product binds to the active site of fumarate hydratase, fumarate as a reactant cannot also bind because malate and fumarate share the same carbon skeleton. Moreover, the hydroxyl group of malate shares the same oxygen as the other reactant, water, and when malate is bound to the active site it should also prevent the binding of that molecule of water. This latter consideration illustrates the fact that at least one of the products of an enzymatic reaction that has two or more reactants usually contains atoms derived from two or more of the reactants and will exclude both or all of those reactants from the active site when it is bound there. Therefore, the structures of the products and the reactants will dictate which products exclude which reactants from the active site. Because the products are directly derived from the reactants and relationships between reactants and products are more readily established, the inhibition displayed by a product is less equivocal than the inhibition displayed by an inhibitor that resembles only the reactant.

Another feature than can be exploited when using products as inhibitors is that they can reverse the reaction, rather than forming simple dead-end complexes. For example, the first product to be released from the active site in an ordered mechanism can convert the complex from which it dissociates back to a complex between enzyme and reactants. In Mechanism 3–75, if product P is the first product to dissociate, its presence can reverse the reaction by promoting the formation of complex E·A·B. In this way product P is a noncompetitive inhibitor with respect to reactant A at subsaturating concentrations of reactant B and an uncompetitive inhibitor with respect to reactant A at saturating concentrations of reactant B. A dead-end inhibitor combining, as does product P, with complex E·Q would be an uncompetitive inhibitor only with respect to reactant A.

Consider an enzymatic reaction in which the reactants, A and B, bind to the active site of the enzyme in a particular order and the products, P and Q, are released in a particular order. If all of these substrates are present initially in the solution at subsaturating concentrations, none of the steps of the reaction is formally irreversible because none of the substrates is missing. In this situation there is no irreversible step in the kinetic mechanism of the reaction.

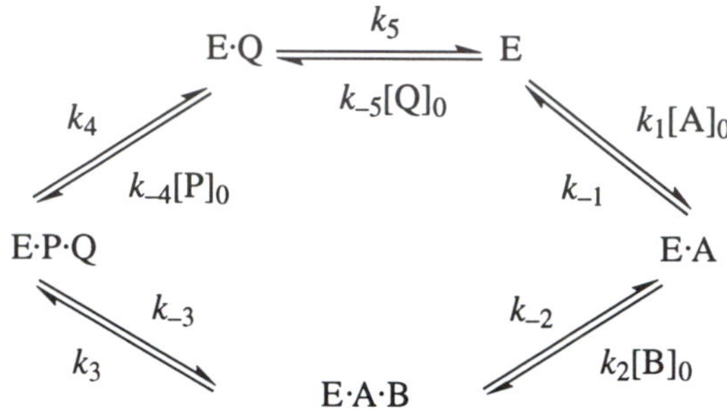


(3–108)

The direction of net flux around the circle will be determined only by the standard free energy change of the reaction and the concentration of the four substrates. If the substrate P is not present initially, the net flux must be in a clockwise direction because step 4 is forced by this choice to be formally irreversible in that direction, and the kinetic mechanism becomes

$$E \cdot Q \underset{k_{-5}[Q]_0}{\overset{k_5}{\rightleftharpoons}} E \underset{k_{-1}}{\overset{k_1[A]_0}{\rightleftharpoons}} E \cdot A \underset{k_{-2}}{\overset{k_2[B]_0}{\rightleftharpoons}} E \cdot A \cdot B \underset{k_{-3}}{\overset{k_3}{\rightleftharpoons}} E \cdot P \cdot Q \overset{k_4}{\rightarrow} E \cdot Q + P \tag{3–109}$$

Because $k_3$, $k_{-3}$, and $k_4$ do not involve entry of substrates into the active site, they can be combined as a composite rate constant by considering the two steps

$$E \cdot A \cdot B \underset{k_{-3}}{\overset{k_3}{\rightleftharpoons}} E \cdot P \cdot Q \overset{k_4}{\rightarrow} E \cdot Q + P \tag{3–110}$$

as one step

$$E \cdot A \cdot B \overset{k_3^{\text{app}}}{\rightarrow} E \cdot Q + P \tag{3–111}$$

In analogy with Equation 3–54

$$k_3^{\text{app}} = \left( \frac{k_4 k_3}{k_{-3} + k_4} \right) \tag{3–112}$$

and Mechanism 3–109 becomes

$$\mathrm{E{\cdot}Q} \underset{k_{-5}[\mathrm{Q}]_0}{\overset{k_5}{\rightleftharpoons}} \mathrm{E} \underset{k_{-1}}{\overset{k_1[\mathrm{A}]_0}{\rightleftharpoons}} \mathrm{E{\cdot}A} \underset{k_{-2}}{\overset{k_2[\mathrm{B}]_0}{\rightleftharpoons}} \mathrm{E{\cdot}A{\cdot}B} \overset{k_3^{\mathrm{app}}}{\rightarrow} \mathrm{E{\cdot}Q+P} \qquad (3\text{–}113)$$

If product Q and reactant A share atoms, so that they cannot bind simultaneously to the enzyme, and if $k_3^{\mathrm{app}} << k_{-5}[\mathrm{Q}]_0$, then Mechanism 3–113 is formally equivalent to Mechanism 3–91. From the conclusions drawn from Mechanism 3–91, it follows that when product Q is added to the reaction, it will behave as a competitive inhibitor with respect to reactant A (Equation 3–98) and a noncompetitive inhibitor with respect to reactant B when the concentration of reactant A is not saturating. Therefore, if product Q is a competitive inhibitor with respect to reactant A and a noncompetitive inhibitor with respect to reactant B, then the conclusion that reactant A must bind to the enzyme before reactant B and product P must be released by the enzyme before product Q is consistent with these observations. On the basis of similar arguments, it can be shown that when product Q is not present, so that Mechanism 3–108 becomes

$$\mathrm{E} \underset{k_{-1}}{\overset{k_1[\mathrm{A}]_0}{\rightleftharpoons}} \mathrm{E{\cdot}A} \underset{k_{-2}}{\overset{k_2[\mathrm{B}]_0}{\rightleftharpoons}} \mathrm{E{\cdot}A{\cdot}B} \underset{k_{-3}}{\overset{k_3}{\rightleftharpoons}} \mathrm{E{\cdot}P{\cdot}Q} \underset{k_{-4}[\mathrm{P}]_0}{\overset{k_4}{\rightleftharpoons}} \mathrm{E{\cdot}Q} \overset{k_5}{\rightarrow} \mathrm{E+Q} \qquad (3\text{–}114)$$

Product P will be a noncompetitive inhibitor with respect to reactant B, an uncompetitive inhibitor with respect to reactant A when saturating initial concentrations of reactant B are present, and a noncompetitive inhibitor with respect to reactant A when reactant B is not at saturating concentrations.[39]

This example illustrates the way in which patterns of product inhibition are used to determine the order in which reactants bind to an enzyme. The relationships between the order in which substrates are bound and released from the enzyme and the patterns of product inhibition have been derived and tabulated (Table 3–1). It should be emphasized that a given pattern of product inhibition is consistent only with a particular mechanism in which a particular order of entry and exit of substrates is postulated and cannot prove that this is the actual order.

If the reactants and products associate and dissociate so rapidly from the active site that they are always in rapid equilibrium with it, and if they bind at random to the active site, then the patterns of product inhibition (Table 3–1) incorporate a specific pair of competitive inhibitions. Usually, one of the products of an enzymatic reaction resembles one of the reactants closely, and the other product resembles the other reactant closely. If the substrates add to the active site at random, any one of them, reactant A, reactant B, product P, or product Q, can associate with the empty active site. If this is the case, the reactants and the products that resemble each other must compete with each other for the empty active site, and the product that resembles a given reactant will be a competitive inhibitor of the enzymatic reaction relative to that reactant.

For example, nitrate reductase catalyzes the reaction

$$\mathrm{H^+ + NADH + NO_3^-} \rightleftharpoons \mathrm{NAD^+ + NO_2^- + H_2O} \qquad (3\text{–}115)$$

The product $\mathrm{NAD^+}$ closely resembles the reactant NADH, and the product nitrite closely resembles the reactant nitrate. The two facts[40] that $\mathrm{NAD^+}$ is a competitive inhibitor with respect to NADH and that nitrite is a competitive inhibitor with respect to nitrate are consistent with a kinetic mechanism in which NADH and nitrate add to the active site of the enzyme at random and nitrite and $\mathrm{NAD^+}$ depart from the active site at random. If all four substrates, NADH, $\mathrm{NAD^+}$, nitrate, and nitrite, can bind to the free enzyme but NADH cannot bind to the active site while $\mathrm{NAD^+}$ is bound and nitrate cannot bind while nitrite is bound, this would explain the observed inhibition. Nitrite, however, is a noncompetitive inhibitor with respect to NADH and $\mathrm{NAD^+}$ is a noncompetitive inhibitor with respect to nitrate.[40] The simplest explanation for this result is that nitrite can bind to both the unoccupied active site of the enzyme and the active site when it is occupied by NADH and that $\mathrm{NAD^+}$ can bind to both the unoccupied active site and the active site occupied by nitrate

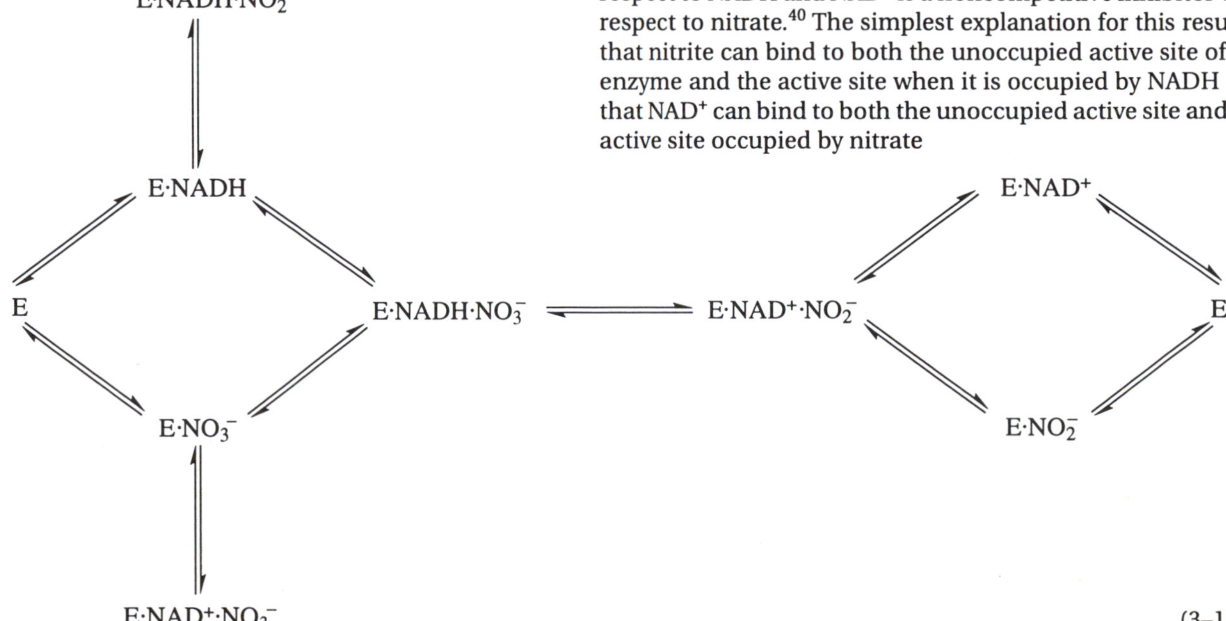


(3–116)

**Table 3–1: Some Patterns of Product Inhibition[a]**

| mechansim | product inhibitor | variable substrate A: unsaturated with B | A: saturated with B | B: unsaturated with A | B: saturated with A |
|---|---|---|---|---|---|
| A must add before B, and P must be released before Q | P<br>Q | NC[b]<br>C[b] | UC[b]<br>C | NC<br>NC | NC<br>NI[c] |
| A and B add at random, P and Q add at random, and both the complexes E·A·Q and E·B·P can form[d] | P<br>Q | C<br>NC | C<br>NI | NC<br>C | NI<br>C |
| A and B add at random, P and Q add at random, B and Q share no common atoms, and B and Q can bind simultaneously to form E·B·Q, but the complex E·A·P cannot form[d] | P<br>Q | C<br>C | NI<br>C | C<br>NC | NI<br>NI |
| P is released after A binds but before B binds, and Q is released after B binds but before A binds[e] | P<br>Q | NC<br>C | NI<br>C | C<br>NC | C<br>NI |

[a] These are patterns expected for product inhibition for an enzymatic reaction with formally two reactants and two products. For a more complete list see ref 39, which is the source for this table. [b] Abbreviations: C, competitive; NC, noncompetitive; UC, uncompetitive. [c] NI indicates no inhibition. In all of these situations, in theory, the saturation of the active site with the indicated reactant would block access to the active site for the product that shares the same atoms. In practice, the reactant, because its concentration cannot be infinite, simply increases the apparent dissociation constant for the inhibitory product by a finite factor. [d] Reactants and products must associate and dissociate so rapidly that they are always at equilibrium with the active site. [e] Refers only to a mechanism confined to one active site, not cases where intermediates are moved between two or more different active sites, as in transcarboxylase.

If this is the mechanism of the enzyme, it would be consistent with the observed behavior.

It is usually the case that two dead-end complexes can form, one between reactant A and the product that shares none or only a few atoms with it and one between reactant B and the product that shares none or only a few atoms with it. If, however, a significant portion of one reactant is transferred to the other reactant during the reaction, the product that has gained the portion may be large enough to exclude the reactant that loses the portion. Pyruvate kinase catalyzes the reaction[41]

$$\text{phospho}\textit{enol}\text{pyruvate} + \text{ADP} + \text{Mg}^{2+} \rightleftharpoons \text{pyruvate} + \text{MgATP} \qquad (3\text{–}117)$$

If the concentration of magnesium cation is held at saturating levels, the enzymatic reaction becomes formally one that has only two reactants, phosph*enol*pyruvate and ADP. The product pyruvate is a competitive inhibitor with respect to the reactant, phospho*enol*pyruvate (Figure 3–10A),[41] and the product MgATP is a competitive inhibitor with respect to ADP (Figure 3–10B). These results are consistent with a kinetic mechanism in which ADP and phospho*enol*pyruvate bind to the active site of the enzyme at random and MgATP and pyruvate are released from the active site at random. The other two product inhibitions that are observed are peculiar, but they make sense. Magnesium adenosine triphosphate is a competitive inhibitor relative to phospho*enol*pyruvate (Figure 3–10C) because it carries the phosphate that is transferred from phospho*enol*pyruvate, and phospho*enol*pyruvate cannot occupy its location in the active site when MgATP is bound. Pyruvate, however, is a noncompetitive inhibitor with respect to ADP. If ADP can bind to the active site with the same dissociation constant regardless of whether pyruvate is bound or not, this would be consistent with the noncompetitive inhibition observed. Because neither ADP nor pyruvate has the phosphate that is transferred, the binding of pyruvate cannot sterically interfere with the binding of ADP.

Even though a significant portion of one reactant is transferred to the other reactant during an enzymatic reaction, the product and the reactant that overlap may not exclude each other from the active site, especially if they are both large molecules. For example, citrate synthase catalyzes the reaction

$$\text{citrate} + \text{HSCoA} \rightleftharpoons \text{oxaloacetate} + \text{acetyl-SCoA} \qquad (3\text{–}118)$$

The reactants enter the active site at random, and the products leave the active site at random,[36] yet the product acetyl-SCoA is a noncompetitive inhibitor with respect to the reactant citrate,[36] even though both of these substrates share the acetyl functionality that is transferred. Presumably, the acetyl-SCoA is so large that it can bind almost as well with its acetyl group displaced from its usual position in the active site as when the acetyl group occupies its normal position, and this allows citrate and acetyl -SCoA to bind simultaneously to the enzyme.

Although unlikely, in some instances the binding of either product to the active site seems to be sufficient to exclude either reactant. Glycine dehydrogenase (decarboxylating) is

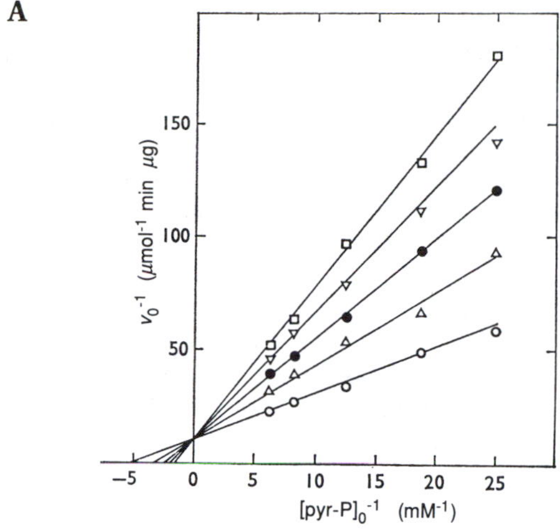


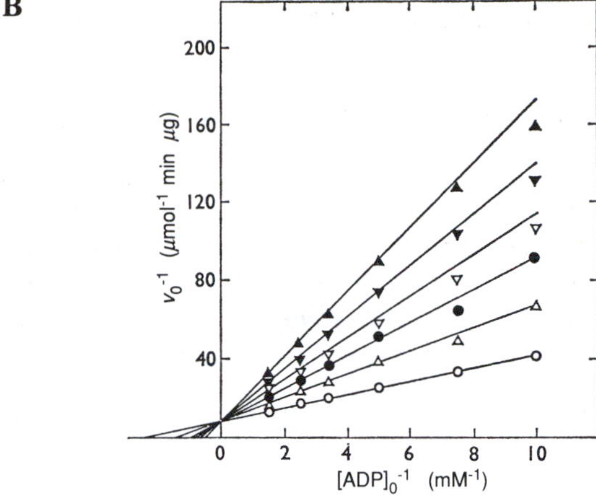


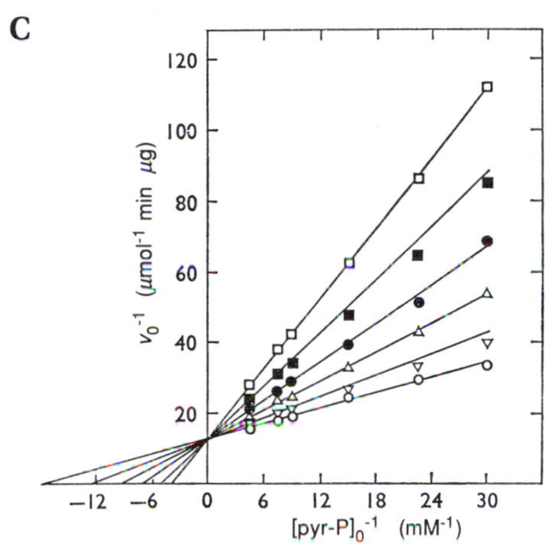


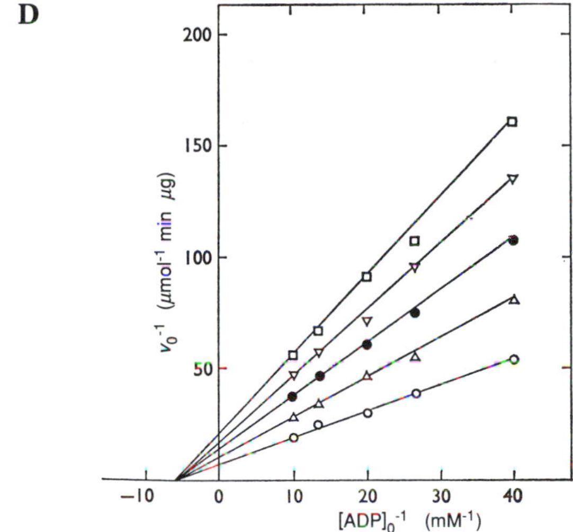


an enzyme containing pyridoxal phosphate and lipoamide that catalyzes the reaction

$$\text{glycine} + \text{protein}\text{–(lipoyl, S–S)} + H^+ \rightleftharpoons CO_2 + \text{protein}\text{–CH}_2\text{CH}_2\text{CH(SH)CH}_2\text{CH}_2\text{–S–CH}_2\text{–}\overset{+}{N}H_3 \qquad (3\text{–}119)$$

Either product, $CO_2$ or the methyleneaminolipoylprotein, is a competitive inhibitor with respect to either reactant, glycine or the oxidized lipoylprotein.[42] This suggested that neither product could bind to the active site to form a dead-end complex when either reactant was bound, and this conclusion was supported by isotope exchange studies. One explanation for these exclusions, which cannot be entirely steric, is that the enzyme itself undergoes a reversible isomerization between release of products and binding of reactants. In one conformation the active site can bind only products and in the other only reactants.

**Figure 3–10:** Double-reciprocal plots for the reversible inhibition by the products of the conversion of phospho*enol*pyruvate and MgADP to pyruvate and MgATP catalyzed by pyruvate kinase.[41] Solutions containing MgADP, phospho*enol*pyruvate, MgATP, and pyruvate were prepared at pH 6.2. The concentration of uncomplexed $Mg^{2+}$ was fixed at 10.0 mM, and that of KCl, at 0.10 M. The initial velocities were followed by coupling the production of pyruvate to the oxidation of NADH catalyzed by lactate dehydrogenase. The initial rate of decrease in $A_{340}$ was used to calculate initial velocities [$v_0$ in micromoles minute$^{-1}$ (microgram of enzyme)$^{-1}$]. Reciprocals of the initial velocities ($v_0^{-1}$) are presented as a function of either (A and C) the reciprocals of the initial concentrations of phospho*enol*pyruvate ([pyr-P]$_0^{-1}$ in millimolar$^{-1}$) or (B and D) the reciprocals of the initial concentrations of ADP ([ADP]$_0^{-1}$ in millimolar$^{-1}$). The inhibition of the reaction by the products, MgATP or pyruvate, are described. (A) The concentrations of pyruvate were 0.0 mM (○), 5.0 mM (△), 10.0 mM (●), 15.0 mM (▽), and 20.0 mM (□). (B) The concentrations of MgATP were 0.0 mM (○), 1.0 mM (△), 2.0 mM (●), 3.0 mM (▽), 4.0 mM (▼), and 5.0 mM (▲). (C) The concentrations of MgATP were 0.0 mM (○), 0.6 mM (▽), 1.25 mM (△), 2.5 mM (●), 4.0 mM (■), and 6.0 mM (□). (D) The concentrations of pyruvate were 0.0 mM (○), 5.0 mM (△), 10.0 mM (●), 15.0 mM (▽), and 20.0 mM (□). Adapted with permission from ref 41. Copyright 1973 Biochemical Society.

## *Ping-Pong Mechanisms*

It is also possible for the first product of a bireactant enzymatic reaction to be released from the active site before the second reactant binds. This is referred to as a **ping-pong mechanism**. If the enzymatic reaction occurs at only one active site, rather than at more than one active site as in the case of methylmalonyl-CoA carboxyltransferase,[42a] the first product released competes with the second reactant to bind, and the second product released competes with the first reactant to bind. 3-Oxoacid CoA-transferase catalyzes the reaction

$$\text{succinyl-SCoA} + \text{acetoacetate} \rightleftharpoons \text{succinate} + \text{acetoacetyl-SCoA} \tag{3–120}$$

The facts that acetoacetyl-SCoA (AACoA) is a competitive inhibitor with respect to succinyl-SCoA (SCoA) but a noncompetitive inhibitor with respect to acetoacetate (AA) and that succinate (S) is a competitive inhibitor with respect to acetoacetate but a noncompetitive inhibitor with respect to succinyl-SCoA[43] are consistent (Table 3–1) with the kinetic mechanism

$$\text{E} \underset{k_{-1}}{\overset{k_1[\text{SCoA}]}{\rightleftharpoons}} \text{E·SCoA} \underset{k_{-2}[\text{S}]}{\overset{k_2}{\rightleftharpoons}} \text{E·CoA} \underset{k_{-3}}{\overset{k_3[\text{AA}]}{\rightleftharpoons}} \text{E·CoA·AA} \underset{k_{-4}[\text{AACoA}]}{\overset{k_4}{\rightleftharpoons}} \text{E} \tag{3–121}$$

In this mechanism, succinate in the reverse direction and acetoacetate in the forward direction bind to the same form of the enzyme, E·CoA, and compete with each other for that form. Acetoacetyl-SCoA and succinyl-SCoA bind to the same form of the enzyme, E, and compete with each other for it. These two competitions are consistent with the conclusion that succinate and acetoacetate, in turn, occupy the same locations in the active site and that succinyl-SCoA and acetoacetyl-SCoA, in turn, occupy the same locations in the active site. The fact that in this mechanism succinate must leave the enzyme before acetoacetate can bind requires that the coenzyme A portion of a succinyl-SCoA remain associated with the active site in the form E·CoA.

The patterns of product inhibition are not sufficient, however, to define a ping-pong mechanism (Table 3–1), but the behavior of the initial velocity as a function of the reactants, in addition to the product inhibitions, is sufficient. If an enzyme has a ping-pong mechanism and both products are absent from the solution, there is a kinetically irreversible step between the binding of the first reactant and the binding of the second reactant because the first product is released before the second reactant is bound. Under these circumstances, either $v_0^{-1}$ as a function of $[\text{A}]_0^{-1}$ at several fixed concentrations of reactant B should define a set of parallel lines (Figure 3–11),[43] or in a direct numerical analysis performed to fit the data to Equation 3–71, the term $K_m^{AB}$ should be zero, within the accuracy of the measurements.

The observation of sets of parallel lines or a $K_m^{AB}$ equal to zero, however, does not demonstrate that a ping-pong mechanism is governing the enzymatically catalyzed reaction. All

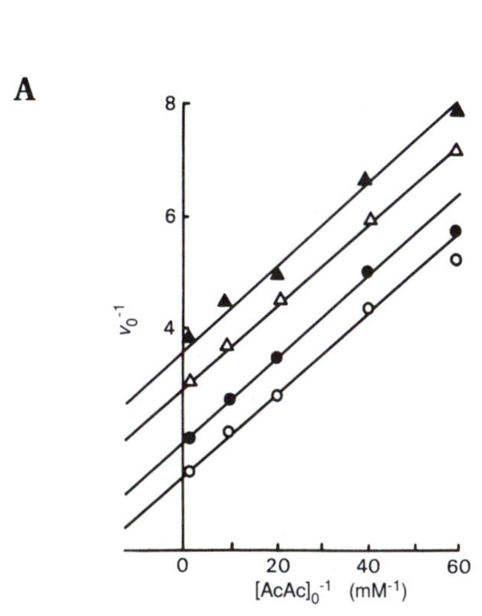


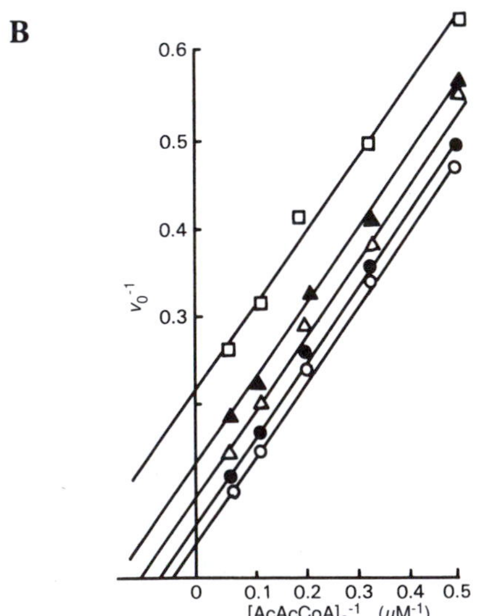


**Figure 3–11:** Double-reciprocal plots of the initial velocity of the reaction catalyzed by 3-oxoacid CoA-transferase from ovine kidney in the two directions.[43] Solutions containing 25 mM $Mg^{2+}$ and either (A) acetoacetate and succinyl-SCoA or (B) acetoacetyl-SCoA and succinate were prepared at pH 8.1 and 30 °C. The initial velocity of each reaction was determined by monitoring the absorbance at 303 nm of the acetoacetyl-SCoA being either formed (A) or lost (B) as the reaction proceeded. Reciprocals of the initial velocities [$v_0^{-1}$ in micromoles$^{-1}$ minute (milligram of enzyme)] are presented either as a function of (A) the reciprocals of the initial concentrations of acetoacetate ($[\text{AcAc}]_0^{-1}$ in millimolar$^{-1}$) at 66 $\mu$M (▲), 100 $\mu$M (△), 200 $\mu$M (●), and 800 $\mu$M (○) succinyl-SCoA or as a function of (B) the reciprocals of the initial concentrations of acetoacetyl-SCoA ($[\text{AcAcCoA}]_0$ in micromolar$^{-1}$) at 1 mM (□), 2 mM (▲), 3 mM (△), 5 mM (●), and 7 mM (○) succinate. Adapted with permission from ref 43. Copyright 1983 Biochemical Society.

that is required for this behavior is that a kinetically irreversible step lie between the binding of the first and the second reactant. In the case of 3-oxoacid CoA-transferase (Figure 3–11), it was the results of the product inhibitions that confirmed that the mechanism was ping-pong. The existence of a ping-pong mechanism can also be confirmed by examining isotope exchange.

## *More Than Two Reactants*

Any enzymatic reaction in which there are more than two reactants or more than two products or both can be turned into a reaction that has formally two reactants and two products by manipulating the concentrations of substrates. For example, consider an enzymatic reaction with four reactants, A, B , C , and D, and three products, P, Q, and R. If $[B]_0$ and $[D]_0$ are kept at saturation and $k_i[B]_0$ and $k_j[D]_0$ are treated formally as first-order rate constants, the reaction can be treated as a reaction with only two reactants, A and C. If no amount of product Q is added, the step in which product Q is released is an irreversible, first-order reaction, and the overall reaction can be treated as a reaction in which two products, P and R, are released. Patterns of product inhibition can be ascertained and the order in which reactants A and C are bound and products P and R are released can be determined by referring to Table 3–1. Different pairs of reactants and products can then be chosen and their order of binding and release determined, respectively. The process can be repeated until the order of all steps is established.[31]

The foregoing discussion has presented two features of steady-state kinetics. First, any step in the mechanism of the enzymatic reaction that involves the binding of a substrate to or the release of a substrate from the enzyme is ultimately responsible for the variations in initial velocity that are observed when the initial concentration of that substrate is varied. Because of this, steady-state kinetics elucidate aspects of the binding and release of substrates. Second, any step following the assembly of all of the reactants on the active site and preceding the release of the first product from the active site cannot be studied in isolation because the rate constants for these steps can be accumulated into composite rate constants without changing the form of the equation for the initial velocity of the other steps of the same type by steady-state kinetics. For these kinetically silent steps, steady-state kinetics provide a value for $V_{max}$. The parameter $V_{max}$ is the overall rate of transformation of reactants into products when the reaction is not limited by the rates at which the reactants enter the active site because, at saturation, the initial concentrations of reactants are so high that their entry into the active site is more rapid than any other step in the mechanism. This is the reason saturation is observed.

## *Suggested Reading*

Cleland, W.W. (1963) The Kinetics of Enzyme-Catalyzed Reactions with Two or More Substrates or Products I. Nomenclature and Rate Equations, *Biochim. Biophys. Acta* 67, 104–137.

**PROBLEM 3–2**

The enzyme fumarate hydratase catalyzes the hydration of fumarate to L-malate (Reaction 3–1). The following data were obtained when fumarate was used as the reactant and the initial velocity of the hydration reaction was measured at pH 5.7, 25 °C, and an enzyme concentration of $2 \times 10^{-6}$ M.

| fumarate (mM) | rate of product formation (mmol $L^{-1}$ $min^{-1}$) |
|---|---|
| 2.0 | 2.5 |
| 3.3 | 3.1 |
| 5.0 | 3.6 |
| 10.0 | 4.2 |

When the same reaction was studied in a buffer containing 0.06 M phosphate, at the same concentration of enzyme, the following data were obtained.

| fumarate (mM) | rate of product formation (mmol $L^{-1}$ $min^{-1}$) |
|---|---|
| 2.0 | 4.0 |
| 3.3 | 5.0 |
| 5.0 | 5.8 |
| 10.0 | 6.6 |

(A) Determine the value of $V_{max}$ and the Michaelis constant, $K_m^F$, for the above conditions. Analyze and discuss briefly any significant differences.

(B) What is the turnover number [moles of product (mole of enzyme)$^{-1}$minute$^{-1}$] for the enzyme in the absence of phosphate?

The Haldane equation relates the kinetic constants $V_{max}$ and $K_m$ for a reaction to the equilibrium constant $K_{eq}$

$$K_{eq} = \frac{({}^{f}V_{max})({}^{r}K_m)}{({}^{r}V_{max})({}^{f}K_m)}$$

where r indicates that the constants are for the reverse reaction, and f, for the forward reaction. The initial rates of the hydration reaction were measured as a function of pH and for various initial concentrations of fumarate. The values obtained, expressed as millimoles of malate liter$^{-1}$ minute$^{-1}$, were

| | fumarate concentration (mM) | | | |
|---|---|---|---|---|
| pH | 10 | 5 | 2.5 | 1.25 |
| 5 | 0.92 | 0.81 | 0.66 | 0.48 |
| 6 | 4.00 | 3.28 | 2.42 | 1.60 |
| 7 | 5.40 | 4.30 | 3.08 | 1.96 |
| 8 | 2.67 | 2.40 | 1.99 | 1.53 |
| 9 | 0.21 | 0.17 | 0.16 | 0.12 |

The values of the maximum initial velocity for the dehydration of L-malate ($V_{max}$), catalyzed by fumarate hydratase, had the following dependence on pH.

| pH | $V_{max}$ (mmol L$^{-1}$ min$^{-1}$) |
|---|---|
| 5 | 0.068 |
| 6 | 0.044 |
| 7 | 3.00 |
| 8 | 5.88 |
| 9 | 2.00 |

(C) The overall equilibrium constant for the reaction, $K_{eq}$, is independent of pH over the range from pH 5 to 9 and has a value of 4.7. Determine values for the Michaelis constant and maximum velocity of the hydration reaction for each value of pH.

(D) Calculate the values of the Michaelis constant for the dehydration reaction for each value of pH. The hydration of fumarate to D,L-malate was measured at pH 5.7 and 25 °C in the absence of enzyme

| fumarate (M) | time (h) |
|---|---|
| 0.0200 | 0 |
| 0.0191 | 20 |
| 0.0183 | 40 |
| 0.0175 | 60 |
| 0.0167 | 80 |

(E) Compare the initial rate constant for this reaction with that for the catalyzed reaction and calculate the factor by which the velocity of the hydration reaction is increased when the reactants are bound to the surface of the enzyme.

**PROBLEM 3–3**

Derive rate equations for the initial velocity of reactions that proceed by the following mechanisms.

$$\begin{array}{ccccc} E + A & \underset{k_{-1}}{\overset{k_1}{\rightleftharpoons}} & EA & & \\ \pm H^+ \ \Big\updownarrow K_{aE} & & \pm H^+ \ \Big\updownarrow K_{aEA} & & \\ EH + A & \underset{k_{-2}}{\overset{k_2}{\rightleftharpoons}} & EHA & \xrightarrow{k_3} & E + P \end{array} \qquad (A)$$

where $K_{aE}$ and $K_{aEA}$ are acid dissociation constants for the active site.

$$E + A \underset{k_{-1}}{\overset{k_1}{\rightleftharpoons}} E{\cdot}A \xrightarrow{k_2} E' + P$$

$$E' + B \underset{k_{-1}}{\overset{k_1}{\rightleftharpoons}} E'{\cdot}B \xrightarrow{k_4} E + Q \qquad (B)$$

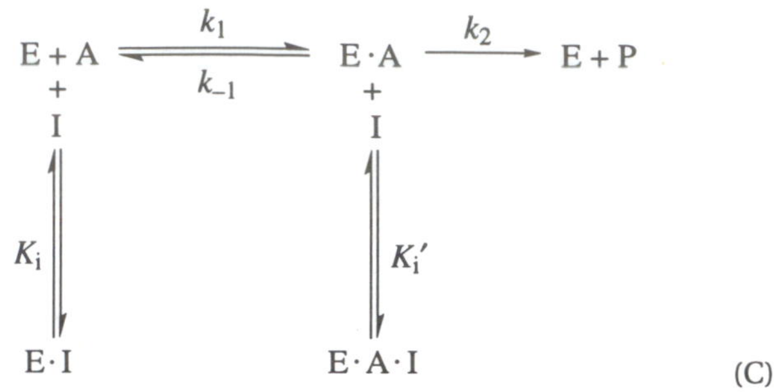


When $K_i' = \infty$, the inhibition is competitive; when $K_i$ and $K_i'$ are finite, the inhibition is noncompetitive

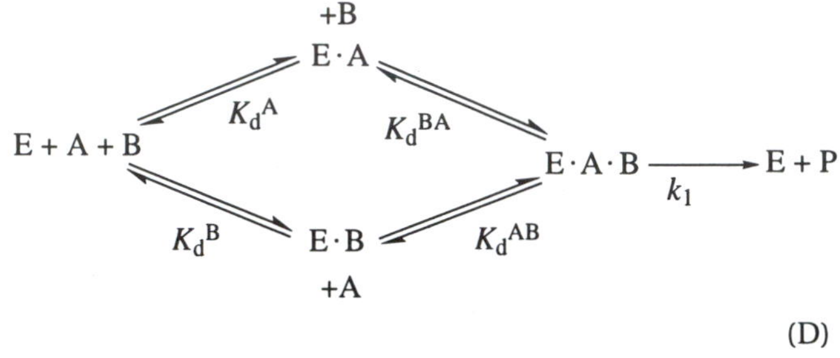


where $K_d^{\,i}$ are dissociation constants.

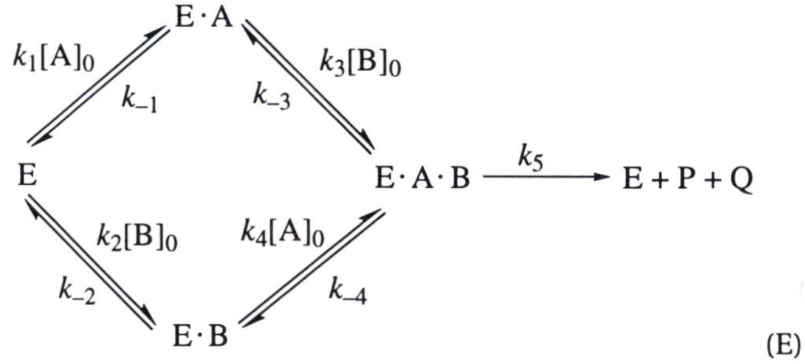


Although not necessary, the method of King and Altman[27,28] can be used to derive this last equation.

**PROBLEM 3–4**

Hexokinase catalyzes the reaction

$$\text{hexose} + \text{MgATP} \rightleftharpoons \text{hexose 6-phosphate} + \text{MgADP}$$

Magnesium cation, $Mg^{2+}$, is required for the enzymatic reaction. It is known that $Mg^{2+}$ forms a chelation complex with ATP

$$Mg^{2+} + ATP \rightleftharpoons Mg^{2+} \cdot ATP$$

and the value for the dissociation constant of this complex has been measured as a function of pH. At a particular pH and at any concentration of $Mg^{2+}$ and ATP, the molar concentrations of MgATP and ATP can be calculated. In this way it is possible to vary the molar concentrations of MgATP and ATP independently. When [ATP] was fixed and [MgATP] was varied systematically, $v_0^{-1}$ was a linear function of [MgATP]$^{-1}$. When [ATP] was changed to a higher fixed level, $v_0$ decreased at every [MgATP] but $V_{max,app}$ for [MgATP] was unchanged (Figure 3–8C). Use these observations to identify the actual reactant in the enzymatic reaction and to identify an inhibitor of the enzymatic reaction.

## PROBLEM 3–5

Fatty-acid synthase, a multienzyme complex, catalyzes the reaction

$$\text{acetyl-SCoA} + n\text{malonyl-SCoA} + 2n\text{NADPH} \rightleftharpoons \text{fatty acid} + 2n\text{NADP}^+ + n\text{CO}_2 + (n+1)\text{HSCoA}$$

When the initial velocity of the reaction at fixed concentrations of malonyl-SCoA and NADPH was followed as a function of the concentration of acetyl-SCoA, the following behavior was observed.[44]

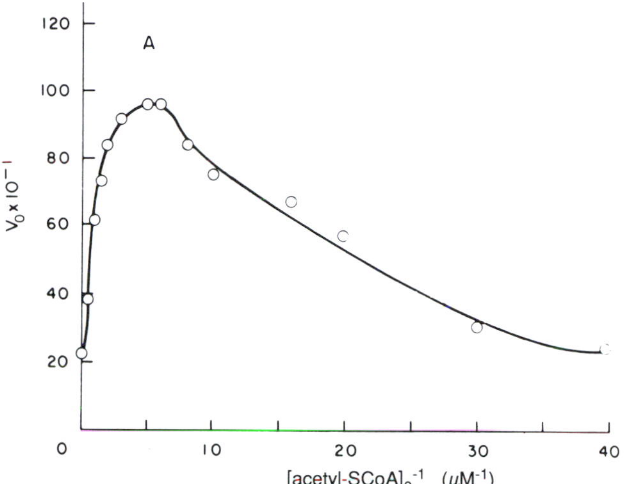


Rate of fatty acid synthesis as a function of acetyl-SCoA concentration at fixed levels of NADPH and malonyl-SCoA in 0.2 M phosphate buffer at pH 7.0.[44] The concentrations of the reactants were 10 $\mu$M malonyl-SCoA and 30 $\mu$M NADPH. Rates were followed spectrophotometrically with a 5-cm cell. Units for initial velocities ($v_0$) on the ordinate are nanomoles of NADPH oxidized minute$^{-1}$ (milligram of protein)$^{-1}$. Adapted with permission from ref 44. Copyright 1974 Academic Press.

The inhibition observed at high concentrations of acetyl-SCoA decreases as the concentration of malonyl-SCoA is raised. Explain all of these observations.

## PROBLEM 3–6

What do the patterns of inhibition of 5′-phosphoribosylformylglycinamidine synthase by 5′-adenylylmethylenediphosphonate (Figure 3–6) state about the order in which the reactants bind to the enzyme?

## PROBLEM 3–7

The enzyme adenine phosphoribosyltransferase catalyzes the reaction

$$\text{adenine} + \text{magnesium 5-phospho-}\alpha\text{-D-ribosyl 1-diphosphate} \rightleftharpoons \text{adenosine 5}'\text{-phosphate} + \text{magnesium pyrophosphate}$$

The double-reciprocal plots of $v_0^{-1}$ as a function of either [adenine]$_0^{-1}$ or [phosphoribosyl diphosphate]$_0^{-1}$ at several fixed concentrations of either phosphoribosyl diphosphate or adenine, respectively, were sets of parallel lines (as in Figure 3–11).

(A) Show that these data fit the function

$$v_0 = \frac{V_{max}[\text{Ad}]_0[\text{PRPP}]_0}{[\text{Ad}]_0[\text{PRPP}]_0 + K_m^{\text{Ad}}[\text{PRPP}]_0 + K_m^{\text{PRPP}}[\text{Ad}]_0}$$

where $[\text{Ad}]_0$ is the molar concentration of adenine and $[\text{PRPP}]_0$ is the molar concentration of magnesium 5-phosphoribosyl 1-diphosphate.

(B) Write the enzymatic mechanism usually associated with this rate equation.

(C) For this mechanism to be correct, what type of inhibition must adenosine 5′-phosphate display toward magnesium phosphoribosyl diphosphate?

(D) For this mechanism to be correct, what type of inhibition must magnesium pyrophosphate display toward adenine?

(E) Adenosine monophosphate is a competitive inhibitor relative to magnesium phosphoribosyl monophosphate, and magnesium pyrophosphate is an uncompetitive inhibitor relative to adenine. Do these observations match your expectations?

(F) Derive the initial velocity equation for the following mechanism

$$\text{E} + \text{PRPP} \underset{k_{-1}}{\overset{k_1}{\rightleftharpoons}} \text{E}\cdot\text{PRPP} + \text{Ad} \underset{k_{-2}}{\overset{k_2}{\rightleftharpoons}}$$

$$\text{E}\cdot\text{PRPP}\cdot\text{Ad} \overset{k_3}{\rightarrow} \text{E} + \text{AMP} + \text{PRPP}$$

(G) Show how this equation can assume the form

$$v_0 \cong \frac{V_{max}[\text{Ad}]_0[\text{PRPP}]_0}{[\text{Ad}]_0[\text{PRPP}]_0 + K_m^{\text{Ad}}[\text{PRPP}]_0 + K_m^{\text{PRPP}}[\text{Ad}]_0}$$

if two reasonable inequalities hold.

## PROBLEM 3–8

The introduction of a kinetically irreversible step between the binding of reactants A and B in an enzymatically catalyzed reaction causes the observed kinetic behavior of the reaction to differ from the case in which only reversible steps separate the binding of reactants A and B. Assume the kinetic mechanism

$$\text{E} \underset{k_{-1}}{\overset{k_1[\text{A}]_0}{\rightleftharpoons}} \text{E}\cdot\text{A} \overset{k_2}{\rightarrow} \text{E}'\cdot\text{A}' \underset{k_{-3}}{\overset{k_3[\text{B}]_0}{\rightleftharpoons}} \text{E}'\cdot\text{A}'\cdot\text{B} \overset{k_4}{\rightarrow} \text{E} + \text{P} + \text{Q}$$

(A) Derive the expression for $v_0$ in terms of $[\text{E}]_{\text{TOT}}$, $[\text{A}]_0$, and $[\text{B}]_0$ for this mechanism.

(B) How does it differ from Equation 3–71 in form?

(C) If $v_0^{-1}$ is plotted against $[\text{A}]_0^{-1}$ at a series of constant values of $[\text{B}]_0$, how will the lines in the set be related to each other?

## PROBLEM 3–9

Procollagen-lysine, 2-oxoglutarate 5-dioxygenase catalyzes the reaction

lysyl peptide + 2-oxoglutarate + $O_2 \rightleftharpoons$
5-hydroxylysylpeptide + succinate + $CO_2$

Ascorbate (Figure 2–32) is also a required addition to maintain the iron in the enzyme in the ferrous form. When the concentration of ascorbate is varied at a fixed concentration of any one of the reactants and saturating concentrations of the other two, $v_0^{-1}$ is a linear function of $[\text{ascorbate}]_0^{-1}$. If the fixed substrate is set at other concentrations, the lines defined by the data are all parallel. For example, if lysyl peptide is the fixed substrate and oxygen and $\alpha$-ketoglutarate are always saturating,[45] the behavior is

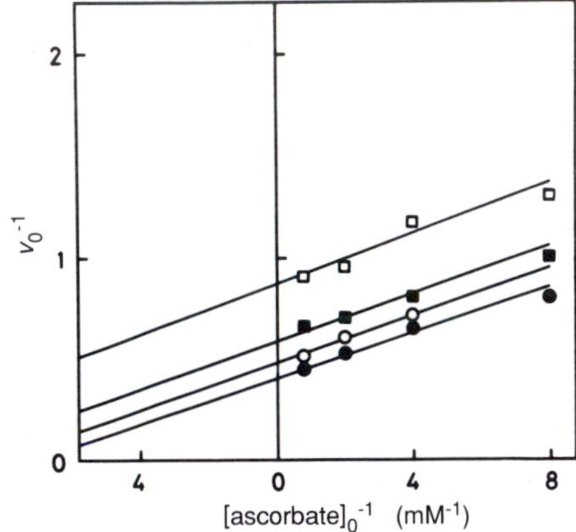


Effect of initial ascorbate concentration ($[\text{ascorbate}]_0^{-1}$ in millimolar$^{-1}$) on the rate of the lysyl hydroxylase reaction at different fixed concentrations of the peptide substrate.[45] The concentrations of the peptide substrate were 0.1 mg mL$^{-1}$ (□), 0.15 mg mL$^{-1}$ (■), 0.2 mg mL$^{-1}$ (○), and 0.75 mg mL$^{-1}$ (●). The concentration of $Fe^{2+}$ was 0.05 mM and that of $\alpha$-ketoglutarate was 0.1 mM. Adapted with permission from ref 45. Copyright 1980 Elsevier Science Publishers.

Explain this behavior on the basis of the role of ascorbate in the enzymatic reaction.

## PROBLEM 3–10

Suppose there is an enzyme that catalyzes the reaction A + B $\rightleftharpoons$ C + D. Suppose reactant A has to bind to the active site before reactant B can bind, and reactant B has to bind before products are formed. Suppose there were an inhibitor I that, like reactant B, could bind to the active site only after reactant A was bound.

(A) Write a kinetic mechanism describing both the enzymatic reaction and its inhibition by the inhibitor I.

(B) Show that the equation for the initial velocity for this kinetic mechanism is

$$v_0 = \frac{V_{max}[A]_0[B]_0}{[A]_0[B]_0 + K_m{}^B\left(1+\frac{[I]}{K_i}\right)[A]_0 + K_m{}^A[B]_0 + K_m{}^{AB}}$$

where $K_i$ is the dissociation constant for the inhibitor from E·A.

(C) If this were the enzymatic mechanism, what would $V_{max}$, $K_m{}^B$, $K_m{}^A$, and $K_m{}^{AB}$ be equal to in terms of the rate constants in your kinetic mechanism in (A)?

(D) What would plots of $v_0^{-1}$ against $[A]_0^{-1}$ at constant $[B]_0$ and several fixed [I] look like?

(E) What would plots of $v_0^{-1}$ against $[B]_0^{-1}$ at constant $[A]_0$ and several fixed [I] look like?

Glycerol kinase catalyzes the reaction

glycerol + MgATP $\rightleftharpoons$ glycerol phosphate + MgADP

(F) Write a chemical mechanism for this reaction.

The chromium complex of ATP, CrATP, cannot be used as a reactant in the glycerol kinase reaction because the chromium–oxygen bonds are too strong, but it is a potent inhibitor of the enzyme.

(G) Examine Figure 3–9 and the following figure and decide which of the reactants for glycerol kinase binds first to the active site. Why?

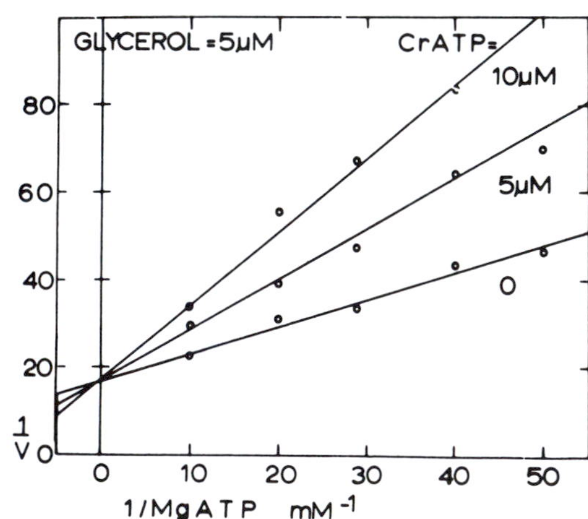


Inhibition of glycerol kinase by CrATP[38] when MgATP is the variable reactant and the concentration of glycerol is constant throughout. The concentrations of CrATP are noted to the right. Reprinted with permission from ref 38. Copyright 1974 *Journal of Biological Chemistry*.

# Binding

Analysis of steady-state kinetics has led to the conclusion that reactants are assembled at an active site on the surface of the enzyme before they are transformed into products. This results from the binding of those reactants to the enzyme within the active site prior to the interconversion between reactants and products. If the association of each substrate can be treated as a discrete step in an enzymatic reaction, then each of these steps should be no different from the binding of any small molecule, or ligand, to a particular site on a protein. It is only what follows that is different. A **ligand** is any molecule that binds to but is not chemically transformed by a protein. It is the lack of chemical transformation that distinguishes a ligand from a substrate. An **agonist** is a ligand that binds to a protein and, upon binding, produces a measurable change in the structure or function of that protein. An **antagonist** is a ligand that binds to a protein, does not elicit the characteristic change in the protein, but blocks the ability of an agonist to effect that change. An **inhibitor** or an **activator** is a ligand that binds to an enzyme and decreases or increases, respectively, the rate of its catalysis. Inorganic anions, inorganic cations, and hydrophobic molecules that bind adventitiously to a protein are also examples of ligands.

It is usually the case that one molecule of a ligand will associate with a particular constellation of amino acids on the surface of a protein and that, because of symmetry, the number of sites on the protein will be equal to the number of protomers it contains. Usually, each of the symmetrically arrayed sites in a homooligomer will bind the ligand independently. In this situation, the molar concentration of sites in a particular solution is equal to the molar concentration of native, properly folded protomers.

The binding of a ligand, L, to a site, ST, on a protein to form a specific complex, ST·L, is quantitatively described by a dissociation constant

$$K_{\mathrm{d}}^{\mathrm{L}} = \frac{a_{\mathrm{ST}} a_{\mathrm{L}}}{a_{\mathrm{ST \cdot L}}} \tag{3–122}$$

where $a_i$ is the activity of component $i$ at equilibrium. In terms of molar concentrations

$$K_{\mathrm{d}}^{\mathrm{L}} = \frac{\gamma_{\mathrm{ST}} \gamma_{\mathrm{L}}}{\gamma_{\mathrm{ST \cdot L}}} \cdot \frac{[\mathrm{ST}]_{\mathrm{eq}}[\mathrm{L}]_{\mathrm{eq}}}{[\mathrm{ST \cdot L}]_{\mathrm{eq}}} \tag{3–123}$$

where $\gamma_i$ is the activity coefficient of component $i$ and the subscript eq refers to concentrations at equilibrium. If the actual molar concentrations of the various participants are low when the measurement is made and if the solution has an ionic strength of 0.1 M or greater, then the ratio of the activity coefficients is usually close enough to unity to be ignored. The dissociation constant, because it is an equilibrium constant, is directly related to the standard free energy change for the binding of the ligand to the protein. As an equilibrium constant, it must be distinguished from a kinetic parameter such as the Michaelis constant, $K_{\mathrm{m}}$, which is rarely directly related to a thermodynamic variable.

The dissociation constant for the binding between a ligand and a site on a molecule of protein can be determined from experimental observations. In any situation in which the population of sites for the ligand is homogeneous, the dissociation is governed by the simple equation

$$K_{\mathrm{d}}^{\mathrm{L}} = \frac{[\mathrm{ST}]_{\mathrm{eq}}[\mathrm{L}]_{\mathrm{eq}}}{[\mathrm{ST \cdot L}]_{\mathrm{eq}}} \tag{3–124}$$

The concentration of bound ligand at equilibrium is equal to

$$[\mathrm{ST \cdot L}]_{\mathrm{eq}} = \frac{[\mathrm{ST}]_{\mathrm{TOT}}[\mathrm{L}]_{\mathrm{eq}}}{K_{\mathrm{d}}^{\mathrm{L}} + [\mathrm{L}]_{\mathrm{eq}}} \tag{3–125}$$

where $[\mathrm{ST}]_{\mathrm{TOT}}$ is the total molar concentration of sites, both occupied and unoccupied. This equation has the form of a function defining a rectangular hyperbola (Equation 3–30) if the concentration of protein, and hence $[\mathrm{ST}]_{\mathrm{TOT}}$, is fixed at a constant value during the experimental measurements. For example, the binding of $[^{125}\mathrm{I}]$-insulin to insulin receptor can be followed by adding increasing amounts of $[^{125}\mathrm{I}]$-insulin to a series of samples each containing the same concentration of protein. Following equilibration, the number of moles of the complex formed between receptor and insulin can be measured[46] by precipitating the complex from solution with poly(ethylene glycol) (Figure 3–12).[47] In this situation the molar concentration of bound ligand in the original sample is a hyperbolic function of the free concentration of insulin at equilibrium.

It is often not possible, however, to fix the concentration of sites at a constant value because unavoidable changes in volume occur during the experimental procedure. It can be assumed, however, that for a given set of sites

$$C_{\mathrm{p}} = \kappa[\mathrm{ST}]_{\mathrm{TOT}} \tag{3–126}$$

where $C_{\mathrm{p}}$ is the concentraion of protein expressed in any convenient units, usually milligrams milliliter$^{-1}$, and $\kappa$ is a constant of proportionality, the units of which are usually grams of protein (mole of sites)$^{-1}$. Parenthetically, it should be noted that this is a very useful number to know. With this assumption, Equation 3–125 becomes

$$\frac{[\mathrm{ST \cdot L}]_{\mathrm{eq}}}{C_{\mathrm{p}}} = \frac{\kappa^{-1}[\mathrm{L}]_{\mathrm{eq}}}{K_{\mathrm{d}}^{\mathrm{L}} + [\mathrm{L}]_{\mathrm{eq}}} \tag{3–127}$$

which defines a rectangular hyperbola in the concentration of L regardless of the variations in $C_{\mathrm{p}}$.

As in the case of data analysis in enzymatic kinetics, the data from a study of binding are submitted to numerical analysis. In this case, the data are a set of measured values of $[\mathrm{ST \cdot L}]_{\mathrm{eq}}/C_{\mathrm{p}}$ corresponding to a set of equilibrium molar concentrations of the free ligand $[\mathrm{L}]_{\mathrm{eq}}$. Again, as in the case of enzymatic kinetics, it is now customary to fit a rectangular hyperbola to the data by weighted, nonlinear least-squares analysis.[48] The two parameters, and their standard deviations, provided by the fitting procedure are $K_{\mathrm{d}}^{\mathrm{L}}$ and $\kappa$. Such nonlinear numerical analysis can be extended to situations where several different sites, at different relative concentrations, are

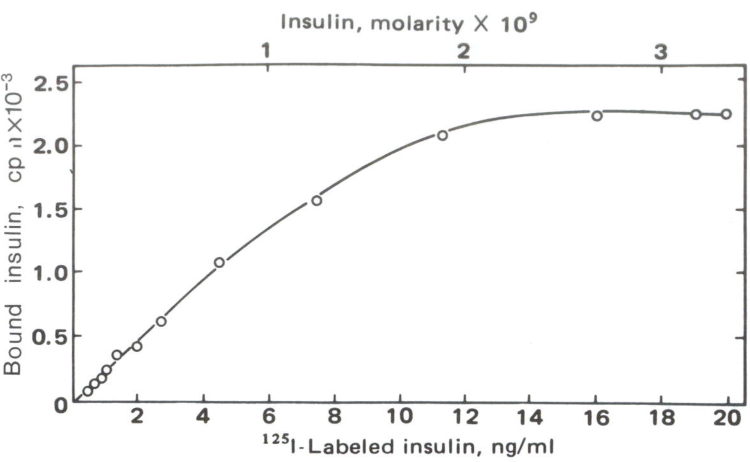


**Figure 3–12:** Binding of [$^{125}$I]-insulin to insulin receptor dissolved in a solution of detergent.[47] Plasma membranes were isolated from rat liver and dissolved in a 2% solution of the detergent Triton X-100. The dissolved receptor was then further purified by affinity adsorption on a matrix to which insulin had been attached covalently. Equal amounts of this purified protein were mixed with solutions (0.2 mL) containing different concentrations of [$^{125}$I]-insulin (100–200 $\mu$Ci $\mu$g$^{-1}$) that had been iodinated on its tyrosines with [$^{125}$I]ICl by electrophilic aromatic substitution. The solutions stood for 30 min at 24 °C to effect equilibrium, and they were then mixed with cold 25% poly(ethylene glycol) (0.5 mL) to precipitate the protein and any bound [$^{125}$I]-insulin. The precipitates were then trapped on filter papers and washed rapidly with 3 mL of an 8% solution of poly(ethylene glycol). The amount of bound [$^{125}$I]-insulin was determined by crystal scintillation counting. The amount of bound [$^{125}$I]-insulin (in counts per minute × $10^{-3}$) is presented as a function of the concentration of insulin (nanograms milliliter$^{-1}$) in the original equilibrium mixture before the poly(ethylene glycol) was added. Reprinted with permission from ref 47. Copyright 1978 National Academy of Sciences.

binding the same ligand, each with its own unique dissociation constant.[49] In either case, the values of $K_d^L$ and $\kappa$ are the **experimental observations** obtained from such analyses.

In the past, the data were first converted to a linear form[50] before the fitting was performed, but this is rarely done anymore because digital computers permit nonlinear numerical analysis to be performed readily. Partly for historical reasons and partly in the interest of clarity, however, when the data are presented graphically for the reader's evaluation, they are usually presented in a linear form. Equation 3–127 can be cast in a linear form either by the inversion[50]

$$\frac{C_p}{[\mathrm{ST\cdot L}]_{eq}} = \kappa + \kappa K_d^L \left( \frac{1}{[\mathrm{L}]_{eq}} \right) \tag{3–128}$$

or by the rearrangement[51]

$$\frac{[\mathrm{ST\cdot L}]_{eq}}{C_p} = \frac{1}{\kappa} - K_d^L \left( \frac{[\mathrm{ST\cdot L}]_{eq}}{C_p[\mathrm{L}]_{eq}} \right) \tag{3–129}$$

Because it was pointed out by Professor Scatchard that the use of Equation 3–128 "gives undue weight to that portion of . . . [the] data which is poorest in experimental precision,"[50] Equation 3–129 is usually used as the linear form of the function. This is in contradistinction to the common use of Equation 3–36 as the linear form of the function for initial velocity. If the quotient of the molar concentration of occupied sites and the concentration of protein, $[\mathrm{ST\cdot L}]_{eq}/C_p$, is measured as a function of the concentration of free ligand, $[\mathrm{L}]_{eq}$, and if these data are plotted as $[\mathrm{ST\cdot L}]_{eq}/C_p$ against $[\mathrm{ST\cdot L}]_{eq}/(C_p[\mathrm{L}]_{eq})$, a straight line should be obtained (Figure 3–13)[52] if there is only one class of sites. The intercept with the ordinate should be $\kappa^{-1}$ and the slope should be $-K_d^L$. Such **Scatchard plots** are routinely used to present graphically the data for the binding of ligands to proteins, agonists and antagonists to receptors, and, under the assumed name of Eadie–Hofstee plots,[53,54] the initial velocities of enzymatic reactions as a function of the concentrations of reactants.

To perform an analysis of the binding of a ligand to a protein, simultaneous measurements must be made at equilibrium of the molar concentration of bound ligand, $[\mathrm{ST\cdot L}]_{eq}$, and the molar concentration of free ligand, $[\mathrm{L}]_{eq}$, in several samples of known protein concentration at a number of different concentrations of free ligand. There are several ways to accomplish this. If the protein is placed on one side of a semipermeable membrane separating two compartments, ligand is added to both compartments, and the system is allowed to reach equilibrium, the molar concentration of the ligand in the compartment without the protein will be equal to $[\mathrm{L}]_{eq}$, and the total molar concentration of the ligand in the compartment containing the protein will be $[\mathrm{L}]_{eq} + [\mathrm{ST\cdot L}]_{eq}$ (Figure 3–13B).[52] The final concentration of protein in the one compartment can be measured independently. This technique is referred to as **equilibrium dialysis**. If a column of beaded, tightly cross-linked dextran is equilibrated with a solution containing a fixed concentration of ligand and a sample of a solution of the protein is passed over the column, the protein, which exits the column at the excluded volume, will bind lig-

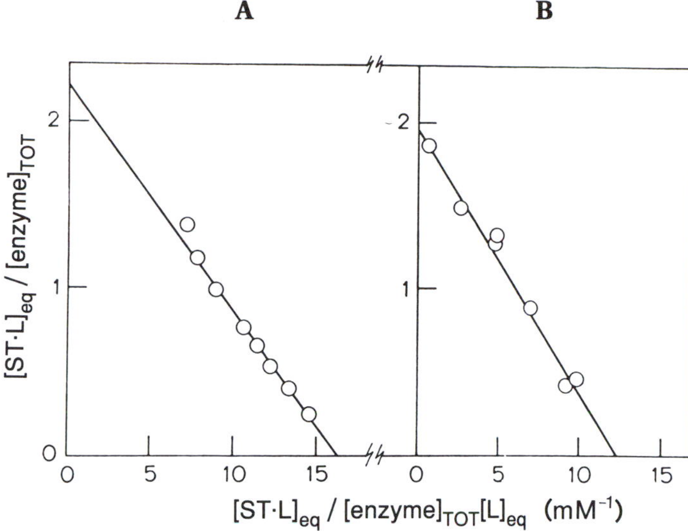


**Figure 3–13:** Scatchard plots for the binding of an equilibrium mixture of [U-$^{14}$C]glucose 6-phosphate and [U-$^{14}$C]fructose 6-phosphate to glucose-6-phosphate isomerase from rabbit muscle.[52] (A) Measurement of free concentration of ligand by rate dialysis. An apparatus separating two chambers by a dialysis membrane was used. The enzyme (6.6 mg $mL^{-1}$) in a buffer at pH 8.0 and at 22 °C was in the upper chamber (1 mL). The lower chamber contained a spiral groove, in contact with the membrane, through which solution was pumped at a steady rate (4 mL $min^{-1}$). [U-$^{14}$C]Glucose 6-phosphate was added at a low concentration to the upper chamber, and the rate at which $^{14}$C appeared in the lower chamber was determined by submitting the constantly flowing solution to scintillation counting. When equilibrium had been reached, the concentration of [U-$^{14}$C]glucose in the upper chamber was increased, and so forth. By calibrating the apparatus in the absence of protein, the rate of appearance of $^{14}$C counts per minute in the lower fluid could be converted to the free concentration of [U-$^{14}$C]glucose 6-phosphate and [U-$^{14}$C]fructose 6-phosphate in the upper solution. From the known total concentration, the concentration of bound substrates could be calculated by difference. (B) Measurement of free and bound concentrations of ligand by equilibrium dialysis. A series of apparatuses separating two chambers by a dialysis membrane was used. In one chamber glucose-6-phosphate isomerase (10.0 mg $mL^{-1}$) was present; in the other, buffer alone. [U-$^{14}$C]Glucose 6-phosphate was added to both sides and allowed to reach equilibrium with fructose 6-phosphate as well as reach equilibrium across the membrane. The concentration of [$^{14}$C]substrates in the side lacking enzyme is the free concentration of substrates in both chambers; the concentration of [$^{14}$C]substrates in the side containing the protein is the sum of the free and the bound concentrations of substrates. The variable $[ST{\cdot}L]_{eq}/[enzyme]_{TOT}$ is plotted as a function of $[ST{\cdot}L]_{eq}/([enzyme]_{TOT}[L]_{eq})$ in units of $millimolar^{-1}$. As the units on $[enzyme]_{TOT}$ were moles of enzyme $liter^{-1}$ and the units on $[ST{\cdot}L]_{eq}$ are moles of sites $liter^{-1}$, the intercept with the ordinate ($\kappa^{-1}$ in Equation 3–129) has the units of moles of sites (mole of $enzyme)^{-1}$, which is also an informative number. The dissociation constants measured by these two techniques were 0.14 M and 0.16 M, respectively. Adapted with permission from ref 52. Copyright 1973 Springer-Verlag.

and as it passes through the beaded dextran until the reaction reaches equilibrium. The excess ligand eluting with the protein should be $[ST{\cdot}L]_{eq}$; and, if the protein has moved far enough to leave behind the deficit it created by binding the ligand, the $[L]_{eq}$ in the region of the excluded volume should be the molar concentration of ligand at which the column was equilibrated.[55] The unidirectional rate at which a ligand passes through a dialysis membrane is directly proportional to its free concentration on the side from which it is leaving. If a known concentration of protein and a known concentration of ligand are placed in a compartment on one side of a dialysis membrane, and the compartment on the other side is flushed continuously with a solution that does not contain the ligand, the rate at which the ligand passes across the membrane will be directly proportional to its free concentration, $[L]_{eq}$, in the compartment containing the ligand and the protein.[56] This rate is determined by monitoring the amount of ligand appearing in the flushing solution as a function of time. If the chamber and the dialysis membrane are calibrated with known concentrations of ligand, the rate can be converted directly into a value for $[L]_{eq}$ and the concentration of bound ligand can be calculated by difference (Figure 3–13A) from the known total concentration of ligand in the compartment with the protein

$$[ST{\cdot}L]_{eq} = [L]_{TOT} - [L]_{eq} \qquad (3\text{–}130)$$

In each of these types of experiments, the values of $[ST{\cdot}L]_{eq}$, $C_p$, and $[L]_{eq}$ can be used to find $K_d{}^L$ and $\kappa$.

When these three techniques were applied separately to the binding at pH 8.0 and 22 °C of the competitive inhibitor 6-phosphogluconate (6-PGA) to glucose-6-phosphate isomerase, values for $K_d^{6\text{-PGA}}$ were 0.08 mM, 0.04 mM, and 0.03 mM, respectively, and values for $\kappa$ were 66,000 g (mol of sites)$^{-1}$, 66,000 g (mol of sites)$^{-1}$, and 63,000 g (mol of sites)$^{-1}$, respectively.[52] These values are indistinguishable from the molar mass of one subunit of this enzyme. At pH 8 and 30 °C, the dissociation constant, determined kinetically for 6-phosphogluconate as an inhibitor of glucose-6-phosphate isomerase, is 0.03 mM,[57] which agrees satisfactorily with the values obtained for its dissociation constant when measured directly.

There is one potential drawback of all of these techniques, referred to as the **Donnan effect**. In each technique, a solution containing the protein and the ligand is physically separated from a solution containing only the ligand. The protein is almost always charged at the pH at which the measurements are made. Even during molecular exclusion chromatography, the dimensions of the two compartments, that containing the protein and that lacking the protein, are greater than the thickness of the ionic double layer around the protein.[57a] If the ligand is also charged, it will act as any other electrolyte in the diffuse layer of dissolved counterions around the protein. It will either be excluded from the layer if it is of the same charge as the protein or enriched in the layer if it is of opposite charge. This has the effect of decreasing or increasing, respectively, the measured concentration of bound ligand in the compartment containing the protein. This source of error can be minimized by performing the binding experiments at a concentration of supporting electrolyte high enough to eliminate the contribution of the ligand itself to the layer of mobile counter-ions. This can be seen in the following way.

Suppose that the solution of protein contains macromolecules, including the molecules of protein themselves, that have a net charge at the pH chosen for the measurements. In the system there are two compartments: the compartment $\alpha$ containing the charged macromolecules and the compartment $\beta$ lacking them. Assume that the only diffusible electrolytes are the ligand, potassium, and chloride, and also assume that the solution of protein contains no binding sites for ligand L. By electroneutrality

$$[\text{charge}]_\alpha = [\text{K}^+]_\beta - [\text{K}^+]_\alpha + [\text{Cl}^-]_\alpha - [\text{Cl}^-]_\beta + z_\text{L}([\text{L}]_\beta - [\text{L}]_\alpha) \tag{3–131}$$

where $[\text{charge}]_\alpha$ is the concentration of nondiffusible charge in compartment $\alpha$ (in equivalents liter$^{-1}$, which can have either a positive or a negative value) and $z_\text{L}$ is the charge on ligand L. This defines the contribution of each ion to the maintenance of electroneutrality. A quantity $f^\text{L}_\text{charge}$ can be defined, which is the fraction of the charge compensation contributed by the ligand L

$$f^\text{L}_\text{charge} = \frac{z_\text{L}([\text{L}]_\beta - [\text{L}]_\alpha)}{[\text{K}^+]_\beta - [\text{K}^+]_\alpha + [\text{Cl}^-]_\alpha - [\text{Cl}^-]_\beta + z_\text{L}([\text{L}]_\beta - [\text{L}]_\alpha)} \tag{3–132}$$

Because at equilibrium

$$\left(\frac{[\text{L}]_\alpha}{[\text{L}]_\beta}\right)^{z_\text{L}} = \frac{[\text{Cl}^-]_\beta}{[\text{Cl}^-]_\alpha} = \frac{[\text{K}^+]_\alpha}{[\text{K}^+]_\beta} \tag{3–133}$$

$$f^\text{L}_\text{charge} = \frac{z_\text{L}\left[\left(\frac{[\text{K}^+]_\beta}{[\text{K}^+]_\alpha}\right)^{\frac{1}{z_\text{L}}} - 1\right][\text{L}]_\alpha}{\left(\frac{[\text{K}^+]_\beta}{[\text{K}^+]_\alpha} - 1\right)[\text{K}^+]_\alpha + \left(\frac{[\text{K}^+]_\beta}{[\text{K}^+]_\alpha} - 1\right)[\text{Cl}^-]_\beta + z_\text{L}\left[\left(\frac{[\text{K}^+]_\beta}{[\text{K}^+]_\alpha}\right)^{\frac{1}{z_\text{L}}} - 1\right][\text{L}]_\alpha} \tag{3–134}$$

This is a complex function but if $z_\text{L} = 1$

$$f^\text{L}_\text{charge} = \frac{[\text{L}]_\alpha}{[\text{K}^+]_\alpha + [\text{Cl}^-]_\beta + [\text{L}]_\alpha} \tag{3–135}$$

or if $z_\text{L} = -1$

$$f^\text{L}_\text{charge} = \frac{[\text{L}]_\beta}{[\text{K}^+]_\alpha + [\text{Cl}^-]_\beta + [\text{L}]_\beta} \tag{3–136}$$

If it is assumed that $[\text{charge}]_\alpha$, by design, is much less than the sum of the concentrations of the diffusible electrolytes, the concentrations of each electrolyte in the two compartments will be approximately the same and

$$f^\text{L}_\text{charge} \cong \frac{[\text{L}]}{[\text{K}^+] + [\text{Cl}^-] + [\text{L}]} \tag{3–137}$$

The difference in concentration of the ligand L between the two compartments due only to the Donnan effect will be

$$z_\text{L}([\text{L}]_\beta - [\text{L}]_\alpha) = f^\text{L}_\text{charge}[\text{charge}]_\alpha \cong \frac{[\text{L}][\text{charge}]_\alpha}{[\text{K}^+] + [\text{Cl}^-] + [\text{L}]} \tag{3–138}$$

This relationship indicates that even if the solution of protein contains no sites for ligand L, it will appear to contain binding sites if the charge of ligand L is opposite that of the nondiffusible charge. The apparent dissociation constant measured for ligand L will be $[\text{K}^+] + [\text{Cl}^-]$. If the Donnan effect is the only reason for the excess ligand in the compartment containing the protein, the apparent dissociation constant should be directly proportional to the concentration of supporting electrolyte. If, however, a concentration of supporting electrolyte is purposely chosen so that it is always much greater than [L], and $[\text{charge}]_\alpha$ is not too large, the amount of ligand retained or excluded from compartment $\beta$ due to the Donnan effect, as opposed to legitimate binding, will be negligible (Equation 3–138). If the charge of ligand L is the same as that of the nondiffusible macromolecules, ligand L will be excluded from the compartment containing the protein. If there are legitimate sites for ligand L, the apparent binding will be less than the actual binding. This effect, however, can also be eliminated by increasing the concentration of the supporting electrolyte. In summary, any apparent binding that displays a significant dependence on ionic strength should be discounted until it is demonstrated that it does not result simply from the Donnan effect.

In all of the many different procedures that have been developed to follow the binding of a ligand to a protein, the three variables that are simultaneously measured are the free concentration of the ligand, $[L]_{eq}$, the total concentration of protein, $C_p$, and the concentration of the complex between protein and ligand, $[ST{\cdot}L]_{eq}$. Each one of a set of solutions whose molar concentrations of protein and ligand are systematically varied is allowed to reach equilibrium and these three concentrations are then measured by techniques that can assess what their values were in the solution when it was at equilibrium. The results from these measurements are the dissociation constant, $K_d^L$, and either the total molar concentration of sites, $[ST]_{TOT}$, in the solution or the grams of protein (mole of sites)$^{-1}$.

It is also possible to use any physical or chemical property of the ligand or of the protein that is altered upon the binding of the ligand to determine a value for the dissociation constant of the ligand. Examples of these properties would be the absorbance of the protein, the fluorescence of the ligand, or the rate of the covalent reaction of a particular amino acid in the protein with a particular chemical reagent. For example, *N*-methylnicotinamide binds to lysozyme to form a complex that absorbs light at 350 nm, a wavelength at which both the ligand and the protein are transparent.[58] In this instance, the absorbance observed, $A_{350}$, should be directly proportional to the molar concentration of the complex between the ligand, *N*-methylnicotinamide, and the protein, lysozyme

$$A_{350} = \varepsilon[ST \cdot L]_{eq} \qquad (3\text{–}139)$$

When Equation 3–139 is combined with Equation 3–129

$$\frac{A_{350}}{C_p} = \varepsilon\kappa^{-1} - K_d^{\,L}\,\frac{A_{350}}{C_p[L]_{eq}} \qquad (3\text{–}140)$$

A plot of $A_{350}/C_p$ as a function of $A_{350}/(C_p[L]_{eq}$, is linear (Figure 3–14)[58] and its slope is $-K_d^L$. The intercept with the ordinate is $\varepsilon\kappa^{-1}$. If it is assumed that there is one site for each mole of protein present and that the protein used is homogeneous

$$\kappa = \frac{C_p}{M_p} \qquad (3\text{–}141)$$

where $M_p$ is the molar mass of the protein. Because $\kappa$ now has a known value, the molar extinction coefficient, $\varepsilon$, of the complex between *N*-methylnicotinamide and lysozyme can be calculated. Nonlinear, weighted, least-squares numerical analysis has also been adapted to analyze directly the behavior of a physical property such as fluorescence or absorbance as a function of the concentration of free ligand.[59]

In the example just described, neither the protein alone nor the ligand alone displayed the physical property mea-

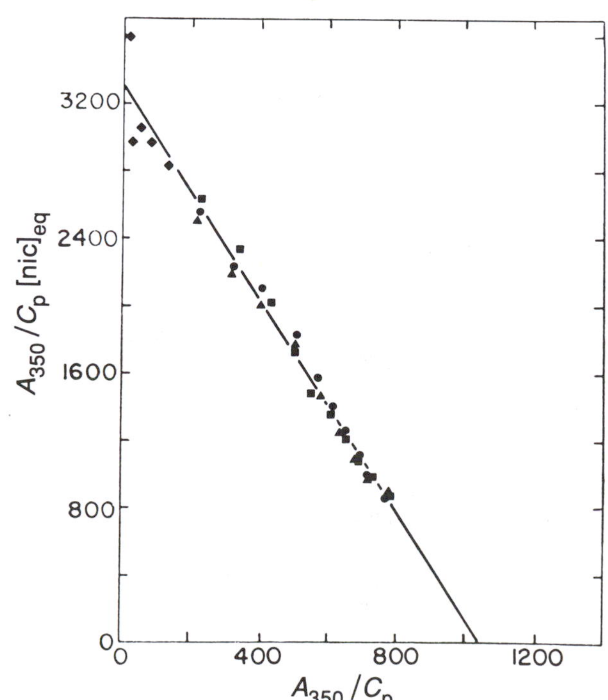


**Figure 3–14:** Binding of *N*-methylnicotinamide to lysozyme[58] followed by the increase in absorbance of the solution at 350 nm. Crystalline lysozyme was dissolved at pH 5 and 22 °C to a final concentration of 10.00 mg mL$^{-1}$ in 2.00 mL. Successive additions of standard solutions of *N*-methylnicotinamide (final concentrations between 5 and 890 mM) were made. After each addition, the absorbance of the solution at 350 nm was measured in a spectrophotometer, and the new concentration of protein resulting from the dilution was calculated. Because the concentration of *N*-methylnicotinamide was always in significant excess over the molar concentration of protein, it was assumed that the total concentration of the *N*-methylnicotinamide was equal to its free concentration, $[\text{nic}]_{eq}$. In this example, the quantity $A_{350}/(C_p[\text{nic}]_{eq})$ is presented as a function of $A_{350}/C_p$. Usually, $[ST{\cdot}L]_{eq}/C_p$ is presented as a function of $[ST{\cdot}L]_{eq}/(C_p[L])$, rather than the other way around. Adapted with permission from ref 58. Copyright 1969 National Academy of Sciences.

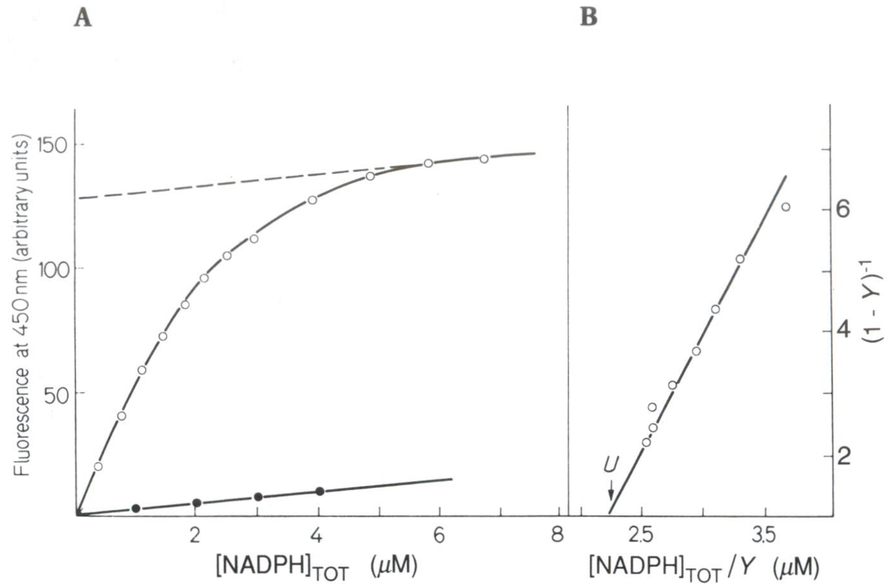


**Figure 3–15:** Binding of NADPH to homoserine dehydrogenase from *E. coli* followed by the enhancement of the fluorescence of NADPH.[60] (A) To a cuvette containing either a 0.7 $\mu$M solution of homoserine dehydrogenase (open circles) or buffer alone (closed circles), successive additions of concentrated solutions of NADPH were made to produce the noted final, total concentrations of ligand. Between each addition, the fluorescence of the solution was determined ($\lambda_{excitation}$ = 285 nm; $\lambda_{emission}$ = 350 nm). The fluorescence (in arbitrary units) is presented as a function of the total concentration of NADPH (in micromolar). Note that saturation occurs at a concentration of NADPH within an order of magnitude of the concentration of enzyme. (B) The fractional saturation of sites for NADPH on homoserine dehydrogenase, $Y$, was calculated with Equations 3–148 and 3–147. The dimensionless quantity $(1 - Y)^{-1}$ is presented as a function of the variable $[\text{NADPH}]_{TOT}/Y$, where $[\text{NADPH}]_{TOT}$ is the total concentration of the ligand (in micromolar). The dissociation constant, $K_d^{L}$ (0.26 $\mu$M), is the reciprocal of the slope of the line. The molar concentration of sites $[\text{ST}]_{TOT}$ (2.2 $\mu$M) is the intercept. Adapted with permission from ref 60. Copyright 1969 Springer-Verlag.

sured, namely, absorbance at 350 nm. Usually, however, one or the other does possess the physical property, and it is merely altered by the association. For example, NADPH displays fluorescence at 450 nm when it is excited by light at a wavelength of 290 nm. Its fluorescence is enhanced about 20-fold when it is bound at the active site of homoserine dehydrogenase.[60] The effect of this enhancement in fluorescence can be observed when the fluorescence of a solution is determined as a function of the total concentration of NADPH (Figure 3–15A).[60] In the absence of the enzyme, the fluorescence of the solution increases linearly with the concentration of NADPH and the fluorescence observed, $F_{free}$, is due entirely to free ligand

$$F_{free} = \alpha[\text{L}]_{eq} \qquad (3\text{–}142)$$

where $\alpha$ is a constant of proportionality. If a fixed concentration of homoserine dehydrogenase is present, however, the fluorescence increases more steeply as NADPH is added because the fluorescence, $F_{bound}$, of the complex between ligand and protein, ST·L, is greater than that of the free NADPH

$$F_{bound} = \beta\,[\text{ST}\cdot\text{L}]_{eq} \qquad (3\text{–}143)$$

where $\beta$ is a constant of proportionality. If $[\text{L}]_{TOT}$ is the total concentration of ligand, both bound and free, and $Y$ is defined as the fraction of the sites occupied, such that

$$Y \equiv \frac{[\text{ST}\cdot\text{L}]_{eq}}{[\text{ST}]_{TOT}} \qquad (3\text{–}144)$$

then the fluorescence observed, $F_{obs}$, at any total concentration of ligand, $[\text{L}]_{TOT}$, will be

$$F_{obs} = F_{free} + F_{bound} \qquad (3\text{–}145)$$

$$F_{obs} = \alpha([\text{L}]_{TOT} - [\text{ST}]_{TOT}Y) + \beta[\text{ST}]_{TOT}Y \qquad (3\text{–}146)$$

After rearrangement

$$Y = \frac{F_{obs} - \alpha[\text{L}]_{TOT}}{(\beta - \alpha)[\text{ST}]_{TOT}} \qquad (3\text{–}147)$$

At saturation, $Y = 1$, and

$$F_{obs} = \alpha[\text{L}]_{TOT} + (\beta - \alpha)[\text{ST}]_{TOT} \qquad (3\text{–}148)$$

This defines the asymptote of the hyperbola observed in the presence of protein. This asymptote will be a line whose slope is the same as the slope of the line in the absence of protein ($\alpha$) and whose intercept at the ordinate will be $(\beta - \alpha)[\text{ST}]_{TOT}$. Using the value of $(\beta - \alpha)[\text{ST}]_{TOT}$ determined from this intercept, values of $Y$ can be obtained for each value of $[\text{L}]_{TOT}$ (Equation 3–147). This approach is a general one for any physical property that is directly proportional to the concentration of the ligand and has significantly different constants of proportionality for free ligand ($\alpha$) and bound ligand ($\beta$), respectively.

These determinations of the fraction of the sites occupied, $Y$, as a function of the total concentration of the ligand, $[\text{L}]_{TOT}$, can be used to determine the dissociation constant for the ligand.[60] If

$$K_d^{\mathrm{L}} = \frac{[\mathrm{ST}]_{\mathrm{eq}}[\mathrm{L}]_{\mathrm{eq}}}{[\mathrm{ST}\cdot\mathrm{L}]_{\mathrm{eq}}} = \frac{([\mathrm{ST}]_{\mathrm{TOT}} - [\mathrm{ST}]_{\mathrm{TOT}}Y)([\mathrm{L}]_{\mathrm{TOT}} - [\mathrm{ST}]_{\mathrm{TOT}}Y)}{[\mathrm{ST}]_{\mathrm{TOT}}Y} \tag{3–149}$$

then

$$\frac{1}{1-Y} = \frac{1}{K_d^{\mathrm{L}}}\left(\frac{[\mathrm{L}]_{\mathrm{TOT}}}{Y} - [\mathrm{ST}]_{\mathrm{TOT}}\right) \tag{3–150}$$

A plot of $(1 - Y)^{-1}$ as a function of $[\mathrm{L}]_{\mathrm{TOT}}/Y$ (Figure 3–15B) has a slope of $(K_d^{\mathrm{L}})^{-1}$.

This example of the binding of NADPH to homoserine dehydrogenase (Figure 3–15) differs from the example of the binding of *N*-methylnicotinamide to lysozyme (Figure 3–14) because the molar concentration of sites was chosen to be within the range of the value of the dissociation constant in the former case, while the molar concentration of sites was far less than the dissociation constant in the latter case. In spectrometric titrations such as these for homoserine dehydrogenase and lysozyme, the molar concentration of bound ligand is not measured directly. To estimate the concentration of binding sites, the effect of the addition of the protein to the solution on the free concentration of the ligand must be monitored (Equation 3–146). To see any effect on the free concentration of the ligand, enough sites must be added to decrease its value significantly. If

$$[\mathrm{ST}]_{\mathrm{TOT}} = [\mathrm{ST}]_{\mathrm{eq}} + [\mathrm{ST}\cdot\mathrm{L}]_{\mathrm{eq}} \tag{3–151}$$

and

$$[\mathrm{L}]_{\mathrm{TOT}} = [\mathrm{L}]_{\mathrm{eq}} + [\mathrm{ST}\cdot\mathrm{L}]_{\mathrm{eq}} \tag{3–152}$$

then by combination with Equation 3–124 and rearrangement

$$K_d^{\mathrm{L}}[\mathrm{L}]_{\mathrm{TOT}} - (K_d^{\mathrm{L}} + [\mathrm{ST}]_{\mathrm{TOT}} + [\mathrm{L}]_{\mathrm{TOT}})[\mathrm{L}]_{\mathrm{eq}} + [\mathrm{L}]_{\mathrm{eq}}^2 = 0 \tag{3–153}$$

This quadratic equation, because it can be solved for the free concentration of ligand, $[\mathrm{L}]_{\mathrm{eq}}$, is useful in its own right, but it also demonstrates that the free concentration of ligand is a function of both the total concentration of ligand and the total concentration of sites, only as long as $[\mathrm{ST}]_{\mathrm{TOT}} \cong K_d^{\mathrm{L}}$. Once the values of $K_d^{\mathrm{L}}$ and $[\mathrm{ST}]_{\mathrm{TOT}}$ have been estimated in preliminary experiments, $[\mathrm{ST}]_{\mathrm{TOT}}$ can be purposely chosen to be of the same magnitude as $K_d^{\mathrm{L}}$, as in the experiments with homoserine dehydrogenase.

Sometimes, however, it is simpler to choose a value of $[\mathrm{ST}]_{\mathrm{TOT}}$ that is much less than $K_d^{\mathrm{L}}$. When $[\mathrm{ST}]_{\mathrm{TOT}} \ll K_d^{\mathrm{L}}$, then

$$([\mathrm{L}]_{\mathrm{eq}} - [\mathrm{L}]_{\mathrm{TOT}})([\mathrm{L}]_{\mathrm{eq}} - K_d^{\mathrm{L}}) \cong 0 \tag{3–154}$$

and

$$[\mathrm{L}]_{\mathrm{eq}} \cong [\mathrm{L}]_{\mathrm{TOT}} \tag{3–155}$$

at all values of [L]. In this situation, exemplified by the binding of *N*-methylnicotinamide to lysozyme (Figure 3–14), where the total molar concentration of sites is small compared to the value of the dissociation constant ($[\mathrm{ST}]_{\mathrm{TOT}} < 1$ mM; $K_d^{\mathrm{L}} = 0.3$ M), the total concentration of ligand is always indistinguishable from its free concentration. Equation 3–149 becomes

$$K_d^{\mathrm{L}} = \frac{[\mathrm{ST}]_{\mathrm{eq}}[\mathrm{L}]_{\mathrm{TOT}}}{[\mathrm{ST}\cdot\mathrm{L}]_{\mathrm{eq}}} = \frac{([\mathrm{ST}]_{\mathrm{TOT}} - [\mathrm{ST}]_{\mathrm{TOT}}Y)[\mathrm{L}]_{\mathrm{TOT}}}{[\mathrm{ST}]_{\mathrm{TOT}}Y} \tag{3–156}$$

and Equation 3–150 becomes

$$\frac{1}{1-Y} = \frac{1}{K_d^{\mathrm{L}}}\left(\frac{[\mathrm{L}]_{\mathrm{TOT}}}{Y}\right) \tag{3–157}$$

In this situation, where $[\mathrm{ST}]_{\mathrm{TOT}} \ll K_d^{\mathrm{L}}$, a plot of $(1 - Y)$ as a function of $[\mathrm{L}]_{\mathrm{TOT}}/Y$ (as in Figure 3–15B) has an intercept indistinguishable from the origin and provides a value for the dissociation constant, $K_d^{\mathrm{L}}$, but no information about the concentration of sites, $[\mathrm{ST}]_{\mathrm{TOT}}$. Although this is a disadvantage, the advantage of choosing $[\mathrm{ST}]_{\mathrm{TOT}} \ll K_d^{\mathrm{L}}$ is that no distinction has to be made between the total concentration of ligand, which can be set by simply mixing up a solution, and the free concentration of the ligand, which would otherwise be a complicated function of the total concentration of the ligand and the total concentration of the sites (Equation 3–153).

In any particular experimental situation, the concentrations of free ligand, $[\mathrm{L}]_{\mathrm{eq}}$, that have been chosen by the investigator for measuring $K_d^{\mathrm{L}}$ must include a range of values less than $10K_d^{\mathrm{L}}$. If it does not, then the site will be saturated at all concentrations within the chosen range, hyperbolic behavior will not be observed, and the equation for a hyperbola cannot be fit to the data. This follows from the facts that

$$\frac{[\mathrm{L}]_{\mathrm{eq}}}{K_d^{\mathrm{L}}} = \frac{[\mathrm{ST}\cdot\mathrm{L}]_{\mathrm{eq}}}{[\mathrm{ST}]_{\mathrm{eq}}} \tag{3–158}$$

and that values of $[\mathrm{ST}\cdot\mathrm{L}]_{\mathrm{eq}}/[\mathrm{ST}]_{\mathrm{eq}}$ in excess of 10 cannot be measured with sufficient accuracy to be useful in defining $K_d^{\mathrm{L}}$. When $[\mathrm{ST}]_{\mathrm{TOT}}$ is chosen so that $[\mathrm{ST}]_{\mathrm{TOT}}$ is of the same order of magnitude as $K_d^{\mathrm{L}}$, values of $[\mathrm{L}]_{\mathrm{eq}}$ required to measure $K_d^{\mathrm{L}}$, and hence the values of $[\mathrm{L}]_{\mathrm{TOT}}$ chosen for the measurement, will be of the same order of magnitude or less than the total concentration of sites. Because $0 < Y < 1$, the term $[\mathrm{ST}]_{\mathrm{TOT}}$ in Equation 3–150 will be significant. In the situation where $[\mathrm{ST}]_{\mathrm{TOT}} \approx K_d^{\mathrm{L}}$, a plot of $(1 - Y)^{-1}$ as a function of $[\mathrm{L}]_{\mathrm{TOT}}/Y$ has an intercept at the ordinate of $[\mathrm{ST}]_{\mathrm{TOT}}$ and the slope of the line is $(K_d^{\mathrm{L}})^{-1}$. A spectrometric titration carried out under these circumstances provides information about both the value of the dissociation constant $K_d^{\mathrm{L}}$ and the total concentration of sites, $[\mathrm{ST}]_{\mathrm{TOT}}$ (Figure 3–15).

Equations 3–124, 3–151, and 3–152 can be combined and rearranged to yield

$$[\mathrm{ST}]_{\mathrm{TOT}}[\mathrm{L}]_{\mathrm{TOT}} - (K_d^{\mathrm{L}} + [\mathrm{ST}]_{\mathrm{TOT}} + [\mathrm{L}]_{\mathrm{TOT}})[\mathrm{ST}\cdot\mathrm{L}]_{\mathrm{eq}} + [\mathrm{ST}\cdot\mathrm{L}]_{\mathrm{eq}}^2 = 0 \tag{3–159}$$

If $K_d^{\mathrm{L}} \ll [\mathrm{ST}]_{\mathrm{TOT}}$

$$([\mathrm{ST}]_{\mathrm{TOT}} - [\mathrm{ST}\cdot\mathrm{L}]_{\mathrm{eq}})([\mathrm{L}]_{\mathrm{TOT}} - [\mathrm{ST}\cdot\mathrm{L}]_{\mathrm{eq}}) \cong 0 \tag{3–160}$$

If $[\mathrm{ST}\cdot\mathrm{L}]_{\mathrm{eq}} \neq [\mathrm{ST}]_{\mathrm{TOT}}$

$$[\mathrm{L}]_{\mathrm{TOT}} \cong [\mathrm{ST}\cdot\mathrm{L}]_{\mathrm{eq}} \tag{3–161}$$

This states that when $[ST]_{TOT} >> K_d^L$, each mole of ligand added to the solution will be bound stoichiometrically by the protein until all sites are occupied. Beyond saturation, where $[ST{\cdot}L]_{eq} = [ST]_{TOT}$, further addition of ligand will resemble the addition of ligand to a solution lacking the protein. In the situation where $[ST]_{TOT} >> K_d^L$, a plot of the physical property responding to the binding of the ligand as a function of the total concentration of ligand provides an estimate of the total concentration of sites but no estimate of the dissociation constant $K_d^L$.

When either NADH or NADPH binds to glutamate dehydrogenase (NAD(P)$^+$), the absorbance of the respective nucleotide at 320 nm decreases by about 15%. If the concentration of enzyme is fixed at 5.30 mg mL$^{-1}$ and increasing concentrations of either NADH or NADPH are added, the rate of increase in the absorbance of the solution is significantly lower than that of the increase when the enzyme is not present; but, beyond a concentration of ligand equal to that of the concentration of enzyme, the rate of increase in absorbance discontinuously assumes the value observed in the absence of enzyme (Figure 3–16).[61] At the point of inflection, the total concentration of ligand must equal the total concentration of sites. Such titrations are used to determine the concentration of sites in a particular solution of protein. For example, the concentration of sites in the 5.30 mg mL$^{-1}$ solutions of glutamate dehydrogenase (NAD(P)$^+$) was 93 $\mu$M; and consequently, there is one site on the enzyme for binding NADH or NADPH for every 57,000 Da of protein. Properly, such a determination should be made at several different concentrations of protein because, as the concentration of protein is increased, the point of inflection must shift proportionally to the right if the total concentration of sites is actually greater than the dissociation constant. This should be demonstrated directly. An experiment of this type is a spectrophotometric titration of the binding sites.

When spectrophotometric studies of binding are performed, a clear distinction among these three regimes must be made. If $[ST]_{TOT} << K_d^L$, changing the concentration of protein will have no effect on the apparent dissociation constant determined using the approximation that $[L]_{TOT} = [L]_{eq}$ a plot of $(1 - Y)^{-1}$ against $Y^{-1}$ will have a zero intercept, and only a value of $K_d^L$ can be obtained. If $[ST]_{TOT} \cong K_d^L$, changing the concentration of protein will change the apparent dissociation constant if it is again assumed, in this case erroneously, that $[L]_{TOT} = [L]_{eq}$, a plot of $(1 - Y)^{-1}$ against $Y^{-1}$ will have an intercept at the ordinate that is greater than zero, and values of both $K_d^L$ and $[ST]_{TOT}$ can be estimated. If $[ST]_{TOT} >> K_d^L$, a titration with an inflection will be observed, the point of inflection will shift proportionally as the concentration of protein is varied, and only a value of $[ST]_{TOT}$ can be estimated.

Often the variation in a physical property of the protein itself upon the binding of the ligand is used to determine the dissociation constant. For example, the flavoenzyme (*S*)-2-hydroxyacid oxidase binds anions such as sulfate. The changes in the optical spectrum of the flavin upon binding of the anion display two isosbestic points, an observation which suggests that only two species are present at equilibrium, the enzyme without sulfate bound and the enzyme with a sulfate

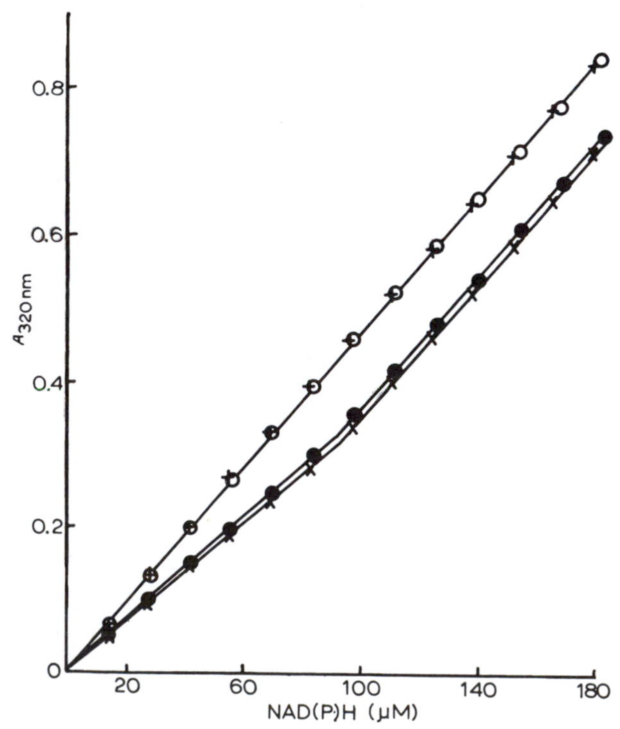


**Figure 3–16:** Spectrophotometric titration[61] of the binding sites on glutamate dehydrogenase (NAD(P)$^+$) for either NADH (●) or NADPH (×). Crystalline glutamate dehydrogenase (NAD(P)$^+$) was dissolved at pH 7.0 in two solutions (3.00 mL) of 0.17 M glutamate. The final concentration of enzyme in each solution was 5.3 mg mL$^{-1}$ based on an extinction coefficient determined by dry weight analysis. Successive additions (5 $\mu$L) of a standard solution of either NADH (●) or NADPH (×) were made to the two solutions, respectively, to establish the noted final concentrations. Two otherwise identical 3.00 mL solutions of 0.17 M glutamate at pH 7.0, to which no enzyme had been added, served as the respective controls. The same successive additions (5 $\mu$L) of either NADH (○) or NADPH (+) were made to the controls. After each successive addition, the absorbance of each solution was measured. The absorbances of the several solutions are presented as a function of the final concentrations of either NADH or NADPH (micromolar). Adapted with permission from ref 61. Copyright 1971 Elsevier Science Publishers.

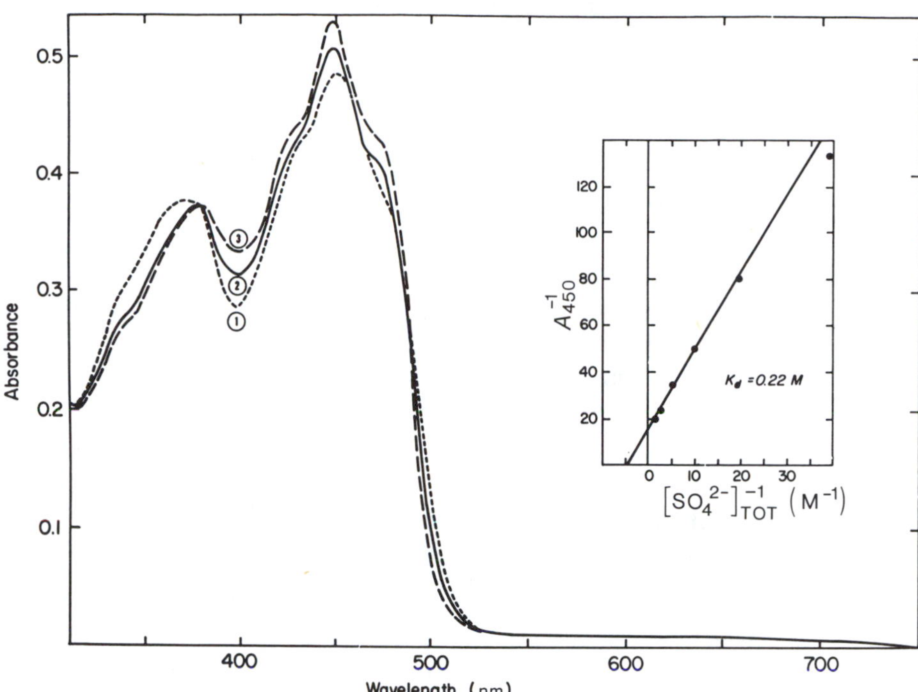


**Figure 3–17:** Binding of sulfate to (*S*)-2-hydroxyacid oxidase followed by the increase in absorbance at 450 nm of the flavin bound to the enzyme.[62] A solution of the crystalline enzyme was prepared at pH 7.6 and 10 °C. The absorption spectra of this solution in the absence of $(NH_4)_2SO_4$ (curve 1) or with $(NH_4)_2SO_4$ added to a final concentration of 0.10 M (curve 2) or 0.37 M (curve 3) are presented. Very similar spectral changes were observed at the same final concentrations of $K_2SO_4$. Isosbestic points can be seen at 380 and 490 nm, and this suggests that only two forms of the enzyme, liganded and unliganded, were present in the solution. Inset: The reciprocal of the absorbance at 450 nm ($A_{450}^{-1}$) is presented as a function of the reciprocal of the total concentration of sulfate ($[SO_4^{2-}]_{TOT}^{-1}$ in $molar^{-1}$). The linear behavior is consistent with a single binding site for sulfate with a dissociation constant ($K_d$) of 0.22 M (Equation 3–166). Adapted with permission from ref 62. Copyright 1971 Elsevier Science Publishers.

bound to a site in the vicinity of the flavin (Figure 3–17).[62] If one wavelength is chosen, the absorbance observed in the absence of sulfate, $A^{uo}$, is

$$A^{uo} = \varepsilon_{uo}[ST]_{TOT} \tag{3–162}$$

where $\varepsilon_{uo}$ is the extinction coefficient of the unoccupied enzyme. The absorbance observed at saturating concentrations of sulfate is

$$A^{sat} = \varepsilon_{occ}[ST]_{TOT} \tag{3–163}$$

where $\varepsilon_{occ}$ is the extinction coefficient of the occupied enzyme. The absorbance, $A^{obs}$, at any intermediate concentration of sulfate is

$$A^{obs} = \varepsilon_{uo}[ST]_{TOT}(1-Y) + \varepsilon_{occ}[ST]_{TOT}Y \tag{3–164}$$

When Equations 3–162 through 3–164 are combined and rearranged

$$Y = \frac{A^{obs} - A^{uo}}{A^{sat} - A^{uo}} = \frac{\Delta A}{\Delta A^{TOT}} \tag{3–165}$$

If the concentration of sulfate is in great excess over the total concentration of sites, Equation 3–157 applies, and, upon combination with Equation 3–165 and rearrangement

$$\frac{1}{\Delta A} = \frac{K_d^{L}}{\Delta A^{TOT}}\left(\frac{1}{[L]_{TOT}}\right) + \frac{1}{\Delta A^{TOT}} \tag{3–166}$$

From the slope of a plot of $\Delta A^{-1}$ as a function of $([L]_{TOT})^{-1}$, the value of $K_d^{L}$ can be calculated (Figure 3–17).[62]

The species within the protein that changes absorbance can also be an amino acid rather than a coenzyme. When succinate, L-malate, or D-malate binds to the active site of aspartate transcarbamylase saturated with carbamyl phosphate, a significant increase in the absorbance at 290 nm occurs due to the perturbation of the spectrum of a tryptophan. This change in absorbance could be used to determine the dissociation constants for these ligands from the active site of the enzyme.[63]

Most of the experimental results and procedures examined so far require that the molar concentration of binding sites be fairly high. For example, in all of the procedures where the bound concentration of the ligand is determined by the difference between the free concentration and the sum of the bound and the free concentrations (Figure 3–13), the bound concentration of the ligand must be of the same order of magnitude as the free concentration. Where differences in the fluorescence of a ligand are being followed (Figure 3–15), unless the increase in quantum yield is enormous, the concentrations of bound and free ligand must also be similar; and, since

$$\frac{[\mathrm{ST\cdot L}]_{eq}}{[\mathrm{L}]_{eq}} = \frac{[\mathrm{ST}]_{eq}}{K_d^{\mathrm{L}}} \tag{3–167}$$

it follows that the concentration of sites must be of the same order of magnitude as the dissociation constant. This was also the stipulation placed on the study of the binding of NADPH to homoserine dehydrogenase (Figure 3–15) if the data were to be used to measure both the molar concentration of sites and the dissociation constant. In the titration of glutamate dehydrogenase by NADH (Figure 3–16) the concentration of sites had to be in excess of the dissociation constant. Finally, in experiments in which either the absorbance of the protein (Figure 3–17) or the absorbance of a complex between the ligand and the protein (Figure 3–14) is monitored, the concentration of the protein must be high enough to produce measurable absorbances. None of these requirements are prohibitive when the protein in question is abundant, soluble, and pure because molar concentrations of 100 $\mu$M–1 mM in sites are easily achieved under these circumstances. In other instances, however, such high concentrations of sites cannot be produced.

Many examples of this problem are encountered in the large group of proteins known as **receptors**. As the name implies, in many cases these proteins are identified only on the basis of their ability to bind a ligand. For example, the insulin receptor is defined as a protein in the plasma membrane of cells that binds insulin (Figure 3–12). Often, for technical reasons, receptors are impure, insoluble, and available only in small quantities. In such instances the procedures described so far do not work. When such problems arise, binding is usually followed by using radiolabeled ligands and rapidly separating the large quantities of unbound ligand from the small amount of bound ligand. For example, the binding of insulin to the insulin receptor (Figure 3–12)[47] was followed by using [$^{125}$I]-insulin of high specific radioactivity (3 × $10^{14}$ cpm $g^{-1}$) and separating the small amount of bound insulin (always < 0.5% of the total insulin present) from the unbound insulin by precipitating the protein with poly(ethylene glycol) and rapidly capturing and washing the precipitate on a filter.[46] The binding of epidermal growth factor to the epidermal growth factor receptor in purified plasma membranes has been followed by using [$^{125}$I]-epidermal growth factor of high specific radioactivity (3 × $10^{13}$ cpm $g^{-1}$) and separating the small amount of bound [$^{125}$I]-epidermal growth factor (always < 5% of the total) from the large amount of unbound [$^{125}$-I]-epidermal growth factor by filtering the membranes directly.[64]

In all of these experiments examining the binding of ligands to receptors, two controls are usually run. It must be shown that the method used to separate physically the bound ligand and the unbound ligand is rapid enough that only insignificant amounts of the ligand dissociate from the site during the procedure. A control for nonspecific binding is also usually performed. In most cases, this involves addition of a high concentration of unradioactive ligand to see if this eliminates binding of the radioactive ligand. A better control, however, is to add a different ligand, known from independent experiments to block completely the binding of the ligand being tested. For example, specific binding of the $\alpha$ toxin from *Naja naja siamensis* to acetylcholine receptor could be defined as the binding that was blocked by a high concentration of carbamyl choline,[65] a structurally different ligand for acetylcholine receptor that competes with $\alpha$ toxin for binding.

Up to this point, the binding between a site on a molecule of protein and a ligand have been discussed with the assumption that the complex that is formed, ST·L, is a dead-end complex. A **dead-end complex** is a complex formed by association of a ligand with a protein in which the ligand is not transformed but dissociates in its original form. Dead-end complexes are defined by a simple dissociation constant. Before they are interconverted, reactants and products bind to the active sites of enzymes as if they were ligands, but their transformation on the active site complicates the picture. The measurement of the dissociation constant for a dead-end complex is a thermodynamic measurement; that for a complex between enzyme and substrate is usually a kinetic measurement.

A direct connection can be made between a dissociation constant, $K_d^{\mathrm{L}}$, and the rate constants governing the kinetics of the association, $k_1$, and dissociation, $k_{-1}$, of the complex. If

$$K_d^{\mathrm{L}} = \frac{[\mathrm{ST}]_{eq}[\mathrm{L}]_{eq}}{[\mathrm{ST\cdot L}]_{eq}} \tag{3–168}$$

and, at equilibrium,

$$k_1[\mathrm{ST}]_{eq}[\mathrm{L}]_{eq} = k_{-1}[\mathrm{ST\cdot L}]_{eq} \tag{3–169}$$

then

$$K_d^{\mathrm{L}} = \frac{k_{-1}}{k_1} \tag{3–170}$$

Equation 3–170, because it contains no concentrations, must be true under any set of circumstances. In particular, even in a situation in which an enzyme and its substrates exist in a steady state, which is not to be confused with existing at equilibrium, Equation 3–170 is valid. If a set of kinetic measurements, made in any situation, provides a numerical value for the ratio $k_{-1}(k_1)^{-1}$, then this numerical value is the numerical value of $K_d^{\mathrm{L}}$.

There is significant confusion between the Michaelis constant for a particular substrate A in an enzymatic reaction, $K_m^{\mathrm{A}}$, and the dissociation constant for the binding of that substrate to the active site, $K_d^{\mathrm{A}}$. This confusion arises from the provocative appeal of the simplest possible mechanism for an enzymatically catalyzed reaction, Mechanism 3–39. In the equation for $v_0$ derived from Mechanism 3–39 (Equation 3–46)

$$K_m^{\mathrm{A}} = \frac{k_{-1}+k_2}{k_1} > \frac{k_{-1}}{k_1} = K_d^{\mathrm{A}} \tag{3–171}$$

The desire is that $k_2 << k_{-1}$ so that

$$K_m^{\mathrm{A}} \cong K_d^{\mathrm{A}} \tag{3–172}$$

If all enzymatic reactions had mechanisms as simple as Mechanism 3–39, and if in all cases the chemical transformation of the reactants were slower than the rate constants for their dissociation from the active site, this approximation would be valid. Unfortunately, most enzymatic reactions have more complicated mechanisms involving multiple steps, and the expression for $K_m$ is more complicated and more distant from the simple expression for a dissociation constant. Even if the mechanism were as simple as Mechanism 3–39, many enzymes are able to perform the chemical transformation more rapidly than the bound reactants are able to dissociate from the active site. For both of these reasons, Equation 3–172 is rarely valid, and if it is to be used, its validity in the particular instance must be established.

This desire to treat a Michaelis constant as if it were a dissociation constant arises from the fact that it is difficult to measure the binding of a substrate to an enzyme because the enzyme will usually transform the substrate while the measurements are being attempted. Often, therefore, kinetic measurements are the only measurements available for determining the dissociation constants between an enzyme and a substrate. Studies with acetylcholinesterase demonstrate how these are dissected.

Acetylcholinesterase catalyzes the reaction

$$(CH_3)_3N^+CH_2CH_2OC(=O)CH_3 + H_2O \rightleftharpoons (CH_3)_3N^+CH_2CH_2OH + HOC(=O)CH_3 \qquad (3\text{–}173)$$

The enzyme also catalyzes the hydrolysis of a number of related acetyl esters, and a series of these esters has been used to study the properties of the active site involved in binding these reactants.[66] It had been shown in earlier studies that the mechanism

$$\text{E-OH} + \text{RO}C(=O)CH_3 \underset{k_{-1}}{\overset{k_1}{\rightleftharpoons}} \text{E-OH}\cdot\text{RO}C(=O)CH_3 \underset{k_{-2}\,[\text{ROH}]_0}{\overset{k_2}{\rightleftharpoons}} \text{E-O}C(=O)CH_3 \xrightarrow{k_3\,[H_2O]} \text{E-OH} + \text{HO}C(=O)CH_3 \qquad (3\text{–}174)$$

where E-OH is a serine in the active site, could explain all of the kinetic observations that had been made with acetylcholinesterase. In this mechanism, the term $k_3[H_2O]$ had the same value for all of the reactants because they were all acetyl esters, and that value had already been determined accurately. The maximum velocity for Mechanism 3–174 when $[ROH]_0$ is zero is

$$V_{max} = \frac{k_2 k_3 [H_2O][E]_{TOT}}{k_2 + k_3[H_2O]} \qquad (3\text{–}175)$$

From the maximum velocities for each reactant and values for $k_3[H_2O]$ and $[E]_{TOT}$, values of the rate constant $k_2$ could be calculated for each reactant. For Mechanism 3–174

$$\frac{V_{max}}{[E]_{TOT}K_m} = \frac{k_2}{\left(\dfrac{k_{-1}+k_2}{k_1}\right)} \qquad (3\text{–}176)$$

With this equation, measured values of $V_{max}/([E]_{TOT}K_m)$, and the values of $k_2$, values of $(k_{-1} + k_2)(k_1)^{-1}$ could be calculated for each substrate. From comparing measured values of $k_2$ to estimated values of $k_{-1}$, it was concluded that, for every reactant except acetylcholine, $k_{-1} >> k_2$ and $(k_{-1} + k_2)(k_1)^{-1}$ was an accurate estimate of the dissociation constant (Equation 3–171) for the binding of each of the reactants to the active site.

Numerical values of the dissociation constants measured for complexes between enzymes and substrates or between enzymes and competitive inhibitors or between proteins and ligands have a number of uses. Because they are directly related to changes in standard free energy, they can often be related to the types of noncovalent interactions between the site and the ligand. In this way, these dissociation constants are often used to probe the structure of the binding site or the active site.

In the case of acetylcholinesterase, the dissociation constants for a series of *N*-methylated 2-aminoethyl acetates [$H_3N^+C_2H_4$–, $(CH_3)H_2N^+C_2H_4$–, $(CH_3)_2HN^+C_2H_4$–, and $(CH_3)_3N^+C_2H_4$–] were compared, respectively, with a series of corresponding alkyl acetates in which nitrogen has been replaced by carbon [$H_3CC_2H_4$–, $(CH_3)H_2CC_2H_4$–, $(CH_3)_2HCC_2H_4$–, and $(CH_3)_3CC_2H_4$–]. All of these acetate esters, both the cations and the alkanes, were reactants in hydrolyses catalyzed by the enzyme (Table 3–2). The fact that little if any difference was observed between the dissociation constants for pairs of substrates having or lacking the cationic charge demonstrates that this feature of the structure of the physiologically relevant reactant, acetylcholine, does not contribute to the favorable change in standard free energy accompanying its binding to the active site. These results suggest that only a hydrophobic effect drives this binding.

Yeast hexokinase binds its substrates with a preference for glucose binding before MgATP.[67,68] Because the substrates associate and dissociate rapidly relative to their turnover and because glucose is usually the first substrate to bind, the Michaelis constant for glucose is approximately equal to the dissociation constant for its binding to the open active site.[68] In such an instance, kinetic values for $K_m$ can be used as estimates of the dissociation constants for the binding of various analogues of glucose to the active site (Table 3–3).[69,70] Most of these derivatives of glucose are phosphorylated at the 6-hydroxyl in a reaction catalyzed by the enzyme.

It is possible to get a feeling for the structure of the active site of hexokinase from the changes in $K_m$ caused by various replacements. The active site of the enzyme seems to be open and unrestrictive at carbon 2 of glucose because epimerization or the replacement of the OH with smaller (hydrogen or fluorine) or larger (chloro or acetamido) groups has little effect on the Michaelis constant. At both carbons 3 and 4, however, epimerization eliminates the ability of the sugar to participate in the reaction. Presumably, there are either amino acids or polypeptide above and below the ring at these loca-

**Table 3–2: Dissociation Constants[66] for the Binding of Reactants to Acetylcholinesterase[a]**

| reactant | dissociation constant (mM) |
|---|---|
| $H_3\overset{+}{N}CH_2CH_2OC(=O)CH_3$ | 27 |
| $H_3CCH_2CH_2OC(=O)CH_3$ | 33 |
| $H_3CH_2\overset{+}{N}CH_2CH_2OC(=O)CH_3$ | 22 |
| $H_3CH_2CCH_2CH_2OC(=O)CH_3$ | 13 |
| $(CH_3)_2H\overset{+}{N}CH_2CH_2OC(=O)CH_3$ | 1.4 |
| $(CH_3)_2HCCH_2CH_2OC(=O)CH_3$ | 2.8 |
| $(CH_3)_3\overset{+}{N}CH_2CH_2OC(=O)CH_3$ | 1.2 |
| $(CH_3)_3CCH_2CH_2OC(=O)CH_3$ | 3.1 |

[a]Dissociation constants were determined from steady-state kinetics at pH 7.8, 25 °C, and $\Gamma/2 = 0.18$ M.

tions in the active site, and anything larger than a hydrogen is sterically excluded. This must also be true beyond the hydroxyl on carbon 3 because methylation at this oxygen produces a substrate that cannot participate in the reaction. A more subtle effect is seen when either the 3-hydroxyl or the 4-hydroxyl is replaced by a fluorine. The 550-fold increase in the Michaelis constant ($\Delta\Delta G° = +15$ kJ mol$^{-1}$) suggests that in each case a noncovalent interaction, such as a hydrogen bond, directed toward the hydroxyl in particular has been prevented from occurring. A similar although 10-fold less dramatic change is seen when either the 6-hydroxyl is replaced by a fluorine or carbon 6 is eliminated entirely (xylose).

**Structure–activity relationships**, such as those presented in Tables 3–2 and 3–3, are often compiled, and other examples could also be cited.[71,72] These studies provide an intuitive picture of the site at which the ligand binds. They also indicate

**Table 3–3: Michaelis Constants[69,70] for Derivatives of Glucose [(2*R*,3*S*,4*R*,5*R*)-2,3,4,5,6-Pentahydroxyhexanal] as Reactants in the Reaction Catalyzed by Hexokinase[a]**

| derivative of glucose | $K_m$ (mM) |
|---|---|
| glucose | 0.14 |
| 2*S* epimer (mannose) | 0.05 |
| 2-deoxy | 0.30 |
| 2-deoxy-(2*R*)-acetamido | 1.0 |
| 2-deoxy-(2*R*)-fluoro | 0.19 |
| 2-deoxy-(2*S*)-fluoro | 0.14 |
| 2-deoxy-(2*R*)-chloro | 2.1 |
| 3*R* epimer (allose) | >100 |
| 3-deoxy-(3*S*)-fluoro | 70 |
| 3-*O*-methyl | >100 |
| 4*S* epimer (galactose) | >50 |
| 4-deoxy-(4*R*)-fluoro | 84 |
| xylose | 10[b] |
| 6-deoxy-(6*R*)-fluoro | 8[b] |

[a]30 °C, pH 7–8.5. [b]Competitive inhibitor, $K_i$ listed.

which portions of the ligand cause it to be recognized and bound at the site upon the protein. Often they are useful in designing derivatives of reactants, activators, or inhibitors that can bind to one protein tightly while being excluded from other proteins.

Measurements of the binding of a reactant to an enzyme in the absence of the other required reactants is often used to confirm or to reject kinetic mechanisms that propose a particular order in which the reactants bind to the active site during enzymatic catalysis. Under these circumstances the only reactant present is usually not transformed by the enzyme, and it binds as if it were a simple ligand. The facts that glucose binds to yeast hexokinase, in the absence of MgATP, with a dissociation constant equal to its Michaelis constant but that MgATP binds much less strongly to the enzyme in the absence of a monosaccharide[68] confirm a kinetic mechanism in which glucose binds first to the active site. Likewise, the facts that isocitrate binds to isocitrate dehydrogenase ($NAD^+$) in the absence of $NAD^+$, with a dissociation constant equal to its Michaelis constant, but that $NAD^+$ does not bind to the enzyme in the absence of isocitrate[73] are consistent with a kinetic mechanism in which isocitrate is required to enter the active site before $NAD^+$.

When reactants enter the active site at random, either reactant will usually bind to the enzyme in the absence of the other. The dissociation constants for either $NAD^+$ (200 $\mu$M) or UDP-glucose (50 $\mu$M) when either binds to UDP-glucose dehydrogenase in the absence of the other[74] are close to the values for their Michaelis constants (500 and 10 $\mu$M, respectively),[75,76] and these observations are consistent with a kinetic mechanism in which the two reactants bind at random in the first step of the two-step oxidation–reduction catalyzed by this enzyme. Likewise, the directly measured dissociation constants for either isoleucine (25 $\mu$M) or ATP (150 $\mu$M) when either binds in the absence of the other[77] to isoleucine-tRNA ligase are close to the values for their dissociation constants calculated from kinetic measurements (4 $\mu$M and 150 $\mu$M, respectively),[78] and these observations are consistent with a kinetic mechanism in which these two reactants enter the active site of the enzyme at random.

There are often ambiguities in measurements of this kind. For example, the kinetic behavior of the inhibition of aspartate transcarbamylase by *N*-(phosphonacetyl)-L-aspartate (Figure 3–7), and studies of product inhibition,[79] as well as other observations, are all consistent with the conclusion that carbamyl phosphate must enter the active site before aspartate does in the reaction leading to products. Aspartate, however, can bind to the unoccupied enzyme with a dissociation constant even less than its dissociation constant when it is productively entering the active site occupied with carbamyl phosphate.[63] The substrate fructose 6-phosphate can bind to six sites on each subunit of 6-phosphofructokinase, but the addition of inorganic phosphate, which eliminates the nonspecific binding attributable to the phosphate on the fructose 6-phosphate, decreases the number of sites to one for each subunit.[80] Presumably, this one remaining site is the active site. Analogues of ATP bind to three sites on every subunit of 6-phosphofructokinase. One of these sites is a site at which fructose 1,6-bisphosphate also binds.[81]

The concentration of active sites in a solution can be determined by a kinetic titration. A **kinetic titration** is a titration of the sites for a tightly and stoichiometrically bound activator, reactant, or inhibitor of an enzyme that is followed by its effect on the kinetics of the enzymatic reaction. For example, a small protein, calmodulin, activates $Ca^{2+}$-transporting ATPase from erythrocyte membranes.[82,83] If the initial velocity of this enzymatic reaction is followed as a function of the concentration of calmodulin at saturation with calcium and MgATP, a monotonic increase in the initial velocity, $\Delta v_0$, is observed that reaches a maximum, $\Delta V_{max}$, at high concentrations of calmodulin. It can be shown[84] that $[CM]_{TOT}^{0.5}$, the total concentration of calmodulin producing an increase in the initial velocial, $\Delta v_0$, equal to half of the maximum increase, $\Delta V_{max}$, is related to the total molar concentration of sites at which calmodulin binds by the equation

$$[CM]_{TOT}^{0.5} = K_{d,app}^{CM} + \frac{[ST]_{TOT}}{2} \tag{3–177}$$

where $K_{d,app}^{CM}$ is the apparent dissociation constant for the binding of calmodulin to $Ca^{2+}$-transporting ATPase. When Equation 3–177 is combined with Equation 3–126

$$[CM]_{TOT}^{0.5} = K_{d,app}^{CM} + \frac{C_P}{2\kappa} \tag{3–178}$$

If the molar concentration of sites can be adjusted to a value in the range of the value of the apparent dissociation constant and initial velocities can be measured at these concentrations, a plot of $[CM]_{TOT}^{0.5}$ as a function of the concentration of the protein (in grams liter$^{-1}$) will have a slope of $(2\kappa)^{-1}$. In this instance, the parameter $\kappa$ is the grams of protein (mole of sites)$^{-1}$. A simple way to understand Equation 3–177 is to consider that the free concentration of calmodulin is less than the total concentration by the amount bound to the enzyme. To effect the same fractional activation as the concentration of enzyme is increased, the total concentration of calmodulin must be increased to adjust for the increase in the bound calmodulin, and the additional calmodulin needed is directly proportional to the concentration of sites.

The same strategy can be used if a tightly bound reactant or inhibitor of the enzyme is available. For example, it is often possible to follow one turnover at concentrations of the enzyme in excess of the dissociation constant for a tightly bound reactant. A titration performed by increasing the concentration of the enzyme and following the increase in the initial rate of that single turnover can be used to estimate the molar concentration of sites for the reactant in a solution of the enzyme.[85] It is also possible to titrate the concentration of sites for an inhibitor if it is stoichiometrically bound by following initial velocities by steady-state kinetics.[86] In this case, the decrease in initial velocity observed is directly proportional to the concentration of bound inhibitor, which is equal to the total concentration of the inhibitor. The molar concentration of inhibitor required to inhibit the enzyme completely is equal to the molar concentration of sites.

## *Suggested Reading*

Hasan, F.B., Cohen, S.G., & Cohen, J.B. (1980) Hydrolysis by Acetylcholinesterase: Apparent Molal Volumes and *Trimethyl* and *Methyl* Subsites, *J. Biol. Chem. 255*, 3898–3904.

### PROBLEM 3–11

Show that Equations 3–175 and 3–176 follow from Mechanism 3–174.

### PROBLEM 3–12

Glucose-6-phosphate isomerase binds glucose 6-phosphate. The following observations[52] were made by equilibrium dialysis at 30 °C and pH 8.0. The concentration of protein was 10.0 g L$^{-1}$.

| [glucose 6-phosphate]$_{free}$ ($\mu$M) | [glucose 6-phosphate]$_{bound}$ ($\mu$M) |
|---|---|
| 46 | 32 |
| 46 | 34 |
| 126 | 66 |
| 260 | 95 |
| 270 | 100 |
| 560 | 112 |
| 2810 | 141 |

(A) Display these data on a Scatchard plot and determine the dissociation constant and the grams of protein (mol of sites)$^{-1}$ at saturation.

(B) The molar mass of glucose-6-phosphate isomerase is 132,000 g mol$^{-1}$. If the solution was monodisperse, how many moles of sites are there for every mole of enzyme?

### PROBLEM 3–13

Adenine phosphoribosyltransferase catalyzes the reaction

$$\text{adenosine monophosphate + pyrophosphate} \rightleftharpoons \text{adenine + 5-phospho-}\alpha\text{-D-ribose 1-diphosphate}$$

Various phosphates have been found to be inhibitors of adenine phosphoribosyltransferase.[87] Kinetic results obtained with one of these compounds are displayed in panel A of the figure on the next page. Values for the dissociation constants (determined as inhibition constant, $K_i$) between various inhibitors and the enzyme are shown in the table on the next page and some are plotted in panel B of the figure on the next page.

(A) What is the difference in free energy of binding between a diol diphosphate that fits the active site well and D-ribose 1,5-bisphosphate?

(B) Assume that the main difference between a diol diphosphate and D-ribose 1,5-bisphosphate is the hydroxyls of the ribose, and discuss this difference of free energy in terms of the enthalpies of formation of the hydrogen bonds and the entropy of approximation. Recall that the active site is constructed for a ribose.

(C) Explain the table and panel B of the figure.

| compound | $K_i$ | $\frac{K_i \text{ analogue}}{K_i \text{PP-ribose-P}}$ |
|---|---|---|
| PP-ribose-P | 0.0002 | 1 |
| D-ribose | >50 | >250,000 |
| α-D-ribose 1-phosphate | >50 | >250,000 |
| β-D-ribose 1-phosphate | >50 | >250,000 |
| D-ribose 5-phosphate | 3.3 | 16,500 |
| D-fructose 6-phosphate | 5.1 | 25,500 |
| D-glucose 6-phosphate | 5.1 | 25,500 |
| D-ribose 1,5-bisphosphate | 0.12 | 600 |
| D-ribulose 1,5-bisphosphate | 0.41 | 2050 |
| D-fructose 1,6-bisphosphate | 0.20 | 1000 |
| D-glucose 1,6-bisphosphate | 0.42 | 2100 |
| 6-phosphogluconic acid | >50 | >250,000 |
| ethylene glycol bisphosphate | 24 | 120,000 |
| 1,3-propanediol bisphosphate | 1.2 | 6000 |
| 1,4-butanediol bisphosphate | 0.45 | 2250 |
| 1,5-pentanediol bisphosphate | 0.39 | 1950 |
| 1,6-hexanediol bisphosphate | 0.38 | 1900 |
| 2-pentanone 1,5-bisphosphate | 0.26 | 1300 |

**A**

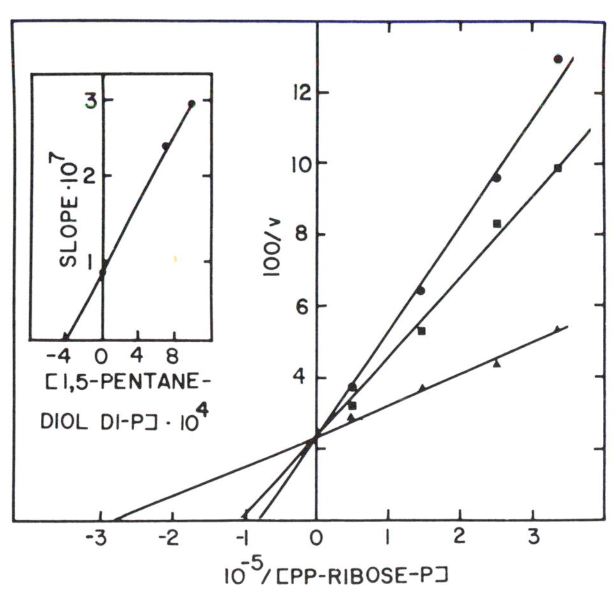


**B**

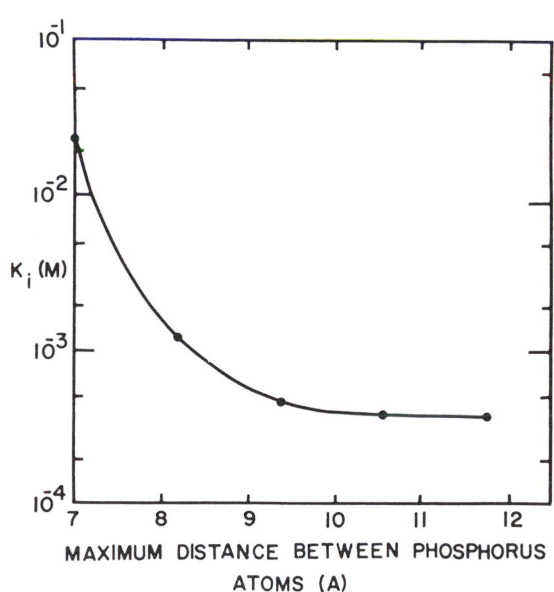


(A) Inhibition of adenine phosphoribosyltransferase by 5-pentanediol diphosphate.[87] Double-reciprocal plot of initial velocity against PP-ribose-P concentration at an adenine concentration of $9 \times 10^{-6}$ M. The concentrations of 1,5-pentanediol diphosphate are 0 (▲), $7 \times 10^{-4}$ M (■), and $10^{-3}$ M (●). The initial velocity ($v$) is expressed as picomoles of AMP per min. The inset represents the secondary plot of slopes against molar concentration of 1,5-pentanediol diphosphate. (B) Relationship between inhibition constants and maximum distance between phosphorus atoms of aliphatic diol bisphosphates.[87] Reprinted with permission from ref 87. Copyright 1970 *Journal of Biological Chemistry.*

# Nonideal, Superstoichiometric, and Substoichiometric Behavior

The data presented in Figure 3–2 display **ideal behavior** when $v_0^{-1}$ is plotted as a function of $[A]_0^{-1}$, as expected for a situation in which the solution contains a homogeneous population of independent active sites with which the reactants associate and at which the bound reactants are transformed to products (Mechanism 3–39). The data presented in Figure 3–13 display ideal behavior when $[ST{\cdot}L]_{eq}/C_p$ is plotted as a function of $[ST{\cdot}L]_{eq}/(C_p[L]_{eq})$, as expected for a situation in which the solution contains a homogeneous population of independent binding sites with which the ligand can associate. It is also the case that, in the experiment described by Figure 3–13, the molar concentration of binding sites determined from the intercept of the line with the ordinate agrees closely with the molar concentration of subunits present in the solution, when this is calculated from the concentration of protein and the length of the polypeptide from which each subunit is composed, even though glucose-6-phosphate isomerase is a dimeric protein. All of the results discussed so far serve as paradigms of ideal, stoichiometric kinetic or thermodynamic behavior. In many instances, however, when a solution of a particular protein is assayed, the results obtained from steady-state kinetics or measurements of binding deviate from such ideal, stoichiometric behavior.

Deviations from ideal behavior in either a plot of reciprocal initial velocities or a Scatchard plot of the binding of a ligand are of two types: positive or negative (Figure 3–18). These two terms are defined on the basis of how the actual behavior of the initial velocity (Figure 3–1B) or the actual concentration of bound ligand (Figure 3–12), as a function of the initial concentration of reactant or the free concentration of ligand, respectively, deviate from ideal behavior. Within any set of measurements of the turnover of a reactant or the binding of a ligand performed on a solution of a protein, the monotonically increasing values of either the initial velocity or the concentration of bound ligand gathered at the lowest concentrations of reactant or ligand, respectively, can be used to define the initial slope of a rectangular hyperbola ($c^{-1}$ in Equation 3–33), and the invariant value of either the initial velocity or concentration of bound ligand measured at the highest concentrations of reactant or ligand, respectively, can be used to define the horizontal asymptote of that rectangular hyperbola ($d^{-1}$ in Equation 3–33). If the initial velocities or the concentrations of bound ligand measured at concentrations of reactant or ligand between these extremes are greater than those predicted by this ideal hyperbola, the behavior is considered to be a **positive deviation**; if they are less than those predicted by this ideal hyperbola, the behavior is considered to be a **negative deviation**.

Superstoichiometric or substoichiometric behavior is also observed in many situations. The molar concentration of protomers present in the solution of a purified protein can be calculated from the amino acid sequence or amino acid sequences of the polypeptide or polypeptides comprising one protomer and the total concentration of protein in the solution. When the molar concentration of bound ligand at saturation is greater than the molar concentration of protomers, this observation is considered to represent **superstoichiometric behavior;** when it is less than the molar concentration of protomers, this observation is considered to represent **substoichiometric behavior**.

## *Superstoichiometric Behavior*

Although there are exceptional circumstances where linear, superstoichiometric binding of a ligand to a protein has been observed,[89] it is usually the case that superstoichiometric behavior displays negative deviation from ideal behavior. This is due to the fact that if more than one binding site for a ligand is present for every protomer, which would require that each of the different binding sites is at a different location on the surface of the protomer, each of these unique sites on the protomer should display a different dissociation constant. Because sites are occupied in proportion to their affinities, the site with highest affinity being occupied preferentially at the lowest concentrations, the behavior observed will be that of a sum of independent associations each with progressively lower affinity (lower curve in Figure 3–18A). Under these circumstances the system behaves as if there were two or more populations of sites in the solution, each present in a molar concentration equal to the molar concentration of protomer. For example, suppose each protomer has three different sites: ST1, ST2, and ST3. These sites have the respective dissociation constants $K_{d1}$, $K_{d2}$, and $K_{d3}$. The values of these constants are different from each other. The three independent reactions occurring at the three independent sites are

$$ST1{\cdot}L \underset{}{\overset{K_{d1}}{\rightleftharpoons}} ST1 + L \tag{3–179}$$

$$ST2{\cdot}L \underset{}{\overset{K_{d2}}{\rightleftharpoons}} ST2 + L \tag{3–180}$$

$$ST3{\cdot}L \underset{}{\overset{K_{d3}}{\rightleftharpoons}} ST3 + L \tag{3–181}$$

The concentration of ligand bound to all sites is

$$[ST{\cdot}L]_{TOT} = [ST1{\cdot}L]_{eq} + [ST2{\cdot}L]_{eq} + [ST3{\cdot}L]_{eq} \tag{3–182}$$

and

$$\begin{aligned}[ST1]_{eq} + [ST1{\cdot}L]_{eq} &= [ST2]_{eq} + [ST2{\cdot}L]_{eq} \\ &= [ST3]_{eq} + [ST3{\cdot}L]_{eq} = \tfrac{1}{3}[ST]_{TOT}\end{aligned} \tag{3–183}$$

where $[\text{ST}]_{\text{TOT}}$ is the total concentration of sites. From Equations 3–182 and 3–183 and the definitions of the three dissociation constants, it follows that

$$[\text{ST}\cdot\text{L}]_{\text{TOT}} = \frac{[\text{S}]_{\text{TOT}}}{3}\left\{\frac{3([\text{L}]_{\text{eq}})^3 + 2([\text{L}]_{\text{eq}})^2(K_{\text{d1}}+K_{\text{d2}}+K_{\text{d3}}) + [\text{L}]_{\text{eq}}(K_{\text{d1}}K_{\text{d2}}+K_{\text{d2}}K_{\text{d3}}+K_{\text{d1}}K_{\text{d3}})}{([\text{L}]_{\text{eq}})^3 + ([\text{L}]_{\text{eq}})^2(K_{\text{d1}}+K_{\text{d2}}+K_{\text{d3}}) + [\text{L}]_{\text{eq}}(K_{\text{d1}}K_{\text{d2}}+K_{\text{d2}}K_{\text{d3}}+K_{\text{d1}}K_{\text{d3}}) + K_{\text{d1}}K_{\text{d2}}K_{\text{d3}}}\right\} \tag{3–184}$$

At low concentrations of ligand

$$\lim_{[\text{L}]\to 0}[\text{ST}\cdot\text{L}]_{\text{TOT}} = [\text{L}]_{\text{eq}}\frac{[\text{ST}]_{\text{TOT}}(K_{\text{d1}}K_{\text{d2}}+K_{\text{d2}}K_{\text{d3}}+K_{\text{d1}}K_{\text{d3}})}{3K_{\text{d1}}K_{\text{d2}}K_{\text{d3}}} \tag{3–185}$$

and this linear equation defines the initial slope of an ideal hyperbola. At saturating concentrations of ligand

$$[\text{ST}\cdot\text{L}]_{\text{TOT}} = [\text{ST}]_{\text{TOT}} \tag{3–186}$$

and this defines the asymptote of an ideal hyperbola. Between these extremes the actual curve defined by Equation 3–184 falls below this ideal hyperbola, and the behavior is one of negative deviation.

If there are only a few, distinct populations of independent sites in the solution, the resulting behavior of the concentration of bound ligand as the concentration of free ligand[49] or the resulting nonlinear Scatchard plot[90] can be resolved by numerical analysis into its several components, each describing the binding of the ligand to sites of successively lower affinity. For example, $\text{Na}^+/\text{K}^+$-transporting ATPase binds ATP tightly at its active site, and there is one active site on every protomer of the protein.[91] Under certain circumstances, however, the enzyme binds more than one mole of adenylyl imidodiphosphate, an analogue of ATP, for each mole of active sites. The behavior is such that the tight binding of the ligand to the active site can be distinguished from much weaker binding at one or more additional sites (Figure 3–19).[92]

There is significant difficulty involved in interpreting results of this type. Proteins often associate weakly with ligands at several biologically irrelevant sites upon their surface when those ligands are present in solution at high concentrations, so it is seldom the case that true saturation is reached when studies of binding are performed. The Donnan effect (Equation 3–138) further complicates the interpretation. Usually, the concentration of bound ligand gradually increases as the concentration of free ligand is increased following the early and obvious occupation of any sites of high affinity. Therefore, it is difficult to define a unique site that has a low affinity for a particular ligand from studies of binding alone, unless it can be shown unambiguously that saturation has been reached and the Donnan effect is not the cause of this behavior.

## *Negative Deviations from Ideal Behavior*

Negative deviations from ideal behavior in either steady-state kinetics or the binding of ligands are sometimes observed even when the number of bound ligands at saturation equals the number of protomers. When such negative deviation is displayed by an oligomeric protein, an extensive theoretical explanation,[93] referred to as **negative cooperativity**,[94] is often invoked to describe it. This theory derives its explanation for the observed negative deviation from the proposition that intramolecular interactions among the sites for binding the ligand on the several protomers of the same molecule of protein are responsible for it. In other words, each individual molecule of the oligomeric protein possesses within itself the origin of the negative deviation; and, if only one oligomeric molecule were present in the solution, the observed properties would be the same. Associated with this theoretical treatment, therefore, is the irrevocable assumption that the solution contains a completely homogeneous population of either the protein binding the ligand or the enzyme transforming the substrates. If the population is not homogeneous but heterogeneous, and different subpopulations of the protein or the enzyme display different dissociation constants for the ligand or kinetic rate constants for the transformation of the reactants, respectively, negative deviations from ideal behavior will necessarily result, not as a consequence, however, of negative cooperativity. This is due to the fact that in a heterogeneous solution either the observed concentrations of bound ligand or the observed initial velocities are summations of the individual values for each subpopulation, and the observed function describing binding or initial velocity is the sum of several individual rectangular hyperbolas, each with different initial slopes and different levels of saturation, as a function of the concentration of free ligand or free reactant. Because the sum of two or more rectangular hyperbolas necessarily displays negative deviation, heterogeneous solutions of a protein or of an enzyme often display negative deviations from ideal behavior simply because they are heterogeneous.

Because solutions of purified protein are often heterogenous, nonideal behavior arising for this reason is common but uninteresting. For example, impure preparations of insulin receptor displayed negative deviations from ideality in their binding of insulin,[47,95] and this was explained in terms of negative cooperativity.[96] When the protein was further purified, however, this nonideal behavior disappeared (Figure 3–12).[47] It could be regained by adding back a solution of glycoproteins that had been separated during the purification. In this case, negative deviations from ideal behavior resulted from the heterogeneity of the impure preparation, perhaps from the random association between molecules of another protein and some of the insulin receptors. In any case, these results demonstrate that negative deviations are not an intrinsic property of insulin receptor itself. At the moment, it is unclear whether or not there are any homogeneous solutions of particular proteins that nevertheless display negative deviations from ideal behavior and, if so, whether or not these deviations require the theoretical treatment of negative cooperativity for their explanation.

It should also be emphasized that there is considerable confusion surrounding the use of the term negative coopera-

A

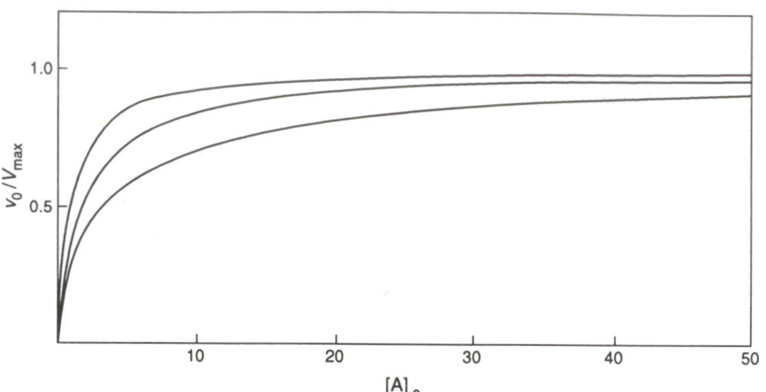


B

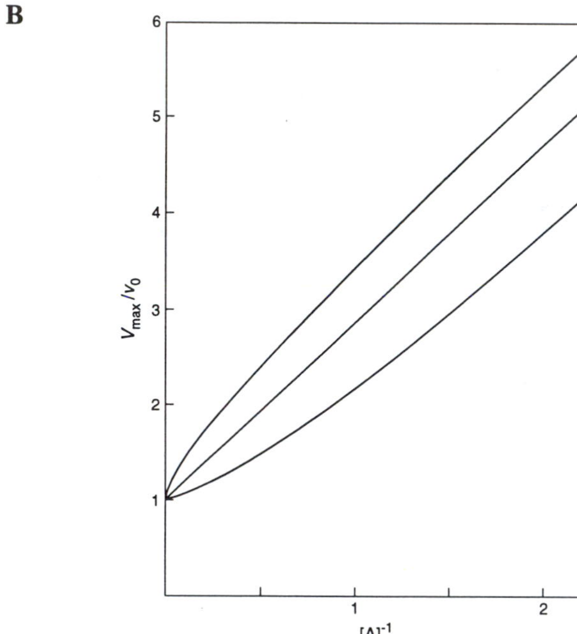


C

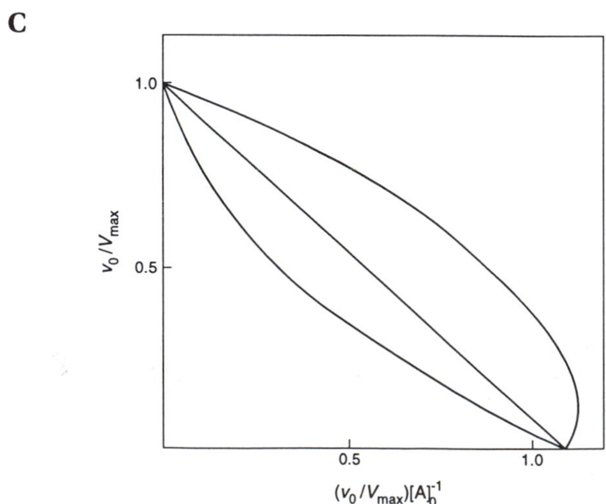


**Figure 3–18:** Comparison of normal hyperbolic behavior with positive and negative deviations from ideal behavior. Arbitrarily, initial velocity as a function of the concentration of reactant A was chosen to illustrate these properties rather than binding of ligands. (A) The normalized initial velocity ($v_0/V_{max}$) is presented as a function of the initial concentration of reactant A ($[A]_0$), (B) the reciprocal of the normalized initial velocity, $V_{max}/v_0$, is presented as a function of the reciprocal of the initial concentration of reactant A ($[A]_0^{-1}$), or (C) the normalized initial velocity ($v_0/V_{max}$) is presented as a function of the quotient of the normalized initial velocity and the initial concentration of reactant A [$(v_0/V_{max})[A]_0^{-1}$]. The three curves in panel A were constructed so that their initial slopes were each equal to 1.1 and their horizontal asymptotes were each equal to 2. The middle curve in all three panels is for the rectangular hyperbola defined by the equation

$$v_0 = \frac{V_{max}[A]_0}{K_m + [A]_0}$$

where $V_{max} = 2$ and $K_m = 1.818$. The lower curve in panel A, chosen to illustrate negative deviation, is the curve defined by the equation

$$v_0 = \frac{V_{max,1}[A]_0}{K_{m,1} + [A]_0} + \frac{V_{max,2}[A]_0}{K_{m,2} + [A]_0}$$

where $V_{max,1} = V_{max,2} = 1$, $K_{m,1} = 1$, and $K_{m,2} = 10$. This would be the equation for a solution containing two independent forms of the enzyme with different values of $K_m$ and $V_{max}$. The upper curve in panel B and the lower curve in panel C are the curves that are also defined by this equation for negative deviation with the same values of the parameters. The upper curve in panel A, chosen to illustrate positive deviation, is the curve defined by the equation

$$v_0 = \frac{V_{max}\left[K_{RT}\frac{[A]_0}{K_d^{TA}}\left(1+\frac{[A]_0}{K_d^{TA}}\right)+\frac{[A]_0}{K_d^{RA}}\left(1+\frac{[A]_0}{K_d^{RA}}\right)\right]}{K_{RT}\left(1+\frac{[A]_0}{K_d^{TA}}\right)^2+\left(1+\frac{[A]_0}{K_d^{RA}}\right)^2}$$

where $V_{max} = 2$, $K_{RT} = 10$, $K_d^{TA} = 3$, and $K_d^{RA} = 0.368$. This is the equation describing the initial velocity of an enzyme with two symmetric, positively cooperative active sites by the formalism of Monod, Wyman, and Changeux.[88] The lower curve in panel B and the upper curve in panel C are the curves that are also defined by this equation for positive deviation with the same values of the parameters. The initial slopes and horizontal asymptotes of all of the curves in panel A are the same.

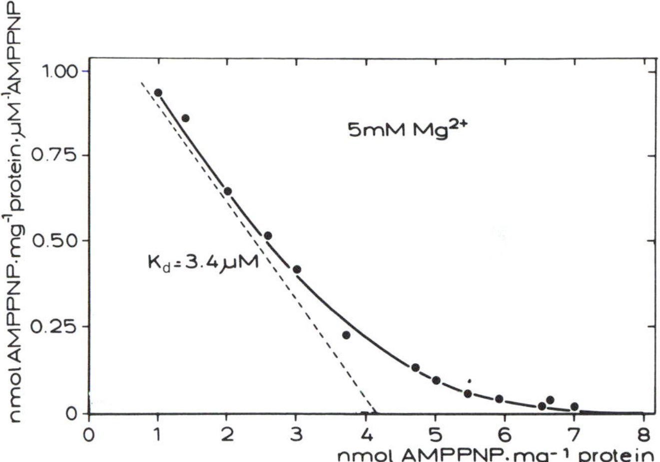


**Figure 3–19:** Binding of adenylyl imidodiphosphate to $Na^+/K^+$-transporting ATPase from rabbit kidney followed by a filtration assay. Fragments of membrane containing $Na^+/K^+$-transporting ATPase as their only protein were suspended at a final concentration of 1 mg of protein $mL^{-1}$ in 5 mM $MgCl_2$ at pH 7.0 and 22 °C in a number of separate 100-μL samples. [$^3$H]Adenylyl imidodiphosphate was added to the several samples at different final concentrations. [$^{14}$C]Sucrose was added to all samples at the same concentration. After 5 min, each sample was passed through a separate filter, which trapped the fragments of membrane and some of the fluid in which they had been suspended. The trapped membranes and trapped fluid were dissolved from the filters and the counts per minute of trapped $^3$H and $^{14}$C were determined by scintillation counting. The amount of fluid trapped by the filter could be determined by the amount of [$^{14}$C]sucrose that was trapped. The amount of [$^3$H]adenylyl imidodiphosphate dissolved in the trapped fluid could then be subtracted from the total [$^3$H]adenylyl imidodiphosphate trapped by the filter. The difference is the bound ligand. The quotient of the amount of bound ligand and the free concentration of adenylyl imidodiphosphate [in nanomoles of AMPPNP (milligrams of protein)$^{-1}$ (micromolar AMPPNP)$^{-1}$] is presented as a function of the amount of bound ligand [in nanomoles of AMPPNP (milligram of protein)$^{-1}$]. The dissociation constant ($K_d$) for the binding of ligand to the sites of highest affinity was 3.4 μM. Adapted with permission from ref 92. Copyright 1981 Elsevier Science Publishers.

tivity. This confusion arises from a common mistake in experimental science, that of confounding the observations and the explanation for those observations. In this instance the error is particularly easy to see. The observations are negative deviations from ideal behavior. These observations exist even in the absence of any explanation of them. Negative cooperativity is one explanation for the observation of negative deviation from ideal behavior. It is not the only explanation, and it is probably not the correct explanation in the majority if not all of the cases in which negative deviation from ideal behavior is observed. The unnoticed sleight of hand is referring to the observations by the term negative cooperativity. In this way a particular explanation is inadvertently forced onto a set of observations.

## *Fractional Site Behavior*

An extreme manifestation of negative cooperativity would be **fractional site behavior**. Consider an oligomeric protein that participates in negative cooperativity. According to the theory, the successively occupied sites on the same molecule of the oligomeric protein display successively lower affinities for the ligand, and it is these different affinities that lead to negative deviations from ideal behavior. In the limit one could argue that some fraction of the unoccupied sites on the protein would have an affinity so low that they would be unable to bind the ligand within the range of concentrations examined. Because almost all oligomeric proteins are dimers or dimers of dimers, it is most likely that an occupied site of high affinity would be paired with an unoccupied site unable to bind the ligand and **half-of-sites behavior** would result. There are elaborate theories to explain why such behavior would be advantageous.[97–99]

The difficulty with behavior resembling half-of-sites behavior or any other type of substoichiometric behavior is that it can so easily arise artifactually. If a certain fraction of the protein of interest in the solution is inactive but otherwise intact or if the solution of the purified protein of interest is contaminated with other proteins, substoichiometric behavior will necessarily be observed. It is always difficult or impossible to determine independently what fraction of the molecules of protein in a solution are properly folded in their native structure and what fraction are not in their native structure and are enzymatically inactive or incapable of binding a ligand, even if all of the polypeptides in the solution are covalently identical to each other. If a significant fraction of the molecules of protein are adventitiously inactive, the molar concentration of active sites or of sites for the binding of a ligand will be less than the molar concentration of protomers, even though there is one site for each properly folded polypeptide in an active molecule of the protein. For example, impure preparations of yeast glyceraldehyde-3-phosphate dehydrogenase display substoichiometric behavior that is not displayed when the enzyme is purified rapidly and more completely.[100] The substoichiometric behavior of solutions of this enzyme has been explained by recourse to an elaborate and appealing theory,[101] even though it probably arises from heterogeneity. It is remarkable how such measurements of the concentration of sites made on solutions of the enzyme, which would be expected to give various degrees of fractional saturation, nevertheless give half-of-sites behavior; the molar concentration of sites is half that of the estimated molar concentration of subunits (Figure 3–20).[101]

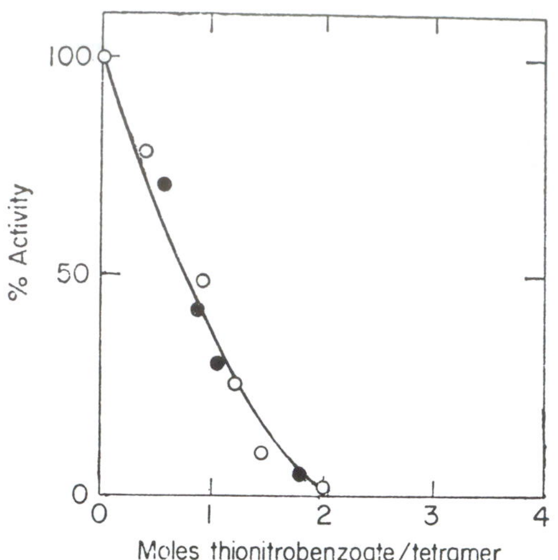


**Figure 3–20:** Inactivation of yeast glyceraldehyde-3-phosphate dehydrogenase by reaction with 5,5′-dithiobis-(2-nitrobenzoate). Purified glyceraldehyde-3-phosphate dehydrogenase was mixed with dithiothreitol at a final concentration of 1 mM for 15 min and then passed over a molecular exclusion column to remove the dithiothreitol. The final concentration of enzyme was $5 \times 10^{-6}$ M in tetramers dissolved at pH 8.5 in 2 mL. Successive additions of 5,5′-dithiobis(2-nitrobenzoate) were made. This reagent was used in the hope of modifying Cysteine 149 in the active site of the enzyme by disulfide interchange. The enzymatic activity was measured after each addition. The amount of 5-thio-2-nitrobenzoate bound covalently to the enzyme after each addition was determined either by using [$^{14}$C]-5,5′-dithiobis(2-nitrobenzoate) and submitting protein precipitated with trichloroacetic acid to scintillation counting (○), or by passing the protein over a molecular exclusion column and releasing the bound 2-carboxy-3-nitrobenzene thiolate anion with dithiothreitol and determining its concentration by absorbance at 412 nm (●). The enzymatic activity (percent) is presented as a function of the moles of thionitrobenzoate incorporated for every mole of protein. Reprinted with permission from ref 101. Copyright 1973 Academic Press.

An additional problem arises when a comparison is made between the concentration of sites and the concentration of protomers. It is not possible to measure the absolute concentration of protein in a solution with a precision of greater than about ±20%. Inaccurate measurements of protein concentration can increase the degree of the observed substoichiometric behavior. It was at one time proposed that $Na^+/K^+$-transporting ATPase displayed half-of-sites behavior until it was demonstrated that, when the concentration of protein was assessed by total amino acid analysis, solutions of the purest preparations of the enzyme contained as high as 0.85 mol of active sites for every mol of subunit.[91]

Observations that do not rely upon measurements of the molar concentration of sites or the absolute concentration of protein can often rule out the existence of fractional site behavior definitively. For example, from measurements of the concentration of phosphorylated active sites and the concentration of protein, it was initially concluded that alkaline phosphatase from *E. coli* displayed half-of-sites behavior. Phosphate is incorporated covalently into the active site of this enzyme as an intermediate in the catalytic reaction. When a phosphate anhydride or phosphate ester with an excellent leaving group, such as AMP, pyrophosphate, or nitrophenyl phosphate, is used as a reactant, the active site can be phosphorylated more rapidly than it dephosphorylates, and the phosphorylated intermediate accumulates. The initial conclusion that the enzyme displayed half-of-sites behavior was based on the observation of a plateau in the level of phosphorylation at higher values of pH, even though this plateau occurred at values significantly less than 1 (Figure 3–21A).[102] These early measurements were based on assays of radioactive phosphorus precipitated with protein by hydrochloric acid. When the relative amounts of phosphorylated and unphosphorylated subunits were measured directly by electrophoresis,[103] a similar decrease in the amount of phosphorylated subunits was observed at high pH, but no plateau was evident. The relative amount of phosphorylated subunits decreased monotonically to less than 10% of the total subunits (Figure 3–21B).[103] As this was a more direct measurement of phosphorylated active sites, it was concluded that the plateau observed in Figure 3–21A had been artifactual. Because the experimental observation supporting half-of-sites behavior was not the fact that a plateau was observed but the fact that the plateau occurred at a value of about 1 (Figure 3–21A), these later results ruled out half-of-sites behavior for alkaline phosphatase.

Both proposals, that for the existence of negative cooperativity and that for the existence of half-of-sites behavior, assume that the nonideal behavior with negative deviation or the substoichiometric behavior is an intrinsic property of the oligomer that results from interactions among the binding sites or active sites on each of its subunits. As such, both theories rely on the ability of the binding of a ligand to the binding site or a substrate to the active site on one protomer to influence the binding of the ligand or substrate to another site on a different protomer in the same oligomer. Because this property is an essential feature of these theories, it can be used as a test for their relevance to a particular example of nonlinear or substoichiometric behavior. This is done by producing hybrid oligomers. A solution is produced containing a homogeneous population of hybrid molecules in each of which half of the binding sites or half of the active sites are normal but the other half have been rendered inactive. For example, this can be done with $\alpha_2$ dimers. A homogeneous population of heterodimers is produced in which one subunit is inactive and the other subunit is normal. If the nonideal or apparently substoichiometric behavior of the normal subunit is unaffected by having a defunct subunit associated with it, then

A

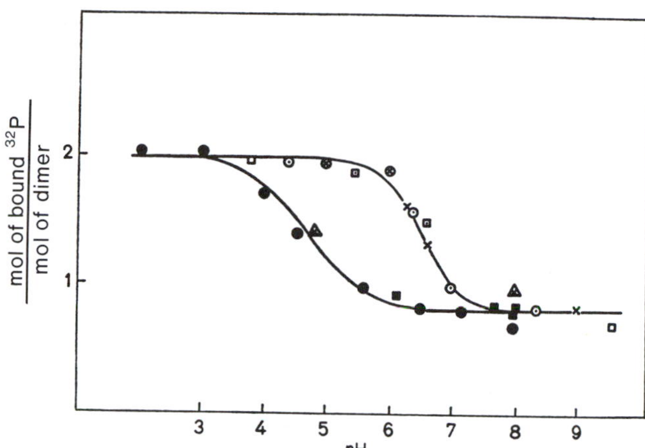

B

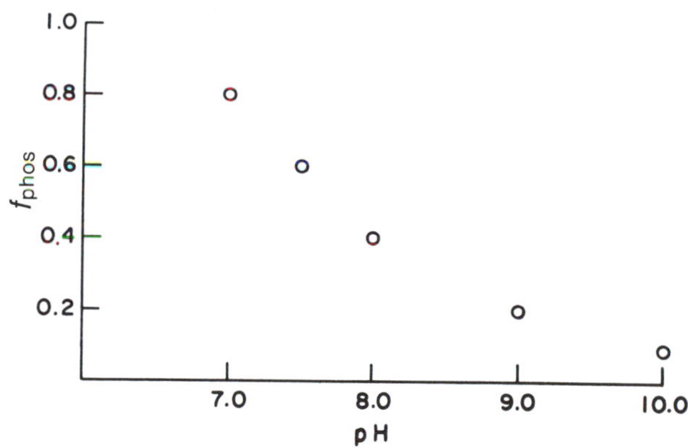


**Figure 3–21:** Dependence on pH of the phosphorylation of the active site of alkaline phosphatase from *E. coli* during its catalysis of the hydrolysis of phosphate esters and anhydrides. (A) Purified enzyme in which the active-site metal was $Zn^{2+}$(⊙, ×, ⊡) or purified enzyme in which the naturally occurring $Zn^{2+}$ had been replaced by $Co^{2+}$ (●, ■, ◬) were dissolved in solutions at the noted values of pH at 0 °C. The enzymes were phosphorylated with either [$^{32}$P]AMP (⊙, ●), [$^{32}$P]ATP (×), or [$^{32}$P]pyrophosphate (⊡, ■). After a short time (4–10 s) to allow steady state to be reached, the protein was precipitated with hydrochloric acid. The precipitate was dissolved in 8 M urea, and noncovalently bound [$^{32}$P]phosphate was removed by molecular exclusion chromatography. The amount of covalently bound [$^{32}$P]phosphate was determined by scintillation counting, and this value (moles of bound $^{32}$P) was divided by the moles of dimeric enzyme (mole of dimer) in the initial mixture. The quantity, moles of bound $^{32}$P (mole of dimer)$^{-1}$, is presented as a function of pH.[102] The release of *p*-nitrophenolate anion (◬) in a burst after addition of *p*-nitrophenyl phosphate to a solution of the enzyme, relative to the molar concentration of dimer, is also presented for comparison. Panel A adapted with permission from ref 102. Copyright 1971 Springer-Verlag. (B) Alkaline phosphatase ($4 \times 10^{-7}$ M) was mixed with various concentrations of *p*-nitrophenyl phosphate (0.1–10 mM) at a particular pH. After the 1–2 s required to reach steady state, the reaction was quenched by adding $Cl_3CCOOH$. The precipitated protein was gathered, dissolved in 4 M urea, and submitted to electrophoresis. Phosphorylated polypeptides had different mobilities from unphosphorylated polypeptides, and the amount of each form could be measured directly from the gel. The fraction ($f_{phos}$) of the total polypeptide that was phosphorylated at steady state is presented as a function of the pH of the solution.[103] Panel B adapted with permission from ref 103. Copyright 1980 American Chemical Society.

ideal or substoichiometric behavior cannot result from interactions between the two sites on the same oligomer because there can be no interactions in this situation.

A heterodimer containing one normal subunit and one mutationally inactivated subunit of alkaline phosphatase from *E. coli* was constructed.[104] The single, normal active site in the hybrid dimer displayed all of the nonideal kinetic behavior and substoichiometric behavior displayed by an active site in a native $\alpha_2$ homodimer even though it was paired with an active site that was unable to catalyze the reaction. The only difference was that all of the values for the rates of the reactions and yields of phosphorylated intermediates were half the values of those same properties for the wild type when they were based on the molar concentrations of dimer rather than on the molar concentration of active subunits. Therefore, interactions between the sites must not be responsible for the peculiarities seen with the native protein and they must be intrinsic to the individual active sites. Likewise, a heterodimer of horse liver alcohol dehydrogenase was constructed in which one of the two active sites was occupied by a tightly bound inhibitor. The uninhibited active site in the hybrid displayed the same nonideal kinetic behavior displayed by an active site in a native $\alpha_2$ homodimer.[105] Again, these results demonstrate that these properties cannot result from negative cooperativity or half-of-sites behavior. It has been pointed out that an acceptable explanation of the nonlinear kinetics observed with horse liver alcohol dehydrogenase, based on the assumption that each active site in the dimer of horse liver alcohol dehydrogenase is independent of the other, can be formulated.[106] Experiments of this type should act to discourage the tendency to explain any peculiar behavior of an oligomeric enzyme in terms of interactions among its active sites rather than in terms of the properties of each independent active site within the oligomer.

The enzyme that has been most extensively studied with regard to the substoichiometric binding of substrates is prokaryotic tyrosine-tRNA ligase. This enzyme catalyzes the reaction

$$\text{tyrosine} + \text{MgATP} + \text{tRNA}^{\text{Tyr}} \rightleftharpoons \text{tyrosyl-tRNA}^{\text{Tyr}} + \text{MgOPO}_3\text{PO}_3\text{H}_2 + \text{AMP} \qquad (3\text{–}187)$$

where $^{2-}OPO_3PO_3H_2$ is inorganic pyrophosphate. The enzyme is an $\alpha_2$ dimer ($n_{aa}$ = 420)[107] for which a crystallographic molecular model is available.[108] In the crystallographic molecular model, the two subunits are indistinguishable and each subunit is the asymmetric unit in a $P3_121$ space group. Therefore, the two subunits in a given $\alpha_2$ dimer in the crystal are related to each other by a crystallographic 2-fold rotational axis of symmetry and must be of identical molecular structure.

Nevertheless, when the binding of tyrosine in a solution of the enzyme as it is normally prepared is assessed, only about 1.0 mol of tyrosine will bind for every 840 mol of amino acids in the solution.[109] This substoichiometric binding was thought to be evidence of half-of-sites behavior.[110] Three observations, however, have contradicted this conclusion. First, two widely separated (3.2 nm), symmetrically displayed binding sites for tyrosine, which bind tyrosine equivalently, can be observed in the crystallographic molecular model of a dimer of the enzyme.[111] Second, when the crystalline enzyme is redissolved, it is able to bind 1.0 mol of tyrosine for every 520 amino acids in the solution.[111] On the basis of this increase in the concentration of sites, it was proposed that the crystallization had produced a preparation of the enzyme of higher purity than that obtained by the usual methods of purification. Third, experiments performed with hybrid heterodimers formed from two differently modified subunits of tyrosine-tRNA ligase are also consistent with the conclusion that the two active sites in a dimer of the enzyme are independent from and unrelated to each other.[112]

During the reaction catalyzed by tyrosine-tRNA ligase, tyrosyladenylate (AMP-Tyr) is formed as an intermediate. This intermediate is a mixed anhydride between the carboxylic acid of tyrosine and the phosphoric acid of adenosine monophosphate

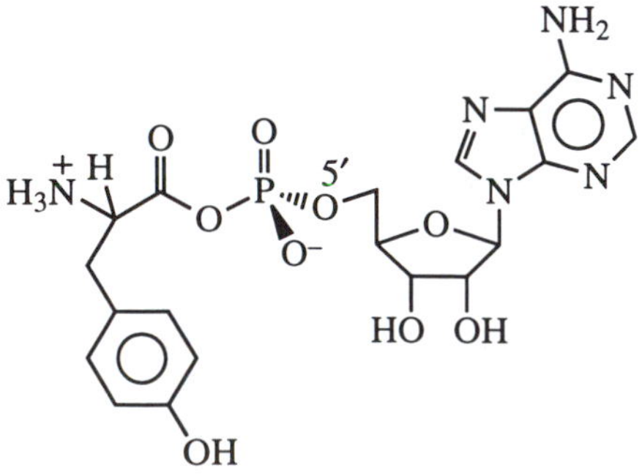

AMP-Tyr

**3–7**

In this way, the anionic oxygen of the carboxylate is replaced by an alkyl phosphate. This increases the acidity of the leaving group significantly and activates the carboxylate of the amino acid so that it can react with the alcoholic oxygen of the tRNA to produce the ester found within the product. The tyrosyladenylate is formed upon the active site of the enzyme from MgATP and tyrosine. About 1.0 mol of tyrosyladenylate is formed for every 840 amino acids in a solution of the enzyme, as it is usually prepared.[112]

The tyrosyladenylate divides the enzymatic reaction into two steps, activation and acyl transfer

$$\text{E} + \text{MgATP} + \text{Tyr} \rightleftharpoons \text{E·AMP-Tyr} + \text{MgOPO}_3\text{PO}_3\text{H}_2 \qquad (3\text{–}188)$$

$$\text{E·AMP-Tyr} + \text{HO-tRNA} \rightleftharpoons \text{E} + \text{AMP} + \text{Tyr-O-tRNA} \qquad (3\text{–}189)$$

It is possible to block the second step of the reaction by removing a segment of the polypeptide at the carboxy-terminal end of the enzyme molecule. This produces an enzyme that can catalyze the first step of the reaction (Reaction 3–188) but not the second (Reaction 3–189), and the overall reaction ceases with a tightly bound tyrosyladenylate in the active site.

A hybrid heterodimer of tyrosine-tRNA ligase can be produced in which one active site has associated with it a truncated carboxy terminus and the other active site has associated with it a normal carboxy terminus. This heterodimer, because all the active sites can still form tyrosyladenylate, displays the same substoichiometric behavior of an unmodified homodimer. Only 1.0 mol of tyrosine is bound for every 840 mol of amino acids in a solution of the heterodimer. When the complex between this heterodimer and tyrosyladenylate is produced at this stoichiometry, only half of the tightly bound tyrosyladenylate can transfer its tyrosine to subsequently added tRNA to produce Tyr-tRNA, because only half of the active sites are complete.[112] Tyrosyladenylate remains tightly bound to the other half of the active sites, making them inactive. If each heterodimer of tyrosine-tRNA ligase in the solution were able to bind only one tyrosyladenylate, half of the individual molecules of protein in the population of heterodimers would have been rendered inactive after this step.

It has been pointed out[112] that, if every heterodimer of tyrosine-tRNA ligase were able to produce only one tyrosyladenylate, the entirely unoccupied heterodimers, released after the transfer of tyrosine to $\text{tRNA}^{\text{Tyr}}$ had occurred at a competent active site, would then bind tyrosine and MgATP at random to only one of its two active sites, just as the unoccupied enzyme did the first time around. Therefore, after the second round, half of these heterodimers would end up with tyrosyladenylate tightly bound to the crippled active sites and they would become inactive. Therefore, if tyrosine-tRNA ligase were to display half-of-sites behavior, half of the remaining active enzyme would have to be inactivated after each cycle because it would end up with tyrosyladenylate bound to a crippled active site and be unable to produce tyrosyladenylate at its normal active site. If this were the case, then a solution of these heterodimers should be incapable of sustained production of tyrosyl-tRNA. Because a solution of these heterodimers, however, is capable of sustained production of tyrosyl-tRNA at only half the rate of the unaltered enzyme, tyrosine-tRNA ligase cannot display half-of-sites behavior.

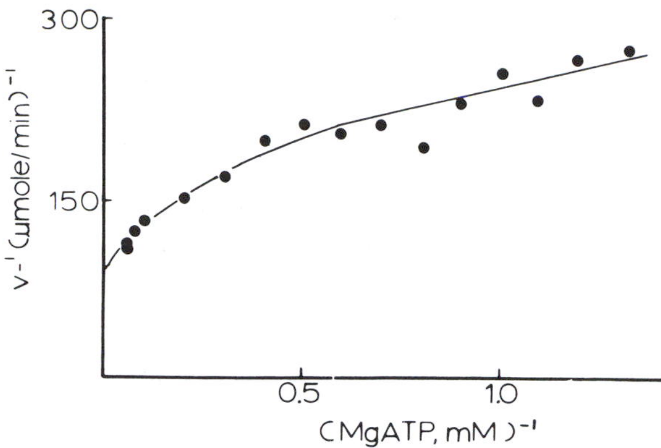


**Figure 3–22:** Double-reciprocal plot of the initial velocity of the phosphorylation of glucose catalyzed by yeast hexokinase.[113] Solutions containing MgATP at various concentration were prepared at pH 6.75 and 25 °C. The concentration of uncomplexed $Mg^{2+}$ cation was fixed at 8 mM and the assay mixture also contained 15 $\mu$M citrate. The concentration of glucose was 20 mM. The reaction was initiated by adding enzyme (to a final concentration of 42 ng $mL^{-1}$) to each solution. Initial velocity was followed by coupling the production of glucose 6-phosphate to the reduction of $NAD^+$ with glucose-6-phosphate dehydrogenase, and the increase in NADH was monitored continuously by its absorbance at 340 nm. Following an approach to steady state, lasting about 2 min, the production of glucose 6-phosphate reached a steady rate that was considered to be its initial velocity. The reciprocal of that initial velocity (minutes $micromole^{-1}$) is plotted as a function of the reciprocal of the concentration of MgATP ($millimolar^{-1}$). Note that the behavior is that expected for a negative deviation from linear behavior (Figure 3–18). Adapted with permission from ref 113. Copyright 1975 *Journal of Biological Chemistry*.

This conclusion is reinforced by other experiments with more complicated hybrids of tyrosine-tRNA ligase.[112] When Histidine 45, an amino acid within the active site, is modified, the rate at which tyrosyladenylate is produced (Reaction 3–188) decreases by a factor of $10^4$, while the rate of transfer of tyrosine from the tyrosyladenylate to tRNA (Reaction 3–189) is unaffected.[112] Four different active sites can be produced: the native active site, an active site in which Histidine 45 has been modified, an active site in which the carboxy terminus has been deleted, and an active site in which both Histidine 45 has been modified and the carboxy terminus has been deleted. When homodimers containing the same two active sites or heterodimers containing two different active sites were produced, all of the numerical values for the enzymatic activities and the rates of the partial reactions that were measured[112] were consistent with the conclusion that the two active sites in these dimers were completely independent of each other. Again, the most reasonable conclusion is that tyrosine-tRNA ligase does not display half-of-sites behavior. It is unclear why the enzyme as it is usually prepared can bind only one tyrosine or one tyrosyladenylate for every 840 amino acids in the solution.

There are at least two oligomeric enzymes that seem to display intrinsic nonideal behavior resulting from the fact that the active sites are gathered into an oligomer. Yeast hexokinase displays negative deviations from ideal behavior in its steady-state kinetics (Figure 3–22).[113] This behavior may arise from the fact that under these circumstances the enzyme is a dimer, rather than a monomer, and the dimer is composed of two subunits arrayed about a screw axis rather than a rotational axis of symmetry.[113a] In the dimer, the two active sites are structurally distinct and therefore could have different affinities for substrates.[114] This educational example, however, is an isolated exception to the rule that oligomeric enzymes are built around rotational axes of symmetry. This property causes each subunit in the usual oligomeric enzyme to be the same as all of the other subunits in the same oligomer and, by extension, each active site to be indistinguishable from every other active site.

Glyceraldehyde-3-phosphate dehydrogenase from *Bacillus stearothermophilus* is an $\alpha_4$ dimer of dimers. Crystallographic molecular models are available for both the unliganded enzyme and the enzyme to which four $NAD^+$ are bound.[115,116] The structures of the subunits in these two respective forms differ from each other by a 4° rigid-body rotation of two hinged domains relative to each other. This movement opens and closes the active site. In neither crystal are any of the three orthogonal 2-fold rotational axes of symmetry that run through the $\alpha_4$ tetramer crystallographic axes of symmetry, and the four subunits within each of these crystallographic molecular models, respectively, do not have identical structures. Nevertheless, the differences in structure among the subunits in a given $\alpha_4$ tetramer appear to result entirely from asymmetry induced by crystal packing. In solution, the protein in either the unliganded or the fully liganded form is presumed to be constructed of four identical subunits related by three orthogonal 2-fold rotational axes of symmetry.

In spite of this presumption, solutions of the enzyme display nonideal behavior with negative deviation in the binding of $NAD^+$.[117] The first two moles of $NAD^+$ for every mole of the $\alpha_4$ tetramer are bound very tightly ($K_{d1}^{NAD^+} < 0.1\ \mu$M), while the second two molecules of $NAD^+$ for every mole of the $\alpha_4$ tetramer are bound much less tightly ($K_d^{NAD^+} = 2\ \mu$M).

It has been possible to crystallize the enzyme with only one of its four active sites occupied by $NAD^+$.[115] In the crystallographic molecular model, the subunit to which the $NAD^+$ has been bound has assumed a structure indistinguishable from that of the subunits in the fully liganded enzyme, while each of the three unoccupied subunits resembles a subunit in the unliganded enzyme. The fact that the liganded subunit has assumed a different conformation from that of the other three, however, cannot explain the negative deviations from linear behavior because this outcome should produce either a positive deviation or no deviation from ideal behavior. There are, however, structural changes that occur in the neighboring subunit containing the unoccupied active site closest to the occupied active site. These small changes may interfere sterically with the binding of $NAD^+$ to this adjacent active site and explain the nonideal behavior of the binding of $NAD^+$ to the enzyme.

The results from the crystallographic molecular model of glyceraldehyde-3-phosphate dehydrogenase from *B. stearothermophilus* suggest that some of the negative deviations from ideal behavior encountered with oligomeric enzymes may be nothing more than steric effects encountered when several successive associations occur on the same molecule. Because $NAD^+$ is such a large substrate, it would not be surprising if the occupation of one active site would produce steric effects in an adjacent subunit that would decrease the access to the active site in that subunit.

Unlike the situation of negative deviations from ideal behavior, it is not possible to explain positive deviations from ideal behavior as the result of heterogeneity in the population of the protein in the solution. Satisfactory explanations of such positive behavior are usually but not always based on interactions among the subunits in an oligomeric protein and global conformational changes that spread over the whole oligomer.

## *Suggested Reading*

Bloch, W., & Schlesinger, M.J. (1974) Kinetics of Substrate Hydrolysis by Molecular Variants of *Escherichia coli* Alkaline Phosphatase, *J. Biol. Chem. 249*, 1760–1768.

## PROBLEM 3–14

(A) Show that, for the mechanism

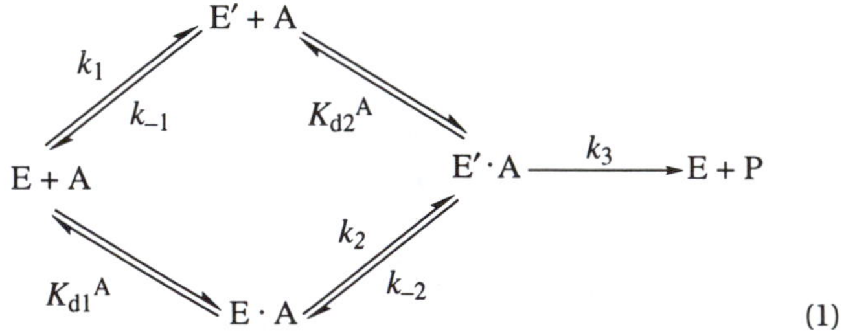


(1)

where E and E′ are two conformations of the enzyme and $K_{d1}{}^A$ and $K_{d2}{}^A$ are dissociation constants

$$\frac{v_0}{[E]_{TOT}} = \frac{k_3(k_1 K_{d1}{}^A + k_2[A]_0)[A]_0}{(k_{-1}+k_1)K_{d1}{}^A K_{d2}{}^A + [(k_{-1}+k_2)K_{d2}{}^A + (k_1+k_{-2}+k_3)K_{d1}{}^A][A]_0 + (k_2+k_{-2}+k_3)([A]_0)^2} \quad (2)$$

The two limits are

$$\lim_{[A]\to\infty} \frac{v_0}{[E]_{TOT}} = \frac{k_3 k_2}{k_2 + k_{-2} + k_3} \quad (3)$$

and

$$\lim_{[A]\to 0} \frac{v_0}{[E]_{TOT}} = \frac{k_3 k_1 [A]_0}{(k_{-1}+k_1)K_{d2}{}^A} \quad (4)$$

(B) What would be the equation for a rectangular hyperbola

$$\frac{v_0}{[E]_{TOT}} = \frac{k_{cat}[A]_0}{K_m + [A]_0} \quad (5)$$

that had the same two limits? In other words, what would be the expressions for $k_{cat}$ and $K_m$ in terms of the rate constants in Equations 3 and 4?

(C) The four equilibria connecting E + A and E′·A in the mechanism are the sides of a linkage box. Write an equation relating $K_{d1}{}^A$, $K_{d2}{}^A$, $k_1$, $k_{-1}$, $k_2$, and $k_{-2}$ that results from the fact that the free energy between these two points is independent of path.

Suppose that $K_{d1}{}^A$ = 100 $\mu$M, $K_{d2}{}^A$ = 1 $\mu$M, $k_1$ = 100 $s^{-1}$, $k_{-1}$ = 1 $s^{-1}$, $k_2$ = 1000 $s^{-1}$, $k_{-2}$ = 0.1 $s^{-1}$, $k_3$ = 100 $s^{-1}$.

(D) Show that these values satisfy the linkage relationship of part C.

(E) At low concentrations of reactant A ($[A]_0 < 1$ $\mu$M), which path around the box will dominate and which are the rate-determining steps? At high concentrations of reactant ($[A]_0 > 100$ $\mu$M), which path around the box will dominate and what is the rate-limiting step?

(F) Make a table containing values for $v_0/[E]_{TOT}$ calculated from Equation 5 for values of $[A]_0$ of 0.01, 0.02, 0.03, 0.10, 1.0, 10, 100, and 1000 $\mu$M. Don't forget the units. Why is the behavior described by Equation 2 negative deviation from ideal behavior?

# Contributions of Binding to Catalysis

During a simple one-step chemical reaction or during one of the discrete steps of a multistep reaction, the nuclei of the atoms involved in the reaction change their mean positions relative to each other. The relative positions of the atomic nuclei in the reactants are different from their relative positions in the products. If there are $n$ nuclei in the reactants, and therefore $n$ nuclei in the products, $3n - 6$ coordinates are required to define the relative positions of these nuclei in space when they are close enough together to participate in the reaction. Assigning a numerical value to each of these $3n - 6$ coordinates defines a relative distribution of the nuclei in space, and the associated electrons automatically assume a distribution in response to that distribution of nuclei. This particular distribution of nuclei and electrons has a specific potential energy, $U$, associated with it. The atomic coordinates and the associated potential energies together create a **potential energy surface** in $(3n - 5)$-dimensional space.

The initial state in a unimolecular reaction in which a bond will be formed between two atoms in the reactant is any conformation of a molecule of the reactant in which the two atoms to be joined are separated sufficiently that they do not affect their respective behavior significantly but are approaching each other on a trajectory that will lead to intramolecular collision and formation of the bond. The final state in such a reaction is the product containing the new bond in its ground state. The initial state in an intramolecular reaction in which a bond will be broken is the ground state of the molecule containing the susceptible bond. The final state in such a reaction is the molecules of product in which the two atoms previously bonded have separated from each other sufficiently so that they no longer affect each other's behavior significantly. The initial state in a multimolecular reaction is the separated molecules of the reactants, too distant from each other to affect their individual free energies but on trajectories that will lead to a productive collision. The final state is the separated molecules of product departing from the productive collision. In any given reaction, the initial state and the final state can be assigned to regions on the potential energy surface describing that reaction.

Any process that converts reactants into products must follow a path along the potential energy surface between the region occupied by the reactants and the region occupied by the products. Each of these paths must pass through a point that has the highest potential energy for that particular path. Within the set of the infinite number of paths between reactant and product, there will be a saddlepoint. The **saddlepoint** is the highest point on a subset of these paths that nevertheless has a lower potential energy than the highest points on all of the other paths. This saddlepoint is analogous to the pass of lowest altitude through a range of mountains separating two geographical points. Because this saddlepoint represents the lowest barrier of potential energy between reactants and products, a chemical reaction connecting reactants and products proceeds on a path through this saddlepoint. If a local minimum of potential energy is encountered along the paths leading through a saddlepoint so that they all have two maxima of potential energy, the reaction proceeds through an **intermediate**. In this instance, the reaction is divided into two steps, and each step is considered as a separate reaction. Consequently, a path through the saddlepoint between reactants and products on the potential energy surface for each step has only one maximum of potential energy, which is the saddlepoint itself. A **reaction coordinate** is a path through this saddlepoint.

A reaction coordinate is presented figuratively (Figure 3-23)[118] as a curve that begins on a plateau whose potential energy is that of the separated reactants and that ends on a plateau whose potential energy is that of the separated products. The abscissa is the distance along the reaction coordinate over the potential energy surface and the ordinate is the potential energy associated with a given distance along the reaction coordinate. Each point on the reaction coordinate defines a distribution of atomic nuclei in space. As the reaction progresses along the reaction coordinate, the nuclei participating in the reaction are changing their relative positions as reactants continuously are transformed into products.

The relative positions of the nuclei in space at the point of highest potential energy, the saddlepoint, is the **transition**

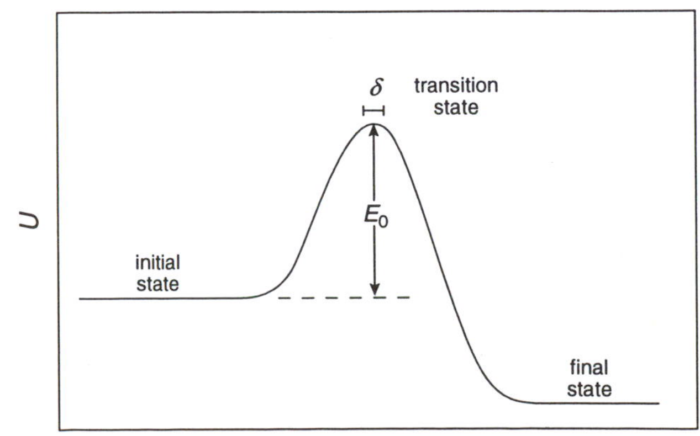


**Figure 3–23:** Variation in potential energy, $U$, as function of the distance along a reaction coordinate.[118] The plateaus on the two sides are the potential energies of separated reactants and separated products. The potential energy at the peak is the potential energy of the transition state. The width, $\delta$, is the portion of the reaction coordinate occupied by the transition state. The height, $E_0$, is the difference in potential energy between reactants and transition state.

**state**. The transition state, like any stable molecule, is a collection of nuclei distributed in space in association with a particular number of electrons, which are confined to the atomic and molecular orbitals dictated by that distribution of nuclei. The only difference between a transition state and either the reactants or the products is that the transition state is directly unstable with respect to both and is immediately transformed with equal probability into either.

For reactants to be transformed into products, sufficient energy must be available for them to reach the potential energy of the transition state. This energy is provided by the kinetic energy of the collision in a multimolecular reaction or by fluctuations in internal energy in a unimolecular reaction, but any other source of energy is as acceptable.

According to absolute rate theory,[118,119] the rate constant of a chemical reaction, $k_r$, is

$$k_r = \kappa_T \frac{k_B T}{h} K^{\ddagger\prime} \quad (3\text{–}190)$$

where $\kappa_T$ is the transmission coefficient and $k_B$ is Boltzmann's constant. The transmission coefficient is the fraction of the transition states successfully reached by the reactants that decompose into products, and it is thought to be close to unity under normal circumstances.[119] The constant $K^{\ddagger\prime}$ is defined as

$$K^{\ddagger\prime} = \frac{Q_\ddagger^{0\prime}}{Q_A^0 \, Q_B^0 \ldots} e^{-E_0/RT} \quad (3\text{–}191)$$

where $Q_\ddagger^{0\prime}$ is the partition function of the transition state calculated from all degrees of freedom except motion along the reaction coordinate, $Q_i^0$ is the partition function of reactant $i$, and $E_0$ is the difference in potential energy between the reactants and the transition state (Figure 3–23). The zeroes as superscripts denote partition functions for the same standard volume.

Although $K^{\ddagger\prime}$ is formally determined only by partition functions and potential energies, it is mathematically indistinguishable from the statistical mechanical expression for an equilibrium constant, and it is possible to define a **standard free energy of activation** $\Delta G^{\circ\ddagger\prime}$ by the equation

$$\Delta G^{\circ\ddagger\prime} \equiv -RT \ln K^{\ddagger\prime} = -RT \ln\left(\frac{hk_r}{\kappa_T k_B T}\right) \quad (3\text{–}192)$$

It should be noted that this equation is a definition of $\Delta G^{\circ\ddagger\prime}$ and tabulations of $\Delta G^{\circ\ddagger\prime}$ are nothing more than logarithmic tabulations of particular rate constants, usually with the assumption that $\kappa_T = 1.0$.

In the absolute rate theory, $Q_\ddagger^{0\prime}$ is the quotient of the actual partition function for the transition state, $Q_\ddagger^0$, and the factor $(2\pi m^\ddagger k_B T)^{1/2}\delta h^{-1}$, where $m^\ddagger$ is the mass of the transition state and $\delta$ is the distance along the reaction coordinate occupied by the transition state (Figure 3–23). This factor is the partition function for translation of the transition state along the reaction coordinate. Therefore,

$$k_r = \frac{\kappa_T}{\delta}\left(\frac{k_B T}{2\pi m^\ddagger}\right)^{1/2}\left[\frac{Q_\ddagger^0}{Q_A^0 \, Q_B^0 \ldots}\right] e^{-E_0/RT} \quad (3\text{–}193)$$

The same reaction can be performed at the same temperature but under two separate situations. For example, it could be performed in solution and on the active site of an enzyme, or on the respective active sites of the same enzymes from two different species, or on the wild type and a mutant form of the same enzyme. If it is assumed that $\kappa_T$, $\delta$, and $m^\ddagger$ are the same in the two situations, then, because

$$\mu_i{}^\circ = -RT \ln Q_i{}^0 \quad (3\text{–}194)$$

where $\mu_i^\circ$ is the standard chemical potential of species $i$

$$-RT \ln\left(\frac{k_1}{k_2}\right) = \Delta E_0 + \left(\mu_{\ddagger 1}{}^\circ - \mu_{\ddagger 2}{}^\circ\right) - \left(\mu_{A1}{}^\circ - \mu_{A2}{}^\circ\right) - \left(\mu_{B1}{}^\circ - \mu_{B2}{}^\circ\right)\ldots \quad (3\text{–}195)$$

and

$$-RT \ln\left(\frac{k_1}{k_2}\right) = \Delta\Delta G^{\circ\ddagger} \quad (3\text{–}196)$$

In Equation 3–196 the difference in standard free energies of activation, $\Delta\Delta G^{\circ\ddagger}$, is the difference in the actual standard free energies of activation for the reactions under the two circumstances. The actual standard free energies of activation themselves, $\Delta G_i^{\circ\ddagger}$, are the differences between the standard chemical potentials of the reactions and the chemical potentials of the respective transition states. Therefore, any difference between the two situations that increases the standard chemical potential of the reactants or decreases the standard chemical potential of the transition state or does both will cause the reaction to have a greater rate in the second situation than in the first. The observed difference in rate between the two situations can be used to calculate the numerical value of the change in the standard chemical potential of the reactants or the standard chemical potential of the transition state or the sum of these two differences. The three assumptions made in deriving Equation 3–196, however, should not be forgotten.

One of the more remarkable advantages of enzymatically catalyzed reactions is the increase in rate accomplished by the enzyme. Comparisons of enzymatically catalyzed rates of reaction and uncatalyzed rates of reaction are usually made with the turnover number of the enzyme, $V_{max}/[\mathrm{E}]_{TOT}$, and the rate constant for the same reaction in aqueous solution at the same temperature. The turnover number, or $k_{cat}$, of an enzymatic reaction is the rate constant for that reaction as it occurs on the active site after all reactants have bound. The rates of association of the reactants with the active site have been eliminated from consideration by adding saturating concentrations of all reactions, as indicated by the choice of $V_{max}$ in the definition. Therefore, a turnover number is always the rate constant for a unimolecular reaction.

If the reaction of interest is a bimolecular or multimolecular reaction, as almost all enzymatic reactions are, the **standard entropy of approximation** realized as the reactants bind to the active site (see Appendix, pp. 251–255) decreases their intrinsic standard entropy and brings their standard chemical potential closer to the standard chemical potential of the transition state.[120] As a result, it is misleading to compare the turnover number of an enzymatic reaction and the rate constant of the respective intermolecular reaction because the change in standard entropy of approximation has already occurred in the former instance. If the comparison were to be a valid one, the change in standard free energy attributable to this standard entropy of approximation would have to be accounted for to obtain the actual difference in the standard free energy of activation, $\Delta\Delta G^{\circ\ddagger}$, from the apparent difference in standard free energy of activation $\Delta G^{\circ\ddagger}_{\text{app}}$

$$-RT\ln\left(\frac{k_{\text{cat}}}{k_{\text{uncat}}}\right) = \Delta\Delta G^{\circ\ddagger}_{\text{app}} \tag{3–197}$$

$$\Delta\Delta G^{\circ\ddagger} - \Delta\Delta G^{\circ\ddagger}_{\text{approx}} = \Delta\Delta G^{\circ\ddagger}_{\text{app}} \tag{3–198}$$

$$\Delta\Delta G^{\circ\ddagger} = \Delta\Delta G^{\circ\ddagger}_{\text{app}} - T\Delta S^{\circ\ddagger}_{\text{approx}} \tag{3–199}$$

From the discussion presented in the Appendix (pp. 251–255), it can be seen that this can be a large correction. Even in simple organic reactions involving no enzymes, accelerations of $10^6$ N (units of mole fraction) have been observed when bimolecular reactions are converted to intramolecular reactions. The theoretical acceleration expected from simply bringing the two reactants in a bimolecular reaction together on an active site and holding them tightly in the proper orientation is a factor of $10^9$ N if the rate of the bimolecular reaction is expressed in units of (mole fraction)$^{-1}$ second$^{-1}$ and the standard entropy of approximation is $-170$ J K$^{-1}$ mol$^{-1}$ based on standard states of one mole fraction. If it is further considered that the active sites of enzymes also gather together catalytic amino acids that replace the acids and bases in an aqueous solution and that standard entropies of approximation should apply also to these catalysts, the issue of how enzymes can realize such large increases in rate seems moot. The question seems rather to be why the rate increases are not larger.

It is during the binding of the reactants to the active site that the standard entropy of approximation is overcome. This decrease in the standard entropy of the reactants caused by the disappearance of their respective translational and rotational degrees of freedom relative to each other represents an increase in intrinsic standard free energy that is paid for by favorable noncovalent interactions between each of the reactants and the active site. The consequent elimination of the standard entropy of approximation brings the standard chemical potentials of the reactants closer to the standard chemical potential of the transition state. For such a strategy to work, the associations between the enzyme and the reactants must be so fast that they do not become rate-limiting, even though a considerable decrease in translational and rotational entropy occurs during these associations. The association rate constants between active sites and reactants are quite large [$(0.01–10) \times 10^8$ M$^{-1}$s$^{-1}$],[119] and approach values expected for the diffusion-controlled limit.[121] Such rapid rates of association are actually more remarkable than the enhanced rates of the chemical reactions once the substrates are bound.

These considerations can be summarized diagrammatically (Figure 3–24). In the absence of the enzyme, reactant A collides with reactant B to produce a transition state of high standard free energy relative to the standard free energy of the reactants. The standard chemical potential of the transition state is higher than the standard chemical potential of the separated reactants by the standard free energy of activation of the uncatalyzed reaction, $\Delta G_u^{\circ\ddagger}$. If the enzyme were catalyzing the reaction only by approximation, the two steps encompassing the association of the reactants, both proceeding rapidly with small standard free energies of activation,

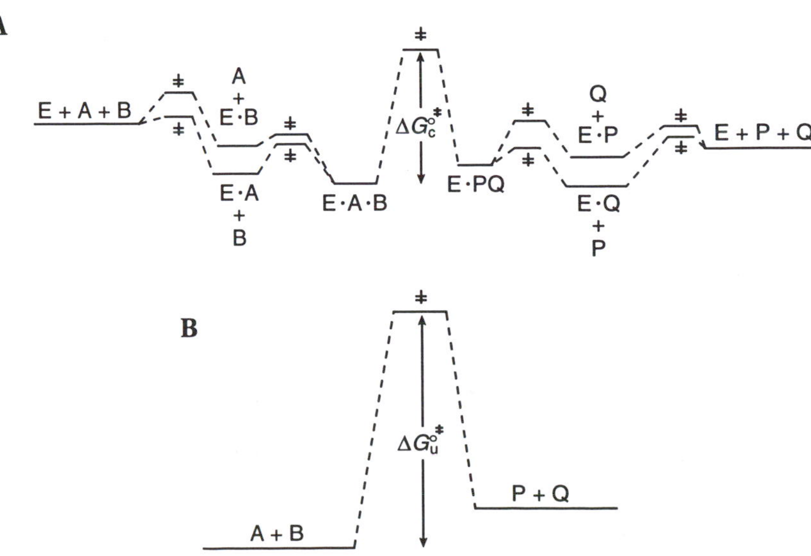


**Figure 3–24:** Diagrammatic representations of the difference in standard free energy of activation between a simple single-step reaction (B) and the steps involved when that reaction occurs at the active site of an enzyme. (A) The steps involved in the reaction at the active site of the enzyme are the binding of reactants A and B to form the ternary complex, E·A·B, the conversion ($\Delta G_c^{\circ\ddagger}$) of the complex with the reactants into the ternary complex of products and the enzyme, E·P·Q, and the dissociation of the products. (B) In the respective bimolecular reaction uncatalyzed by the enzyme, the reactants pass through only one transition state ($\Delta G_u^{\circ\ddagger}$) rather than the five transition states for the enzymatically catalyzed reaction.

would bring the reactants into juxtaposition and proper alignment. The standard free energy of activation for the chemical transformation on the active site of the enzyme, $\Delta G_c^{\circ\ddagger}$, would be decreased relative to that of the bimolecular reaction because of the consequent elimination of the standard entropy of approximation

$$\Delta G_c^{\circ\ddagger} = \Delta G_u^{\circ\ddagger} + T\Delta S_{approx}^{\circ\ddagger} \qquad (3\text{–}200)$$

This decrease in the standard free energy of activation ($\Delta S_{approx}^{\circ\ddagger}$ is always negative) results from the fact that the standard chemical potential of the reactants has increased. This increase results from the fact that a certain fraction of the standard free energy of noncovalent interaction between reactant and active site has been expended to compensate for the loss of the translational and rotational entropy of the free reactants.

When an enzyme catalyzes a unimolecular reaction, differences in standard entropy of approximation should also be important. In this case, the active site of the enzyme, by binding the reactant in the conformation in which the reactive locations are juxtaposed, eliminates the entropy of internal rotations that decreases the probability that these locations will collide when the molecule is free in solution. For example, for chorismate to be converted into prephenate (Reaction 1–99), the distal carbon of the olefin must collide with the proper carbon of the cyclohexadiene (Figure 1–16). Between these two locations are two bonds about which full rotation can occur. If these were unhindered, the standard entropy of internal rotation should be about +40 J $K^{-1}$ $mol^{-1}$. The elimination of this amount of entropy of internal rotation would translate into a modest increase in the rate of the reaction of 100-fold.

It is also necessary, however, for the cyclohexadiene of chorismate to assume the conformation placing the substituent containing the exocyclic olefin in an axial orientation.[122] The equilibrium constant[123] for the isomerization between the equatorial and axial conformers at this location is 0.14 at 25 °C in aqueous solution. If the axial conformer is the only one that can fit into the active site, then an apparent acceleration of 10-fold should be observed when the turnover number of the enzyme is compared to the rate constant of the uncatalyzed reaction. The turnover number of the enzymatically catalyzed reaction is $2 \times 10^6$ times that of the nonenzymatic reaction in water at pH 7.5 and 37 °C. Together, the freezing of internal rotations and the enforcement of the axial conformation should only provide an acceleration of $10^3$, and a factor of only $10^3$ remains unaccounted for.

If, during the binding of the reactants to an enzyme, they not only are oriented properly to achieve the configuration of the transition state but also are physically contorted into that configuration, catalysis by **strain** results. In this situation, an additional fraction of the standard free energy of association is used to provide the energy necessary to contort the reactants toward the transition state. This strategy was originally defined in detail by Pauling in 1948 in the following credo:[124]

> I believe that an enzyme has a structure closely similar to that found for antibodies, but with one important difference, namely, that the surface configuration of the enzyme is not so closely contemporary [*sic*] to its specific substrate as is that of an antibody to its homologous antigen, but is instead complementary to an unstable molecule with only transient existence—namely, the "activated complex" for the reaction that is catalyzed by the enzyme. The mode of action of an enzyme would then be the following: the enzyme would show a small power of attraction for the substrate molecule or molecules, which would become attached to it in its active surface region. This substrate molecule, or these molecules, would then be strained by the forces of attraction to the enzyme, which would tend to deform it into the configuration of the activated complex, for which the power of attraction by the enzyme is the greatest. The activated complex would then, under the influence of ordinary thermal agitation, either reassume the configuration corresponding to the reactants, or assume the configuration corresponding to the products. The assumption made above that the enzyme has a configuration complementary to the activated complex, and accordingly has the strongest power of attraction for the activated complex, means that the activation energy for the reaction is less in the presence of the enzyme that in its absence, and accordingly that the reaction would be speeded up by the enzyme. My colleague Professor Carl Niemann and I are carrying out experiments on inhibition of enzyme activity designed to test this postulate, by the search for inhibitors that have a greater power of combination with the enzyme than have the substrate molecules themselves.

The crystallographic molecular model of lysozyme has provided experimental evidence to support this hypothesis. Lysozyme catalyzes the hydrolysis of the hexasaccharide ($\beta$1,4)hexa-*N*-acetylglucosaminopyranose (Figure 3–25). Within the hexasaccharide, the cleavage occurs between the fourth and the fifth *N*-acetylglucosamine.[125] When lysozyme was crystallized with ($\beta$1,4)tri-*N*-acetylglucosaminopyranose occupying the active site, a crystallographic molecular model incorporating reasonable hydrogen bonds and hydrophobic interactions could be constructed for the complex between the trisaccharide and the enzyme from the resulting map of electron density.[126] When three additional skeletal models of the proper monosaccharides were added to the reducing end of the trisaccharide in the molecular model, these last three monosaccharides could be placed in the active site and linked in a similar array of hydrogen bonds and hydrophobic interactions only if the skeletal model of the hexa-*N*-acetylglucosaminopyranose was contorted from the all-chair conformation. The contortion required to fit the skeletal models of the last two monosaccharides into the skeletal model of the active site required a flattening of monosaccharide D into the sofa configuration (Figure 3–26)[126,127] in which carbon 5, oxygen 5, carbon 1, and carbon 2 of monosaccharide D were all in the same plane

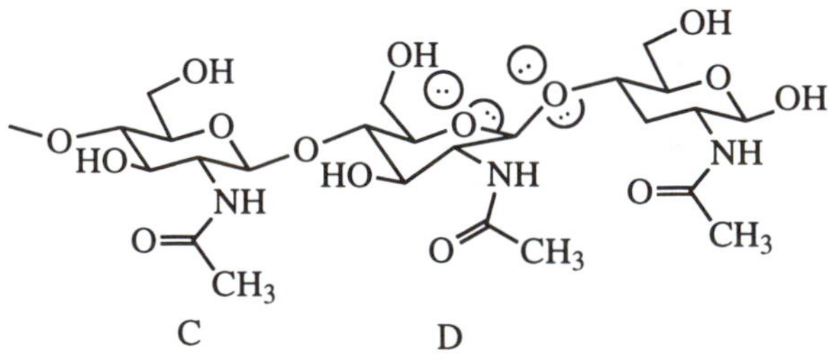

C D

**3–8**

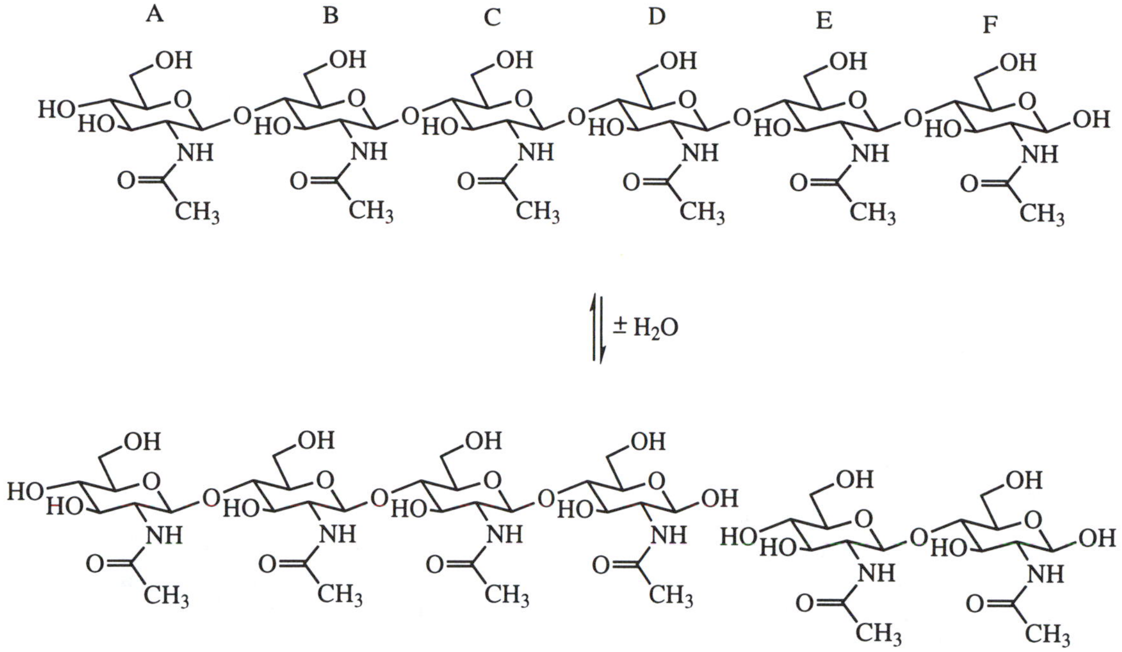

**Figure 3–25:** Hydrolysis of ($\beta$1,4)hexa-*N*-acetylglucosaminopyranose to a molecule of ($\beta$1,4)di-*N*-acetylglucosaminopyranose and a molecule of ($\beta$1,4)tetra-*N*-acetylglucosaminopyranose.

**Figure 3–26:** Molecular model of ($\beta$1,4)hexa-*N*-acetylglucosaminopyranose built into the crystallographic molecular model of the active site of lysozyme.[127] A crystallographic molecular model of lysozyme was constructed from a map of electron density calculated from a data set and phases from multiple isomorphous replacement. The location at which ($\beta$1,4)tri-*N*-acetylglucosaminopyranose was bound by the enzyme was determined from a difference map of electron density calculated from the initial data set and the data set for crystals that had been soaked in the trisaccharide. A molecular model of the trisaccharide was then built into this location in the crystallographic molecular model by making reasonable hydrogen bonds and van der Waals contacts. The model was then extended by hand to complete a model of the hexasaccharide bound in the active site. The hexasaccharide is represented by the heavy lines; the amino acids surrounding the active site in the crystallographic molecular model are also included. Adapted with permission from ref 127. Copyright 1974 Academic Press.

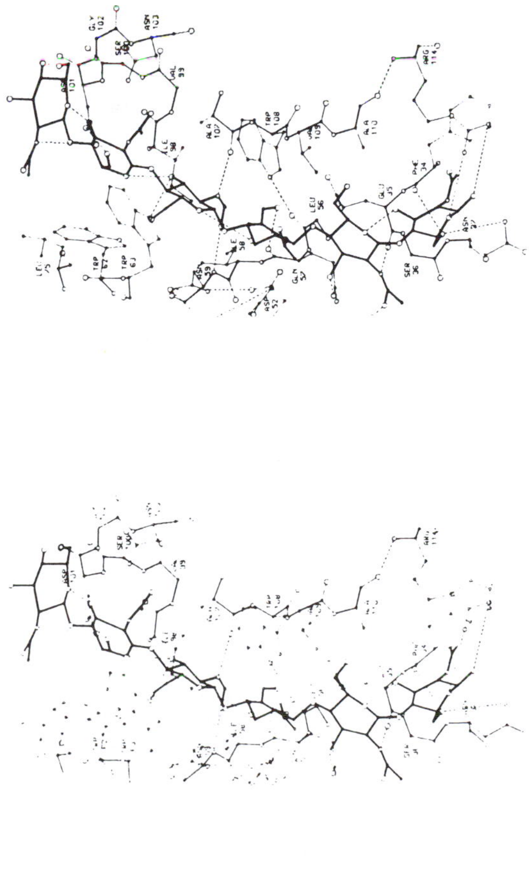

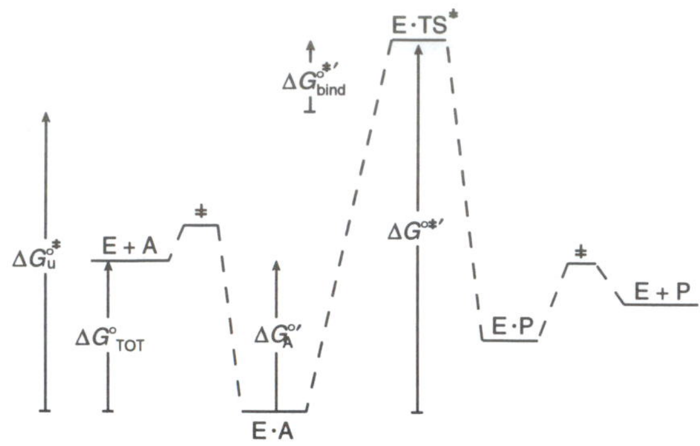


enyme that binds reactant efficiently

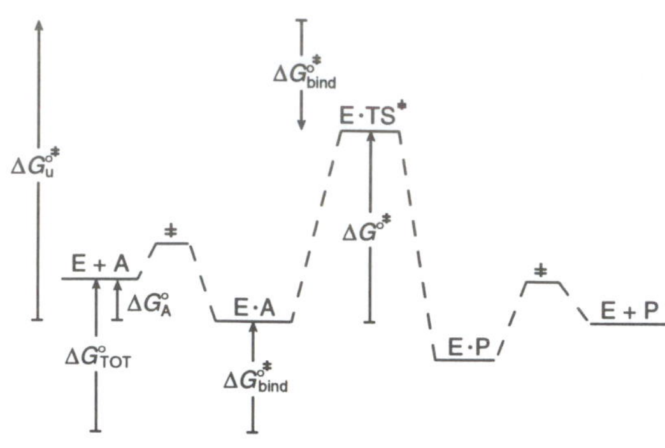


enyme that binds transition state efficiently

**Figure 3–27:** Diagrammatic representation of the effect of binding the transition state more efficiently than the ground state on the standard free energy of activation. (A) An enzyme whose active site binds the reactant most efficiently will have a high affinity ($\Delta G_A{}^{\circ\prime} >> 0$) for the reactant, but it will bind the transition state poorly because noncovalent interactions are lost ($\Delta G^{\circ\ddagger\prime}_{bind} > 0$) as the reactant is transformed to the transition state. The lowering of the standard free energy of the E·A complex and the raising of the energy of the E·TS$^{\ddagger}$ complex causes the standard free energy of activation ($\Delta G^{\circ\ddagger\prime} >> 0$) to be larger than the standard free energy of activation of the uncatalyzed reaction ($\Delta G_u{}^{\circ\ddagger}$), corrected for standard entropy of approximation. (B) An enzyme whose active site binds the transition state most efficiently will have a low affinity ($\Delta G_A{}^{\circ} > 0$) for the reactant, but it will bind the transition state with high affinity because noncovalent interactions are gained ($\Delta G^{\circ\ddagger}_{bind} < 0$)as the reactant is transformed to the transition state. The raising of the standard free energy of the E·A complex and the lowering of the standard free energy of the E·TS$^{\ddagger}$ complex causes the standard free energy of activation ($\Delta G^{\circ\ddagger} > 0$) to be smaller than the standard free energy of activation of the uncatalyzed reaction ($\Delta G_u{}^{\circ\ddagger}$) corrected for the standard entropy of approximation.

On the basis of this model-building study, the crystallographers concluded that one way lysozyme catalyzes the hydrolysis is to use the energy of its association to contort the hexasaccharide into a form that resembles the planar oxonium cation that is the intermediate in this acetal hydrolysis.

**3–9**

Such an oxonium cation is an obligatory intermediate of high energy in the hydrolysis of all acetals and ketals. By the Hammond postulate,[128] the transition state between a reactant and an intermediate of high energy should resemble the intermediate more closely than the reactant. Therefore, the conclusion was reached that lysozyme was catalyzing the reaction by contorting the substrate into the transition state.

This process can be viewed dynamically.[124] As the nuclei around the bond to be broken move in the direction of the configuration they must assume in the transition state, the hexa-*N*-acetylglucosamine fits the active site more effectively, and the free energy of formation for the noncovalent forces decreases accordingly. As the nuclei move away from the configuration they must assume in the transition state, the hexa-*N*-acetylglucosamine fits the active site less effectively, and the standard free energy of formation for the noncovalent forces increases accordingly. Because the standard free energy of formation for the noncovalent forces is decreasing continuously as the reactant moves along the reaction coordinate toward the transition state, the standard free energy of activation is less than it would have been if the reaction were occurring in free solution, and the rate of the reaction is greater. The consequence of this process is that the standard chemical potential of the transition state is decreased by the additional standard free energy of noncovalent interaction realized along the trajectory.

These effects can be represented on free energy diagrams (Figure 3–27).[129] It is usually the case that the reactant and the active site have complementary hydrophobic surfaces and donors and acceptors of hydrogen bonds. When all of these noncovalent interactions are in effect, the maximum standard free energy of dissociation, $\Delta G^{\circ}_{TOT}$ (always a positive number), is exerted. Suppose the enzyme catalyzing the reaction had an active site constructed so that all of these interactions were at their maximum strength when the reactant was bound. This would cause the reactant to be bound tightly, and its apparent standard free energy of dissociation, $\Delta G_A{}^{\circ\prime}$, would equal $\Delta G^{\circ}_{TOT}$. During the approach to the transition state, the strength of these interactions would have to decrease as the

**Table 3–4: Effect of Changes in Structure on Kinetic Parameters of Thermolysin[a]**

| peptide[b] | $K_m$[c] (mM) | $k_{cat}$[d] ($s^{-1}$) |
|---|---|---|
| Cbz[e]-Gly-Leu-D-Ala | 16 | 5 |
| Cbz-Gly-Leu-amide | 21 | 105 |
| Cbz-Gly-Leu-Gly-D-Ala | 13 | 25 |
| Cbz-Gly-Leu-Gly | 11 | 65 |
| Cbz-Gly-Leu-Gly-Phe | 11 | 130 |
| Cbz-Gly-Leu-Phe | 2.4 | 120 |
| Cbz-Gly-Leu-Leu | 2.6 | 370 |
| Cbz-Gly-Leu-Ala | 11 | 780 |
| Cbz-Gly-Leu-Gly-Ala | 14 | 420 |
| Cbz-Phe-Gly-Leu-Ala | 1.0 | 450 |
| Cbz-Gly-Leu-Ala | 11 | 780 |
| Cbz-Ala-Gly-Leu-Ala | 9 | 5200 |

↑ point of cleavage

[a]Determined from measurements of initial velocities of hydrolysis, as determined by colorimetric assay for $\alpha$-amino group, at 40 °C and pH 7.0. [b]Peptide used as reactant. [c]From double-reciprocal plots. [d]$V_{max}/[E]_{TOT}$. [e]Cbz, benzyloxycarbonyl.

nuclei changed their configuration from the one they had in the reactants. In this situation, the standard free energy of activation, $\Delta G^{\circ\ddagger\prime}$ would be increased relative to the standard free energy of activation of the uncatalyzed reaction, $\Delta G_u^{\circ\ddagger}$ (corrected for entropy of approximation) by the amount of standard free energy of association lost during this transition, $\Delta G^{\circ\ddagger}_{bind}$, and the enzyme would decrease the rate of the reaction. Suppose, however, that the active site were constructed so that all of these interactions were at their maximum only when the transition state was reached. In this case, the reactant, because it is not the transition state, would bind less well than it did in the first active site, and the observed standard free energy of dissociation, $\Delta G_A{}^\circ$, would be less than $\Delta G^\circ_{TOT}$ by a certain amount, $\Delta G^{\circ\ddagger}_{bind}$. During the approach to the transition state, the strength of these associative interactions would be increasing in magnitude until the total standard free energy of noncovalent interactions was expressed as the transition state was reached, and the standard free energy of activation, $\Delta G^{\circ\ddagger}$, would be less than the standard free energy of activation of the uncatalyzed reaction by the amount of free energy of association gained during this transition, $\Delta G^{\circ\ddagger}_{bind}$. In this case, the reaction is accelerated by the enzyme.

These arguments demonstrate how standard free energy of dissociation is interchangeable with standard free energy of activation.[130] They explain why changes in the structure of a reactant distant from the location of the chemical transformation catalyzed by an enzyme can have little effect on $K_m$ while changing $k_{cat}$ dramatically.[130] For example, a series of peptides was synthesized in which the substituents on either side of a glycylleucine dipeptide were changed at random. When these peptides were used as substrates for thermolysin, the turn-over number of the enzyme for cleavage of the glycylleucyl peptide bond varied by a factor of 1000 while the $K_m$ varied only by a factor of 16 (Table 3–4).[131] An explanation for this behavior is that standard free energy of dissociation is used to contort the reactants toward the transition state. The more noncovalent interactions complementary to the active site that are provided by the reactant, the more total standard free energy of dissociation, $\Delta G^\circ_{TOT}$, is available for this diversion.

The interconvertibility of standard free energy of dissociation and standard free energy of activation at the active site of an enzyme is usually expressed as a ratio between the dissociation constant for the reactant, $K_d{}^A$, and the turnover number, $k_{cat}$. The total standard free energy of dissociation available from noncovalent forces, $\Delta G^\circ_{TOT}$, is related to the observed standard free energy of dissociation, $\Delta G_A{}^\circ$, and the standard free energy of association diverted, $\Delta G^{\circ\ddagger}_{bind}$, by

$$\Delta G^\circ_{TOT} = \Delta G_A{}^\circ - \Delta G^{\circ\ddagger}_{bind} \qquad (3\text{–}201)$$

The observed standard free energy of activation, $\Delta G^{\circ\ddagger}_{strain}$, is related to the standard free energy of activation of the same reaction occurring at the active site with a reactant that is not strained, $\Delta G_{us}{}^{\circ\ddagger}$, by

$$\Delta G^{\circ\ddagger}_{strain} = \Delta G_{us}{}^{\circ\ddagger} + \Delta G^{\circ\ddagger}_{bind} \qquad (3\text{–}202)$$

and, by combination

$$\Delta G^\circ_{TOT} = -\Delta G^{\circ\ddagger}_{strain} + \Delta G_{us}{}^{\circ\ddagger} + \Delta G_A{}^\circ \qquad (3\text{–}203)$$

If $K^A_{d,TOT}$ is the dissociation constant that would have been observed if the active site fit the reactant exactly, $K_d{}^A$ is the observed dissociation constant for the reactant, $k_{cat}$ is the observed turnover number for the strained reaction, and $k_{us}$ is the turnover number for the unstrained reaction

$$RT \ln K^A_{d,TOT} = RT \ln\left(\frac{k_{us}}{k_{cat}}\right) + RT \ln K_d{}^A \qquad (3\text{–}204)$$

or

$$K^A_{d,TOT} = \left(\frac{K_d{}^A}{k_{cat}}\right) k_{us} = \exp\left(-\frac{\Delta G^\circ_{TOT}}{RT}\right) \qquad (3\text{–}205)$$

If a series of reactants is examined, all of whose rate constants at the active site for an unstrained reactant, $k_{us}$, would be the same, the quantity $K_d{}^A/k_{cat}$ is directly proportional to the exponent of the total standard free energy of dissociation that can be effected by all of the available noncovalent forces when they are fully engaged in the complex between the active site and the transition state. The ratio $K_d{}^A/k_{cat}$ is widely used to compare different enzymes that catalyze the same reaction or different reactants for the same enzyme. It accounts for the conversion of standard free energy of dissociation into standard free energy of activation and is a direct measurement of total standard free energy of the available noncovalent forces.

The conversion of standard free energy of dissociation and standard free energy of activation in lysozyme can be roughly quantified. ($\beta$1,4)Tri-*N*-acetylglucosaminopyranose,

$(GlcNAc)_3$, has a dissociation constant from the first three sites in the active site of lysozyme of $2 \times 10^{-7}$ mole fraction,[132] corresponding to a standard free energy of dissociation of +38 kJ mol$^{-1}$. ($\beta$1,4)Hexa-*N*-acetylglucosaminopyranose [$(GlcNAc)_6$] has a dissociation constant, when bound productively as a substrate, of $1 \times 10^{-7}$ mole fraction[132] or a standard free energy of dissociation of +40 kJ mol$^{-1}$. The hexasaccharide, because it is twice as long as the trisaccharide, should have a total standard free energy of noncovalent interaction about twice that of the trisaccharide, but the observed standard free energy of dissociation is the same. $\beta$-Ethyl-*N*-acetylglucosamine, $GlcNAcC_2H_5$

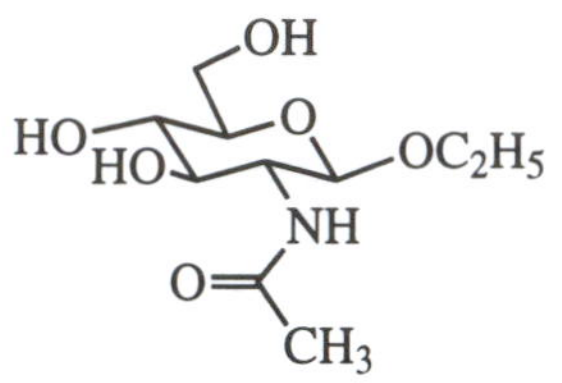

OH HO HO O $OC_2H_5$ NH O= $CH_3$

**3–10**

is a reactant for the hydrolysis catalyzed by lysozyme. It is so small that it is hard to see how the active site could contort it; thus, its rate of hydrolysis can be used as an estimate of the rate of an unstrained reaction at the active site. A comparison of the two turnover numbers

$$\frac{k_{cat}(GlcNAc)_6}{k_{cat}(GlcNAcC_2H_5)} = 3 \times 10^4 \qquad (3\text{–}206)$$

gives a difference in the standard free energies of activation of +25 kJ mol$^{-1}$. This is about the amount of standard free energy of dissociation that seems to be missing from the productive binding of the hexasaccharide to the active site—free energy of dissociation that has been diverted to lower the free energy of activation.

As well as changing the difference in free energy between reactants and transition state, the free energy of binding can be used to alter the difference in free energy between reactants and products while they are bound at the active site. The equilibrium constant for the reaction catalyzed by lactate dehydrogenase

$$\text{NAD}^+ + \text{lactate} \rightleftharpoons \text{pyruvate} + \text{NADH} \qquad (3\text{–}207)$$

at pH 7 in free solution is $4 \times 10^{-5}$, but when all of the substrates are within the active site of the enzyme, the equilibrium constant for the same reaction is 0.25.[133] This represents a shift in the standard free energy change for the reaction as written of −22 kJ mol$^{-1}$ when it takes place on the active site. This shift can be explained with the following linkage box

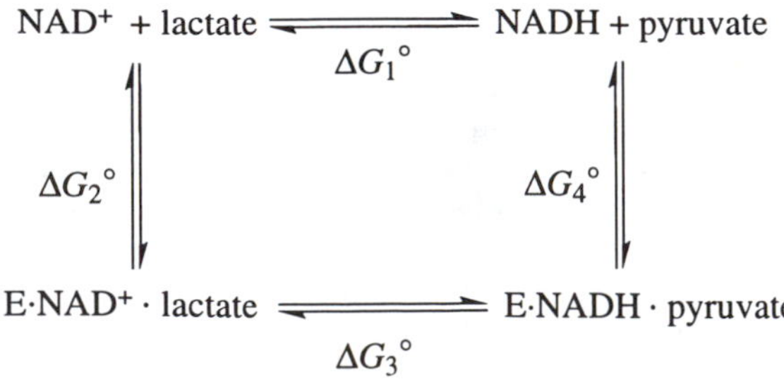

$$\begin{array}{ccc} \text{NAD}^+ + \text{lactate} & \xrightleftharpoons[\;]{\Delta G_1^\circ} & \text{NADH} + \text{pyruvate} \\ \Delta G_2^\circ \updownarrow & & \updownarrow \Delta G_4^\circ \\ \text{E·NAD}^+ \cdot \text{lactate} & \xrightleftharpoons[\Delta G_3^\circ]{} & \text{E·NADH} \cdot \text{pyruvate} \end{array} \qquad (3\text{–}208)$$

where the two binding reactions are considered as dissociations. By observation

$$\Delta G_1^\circ - 22 \text{ kJ mol}^{-1} = \Delta G_3^\circ \qquad (3\text{–}209)$$

It necessarily follows that

$$\Delta G_2^\circ = \Delta G_4^\circ - 22 \text{ kJ mol}^{-1} \qquad (3\text{–}210)$$

or that the products dissociate from the enzyme less readily, by +22 kJ mol$^{-1}$, than the reactants. A standard free energy change of −22 kJ mol$^{-1}$ must occur in the noncovalent forces holding the substrates in the active site as reactants are turned into products.

A similar effect is observed with isoleucine-tRNA ligase. When the substrates are on the active site,[134] the standard free energy change for the adenylation of the amino acid (Reaction 3–188) is +4 kJ mol$^{-1}$; but, in solution,[134,135] the standard free energy change for the same reaction has been estimated to be[134,135] between +25 and +40 kJ mol$^{-1}$.

Such shifts in the standard free energy change for an enzymatically catalyzed reaction, brought about by the preferential binding of reactants or products to the active site, may be advantageous. It has been argued that when the standard free energy change for the reaction catalyzed by an enzyme is shifted sufficiently so that it has a value of 0 within the active site, the enzyme can catalyze the reaction with the highest efficiency.[136] This is supposed to be essential when the role of that reaction in metabolism is to perform the net transformation of substrates either in one direction or in the other direction, depending on the demands of the moment.[137]

Another way to present the effects of binding upon the rate of an enzymatic reaction is to consider the consequences associated with the **binding of the transition state** to the active site of an enzyme.[138,139] In the simple case of an enzyme that operates on two reactants, the following linkage box can be constructed[139]

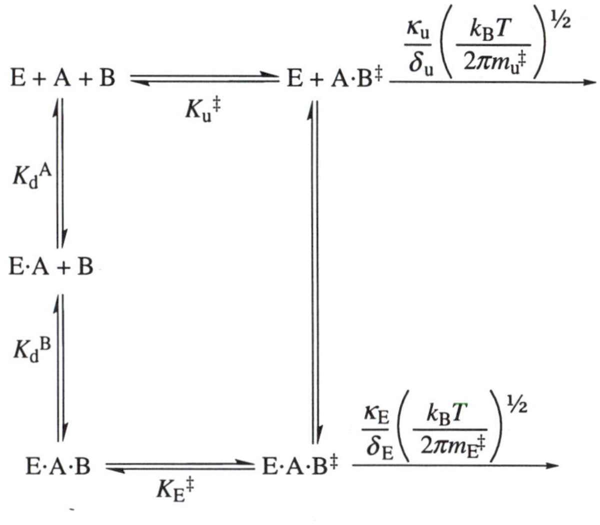

$$\begin{array}{ccc} \text{E + A + B} & \xrightleftharpoons[K_u^\ddagger]{} & \text{E + A·B}^\ddagger \xrightarrow{\frac{\kappa_u}{\delta_u}\left(\frac{k_B T}{2\pi m_u^\ddagger}\right)^{1/2}} \\ K_d^A \updownarrow & & \\ \text{E·A + B} & & \updownarrow \\ K_d^B \updownarrow & & \\ \text{E·A·B} & \xrightleftharpoons[K_E^\ddagger]{} & \text{E·A·B}^\ddagger \xrightarrow{\frac{\kappa_E}{\delta_E}\left(\frac{k_B T}{2\pi m_E^\ddagger}\right)^{1/2}} \end{array} \qquad (3\text{–}211)$$

where $K_d^A$, $K_d^B$, and $K_d^{TS}$ are the dissociation constants for reactant A, reactant B, and the transition state, respectively; A·B$^\ddagger$ is the transition state; $\kappa_u$, $\delta_u$, and $m_u^\ddagger$ and $\kappa_E$, $\delta_E$, and $m_E^\ddagger$ are the transmission coefficient, length of the reaction coordinate at the transition state, and mass of the transition state for the uncatalyzed and enzymatically catalyzed reactions, respectively; and $K_u^{\ddagger}$ and $K_E^{\ddagger}$ are the equilibrium constants between the respective reactants and transition states. If the expressions containing the transmission coefficients, the lengths of the reaction coordinates at the transition state, and the masses of the transition states for the uncatalyzed reaction and the enzymatically catalyzed reaction are approximately equal, then it follows that

$$\frac{K_d^{\mathrm{B}} K_d^{\mathrm{A}}}{K_d^{\mathrm{TS}}} \cong \frac{k_E}{k_u} \qquad (3\text{–}212)$$

This expression states that the ratio between the product of the dissociation constants for the reactants and the dissociation constant for the transition state approximately determines the ratio between the rates of the enzymatically catalyzed reaction and the uncatalyzed reaction. Therefore, to catalyze a reaction, an enzyme must bind the transition state more tightly than the reactants, and the degree of preference it displays for the transition state relative to the reactants determines the increase in rate that is realized.

This statement alone, however, because it is indisputable,[119] is uninformative. The informative observations are the details of the explicit strategies used to bind the transition state preferentially. One strategy is to design an active site in which the noncovalent forces increase as the configuration of the atoms in the reactants moves toward that of the transition state. Another strategy is to provide appropriately located acids and bases. The transition states in most biochemical reactions involve protons in transit. A bare proton leaving a shrinking lone pair of electrons or a bare proton being attracted to an enlarging lone pair of electrons are protons in transit. They are structural features of the transition state that must be bound by the enzyme if the transition state is to be bound more tightly than the reactants, in which these peculiar situations do not exist. To stabilize and provide these protons in transit, acids and bases must be located appropriately within the active site.

Another misunderstanding of this statement that the enzyme must bind the transition state more tightly is the impression it leaves that the enzyme actually binds transition states present in the solution and that this ability would explain its catalysis. An enzyme binds otherwise unassociated and unactivated reactants and the transition states are achieved within the active site. If an enzyme only preferentially bound transition states formed in solution, it could only decrease the rate of the reaction because by definition a transition state decomposes to products instantaneously. That the enzyme must bind the transition state more strongly than the ground state is only a thermodynamic argument, not something that occurs.

If it is necessarily the case that a transition state is bound more tightly to an active site than the substrates, it should be possible to design molecules whose structure resembles the structure of the transition state and to demonstrate that they bind to the enzyme far more tightly than do the substrates.[124] In almost every instance, transition states involve bond lengths and bond angles that cannot be reproduced in a stable molecule. By the Hammond postulate,[128] however, an intermediate of high energy encountered along a reaction coordinate resembles in its structure the transition states on either side of its well of potential energy more closely than do either the immediate reactant or the immediate product, respectively. Because one is unable to synthesize analogues of transition states, **analogues of high-energy intermediates** have been synthesized and shown to be potent inhibitors of enzymes. In spite of the inaccuracy of the phrase, analogues of high-energy intermediates are often referred to as **transition-state analogues.**

The trisaccharide *N*-acetylglucosaminopyranosyl-($\beta$1,4)-*N*-acetylmuramopyranosyl-($\beta$1,4)-*N*-acetylglucosamine binds to the active site of lysozyme in subsites A, B, and C with a dissociation constant of 4 $\mu$M, while the tetrasaccharide *N*-acetylglucosaminopyranosyl-($\beta$1,4)-*N*-acetylmuramopyranosyl-($\beta$1,4)-*N*-acetylglucosaminopyranosyl-($\beta$1,4)-*N*-acetylmuramate binds with a dissociation constant of 500 $\mu$M.[140] This increase in the dissociation constant for the tetrasaccharide relative to the trisaccharide is thought to be due to the fact that the more stable chair conformation of the pyranose of the additional *N*-acetylmuramate in the fourth position of the tetrasaccharide cannot fit into the binding site for the D pyranose in the active site of the enzyme (Figure 3–26), and this incompatibility is reflected in the dissociation constants. The $\delta$-lactone of ($\beta$1,4)-tri-*N*-acetylglucosaminpyranosyl-2-acetamido-2-deoxygluconate

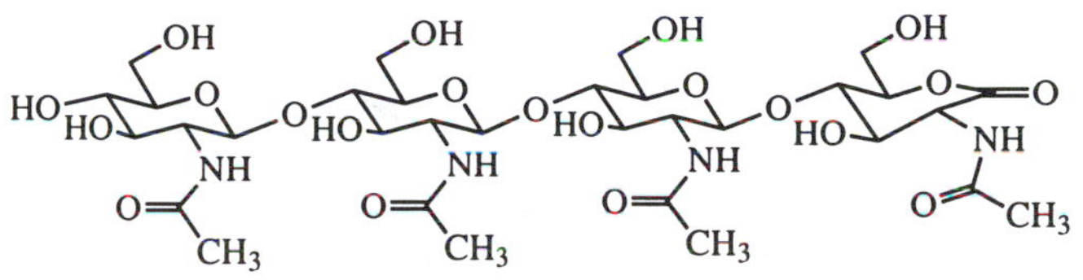

**3–11**

however, binds to the active site of lysozyme with a dissociation constant of 0.2 $\mu$M, while the trisaccharide ($\beta$1,4)-tri-*N*-acetylglucosaminopyranose binds to the active site with a dissociation constant of 5 $\mu$M.[141] This result indicates that a planar conformation rather than a chair conformation at the reducing end of the monosaccharide that occupies the binding site for the D pyranose is the structure compatible with this binding site. On the basis of these observations it was proposed that the $\delta$-lactone resembles the intermediate oxonium cation the active site has evolved to stabilize.

In the crystallographic molecular model obtained from a crystal of lysozyme in which the active site was occupied by tetrasaccharide **3–11**, the lactone was found to occupy the binding site for the D pyranose at the reducing end of the tetrasaccharide, and it was bound in the planar, sofa conformation.[127] Because the terminal $\delta$-lactone definitely occupies this site when tetrasaccharide **3–11** is bound even though a pyranose in the chair conformation at this position in a tetrasaccharide might not occupy the same site, only an upper limit to the difference in the standard free energy of dissociation between lactone and pyranose to the D site can be calculated. This upper limit for $\Delta\Delta G°$ is –20 kJ mol$^{-1}$.

Many analogues of high-energy intermediates have been described. Phosphoglycolate **(3–12)**

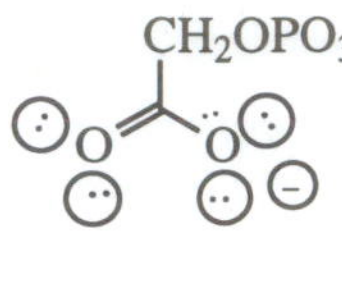

3–12 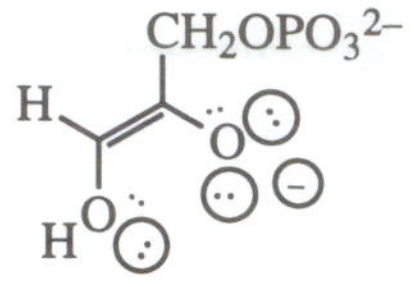 3–13

is a competitive inhibitor of triosephosphate isomerase. It displays a dissociation constant (7 $\mu$M)[138] that is significantly less than the dissociation constants for either of the substrates ($\geq$100 $\mu$M),[142,143] and it has been proposed that this is due to the resemblance between phosphoglycolate (**3–12**) and the intermediate *cis*-enediolate (**3–13**) in the mechanism of the isomerization. The planar compounds pyrrole-2-carboxylate (**3–14**)[144] and 1-pyrroline-2-carboxylate (**3–15**)[145]

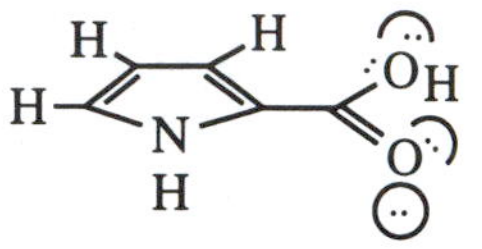

3–14 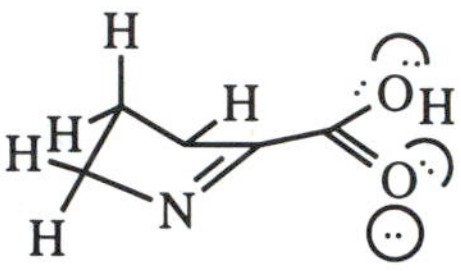 3–15

bind as inhibitors to proline racemase with apparent dissociation constants 100 times smaller than the Michaelis constants for either isomer of proline. It has been suggested[146] that these planar compounds resemble the *gem*-enediol **3–16**

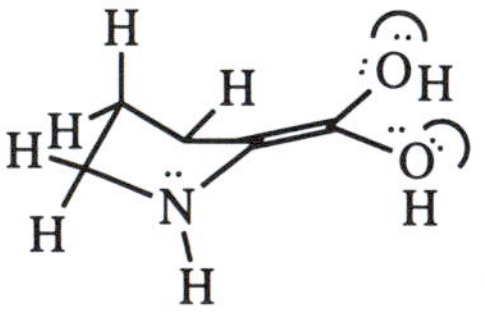

3–16

that is the intermediate in the reaction.[147]

The tetravalent intermediate in acyl exchange reactions

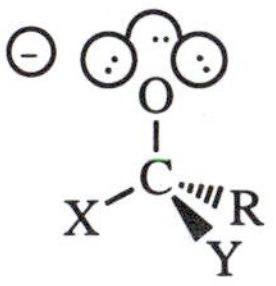

3–17

where X and Y are heteroatomic substituents, has been mimicked with sulfoximines (**3–18**),[148] sulfonamides (**3–19**),[149] thiohemiacetals (**3–20**),[150] carbinolamines (**3–21**),[138,151,152] hemiketals (**3–22**),[153] alkylboronic acids (**3–23**),[154,155] and phosphonamidates (**3–24**)[156–158]

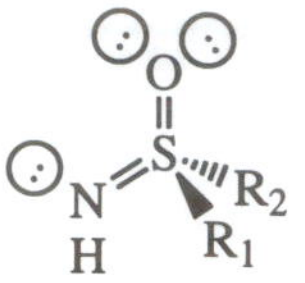

3–18 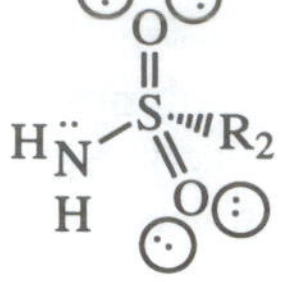 3–19

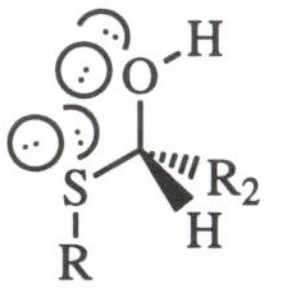

3–20 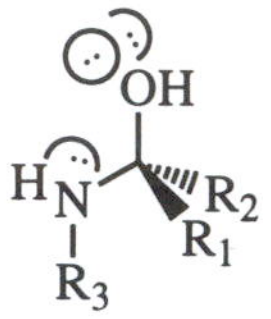 3–21

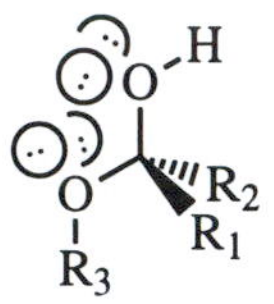

3–22 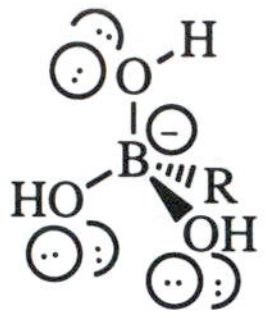 3–23 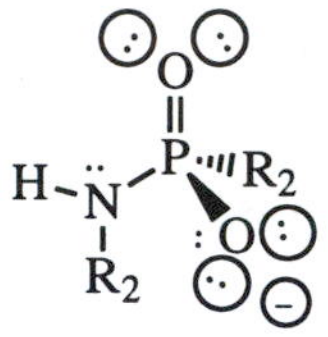 3–24

Analogues of peptides have been synthesized that incorporate a phosphonamidate (**3–24**) in place of a peptide bond. For example, the phosphonamidate

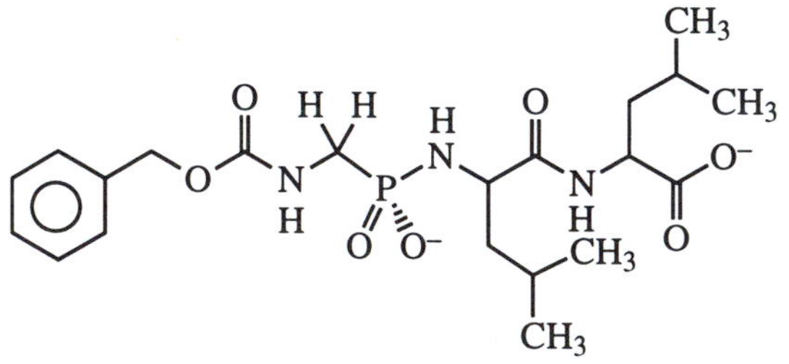

3–25

is identical to the peptide Cbz-Gly-Leu-Leu (Table 3–4) with the exception that one of the acyl carbons has been replaced by the phosphorus of the phosphonamidate. This compound inhibits thermolysin strongly as a competitive inhibitor ($K_i$ = 9 nM).[157]

Thermolysin is an enzyme that effectively diverts standard free energy of dissociation to lower the free energy of activation (Table 3–4). A series of phosphonamidates was synthesized, and each phosphonamidate was shown to be a potent competitive inhibitor of the hydrolytic reactions catalyzed by the enzyme.[157] If the phosphonamidates are bound to the

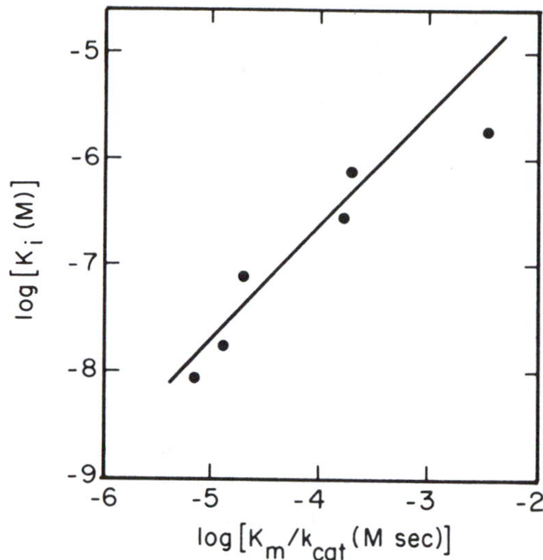


**Figure 3–28:** Correlation between the dissociation constant, $K_i$ (molar), for a set of phosphonamidates from the active site of thermolysin and the values of $K_m/k_{cat}$ (molar second) for the homologous reactants.[157] The reactants (Table 3–4) for which phosphonamidate analogues (**3–24**) were synthesized were, in order of increasing $K_m/k_{cat}$, Cbz-Gly-Leu-Leu, Cbz-Gly-Leu-Ala, Cbz-Gly-Leu-Phe, Cbz-Gly-Leu-Gly, Cbz-Gly-Leu-$NH_2$, and Cbz-Gly-Leu-D-Ala. Their $K_m$ and $k_{cat}$ values at 40 °C and pH 7 are listed in Table 3–4. The phosphonamidate replaced the acyl carbon and oxygen on each glycine (for example, phosphonamidate **3–25**). The resulting compounds were all inhibitors of the enzyme at 25 °C when concentration of the enzyme was $10^{-8}$ M and the reactant being hydrolyzed was Phe-Gly-Leu-Ala at 2 mM. The values of $K_i$ were estimated from the inhibition of the initial velocities of the enzymatic activity as a function of the concentrations of the inhibitors. Reprinted with permission from ref 157. Copyright 1983 American Chemical Society.

active site of thermolysin as if they were tetravalent intermediates in the normal catalytic mechanism, then they should be bound with almost the full standard free energy of the available noncovalent interactions between active site and transition state, and the values of $K_i$ for each competitive inhibitor should be approximately equal to $K_{d,TOT}^{pep}$ (Equation 3–205) for the corresponding peptide. If this is the case, and $k_{us}$ is the same for all of the peptides in the series, the values of log $K_i$ should be a linear function of the respective values of log $(K_d^{pep}/k_{cat})$ with a slope of 1. This was found to be so (Figure 3–28, actual slope = 1.07).[157] Such a correlation simultaneously provides evidence, in the case of thermolysin, for both the diversion of standard free energy of association to lower the standard free energy of activation and the ability of a phosphonamidate to substitute for the tetravalent intermediate (Figure 1–7) in the enzymatically catalyzed hydrolysis of an amide.

Thermolysin catalyzes the hydrolysis of a peptide bond by activating water and catalyzing its direct addition to the amide. Another whole group of proteinases, peptidases, and esterases, such as trypsin, chymotrypsin, elastase, and acetylcholinesterase, catalyzes hydrolyses of acyl compounds by forming covalent intermediates between serines in the active sites of the enzymes and the acyl carbons of the reactants. In the case of acetylcholinesterase, the covalent intermediate is an ester formed between Serine 200 in the active site and the acetyl moiety of the reactant (Mechanism 3–174). Several trifluoromethyl ketones were synthesized to act as precursors to the respective hemiacetals formed from the serine in the active site[159]

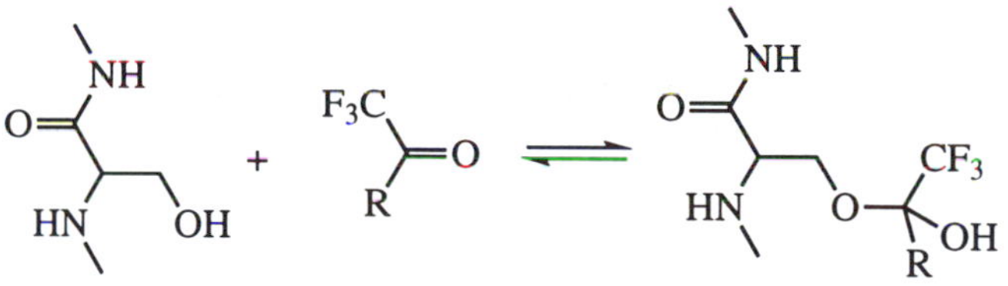

(3–213)

These hemiacetals, because they mimic the tetravalent intermediate of the acyl exchange reaction

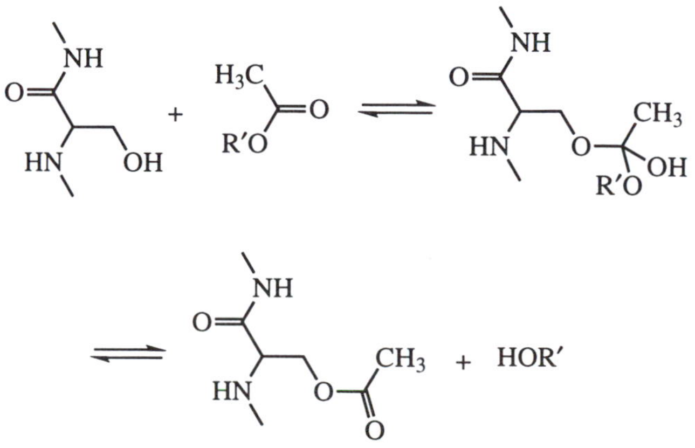

(3–214)

are examples of intermediate analogues formed on the enzyme itself (Equation 3–213) from a precursor. Because they can form these adducts, these trifluoromethyl ketones are potent inhibitors of the enzymatic reaction. For example, the analogue of acetylcholine [R = $-(CH_2)_3N^+(CH_3)_3$] had a dissociation constant from the enzyme, $K_i$, of 0.06 nM.

One interesting observation, however, was that the uncharged alkyl analogue of acetylcholine [R = $-(CH_2)_3C(CH_3)_3$] had a much larger dissociation constant ($K_i$ = 12 nM). This was unexpected because the dissociation constants for the two analogous reactants are almost identical (Table 3–2). One explanation for this difference is that the structure of the tetravalent intermediate differs from that of the reactant (Equation 3–214). If the methyl group, the serine, and the acyl oxygen are held tightly in the active site by covalent and noncovalent interactions, the choline (R′O in Equation 3–214) must move to accommodate the conversion of the trigonal carbon to a tetrahedral carbon. If, during this movement, the quaternary ammonium cation comes closer to a carboxylate in the active site, this electrostatically favorable juxtaposition would lower the free energy of the transition state, relative to the ground state.

The active site in the crystallographic molecular model of acetylcholinesterase[160] is lined with aromatic amino acids, and it is known from model studies that neutral aromatic functional groups can form highly specific binding sites for quaternary ammonium cations.[161] The remarkable feature of the active site is the number of aromatic groups it contains, and it is possible that they create at least two adjacent pockets for the quaternary ammonium cation, one for the cation in the reactant and one for the cation in the tetravalent intermediate. On the edge of the pocket in which the quatermary ammonium cation of the reactant binds is Glutamate 199. It is possible that as the tetravalent intermediate forms, the quaternary ammonium cation moves toward this glutamate, and the resulting electrostatic interaction lowers the free energy of the transition state.

The inhibition effected by analogues of high-energy intermediates has also been used to distinguish between alternatives for the structure of the transition state in a particular enzymatic reaction. For example, the transition state in the reaction catalyzed by chorismate mutase (Reaction 1–99) could proceed through either a chair form of the transition state or a boat form. The fact that *exo*-6-hydroxybicyclo[3.3.1]nonane-1, *exo*-3-dicarboxylate (**3–26**) was a reasonable competitive inhibitor of chorismate mutase ($K_i$ = 400 $\mu$M) while *exo*-6-hydroxybicyclo[3.3.1]nonane-1, *endo*-3-dicarboxylate (**3–27**) was not ($K_i$ > 10 mM)

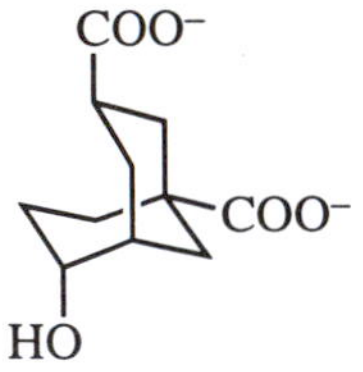

COO⁻ COO⁻ HO

**3–26**

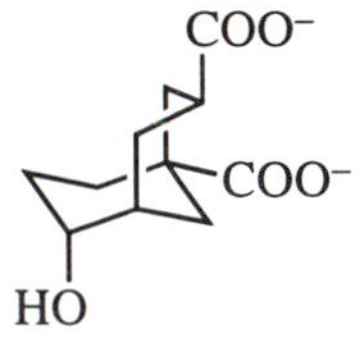

COO⁻ COO⁻ HO

**3–27**

has been used as an argument that the transition state in the enzymatic reaction is in the chain conformation.[162] Likewise, the fact that *N*-(phosphonomethyl)glycine

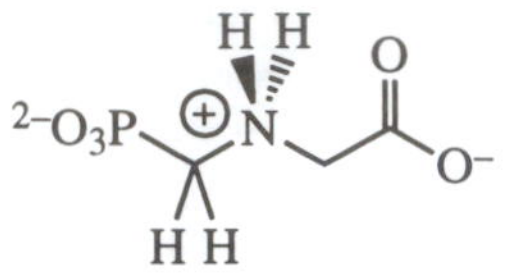

$^{2-}O_3P$ H H ⊕N O O⁻ H H

**3–28**

is a competitive inhibitor that binds tightly to 3-phosphoshikimate 1-carboxyvinyltransferase ($K_i$ = 1 $\mu$M at pH 6.8) has been used as part of the evidence for a carbocation as an intermediate in the mechanism of the enzymatic reaction.[163] The fact that L-methionine (*S*)-sulfoximine *N*-phosphate (**3–29**)[148,164]

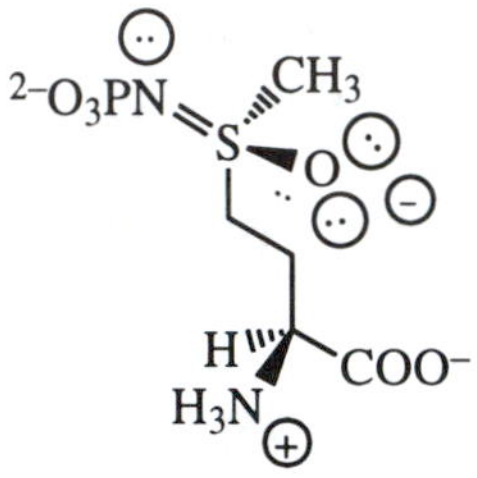

$^{2-}O_3PN$ S $CH_3$ O H COO⁻ $H_3N$⊕

**3–29**

$^{2-}O_3PO$ C $NH_3$ O H COO⁻ $H_3N$⊕

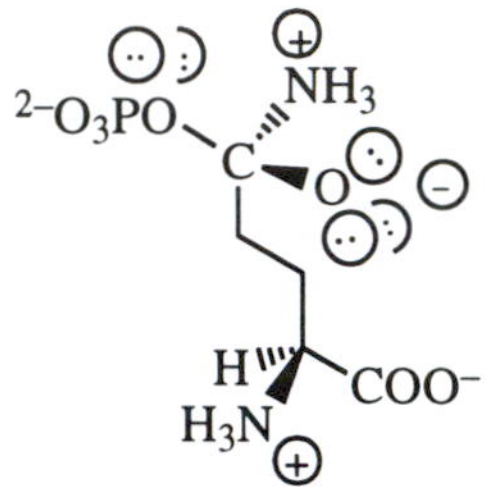

**3–30**

forms an extremely tight complex with glutamate–ammonia lyase when the active site is also occupied by ADP[164] can be used as evidence for the tetrahedral species **3–30** as an intermediate of high energy in the enzymatic reaction. To provide evidence for the existence of the anionic intermediate **3–31**

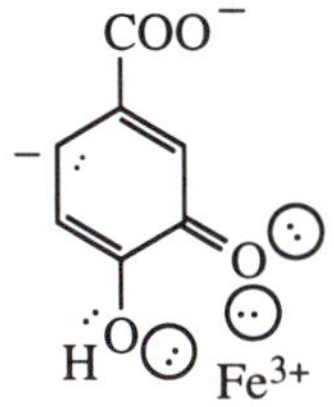

COO⁻ O H O $Fe^{3+}$

**3–31**

COO⁻ N O H O

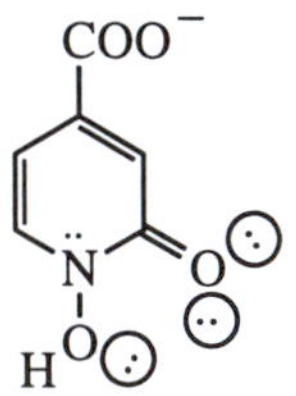

**3–32**

the hydroxamate **3–32** was synthesized and shown to be a potent inhibitor ($K_i < 0.01\ K_d^{protocatechuate}$) of protocatechuate 3,4-dioxygenase.[165]

It has been pointed out by Jencks[166] that if an active site catalyzes a reaction by binding the transition state or an intermediate of high energy more tightly than it does the substrates,[124] an antibody raised against an analogue of either a transition state or a high-energy intermediate, because it would bind the corresponding transition state or high-energy intermediate tightly, should catalyze the corresponding reaction. Such catalytic antibodies have been produced. For example, monoclonal immunoglobulins were raised against an analogue[167]

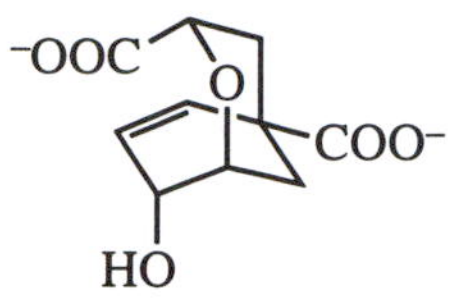

$^{-}$OOC ... O ... COO$^{-}$ ... HO

**3–33**

of the transition state in the chorismate mutase reaction ($K_i$ = 0.1 $\mu$M) that had been attached as a hapten to a protein carrier through tyrosine[168]

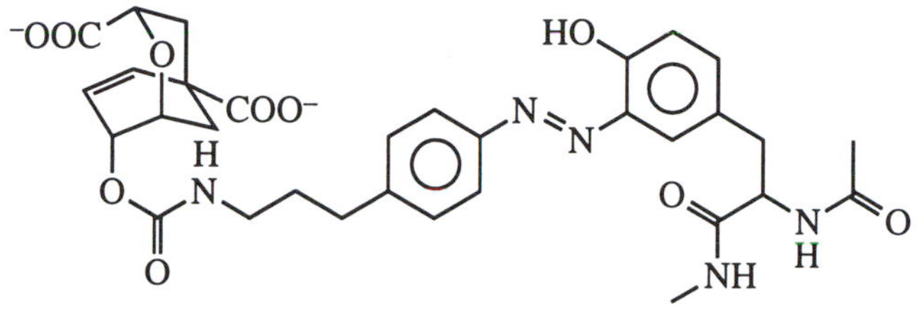

**3–34**

One of the monoclonal immunoglobulins produced was able to catalyze enzymatically ($k_{cat}$ = 2.7 min$^{-1}$, $K_m$ = 260 $\mu$M at 10 °C) the chorismate mutase reaction. Although this monoclonal immunoglobulin was not so efficient as chorismate mutase itself ($k_{cat}$ = 820 min$^{-1}$, $K_m$ = 290 $\mu$M at 10 °C), its turnover number was $10^4$ times that of the first-order, uncatalyzed reaction under the same conditions. The same analogue of the transition state was used in other experiments[169] and a similar monoclonal antibody was isolated.

Analogues of high-energy intermediates should be distinguished from **bisubstrate analogues.**[120] When two of the substrates of an enzyme have been attached together synthetically, the free energy of dissociation for the conjugate must be greater than the sum of the standard free energies of dissociation for each separate reactant for the hybrid to be considered an analogue of a high-energy intermediate in the reaction. For example, *N*-(phosphonacetyl)-L-aspartate (**3–4**), a competitive inhibitor of aspartate transcarbamylase, combines the structure of carbamyl phosphate (**3–5**) and aspartate (**3–6**). The dissociation constant for carbamyl phosphate from the empty active site of the enzyme is 27 $\mu$M, and the dissociation constant for aspartate from the empty active site is 11 mM.[63] The respective standard free energies of dissociation, based on a standard state of one mole fraction, are +36 and +21 kJ mol$^{-1}$. The dissociation constant ($K_i$) for *N*-(phosphonacetyl)-L-aspartate, when it is acting as a competitive inhibitor binding to the empty active site of aspartate transcarbamylase, is 30 nM, from which a standard free energy of dissociation of +53 kJ mol$^{-1}$ can be calculated. Because the standard free energy of dissociation for the bisubstrate analogue is less than the sum of the free energies of dissociation for the two substrates, it is probably not an analogue of a high-energy intermediate in the enzymatic reaction even though its dissociation constant is quite small.

## *Suggested Reading*

Jencks, W.P. (1975) Binding Energy, Specificity, and Enzymic Catalysis, *Adv. Enzymol. Relat. Areas Mol. Biol. 43*, 219–410.

### PROBLEM 3–15

The catalytic subunit of aspartate transcarbamylase catalyzes the following reaction:

$$\text{aspartate} + \text{carbamyl phosphate} \rightleftharpoons \text{carbamyl aspartate} + \text{phosphate}$$

(A) Write a mechanism for this reaction, involving a tetravalent intermediate, as it would occur in the active site of the enzyme. Indicate the removals and additions of protons by active site acids and bases.

(B) Succinate is an inhibitor of the enzyme that competes with aspartate for the same corner of the active site. Draw the structures of succinate and aspartate side by side in the same orientation. What is the difference between the two?

When the catalytic subunit of aspartate transcarbamylase is mixed with saturating concentrations of carbamyl phosphate and succinate, a significant conformational change occurs in the enzyme that is associated with an increase in the absorbance of the protein at 290 nm[169a] and a 1.8% increase in the sedimentation coefficient.[169b] Neither carbamyl phosphate nor succinate alone causes this change in conformation; nor does it occur when the carbamyl phosphate corner of the active site is occupied by $HOPO_3^{2-}$(phosphate) and either aspartate or succinate occupies its corner. The following are dissociation constants for succinate and aspartate from the enzyme when it is saturated with either phosphate or carbamyl phosphate[169a]

| ligand | phosphate | carbamyl phosphate |
|---|---|---|
| aspartate | $6.7 \times 10^{-3}$ M | $>20 \times 10^{-3}$ M |
| succinate | $33 \times 10^{-3}$ M | $0.74 \times 10^{-3}$ M |

The conclusion that can be drawn from these measurements is that carbamyl phosphate increases the affinity of the enzyme for succinate 50-fold but decreases the affinity of the enzyme for aspartate 3-fold.

(C) Examine closely your mechanism for the enzymatic reaction and the structural difference between aspartate and succinate and explain the values of the dissociation constants tabulated above in terms of strain.

(D) Assume that the same noncovalent forces are involved in the binding of either aspartate or succinate to the enzyme and that the available noncovalent free energy for aspartate changes by the same amount as that for succinate when carbamyl phosphate is substituted for phosphate. How much free energy of association is used by the enzyme to lower the free energy of activation for the reaction?

## PROBLEM 3–16

(A) Write the chemical mechanism for the hydrolysis of a peptide bond, RCONHR, when it is not catalyzed by an enzyme. Include all proton transfers explicitly. Draw the reactants, all intermediates, and the products with proper stereochemistry.

A series of inhibitors has been synthesized for thermolysin[157]

| inhibitor | $K_i$ (nM) |
|---|---|
| $CbzNHCH_2PO_2$-L-Leu-D-Ala | 1700 |
| $CbzNHCH_2PO_2$-L-Leu-amide | 760 |
| $CbzNHCH_2PO_2$-L-Leu-Gly | 270 |
| $CbzNHCH_2PO_2$-L-Leu-L-Phe | 78 |
| $CbzNHCH_2PO_2$-L-Leu-L-Ala | 78 |
| $CbzNHCH_2PO_2$-L-Leu-L-Leu | 9.1 |

(B) By drawing the appropriate structures explain why these inhibitors bind so much more tightly to the enzyme than their analogous substrates (Table 3–4).

One way to write the mechanism of the reaction catalyzed by thermolysin and compare it to the mechanism of the nonenzymatic reaction is

(see figure below)

where Int is the central intermediate in peptide hydrolysis, Th is thermolysin

$$K_{int}^{unc} = \frac{[(\text{Int})]}{[\text{RCONHR}'][\text{H}_2\text{O}]}; \quad K_{int}^{cat} = \frac{[\text{Th}\cdot(\text{Int})]}{[\text{Th}\cdot\text{RCONHR}'\cdot\text{H}_2\text{O}]}$$

and all the other constants are dissociation constants.

(C) It has been shown that the values for $K_m$ (Table 3–4) for the reactants analogous to these inhibitors are approximately equal to their dissociation constants from the active site. These tabulated values are presumably apparent dissociation constants

$$K_{d,app}^{pep} = \frac{([\text{Th}]+[\text{Th}\cdot\text{H}_2\text{O}])[\text{RCONHR}']}{[\text{Th}\cdot\text{RCONHR}']+[\text{Th}\cdot\text{RCONHR}'\cdot\text{H}_2\text{O}]}$$

Show that

$$K_{d,app}^{pep} = \frac{K_d^{\text{H}_2\text{Opep}}\left(1+\dfrac{K_d^{\text{H}_2\text{O}}}{[\text{H}_2\text{O}]}\right)}{\left(1+\dfrac{K_d^{\text{pepH}_2\text{O}}}{[\text{H}_2\text{O}]}\right)}$$

(D) Show that

$$K_d^{int} = \frac{K_d^{\text{H}_2\text{O}}\,K_{int}^{unc}\,K_{d,app}^{pep}\left(1+\dfrac{K_d^{\text{pepH}_2\text{O}}}{[\text{H}_2\text{O}]}\right)}{K_{int}^{cat}\left(1+\dfrac{K_d^{\text{H}_2\text{O}}}{[\text{H}_2\text{O}]}\right)}$$

(E) Show that there should be a correlation among the values for $K_i$ for the six inhibitors and the values for $K_d^{pep}$ and $k_{cat}$ for the respective substrates if the following assumptions are made.

(a) The initial velocity $v_0 = k_{cat}^{app}[\text{Th}\cdot\text{RCONHR}'\cdot\text{H}_2\text{O}] = k_{cat}[\text{Th(Int)}]$ and tabulated $k_{cat}^{app} = k_{cat}K_{int}^{cat}$
(b) The rate constant $k_{cat}$ is the same for all substrates.
(c) Both the enzymatic reaction and the uncatalyzed reaction proceed through the same intermediate.
(d) The equilibrium constant, $K_{int}^{unc}$, is the same for all the substrates because the differences among them are distant from the point of hydrolysis.
(e) The dissociation constant, $K_d^{\text{pepH}_2\text{O}}$, has the same value for all of the substrates because the $H_2O$ is binding to the enzyme distant from the differences among the substrates. (Note that this assumption is not required if $[\text{H}_2\text{O}] >> K_d^{\text{pepH}_2\text{O}}$.)
(f) The equilibrium constants, $K_i$ and $K_d^{int}$, vary in direct proportion to each other as the substrates and their analogous inhibitors are varied in structure.

(F) Show that the results display this correlation.

**Problem 3–16 (B)**

$$\text{Th} + \text{RCONHR}' + \text{H}_2\text{O} \overset{K_{int}^{unc}}{\rightleftharpoons} \text{Th} + (\text{Int}) \xrightarrow{k_{unc}} \text{RCOOH} + \text{H}_2\text{NR}'$$

$K_d^{\text{H}_2\text{O}}$: Th + RCONHR′ + $H_2O$ ⇌ RCONHR′ + Th·$H_2O$

$K_d^{pep}$: Th + RCONHR′ + $H_2O$ ⇌ Th·RCONHR′ + $H_2O$

$K_d^{\text{H}_2\text{Opep}}$: RCONHR′ + Th·$H_2O$ ⇌ Th·RCONHR′·$H_2O$

$K_d^{\text{pepH}_2\text{O}}$: Th·RCONHR′ + $H_2O$ ⇌ Th·RCONHR′·$H_2O$

$K_d^{int}$: Th + (Int) ⇌ Th·(Int)

$$\text{Th}\cdot\text{RCONHR}'\cdot\text{H}_2\text{O} \overset{K_{int}^{cat}}{\rightleftharpoons} \text{Th}\cdot(\text{Int}) \xrightarrow{k_{cat}} \text{RCOOH} + \text{H}_2\text{NR}'$$

**PROBLEM 3–17**

The enzyme inorganic pyrophosphatase catalyzes the reaction

$$MgO_3POPO_3^{2-} + H_2O \rightleftharpoons MgO_3PO^- + HOPO_3^{2-} + H^+$$

where $MgO_3POPO_3^{2-}$ is the magnesium complex of inorganic pyrophosphate, $HOPO_3^{2-}$ is inorganic phosphate, and $MgO_3PO^-$ is the magnesium complex of inorganic phosphate. From the following equilibrium constants[169c] for the species at pH 7, calculate the equilibrium constant for this reaction as written

$$\frac{[HOPO_3^{2-}]^2}{[^{2-}O_3POPO_3H_2]} = 2.7 \times 10^4 \text{ M}$$

$$\frac{[Mg][HOPO_3^{2-}]}{[MgO_3PO^-]} = 5 \times 10^{-2} \text{ M}$$

$$\frac{[Mg][^{2-}O_3POPO_3H_2]}{[MgO_3POPO_3^{2-}]} = 2.7 \times 10^{-4} \text{ M}$$

On the active site, the equilibrium constant at pH 7 for the reaction catalyzed by the enzyme is 5. Compare the two equilibrium constants, the one in solution with the one on the active site. Discuss their relative magnitudes in terms of approximation and strain.

**PROBLEM 3–18**

The competitive inhibitors in the opposite table were designed as analogues of high-energy intermediates for the respective enzymes. Draw the high-energy intermediate in the reaction that they were designed to mimic in the same orientation as the drawing of the compound.

**PROBLEM 3–19**

Is the magnesium complex of $P^1,P^5$-di(adenosine-5′)pentaphosphate a bisubstrate analogue or an analogue of a high-energy intermediate for adenylate kinase

$$AMP + MgATP \rightleftharpoons ADP + MgADP$$

if its dissociation constant as a competitive inhibitor[170j] is 3 nM while the dissociation constants from the active site for AMP and MgATP are 0.1 and 0.6 mM?

| enzyme[170] | compound | $K_i$ |
|---|---|---|
| aconitate hydratase[170a] (4.2.1.3) | 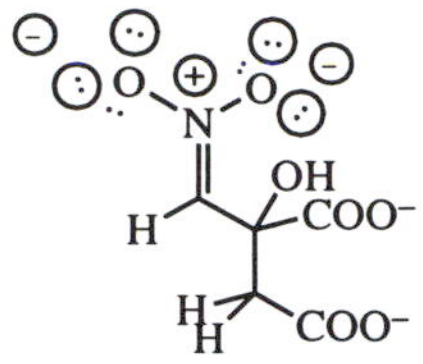 | 60 nM |
| isocitrate lyase[170b] (4.1.3.1) | 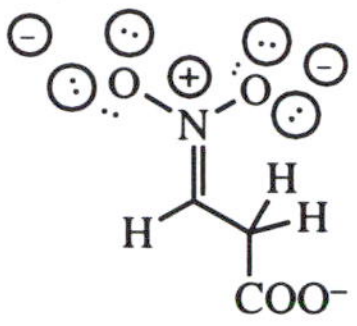 | <20 $\mu$M |
| orotodine-5′-phosphate[170c] decarboxylase (4.1.1.23) | 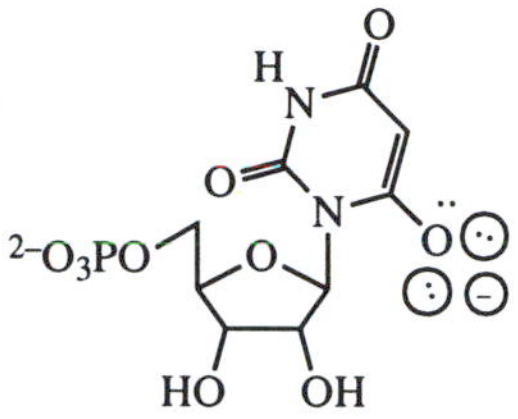 | 10 pM |
| adenosine deaminase[170d] (3.5.4.4) | 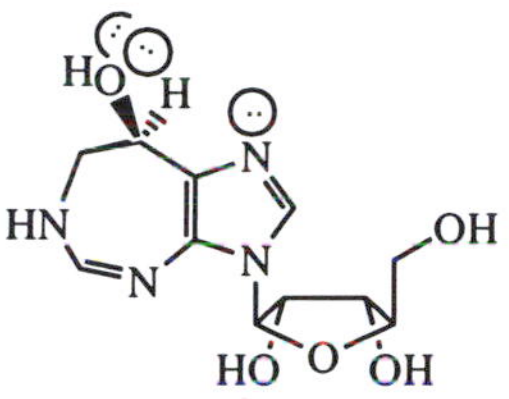 | 3 pM |
| acetylcholinesterase[170e] (3.1.1.7) | 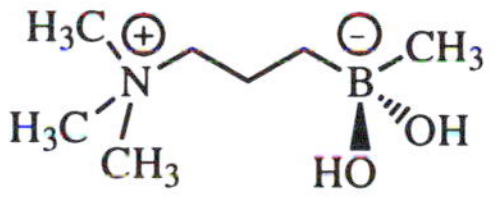 | 30 nM |
| glutamine synthase[170f] (1.4.1.14) | 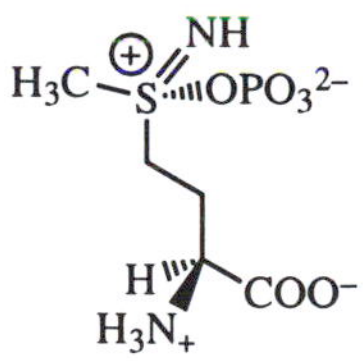 | <4 $\mu$M |
| glucose-6-phosphate[170g] isomerase (5.3.1.9) | 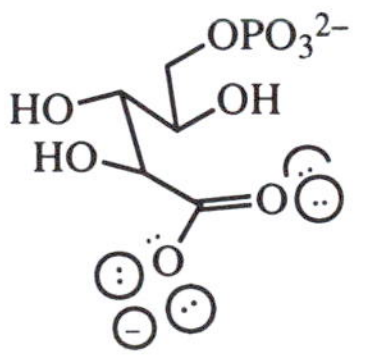 | 0.3 $\mu$M |
| enolase[170h] (4.2.1.11) | 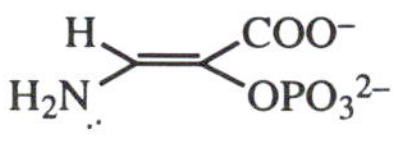 | 0.1 $\mu$M |
| adenosine deaminase[170i] (3.5.4.4) | 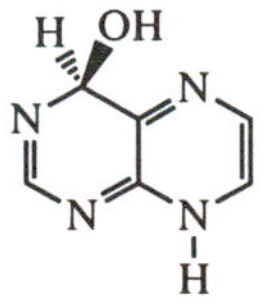 | 25 $\mu$M |

# Crystallographic Studies of Active Sites

It is possible to identify, in crystallographic molecular models of enzymes, the active sites at which the reactants bind and at which they are converted to products. This active site will usually be a crevice[170k] or tubular hole (Figure 3–29)[171] deep enough, wide enough, and long enough to accommodate all of the reactants when they are bound to the enzyme simultaneously. Its walls and floor are formed from strands of the polypeptide forming the protein.

In a monomeric enzyme, each molecule of the protein has an active site at a unique location on its surface. In an oligomeric enzyme, each molecule of the oligomer usually contains as many active sites as there are protomers. The active sites are all identical to each other, and they are distributed over the surface of the protein by the same symmetry relationships that govern the arrangements of the protomers themselves. Although it is usually the case that each protomer contains one complete active site, there are many examples in which an active site in an oligomeric protein is constructed from strands of polypeptide from two adjacent protomers.

Ribulose-1,5-bisphosphate carboxylase[171] is composed of eight copies of a folded $\alpha$ polypeptide ($n_{aa}$ = 475) and eight copies of a folded $\beta$ polypeptide ($n_{aa}$ = 123). The octamer contains four $\alpha_2$ dimers arranged around a 4-fold rotational axis of symmetry, and the two 2-fold rotational axes of symmetry relating the two subunits within the $\alpha_2$ dimer are normal to the 4-fold rotational axis of symmetry. The $\beta$ subunits link the four $\alpha_2$ dimers together around the 4-fold axis of symmetry, so that each $\alpha_2$ dimer is a unit of the higher structure, much as each catalytic trimer is a subunit of the overall structure of aspartate transcarbamylase.[171a] Each $\alpha_2$ dimer of ribulose-bisphosphate carboxylase contains two active sites symmetrically displayed around the 2-fold rotational axis of symmetry. These two active sites lie at the interface between the two $\alpha$ subunits. The amino acids that form each of the two active sites (Figure 3–29) are Threonine 173, Lysine 175, Lysine 177, Threonine 201, Aspartate 203, Glutamate 204, Histidine 294, Arginine 295, Histidine 327, Lysine 334, Leucine 335, Serine 379, Glycine 380, Glycine 381, Glycine 403, and Glycine 404 from one subunit and Glutamate 60, Threonine 65, Tryptophan 66, and Asparagine 123 from the adjacent subunit across the interface.[171]

This catalogue of the amino acids forming the active site of ribulose-bisphosphate carboxylase illustrates the fact that an active site is formed from a cluster of neighboring amino acids on the surface of a protein, regardless of their covalent affiliation. The purpose for which the polypeptide has folded to form a globular structure is to gather these amino acids in proper alignment to catalyze the enzymatic reaction. Because the polypeptide folds into a complexly ordered, globular structure, amino acids distant in the amino acid sequence or from different polypeptides are brought together opportunistically to form an active site.

One of the more unusual active sites is that in the aspartic proteinase from Rous sarcoma virus (Figure 3–30).[172] The protein is an $\alpha_2$ dimer composed of two identical polypeptides ($n_{aa}$ = 124). It is a member of a family of aspartic proteinases that includes the pepsins. Most of the other members of the family, however, are composed of folded polypeptides at least twice as long as that of the aspartic proteinase from Rous sarcoma virus. These larger proteins all have an internal duplication, and the two internally repeating domains are arranged around a 2-fold rotational axis of pseudosymmetry, like that relating the two domains in thiosulfate sulfurtransferase.[172a] In the large aspartic proteinases, this axis of pseudosymmetry passes through the one active site formed by the single folded polypeptide. In the aspartic proteinase from Rous sarcoma virus, however, each of the two identically folded polypeptides assumes the position of one of the two domains in a larger aspartic proteinase, the protein is an $\alpha_2$ dimer containing only one active site, the two folded polypeptides are arranged around a 2-fold axis of symmetry, and the 2-fold axis of symmetry passes through the single active site (Figure 3–30).

The initial step in a crystallographic study of the active site of an enzyme is to construct a refined crystallographic molecular model of the enzyme in one of its forms by the usual procedures of crystallization, diffraction of X-rays, multiple isomorphous replacement, calculation of a map of electron density, construction of a molecular model, and refinement. The active site is usually unoccupied by substrates or inhibitors in this initial molecular model because neither substrates nor inhibitors were present in the crystal. Sometimes the occupied form of the enzyme is the form chosen for the initial crystals, but small ligands are hard to find in the initial map of electron density in any case. In most instances, it is the crystals of the unoccupied enzyme, the data sets from those crystals, both $|F_o|$ and $|F_c|$, the phases for the reflections, both $\alpha_o$ and $\alpha_c$, the refined map of electron density, and the molecular model itself that are the starting points in the study of the active site. If, however, crystals of a liganded form of the enzyme provided the initial map of electron density, the same parameters from these crystals are the starting points in further studies. For the sake of argument it will be assumed that the crystallographic molecular model of the unliganded enzyme was the first to be solved.

Crystals of an enzyme whose active site is occupied by one or more substrates or competitive inhibitors can be prepared in at least two ways. Crystals of the enzyme with an unoccupied active site can be transferred to a solution containing one or more of these ligands, which diffuse into the crystal and occupy the active sites. Often this procedure causes the crystals to shatter as adjustments in the structure of the protein occur in response to the occupation of the active site. Sometimes this shattering can be prevented by cross-linking the crystals. Alternatively, the ligand can be present at saturating concentrations during crystallization and become directly incorporated into the crystal while on the active site of the enzyme. This strategy, however, often leads to crystals of a different space group from the crystals of the unoccupied enzyme simply because the enzyme is a different molecule when one or more ligands are bound than when one is not. For example, while the unliganded $\alpha_2$ dimer of alcohol dehydrogenase crystallizes in the $C222_1$ space group with a single protomer as the asymmetric unit, the form of the enzyme in which the active site is occupied by NADH crystallizes in the $P1$ space group with the entire $\alpha_2$ dimer necessarily the asymmetric unit.[173]

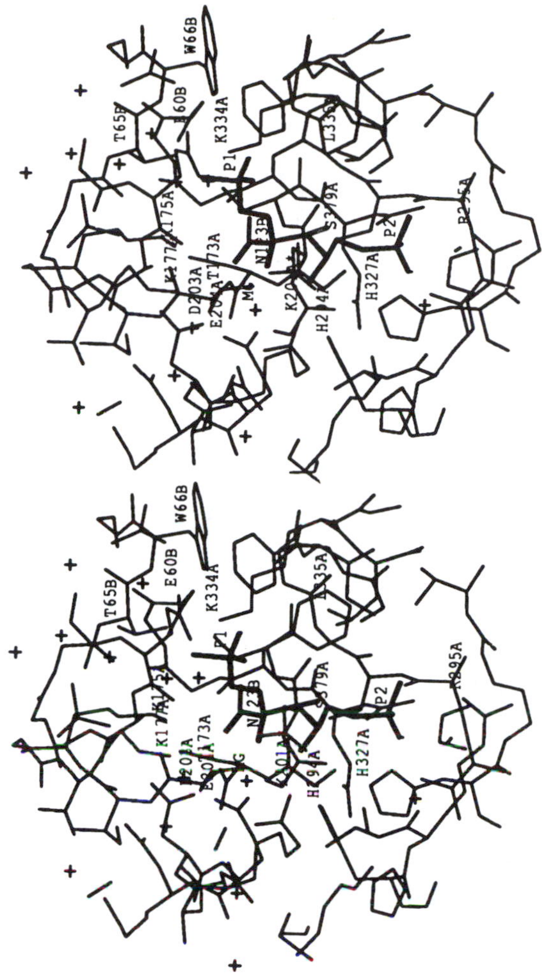


**Figure 3–29:** Crystallographic molecular model of the active site of ribulose-bisphosphate carboxylase occupied with 2-carboxyarabinitol 1,5-bisphosphate, an analogue of a high-energy intermediate in the enzymatic reaction.[171] The complex between ribulose-bisphosphate carboxylase and 2-carboxyarabinitol 1,5-bisphosphate was crystallized from a solution of ammonium sulfate (space group $C222_1$). The phases of the reflections were determined by multiple isomorphous replacement with three derivatives. A model of the polypeptide was inserted into the map of electron density, and the resulting molecular model was used to initiate refinement. After several cycles of refinement, a molecular model of 2-carboxyarabinitol 1,5-bisphosphate was inserted into the difference map of electron density into a feature of positive electron density that was unmistakably the correct size and shape. Refinement then proceeded to a final *R*-factor of 0.24. The view is of the analogue (in darker lines) surrounded by the amino acids creating the active site. In the final crystallographic molecular model the active site is a blind tube, open at one end to the solution, into which the 2-carboxyarabinitol 1,5-bisphosphate has inserted. The position of the essential $Mg^{2+}$ cation is indicated. Reprinted with permission from ref 171. Copyright 1990 Academic Press.

If a crystal of the occupied enzyme has a different space group from a crystal of the unoccupied enzyme, it necessarily presents an entirely new crystallographic problem. In most cases this new problem can be solved rapidly using the method of **molecular replacement.** Methods are available to rotate and translate the crystallographic molecular model of the unoccupied enzyme, derived in the earlier study, within the new unit cell until the *R*-factor between the observed data set from the new crystal and the data set calculated from the old crystallographic molecular model within the new unit cell reaches a minimum. At this point, it is assumed that the position and orientation of the old molecular model is the same as the orientation of the molecule of protein in the actual unit cell of the new crystal. This provisional molecular model positioned in the unit cell of the new crystal is then refined against the data set for the new crystal. The ligand is incorporated at the stage of the refinement at which a feature of positive difference electron density is definitive enough to define its position and orientation in the active site.

If molecular replacement fails, the phases of the individual reflections in the new data set must be established by multiple isomorphous replacement, and the elucidation of the structure becomes as involved and laborious as any new crystallographic investigation. Once the map of electron density is obtained, the folded polypeptide, because it is very similar or identical in structure to the folded polypeptide of the unoccupied enzyme, can be recognized and accounted for immediately and any extra electron density can usually be assigned unambiguously to the molecule or molecules of the ligand.

If crystals of the occupied enzyme can be obtained that are **isomorphous** to crystals of the unoccupied enzyme, the phases of the reflections that are already available from the crystallographic study of the unoccupied enzyme can be used. This significant advantage puts a premium on obtaining isomorphous crystals even though those isomorphous crystals may not have the enzyme in the proper conformation for catalysis. An observed data set, $|F_{o,le}|$, is collected from the isomorphous crystals of the liganded enzyme (le). Were the phases of all of these reflections known precisely, an unambiguous map of electron density could be calculated from this data set immediately, but they are not known because the presence of the ligand has produced a different set of reflections from those collected from crystals of the unliganded enzyme (ule). To solve this new structural problem, the information gathered from the crystals of the unliganded enzyme is relied upon as heavily as possible.

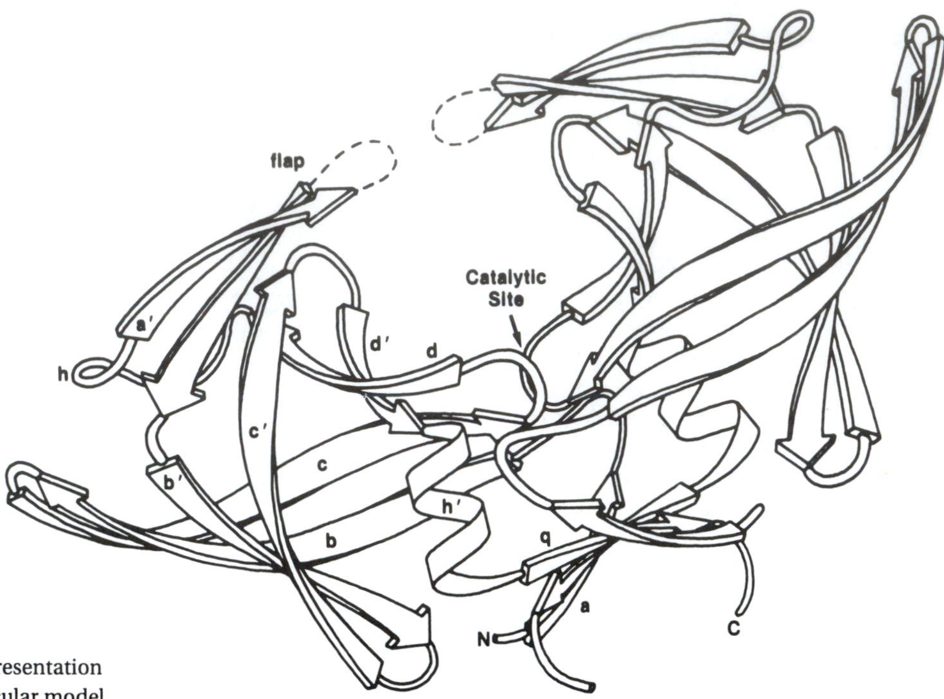


**Figure 3–30:** Schematic representation of the crystallographic molecular model of the $\alpha_2$ dimer of the aspartic proteinase from Rous sarcoma virus.[172] Crystals of the aspartic proteinase from Rous sarcoma virus were formed in a solution of ammonium sulfate (space group $P3_121$). The phases of the reflections to 0.3 nm were determined by multiple isomorphous replacement using four derivatives. The electron density of each of the two protomers of identical amino acid sequence forming the $\alpha_2$ dimer could be identified, and a molecular model of the polypeptide could be inserted into each protomer. The 2-fold rotational axis of symmetry relating the two protomers did not coincide with any of the crystallographic 2-fold rotational axes of symmetry and the asymmetric unit was the dimer itself. Refinement was performed until the final $R$-factor was 0.144. Following refinement, it could be concluded that the two protomers are related by a "nearly perfect"[172] 2-fold rotational axis of symmetry. As this enzyme is obviously homologous to other aspartic proteinases, the active sites of which have been defined crystallographically, the active site in the crystallographic molecular model of the aspartic proteinase from Rous sarcoma virus (catalytic site) was assumed to occupy a homologous location. This placed it upon the 2-fold rotational axis of symmetry. Reprinted with permission from ref 172. Copyright 1990 American Chemical Society.

The first step in the process is to construct a **map of difference electron density.**[173a] To perform this calculation, the observed data set, $|F_{o,le}|$, for the liganded enzyme is considered to be the fundamental data set. Decisions are made both about the phases to be used and about the other data set to be subtracted. The choice of phases, although critical, is arbitrary. The crystallographer bases her decision on whether she believes the observed phases for the unliganded enzyme, $\alpha_o^{ule}$, calculated from the original isomorphous replacements or the calculated phases, $\alpha_c^{ule}$, obtained from the Fourier transform of the refined molecular model are the more reliable set of phases for the reflections from the crystal of the unliganded enzyme (ule).

The calculation of the map of difference electron density is referred to by the shorthand $(nF_i–F_j)$, where $n$ is either the integer 1 or the integer 2 denoting the multiplication of the individual amplitudes of the data set $|F_i|$ that has been performed and $F_j$ refers to the data set $|F_j|$. Each of the normalized amplitudes in the data set $|F_i|$ is multiplied by $n$, and the normalized amplitude with the same index in the data set $|F_j|$ is subtracted from it to form a difference data set $|nF_i–F_j|$. The first map of difference electron density usually calculated is $(F_{o,le}–F_{o,ule})$ using the chosen set of phases.[173a]

An example of such a map of difference electron density is that obtained by subtracting the data set for unliganded thermolysin from the data set for thermolysin to which the competitive inhibitor $N$-(3-phenylpropionyl)-L-phenylalanine

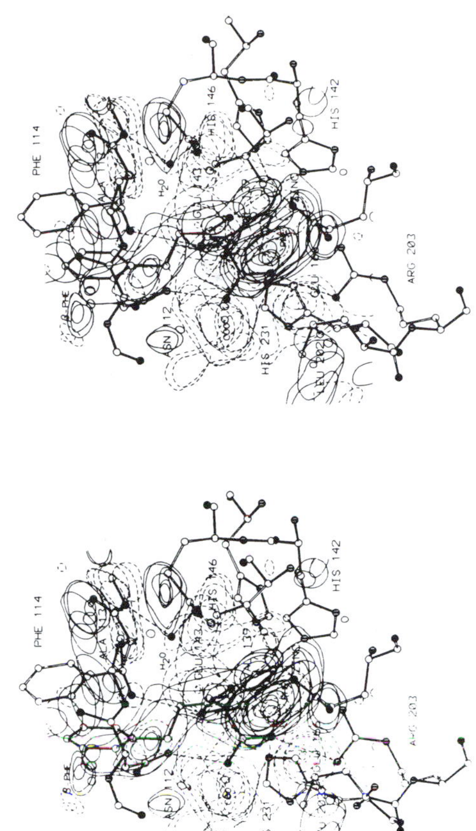


**Figure 3–31:** Map of difference electron density between unliganded thermolysin and thermolysin to which *N*-(3-phenylpropionyl)-L-phenylalanine (**3–35**) had been bound at the active site.[174] Thermolysin was crystallized from mixtures of dimethyl sulfoxide and water 1.4 M in calcium acetate. A data set, $F_{nat}$, was gathered to spacings of 0.23 nm and phases were determined by isomorphous replacement with five derivatives. This provided the initial map of electron density into which a molecular model of the polypeptide was inserted. This initial molecular model was submitted to refinement to produce the crystallographic molecular model represented by the molecular skeleton in the figure. Crystals were then soaked in mother liquor containing *N*-(3-phenylpropionyl)-L-phenylalanine for several days and a new data set, $F_{deriv}$, was then gathered. A map of difference electron density[173a] was calculated from the difference ($F_{deriv}$–$F_{nat}$). Positive electron density (solid lines) identifies features present in the liganded enzyme but not the unliganded enzyme; negative density (broken lines), the reverse. The large uncompensated peak of positive electron density marks the location of the inhibitor. A skeletal model of *N*-(3-phenylpropionyl)-L-phenylalanine has been inserted at this location. Its phenyl ring occupies the major feature of positive electron density and its carboxyl group is shown as hydrogen-bonded to Asparagine 112 and Arginine 203. Regions of positive electron density paired with negative electron density arise from shifts in the molecule of protein upon binding of the inhibitor. Reprinted with permission from ref 174. Copyright 1977 American Chemical Society.

**3–35**

has been bound (Figure 3–31).[174] Wherever there is more electron density within the unit cell of the liganded enzyme than there is electron density within the unit cell of the unliganded enzyme, positive electron density (denoted by a solid line) is present in the difference map. The inhibitor itself produces a significant feature of positive electron density at the active site (center of Figure 3–31).[174] Wherever there is less electron density in the unit cell of the liganded enzyme, negative electron density (denoted by a broken line) is present in the difference map.

Negative electron density can arise in two ways. If a water molecule or some other small molecule or several of these molecules, bound to the unliganded enzyme near the location occupied by the ligand, are displaced when the ligand is bound, this displacement will register as negative electron density. A large region of negative density, presumably due to such a displacement, appears adjacent to the carboxylate of the ligand in Figure 3–31. (The carboxylate is hydrogen-bonded by Asparagine 112 and Arginine 203.) If one of the amino acids or a segment of the polypeptide in the enzyme moves when the ligand binds, the location it vacated will register as negative electron density and the new location of the realigned portion will register as positive electron density. Leucine 202, Glutamate 143, and a portion of the polypeptide including Alanine 113 and Phenylalanine 114 display such shifts in Figure 3–31. These realignments can be recognized as pairs of adjacent positive and negative features of approximately equal magnitude.

This initial ($F_{o,le} - F_{o,ule}$) map of difference electron density locates the active site within the unit cell, but it has a major shortcoming associated with it. An active site is required to orient substrates so that they can participate as constituents in a formally intramolecular reaction. The noncovalent interactions almost always employed to perform this orientation are hydrogen bonds. The donors and acceptors of hydrogen

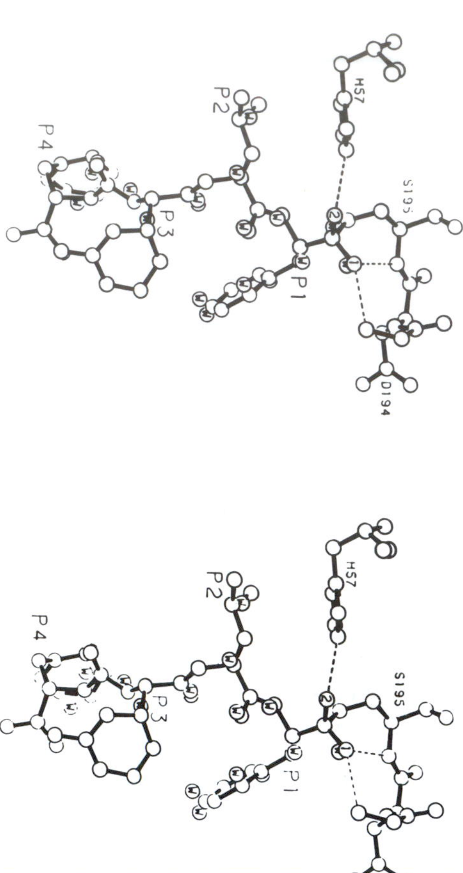


**Figure 3–32:** Crystallographic molecular model of the inhibitor chymostatin bound within the active site of *S. griseus* proteinase A, including the molecules of water displaced from the active site by the inhibitor.[175] Proteinase A was crystallized from 1.3 M sodium phosphate, pH 4.1 (space group $P4_2$). A data set, $F_{nat}$, was collected to spacings of 0.18 nm. The phases of these reflections were estimated from four isomorphous replacements, and the initial crystallographic molecular model was refined to an *R*-factor of 0.14. This refined crystallographic molecular model contained 200 molecules of water. Crystals were then soaked in mother liquor containing chymostatin, and a new data set, $F_{chymo}$, was then collected. The map of difference electron density ($F_{chymo} - F_{nat}$) was then calculated to locate the position of the chymostatin in the unit cell. The inhibitor was found to overlap 18 of the water molecules in the refined crystallographic molecular model of the unoccupied enzyme. These molecules of water were removed from the crystallographic molecular model, and a data set, $F_c$, was then calculated from it. The difference map of electron density ($F_{chymo} - F_c$) was then used to position the chymostatin as accurately as possible. This resulting molecular model was then refined against the data set $F_{chymo}$. The stereo drawing presents the final configuration of the chymostatin (P1–P4) and three of the amino acids in the active site (H57, S195, and D194). Fourteen of the 15 waters that were located within the active site of the unliganded crystallographic molecular model at locations overlapping that of the chymostatin and that are presumably displaced by the chymostatin as it associates are indicated as W. Reprinted with permission from ref 175. Copyright 1985 Academic Press.

bonds in an active site unoccupied by substrates form hydrogen bonds with more or less rigidly held molecules of water.[170a] In addition, large pockets intended for purely hydrophobic portions of the substrates, even though they themselves are lined by hydrophobic amino acids, do not remain empty in the absence of substrates but are occupied by solvent. When a ligand binds to the active site by occupying the donors and acceptors for these hydrogen bonds and the hydrophobic pockets, the molecules of water occupying these locations are displaced. For example, in the active site of proteinase A from *Streptomyces griseus* there are 15 rigidly held molecules of water that are displaced when the competitive inhibitor chymostatin binds to the active site (Figure 3–32).[175] The negative electron density associated with such displacements cancels a significant fraction of the positive electron density associated with the ligand. In the case of the binding of chymostatin to proteinase A, the negative electron density from the 15 molecules of water (150 electrons) present in the active site of the crystallographic molecular model of the unliganded enzyme would have canceled a significant fraction of the electron density of the chymostatin (324 electrons) bound at the active site in the liganded enzyme had these molecules of water not been subtracted from the crystallographic molecular model before the map of difference electron density was calculated.

Often the cancelation is obvious and can be accounted for immediately. For example, in the ($F_{o,le} - F_{o,ule}$) difference map of electron density for the binding of lactone **3–11** to the active site of lysozyme, there is no positive electron density at the location of the glycosidic oxygen connecting the *N*-acetylglucosaminopyranose at subsite A to the *N*-acetylglucosaminopyranose at subsite B because a molecule of water occupies this location in the unliganded enzyme (Figure 3–33).[127] The exchange of one oxygen for another oxygen causes the difference map of electron density to be blank at this location.

One way to avoid the shortcomings associated with the displacement of the water molecules is to displace them first computationally,[174,176] as was done in the study of chymostatin (Figure 3–32). Once the boundaries of the ligand within the active site have been defined by the ($F_{o,le} - F_{o,ule}$) map of difference electron density, each of the water molecules in the refined molecular model intersecting or located within those boundaries can be removed to produce a dehydrated (dh) active site. A data set of reflections $|F_{calc,dhule}|$ can be calculated for the dehydrated, unliganded enzyme. The ($F_{o,le} - F_{calc,dhule}$) map of difference electron density will display the full positive difference electron density associated with the bound ligand (Figure 3–34A).[176]

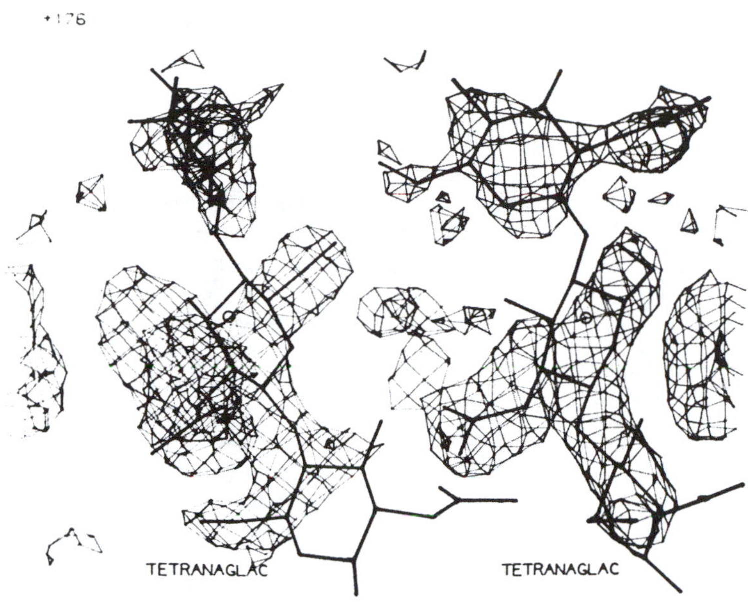


**Figure 3–33:** Map of difference electron density between unliganded lysozyme and lysozyme to which lactone **3–11** has been bound at the active site.[127] Lysozyme was crystallized from 1 M NaCl at pH 4.7. A data set, $F_{nat}$, was gathered to spacings of 0.25 nm, and phases were estimated with three different isomorphous replacements. This provided an initial map of electron density and a crystallographic molecular model. Crystals were then soaked in solutions of lactone **3–11** for 24 h, and a new data set, $F_{lactone}$, was then gathered. A map of difference electron density was calculated from the difference ($F_{lactone} - F_{nat}$) and the observed phases from the earlier isomorphous replacements. Two orthogonal views of that map of difference electron density are presented for the positive electron density located in subsite A (upper) and subsite B (lower) of the active site of the enzyme. A skeletal model of lactone **3–11** was inserted into the positive electron density in the difference map, and this skeletal model is represented by the structure drawn within the electron density. The three pyranoses in subsites A, B, and C are presented. Reprinted with permission from ref 127. Copyright 1974 Academic Press.

A molecular model of the ligand is inserted into the envelope of electron density assigned to the ligand in the map of difference electron density (Figure 3–34A). This positions the ligand in the unit cell. Because a refined molecular model of the unliganded protein has previously been positioned in the same unit cell, this operation inserts the model of the ligand into the model of the active site. Often this job is easier if the ($2F_{o,le} - F_{calc,dhule}$) map of difference electron density is used.[176] In this map the electron density of the entire protein is present, but the electron density of the ligand has twice the intensity of the electron density of the protein.

These operations create a crystallographic molecular model of the enzyme with the ligand occupying the active site. An observed data set, $|F_{o,le}|$, containing indexed reflections from a crystal of the liganded enzyme is also available. An initial molecular model and an observed data set are all that are required to initiate a refinement of the molecular model of the liganded enzyme. The refinement produces a refined crystallographic molecular model of the liganded active site. As with all refinements, the reliability of the final molecular model depends entirely on the validity of the manual insertion of the ligand into the original difference map of electron density (Figure 3–34A).

Because a crystallographic molecular model of an active site occupied by a ligand is a static structure and an enzymatically catalyzed reaction is a rapid kinetic process, the information gained about the catalytic mechanism of the enzyme from the crystallographic molecular model depends entirely upon the **choice of the ligand**. Almost every enzyme catalyzes a reaction involving two or more reactants. In such a situation, the active site can be located by occupying it with only one of the two reactants. For example, both MgADP and MgATP were bound in separate experiments at the active site of crystalline phosphoglycerate kinase by soaking crystals in solutions of the respective ligand. Both of these substrates were bound at the same location and in the same orientation on the surface of the enzyme, and this observation both located the active site and defined some of the noncovalent interactions between the active site and an adenine nucleotide.[177] Most of the dehydrogenases have been crystallized with either $NAD^+$ or NADH occupying their active sites[173,178,179] in the absence of the other respective reactant[179] or in the presence of an inhibitor that is competitive with the other reactant.[173] In most instances, this has permitted the description of the networks of hydrogen bonds that form between the amino acids of the active site and the $NAD^+$ or NADH (Figure 3–35).[173]

This practice of using just one of two or three reactants, however, is often misleading. For example, crystals of the unliganded citrate synthase were soaked in solutions of citrate and isomorphous crystals of the enzyme to which citrate is bound are obtained. When crystals were formed from enzyme

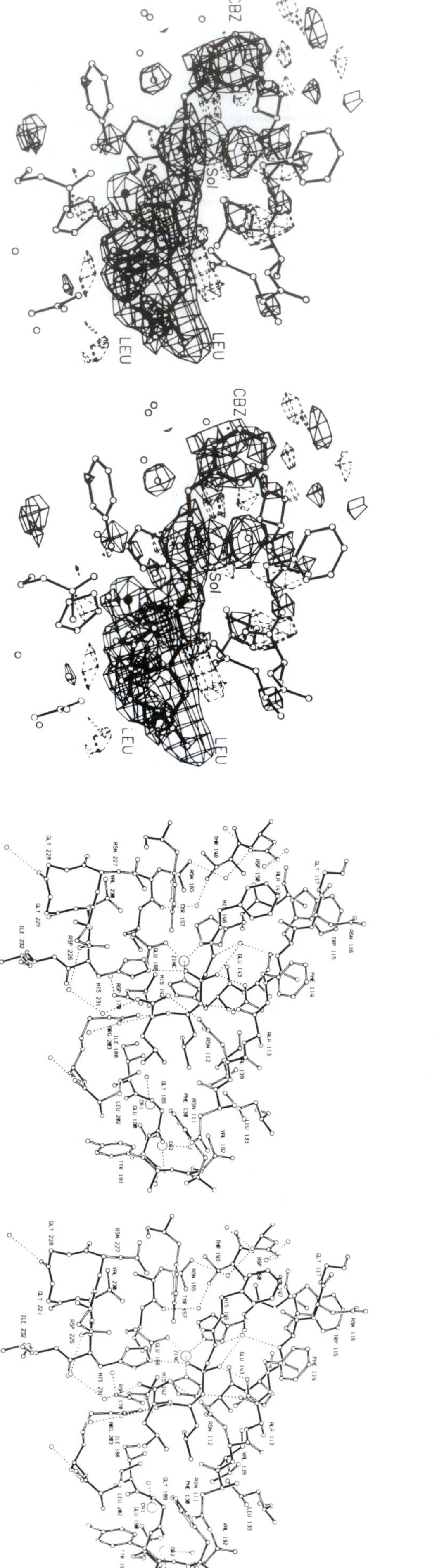


**Figure 3–34:** Map of difference electron density between unliganded thermolysin and thermolysin to which phosphonamidate **3–25** had been bound at the active site.[176] A data set from crystals of unliganded thermolysin (see Figure 3–31) gathered to spacings of 0.16 nm and a crystallographic molecular model refined against this data set served as the starting point of the study. Crystals of thermolysin were soaked in a 10 mM solution of phosphonamidate **3–25** for several days. A data set, $|F_{\text{complex}}|$, was gathered to spacings of 0.16 nm from these crystals. From maps of difference electron density, the general location occupied by the phosphonamidate in the active site could be defined. All molecules of water overlapping this location in the refined crystallographic molecular model were removed, and a data set, $|F_{\text{nat,calcd}}|$, and a set of phases, $\alpha_{\text{nat,calcd}}$, was calculated from this dehydrated model to spacings of 0.16 nm. The map of difference electron density $(F_{\text{complex}} - F_{\text{nat,calcd}})$ was calculated from these two data sets and the set of phases. (A) Features of this map of difference electron density in the location of the active site occupied by phosphonamidate **3–25**. A few of the side chains from the refined crystallographic model of the unliganded enzyme are included, and a skeletal model of phosphonamidate **3–25** has been inserted into the large feature of positive electron density. (B) The initial crystallographic molecular model of phosphonamidate in the active site (panel A) was refined against the data set $|F_{\text{complex}}|$, and the final refined crystallographic molecular model is presented. Reprinted with permission from ref 176. Copyright 1987 American Chemical Society.

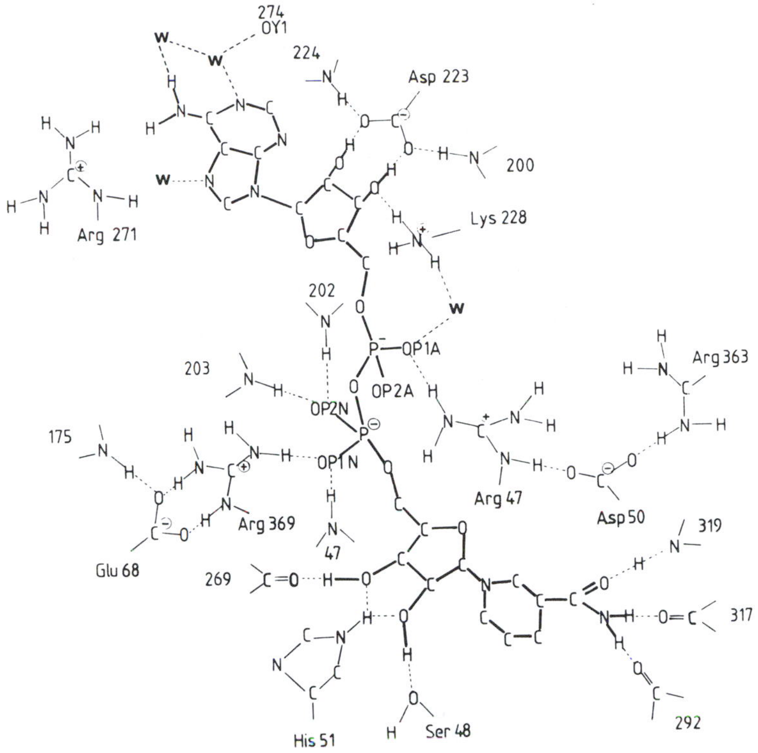

**Figure 3–35:** Diagrammatic representation of the network of hydrogen bonds between NADH and amino acid side chains in the polypeptide surrounding the active site of horse liver alcohol dehydrogenase.[173] The ternary complex between alcohol dehydrogenase, NADH, and dimethyl sulfoxide was crystallized from 25% 2-methyl-2,4-pentanediol at pH 7.0 (space group *P*1). The data set was gathered to Bragg spacings of 0.29 nm, and phases were determined by isomorphous replacement with four derivatives. The initial crystallographic molecular model of the liganded enzyme was built with the assistance of a previously refined crystallographic molecular model of the unliganded enzyme from a data set at spacings of 0.24 nm. Unfortunately, the unliganded enzyme crystallizes in a different space group ($C222_1$). Nevertheless, details of the atomic structure of the liganded enzyme could be based on this more accurate model of the unliganded enzyme. All amino acid side chains and peptide functional groups forming hydrogen bonds to the NADH in the final refined crystallographic molecular model of the ternary complex are presented. Reprinted with permission from ref 173. Copyright 1984 American Chemical Society.

to which both citrate and coenzyme A had been bound, however, the enzyme changed its conformation, causing the active site to rearrange and the details of the interactions between citrate and the amino acids around it to change significantly.[180]

If an enzyme has only one reactant and one product or if one of the two reactants is water itself, competitive inhibitors are usually used to define the active site. For example, thymidine 3′,5′-bisphosphate, a competitive inhibitor of deoxyribonuclease I, has been used to identify the active site in the crystallographic molecular model of the enzyme.[181]

Sometimes, a reactant in a formally unimolecular reaction is used as a ligand in crystallographic studies even though it is transformed by the enzyme. *N*-Formyltryptophan is a reactant in an oxygen exchange reaction catalyzed by chymotrypsin at its carboxylate[182]

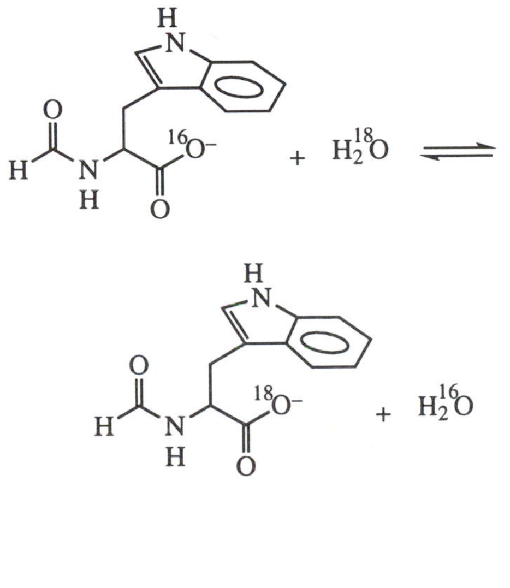

$$+\ H_2{}^{18}O \rightleftharpoons \quad +\ H_2{}^{16}O$$

(3–215)

The reaction is slow and its product is chemically equivalent to the reactant. The crystallographic molecular model of α-chymotrypsin occupied by *N*-formyltryptophan has been used to examine the interactions between this reactant and the amino acids in the active site.[183] The reactant and the product in the reaction catalyzed by triosephosphate isomerase are present in an almost equimolar mixture when they are bound to the active site,[184] and they resemble each other quite closely

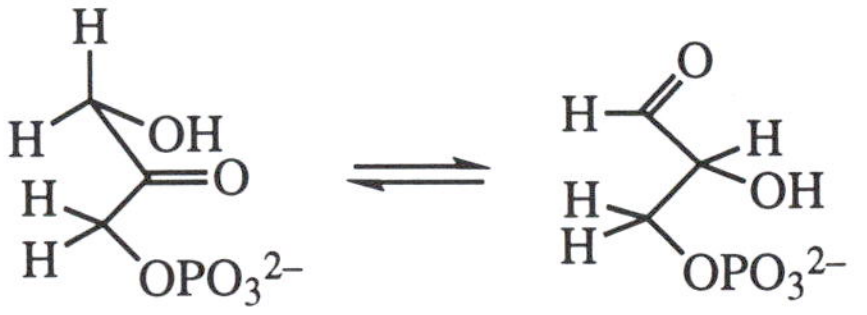

$$\rightleftharpoons$$

(3–216)

although not so closely as do the reactant and product in Reaction 3–215. Nevertheless, crystals of triosephosphate isomerase were soaked in a solution of dihydroxyacetone phosphate, and the difference map of electron density had a large positive feature about the size of one of the substrates[185] located adjacent to Glutamate 165, an amino acid that had been identified chemically in earlier studies as a participant in the active site.

Bisubstrate analogues can also be used as ligands to identify catalytic amino acids in the active site of an enzyme. The bisubstrate analogue (3*S*)-5-(3-carboxy-3-hydroxypropyl)-NAD$^+$

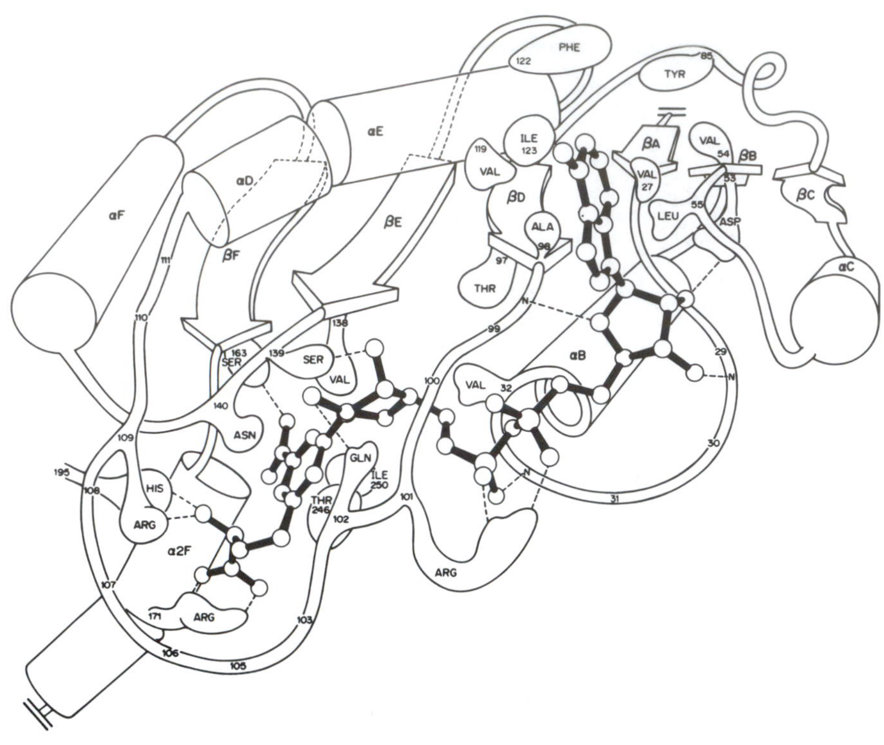


**Figure 3–36:** Diagrammatic representation of the complex between pig lactate dehydrogenase and bisubstrate analogue **3–36**.[186] The complex between enzyme and substrate crystallizes in the $P3_221$ space group. From these crystals a data set was gathered to Bragg spacings of 0.27 nm. A refined crystallographic molecular model of the complex between dogfish lactate dehydrogenase, $NAD^+$, and pyruvate was already available. This other enzyme complexed with these other substrates, however, had crystallized in the $P42_12$ space group. Consequently, the crystals of the complex for which a refined crystallographic molecular model was available were not isomorphous with the crystals containing the complex between pig lactate dehydrogenase and bisubstrate analogue **3–36**. Therefore, the method of molecular replacement was used to obtain an initial crystallographic molecular model for refinement against the unphased data set from the $P3_221$ crystals. The available crystallographic molecular model was positioned along the crystallographic 2-fold rotational axis of symmetry in the $P3_221$ unit cell by rotation and translation until the *R*-factor reached a minimum value. The amino acids in the available crystallographic molecular model were changed so that it had the amino acid sequence of the pig enzyme rather than the dogfish enzyme and the 3-carboxy-3-hydroxypropyl group was inserted into the electron density found in the map of electron density adjacent to the 5 position on the nicotinamide ring. The molecular model was then refined to an *R*-factor of 0.20 at spacing of 0.27 nm. The diagram presents an exact skeletal model of the conformation of analogue **3–36** in the final molecular model and a diagram of the amino acid side chains that form contact with it around the active site. Reprinted with permission from ref 186. 

**3–36**

is a bisubstrate analogue for lactate dehydrogenase, which catalyzes the reaction

$$\mathrm{H{-}C(OH)(CH_3){-}COO^-} + \mathrm{NAD^+} \rightleftharpoons \mathrm{O{=}C(CH_3){-}COO^-} + \mathrm{NADH} \qquad (3\text{–}217)$$

When this bisubstrate analogue is added to a solution of lactate dehydrogenase, the hydride is transferred back and forth at the active site intramolecularly between carbon 3 of the propyl side chain and carbon 4 of the nicotinamide, an observation demonstrating that the one analogue effectively mimics the two substrates. A crystallographic molecular model of the bisubstrate analogue occupying the active site of the enzyme has been constructed (Figure 3–36).[186]

Unquestionably, the most informative ligands that can be chosen for crystallographic studies of the active site of an enzyme are analogues of high-energy intermediates. One of the phosphonamidate inhibitors of thermolysin, phosphonamidate **3–25**, is based on the reactant Cbz-Gly-Leu-Leu (Table 3–4). Crystals of thermolysin soaked in a solution of phosphonamidate **3–25** were used to generate a difference map of electron density (Figure 3–34A) into which a skeletal model of the phosphonamidate was inserted. A refined crystallographic molecular model of the active site occupied by this analogue of the tetravalent intermediate was then produced (Figure 3–34B).[176] Aside from identifying all of the important catalytic residues, the crystallographic molecular model defines the noncovalent interactions holding the ana-

**Figure 3–37:** Mechanism by which serine proteinases catalyze the hydrolysis of a peptide bond. Attack of the serine on the acyl carbon produces the tetravalent intermediate that ejects the amine upon protonation of the nitrogen to produce the acyl intermediate. Water then attacks, and protonation of the oxygen of the serine in the resulting tetravalent intermediate leads to its ejection to complete the reaction.

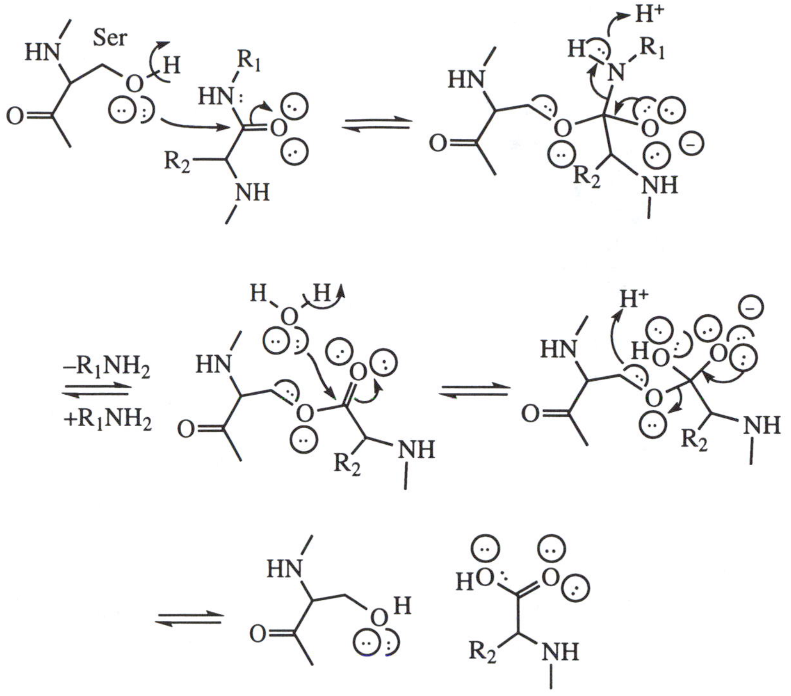

logue in the active site along its entire length. It was also observed that the binding of this analogue of the tetravalent intermediate was accompanied by almost no change in the structure of the active site, and these observations implied that the active site of thermolysin, even when it is unoccupied, is built to bind the tetravalent intermediate efficiently. This would be expected if strain were operating in the catalytic mechanism of this enzyme.

An even more graphic example of strain has been observed in crystallographic studies of serine proteolytic enzymes. In this class of enzymes, a peptide bond is hydrolyzed by first forming a transient ester with a serine within the active site, which in turn is hydrolyzed (Figure 3–37). The substrates and the ester at the serine are all flat, planar, and trigonal at the acyl carbon, but the two tetravalent intermediates are tetrahedral at the acyl carbon and have an anionic oxygen atom as one of their substituents. Phenylboronic acid (**3–23**, R = phenyl) is an inhibitor of several of these enzymes[154] because of its ability to form a tetravalent adduct with the serine

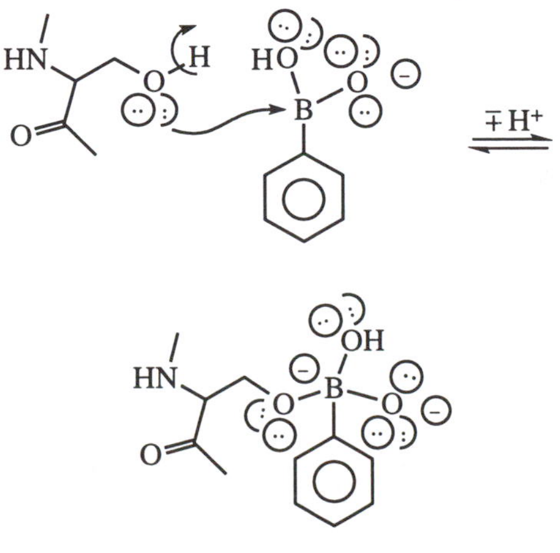

(3–218)

In the crystallographic molecular model of this adduct with subtilisin, the anionic oxygen ended up in a pocket, referred to as the oxyanion hole, surrounded by two appropriately positioned hydrogen-bond donors (Figure 3–38).[155] The donors are the amide nitrogen of Asparagine 155 and the amide nitrogen from Serine 221 in the polypeptide backbone. Because these noncovalent interactions do not occur between the acyl oxygen and these two hydrogen-bond donors in the crystallographic molecular models of substrates and inhibitors bound in the active site, it is assumed that they come into being as the transition state is reached. In this way, the free energy of formation of these two hydrogen bonds can be used to lower the free energy of activation for the reaction.

A similar pair of hydrogen bonds is seen in the crystallographic molecular model of chymostatin within the active site of proteinase A from *S. griseus* (Figure 3–32). Chymostatin is an aldehyde that forms a hemiacetal anion with the serine in the active site

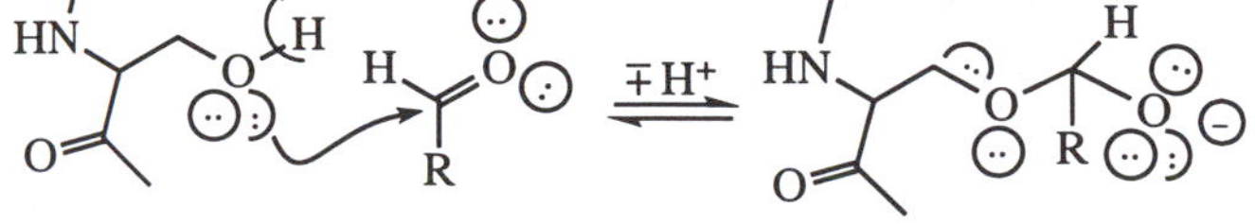

(3–219)

Both the *S*-hemiacetal and the *R*-hemiacetal are formed in equimolar ratio (oxygen 1 and oxygen 2 in Figure 3–32, respectively). It is the oxygen of the *S*-hemiacetal that ends up in the oxyanion hole. In this instance, the two donors forming the oxyanion hole are both amide nitrogens from the polypeptide backbone.

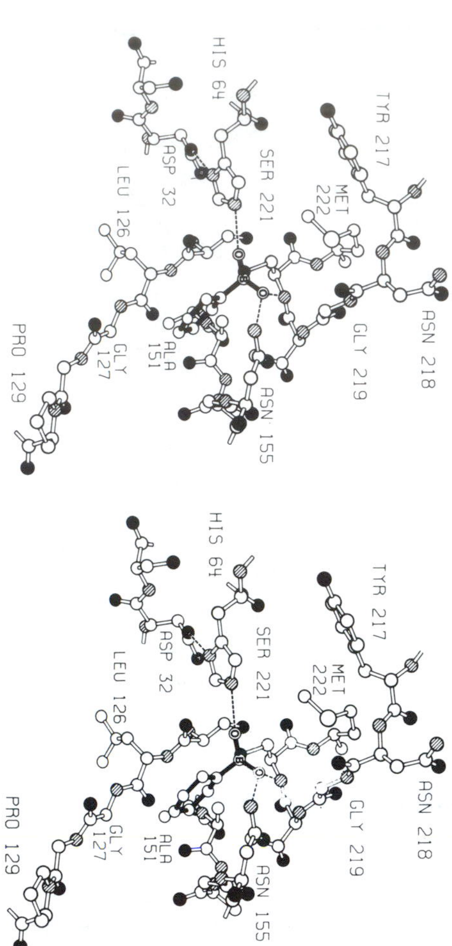


**Figure 3–38:** Crystallographic molecular model of phenylboronic acid in a tetravalent adduct with Serine 221 in the active site of subtilisin.[155] Crystals of unliganded subtilisin grown in ammonium sulfate at pH 5.9 provided a data set to Bragg spacing of 0.2 nm; and, after isomorphous replacement, a refined crystallographic molecular model ($R = 0.23$). From this refined crystallographic molecular model, a set of phases, $\alpha_c$, was calculated. Isomorphous crystals of unliganded subtilisin were formed in ammonium sulfate at pH 7.5 and provided a data set, $|F_{nat,7.5}|$, to Bragg spacing of 0.25 nm. These crystals were then soaked with a saturated solution of phenylboronic acid dissolved in mother liquor. From these crystals a third data set, $|F_{deriv}|$, was collected to Bragg spacing of 0.25 nm. The map of difference electron density, $(F_{nat,7.5} - F_{deriv})$, was calculated using $\alpha_c$. A molecular model of phenylboronic acid was inserted into the large feature of positive electron density adjacent to Serine 221 in the refined crystallographic molecular model of the unliganded enzyme in such a way that a covalent bond was formed between the oxygen of the serine and the boron. This provided a molecular model of the covalent complex between the enzyme and phenylboronic acid. This molecular model was then refined against the data set $|F_{deriv}|$ to produce the refined crystallographic molecular model presented in the figure. Reprinted with permission from ref 155. Copyright 1975 *Journal of Biological Chemistry*.

A difference map of electron density between lysozyme with an unoccupied active site and lysozyme with an active site occupied by lactone **3–11**, the analogue of the oxonium cation, has been prepared from crystals of the enzyme soaked in the analogue.[127] Although $\delta$-lactones of this type are supposed to assume the half-chair conformation as the conformation of lowest energy in solution

**3–37**

this conformation did not fit the volume of positive electron density. Either a sofa conformation (**3–38**) or a [3,0] boat conformation (**3–39**), however, did fit the electron density satisfactorily

**3–38**

**3–39**

It has been concluded that one or both of these two conformations of the lactone are enforced by strain at the active site. The advantages of either of these conformations to the reaction catalyzed by the enzyme are obvious. For example, in the sofa conformation of the oxonium intermediate

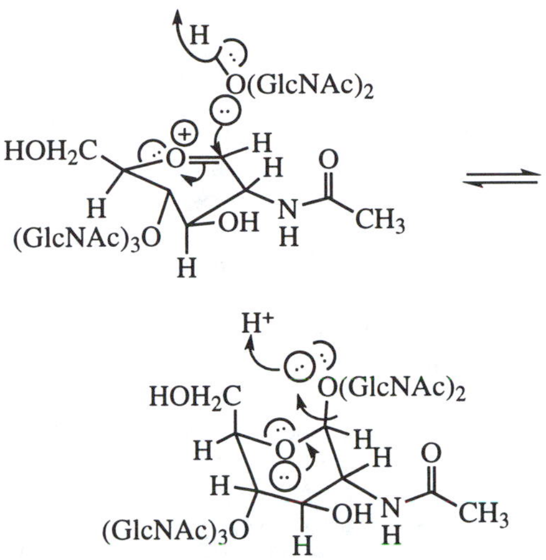

(3–220)

oxygen 5 and carbon 1 of the oxonium cation and their substituents would all lie in the same plane, consistent with the existence of a $\pi$ bond between oxygen 5 and carbon 1. In the [3,0] boat conformation of the reactant, a lone pair of electrons on oxygen 5 is antiperiplanar to the $\sigma$ bond to be broken, and it should assist in the reaction by providing push.[187]

Analogues of the pentacoordinate intermediate of high energy in phosphate transfer[188,189] reactions have also been studied crystallographically. In aqueous solutions, oxovanadium(IV) usually forms octahedral complexes with six covalently bound oxygen ligands, but complexes in which only five oxygens surround the vanadium(IV) have been observed. A complex between uridine and oxovanadium(IV) was found to inhibit ribonuclease competitively ($K_i$ = 10 $\mu$M).[190] Ribonuclease, as part of its normal mechanism, catalyzes the hydrolysis of 2′, 3′-cyclic nucleoside phosphates

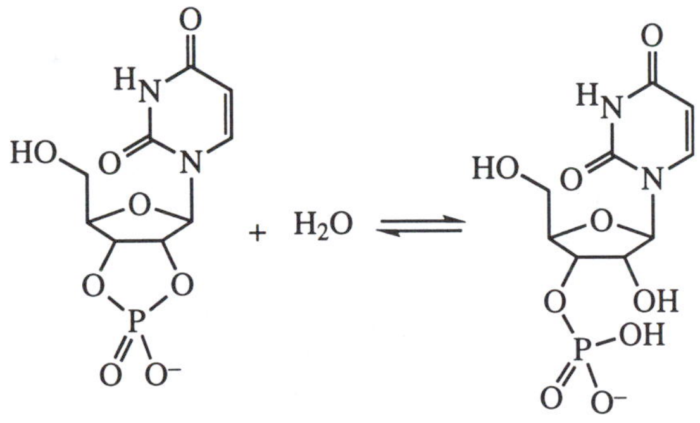

$$+ \; H_2O \rightleftharpoons$$

(3–221)

In this reaction, pentavalent intermediate **3–40** should be formed in which the leaving group, oxygen 2′, should be at an apical position in a trigonal bipyramid

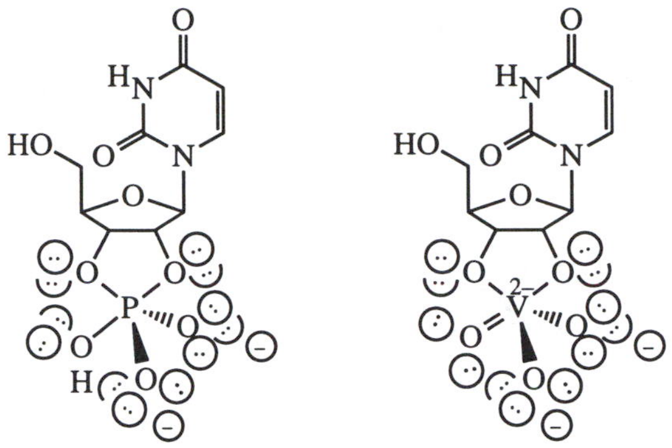

**3–40** **3–41**

It was proposed that the inhibition observed with the complex between oxovanadium(IV) and uridine results from the binding to the active site of the pentacoordinate complex **3–41** between uridine, vanadate, and water. In the crystallographic molecular model, obtained from neutron diffraction of crystals of ribonuclease soaked in solutions of uridine and oxovanadium(IV), such a pentacoordinate complex was observed. In this complex, the oxygens around the vanadium assumed a distorted trigonal bipyramid (Figure 3–39).[191] In this refined crystallographic molecular model, because it was the result of neutron diffraction, hydrogens and deuteriums are readily observed. In the model an unambiguously defined hydrogen bond exists between Lysine 41 of ribonuclease and oxygen 2′ of the ribose. In crystallographic molecular models of substrates within the active site of ribonuclease, Lysine 41 is not close enough to any of the oxygens around the phosphorus to form a hydrogen bond.[192] An inference drawn from these two observations is that as the pentacoordinate trigonal bipyramid is achieved in the transition state, the movement of oxygen 2′ is sufficient to permit it to form a hydrogen bond with Lysine 41 and that the free energy of formation of this hydrogen bond lowers the free energy of activation for the reaction.

The **noncovalent forces** that provide the positive free energy of dissociation necessary to bind the substrates with sufficiently small dissociation constants are apparent in the crystallographic molecular models of inhibitors, substrates, bisubstrate analogues, and analogues of high-energy intermediates bound within active sites. The major contribution to the positive standard free energy of dissociation from these complexes is from the **hydrophobic force**. The juxtaposition of hydrophobic surfaces to the exclusion of water (Figures 3–31, 3–32, 3–34, 3–36, and 3–38) is often observed in a crystallographic molecular model of a ligand bound at an active site. For example, the two leucines at the carboxy-terminal end of phosphonamidate **3–32** are bound to thermolysin in a pocket formed from leucine, valine, isoleucine, and phenyl-

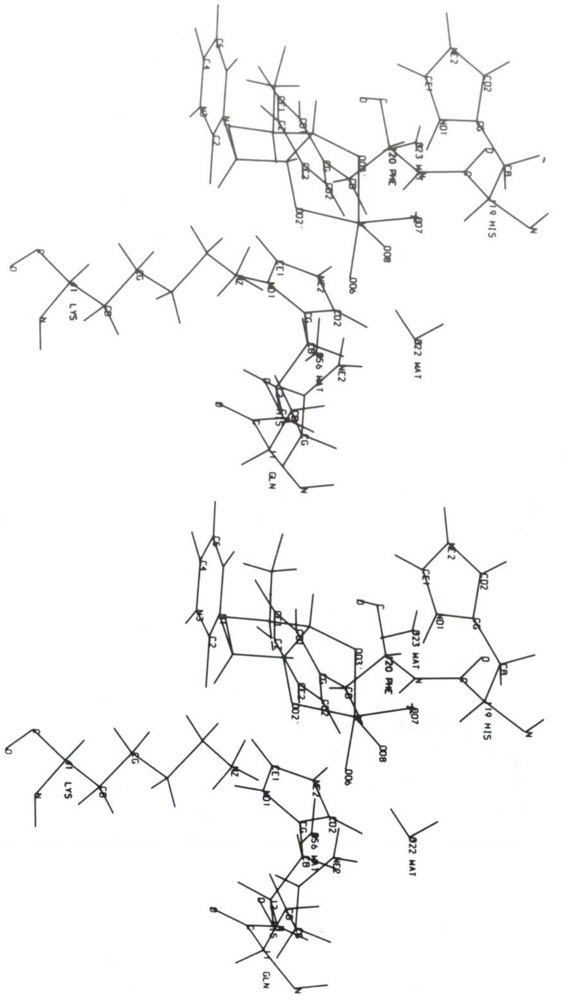

**Figure 3–39:** Crystallographic molecular model of the complex between bovine ribonuclease and uridyl vanadate **3–41**.[191] Crystals of ribonuclease A were grown at pH 5.3 in 43% *tert*-butyl alcohol. The crystals were soaked in a solution of uridyl vanadate in 50% $C^2H_3O^2H$–50% $^2H_2O$ both to produce the complex between enzyme and inhibitor and to exchange hydrogen for deuterium at all exchangeable positions. From one crystal, a data set, $|F_{complex,neut}|$, was collected to Bragg spacing of 0.2 nm by neutron diffraction. The available refined crystallographic molecular model of ribonuclease, obtained from crystals of the same space group, could be used to calculate a data set, $|F_{calc,neut}|$, and a set of phases, $\alpha_{calc,neut}$, for neutron diffraction from this unit cell. In the map of difference neutron scattering density ($F_{complex,neut} - F_{calc,neut}$) calculated with $\alpha_{calc,neut}$, the large feature of positive neutron scattering density corresponding to uridyl vanadate **3–41** could be readily located within the active site of the refined crystallographic molecular model of the unliganded enzyme. A molecular model of uridyl vanadate was inserted into this feature to produce an initial molecular model. This initial molecular model was refined against the data set $|F_{complex,neut}|$ to produce a refined neutron crystallographic molecular model ($R$ = 0.21 at 0.2 nm). Along the way, the positions in the molecule at which deuterons had exchanged for protons could be easily identified because of the large difference in neutron scattering capacity of these two nuclei. The final refined neutron crystallographic molecular model shown in the figure contains hydrogens and deuteriums as observed features because it is derived from neutron diffraction. Reprinted with permission from ref 191. Copyright 1983 National Academy of Sciences.

alanine (Figure 3–34B). In the crystallographic molecular model of the active site of lactate dehydrogenase (Figure 3–36), the planar $\pi$ system of the adenine ring of $NAD^+$ is sandwiched between an alanine, a valine, and a leucine on one side and a valine and a leucine on the other.

Although this fact is seldom emphasized, monosaccharides and polysaccharides have hydrophobic axial surfaces.[193] For example, cyclodextrin, a cyclic, cylindrical polymer of glucose, has a structure that turns an axial surface of each of its monomers toward the axis of the cylinder, and this forms a hydrophobic pocket. In the $\beta$ pyranose of arabinose, one of the two axial surfaces is more hydrophobic than the other

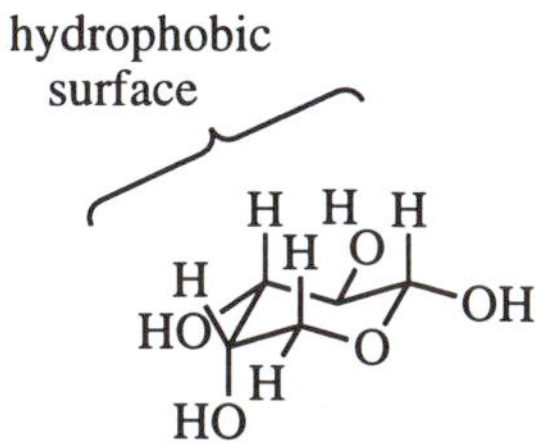

hydrophobic surface

**3–42**

because of the one axial hydroxyl. In the refined crystallographic molecular model of the complex between arabinose and the arabinose binding protein, this surface of the monosaccharide is flush against the $\pi$ system of a tryptophan and the edge of a phenylalanine.[194] The monosaccharide bound in the B site in lysozyme has one of its axial faces flush against the $\pi$ system of Tryptophan 62 (Figure 3–26). Because of the high frequency with which hydrophobic amino acids and the hydrophobic portions of hydrophilic amino acids line the surfaces of enzymatic active sites in crystallographic molecular models, it is assumed that significant hydrophobic forces are involved in the association of any ligand, other than an inorganic ion, to the active site of an enzyme.

A more obvious feature, however, of the crystallographic molecular model of ligands bound to active sites is the networks of **hydrogen bonds** ensnaring them (Figures 3–34, 3–35, and 3–36). An interesting further example of such a network of hydrogen bonds is the array of donors and acceptors on a loop of polypeptide in the active site of citrate (*si*) synthase that enfold the acceptors and donors around the edge of the adenine ring of the coenzyme A (Figure 3–40).[180] One of the more extensive examples, however, of hydrogen bonding between a

**Figure 3–40:** Crystallographic molecular model of the hydrogen-bond donors surrounding the adenine ring of coenzyme A bound within the active site of citrate (*si*) synthase.[180] Monoclinic crystals of citrate (*si*) synthase from chicken heart were produced at 1.2 M citrate, pH 6, in the presence of coenzyme A. A data set was gathered to Bragg spacings of 0.17 nm. The building of the crystallographic molecular model of this liganded chicken heart enzyme was initiated by molecular replacement using a previous crystallographic molecular model of the unliganded enzyme from pig heart obtained from tetragonal crystals by isomorphous replacement. The tetragonal crystallographic molecular model was positioned in the monoclinic unit cell by rotating and translating it until the *R*-factor was at a minimum. This initial molecular model was then refined against the monoclinic data set for the liganded chicken heart enzyme. During the refinement, the amino acid sequence of the pig heart model was changed to that of the chicken heart enzyme, and models of citrate and co-enzyme A were inserted into the appropriate features of positive electron density that appeared in maps of difference electron density. In the portion of the final refined crystallographic molecular model presented in the figure, the amide NH of Valine 315, the acyl oxygens of Valine 315 and Tyrosine 318, and the hydroxyl oxygen of the pantetheine of the coenzyme A form hydrogen bonds to four of the five positions around the adenine ring. Reprinted with permission from ref 180. Copyright 1982 Academic Press.

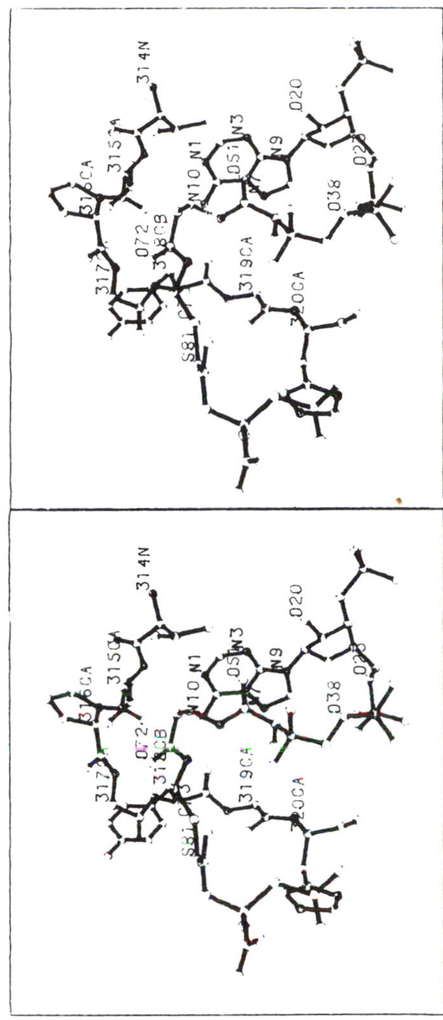

ligand and its site is found in the crystallographic molecular model of the complex between L-arabinose and the L-arabinose binding protein (Figure 3–41).[194] Of the 14 donors and acceptors of hydrogen bonds on the L-arabinose itself, eight are occupied by acceptors and donors from the amino acids surrounding the site and two are occupied by donors and acceptors of tightly bound molecules of water. Every donor of a hydrogen bond on the L-arabinose is associated with an acceptor. All of these observations are consistent with the fact that whenever either a donor or an acceptor on the ligand that is occupied by water in free solution is vacated upon the binding of the ligand to a site and is not replaced by either an acceptor or a donor from the site itself, there is a positive change in free energy, and they are also consistent with the fact that donors outnumber acceptors in a solution of protein. From these two facts it follows that the maximum possible number of hydrogen bonds will form between the ligand and the binding site and very few, if any, of the donors on the ligand will be vacant in the complex.

The formation of a hydrogen bond between an acceptor on the active site and a donor on the ligand or an acceptor on the ligand and a donor on the active site should be an almost isothermic reaction[195,196] because the concentration of hydrogen bonds in the solution does not increase during the reaction.[197,197a] It is possible, however, that the formation of a hydrogen bond between an acceptor or a donor on an otherwise rigidly held ligand and a rigidly fixed donor or acceptor, respectively, on the protein involves significant standard entropy of approximation. Two series of analogues of the tetravalent intermediate in the proteolytic reaction catalyzed by thermolysin have been synthesized. The various analogues in the first series were phosphonates (**3–43**) and the various analogues in the second series were the corresponding phosphonamidates (**3–44**), homologous, respectively, to the same set of peptides.

**3–43**

**3–44**

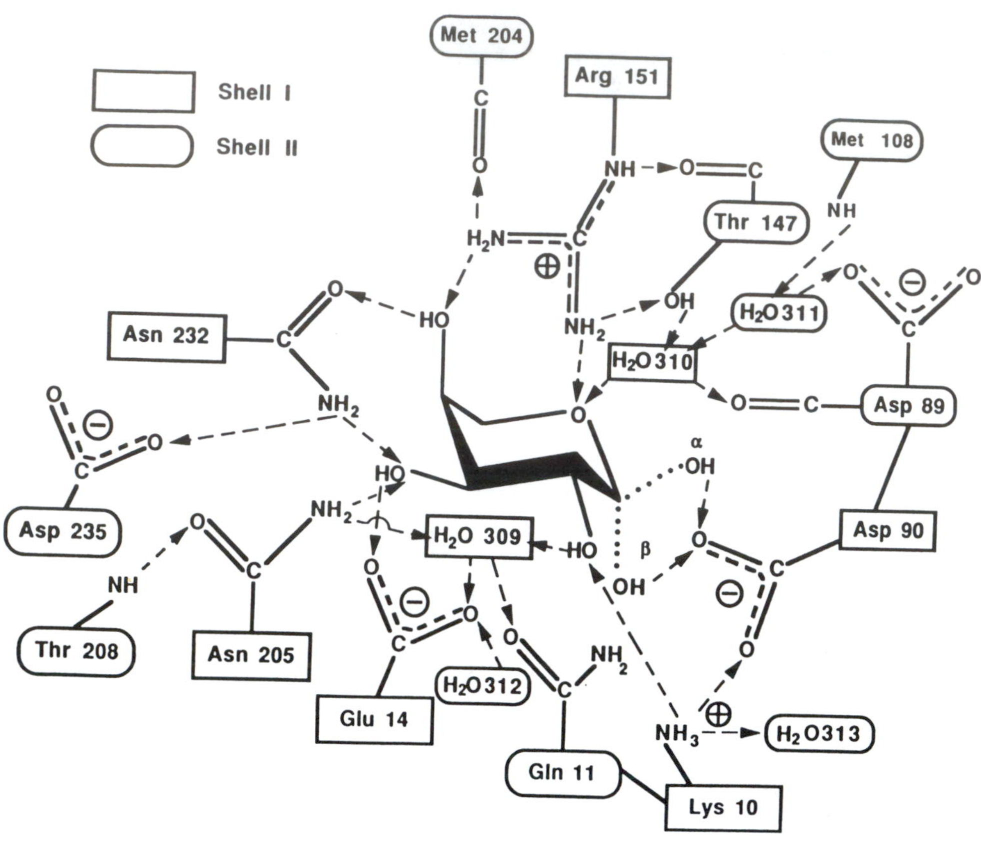


**Figure 3–41:** Schematic diagram of the complex networks of hydrogen bonds formed between L-arabinose and the various donors and acceptors surrounding it in the binding site of a crystallographic molecular model of the complex between L-arabinose and L-arabinose binding protein from *E. coli.*[194] The purified protein contains one molecule of tightly bound L-arabinose. Crystals of arabinose binding protein ($P2_12_12_1$) provided a data set to Bragg spacings of 0.17 nm. A map of electron density was obtained by isomorphous replacement and refined to an *R*-factor of 0.14. The molecular model inserted into the initial map of electron density included both the polypeptide and a model of L-arabinose. Electron density for the O1 oxygen of both the $\alpha$ and $\beta$ anomers of L-arabinose was observed, indicating that both anomers were bound to the site. Hydrogen bonds were assumed to exist between any donor and acceptor the heteroatoms of which were less than 0.31 nm apart and had reasonable angular dispositions. All of the hydrogen bonds involving donors and acceptors on the arabinose are considered primary hydrogen bonds, and all other donors and acceptors in networks involving the primary donors and acceptors are considered secondary hydrogen bonds. All of the primary and secondary hydrogen-bonded atoms are represented. Reprinted with permission from ref 194. Copyright 1984 Macmillan Magazines Limited.

When the dissociation constants ($K_i$) for each phosphonate and its corresponding phosphonamidate, all acting as competitive inhibitors of thermolysin, were compared, a consistent difference, equivalent to 16.7 ± 0.4 kJ mol$^{-1}$, was observed.[198] Each phosphonamidate had the larger free energy of dissociation (the smaller dissociation constant) relative to the corresponding phosphonate.

Between the refined crystallographic molecular model of the complex between a phosphonate and the enzyme and the crystallographic molecular model of the corresponding phosphonamidate and the enzyme, no difference in the orientations or the positions of the two respective inhibitors or amino acids within the active site could be observed.[199] In the crystallographic molecular model of the phosphonamidate bound in the active site, the nitrogen of the phosphonamidate is 0.30 nm from the acyl oxygen of the peptide bond of Alanine 113 with a reasonable orientation for a hydrogen bond (Figure 3–34B). One difference between the complex of a phosphonate and the active site and the complex of the corresponding phosphonamidate and the active site is the existence of this hydrogen bond. If this were the only difference, the free energy of formation of this rigidly oriented hydrogen bond, relative to the hydrogen bond between the same nitrogen–hydrogen and a molecule of water when it was free in solution, would be –17 kJ $mol^{-1}$. There are, however, at least two other differences.

First, the ester oxygen of a phosphonate is forced to reside next to the oxygen of Alanine 113 in the complex between a phosphonate and the active site, and this should engender an electron repulsion. To test this possibility, two competitive inhibitors of thermolysin, both of which were phosphonates

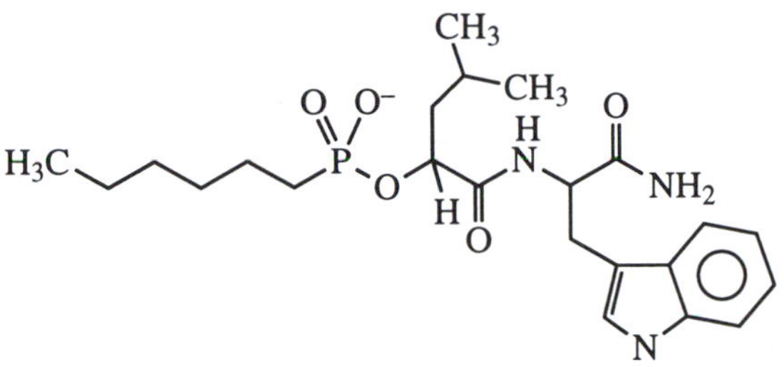

**3–45**

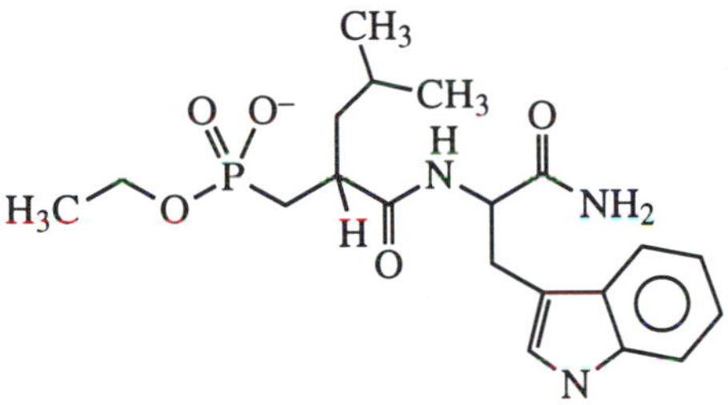

**3–46**

were synthesized.[200] The difference between the two was that the oxygen and the carbon of the phosphonate were interchanged. At the active site this would have replaced the oxygen with a carbon and a hydrogen, yet no difference in dissociation constant between these two competitive inhibitors was observed. The advantage of this experiment was that solvation of the two inhibitors in free solution should have been the same around the phosphorus but in the active site the oxygen would be replaced by the methylene. This result suggests that electron repulsion is insignificant, but there may be steric repulsion in the case of the methylene equivalent in magnitude to the electron repulsion of the oxygen.

Second, the two free anionic oxygens of a phosphonate are less basic than the anionic oxygens of a phosphonamidate and should bind less strongly to the zinc in the active site. This could account for some or all of the difference in free energy of dissociation that was observed. To test this possibility, two sets of phosphonamidates (**3–44**), phosphinates (**3–43**), and phosphonates (**3–47**)

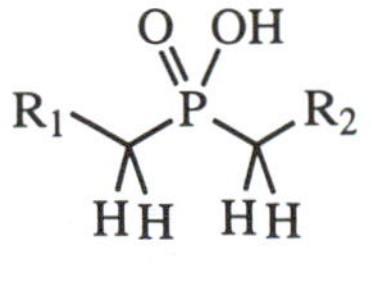

**3–47**

each set analogous to a peptide cleaved by thermolysin, were synthesized.[200] The phosphinates had the same $pK_a$ as the phosphonamidates, but the values for the $pK_a$ of the analogous phosphonates were about 1.65 units lower. When the dissociation constant of the phosphonamidates and phosphinates were compared, the free energies of dissociation differed by only 6 kJ $mol^{-1}$, while the free energies of dissociation between phosphonamidates and phosphonates differed, as before, by 17 kJ $mol^{-1}$. When comparisons between phosphinates and phosphonamidates were made with the same series of peptide analogues used in the original comparisons between phosphonamidates and phosphonates,[201] the differences in free energy of dissociation between phosphonamidates and phosphinates was even less, 0.5 kJ $mol^{-1}$. From these observations it might be concluded that the majority of the difference in free energy of dissociation between phosphonamidate and phosphonate is due not to the hydrogen bond but to the difference in basicity, which causes the phosphoryl oxygens of the phosphonamidate to bind more tightly to the zinc. When a series of arylphosphonate inhibitors of thermolysin that differed in the basicity of the phosphonate were examined, however, no differences in dissociation constant were observed.[201]

From comparisons of the dissociation constants of all of these phosphonamidates, phosphonates, and phosphinates, it can be concluded that the differences in observed free energies of dissociation resulted from some combination of several factors: (1) the difference in hydrogen bonding in free solution between water and the nitrogen–hydrogen of the phosphonamidate and the oxygen of the phosphonate; (2) the difference in hydrophobic effect of the methylene in the phosphinates and the oxygen of the phosphonates or the nitrogen–hydrogen of the phosphonamidates; (3) the difference in electron repulsion or steric repulsion between the acyl oxygen of Alanine 113 and the oxygen of the phosphonates, the nitrogen–hydrogen of the phosphonamidates, or the methylene of the phosphinates; (4) the difference in basicity between the phosphinates or the phosphonamidates and the phosphonates; and (5) the hydrogen bond itself. These experiments, at face value so simple, illustrate the difficulty of separating the various contributions to binding whenever structure–activity comparisons are performed.

The results of these experiments suggest at least that the difference in free energy between the hydrogen bond in the complex of a phosphonamidate and the active site of thermolysin and the hydrogen bond of the same phosphonamidate with water in free solution is greater than –17 kJ $mol^{-1}$. The two states that differ by this rather small amount of free energy of formation are the nitrogen–hydrogen bond of the phosphonamidate hydrogen-bonded as a donor intramolecularly to the acyl oxygen of Alanine 113 and the nitrogen–hydrogen bond of the phosphonamidate hydrogen-bonded as a donor intermolecularly to water. As the values for the $pK_a$ of the oxygen of water and the oxygen of a peptide bond are essentially the same, no difference in enthalpy should exist between these two states. The lone pair of electrons on the peptide oxygen fixed and oriented adjacent to the NH of the bound phosphonamidate, however, acts as if its entropy of approximation is no more negative than –50 J $K^{-1}$ $mol^{-1}$ relative to the entropy of approximation associated with the one mole fraction of water surrounding the unbound

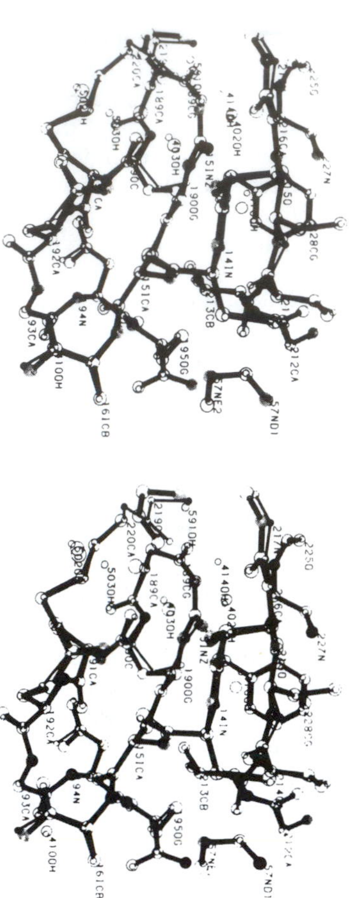

**Figure 3–42:** Crystallographic molecular models of the binding of either lysine (narrow lines and small dots) or arginine (wide lines and large dots) to the portion of the active site of trypsin, the specificity pocket, that determines its specificity for lysines and arginines.[205] The two crystallographic molecular models are superposed to show that the same conformation of the specificity pocket accommodates both side chains and that the side chains themselves superpose. The two crystallographic molecular models from which these regions come are those of the complex between native bovine pancreatic trypsin inhibitor ($n_{aa}$ = 58) and trypsin, in which a lysine (Lysine 15) of the inhibitor occupies the specificity pocket, and the complex between trypsin and bovine pancreatic trypsin inhibitor, in which Lysine 15 has been enzymatically replaced with an arginine. In the models, the side chain of the lysine occupies exactly the same location as that of arginine with the ε-amino group of the lysine occupying the location of the guanidine carbon of the arginine. Water 414, Water 416, and the acyl oxygen of Serine 190 form a triangular array of hydrogen-bond acceptors for the three donors on the lysine. Aspartate 189, Water 403, Water 416, and the acyl oxygen of Glycine 219 form a planar array of hydrogen-bond acceptors for the five donors on the arginine. Reprinted with permission from ref 205. 

inhibitor. In other, more equivocal experiments, the free energies of formation of single hydrogen bonds between amino acids in the active site of tyrosine-tRNA ligase and donors and acceptors on a bound substrate was estimated[197] to be about $-4$ kJ mol$^{-1}$. In these experiments, amino acids in the active site were replaced systematically by site-directed mutation, a procedure that involves significant and unpredictable steric effects. In any case, there is agreement that entropy of approximation can be of sufficient magnitude to overcome the competition by water and provide a favorable, but not extraordinarily favorable, free energy of formation to one or more but probably not the majority of the hydrogen bonds in a complex between a ligand and a site on a protein.

The role of **ionic interactions** in the formation of complexes between substrates and active sites is difficult to assess. Again this is due to the fact that there are almost no fixed charges in biochemical metabolites. The charged molecules in biochemistry are almost always cationic acids or anionic bases that can be neutralized in acid–base reactions and that have multiple donors and acceptors of hydrogen bonds. It is difficult to distinguish a hydrogen bond between a donor and an acceptor that happen to be oppositely charged from an interaction due solely to electrostatic attraction. An example of this difficulty is the charged hydrogen bonds between the two arginines in the active site that bind the two carboxylates on the substrates of aspartate aminotransferase.[202] Are these hydrogen bonds or electrostatic interactions? In the one instance that had always been presented as a purely electrostatic interaction,[203] that between acetylcholine and acetylcholinesterase, significant doubt has been cast upon this interpretation.[66] At the moment, there seems to be little evidence for the existence of anything so simple as an anion within the active site of this enzyme to balance the quaternary ammonium cation of the reactant. An example of an active site in which no cationic amino acid is located adjacent to a negative charge on a substrate is found with tryptophan synthase. In the crystallographic molecular model of the active site occupied by the competitive inhibitor propane phosphate, the phosphate is surrounded only by uncharged donors and acceptors of hydrogen bonds.[204] Most of the substrates of enzymatic reactions are charged at neutral pH and probably enter the active site as charged molecules. In crystallographic molecular models of these complexes, however, there is no obvious pattern in which each charge on the substrate is exactly balanced by an amino acid bearing the opposite charge in the site. Rather the distributions of charges over occupied active sites is much more disorganized and unpredictable (Figures 3–34B, 3–35, and 3–36).

There is, however, one example of a complex of a positively charged substrate with a negatively charged amino acid in the active site that is usually cited as a simple electrostatic interaction. Trypsin is a proteolytic enzyme that cleaves peptide bonds to the carboxy-terminal side of lysines and arginines exclusively. The enzyme is confined to cleave at these locations because, in the region of its active site where the side chain of the amino acid to the amino-terminal side of the susceptible bond is bound, there is a strategically placed aspartate, Aspartate 189.[183] Before refined crystallographic molecular models of complexes between trypsin and inhibitors were available, the interaction between trypsin and the lysines or arginines on its reactants was thought to be ionic.

Refined crystallographic molecular models are now available for complexes between trypsin and small inhibitory proteins in which either lysine or arginine occupies the pocket for the targeted amino acid in the reactant (Figure 3–42).[205] In these crystallographic models, it is clear that either the lysine or the arginine, in turn, participates in a complex network of hydrogen bonds rather than participating only in an ionic interaction. This observation is remarkable because the arrangement of the hydrogen-bond donors in lysine and the hydrogen-bond donors in arginine are so different. Nevertheless, hydrogen-bond acceptors are arranged within the binding site with great precision so as to accommodate either side chain. When lysine is bound, its three tetrahedrally arrayed hydrogen-bond donors find three tetrahedrally arrayed hydrogen-bond acceptors: Water 416, Water 414, and the acyl oxygen of Serine 190 within the polypeptide (Figure 3–42). The carboxylate of Aspartate 189 is not directly bound to lysine because its side chain is too short, but it is hydrogen-bonded to Water 414. Arginine, however, is longer, by one atom, than lysine, and arginine forms a bis-hydrogen bond with Aspartate 189 directly (Figure 3–42). The other three hydrogen-bond donors of the planar guanidinium in the arginine form hydrogen bonds with the acyl oxygen of Glycine 219, Water 416, and Water 403. Only one of these five hydrogen-bond acceptors, Water 416, is also used in the complex with lysine. When arginine is bound, Water 414 is displaced and the hydrogen-bond acceptor on the acyl oxygen of Serine 190 is vacant. When lysine is bound, the hydrogen-bond acceptor on Water 403 and the lone pair on Aspartate 189 are vacant. The vacant hydrogen-bond acceptors in a given complex that are used to bind the other reactant in the other complex are energetically inconsequential because the concentration of hydrogen-bond donors is in excess over the concentration of hydrogen-bond acceptors in the solution. It is noteworthy, however, that every hydrogen-bond donor on the respective reactants ends up in a hydrogen bond when it is bound at the active site.

The magnitude of the ionic interaction between Aspartate 189 and the lysine or the arginine has not been established because the local dielectric constant is unknown. To the extent that water of hydration had to be displaced from either the primary ammonium or the guanidinium or Aspartate 189 itself, the favorable ionic interaction, regardless of its magnitude, between the cationic and anionic groups should compensate for this expenditure of standard free energy. If, for example, a methionine in the reactant polypeptide were to bind to the pocket instead of a lysine, it would displace waters of hydration around Aspartate 189 without a compensating electrostatic interaction. This lack of compensation would explain why trypsin does not cleave adjacent to methionine but does cleave adjacent to lysine. The exclusion of most of the other amino acids from the pocket, however, is probably due to steric effects with the exception of serine, cysteine, glycine, and alanine. Why trypsin fails to cleave adjacent to these small amino acids could be explained by assuming that the hydrophobic force between the side chain of lysine or arginine and the walls of the pocket is required for catalysis.

## *Suggested Reading*

Quiocho, F.A., & Vyas, N.K. (1984) Novel Stereospecificity of the L-Arabinose-Binding Protein, *Nature 310*, 381–386.

Holden, H.M., Tronrud, D.E., Monzingo, A.F., Weaver, L.H., & Matthews, B.W. (1987) Slow- and Fast-Binding Inhibitors of Thermolysin Display Different Modes of Binding: Crystallographic Analysis of Extended Phosphonamidate Transition-State Analogues, *Biochemistry 26*, 8542–8553.

### PROBLEM 3–20

Phenobarbital is a drug used as an anticonvulsant and a sedative

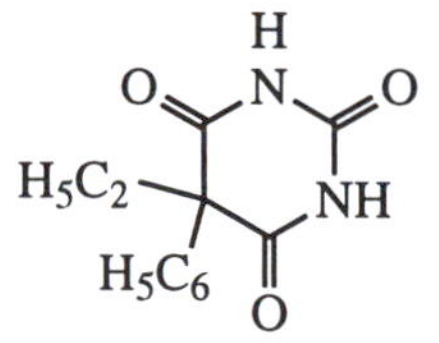

Its action as a drug results from its reversible binding to the active sites of target proteins.

(A) Draw the correct three-dimensional structure of phenobarbital and indicate clearly all hybridizations and $\sigma$ lone pairs. Show all $\pi$ electrons, labeled as $\pi$ electrons. Use correct angles for all bonds and lone pairs. How many $\pi$ electrons are there?

(B) Draw a hypothetical structure for phenobarbital in the binding site of a target protein. Provide appropriate hydrogen-bond donors and acceptors from among the amino acid side chains, and cover hydrophobic surfaces with appropriate hydrophobic amino acid side chains.

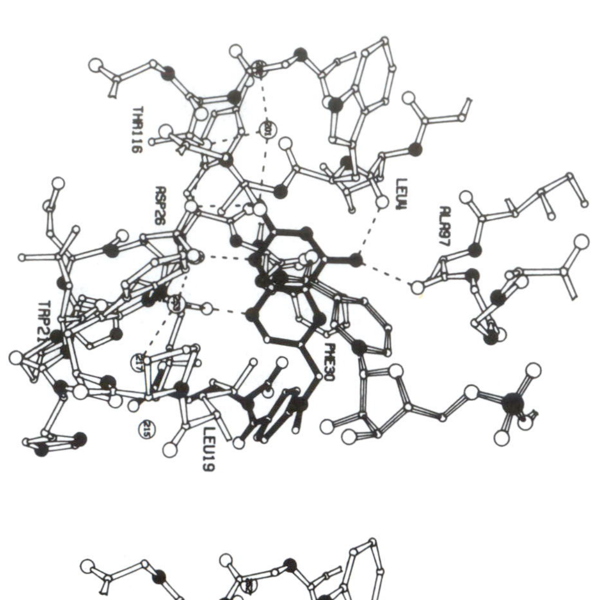


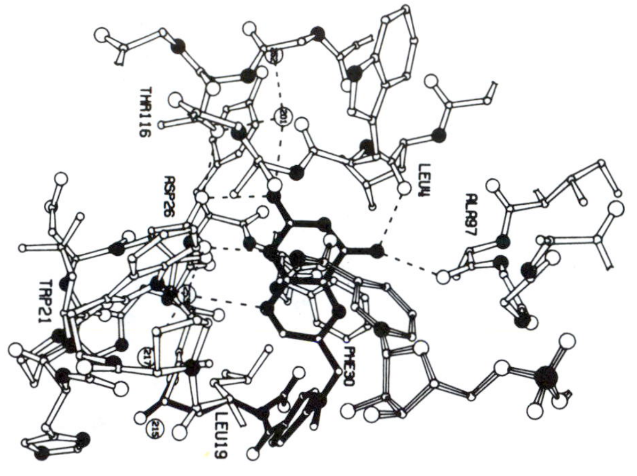


Pteridine-binding site of *Lactobacillus casei* dihydrofolate reductase.[206] Methotrexate is indicated by solid bonds, protein by open bonds, and a portion of the NADPH molecule by striped bonds. Carbon atoms are represented by smaller open circles, oxygen atoms by larger open circles, and nitrogen atoms by blackened circles. Large numbered circles represent fixed solvent molecules. Reprinted with permission from ref 206. Copyright 1982 *Journal of Biological Chemistry.*

**PROBLEM 3–21**

Methotrexate is a strong, competitive inhibitor of dihydrofolate reductase

A view of the crystallographic molecular model of the complex between methotrexate and the active site of the enzyme is presented in the figure in the upper left. There are six hydrogen bonds, indicated with dashed lines, between the inhibitor and the active site. The donors and acceptors from the active site are Water 253, Aspartate 26, Water 201, the acyl oxygen of Leucine 4, and the acyl oxygen of Alanine 97.

(A) Draw the $\sigma$ structure of only the pteridine ring of methotrexate. Include all $\sigma$ lone pairs of electrons and acidic protons in their proper orientations. As an afterthought, to make your drawing more familiar, add $\pi$ electrons distributed as they would be in one of the resonance structures.

(B) Around the pteridine ring in your drawing, place the functional groups that form the hydrogen bonds seen in the figure. Draw each functional group fully enough so that it is clear what it is. Draw each functional group with the proper size to match your drawing of the pteridine ring. Orient each functional group so that an ideal hydrogen bond is formed in each instance. Put $\sigma$ lone pairs on the acceptors of hydrogen bonds in the proper orientations. Put acidic hydrogens on the donors of hydrogen bonds in the proper orientations.

(C) Two of the hydrogen bonds as they are formed in the crystallographic molecular model are not in the ideal configuration. Which two are they and why are they not ideal? Why does it not matter?

The following figure is a view of the active site in the crystallographic molecular model of chymotrypsin[207] (reprinted with permission from ref 207; copyright 1976 American Chemical Society).

The narrow lines are supposed to represent hydrogen bonds. Focus your attention on the 10 hydrogen bonds involving water molecules W6 and W7, the amide nitrogen of Isoleucine 16, the acyl oxygen of Glycine 193 (note that the $\alpha$-carbon and amide nitrogen of Glycine 193 are identified incorrectly), and the side chains of Aspartate 194, Glutamine 30, Histidine 40, and Serine 32.

(D) Attempt to draw out each of these hydrogen bonds with the care you exercised in your answer to part B.

(E) If you have done your work carefully enough you will have found at least three hydrogen bonds indicated in the figure that are formally impossible for electronic or steric reasons. Identify each formally impossible hydrogen bond and describe why each is impossible.

(F) Assume that the crystallographer knew these were impossible as written and meant to indicate a different arrangement from single hydrogen bonds. Provide a reasonable explanation for what may be happening at each of these formally impossible locations.

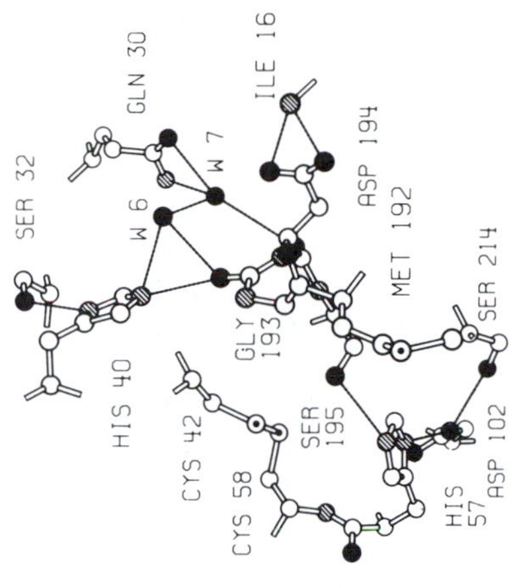


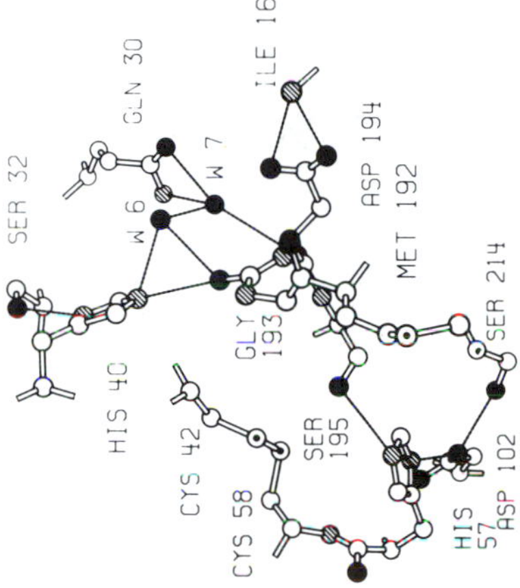


# Induced Fit

From crystallographic studies, it has become apparent that the active sites of enzymes display a spectrum of structural responses to the binding of substrates. At one extreme of the spectrum are active sites whose structures are affected very little when a substrate binds. An example of such an active site is that of thermolysin. The difference map of electron density of the protein (Figure 3–34A) in the region of the active site of this enzyme displays only insignificant reorientations of the amino acids surrounding and creating the active site upon the binding of an analogue of the tetravalent intermediate in the enzymatic reaction.[158] At the other extreme of the spectrum are active sites that close around the substrates as they bind so that the amino acids forming the active site are assembled and properly oriented only after the substrates have bound. An example of such an active site is that of citrate (*si*) synthase.[180] When the two substrates of citrate (*si*) synthase, citrate and coenzyme A, are bound to the enzyme, its structure changes so significantly that the unliganded enzyme and the liganded enzyme form crystals with completely different space groups, $P4_12_12$ and $C2$, respectively. This difference in structure results from a rigid-body movement between two hinged domains within the same subunit that closes the active site, which lies in the crevice between the domains around the bound substrates. In the process, amino acids participating in catalysis and those binding the substrates are considerably reoriented to form a functional active site in the closed form where none existed in the open form. Similar closings of the active site following the binding of substrates have been observed crystallographically in hexokinase (Figure 3–43),[208] lactate dehydrogenase,[186] adenylate kinase,[209] and alcohol dehydrogenase.[173] Between the extremes represented by thermolysin and citrate (*si*) synthase, examples of a variety of intermediate structural responses can be found.

A large structural change such as the one accompanying the binding of substrates to citrate (*si*) synthase or hexokinase is referred to as an **induced fit.**[210] That such alterations in the structure of the protein could occur in response to the binding of a substrate or substrates was proposed before any such changes were verified crystallographically. The proposal provided an explanation for the specificity that enzymes displayed toward certain substrates. For example, 5′-nucleo-

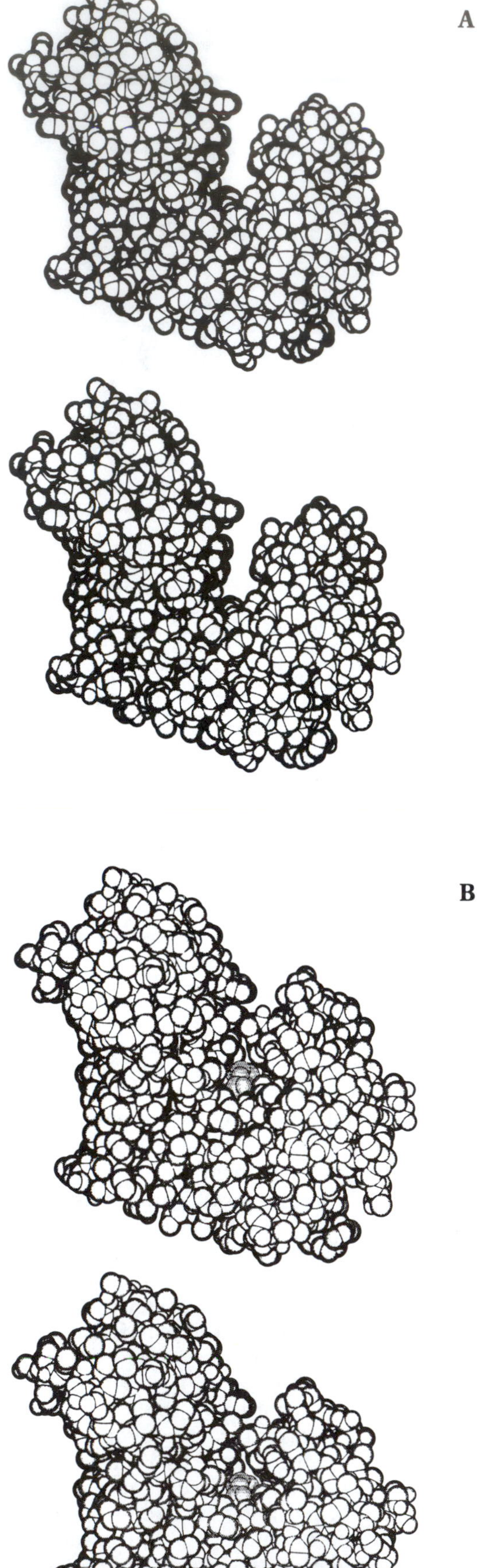


**Figure 3–43:** Space-filling model of the two conformations of a subunit of hexokinase: the form in which the active site is open (A) and the form in which the active site is occupied by glucose and the protein has closed around the substrate (B).[208] The crystallographic molecular model of the open form of the protein was constructed from a map of electron density calculated from the reflections of a crystal of hexokinase with space group $P2_12_12_1$. The phases were determined by isomorphous replacement. The crystallographic molecular model of the closed form was obtained from crystals of the enzyme complexed with glucose. Although these crystals were, by chance, also of the space group $P2_12_12_1$, the dimensions of the unit cell and the packing of the protein were completely different. The phases of the reflections from this nonisomorphous crystal were also determined by multiple isomorphous replacement independently of the previous crystallographic study of the open form. The active site lies in the deep cleft between the two lobes of the open form (panel A). When glucose is bound in this cleft, the lobes close around the active site to enclose the molecule of glucose (atoms highlighted with zigzag pattern in panel B). Reprinted with permission from ref 208. Copyright 1979 American Association for the Advancement of Science.

tidase catalyzes the hydrolysis of the phosphate ester of adenosine 5′-monophosphate 100-fold more rapidly than it does that of ribose 5-phosphate. One explanation for this kinetic preference is that the larger substrate causes an induced fit to occur more effectively.

Induced fit can be used only to improve the specificity of an enzyme at the expense of its catalytic efficiency.[130] This is due to the fact that a portion of the change in free energy accompanying the binding of the reactant must be expended in shifting the equilibrium between the open and closed forms of the enzyme. Assume, as is reasonable, that the amino acid sequence of the polypeptide folded to form the native structure of a subunit of the enzyme contains the information necessary to specify two distinct native structures, usually referred to as two distinct **conformations** of the enzyme. These two conformations could be, for example, the open conformation ($E_o$) and the closed conformation ($E_c$) observed in the two crystallographic molecular models of citrate (*si*) synthase. Assume also, as is observed,[180] that in the absence of substrates the open conformation is more stable than the closed but that when the substrates are bound the closed conformation is more stable than the open. The equilibria describing the situation are

$$\begin{array}{ccc} E_o + S & \underset{\Delta G_4^\circ}{\rightleftharpoons} & E_o{\cdot}S \\ \Delta G_2^\circ \updownarrow & & \updownarrow \Delta G_3^\circ \\ E_c + S & \underset{\Delta G_1^\circ}{\rightleftharpoons} & E_c{\cdot}S \end{array} \quad (3\text{–}222)$$

The standard free energy changes for binding of the substrate to the two respective forms of the enzyme are for the directions of dissociation, from right to left, and those for the conformational changes are for the directions from closed to open, upward. Because free energy is a state function and independent of path

$$\Delta G_1^\circ + \Delta G_2^\circ = \Delta G_3^\circ + \Delta G_4^\circ \quad (3\text{–}223)$$

Because the enzyme is open in the absence of substrates and closed in the presence of substrates

$$\Delta G_2^\circ < 0 \quad (3\text{–}224)$$

and

$$\Delta G_3^\circ > 0 \quad (3\text{–}225)$$

from which it follows that

$$\Delta G_1^\circ >> \Delta G_4^\circ \quad (3\text{–}226)$$

These **linkage relationships** state that if and only if the substrates bind more tightly to the closed form than the open form, as is reasonable, will the binding of substrates be able to invert the equilibrium between the open form and the closed form of the enzyme. It is this difference in the standard free energies of dissociation that tips the balance. If almost all of the unoccupied enzyme is in the open conformation to begin with, the total change in standard free energy for dissociation of substrates from the active complex, $E_c{\cdot}S$, is either $\Delta G_1^\circ + \Delta G_2^\circ$ or $\Delta G_3^\circ + \Delta G_4^\circ$, but either of these sums is always more negative than $\Delta G_1^\circ$, the standard free energy of dissociation of substrates from the closed form. Therefore, the substrates bind less tightly to the enzyme than they would if only the closed conformation were present from the beginning. Therefore, tight binding is sacrificed to perform the conformational change.

In return for this sacrifice, specificity is increased. Because the active site assembles only after all of the proper substrates are present, the enzyme avoids catalyzing a reaction for which it was not intended. If the enzyme were always present in the closed conformation, the substrates would bind more tightly, the Michaelis constant would be smaller, and the enzyme could be a better catalyst. Nothing, however, could prevent molecules that contained only a portion of one of the substrates from binding and participating in the normal reaction mechanism.

The smaller molecule of most concern is water, and the reaction catalyzed by yeast hexokinase illustrates this problem most graphically. Hexokinase transfers the $\gamma$-phosphate from MgATP to the primary alcohol on carbon 6 of glucose. Water is also a primary alcohol and should be able to occupy the same location in the active site occupied by the primary alcohol of glucose. If the active site in hexokinase were properly configured in the absence of the glucose, MgATP could bind, water would be present at the alcoholic site, and the phosphate would be transferred to the water. In this instance, hexokinase would catalyze the hydrolysis of ATP.

Yeast hexokinase does catalyze the hydrolysis of MgATP in the absence of glucose, but the $K_m$ for MgATP (4 mM) and the $V_{max}$ (4 $\mu$mol min$^{-1}$ mg$^{-1}$) are higher and lower, respectively, than the $K_m$ for MgATP (0.2 mM) and the $V_{max}$ (150 $\mu$mol min$^{-1}$ mg$^{-1}$) for the transfer of phosphate to the primary hydroxyl of glucose. From what is known about the conformational changes in which yeast hexokinase engages (Figure 3–43), one explanation for this counterproductive hydrolytic activity would be that occasionally the active site closes when MgATP is bound but glucose is not bound. Because water would necessarily be occupying the region of the active site at which the primary hydroxyl of glucose would normally be located, the hydrolysis of MgATP would result as soon as the proper active site was assembled.

If this is the explanation for the observed hydrolysis of MgATP, it should be possible to amplify the error. When D-lyxose

**3–48**

the five-carbon analogue of D-mannose missing carbon 6, is added at a concentration of 0.1 M ($K_m^{lyx}$ = 5 mM) the hydrolysis of MgATP is remarkably accelerated to levels approaching the normal levels of hexokinase activity [$K_m^{MgATP}$ = 0.1 mM and $V_{max}$ = 70 $\mu$mol min$^{-1}$ mg$^{-1}$].[211] D-Xylose, the five-carbon analogue of D-glucose, also accelerates the hydrolysis of MgATP. Because both mannose and glucose are excellent substrates for hexokinase (Table 3–3), these results suggest that lyxose or xylose, respectively, instead of mannose or glucose, induces the closing of the active site when MgATP is also bound but that a water molecule is trapped at the position usually occupied by a hydroxymethyl group and acts as the acceptor.

In a similar set of experiments, it could be shown that 3-phosphoshikimate 1-carboxyvinyltransferase would catalyze hydrogen exchange at carbon 3 of phospho*enol*pyruvate only when 4,5-dideoxyshikimate 3-phosphate was present.[211a] This analogue is missing the 5-hydroxyl to which the enzyme normally transfers the carboxyvinyl group of phospho*enol*pyruvate, but its ability to activate the catalysis of a partial reaction indicates that it can nevertheless induce the active site to close.

Under certain circumstances, the process of induced fit would not be able to accomplish the increase in specificity that has just been described. Consider the closed form of the enzyme, $E_c$. Suppose that both the proper reactant, glucose in the case of hexokinase, and an improper reactant, water in the case of hexokinase, could equilibrate so rapidly with the active site in its closed conformation that both the association and dissociation of each of these reactants with and from the closed conformation were all faster than the turnover of the enzyme. If this were the case, adding the ability of the enzyme to open and close would not change the ratio of the rates for the participation of the two reactants in the enzymatic reaction. As soon as the enzyme closed, these two competing reactants would equilibrate on the active site, and the fact that the enzyme had just closed would be irrelevant. Therefore, if the proper and improper reactant were able to enter and leave the closed form of the enzyme rapidly, induced fit would not improve the specificity of the enzyme. Induced fit can increase the specificity of an enzyme only if the active site closes so tightly on the proper reactant that an improper reactant cannot displace it before turnover occurs. This seems to be the case with hexokinase (Figure 3–43). It is unlikely that water could displace a molecule of glucose upon which the enzyme has closed. This consideration suggests that induced fit will usually involve the closing of the active site around a reactant as in hexokinase.

It must also be the case that the closing of the enzyme around an improper reactant be significantly slower than both the closing of the enzyme around the proper reactant and the turnover of the complex of the closed enzyme and the proper reactant. The closing of the enzyme around an improper reactant must be so slow as to be inconsequential. It has been pointed out by Fersht[212] that if all forms of the unliganded enzyme and all of the complexes of the enzyme and the reactants, both proper and improper, were to come to equilibrium faster than these reactants were converted to products, the enzyme would be no better off than it would have been if the active site had not undergone a change in conformation and were simply a rigid template with properly arrayed catalytic groups at all times. In other words, induced fit does not improve the specificity of the enzyme if the steps governing the binding of both proper and improper reactants are under thermodynamic control. The ability of induced fit to improve specificity is manifested only when the enzyme can close to the active conformation rapidly around the proper reactant but only slowly around the improper reactant. In other words, the steps governing the binding of proper and improper reactants must be under kinetic control. For example, if it is the case that induced fit defines the specificity of hexokinase, then the closing of the enzyme around MgATP and water is the rate-limiting step of the slow ATPase activity of the enzyme.

When two reactants of a bimolecular reaction that is catalyzed by an enzyme induce together the closing of the active site around themselves, they bind to the enzyme synergistically.[213] The reason for this is most easily understood by examining the linkage cube for this reaction

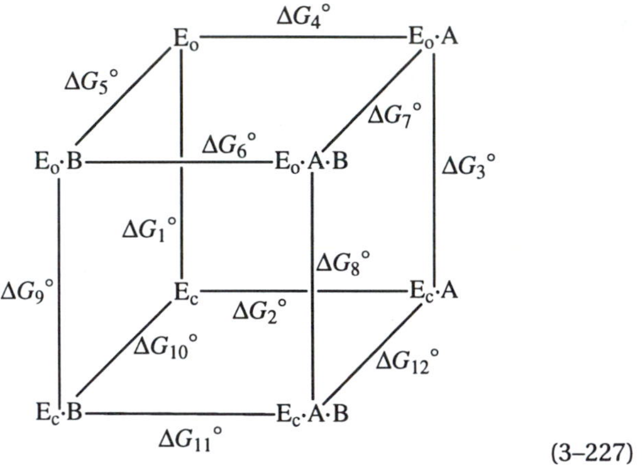


(3–227)

In this representation, the changes in standard free energies $\Delta G_2°$, $\Delta G_4°$, $\Delta G_5°$, $\Delta G_6°$ $\Delta G_7°$, $\Delta G_{10}°$ $\Delta G_{11}°$, and $\Delta G_{12}°$ are standard free energies of dissociation and $\Delta G_1°$, $\Delta G_3°$, $\Delta G_8°$, and $\Delta G_9°$ are standard free energies of isomerization in the direction from closed to open, as in Equation 3–222. Double arrows are replaced with lines for simplicity. If it is assumed, to simplify matters, that each of the two reactants, A and B, binds with the same affinity to the open conformation of the enzyme whether the other is bound or not, then $\Delta G_5° = \Delta G_7°$ and $\Delta G_4° = \Delta G_6°$. If it is assumed that the enzyme does not close until both reactants are present, then $\Delta G_1° < 0$, $\Delta G_3° < 0$, and $\Delta G_9° < 0$. When both reactants are bound, the conformational change takes place, and therefore, $\Delta G_8° > 0$. When A binds to the enzyme in the absence of B to form $E_o$·A, the standard free energy of dissociation is $\Delta G_4°$. When A binds to the enzyme after B has been bound to form $E_c$·A·B, the standard free energy of dissociation is $\Delta G_6° + \Delta G_8°$ or $\Delta G_9° + \Delta G_{11}°$, both of which must be more positive than $\Delta G_4°$. When B binds to the enzyme in the absence of A to form $E_o$·B, the standard free energy of dissociation is $\Delta G_5°$. When B binds to the enzyme after A has been bound to form $E_c$·A·B, the standard free energy of dissociation is $\Delta G_7° + \Delta G_8°$ or $\Delta G_3° + \Delta G_{12}°$, both of which must be more positive than $\Delta G_5°$. Therefore, each reactant will bind more tightly when the other is already bound.

This synergy can be observed in the binding of reactants to yeast hexokinase. The dissociation constant for the binding of MgATP in the absence of monosaccharide to the empty, open active site is 4 mM. When lyxose binds along with MgATP, the Michaelis constant for MgATP is 0.1 mM, and when mannose binds along with MgATP, the Michaelis constant for MgATP is 28 $\mu$M.[213]

In the instances of hexokinase and citrate (*si*) synthase, the clefts between the two hinged domains appear to close sufficiently that entry and exit of substrates from the closed conformation should be very difficult. In these cases it could be argued that the closed form could not by itself be a more efficient enzyme. Even though substrates bind more tightly to the closed form, their entry into it and exit from it should be too slow for it to be effective. If this is the case, induced fit can also be viewed as a way to create an active site that surrounds the substrates on all sides but that can nevertheless bind and release reactants and products rapidly.

### *Suggested Reading*

DelaFuente, G., Lagunas, R., & Sols, A. (1970) Induced Fit in Yeast Hexokinase, *Eur. J. Biochem. 16*, 226–233.

# Association of Reactants and Dissociation of Products

The second-order rate constants for the associations of reactants with the active site of an enzyme usually have values that are within 1 or 2 orders of magnitude of the rate constant predicted for a diffusion-controlled reaction between a small molecule and a macromolecule,[121,214] which is about $10^9$ M$^{-1}$ s$^{-1}$ at 25 °C. The concentrations of most substrates in the cytoplasm, however, are usually less than 1 mM, which means that the apparent first-order rate constants of association between reactants and enzymes are usually significantly less than $10^6$ s$^{-1}$. With this in mind, it should come as no surprise that some enzymes have turnover numbers that are equal to or greater than the apparent first-order rate constant with which the reactants, when present at physiological concentrations, associate with the active site. This leads to the possibility that the association of the reactants with the active site, rather than the turnover of the bound reactants, can be the rate-limiting step in an enzymatic reaction under physiological conditions.

This possibility can be realized, however, only if the reactants are sticky reactants. A **sticky reactant** is a reactant that is converted to products faster than it can dissociate from the active site. If the reactants are not sticky and dissociate from the active site faster than they can be converted into products, then a **rapid preequilibrium** is established between the enzyme and the reactants, and turnover must be the rate-limiting step.[215] If, however, the reactants are sticky because the turnover number of the enzyme is greater than either of the first-order rate constants for their dissociation, then most of the time, molecules of bound reactant are converted to product before they can dissociate. In this situation, association of the reactants with the active site can be the rate-limiting step.[215]

If the reactants in a particular enzymatic reaction are sticky reactants because the turnover number of the enzyme is very large, then two situations can be distinguished. If the concentrations of reactants are at saturation, then as each active site releases products, it is immediately filled with reactants, and the rate-limiting step of the reaction must be the turnover of the bound reactants. If the concentration of one of the reactants is so low that the rate of its association with the active site becomes less than the rate of turnover, and if that reactant is sticky so that almost every molecule is converted to product once it associates with the active site, then the rate-limiting step in the reaction will become the association of the reactant with the active site. It follows that when the reactants in an enzymatically catalyzed reaction are sticky, a change in rate-limiting step must occur when the concentration of one of them is decreased below a certain level.

Coincident with this change in rate-limiting step is a change in reaction order. At high concentrations of the reactant, turnover, a first-order process, is rate-limiting; at low concentrations of the reactant, association, a second-order process, is rate-limiting. As a result, the behavior observed is an increase in initial velocity that is proportional to the concentration of reactant at low concentrations, followed by an approach to a horizontal asymptote at higher concentrations of reactant. This apparent saturation, however, is not caused by saturation of the active site but by a change in rate-limiting step. This behavior will be indistinguishable from the behavior of an enzyme that does not have sticky reactants. In each instance, the initial velocity is a hyperbolic function of the concentration of reactant. In the case of a rapid preequilibrium, where reactants are not sticky, the values of $K_m$ are determined in part by the dissociation constants of the reactants from the active site. In a situation where the reactants are sticky, the values of $K_m$ are determined by the ratio between the rate constants for the associations of the reactants and the turnover number of the enzyme.

This general conclusion can be most easily seen for the simplest kinetic mechanism (Mechanism 3–39). When the reactants are sticky, $k_2 >> k_{-1}$, and the equation for the initial velocity (Equation 3–46) becomes

$$v_0 = \frac{k_2[\mathrm{E}]_{\mathrm{TOT}}[\mathrm{A}]_0}{\dfrac{k_2}{k_1} + [\mathrm{A}]_0} \tag{3–228}$$

In this equation, $K_m$ is $k_2(k_1)^{-1}$. If the concentration of reactant is so low that its apparent rate constant of association becomes rate-limiting ($k_1[\mathrm{A}]_0 < k_2$), then $v_0$ is equal to $k_1[\mathrm{E}]_{\mathrm{TOT}}[\mathrm{A}]$, the rate law for a second-order process. In these circumstances, $k_{cat}/K_m$ is equal to $k_1$, the second-order rate constant for association of reactant and active site. This is the slope of the linear relationship between initial velocity and the concentration of reactant at low concentration. When $[A]_0$ is at saturation, $v_0 = k_2\ [E]_{TOT}$, which is the rate equation for a first-order process with a rate constant of $k_2$.

If the turnover number of an enzyme is so fast that the reactants are sticky and if the rate of association of one or more reactants, when they are at their physiological concentrations, is the rate-limiting step in the reaction, then the enzyme will have evolved to a state where increase in its turnover number cannot increase the rate of the enzymatic reaction as it occurs in the cytoplasm, and natural selection will no longer operate upon the turnover number. At this state the enzyme will have become a perfect catalyst because there is no way to increase the rate of diffusion of a reactant to the active site, short of increasing its cellular concentration.[121,216]

If the rate of diffusion of one of the reactants to the active site is rate-limiting at its physiological concentrations, then the $K_m$ for that reactant will be greater than its physiological concentration, the initial velocity at the physiological concentration of that reactant will be proportional to its concentration, and the constant of that proportionality, $k_{cat}/K_m$, will be equal to the second-order rate constant for its association with the enzyme. This fact can be used to distinguish a mechanism involving a sticky reactant from a mechanism involving a rapid preequilibrium. The addition of solutes such as glycerol, sucrose, poly(ethylene glycol), or Ficoll to an aqueous solution increases its viscosity. Rates of diffusion are inversely proportional to the viscosity of the solution,[216a] and the rates of bimolecular reactions, which depend upon the frequency with which the reactants collide in solution, are decreased by increases in the viscosity of the solution. If the rate-limiting step in an enzymatic reaction, at subsaturating concentrations of a reactant, is the association of that reactant with the active site, then the rate of the enzymatic reaction should decrease when the viscosity of the solution is increased.

At concentrations of reactants that saturate an enzyme, none of the pseudo-first-order rate constants for the associations between an enzyme and its reactants can be rate-limiting. This is because the concentrations of reactants, and consequently their pseudo-first-order rate constants of association, have been purposely increased to a level at which no further increases in the rate of the enzymatic reaction are observed. At concentrations of a reactant less than the numerical value for its $K_m$, however, the initial velocity of an enzymatic reaction is usually directly proportional to the concentration of that reactant (Equation 3–72). If the concentration of one reactant is less than its $K_m$, if the concentrations of all other reactants are saturating, and if the association of the one reactant is the rate-limiting step in the reaction, then its pseudo-first-order rate constant of association is $k_{cat}/K_m$. If the association of the reactant is the rate-limiting step of the reaction at initial velocities less than $V_{max}$, then $k_{cat}/K_m$ should be inversely proportional[217] to the viscosity of the solution, $\eta$

$$\frac{K_m}{k_{cat}} = \frac{1}{k_1^{\,0}} \frac{\eta}{\eta_0} \qquad (3\text{–}229)$$

where $k_1^{\,0}$ is the bimolecular rate constant for association of the subsaturating reactant with the enzyme at the reference viscosity $\eta_0$, usually that of water alone, and $K_m$ is the Michaelis constant for that reactant.

The parameter $K_m/k_{cat}$ for the hydrolysis of *N*-(methoxycarbonyl)-L-tryptophan *p*-nitrophenyl ester by chymotrypsin increases as the relative viscosity, $\eta/\eta_0$, of the solution increases (Figure 3–44A).[217] The increase is less than the direct proportion predicted by Equation 3–229 between $K_m/k_{cat}$ and the viscosity, so the rate of association of the reactants is not the rate-limiting step in the reaction but is a rate-determining step. The order of magnitude of the rate constant for the rate-limiting step in the reaction must be the same as that of the pseudo-first-order rate constant for association of the reactant with the active site at these concentrations, or the rate constant of the reaction would have been invariant with the viscosity. In the case of the reaction catalyzed by triosephosphate isomerase, the increase in the parameter $K_m/k_{cat}$, when glyceraldehyde 3-phosphate is the reactant, is directly proportional to the viscosity of the solution (Figure 3–44B).[121] This result is consistent with a kinetic mechanism in which the association between glyceraldehyde 3-phosphate and the enzyme is completely rate-limiting at low concentrations of the reactant. If this is the case, the only reason the initial velocity of the reaction decreases below $V_{max}$ as the concentration of glyceraldehyde 3-phosphate is decreased is that a change in rate-limiting step occurs from turnover, at high concentrations, to association, at low concentrations. Therefore, the Michaelis constant for this reaction cannot be related to the dissociation constant between enzyme and reactant.

There is always the possibility that the change in viscosity produced by adding these agents affects the enzymatic reaction indirectly, for example, by altering the structure of the protein. If this were the case, the turnover number itself should be decreased by the increase in viscosity. Contrarily, if only association rates are being affected by the increase in viscosity, the turnover number cannot decrease. In the case of the chymotryptic hydrolysis of *N*-(methoxycarbonyl)-L-tryptophan *p*-nitrophenyl ester, $k_{cat}$ actually increased by about 10% over the interval that $k_{cat}/K_m$ decreased by a factor of 2.

If the viscosity of the medium were affecting the structure of the protein, then these effects should persist even after turnover is drastically reduced to the point where it becomes completely rate-limiting. *N*-Acetyl-L-tryptophan *p*-nitroanilide could be used as such a control reactant for chymotrypsin, because its turnover number is $10^5$ times slower than that of *N*-(methoxycarbonyl)-L-tryptophan *p*-nitrophenyl ester. With this slower reactant, changes in the viscosity of the solution had no effect on the parameter $K_m/k_{cat}$ (Figure 3–44A). Rather than using an alternate reactant to decrease the turnover number, the enzyme can be changed. Triosephosphate isomerase in which Glutamate 165 had been changed to aspartate by site-directed mutation, which decreases its turnover by $10^3$, catalyzes the isomerization of glyceraldehyde 3-phosphate in a reaction that is insensitive to the viscosity of the solution (Figure 3–44B). These controls are consistent with the conclusion that the fastest enzymes are so fast that the diffusion of reactants to their active site is the rate-limiting step of the reaction below saturation.

Because the dissociation constants between enzymes and substrates are often in the range of $10^{-6}$ M and the respective second-order association rate constants are usually less than $10^8\ M^{-1}\ s^{-1}$, it follows that the rate of dissociation of a product from the active site may often be slow enough to be the rate-limiting step in an enzymatically catalyzed reaction even

A

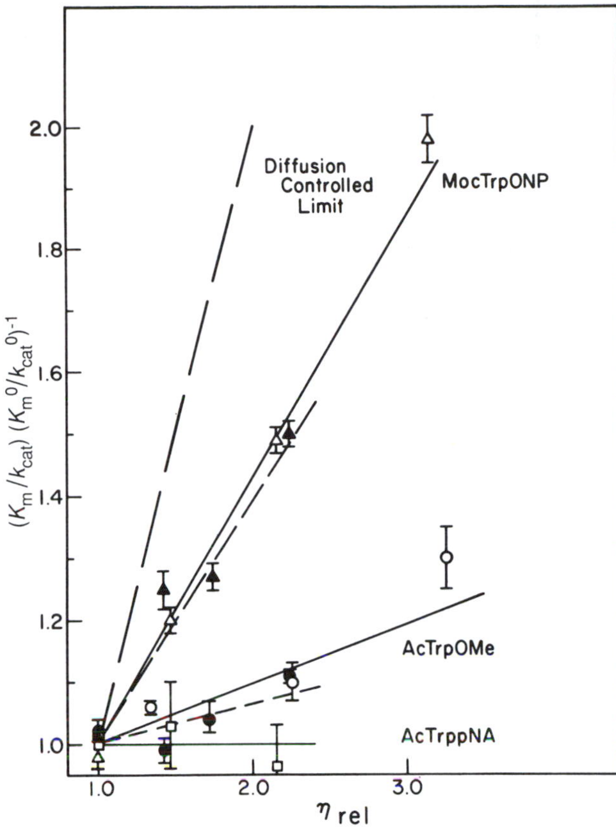


B

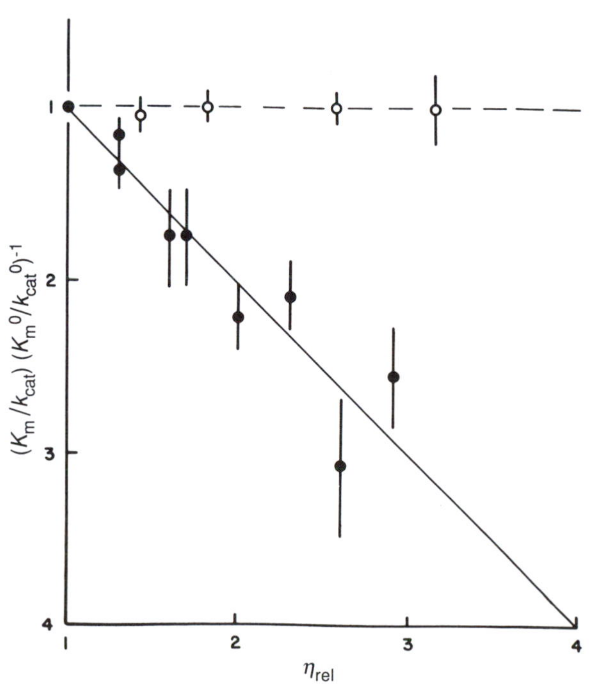


**Figure 3–44:** Effect of viscosity of the solution on the rate of the hydrolysis of an anilide and two esters of tryptophan by chymotrypsin (A)[217] and on the rate of the conversion of glyceraldehyde 3-phosphate to dihydroxyacetone phosphate catalyzed by triosephosphate isomerase (B).[121] (A) The initial rates of hydrolysis catalyzed by chymotrypsin were measured for the *p*-nitroanilide (□, AcTrppNA) and the methyl ester (○, ●, AcTrpOMe) of *N*-acetyltryptophan and the *p*-nitrophenyl ester of *N*-(methoxycarbonyl)-L-tryptophan (Δ, ▲, MocTrpONP) at pH 8.0, 25 °C, and an ionic strength of 0.5 M in solutions containing various concentrations of sucrose (open symbols) or Ficoll (closed symbols) to increase the viscosity. From measurements of initial velocities at several concentration of each reactant, $k_{cat}$ and $K_m$ could be determined for each set of conditions. The relative rates $(K_m/k_{cat})\ (K_m^0/k_{cat}^0)^{-1}$, where $K_m$ and $k_{cat}$ refer to the solutions of increased viscosity and $K_m^0$ and $k_{cat}^0$ refer to the solutions in the absence of sucrose or Ficoll, are presented as a function of the relative viscosity $\eta_{rel}$, which is the viscosity of the particular solution, $\eta$, divided by the viscosity of the solution in the absence of sucrose or Ficoll, $\eta_0$. The dashed line labeled "Diffusion Controlled Limit" is the line defining a direct proportionality between relative rate and relative viscosity as predicted by Equation 3–229. Panel A adapted with permission from ref 217. Copyright 1982 American Chemical Society. (B) The initial rates of conversion of D-glyceraldehyde 3-phosphate to dihydroxyacetone phosphate catalyzed by wild-type triosephosphate isomerase (closed symbols) or a mutant in which Glutamate 165 had been changed to aspartate (open symbols) were measured at pH 7.6 and 30 °C in solutions containing various concentrations of glycerol to increase the viscosity. The initial velocities at several concentrations of glyceraldehyde 3-phosphate were used to determine $k_{cat}$ and $K_m$, and the data are presented as in panel A. To imply that increases in viscosity inhibit the enzyme, the ordinate is inverted in panel B. Panel B adapted with permission from ref 121. Copyright 1988 American Chemical Society.

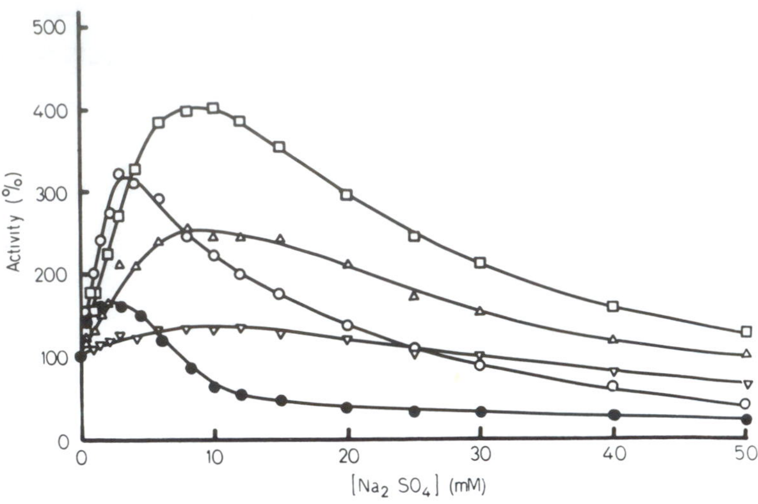


**Figure 3–45:** Effect of sulfate on the enzymatic activity of yeast phosphoglycerate kinase.[219] Samples were prepared that contained phosphoglycerate kinase in 5 $\mu$M (●), 50 $\mu$M (○), 0.5 mM (□), 1.5 mM(△), or 5.0 mM (▽) MgATP and 1.0 mM free $Mg^{2+}$, 50 mM 3-phosphoglycerate, and various concentrations of sodium sulfate ([$NaSO_4$] in millimolar). The initial velocity of the production of 1,3-bisphosphoglycerate was monitored by the decrease in the concentration of NADH in a coupled assay involving glyceraldehyde-3-phosphate dehydrogenase, creatine kinase, creatine phosphate, and triosephosphate isomerase. Initial rates of decrease in the fluorescence of the NADH were measured continuously for each sample. The activity of the enzyme (percent of the initial velocity in the absence of phosphate) is presented as a function of the concentration of sulfate (millimolar). Adapted with permission from ref 219. Copyright 1978 Springer-Verlag.

though its rate of association is fast. Many enzymatic reactions in which the dissociations of the products are rate-limiting have been identified. Therefore, the rate of the release of products from the active site is an important factor in the overall rate of the enzymatic reaction.

Now that the molecular details of the noncovalent interactions between substrates and active sites are appreciated, it becomes possible to imagine the steps in the dissociation of a product. Because almost every substrate is held in the active site of an enzyme by many localized hydrogen bonds and hydrophobic interactions and because covalent bonds about which free rotation can occur are located between these individual points of contact within most substrates, the release of a product can be viewed as the culmination of a series of steps taken in a random order.[218] A molecule of a given product can be divided into segments, between each of which is one or more covalent bonds about which rotation can occur. Even though the configurational flexibility of a bound or partially bound product is likely to be much less than what it displays in solution, it is possible to imagine each of these segments independently lifting out of the active site momentarily as the subset of the noncovalent interactions holding that segment in the active breaks and, albeit hindered, rotation occurs around the covalent bonds connecting that segment to the rest of the substrate. Just as quickly, however, that segment reestablishes its connection with the active site before the neighboring segments can break theirs. These local dissociations and reassociations from subsites of the active site occur randomly over the molecule of product. Eventually, at some instant, all of the local contacts are broken at the same time, and the molecule of product departs.

A simple example of such a process may occur within the active site of yeast phosphoglycerate kinase,[218] the enzyme catalyzing the reaction

$$\text{3-phosphoglycerate} + \text{MgATP} \rightleftharpoons \text{1,3-bisphosphoglycerate} + \text{MgADP} \qquad (3\text{–}230)$$

As the reaction is written, the product is 1,3-bisphosphoglycerate. At high concentrations, sulfate is a competitive inhibitor with respect to MgATP in the forward reaction and with respect to 1,3-bisphosphoglycerate, but not with respect to MgADP, in the reverse reaction.[218] These results suggest that sulfate can bind to the same location in the active site where the phosphate that is transferred between the substrates is bound. At low concentrations, however, sulfate and several other polyanions activate the enzymatic reaction (Figure 3–45).[219] A possible explanation for both the activation and the inhibition observed in the respective ranges of concentration is that during the instant that the 1-phosphate of the product 1,3-bisphosphoglycerate has momentarily dissociated and before it can reassociate, sulfate can seize the region of the active site it normally occupies (Figure 3–46). The sulfate prevents the reassociation of the 1-phosphate, and the 1,3-bisphosphoglycerate is forced to depart as soon as the 3-phosphate breaks away. In this view of the process of dissociation, the 3-phosphate in the absence of sulfate would usually break away after the 1-phosphate had already reassociated, and the 3-phosphate would have to dissociate many times before it happened to dissociate at the same time that the 1-phosphate was also unattached. The sulfate, however, at concentrations high enough to occupy the active site effectively but not high enough to occupy the active site continu-

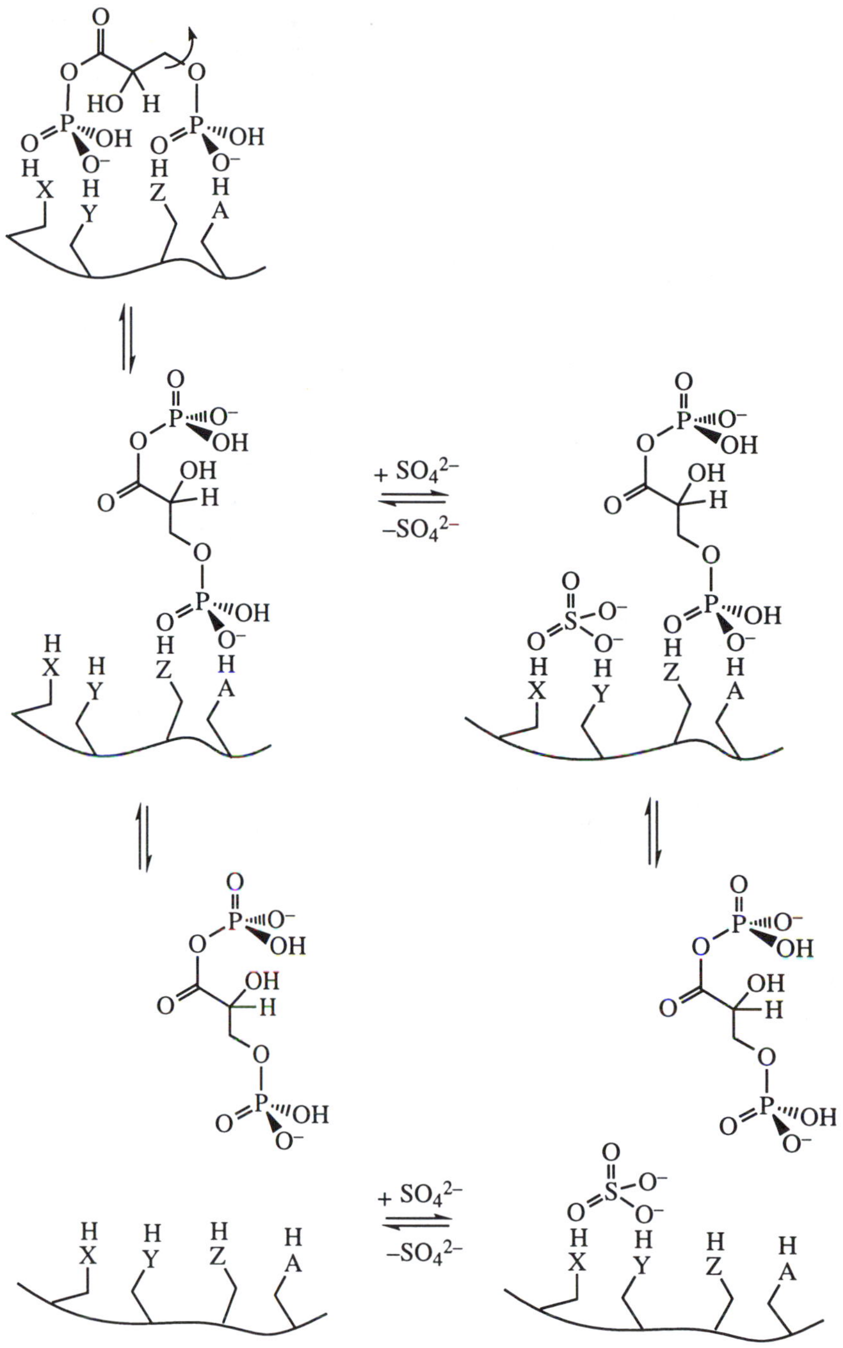

**Figure 3–46:** Schematic presentation of the possibility that sulfate may accelerate the dissociation of 1,3-bisphosphoglycerate from the active site of yeast phosphoglycerate kinase.[219] The dissociation can be considered to occur in at least two steps: the release of the 1-phosphate followed by the release of the 3-phosphate. It is possible that sulfate occupies the subsite for the 1-phosphate, preventing its reassociation and hastening the dissociation of the whole molecule.

ously, would block the reassociation of the 1-phosphate and activation rather than inhibition would result. For this to be the explanation, the dissociation of the product must be a rate-limiting step in the enzymatic reaction.

If this explanation is correct, phosphoglycerate kinase serves as an example not only of stepwise dissociation of a product but also of a mechanism by which reactants or other ligands could accelerate the dissociation of a product from the active site. The individual sites of noncovalent interaction distributed over the reactants in an enzymatic reaction are usually found distributed over the products, albeit in a different order. For example, the $\gamma$-phosphate of MgATP is the same phosphate as the 1-phosphate of 1,3-bisphosphoglycerate in the reaction catalyzed by phosphoglycerate kinase, but the 3-phosphate of 3-phosphoglycerate is the 3-phosphate of 1,3-bisphosphoglycerate. It follows that a reactant usually has a substituent that can occupy the region of the active site occupied by the respective substituent on a product. In this view, a molecule of reactant should always have the potential to hasten the departure of one of the molecules of product that immediately precedes it upon the active site. In the particular case of phosphoglycerate kinase, MgATP, the $\gamma$-phosphate of which would be a polyanion that can bind at the site for the 1-phosphate of 1,3-bisphosphoglycerate, should be able to accelerate the dissociation of the 1,3-bisphosphoglycerate just as sulfate was able to do. As it turns out, MgATP does display kinetic behavior referred to as **substrate activation**,[219] consistent with an ability to perform this acceleration of the reaction by displacement of the product.

Consider an enzyme whose product, C–D, was bound at the active site by separate subsets of noncovalent interactions to segments C and D, respectively. Assume that the reactant, A–B, is bound by the same subsets of noncovalent interactions to segments A and B, respectively. Under initial conditions in which the concentration of the product C–D is zero, the replacement of product C–D by reactant A–B during the kinetic mechanism could occur in the following sequence

$$\mathrm{E}\overset{\mathrm{C}}{\underset{\mathrm{D}}{|}} \underset{k_{-1}}{\overset{k_1}{\rightleftharpoons}} \mathrm{E}^{\mathrm{C-D}} \xrightarrow{k_3} \mathrm{E} \underset{k_{-2}}{\overset{k_2[\mathrm{A-B}]}{\rightleftharpoons}} \mathrm{E}_{\mathrm{B-A}} \underset{k_{-4}}{\overset{k_4}{\rightleftharpoons}} \mathrm{E}\overset{\mathrm{A}}{\underset{\mathrm{B}}{|}}$$

$$\mathrm{E}^{\mathrm{C-D}} \underset{k_{-2}}{\overset{k_2[\mathrm{A-B}]}{\rightleftharpoons}} \mathrm{E}^{\mathrm{C-D}}_{\mathrm{B-A}} \xrightarrow{k_3} \mathrm{E}_{\mathrm{B-A}}$$

(3–231)

In this scheme, the binding of segment B of the reactant blocks the rebinding of segment D of the product. The rate of the dissociation of the product C–D from the complex $\mathrm{E}\overset{\mathrm{C}}{\underset{\mathrm{D}}{|}}$ between enzyme and product is

$$\frac{\mathrm{d[C-D]}}{\mathrm{d}t} = k_3\,([\mathrm{E^{CD}}]+[\mathrm{E^{CD}_{BA}}]) \tag{3–232}$$

From the various steady-state relationships it follows that

$$\frac{\mathrm{d[C-D]}}{\mathrm{d}t} = k_1^{\mathrm{app}}[\mathrm{E}\overset{\mathrm{C}}{\underset{\mathrm{D}}{|}}] \tag{3–233}$$

where

$$k_1^{\mathrm{app}} = k_1 \frac{k_3(k_3+k_{-2})+k_3k_2[\mathrm{A-B}]}{(k_3+k_{-2})(k_3+k_{-1})+k_3k_2[\mathrm{A-B}]} \tag{3–234}$$

The logarithm of the apparent first-order dissociation constant $k_1^{\mathrm{app}}$ for the dissociation of product can be plotted as a function of the logarithm of the concentration of reactant, [A–B] (Figure 3–47). Because $k_3k_1(k_3+k_{-1})^{-1}$ is always less than $k_1$, the addition of reactant A–B will always accelerate the dissociation of product C–D.

Mechanism 3–231 is branched and contains two paths between $\mathrm{E}\overset{\mathrm{C}}{\underset{\mathrm{D}}{|}}$ and $\mathrm{E}\overset{\mathrm{A}}{\underset{\mathrm{B}}{|}}$, and the fluxes through the two paths shift in relative magnitude as [A–B] is increased in concentration. Such a situation is known to produce negative deviations from linearity,[91] and it is possible that many of the instances of such behavior result from the process of reactants displacing products.

The displacement of products by reactants at the active site seems so unavoidable, however, that it is possible it occurs far more often than the frequency of negative deviations would suggest. Because the displacement of a product from the active site by a reactant is usually a step in the normal association of the reactant, it should be able to occur even when obvious deviations from linear kinetic behavior are not observed. It can be recognized, however, by other anomalies.

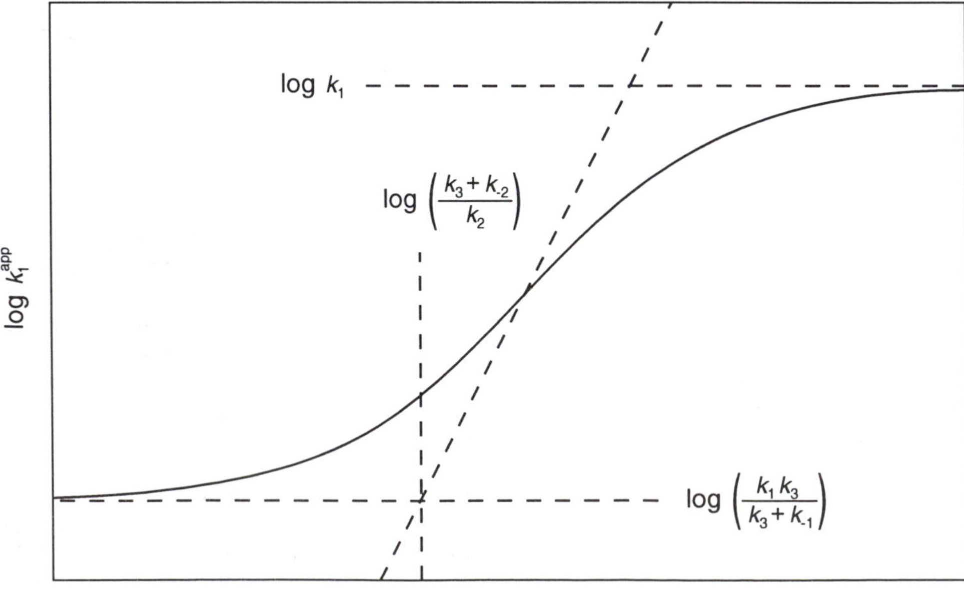


**Figure 3–47:** Graphical representation of the function defined by Equation 3–234.

For example, aconitate hydratase, which catalyzes the three reactions[220]

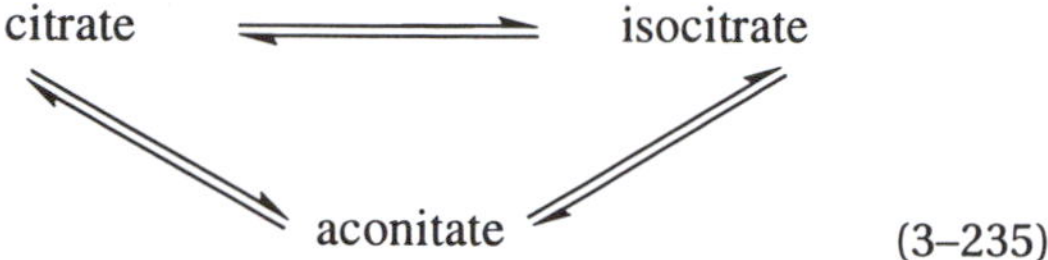


(3–235)

displays linear and uncomplicated initial rate kinetics with each of the substrates as reactants.[221] The unassisted release of [$^{14}$C]citrate from the active site in the absence of any other substrates is far too slow, however, for the unassisted release of citrate to be a step in the reaction

$$\text{aconitate} \rightleftharpoons \text{citrate} \qquad (3\text{–}236)$$

If the release of [$^{14}$C]citrate is measured in the presence of unlabeled citrate, however, its release becomes fast enough.[221] In addition, the effects of pH on the rate of the enzymatically catalyzed reaction are inconsistent with the effects of pH on the inhibition shown by analogues of the substrates. It has been concluded that as one or the other of the carboxyl arms of a molecule of a product bound at the active site of aconitate hydratase momentarily pulls away from the active site, it can either be trapped by a proton or its subsite can be seized by the respective carboxyl arm of a molecule of the reactant in the solution. Either of these events would prevent the carboxyl arm of the product from reassociating and would hasten the rate of dissociation of the product. If this is the case, aconitate hydratase would be an example of an enzyme the kinetics of which are unremarkable but that nevertheless uses reactants to displace products.

There are several factors that would limit the effectiveness of such mechanisms for the acceleration of the dissociation of product and hence the rate of the enzymatic reaction. First, in the conformation in which only a portion of the product has dissociated, there will likely remain considerable steric hindrance to the approach of the displacing ligand. This would be particularly acute if the displacing ligand was large and could only associate with the momentarily exposed portion of the active site in one particular orientation. For example, a rigid molecule like the pyranose of D-glucose acting as a reactant may only be able to associate with an active site in one specific orientation which could not be accommodated if any portion of the product was still bound. Second, not only such steric effects but also charge repulsion could prevent a molecule of the displacing ligand from associating even partially before the entire molecule of product has dissociated. For example, the association of sulfate with the subsite on the active site momentarily vacated by the 1-phosphate of 1,3-bisphosphoglycerate should be hindered by the negative charge of the nearby 1-phosphate (Figure 3–46). Third, if the mechanism of the enzyme involves induced fit, the active site may be continuously changing its conformation as the product dissociates and this might preclude the association of any other ligand until it was completely vacated. Finally, the ligand-promoted dissociations of products are often observed only at high, unphysiological concentrations of the displacing ligand, and such examples would be irrelevant biologically. Nevertheless, the promotion of the dissociation of a molecule of the product from the active site by a molecule of the reactant or another ligand may be a common feature of enzymatically catalyzed reactions.

### *Suggested Reading*

Scopes, R.K. (1978) Binding of Substrates and Other Anions to Yeast Phosphoglycerate Kinase, *Eur. J. Biochem. 91*, 119–129.

## Transfer of Substrates among Active Sites

In most reactions that occur sequentially within a metabolic pathway, the product of the preceding reaction in the pathway is released into the cytoplasm from the one active site, it diffuses freely, and it is bound from the cytoplasm to the next active site as the reactant in the next reaction. There are, however, instances where products and reactants are transferred between active sites without being released into the cytoplasm as free solutes. The most commonly cited examples of this process are reactions catalyzed by particular multienzyme complexes.

One of the most obvious examples of such transfer occurs in the multienzyme complex responsible for the pyruvate dehydrogenase reaction

$$\text{pyruvate} + \text{CoASH} + \text{NAD}^+ \rightleftharpoons \text{acetyl-SCoA} + CO_2 + \text{NADH} \qquad (3\text{–}237)$$

The core protein of the complex from *E. coli* is an octahedral oligomeric protein[221a] composed of 24 identical, folded polypeptides. The 316 amino-terminal amino acids of the sequence of each of these core polypeptides[222] consists of three internally repeating domains about 100 amino acids in length. Each of these domains contains a lysine to which a lipoamide has been covalently attached,[223] and following each of these three domains there are short segments about 10 amino acids in length containing only alanines and prolines. It is thought that these short segments act as hinges of random coil and that the three domains and their attached lipoamides form a flexible, jointed arm extending out from each protomer in the core. The purpose of this flexibility is to permit the lipoamide to move among the three independent active sites in the complex.

The pyruvate dehydrogenase complex can be resolved into three separate proteins, each of which carries the active site responsible for one of the steps in the reaction. These three independent active sites are all represented in the complete complex. The active site of pyruvate decarboxylase

$$\text{pyruvate} + \text{lipoamide} \rightleftharpoons \text{acetyldihydrolipoamide} + CO_2 \qquad (3\text{–}238)$$

transfers the acetyl group of pyruvate to a lipoamide. The acetyldihydrolipoamide then swings into the active site of dihydrolipoamide transacetylase

acetyldihydrolipoamide + CoASH ⇌

acetyl-SCoA + dihydrolipoamide (3–239)

and the acetyl group is transferred to coenzyme A. The dihydrolipoamide then swings into the active site of lipoamide dehydrogenase

$$\text{dihydrolipoamide} + NAD^+ \rightleftharpoons NADH + \text{lipoamide} + H^+ \quad (3\text{–}240)$$

and the dihydrolipoamide is converted to lipoamide. The lipoamide then swings back to the active site of pyruvate decarboxylase.

In this scheme, the impression is left that the pyruvate dehydrogenase complex can be divided into 24 independent sectors in which these cycles proceed in isolation. It has been demonstrated, however, that the domains bearing the lipoamides are so flexible that all of the active sites within a given complex are chemically interconnected. A set of pyruvate dehydrogenase complexes containing decreasing ratios of the active site for pyruvate decarboxylase can be produced either by reconstituting the complex with less than stoichiometric amounts of this particular protein[224,225] or by adding an inhibitor of such high affinity that it binds stoichiometrically to a portion of these active sites.[225] The overall enzymatic activity decreases in direct proportion to the loss of these active sites, and the rate at which lipoamide is acetylated by [2-$^{14}$C]pyruvate in the absence of coenzyme A and NADH also decreases in direct proportion to the loss of these active sites (Figure 3–48).[226] All of the lipoamides in the complex, however, still become acetylated in a simple first-order reaction (Figure 3–48), even when as little as 10% of the active sites for pyruvate decarboxylase are present.[225] These results state that acetyl groups can enter the complex at any active site for pyruvate decarboxylase and then be transferred among all of the lipoamides. Presumably this occurs by uncatalyzed thiol exchange–acyl exchange[225] in the center of the complex

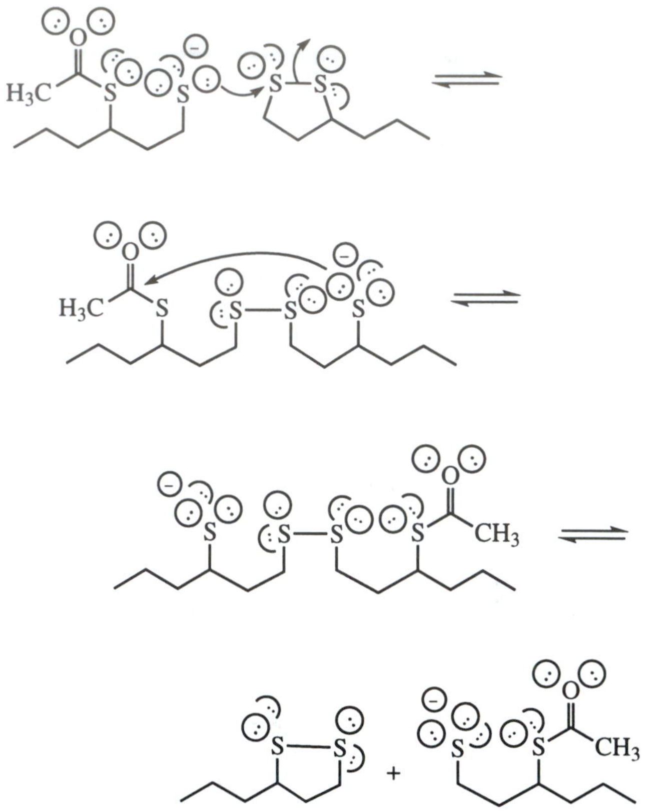

(3–241)

Therefore, not only are acetyl groups transferred among active sites in the complex, but they are also transferred among all the lipoamides attached to the core of the complex. As a result, complexes that are missing a portion of their components are still able to function efficiently.

Most multienzyme complexes, however, do not transfer a substrate among their active sites in the form of a covalent adduct like acetyldihydrolipoamide. In many multienzyme complexes, even though all the active sites are on the same protein, substrates are not transferred preferentially among the active sites in the same cmplex but pass among active sites through the solution. Chorismate mutase–prephenate dehydratase from *E. coli* is a multienzyme complex catalyzing two sequential reactions in the biosynthetic pathway for phenylalanine

chorismate ⇌ prephenate (3–242)

$$\text{prephenate} \rightleftharpoons \text{phenylpyruvate} + H_2O + CO_2 \quad (3\text{–}243)$$

These two reactions are catalyzed by two distinct active sites[227] that reside on the same folded polypeptide.[228]

Prephenate, the product of the first reaction, is the exclusive reactant in the second reaction. The progress of the overall reaction can be followed by assaying for the concentrations of the three substrates as a function of time (Figure 3–49A).[227] As the reaction progresses, the concentration of prephenate in the solution increases, reaches a maximum, and declines as all of the substrate is converted to phenylpyruvate. The important question is whether all of the prephenate passes

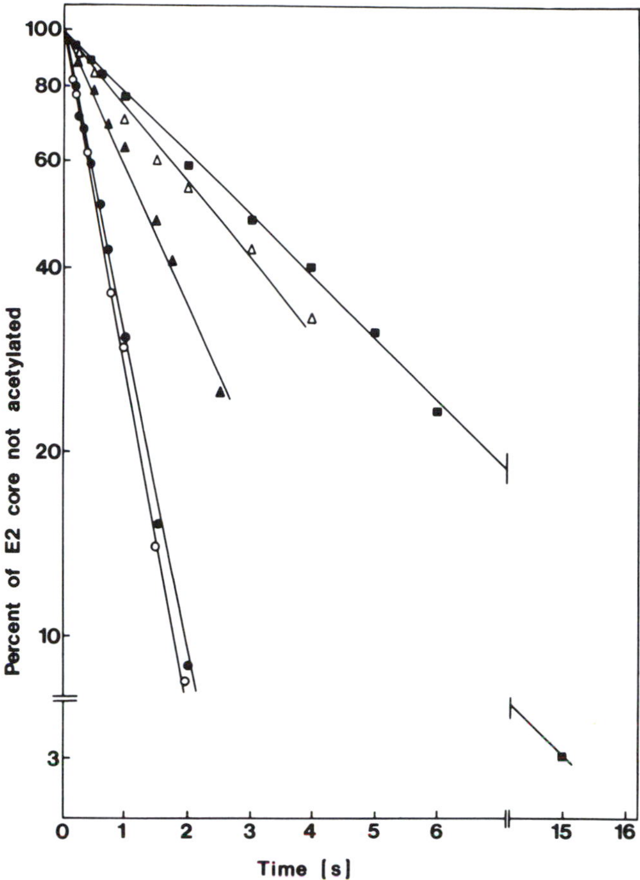


**Figure 3–48:** Incorporation of [2-$^{14}$C]pyruvate into a series of partly assembled pyruvate dehydrogenase complexes.[226] The pyruvate dehydrogenase complexes were reassembled from purified pyruvate decarboxylase and a purified form of the complex between dehydrolipoamide transacetylase and lipoamide dehydrogenase with a stoichiometry of 0.88 polypeptides of the dehydrogenase for every polypeptide of the transacetylase. Various ratios of these two purified proteins were mixed and allowed to reassemble and the products were purified. This produced complexes with 0.21 (■), 0.39 (△), 0.71 (▲), and 1.29 (○) polypeptides of the decarboxylase for every polypeptide of the transacetylase. The native protein (●) has 1.24 polypeptides of the decarboxylase for every polypeptide of the transacetylase. The transfer of the acetyl moiety from [2-$^{14}$C]pyruvate (18 $\mu$M) to the lipoamides of the transacetylase was monitored as a function of time at 0.7 mM thiamin pyrophosphate, 10 mM $MgCl_2$, and 1.0 mM $NAD^+$, pH 7.0, 25 °C. The reaction was quenched at various times by mixing with trichloroacetic acid, and the radioactivity bound to protein was assayed. The logarithm of the percent lipoamide not acetylated is presented as a function of time (seconds). Notice that the lipoamide became acetylated in each case in a simple first-order reaction. Reprinted with permission from ref 226. Copyright 1978 National Academy of Sciences.

through the solution in going from the first active site to the second active site or if some of it passes between the active sites without leaving the enzyme. To examine this question, [$^{14}$C]chorismate and nonradioactive prephenate were mixed with the enzyme, and the production of [$^{14}$C]prephenate was monitored (Figure 3–49C). The production of [$^{14}$C]prephenate followed the calculated curve (solid line in Figure 3–49C) expected if all of the prephenate produced at the active site of chorismate mutase had to pass through the solution to reach the active site of prephenate dehydratase. If all of the prephenate produced at the first active site had been transferred to the second active site without leaving the enzyme, the production of [$^{14}$C]phenylpyruvate would have begun at a steady rate (dashed line in Figure 3–49C). If, however, all of the [$^{14}$C]prephenate had to enter the solution before it entered the second active site, it should have been diluted by the unradioactive prephenate, and there should have been a lag in production as [$^{14}$C]prephenate was built up, as was observed.

Unlike chorismate mutase–prephenate dehydratase, tryptophan synthase is an enzyme in which the intermediate product does not seem to have to enter the solution to pass between the two active sites. Tryptophan synthase is an $\alpha_2\beta_2$ heterodimer that catalyzes the reaction

$$\text{serine} + \text{1-(indol-3-yl)glycerol 3-phosphate} \rightleftharpoons \text{tryptophan} + \text{glyceraldehyde 3-phosphate} \quad (3\text{–}244)$$

The isolated $\alpha$ subunits have an active site that catalyzes the reaction[229]

$$\text{1-(indol-3-yl)glycerol 3-phosphate} \rightleftharpoons \text{glyceraldehyde 3-phosphate} + \text{indole} \quad (3\text{–}245)$$

and the isolated $\beta_2$ subunits have an active site that catalyzes the reaction[229]

$$\text{indole} + \text{serine} \rightleftharpoons \text{tryptophan} + H_2O \quad (3\text{–}246)$$

In the crystallographic molecular model of the enzyme,[204] the active site for Reaction 3–245 has been located on the $\alpha$ subunit by soaking crystals in a solution of the competitive inhibitor 3-(indol-3-yl)propanol phosphate, and the active site for Reaction 3–246 has been located on the $\beta$ subunit by locating the pyridoxal phosphate present in the enzyme as it was originally crystallized. The distance in the crystallographic molecular model between an active site on an $\alpha$ subunit and the closest active site on a $\beta$ subunit is 2.5–3.0 nm.

In the absence of serine, intact tryptophan synthase produces indole from 1-(indol-3-yl)glycerol 3-phosphate but the

A 

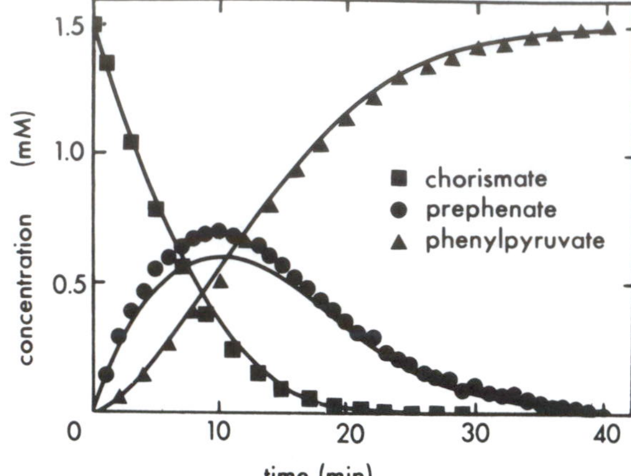


B

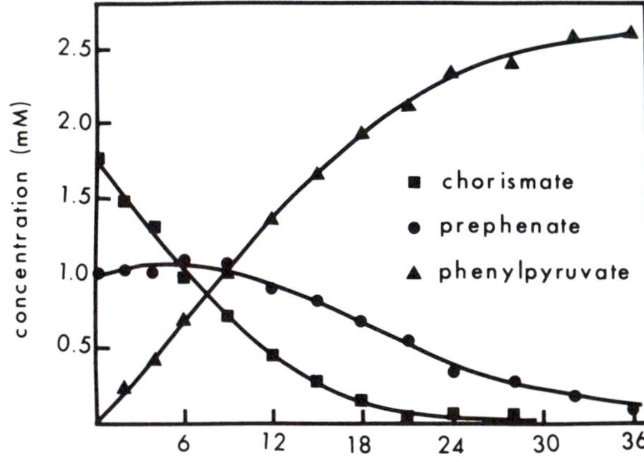


C

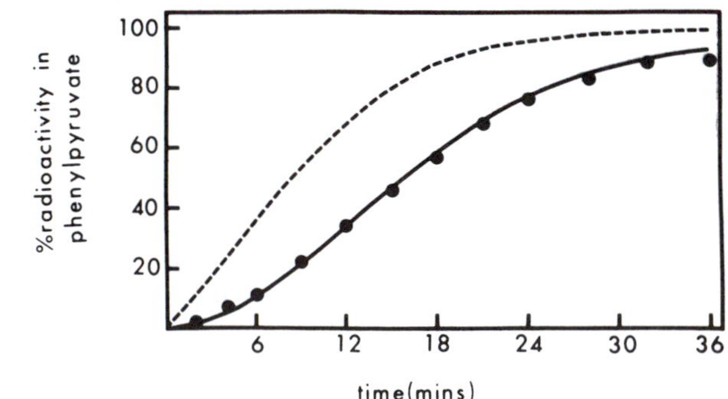


**Figure 3–49:** Changes of concentrations of substrates with time in solutions containing *E. coli* chorismate mutase–prephenate dehydratase.[227] (A) Chorismate mutase–prephenate dehydratase was mixed with 1.5 mM chorismate at pH 7.5 and 37 °C. At the noted times (in minutes), samples were removed and assayed for chorismate (■), prephenate (●) and phenylpyruvate (▲). The three smooth curves are for theoretical functions derived from the kinetic constants ($K_m$ and $V_{max}$) for the two active sites and from the assumption that all products and reactants dissociate into the solution and associate from the solution. Any preferential channeling between active sites on the same protein would have caused the observed concentrations of prephenate to be less than the theoretical function. (B) Chorismate mutase–prephenate dehydratase was mixed under the same conditions with 1.76 mM [U-$^{14}$C]chorismate (12,000 cpm $\mu$mol$^{-1}$) and 1.0 mM prephenate and the concentrations of the three substrates, disregarding the radioactivity, were again monitored as a function of time. The curves are calculated with the same assumptions and the same functions as were used for the curves in panel A. (C) In each of the samples from the experiment in panel B, the percent of the total $^{14}$C radioactivity in phenylpyruvate was determined as a function of time (minutes), disregarding the total molar concentration of the phenylpyruvate. The solid line represents the predictions of the functions used in panels A and B that were derived on the assumption that no channeling occurs The dashed line is for a function derived on the assumption that prephenate is completely channeled between the active sites. Reprinted with permission from ref 227. Copyright 1978 American Chemical Society.

rate at which indole is released into the solution is 50 times slower than the rate at which tryptophan is produced from 1-(indol-3-yl)glycerol 3-phosphate and serine under the same circumstances.[230,231] When serine is present, indole production from 1-(indol-3-yl)glycerol 3-phosphate ceases (< 20%).[230,231] Furthermore, during a reaction in which 100 nmol of tryptophan was produced from 1-(indol-3-yl)glycerol 3-phosphate and serine, no indole (< 2 nmol) was released into the solution.

These results have been interpreted as evidence for channeling of indole between an active site on an $\alpha$ subunit and an active site on a $\beta$ subunit. **Channeling** is the noncovalent transfer of a substrate between two active sites by a process that avoids the dissociation of the substrate into the solution from the one active site and its association from the solution with the other active site. Channeling in the specific case of tryptophan synthase would be the transfer of a molecule of indole produced in Reaction 3–245 on the active site of the $\alpha$ subunit to the active site of the $\beta$ subunit without permitting the molecule of indole to dissociate into the solution. When serine is present, all of the channeled indole would be immediately transformed into tryptophan, but when serine is absent, the unconsumed indole would slowly dissociate from the enzyme, and its dissociation would be the rate-limiting step in Reaction 3–225. If this were the explanation, Reaction 3–245 would be a side reaction not on the main pathway for Reaction 3–244.

In the crystallographic molecular model of tryptophan synthase there is a tunnel passing through the protein 2.5 nm in length and connecting the active site for Reaction 3–245 on the $\alpha$ subunit with the active site for Reaction 3–246 on the $\beta$ subunit.[204] The tunnel is enclosed along its length on all sides by protein. It is wide enough to permit an indole to pass through. In the crystallographic molecular model of the enzyme with 3-indolylpropanol phosphate occupying the active site in the $\alpha$ subunit, the indolyl is in the bottom of that active site as far from the solution as possible and sitting at the opening of the tunnel. Upon cleavage of the normal substrate, 1-(indol-3-yl)glycerol 3-phosphate, the indole would be released at this location in the active site, and it would be hindered from being released into the solution. Because of these details it is believed that the indole feeding the active site on the $\beta$ subunit during the overall reaction (Reaction 3–244) is transferred between the two active sites through this tunnel.

This mechanism provides an explanation for an earlier observation. When 1-([2-$^{14}$C]-indol-3-yl)glycerol 3-phosphate is used as a reactant in the presence of serine, [2′-$^{14}$C]tryptophan is produced by tryptophan synthase and no [2-$^{14}$C]indole is released into the solution.[232] If, however, unradioactive indole is added to the solution, the rate of production of [2′-$^{14}$C]tryptophan decreases remarkably, and significant amounts of [2-$^{14}$C]indole are released into the solution. As the concentration of unradioactive indole is increased, the rate of production of [2′-$^{14}$C]tryptophan decreases further and the rate of production of [2-$^{14}$C]indole increases. One explanation for these observations is that free indole can enter the active site for Reaction 3–246 on the $\beta$ subunit from the solution in preference to indole arriving through the tunnel. If the flux of indole entering from the solution is sufficient to block movement through the tunnel, the [2-$^{14}$C]indole produced at the active site for Reaction 3–245 on the $\alpha$ subunit would dissociate into the solution because it is unable to move through the tunnel.

## *Suggested Reading*

De Moss, J.A. (1962) Studies on the Mechanism of the Tryptophan Synthetase Reaction, *Biochim. Biophys. Acta 62*, 279–293.

## *References*


1. Bloch, W., MacQuarrie, R.A., & Bernhard, S.A. (1971) *J. Biol. Chem. 246*, 780–790.
2. Bates, D.J., & Frieden, C. (1973) *J. Biol. Chem. 248*, 7878–7884.
3. Samuel, D., Estroumza, J., & Ailhoud, G. (1970) *Eur. J. Biochem. 12*, 576–582.
4. Kivirikko, K.I., Shudo, K., Sakakibara, S., & Prockop, D.J. (1972) *Biochemistry 11*, 122–129.
5. Hagemann, R.V., Orme-Johnson, W.H., & Burris, R.H. (1980) *Biochemistry 19*, 2333–2342.
6. Trentham, D.R., McMurray, C.H., & Pogson, C.I. (1969) *Biochem. J. 114*, 19–24.
7. Koerner, T.A.W., Cary, L.W., Bhacca, N.S., & Younathan, E.S. (1973) *Biochem. Biophys. Res. Commun. 51*, 543–549.
8. Benkovic, P.A., Bullard, W.P., de Maine, M., Fishbein, R., Schrat, K.J., Steffans, J.J., & Benkovic, S.J. (1974) *J. Biol. Chem. 249*, 930–931.
9. Hall, R.L., Vennesland, B., & Kezdy, F. (1969) *J. Biol. Chem. 244*, 3991–3998.
10. Hafner, E.W., & Wellner, D. (1971) *Proc. Natl. Acad. Sci. U.S.A. 68*, 987–991.
11. Hayzer, D.J., & Leisinger, T. (1981) *Biochem. J. 197*, 269–274.
12. Alberts, A.W., & Vagelos, P.R. (1968) *Proc. Natl. Acad. Sci. U.S.A. 59*, 561–568.
13. Fall, R.R., & Vagelos, P.R. (1972) *J. Biol. Chem. 247*, 8005–8015.
14. Alberts, A.W., Gordon, S.G., & Vagelos, P.R. (1971) *Proc. Natl. Acad. Sci. U.S.A. 68*, 1259–1263.
15. Paoletti, F. (1983) *Arch. Biochem. Biophys. 222*, 489–496.
16. Libertini, L.J., & Smith, S. (1978) *J. Biol. Chem. 253*, 1393–1401.
17. Abdinejad, A., Fisher, A.M., & Kumar, S. (1981) *Arch. Biochem. Biophys. 208*, 135–145.
18. Sasaki, R., Ikura, K., Sugimoto, E., & Chiba, H. (1975) *Eur. J. Biochem. 50*, 581–593.
19. Albright, F., & Schroepfer, G.J. (1970) *Biochem. Biophys. Res. Commun. 40*, 661–666.
20. Harting, J., & Velick, S. (1954) *J. Biol. Chem. 207*, 857–865.
21. Massey, V. (1953) *Biochem. J. 53*, 67–71.
22. Cleland, W.W. (1979) *Methods Enzymol. 63*, 103–138.
23. Williams, A. (1970) *Biochemistry 9*, 3383–3390.
24. Purich, D.L., & Fromm, H.J. (1971) *J. Biol. Chem. 246*, 3456–3463.
25. Villet, R.H., & Dalziel, K. (1972) *Eur. J. Biochem. 27*, 251–258.
26. Knight, W.B., & Cleland, W.W. (1989) *Biochemistry 28*, 5728–5734.
27. King, E.L., & Altman, C. (1956) *J. Phys. Chem. 60*, 1375–1378.
28. Segel, I.H. (1975) *Enzyme Kinetics: Behavior and Analysis of Rapid Equilibrium and Steady-State Enzyme Systems*, pp 506–523, John Wiley & Sons, New York.
29. Mahler, H.R., & Cordes, E.H. (1971) *Biological Chemistry: Second Edition*, pp 295–299, Harper & Row, New York.

30. Orsi, B.A., & Cleland, W.W. (1972) *Biochemistry 11*, 102–109.
31. Li, H., & Buchanan, J.M. (1971) *J. Biol. Chem. 246*, 4720–4726.
32. De Leo, A.B., Dayan, J., & Sprinson, D.B. (1973) *J. Biol. Chem. 248*, 2344–2353.
33. Lueck, J.D., & Fromm, H.J. (1974) *J. Biol. Chem. 249*, 1341–1347.
34. Collins, K.D., & Stark, G.R. (1971) *J. Biol. Chem. 246*, 6599–6605.
35. Hoogenraad, N.J. (1974) *Arch. Biochem. Biophys. 161*, 76–82.
36. Matsuoka, Y., & Sreve, P.A. (1973) *J. Biol. Chem. 248*, 8022–8030.
37. Noat, G., Ricard, J., Borel, M., & Got, C. (1970) *Eur. J. Biochem. 13*, 347–363.
38. Janson, C.A., & Cleland, W.W. (1974) *J. Biol. Chem. 249*, 2562–2566.
39. Cleland, W.W. (1963) *Biochim. Biophys. Acta 67*, 104–137.
40. Howard, W.D., & Solomonson, L.P. (1981) *J. Biol. Chem. 256*, 12725–12730.
41. Ainsworth, S., & MacFarlane, N. (1973) *Biochem. J. 131*, 223–236.
42. Fujiwara, K., & Motokawa, Y. (1983) *J. Biol. Chem. 258*, 8156–8162.
42a. Kyte, J. (1995) *Structure in Protein Chemistry*, p 305, Garland Publishing, New York.
43. Sharp, J.A., & Edwards, M.R. (1983) *Biochem. J. 213*, 179–185.
44. Katiyar, S.S., & Porter, J.W. (1974) *Arch. Biochem. Biophys. 163*, 324–331.
45. Puistola, U., Turpeenniemi-Hujanen, T.M., Myllylä, R., & Kivirikko, K.I. (1980) *Biochim. Biophys. Acta 611*, 40–50.
46. Cuatrecasas, P., & Hollenberg, M.D. (1976) *Adv. Protein Chem. 30*, 251–451.
47. Maturo, J.M., & Hollenberg, M.D. (1978) *Proc. Natl. Acad. Sci. U.S.A. 75*, 3070–3075.
48. deLean, A., Munson, P.J., & Rodbard, D. (1978) *Am. J. Physiol. 235*, E97–E102.
49. Munson, P.J., & Rodbard, D. (1980) *Anal. Biochem. 107*, 220–229.
50. Klotz, I.M., Walker, F.M., & Pivan, R.B. (1946) *J. Am. Chem. Soc. 68*, 1486–1490.
51. Scatchard, G. (1949) *Ann. N.Y. Acad. Sci. 51*, 660–672.
52. Bruch, P., Schnackerz, K.D., Chirgwin, J.M., & Noltmann, E.A. (1973) *Eur. J. Biochem. 36*, 564–568.
53. Eadie, G.S. (1942) *J. Biol. Chem. 146*, 85–93.
54. Hofstee, B.H.J. (1959) *Nature 184*, 1296–1298.
55. Hummel, J.P., & Dreyer, W.J. (1962) *Biochim. Biophys. Acta 63*, 530–532.
56. Colowick, S.P., & Womack, F.C. (1969) *J. Biol. Chem. 244*, 774–777.
57. Dyson, J.E.D., & Noltmann, E.A. (1968) *J. Biol. Chem. 243*, 1401–1414.
57a. Kyte, J. (1995) *Structure in Protein Chemistry*, p 29, Garland Publishing, New York.
58. Deranleau, D.A., Bradshaw, R.A., & Schwyzer, R. (1969) *Proc. Natl. Acad. Sci. U.S.A. 63*, 885–889.
59. Mackay, D., Panjehshahin, M.R., & Bowmer, C.J. (1991) *Biochem. Pharmacol. 41*, 2011–2018.
60. Janin, J., van Rapenbusch, R., Truffa-Bachi, P., & Cohen, G.N. (1969) *Eur. J. Biochem. 8*, 128–138.
61. Egan, R.R., & Dalziel, K. (1971) *Biochim. Biophys. Acta 250*, 47–50.
62. Schuman, M., & Massey, V. (1971) *Biochim. Biophys. Acta 227*, 500–520.
63. Collins, K.D., & Stark, G.R. (1969) *J. Biol. Chem. 244*, 1869–1877.
64. Carpenter, G., King, L., & Cohen, S. (1979) *Proc. Natl. Acad. Sci. U.S.A. 254*, 4884–4891.
65. Sine, S., & Taylor, P. (1980) *J. Biol. Chem. 255*, 10144–10156.
66. Hasan, F.B., Cohen, S.G., & Cohen, J.B. (1980) *J. Biol. Chem. 255*, 3898–3904.
67. Rudolph, F.B., & Fromm, H.J. (1971) *J. Biol. Chem. 246*, 6611–6619.
68. Colowick, S.P., Womack, F.C., & Nielsen, J. (1969) in *The Role of Nucleotides for the Function and Conformation of Enzymes* (Kalckar, H.M., Klenow, H., Munch-Petersen, A., Ottesen, M., & Thaysen, J.H., Eds.) pp 15–43, Munksgaard, Copenhagen.
69. Bessell, E.M., Foster, A.B., & Westwood, J.H. (1972) *Biochem. J. 128*, 199–204.
70. Sols, A., de la Fuente, G., Villar-Palasi, C., & Asensio, C. (1958) *Biochim. Biophys. Acta 30*, 92–101.
71. Holler, E., Rainey, P., Orme, A., Bennett, E.L., & Calvin, M. (1973) *Biochemistry 12*, 1150–1159.
72. Majamaa, K., Hanauske-Abel, H.M., Günzler, V., & Kivirikko, K.I. (1984) *Eur. J. Biochem. 138*, 239–245.
73. Kuehn, G.D., Barnes, L.D., & Atkinson, D.E. (1971) *Biochemistry 10*, 3945–3951.
74. Franzen, J.S., Kuo, I., Eichler, A.J., & Feingold, D.S. (1973) *Biochem. Biophys. Res. Commun. 50*, 517–523.
75. Maxwell, E.S., Kalckar, H.M., & Strominger, J.L. (1956) *Arch. Biochem. Biophys. 65*, 2–10.
76. Salitis, G., & Oliver, I.T. (1964) *Biochim. Biophys. Acta 81*, 55–60.
77. Penzer, G.R., Bennett, E.L., & Calvin, M. (1971) *Eur. J. Biochem. 20*, 1–13.
78. Cole, F.X., & Schimmel, P.R. (1970) *Biochemistry 9*, 480–489.
79. Porter, R.W., Modebe, M.O., & Stark, G.R. (1969) *J. Biol. Chem. 244*, 1846–1859.
80. Lorenson, M.Y., & Mansour, T.E. (1969) *J. Biol. Chem. 244*, 6420–6431.
81. Liou, R.S., & Anderson, S.R. (1978) *Biochemistry 17*, 999–1004.
82. Gopinath, R.M., & Vincenzi, F.F. (1977) *Biochem. Biophys. Res. Commun. 77*, 1203–1209.
83. Jarrett, H.W., & Penniston, J.T. (1977) *Biochem. Biophys. Res. Commun. 77*, 1210–1216.
84. Jarrett, H.W., & Kyte, J. (1979) *J. Biol. Chem. 254*, 8237–8244.
85. Meighen, E.A., & Hastings, J.W. (1971) *J. Biol. Chem. 246*, 7666–7674.
86. Kyte, J. (1972) *J. Biol. Chem. 247*, 7634–7641.
87. Gadd, R.E.A., & Henderson, J.F. (1970) *J. Biol. Chem. 245*, 2979–2984.
88. Monod, J., Wyman, J., & Changeux, J. (1965) *J. Mol. Biol. 12*, 88–118.
89. Teresi, J.D., & Luck, J.M. (1948) *J. Biol. Chem. 174*, 653–661.
90. Spears, G., Sneyd, J.G.T., & Loten, E.G. (1971) *Biochem. J. 125*, 1149–1151.
91. Moczydlowski, E.G., & Fortes, P.A.G. (1981) *J. Biol. Chem. 256*, 2346–2356.
92. Schuurmans-Stekhoven, F.M.A.H., Swarts, H.G.P., dePont, J.J.H.H.M., & Bonting, S.L. (1981) *Biochim. Biophys. Acta 649*, 533–540.
93. Koshland, D.E., Nemethy, G., & Filmer, D. (1966) *Biochemistry 5*, 365–385.
94. Levitzki, A., & Koshland, D.E. (1969) *Proc. Natl. Acad. Sci. U.S.A. 62*, 1121–1128.
95. Kahn, C.R., Freychet, P., Roth, J., & Neville, D.M. (1974) *J. Biol. Chem. 249*, 2249–2257.
96. deMeyts, P., Bianco, A.R., & Roth, J. (1976) *J. Biol. Chem. 251*, 1877–1888.
97. Lazdunski, M. (1972) *Curr. Top. Cell. Regul. 6*, 267–310.
98. Seydoux, F., Malhotra, O.P., & Bernhard, S.A. (1974) *CRC Crit. Rev. Biochem. 2*, 227–257.
99. Schwendimann, B., Ingbar, D., & Bernhard, S.A. (1976) *J. Mol. Biol. 108*, 123–138.
100. Gennis, L.S. (1976) *Proc. Natl. Acad. Sci. U.S.A. 73*, 3928–3932.
101. Stallcup, W.B., & Koshland, D.E. (1973) *J. Mol. Biol. 80*, 41–62.

102. Lazdunski, M., Petitclerc, C., Chappelet, D., & Lazdunski, C. (1971) *Eur. J. Biochem. 20*, 124–139.
103. Cocivera, M., McManaman, J., & Wilson, I.B. (1980) *Biochemistry 19*, 2901–2907.
104. Bloch, W., & Schlesinger, M.J. (1974) *J. Biol. Chem. 249*, 1760–1768.
105. Anderson, D.C., & Dahlquist, F.W. (1982) *Biochemistry 21*, 3578–3584.
106. Andersson, P., & Petterson, G. (1982) *Eur. J. Biochem. 122*, 559–568.
107. Winter, G., Koch, G.L.E., Hartley, B.S., & Barker, D.G. (1983) *Eur. J. Biochem. 132*, 383–387.
108. Irwin, M.J., Nyborg, J., Reid, B.J., & Blow, D.M. (1976) *J. Mol. Biol. 105*, 577–586.
109. Mulvey, R.S., & Fersht, A.R. (1977) *Biochemistry 16*, 4005–4013.
110. Fersht, A.R. (1975) *Biochemistry 14*, 5–12.
111. Montheilhet, C., & Blow, D.M. (1978) *J. Mol. Biol. 122*, 407–417.
112. Ward, W.H.J., & Fersht, A.R. (1988) *Biochemistry 27*, 1041–1049.
113. Shill, J.P., & Neet, K.E. (1975) *J. Biol. Chem. 250*, 2259–2268.
113a. Kyte, J. (1995) *Structure in Protein Chemistry*, p 320, Garland Publishing, New York.
114. Anderson, W.F., & Steitz, T.A. (1975) *J. Mol. Biol. 92*, 279–287.
115. Leslie, A.G.W., & Wonacott, A.J. (1984) *J. Mol. Biol. 178*, 743–772.
116. Biesecker, G., Harris, J.I., Thierry, J.C., Walker, J.E., & Wonacott, A.J. (1977) *Nature 266*, 328–333.
117. Allen, G., & Harris, J.I. (1975) *Biochem. J. 151*, 747–749.
118. Frost, A.A., & Pearson, R.G. (1961) *Kinetics and Mechanism: A Study of Homogeneous Chemical Reactions*, pp 77–102, John Wiley & Sons, New York.
119. Kraut, J. (1988) *Science 242*, 533–540.
120. Jencks, W.P. (1975) *Adv. Enzymol. Relat. Areas Mol. Biol. 43*, 219–410.
121. Blacklow, S.C., Raines, R.T., Lim, W.A., Zamore, P.D., & Knowles, J.R. (1988) *Biochemistry 27*, 1158–1167.
122. Andrews, P.R., Smith, G.D., & Young, I.G. (1973) *Biochemistry 12*, 3492–3498.
123. Copley, S.D., & Knowles, J.R. (1987) *J. Am. Chem. Soc. 109*, 5008–5013.
124. Pauling, L. (1948) *Am. Sci. 36*, 50–58.
125. Rupley, J.A. (1967) *Proc. R. Soc. London, B 167*, 416–428.
126. Blake, C.C.F., Johnson, L.N., Mair, G.A., North, A.C.T., Phillips, D.C., & Sarma, V.R. (1967) *Proc. R. Soc. London, B 167*, 378–388.
127. Ford, L.O., Johnson, L.N., Machin, P.A., Phillips, D.C., & Tjian, R. (1974) *J. Mol. Biol. 88*, 349–371.
128. Hammond, G.S. (1955) *J. Am. Chem. Soc. 77*, 334–338.
129. Jencks, W.P. (1969) *Catalysis in Chemistry and Enzymology*, pp 605–614, McGraw-Hill, New York.
130. Jencks, W.P. (1969) *Catalysis in Chemistry and Enzymology*, pp 282–320, McGraw-Hill, New York.
131. Morihara, K., & Tsuzuki, H. (1970) *Eur. J. Biochem. 15*, 374–380.
132. Rossi, G.L., Holler, E., Kumar, S., Rupley, J.A., & Hess, G.P. (1969) *Biochem. Biophys. Res. Commun. 37*, 757–766.
133. Gutfreund, H., & Trentham, D.R. (1975) in *Energy Transformation in Biological Systems: Ciba Foundation Symposium 31*, pp 69–86, Elsevier, Amsterdam.
134. Holler, E., & Calvin, M. (1972) *Biochemistry 11*, 3741–3752.
135. Wells, T.N.C., & Fersht, A.R. (1986) *Biochemistry 25*, 1881–1886.
136. Knowles, J.R. (1980) *Annu. Rev. Biochem. 49*, 877–919.
137. Burbaum, J.J., & Knowles, J.R. (1989) *Biochemistry 28*, 9306–9317.
138. Wolfenden, R. (1969) *Nature 223*, 704–705.
139. Lienhard, G.E. (1973) *Science 180*, 149–154.
140. Chipman, D.M., Grisaro, V., & Sharon, N. (1967) *J. Biol. Chem. 242*, 4388–4394.
141. Secemski, I., Lehrer, S.S., & Lienhard, G.E. (1972) *J. Biol. Chem. 247*, 4740–4748.
142. Albery, W.J., & Knowles, J.R. (1976) *Biochemistry 15*, 5627–5631.
143. Burton, P.M., & Waley, S.G. (1968) *Biochim. Biophys. Acta 151*, 714–715.
144. Cardinale, G.J., & Abeles, R.H. (1968) *Biochemistry 7*, 3970–3978.
145. Keenan, M.V., & Alworth, W.L. (1974) *Biochem. Biophys. Res. Commun. 57*, 500–504.
146. Jencks, W.P. (1966) in *Current Aspects of Biochemical Energetics* (Kaplan, N.O., & Kennedy, E.P., Eds.) pp 273–298, Academic Press, New York.
147. Rudnick, G., & Abeles, R.H. (1975) *Biochemistry 14*, 4515–4522.
148. Manning, J.M., Moore, S., Rowe, W.B., & Meister, A. (1969) *Biochemistry 8*, 2681–2685.
149. Masters, D.S., & Meister, A. (1982) *J. Biol. Chem. 257*, 8711–8715.
150. Westerik, J.O., & Wolfenden, R. (1972) *J. Biol. Chem. 247*, 8195–8197.
151. Evans, B.E., & Wolfenden, R.V. (1973) *Biochemistry 12*, 392–397.
152. Evans, B.E., Mitchell, G.N., & Wolfenden, R. (1975) *Biochemistry 14*, 621–624.
153. Poulos, T.L., Alden, R.A., Freer, S.T., Birktoft, J.J., & Kraut, J. (1976) *J. Biol. Chem. 251*, 1097–1103.
154. Koehler, K.A., & Lienhard, G. (1971) *Biochemistry 10*, 2477–2483.
155. Matthews, D.A., Alden, R.A., Birktoft, J.J., Freer, S.T., & Kraut, J. (1975) *J. Biol. Chem. 250*, 7120–7126.
156. Kam, C., Nishino, N., & Powers, J.C. (1979) *Biochemistry 18*, 3032–3038.
157. Bartlett, P.A., & Marlowe, C.K. (1983) *Biochemistry 22*, 4618–4624.
158. Tronrud, D.E., Monzingo, A.F., & Matthews, B.W. (1986) *Eur. J. Biochem. 157*, 261–268.
159. Allen, K.N., & Abeles, R.H. (1989) *Biochemistry 28*, 8466–8473.
160. Sussman, J.L., Harel, M., Frolow, F., Oefner, C., Goldman, A., Toker, L., & Silman, J. (1991) *Science 253*, 872–879.
161. Doughtery, D.A., & Stauffer, D.A. (1990) *Science 250*, 1558–1560.
162. Andrews, P.R., Cain, E.N., Rizzardo, E., & Smith, G.D. (1977) *Biochemistry 16*, 4848–4852.
163. Steinrücken, H.C., & Amrhein, N. (1984) *Eur. J. Biochem. 143*, 351–357.
164. Rowe, W.B., Ronzio, R.A., & Meister, A. (1969) *Biochemistry 8*, 2674–2680.
165. Whittaker, J.W., & Lipscomb, J.D. (1984) *J. Biol. Chem. 259*, 4476–4486.
166. Jencks, W.P. (1969) *Catalysis in Chemistry and Enzymology*, p 288, McGraw-Hill, New York.
167. Bartlett, P.A., & Johnson, C.R. (1985) *J. Am. Chem. Soc. 107*, 7792–7793.
168. Jackson, D.Y., Jacobs, J.W., Sugasawara, R., Reich, S.H., Bartlett, P.A., & Schultz, P.G. (1988) *J. Am. Chem. Soc. 110*, 4841–4842.
169. Hilvert, D., Carpenter, S.H., Nared, K.D., & Auditor, M.T. (1988) *Proc. Natl. Acad. Sci. U.S.A. 85*, 4953–4955.
169a. Collins, K.D., & Stark, G.R. (1969) *J. Biol. Chem. 244*, 1869–1877.
169b. Gerhardt, J.C., & Schachman, H.K. (1968) *Biochemistry 7*, 538–552.
169c. Springs, B., Welsh, K.M., & Cooperman, B.S. (1981) *Biochemistry 20*, 6384–6391.
170. International Union of Biochemistry, Nomenclature Committee (1984) *Enzyme Nomenclature*, Academic Press, Orlando, FL.
170a. Schloss, J.V., Porter, D.J.T., Bright, H.J., & Cleland, W.W. (1980) *Biochemistry 19*, 2358–2362.

170b. Schloss, J.V., & Cleland, W.W. (1982) *Biochemistry 21*, 4420–4427.

170c. Levine, H.L., Brody, R.S., & Westheimer, F.H. (1980) *Biochemistry 19*, 4993–4999.

170d. Frick, L., Wolfenden, R., Smal, E., & Baker, D.C. (1986) *Biochemistry 25*, 1616–1621.

170e. Koehler, K.A., & Hess, G.P. (1974) *Biochemistry 13*, 5345–5350.

170f. Ronzio, R.A., Rowe, W.B., & Meister, A. (1969) *Biochemistry 8*, 1066–1075.

170g. Chirgwin, J.M., & Noltmann, E.A. (1975) *J. Biol. Chem. 250*, 7272–7276.

170h. Spring, T.G., & Wold, F. (1971) *Biochemistry 10*, 4655–4660.

170i. Evans, B.E., & Wolfenden, R.V. (1973) *Biochemistry 12*, 392–398.

170j. Lienhard, G.E., & Secemski, I.I. (1973) *J. Biol. Chem. 248*, 1121–1123.

170k. Kyte, J. (1995) *Structure in Protein Chemistry*, p 218, Garland Publishing, New York.

171. Knight, S., Andersson, I., & Branden, C. (1990) *J. Mol. Biol. 215*, 113–160.

171a. Kyte, J. (1995) *Structure in Protein Chemistry*, p 332, Garland Publishing, New York.

172. Jaskólski, M., Miller, M., Rao, J.K.M., Leis, J., & Wlodawer, A. (1990) *Biochemistry 29*, 5889–5898.

172a. Kyte, J. (1995) *Structure in Protein Chemistry*, p 329, Garland Publishing, New York.

173. Eklund, H., Samana, J.P., & Jones, T.A. (1984) *Biochemistry 23*, 5982–5996.

173a. Kyte, J. (1995) *Structure in Protein Chemistry*, p 141, Garland Publishing, New York.

174. Kester, W.R., & Matthews, B.W. (1977) *Biochemistry 16*, 2506–2516.

175. Delbaere, L.T.J., & Brayer, G.D. (1985) *J. Mol. Biol. 183*, 89–103.

176. Holden, H.M., Tronrud, D.E., Monzingo, A.F., Weaver, L.H., & Matthews, B.W. (1987) *Biochemistry 26*, 8542–8553.

177. Blake, C.C.F., & Evans, P.R. (1974) *J. Mol. Biol. 84*, 585–601.

178. Hill, E., Tsernoglou, D., Webb, L., & Banaszak, L.J. (1972) *J. Mol. Biol.* 72, 577–591.

179. Birktoft, J.J., Fernley, R.T., Bradshaw, R.A., & Banaszak, L.J. (1982) in *Molecular Structure and Biological Activity* (Griffin, J.F., & Duax, W.L., Eds.) pp 37–55, Elsevier, New York.

180. Remington, S., Wiegand, G., & Huber, R. (1982) *J. Mol. Biol. 158*, 111–152.

181. Suck, D., Oefner, C., & Kabsch, W. (1984) *EMBO J. 3*, 2423–2430.

182. Bender, M.L., & Kemp, K.C. (1957) *J. Am. Chem. Soc. 79*, 116–117.

183. Steitz, T.A., Henderson, R., & Blow, D.M. (1969) *J. Mol. Biol. 46*, 337–348.

184. Rose, I.A. (1981) *Philos. Trans. R. Soc. London, B 293*, 131–143.

185. Alber, T., Banner, D.W., Bloomer, A.C., Petsko, G.A., Phillips, D., Rivers, P.S., & Wilson, I.A. (1981) *Philos. Trans. R. Soc. London, B 293*, 159–171.

186. Grau, U.M., Trommer, W.E., & Rossmann, M.G. (1981) *J. Mol. Biol. 151*, 289–307.

187. Gorenstein, D.G., Findlay, J.B., Luxon, B.A., & Kar, D. (1977) *J. Am. Chem. Soc. 99*, 3473–3478.

188. Cantley, L.C., Josephson, L., Warner, R., Yanagisawa, M., Lechene, C., & Guidotti, G. (1977) *J. Biol. Chem. 252*, 7421–7423.

189. Cantley, L.C., Cantley, L.G., & Josephson, L. (1978) *J. Biol. Chem. 253*, 7361–7368.

190. Lindquist, R.N., Lynn, J.L., & Lienhard, G.E. (1973) *J. Am. Chem. Soc. 95*, 8762–8768.

191. Wlodawer, A., Miller, M., & Sjölin, L. (1983) *Proc. Natl. Acad. Sci. U.S.A. 80*, 3628–3631.

192. Alber, T., Gilbert, W.A., Ponzi, D.R., & Petsko, G.A. (1982) *Ciba Foundation Symp. 93*, 4–24.

193. Jencks, W.P. (1969) *Catalysis in Chemistry and Enzymology*, pp 409–411, McGraw-Hill, New York.

194. Quiocho, F.A., & Vyas, N.K. (1984) *Nature 310*, 381–386.

195. Hine, J. (1972) *J. Am. Chem. Soc. 94*, 5766–5771.

196. Stahl, N., & Jencks, W.P. (1986) *J. Am. Chem. Soc. 108*, 4196–4205.

197. Fersht, A.R., Shi, J., Knill-Jones, J., Lowe, D.M., Wilkinson, A.J., Blow, D.M., Brick, P., Carter, P., Waye, M.M.Y., & Winter, G. (1985) *Nature 314*, 235–238.

197a. Kyte, J. (1995) *Structure in Protein Chemistry*, p 167, Garland Publishing, New York.

198. Bartlett, P.A., & Marlowe, C.K. (1987) *Science 235*, 569–571.

199. Tronrud, D.E., Holden, H.M., & Matthews, B.W. (1987) *Science 235*, 571–574.

200. Grobelny, D., Goli, U.B., & Galardy, R.E. (1989) *Biochemistry 28*, 4948–4951.

201. Morgan, B.P., Scholtz, J.M., Ballinger, M.D., Zipkin, I.D., & Bartlett, P.A. (1991) *J. Am. Chem. Soc. 113*, 297–307.

202. Kirsch, J.F., Eichele, G., Ford, G.C., Vincent, M.G., & Jansonius, J.N. (1984) *J. Mol. Biol. 174*, 497–525.

203. Wilson, I.B., & Bergmann, F. (1950) *J. Biol. Chem. 185*, 479–489.

204. Hyde, C.C., Ahmed, S.A., Padlan, E.A., Miles, E.W., & Davies, D.R. (1988) *J. Biol. Chem. 263*, 17857–17871.

205. Bode, W., Walter, J., Huber, R., Wenzel, H.R., & Tschesche, H. (1984) *Eur. J. Biochem. 144*, 185–190.

206. Bolin, J.T., Filman, D.J., Matthews, D.A., Hamlin, R.C., & Kraut, J. (1982) *J. Biol. Chem. 257*, 13650–13662.

207. Birktoft, J.J., Kraut, J., & Freer, S.T. (1976) *Biochemistry 15*, 4481–4485.

208. Anderson, C.M., Zucker, F.H., & Steitz, T.A. (1979) *Science 204*, 375–380.

209. Pai, E.F., Sachsenheimer, W., Schirmer, R.H., & Schulz, G.E. (1977) *J. Mol. Biol. 114*, 37–45.

210. Koshland, D.E. (1958) *Proc. Natl. Acad. Sci. U.S.A. 44*, 98–104.

211. de la Fuente, G., & Sols, A. (1970) *Eur. J. Biochem. 16*, 226–233.

211a. Anton, D.L., Hedstrom, L., Fish, S.M., & Abeles, R.H. (1983) *Biochemistry 22*, 5903–5908.

212. Fersht, A. (1985) *Enzyme Structure and Mechanism. Second Edition*, pp 331–333, W.H. Freeman and Company, New York.

213. Viola, R.E., Raushel, F.M., Rendina, A.R., & Cleland, W.W. (1982) *Biochemistry 21*, 1295–1302.

214. Hammes, G.G., & Schimmel, P.R. (1970) in *The Enzymes, Volume II: Kinetics and Mechanism* (Boyer, P.D., Ed.) pp 67–114, Academic Press, New York.

215. Rocek, J., Westheimer, F.H., Eschenmoser, L., Moldovanyi, L., & Schreiber, J. (1962) *Helv. Chim. Acta 45*, 2554–2567.

216. Albery, W.J., & Knowles, J.R. (1976) *Biochemistry 15*, 5631–5640.

216a. Kyte, J. (1995) *Structure in Protein Chemistry*, p 27, Garland Publishing, New York.

217. Brouwer, A.C., & Kirsch, J.F. (1982) *Biochemistry 21*, 1302–1307.

218. Scopes, R.K. (1978) *Eur. J. Biochem. 91*, 119–129.

219. Scopes, R.K. (1978) *Eur. J. Biochem. 85*, 503–516.

220. Rose, I.A., & O'Connell, E.L. (1967) *J. Biol. Chem. 242*, 1870–1879.

221. Schloss, J.V., Emptage, M.H., & Cleland, W.W. (1984) *Biochemistry 23*, 4572–4580.

221a. Kyte, J. (1995) *Structure in Protein Chemistry*, p 342, Garland Publishing, New York.

222. Stephens, P.E., Darlison, M.G., Lewis, H.M., & Guest, J.R. (1983) *Eur. J. Biochem. 133*, 481–489.

223. Daigo, K., & Reed, L.J. (1962) *J. Am. Chem. Soc. 84*, 666–671.

224. Bates, D.L., Danson, M.J., Hale, G., Hooper, E.A., & Perham, R.N. (1977) *Nature 268*, 313–316.
225. Collins, J.H., & Reed, L.J. (1977) *Proc. Natl. Acad. Sci. U.S.A. 74*, 4223–4227.
226. Danson, M.J., Fersht, A.R., & Perham, R.N. (1978) *Proc. Natl. Acad. Sci. U.S.A. 75*, 5386–5389.
227. Duggleby, R.G., Sneddon, M.K., & Morrison, J.F. (1978) *Biochemistry 17*, 1548–1554.
228. Gething, M.H., & Davidson, B.E. (1976) *Eur. J. Biochem. 71*, 327–336.
229. Creighton, T.E. (1970) *Eur. J. Biochem. 13*, 1–10.
230. DeMoss, J.A. (1962) *Biochim. Biophys. Acta, 62*, 279–293.
231. Yanofsky, C., & Rachmeler, M. (1958) *Biochim. Biophys. Acta 28*, 640–641.
232. Matchett, W.H. (1974) *J. Biol. Chem. 249*, 4041–4049.


# Appendix: Molecularity and Approximation

## *Intramolecular and Intermolecular Processes*

**Intramolecular** chemical reactions often occur at rates much faster than equivalent **intermolecular** reactions, and intramolecular associations often occur with association equilibrium constants much larger than those of equivalent intermolecular associations. A particularly informative series illustrating such effects can be gathered[1] from among the reactions involving intramolecular nucleophilic catalysis of the hydrolysis of phenyl esters by the carboxylate anion. The mechanism for this nucleophilic catalysis has been shown to involve the formation of an intermediate anhydride which in the intramolecular examples would be cyclic (Figure 3–50). These reactions should involve a preequilibrium in which the tetrahedral intermediate is formed from which the phenolate is eventually ejected. It is the ejection of the phenolate that is monitored as the reaction progresses. The rate constant, in units of reciprocal seconds, for the appearance of phenolate would be equal to $K_{eq}^{TH}k_E$, where $K_{eq}^{TH}$ is the equilibrium constant for the formation of the tetrahedral intermediate. If it is assumed that for all compounds in the series the rate con-

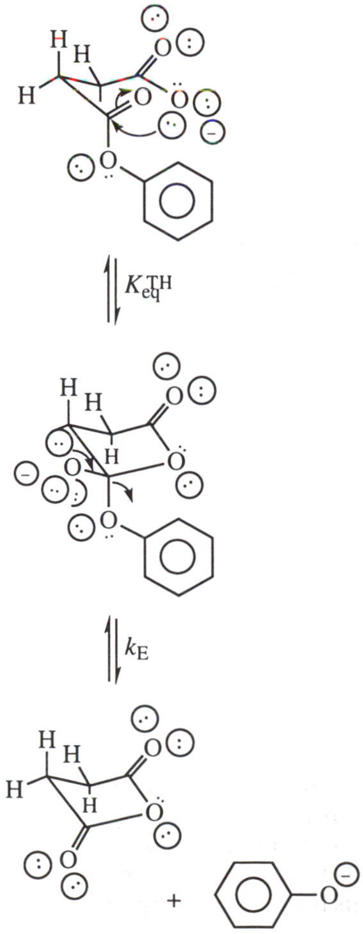

**Figure 3–50:** Mechanism for the formation of the intramolecular anhydride that is the intermediate in the hydrolysis of phenyl succinate by intramolecular nucleophilic catalysis. The tetrahedral intermediate that leads to the anhydride has both a phenolate and a carboxylate as potential leaving groups; and, because the latter is the better, it should decompose to reactant much more frequently than to anhydride.

stant $k_E$, which in all cases is for a chemically equivalent first-order relaxation, has the same value, then the increases observed in the rates result from increases in $K_{eq}^{TH}$, an equilibrium constant. It is likely, however, that these increases in the equilibrium constant are due almost entirely to increases in the respective rate constants for the formation of the intermediate rather than decreases in the rate constants for its dissociation.

A comparison of the first-order rate constants of phenolate release for a series of intramolecular reactions,[1] to the estimated pseudo first-order rate constant for the intermolecular reaction between phenyl acetate and acetate anion, when the catalysis by acetate anion proceeds through acetic anhydride as an intermediate,[2,3] indicates how large these intramolecular accelerations of rate can be (Table 3–5). The increase in $K_{eq}^{TH}$ of $2.2 \times 10^5$ in the case of phenyl succinate compared to that for 1 M acetate is similar in magnitude to the $3 \times 10^5$ increase in the association equilibrium constant for the intramolecular formation of succinic anhydride compared to the intermolecular formation of acetic anhydride at 1 M acetic acid,[4] and this suggests that the changes seen are representative. An enhancement of $5 \times 10^5$, when compared to nucleophilic attack by 1 M hydroxide, was seen in the intramolecular attack of alkoxide in the intramolecular $S_N2$ reaction[5]

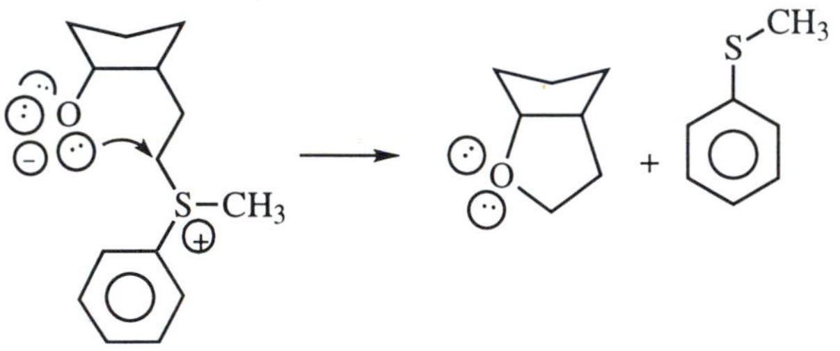

(3–247)

Again, this can be compared with the enhancement of $2 \times 10^5$ for the intramolecular nucleophilic enhancement in phenyl succinate (Table 3–5), also a situation in which a five-membered ring is produced.

Many other examples of intramolecular accelerations of rates have been reported,[3,6] but the accelerations presented in Table 3–5 are among the largest that have been noted for each particular size of ring formed during each of these intramolecular reactions. In some of the other instances, electronic effects are difficult to separate from the effects of approximation. For example, in the intramolecular reaction

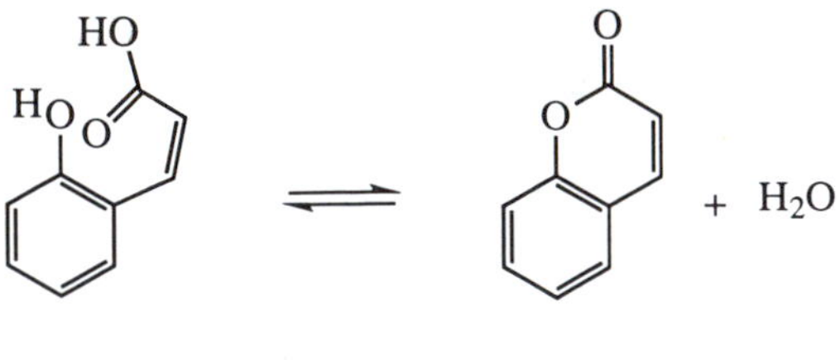

$$\rightleftharpoons \quad + \; H_2O \qquad (3\text{–}248)$$

the two reactants are connected electronically by the $\pi$ system in addition to being juxtaposed by the $\sigma$ system.

At one time it was fashionable to refer to these accelerations in rate as due to an increase in the **effective molarity** of one of the reactants brought about by attaching it covalent-

**Table 3–5: Relative Rate Constants[1,3] for Anhydride Formation[a]**

| phenyl ester | relative rate of anhydride formation[b] | $T\Delta S^{o\ddagger}_{approx}$[c] (kJ mol$^{-1}$) |
|---|---|---|
| 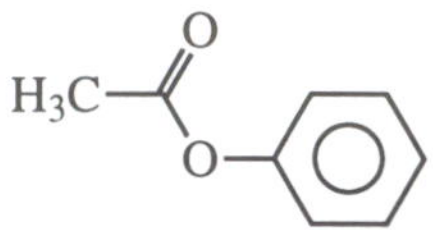 phenyl acetate | 1.0 (1 M $CH_3COO^-$) | |
| 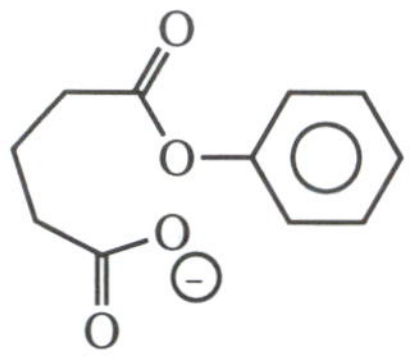 phenyl glutarate | $1 \times 10^3$ | < −17 |
| 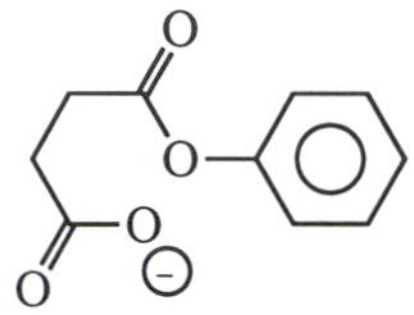 phenyl succinate | $2 \times 10^5$ | < −30 |
| 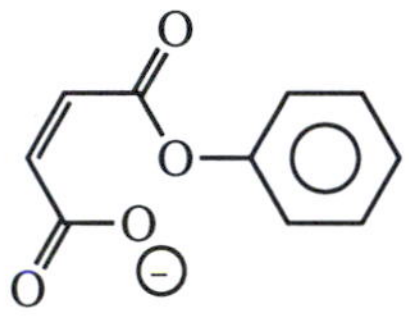 phenyl maleate | $1 \times 10^7$ | < −40 |
| 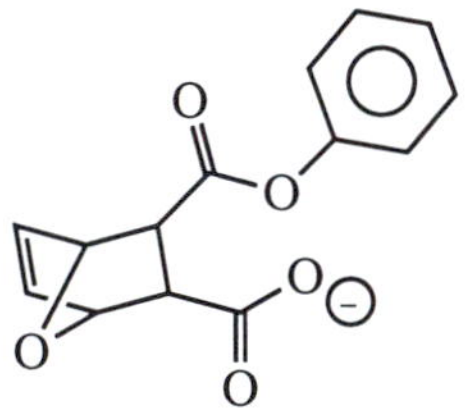 bicyclic (oxanorbornene) phenyl ester | $5 \times 10^7$ | < −44 |

[a]The first-order rate constants for the hydrolysis of the various phenyl monoesters of dicarboxylic acids were determined[1] as a function of pH in the range pH 4–8. From the pH–rate behavior of each of these rate constants, the first-order rate constant for intramolecular nucleophilic catalysis of the respective hydrolysis by the appended carboxylate could be calculated. From these values, the first-order rate constants for intramolecular anhydride formation could be calculated. These rate constants were determined for each phenyl monoester at 25, 30, or 35 °C. The values of these rate constants were adjusted to the same temperature and presented relative to the first-order rate constant for intramolecular anhydride formation from phenyl glutarate.[1] These first-order rate constants could be related to the pseudo first-order rate constant for the intermolecular formation of acetic anhydride from excess acetate anion and phenyl acetate during the intermolecular nucleophilic catalysis of the hydrolysis of phenyl acetate by acetate anion.[3] [b]First-order rate constants for anhydride formation are presented relative to the calculated[3] pseudo first-order rate constant for the formation of acetic anhydride from phenyl acetate and 1.0 M acetate anion, which is arbitrarily assigned the value of 1.0. [c]The standard entropy of approximation calculated by Equation 3–253 assuming $\Delta\Delta H^{o\ddagger} \cong 0$ was multiplied by 298 K. The standard state was 1 M for acetate anion.

ly to the other. As more exaggerated examples of this phenomenon were reported, however, the unreality of discussing concentrations of $5 \times 10^7$ M became apparent,[4] and a more sophisticated view of the situation was required.

In all instances in which an intramolecular reaction is compared to an intermolecular reaction, the difference observed in the two rate constants $k_{intra}$ and $k_{inter}$ is due in large part to an increase in the entropy of activation caused simply by the fact that a unimolecular reaction is being compared to a multimolecular reaction. This increase in the entropy of activation for the reaction results from the fact that the standard entropy of approximation is missing from the standard free energy change for the intramolecular reaction. The **standard entropy of approximation**, $\Delta S^{\circ\ddagger}_{approx}$, is the change in standard entropy due solely to bringing the separate reactants together into the same complex prior to the reaction. It is a negative number because the intrinsic entropy of two separate reactants relative to the transition state is greater than the intrinsic entropy of one molecule containing the two reactants or of a complex into which the two reactants have been assembled relative to the transition state. Because the standard entropy of approximation is missing from the standard entropy of activation for the intramolecular reaction, owing to the fact that approximation has already been accomplished synthetically, the standard entropy of activation for the intramolecular reaction is greater than the standard entropy of activation for the intermolecular reaction.

These relationships can be expressed in equations. The standard entropy of activation, $\Delta S^{\circ\ddagger}_{intra}$, of the intramolecular reaction should be related to the standard entropy of activation, $\Delta S^{\circ\ddagger}_{inter}$, of the intermolecular reaction by

$$\Delta S^{\circ\ddagger}_{inter} = \Delta S^{\circ\ddagger}_{intra} + \Delta S^{\circ\ddagger}_{approx} \qquad (3\text{–}249)$$

if the same reaction with the same change in standard enthalpy is occurring once the reactants have been approximated. If this is an adequate description of the situation, then

$$R \ln \frac{k_{intra}}{k_{inter}} = \Delta S^{\circ\ddagger}_{intra} - \Delta S^{\circ\ddagger}_{inter} = -\Delta S^{\circ\ddagger}_{approx} \qquad (3\text{–}250)$$

Because $\Delta S^{\circ\ddagger}_{approx} < 0$, $k_{intra} > k_{inter}$.

The magnitude of the standard entropy of approximation is determined by the difference between two other standard entropy changes, the standard entropy of molecularity and the standard entropy of rotational restraint.[4] The formation of a transition state during an intermolecular reaction requires that two or more independent molecules become one molecule, and this involves a considerable decrease in standard entropy. The standard entropy change responsible for this decrease, the **standard entropy of molecularity**, $\Delta S^{\circ\ddagger}_{molec}$, has a negative value and is a major, unavoidable, unfavorable term in the standard free energy of activation in any intermolecular reaction. In an intramolecular reaction, however, the decrease in standard entropy due to the standard entropy of molecularity does not occur, because reactants are already on only one molecule, and this has the effect of increasing dramatically the standard entropy of activation for the intramolecular reaction relative to the intermolecular reaction and hence increasing its rate.

There is affiliated with an intramolecular reaction, however, a **standard entropy of rotational restraint**, which, conversely, is irrelevant to an intermolecular reaction. It arises from the fact that the formation of the transition state during an intramolecular reaction requires that a portion of the rotational entropy in the molecule be eliminated because only a fraction of the accessible rotational isomers can participate in the reaction productively. The standard entropy change accompanying this decrease in the number of rotational isomers, the standard entropy of rotational restraint, $\Delta S^{\circ\ddagger}_{rot}$, has a negative value and its inclusion caused the standard entropy of activation for the reaction to be smaller than it would be if no rotation occurred and only a productive rotational isomer were present.

The relationships between the standard entropy of approximation and the standard entropy of molecularity and standard entropy of rotational restraint are

$$\Delta S^{\circ\ddagger}_{approx} = \Delta S^{\circ\ddagger}_{molec} - \Delta S^{\circ\ddagger}_{rot} \qquad (3\text{–}251)$$

The magnitudes of each of these terms can be discussed in turn.

The standard entropy of molecularity is the decrease in standard entropy that should accompany the change of an intermolecular reaction to a rigidly oriented intramolecular reaction.[4] In the specific case of a bimolecular reaction, the two independent reactants have six translational and six rotational degrees of freedom, but the one transition state, formed by the association of the two others, should have only three translational and three rotational degrees of freedom. The standard entropy change associated with the loss of the three translational and three rotational degrees of freedom during this inescapable association, calculated for the situation in which the two reactants and the transition state or product are dissolved in a solution at 25 °C with a standard state of 1 M in solutes, has been estimated[4] to be between −190 and −210 J K$^{-1}$ mol$^{-1}$. This estimate can be compared to the standard entropy change observed for a simple bimolecular reaction, such as the dimerization of cyclopentadiene in the liquid phase, during which the standard entropy change is −130 to −170 J K$^{-1}$ mol$^{-1}$ with the same choice of standard state. The difference between the calculated standard entropy change and the observed standard entropy change in the particular instance of cyclopentadiene can be completely accounted for by the presence of low frequency vibrations in the dimer that could not be present in the two monomers because of their smaller size. It has been concluded[4] that −190 to −210 J K$^{-1}$ mol$^{-1}$ is an adequate estimate for $\Delta S^{\circ\ddagger}_{molec}$, the maximum decrease in standard entropy change expected from converting a bimolecular reaction into a unimolecular reaction, when molarities are used as units of concentration for standard states. If mole fractions are used as units of concentration for standard states and the concentration of solvent is about 50 M, which is about the concentration of water, the same range for expected standard entropy of approximation for a bimolecular reaction would be −150 to −180 J K$^{-1}$ mol$^{-1}$.

As the example of the dimerization of cyclopentadiene illustrates, an intramolecular reaction, because it usually involves a larger and more flexible molecule than any of the reactants in an intermolecular reaction, can never realize all of this favorable standard entropy of molecularity. Major factors in decreasing the portion of the standard entropy of molecularity that an intramolecular reaction will enjoy are the internal rotations within the intramolecular reactant itself. These rotations decrease the probability that the necessary juxtaposition of reactants will occur. For example, in the case of phenyl succinate (Figure 3–50) the rotational orientation around three carbon–carbon bonds must be appropriate if the carboxyl oxygen is to be placed adjacent to the acyl carbon. It has been estimated[4] from the results of thermodynamic and kinetic measurements from a number of intramolecular reactions that the entropy of rotational restraint decreases by about 20 J $K^{-1}$ $mol^{-1}$ for every bond that lies between the two atoms participating directly in the reaction and about which free rotation can occur.

When two similar intramolecular reactions are compared, for which it is assumed that differences between their standard enthalpies of formation are negligible[1]

$$\Delta\Delta S^\circ = R \ln\left(\frac{K_{eq1}}{K_{eq2}}\right) \tag{3–252}$$

where $\Delta\Delta S^\circ$ is the difference between their standard entropy changes and $K_{eq1}$ and $K_{eq2}$ are the respective association equilibrium constants. The change in relative rate of phenoxide release in going from phenyl glutarate to phenyl succinate (Table 3–5) is 230-fold and in going from phenyl succinate to the phenyl ester of the fused ring is also 230-fold. In each comparison, one less carbon–carbon bond around which free rotation is allowed is found in the more constrained member of the pair. The changes in rate, presumably reflecting differences in $K_{eq}^{TH}$, are equivalent in each comparison to 45 J $K^{-1}$ $mol^{-1}$. It has been noted, however, that the case of the nucleophilic catalysis of the hydrolysis of phenyl esters provides the largest standard entropy of rotational restraint (carbon–carbon bond)$^{-1}$ yet observed.[4] One explanation for the unusually large change in the relative rates as axes of rotation are eliminated in this particular case would be that the differences are not entirely entropic in origin but also involve favorable decreases in the standard enthalpy of formation of the tetravalent intermediate.

With small molecules, the effect of approximation on the rate constant is usually significant only when the two central atoms that participate in the reaction become involved in a five-membered or six-membered ring in the transition state. A four-membered ring is usually too strained, because of the normal bond angles of commonly encountered molecules, to provide any favorable approximation. A seven-membered ring, if there is free rotation about every bond, has too small a value of $\Delta S_{rot}^{\circ\ddagger}$ to exhibit a $\Delta S_{approx}^{\circ\ddagger}$ small enough to have a noticeable effect on rates or equilibrium. For example, even in the intramolecular nucleophilic catalysis of phenyl ester hydrolysis, a series of reactions unusually prone to intramolecular catalysis, phenyl adipate would show a rate of phenolate release due to nucleophilic catalysis only 4-fold greater than that for the same reaction of phenyl acetate in 1.0 M sodium acetate. Large, rigid molecules in which the two atoms that must react are more than six atoms apart yet close enough to collide are difficult to construct, with the relevant exception of a molecule of protein or nucleic acid.

The magnitude of the actual difference in standard entropy of activation between a given intramolecular reaction and the corresponding intermolecular reaction will be less than the magnitude of the standard entropy of approximation because vibrational degrees of freedom, unavailable to the reactants in the intermolecular reaction, are available to the necessarily larger reactant in the intramolecular reaction and because steric effects that do not apply to the intermolecular reaction are often unavoidable consequences of designing the intramolecular reactant. If the magnitude of $\Delta\Delta S^{\circ\ddagger}$, the actual difference between the standard entropies of activation, must be less than the magnitude of $\Delta\Delta S_{approx}^{\circ\ddagger}$, then

$$T\Delta S_{approx}^{\circ\ddagger} < -RT \ln\left(\frac{k_{intra}}{k_{inter}}\right) - \Delta\Delta H^{\circ\ddagger} \tag{3–253}$$

where $k_{intra}$ and $k_{inter}$ are the intramolecular and intermolecular rate constants. If the differences in the actual standard enthalpies of activation, $\Delta\Delta H^{\circ\ddagger}$, are known, the estimates of the upper limits for $\Delta S_{approx}^{\circ\ddagger}$ can incorporate them. If they are unknown, they can be assumed to be zero, for the sake of argument, but such an assumption can be misleading.

In several reactions displaying large increases in the rate of the reaction due to covalent approximation of the reactants, the major effect of the approximation is on the standard enthalpy of activation for the reaction.[6] For example,[7,8] 2,2,3,3-tetra-methylsuccinanilide displays a rate of aniline release at pH 5, 1200 times greater than that of succinanilide itself. This increase in rate, however, which is equivalent to a change in the free energy of activation of −18 kJ $mol^{-1}$, is accompanied by a change in the standard enthalpy of activation, $\Delta\Delta H^{\circ\ddagger}$, of −25 kJ $mol^{-1}$. Therefore, in this case the standard entropy of activation actually decreases as the rate of the reaction is enhanced by approximation. An explanation for this unusual behavior may be the compression that must occur in either the transition state or the tetrahedral intermediate of the tetramethyl analogue because of the vicinal dimethyls on the other side of the ring

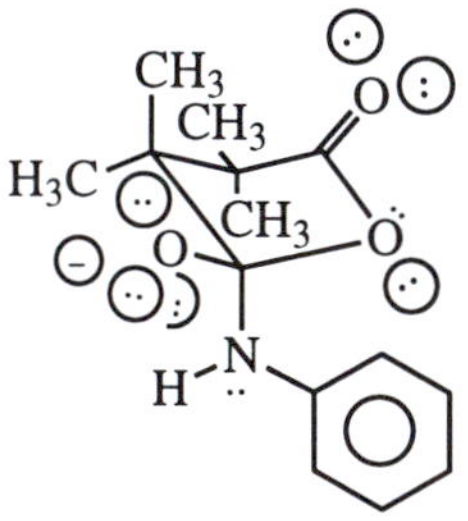

**3–49**

The steric forces on the back of the ring, because they involve van der Waals repulsions among the methyl groups, would be enthalpic effects. They will be the forces that either compress the attacking oxygen against the acyl carbon and lower the standard enthalpy of the transition state or lock the tetravalent intermediate into the ring and thereby enhance exocylic ejection of aniline.

A similar enthalpic effect could explain the enhancement in the rate of acid-catalyzed lactonization of 3-(*o*-hydroxyphenyl)propionate when three methyls are incorporated into its structure to produce 3-[*o*-hydroxy-*o*′-(methylphenyl)]-2,2-dimethylpropionate[9]

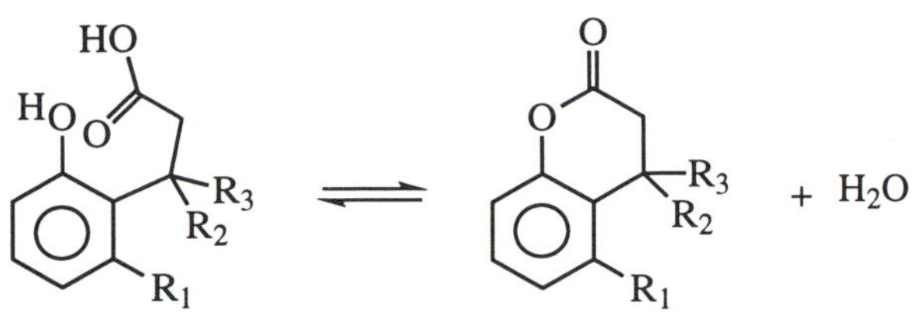

(3–254)

where $R_1 = R_2 = R_3$ and all are either hydrogen or methyl. The observed difference in the rates of acid-catalyzed lactonization for the trimethyl compound relative to the trihydro compound of $5 \times 10^{10}$ could arise in large part from steric effects that decrease the standard enthalpy of activation rather than increasing the standard entropy of activation. Unfortunately, no standard enthalpies of activation were measured in this case.

When the upper limits of $T\Delta S^{\circ\ddagger}_{approx}$ are calculated from the relative rates of the intramolecular and bimolecular nucleophilic catalysis of the hydrolysis of phenyl esters, on the assumption that $\Delta\Delta H^{\circ\ddagger}$ is equal to zero,[1] they are all equal to or greater than −44 kJ mol$^{-1}$ (Table 3–5). If it is assumed that the fused ring retains two rotational axes, which is a generous assumption, $\Delta S^{\circ\ddagger}_{approx}$ should be about −150 J K$^{-1}$ mol$^{-1}$ and $T\Delta S^{\circ\ddagger}_{approx}$ about −44 kJ mol$^{-1}$. The agreement between experiment and theory is certainly fortuitous.

At least three points are illustrated by this exercise. First, the largest intramolecular increases in rate yet measured, with the educational exceptions of the cases involving severe, compressive steric effects in the transition states, are of a magnitude less than that expected for the transformation of an intermolecular reaction into a fully constrained intramolecular reaction. It should be noted, however, that an enzymatically catalyzed reaction may be severely constrained. Second, the maximum decrease in the standard free energy of activation to be expected when a bimolecular reaction is turned into an intramolecular reaction, as it is on the active site of an enzyme, is −60 kJ mol$^{-1}$, which would produce a rate enhancement of $3 \times 10^{10}$ M. Third, the larger the number of bonds about which rotation can occur between the atoms that must collide during the reaction, the smaller will be the decrease in standard free energy of activation to be expected when an intermolecular reaction becomes an intramolecular reaction. Therefore, it should be the purpose of the enzyme to confine such rotation. Considering all of these points, the rates of enzymatic reactions seem to be too slow.

## *Suggested Reading*

Page, M.I., & Jencks, W. (1971) Entropic Contributions to Rate Accelerations in Enzymic and Intramolecular Reactions and the Chelated Effect, *Proc. Natl. Acad. Sci. U.S.A. 68*, 1678–1683.

## *References*

1. Bruice, T.C., & Pandit, U.K. (1960) *J. Am. Chem. Soc. 82*, 5858–5865.
2. Bruice, T.C., & Turner, A. (1970) *J. Am. Chem. Soc. 92*, 3422–3428.
3. Bruice, T.C. (1970) in *The Enzymes: Kinetics and Mechanism* (Boyer, P.D. Ed.) 3rd ed., Vol II, pp 217–279, Academic Press, New York.
4. Page, M.I., & Jencks, W.P. (1971) *Proc. Natl. Acad. Sci. U.S.A. 68*, 1678–1683.
5. Coward, J.K., Lok, R., & Takagi, O. (1976) *J. Am. Chem. Soc. 98*, 1057–1058.
6. Jencks, W.P. (1969) *Catalysis in Chemistry and Enzymology*, McGraw-Hill, New York.
7. Taylor, R., Kennard, O., & Versichel, W. (1984) *J. Am. Chem. Soc. 106*, 244–248.
8. Higuchi, T., Eberson, L., & Herd, A.K. (1966) *J. Am. Chem. Soc. 88*, 3805–3808.
9. Milstein, S., & Cohen, L.A. (1972) *J. Am. Chem. Soc. 94*, 9158–9165.

*Chapter* **FOUR**

# Transfer of Protons

## General Acid Catalysis and General Base Catalysis

Proton transfers are involved in almost all biochemical reactions. The accepted chemical mechanisms that would be written for the various reactions involved in metabolism, both those that involve coenzymes (Chapter 2) and those that do not (Chapter 1), require that protons be provided for existing or incipient lone pairs of electrons and that protons be removed from diminishing or disappearing pairs of electrons. Although mechanistic shorthand allows protons to be written as $H^+$, there is no such thing in aqueous solution as a bare proton. A proton either occupies one pair of electrons or is in the process of being transferred intimately between two pairs of electrons. If a proton is provided to a reactant, it is provided residing upon a pair of electrons of an acid. If a proton is removed from a reactant, it is removed by being transferred to the lone pair of electrons of a base intimately associated with the reactant. The requirements for these participating acids and bases are demonstrated by studying systematically the rate of the reaction as a function of the pH of the solution and of the concentrations of added acids and bases.

As the pH of a solution is decreased, the concentration of hydronium ion increases. If the rate of a reaction increases as the pH of the solution is decreased, in the absence of significant concentrations of added acids or bases, it is usually assumed that it is only the increase in the concentration of hydronium ion that is responsible for the increase in rate. **Hydronium ion catalysis** is the increase in the rate of a reaction produced by an increase only in the concentration of hydronium ion. The existence of hydronium ion catalysis is established by measuring the rate of the reaction at a series of different values of pH. Although one can use a machine known as a pH stat, the pH of the solution is usually held constant for each individual rate measurement by adding a buffer of sufficient strength to prevent any change in pH due to the reaction itself. At this constant pH, the rate constant for the reaction is measured and tabulated. If this method is used, however, one must be aware that the buffer itself, because it is necessarily composed of an acid and its conjugate base, often affects the rate of the reaction (Figure 4–1).[1] To correct for this effect, the concentration of buffer is varied at constant pH, which is established only by the ratio of the acid to the conjugate base in the buffer, and the value of the rate constant in the absence of buffer is obtained by extrapolation to zero concentration. For example, the pseudo-first-order rate constant, $k_{obs}$, for the hydrolysis of the enol ether

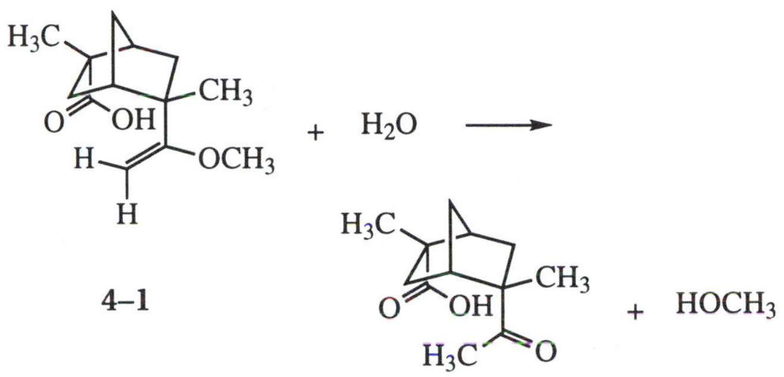

4–1 + $H_2O$ → [product] + $HOCH_3$ (4–1)

was measured at several values of pH and varying concentrations of phosphate to estimate, by extrapolation, the behavior of the rate constant as a function of pH in the absence of any phosphate (Figure 4–1). Because the pseudo-first-order rate constants in the absence of the buffer (intersections of the lines with the ordinate) increase as the pH decreases, hydronium ion catalysis is occurring.

The rate of a particular reaction may increase as the pH is increased, in the absence of significant concentrations of added acids or bases. Again the values of the rate constants in the absence of buffer are estimated by extrapolation, but in this instance the rate constants in the absence of buffer will increase as the pH increases rather than decreasing as they do in Figure 4–1.

Formally, this behavior could be referred to as hydroxide ion catalysis, and mathematically it is often treated as such. For example, the increase in a pseudo-first-order rate constant with an increase in pH can usually be presented as a second-order rate constant times the molar concentration of hydroxide ion, $k_{OH^-}[OH^-]$. It should be remembered, however, that, because of the water constant, $K_w$

$$k_{OH^-}[OH^-] = \frac{k_{OH^-}K_w}{[H^+]} \quad (4\text{–}2)$$

As this relationship demonstrates, it is impossible to distinguish in the absence of other information whether the increase in rate is due to the increase in the concentration of hydroxide ion or to the decrease in the concentration of hydronium ion. Mechanistically, this is an important distinction because hydronium ion and hydroxide ion, although

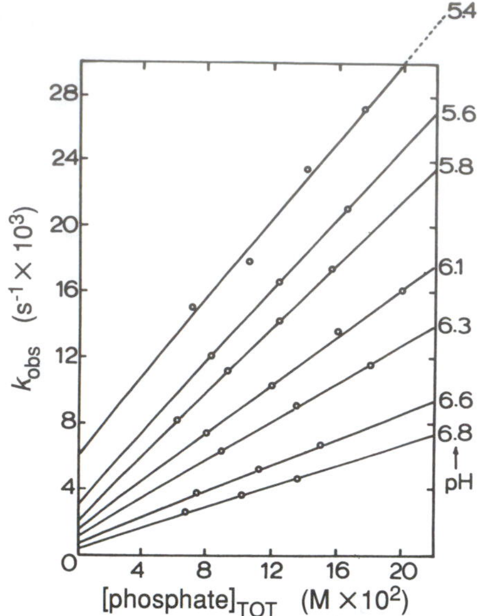


**Figure 4–1:** Catalysis of the hydrolysis of methyl enol ether **4–1** by phosphate anion at various pH values.[1] Solutions of the methyl enol ether **4–1** (0.4 mM) were prepared in 5% ethanol and 1 M KCl, with the noted concentrations of total potassium phosphate at the noted pH at 25 °C. The hydrolysis of the methyl enol ether (Reaction 4–1) was followed spectrophotometrically by the decrease in absorbance at 230 nm from the vinyl ether and the increase in absorbance at 280 nm from the ketone. In all instances the hydrolysis was a first-order reaction in the concentration of the methyl enol ether. The pseudo-first-order rate constant was obtained from each kinetic run by nonlinear least-squares analysis of the curves. These pseudo-first-order rate constants, $k_{obs}$ (seconds$^{-1} \times 10^3$), are plotted as a function of the total concentration of phosphate, [phosphate]$_{TOT}$ (molar $\times 10^2$), at a series of pH values. Adapted with permission from ref 1. Copyright 1976 American Chemical Society.

mathematically symmetrical, are chemically distinct. On the one hand, the hydroxide ion itself, acting as a nucleophile or directly as a base, could be a reactant. On the other hand, an acid dissociation on one of the reactants could be a step in the mechanism of the reaction. As the pH is increased but remains less than the p$K_a$ of an acid, RAH, on a reactant, the concentration of the conjugate base of that acid, [RA⊙], increases in inverse proportion to the concentration of hydronium cation. If the only form of the reactant contributing significantly to the reaction is the unprotonated form, then, at values of pH less than the p$K_a$ of the acid, where RA⊙ is the limiting reactant

$$\text{rate} = k_b'[\text{RA}^\odot] \cong \frac{k_b' K_a^{\text{RAH}}}{[\text{H}^+]}[\text{RAH}]_{\text{TOT}} \quad (4\text{–}3)$$

where [RAH]$_{TOT}$ is the total concentration of the reactant in all of its forms and $k_b'$ is a pseudo-first-order rate constant. It is also the case that

$$k_{obs} = \frac{k_b' K_a^{\text{RAH}}}{[\text{H}^+]} = \frac{k_b' K_a^{\text{RAH}}}{K_w}[\text{OH}^-] = k_b[\text{OH}^-] \quad (4\text{–}4)$$

Therefore, at values of pH below the p$K_a$ of the reactant, a requirement for an acid on a reactant to be unprotonated cannot be distinguished from the direct involvement of the basic, nucleophilic hydroxide anion as a reactant because if the pseudo-first-order rate constant of a reaction is inversely proportional to the concentration of hydronium cation (Equation 4–3), it is necessarily directly proportional to the concentration of hydroxide anion (Equation 4–4). All of these considerations affect the interpretation of the effect of pH on the rate of a reaction the rate of which increases with the basicity of the solution.

A **pH–rate profile** is a plot of the estimated rate constant for a reaction, usually converted to its logarithm, as a function of pH at constant temperature and ionic strength in the absence of significant concentrations of added acids and bases. The pH–rate profile for the observed pseudo-first-order rate constant for the hydrolysis of *N*-acetylimidazole, AcIm

$H_3C$–C(=O)–N(imidazole)N + $H_2O$ $\xrightarrow{k_{obs}}$ $H_3C$–C(=O)–OH + HN(imidazole)N (4–5)

extrapolated to zero buffer concentration, has regions in which the rate constant decreases with increases in pH and regions in which it increases with increases in pH (Figure 4–2).[2] The criterion establishing this as a reaction first-order in acetylimidazole was that

$$\text{rate} = k_{obs}[\text{AcIm}]_{\text{TOT}} \quad (4\text{–}6)$$

where [AcIm]$_{TOT}$ is the total concentration of *N*-acetylimidazole in all of its forms at any time during the reaction. The behavior of the observed pseudo-first-order rate constant, $k_{obs}$, as a function of pH can be fully explained by the equation[2]

$$k_{obs} = \frac{k_{\text{H}^+} K_{a1}^{\text{AcIm}}[\text{H}^+] + k_0 K_{a1}^{\text{AcIm}} + k_{\text{OH}^-} K_{a1}^{\text{AcIm}}[\text{OH}^-]}{K_{a1}^{\text{AcIm}} + [\text{H}^+]} \quad (4\text{–}7)$$

where $K_{a1}^{\text{AcIm}}$ is the first acid dissociation constant of the *N*-acetylimidazole (p$K_{a1}^{\text{AcIm}}$ = 3.6)

$H_3C$–C(=O)–N(imidazole)$NH^+$ $\overset{K_{a1}^{\text{AcIm}}}{\rightleftharpoons}$ $H_3C$–C(=O)–N(imidazole)N + $H^+$ (4–8)

In terms of species actually present in the solution, Equations 4–6 and 4–7 can be rewritten as

$$\text{rate} = k_{\text{H}^+}' K_{a1}^{\text{AcIm}}[\text{H}_2\text{O}][\text{AcImH}^+] + k_0'[\text{H}_2\text{O}][\text{AcIm}] + k_{\text{OH}^-}[\text{OH}^-][\text{AcIm}] \quad (4\text{–}9)$$

where AcIm is the free base of *N*-acetylimidazole, $AcImH^+$ is its conjugate acid, $k_{H^+}' = k_{H^+}[H_2O]^{-1}$, and $k_0' = k_0[H_2O]^{-1}$ (see Equation 4–7). This rate equation implies that water collides productively with the conjugate acid of *N*-acetylimidazole, water collides productively with unprotonated *N*-acetylimidazole, and hydroxide anion collides productively with unprotonated *N*-acetylimidazole, and each of these collisions has its respective rate constant. Because

$$k_0'[H_2O][AcIm] = \frac{k_0 K_a^{AcIm}}{K_w}[OH^-][AcImH^+] \qquad (4\text{–}10)$$

it is possible that the second term describes a reaction between hydroxide anion and the conjugate acid of *N*-acetylimidazole or the sum of a reaction between water and *N*-acetylimidazole and a reaction between hydroxide ion and the conjugate acid. Because

$$k_{OH^-}[OH^-][AcIm] = \frac{k_{OH^-}K_w}{K_{a2}^{AcIm}}[AcIm^-] \qquad (4\text{–}11)$$

where $AcIm^-$ is the enolate anion of *N*-acetylimidazole and $K_{a2}^{AcIm}$ is the acid dissociation producing this enolate

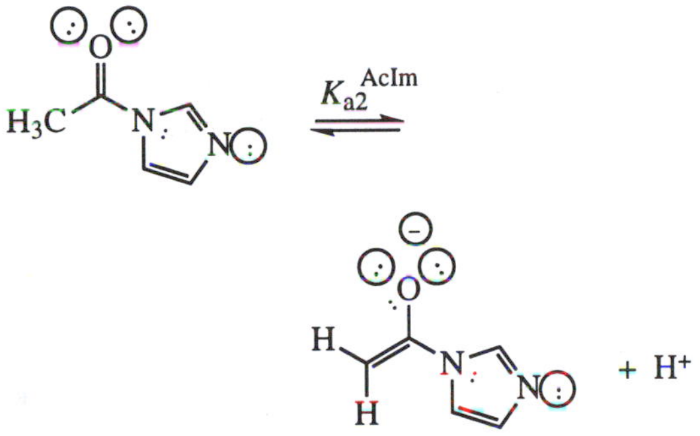

$$\xrightleftharpoons{K_{a2}^{AcIm}} \; + H^+ \qquad (4\text{–}12)$$

it is possible that the third term refers to a reaction involving this enolate. Mechanisms that can be written involving the enolate, however, seem unrealistic.

At values of pH greater than $pK_{a1}^{AcIm}$ (3.6), where $[H^+] < K_{a1}^{AcIm}$, Equation 4–7 simplifies to

$$k_{obs} = k_{H^+}[H^+] + k_0 + k_{OH^-}[OH^-] \qquad (4\text{–}13)$$

At these higher values of pH, $[AcIm]_{TOT} = [AcIm]$, and these three terms can be thought of as arising from a transition state ($k_{H^+}[H^+]$) containing as its component parts a neutral *N*-acetylimidazole, a proton, and any number of molecules of water; a transition state ($k_0$) containing as its component parts a neutral *N*-acetylimidazole and any number of molecules of water; and a transition state ($k_{OH^-}[OH^-]$) containing as its component parts a neutral *N*-acetylimidazole, a hydroxide anion, and any number of molecules of water, respectively.

When none of the $pK_a$ values for the acid–bases on the reactant fall in the range of pH examined, the behavior of the observed pseudo-first-order rate constant takes the form of Equation 4–13 or some subset of the terms in Equation 4–13. For example the pH–rate profile for the ketonization of vinyl alcohol

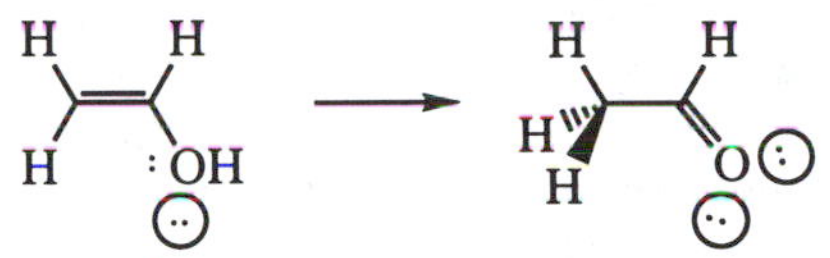

$$(4\text{–}14)$$

is strictly consistent with Equation 4–13 from pH 2 to 6 (Figure 4–3).[3] It also displays a much broader range governed by the uncatalyzed term $k_0$ (middle region of the curve).

It is often the case that the addition of an acid–base to the solution increases the rate constant of a reaction (Figure 4–1). **General acid–base catalysis** is the increase in the rate of a

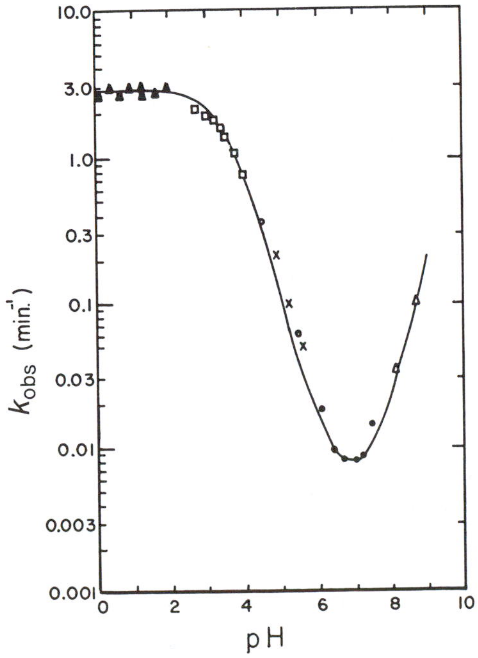

**Figure 4–2:** pH–Rate profile for the hydrolysis of *N*-acetylimidazole at 25 °C and an ionic strength of 0.2 M.[2] *N*-Acetylimidazole was added to a series of solutions adjusted to the noted pH values with HCl (▲) or maintained at the noted pH values with sodium formate buffer (□), sodium acetate buffer (○), sodium succinate buffer (×), imidazolium chloride buffer (●), or tris(hydroxymethyl)aminomethane hydrochloride buffer (△). The ionic strength was maintained by the appropriate concentration of sodium chloride. The values of the pseudo-first-order rate constants at each pH were extrapolated to zero buffer concentration. The progress of each reaction was followed by the decrease in absorbance at 245 nm resulting from the loss of *N*-acetylimidazole. All reactions were first-order in *N*-acetylimidazole, and pseudo-first-order rate constants were obtained by plotting the logarithm of the extent of the reaction as a function of time. The logarithms of the estimated pseudo-first-order rate constants in the absence of buffer, $k_{obs}$ (minutes$^{-1}$), are plotted as a function of the pH. The solid line has been fit to the data and is the line for the equation $k_{obs} = 2.8\ min^{-1}[AcImH^+] + 0.005\ min^{-1}[AcIm] + 0.005\ min^{-1}[AcIm][OH^-]$ (Equation 4–9). Adapted with permission from ref 2. Copyright 1959 *Journal of Biological Chemistry.*

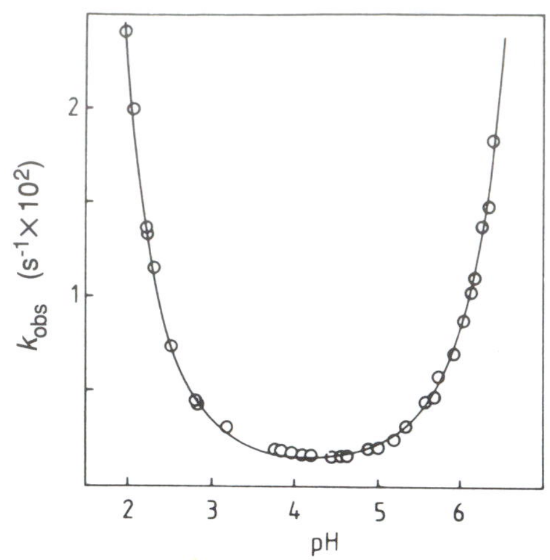


**Figure 4–3:** pH–Rate profile for the ketonization of vinyl alcohol at 15 °C at an ionic strength of 1 M.[3] Methoxy(vinyloxy)methyl acetate was added to a series of solutions of 1 M KCl in which the pH had been adjusted to different values with HCl or by adding low concentrations (0.1 mM) of buffer. The methoxy(vinyloxy)methyl acetate in all instances hydrolyzed immediately to vinyl alcohol, methanol, formaldehyde, and acetic acid. The vinyl alcohol then slowly tautomerized to acetaldehyde (Reaction 4–14) in a reaction that could be followed to greater than 90% completion by the decrease in absorbance at 210 nm due to the vinyl chromophore. Because this reaction proceeds with neither the production nor the consumption of protons, only weak buffering was required, and it was independently established that the buffer had no effect on the rate of the reaction when it was present at these low concentrations. The pseudo-first-order decreases in $A_{210}$ were fit by a nonlinear least-squares program to obtain values of the pseudo-first-order rate constant. This rate constant, $k_{obs}$ (seconds$^{-1}$ × 10$^2$), is plotted as a function of pH. Adapted with permission from ref 3. Copyright 1982 American Chemical Society.

reaction produced by adding an acid–base to the solution. Because it is impossible to add an acid to water without simultaneously producing its conjugate base or to add a base without simultaneously producing its conjugate acid, it must be established which of the two forms is responsible for the acceleration observed. This is done by using the conjugate acid and conjugate base in constant ratio as the buffer. As the concentration of the acid–base is increased at constant ratio of conjugate acid to conjugate base, the pseudo-first-order rate constant for the reaction increases (Figures 4–1 and 4–4).[1,4] This behavior can be described by the relationship

$$k_{obs} = k_{pH} + k_{gen}[\text{buffer}]_{TOT} \tag{4–15}$$

where $k_{pH}$ is a composite rate constant containing the terms for hydronium ion catalysis, no catalysis, and hydroxide ion catalysis (Equation 4–13) and $k_{gen}$ is a composite rate constant reflecting the effects of the added acid and base. Its kinetic order is one greater than $k_{obs}$. This linear behavior states that the term $k_{gen}[\text{buffer}]_{TOT}$ producing an increase in the rate results from a reaction whose transition state is composed of the reactant and either the conjugate acid or the base of the buffer. In other words, the acid or the base, when it participates directly in the reaction, causes it to proceed more rapidly.

If the ratio between the acid and the base in the buffer is changed, and inescapably the pH as well, and the experiment is repeated, the slope of the line, $k_{gen}$, will usually change (Figures 4–1 and 4–4). If the composite rate constant, $k_{gen}$, increases as the ratio between acid and base is increased, the conjugate acid of the buffer is the more important participant (Figure 4–1). If it decreases, the base is the more important participant.

The term, $k_{gen}[\text{buffer}]_{TOT}$, is itself often a sum of two separate terms

$$k_{gen}[\text{buffer}]_{TOT} = k_{ga}[\text{HA}] + k_{gb}[\text{A}] \tag{4–16}$$

where HA is the acid and A is the conjugate base of the buffer. The numerical values for these two terms are obtained by extrapolation. If $k_{gen}$ is plotted as a function of the fraction of the total buffer that is the conjugate acid, the extrapolated value for the point at which all of the buffer is the conjugate acid, $[\text{buffer}]_{TOT}$ = [HA] and [A] = 0, is $k_{ga}$ and the extrapolated value for the point at which all of the buffer is the conjugate base, $[\text{buffer}]_{TOT}$ = [A] and [HA] = 0, is $k_{gb}$ (inset to Figure 4–4). If the reactants other than the general acid–bases contain no acid–bases with values of p$K_a$ within the range of pH studied, Equations 4–16, 4–13, and 4–15 can be combined to obtain a general expression for acid–base catalysis

$$k_{obs} = k_{H^+}[H^+] + k_0 + k_{OH^-}[OH^-] + k_{ga}[\text{HA}] + k_{gb}[\text{A}] \tag{4–17}$$

In each particular reaction, the two terms in Equation 4–17 governing the particular general acid–base catalyst have relative degrees of importance reflected in the numerical values of the rate constants. The hydrolysis of the ethyl hemiacetal of formaldehyde

$$H_3C\text{–}CH_2\text{–}O\text{–}CH_2\text{–}OH + H_2O \longrightarrow CH_3CH_2OH + HO\text{–}CH_2\text{–}OH \tag{4–18}$$

in a solution buffered with chloroacetate (Figure 4–4), has a pseudo-first-order rate constant the equation for which contains two terms of similar magnitude (20 × 10$^{-3}$ M$^{-1}$ s$^{-1}$ and 2 × 10$^{-3}$ M$^{-1}$ s$^{-1}$), one directly proportional to the concentration of chloroacetic acid, [$ClCH_2COOH$], and the other directly proportional to the concentration of chloroacetate anion, [$ClCH_2COO^-$], respectively

$$k_{obs} = k_{ph} + k_{ga}[ClCH_2COOH] + k_{gb}[ClCH_2COO^-] \tag{4–19}$$

The hydrolysis of enol ether **4–1** (Reaction 4–1) , however, has a pseudo-first-order rate constant the equation for which, within experimental error, contains no term that is directly proportional to the concentration of the conjugate base of the phosphate buffer, $HOPO_3^{2-}$, even though it contains a term that is directly proportional to the conjugate acid of the phosphate buffer, $HOPO_3H^-$ (Figure 4–1). The hydrolysis of *N*-

acetylimidazole (Reaction 4–5), in a solution buffered with imidazole, has a pseudo-first-order rate constant the equation for which, within experimental error, has no term directly proportional to the cationic conjugate acid of imidazole, $H_2Im^+$, even though it contains a significant term directly proportional to the concentration of the neutral free base, HIm.[5]

Each of the terms in the equation for the composite rate constant (Equation 4–17), because it appears as the participant in a sum, refers to an independent route between reactants and products through a different transition state composed respectively from the molecules the concentrations of which appear in that particular term in the rate law as well as an unknown number of water molecules. If it has been observed that the term in Equation 4–17 proportional to the concentration of the conjugate acid of the buffer, $k_{ga}$[HA], is significant, the reaction pathway represented by this term is referred to as **general acid catalysis.** If it has been observed that the term in Equation 4–17 proportional to the concentration of the conjugate base of the buffer, $k_{gb}$[A], is significant, the reaction pathway represented by this term is referred to as **general base catalysis.**

There are many reactions that seem to display neither general acid catalysis nor general base catalysis even though their rates are functions of the pH of the solution. For example, the pseudo-first-order rate constant, $k_{obs}$, for the hydrolysis of formaldehyde acetal **4–2**

**4–2**

$$\text{Cl–C}_6\text{H}_4\text{–O–CH}_2\text{–OCH}_3 \longrightarrow \text{Cl–C}_6\text{H}_4\text{–OH} + \text{HOCH}_2\text{OH} + \text{HOCH}_3 \qquad (4\text{–}20)$$

was unaffected by the concentration of buffer (sodium formate), but the pseudo-first-order rate constant was directly proportional to the concentration of hydronium cation.[6] The relationship at 56 °C at an ionic strength of 1.0 M in $H_2O$ is

$$k_{obs} = 1.06\ \text{M}^{-1}\ \text{min}^{-1}\ [\text{H}^+] \qquad (4\text{–}21)$$

In this case, as in the case of the hydrolysis of almost all acetals, the rate constant for the reaction is a function only of the hydronium ion concentration, it remains proportional to the hydronium ion concentration even at relatively high values of pH, and it is unaffected by the addition of any other acids or bases to the solution.[7] As such, the hydrolysis of an acetal serves as an example of a reaction that displays specific acid catalysis. **Specific acid catalysis** is the increase in the

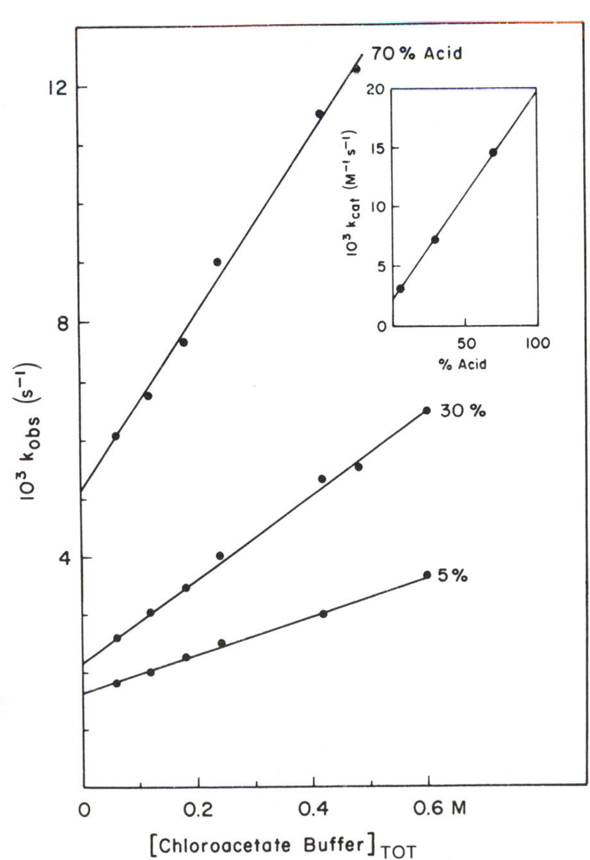


**Figure 4–4:** Catalysis of the hydrolysis of formaldehyde ethyl hemiacetal by chloroacetate buffers at 25 °C and an ionic strength of 1.00 M.[4] Formaldehyde ethyl hemiacetal was formed by mixing formaldehyde with ethyl alcohol and allowing the mixture to stand for 3 h before use. The formaldehyde ethyl hemiacetal was added to a final concentration of 0.1 mM to a solution of 1.00 M KCl containing the noted total concentration of potassium chloroacetate plus chloroacetic acid mixed in the noted ratios (95:5, 70:30, and 30:70). The decomposition of the formaldehyde ethyl hemiacetal was followed by trapping the formaldehyde produced with thiosemicarbazide (0.01 M). The increase in absorbance at 275 nm due to the formation of the thiosemicarbazone was followed with time. For each individual run, pseudo-first-order rate constants were obtained from the slopes of plots of ln ($A_{275} - A_{275}^{\infty}$) against time. The pseudo-first-order rate constants, $k_{obs}$ (seconds$^{-1}$ × $10^3$) are plotted as a function of the total molar concentration of chloroacetate anion plus chloroacetic acid, ([Chloroacetate Buffer]$_{TOT}$). The slopes of these lines, $k_{cat}$ (molar$^{-1}$ seconds$^{-1}$ × $10^3$), are plotted in the inset as a function of the percent of the total chloroacetate that was chloroacetic acid (% Acid). Adapted with permission from ref 4. Copyright 1978 American Chemical Society.

rate of a reaction produced by decreasing the pH when no general acid catalysis can be observed. **Specific base catalysis** is the increase in the rate of a reaction produced by increasing the pH when no general base catalysis can be observed. The fact that these terms have negative definitions makes them always equivocal. It is always possible that general acid–base catalysis has not yet been observed.

The terms *general acid catalysis, general base catalysis, specific acid catalysis,* and *specific base catalysis* should be reserved to describe observations. As in the kinetics of enzymatic reactions, it is important to make a distinction between the observation and an explanation for that observation. The importance of this distinction between observation and explanation is illustrated by the fact that there are always several explanations for the observation of either general acid catalysis or general base catalysis. For example, the hydrolysis of ethyl dichloroacetate, $EtOCOCHCl_2$, in a solution buffered by tris(hydroxymethyl)aminomethane has a pseudo-first-order rate constant that, within experimental error, has no term proportional to the concentration of the conjugate acid of tris(hydroxymethyl)aminomethane, $TrisH^+$, even though it has a term proportional to the concentration of the free base of this buffer, Tris.[8] In the rate law for the reaction, however, the respective term can be written in two equivalent ways

$$k_{gb}'[\mathrm{Tris}][\mathrm{EtOCOCHCl_2}][\mathrm{H_2O}]$$

$$= \frac{k_{gb}' K_a^{\mathrm{Tris}}}{K_a^{\mathrm{H_2O}}}[\mathrm{TrisH^+}][\mathrm{EtOCOCHCl_2}][\mathrm{OH^-}] \qquad (4\text{–}22)$$

where $k_{gb}'$ is $k_{gb}[H_2O]^{-1}$, $K_a^{\mathrm{Tris}}$ is the acid dissociation constant of the conjugate acid of tris(hydroxymethyl)aminomethane, and $K_a^{\mathrm{H_2O}}$ is the acid dissociation constant of water. The term to the left in Equation 4–22 implies that the free base of the catalyst combines with a molecule of the ester and a molecule of water to form the transition state. The term to the right in Equation 4–22 implies that a molecule of the cationic conjugate acid of the catalyst combines with a molecule of the ester and a hydroxide anion to form the transition state. At least three different arrangements of constituents

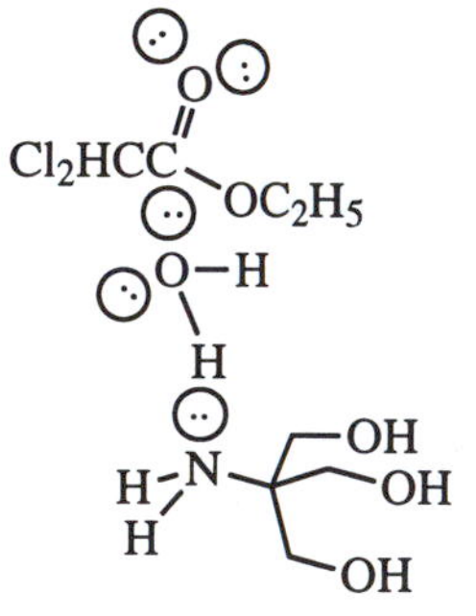

**4–3** 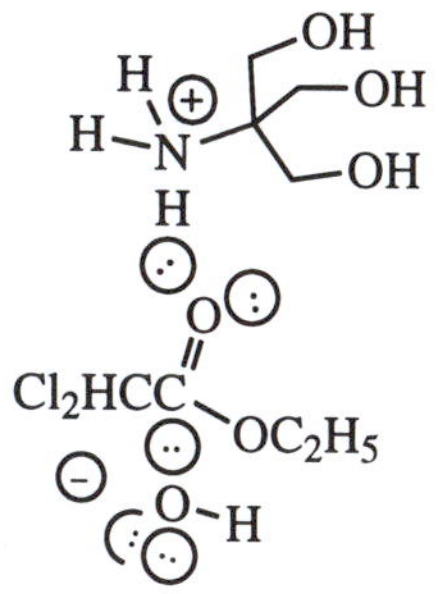 **4–4**

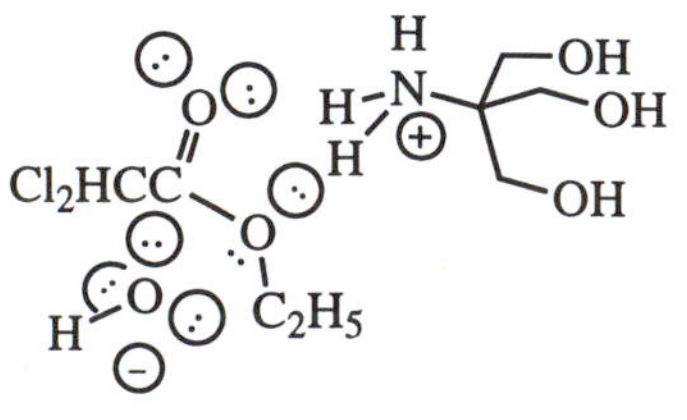

**4–5**

within the transition state are consistent with the terms of Equation 4–22.[8]

To decide among the various alternatives, other independent experiments must be performed. For example, from the results of more extensive studies, it was concluded that the term directly proportional to the concentration of chloroacetic acid, $ClCH_2COOH$, in the composite pseudo-first-order rate constant for the hydrolysis of the ethyl hemiacetal of formaldehyde (Reaction 4–18) arose from arrangement **4–6** in the transition state while the term directly proportional to the concentration of chloroacetate anion, $^-OAcCl$, arose from arrangement **4–7**

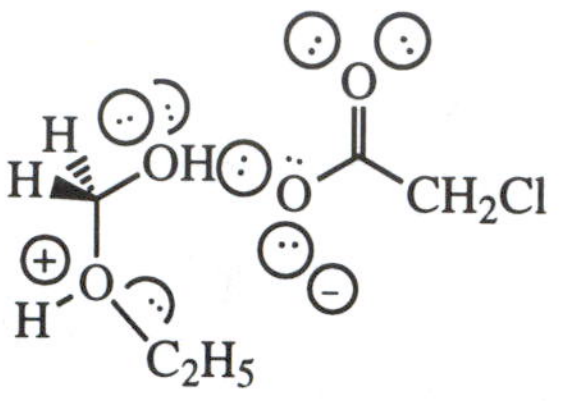

**4–6** $k_{ga}[ClCH_2COOH]$ **4–7** $k_{gb}[ClCH_2COO^-]$

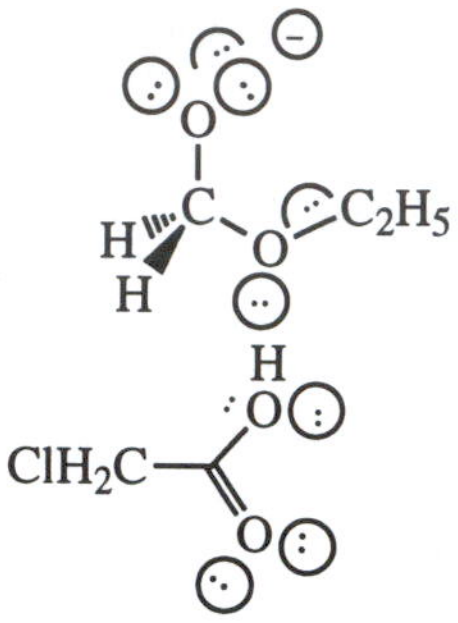

Ironically, in this instance it was concluded that neither term referred to the most straightforward possibility. It should be noted that all protons present in the species participating in each of the terms in the rate laws are present in the respective transition state, but they are sometimes present in different locations. This simply means that the rate law can make no distinction among the tautomers of the transition state.

The distinction between observation and explanation is also critical when considering specific acid or base catalysis because there are also at least two different explanations for these observations. A preequilibrium may occur involving an acid–base reaction followed by a rate-limiting step involving bond-breaking or bond-making, or hydronium ion may participate as a general acid or hydroxide ion as a general base in a reaction with a Brønsted coefficient near unity.

In many reactions, it is only the conjugate base or only the conjugate acid of one of the reactants that participates significantly. Because acid–base reactions are so rapid, the equilibrium between the reactive species and its unreactive conjugate acid or conjugate base may be established prior to a rate-limiting step that does not involve any proton transfers. For example, one explanation for the observation of specific acid catalysis in the hydrolysis of formaldehyde acetal **4–2** would be that the protonation of one of the acetal oxygens in a rapid acid–base equilibrium necessarily precedes the rate-limiting departure of the leaving group

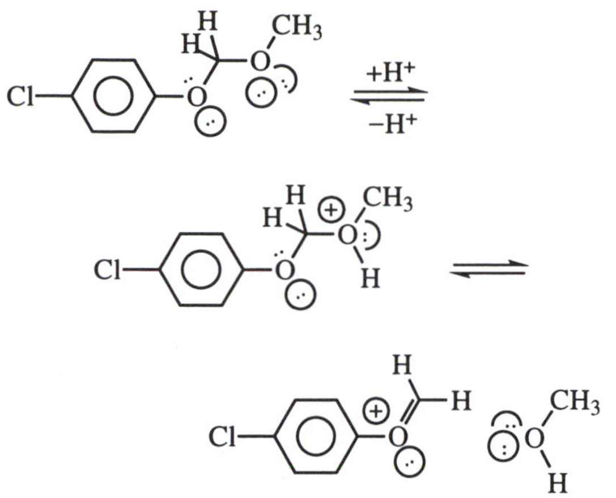

(4–23)

Because the concentration of the protonated acetal should be directly proportional to the concentration of hydronium cation at values of pH significantly greater than the p$K_a$ of the oxygen, which is less than zero, this mechanism is consistent with the observation of specific acid catalysis (Equation 4–21). Originally, it was hoped that the distinction between general and specific acid–base catalysis could distinguish such a pre-equilibrium mechanism from a mechanism in which a proton must be transferred in the transition state of the rate-limiting step. Unfortunately, an alternative explanation that is also consistent with the observation of specific acid catalysis would be that the protonation of the acetal oxygen by either an added acid or by hydronium cation occurs simultaneously with the cleavage of the carbon–oxygen bond but that general acid catalysis is not observed because the Brønsted coefficient for the reaction is near unity.

A related problem that can be discussed simultaneously with that of the explanations for specific acid–base catalysis is the relative magnitudes of the rate of general acid or base catalysis and the rate of the same reaction in the absence of any added acid–base. Even when general acid–base catalysis is observed, it rarely results in increases in rate of more than about a factor of 10 at concentrations of buffer less than 1.0 M (Figures 4–1 and 4–4). Specific acid–base catalysis could simply be viewed as the extreme where the magnitude of general acid–base catalysis was so small that it could not be detected experimentally above the rate of the reaction in the absence of added acid–bases. In this alternative explanation, the reaction is susceptible to general acid–base catalysis, but it simply cannot be detected. Because general acid–base catalysis is usually so unremarkable, and may often be undetectable, the impression might be left that such catalysis may be expendable in enzymatically catalyzed reactions. To understand the fallacy of this impression as well as the alternative explanation for the observation of specific acid–base catalysis, the reasons that general acid–base catalysis is so unremarkable when reactions are performed in aqueous solution must be appreciated. The basis of this appreciation is the Brønsted relationship.

The numerical values of the rate constants for either the general acid catalysis or the general base catalysis of a certain reaction by a series of acids or bases with different p$K_a$ values are usually found to be a function of those p$K_a$ values. The function relating the rate constants for general acid or base catalysis and the p$K_a$ values of the catalysts is the **Brønsted relationship.**

If the catalysis observed has been designated as general acid catalysis because there is a term, $k_{ga}$[HA], that is directly proportional to the concentration of the conjugate acid of the buffer in the expression for the observed rate constant, $k_{obs}$, and if a series of acids all with the same atom to which the proton is attached and with the same charge is chosen, the Brønsted relationship usually observed is

$$\log\left(\frac{k_{ga}}{p}\right) = \log G_A + \alpha \log\left(\frac{qK_a}{p}\right) \tag{4–24}$$

where $p$ and $q$ are integers and log $G_A$ is a parameter. The other parameter, $\alpha$, is referred to as a **Brønsted coefficient.** The term $k_{ga}/p$ is the observed general acid rate constant, $k_{ga}$, divided by the number of equivalent protons, $p$, on the acid. This division yields the rate constant in terms of the molar concentration of acidic protons. The term $qK_a/p$ is the observed acid dissociation constant $K_a$ corrected statistically[9] for the number of equivalent protons on the acid, $p$, and the number of equivalent lone pairs on the conjugate base, $q$. After this correction, the dissociation constant is expressed in terms of the molar concentration of individual protons on the conjugate acid and the molar concentration of individual lone pairs on the conjugate base. If log ($k_{ga}/p$) is plotted as a function of p$K_a$ + log ($p/q$) a linear relationship is usually observed with a slope of $-\alpha$. For example, when a series of acids were chosen as general acid catalysts for the cleavage of *N*-(*p*-nitrophenyl)carbamate

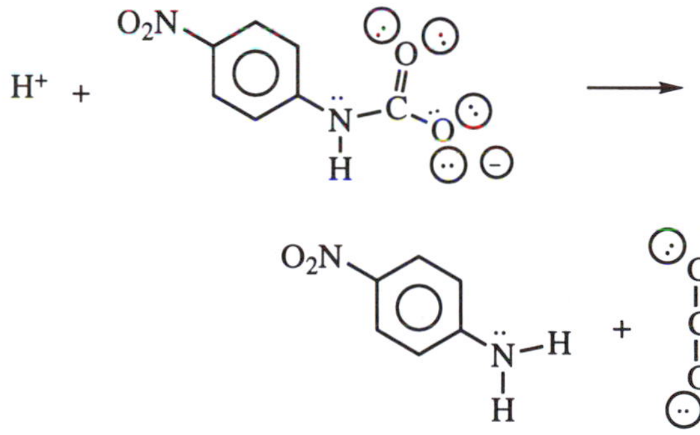

(4–25)

the rate constants for the respective observed general acid catalyses were related to the values of the p$K_a$ of the acids by a Brønsted relationship (Equation 4–24) with $\alpha$ = 0.84 (Figure 4–5A).[10] The points for the oxygen acids and the points for the nitrogen acids defined two distinguishable lines of the same slope.

If the catalysis observed has been designated as general base catalysis because there is a term, $k_{gb}$[A], that is directly proportional to the concentration of the conjugate base of the buffer in the expression for the observed rate constant, $k_{obs}$, and if a series of bases all with the same atom on which the basic lone pairs are situated and with the same charge is chosen, the Brønsted relationship usually observed is

$$\log\left(\frac{k_{gb}}{q}\right) = \log G_B - \beta \log\left(\frac{qK_a}{p}\right) \tag{4–26}$$

where log $G_B$ is a parameter and $K_a$ is the acid dissociation constant of the conjugate acid of the catalytic general base.

A

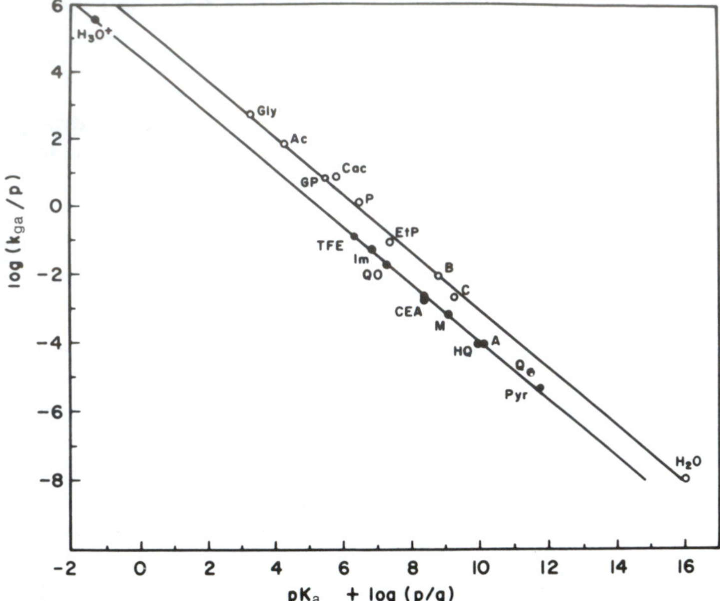

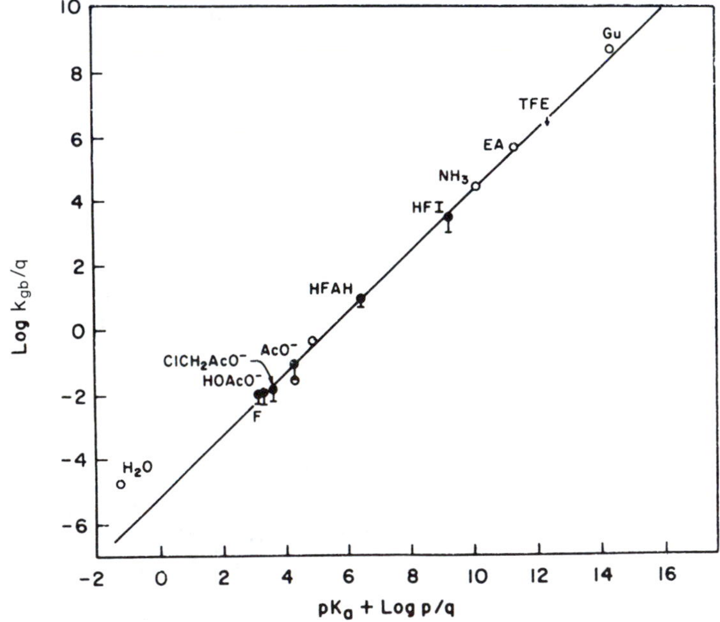


**Figure 4–5:** Brønsted plots for general acid catalysis (A) and general base catalysis (B). (A) General acid catalysis for the cleavage of *N*-(*p*-nitrophenyl)carbamate (Reaction 4–25).[10] Solutions of *N*-(*p*-nitrophenyl)carbamate were prepared at 1.0 M ionic strength (maintained with KCl) and 25 °C. The solutions were buffered with the respective general acid catalyst and its conjugate base. The cleavage was followed spectrophotometrically at 405 nm, the absorption maximum of the product, *p*-nitrophenylaniline. Rate constants for general acid catalysis, $k_{ga}$, were determined, and the logarithms of their statistically corrected values, $k_{ga}/p$ (molar$^{-1}$ second$^{-1}$), are plotted as a function of the negative logarithms of the statistically corrected values of the acid dissociation constant for each catalyst, $K_a(q/p)$. Oxygen acids (○) glycine, acetic acid, phosphate, ethyl phosphate, boric acid, bicarbonate, and water are plotted separately from nitrogen acids (●) trifluorethylammonium, imidazolium, 3-quinuclidone, cyanoethylammonium, morpholinium, 3-hydroxyquinuclidinium, ammonium, quinuclidinium, and pyrrolidinium. (B) General base catalysis for the dissociation of *p*-methoxyacetophenone bisulfite (Reaction 4–27).[11] Solutions of *p*-methoxyacetophenone bisulfite were prepared at 1.0 M ionic strength [maintained with either KCl (closed symbols) or $(CH_3)_4NCl$ (open symbols)] and 25 °C. The solutions were buffered with the general base catalyst and its conjugate acid. The dissociation was followed spectrophotometrically at 285 nm, the absorption maximum of the product, *p*-methoxyacetophenone. Rate constants for general base catalysis, $k_{gb}$, were determined, and the logarithms of their statistically corrected values, $k_{gb}/q$ (molar$^{-1}$ second$^{-1}$), are plotted as a function of the negative logarithms of the statistically corrected values of the acid dissociation constants of the respective conjugate acids $K_a(q/p)$. The general bases were water, formate, glycolate, chloroacetate, acetate, hexafluoroacetone hydrate anion, ammonia, ethylamine, trifluoroethanol anion, and guanidine. Adapted with permission from refs 10 and 11. Copyright 1980 and 1977 American Chemical Society.

The other parameter, $\beta$, is also referred to as a Brønsted coefficient. The integers $p$ and $q$ serve the same purpose as before. The rate constant for general base catalysis is divided by the numbers of equivalent lone pairs $q$ to yield the rate constant in terms of the molar concentration of those lone pairs. The general base catalysis of the dissociation of $p$-methoxyacetophenone bisulfite

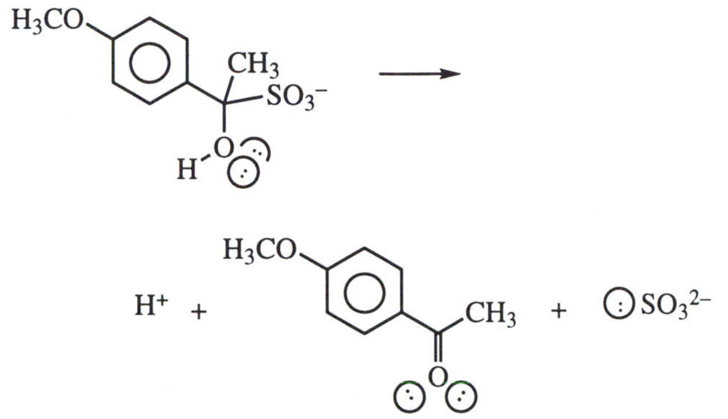

(4–27)

follows the Brønsted relationship with $\beta = 0.94$ (Figure 4–5B).[11]

The Brønsted relationships demonstrate that the acidity of the general acid or the basicity of the general base determines the effect that the catalyst will have on the rate of the reaction. The more acidic the general acid, the greater will be the rate constant for general acid catalysis. The more basic the general base, the greater will be the rate constant for general base catalysis. The magnitude of the values for $\alpha$ and $\beta$, however, will determine the magnitude of these effects. If the value for $\alpha$ or $\beta$ is small, the acidity or basicity of the catalyst has little effect on the respective rate constant; but, if $\alpha$ or $\beta$ is large, the acidity or basicity of the catalyst will have a large effect. It is believed that if the mechanism involves the general acid as a proton donor in the transition state, a small value for $\alpha$ indicates that the bond between the general acid and its proton is only marginally broken at the transition state and that if the mechanism involves the general base as a proton acceptor in the transition state, a small value for $\beta$ indicates that the bond between general base and the proton is only marginally formed at the transition state. Under these circumstances, any source of a proton or any lone pair, respectively, would be about the same. If $\alpha$ has a value near 1.0 and if the general acid acts as a proton donor in the transition state, it is generally assumed that the bond between the general acid and the proton is largely broken at the transition state; and, if $\beta$ has a value near 1.0 and the general base acts as a proton acceptor in the transition state, it is generally assumed that the bond between the general base and the proton is largely formed in the transition state. Under these circumstances the strength of the acid or the strength of the base, respectively, should be fully manifested.

It has already been noted that if a reaction actually has a mechanism requiring that a proton transfer occur before the rate-limiting step and if that proton transfer is so rapid as to be at equilibrium, specific acid–base catalysis should be observed. It is also possible, however, that a reaction displays specific acid–base catalysis because the magnitude of the existing general acid–base catalysis is too small to be observed. It is the Brønsted relationships that can cause general acid or base catalysis to be weak or immeasurable. Consider a reaction susceptible to general acid catalysis. Assume for the moment that the hydronium cation, $H_3O^+$, behaves as a typical general acid in the set. From the Brønsted relationship for the reaction of interest (Equation 4–24)

$$k_{ga}{}^{HA} = \left(\frac{q_{HA}}{2}\right)^{\alpha}\left(\frac{p_{HA}}{3}\right)^{1-\alpha}\left(\frac{K_a{}^{HA}}{K_a{}^{H_3O^+}}\right)^{\alpha} k_{ga}{}^{H_3O^+} \qquad (4\text{–}28)$$

where HA is any general acid in the series tested as catalysts. If it is also assumed that terms involving $q_{HA}$ and $p_{HA}$ are together nearly equal to 1, then

$$k_{ga}{}^{HA} \cong \left(\frac{K_a{}^{HA}}{K_a{}^{H_3O^+}}\right)^{\alpha} k_{ga}{}^{H_3O^+} \qquad (4\text{–}29)$$

If only acid catalysis is important, then

$$k_{obs} = k_0 + k_{H^+}[H^+] + k_{ga}{}^{HA}[HA] \qquad (4\text{–}30)$$

If

$$k_{H^+}[H^+] = k_{ga}{}^{H_3O^+}[H^+] \qquad (4\text{–}31)$$

then

$$k_{obs} \cong k_0 + k_{H^+}[H^+] + \left(\frac{K_a{}^{HA}}{K_a{}^{H_3O^+}}\right)^{\alpha} k_{H^+}[HA] \qquad (4\text{–}32)$$

$$k_{obs} \cong k_0 + k_{H^+}[H^+] + \left(\frac{[A][H^+]}{[H_2O][HA]}\right)^{\alpha} k_{H^+}[HA] \qquad (4\text{–}33)$$

$$k_{obs} \cong k_0 + k_{H^+}[H^+] + \left(\frac{[A]}{[H_2O]}\right)^{\alpha}\left(\frac{[HA]}{[H^+]}\right)^{1-\alpha} k_{H^+}[H^+] \qquad (4\text{–}34)$$

Because $[A] \ll [H_2O]$ under all possible circumstances, the term $([HA]/[H^+])^{1-\alpha}$ must be sufficiently large to counteract the effect of the term $([A]/[H_2O])^{\alpha}$ for general acid catalysis to be significant. In particular, if $\alpha$ is near a value of 1.0, the disparity between the concentration of the conjugate base of the general acid and the concentration of water will dominate and no general acid catalysis will be observed. If, however, $\alpha$ is sufficiently small and the acid is sufficiently weak that concentrations of its conjugate acid much greater than the concentration of hydronium ion can be achieved, then general acid catalysis will be significant. A similar explanation can be formulated for the competition between hydroxide ion catalysis and general base catalysis.

From all of these considerations, it is clear that the reason that both general acid and base catalysis are so unremarkable in any aqueous solution is that water is the predominant acid–base in that solution. Because the concentration of water is so high relative to any added base or acid, the concentrations of its conjugate acid or its conjugate base are anomolously high relative to those of an added base or acid, and they act preferentially as the general acids and general bases required in the reaction. Another way of looking at the situation is that $H_3O^+$ is such a strong acid and $OH^-$ is such a strong base that, even when their concentrations are small relative to an added general acid or general base, they can compete successfully. Most reactions of biochemical significance have an absolute, inescapable requirement for catalytic acids or bases or both. It is just that this requirement is masked either partially or completely by the solvent. With this in mind, it becomes of interest to contrast the situation of a reactant within an active site and the situation encountered by the same reactant in solution.

The active site of an enzyme binds substrates specifically and in the proper orientation because **aligned hydrogen bonds** form between donors and acceptors on the amino acids lining the active site and acceptors and donors on the substrates. The donors and acceptors of hydrogen bonds involved in the recognition of substrates can automatically become general acids and general bases required in the reaction. For example, in the active site of triosephosphate isomerase, the carbonyl oxygen of dihydroxyacetone phosphate is recognized by being hydrogen-bonded to the donors from Lysine 13 and Histidine 95.[12] Because the imidazole of the histidine is in its neutral form in the complex,[13] it is the alkylammonium cation of the lysine that is the ultimate general acid providing the proton necessary to enolize the reactant as Glutamate 165 removes the proton

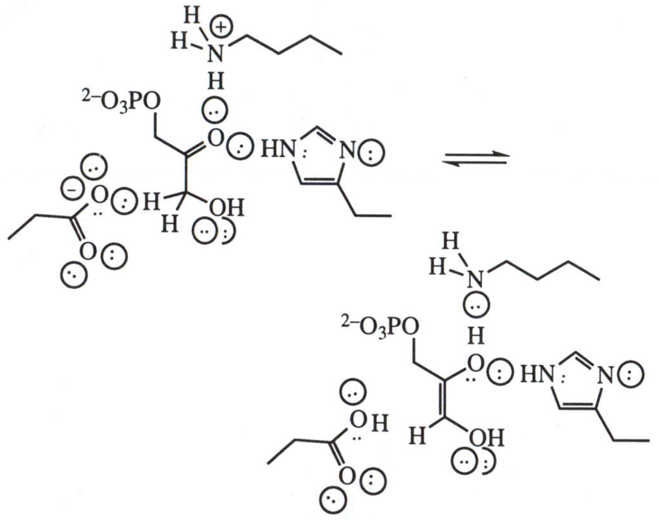

(4–35)

If the mechanism of a reaction that occurs in solution requires the participation of a general acid or a general base in the transition state of the rate-limiting step and the only general acids and general bases present are other molecules in solution, a molecule of the reactant must collide with a molecule of the catalyst. Even if the catalyst is a hydronium cation, a molecule of water, or a hydroxide anion, its participation in the reaction is intermolecular. In the active site of an enzyme, however, the reaction between the bound reactant and the general acid or general base is a unimolecular reaction, and the **entropy of approximation** gained by converting an intermolecular reaction into an **intramolecular reaction** increases the efficiency of the catalyst.

The hydrolysis of acetals (Reaction 4–20) is normally a specific acid-catalyzed reaction. It is possible, however, to provide an intramolecular general acid catalyst in the form of the *o*-carboxylic acid on phenyl methyl acetal **4–8**

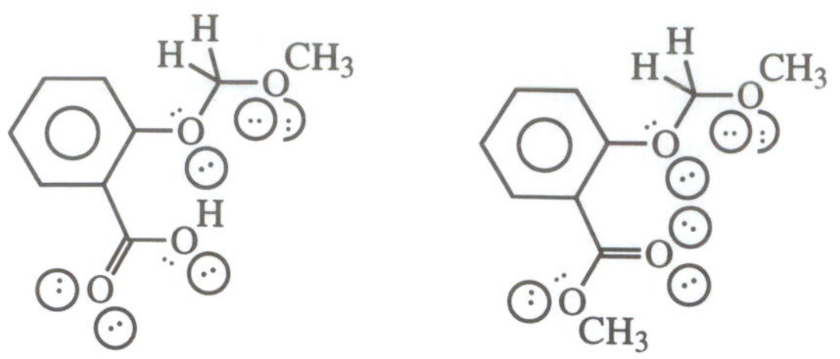

**4–8** **4–9**

The methyl ester **4–9** of acetal **4–8** serves as a control that incorporates the electronic structure without the ability to provide a proton. The pH–rate profiles for the hydrolysis of acetal **4–8** and acetal **4–9** differ remarkably (Figure 4–6).[6] The hydrolysis of acetal **4–9**, as with other acetals, displays only hydroniun ion catalysis. The hydrolysis of acetal **4–8**, however, displays a plateau in its pH–rate profile such that at values of pH greater than 5, in the neutral range of pH, the rate of its hydrolysis is 270-fold greater than the rate of hydrolysis of the methyl ester.

The behavior of the hydrolysis of acetal **4–8** as a function of pH can be explained satisfactorily in the following way.[6] Below pH 1, the concentration of hydronium ion is so high that hydronium ion catalysis dominates, and

$$k_{obs} = k_{H^+}[H^+] \qquad (4\text{–}36)$$

Above pH 1, the concentration of hydronium ion becomes so small that intramolecular general acid catalysis dominates

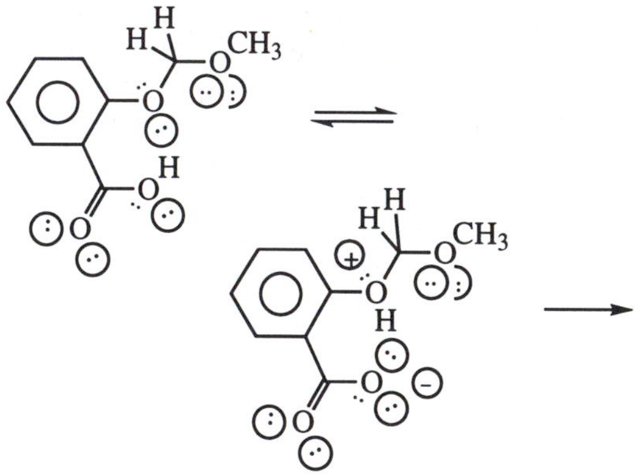

(4–37)

Until the p$K_a$ of the *o*-carboxylic acid is reached, the rate of the reaction is invariant with pH. At values of pH in excess of the p$K_a$, the concentration of the active carboxylic acid decreases by a factor of 10 for every increase of 1 unit in the pH. This example demonstrates that intramolecular general acid catalysis can be observed in cases where intermolecular general acid catalysis cannot, and it serves as an example of the effect of entropy of approximation on acid–base catalysis.

Examples of intramolecular general acid and base catalysis are common. The observed pH–rate profile for the hydrolysis of enol ether **4–1** (Reaction 4–1) displays a plateau very similar to that observed in Figure 4–6, which suggests that the *endo*-carboxylic acid acts as a general acid protonating intramolecularly the peripheral carbon of the enol to form the oxonium cation[1]

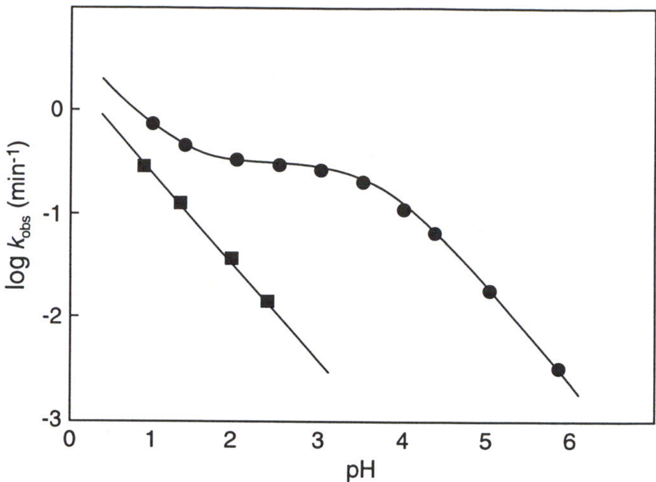


**Figure 4–6:** Intramolecular general acid catalysis of the hydrolysis of the *o*-carboxyphenyl methyl acetal of formaldehyde.[6] Solutions of either *o*-carboxyphenyl methyl formaldehyde acetal **4–8** (○) or *o*-(methoxycarbonyl)phenyl methyl formaldehyde acetal **4–9** (□) were prepared at ionic strength of 1.0 M (maintained with KCl) and 56 °C. Solutions were either unbuffered or buffered with sodium formate or potassium acetate. All values of observed rate constants were those obtained by extrapolation to eliminate any effect of buffer on the rate. In most instances, the buffer had no significant effect on the rate of the reaction. The release of the respective phenol as product was followed by the increase in absorbance at 302 nm, and pseudo-first-order rate constants were obtained from slopes of plots of the logarithms of the absorbances as a function of time. The pH–rate profiles presented are the logarithms of the tabulated pseudo-first-order rate constants for the hydrolysis of the respective acetals, $k_{obs}$ (minute$^{-1}$), plotted as functions of pH.

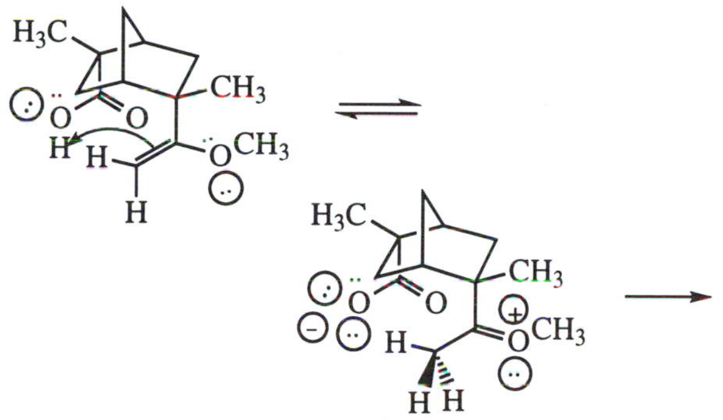

(4–38)

which is then hydrated.

The observed rate expression for the aminolysis of *N*-acetylimidazole, AcIm, by aliphatic amines, $RNH_2$

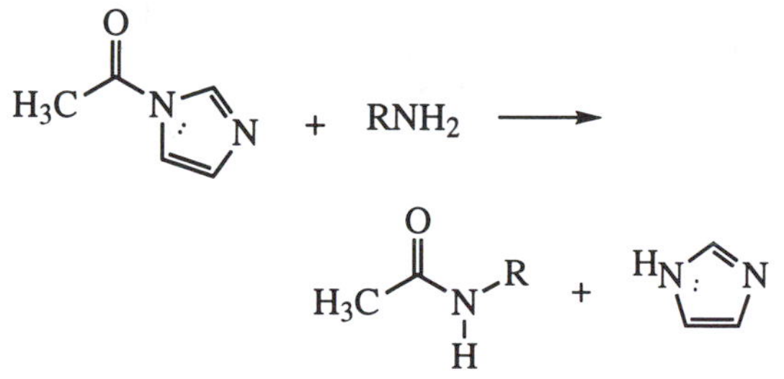

(4–39)

has a term, $k_B[RNH_2][AcIm]$, that is assumed to describe the attack of the unprotonated amine on the unprotonated *N*-acetylimidazole.[14] The rate constant, $k_B$, for neutral unprotonated 1,3-diaminopropane is 75 times greater than the respective rate constant for neutral 1-aminopropane even though the conjugate monoacids of each of these compounds have similar values for their acid dissociation constants. This is believed to be due to the ability of the second amino group of 1,3-diaminopropane to remove intramolecularly a proton from the tetravalent intermediate

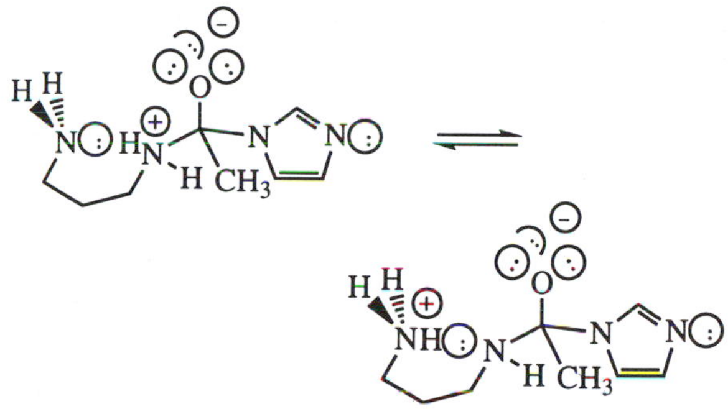

(4–40)

and thereby lock the distal nitrogen onto the acyl carbon. Intermolecular general base catalysis of the reaction between monoamines and *N*-acetylimidazole is also readily observed, and 1,3-diaminopropane participating as an intramolecular general base (Equation 4–40) is only as efficient as a 1 M concentration of a second molecule of 1-aminopropane acting as an unattached general base catalyst of its own aminolysis of *N*-acetylimidazole. Therefore, the entropy of approximation exerted in this reaction is not significant.

The rate expression for the enolization of aminoketone **4–10**, $H_2NKet$

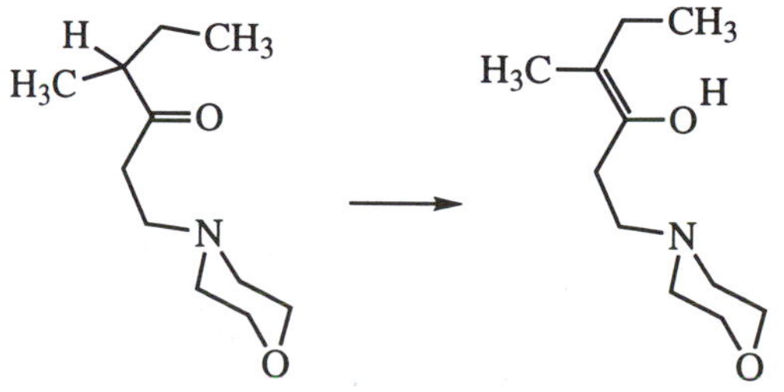

**4–10**

(4–41)

contains a term, $k_{OH^-}[OH^-][H_3N^+Ket]$.[15] This term is believed to reflect the ability of the tertiary ammonium cation to act as an intramolecular general acid

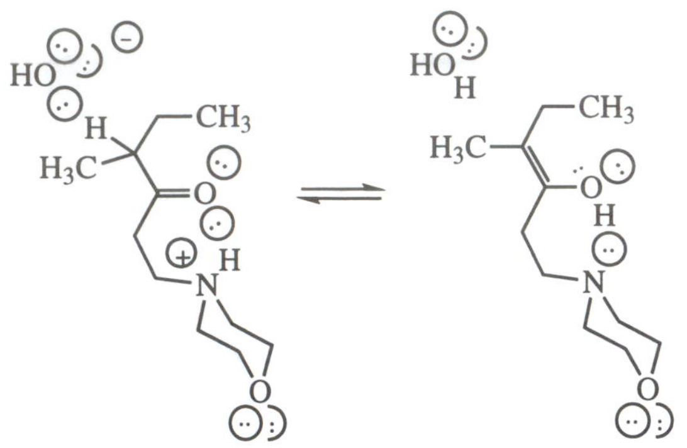

(4–42)

in concert with hydroxide ion catalysis, although the kinetically indistinguishable mechanism in which the basic nitrogen removes the $\alpha$-proton directly could not be ruled out. If the mechanism is as shown in Reaction 4–42, the rate of this intramolecular reaction is at least 100 times greater than the rate of the reaction catalyzed by hydroxide anion alone when the nitrogen of aminoketone **4–10** is unprotonated.

The incorporation of a second intramolecular catalyst can lead to a further increase in the rate of a reaction. The *endo*-imidazole in norbornyl ester **4–11**

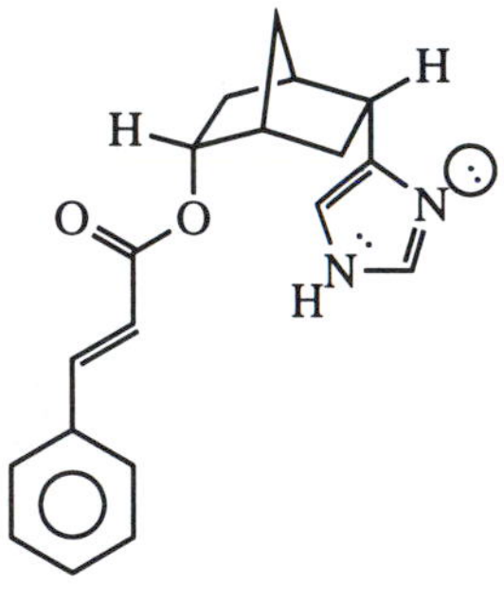

**4–11** 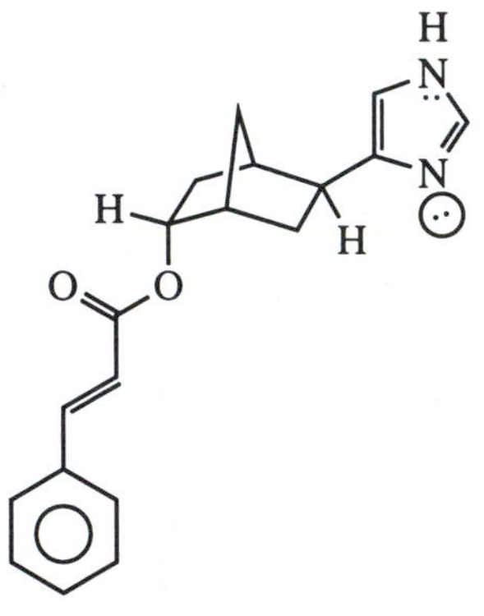 **4–12**

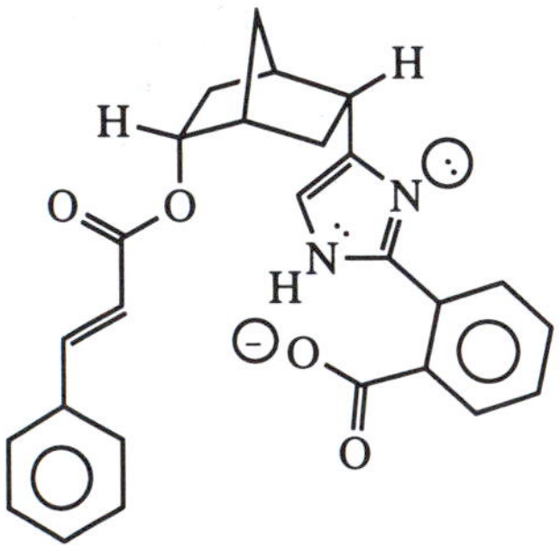

**4–13**

acts as an intramolecular catalyst of its own hydrolysis.[16] As the pH of the solution is decreased, the pH–rate profile for the hydrolysis of norbornyl ester **4–11** displays a plateau consistent with a change in mechanism from hydroxide ion catalysis to intramolecular general base catalysis (Figure 4–7),[16] the reciprocal behavior to that observed in Figure 4–6. Consistent with this conclusion is the observation that the *exo* analogue **4–12** of norbornyl ester **4–9** displays no plateau. At values of pH less than the $pK_a$ of the imidazole, the reaction involving intramolecular general base catalysis is only 6 times faster than the reaction involving only hydroxide ion catalysis. If, however, an *o*-carboxyphenyl group is attached to the *endo*-imidazole of norbornyl ester **4–11** to form norbornyl ester **4–13**, the rate of hydrolysis of the ester at pH 7 increases[17] by an additional factor of $4 \times 10^4$. The effective concentration of the intramolecular *o*-carboxyphenyl group relative to intermolecular benzoate is 3.6 mole fraction, consistent with an entropy of approximation of 11 J $K^{-1}$ $mol^{-1}$.

A feature of most active sites, in addition to the conversion of donors and acceptors of hydrogen bonds to general acids and bases and entropy of approximation, is that after the reactants have bound and, in the process, displaced the waters of hydration within the empty site, the active site has become **anhydrous.** Therefore, the lone pairs and protons on the reactants do not have access to molecules of water, hydronium ions, or hydroxide ions. In the absence of access to water, other general acids and general bases must be provided.

In benzene, the epimerization of 2,3,4,6-tetramethylglucose (TMG)

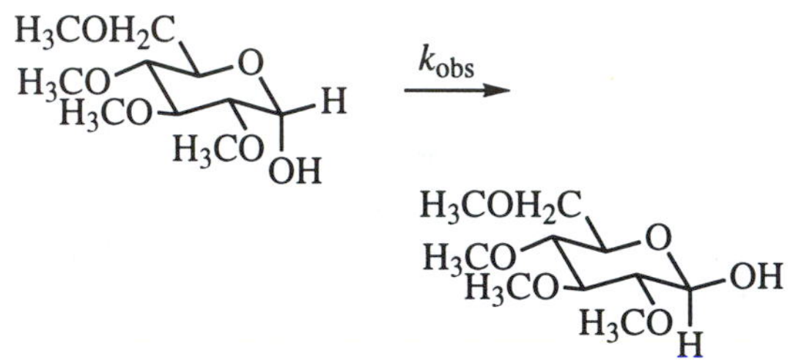

$$\xrightarrow{k_{obs}}$$

(4–43)

is very slow and variable in rate, suggesting that what little reaction is observed has an absolute requirement for traces of general acids or general bases contaminating the solvents.[18] When pyridine (Pyr) is added as a catalyst, the rate law contains, in addition to the negligible term for the uncatalyzed reaction, a term of the form $k_{pygl}$[Pyr][TMG]$^2$; and, when both pyridine and phenol (Phe) are added, the rate law contains an additional term of the form $k_{pyph}$[Pyr][Phe][TMG]. In each of these two instances, the individual terms are termolecular, and it was originally concluded[18] that, in benzene, the epimerization of tetramethylglucose has an absolute requirement for a general acid and a general base simultaneously

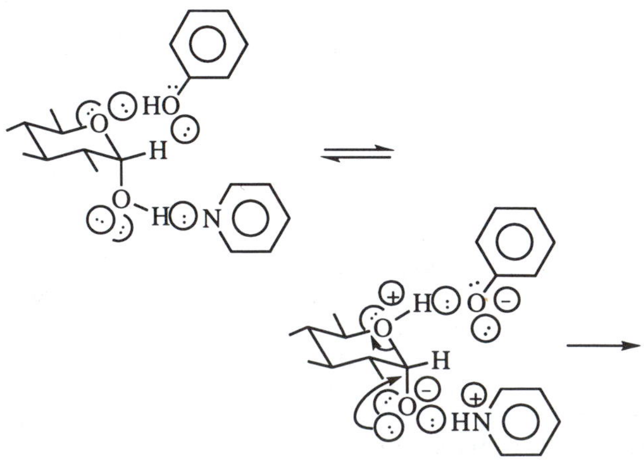

(4–44)

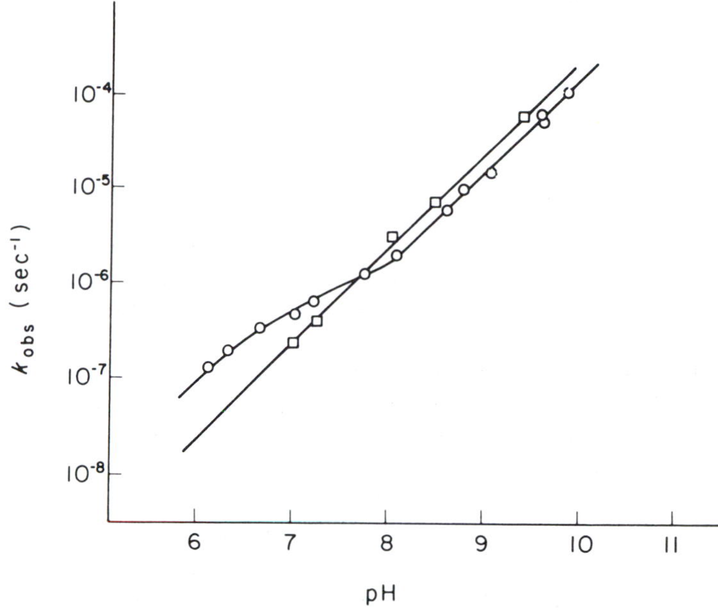


**Figure 4–7:** Intramolecular general base catalysis of the hydrolysis of an ester by the imidazolyl group.[16] Solutions of either *endo*-norbornyl ester **4–11** (○) or *exo*-norbornyl ester **4–12** (□) were prepared at ionic strength of 0.1 M (maintained with KCl) and 60 °C. The solutions were buffered with carbonate or *N,N*-bis(2-hydroxyethyl)glycine. The progress of the reaction was followed by monitoring the decrease in the absorbance of the ester at 310 nm. Specific pseudo-first-order rate constants, $k_{obs}$ (second$^{-1}$) are plotted as a function of the pH. Adapted with permission from ref 16. Copyright 1977 National Academy of Sciences.

It has since been shown that some of the catalysis observed, if not all of it, is due to general base catalysis by either glucosate anion or by phenoxide anion acting alone. The concentrations of these strongly basic anions are directly proportional either to [Pyr][TMG] or to [Pyr][Phe], respectively. General base catalysis by other strong bases, acting alone, also is sufficient[19] for mutarotation

(4–45)

Nevertheless, all of the results agree on the absolute necessity for added catalyst when the epimerization is performed in anhydrous benzene.

One of the most effective catalysts of the mutarotation of 2,3,4,6-tetramethylglucose in benzene is 2-hydroxypyridine[20]

(4–46)

**4–14**

a compound chosen on the assumption that it would be a bifunctional catalyst

(4–47)

At a concentration of 0.05 M in benzene, 2-hydroxypyridine catalyzes the epimerization of tetramethylglucose 200 times more rapidly than 0.05 M phenol and 0.05 M pyridine. Yet, in aqueous solution, the mutarotation of glucose is not significantly catalyzed by the addition of 0.1 M 2-hydroxypyridine because the water itself and the hydronium cations and hydroxide anions unavoidably present in it are even more efficient catalysts than 2-hydroxypyridine. The implication of these observations is that in an active site, which does not have access to the aqueous solution, the general acids and general bases provided by the various amino acids become essential to catalysis even if they are unnecessary when the reactants are dissolved in solution.

## *Suggested Reading*

Jencks, W.P., & Carriulo, J. (1961) General Base Catalysis of Ester Hydrolysis, *J. Am. Chem. Soc. 83*, 1743–1750.

## PROBLEM 4–1

(A) Write a step-by-step mechanism for the following reaction

$$H^+ + O{=}C(CH_3)_2 + H_2NR \rightleftharpoons H_2O + (H_3C)_2C{=}\overset{\oplus}{N}HR$$

Hydroxylamine reacts with the iminium cation to yield the hydroxamate very rapidly, but it reacts only very slowly with acetone. Therefore, it is possible to follow the rate at which the iminium cation is formed when acetone is mixed with various amines by trapping the iminium cation with hydroxylamine as it is formed. It has been shown that the reaction in the direction written, that of iminium cation formation, is catalyzed by general acids.

(B) Choose the step in your overall mechanism where a general acid would participate directly and draw a picture of the general acid (AH) donating its proton to the lone pair on the intermediate compound present at that step. Remember, you are considering only the reaction proceeding from left to right. The reverse reaction, the hydrolysis of the iminium cation, has been prevented by trapping the iminium cation as it is formed.

The rates at which each of several primary amines $RCH_2NH_2$ react with acetone to form the iminium cation were determined at a series of pH values. In addition, the $pK_a$ values for each of the amino nitrogens in each of the reactants were determined. With these $pK_a$ values, the amount of every form of the primary amine present at a given pH could be calculated, as well as the $pK_a$ of the primary amino group when the other amino group is either neutral or protonated. From the behavior of the rate constants as a function of pH and a knowledge of the concentrations of the various protonated species at each pH, the rate constants ($k_{am}$) for the reaction of each form of a given amine with acetone could be calculated.

| amine | $pK_a$ | $10^3 k_{am}$ ($M^{-1}\ s^{-1}$) |
|---|---|---|
| $n$-$C_4H_9NH_2$ | 10.33 | 94.7 |
| $CH_3O(CH_2)_2NH_2$ | 9.09 | 15.8 |
| $CH_3O(CH_2)_3NH_2$ | 9.82 | 47.8 |
| $(CH_3)_3N^+(CH_2)_2NH_2$ | 6.7 | 0.9 |
| $(CH_3)_2N^+H(CH_2)_2NH_2$ | 6.71 | 797 |
| $(CH_3)_2N(CH_2)_2NH_2$ | 9.27 | 21 |
| $(CH_3)_2N^+H(CH_2)_3NH_2$ | 8.50 | 108 |
| $(CH_3)_2N(CH_2)_3NH_2$ | 9.75 | 51.5 |
| $(CH_3)_2N^+H(CH_2)_4NH_2$ | 9.23 | 69.4 |
| $(CH_3)_2N(CH_2)_4NH_2$ | 10.00 | 66.7 |
| $(CH_3)_2N^+H(CH_2)_5NH_2$ | 10.06 | 159 |
| $(CH_3)_2N(CH_2)_5NH_2$ | 10.34 | 84.4 |
| $(CH_3)_2N^+HC_5H_8CH_2NH_2$[a] | 8.33 | 388 |
| $(CH_3)_2NC_5H_8CH_2NH_2$[a] | 9.61 | 19 |

[a]The tertiary protonated and unprotonated forms of *trans*-2-[(dimethylamino)methyl]cyclopentylamine. Data from ref 21.

The rate constant is the second-order rate constant for the reaction between acetone and the primary amine noted in the left-hand column. The $pK_a$ is for the primary amino group. The primary amine must be unprotonated to react, but a diamine can be protonated or unprotonated at the other nitrogen.

(C) Make a Brønsted plot of all of the experimental results. Draw a line through the values for all of the neutral compounds and the quaternary ammonium cation and determine the slope of the line.

All of the cationic compounds, except the quaternary ammonium, deviate in their behavior from that expected from the $pK_a$ of their primary amino group.

(D) Explain this deviation in terms of intramolecular general acid catalysis of the formation of the iminium cation by drawing several structures, one for each of the cationic primary amines, that explicitly incorporate this catalytic step and that show the proton transfer unambiguously.

(E) Explain the two pH–rate profiles displayed in the figure.[21]

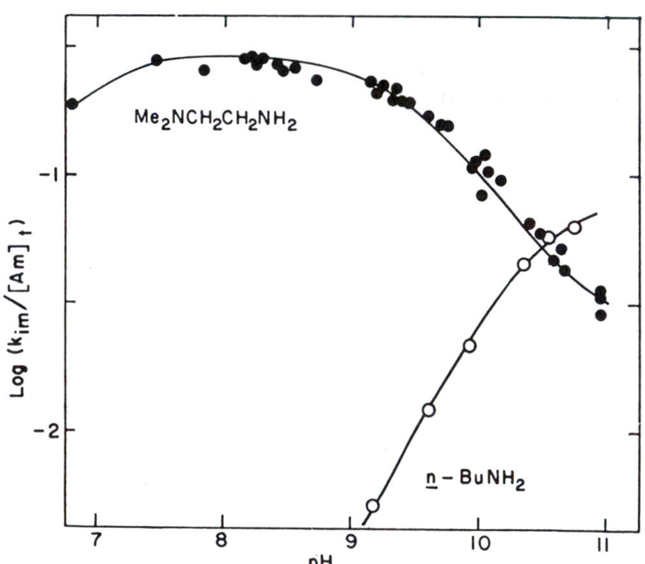


Plot of log ($k_{im}$/ $[Am]_t$) against pH for (●) 2-(dimethylamino)ethylamine and (○) $n$-butylamine. The rate constant $k_{im}$ is for the total rate of iminium cation formation from all forms of the amine, and $[Am]_t$ is the total concentration of primary amine in all forms. First explain the behavior of $n$-butylamine, and then provide a detailed explanation of the increase in rate, followed by the plateau, followed by the decrease in the curve for 2-(dimethylamino)ethylamine as the pH is decreased. Reprinted with permision from ref 21. Copyright 1973 American Chemical Society.

The difference in standard free energy of activation between two reactions is related to the rate constants by

$$\Delta\Delta G^{\circ\ddagger} = -RT\ln(k_1/k_2)$$

(F) Use the rate of the reaction, $k_{am}$, of the 5-(*N,N*-dimethylammonium)pentylamine, $Me_2N^+H(CH_2)_5NH_2$, as the example of the reaction showing the least intramolecular general acid catalysis. With the Brønsted relationship, correct the rate constants of all of the other reactions showing intramolecular general acid catalysis for the differences in the p$K_a$ values of their primary amino groups from that of $(CH_3)_2N^+H(CH_2)_5NH_2$ so that they can be compared directly to the rate constant for $(CH_3)_2N^+H(CH_2)_5NH_2$. Calculate $\Delta\Delta G^{o\ddagger}_{corr}$ for each of these corrected values relative to $(CH_3)_2N^+H(CH_2)_5NH_2$. Assume that entropy of approximation is responsible for all of this corrected difference in free energy of activation, $\Delta\Delta G^{o\ddagger}_{corr} = -T\Delta\Delta S^{o\ddagger}_{approx}$, and determine $\Delta\Delta S^{o\ddagger}_{approx}$ for $(CH_3)_2N^+H(CH_2)_5NH_2$, $(CH_3)_2N^+H(CH_2)_4NH_2$, $(CH_3)_2N^+H(CH_2)_3NH_2$, *cis*-$(CH_3)_2N^+HC_6H_{10}NH_2$, and $(CH_3)_2N^+H(CH_2)_2NH_2$.

(G) Comment on the trend in $\Delta\Delta S^{o\ddagger}_{approx}$.

**PROBLEM 4–2**

The hydrolysis of the methyl enol ether of cyclohexanone in water is catalyzed by general acids; the rate-limiting step is protonation at the $\alpha$ position of the enol ether

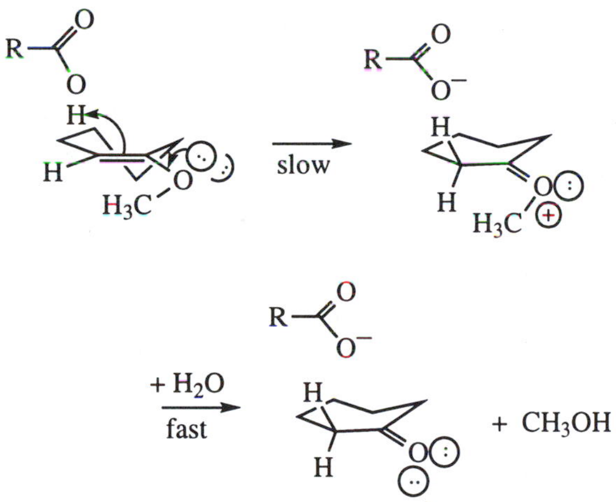

The Brønsted plot for catalysis of the hydrolysis of the methyl enol ether of cyclohexanone by carboxylic acids is strictly linear, except for the deviation of pivalic acid[22]

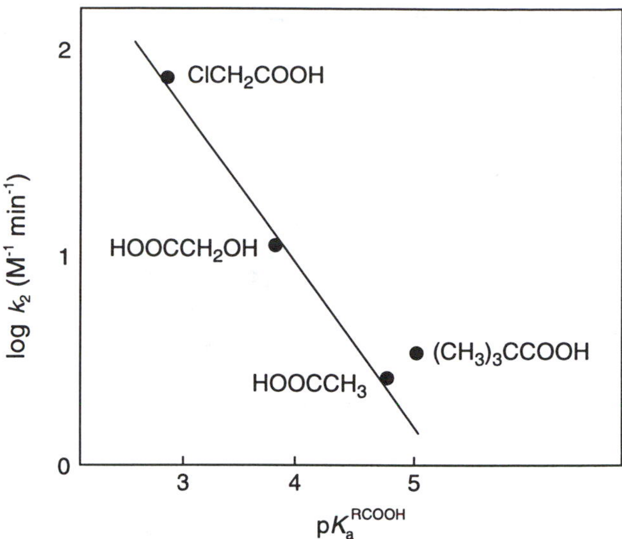


where

$$\frac{-d[\text{enol ether}]}{dt} = k_2[\text{enol ether}][\text{RCOOH}]$$

Suggest a hypothesis to explain this deviation of pivalic acid. How could you test your hypothesis?*

**PROBLEM 4–3**

The tautomerization of certain azomethines related to pyridoxal phosphate is catalyzed by aqueous imidazole buffers

$$R_1R_2C{=}NCHR_3R_4 \rightarrow R_1R_2CHN{=}CR_3R_4$$

The rate law for the reaction is

$$v = k[\text{Im}][\text{ImH}^+][\text{azomethine}]$$

where Im is imidazole base and $ImH^+$ is the conjugate acid. Write a mechanism consistent with the rate law.*

**PROBLEM 4–4**

The hydration of 1,3-dichloroacetone is catalyzed by imidazole buffers. The rate law for the reaction contains a term, $k[\text{Im}]^2[H_2O][ClCH_2COCH_2Cl]$, where Im is the free base of imidazole. Write a mechanism consistent with this term that does not involve addition of imidazole to the carbonyl group of the 1,3-dichloroacetone.*

# Enzymatic pH–Rate Profiles

The rates of all enzymatic reactions are sensitive to pH. The effects of pH on the activity of an enzyme, measured by the standard assay used to purify the enzyme, are theoretically meaningless, and the often-encountered pH–activity relationships have only practical value.[23] If, however, the effects of pH on the underlying parameters governing the initial velocity of an enzymatic reaction, namely the maximum initial velocity, $V_{max}$, or the turnover number, $k_{cat}$, and the Michaelis constants, $K_m$, are assessed, the observed behavior is often informative. Because the $V_{max}$ of an enzymatic reaction usually varies dramatically with the pH of the solution, different concentrations of enzyme are often used for the measurements at

*G.E. Lienhard, Department of Biochemistry, Dartmouth School of Medicine, formulated these problems.

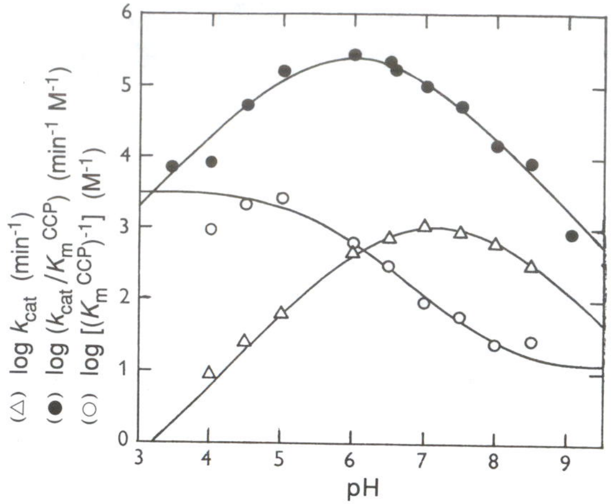


**Figure 4–8:** pH–Rate profiles for the hydrolysis of cytidine 2′,3′-cyclic phosphate catalyzed by pancreatic ribonuclease.[24] Solutions of cytidine 2′,3′-cyclic phosphate (25–30 $\mu$M) were prepared at various values of pH (4.0–9.0) buffered with sodium formate, sodium acetate, sodium malonate, disodium pentane-3,3-dicarboxylate, *N*-ethylmorpholinium chloride, or sodium borate at 25 °C and a final ionic strength of 0.2 M (maintained with sodium chloride). None of the buffers was found to affect the rate of the enzymatic reaction. The reaction was followed spectrophotometrically in these buffered solutions or with a pH stat in unbuffered solution. The reaction was initiated by adding enzyme, and initial velocities were measured over 1–2 min for each solution. Values for $V_{max}$ and $K_m^{CCP}$ were obtained for each pH from a set of measurements of initial velocity as a function of the concentration of cytidine 2′,3′-cyclic phosphate. The value of the turnover number, $k_{cat}$, was obtained at each pH by dividing $V_{max}$ by the molar concentration of enzyme in the solution for that particular set of measurements. The logarithms of $k_{cat}$ (△, minute$^{-1}$), $k_{cat}/K_m^{CCP}$ (●, minute$^{-1}$ molar$^{-1}$), and $(K_m^{CCP})^{-1}$ (○, millimolar$^{-1}$) are plotted as a function of pH. Adapted with permission from ref 24. Copyright 1962 Biochemical Society.

different values of pH to keep the initial velocities in a manageable range. Consequently, it is the turnover number of the enzyme, $k_{cat}$—which is maximum velocity, $V_{max}$, divided by the molar concentration of active sites, $[E]_{TOT}$—that is routinely used in such an analysis rather than $V_{max}$.

Pancreatic ribonuclease catalyzes the hydrolysis of cytidine 2′,3′-cyclic phosphate (Reaction 3–221). This is a simple reaction with one formal reactant, the cyclic phosphate ester. The Michaelis constant, $K_m^{CCP}$, for cytidine 2′,3′-cyclic phosphate and the maximum initial velocity of the reaction, $V_{max}$, can be determined at a given pH by running the reaction at various initial concentrations of cytidine 2′,3′-cyclic phosphate. The values for $K_m^{CCP}$ and $V_{max}$ for a series of different values of pH have been gathered.[24] The turnover numbers, $k_{cat}$, could be calculated from the various values of $V_{max}$ and $[E]_{TOT}$. The effects of pH on the reaction were evaluated by plotting the logarithms of $k_{cat}$, $K_m^{CCP}$, and $k_{cat}/K_m^{CCP}$ as functions of the pH (Figure 4–8).[24]

What was observed with ribonuclease is typical of what is observed with most enzymes. Below certain values of pH, the logarithms of $k_{cat}$ and $k_{cat}/K_m$, respectively, decrease as a linear function of pH with a slope of +1. Above certain higher values of pH, the logarithms of $k_{cat}$ and $k_{cat}/K_m$, respectively, decrease as a linear function of pH with a slope of –1. The logarithm of $K_m^{-1}$, however, is usually a complex function of pH showing several inflections.

The most reasonable explanation for the observation of lines with unit slopes in these pH–rate profiles is that only one general base or acid in the active site of the enzyme is titrating in these ranges of pH. If two or more general bases or acids that were essential for the enzymatic reaction were titrating in the range of pH examined, the slopes of the lines should have been greater than or equal to +2 or less than or equal to –2, respectively. If such a steep slope is observed in a pH–rate profile, however, one should suspect that it is only the unfolding of the enzyme at the extremes of pH that is being monitored, a reaction that characteristically displays a high order in either proton or hydroxide ion concentration.

In many enzymatic reactions, the logarithmic decreases on the two sides of the maximum in the plot of $k_{cat}$ or $k_{cat}/K_m$ against pH are separated by a plateau over which the value of the particular parameter is invariant with pH (Figure 4–9A).[25] The region of pH over which measurements of enzymatic activity can be made is usually confined by the unfolding of the protein at both high and low pH. Often the plateau extends beyond this range of permitted measurements and only one of the regions in which the value of the parameter decreases logarithmically with pH is observed (Figure 4–9B)[26] or no effect of pH on the parameter is observed. Whether or not the values of the parameters would eventually begin to decrease at higher or lower values of pH if the enzyme did not unfold is unknown. It is not necessary for $k_{cat}$ to decrease on one or both sides of its maximum if $k_{cat}/K_m$ decreases on both sides, and it is not necessary for $k_{cat}/K_m$ to decrease on one or both sides of its maximum if $k_{cat}$ decreases on both sides. For example, in the case of aspartate aminotransferase, $k_{cat}/K_m$ varies in the usual manner with pH while $k_{cat}$ is invariant with pH (Figure 4–10).[27]

From an examination of all these curves and others like them, several conclusions can be reached. If the value of either $k_{cat}$ or $k_{cat}/K_m$ decreases as the pH is decreased, there must be a base that participates directly in the enzymatic reaction or whose titration causes the polypeptide forming the active site to shift into an inactive conformation. In either case, when the conjugate acid of that base is formed by addition of a proton, the enzymatic reaction cannot proceed effectively. The region of the pH–rate profile where the value of the parameter decreases by a factor of 10 for every increase in the proton concentration of a factor of 10 must occur at values of pH less than the value of the p$K_a$ for the conjugate acid of this base. If the value of either $k_{cat}$ or $k_{cat}/K_m$ decreases as the pH is increased, there must be an acid that participates directly in the enzymatic reaction or whose titration causes the polypeptide forming the active site to shift into an inactive conformation. In either case, when the conjugate base of that acid is formed by the removal of a proton, the enzymatic reaction

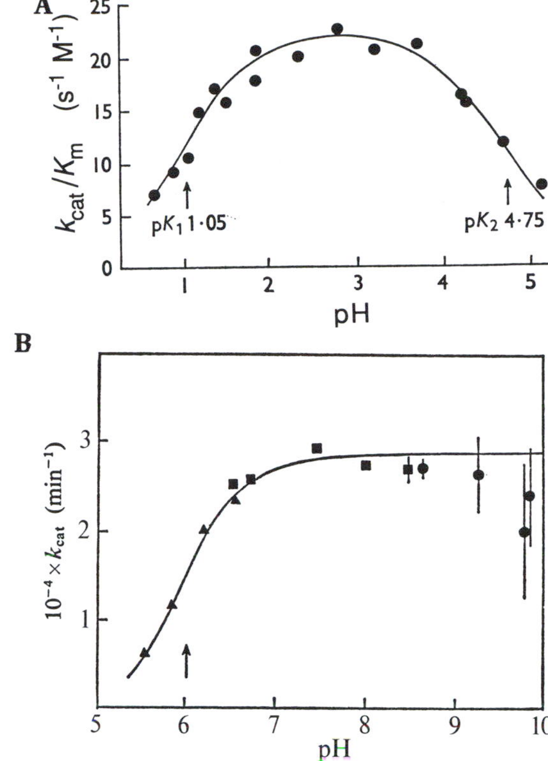


**Figure 4–9:** pH–Rate profiles for the reactions catalyzed by pepsin (A) and triosephosphate isomerase (B). (A) Effect of pH on the parameter $k_{cat}/K_m$ for the hydrolysis of *N*-acetyl-L-phenylalanyl-L-phenylalanyl amide at 37 °C.[25] Solutions of the dipeptide were prepared in 0.1 M sodium citrate and brought to the noted pH with HCl. The reaction was initiated by the addition of enzyme, and the initial velocity was monitored by the appearance of ninhydrin-positive amine. Initial velocities were estimated by eye from plots of $A_{570}$ as a function of time. Values of $k_{cat}$ and $K_m$ were obtained from the behavior of the initial velocity as a function of the concentration of dipeptide at a given pH and the molar concentration of active sites. The parameter $k_{cat}/K_m$ (second$^{-1}$ molar$^{-1}$) is plotted as a function of pH. Values of 1.05 and 4.75 were obtained for the two apparent p$K_a$ values governing the behavior. (B) Effect of pH on the parameter $k_{cat}$ for the isomerization of dihydroxyacetone phosphate by triosephosphate isomerase at 30 °C.[26] Solutions of dihydroxyacetone phosphate were prepared at 0.1 M ionic strength (maintained with NaCl), and the pH was buffered at the noted values with sodium 2,2-dimethylglutarate (▲), triethanolammonium chloride (■), or sodium carbonate (●). The reaction was initiated by adding enzyme, and the initial velocity was monitored by coupling the production of glyceraldehyde 3-phosphate to the oxidation of NADH (loss of $A_{340}$) by glyceraldehyde-3-phosphate dehydrogenase. From the initial velocities at a particular pH and molar concentration of enzyme, values of $k_{cat}$ for each value of pH could be obtained. These values of $k_{cat}$ (minute$^{-1}$ × $10^{-4}$) are plotted as a function of pH. The arrow marks an apparent p$K_a$ of 6.0. Adapted with permission from refs 25 and 26. Copyright 1969 and 1972 Biochemical Society.

cannot proceed effectively. The region in which the value of the parameter decreases by a factor of 10 for every decrease in the proton concentration of a factor of 10 must occur at values of pH greater than the value of the p$K_a$ for this acid. Because the inflections in the curve for $k_{cat}$ usually occur at different values of pH than the inflections for the curve of $k_{cat}/K_m$ (Figures 4–8 and 4–10), either the acids and bases involved in controlling these parameters are different, or the values of the p$K_a$ for the same acids and bases have changed during the reaction, or the effects of these acids and bases on these parameters differ in their magnitude, or these inflections result from a change in the rate-limiting step of the enzymatic reaction.

It was originally noted[28–30] that the behavior of each of the kinetic parameters governing the initial velocity of an enzymatic reaction as a function of pH resembles the behavior of the molar concentration of one of the individual ionization states of a polyprotic acid as a function of pH. Because the active site of an enzyme is a polyprotic acid, the pH–rate profile of an enzyme can be explained if it is assumed that only one of the ionization states of its active site is competent for catalysis.

Suppose, as is observed crystallographically, that the active site of an enzyme contains several acid–bases. If so, an active site, as a polyprotic acid, must display several macroscopic acid dissociation constants, each successively defining the loss of a proton

$$EH_n \overset{K_{a1}^{E}}{\rightleftharpoons} EH_{(n-1)} + H^+ \overset{K_{a2}^{E}}{\rightleftharpoons} \cdots \overset{K_{a(n-1)}^{E}}{\rightleftharpoons} EH + (n-1)H^+ \overset{K_{an}^{E}}{\rightleftharpoons} E + nH^+ \tag{4–48}$$

where the protons considered are only those that originate on acids within the active site. Because the slopes at which the logarithms decrease on either side of the maxima in the observed behavior of enzymatic reactions are usually either +1 or −1 (Figure 4–8), at most only two of these ionizations must be important to the enzymatic reaction in the measurable range. These would be the loss of one proton from an inactive form of the active site to yield the active form of the active site and the loss of the next proton from the active form of the active site to yield an inactive form of the active site

$$EH_{(n-m+1)} + (m-1)H^+ \overset{K_{am}^{E}}{\rightleftharpoons} EH_{(n-m)} + mH^+ \overset{K_{a(m+1)}^{E}}{\rightleftharpoons} EH_{(n-m-1)} + (m+1)H^+ \tag{4–49}$$

where $EH_{(n-m)}$ is the active form of the active site. If only these three states are present at significant concentrations in the range of pH chosen

$$[EH_{(n-m)}] = \alpha_m^{E}[E]_{TOT} \tag{4–50}$$

where $[E]_{TOT}$ is the total concentration of the three forms of the free enzyme directly coupled by the two ionizations

$$[E]_{TOT} = [EH_{(n-m+1)}] + [EH_{(n-m)}] + [EH_{(n-m-1)}] \tag{4–51}$$

and

$$\alpha_m^{E} = \left( \frac{[H^+]}{K_{am}^{E}} + 1 + \frac{K_{a(m+1)}^{E}}{[H^+]} \right)^{-1} \tag{4–52}$$

Again, the term $\alpha_m^{E}$ is the fraction of the enzyme in these three forms present as the form $EH_{(n-m)}$ at any given pH.

If it is assumed[30] that the kinetic mechanism of a particular enzymatic reaction is

$$\mathrm{EH}_{(n-m+1)} \overset{K_{am}{}^{\mathrm{E}}}{\rightleftharpoons} \mathrm{EH}_{(n-m)} + \mathrm{H}^+ \overset{K^{\mathrm{E}}_{\mathrm{a}(m+1)}}{\rightleftharpoons} \mathrm{EH}_{(n-m-1)} + 2\mathrm{H}^+$$

$$k_{-1} \Big\uparrow\Big\downarrow k_1[\mathrm{A}]$$

$$\mathrm{A{\cdot}EH}_{(n-m+1)} \overset{K_{am}{}^{\mathrm{EA}}}{\rightleftharpoons} \mathrm{A{\cdot}EH}_{(n-m)} + \mathrm{H}^+ \overset{K^{\mathrm{EA}}_{\mathrm{a}(m+1)}}{\rightleftharpoons} \mathrm{A{\cdot}EH}_{(n-m-1)} + 2\mathrm{H}^+$$

$$\Big\downarrow k_2$$

$$\mathrm{E+P} \tag{4–53}$$

it can be shown, by solving for $v_0$, that

$$v_0 = \frac{V_{\max}[\mathrm{A}]_0}{[\mathrm{A}]_0 + K_{\mathrm{m}}{}^{\mathrm{A}}} \tag{4–54}$$

In Equation 4–54

$$V_{\max} = k_2 \alpha_m{}^{\mathrm{EA}}[\mathrm{E}]_{\mathrm{TOT}} \tag{4–55}$$

and

$$K_{\mathrm{m}}{}^{\mathrm{A}} = \left(\frac{k_{-1}+k_2}{k_1}\right)\left(\frac{\alpha_m{}^{\mathrm{EA}}}{\alpha_m{}^{\mathrm{E}}}\right) \tag{4–56}$$

The parameter $[\mathrm{E}]_{\mathrm{TOT}}$ is the concentration of the enzyme in all six of its forms, and $\alpha_m{}^{\mathrm{EA}}$ is the fraction of the complex between enzyme and reactant in all of the three forms that has $m$ protons in its active site

$$\alpha_m{}^{\mathrm{EA}} = \left(\frac{[\mathrm{H}^+]}{K_{am}{}^{\mathrm{EA}}} + 1 + \frac{K^{\mathrm{EA}}_{\mathrm{a}(m+1)}}{[\mathrm{H}^+]}\right) \tag{4–57}$$

In this specific kinetic mechanism, it has been assumed that the acid dissociations are so rapid as to be always at equilibrium rather than at steady state, that the only form of the enzyme capable of binding reactant is $\mathrm{EH}_{(n-m)}$, and that the only form of the enzyme capable of yielding product is $\mathrm{A{\cdot}EH}_{(n-m)}$. In other words, "there is only one state of ionization of the active site that is capable of catalyzing the interconversion of [reactant] and product."[23] If these assumptions are accurate, then the values of the macroscopic acid dissociation constants for the active site of the free enzyme can be obtained from the behavior of $k_{\mathrm{cat}}/K_{\mathrm{m}}$ as a function of pH and those for the complex between enzyme and reactant can be obtained from the behavior of $k_{\mathrm{cat}}$ as a function of pH, respectively.

Consider $k_{\mathrm{cat}}$ (Figure 4–11).[31] If Mechanism 4–53 describes the actual situation, then

$$\log k_{\mathrm{cat}} = \log \alpha_m{}^{\mathrm{EA}} + \log k_2 \tag{4–58}$$

This will be the sum of a constant term, $\log k_2$, and a term that varies with pH, $\log \alpha_m{}^{\mathrm{EA}}$. The logarithm of $\alpha_m{}^{\mathrm{EA}}$ varies with pH because

$$\log \alpha_m{}^{\mathrm{EA}} = -\log\left(\frac{[\mathrm{H}^+]}{K_{am}{}^{\mathrm{EA}}} + 1 + \frac{K^{\mathrm{EA}}_{\mathrm{a}(m+1)}}{[\mathrm{H}^+]}\right) \tag{4–59}$$

If $K_{am}{}^{\mathrm{EA}} \gg K^{\mathrm{EA}}_{\mathrm{a}(m+1)}$, then when $[\mathrm{H}^+] > K_{am}{}^{\mathrm{EA}}$

$$\log \alpha_m{}^{\mathrm{EA}} = \mathrm{pH} - \mathrm{p}K_{am}{}^{\mathrm{EA}} \tag{4–60}$$

$$\log k_{\mathrm{cat}} = \mathrm{pH} - \mathrm{p}K_{am}{}^{\mathrm{EA}} + \log k_2 \tag{4–61}$$

and $\log k_{\mathrm{cat}}$ increases as a function of pH with a slope of +1. When $K_{am}{}^{\mathrm{EA}} > [\mathrm{H}^+] > K^{\mathrm{EA}}_{\mathrm{a}(m+1)}$

$$\log \alpha_m{}^{\mathrm{EA}} = 0 \tag{4–62}$$

and

$$\log k_{\mathrm{cat}} = \log k_2 \tag{4–63}$$

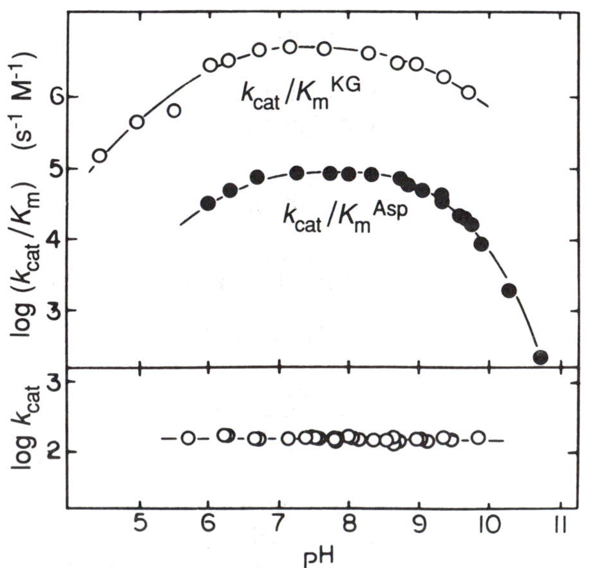


**Figure 4–10:** pH–Rate profiles for the transamination of aspartate and 2-ketoglutarate catalyzed by aspartate aminotransferase.[27] Solutions of aspartate and 2-ketoglutarate at various concentrations were prepared at 25 °C at values of pH buffered with *N*,*N*′-bis(2-hydroxyethyl)piperazine, 2-(*N*-morpholino)ethanesulfonic acid (MES), *N*-(2-hydroxyethyl)piperazine, 3-[[tris(hydroxymethyl)methyl]amino]propanesulfonic acid (TAPS), 2-(*N*-cyclohexylamino)ethanesulfonic acid (CHES), or 3-(cyclohexylamino)-1-propanesulfonic acid (CAPS). The reaction was initiated by adding enzyme, and the initial velocities were monitored by coupling the production of oxaloacetate to the oxidation of NADH (decrease in $A_{340}$) with malate dehydrogenase. From a set of initial velocities at a given pH as a function of the concentrations of both aspartate and 2-ketoglutarate and from a knowledge of the molar concentration of the enzyme, values of the turnover number, $k_{\mathrm{cat}}$, the Michaelis constant for aspartate, $K_{\mathrm{m}}{}^{\mathrm{asp}}$, and the Michaelis constant for 2-ketoglutarate, $K_{\mathrm{m}}{}^{\mathrm{KG}}$ (Equation 3–71) could be determined. The logarithm of $k_{\mathrm{cat}}/K_{\mathrm{m}}{}^{\mathrm{Asp}}$ (●, second$^{-1}$ molar$^{-1}$) is plotted as a function of pH; the solid line is calculated for apparent p$K_{\mathrm{a}}$ values of 6.4 and 9.2. The logarithm of $k_{\mathrm{cat}}/K_{\mathrm{m}}{}^{\mathrm{KG}}$ (○, second$^{-1}$ molar$^{-1}$) is plotted as a function of pH; the solid line is calculated for apparent p$K_{\mathrm{a}}$ values of 5.8 and 9.2. The logarithm of $k_{\mathrm{cat}}$ (second$^{-1}$) is invariant with pH. Adapted with permission from ref 27. Copyright 1983 American Chemical Society.

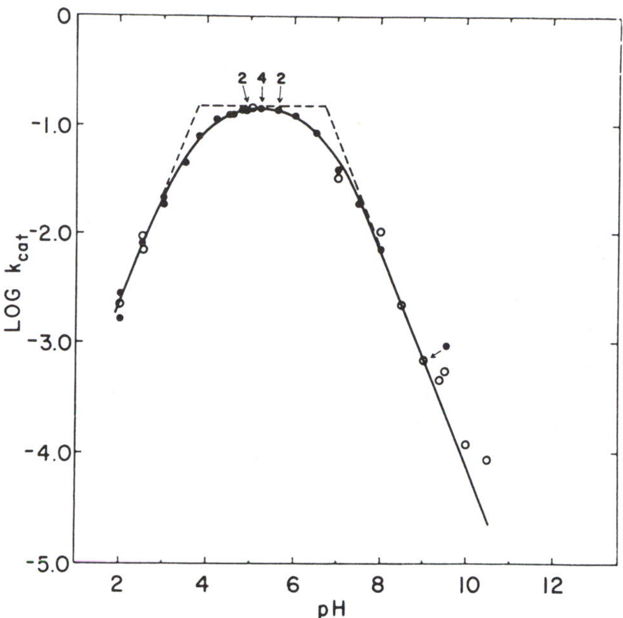


**Figure 4–11:** pH–Rate profile for the turnover number, $k_{cat}$ (second$^{-1}$), for the hydrolysis of ($\beta$1,4)hexa-$N$-acetylglucosamine (Figure 3–25) by lysozyme.[31] Solutions of the ($\beta$1,4)hexasaccharide of $N$-acetylglucosamine were prepared at 40 °C and 0.1 M ionic strength (maintained with NaCl) and at various pH values buffered with sodium acetate, sodium phosphate, or sodium carbonate. The hydrolysis of the hexasaccharide was followed either by monitoring the increase in reducing sugar with $K_3(Fe(CN)_6)$ or by separating the products of the reaction chromatographically. From measurements of the initial velocity of the reaction at a series of concentrations of the hexasaccharide at a constant concentration of enzyme and a constant pH, values for $V_{max}$ and $K_m$ at that pH could be obtained. The values of $k_{cat}$ were calculated from values of $V_{max}$ and $[E]_{TOT}$. The logarithm of $k_{cat}$ (second$^{-1}$) is presented as a function of pH. The two intersections of the three asymptotes dashed lines) define apparent p$K_a$ values of 3.8 and 6.7. Reprinted with permission from ref 31. Copyright 1973 *Journal of Biological Chemistry.*

When $K_{a(m+1)}^{EA} > [H^+]$

$$\log \alpha_m{}^{EA} = pK_{a(m+1)}^{EA} - pH \tag{4–64}$$

$$\log k_{cat} = pK_{a(m+1)}^{EA} - pH + \log k_2 \tag{4–65}$$

and log $k_{cat}$ decreases as a function of pH with a slope of –1. If Mechanism 4–53 describes the actual situation, Equations 4–61, 4–63, and 4–65 define the asymptotes (dashed lines in Figure 4–11) of log $k_{cat}$ as a function of pH. By solving the two pairs of these three simultaneous equations (Equations 4–61 and 4–63 and Equations 4–63 and 4–65), it is easily shown that the line defined by Equation 4–61 intersects the line defined by Equation 4–63 when pH equals p$K_{am}{}^{EA}$ and the line defined by Equation 4–63 intersects the line defined by Equation 4–65 when pH equals p$K_{a(m+1)}^{EA}$.

If Mechanism 4–53 describes the actual situation

$$\log\left(k_{cat}/K_m\right) = \log \alpha_m{}^{E} + \log\left(\frac{k_1 k_2}{k_{-1} + k_2}\right) \tag{4–66}$$

which is formally equivalent to Equation 4–58. It follows that if $K_{am}{}^{E} >> K_{a(m+1)}^{E}$, there will be three asymptotes, one of slope +1, one of slope 0, and one of slope –1, and the two points of intersection of these three lines will occur where pH equals p$K_{am}{}^{E}$ and pH equals p$K_{a(m+1)}^{E}$. Therefore, if Mechanism 4–53 did describe the actual situation and if the two successive values in each pair of acid dissociation constants, $K_{am}{}^{E}$ and $K_{a(m+1)}^{E}$ and $K_{am}{}^{EA}$ and $K_{a(m+1)}^{EA}$, are sufficiently different from each other, the values for the macroscopic acid dissociation constants of the free enzyme or those of the complex between enzyme and reactant could be read from the respective graph.

The argument so far has assumed that Mechanism 4–53 is the actual mechanism of the enzymatic reaction, which is unlikely in most cases.[32] To be safe, in the absence of independent verification, it is best to refer to the observed intersections of the asymptotes as apparent p$K_a$ values. These are generally identified as p$K_{a1}{}^{E}$, p$K_{a2}{}^{E}$, p$K_{a1}{}^{EA}$, and p$K_{a2}{}^{EA}$. In practice, the intersections of the asymptotes in a plot of log $k_{cat}$ against pH, such as the one displayed in Figure 4–11, should be referred to as apparent acid dissociation constants of the complex between enzyme and reactant, p$K_{a1}{}^{EA}$ and p$K_{a2}{}^{EA}$. The intersections of the asymptotes in a plot of log ($k_{cat}/K_m$) against pH, such as those in Figure 4–10, should be referred to as apparent acid dissociation constants of the free enzyme, p$K_{a1}{}^{E}$ and p$K_{a2}{}^{E}$. If the two successive values of any pair of apparent dissociation constants are not sufficiently different from each other, then no plateau occurs in the curves (Figure 4–8) and the apparent acid dissociation constants must be estimated by numerical analysis.[30]

Treatments of rate data by these analytical algorithms to obtain estimates of the actual values of the macroscopic acid dissociation constants for free enzyme and the complex between enzyme and reactant, respectively, always assume, either explicitly or tacitly, that Mechanism 4–53 describes the enzymatic reaction. There is also one simple variant of Mechanism 4–53 that leads to the same expression of $v_0$, namely

$$\begin{array}{ccccc}
EH_{(n-m+1)} & \underset{}{\overset{K_{am}{}^{E}}{\rightleftharpoons}} & EH_{(n-m)} & \overset{K_{a(m+1)}^{E}}{\rightleftharpoons} & EH_{(n-m-1)} \\
K_1 \updownarrow & & K_2 \updownarrow & & K_3 \updownarrow \\
A{\cdot}EH_{(n-m+1)} & \overset{K_{am}{}^{EA}}{\rightleftharpoons} & A{\cdot}EH_{(n-m)} & \overset{K_{a(m+1)}^{EA}}{\rightleftharpoons} & A{\cdot}EH_{(n-m-1)} \\
 & & \downarrow k_2 & &
\end{array} \tag{4–67}$$

In this mechanism the rate constant for the catalytic step $k_2$ is slow and all other rate constants are so fast that everything else is in equilibrium. In this case the intersections of the asymptotes also define values for $pK_a^{E}$ and $pK_a^{EA}$.

It has been pointed out, however, that both Mechanisms 4–53 and 4–67 are too simple to describe the kinetic mechanisms of most if not all enzymes[23] and that any increase in the complexity of the mechanism causes $k_{cat}$ and $k_{cat}/K_m$ no longer to be simple products of $\alpha_m^{EA}$ and $\alpha_m^{E}$, respectively.[32] Therefore, in most cases, the intersections of the asymptotes are not values for $pK_a^{EA}$ and $pK_a^{E}$, respectively. Any statement that they are must be accompanied by independent evidence of this claim. The situation, however, is nothing more serious than an instance of what is true of all kinetic mechanisms: they may be consistent with the observations but cannot be proven by the observations. If, however, the kinetic mechanism of a particular enzymatic reaction has been established independently and it differs from Mechanisms 4–53 and 4–67, then the meaning of the intersections of the asymptotes, whether or not they define the macroscopic acid dissociation constants, can be stated explicitly because an equation for the initial velocity can be derived by the usual formalism.[32]

To illustrate the fact that there is no requirement for the apparent $pK_a$ values to be the actual macroscopic $pK_a$ values for an active site, a simple example of a situation in which an apparent $pK_a$ is not an actual macroscopic $pK_a$ can be formulated.[33] Suppose that an enzymatic reaction has two steps following the binding of reactants, a slow, pH-independent step followed by a step that requires an unprotonated base. Suppose also that at high pH, above the $pK_a$ of the conjugate acid of the base, the second step is much faster than the first. As the pH is lowered, flux through the two steps would remain constant even as the pH dropped below the actual $pK_a$ of the base because the second step would remain faster than the first. Below a certain value of pH, however, there would be a change in the rate-limiting step as the concentration of unprotonated base became so small that the rate of the second step necessarily became slower than the first. Below this value of pH, the overall rate, now limited by the second step rather than the first, would decrease by a factor of 10 for every decrease of 1 unit in pH. Therefore, the pH at which the change in the rate-limiting step occurs only appears to define one of the acid dissociation constants of the active site; the actual value of the $pK_a$ of the conjugate acid of the base controlling the rate of the second step is larger than the pH at which the asymptotes intersect. Therefore, the apparent values of $pK_{a1}^{EA}$ and $pK_{a2}^{EA}$, respectively, are always less than or equal to the actual macroscopic $pK_a$ values for the unoccupied or occupied active site. An argument similar to that just made for a base can be made for an acid in an active site. The conclusion drawn from this argument is that the apparent values of $pK_{a2}^{E}$ and $pK_{a2}^{EA}$ are always greater than or equal to the actual values for these constants. These two conclusions are valid in all situations and can be related as four inequalities: $pK_{a1}^{E} \le pK_{am}^{E}$, $pK_{a1}^{EA} \le pK_{am}^{EA}$, $pK_{a2}^{E} \ge pK^{E}_{a(m+1)}$, and $pK_{a2}^{EA} \ge pK^{EA}_{a(m+1)}$. In these inequalities $pK_{a1}$ and $pK_{a2}$ are the observed parameters (Figures 4–10 and 4–11), and $pK_{am}$ and $pK_{a(m+1)}$ are the actual macroscopic $pK_a$ values for the active site (Equation 4–48).

Aside from the indiscriminate and unsubstantiated use of pH–rate profiles to gather what only appear to be macroscopic $pK_a$ values of the active site, there are examples of more informative uses of pH–rate profiles. Perhaps the least equivocal examples are cases in which an apparent $pK_a$ obtained from kinetic measurements can be correlated with a $pK_a$ measured directly. The two amino acids Glutamate 35 and Aspartate 52 cause the active site of lysozyme (Figure 3–26) to be at least a dicarboxylic acid. By difference titration between the free enzyme and the enzyme esterified at Aspartate 52, the two successive macroscopic acid dissociation constants, $pK_{a1}^{E}$ and $pK_{a2}^{E}$, for the two ionizations

$$EH_2 \underset{}{\overset{K_{a1}}{\rightleftharpoons}} EH + H^+ \underset{}{\overset{K_{a2}}{\rightleftharpoons}} E + 2H^+ \tag{4–68}$$

have been found to be 4.3 and 6.0, respectively, at 40 °C in 0.15 M KCl.[34] When the behavior of the logarithm of $k_{cat}/K_m$ as a function of pH at 40 °C in 0.1 M NaCl was determined independently,[31] it was found to be governed by two apparent macroscopic acid dissociation constants with $pK_{a1}^{E} = 4.2$ and $pK_{a2}^{E} = 6.1$. On the basis of this correlation and the patent involvement of these two amino acids in the active site of the crystallographic molecular model of the enzyme, it can be concluded that the apparent $pK_a$ values are actual macroscopic $pK_a$ values of the active site and that the free enzyme must be in the form in which only one proton is occupying these two carboxylic acids to be capable of binding the reactant in a productive manner.

The example of the carboxylates in the active site of lysozyme presents another opportunity to distinguish between macroscopic and microscopic acid dissociation constants.[35] The macroscopic dissociation constants for the dicarboxylic acid that is in the active site of lysozyme refer only to the total number of protons on each species, not to their individual locations. In the particular case of the acitve site of lysozyme

EH⁻ (35: $CO_2^-$; 52: $CO_2H$)

HEH (35: $CO_2H$; 52: $CO_2H$) $\xrightleftharpoons{K_a^{35}}$ EH⁻ $\xrightleftharpoons{K_a^{52-}}$ $E^{2-}$ (35: $CO_2^-$; 52: $CO_2^-$)

HEH $\xrightleftharpoons{K_a^{52}}$ HE⁻ $\xrightleftharpoons{K_a^{35-}}$ $E^{2-}$

EH⁻ $\xrightleftharpoons{K_T}$ HE⁻

HE⁻ (35: $CO_2H$; 52: $CO_2^-$)

(4–69)

there are two tautomers, $EH^-$ and $HE^-$, of the state with only one proton. The form $EH^-$ has the proton on Aspartate 52 and the form $HE^-$ has the proton on Glutamate 35. The two macroscopic acid dissociation constants are defined as

$$K_{a1}^{E} = \frac{[H^+]([HE^-]+[EH^-])}{[HEH]} \tag{4–70}$$

and

$$K_{a2}{}^{E} = \frac{[H^+][E^{2-}]}{[HE^-]+[EH^-]} \quad (4\text{–}71)$$

The microscopic dissociation constants are

$$K_a{}^{35} = \frac{[H^+][EH^-]}{[HEH]} \quad (4\text{–}72a) \qquad K_a{}^{52} = \frac{[H^+][HE^-]}{[HEH]} \quad (4\text{–}72b)$$

$$K_a{}^{52-} = \frac{[H^+][E^{2-}]}{[EH^-]} \quad (4\text{–}73a) \qquad K_a{}^{35-} = \frac{[H^+][E^{2-}]}{[HE^-]} \quad (4\text{–}73b)$$

Each microscopic constant deals with a specific dissociation from a specific location in a particular ionization state of the active site. The relationship between the macroscopic and microscopic constants are

$$K_{a1}{}^{E} = K_a{}^{35} + K_a{}^{52} \quad (4\text{–}74)$$

and

$$\frac{1}{K_{a2}{}^{E}} = \frac{1}{K_a{}^{35-}} + \frac{1}{K_a{}^{52-}} \quad (4\text{–}75)$$

The macroscopic constants, responding only to the number of protons bound, govern any measurement in which the individual acid–bases are not examined separately. Such measurements would be acid–base titrations, pH–rate profiles of kinetic parameters, or pH effects on binding of ligands. The microscopic constants govern the molecular events more fundamentally, but they can be measured only indirectly. For example, when Aspartate 52 in the active site of lysozyme was esterified, $K_a{}^{35}$ in the unesterified enzyme could be measured by a difference titration because Aspartate 52 could not ionize.[36] From these results, it was estimated that in the unesterified native enzyme, $pK_a{}^{35} = 5.2$, $pK_a{}^{35-} = 6.0$, $pK_a{}^{52} = 4.4$, and $pK_a{}^{52-} = 5.2$ at 25 °C and 0.15 M KCl. These should be compared with $pK_{a1}{}^{E} = 4.3$ and $pK_{a2}{}^{E} = 6.1$. The first macroscopic constant is close to $pK_a{}^{52}$ and the second is close to $pK_a{}^{35-}$ as expected. Also as expected from the proximity of these two amino acids, the ionization of one raises the microscopic $pK_a$ of the other.

In Mechanism 4–69, the form of the active site containing only one proton has two states: one in which the proton is on Glutamate 35 and one in which the proton is on Aspartate 52. No pH–rate profile, because it detects only macroscopic acid dissociation constants, can distinguish between these two tautomers. It is undoubtedly the case that only one of the two tautomers is enzymatically active, and the fraction of the molecules that exist in this tautomer will determine the ultimate enzymatic activity. Because the equilibrium constant, $K_T$, relating the two tautomers cannot be a function of pH, the ratio between the two tautomers does not vary with pH.

The dissociation of the proton from the aldimine nitrogen on the lysylpyridoximine in the active site of aspartate aminotransferase

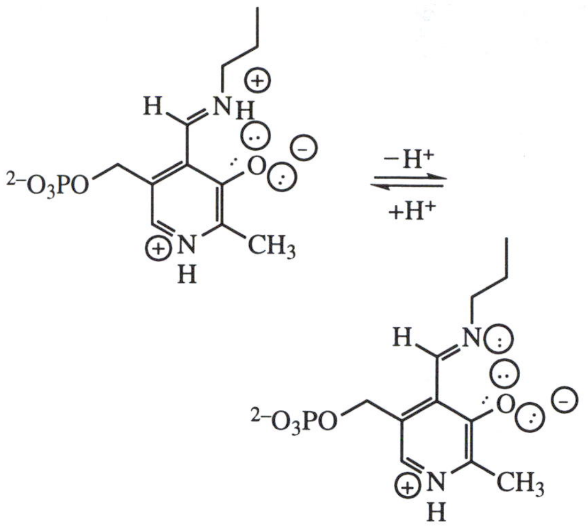

(4–76)

can be assessed by following the absorbance of the holoenzyme, which shifts to lower wavelengths as the dipolar $\pi$ system is shortened upon departure of the proton.[37] The $pK_a$ for this acid dissociation (Reaction 4–76) at 25 °C is 6.3. In the initial velocity kinetics of aspartate aminotransferase,[27] an inflection is observed in the plot of log $(k_{cat}/K_m{}^{Asp})$ as a function of pH (Figure 4–10) that is consistent with a $pK_a{}^{E}$ of 6.4. Above pH 6.4, log $(k_{cat}/K_m)$ is invariant with pH; and, below pH 6.4, it decreases with decreasing pH. If this $pK_a{}^{E}$ has been assigned correctly to the ionization of the coenzyme (Reaction 4–76) on the basis of the identical numerical values, then when the conjugate acid of the coenzyme is present, the active site cannot bind the reactant productively, but when the conjugate base is present it can. A chemical explanation for this observation[27] would be that, because aspartate enters the active site with its $\alpha$ amino group protonated, the first step in the reaction would be for the phenolate of the coenzyme to remove that proton (Figure 2–2). This would require that the phenolate be free of the hydrogen bond to the iminium nitrogen and available to act as a base.

The kinetic parameters defining the initial velocity of triosephosphate isomerase display inflections at both high and low pH.[26] The plot of log $k_{cat}/K_m{}^{G3P}$ for the initial velocity of the conversion of D-glyceraldehyde 3-phosphate to dihydroxyacetone phosphate displays an apparent $pK_a$ of 6.0 at the acidic side of the maximum. The $pK_a$ for the acid dissociation of the phosphate in the reactant itself

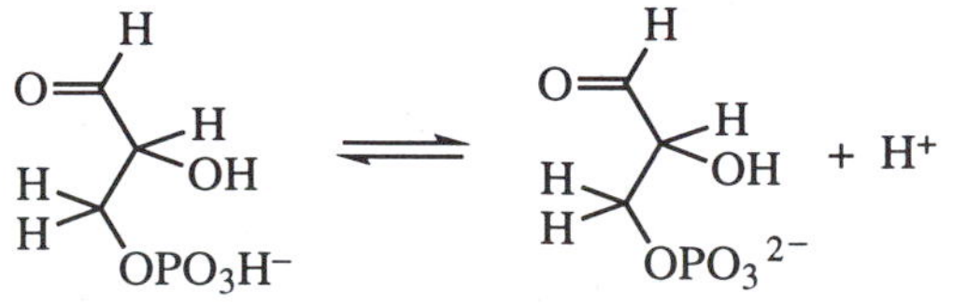

(4–77)

is 6.3. The correspondence of these two numerical values suggested that $k_{cat}/K_m{}^{G3P}$ is reflecting only the acid dissociation of the reactant and not an acid dissociation of the enzyme.

Suppose that an enzyme binds only one ionization state of a reactant and, once the reactant is bound, it is held in the active site at the particular acid–base responsible for that ionization so intimately that the bound reactant does not have an acid dissociation constant within the range of measurement. If this is the situation, then

$$\begin{array}{l} E + AH_{(n-m+1)} \\ \quad \Big\Updownarrow K_{am}^{A} \\ E + AH_{(n-m)} \underset{k_{-1}}{\overset{k_1}{\rightleftharpoons}} E{\cdot}AH_{(n-m)} \xrightarrow{k_2} \\ \quad \Big\Updownarrow K_{a(m+1)}^{A} \\ E + AH_{(n-m-1)} \end{array} \qquad (4\text{–}78)$$

If, for the sake of simplicity, only two ionizations are considered, then the initial concentration of the only form of the reactant recognized by the enzyme, namely $AH_{(n-m)}$, will be

$$[AH_{(n-m)}]_0 = \alpha_m{}^{A}[A]_{0,TOT} \qquad (4\text{–}79)$$

where $\alpha_m{}^{A}$ is the fraction of the reactant in the proper state and $[A]_{0,TOT}$ is the sum of the initial concentrations of the reactants in all states. Because the initial velocities of most enzymatic reactions satisfy the same equation

$$v_0 = \frac{V_{max}[AH_{(n-m)}]_0}{[AH_{(n-m)}]_0 + K_m} \qquad (4\text{–}80)$$

it necessarily follows that

$$v_0 = \frac{V_{max}[A]_{0,TOT}}{[A]_{0,TOT} + \dfrac{K_m}{\alpha_m{}^{A}}} \qquad (4\text{–}81)$$

The apparent Michaelis constant, $K_m{}^{app}$, is equal to $K_m/\alpha_m{}^{A}$, and it is only this apparent Michaelis constant that will be the parameter derived from the experimental observations. If so, then

$$k_{cat}/K_m{}^{app} = \alpha_m{}^{A}k_{cat}/K_m \qquad (4\text{–}82)$$

If $K_m$ and $k_{cat}$ themselves are invariant with pH within the range of measurement, $k_{cat}/K_m{}^{app}$ will reflect only the acid dissociation constants of the free reactant, not those of the free enzyme.

It was possible that the apparent acid dissociation constant of $pK_a = 6.0$ governing the behavior of $k_{cat}/K_m{}^{G3P}$ in the reaction catalyzed by triosephosphate isomerase was that for the acid dissociation of the phosphate of the glyceraldehyde 3-phosphate rather than that for the free enzyme. That it is the ionization of the reactant was shown in two ways.[38] 2-Hydroxy-4-phosphonobutyraldehyde (**4–15**)

**4–15**

is a substrate for triosephosphate isomerase, but the $pK_a$ of the phosphonate is 7.6 rather than 6.3. The $pK_a$ governing $k_{cat}/K_m$ for 2-hydroxy-4-phosphonobutyraldehyde is 7.5 rather than 6.0. Therefore, the apparent $pK_a$ governing the kinetics shifts as expected when one reactant is substituted with another. If the apparent $pK_a$ observed in the normal behavior of $k_{cat}/K_m$ is that for the phosphate of the reactant, then it must be the dianions (Equation 4–77) that are the only forms of the two substrates, glyceraldehyde and butyraldehyde, bound by the enzyme, and the monoanions must be ignored by it. If this is the case, dihydroxyacetone sulfate, which can exist only as the monoanion, should not be a reactant in the reverse reaction and it should not be an inhibitor of the forward reaction. It is neither.

If the acid dissociation constants for the reactant are known and the form of the reactant that binds to the enzyme can be determined independently, as in these last two experiments, $\alpha_m{}^{A}$ can be calculated for each value of pH and $K_m$ can be calculated from the observed Michaelis constant, $K_m{}^{app}$, for each value of pH (Equation 4–82). This eliminates the acid dissociation constants of the reactant from the data and leaves only the acid dissociation constants of the enzyme. Because the effect of the acid dissociation constants is only on the free concentration of the active form of the reactant, this correction is not mechanism-dependent. For example, from the kinetic behavior of the reaction, it was concluded that only zwitterionic ornithine

**4–16**

was a substrate for ornithine carbamoyltransferase. Total concentrations of ornithine were corrected with an equation for $\alpha_2{}^{Orn}$ so that $K_m$, rather than $K_m{}^{app}$, could be tabulated at each value of pH.[39]

Although there is some disagreement on this point, one of the ways to determine the $pK_a$ of a particular amino acid in a protein is to monitor the rate of its modification by a reagent that can react with only one of its states of ionization. Usually an electrophilic reagent is used to modify the amino acid so that only the conjugate base of the amino acid will react. For example, the lysine in the active site of acetoacetate decarboxylase can be acetylated with 2,4-dinitrophenyl propionate

(4–83)

The rate of the acylation as a function of pH provided the p$K_a$ of this lysine, which is 5.9 (Figure 4–12).[32] This value, as with the values for the p$K_a$ of many glutamates, aspartates, tyrosines, and other lysines found in active sites, is shifted into the neutral range from the value it would have in a fully exposed location.

The value of an acid dissociation constant observed in such a chemical modification of the active site of an enzyme can often be correlated with the value for an apparent acid dissociation constant observed in a kinetic measurement. In the crystallographic molecular model of chymotrypsin,[40] Histidine 57 is located in the active site adjacent to Serine 195, which bears the hydroxyl at which the acyl enzyme is formed (Figure 3–37).[41] The p$K_a$ of Histidine 57 in the unoccupied enzyme has been found to be 6.8 on the basis of the rate of its arylation with 1-fluoro-2,4-dinitrobenzene.[42] This measured value of the p$K_a$ may be in error if the reaction of Histidine 57 with 1-fluoro-2,4-dinitrobenzene involves a step in which the 1-fluoro-2,4-dinitrobenzene binds to the active site before the arylation. If, however, the measurement is not complicated by such prereaction binding, this value for the p$K_a$ is somewhat surprising because it is known from crystallographic molecular models that the lone pair of the conjugate base of this imidazole forms a hydrogen bond with the hydroxyl of Serine 195 (Figure 4–13),[40] while in the conjugate acid of Histidine 57, the nitrogen–hydrogen bond on the opposite nitrogen forms a hydrogen bond with the anionic carboxylate.[43] Therefore, the p$K_a$ measured is for the acid–base reaction

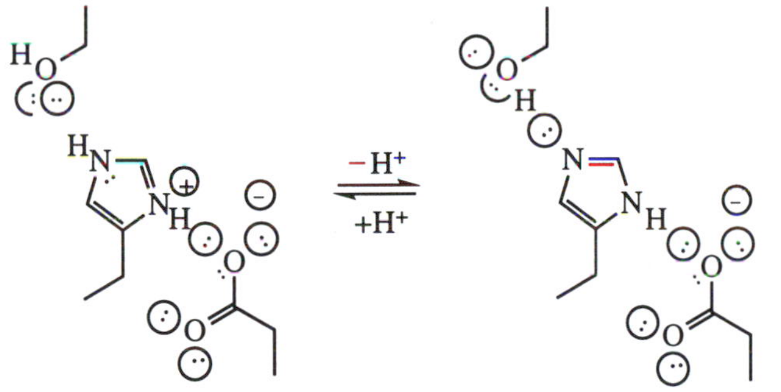

(4–84)

and it is this constellation of amino acids that displays the macroscopic p$K_a$ of 6.8. Presumably, the increase in p$K_a$ necessarily caused by forming a hydrogen bond between the conjugate acid of the histidine and the anionic carboxylate is fortuitously canceled by the decrease in p$K_a$ necessarily caused by forming a hydrogen bond between the conjugate base and Serine 195 because the observed p$K_a$ is about the same as an undistracted imidazole on the surface of a protein.[43a]

The maximum initial velocity, $V_{max}$, for the acylation of Serine 195 in the active site (Figure 4–13) of chymotrypsin by nitrophenyl esters, the first step in the hydrolysis of these esters catalyzed by the enzyme, is a function of pH and decreases at low pH through an apparent p$K_a$ of 6.7.[44] The coincidence of this p$K_a$ with that observed directly for Histidine 57 suggests that the p$K_a$ governing the pH–rate profile of the $V_{max}$ of the acylation reaction is that of the complex acid–base formed from Aspartate 102, Histidine 57, and Serine 195. If this is the case, then this constellation must be present in the initial complex between the enzyme and the nitrophenyl ester as the conjugate base (Reaction 4–84) for the active site to be capable of catalyzing the hydrolysis of this substrate. Because a proton must be removed from Serine 195 either before or after the tetravalent intermediate is formed in the acylation reaction, the role of Histidine 57 in the constellation must be to act as the general base during the acylation.

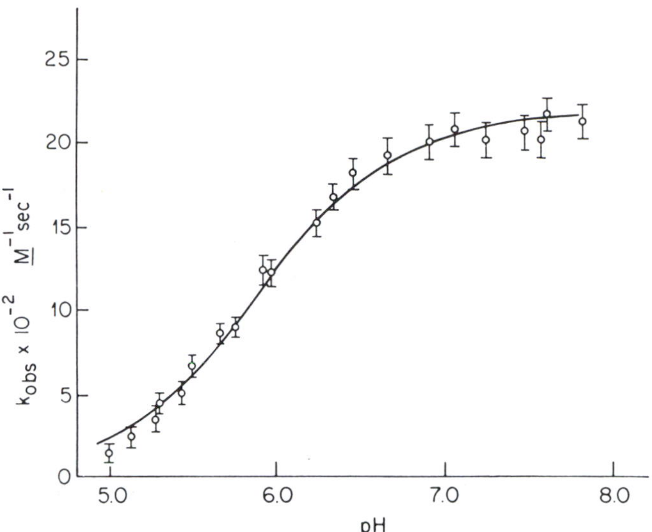


**Figure 4–12:** pH–Rate profile for the reaction of 2,4-dinitrophenyl propionate with acetoacetate decarboxylase.[32] Solutions of 2,4-dinitrophenyl propionate were prepared at various values of pH buffered with lutidine sulfate, *N,N,N′N′*-tetramethylethylenediammonium sulfate, or 4-[(2-dimethylamino)ethyl]morpholinium sulfate at 30 °C. The spontaneous rate of hydrolysis was recorded, and then a sample of acetoacetate decarboxylase was added. This caused a burst of 2,4-dinitrophenol production. The rate of the spontaneous hydrolysis of the 2,4-dinitrophenol in the absence of enzyme was subtracted from the rate of this burst, and a pseudo-first-order rate constant was calculated from the rate at which the enzyme reacted with the 2,4-dinitrophenyl propionate for each sample. This pseudo-first-order rate constant was divided by the concentration of 2,4-dinitrophenyl propionate to obtain the apparent second-order rate constant, $k_{obs}$ (molar$^{-1}$ second$^{-1}$), for the reaction of acetoacetate decarboxylase with the 2,4-dinitrophenyl propionate, and this parameter is plotted as a function of pH. In separate experiments it was shown that the amplitude of the entire burst of 2,4-dinitrophenol was equal to the molar concentration of active sites, that after the burst was complete, the enzyme had lost greater than 98% of its activity, that the loss of activity was coincident with the burst, that enzyme acetylated at the lysine within its active site showed no burst, and that occupation of the active site with acetopyruvate also eliminated the burst. The curve drawn through the points is that for an acid–base titration with a p$K_a$ of 5.9. Reprinted with permission from ref 32. Copyright 1971 American Chemical Society.

In the binary complex between NADH and lactate dehydrogenase, there is a histidine the preferential reaction of which with diethyl pyrocarbonate inactivates the enzyme. The rate of its reaction with diethyl pyrocarbonate as a function of

**Figure 4–13:** Catalytic acid–base in the active site of $\alpha$-chymotrypsin.[40] $\alpha$-Chymotrypsin was crystallized from 48% $(NH_4)_2SO_4$ and 0.2 M sodium citrate, pH 4.2. Amplitudes of reflections were collected to a spacing between the diffracting planes of 0.168 nm. An earlier crystallographic molecular model was used to provide the starting coordinates for the refinement. When the refinement was complete, the $R$-factor was 0.23. The complex acid–base formed by the cluster of Aspartate 102, Histidine 57, and Serine 195, the serine acetylated during the catalytic reaction, was located in the final crystallographic molecular model of the unoccupied enzyme. The electron density displayed is a difference map of electron density for amplitudes $(2F_o - F_c)$ and phases of $\alpha_c$, where $F_o$ are the observed amplitudes and $F_c$ are the calculated amplitudes for the final model, and $\alpha_c$ are the calculated phases for the final model. The skeletal structures are the final refined crystallographic molecular model in this region. Reprinted with permission from ref 40. Copyright 1985 Academic Press.

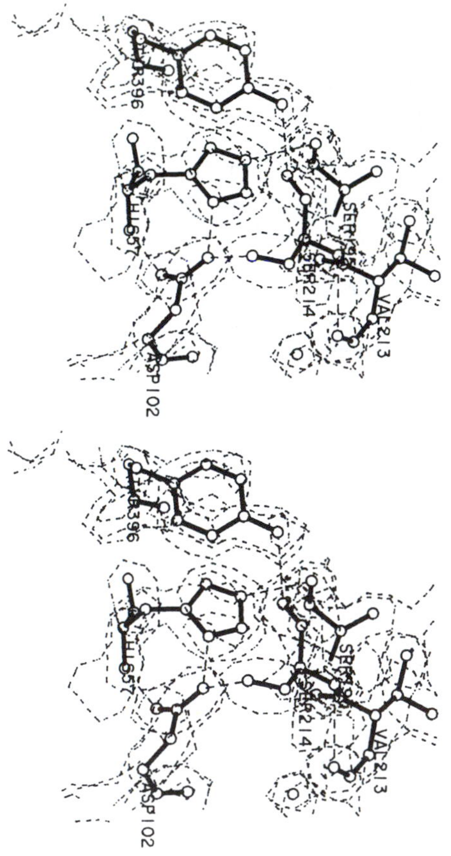


pH is governed by an apparent p$K_a$ of 6.7.[45] The dissociation constant between the binary complex of enzyme and NADH and $o$-nitrophenylpyruvate, a reactant for which hydride transfer is very slow, increases at high pH as the result of an apparent p$K_a$ of 6.8. If this p$K_a$ is the p$K_a$ of the histidine that reacts with diethyl pyrocarbonate, then the substrate is bound only when this histidine is present as the conjugate acid. In the crystallographic molecular model of a bisubstrate analogue within the active site (Figure 3–36) there is a histidinium, Histidinium 195, hydrogen-bonded as a donor to the carbonyl oxygen of the inhibitor as an acceptor. It is this histidinium that should be the acid adding a proton to the keto oxygen of pyruvate, and its conjugate base should be the base removing a proton from the hydroxyl oxygen of lactate during the enzymatic reaction.

For several reasons, the kinetic observations made at initial velocity of pH–rate profiles are almost always ambiguous. Only macroscopic dissociation constants rather than microscopic acid dissociation constants are measured, effects of changes of the rate-limiting step cannot be distinguished from inflections due to a macroscopic acid dissociation constant, and an explicit kinetic mechanism involving a particular acid–base reaction can only be consistent with observations and cannot be required by the observations. The situation, however, can be made less ambiguous in two other ways. If a single step in the enzymatic reaction can be studied in isolation from the others, then the effect of pH on that one step may be susceptible to interpretation. If the binding of inhibitors to the enzyme is a function of pH, then it is possible to assign values of acid dissociation constants on the basis of the assumptions that only the dissociation constants of the inhibitors are being measured and that there are no complications resulting from covalent transformation of the bound inhibitor. In such instances, an acid dissociation constant extracted from the behavior of the dissociation constant of the inhibitor as a function of pH can be either for the enzyme or for the inhibitor itself. If, however, the acid dissociation constants for the inhibitor have been measured in separate experiments, these can be distinguished numerically from the acid dissociation constants of the enzyme.

Because the absorbance of $NAD^+$ at 340 nm increases significantly at the instant that it accepts a hydride, monitoring an enzymatic reaction involving this substrate at this wavelength follows only this one explicit chemical step. Ethanol and $NAD^+$ can be added rapidly to a solution containing alcohol dehydrogenase and the burst of absorbance at 340 nm for only one transfer of a hydride for each active site can be monitored. If the concentrations of these two reactants are high enough, all of the active sites in the solution will be filled much faster than the rate of the transfer of the hydride

(4–85)

and only this one step should contribute to the observed rate constant. Under these circumstances, the observed rate constant for hydride transfer is a function of pH and displays an apparent p$K_a$ of 6.4 (Figure 4–14).[46] Because only a single p$K_a$ appears to be controlling the behavior, it was assumed that it

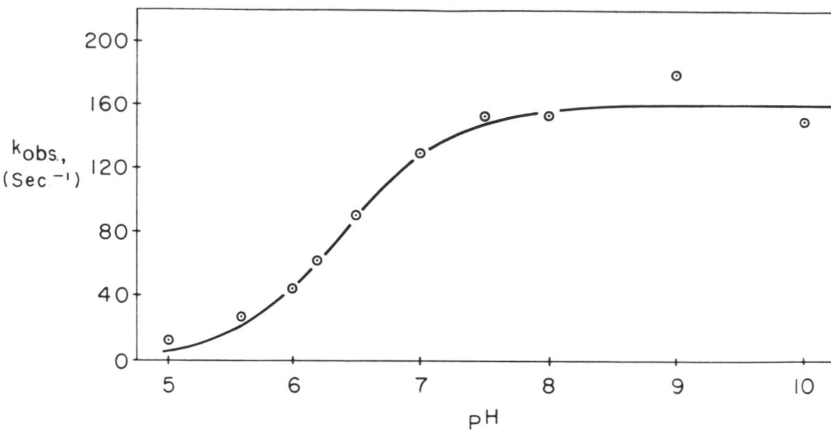


**Figure 4–14:** pH–Rate profile for the transfer of a hydride from ethanol to $NAD^+$ at the active site of liver alcohol dehydrogenase.[46] Solutions of 200 mM ethanol and 4 mM $NAD^+$ were prepared at various values of pH buffered with sodium phosphate or sodium glycinium phosphate at 25 °C and at an ionic strength of 0.1 M. Each of these solutions was mixed in turn with an equal volume of alcohol dehydrogenase at a 30 mM concentration of active sites. The mixing was performed in a stopped-flow apparatus, and the change in absorbance at 340 nm was monitored by an oscilloscope. A burst of absorbance at 340 nm was observed corresponding to a single hydride transfer for each active site. The pseudo-first-order rate constant for this burst was measured from each kinetic trace. The value of this rate constant, $k_{obs}$ (second$^{-1}$), is plotted as a function of the pH. The line drawn is for an acid–base titration with a p$K_a$ of 6.4. Reprinted with permission from ref 46. Copyright 1972 *Journal of Biological Chemistry.*

was the microscopic p$K_a$ of a single acid–base in the active site. This assumption is valid only if all of the acid–base reactions involved in binding the reactants are so much faster than the hydride transfer that they reach equilibrium before any hydride transfer can occur.

It is usually assumed that the dissociation constant measured for an inhibitor of an enzymatic reaction is a simple equilibrium constant between the active site and a ligand. If a protein binds a ligand at a particular site, the number of protons in the site and their distribution will usually affect the observed dissociation constant for that ligand. The apparent dissociation constant that is measured will be a complex function of the pH. Consider the simplest situation in which only forms of the site containing either $n - m + 1$ protons, $STH_{(n-m+1)}$, or $n - m$ protons, $STH_{(n-m)}$, exist in significant concentration within the range of pH covered. The ligand, L, can bind to either form of the site, and protons can enter or leave the occupied site

$$\begin{array}{ccc} STH_{(n-m+1)} + L & \underset{K_{d1}{}^{L}}{\rightleftharpoons} & L{\cdot}STH_{(n-m+1)} \\ \Big\updownarrow K_{am}{}^{ST} & & \Big\updownarrow K_{am}{}^{STL} \\ STH_{(n-m)} + L & \underset{K_{d2}{}^{L}}{\rightleftharpoons} & L{\cdot}STH_{(n-m)} \end{array} \qquad (4\text{–}86)$$

In this mechanism, the equilibrium constants $K_{am}{}^{ST}$ and $K_{am}{}^{STL}$ are acid dissociation constants and the equilibrium constants $K_{d1}{}^{L}$ and $K_{d2}{}^{L}$ are the dissociation constants for the ligand from the two ionization states of the site. It can be shown by inspection that

$$K_{am}{}^{ST}K_{d1}{}^{L} = K_{am}{}^{STL}K_{d2}{}^{L} \qquad (4\text{–}87)$$

If the apparent dissociation constant for ligand L is defined as

$$K^{L}_{d,app} = \frac{([STH_{(n-m+1)}]+[STH_{(n-m)}])[L]}{[L{\cdot}STH_{(n-m+1)}]+[L{\cdot}STH_{(n-m)}]} \qquad (4\text{–}88)$$

then it follows that

$$K^{L}_{d,app} = \frac{\left(1+\dfrac{[H^+]}{K_{am}{}^{ST}}\right)}{\left(1+\dfrac{[H^+]}{K_{am}{}^{STL}}\right)}K_{d2}{}^{L} \qquad (4\text{–}89)$$

and

$$pK^{L}_{d,app} = \log\left(1+\frac{[H^+]}{K_{am}{}^{STL}}\right) - \log\left(1+\frac{[H^+]}{K_{am}{}^{ST}}\right) + pK_{d2}{}^{L} \qquad (4\text{–}90)$$

If the ligand binds more tightly to the site containing $(n - m + 1)$ protons than the site with $n - m$ protons ($K_{d1}{}^{L} < K_{d2}{}^{L}$), then, because of Equation 4–87

$$K_{am}{}^{ST} > K_{am}{}^{STL} \qquad (4\text{–}91)$$

When $[H^+] > K_{am}{}^{ST} > K_{am}{}^{STL}$

$$pK^{L}_{d,app} \cong pK_{d2}{}^{L} + pK_{am}{}^{STL} - pK_{am}{}^{ST} = pK_{d1}{}^{L} \qquad (4\text{–}92)$$

which is invariant with pH. When $K_{am}^{ST} > [H^+] > K_{am}^{STL}$

$$pK^{L}_{d,app} \cong pK_{am}^{STL} + pK_{d2}^{L} - pH \quad (4\text{–}93)$$

which decreases as pH is increased. When $K_{am}^{ST} > K_{am}^{STL} > [H^+]$

$$pK^{L}_{d,app} \cong pK_{d2}^{L} \quad (4\text{–}94)$$

which is invariant with pH. The asymptote defined by Equation 4–92 designates $pK_{d1}^{L}$, the asymptote defined by Equation 4–94 designates $pK_{d2}^{L}$, the intersection of the asymptotes defined by Equations 4–92 and 4–94 occurs when $pH = pK_{am}^{ST}$, and the intersection of the asymptotes defined by Equations 4–93 and 4–94 occurs when $pH = pK_{am}^{STL}$.

The reaction catalyzed by aspartate aminotransferase

$$\text{aspartate} + \text{2-ketoglutarate} \rightleftharpoons \text{glutamate} + \text{oxaloacetate} \quad (4\text{–}95)$$

involves the coenzyme pyridoxal phosphate. This transamination can be divided into two steps

$$\text{aspartate} + \text{lysylpyridoximine phosphate} \rightleftharpoons \text{oxaloacetate} + \text{pyridoxamine phosphate} + \text{lysine} \quad (4\text{–}96)$$

and

$$\text{lysine} + \text{2-ketoglutarate} + \text{pyridoxamine phosphate} \rightleftharpoons \text{glutamate} + \text{lysylpyridoximine phosphate} \quad (4\text{–}97)$$

where the lysine is a lysine side chain from the enzyme itself. If 2-ketoglutarate enters the active site when the coenzyme is the lysylpyridoximine phosphate, no further reaction can occur. Therefore, 2-ketoglutarate is a competitive inhibitor with respect to aspartate. The observed inhibition constant of 2-ketoglutarate is a function of pH and obeys Equation 4–90 (Figure 4–15).[27] By fitting Equation 4–90 to the data it was found that $K_{d1}^{KG} = 0.8$ mM, $K_{d2}^{KG} = 27$ mM, $pK_{am}^{ST} = 6.5$, and $pK_{am}^{STKG} = 8.2$.

These values state that 2-ketoglutarate binds more tightly to the more protonated form of the enzyme and, when bound at the site, it raises the $pK_a$ of the enzyme. The $pK_a$ of 6.5 is for the iminium nitrogen of the lysylpyridoximine (Equation 4–76). If 2-ketoglutarate, an anion, binds adjacent to this imine, it should bind more tightly when the imine is positive rather than zwitterionic and the presence of the 2-ketoglutarate should cause the imine to be a weaker acid than it is in the absence of 2-ketoglutarate.

To explain the observed behavior of the dissociation constant for an inhibitor, $K_i$, as a function of pH, it is often assumed, in analogy with the explanation of the behavior of $k_{cat}$ and $k_{cat}/K_m$, that the inhibitor can bind to only one form of the enzyme. For example, this would be equivalent to making the assumption in Mechanism 4–86 not only that $K_{d2}^{L}$ was greater than $K_{d1}^{L}$ but that $K_{d2}^{L}$ was infinitely large. For Equation 4–87 to hold in the limit, as it must, this would require that $K_{am}^{SL}$ be zero if $K_{d1}^{L}$ and $K_{am}^{ST}$ are to remain finite. Therefore, if the ligand can bind only to the protonated state, $STH_{(n-m+1)}$, of the site, it necessarily follows that in the occupied site the protonated acid can no longer dissociate. This can be generalized to obtain the statement that if the ligand can bind only to one protonation state of the site, then the acid–bases within the site cannot participate in proton exchanges with the solution once the site is occupied. If it is assumed that the inhibitor can bind to only one form of the active site, $EH_m$, then

$$K_{i,app} = K_i / \alpha_m^{E} \quad (4\text{–}98)$$

where $\alpha_m^{E}$ is defined by Equation 4–52 and $K_i$ is the dissociation constant between the inhibitor and the form of the enzyme, $EH_{n-m}$, with the correct number of protons in the active site.

The apparent dissociation constant for 2,4-diamino-6,7-dimethylpteridine, acting as a competitve inhibitor of dihydrofolate reductase, varies with pH.[47] It increases below an apparent $pK_a$ of 5.9 and increases above an apparent $pK_a$ of 7.9. The parameter $k_{cat}/K_m^{DHF}$ for the reaction catalyzed by dihydrofolate reductase

$$\text{7,8-dihydrofolate} + \text{NADPH} \rightleftharpoons \text{5,6,7,8-tetrahydrofolate} + \text{NADP}^+ \quad (4\text{–}99)$$

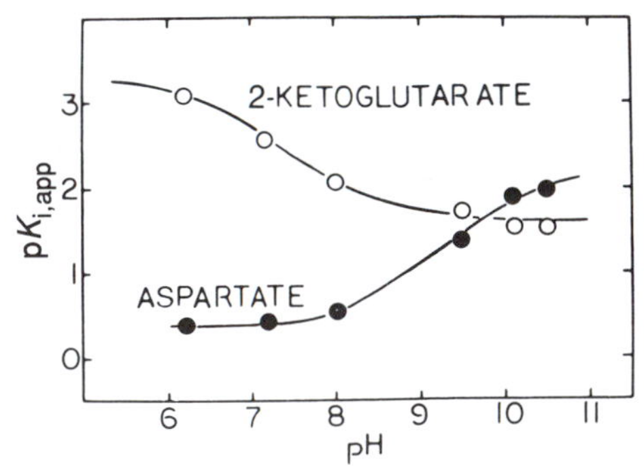


**Figure 4–15:** Dependence on the pH of the competitive inhibition of the transamination catalyzed by aspartate aminotransferase (Reaction 4–95) by either 2-ketoglutarate (○) or aspartate (●).[27] Solutions containing various concentrations of aspartate and 2-ketoglutarate at various values of pH (see Figure 4–10) were prepared at 25 °C. The reaction was initiated by adding aspartate aminotransferase, and the production of oxaloacetate was followed by coupling it to the oxidation of $NAD^+$ with malate dehydrogenase. The increase in $A_{340}$ with time was used to measure initial velocities. Both 2-ketoglutarate and aspartate are reactants in the reaction catalyzed by aspartate aminotransferase. Each of them, however, at high concentration, higher than that necessary to obtain their respective maximum velocities, becomes an inhibitor of the enzyme; each is a competitive inhibitor with respect to the other. From the competitive inhibition observed, values for the apparent dissociation constant for each reactant, $K_{i,app}^{KG}$ and $K_{i,app}^{Asp}$, were calculated at each pH. The negative logarithms of these apparent dissociation constants (in units of molarity), $pK_{i,app}$, are plotted as functions of pH. The curves drawn are fits of Equation 4–90 to the respective data. Adapted with permission from ref 27. Copyright 1983 American Chemical Society.

where $K_m^{DHF}$ is the Michaelis constant for 7,8-dihydrofolate, decreases above a p$K_a$ of 8.1. Because 2,4-diamino-6,7-dimethylpteridine resembles 7,8-dihydrofolate and is a competitive inhibitor with respect to 7,8-dihydrofolate, the coincidence of these two apparent p$K_a$ values (p$K_a$ = 7.9 and p$K_a$ = 8.1), each of which should be for the free enzyme, suggests that they reflect the same acid dissociation. The apparent p$K_a$ associated with the binding of the inhibitor, because it governs an equilibrium between a protein and ligand, should be a simple macroscopic p$K_a$ for the free enzyme rather than a complex kinetic parameter. Therefore, the apparent p$K_a$ observed in the behavior of $k_{cat}/K_m^{DHF}$ should also be a simple macroscopic p$K_a$ for the free enzyme.

Azide anion is a competitive inhibitor with respect to formate of formate dehydrogenase, which catalyzes the reaction

$$NAD^+ + HCO_3^- \rightleftharpoons NADH + CO_2 \qquad (4\text{–}100)$$

The inhibition constant for azide varies with pH[48] according to Equation 4–98 with p$K_{a(m+1)}^E$ = 6.4 and p$K_{am}^E$ = 8.3. If the inhibition constant for azide is detecting only the macroscopic acid dissociation constants for the free enzyme, and if the parameter $k_{cat}/K_m^F$, where $K_m^F$ is the Michaelis constant for formate, is also governed only by the acid dissociation constants for the free enzyme, as it is usually assumed to be, the respective values for the p$K_a$ should coincide. For the reaction catalyzed by formate dehydrogenase, the parameter $k_{cat}/K_m^F$ decreases below a p$K_a$ of 6.1 and decreases above a p$K_a$ of 8.5. Although the differences between the two measurements may be within experimental error, it is of note that the p$K_a$ values determined from the behavior of $k_{cat}/K_m^F$ are less than and greater than, respectively, the p$K_a$ values determined from the dissociation constants for the inhibitor, as expected if the step involving this acid–base were not the rate-limiting step at the optimum pH.

The enthalpy of dissociation is the enthalpy change for an acid dissociation. The acid–bases on the various amino acids when they are in free solution have characteristic standard enthalpies of dissociation.[49] Carboxylic acids have standard enthalpies of dissociation near 0 (–4 to +4 kJ mol$^{-1}$), imidazoles and phenols have standard enthalpies of dissociation in an intermediate range (+25 to +35 kJ mol$^{-1}$), and ammonium cations have the highest standard enthalpies of dissociation (+40 to +55 kJ mol$^{-1}$). It is unclear, however, whether or not these characteristic standard enthalpies of dissociation are altered by the same environmental aspects of the active site that so dramatically shift the values of the acid dissociation constants themselves. In the absence of any evidence concerning this question, the use of standard enthalpies of dissociation to identify the type of amino acid[50] responsible for a particular p$K_a$ may be misleading. For example, the standard enthalpy of dissociation for an amino acid governing the pH dependence of papain at low pH was measured to be 0. On this basis the apparent p$K_a$ was assigned as a carboxylate.[51] It was later shown that the group responsible for this p$K_a$ was the imidazole of a histidine.

The acid dissociation constants of simple acid–bases free in aqueous solution are also affected by the addition of miscible organic solvents that lower the dielectric constant.[52,53] At one time it was thought that these characteristic shifts in p$K_a$ could be used to identify the type of amino acid responsible for the acid–base catalysis.[50,54] This method, however, has also been shown to be equivocal.[53]

## *Suggested Reading*

Kiick, D.M., & Cook, P.F. (1983) pH Studies toward the Elucidation of the Auxiliary Catalyst for Pig Heart Aspartate Aminotransferase, *Biochemistry 22*, 375–382.

### PROBLEM 4–5

Show that the kinetic mechanism

$$\mathrm{EH + H^+} \underset{}{\overset{}{\rightleftharpoons}}_{K_a^E} \mathrm{E + 2H^+}$$

$$\mathrm{E + 2H^+ + A} \underset{k_{-1}}{\overset{k_1}{\rightleftharpoons}} \mathrm{E{\cdot}X + 2H^+}$$

$$\mathrm{E{\cdot}XH + H^+} \overset{K_a^{EX}}{\rightleftharpoons} \mathrm{E{\cdot}X + 2H^+}$$

$$\mathrm{E{\cdot}XH + H^+} \overset{k_2}{\longrightarrow} \mathrm{E{\cdot}YH + H^+}$$

$$\mathrm{E{\cdot}YH_2} \overset{K_a^{EY}}{\rightleftharpoons} \mathrm{E{\cdot}YH + H^+}$$

$$\mathrm{E{\cdot}YH + H^+} \overset{k_3[H^+]}{\longrightarrow} \mathrm{E{\cdot}ZH_2 + H^+}$$

$$\mathrm{E{\cdot}ZH_2 + H^+} \overset{K_a^{EZ}}{\rightleftharpoons} \mathrm{E{\cdot}ZH + 2H^+}$$

$$\mathrm{E{\cdot}ZH + 2H^+} \overset{k_4}{\longrightarrow} \mathrm{E + P}$$

gives a rate equation of the form

$$v_0 = \frac{k_{cat}'[E]_{TOT}[A]_0}{K_m' + [A]_0}$$

where

$$k_{cat}' = \frac{k_{cat}}{\dfrac{K_{a1}}{[H^+]} + 1 + \dfrac{[H^+]}{K_{a2}}}$$

and

$$\frac{k_{cat}'}{K_m'} = \frac{\dfrac{k_{cat}}{K_m}}{\dfrac{K_{a3}}{[H^+]} + 1 + \dfrac{[H^+]}{K_{a4}}}$$

The primed constants vary with pH and the unprimed constants do not.

Write the expressions defining each of the four observed apparent acid dissociation constants, $K_{a1}$, $K_{a2}$, $K_{a3}$, and $K_{a4}$. Are any of them true macroscopic acid dissociation constants?

### PROBLEM 4–6

Show that Equation 4–89 follows from Equation 4–88.

# Kinetic Isotope Effects

Reactions in which hydrogen is transferred as a proton from a reactant to a general base or from a general acid to a reactant in a rate-limiting step proceed more slowly when the protium is replaced by deuterium. For example, the removal of deuterium as a deuteron from 4-(4-nitrophenoxy)-2-[3,3-$^2H_2$]butanone by phenoxide anion acting as a general base in aqueous solution

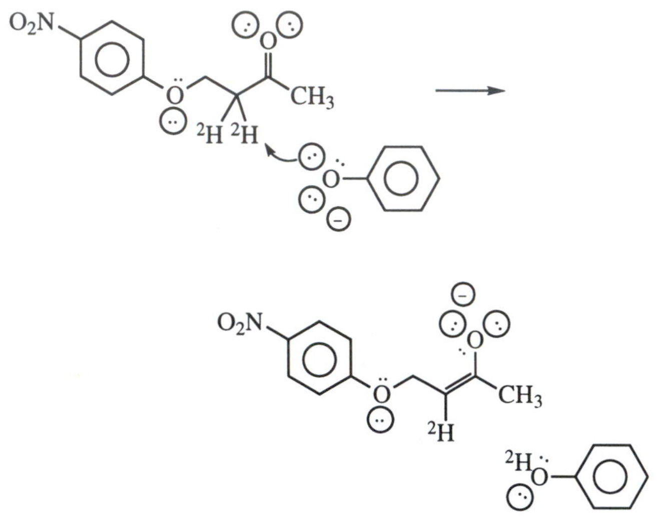

(4–101)

occurs 6.1 times more slowly than the removal of protium as a proton from the same position by the same base.[55] The decrease in rate produced by replacing a protium with a deuterium is an example of a kinetic isotope effect. A **kinetic isotope effect** is a change in the rate of a reaction produced by changing the isotope at one of the atoms in one of the reactants.

The appearance of a deuterium kinetic isotope effect provides experimental evidence that the removal of that particular hydrogen, usually as a proton but sometimes as a hydride or a hydrogen radical, is involved in a rate-limiting step of an enzymatic reaction. For example, the fact that substitution of protium with deuterium at hydrogen 5 on glucose 6-phosphate decreased the rate of the reaction catalyzed by *myo*-inositol-1-phosphate synthase was used as evidence that this hydrogen is removed in one of the steps of the enzymatic reaction even though it is not removed in the overall reaction[56]

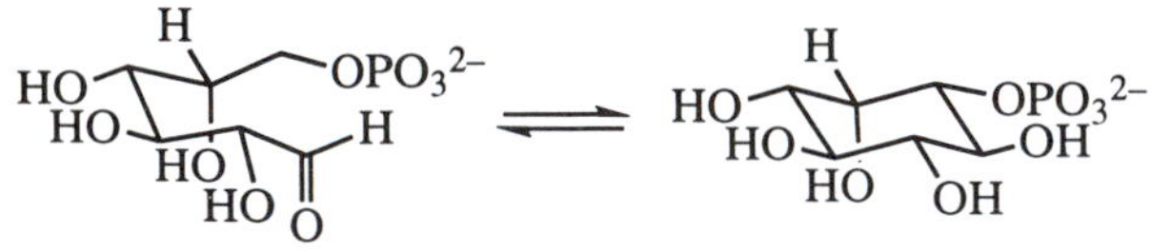

(4–102)

Kinetic isotope effects are also observed in enzymatic reactions when protium is replaced by tritium or when any one of the natural isotopes of the lighter elements such as $^{12}$carbon, $^{14}$nitrogen, or $^{16}$oxygen is replaced by a heavier isotope. In these latter instances, involving elements of the second row, the observation of a kinetic isotope effect provides evidence that one of the bonds to that particular atom is broken during a rate-determining step.

A deuterium or a tritium can be substituted synthetically for a protium on carbon, and it can be expected to remain on that carbon long enough to perform an experiment. Deuteriums on sulfur, oxygen, or nitrogen, however, exchange rapidly with the hydrogens of the water. If deuterium oxide is used in place of water to ensure that deuterium is present on a particular oxygen or nitrogen, all of the oxygens and nitrogens accessible to the solution will have their hydrogens replaced with deuterium. Because it is not possible to confine deuterium to just one oxygen, nitrogen, or sulfur in either a substrate or the enzyme, most of the hydrogen kinetic isotope effects reported in studies of enzymes are kinetic isotope effects resulting from the replacement of hydrogen on carbon by deuterium or tritium.

The usual explanation for a **deuterium kinetic isotope effect** is based on an examination of vibrational energy levels.[57] In the reactant, the carbon–hydrogen bond in the ground state participates in a stretching vibration. In a linear transition state in which the hydrogen is being transferred as a proton from carbon to a lone pair of electrons on a general base, all of this vibrational motion is converted into translational motion along the reaction coordinate. The hydrogen in that transition state no longer has stretching vibrational energy because its vibrational energy has been converted entirely into this translational energy, which is part of the free energy of activation. It is this conversion of vibrational energy into free energy of activation that causes the kinetic isotope effect.

In either a molecule or a transition state, the arrangement of nuclei and the accompanying electrons in space creates a potential energy surface. If one nucleus is moved with respect to another nucleus, the potential energy varies. If, for example, only the hydrogen in a carbon–hydrogen bond in the molecule is moved along the axis of the bond, the potential energy function for stretching that bond is generated (Figure 4–16). Because the nucleus of a deuterium atom has the same charge as the nucleus of a protium atom, the potential function for a carbon–deuterium bond is identical to that for a carbon–protium bond.

Stretching vibration along a carbon–hydrogen bond is controlled by this potential function, but it is quantized. The relationship between the vibrational energy, in excess of the potential energy at the bottom of the well and the observed wavelength of infrared absorption, is

$$E_v = (n_v + {}^1/_2)h\nu_0 \qquad (4\text{–}103)$$

where $n_v$ is the vibrational quantum number and $\nu_0$ is the frequency of a photon whose energy matches that of the steps between the vibrational energy levels. The difference in energy between the bottom of the potential well and the ground state is the zero-point energy, $\Delta\varepsilon$, and

$$\Delta\varepsilon = {}^1/_2\, h\nu_0 \qquad (4\text{–}104)$$

The infrared absorption maximum for the stretching vibration of a carbon–hydrogen bond in normally encountered molecules is around 2900 cm$^{-1}$. Therefore, the zero-point energy

**Figure 4–16:** Schematic drawing of the potential energy for the stretching of a carbon–hydrogen bond. The potential energy is plotted as a function of the distance between the nuclei of the carbon and the hydrogen atoms. When that distance is short, the nuclei repel and the potential energy rises dramatically. When the distance is long, the atoms do not interact and the potential energy is near zero. At intermediate distances, electron overlap creates a covalent bond, which is a favorable interaction. Because the stretching vibrations within the bottom of the potential well are quantized, they can only take on certain energies. The zero-point energy for a carbon–protium bond, $\Delta\varepsilon_{CH}$, is higher than that for a carbon–deuterium bond, $\Delta\varepsilon_{CD}$, because of the difference in mass between a proton ($1.67 \times 10^{-24}$ g) and a deuteron ($2.34 \times 10^{-24}$ g). The drawing is not to scale. The ground-state energy of the carbon–hydrogen bond should be about $-340$ kJ mol$^{-1}$ below the origin, while $\Delta\varepsilon_{CH}$ should be only $+17.4$ kJ mol$^{-1}$.

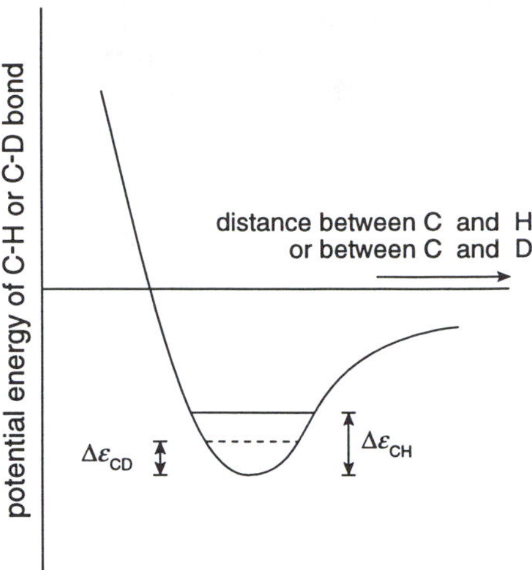


for a carbon–protium bond is about +17.4 kJ mol$^{-1}$. The infrared absorption maximum for the stretching vibration of a carbon–deuterium bond is around 2200 cm$^{-1}$. Therefore, the zero-point energy for a carbon–deuterium bond is about +12.6 kJ mol$^{-1}$. In the transition state all of this vibrational energy is converted into translational energy along the reaction coordinate; and, if the transition state were symmetrical and linear, a carbon–protium bond would require 4.8 kJ mol$^{-1}$ less free energy of activation to reach the transition state than a carbon–deuterium bond. This would lead to a rate enhancement of the reaction when protium is present rather than deuterium because

$$\frac{k^{CH}}{k^{CD}} = e^{-\Delta\Delta G^{\circ\ddagger}/RT} = 7 \qquad (4\text{–}105)$$

Extensive tabulations are available of the observed deuterium kinetic isotope effects for the removal of hydrogen from various carbons as a proton by various general bases.[55] When the p$K_a$ of the conjugate acid of the general base is equal to the p$K_a$ of the carbon acid, the values observed lie between 6 and 12. Because such transfers should proceed through symmetric transition states, they should represent the maximum values that can be obtained.[57] That values in excess of 7 are common has usually been explained by assuming that energy from the bending vibrations of the carbon–hydrogen bond as well as energy from the stretching vibrations is incorporated into the translational energy along the reaction coordinate. Recently, however, these larger kinetic isotope effects have been explained as the results of tunneling of a proton through the energy barrier for transfer.[58] Both coupling of bending modes to movement along the reaction coordinate and tunneling through the barrier could occur simultaneously in some situations because these two processes are not mutually exclusive.

If a sufficient number of the values for the simple deuterium kinetic isotope effects for the removal of protons from carbon by general bases are plotted as a function of the difference in p$K_a$ between the carbon acid and the conjugate acid of the general base, there seems to be a maximum when the two values of the p$K_a$ are the same.[55] As the p$K_a$ of the conjugate acid of the general base becomes larger or smaller than the p$K_a$ of the carbon acid, the deuterium kinetic isotope effect decreases monotonically. When the difference in p$K_a$ is 10, about half of the kinetic isotope effect has been lost. For example, the deuterium kinetic isotope effect for the removal of a proton from sulfonatoacetone[59] by hydroxide anion ($\Delta$p$K_a$ = −2.0) is 7.4, but the deuterium kinetic isotope effect for the removal of the same proton by acetate anion ($\Delta$p$K_a$ = 9.6) is 3.8. These decreases in the kinetic isotope effect on either side of the maximum are thought to result from the increasing asymmetry of the transition state due to the changes in the relative attractive forces of the two lone pairs for the proton that sits between them.[57] The stretching frequency of the bond between hydrogen and the carbon of the acid is completely converted into translational motion along the reaction coordinate, and as a result this vibrational mode disappears. The transition state, however, sits at a saddlepoint on the potential energy surface. Because the potential energy surface in the vicinity of the transition state has literally the shape of a saddle, perpendicular to the potential maximum of the reaction coordinate is a potential well. This potential well has a symmetical vibrational mode associated with it. If the carbon and the base removing the proton have equal affinity for the proton, then the proton remains stationary as vibration occurs in this perpendicular symmetrical vibrational mode of the transition state and there is no difference in energy between the transition state with protium and the one with deuterium. If, however, the affinities of the carbon and the base differ from each other, then the proton moves in concert with the atom that has the higher affinity and the transition state containing deuterium will have a lower energy than that containing protium. This deuterium isotope effect on the transition state therefore must decrease the deuterium isotope effect on the reaction. As a result, the greater the difference in p$K_a$, regardless of its sign, between the carbon and the conjugate acid of the base, the smaller will be the deuterium kinetic isotope effect on the rate of the proton transfer.

The maximum magnitudes expected for other classes of kinetic isotope effects depend on the masses of the two atoms participating in the bond. The differences in zero-point energies between protium and deuterium for nitrogen–hydrogen and oxygen–hydrogen bonds[55] are 5 and 6 kJ mol$^{-1}$, so the maximum kinetic isotope effects seen with these bonds are in the same range as those for carbon–hydrogen bonds. From a semiclassical treatment of vibrational energies, it has been proposed that, on the basis of the differences in mass among hydrogen, deuterium, and tritium, the following relationship should hold[60]

$$\frac{{}^{\mathrm{H}}k}{{}^{\mathrm{T}}k} = \left(\frac{{}^{\mathrm{H}}k}{{}^{\mathrm{D}}k}\right)^{1.44} \tag{4–106}$$

This relationship is consistent with experimental values for the deuterium and tritium kinetic isotope effects for various reactions, and it is used widely as an established fact.[61] As an example, if a deuterium kinetic isotope effect in a given reaction were 7, then the tritium kinetic isotope effect for the same reaction should be 16.

Because the fractional difference in mass is much smaller between $^{13}$carbon and $^{12}$carbon, between $^{14}$carbon and $^{12}$carbon, or between $^{16}$oxygen and $^{18}$oxygen than between $^{1}$hydrogen and $^{2}$hydrogen, carbon and oxygen kinetic isotope effects are much smaller. In a reaction in which a carbon–carbon bond is broken during the rate-limiting step, if one of the $^{12}$carbons is replaced with a $^{13}$carbon, the largest kinetic isotope effects $^{12}k/^{13}k$ that have been observed[62] are about 1.06. If $^{12}$carbon is replaced with $^{14}$carbon instead, carbon kinetic isotope effects $^{12}k/^{14}k$ in reactions with symmetrical transition states[63] are between 1.09 and 1.15. The oxygen kinetic isotope effects $^{16}k/^{18}k$ observed when a carbon–oxygen bond is broken in the rate-limiting step[64] and $^{16}$oxygen is replaced with $^{18}$oxygen can be as high as 1.08.

The effect that substituting deuterium will have on the equilibrium constant of a reaction depends upon the stability of the $\sigma$ bond between the product and the deuterium relative to that between the product and protium and the stability of the $\sigma$ bond between the reactant and the deuterium relative to that between the reactant and protium. When the deuterium is transferred from carbon as an acid to either nitrogen or oxygen acting as a base, the deuterium equilibrium isotope effect should cause the equilibrium constant to shift in favor of the carbon acid upon deuterium substitution, because the difference in zero-point energies is smaller between a carbon–protium bond and a carbon–deuterium bond ($\Delta\varepsilon$ = 4.8 kJ mol$^{-1}$) than it is between a nitrogen–hydrogen bond and a nitrogen–deuterium bond ($\Delta\varepsilon$ = 5 kJ mol$^{-1}$). If only the differences in zero-point energies for stretching vibrations were important, the maximum deuterium equilibrium isotope effect ($^{\mathrm{H}}K_{\mathrm{eq}}/^{\mathrm{D}}K_{\mathrm{eq}}$) would be 1.4 for transfer from carbon to oxygen. The deuterium equilibrium isotope effect, however, is also affected by the other substituents around the central atom. For example, in transfers of hydrogen between two carbons, deuterium equilibrium isotope effects ($^{\mathrm{H}}K_{\mathrm{eq}}/^{\mathrm{D}}K_{\mathrm{eq}}$) as large as 1.3 are observed when the transfer is from a carbon substituted with two oxygens to a carbon substituted only with carbons and hydrogens.[65]

Simple deuterium kinetic isotope effects are readily observed in enzymatically catalyzed reactions. For example, the reaction catalyzed by *cis,cis*-muconate cycloisomerase

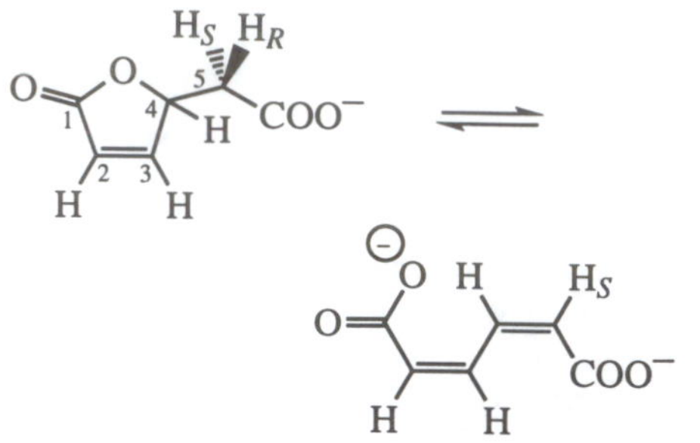

(4–107)

involves the removal, as a proton, of the *pro-R* hydrogen on carbon 5 of the reactant during the elimination. When protium is replaced by a deuterium at this position,[66] the maximum initial velocity of the reaction, $V_{max}$, decreases by a factor of 2.5. This result demonstrates that the removal of this hydrogen as a proton, presumably by a general base in the active site, occurs during a rate-determining step in the reaction.

It is possible that information about the structure of the transition state or the degree of symmetry of the proton transfer could be inferred from the magnitude of a deuterium kinetic isotope effect. This can be accomplished, however, only if the magnitude of the deuterium kinetic effect at the step in which the carbon–hydrogen bond is broken can be determined. The **intrinsic kinetic isotope effect** is the kinetic isotope effect for the specific step in an enzymatically catalyzed reaction at which the isotopically substituted bond is broken by passing through a single transition state. Most enzymatically catalyzed reactions proceed through many steps, and the rate constants for several steps are usually consequential to the composite rate constant for the overall reaction. Usually, however, only one of these steps can display an intrinsic kinetic isotope effect. It necessarily follows that the observed deuterium kinetic isotope effect for an enzymatic reaction is always less than or equal to the intrinsic deuterium kinetic isotope effect for the step in which the hydrogen is transferred.

Consider an enzymatic reaction that proceeds through a series of steps

$$\mathrm{E_1} \underset{k_{-1}}{\overset{k_1[\mathrm{A}]}{\rightleftharpoons}} \mathrm{E_2} \underset{k_{-2}}{\overset{k_2[\mathrm{B}]}{\rightleftharpoons}} \mathrm{E_3} \underset{k_{-3}}{\overset{k_3}{\rightleftharpoons}} \mathrm{E_4} \underset{k_{-4}}{\overset{k_4}{\rightleftharpoons}} \mathrm{E_5} \underset{k_{-5}}{\overset{k_5}{\rightleftharpoons}} \mathrm{E_6} \underset{k_{-6}[\mathrm{P}]}{\overset{k_6}{\rightleftharpoons}} \mathrm{E_7} \underset{k_{-7}[\mathrm{Q}]}{\overset{k_7}{\rightleftharpoons}} \mathrm{E_1} \tag{4–108}$$

If neither product P nor product Q is present and the initial velocities are being measured, the conversion of $E_6$ to $E_7$ is irreversible. If the reactant that is deuterated is reactant B, if the step at which the carbon–hydrogen bond is broken is step 4, if reactant A is present at saturation, and if all forms of the enzyme are in steady state, Mechanism 4–108 can be rewritten as

$$\mathrm{E_2} \underset{k_{-23}'}{\overset{k_{23}'[\mathrm{B}]}{\rightleftharpoons}} \mathrm{E_4} \underset{k_{-4}}{\overset{k_4}{\rightleftharpoons}} \mathrm{E_5} \xrightarrow{k_{56}'} \mathrm{E_7} \xrightarrow{k_7} \mathrm{E_2} \tag{4–109}$$

where the composite rate constants are defined as[67]

$$k_{23}'[\mathrm{B}] = \frac{k_3 k_2[\mathrm{B}]}{k_{-2} + k_3} \tag{4–110}$$

and

$$k_{-23}' = \frac{k_{-2} k_{-3}}{k_3 + k_{-2}} \tag{4–111}$$

and

$$k_{56}' = \frac{k_6 k_5}{k_{-5} + k_6} \tag{4–112}$$

If the commitment forward, $c_f$, is defined[68] as the ratio between the forward rate constant ($k_4$) for the step in which the carbon–hydrogen bond is broken and the composite rate constant for reversal from the form of the enzyme, $E_4$, immediately preceding this step to the step at which the deuterated reactant is released back into the solution, then

$$c_f = \frac{{}^{\mathrm{H}}k_4}{k_{-23}'} = \frac{{}^{\mathrm{H}}k_4(k_3 + k_{-2})}{k_{-2}k_{-3}} \tag{4–113}$$

If the commitment in reverse, $c_r$, is defined as the ratio between the reverse rate constant ($k_{-4}$) for the step in which the carbon–hydrogen bond is broken and the composite rate constant for forward flux from the form of the enzyme immediately following this step, $E_5$, to the first irreversible step, then

$$c_r = \frac{{}^{\mathrm{H}}k_{-4}}{k_{56}'} = \frac{{}^{\mathrm{H}}k_{-4}(k_{-5} + k_6)}{k_6 k_5} \tag{4–114}$$

Almost any mechanism for an enzymatic reaction, no matter how complex, can be rewritten in the form of Mechanism 4–109 by using composite rate constants.[67] Once this has been done, all of the mechanisms become formally equivalent. It can be shown[68] that, for any mechanism written in this form, if the isotopic substitution affects only the one step, then

$$\frac{{}^{\mathrm{H}}V_{\max} / K_{\mathrm{mH}}{}^{\mathrm{B}}}{{}^{\mathrm{D}}V_{\max} / K_{\mathrm{mD}}{}^{\mathrm{B}}} = \frac{\dfrac{{}^{\mathrm{H}}k_4}{{}^{\mathrm{D}}k_4} + c_f + c_r \dfrac{{}^{\mathrm{D}}k_{-4}{}^{\mathrm{H}}k_4}{{}^{\mathrm{H}}k_{-4}{}^{\mathrm{D}}k_4}}{1 + c_f + c_r} \tag{4–115}$$

where ${}^{\mathrm{H}}V_{\max}$ and ${}^{\mathrm{D}}V_{\max}$ are the maximum initial velocities for the reactants containing hydrogen and deuterium, respectively, and $K_{\mathrm{mH}}{}^{\mathrm{B}}$ and $K_{\mathrm{mD}}{}^{\mathrm{B}}$ are the Michaelis constants for the reactant B when it contains a protium or a deuterium, respectively.

If, as is usually the case, the deuterium equilibrium isotope effect, ${}^{\mathrm{D}}k_{-4}{}^{\mathrm{H}}k_4/{}^{\mathrm{H}}k_{-4}{}^{\mathrm{D}}k_4$ is close to 1 ,the fraction of the intrinsic kinetic isotope effect reflected in the measured kinetic parameters will depend on the magnitudes of the commitment forward and the commitment in reverse, $c_f$ and $c_r$, respectively. If $c_f$ is large, so that $E_4$ is always converted to $E_5$ without reversal from $E_4$ to free reactant, and $c_r$ is small, so that $E_5$ is always converted to products without reversal of the step in which the substituted bond is broken, then no isotope effect will be observed (Figure 4–17A). If $c_r$ is large, so that $E_5$ often reverses to $E_4$ before the first product is released, and $c_f$ is small, so that $E_4$ often reverses to release reactant B, the isotope effect observed will only be the deuterium equilibrium isotope effect of the affected step

$$\frac{{}^{\mathrm{D}}k_{-4}{}^{\mathrm{H}}k_4}{{}^{\mathrm{D}}k_4{}^{\mathrm{H}}k_{-4}} = \frac{{}^{\mathrm{H}}K_{\mathrm{eq}}}{{}^{\mathrm{D}}K_{\mathrm{eq}}} \tag{4–116}$$

and this can be less than 1 or greater than 1 (Figure 4–17B). The full intrinsic isotope effect will be seen only if $c_f$ is small, so that the form of the enzyme immediately preceding the step in which the bond is broken is in equilibrium with free reactant B, and if $c_r$ is small, so that once the form of the enzyme following the step in which the bond is broken is produced, it rapidly yields product (Figure 4–17C)

$$\frac{{}^{\mathrm{H}}V_{\max}/K_{\mathrm{mH}}{}^{\mathrm{B}}}{{}^{\mathrm{D}}V_{\max}/K_{\mathrm{mD}}{}^{\mathrm{B}}} = \frac{{}^{\mathrm{H}}k_4}{{}^{\mathrm{D}}k_4} \tag{4–117}$$

This almost never happens.

It should be noted that formally equivalent circumstances cause the values for the observed acid dissociation constants in the pH–rate profile of an enzymatic reaction to differ from the intrinsic macroscopic acid dissociation constants for the active site. The only difference is that a kinetic isotope effect changes an intrinsic rate constant discontinuously while variation of the pH changes an intrinsic rate constant continuously.

An example of the masking of an intrinsic deuterium isotope effect by another step in the reaction occurs with alcohol dehydrogenase.[69] With the native enzyme, the deuterium kinetic isotope effect on $V_{\max}$, ${}^{\mathrm{H}}V_{\max}/{}^{\mathrm{D}}V_{\max}$, is only 1.3 for deuterioethanol as a reactant because the dissociation of NADH as a product, which is a step subsequent to hydride transfer, is the rate-limiting step in the reaction. When the enzyme is modified at several of the lysines within its active site by amidination with the methyl imidate of 2-pyridinecarboxylic acid, the dissociation rate of NADH is increased more than 30-fold. With the modified enzyme, the deuterium kinetic isotope effect on $V_{\max}$ is 5. The increase observed in the deuterium kinetic isotope effect arises because the step in which the hydride is transferred has become rate-limiting rather than the subsequent dissociation of NADH.

Because numerical values for the commitments, $c_f$ and $c_r$, are almost never available, a method for determining the intrinsic kinetic isotope effect has been proposed,[61] which is based on Equation 4–106. If the equilibrium constant for the enzymatically catalyzed reaction is not affected significantly by the substitution of deuterium for hydrogen or tritium for hydrogen, then

$$\frac{\dfrac{{}^{\mathrm{H}}V_{\max}/K_{\mathrm{mH}}{}^{\mathrm{B}}}{{}^{\mathrm{D}}V_{\max}/K_{\mathrm{mD}}{}^{\mathrm{B}}} - 1}{\dfrac{{}^{\mathrm{H}}V_{\max}/K_{\mathrm{mH}}{}^{\mathrm{B}}}{{}^{\mathrm{T}}V_{\max}/K_{\mathrm{mT}}{}^{\mathrm{B}}} - 1} = \frac{\dfrac{{}^{\mathrm{H}}k}{{}^{\mathrm{D}}k} - 1}{\left(\dfrac{{}^{\mathrm{H}}k}{{}^{\mathrm{D}}k}\right)^{1.44} - 1} \tag{4–118}$$

Even though very accurate values for the deuterium and tritium isotope effects are required, this equation has been used successfully to determine, among others, the intrinsic kinetic isotope effect in the reaction catalyzed by dopamine $\beta$-monooxygenase[70]

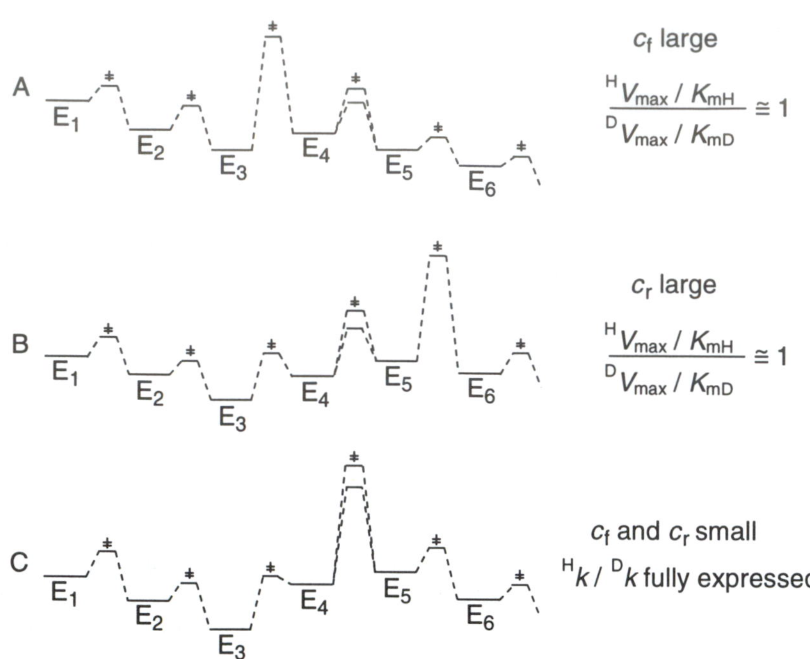


**Figure 4–17:** Diagrammatic representations of various possible reaction coordinates for Mechanism 4–108. (A) Example of a reaction coordinate in which the commitment forward is large because the free energy of activation of the step ($E_3 \rightarrow E_4$) preceding the step in which the intrinsic isotope effect is exerted ($E_4 \rightarrow E_5$) is large. (B) Example of a reaction coordinate in which the commitment in reverse is large because the free energy of activation of the step ($E_5 \rightarrow E_6$) following the step in which the intrinsic isotope effect is exerted ($E_4 \rightarrow E_5$) is large. (C) Example of a reaction coordinate in which the free energy of activation of the step in which the intrinsic isotope effect is exerted ($E_4 \rightarrow E_5$) is so large that both the commitment forward and the commitment in reverse are small.

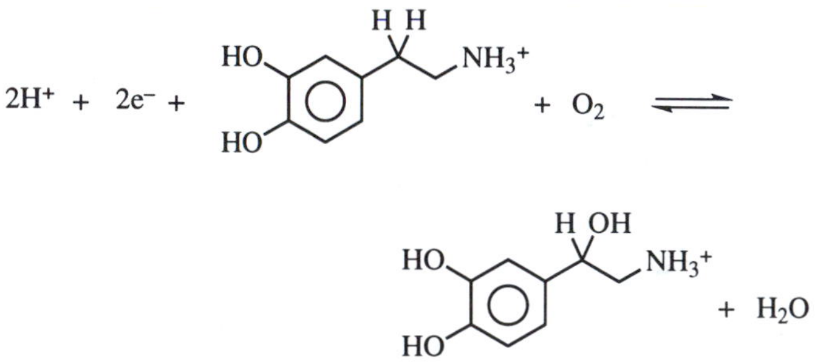

$2H^+$ + $2e^-$ + (HO, HO, H H, $NH_3^+$) + $O_2$ $\rightleftharpoons$

(HO, HO, H OH, $NH_3^+$) + $H_2O$

(4–119)

Even though the observed deuterium kinetic isotope effect on $V_{max}/K_m{}^{DAm}$ when [2,2-$^2H_2$]dopamine (DAm) was used as a reactant was only 2.8 and the observed tritium kinetic isotope effect on $V_{max}/K_m{}^{DAm}$ when [2-$^3$H]dopamine was used as a reactant was only 6.0, the intrinsic deuterium isotope effect calculated with Equation 4–118 is 9. The magnitude of this value was presented as evidence that the transition state in the step at which the hydrogen is removed is symmetric and the equilibrium constant for this step is close to one.

Malate dehydrogenase (oxaloacetate-decarboxylating) (NADP) catalyzes the reaction

$$\text{malate} + \text{NADP}^+ \rightleftharpoons \text{pyruvate} + CO_2 + \text{NADPH} \qquad (4\text{–}120)$$

in which a hydride is removed from malate and carbon dioxide is removed from the intermediate oxaloacetate. By measuring deuterium, tritium, and $^{13}$carbon kinetic isotope effects and applying both Equations 4–115 and 4–118 as well as other kinetic measurements, it was possible to calculate[71] an intrinsic deuterium isotope effect for the hydride transfer from malate to $NADP^+$ of 6. This value is in the range expected for an isoergonic hydride transfer with a symmetric transition state if bending vibrational modes do not contribute significantly to the free energy of activation. The directly observed deuterium kinetic isotope effect on $V_{max}/K_m{}^{malate}$, however, is only 1.5 at all values of pH,[68] because the hydride transfer is not the only slow step in the overall reaction.

The magnitude of an intrinsic deuterium kinetic isotope effect has been used as an indicator of alterations in the transition state caused by variation in the substrates.[72] The reduction potential of the hydride acceptor in the oxidation–reduction of benzyl alcohol catalyzed by alcohol dehydrogenase was changed from –0.32 to –0.26 V by using an analogue of $NAD^+$, and the intrinsic deuterium kinetic isotope effect observed with [1,1-$^2H_2$]benzyl alcohol increased from 4 to 6.5. This increase was presented as evidence that as the reduction potential became more positive, the transition state for hydride transfer became more symmetric. As an increase in reduction potential should cause the transition state to occur earlier, the change is proposed to be between a late transition state and a symmetric transition state.

As with the effects of pH on the kinetics of an enzymatic reaction, the interpretation of a kinetic isotope effect is more straightforward if the step in the mechanism at which the hydrogen is removed can be studied directly. When a high molar concentration of amine oxidase (copper-containing), in the oxidized form, was mixed with a high concentration of dopamine, the transfer of the hydride from the dopamine to the coenzyme on the protein[73] could be followed directly by its change in absorbance.[74] The rate of the reaction was extrapolated to saturation with dopamine to eliminate the contributions of the rate of association of the dopamine with the active site. The deuterium kinetic isotope effect upon the rate of reduction observed when [1,1-$^2H_2$]dopamine was used as reactant was 13. In the mechanism usually written for this reaction the hydrogen in question is removed as a proton by a base from the active site,[73] and the observed kinetic isotope effect is assumed to be the intrinsic deuterium kinetic isotope effect for this step. When $V_{max}/K_m^{DAm}$ was assessed, however, the deuterium kinetic isotope effect was only 6. When the same experiments were performed with benzylamine,[75] however, the deuterium kinetic isotope effect observed directly (13) was the same as the deuterium kinetic isotope effect (12) on $V_{max}/K_m$. These results reiterate the fact that deuterium kinetic isotope effects on $V_{max}/K_m$ can be influenced by steps in the mechanism other than the acid–base reaction of interest.

**Solvent kinetic isotope effects** are the effects on the rate of a reaction that result when an isotopic form of the solvent is substituted for the naturally occurring isotopic form. The most common solvent kinetic isotope effect is that observed when $^2H_2O$ is used as solvent instead of $^1H_2O$. It is possible to use the existence of a solvent kinetic deuterium isotope effect on the kinetic parameters $V_{max}$ or $V_{max}/K_m$ or both as evidence for a contribution of proton transfer to a step in the enzymatic mechanism.[76] The quantitative interpretation of the observed effects is similar to that for a specific deuterium isotope effect (Mechanism 4–108 and Equation 4–115) except that the proton being transferred cannot be pinpointed and it is possible for the rates of several steps in the mechanism to be affected rather than just one. These drawbacks make the interpretation far less informative.

In simple situations, it is also possible to use solvent deuterium isotope effects, if they exist, to determine how many protons are being transferred in the transition state of the step involving the proton transfer.[77] If there is only one proton being transferred in the transition state of one of the slow steps of the reaction and if, as a result, the reaction displays a solvent deuterium kinetic isotope effect, there should be a linear relationship between the fraction of the hydrogen in the solvent that has been replaced by deuterium and the numerical value of the parameter that displays the isotope effect. As hydrogen is replaced by deuterium, the rate should decrease linearly in concert. If, however, two protons are being transferred in the transition state of the slow step in the reaction producing the solvent kinetic isotope effect, the numerical value of the parameter that displays the isotope effect should be a quadratic function of the fraction of the hydrogen that has been replaced with deuterium (Figure 4–18).[78]

For such an interpretation to be valid in a complex, multistep reaction such as that catalyzed by an enzyme, only one step can be responsible for the entire observed solvent deuterium kinetic isotope effect. If the observed kinetic isotope effect is the combination of kinetic isotope effects on several steps, simple interpretations of the effect of the fraction of deuterium in the solvent on the observed kinetic parameter are not possible. For example, it was necessary to conclude that the solvent deuterium kinetic isotope effect on $k_{cat}$ that was observed with the proteinase from human immunodeficiency virus 1 was entirely the result of an effect on the step in which the tetravalent intermediate in the hydrolysis of the peptide bond decomposed to amine and carboxylic acid (Figure 1–7). Otherwise, it could not have been concluded that the quadratic behavior of this kinetic isotope effect as a function of the mole fraction of deuterium oxide in the solvent (Figure 4–18) was evidence that two protons were being transferred in the transition state of this step.[78]

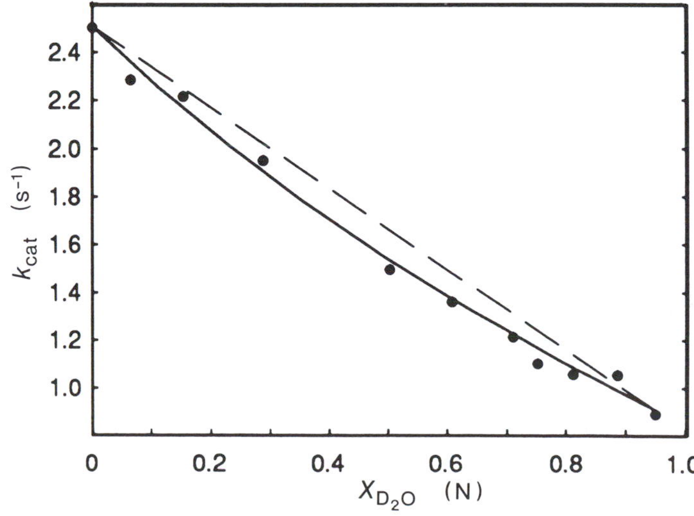


**Figure 4–18:** Variation in the numerical value of the turnover number of the proteinase from human immunodeficiency virus 1 with changes in the mole fraction ($X_{D_2O}$) of $^2H_2O$ in the solvent.[78] Solutions containing various concentrations of the peptide acetyl-RASQNYPVV-amide were prepared at pH of 6 and 37 °C at an ionic strength of 0.2 M. The solutions had various mole fractions of $^2H_2O$ in $^1H_2O$. The hydrolysis of the peptide at the tyrosine was initiated by adding the enzyme, and the reaction was quenched at successive time intervals by mixing samples with trifluoroacetic acid. The rate of product formation was assessed directly by high-pressure liquid chromatography. For each mole fraction of $^2H_2O$, the initial velocities at several concentrations of acetyl-RASQNYPVV-amide were used to obtain $V_{max}$ and, from $V_{max}$, $k_{cat}$. The turnover number, $k_{cat}$ (second$^{-1}$), is presented as a function of the mole fraction ($X_{D_2O}$) of $^2H_2O$. The curve that has been fit to the data is for the quadratic equation $k_{cat} = 2.5(1 - 1.57X_{D_2O})^2$. The dashed line is drawn arbitrarily between the data point at $X_{D_2O} = 0$ and the data point at $X_{D_2O} = 0.96$ to demonstrate the deviation of the data points from linear behavior. Adapted with permission from ref 78. Copyright 1991 American Chemical Society.

## *Suggested Reading*

Farnum, M., Palcic, M., & Klinman, J.P. (1986) pH Dependence of Deuterium Isotope Effects and Tritium Exchange in the Bovine Plasma Amine Oxidase Reaction: A Role for Single Base Catalysis in Amine Oxidation and Imine Exchange, *Biochemistry 25*, 1898–1904.

### PROBLEM 4–7

Enolase catalyzes the reaction

2-phosphoglycerate ⇌ phospho*enol*pyruvate

(A) Write a mechanism for this reaction indicating all removals and additions of protons. Incorporate acid (–AH) and base (–B⊖) groups on the enzyme into your mechanism.

2-Phospho[2-$^2$H]glycerate was synthesized as a reactant for enolase. The following data were obtained under the same set of conditions (0.9 $\mu$g of enolase mL$^{-1}$, 30 °C, pH 7.81) for the two reactants.[79]

| 2-phospho[2-$^2$H]glycerate | | 2-phospho[2-$^1$H]glycerate | |
|---|---|---|---|
| [reactant] (mM) | $v_0$ ($\Delta A_{230}$ min$^{-1}$) | [reactant] (mM) | $v_0$ ($\Delta A_{230}$ min$^{-1}$) |
| 0.25 | 0.480 | 0.25 | 0.228 |
| 0.10 | 0.369 | 0.10 | 0.169 |
| 0.05 | 0.274 | 0.05 | 0.123 |
| 0.025 | 0.180 | 0.025 | 0.086 |

where $v_0$ is the initial velocity of the dehydration reaction.

(B) Determine $V_{max}$ and $K_m$ for each substrate under these conditions.

The parameter $V_{max}$ for the enzymatic reaction was measured in the direction of dehydration, as a function of pH for the reactant containing protium and the reactant containing deuterium, respectively.[79]

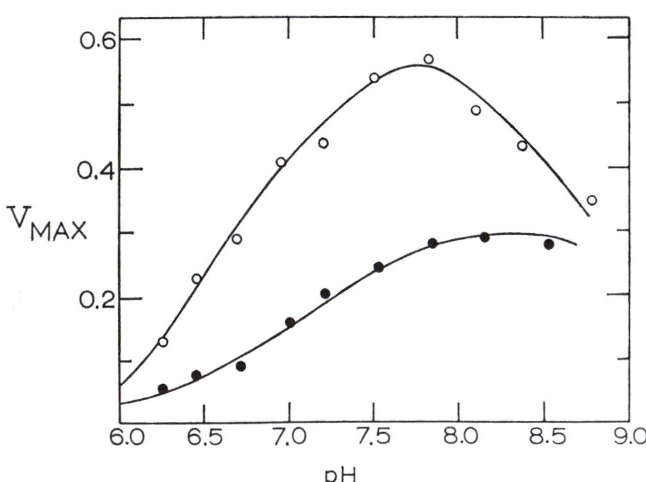


Maximum velocities for protiated (○) and deuterated (●) reactants as a function of pH. Reprinted with permission from ref 79. Copyright 1973 American Chemical Society.

(C) Explain, by a series of acid–base equilibria, why $V_{max}$ for the protiated reactant shows a maximum.

When the ratio of $V_{max}$ for the two reactants is plotted as a function of pH, the following results were obtained[79] (adapted with permission from ref 74; copyright 1973 American Chemical Society).

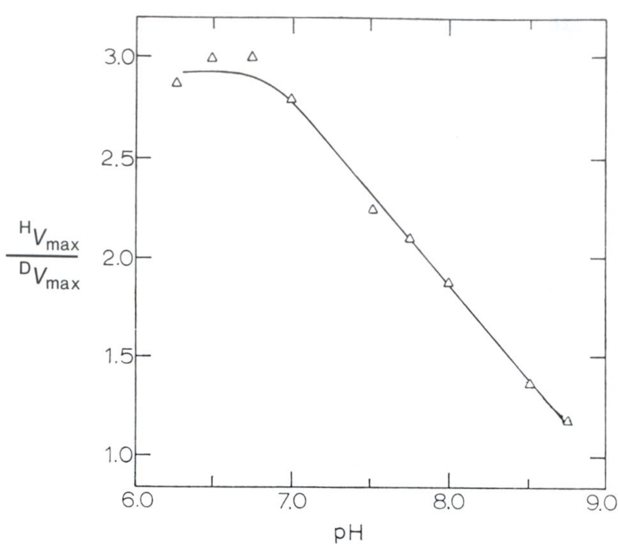


(D) Explain, in terms of the considerations you made in parts A–C, why the importance of the deuterium isotope effect diminishes as the pH is increased into the range where $^{H}V_{max}$ is decreasing.

### PROBLEM 4–8

The enzyme steroid Δ-isomerase catalyzes the following reaction

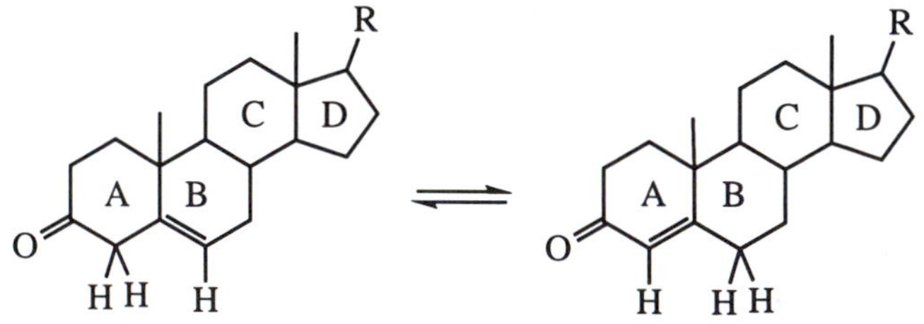

(1)

The enzyme has been purified from *Pseudomonas testosteroni* and is used for the synthesis of steroids in the pharmaceutical trade. The enzyme is rather tolerant about R, using many steroids as substrates. Since the chemical transformation occurs in only the A and B rings, the C and D rings will be ignored.

The same transformation catalyzed by the enzyme occurs spontaneously, albeit at a much reduced rate, under acidic conditions. The mechanism of this nonenzymatic reaction is

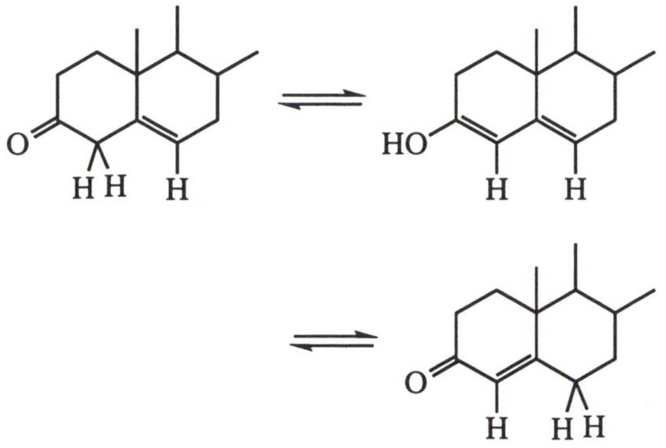

(2)

The sawhorse structure of the A and B rings of a ketosteroid substrate is

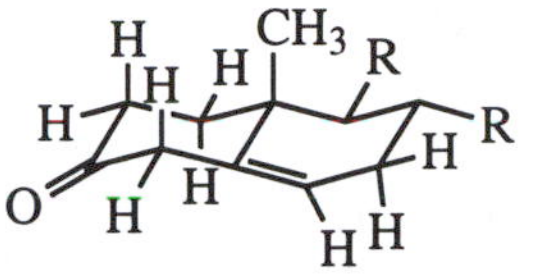

where $R_1$ and $R_2$ indicate the remainder of the steroid molecule. Build just this structure with molecular models.

When the following deuterated reactants were synthesized and mixed with the enzyme in buffered $H_2O$, the noted products were obtained.[79a] The numbers in parentheses below each structure are the moles of deuterium for each mole of the compound.

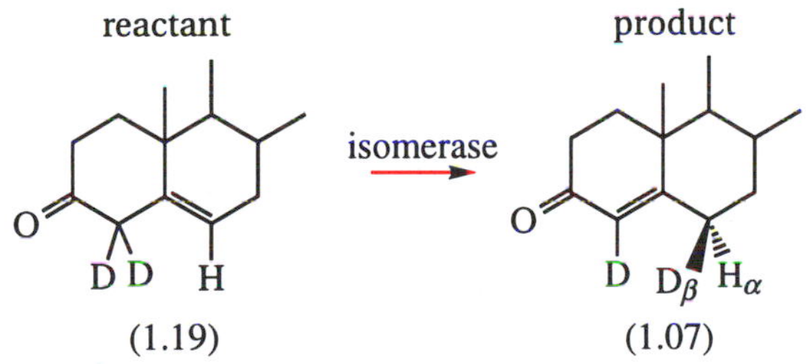

reactant → isomerase → product
(1.19) (1.07) (3)

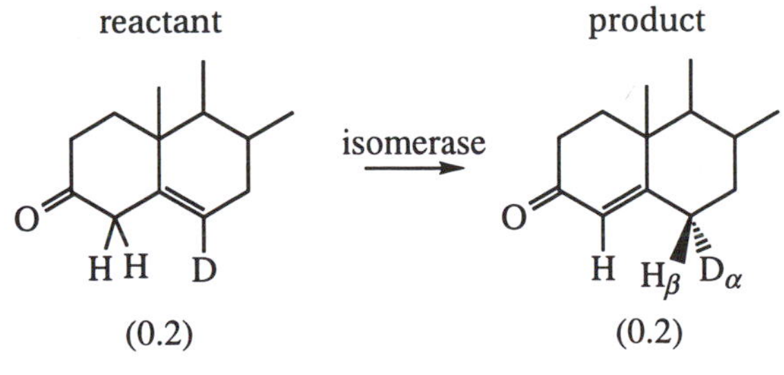

reactant → isomerase → product
(0.2) (0.2) (4)

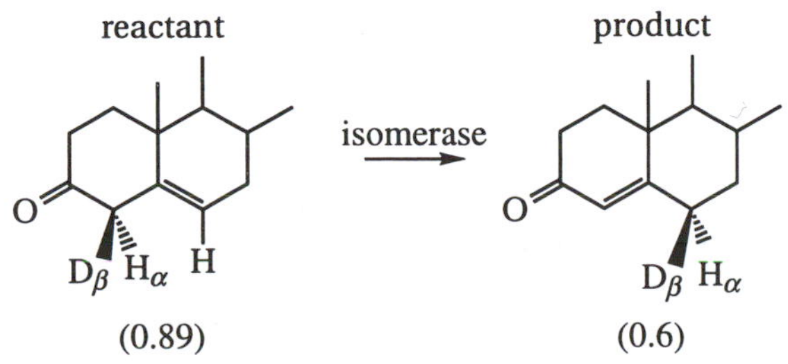

reactant → isomerase → product
(0.89) (0.6) (5)

In each case the reaction is written as if it were irreversible because the equilibrium strongly favors the conjugated product. The absolute configuration of each deuterated product was verified by infrared spectroscopy; and its deuterium content, by mass spectroscopy.

The kinetics of Reaction 3, in which the 4$\beta$-monodeuterated ketosteroid was used as substrate, were compared with those of the reaction in which the undeuterated ketosteroid of the same structure was used as substrate. It was determined that

$$\frac{^{\mathrm{H}}V_{\mathrm{max}}}{^{\mathrm{D}}V_{\mathrm{max}}} = 5.3$$

(A) Write out a detailed enzymatic mechanism that explains the above observations. Choose specific amino acid side chains to perform various catalytic functions. It is unknown which amino acid side chains actually perform these roles, but you should pick appropriate ones. Start with the reactant bound to the active site and finish with the product bound to the active site. Draw the structure of substrate, intermediates, and product in sawhorse configuration and indicate clearly all stereochemistry. Your mechanism must explain the absolute configuration of the deuterated products and the retention of the deuterium in the product.

(B) When the enzymatic reaction was performed in deuterium oxide ($^2H_2O$) with an undeuterated reactant, 0.12 mol of deuterium was incorporated at the 6$\beta$ position for every mole of product produced. As Reactions 3 and 5 proceed, some deuterium is lost to the solvent. Explain all of these observations by referring to your mechanism in part A. At which step is the deuterium gained from or lost to the solvent and how does this happen? Present your answer in a series of acid–base equilibria.

The magnitude of the deuterium isotope effect observed in $V_{\mathrm{max}}$ indicates that the breaking of the carbon–hydrogen bond is the rate-limiting step in the reaction. If this is the case, all subsequent steps are very rapid and can be ignored. Convince yourself that, under these circumstances, the following kinetic mechanism is appropriate.

$$\mathrm{E} + \mathrm{A} \underset{k_{-1}}{\overset{k_1}{\rightleftharpoons}} \mathrm{E{\cdot}A} \xrightarrow{k_2} \mathrm{E} + \mathrm{P}$$

The $K_{\mathrm{m}}$ for monodeuterated reactant in Reaction 5 is $^{\mathrm{D}}K_{\mathrm{m}} = 1.4 \times 10^{-4}$ M, while that for the same reactant when it is undeuterated is $^{\mathrm{H}}K_{\mathrm{m}} = 3.1 \times 10^{-4}$ M. The turnover number for the undeuterated substrate, $k_{\mathrm{cat}}$, is $17 \times 10^6$ min$^{-1}$.

(C) Make the reasonable assumption that the presence of a deuterium at the 4$\beta$ position has no effect on the binding of substrate to the enzyme

$$^{\mathrm{D}}k_1 = {}^{\mathrm{H}}k_1 = k_1 \qquad {}^{\mathrm{D}}k_{-1} = {}^{\mathrm{H}}k_{-1} = k_{-1}$$

Explain mathematically why $^{\mathrm{H}}K_{\mathrm{m}} \neq {}^{\mathrm{D}}K_{\mathrm{m}}$, even though binding is unaltered, and determine definite values for $k_1$, $k_{-1}$, and $^{\mathrm{D}}k_2$.

(D) Draw the sawhorse structure of the A and B rings of a ketosteroid; and, on your drawing, indicate how the enzyme could strain it and cause it to resemble the intermediate in the enzymatic reaction.

## PROBLEM 4–9

The enzyme adenine phosphoribosyltransferase catalyzes the following reaction

adenine + magnesium 5-phospho-$\alpha$-D-ribose 1-pyrophosphate
$\rightleftharpoons$ adenosine 5′-phosphate + magnesium pyrophosphate

(A) Draw a chemical mechanism for the enzymatic reaction as it would occur within the active site. Use specific amino acid residues as general acids and bases. Draw the reaction with proper stereochemistry at carbon 1 of the ribose throughout.

The kinetics of the enzymatic reaction have been examined as a function of pH (Figure A)[80] and a solvent deuterium kinetic isotope effect has been observed when the enzymatic reaction is carried out in $^2H_2O$ (Figure B).[80]

(B) What appears to be the p$K_a$ of the group in the active site whose ionization causes the behavior observed in Figure A? Is it a general acid or a general base in the mechanism of the enzymatic reaction? What form of the enzyme is titrating? What side chain in your mechanism could be the group that is titrating?

The possibility can now be examined that the deuterium isotope effect seen in Figure B could arise from the removal of a deuterium atom from $N^6$,$N^6$,$N^9$-trideuteroadenine during a rate-limiting step in the enzymatic reaction. In $^2H_2O$ the only form of adenine present is $N^6$,$N^6$,$N^9$-trideutero-adenine.

(C) From the value for the equilibrium constant for the following acid–base reaction

$$B^{\ominus} + \text{(adenine, N9–H; NH}_2\text{)} \underset{K_{eq}}{\rightleftharpoons} BH + \text{(adenine anion, N9}^{\ominus}\text{; NH}_2\text{)}$$

where BH has a p$K_a$ = 6.0 and adenine has a p$K_a$ = 9.8 and, with the assumption that proton transfer in a favored direction is diffusion-controlled

$$BH + \text{(adenine anion, N9}^{\ominus}\text{; NH}_2\text{)} \xrightarrow{10^{10}\ M^{-1}s^{-1}} B^{\ominus} + \text{(adenine, N9–H; NH}_2\text{)}$$

calculate the rate constant $k_B$ for the reaction

**A**

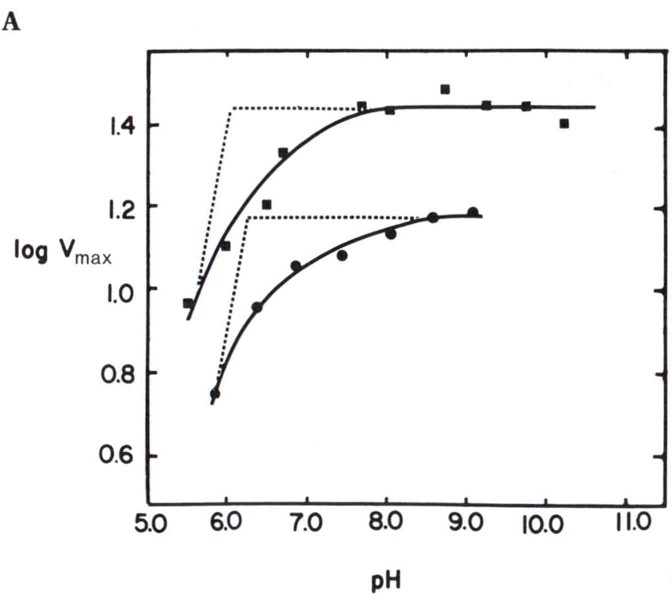


**Figure A:** Effect of pH and temperature on the logarithm of the maximum velocity.[80] Initial velocity was measured as a function of phosphoribose pyrophosphate concentration and of adenine concentration at a series of fixed concentrations of the other reactant in 0.15 M histidine–Tris–glycine buffers and 1 mM $MgSO_4$ at 10 °C (●), and 30 °C (■). Different enzyme concentrations were used at each temperature. Maximum velocity was obtained from replots of ordinates of double-reciprocal plots of these data. Adapted with permission from ref 80. Copyright Academic Press.

**B**

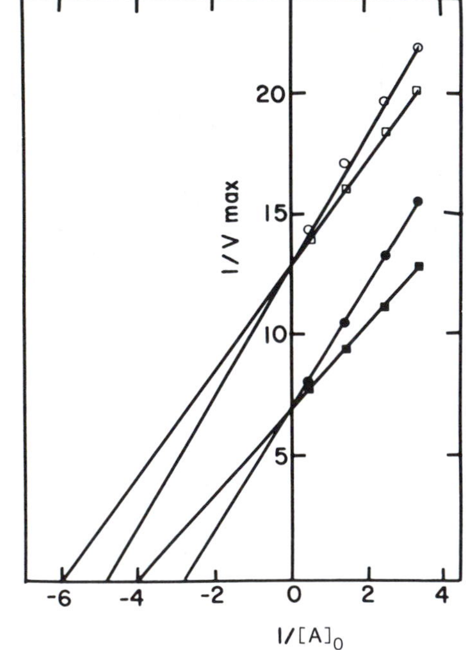


**Figure B:** Effect of $^2H_2O$ on the maximum velocity and the Michaelis constants.[80] Initial velocity was measured as a function of phosphoribose pyrophosphate concentration and of adenine concentration at a series of fixed concentrations of the other reactant in 0.1 M Tris buffer, pH 8.1, and 1 mM $MgSO_4$, in $^1H_2O$ and $^2H_2O$. Ordinates of double-reciprocal plots of these data are replotted against the reciprocal of the reactant (A) concentration. Phosphoribose pyrophosphate is the variable reactant in $H_2O$ (■) and $^2H_2O$ (□); adenine is the variable reactant in $H_2O$ (●) and $^2H_2O$ (○). Adapted with permission from ref 80. Copyright Academic Press.

$$B^{\ominus} + \text{adenine (N-H)} \xrightarrow{k_B} BH + \text{adenine anion (N}^{\ominus})$$

(D) Considering the effect of approximation, what would you expect to be the first-order rate constant for proton removal from adenine on the active site?

(E) The turnover number of adenine phosphoribosyltransferase is 7 $s^{-1}$. To what is the deuterium isotope effect not due, and to what else could it be due?

# Stereochemistry

When you observe the skeletal model of a substrate such as (*S*)-malate

COO⁻ / H, OH / H, H / COO⁻

**4–17**

you can readily distinguish the two hydrogens on carbon 3 because you are a diastereomeric object. If carbon 2 is placed behind and carbon 4 is placed in front and down, as in the view of Structure **4–17**, one of the hydrogens is to the right and the other is to the left. No matter how the skeletal model was rotated in space or about its bonds you could always identify the right hydrogen and the left hydrogen. A **prochiral atom** in a molecule is an atom that is achiral but that could become chiral with a single substitution or addition. As with the term *chiral*, *prochiral* refers to an atom in the molecule and not to the molecule itself. Carbon 3 of (*S*)-malate is prochiral because if one of the hydrogens were substituted with a methyl group, carbon 3 would become chiral. This substitution is formally equivalent to distinguishing the left hydrogen from the right hydrogen. Any tetrahedral carbon on which reside any two chemically identical substituents, which can nevertheless be distinguished as unique by a diastereomeric observer, is a prochiral carbon.

Properly, the prochiral, tetrahedral carbon to which the substituents that make it prochiral are attached is distinguished from the substituents themselves. The substituents are classified as enantiotopic or diastereotopic. If the replacement of one of the two identical substituents on a prochiral carbon with an achiral functional group different from all of the other substituents would create an enantiomeric product, then the two identical substituents are referred to as **enantiotopic substituents**. If the replacement of one of the two identical substituents on a prochiral carbon with an achiral substituent different from all the other substituents would create a diastereomeric product, then the two identical substituents are referred to as **diastereotopic substituents**. The substituents should not be referred to as prochiral substituents, for example, prochiral hydrogens, because this is the term reserved for identifying the central carbon.

The two identical but distinguishable stereoheterotopic substituents on a prochiral, tetrahedral carbon can be unambiguously designated as *pro-R* or *pro-S*.[81] Choose one of the two identical substituents and assume that its priority is greater than the other. If this renders the carbon *S*, the elevated substituent is the *pro-S* substituent; if this renders the carbon *R*, the elevated substituent is the *pro-R* substituent. In the case of (*S*)-malate, the *pro-R* and *pro-S* hydrogens can be designated with subscripts

COO⁻ / H, OH / $H_R$, $H_S$ / COO⁻

**4–18**

They are diastereotopic hydrogens.

Any planar, trigonal carbon that has two chemically identical faces, which can nevertheless be distinguished as unique by a diastereomeric observer, is also a prochiral carbon. The two faces themselves are referred to as enantiotopic faces or diastereotopic faces to distinguish them from the central carbon itself, which is prochiral. Again, if addition to the prochiral carbon produces an enantiomer, the faces are enantiotopic; if addition produces a diastereomer, the faces are diastereotopic. The two stereoheterotopic faces are designated *re* and *si*.[81] If the order of substituents around the trigonal carbon is clockwise, then you are viewing the *re* face. If the order is counterclockwise, then you are viewing the *si* face. The *re* face of oxaloacetate is

$^{-}OOC-C(=O)-CH_2COO^{-}$

**4–19**

The *si* face is on the opposite side. These two faces, *re* and *si*, are the enantiotopic faces of oxaloacetate.

The active site of an enzyme is also a diastereomeric object. When (*S*)-malate is bound as a reactant to fumarate hydratase, it will be bound with the two carboxylates in the *trans* configuration because the product, fumarate, is the *trans* acid. The two carboxylates will be hydrogen-bonded to particular amino acids at the proper locations, and this will pin the substrate in the *trans* configuration with the carboxylates in the proper orientation to achieve maximal overlap between the $\pi$ molecular orbital systems of the carboxylates and the developing $\pi$ bond between carbons 2 and 3. If there is, within the active site, a general base responsible for removing the proton from carbon 3 during the elimination, it will be positioned so that it can remove only one of the two prochiral hydrogens because, as with most proton transfers, the hydrogen removed must be on the line of centers between carbon 3 and the heteroatom of that general base. During the enzymatic reaction, the hydrogen removed from carbon 3 of every (*S*)-malate will always be the same one, either the *pro-R* or the *pro-S*, because every active site in the solution is stereochemically identical.

Any isotropic solution of achiral molecules, for example, an aqueous solution containing salts and achiral general acids and achiral general bases, will usually make little distinction between two stereoheterotopic hydrogens on a prochiral carbon. For example, if a hydrogen is removed from carbon 3 of (*S*)-malate by an achiral general base during an elimination in free solution, the two hydrogens will usually have similar probabilities of being removed because the preference observed in their removal has its origins only in the fact that (*S*)-malate itself is enantiomeric. It is only because chemists usually perform reactions in isotropic solution that the absolute stereochemistry enforced by an enzyme is considered peculiar.

It is also peculiar because it is arbitrary but absolute. The hydrogen removed or added in a reaction catalyzed by an enzyme is the one that happens to end up in line between a general base in the active site and the carbon of the substrate. This is usually an arbitrary outcome of evolution by natural selection unless there are overriding chemical advantages of one choice over the other. Regardless, however, of which hydrogen happens to end up between the carbon and the general base, once that alignment has occurred and been fixed in the population, the active site has been designed to remove or add only that one stereoheterotopic hydrogen every time the enzyme turns over. For example, every time a molecule of (*S*)-malate is converted to fumarate by fumarate hydratase, the *pro-R* hydrogen is removed,[82,83] not the *pro-S*, because that is the side of the reactant on which the general base in the active site happened to end up as evolution by natural selection proceeded.

If it can be shown that only one of the two stereoheterotopic hydrogens on a chiral carbon is removed or added during an enzymatic reaction, then it is evident that there is a general base in the active site that always removes that proton or that there is a general acid in the active site that always adds that proton, respectively. It should be mentioned that the general base or the general acid could be a molecule of water confined by the active site to act at only one stereoheterotopic position. If, however, the step in the reaction in which the proton is removed or added had occurred in solution, such an absolute preference would not have been displayed. In addition, the identification of the hydrogen operated on by the enzyme positions the general acid–base within the active site.

Diaminopimelate decarboxylase catalyzes the decarboxylation of *meso*-(2*S*,6*R*)-2,6-diaminopimelate. Because the normally encountered (*S*)-lysine is the product, the enzyme must decarboxylate the *R* end of the reactant. When the enzymatically catalyzed reaction is carried out in $^{2}H_2O$

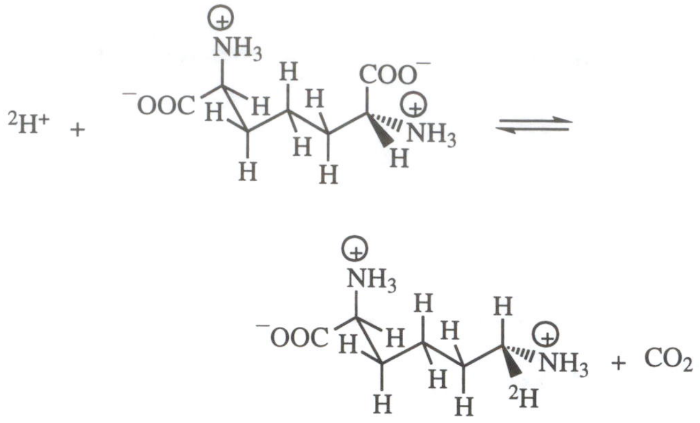

(4–121)

the deuteron that replaces the carbon dioxide is added within the active site so that only the (2*S*,6*R*)-[6-$^{2}$H]lysine is produced.[84] Therefore, in the normal enzymatic reaction carried out in $^{1}H_2O$, the proton that replaces the carbon dioxide always becomes the *pro-R* hydrogen at carbon 6 of the product.

This reaction is catalyzed with participation of pyridoxal phosphate as a coenzyme within the active site. The base to which the proton is added by the general acid within the active site is enamine **4–20**

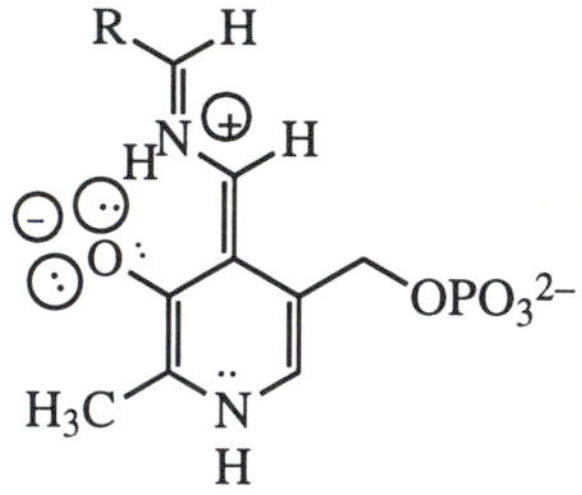

**4–20**

and to explain the observed stereochemistry, the general acid in the active site adding the proton must sit above the plane of the enamine in the view presented. Therefore it adds the proton to the *re* face of the enamine. Therefore, the determination of the stereochemistry of this reaction has revealed the orientation of the general acid within the active site. If the same achiral enamine were formed in free solution, it would be protonated with almost equal probability from either the *si* face or the *re* face by a hydronium cation, a molecule of water, or an achiral general acid because both faces are almost the same to any achiral acid. In free solution in $^{2}H_2O$, both (2*S*,6*S*)-6-deuterolysine and (2*S*,6*R*)-6-deuterolysine would be produced, and the product would be almost completely racemic at carbon 6.

The conclusion that a particular reaction, if carried out in free solution, would lead to a racemic product cannot be drawn unambiguously if the reactant itself, unlike enamine **4–20**, has a chiral center in the vicinity of the site of reaction. For example, (*S*)-5,6-dihydroorotate undergoes hydrogen exchange in free solution in $^{2}H_2O$ 10-fold more rapidly at the *pro-R* position at carbon 5 than at the *pro-S* position[85]

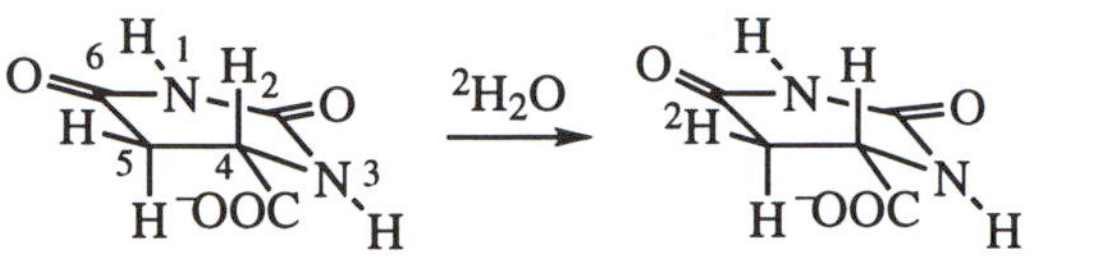

(4–122)

presumably because of the proximity of the chirality at carbon 4. In this instance, however, the enzyme, dihydroorotate dehydrogenase, removes the *pro-S* hydrogen exclusively from carbon 5. This observation demonstrates that the intermediate enol in the enzymatic reaction is generated by the removal of a proton by a general base in the active site positioned by chance to remove the less acidic proton of the reactant. This example reemphasizes the point that the final choice of which hydrogen is removed is unpredictable. Nevertheless, it may be the case that in some instances differences in acidity caused by differences in the electronic configurations in the two possible transition states decide which proton is chosen by natural selection.

The enzymatic reaction catalyzed by diaminopimelate decarboxylase proceeds with inversion of configuration at carbon 6 (Reaction 4–121). When a reaction occurs with **retention of configuration**, the substituent replacing the substituent that was removed occupies the same position relative to the other three substituents on the same carbon. When a reaction proceeds with **inversion of configuration**, the new substituent occupies a position in which the other three substituents have the opposite relative stereochemistry in the product from their relative stereochemistry in the reactant. Many enzymes, using pyridoxal phosphate as a coenzyme, catalyze decarboxylations of amino acids formally equivalent to the decarboxylation catalyzed by diaminopimelate decarboxylase. Usually, each has as its reactant the respective naturally occurring (*S*)-amino acid, and the hydrogen added is the *pro-R* hydrogen in the product.[84] Therefore, these other enzymes usually have, as has diaminopimelate decarboxylase, general acids in their active sites on the *re* face of the respective enamine **4–20**. These other enzymes, however, usually proceed with retention of configuration, while diaminopimelate decarboxylase proceeds with inversion of configuration. Therefore, in these other reactions the general acid–bases catalyzing the decarboxylation step must reside on the same face, the *re* face, as the general acid protonating the enamine in the next step. In the reaction catalyzed by diaminopimelate decarboxylase, however, they must reside on the opposite face, the *si* face, from the general acid protonating the enamine. This comparison illustrates that it is the absolute stereochemistry of the reaction that identifies the absolute position within the active site of the general acid–bases responsible for that particular step.

The designation of retention or inversion, however, elucidates the relative orientation of the catalytic groups. The enzymatic reaction catalyzed by propionyl-CoA carboxylase

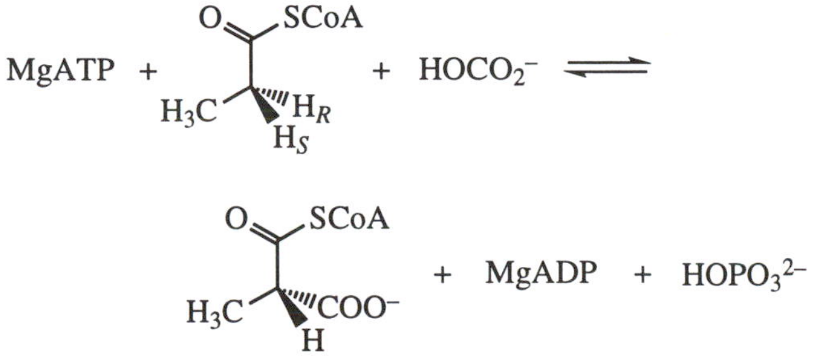

(4–123)

proceeds with retention of configuration at carbon 2. This conclusion is based on two observations. When either deuterium[86] or tritium[87] is substituted for the *pro-S* hydrogen, each is retained, respectively, in the product. When deuterium is substituted for the *pro-R* hydrogen, it is not retained in the product.[87] Therefore, the general base in the active site that removes the proton to produce the enol at carbon 2 must be on the same side of the reactant as the *N*-carboxybiotin that delivers the activated carbonic acid to that same carbon in the enol.

Just as retention or inversion of configuration at a particular chiral or prochiral carbon indicates the relative orientation of the catalytic functional groups within the active site, the same information is gained by a determination of whether an addition–elimination catalyzed by an enzyme proceeds in a *syn* sense or an *anti* sense. In the elimination–addition catalyzed by enolase

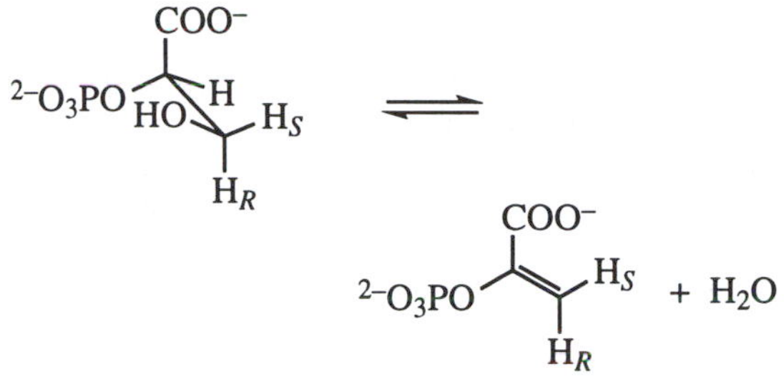

(4–124)

the reaction proceeds with the *anti* elimination of water, because the *pro-R* hydrogen of the 2-phosphoglycerate is found *cis* to the phosphate in the phospho*enol*pyruvate.[88] Therefore, in the active site of the enzyme, the general base that removes the hydrogen from carbon 2 is on the opposite side of the bound reactant from the general acid that activates the hydroxyl.

In aqueous solution, addition–elimination reactions following an E2 mechanism usually proceed in an *anti* sense, with the general base removing the proton from the side of the reactant opposite the side from which the leaving group departs.[89] This preference may be due entirely or only in part to steric effects. Many enzymatic reactions also show this *anti* preference. Histidase catalyzes the *anti* elimination of ammonium from histidine[90,91]

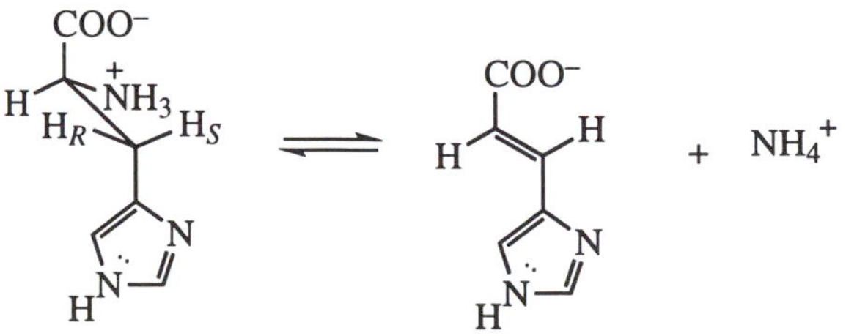

(4–125)

because the *pro-R* hydrogen is removed during the reaction.[90,91] The addition of water to the aliphatic olefin in oleic acid, catalyzed by an enzyme from a pseudomonad, also proceeds in an *anti* sense

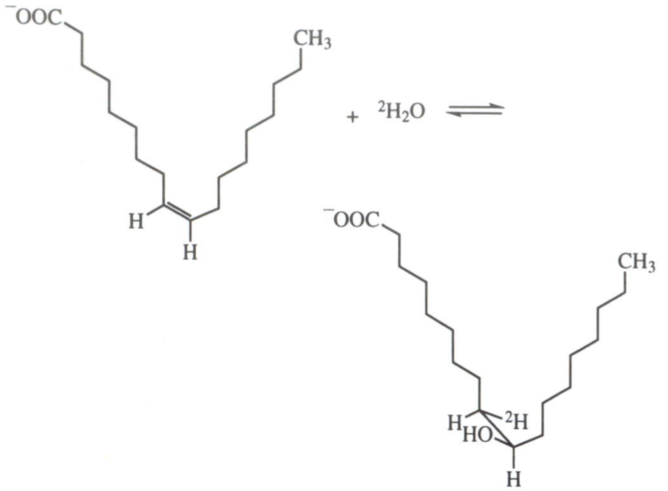

(4–126)

because deuterium is incorporated at the *pro-R* position when the reaction is carried out in $^2H_2O$.[92]

It has been found that certain eliminations, in particular, eliminations proceeding through an E1cB mechanism or E2 eliminations involving ion pairing between the general base and the leaving group, can yield variable percentages of the products of *syn* elimination.[89] A particularly relevant example for the present purposes is the elimination of acetic acid from (2*R*,3*R*)-*S*-*tert*-butyl-3-acetoxy[2-$^2$H]butanethioate[93]

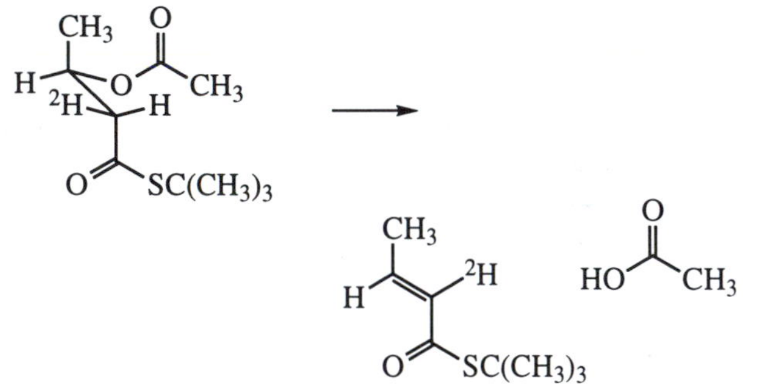

(4–127)

The yield of deuterated product (43% of the total yield of the *E* isomer) was used to calculate that when carbon 2 has two hydrogens and no deuterium isotope effect is operating, the yield of *syn* product would be 14%. This reaction probably proceeds through an E1cB mechanism

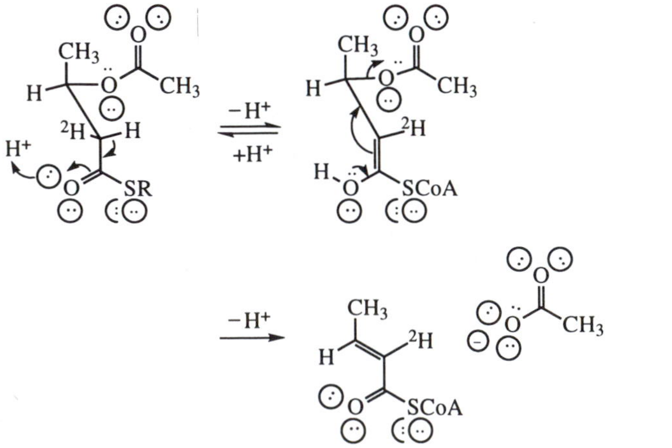

(4–128)

In any event, the product ratios observed suggest that the difference in free energy of activation between the *syn* and *anti* elimination–additions is probably fairly small and can be influenced considerably by the conditions.

*Syn* elimination–additions adjacent to carbonyl groups or acyl groups have often been observed in enzymatically catalyzed reactions. The dehydration of (2*R*,3*S*)-3-hydroxy[2-$^3$H]butyryl-SCoA to produce crotonyl-SCoA, catalyzed by enoyl-CoA hydratase

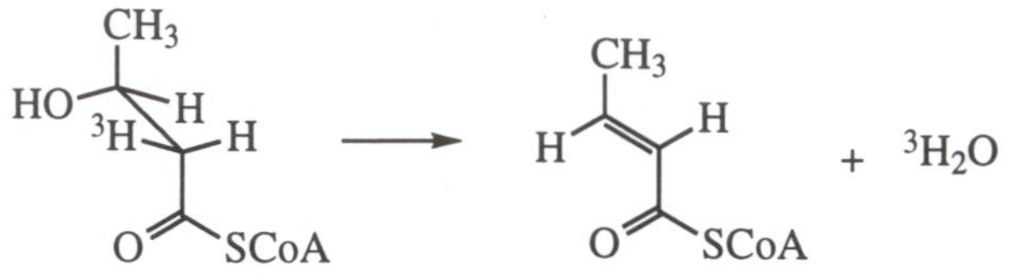

(4–129)

produces a product that has only 20% the specific radioactivity of the reactant. Therefore, even though the tritium kinetic isotope effect would be against the reaction as it is found to occur, the enzyme nevertheless removes the triton because it is in the *pro-R* position and aligned with the general base in the active site. The hydrogen in the *pro-S* position is not removed, even though the carbon–hydrogen bond is easier to break, because it is on the opposite side of the reactant as it is held in the active site with the methyl and acyl carbons in the necessary *trans* configuration. The small amount of tritium remaining in the product probably arises from nonenzymatic racemization of the reactant during its synthesis or during the incubation

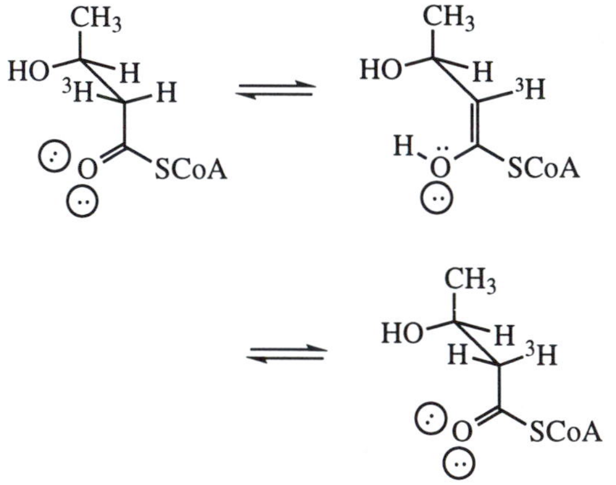

(4–130)

This is a common problem in stereochemical experiments.

Enzymatically catalyzed dehydrogenation–hydrogenations, as opposed to dehydration–hydrations, usually proceed with *syn* stereochemistry.[94] The dehydrogenation of 5-$\alpha$-cholestan-3$\beta$-ol to $\Delta^7$-5$\alpha$-cholestan-3$\beta$-ol, catalyzed by an enzyme in cockroaches

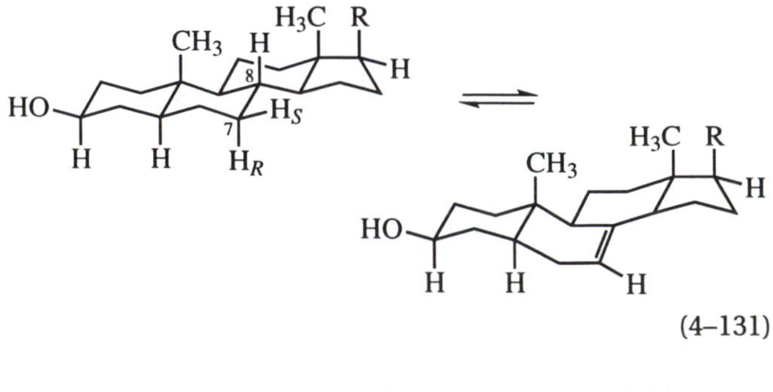

(4–131)

proceeds with the removal of the methine hydrogen at carbon 8 and the *pro-S* hydrogen at carbon 7 because when (7*S*)-[7-$^3$H]-5$\alpha$-cholestan-3$\beta$-ol was fed to roaches, the tritium was removed (>90%), but when (7*R*)-[7-$^3$H]-5$\alpha$-cholestan-3$\beta$-ol was fed to roaches, the tritium was retained (>85%) in the product.[95] There is, however, at least one dehydrogenation that proceeds with *anti* stereochemistry. Acyl-CoA dehydrogenase converts butyryl-SCoA to crotonyl-SCoA by removing the *pro-R* hydrogen[96] from carbon 3 and the *pro-R* hydrogen[97] from carbon 2

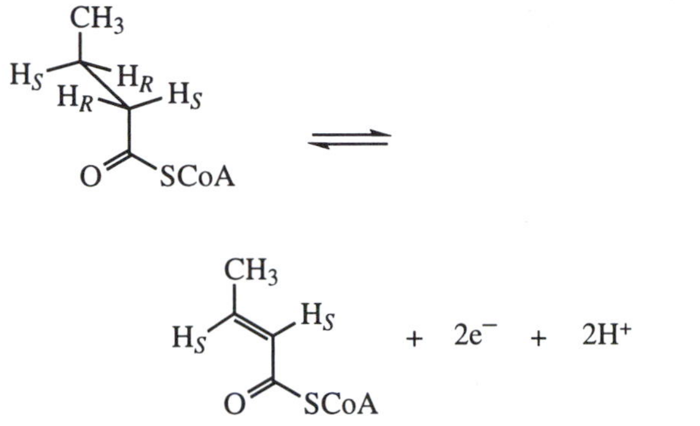

(4–132)

which is an *anti* dehydrogenation. Examples of both of the stereochemical outcomes in both elimination–addition and dehydrogenation–hydrogenation reactions again illustrate the fact that the stereochemical outcome of an enzymatic reaction often depends on the adventitious positioning of the general acids and bases and the coenzymes within the active site relative to the bound reactant, rather than the innate electronic preference of the reaction.

The isomerizations catalyzed by a set of related enzymes, glucose-6-phosphate isomerase, ribose-5-phosphate isomerase, mannose-6-phosphate isomerase, triosephosphate isomerase, xylose isomerase, L-arabinose isomerase, and L-fucose isomerase, are addition–elimination reactions whose common mechanism has been demonstrated by stereochemical studies. The natural aldose substrate in each of these reactions is chiral at carbon 2, but both possible chiralities, *S* and *R*, are represented within the set. The ketose substrate has a prochiral carbon where the aldehyde of the corresponding aldose is located, and the identity of the added hydrogen at this prochiral carbon of the ketose has been determined in each case. When the aldose is *R* at carbon 2 (D-xylose, D-glyceraldehyde 3-phosphate, D-glucose 6-phosphate, D-ribose 5-phosphate, and L-arabinose), the ketose produced in the enzymatic reaction has the added hydrogen at the *pro-R* position

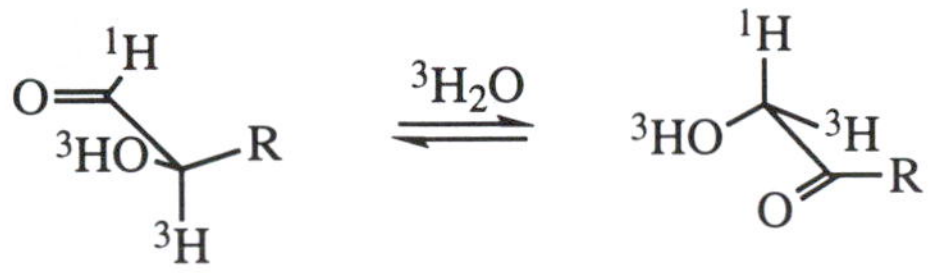

(4–133)

When the aldose is *S* at carbon 2 (D-mannose 6-phosphate, L-fucose), the ketose produced has the added hydrogen at the *pro-S* position

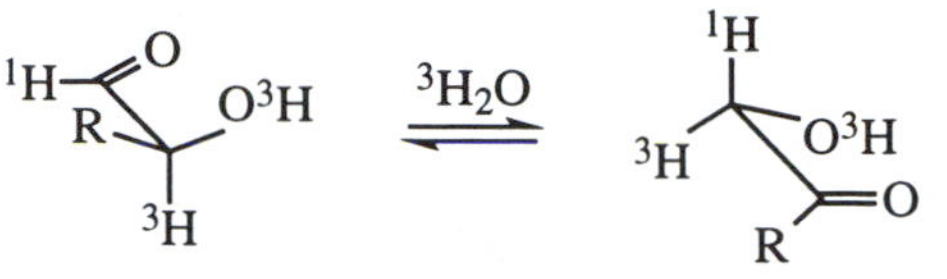

(4–134)

These conclusions were reached by bringing the reactions to equilibrium in $^3H_2O$, isolating the products, and determining their absolute stereochemistries.[98]

This set of enzymes also displays various degrees of internal hydrogen transfer between carbons 2 and 1 during the isomerization. For example, when the product of the reaction is trapped by glucose-6-phosphate dehydrogenase, glucose-6-phosphate isomerase will transfer a tritium from carbon 1 of D-fructose 6-phosphate to carbon 2 of D-glucose 6-phosphate

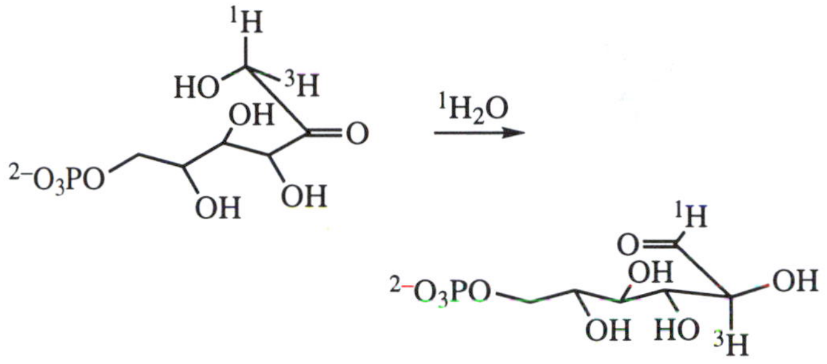

(4–135)

when the reaction is run in $^1H_2O$.[99] The ratio between transfer of the tritium intramolecularly and its loss to solvent while it is briefly located on the general base responsible for removing it and then adding it back varies from 5 to 0.3 depending on the conditions. Similar results have been observed with D-mannose 6-phosphate and mannose-6-phosphate isomerase;[100] and, in the case of xylose isomerase, the transfer of a proton back and forth between carbons 1 and 2 occurs with no loss to or incorporation from the solvent.[98] All of these facts have led to the conclusion that a single respective general base–acid in the active site of each of these isomerases removes the proton to form a *cis*-enediol, moves across the face of the planar *cis*-enediol and adds the proton back to the *cis*-enediol from the same face from which it was removed

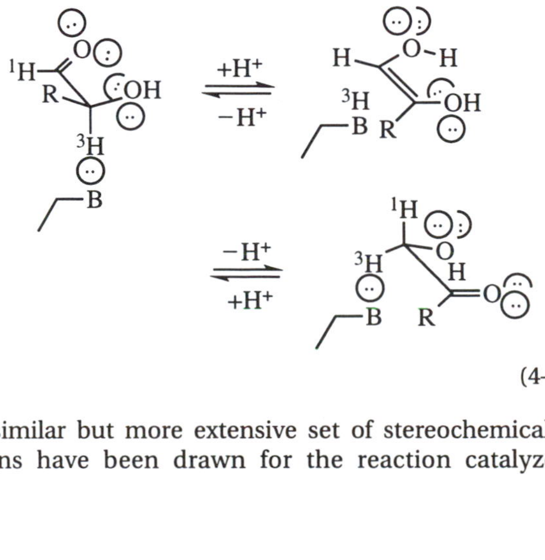

(4–136)

A similar but more extensive set of stereochemical conclusions have been drawn for the reaction catalyzed by

*O*-succinylhomoserine (thiol)-lyase[101,102] (Reaction 2–19). The stripped intermediate in the reaction catalyzed by this enzyme is an adduct between pyridoxal phosphate and vinylglycine that has lost the $\alpha$-hydrogen. In the normal reaction, this intermediate is formed from *O*-succinylhomoserine by removal of the $\alpha$-hydrogen, protonation of the aldehyde carbon of the pyridoxal phosphate, removal of the $\beta$-hydrogen, and expulsion of the leaving group. The enzyme, however, will also accept vinylglycine as a reactant and produce cystathionine

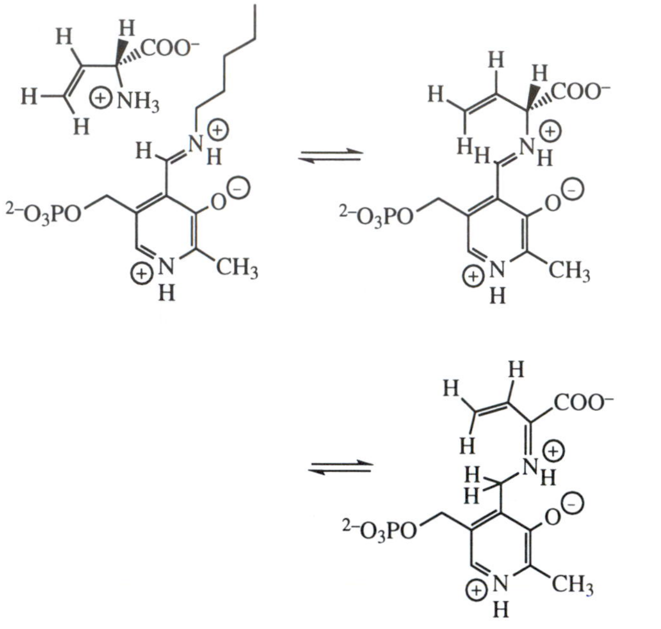

(4–137)

Both (*E*)-[3,4-$^2$H]vinylglycine and (*Z*)-[4-$^2$H]vinylglycine were synthesized chemically, and the absolute stereochemistries of the cystathionine produced in the normal reaction and the $\alpha$-ketobutyrate produced in an abortive reaction

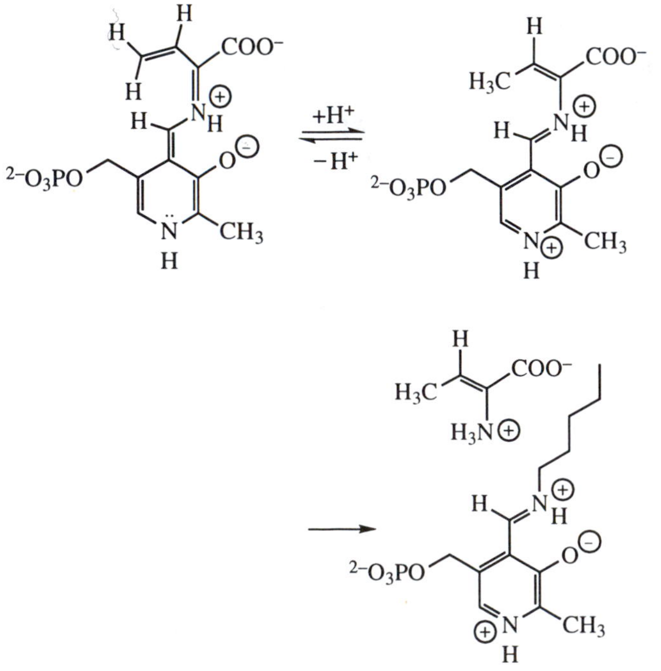

(4–138)

were determined. From all of these results, it was concluded that the $\alpha$-hydrogen, the $\beta$-hydrogen, and the leaving group were all removed from *O*-succinylhomoserine from the same face.[101] In theory, the same flexible general acid–base, confined to this face, could be responsible for all of these steps. It would first remove the $\alpha$-hydrogen as a proton from the pyridoxal adduct of *O*-succinylhomoserine, add that proton to the aldehyde carbon of the pyridoxal phosphate, remove the $\beta$-hydrogen as a proton, and use that proton to protonate the oxygen of the leaving group. The reaction would be completed by reversing these steps after the nucleophile has added. The most logical candidate for this general base is the lysine of the lysylpyridoximine.

Stereochemical studies of enzymatic reactions should always be designed so that both stereochemical outcomes at a prochiral carbon are examined simultaneously. For example, in the stereochemical study of the reaction catalyzed by diaminopimelate decarboxylase, (2*S*,6*R*)-2,6-diamino[2,6-$^2H_2$]pimelate was converted to (2*S*)-[2,6-$^2H_2$]lysine in $H_2O$

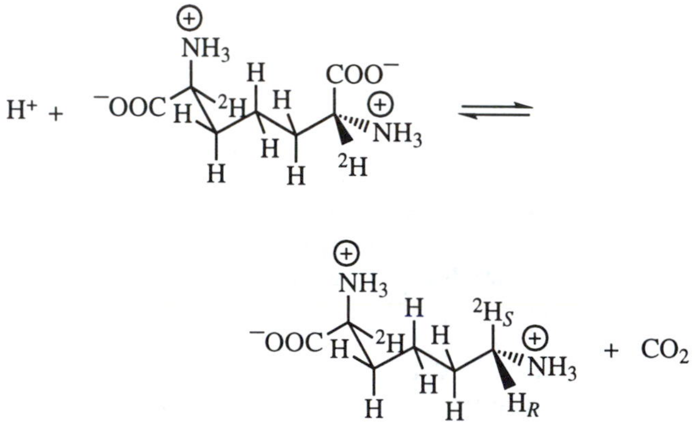

(4–139)

and (2*S*,6*R*)-2,6-diaminopimelate was converted to (2*S*)-[6-$^2$H]lysine in $^2H_2O$ (Reaction 4–121). Each product, which should have the opposite stereochemistry from the other, was separately converted to the respective phthalimide **4–21**

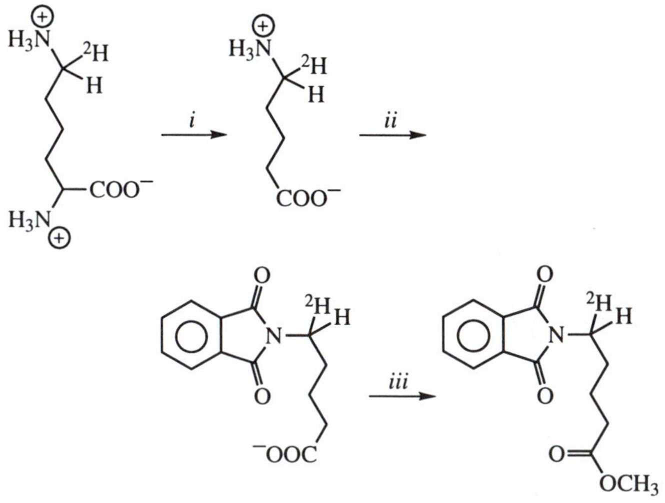

**4–21**

(4–140)

with L-lysine oxidase (*i*), phthalic anhydride (*ii*), and diazomethane (*iii*).[84] The two products had equal but opposite specific rotations at 589 nm. The specific rotation of the product from Reaction 4–139 was +0.84°, and that of the product from Reaction 4–121 was –0.92°. The specific rotation of authentic (*R*)-5-phthalimido[5-$^2$H]valerate, synthesized from L-glutamate, had a specific rotation of –0.95. This observation demonstrated that the product from Reaction 4–121 had the *R* stereochemistry.

In the foregoing example, the stereochemistry of the product of the enzymatic reaction was inverted by changing the isotopic composition of two different reactants, the water and the diaminopimelate, to produce the two opposite stereochemical outcomes. It is also possible to use two identical reactants of opposite stereochemical labeling. Again, the occurrence of the reciprocal outcomes in the experiment provides reassurance that the observations resulted from only the stereochemistry of the active site rather than some anomalous kinetic isotope effect. The two diastereomers of [6-$^2$H]shikimate with the opposite stereochemistry at carbon 6 were synthesized (Figure 4–19)[103] by a Diels–Alder reaction (*i*), followed by *cis*-dihydroxylation of the double bond with osmium tetroxide *anti* to the acetoxy groups (*ii*), protection of the resulting vicinal diol as the dimethyl ketal (*iii*), elimination of acetic acid (*iv*), and hydrolysis in acid (*v*). The stereochemistry of the final product at carbon 6 is dictated by the *endo* preference of the Diels–Alder reaction that places the methylcarboxy group *cis* to the two acetoxy groups and, hence, $H_Z$ also *cis* to the two acetoxy groups.[104] The two 6-deuterioshikimates of known stereochemistry were synthesized from the two appropriate deuterioacrylates. Methyl (*Z*)-[3-$^2$H]acrylate yields (3*R*,4*S*,5*R*,6*R*)-[6-$^2$H]shikimate, and methyl (*E*)-[2,3-$^2H_2$]acrylate yields (3*R*,4*S*,5*R*,6*S*)-[6-$^2$H]shikimate.[104] When these two isomers were separately incubated with the bacterium *Escherichia coli*, the (3*R*,4*S*,5*R*,6*S*)-[6-$^2$H]shikimate was the precursor to deuteriophenylalanine and the (3*R*,4*S*,5*R*,6*R*)-[6-$^2$H]shikimate was the precursor to undeuterated phenylalanine. Between shikimate and phenylalanine the hydrogen at carbon 6 of the shikimate is removed during the elimination catalyzed by chorismate synthase

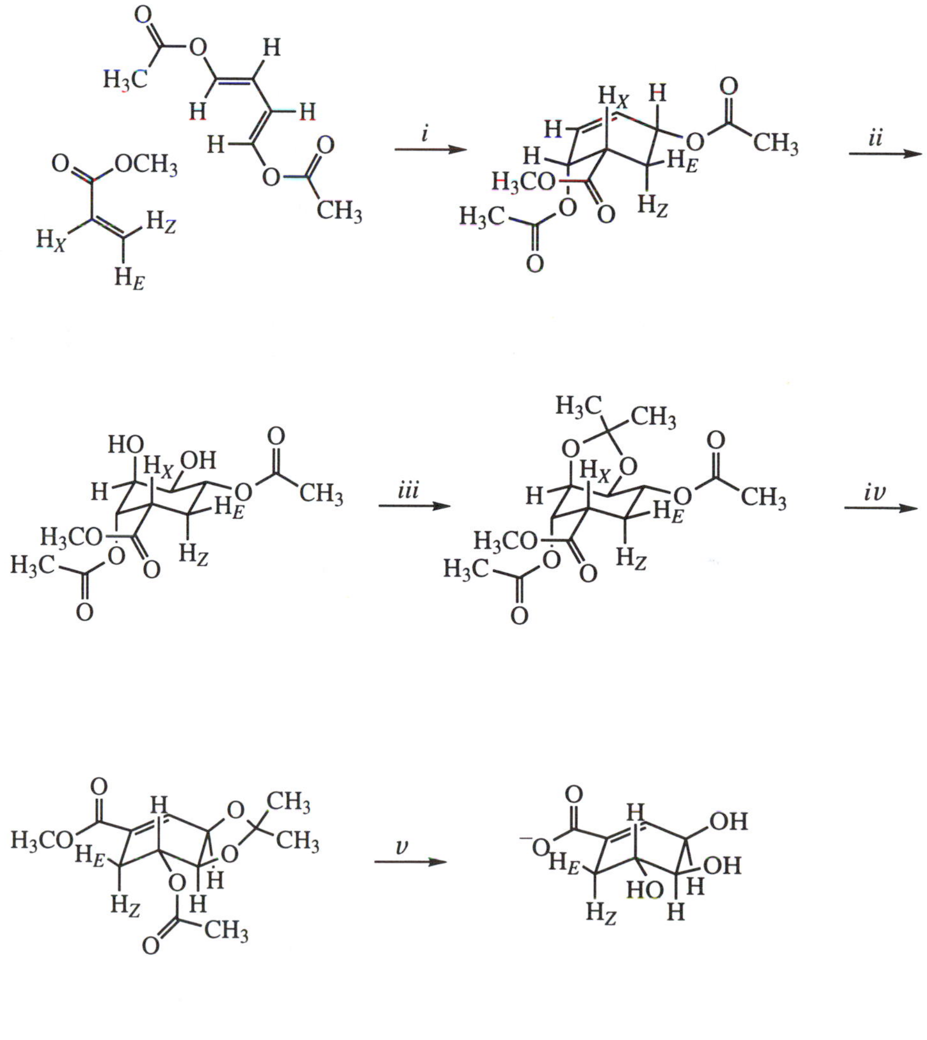

**Figure 4–19:** Synthesis of D-shikimate from methyl acrylate and *trans,trans*-1,4-diacetoxy-1,3-butadiene.[103,104] (*i*) The diene and dienophile were condensed in anhydrous xylene at reflux in a Diels–Alder reaction that proceeded with the usual stereochemical consequences. The two outwardly directed functional groups on the diene ended up *cis* to each other in the cyclohexene. The two enantiomers of the *endo* product were produced in this reaction, only one of which is shown. (*ii*) Osmium tetroxide in tetrahydrofuran/pyridine was used to perform a *cis*-hydroxylation of the olefin. The hydroxylation proceeded *trans* to the two adjacent acetoxy groups because of their bulk and the bulk of the osmium tetroxide. (*iii*) The two vicinal *cis*-hydroxyls in the product formed an acetal with anhydrous acetone under a stream of hydrochloric acid. (*iv*) The elimination of acetic acid by concerted loss of the acidic proton, $H_x^+$, and the good leaving group, $CH_3COO^-$, was accomplished by pyrolysis in a sealed tube at 285 °C. (*v*) The acetal was hydrolyzed in 60% acetic acid to yield racemic shikimic acid. The D enantiomer of shikimate was isolated as the crystalline salt with (–)-1-amino-1-phenylethane.

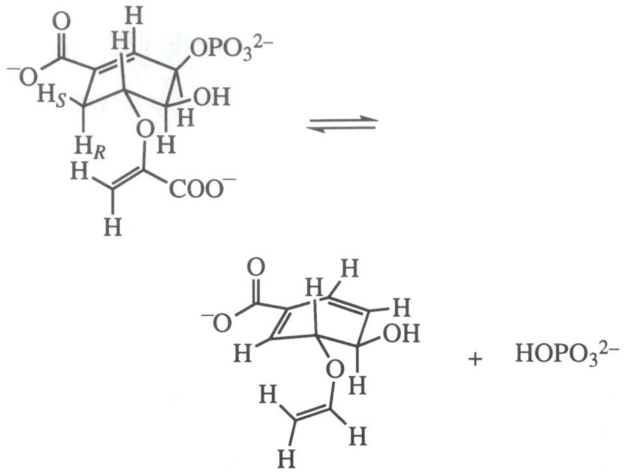

(4–141)

Therefore, because the general base in the active site removes the *pro-R* proton during the elimination, it must reside on the opposite side of the substrate from the general acid that protonates the phosphate to decrease its negative charge and make it a better leaving group.

The ultimate assignment of the absolute stereochemistry of the pair of reactants presented separately to an enzyme or the pair of products produced separately by an enzyme is usually based on an unambiguous stereochemical synthesis of one or both members of the labeled pair of substrates, as was the case with [6-$^2$H]shikimate. The need for this custom is illustrated by the history of the stereochemical studies of the reaction catalyzed by fumarate hydratase. The (2*S*)-[3-$^2$H]malate produced from fumarate by fumarate hydratase in $^2H_2O$ was shown to be stereochemically homogeneous.[105] On the basis of the assumption that in the crystalline solid malate would adopt a *trans* configuration, a theoretical calculation of the line broadening of the nuclear magnetic resonance absorption for the [3-$^2$H]malates in the solid state predicted a numerical value (3.6 ± 0.6 G) for (2*S*,3*S*)-[3-$^2$H]malate much closer to the value observed (3.5 ± 0.2 G) than the value predicted (2.3 ± 0.5 G) for (2*S*,3*R*)-[3-$^2$H]malate.[106] From these theoretical calculations and the direct observations, it was concluded that fumarate hydratase catalyzed the *syn* addition of water to fumarate.

Soon afterward, a synthesis of the racemic mixture of (2*R*,3*R*)-[3-$^2$H]malate and (2*S*,3*S*)-[3-$^2$H]malate was reported.[82,83] The racemate was synthesized from *trans,trans*-2,5-dimethoxy-3,4-epoxytetrahydrofuran

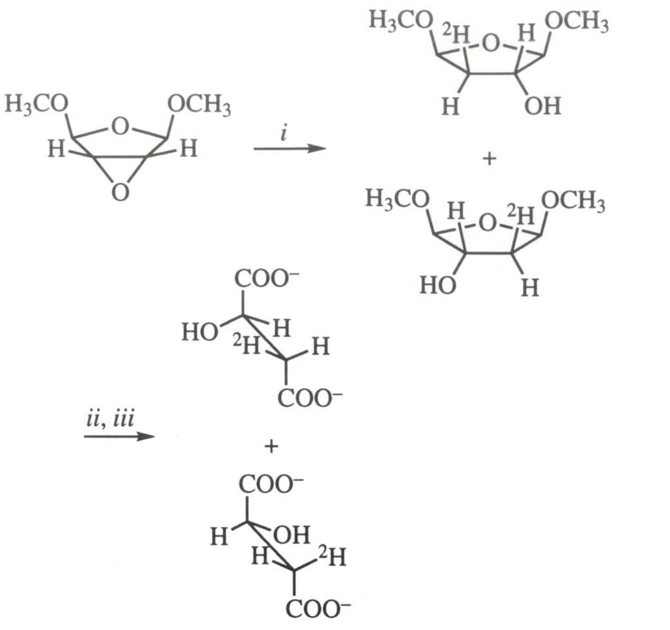

(4–142)

by treatment with lithium aluminum deuteride (*i*), a reagent known to proceed in a substitution reaction by inversion of configuration, followed by acid hydrolysis (*ii*) of the diacetal and oxidation (*iii*) of the dialdehyde. One of these two enantiomers, the (2*S*,3*S*) enantiomer, was the diastereomer then thought to be the product of the hydration catalyzed by fumarate hydratase in $^2H_2O$. It was a surprise when the physical properties of the synthetic product did not match those of the enzymatic product, which was definitely a (2*S*)-[3-$^2$H]malate. They did, however, match the physical properties of a (2*R*)-[3-$^2$H]malate known to have the same absolute stereochemistry at carbon 3 as the (2*S*)-[3-$^2$H]malate obtained in the fumarate hydratase reaction. This (2*R*)-[3-$^2$H]malate, therefore, must have been (2*R*,3*R*)-[3-$^2$H]malate, and (2*S*,3*R*)-[3-$^2$H]malate must be the product of the hydration of fumarate by fumarate hydratase in $^2H_2O$. It follows that fumarate hydratase must perform an *anti* addition–elimination reaction

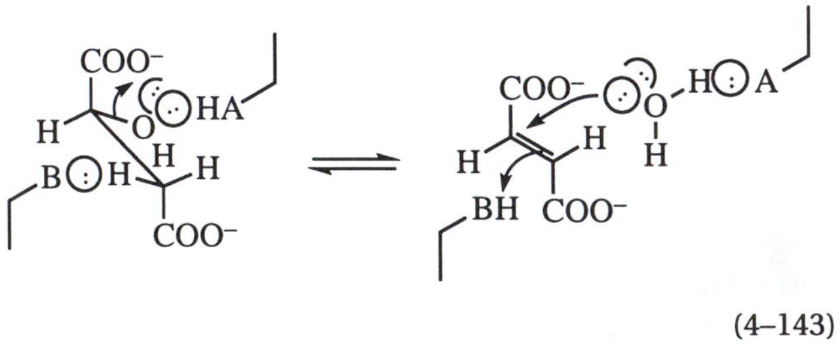

(4–143)

that the general base–acid removing or adding the proton from or to carbon 3, respectively, must be on the opposite side of the bound substrate from the general acid–base protonating the oxygen of malate or removing a proton from the water, and that the theoretical calculations had been unreliable.

One of the more informative stereochemical tools in studying the orientation of general acid–bases within the active sites has been the chiral methyl group.[107,108] The most widely used compound with a chiral methyl group is [2,2-$^3$H,$^2$H]acetate in its two enantiomeric forms

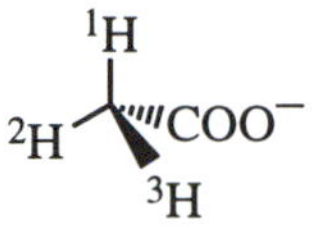

4–22 ($S$) 4–23 ($R$)

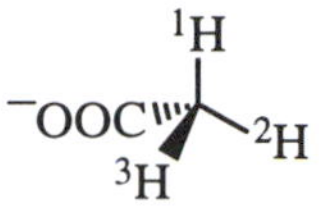

Enantiomeric acetate can be synthesized from [2-$^3$H]glyoxylate (Figure 4–20)[108] by reduction with NAD$^1$H and lactate dehydrogenase (*i*), reduction of the carboxylate to the alcohol with lithium aluminum hydride (*ii*), activation for nucleophilic substitution of one of the two equivalent hydroxyls of the ethylene glycol at random with *p*-bromobenzenesulfonyl chloride (*iii*), nucleophilic substitution with a hydride from lithium aluminum deuteride with inversion of configuration (*iv*), and oxidation of the alcohol to the carboxylic acid with oxygen over platinum (*v*). The synthesis provides an equimolar mixture of [2-$^2$H]acetate and (*S*)-[2,2-$^2$H,$^3$H]acetate. The stereochemistry is assigned directly as *S* because lactate dehydrogenase is known to produce the *S* isomer of glycolate from [$^3$H]glyoxylate and unlabeled NADH and the nucleophilic substitution with deuteride proceeds with inversion of configuration. The (*R*)-acetate can be obtained by the same sequence of reactions with the exception that glyoxylate reductase from spinach is used in the first step to produce stereochemically pure (*R*)-[2-$^3$H]glycolate rather than (*S*)-[2-$^3$H]glycolate. The two enantiomeric acetates can be converted readily to the acetyl coenzymes A with acetate kinase and phosphate acetyltransferase.[107]

The synthesis just described was one of the first for enantiomeric acetate, and it will illustrate the combined use of enzymatic and synthetic steps. There are several more recently developed methods for synthesizing enantiomeric acetate that are more useful.[109,110] In one of these methods,[110] a racemic mixture of the two enantiomers of (*E*)-2-deuterio-1-phenyloxirane **4–24** is resolved into the pure 2*R* and 2*S* isomers by reacting the epoxide with (*S*)-(1-phenyl)ethylamine, separating the two diastereomeric aminohydrins, and regenerating separately the (2*R*)-2-deuterio-1-phenyloxirane and the (2*S*)-2-deuterio-1-phenyloxirane by exhaustive methylation of the amine and intramolecular nucleophilic substitution in strong base. The two resolved enantiomers of (*E*)-2-deuterio-1-phenyloxirane are then separately reduced with Li$^3$HB(C$_2$H$_5$)$_3$ and oxidized to the respective [2,2-$^2$H,$^3$H]phenyl ketones with CrO$_3$ followed by treatment with peroxytrifluoroacetic acid to produce Baeyer–Villiger rearrangement

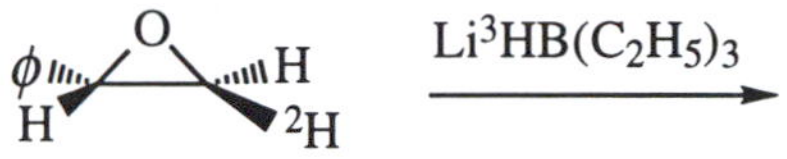

**4–24** $\xrightarrow{\text{Li}^3\text{HB(C}_2\text{H}_5)_3}$

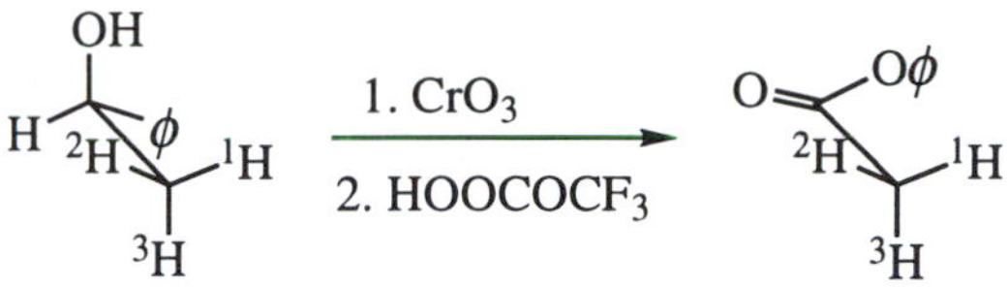

$\xrightarrow[\text{2. HOOCOCF}_3]{\text{1. CrO}_3}$ (4–144)

The two enantiomeric [2,2-$^2$H,$^3$H]phenyl acetates are then saponified and collected as the respective sodium acetates. They can then be converted to the respective acetyl coenzymes A.

The use of enantiomeric acetyl-SCoA to answer a stereochemical question relies upon the kinetic isotope effect and the planar, rigid structure of an enol or enolate anion. Consider (*R*)-[2,2-$^2$H,$^3$H]acetyl-SCoA bound to an active site (Figure 4–21). The first step in almost every enzymatic reaction using acetyl-SCoA as a reactant is the removal of a proton to create an enol. The acetyl-SCoA will be bound unavoidably to the active site through its coenzyme A and should be hydrogen-bonded at its acyl oxygen by at least the general acid necessary to catalyze formation of the enol. The chiral methyl

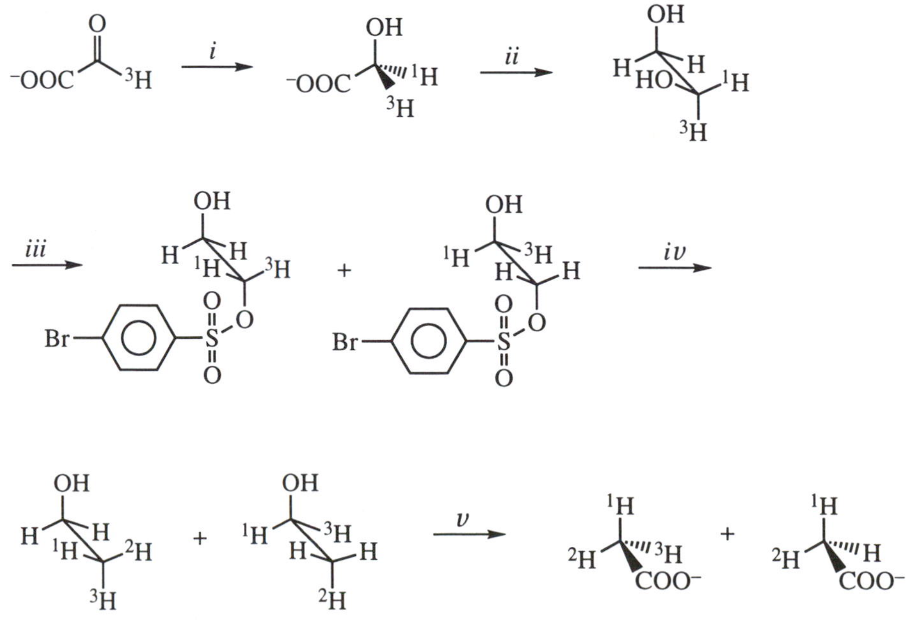

**Figure 4–20:** Synthesis of enantiomeric acetate from [2-$^3$H]glyoxylate.[108] (*i*) [2-$^3$H]Glyoxylate was reduced with NADH and lactate dehydrogenase to produce (2*S*)-[2-$^3$H]glycolate. (*ii*) The methyl ester of (2*S*)-[2-$^3$H]glycolate was reduced chemically with LiAlH$_4$ to produce (1*S*)-[1-$^3$H]ethylene glycol. (*iii*) A mixture of the two brosylates of the ethylene glycol was produced by sulfonylation with *p*-bromobenzenesulfonyl chloride. (*iv*) The leaving ability of the brosylate group was exploited to promote the reduction of the two regioisomers with LiAl$^2$H$_4$ by substitution with hydride with inversion of configuration. (*v*) The resulting mixture of (2*S*)-[2,2-$^2$H,$^3$H]ethanol and (2*S*)-1-tritio-2-deuterioethanol was oxidized with O$_2$ over a platinum catalyst producing a mixture of (*S*)-[2,2-$^2$H,$^3$H]acetate and [2-$^2$H]acetate.

(*S*)-[2,2-$^2$H, $^3$H]acetate

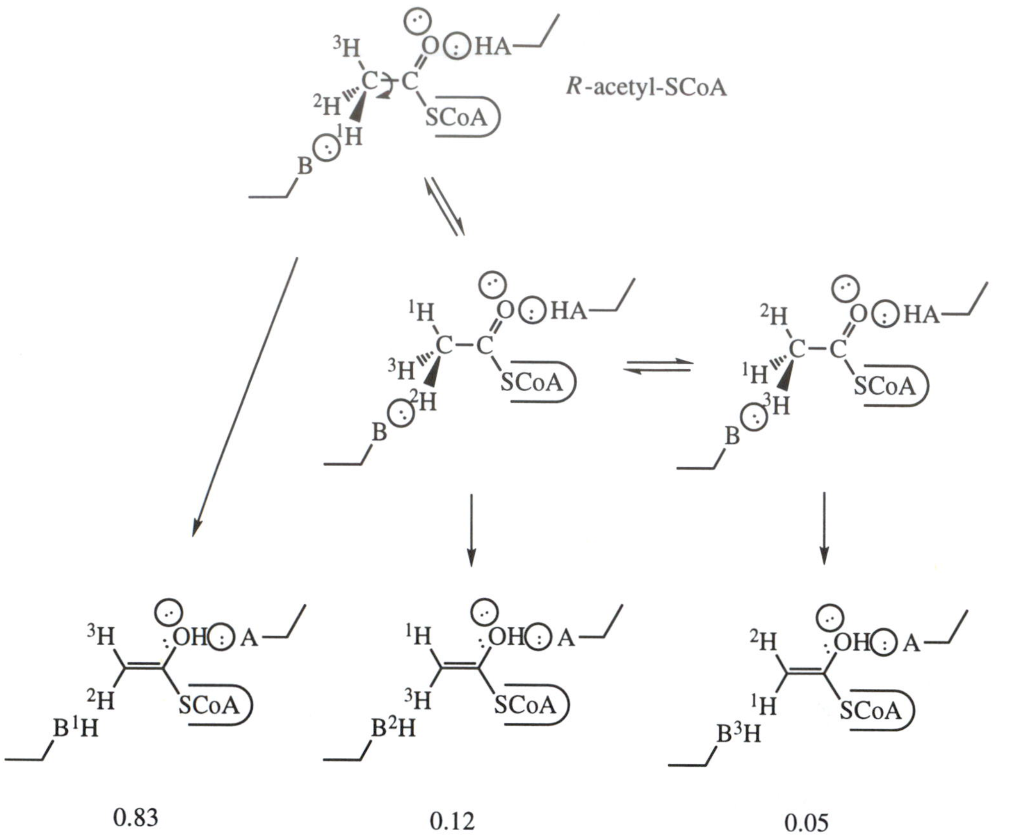

*R*-acetyl-SCoA

0.83 0.12 0.05

**Figure 4–21:** Production of the three respective enols of (*R*)-[2,2-$^2$H,$^3$H]acetyl-SCoA at the active site of an enzyme. In the active site of the enzyme, the coenzyme A of the reactant will be extensively immobilized by noncovalent interactions and the acyl oxygen will be hydrogen-bonded by at least the general acid (HA$^-$) responsible for protonation of the enolate. The methyl group, however, cannot be immobilized by the active site and must be free to rotate. As each of the hydrogens spins by the general base (–B⊙) responsible for its removal, the probability that it will be removed should be directly proportional to the appropriate kinetic isotope effect. If the kinetic isotope effects for this proton removal were $^{H}k/^{D}k = 7$ and $^{H}k/^{T}k = 16$, then the probability that the base will remove the protium is 0.83; the deuterium, 0.12; and the tritium, 0.05. The products of each of these reactions are planar and rigid, and they are pinned to the active site.

group, because it cannot be hydrogen-bonded, will be free to rotate. Assume that the base removing the hydrogen as a proton is above the plane of the acyl carbon and that normal kinetic isotope effects of $^{H}k/^{D}k = 7$ and $^{H}k/^{T}k = 16$ govern the removal of the proton by the base. If this were the case, 83% of the enol formed would be the (*E*)-[2,2-$^2$H,$^3$H]enol and 5% would have lost its $^3$H altogether. Rotation about the carbon–carbon double bond of the enol cannot occur. The electrophile added to the enol in the second step of the reaction will approach either from below the plane of the enol as pictured, in which case inversion of configuration will occur, or from above the plane of the enol, in which case retention of configuration will occur.

In the reaction catalyzed by malate synthase

$$H_2O + \text{acetyl-SCoA} + \text{glyoxylate} \rightleftharpoons (S)\text{-malate} + \text{HSCoA} \quad (4\text{–}145)$$

the electrophile is glyoxylate. Because (*S*)-malate is the product of the reaction, the enol must add

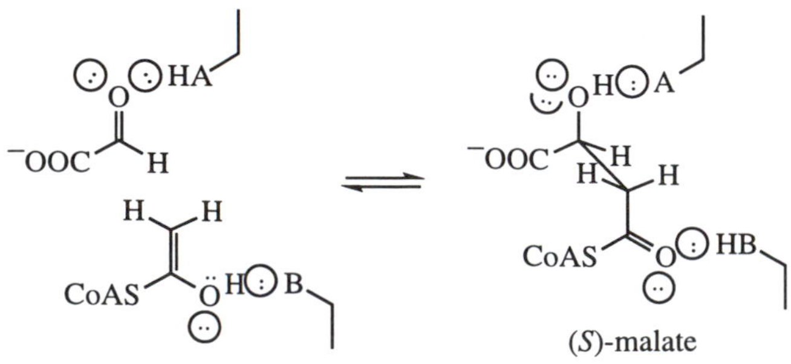

(*S*)-malate

(4–146)

to the *si* face of the glyoxylate. This stereochemical conclusion follows only from a determination of which enantiomer of the malate, *S* or *R*, is formed during the reaction. The stereochemistry at the methyl carbon of the acetyl-SCoA can be assessed by using enantiomeric acetyl-SCoA.

This is most easily understood by making a prediction (Figure 4–22). If the reaction proceeds with retention of configuration at carbon 3 of the malate, (*R*)-acetyl-SCoA should give (2*S*,3*R*)-[3,3-$^2$H,$^3$H]malate as the major product and (*S*)-acetyl-SCoA should give (2*S*,3*S*)-[3,3-$^2$H,$^3$H]malate as the major product. If the reaction proceeds with inversion of configuration at carbon 3 of the malate, (*R*)-acetyl-SCoA should give (2*S*,3*S*)-[3,3-$^2$H,$^3$H]malate as the major product and (*S*)-acetyl-SCoA should give (2*S*,3*R*)-[3,3-$^2$H,$^3$H]malate as the major product.

Fumarate hydratase is used to distinguish between (2*S*,3*R*)-[3,3-$^2$H,$^3$H]malate and (2*S*,3*S*)-[3,3-$^2$H,$^3$H]malate. The dehydration catalyzed by fumarate hydratase is always *anti* regardless of the disposition of the isotopes of hydrogen on carbon 3 of the malate. Fumarate hydratase removes the tritium from (2*S*,3*R*)-[3,3-$^2$H,$^3$H]malate but removes the deuterium from (2*S*,3*S*)-[3,3-$^2$H,$^3$H]malate. If the kinetic isotope effects in the active site of malate synthase had been 16:2.29:1, then 82% of the malate should have been (2*S*)-[3,3-$^2$H,$^3$H]malate; 12%, (2*S*)-[3-$^3$H]malate; and 5%, (2*S*)-[3-$^2$H]malate from a given enantiomer of [2,2-$^2$H,$^3$H]acetyl-SCoA. Each mixture should have lost either 87% or 13% of its tritium, respectively, upon incubation with fumarate hydratase, depending on the stereochemistry of the malate synthase reaction (Figure 4–22). When (*R*)-[2,2-$^2$H,$^3$H]acetyl-SCoA was incubated with glyoxylate and malate synthase, the

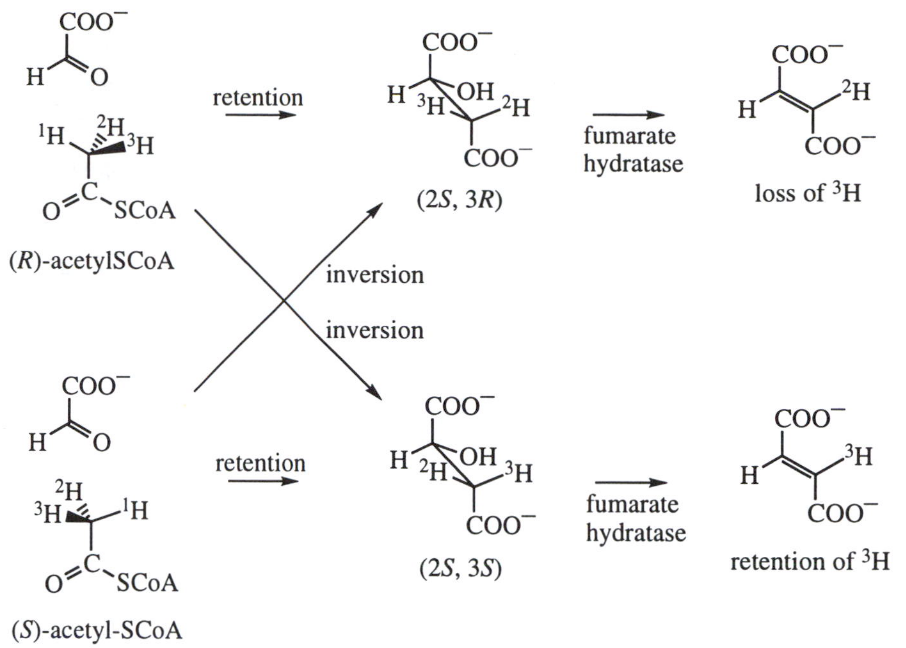

**Figure 4–22:** Predictions for the stereochemical consequences of the malate synthase reaction if it proceeds with either inversion or retention of configuration from either ($R$)-acetyl-SCoA or ($S$)-acetyl-SCoA. The actual stereochemistries of the products are assessed by examining the loss of tritium upon dehydration of the ($S$)-malate with fumarate hydratase.

($R$)-acetylSCoA — retention → (2$S$, 3$R$) — fumarate hydratase → loss of $^3$H

($S$)-acetyl-SCoA — retention → (2$S$, 3$S$) — fumarate hydratase → retention of $^3$H

inversion

inversion

[$^3$H]malate produced lost only 31% of its tritium upon conversion to fumarate,[107] so the major isomer must have been (2$S$,3$S$)-[3,3-$^2$H,$^3$H]malate. When ($S$)-[2,2-$^2$H,$^3$H]acetyl-SCoA was incubated with glyoxylate and malate synthase, the [$^3$H]malate produced lost 69% of its tritium upon conversion to fumarate,[107] so the major isomer must have been (2$S$,3$R$)-[3,3-$^2$H,$^3$H]malate. Therefore, malate synthase catalyzes the condensation with inversion of configuration at the methyl carbon of acetyl-SCoA. In independent experiments,[111] it was determined that the intrinsic kinetic isotope effect, $^{H}k/^{D}k$ for malate synthase is 3.7, an observation which explains, in part, the greater than theoretical and less than theoretical percentages of tritium loss, respectively.

If the reaction catalyzed by malate synthase proceeds by inversion of configuration, the general base within the active site that removes the proton from the methyl group to form the enol must be on the opposite side of the bound acetyl-SCoA from the position at which the glyoxylate is held within the active site with its *si* face directed toward the methyl carbon of the bound acetyl-SCoA. It should be noted, however, that in this instance the absolute stereochemistry of the reaction cannot be determined. There is no way to learn whether the general base in the active site is on the *si* face of the bound acetyl-SCoA as shown in Figure 4–21 or on the *re* face. Only the relative stereochemistry of the reaction is assessed.

Once it was understood that malate synthase proceeds with inversion of configuration at the methyl carbon of acetyl-SCoA, a routine assay for the stereochemistry of any [2,2-$^2$H,$^3$H]acetate derived from an experiment became available.[112] ($R$)-[2,2-$^2$H,$^3$H]Acetate is the enantiomeric acetate which, when converted to malate by malate synthase, produces [$^3$H]malate that retains most of its tritium upon incubation with fumarate hydratase. ($S$)-[2,2-$^2$H,$^3$H]Acetate is the enantiomeric acetate which, when converted to malate by malate synthase, produces [$^3$H]malate that loses most of its tritium upon incubation with fumarate hydratase. The assay can be expanded to assess the absolute stereochemistry of any [2,2-$^2$H,$^3$H]methyl group in a molecule if that methyl group can be converted into the methyl group of acetate without racemization. This is usually done by chemically oxidizing the immediate product of the reaction.

Isopentenyl-diphosphate Δ-isomerase

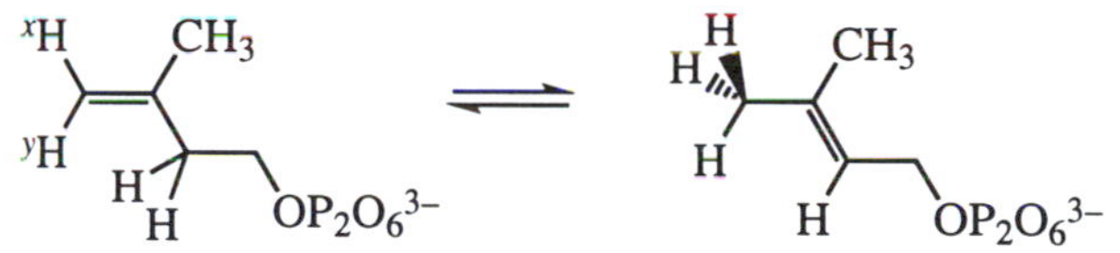

**4–25** ⇌ **4–26**

(4–147)

catalyzes the 1,2-shift of a carbon–carbon double bond. Both ($E$)-[4-$^3$H]-3-methyl-3-butenyl pyrophosphate (Structure **4–25**; $x$ = 3, $y$ = 1) and ($Z$)-[4-$^3$H]-3-methyl-3-butenyl pyrophosphate (Structure **4–25**; $x$ = 1, $y$ = 3) were synthesized enzymatically from the appropriate [2-$^3$H]mevalonic acids.[113] Each was separately incubated with isopentenyl-diphosphate Δ-isomerase in $^2H_2O$ to generate the two dimethylallyl pyrophosphates **4–26** with chiral ($E$)-methyl groups of opposite stereochemistry at carbon 4. The two dimethylallyl pyrophosphates were converted to the two farnesols which contained the respective chiral methyl groups at their aliphatic ends. These vinyl methyl groups were converted to the methyl groups of acetate by ozonolysis and the iodoform reaction. The product from ($E$)-[4-$^3$H]-3-methyl-3-butenyl pyrophosphate gave the ($S$)-acetate (64% loss of $^3$H), and the product from ($Z$)-[4-$^3$H]-3-methyl-3-butenyl pyrophosphate gave the ($R$)-acetate (64% retention of $^3$H). From these results, it could be concluded that the general acid in the active site that adds the proton to carbon 4 of 3-methyl-3-butenyl pyrophosphate must be on the *re* face of the olefin when it is bound in the active site. It has also been shown that the *pro-2R* hydrogen[114] is removed by the respective general base on the active site of the enzyme during the isomerization

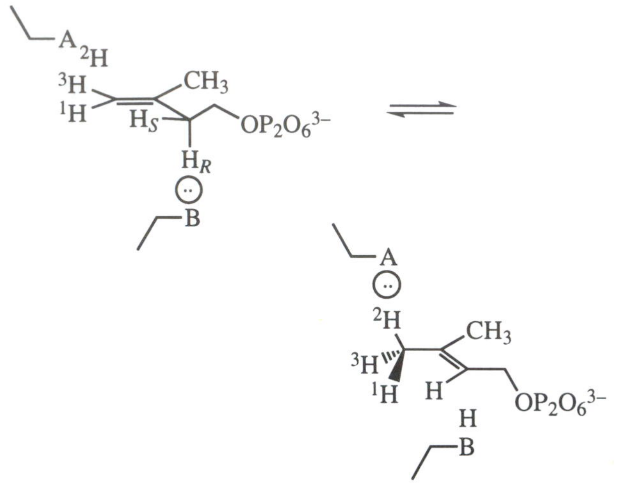

(4–148)

The general acid–base operating on carbon 4 therefore must be on the opposite side of the bound substrate in the active site from the general base–acid operating on carbon 2, and the same general acid–base cannot fulfill both roles.

Pyruvate kinase also forms a methyl group during its enzymatic reaction

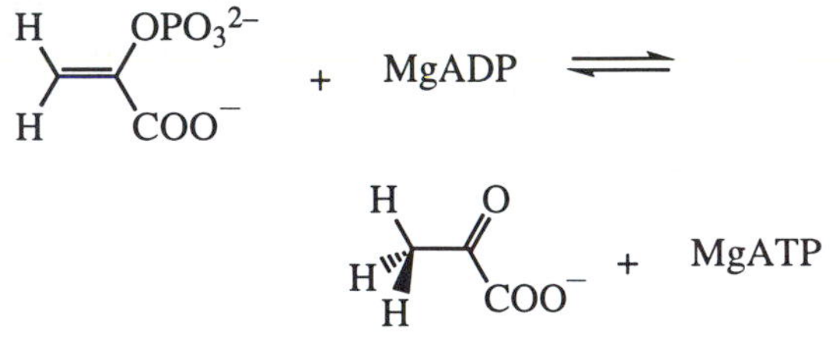

(4–149)

Both (*E*)-[3,3-$^2$H,$^3$H]phospho*enol*pyruvate and (*Z*)-[3,3-$^2$H,$^3$H]phospho*enol*pyruvate were synthesized enzymatically from [1-$^3$H]glucose in $^2H_2O$ and [1-$^2$H]glucose in $^3H_2O$. Each was separately incubated with pyruvate kinase in $^1H_2O$ to generate the two respective enantiomers of pyruvate. These enantiomers of pyruvate were oxidized with $H_2O_2$ to produce the respective enantiomeric acetates. The [3,3-$^2$H,$^3$H]pyruvate from (*E*)-[3,3-$^2$H,$^3$H]phospho*enol*pyruvate gave (*S*)-[2,2-$^2$H,$^3$H]acetate (84% loss of $^3$H), and the [3,3-$^2$H,$^3$H]pyruvate from (*Z*)-[3,3-$^2$H,$^3$H]phospho*enol*pyruvate gave (*R*)-[2,2-$^2$H,$^3$H]acetate (68% retention of $^3$H).[115] Therefore, the general acid that adds the proton to phospho*enol*pyruvate during the enzymatic reaction is on the *si* face

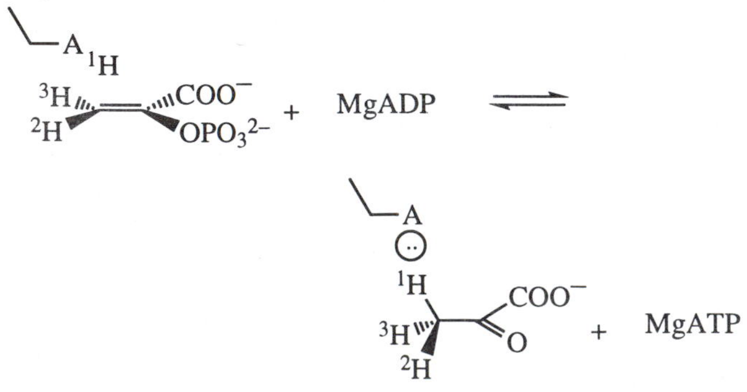

(4–150)

when the phospho*enol*pyruvate is bound within the active site.

These enantiomeric pyruvates of known absolute stereochemistry (Reaction 4–150) can be used to follow stereochemistry in reactions involving pyruvate as a reactant. 4-Hydroxy-2-oxoglutarate aldolase

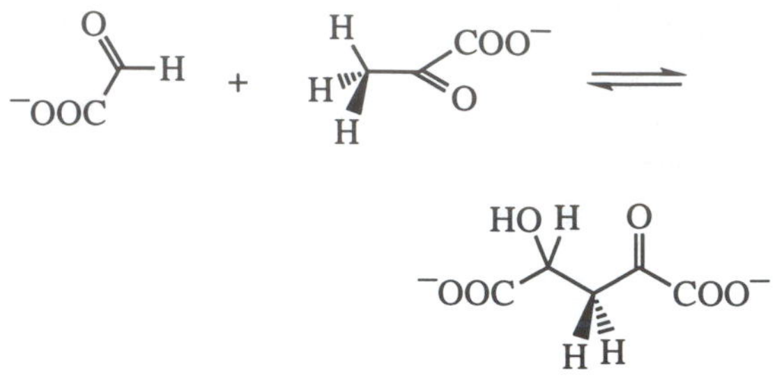

(4–151)

removes a proton from carbon 3 of pyruvate and then condenses the enol with glyoxylate. The 4-hydroxy-2-oxoglutarate produced can be converted to malate by treatment with $H_2O_2$. The [$^3$H]malate produced from (3*R*)-[3,3-$^2$H,$^3$H]pyruvate lost 81% of its tritium upon incubation with fumarate hydratase, and the [$^3$H]malate produced from (3*S*)-[3,3-$^2$H,$^3$H]pyruvate retained 63% of its tritium upon incubation with fumarate hydratase.[116] Therefore, 4-hydroxy-2-oxoglutarate aldolase proceeds with retention of configuration at carbon 3 of the pyruvate

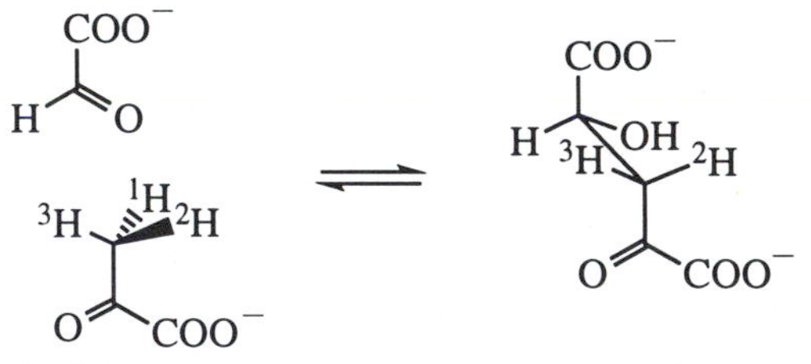

(3*R*)-[3,3-$^2$H,$^3$H]pyruvate (2*S*,3*R*)-4-hydroxy-2-oxoglutarate

(4–152)

In this case, it is possible that the general base removing the proton to form the enol of pyruvate then acts as a general acid to protonate the carbonyl oxygen of glyoxalate because the proton is removed from the same side of the pyruvate to which the aldehyde adds.

As with reactions involving enantiomeric acetyl-SCoA, reactions involving enantiomeric pyruvate depend upon the inability of the enol of pyruvate to undergo rotation between carbons 2 and 3. The observation that chirality is retained in the products of these reactions states that the intermediate in these reactions in which the formerly chiral carbon has lost the protium and still retains the tritium and the deuterium cannot rotate around the bond between that carbon and the carbon to which it is attached. The stereochemical outcome of an enzymatic reaction can also demonstrate that the intermediate in that reaction is able to rotate around adjacent carbon–carbon bonds during its lifetime.

When either (*R*)-[2,2-$^2$H,$^3$H]ethanolamine or (*S*)-[2,2-$^2$H,$^3$H]ethanolamine has been converted to acetaldehyde

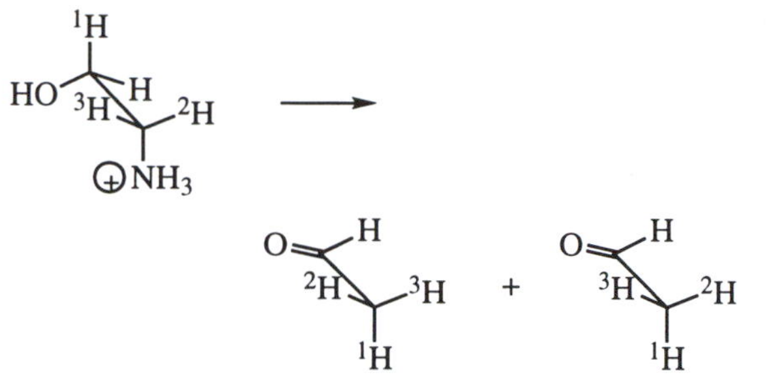

(4–153)

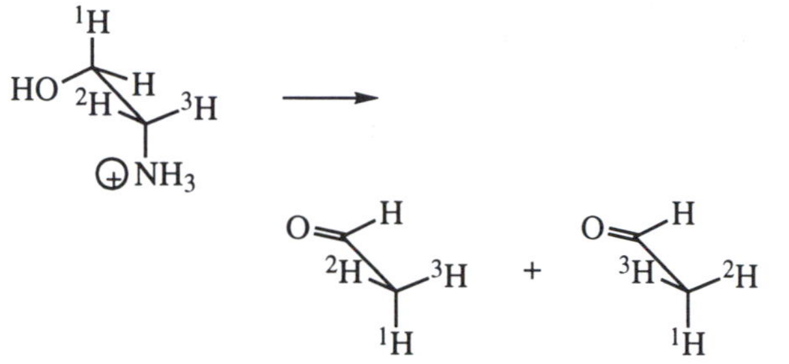

(4–154)

by ethanolamine ammonia-lyase, an enzyme using coenzyme $B_{12}$ in its active site, the methyl group on the resulting acetaldehyde is racemic.[117] This is not due to the release of an achiral but rigid intermediate that is then rendered racemic by further reaction with the solvent, because when either (2*S*)-2-aminopropanol or (2*R*)-2-aminopropanol is used as a reactant, the products are enantiomeric. When (2*S*)-2-aminopropanol is converted to propionaldehyde with ethanolamine ammonia-lyase containing [5′-$^{3}$H]adenosylcobalamine, (2*S*)-[2-$^{3}$H]propionaldehyde is the product; but, when (2*R*)-2-aminopropanol is the reactant, (2*S*)-[2-$^{3}$H]propionaldehyde is again the product[118]

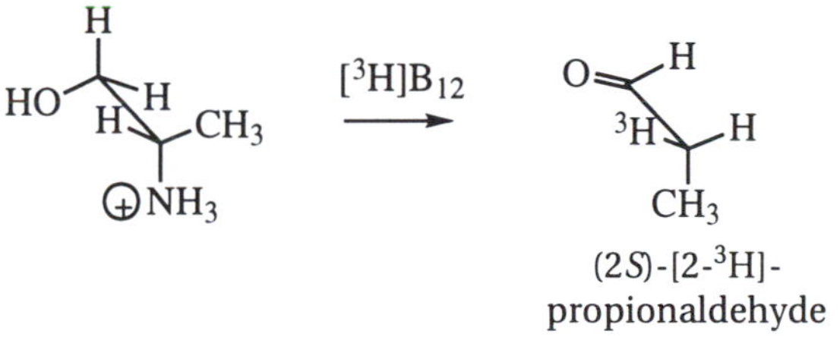

(2*S*)-[2-$^{3}$H]-propionaldehyde

(4–155)

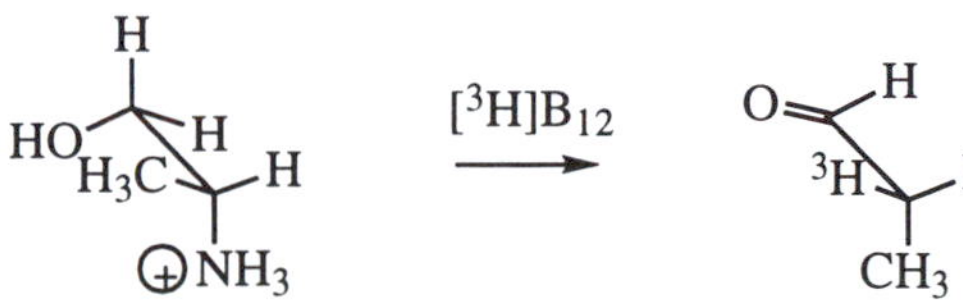

(4–156)

All of these results can be explained if the intermediate formed in the reaction catalyzed by ethanolamine ammonia-lyase is free to rotate about its carbon–carbon bond while bound in the active site. For example, if the intermediate were a carbocation, produced by removal of hydride from the *pro-S* position[118] of carbon 1 followed by amino migration, then the resulting carbocation at carbon 2 derived from (2*S*)-2-aminopropanol could rotate before it was quenched by hydride transfer from the [5′-$^{3}$H]-5′-deoxyadenosine

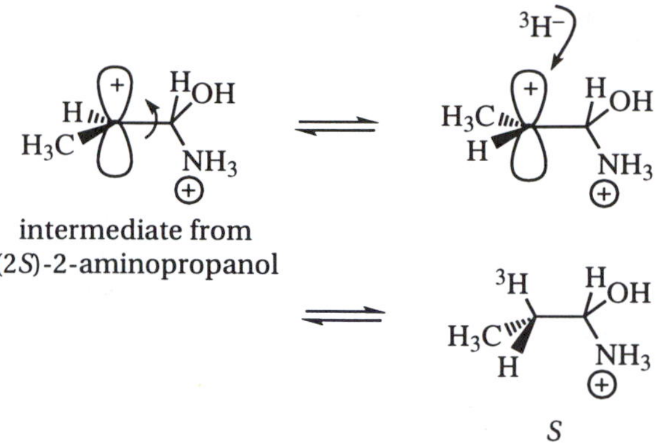

intermediate from (2*S*)-2-aminopropanol

*S*

(4–157)

A similar argument could be made if the intermediate were a radical or a carbanion rather than a carbocation. The intermediate produced directly from (2*R*)-2-aminopropanol would be the intermediate to the right in the first step of Reaction 4–157. If that rotamer were the more stable one because of steric interactions in the active site, then both (2*S*)-2-aminopropanol and (2*R*)-2-aminopropanol would give the same product, (2*S*)-[2-$^{3}$H]propionaldehyde, as is observed. Rotation of the type described would also explain the racemization observed when the two substituents on carbon 2 are merely isotopes of hydrogen which cannot be sterically distinguished. In this instance, stereochemistry has provided information about one of the properties of the intermediate in the reaction, namely, that it is capable of dihedral rotation within its lifetime.

The wide use of enantiomeric acetate, in which three otherwise identical hydrogen atoms are distinguished by using the three isotopes protium, deuterium, and tritium, inspired the synthesis and use of enantiomeric phosphate monoesters and anhydrides, in which three otherwise identical oxygen atoms are distinguished again by using the three isotopes $^{16}$O, $^{17}$O, and $^{18}$O. Enantiomeric phosphate can be prepared synthetically from specific enantiomers of vicinal bisnucleophiles, such as 2-(methylamino)-1-phenyl-1-propanol **4–27**[119] or 1,2-diphenyl-ethylene glycol **4–28**[120]

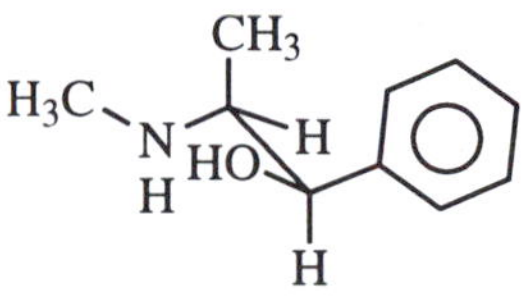

**4–27** 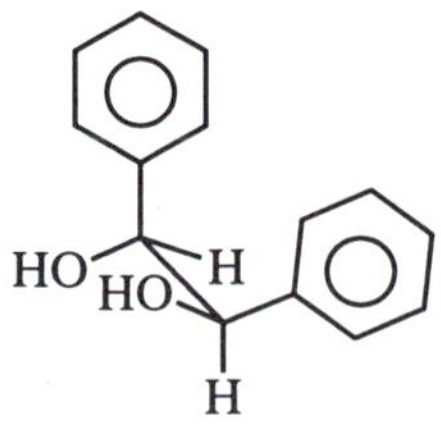 **4–28**

Enantiomers of these compounds of known absolute stereochemistry determine the stereochemistry of the ultimate phosphate product. (*R*)-[$^{16}$O,$^{17}$O,$^{18}$O]-1-Phosphono-(2*S*)-propane-1,2-diol[119] or adenosine $\gamma$-(*S*)-[$^{16}$O,$^{17}$O,$^{18}$O]triphosphate[121] can be synthesized from (1*R*,2*S*)-2-(methylamino)-1-phenylpropanol (Figure 4–21). (1*R*,2*S*)-2-(Methylamino)-1-phenylpropanol was converted to the cyclic phosphorylchloride **4–29**. Cyclic phosphorylchloride **4–29** was reacted with 2-benzyl-(*S*)-propane-1,2-diol, and the two epimers at phosphorus were resolved chromatographically. The major isomer was hydrolyzed in acidic $H_2{}^{18}O$, and the phosphate diester was reduced with $H_2$ over palladium on carbon to produce a stereochemically pure monoester.

The synthesis itself did not define the stereochemistry of the final monoester, but it could be assigned unambiguously as *R* by metastable ion mass spectroscopy.[119] The $[^{16}O,^{17}O,^{18}O]$-1-phosphono-(2*S*)-propane-1,2-diol was converted by cyclization and methylation to a mixture of three isotopically distinct *syn* (4–32) and three isotopically distinct *anti* (4–33) cyclic triesters

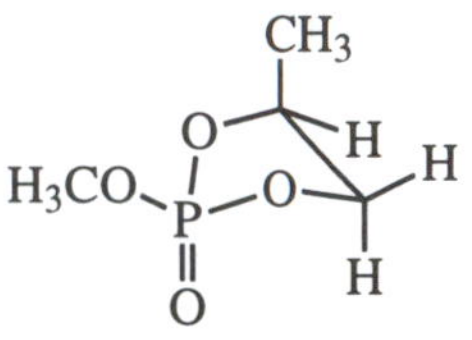

**4–32** 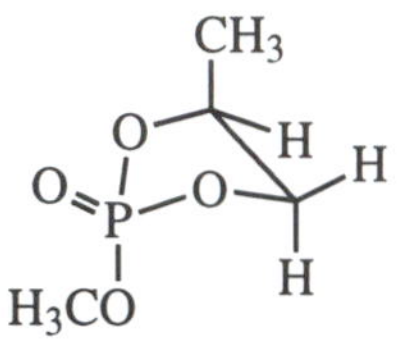 **4–33**

This mixture of diastereomers could be separated into a mixture of the three *syn* compounds and a mixture of the three *anti* compounds. These two mixtures were then submitted separately to methanolysis to produce the trimethyl phosphates. Each of these two mixtures of trimethyl phosphates derived from the original $[^{16}O,^{17}O,^{18}O]$-1-phosphono-(2*S*)-propane-1,2-diol contained three isotopically distinct products. The three products present in each of these mixtures would be different depending on whether the original phosphonopropanediol had been *R* or *S* at phosphorus.

The three products in each of the two mixtures of trimethyl phosphates would have had molecular masses of 141, 142, and 143 daltons, regardless of whether the original phosphonopropanediol were *R* or *S* at phosphorus, so direct mass spectroscopy could not have distinguished between the two possibilities. It just so happens, however, that the positive ion of trimethyl phosphate is a metastable ion that loses a neutral molecule of formaldehyde, producing a positive ion of mass (M–30), while passing through a mass spectrometer. The pattern of these daughter ions does differ depending on whether the original phosphonopropanediol was *R* or *S* at phosphorus. In a double-focusing mass spectrometer, it is possible, by varying systematically the electrostatic field, to determine what pattern of daughter ions arises from each of the parent ions of mass 141, 142, and 143. From the patterns of daughter ions observed, it could be shown that the $[^{16}O,^{17}O,^{18}O]$-1-phosphono-(2*S*)-propanediol **4–30** had the *R* stereochemistry at phosphorus.

It is this assignment that has served as the basis for the stereochemical assignments of the other enantiomeric phosphates. The reduction with $H_2$ over palladium on carbon does not disturb the stereochemistry at phosphorus, hydrolysis in acidic $H_2{}^{18}O$ proceeds with inversion of configuration, and nucleophilic substitution on the cyclic phosphorylchloride **4–29** proceeds with retention of configuration.[119] Therefore, the major epimer of the cyclic $[^{17}O]$phosphorylchloride **4–29** must have had the *R* configuration at phosphorus. This *R*-epimer of cyclic phosphorylchloride **4–29** was separated from the *S*-epimer chromatographically, and it was hydrolyzed with $Li^{18}OH$ in $H_2{}^{18}O$ and dioxane with retention of configuration. The product of this reaction was submitted to nucleophilic substitution with ADP with inversion of configuration, and adenosine $\gamma$-(*S*)-$[^{16}O,^{17}O,^{18}O]$triphosphate was released by reduction with $H_2$ over palladium on carbon.

The absolute stereochemistry of the enantiomeric $[^{16}O,^{17}O,^{18}O]$phosphate in either the reactant or the product of an enzymatic reaction can be assessed by phosphorus nuclear magnetic resonance spectroscopy. Usually, the phosphate in any particular product is transferred enzymatically by reactions of known stereochemistries to a reference diol, either *S*-propane-1,2-diol (Figure 4–23)[122] or glucose.[123] The phosphate monoester is then cyclized to the cyclic diester with inversion of configuration. Because the presence of $^{17}O$ broadens the absorbances of phosphorus in a nuclear magnetic resonance spectrum, signals from cyclic diesters containing only $^{18}O$ and $^{16}O$ are all that are observed. Because $^{18}O$ causes shifts in the absorbance of phosphorus that depend on geometry, the various diastereomers can be separately distinguished[122,123] and related back to a reference compound of known stereochemistry.

Another approach to enantiomeric phosphate has been to use two isotopes of oxygen, $^{16}O$ and $^{18}O$, and replace the third oxygen with sulfur.[124] For example, a synthesis of adenosine (*R*)-$[\gamma,\gamma\text{-}^{16}O,^{18}O]$-$\gamma$-thiotriphosphate has been developed[125] and a sequence of reactions, based on the absolute stereochemical preferences of acetate kinase and pyruvate kinase, has been developed to determine the absolute stereochemistry of any $[^{18}O]$thiophosphate that can be transferred to become the $\beta$-phosphate on ADP.[126]

Enantiomeric phosphate has been used to examine the stereochemistry of the phosphonotransfer reaction of the type

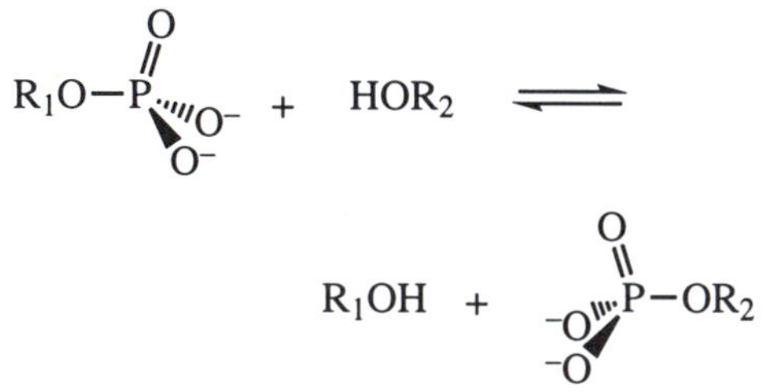

$$R_1O\text{–}PO_3^{2-} + HOR_2 \rightleftharpoons R_1OH + {}^{2-}O_3P\text{–}OR_2 \qquad (4\text{–}158)$$

catalyzed by a large collection of enzymes. In all cases, the enzymes catalyze the transfer of a phosphonate group from an atom of oxygen in one compound to an atom of oxygen in another compound. The significant majority of these enzymatic reactions proceed with inversion of configuration at phosphorus. For example, glycerol kinase proceeds with inversion of configuration at phosphorus[121]

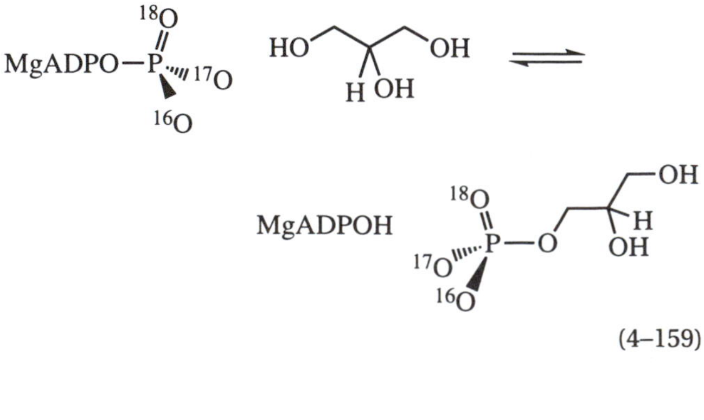

$$\text{MgADPO–P}(^{18}O)(^{17}O)(^{16}O) + \text{glycerol} \rightleftharpoons \text{MgADPOH} + (^{18}O)(^{17}O)(^{16}O)\text{P–O–glycerol} \qquad (4\text{–}159)$$

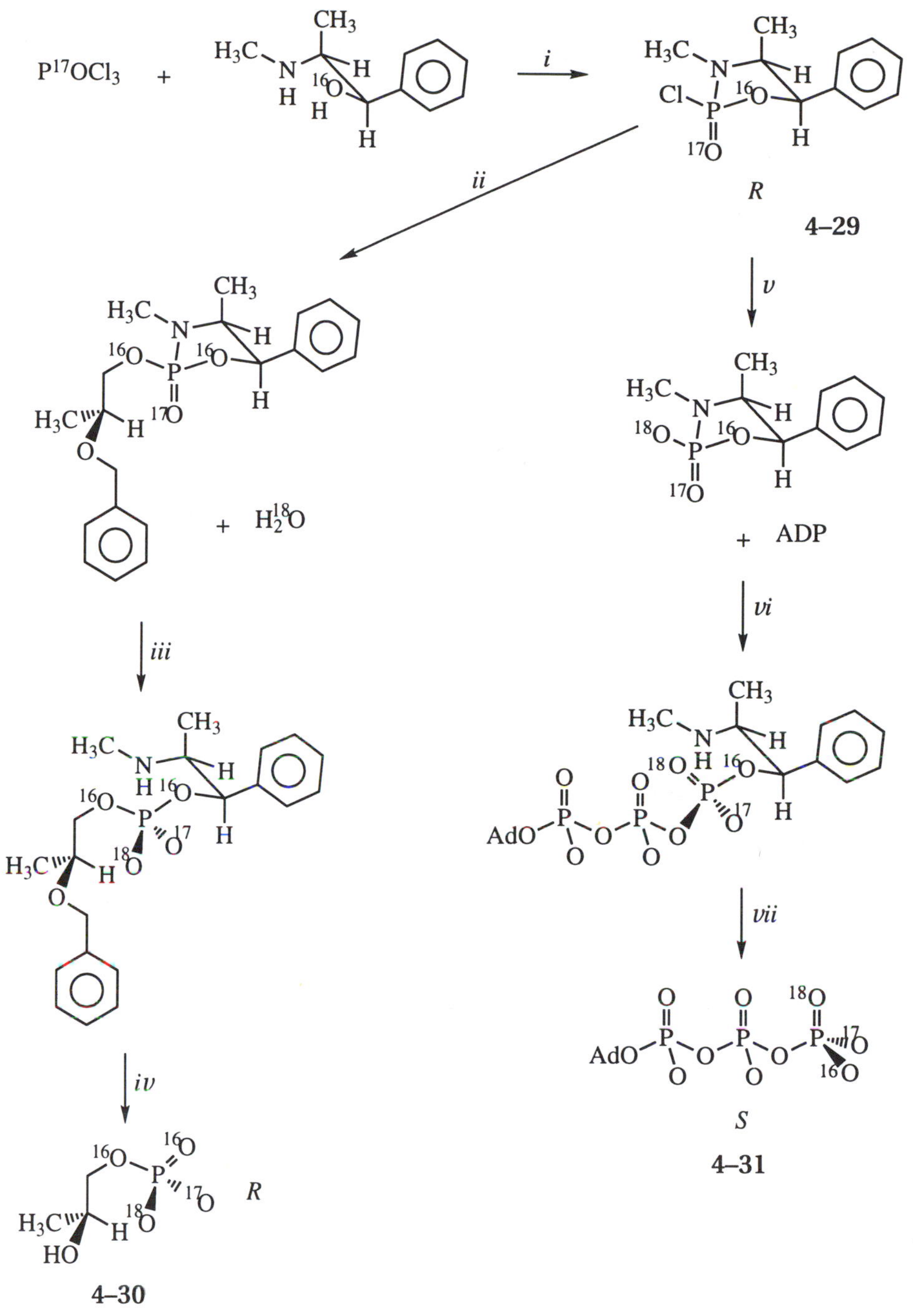

**Figure 4–23:** Synthesis of either ($R$)-[$^{16}$O,$^{17}$O,$^{18}$O]-1-phosphono-(2$S$)-propane-1,2-diol (**4–30**)[119] or adenosine $\gamma$-($S$)-[$^{16}$O,$^{17}$O,$^{18}$O]triphosphate (**4–31**)[121] from $P^{17}OCl_3$ and (1$R$,2$S$)-2-(methylamino)-1-phenylpropanol (**4–30**). (1$R$,2$S$)-2-(Methylamino)-1-phenylpropanol was converted to the cyclic phosphorylchloramidate **4–29** with $P^{17}OCl_3$ in ether using triethylamine as the base (*i*). Cyclic phosphorylchloramidate (**4–29**) was submitted to substitution with retention of configuration by 2-benzyl-($S$)-propane-1,2-diol, again using triethylamine as the base, and the two epimers at phosphorus were resolved chromatographically (*ii*). The major epimer was hydrolyzed in acidic $H_2{}^{18}O$ (*iii*), and the phosphate amidoester was reduced with $H_2$ over palladium on carbon to produce the monoester (*iv*). The two epimers of cyclic phosphorylchloramidate **4–29** were separated chromatographically, and the major one was hydrolyzed with $Li^{18}OH$ in $H_2{}^{18}O$ and dioxane with retention of configuration (*v*). The product of this reaction was submitted to nucleophilic substitution with ADP in dimethyl sulfoxide with inversion of configuration (*vi*), and adenosine $\gamma$-($S$)-[$^{16}$O,$^{17}$O,$^{18}$O]triphosphate was released by reduction with $H_2$ over palladium on carbon in ethanol–water (*vii*).



These observations suggest that such phosphonotransfer reactions are simple in-line transfers. The pentavalent intermediate formed from entry of the acceptor, which necessarily has the oxygen of the acceptor at one axial position, has the oxygen of the donor at the other axial position, and no pseudorotation takes place. A mechanism in which the donor oxygen resides at an equatorial position in the initial pentavalent intermediate and then assumes an axial position by pseudorotation prior to departure would proceed with retention of configuration. In the active sites of those enzymes that catalyze phosphonotransfer reactions that proceed with inversion of configuration, the donor and acceptor must be held in such a way that the oxygen of the acceptor lies upon the opposite side of the phosphorus on the line of the phosphorus–oxygen bond of the donor.

There are several enzymatic phosphonotransfers, however, that proceed with retention of configuration at phosphorus. In each of these reactions, there is independent evidence that a phosphorylated form of the enzyme acts as an intermediate in the enzymatic mechanism. Alkaline phosphatase catalyzes the general phosphonotransfer reaction (Reaction 4–158) in which the donor and acceptor can be any two alcohols, including water, which is the alcohol where R is H. Because water is usually present at high concentration, it is usually the acceptor when any phosphate monoester is the reactant. For this reason, the reaction usually catalyzed by alkaline phosphatase is the hydrolysis of a phosphate monoester to the alcohol and $HOPO_3{}^{2-}$

$$ROPO_3{}^{2-} + H_2O \rightleftharpoons ROH + HOPO_3{}^{2-} \qquad (4\text{–}160)$$

If, however, an alcohol other than water is present in the solution at high concentration, the phosphate will be transferred from the donor to that alcohol as well. Alkaline phosphatase transfers phosphate from phenylphosphate to (*S*)-propane-1,2-diol with retention of configuration[127]

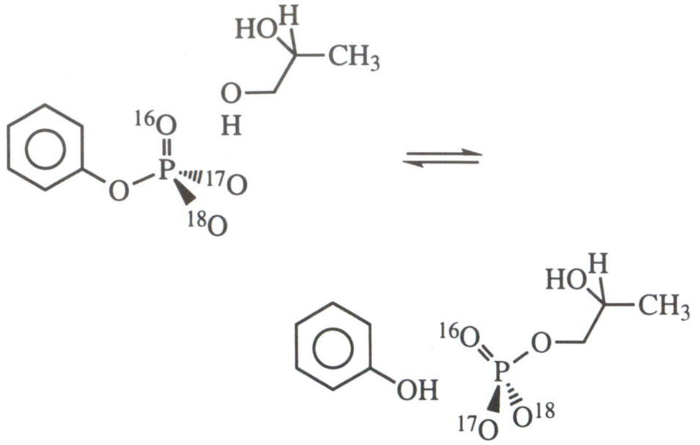

(4–161)

It has been established independently that, in the first step of the enzymatic reaction, the donor transfers the phosphate to Serine 102 within the active site of the enzyme.[128] The donor then leaves the active site and is replaced by the acceptor, and the phosphate is transferred from Serine 102 to the acceptor. The detailed structure of the active site[128] is consistent with both transfers occurring with inversion of configuration for net retention of configuration.

Nucleoside diphosphate kinase

$$\text{NDP} + \text{MgATP} \rightleftharpoons \text{MgNTP} + \text{ADP} \qquad (4\text{–}162)$$

catalyzes the phosphorylation of any nucleoside diphosphate (NDP), either a ribonucleotide or deoxyribonucleotide. It also proceeds with retention of configuration at phosphorus,[129] and independent results[130] are consistent with a mechanism in which the nucleoside triphosphate phosphorylates an amino acid within the active site, leaves as the nucleoside diphosphate, and is replaced by the other nucleoside diphosphate, which accepts the phosphate from that amino acid. The results with nucleoside diphosphate kinase can be contrasted with those obtained with adenylate kinase

$$\text{AMP} + \text{MgATP} \rightleftharpoons \text{ADP} + \text{MgNTP} \qquad (4\text{–}163)$$

an enzyme that catalyzes a very similar reaction. Adenylate kinase proceeds with inversion of configuration at phosphorus,[125] but it binds both substrates simultaneously to the active site to catalyze a direct transfer between the donor and the acceptor.[131] Presumably, this is accomplished by aligning the oxygen of the acceptor with the oxygen–phosphorus bond of the leaving group in the donor and proceeding with an in-line transfer with inversion of configuration and no pseudorotation. The difference between the reactions catalyzed by nucleoside diphosphate kinase and adenylate kinase is that the phosphoenzyme in the former reaction always reacts with a nucleoside diphosphate, including ADP, but if there were a phosphoenzyme in the latter it would have to react with both a nucleoside monophosphate, AMP, and a nucleoside diphosphate, ADP, and still be able to recognize each as an adenine nucleotide, which would be difficult if not impossible.

The three isotopes of oxygen have been used along with sulfur to produce enantiomeric inorganic thiophosphate[124]

*S*
**4–34**

*R*
**4–35**

and a method has been developed to distinguish the two enantiomers by examination of the nuclear magnetic resonance spectrum of adenosine [$\beta,\beta$-$^{17}$O, $^{18}$O]-$\beta$-thiotriphosphate produced enzymatically from each.[132] The production of chiral thiophosphate has been used to determine the stereochemical consequences of reactions in which MgATP is hydrolyzed to provide energy or synthesized to store energy[133]

$$H_2O + \text{MgATP} \rightleftharpoons \text{MgADP} + HOPO_3^{2-} \qquad (4\text{–}164)$$

When adenosine (*S*)-[$\gamma,\gamma$-$^{17}$O,$^{18}$O]-$\gamma$-thiotriphosphate is hydrolyzed in $H_2{}^{16}O$ by $Ca^{2+}$-transporting ATPase, an enzyme responsible for the transport of calcium across cellular membranes, the [$^{17}$O,$^{18}$O]thiophosphate produced has the *S* configuration.[134] The retention of configuration that was observed is due to the existence of a phosphorylated intermediate in the mechanism of the enzymatic reaction. The $\gamma$-phosphate of MgATP is transferred to an aspartate in the active site to produce a mixed anhydride,[135] which is then hydrolyzed by water. The two successive steps, each proceeding with inversion of configuration, cause the net result to be retention of configuration. When adenosine (*S*)-[$\gamma$-$^{18}$O]-$\gamma$-thiotriphosphate, however, is hydrolyzed in $H_2{}^{17}O$ by the ATPase of myosin, (*S*)-[$^{17}$O,$^{18}$O]thiophosphate is produced.[136] The inversion of configuration that was observed in this case has been used as evidence that water directly hydrolyzes the MgATP on the active site of myosin through an in-line nucleophilic substitution.

## *Suggested Reading*

Willadsen, P., & Eggerer, H. (1975) Substrate Stereochemistry of the Enoyl-CoA Hydratase Reaction, *Eur. J. Biochem. 54*, 247–252.

Willadsen, P., & Eggerer, H. (1975) Substrate Stereochemistry of the Acetyl-CoA Acetyl Transferase Reaction, *Eur. J. Biochem. 54*, 253–258.

### PROBLEM 4–10

In the enzymatic reaction catalyzed by arginase

$$\text{arginine} + H_2O \rightleftharpoons \text{ornithine} + \text{urea}$$

how does a general acid–base in the active site enforce the regiospecificity of the reaction so that ornithine is formed rather than citrulline? Write a mechanism explaining this regiospecificity. The answer to this question is an example of a general strategy in enzymatic regioselectivity.

### PROBLEM 4–11

When dihydroxyacetone phosphate is produced from glyceraldehyde 3-phosphate in $^2H_2O$ by triosephosphate isomerase, 1.0 deuterium atom is incorporated at carbon 1 for every molecule of product formed. If this monodeuterated product is isolated and incubated again with the enzyme in $^2H_2O$, and the glyceraldehyde 3-phosphate formed is isolated, it contains a deuterium attached to carbon 2 but carbon 1 has only a protium, its deuterium having been lost. What do these results suggest about the stereochemical properties of this reaction when it is catalyzed by the enzyme?

## PROBLEM 4–12

(A) Write the steps in the mechanism of the reaction catalyzed by D-serine dehydratase

$$\text{D-serine} \rightleftharpoons \text{pyruvate} + NH_4^+$$

The central intermediates in the mechanism are the iminium cation of pyruvate, $CH_3CNH_2^+COO^-$, and its enamine. If you happen to look up this enzyme, you will learn that it contains pyridoxal phosphate as a catalyst. Forget this fact for the moment, and do not use pyridoxal phosphate anywhere in your mechanism, but do use acids, HA, and bases, B, to move protons.

In this reaction the oxygen on carbon 3 of serine is replaced by a hydrogen. This problem will examine the stereochemistry of this substitution.[136a] D-Serine dehydratase utilizes only D-serine [(2*R*)-serine] as a substrate, never L-serine [(2*S*)-serine]. Likewise, the enzyme D-amino acid oxidase turns only D-serine [(2*R*)-serine] into 3-hydroxypyruvate, never L-serine. Therefore, in a mixture of (2*R*)-serine and (2*S*)-serine, only the (2*R*)-serine is a substrate for either enzyme and the (2*S*)-serine is untouched.

(*E*)-[3-$^3$H]Propenoic acid was dissolved in methanol, and solid $Ag(CH_3COO)_2$ was added, followed by $Br_2$. In this reaction an enantiomeric pair of [3-$^3$H]-2-bromo-3-methoxypropanoic acids was formed by *trans* addition to the double bond. Ammonia ($NH_3$) was then added to the solution to displace the Br in an $S_N2$ reaction. After several hours, the whole mixture was treated with HBr to remove the methyl group from the oxygen by an $S_N2$ reaction at that methyl group. After purification, an enantiomeric pair of diasteriomeric [$^3$H]serines was obtained from the overall reaction.

(B) Draw, in the sawhorse convention, the absolute structures of the two enantiomeric serines in the enantiomeric mixture and name each in the *R,S* convention.

A portion of this enantiomeric mixture was treated with D-amino acid oxidase, and the [$^3$H]-3-hydroxypyruvate that resulted was isolated. This was treated with $H_2O_2$ to produce glycolate that was then turned into glyoxylate by (*S*)-2-hydroxy-acid oxidase. The glyoxylate contained 80% of the tritium originally in the pyruvate; the other 20% was lost as $^3H_2O$.

A portion of the original enantiomeric mixture of the two serines was also dissolved in $^2H_2O$ and treated with a mixture of D-serine dehydratase, pyruvate carboxylase, $CO_2$, MgATP, malate dehydrogenase, and NADH, all in $^2H_2O$. The resulting [3,3-$^2$H,$^3$H]malate was purified from the reaction. Pyruvate carboxylase replaces a proton on pyruvate with $CO_2$ added with retention of configuration, and the enzyme prefers to remove a proton by a factor of 3 over a deuteron. Malate dehydrogenase only produces L-malate [(*S*)-malate] from oxaloacetate. The [3,3-$^2$H,$^3$H]malate was incubated with fumarate hydratase, and 41% of the $^3$H was found in the fumarate and 58% was recovered as $^3H_2O$.

The enantiomeric pair of serines of part B was acetylated on their amino nitrogens to produce an enantiomeric pair of [3-$^3$H]-*N*-acetylserines. This mixture was treated with hog kidney acylase, an enzyme that hydrolyzes acetyl groups only from *N*-acetyl-L-amino acids [(2*S*)-*N*-acetyl-L-amino acids]. The optically pure [3-$^3$H]serine that was formed by this enzyme was purified and this optically pure compound was racemized only at carbon 2 to give an equal mixture of (2*R*)- and (2*S*)-[3-$^3$H]serines. This mixture contained a pair of diastereomers.

(C) Draw, in the sawhorse convention, the absolute structure of the two stereoisomers of [3-$^3$H]serine that were present in this diastereomeric mixture.

(D) When the diastereomeric mixture was treated with D-amino acid oxidase, and the [3-$^3$H]-3-hydroxypyruvate was purified and treated with $H_2O_2$ followed by glycolate oxidase, how much of the tritium in the [3-$^3$H]-3-hydroxypyruvate was retained in the glyoxylate?

When the diastereomeric mixture, dissolved in $^2H_2O$, was incubated with D-serine dehydratase, pyruvate carboxylase, $CO_2$, MgATP, malate dehydrogenase, and NADH in $^2H_2O$, [3,3-$^2$H,$^3$H]malate could be isolated. When this [3,3-$^2$H,$^3$H]malate was incubated with fumarate hydratase, 37% of the $^3$H was released as $^3H_2O$ and 63% of the $^3$H was retained in the fumarate.

(E) Refer to your mechanism for D-serine dehydratase in part A and draw the removal of the –OH from the serine and the addition of the $H^+$ to the eneamine with the proper stereochemistry based on all of these results.

(F) Does the enzyme release into solution the enamine of pyruvate or pyruvate itself as the product? How can you be sure?

## PROBLEM 4–13

Branched chain acyl-CoA dehydrogenase catalyzes the following reaction:

$$RCH_2R'CHCO\text{-}SCoA + FAD \rightleftharpoons$$
$$RCH{=}CR'CO\text{-}SCoA + FADH_2$$

It is generally believed that the two electrons are removed from the acyl-SCoA and transferred onto the flavin either through an extended $\pi$ system or by electron transfer and that each of the two hydrogens is removed as a proton by two general bases, respectively. The enzyme catalyzes the following oxidations[136b]

2-methylpropanoyl-coenzyme A →
2-methylpropenoyl-coenzyme A

(*S*)-2-methylbutanoyl-coenzyme A →
(*E*)-2-methyl-2-butenoyl-coenzyme A

(*R*)-2-methylbutanoyl-coenzyme A →
2-ethylpropenoyl-coenzyme A

(A) Explain these stereochemical results by drawing three pictures of the active site of the enzyme. In each case, the structure of the active site must be the same, but a different one of the three reactants, respectively, will be bound to it. The drawings must clearly indicate the three-dimensional dispositions of general bases and substrate molecules that explain the stereochemical results. Abbreviate the coenzyme A as -SCoA and disregard its detailed structure. Choose two specific general bases.

(B) Explain why the enzyme cannot catalyze the following oxidation because of an absolute stereochemical requirement.

(*R*)-2-methylbutanoyl-coenzyme A →
(*Z*)-2-methyl-2-butenoyl-coenzyme A

The $K_m$ for (*S*)-2-methylbutanoyl-coenzyme A is 20 $\mu$M. Both *n*-butanoyl-coenzyme A and *n*-pentanoyl-coenzyme A inhibit the oxidation of (*S*)-2-methylbutanoyl-coenzyme A. At 100 $\mu$M (*S*)-2-methylbutanoyl-coenzyme A, 100 $\mu$M *n*-butanoyl-coenzyme A inhibits the initial velocity of the oxidation of (*S*)-2-methylbutanoyl-coenzyme A by 58%, and 100 $\mu$M *n*-pentanoyl-coenzyme A inhibits the initial velocity of the oxidation of 2-methylbutanoyl-coenzyme A by 52%.

(C) Assume that *n*-butanoyl-coenzyme A and *n*-pentanoyl-coenzyme A are competitive inhibitors of the enzyme and calculate the $K_i$ for each.

*n*-Butanoyl-coenzyme A is oxidized by the enzyme at a rate only 5% that of the rate for the oxidation of (*S*)-2-methylbutanoyl-coenzyme A when either of these substrates is present at a concentration of 100 $\mu$M. *n*-Pentanoyl-coenzyme A is not oxidized at all by the enzyme.

(D) Provide an explanation, consistent with both your structure of the active site in part A and the dissociation constants calculated in part C, for the fact that *n*-butanoyl-coenzyme A is a poor substrate and *n*-pentanoyl-coenzyme A is not a substrate for the enzyme.

## PROBLEM 4–14

Fatty-acid synthase catalyzes the following sequential conversions

$$\text{H}^+ + \text{acetyl-SCoA} + \text{malonyl-SCoA} \rightleftharpoons \text{CO}_2 + \text{HSCoA} + \text{acetoacetyl-SCoA} \quad (1)$$

$$\text{H}^+ + \text{acetoacetyl-SCoA} + \text{NADPH} \rightleftharpoons (R)\text{-3-hydroxybutyryl-SCoA} + \text{NADP}^+ \quad (2)$$

$$(R)\text{-3-hydroxybutyryl-SCoA} \rightleftharpoons \text{H}_2\text{O} + \text{crotonyl-SCoA} \quad (3)$$

$$\text{H}^+ + \text{crotonyl-SCoA} + \text{NADPH} \rightleftharpoons \text{butyryl-SCoA} + \text{NADP}^+ \quad (4)$$

Crotonyl-SCoA was enzymatically hydrated in $^2H_2O$ to (3*R*)-[2-$^2$H]-3-hydroxybutyryl-SCoA until equilibrium was reached (Reaction 3), and the (3*R*)-[2-$^2$H]-3-hydroxybutyryl-SCoA was purified. The (3*R*)-[2-$^2$H]-3-hydroxybutyryl group was transferred to benzylamine to produce *N*-benzyl-(3*R*)-[2-$^2$H]-3-hydroxybutyramide. The nuclear magnetic resonance spectrum of this product displayed a doublet at 2.4 ppm that was assigned to the hydrogen on carbon 2 of the (3*R*)-[2-$^2$H]-3-hydroxybutyryl group.[137] The doublet integrated with a magnitude of only one proton, which demonstrated that carbon 2 had both a hydrogen and a deuterium attached to it.

(A) What hydrogen is responsible for the splitting of the resonance of the single hydrogen on carbon 2 into a doublet?

The coupling constant for the splitting of the resonance of the hydrogen on carbon 2 was 5.5 Hz. When the enzymatic reaction was brought to equilibrium with (3*R*)-[2,2-$^2H_2$]-3-hydroxybutyryl-SCoA in $H_2O$, the resulting *N*-benzyl-(3*R*)-[2-$^2$H]-3-hydroxybutyramide again displayed a doublet in its nuclear magnetic spectrum at 2.4 ppm, but the coupling constant was 8.4 Hz.

(B) In the spectrum of the first product with a coupling constant of 5.5 Hz, no contribution from a doublet with a coupling constant of 8.4 Hz was detected, and in the spectrum of the second product with a coupling constant of 8.4 Hz, no contribution from a doublet with a coupling constant of 5.5 Hz could be detected. What does this observation demonstrate about the stereochemical purity of the two products?

(C) When (*E*)-2,3-epoxybutyric acid was reduced with sodium borodeuteride (NaB$^2H_4$) in $^2H_2O$, the racemic [2-$^2$H]-3-hydroxybutyric acid produced also displayed a doublet in its nuclear magnetic resonance spectrum at 2.5 ppm. The doublet integrated with a magnitude of one proton.[138] This doublet had a coupling constant of 9.0 Hz. What is the absolute stereochemistry at carbon 2 of the (3*R*)-[2-$^2$H]-3-hydroxybutyric acid in this racemic synthetic product?

(D) Assume that the (3*R*)-[2-$^2$H]-3-hydroxybutyryl group on the *N*-benzyl-(3*R*)-[2-$^2$H]-3-hydroxybutyramide that displayed a doublet with a coupling constant of 8.4 Hz had the same absolute stereochemistry as this synthetic product and that the enzymatic product that displayed a doublet with a coupling constant of 5.5 Hz had the opposite stereochemistry at carbon 2. Write a chemical reaction with all reactants and products in the proper stereochemistry that describes the enzymatic hydration of crotonyl-SCoA. Is this a *syn* addition or an *anti* addition?

(E) Write the chemical mechanism for Reaction 1, in which the malonyl-SCoA contributes carbons 1 and 2 of the acetoacetyl-SCoA and the acetyl-SCoA contributes carbons 3 and 4.

Malonyl-SCoA is produced from acetyl-SCoA in a reaction catalyzed by acetyl-CoA carboxylase that uses biotin as a coenzyme

$$\text{MgATP} + \text{acetyl-SCoA} + \text{HCO}_3^- \rightleftharpoons \text{MgADP} + \text{HOPO}_3^{2-} + \text{malonyl-SCoA}$$

The 2-hydrogen on acetyl-SCoA is replaced by the carboxyl group with retention of configuration.

(F) If ($R$)-[2,2-$^2$H,$^3$H]acetyl-SCoA were carboxylated enzymatically, what would be the stereochemistry of the [2,2-$^2$H,$^3$H]malonyl-SCoA produced?

When ($R$)-[2,2-$^2$H,$^3$H]acetyl-SCoA was incubated with acetyl-CoA carboxylase and fatty acid synthase simultaneously, the butyryl-SCoA produced[139] retained 30.4% of the tritium at carbon 2. When ($S$)-[2,2-$^2$H,$^3$H]acetyl-SCoA was incubated with this mixture of enzymes, the butyryl-SCoA retained 34.3% of the tritium at carbon 2. When racemic [2-$^3$H]acetyl-SCoA was incubated with this mixture of enzymes, the butyryl-SCoA retained 32.3% of the tritium at carbon 2.

(G) The first hydrogen is removed from carbon 2 of the original acetyl-SCoA by acetyl-CoA carboxylase. At which step is the second hydrogen removed? What is the absolute stereochemistry of this step?

(H) Draw, with correct stereochemistry, the reactants and products of Reaction 1 if it were performed with ($R$)-[2-$^3$H]malonyl-SCoA and unlabeled acetyl-SCoA. Does Reaction 1 proceed with retention of configuration, inversion of configuration, *syn* addition, or *anti* addition? Only one answer is correct.

The hydrogen added as a proton from the solvent in Reaction 4 is added to carbon 2 of crotonyl-SCoA and the hydrogen added as a hydride from NADPH is added to carbon 3 of crotonyl-SCoA. When Reaction 4 is carried out in $^2H_2O$, with unlabeled crotonyl-SCoA and unlabeled NADPH, the resonance assigned to hydrogens on carbon 2 in the nuclear magnetic spectrum of the $N$-benzylbutyramide derived from the butyryl-SCoA produced in the reaction integrates with a magnitude of only one hydrogen.

The butyryl-SCoA formed during the enzymatic conversion in $^2H_2O$ was isolated and then incubated with acyl-CoA dehydrogenase (Reaction 4–132 on page 297) in $^1H_2O$. None (<10%) of the deuterium incorporated by fatty acid synthase was lost upon this second incubation. The product formed during conversion of crotonyl-SCoA to butyryl-SCoA with [$^3$H]NADPH (Reaction 4) was isolated and then incubated with acyl-CoA dehydrogenase. Most of the tritium (83%) incorporated by the fatty acid synthase (Reaction 4) was lost during the second incubation.

(I) Draw a mechanism with proper stereochemistry for Reaction 4 if it were carried out in $^2H_2O$ with [$^3$H]NADPH. The mechanism should proceed through a reasonable enol intermediate. Does Reaction 4 proceed with *syn* addition or *anti* addition?

## PROBLEM 4–15

Citrate (*re*) synthase adds the enol of acetyl-SCoA to the *re* face of oxaloacetate to produce citrate in which the *pro-R* carboxymethyl group is derived from the acetyl-SCoA. Citrate (*si*) synthase adds the enol of acetyl-SCoA to the *si* face of oxaloacetate to produce citrate in which the *pro-S* carboxymethyl group is derived from the acetyl-SCoA. Citrate (*pro*-3$S$) lyase cleaves citrate to oxaloacetate and acetate by removing the *pro-S* carboxymethyl group.

(A) Copy the structure of citrate

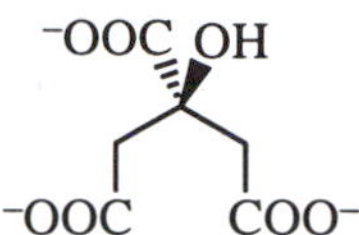


and identify the *pro-S* and *pro-R* carboxymethyl groups.

(B) If [3-$^{14}$C]oxaloacetate were converted to citrate by citrate (*re*) synthase and the [$^{14}$C]citrate produced were converted to oxaloacetate by citrate (*pro*-3$S$) lyase, which final product would be radioactive?

Two specimens of [$^3$H]malate were synthesized. [$^3$H]Malate, specimen 1, was made by incubating fumarate with fumarate hydratase in $^3H_2O$. [$^3$H]Malate, specimen 2, was made by incubating [2,3-$^3H_2$]fumarate with fumarate hydratase in $^1H_2O$. [$^3$H]Malate, specimen 1, was incubated with malate dehydrogenase, NAD$^+$, acetyl-SCoA, and citrate (*re*) synthase. The [$^3$H]citrate produced was incubated with citrate (*pro*-3$S$) lyase in $^2H_2O$, and the [$^3$H,$^2$H]acetate produced was incubated with MgATP, acetate kinase, phosphate acetyltransferase, CoASH, glyoxylate, and malate synthase. The [$^3$H,$^2$H]malate produced in this last reaction retained 61% of its tritium upon incubation with fumarate hydratase in $^1H_2O$. [$^3$H]Malate, specimen 2, was submitted to the same cycle of reactions and the [$^3$H,$^2$H]malate produced lost 56% of its tritium upon incubation with fumarate hydratase in $^1H_2O$.[112]

(C) Write out each of the steps in this cycle that begins with [$^3$H]malate, specimen 1. Show proper stereochemistry at each step unambiguously.

(D) Draw the mechanism of the reaction of citrate (*pro*-3$S$) lyase operating in $^2H_2O$ with the stereochemistry defined by these experiments on the [$^3$H]citrate derived from [$^3$H]malate, specimen 1. Indicate stereochemistry unambiguously in your drawings. Use an enol as the leaving group and protonate the enol stereochemically on carbon while it is in the active site. Does the reaction proceed with inversion or retention of configuration at carbon 2 of the acetate?

(E) Draw the mechanism for the reaction of citrate (*pro*-3$S$) lyase, with proper stereochemistry, operating in $^2H_2O$ on the [$^3$H]citrate derived from the [$^3$H]malate, specimen 2.

($R$)-[2,2-$^2$H,$^3$H]Acetyl-SCoA was incubated with citrate (*si*) synthase and the [$^2$H,$^3$H]citrate produced was cleaved by citrate (*pro*-3$S$) lyase to [2,2-$^2$H,$^3$H]acetate. The [2,2-$^2$H,$^3$H]acetate produced was converted to [$^3$H,$^2$H]malate by incubation with ATP, acetate kinase, phosphate acetyltransferase, CoASH, glyoxylate, and malate synthase. The [$^3$H,$^2$H]malate produced retained 65% of its tritium upon incubation with fumarate hydratase. When the complete cycle was repeated with ($S$)-[2,2-$^2$H,$^3$H]acetyl-SCoA, the [$^3$H,$^2$H]malate produced lost 60% of its tritium upon incubation with fumarate hydratase.[112]

(F) Draw, with proper stereochemistry, the mechanism of the reaction catalyzed by citrate (*si*) synthase operating on ($R$)-[2,2-$^2$H,$^3$H]acetyl-SCoA. Use a stereochemically generated enol as the nucleophile. Does the reaction proceed with retention or inversion of configuration?

## PROBLEM 4–16

Triosephosphate isomerase from rabbit muscle is a dimeric protein composed of two identical subunits, each a polypeptide 248 amino acids in length. It catalyzes the reaction

D-glyceraldehyde 3-phosphate ⇌
dihydroxyacetone phosphate

It is known that the unhydrated, carbonyl forms of the substrates are the only ones recognized by the enzyme, although they are the minor forms present in aqueous solution.

(A,B) Draw, with proper stereochemistry where necessary, two possible chemical mechanisms for this reaction and label them A and B. Mechanism A must involve proton transfers only, while mechanism B must involve a 1,2-hydride shift.

(C) Draw a hypothetical picture of the substrate D-glyceraldehyde 3-phosphate bound to the active site of the enzyme. Employ specific amino acid side chains to form the noncovalent interactions.

When dihydroxyacetone phosphate and glyceraldehyde 3-phosphate were incubated with the enzyme in $^3H_2O$, tritium was incorporated into the dihydroxyacetone phosphate.[139a]

| compounds | isomerase (units) | specific activity of compound (cpm $\mu mol^{-1}$) | specific activity of $^3H_2O$ (cpm $\mu atom^{-1}$) |
|---|---|---|---|
| experiment 1 | | | |
| dihydroxyacetone-P | none | 0 | 699 |
| glyceraldehyde-3-P | none | 0 | 699 |
| experiment 2 | | | |
| dihydroxyacetone-P | 1600 | 303 | 278 |
| experiment 3 | | | |
| dihydroxyacetone-P | 475 | 3150 | 3590 |

When [$^3$H]dihydroxyacetone phosphate (777 cpm $mol^{-1}$) was oxidized to 2-phosphoacetate and formaldehyde with periodate, the formaldehyde was radioactive (835 cpm $mol^{-1}$)

(D) How many hydrogens on which carbon of the dihydroxyacetone phosphate are equilibrated by the enzyme with the $^3H_2O$ during catalysis?

The intention of the next series of experiments[139b] was to establish the stereochemistry of the incorporation of tritium into dihydroxyacetone phosphate. To do this, the absolute stereochemistry of (S)-2-hydroxy-acid oxidase was determined. This enzyme catalyzes the reaction

$$^{-}OOCCH_2OH + O_2 \rightleftharpoons {}^{-}OOCCHO + H_2O_2$$

glycolate glyoxylate

(S)-2-Hydroxy-acid oxidase readily converts L-lactate into pyruvate but fails entirely to catalyze the same reaction with D-lactate. L-Lactate dehydrogenase will convert glyoxylate and NAD$^3$H into NAD$^+$ and [$^3$H]glycolate (78,050 cpm $\mu mol^{-1}$). When this [$^3$H]glycolate is turned into glyoxylate by glycolic acid oxidase, all of the $^3$H ($\leq$100 cpm $\mu mol^{-1}$ remains) is lost. Notice that glycolate and lactate differ only by a methyl group.

(E) Indicate on a stereochemically unambiguous drawing of glycolate which hydrogen is removed by (S)-2-hydroxy-acid oxidase.

A sample of [$^3$H]dihydroxyacetone phosphate (24,000 cpm $\mu mol^{-1}$), prepared as in part D above, was dephosphorylated to dihydroxyacetone, which was oxidized to glycolate, formaldehyde, and $CO_2$ by careful treatment with periodate. The [$^3$H]glycolate from this mixture (10,800 cpm $\mu mol^{-1)}$ was isolated and turned into glyoxylate with (S)-2-hydroxy-acid oxidase. The specific radioactivity of the glyoxylate was 188 cpm $\mu mol^{-1}$.

(F) Why was the specific radioactivity of the glycolate half that of the dihydroxyacetone phosphate?

(G) Indicate, on a stereochemically unambiguous drawing of dihydroxyacetone phosphate, the hydrogen or hydrogens equilibrated with $^3H_2O$ by triosephosphate isomerase.

(H,I) Which mechanism, A or B, is consistent with these observations? Draw the step-by-step mechanism as it occurs on the active site, involving choices of specific amino acid side chains as catalysts.

## PROBLEM 4–17

Write out the chemical equations for the reactions used to convert dimethylallyl pyrophosphate to acetate in the stereochemical study of isopentenyl-diphosphate Δ-isomerase (Reaction 4–147).

## PROBLEM 4–18

[1-$^3$H]Glucose was converted to [1,1-$^3$H,$^2$H]fructose 6-phosphate with hexokinase and glucose-6-phosphate isomerase in $^2H_2O$. [1-$^3$H]Mannose was converted to [1,1-$^3$H,$^2$H]fructose 6-phosphate with hexokinase and mannose-6-phosphate isomerase in $^2H_2O$. The two respective stereoisomers of [1,1-$^3$H,$^2$H]fructose 6-phosphate produced were of opposite stereochemistry at carbon 1.

(A) Assume that the enzymatic mechanisms for triosephosphate isomerase, glucose-6-phosphate isomerase, and mannose-6-phosphate isomerase are all the same. From your knowledge of the active site of triosephosphate isomerase, draw two mechanisms, one for glucose-6-phosphate isomerase and one for mannose-6-phosphate isomerase, that explain why the stereochemistries at carbon 1 of the two stereoisomers of [1,1-$^3$H,$^2$H]fructose 6-phosphate produced in the respective reactions are opposite to each other.

Use the same amino acids as are found in the active site of triosephosphate isomerase in the same roles as general acid–base catalysts.

(B) Explain how the deuterium is introduced at carbon 1 of each fructose 6-phosphate and why, after a sufficient time at equilibrium, the percent deuterium in the proper location on carbon 1 of the fructose 6-phosphate becomes equivalent to the percent deuterium for each atom of hydrogen in the $^2H_2O$.

(C) What are the absolute stereochemistries at carbon 1 of the [1,1-$^3$H,$^2$H]fructose 6-phosphates from [1-$^3$H]glucose and [1-$^3$H]mannose, respectively?

The two stereoisomers of [1,1-$^3$H,$^2$H]fructose 6-phosphate were separately incubated with 6-phosphofructokinase and fructose-bisphosphate aldolase, and the two stereoisomers of [1,1-$^3$H,$^2$H]dihydroxyacetone 1-phosphate were isolated. These stereoisomers of [1,1-$^3$H,$^2$H]dihydroxyacetone 1-phosphate were converted to (*E*)-[$^3$H,$^2$H]phospho*enol*pyruvate and (*Z*)-[$^3$H,$^2$H]phospho*enol*pyruvate, respectively

$^2$H $^3$H $^{2-}O_3PO$ $COO^-$ *E* $^3$H $^2$H $^{2-}O_3PO$ $COO^-$ *Z*

by incubation with triosephosphate isomerase, glyceraldehyde-3-phosphate dehydrogenase in the presence of arsenate, phosphoglycerate mutase, and enolase. Each stereoisomer of [$^3$H,$^2$H]phospho*enol*pyruvate was incubated separately with 3-phosphoshikimate 1-carboxyvinyltransferase and shikimate 3-phosphate

$COO^-$ $^{2-}O_3PO$ OH HO + H H $^{2-}O_3PO$ $COO^-$ ⇌

$COO^-$ H H $^{2-}O_3PO$ O $COO^-$ HO + $HOPO_3^{2-}$

The $O^5$-(1-carboxyvinyl)-3-phosphoshikimate was converted to chorismate

$COO^-$ H H O $COO^-$ HO

with chorismate synthase. The chorismate was converted to lactate by the following sequence of chemical reactions.

$COO^-$ H H O $COO^-$ HO —DMSO, RT, 12 h→ $COO^-$ H H O $COO^-$

—$^1H_2$, Wilkinson's catalyst→ $COO^-$ H H H O H $COO^-$

—Birch reduction→ H H H HO H $COO^-$

Wilkinson's catalyst adds $H_2$ *syn* to a double bond and ultimately produces both a (2*S*)-lactate and a (2*R*)-lactate because it does not distinguish one face of the olefin from the other. The (2*S*)-lactate was eliminated by incubation with (*S*)-lactate dehydrogenase and the stereochemically pure (2*R*)-lactate was submitted to Kuhn–Roth oxidation to produce chiral [$^3$H,$^2$H]acetate.

(D) When the $O^5$-(1–carboxyvinyl)-3-phosphoshikimate derived from (*E*)-[$^3$H,$^2$H]phospho*enol*pyruvate was submitted to this sequence, (*R*)-[$^3$H,$^2$H]acetate was the preferential product. When $O^5$-(1-carboxyvinyl)-3-phosphoshikimate derived from (*Z*)-[$^3$H,$^2$H]phospho*enol*pyruvate was submitted to this sequence, (*S*)-[$^3$H,$^2$H]acetate was the preferential product.[139c] Show by drawing the proper stereochemistry of every product between phospho*enol*pyruvate and acetate, including both lactates, why these two results demonstrate that (*E*)-[$^3$H,$^2$H]phospho*enol*pyruvate yields (*E*)-[5,5-$^3$H,$^2$H]-$O^5$-(1-carboxy-vinyl)-3-phosphoshikimate and (*Z*)-[$^3$H,$^2$H]phospho*enol*pyruvate yields (*Z*)-[5,5-$^3$H,$^2$H]-$O^5$-(1-carboxy-vinyl)-3-phosphoshikimate. Explain why the (*S*)-lactate was eliminated before the Kuhn–Roth oxidation was performed.

(E) Examine the mechanism for 3-phosphoshikimate 1-carboxyvinyltransferase. Assume that the proton is added and removed from the enolpyruvate by the same acid–base in the active site and that, as has been demonstrated, the addition of a proton and the removal of that same proton is in preference to addition and removal of a deuteron or a triton and that this explains why chiral acetate is the final product. Draw the mechanism with correct stereochemistry for the addition of shikimate 3-phosphate to the carbocation and the removal of phosphate to regenerate the carbocation. Your mechanism must explain the observed retention of configuration in the product. Provide specific amino acids of your own choice as general acid–base catalysts of these two steps.

# Modification and Labeling of the Active Site

The general acid–bases catalyzing the reactions that occur within the active site of an enzyme are usually amino acids within the sequence of the folded polypeptide. The particular amino acid in the amino acid sequence acting as a general acid–base in a specific step in the mechanism of the enzyme can be identified by its covalent modification with a reagent designed to bind to the active site and position an electrophile adjacent to that amino acid. Such a reagent is referred to as an active-site label. An **active-site label** is a reagent that has been designed to modify covalently a catalytic acid–base in the active site while it is bound noncovalently at the location normally occupied by a substrate. The electrophile on the active-site label reacts covalently with the lone pair of electrons on the conjugate base of the amino acid, and the modified amino acid is identified by isolating and sequencing a peptide containing it.

From kinetic, isotopic, and stereochemical studies of triosephosphate isomerase, the role of two general acid–bases in its mechanism could be defined

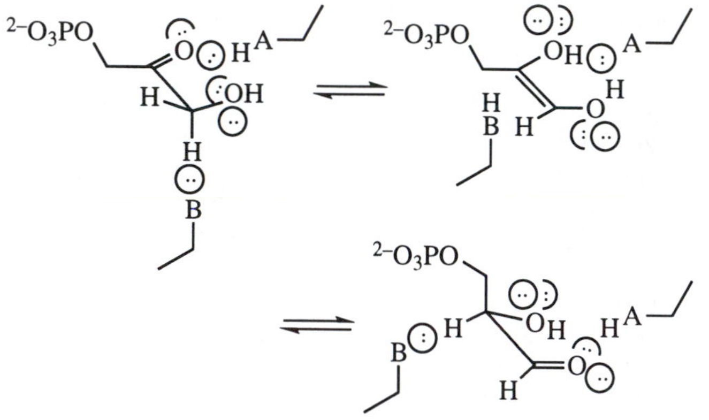

(4–165)

From this information, reagents were designed that would place an electrophilic center adjacent to the general base responsible for removing the acidic proton from carbon.[140–142] The simplest way to accomplish this was to substitute the hydroxyl on carbon 1 of glyceraldehyde 3-phosphate with a halogen[143]

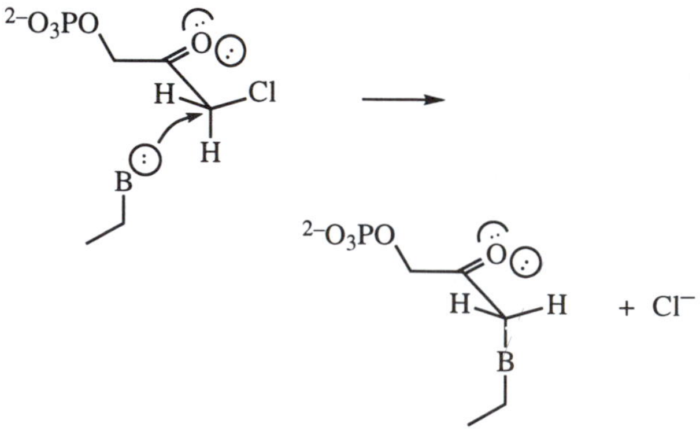

(4–166)

In this way, the amino acid acting as a general base during the enzymatic reaction would become alkylated. The alkyl halide was made radioactive by incorporating a radioactive isotope, so that any peptide that contained the modified amino acid could be identified easily.

3-Iodo-1-hydroxy-2-propanone phosphate, 3-bromo-1-hydroxy-2-propanone phosphate, and 3-chloro-1-hydroxy-2-propanone phosphate were each synthesized. If these compounds are capable of alkylating the general base within the active site, they should have inactivated the enzyme irreversibly, and it was observed that each did (Figure 4–24).[143] 3-Chloro-1-hydroxy-2-propanone [$^{32}$P]phosphate was prepared, and triosephosphate isomerase was modified with this reagent. The enzyme was then denatured, and the denatured, modified polypeptide was digested with trypsin. Only one radioactive peptide was observed upon peptide mapping. This tryptic peptide was purified, and its amino acid sequence was found to be WVLAYEPVWAIGTGK. When the modified peptide was treated with hydroxylamine, the hydroxamate of glutamic acid could be identified as the product

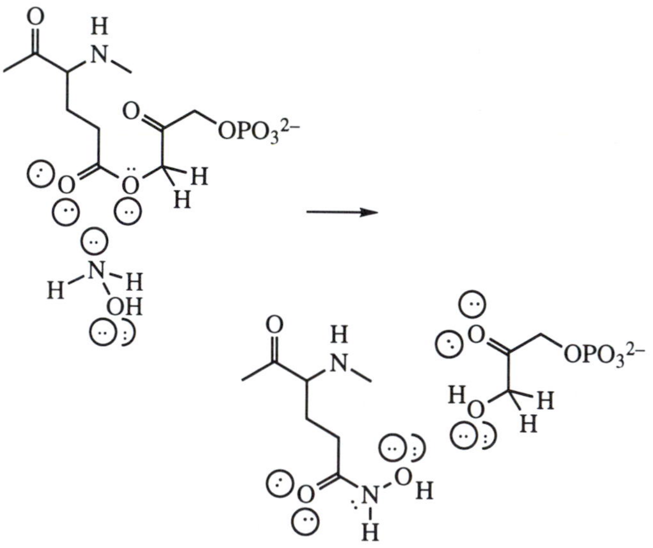

(4–167)

Therefore, the glutamate in this peptide was the nucleophile that had been alkylated by the reagent. The amino acid sequence of the peptide identifies the glutamate as Glutamate 165 in the total amino acid sequence[144] of the polypeptide of triosephosphate isomerase. From the results of several such experiments, Glutamate 165 was deduced to be the base responsible for removing the proton from carbon during the enzymatic reaction (Reaction 4–165). This conclusion was subsequently confirmed by X-ray crystallography[145] and site-directed mutation.[146]

The electrophile used as an active-site label does not have to be a simple alkylating agent. Thymidylate synthase forms a covalent adduct between cysteine and the $\alpha,\beta$-olefin of 2′-deoxyuridine 5′-monophosphate

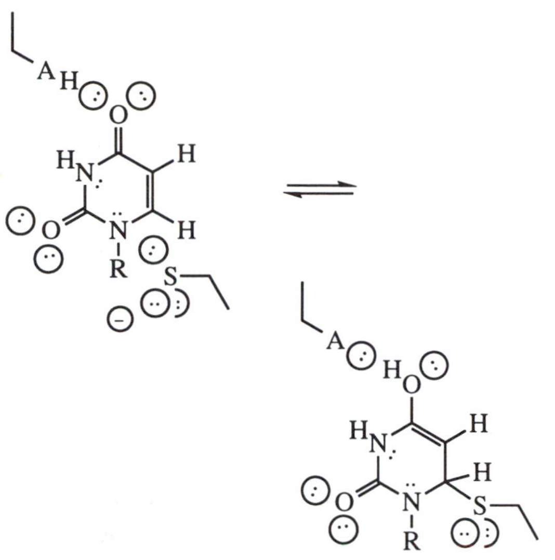

(4–168)

as part of its normal reaction mechanism. 1-($\beta$-D-2′-Deoxyribofuranosyl)-3-azapurin-2-one 5′-monophosphate is an active-site label that capitalizes on the existence of this intermediate[147]

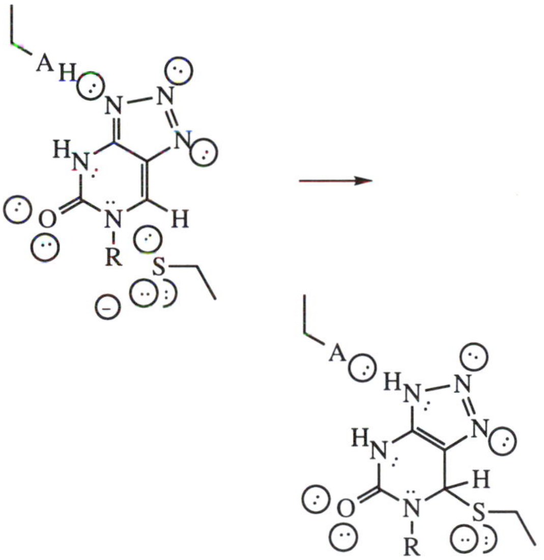

(4–169)

The creation of the aromatic triazole causes the nucleophilic addition to be kinetically irreversible and causes the active-site label to inactivate the enzyme.

There are several **criteria** that should be met before a reagent synthesized as an active-site label can be designated as capable of modifying the enzyme while it is bound in place of a substrate at the active site. First, the rate of the reaction between the enzyme and the active-site label should display saturation as a function of the concentration of the reagent. Second, the natural substrates or competitive inhibitors of the enzyme should competitively inhibit the inactivation produced by the active-site label. Third, the dissociation constants for substrates or other ligands as competitive inhibitors of the inactivation produced by an active-site label should be the same as their dissociation constants from the active site in the absence of the active-site label. Fourth, the active-site label should be either a reactant in a reaction catalyzed by the enzyme or itself a competitive inhibitor of the normal enzymatic reaction with respect to the reactant that it resembles. Fifth, the dissociation constant for the active-site label acting as a reactant in or as a competitive inhibitor of the enzymatic reaction should be the same as its dissociation constant in the inactivation reaction. The first criterion shows that the active-site label binds to the enzyme before it can inactivate the enzyme. The last four criteria connect this binding to the nor-

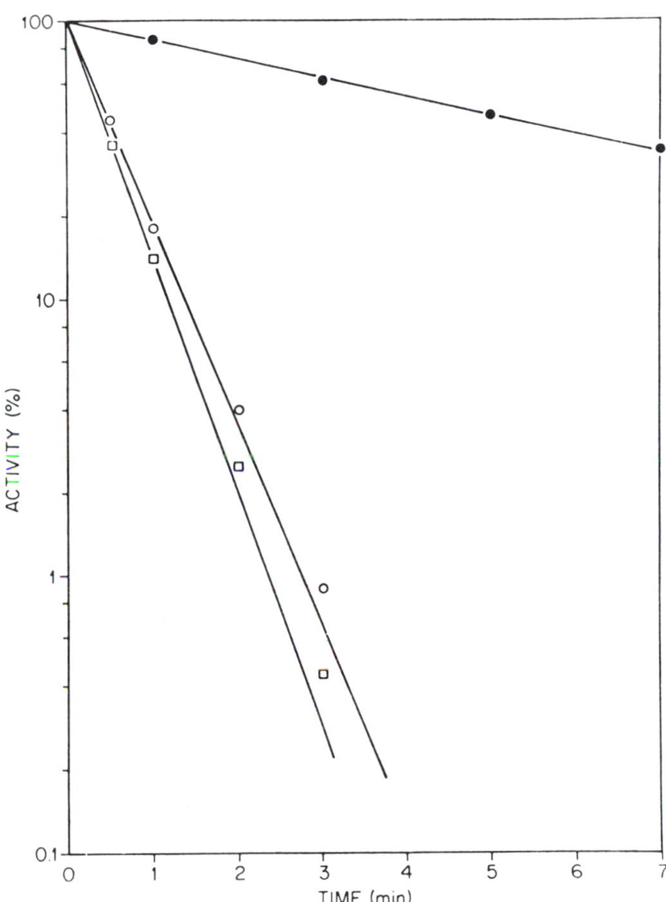


**Figure 4–24:** Inactivation of rabbit muscle triosephosphate isomerase by 3-halo-1-hydroxy-2-propanone phosphates.[143] Mixtures containing 25 $\mu$g mL$^{-1}$ triosephosphate isomerase in 0.1 M imidazolium chloride, pH 6.5, at 2 °C were prepared with either 10 $\mu$M 3-iodo-1-hydroxy-2-propanone phosphate (●), 10 $\mu$M 3-chloro-1-hydroxy-2-propanone phosphate (○), or 10 $\mu$M 3-bromo-1-hydroxy-2-propanone phosphate (□). At the times indicated, 0.1 mL samples from each mixture were removed and diluted into 4.9 mL of 10 mM 2-mercaptoethanol in 0.02 M triethanolammonium chloride, pH 7.9, to discharge rapidly the respective halopropanone. These quenched samples were then assayed for enzymatic activity. The logarithm of the percent activity remaining is plotted as a function of the time (minutes) the enzyme was exposed to the halopropanone before the alkylation of the active site was halted by the quenching solution. Adapted with permission from ref 143. Copyright 1971 American Chemical Society.

mal binding of substrates or competitive inhibitors to the active site.

If an active-site label (ASL) first binds to the active site of an enzyme and then modifies a base within the active site, it follows that

$$\mathrm{E} + \mathrm{ASL} \underset{k_{-1}}{\overset{k_1}{\rightleftharpoons}} \mathrm{E}\cdot\mathrm{ASL} \xrightarrow{k_i} \mathrm{E}-\mathrm{ASL} \qquad (4\text{–}170)$$

where E – ASL is the enzyme that has been modified covalently by the active-site label. If the concentration of active-site label is in large molar excess over the concentration of active sites and if the active-site label is lost from the solution only by reaction with the active site, its concentration will remain essentially constant throughout the reaction. In this case, it can be shown[148] that

$$\ln\left(\frac{[\mathrm{E}]_0 - [\mathrm{E}-\mathrm{ASL}]}{[\mathrm{E}]_0}\right) = -k_{\mathrm{obs}}t \qquad (4\text{–}171)$$

where $[\mathrm{E}]_0$ is the initial concentration of enzyme

$$k_{\mathrm{obs}} = \frac{k_i[\mathrm{ASL}]}{K_m^{\mathrm{ASL}} + [\mathrm{ASL}]} \qquad (4\text{–}172)$$

and

$$K_m^{\mathrm{ASL}} = \frac{k_{-1} + k_i}{k_1} \qquad (4\text{–}173)$$

The term $([\mathrm{E}]_0 - [\mathrm{E}-\mathrm{ASL}])\,[\mathrm{E}]_0^{-1}$ is equal to the fraction of the enzymatic activity remaining at a particular time, if covalent modification by the active-site label is the only reason that enzymatic activity is disappearing. It is the disappearance of the enzymatic activity that is followed in the experiment. The inactiviation of triosephosphate isomerase by 3-halo-1-hydroxy-2-propanone phosphates is a pseudo-first-order reaction (Figure 4–24) as predicted by Equation 4–171. From such first-order behavior, the observed pseudo-first-order rate constant, $k_{\mathrm{obs}}$, can be measured directly.

Because the covalent reaction is usually slow, $k_{-1} >> k_i$ and

$$K_m^{\mathrm{ASL}} \cong \frac{k_{-1}}{k_1} = K_d^{\mathrm{ASL}} = \frac{[\mathrm{ASL}][\mathrm{E}]}{[\mathrm{E}\cdot\mathrm{ASL}]} \qquad (4\text{–}174)$$

where $K_d^{\mathrm{ASL}}$ is the dissociation constant for the complex between the active-site label and the active site.

It is the behavior of the observed pseudo-first-order rate constant of inactivation, $k_{\mathrm{obs}}$, as a function of the molar concentration of active-site label (Equation 4–172) that demonstrates **saturation**. 1-Hydroxybut-3-en-2-one phosphate was synthesized as an active-site label for fructose-bisphosphate aldolase.[149] One of the steps in the reaction catalyzed by this enzyme is the formation of an enamine of dihydroxyacetone phosphate. The enamine is formed from the imine between dihydroxyacetone phosphate and a lysine on the enzyme by removal of a proton from carbon 3

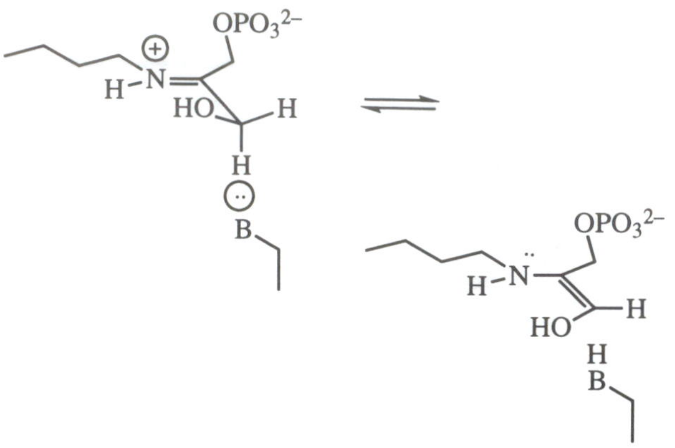

(4–175)

1-Hydroxybut-3-en-2-one phosphate, when bound to the active site as the analogous adduct, should place an $\alpha,\beta$-unsaturated imine adjacent to the amino acid responsible for removing the proton in the normal reaction, and a Michael addition should result

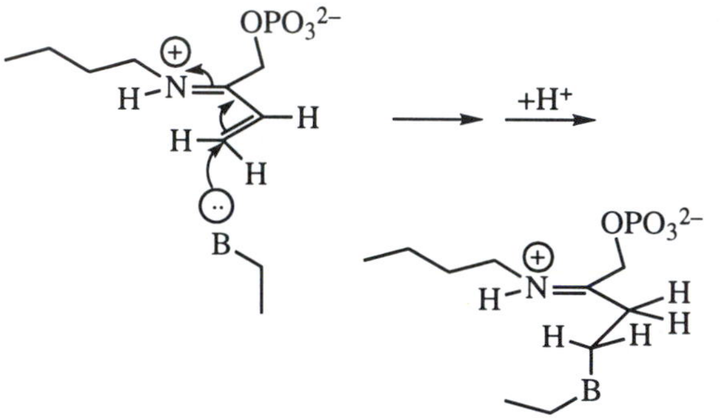

(4–176)

When 1-hydroxybut-3-en-2-one phosphate was added to a solution of fructose-bisphosphate aldolase and samples were withdrawn at various intervals and diluted to dissociate the reversibly bound inhibitor, the enzymatic activity was observed to disappear irreversibly in a pseudo-first-order process.[150] From these measurements, observed pseudo-first-order rate constants of inactivation were obtained using Equation 4–171. When these observed rate constants were plotted as a function of the concentration of the active-site label, saturation was observed (Figure 4–25).[150] The continuous curve drawn in the figure is defined by Equation 4–172 with $k_i = 0.11\ \mathrm{s}^{-1}$ and $K_m^{\mathrm{ASL}} \approx K_d^{\mathrm{ASL}} = 99\ \mu\mathrm{M}$. Therefore, the observed behavior is consistent with a kinetic mechanism in which the active-site label binds to the enzyme before covalently modifying it.

Often the active-site label is an active enough electrophile to react with the water or other bases in the solution. In this instance, the concentration of active-site label decreases systematically during the reaction, and the activity of the enzyme does not decrease in a simple first-order reaction. The initial rate of decrease in the enzymatic activity could be used as $k_{\mathrm{obs}}$ in Equation 4–172, but it is often technically difficult to make this measurement. The situation, however, can be viewed as a competition for the reagent between the nucleophile in the active site and nucleophiles in the solution. Under these circumstances, the final yield of modified enzyme, after all of the active-site label has been destroyed by reacting with the solution, can be expressed as a fraction, $\alpha^{\infty}_{\text{E·ASL}}$ of the amount of enzyme originally present. The yield, $\alpha^{\infty}_{\text{E·ASL}}$, is simply the enzymatic activity at the end of the reaction divided by the enzymatic activity at the start of the reaction.

If the active site cannot recognize the product of the reaction between the active-site label and the solution, it can be shown[151] that

$$\ln\left(1-\alpha^{\infty}_{\text{E·ASL}}\right)=\frac{k_{\text{i}}}{k_{\text{D}}}\ln\left(\frac{K_{\text{d}}^{\text{ASL}}}{[\text{ASL}]_0+K_{\text{d}}^{\text{ASL}}}\right) \quad (4\text{–}177)$$

where $k_D$ is the rate constant for the reaction in which the active-site label is destroyed by reaction with the solution, $k_i$ is the rate at which it reacts with the active site, $[\text{ASL}]_0$ is the initial molar concentration of active-site label, and $K_d^{\text{ASL}}$ is its dissociation constant from the active site. If the active site, however, can bind the product of the reaction between the active-site label and the solution, then as the product builds up, it will compete with the active-site label itself for the active site. If the product has the same dissociation constant, $K_d^{\text{ASL}}$, as the active-site label itself, then it can be shown that

$$\ln\left(1-\alpha^{\infty}_{\text{E·ASL}}\right)=\frac{k_{\text{i}}}{k_{\text{D}}}\left(\frac{[\text{ASL}]_0}{K_{\text{d}}^{\text{ASL}}+[\text{ASL}]_0}\right) \quad (4\text{–}178)$$

From Equation 4–177 or 4–178, it can be ascertained whether or not the reaction between the active site and the active-site label displays saturation[151] and a dissociation constant for the active-site label from the enzyme can be determined.

As must be the case with any reagent that modifies amino acids within an active site, if an active-site label is binding at the active site before modifying the enzyme, then the substrate it was designed to resemble should decrease the rate of the modification by occupying the active site in its stead. For example, the addition of the natural substrate, dihydroxyacetone phosphate, at increasing concentrations causes the rate of inactivation of fructose-bisphosphate aldolase by 1-hydroxybut-3-en-2-one phosphate (Reaction 4–176) to decrease accordingly.[150]

If a natural substrate and the active-site label are competing for the same location within the active site, and if the conversion of the natural substrate at the active site is either prevented by leaving out another required substrate or is simply ignored for the sake of simplicity, then the mechanism should be

$$\text{E·S}\overset{K_{\text{d}}^{\text{S}}}{\rightleftharpoons}\text{S}+\text{E}+\text{ASL}\underset{k_{-1}}{\overset{k_1}{\rightleftharpoons}}\text{E·ASL}\overset{k_{\text{i}}}{\rightarrow}\text{E}-\text{ASL} \quad (4\text{–}179)$$

where $K_d^S$ is the dissociation constant for the substrate from the active site. It can be shown[152] that under these circumstances the observed pseudo-first-order rate constant for irreversible inactivation should be

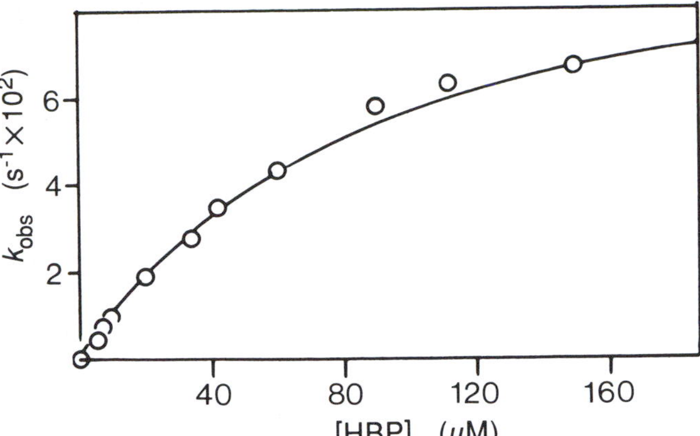


**Figure 4–25:** Behavior of the observed pseudo-first-order rate constant for the inactivation of fructose-bisphosphate aldolase by 1-hydroxybut-3-en-2-one phosphate.[150] Mixtures were prepared containing various concentrations of 1-hydroxybut-3-en-2-one phosphate at pH 7.0 and 37 °C. The reaction was initiated in each case by adding fructose-bisphosphate aldolase. Samples (50 μL) were removed at various times (at about 1 min intervals) and were diluted immediately into an assay mixture (750 μL) for measuring enzymatic activity. The initial velocity of the enzymatic reaction was measured continuously at 340 nm by coupling the production of the triose phosphates to the reduction of NADH with triosephosphate isomerase and glycerol-3-phosphate dehydrogenase. The initial velocity of the enzymatic activity was plotted as a function of the time of exposure to 1-hydroxybut-3-en-2-one phosphate, and an observed pseudo-first-order rate constant of inactivation, $k_{obs}$ (seconds$^{-1}$), was calculated from the time course for each concentration of active-site label. The observed pseudo-first-order rate constants, $k_{obs}$ (seconds$^{-1}\times 10^2$), are plotted as a function of the concentration of 1-hydroxybut-3-en-2-one phosphate, [HBP] (micromolar). Adapted with permission from ref 150. Copyright 1979 American Chemical Society.

$$k_{\text{obs}}=\frac{k_{\text{i}}[\text{ASL}]}{K_{\text{m}}^{\text{ASL}}\left(1+\frac{[\text{S}]}{K_{\text{d}}^{\text{S}}}\right)+[\text{ASL}]} \quad (4\text{–}180)$$

and the **substrate should be a competitive inhibitor of the inactivation** with respect to the active-site label. Equation 4–180 can be rewritten in reciprocal form

$$\frac{1}{k_{\text{obs}}}=\frac{1}{k_{\text{i}}}+\frac{K_{\text{m}}^{\text{ASL}}}{k_{\text{i}}}\left(1+\frac{[\text{S}]}{K_{\text{d}}^{\text{S}}}\right)\frac{1}{[\text{ASL}]} \quad (4\text{–}181)$$

Isocitrate lyase performs an aldol condensation on succinate and glyoxylate to produce (2*R*,3*S*)isocitrate

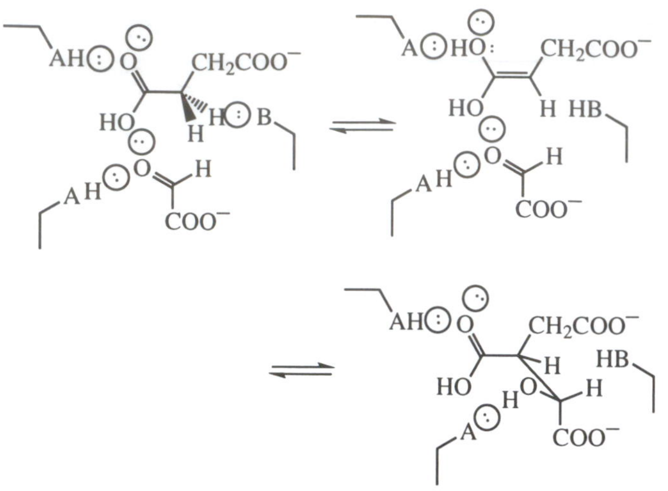

(4–182)

Bromopyruvate was selected[153] as an active-site label for the general base–acid in the active site responsible for removing the proton from carbon

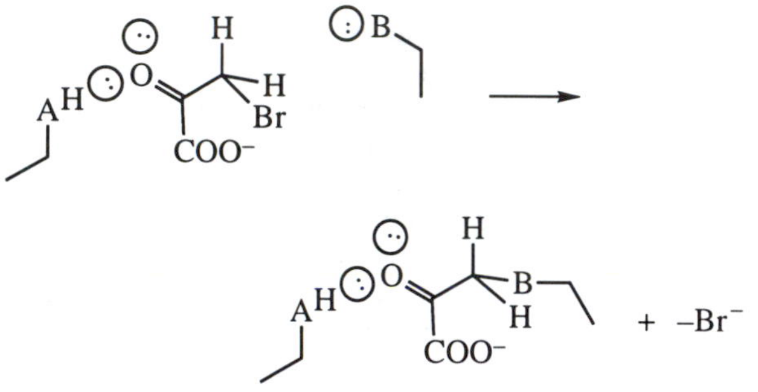

(4–183)

Bromopyruvate (BrP) inactivates isocitrate lyase in a pseudo-first-order reaction, and the observed pseudo-first-order rate constants of inactivation for various concentrations of bromopyruvate and isocitrate were determined. When $k_{obs}^{-1}$ was plotted as a function of $[BrP]^{-1}$ at two fixed concentrations of isocitrate, competitive inhibition of the inactivation was observed (Figure 4–26).[153] The dissociation constant for isocitrate as a competitive inhibitor of the inactivation calculated with Equation 4–181 from the slopes of the lines in Figure 4–26 was 90 $\mu$M, which compares favorably to the Michaelis constant of 110 $\mu$M for isocitrate as a substrate for the enzyme. The **coincidence of these two dissociation constants** suggests that isocitrate binds to the same site on the enzyme when it is a reactant as it does when it is a competitive inhibitor of the irreversible inactivation caused by bromopyruvate and that, therefore, bromopyruvate alkylates the enzyme while it is bound at the active site.

The irreversible inactivation of fructose-bisphosphate aldolase caused by 1-hydroxybut-3-en-2-one phosphate (Reaction 4–176) is inhibited by adding dihydroxyacetone phosphate at various fixed concentrations. The values of the various pseudo-first-order rate constants of inactivation were used with Equation 4–181 to calculate the dissociation constant of the dihydroxyacetone phosphate in competition with the active-site label. The value calculated (1.4 $\mu$M) was close enough to the value for the dissociation constant (4.5 $\mu$M) for dihydroxyacetone phosphate to the active site when it is a substrate to conclude that the inactivation was occurring within the active site.

A competitive inhibitor of the enzymatic reaction can also be used to demonstrate the specificity of an active-site label. *N*-Bromoacetyl-*N*-methylphenylalanine was synthesized as an active-site label for carboxypeptidase A.[154] It inactivated the enzyme in a pseudo-first-order process, and the observed rate constant displayed saturation ($K_m^{BrAMP}$ = 5 mM). A competitive inhibitor of the enzyme, 2-phenylacetate ion, competitively inhibited the irreversible inactivation produced by *N*-bromoacetyl-*N*-methylphenylalanine. The dissociation constant for phenylacetate as a competitive inhibitor of the inactivation (1.1 mM) was in reasonable agreement with its dissociation constant as a competitive inhibitor of the enzymatic activity (0.4 mM), and this was used as evidence that the inactivation occurred while the reagent was bound to the active site of the enzyme. In the normal enzymatic reaction of carboxypeptidase A, the nitrogen of the leaving group must be protonated by a general acid

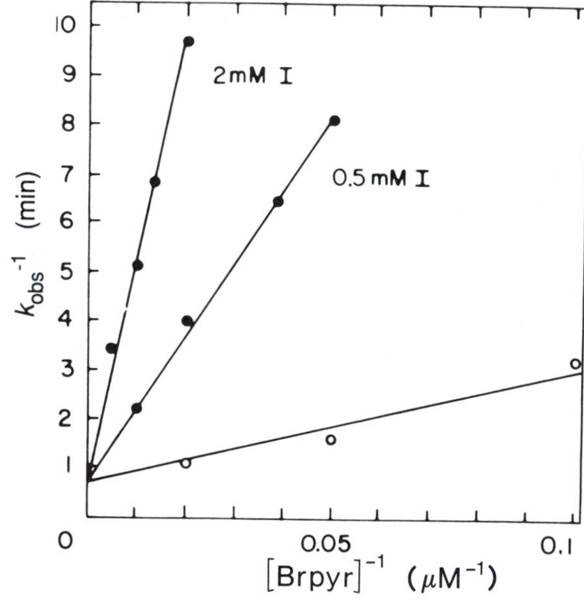


**Figure 4–26:** Competition between isocitrate and bromopyruvate during the inactivation of isocitrate lyase.[153] Mixtures containing various concentrations of bromopyruvate and 0.5 mM isocitrate (●), 2.0 mM isocitrate (●), or no isocitrate (○) were prepared at pH 7.7 and 30 °C. For each mixture, inactivation of the enzyme was followed as a function of time and pseudo-first-order rate constants of inactivation were calculated from the loss of activity over the first half-time of the reaction. The reciprocals of these rate constants, $k_{obs}^{-1}$ (minute$^{-1}$), are plotted as a function of the concentration of bromopyruvate, $[Brpyr]^{-1}$ (micromolar$^{-1}$), in the initial mixture. The lines drawn are those defined by Equation 4–181 for $k_i$ = 1.0 min$^{-1}$, $K_d^{isocit}$ = 90 $\mu$M, and $K_d^{Brpyr}$ = 30 $\mu$M. Adapted with permission from ref 153. Copyright 1969 Academic Press.

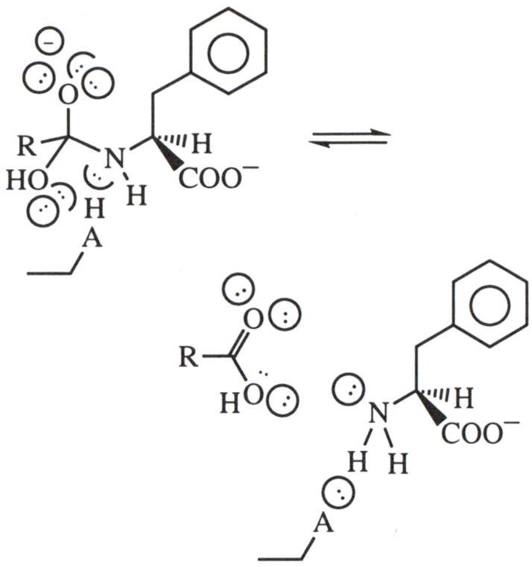

(4–184)

and it may be the conjugate base of that general acid which reacts with the active-site label

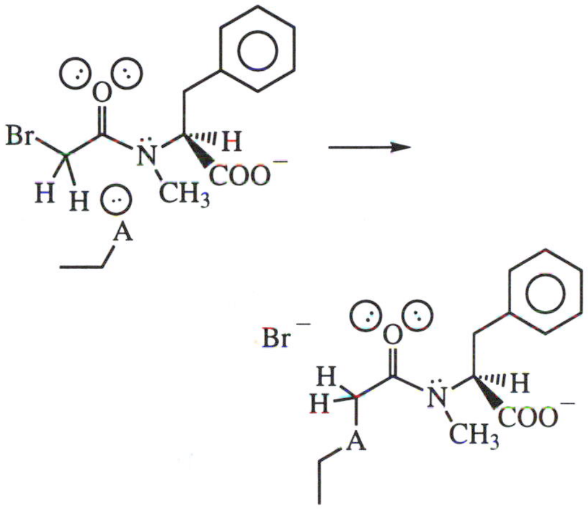

(4–185)

The rate at which an active-site label modifies the targeted amino acid is usually very slow (Figures 4–24 and 4–25) compared to the rates at which it associates and dissociates from the active site and the rates at which the normal substrate is turned over. This means that the initial velocity of the enzymatic reaction can usually be measured in the presence of the reagent over an interval short enough that significant inactivation has not yet occurred. In this way, the ultimately irreversible inhibitor can be assessed for its **ability to inhibit the enzymatic reaction reversibly**. If the active-site label does enter the active site before it reacts with the enzyme, then it should be a competitive inhibitor with respect to the reactant that it resembles.

Amidophosphoribosyltransferase transfers the equivalent of ammonia between glutamine and 5-phosphoribose 1-diphosphate

glutamine + 5-phosphoribose 1-diphosphate + $H_2O \rightleftharpoons$

glutamate + 5-phosphoribosylamine + pyrophosphate

(4–186)

The enzyme probably releases ammonia from glutamine by hydrolysis

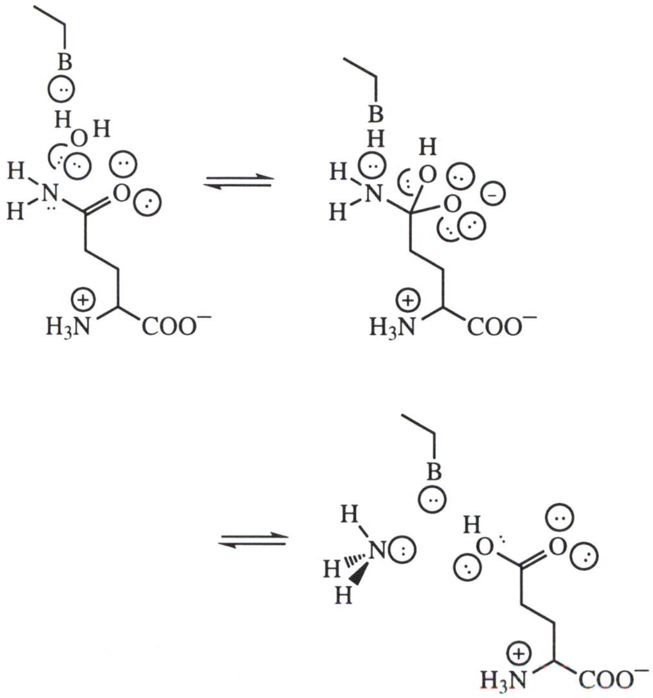

(4–187)

and then transfers it intramolecularly to the phosphoribosyl diphosphate within the active site because free ammonia can substitute, albeit poorly, for glutamine. If this is the case, the enzyme must use at least one general acid–base to protonate the nitrogen of the leaving group. 6-Diazo-5-oxonorleucine was chosen[155] as an active-site label to modify the base protonating the nitrogen during the hydrolysis because it should place a diazoketone adjacent to this base

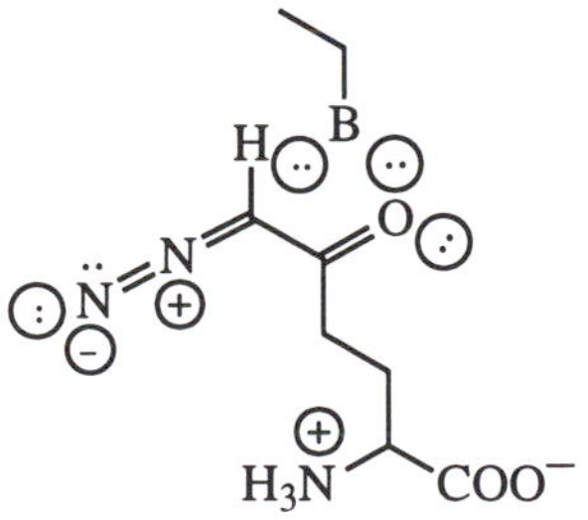

**4–36**

6-Diazo-5-oxonorleucine inactivates amidophosphoribosyltransferase irreversibly; and, when phosphoribosyl diphosphate is present, the rate of inactivation increases 15-fold. 6-Diazo-5-oxonorleucine (Don) at short times is a reversible competitive inhibitor, with respect to glutamine, of the normal enzymatic reaction, and its dissociation constant as a reversible competitive inhibitor ($K_i^{Don}$ = 20 $\mu$M) is close to its dissociation constant as an irreversible active-site label ($K_m^{Don}$ = 30 $\mu$M). This coincidence is evidence that the irreversible modification occurs while the 6-diazo-5-oxonorleucine is occupying the active site in place of glutamine.

Phospho*enol*pyruvate carboxylase catalyzes the carboxylation of the enol of pyruvate

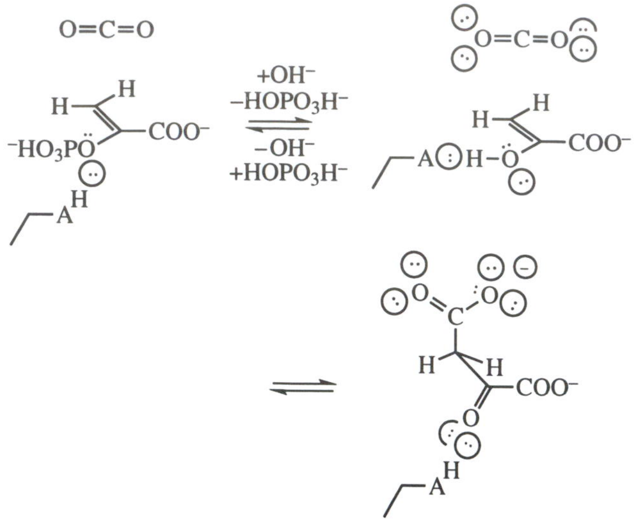

(4–188)

In the presence of $CO_2$, the enzyme is irreversibly inactivated by 3-bromophospho*enol*pyruvate, but this inactivation is not observed in the absence of $CO_2$.[156] It was proposed that the electrophilic reagent 3-bromooxaloacetate is formed on the active site during this reaction

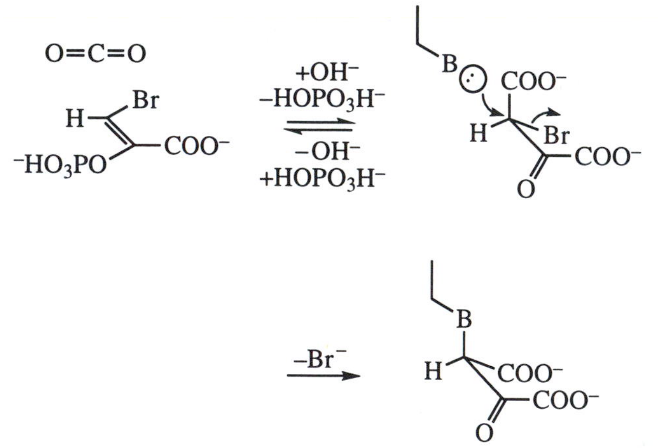

(4–189)

This active-site label would be responsible for the irreversible inactivation, which would arise from the alkylation of one of the general acid–bases within the active site. 3-Bromophospho*enol*pyruvate is a competitive inhibitor of the enzymatic reaction. Its dissociation constant as a competitive reversible inhibitor ($K_i^{BrPEP}$ = 30 $\mu$M) is similar to its dissociation constant ($K_m^{BrPEP}$ = 70 $\mu$M) as an irreversible inhibitor, in the presence of $CO_2$, and this coincidence is presented as evidence that the inactivation occurs from within the active site.

Isocitrate dehydrogenase ($NADP^+$) catalyzes the reductive carboxylation of 2-ketoglutarate

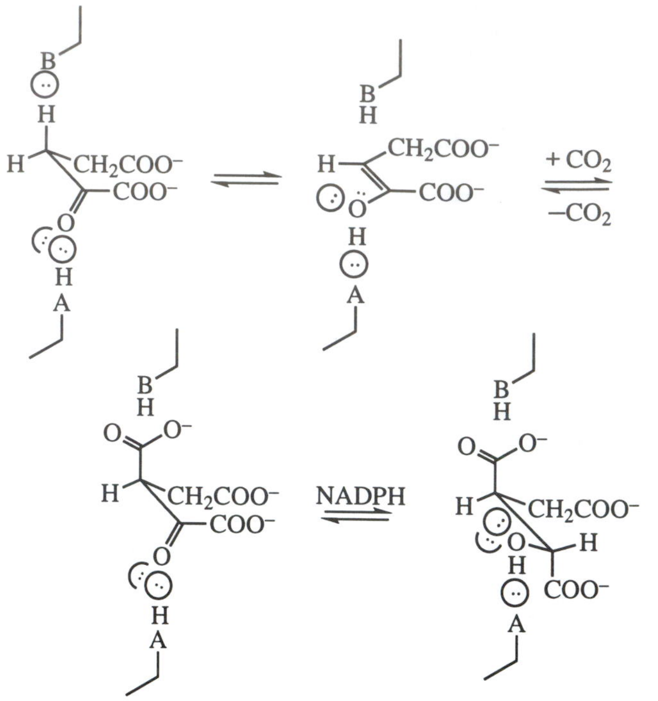

(4–190)

3-Bromo-2-ketoglutarate irreversibly inactivates the enzyme.[157] It was intended that the 3-bromo-2-ketoglutarate react with the base required to form the enol

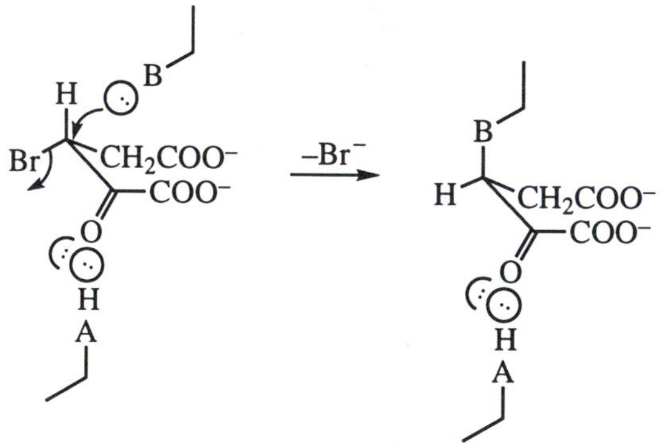

(4–191)

3-Bromo-2-ketoglutarate is a reversible competitive inhibitor with respect to (2*R*,3*S*)-isocitrate of the normal enzymatic reaction in the direction of oxidative decarboxylation (Figure 4–27).[157] The dissociation constant of 3-bromo-2-ketoglutarate for this competitive inhibition ($K_i^{BrKG}$ = 100 $\mu$M) is similar to its dissociation constant ($K_m^{BrKG}$ = 250 $\mu$M) as an active-site label.

3-Bromo-2-ketoglutarate is also a reactant for isocitrate dehydrogenase ($NADP^+$)

3-bromo-2-ketoglutarate + $H^+$ + NADPH ⇌ 3-bromo-2-hydroxyglutarate + $NADP^+$

(4–192)

and its Michaelis constant at saturating NADPH is 250 $\mu$M. Its reduction is competitively inhibited by (2*R*,3*S*)-isocitrate.[157] The dissociation constant for (2*R*,3*S*)-isocitrate as a competitive inhibitor of the reduction of 3-bromo-2-ketoglutarate ($K_i$ = 3 $\mu$M) is the same as its dissociation constant from the active site when it is a substrate. The 3-bromo-2-ketoglutarate, however, is not carboxylated by the enzyme during Reaction 4–192. If the explanation for this fact is that the bromine is mistaken for the carboxylate (as shown in Reaction 4–192), then it would have to be the opposite enantiomer that acts as a substrate from the enantiomer that acts as the irreversible inhibitor if the catalytic base were being alkylated (as shown in Reaction 4–191) because the normal enzymatic reaction proceeds with retention of configuration at carbon 3 (as shown in Reaction 4–190).[158] This puzzling observation was an early indication that the reaction of 3-bromo-2-ketoglutarate with isocitrate dehydrogenase might not be straightforward.

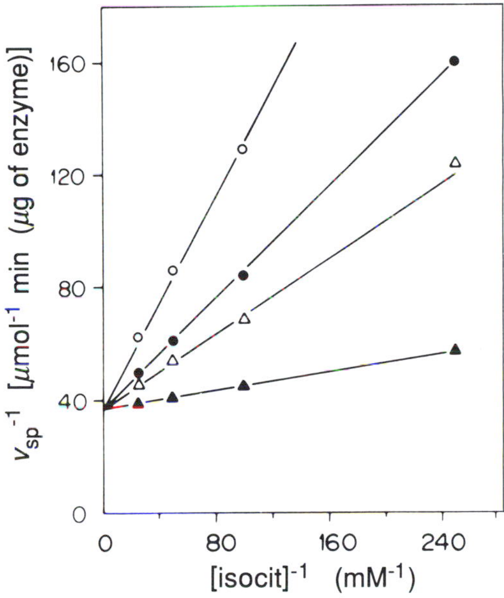


**Figure 4–27:** Competition of the active-site label, 3-bromo-2-ketoglutarate, with the normal reactant, isocitrate, for the active site of isocitrate dehydrogenase ($NADP^+$).[157] Mixtures containing 0.1 mM $NADP^+$ at pH 7.4 and 25 °C and various concentrations of isocitrate were prepared. These solutions also contained 25 $\mu$M (△), 50 $\mu$M (●), or 100 $\mu$M (○) 3-bromo-2-ketoglutarate or no added 3-bromo-2-ketoglutarate (▲). The enzymatic reaction was initiated by adding isocitrate dehydrogenase ($NADP^+$) and the production of NADPH was followed by its absorbance at 340 nm. From these measurements of the initial velocities of the enzymatic reaction, specific activities, $v_{sp}$ [micromoles minute$^{-1}$ (microgram of enzyme)$^{-1}$], were calculated. The reciprocals of these specific activities are presented as a function of the reciprocals of the concentrations of isocitrate (millimolar$^{-1}$). The lines are drawn for $K_i^{BrKG}$ = 100 $\mu$M and $K_m^{isocit}$ = 2 $\mu$M. Adapted with permission from ref 157. Copyright 1981 American Chemical Society.

Aspartate aminotransferase usually catalyzes the transamination

$$\text{aspartate} + \text{2-ketoglutarate} \rightleftharpoons \text{oxaloacetate} + \text{glutamate} \quad (4\text{–}193)$$

but it will also catalyze the reaction[159]

$$\text{aspartate} + \text{3-bromopyruvate} \rightleftharpoons \text{oxaloacetate} + \text{pyruvate} + Br^- + NH_4^+ \quad (4\text{–}194)$$

In Reaction 4–194, the bromopyruvate is thought to react with the pyridoxamine of pyridoxal phosphate, which is formed by transamination from aspartate, and to regenerate by $\beta$-elimination the lysylpyridoxamine in the active site (Figure 4–28). Bromopyruvate, as well as being a reactant for the enzyme, is also an active-site label that produces irreversible inactivation. Its dissociation constant as an active-site label ($K_m^{ASL}$ = 12 mM) was indistinguishable from its Michaelis constant as a reactant ($K_m$ = 15 mM), and this coincidence was presented as evidence that both the $\beta$-elimination and the inactivation proceed through the same intermediate. If this is the case, it is likely that either the imine of the bromopyruvate or the enamine formed by the elimination of $Br^-$ is the electrophile modifying a general base within the active site

B: + enamine–pyridoxaldimine ($^{2-}O_3PO$) → B–$CH_2$–C($COO^-$)=N$^+$H–CH= pyridoxal adduct ($^{2-}O_3PO$)

(4–195)

An additional observation, often cited as a criterion for the specificity of active-site labeling, is the correlation between **incorporation of the reagent** and the **loss of enzymatic activ-**

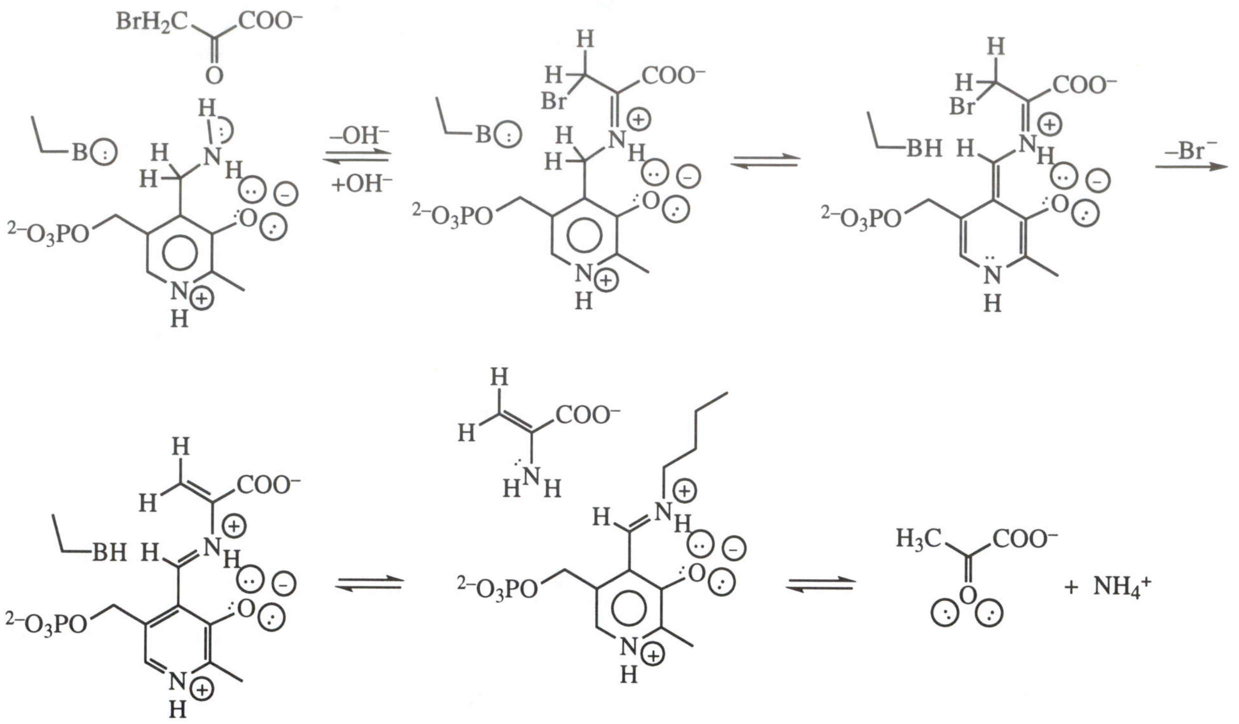

**Figure 4–28:** Participation of bromopyruvate as the second reactant in the enzymatic mechanism of aspartate aminotransferase. The pyridoxamine form of the coenzyme is formed from the first reactant, aspartate. It then reacts with bromopyruvate to form the external pyridoximine. As in the normal next step of the transamination reaction, the general base in the active site removes the benzyl proton from the external pyridoximine. Unlike the normal reaction, however, the 3-bromo-2-imine loses $Br^-$ to produce the internal pyridoximine of pyruvylenamine. Again as in the normal reaction, the enamine is released by transamination, and it decomposes rapidly in the solution to pyruvate and ammonium cation.

**ity**. When amidophosphoribosyltransferase (Reaction 4–184) is irreversibly inactivated by [6-$^{14}$C]-6-diazo-5-oxonorleucine (**4–36**), the covalent incorporation of the radioactive reagent into the protein can be followed as well as the irreversible loss of enzymatic activity. A close correlation between these two properties was observed (Figure 4–29).[160] The extrapolated value for incorporation at full inactivation was 0.85 mol of reagent for every mole of protomer in the solution. The advantage of such behavior is that it provides assurance that the major radioactive peptide derived from digestion of the enzyme will be the peptide containing the amino acid whose modification led to the inactivation of the enzyme.

A stoichiometry in such an experiment of greater than 1, however, or a percent inactivation that is not directly proportional to incorporation does not rule out the possibility that the reagent did inactivate the enzyme while it was bound in place of a substrate within the active site. An incorporation in excess of 1 mol for every mole of active site may only mean that the electrophile is reactive enough to modify other amino acids on the surface of the protein in addition to the targeted amino acid in the active site. The product of the reaction that occurred within the active site can be distinguished from the products of the other reactions that occurred outside the active site by changes in the yield of these products in response to ligands for the active site. The yield of the product of modification from within the active site will decrease significantly when the amino acid involved is protected from modification by adding substrates or competitive inhibitors, but the yield of the products of modification at other amino acids should be unaffected by these additions.

When carboxypeptidase was modified to complete inactivation with *N*-([1-$^{14}$C]bromoacetyl)-*N*-methylphenylalanine (Reaction 4–185), 2 mol of the reagent were incorporated for every mole of the enzyme.[161] Two of the cyanogen bromide fragments of the protein became radioactive during the modification. When the modification was repeated under the same conditions but in the presence of the competitive inhibitor phenylalanine, only 10% of the enzymatic activity was lost and the incorporation of radioactivity into one of the two fragments decreased 10-fold while the incorporation into the other was unchanged. This experiment was sufficient to identify the fragment containing the amino acid modified from within the active site. This ability of competitive inhibitors or substrates to protect a particular peptide against modification is a necessary criterion to demonstrate that the peptide which has been isolated contains the amino acid from the active site whose modification was correlated with the inactivation.

Active-site labels designed to position an electrophile adjacent to a particular general base involved in the mechanism of an enzymatic reaction can be used to assign an **acid dissocia-**

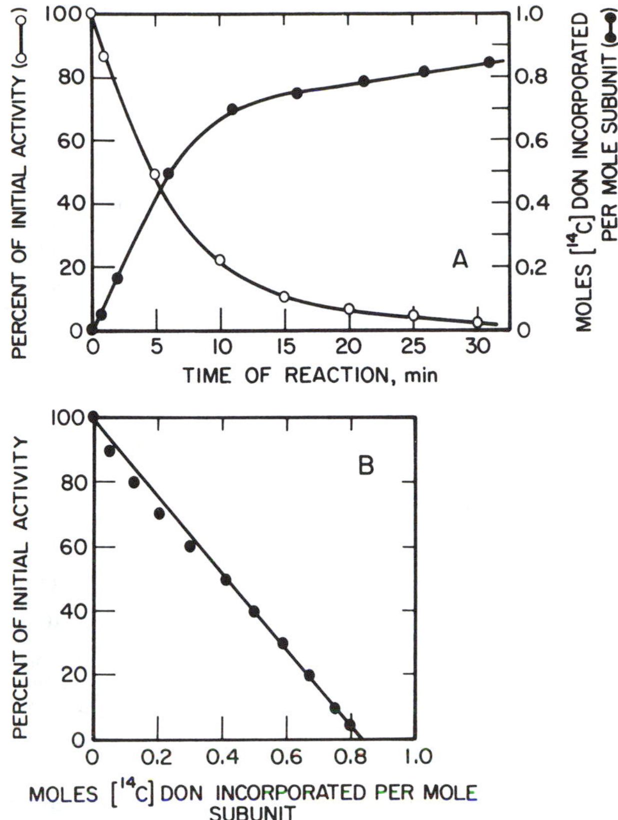


**Figure 4–29:** Incorporation of [6-$^{14}$C]-6-diazo-5-oxonorleucine into amidophosphoribosyltransferase.[160] A solution containing 10 $\mu$M [6-$^{14}$C]-6-diazo-5-oxonorleucine, 7.5 mM $MgCl_2$, and 5 mM 5-phosphoribosyl 1-diphosphate was prepared at pH 7.5 and 37 °C.[145] Amidophosphoribosyltransferase (5 $\mu$M final concentration in protomers) was added to initiate the reaction. At the noted times, samples (10 $\mu$L) were removed for assay of enzymatic activity, and other samples (75 $\mu$L) were removed and quenched with 175 $\mu$L of 10% trichloroacetic acid at 0 °C. Serum albumin (0.1 mg) was added to these latter samples as a coprecipitant. After 4 h, the precipitates were washed with ethanol, dissolved in 1 M NaOH, neutralized with HCl, and submitted to scintillation counting. (A) Enzymatic activity, as percent of initial activity (○), and incorporation of $^{14}$carbon, into protein, as moles of [6-$^{14}$C]-6-diazo-5-oxonorleucine incorporated for each mole of protomer (●) are presented as a function of time (minutes). (B) Enzymatic activity, as percent of initial activity, is plotted as a function of the incorporation of [6-$^{14}$C]-6-diazo-5-oxonorleucine, in moles of incorporation for each mole of protomer. Adapted with permission from ref 160. Copyright 1983 *Journal of Biological Chemistry.*

**tion constant** to the general base. Because the conjugate base of the amino acid is required in the reaction with the electrophile, the behavior of the rate of inactivation as a function of pH provides an estimate for the p$K_a$ of that amino acid in the occupied active site. The rate constant ($k_i$ in Equation 4–172) for the irreversible inactivation of carboxypeptidase by *N*-bromoacetyl-*N*-methylphenylalanine (Reaction 4–185) was determined as a function of the pH.[154] When log $k_i$ was plotted as a function of pH, the inflection assigned to the p$K_a$ of the modified amino acid occurred at pH 7.0. The rate of the reaction decreased at values of pH below the inflection at pH 7.0 and remained constant at values of pH above the inflection at pH 7.0, as expected if the conjugate base of the general acid–base were the reactive species. The modified amino acid was identified as Glutamate 270 in the amino acid sequence of the enzyme,[161] an amino acid that was known from the crystallographic molecular model to be within the active site.[162] In the enzymatic reaction catalyzed by carboxypeptidase, the pH–rate profiles for catalysis of the hydrolysis of several reactants display inflections assigned to an acid dissociation constant with a p$K_a$ of 7.0.[161] Because the p$K_a$ of Glutamate 270 is 7.0 when *N*-bromoacetyl-*N*-methylphenylalanine is bound to the active site, it has been proposed[161] that the apparent values for the p$K_a$ observed in pH–rate profiles are the actual values for the p$K_a$ of Glutamate 270 and that Glutamate 270 is a general base–acid involved directly in the mechanism of the enzyme.

The reaction catalyzed by glucose-6-phosphate isomerase

$$\text{glucose 6-phosphate} \rightleftharpoons \text{fructose 6-phosphate} \qquad (4\text{–}196)$$

proceeds through a *cis*-enediol, as does that catalyzed by triosephosphate isomerase (Reaction 4–165). For this to occur, a general base–acid is necessary to remove and add the proton from and to carbon and a different general acid–base is necessary to add and remove the other proton to and from oxygen. 1,2-Anhydro-D-mannitol 6-phosphate was chosen[163] as an active-site label

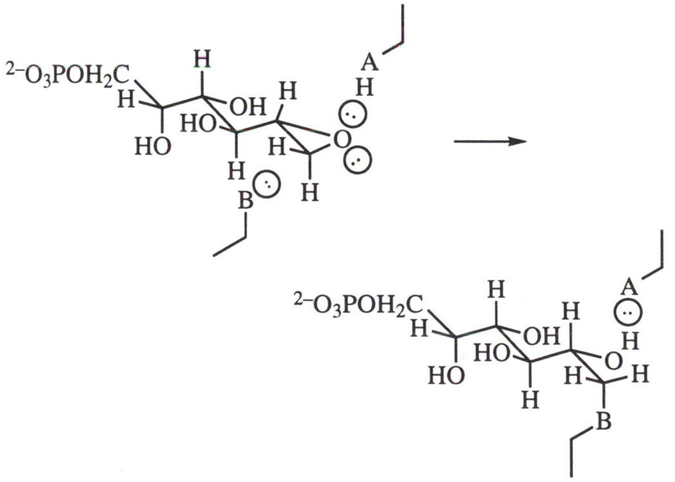

(4–197)

which would alkylate, under catalysis by the general acid responsible for adding the proton to oxygen in the normal reaction, the general base responsible for removing the proton from carbon. The intramolecular rate constants ($k_i$ in Equation 4–172) for the irreversible inactivation caused by the active-site label were determined as a function of pH (Figure 4–30).[163] The two values for p$K_a$ (6.3 and 10.0) are thought to be the values for the microscopic p$K_a$ of the general base–acid operating on the carbons of the substrate and the general acid–base operating on the oxygens, respectively. These acid

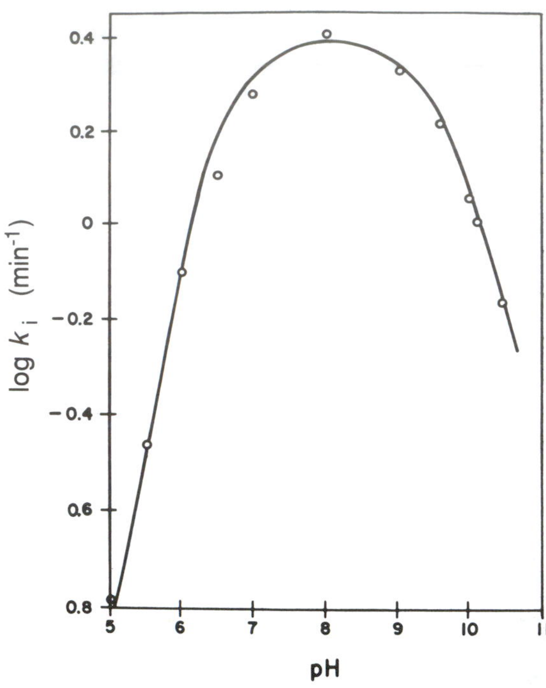


**Figure 4–30:** Dependence on pH of the logarithm of the rate constant for inactivation of glucose-6-phosphate isomerase by 1,2-anhydro-D-mannitol 6-phosphate.[163] Solutions containing 2 or 4 mM (2*R*)-1,2-anhydro-D-mannitol 6-phosphate were prepared at 25 °C and various values of pH buffered with sodium acetate, imidazolium chloride, triethanolammonium chloride, or sodium glycinate. The two concentrations of reagent chosen were known to be saturating for the active site, and the fact that the rates of inactivation were the same for each verified this expectation. The inactivation of the enzyme was initiated by adding yeast glucose-6-phosphate isomerase and was followed by withdrawing samples at appropriate times, diluting 100-fold with cold assay buffer, and performing an assay for enzymatic activity. The pseudo-first-order rate constant of inactivation, $k_i$, for each value of pH and each concentration of epoxide was obtained by plotting the logarithm of the residual enzymatic activity as a function of time. The logarithms of these rate constants, log $k_i$ (minute$^{-1}$), are plotted as a function of the pH. Adapted with permission from ref 163. Copyright 1973 *Journal of Biological Chemistry.*

dissociation constants are uncomplicated by the rate constants of other steps in the catalytic reaction and should be more direct measurements of the actual acid dissociation constants for these amino acids within the active site. They compare favorably to the apparent acid dissociation constants (p$K_a$ = 6.9 and p$K_a$ = 10.6) observed in the behavior of $V_{max}$ as a function of pH.

3-Bromophospho*enol*pyruvate, in its modification of phospho*enol*pyruvate carboxylase (Reaction 4–188), is an example of a subclass of active-site labels. 3-Bromophospho*enol*pyruvate is a poor electrophile because it is a vinyl bromide. Upon carboxylation in the active site, however, it becomes a secondary alkyl bromide, which is a sufficient electrophile. An active-site label that must be chemically altered at the active site, in a reaction resembling the reaction normally catalyzed by the enzyme, before it can modify an amino acid in the active site is referred to as a **mechanism-based inhibitor.**[164] The fact that the enzyme must catalyze the activation of the inhibitor within the active site ensures that the two forms of the inhibitor, unactivated and activated, are able to bind to the active site. It does not, however, ensure that the active inhibitor will remain in the active site long enough to react with an amino acid within the active site. For example, allyl alcohol inactivates alcohol dehydrogenase by being converted by the enzyme to acrolein. The acrolein, however, departs from the active site and then inactivates the enzyme nonspecifically from solution.[165] In such a situation, the activated inhibitor that has been released from the active site into the solution becomes indistinguishable from the same compound if it were added directly to the solution, with the exception that its specificity is more difficult to verify. The reason is that it is difficult to distinguish the effects of competitive inhibitors and substrates on the production of the active inhibitor from their effects on the modification of the enzyme by that active inhibitor. One way to prevent the released electrophile from inactivating the enzyme from solution rather than before it can leave the active site is to add a **scavenger** to the solution that intercepts it as it leaves the active site. For example, one of the observations demonstrating that acrolein was inactivating alcohol dehydrogenase after it had been released from the active site was that dithiothreitol prevented the inactivation of the enzyme by allyl alcohol. Ideally, a purely mechanism-based inhibitor would enter the active site because it was mistaken for the normal reactant for the enzymatic reaction, and it would be converted to an electrophile so reactive that it would modify the enzyme before it could leave the active site. Few electrophiles are so reactive.

The problem can be presented as a simple partition at the step following the production of the electrophile, El, from the reagent, R, at the active site

$$\mathrm{E + R} \underset{k_{-1}}{\overset{k_1}{\rightleftharpoons}} \mathrm{E{\cdot}R} \underset{k_{-2}}{\overset{k_2}{\rightleftharpoons}} \mathrm{E{\cdot}El} \xrightarrow{k_i} \mathrm{E - El}$$
$$\mathrm{E{\cdot}El} \underset{k_{-3}}{\overset{k_3}{\rightleftharpoons}} \mathrm{E + El} \qquad (4\text{–}198)$$

In this scheme, once the electrophile has been produced, either it dissociates ($k_3$) as the product of an enzymatic reaction usually would do or it reacts covalently in the active site ($k_i$) to modify the enzyme irreversibly. If rebinding to the active site ($k_{-3}$) is disregarded as irrelevant to the argument, this partition is governed only by the two rate constants, $k_i$ and $k_3$. As in any partition, the fraction of the reagent covalently modifying the enzyme before being released will be equal to $k_i\,(k_i + k_3)^{-1}$. The smaller this fraction, the more equivalents of the reagent must be added to obtain complete modification if the released electrophile cannot itself reenter or is prevented by a scavenger from reentering the active site. The number of equivalents required will be $(k_i + k_3)k_i^{-1}$. Along with the effects of scavengers, the number of equivalents of the reagent required to inactivate the enzyme is indicative of the mechanism of the inactivation.

3-Decynoyl-*N*-acetylcysteamine behaves as an irreversible, covalent active-site label for 3-hydroxydecanoyl-[acyl-carrier-protein]dehydratase.[166] One of the reactions normally catalyzed by this enzyme is

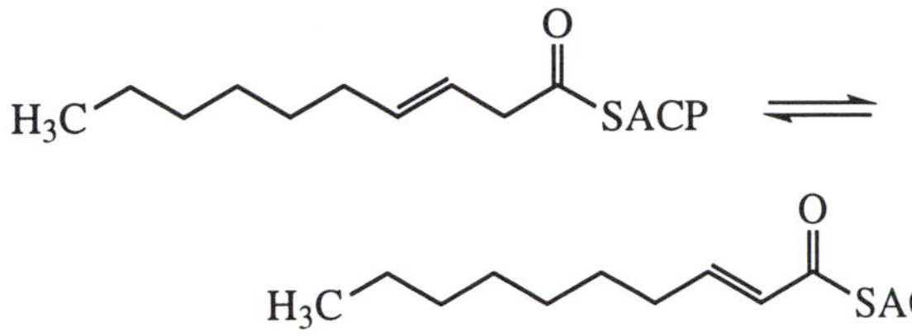

(4–199)

and it has been shown that the way in which the alkyne inactivates the enzyme is first to be converted to the allene

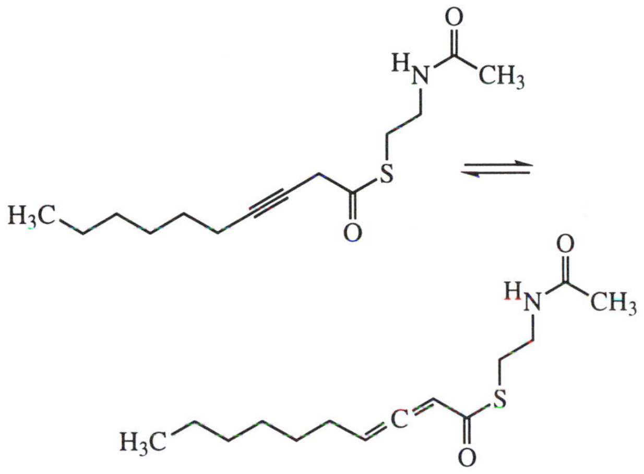

(4–200)

which then modifies the protein. The allene itself is a far superior active-site label to the alkyne. It would be difficult to distinguish whether one molecule of enzyme simply produces an allene as a free product that then inactivates another molecule of the enzyme from solution or an allene, once it is formed on the active site of one molecule of enzyme, alkylates that molecule of enzyme before it departs. Once the enzyme, however, had serendipitously identified the allene, 2,3-decadienoyl-*N*-acetylcysteamine, as the actual active-site label, its specificity could be examined directly.

One way to avoid the ambiguity between mechanism-based inactivation and simple active-site labeling is to design a mechanism-based inhibitor that remains bound tightly to the active site after it has been activated.[165] Acetylenic mechanism-based inhibitors have been used with enzymes employing pyridoxal phosphate as a coenzyme.[167] 2-Amino-4-pentynoic acid[168] is a mechanism-based inhibitor of *O*-succinylhomoserine (thiol)-lyase.[169] The inactivation is thought to result from the formation of the allene

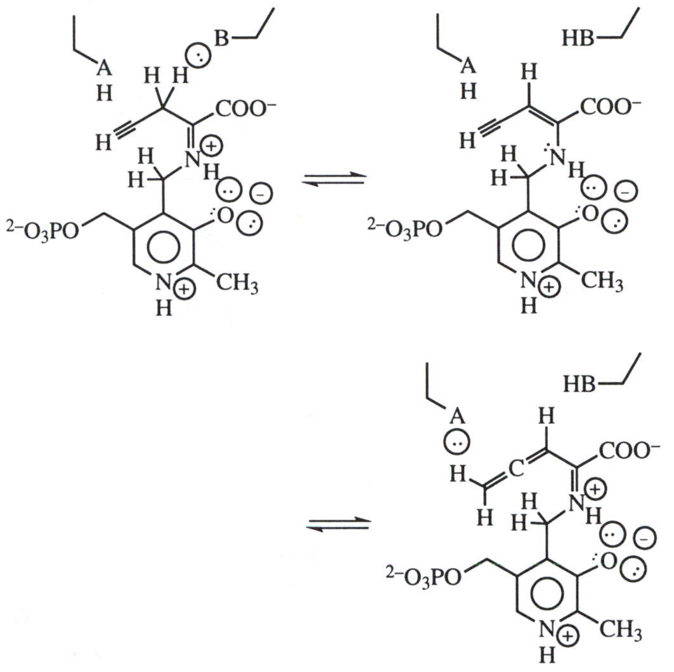

(4–201)

by acid–base reactions analogous to those of the normal mechanism (Reactions 2–16 and 2–17). In this case, the allene that is formed is attached covalently to the pyridoxamine phosphate as an imine through its $\alpha$-carbon. The covalently attached allene can then alkylate one of the general base–acids within the active site.[167] Even so, 4 mol of the 2-amino-4-pentynoic acid were required to inactivate 1 mol of active sites, and it has been shown that after the mechanism-based inhibitor is converted at the active site, the majority of the reagent (75%) is released as a product into the solution before it can inactivate the enzyme directly. Therefore, in this instance, for the partition defined by Equation 4–198, $k_3$ equals $3k_i$ even though the electrophile as it is formed is covalently bound to the enzyme.

Acetylenic active-site labels have also been used to modify enzymes that proceed through enol intermediates.[170] For example, the flavoenzyme butyryl-CoA dehydrogenase catalyzes the oxidation of butyryl-SCoA

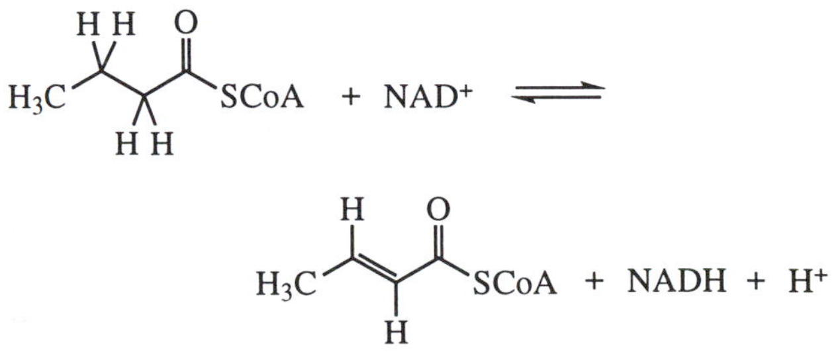

(4–202)

It is believed that this reaction passes through an enol formed by removal of the acidic proton at carbon 2 and that two electrons are then transferred from the enol to flavin. If an enol is the intermediate, locating an alkyne adjacent to carbon 2 should permit the formation of an allene

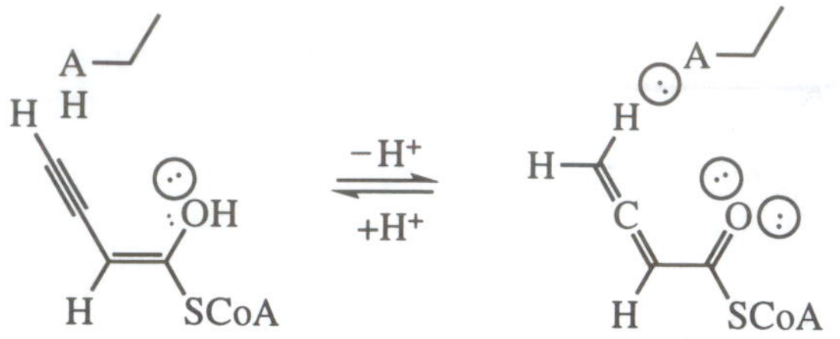

(4–203)

that should be able to react with one of the general base–acids responsible for proton transfers in the normal enzymatic reaction. When (3-pentynoyl)pantetheine was incubated with butyryl-CoA dehydrogenase, the enzyme was irreversibly inactivated coincidently with the covalent incorporation of the active-site label into the protein.[170]

A series of $\alpha$-amino acids that contain leaving groups such as fluorines[171–173] or sulfates[174] attached at the respective $\beta$-carbons have been used as mechanism-based inhibitors of enzymes containing pyridoxal phosphate. An example of this class is D-fluoroalanine, which has been used as a mechanism-based inhibitor of glycine hydroxymethyltransferase. During the normal reaction of this enzyme, an adduct is formed from L-serine, tetrahydrofolate ($RNH_2$), and pyridoxal phosphate. The carbon–carbon bond that is normally broken in this adduct

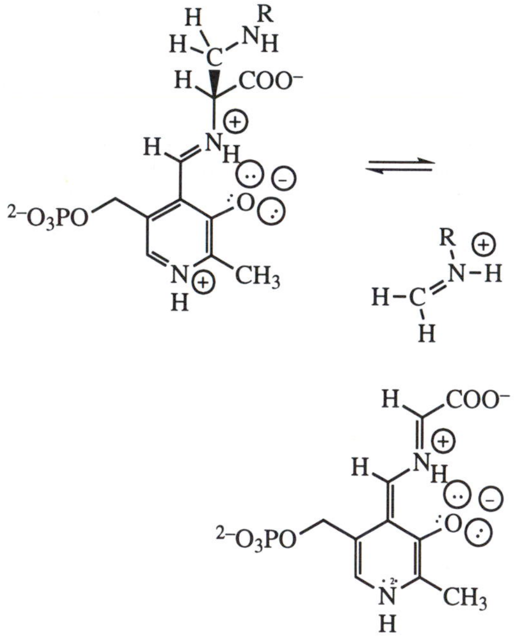

(4–204)

is positioned in parallel with the $\pi$ system of the coenzyme by the stereochemistry of the active site. To ensure that a proton instead was removed in the case of the mechanism-based inhibitor, the D-isomer of fluoroalanine was used[172]

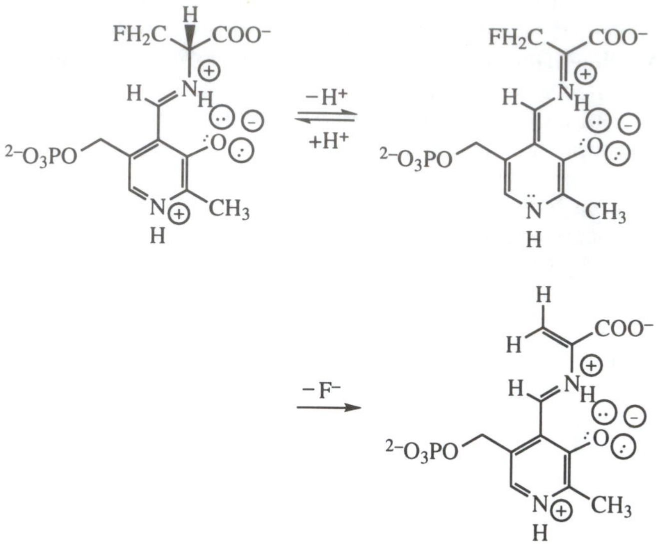

**4–37**

(4–205)

Following removal of the proton, the $\beta$-elimination of a fluoride ion (Reaction 4–205) can occur to produce imine **4–37** formed between an enamine and pyridoxal phosphate. In all of the examples of this class of mechanism-based inhibitors, the imine of the same or the corresponding enamine is formed by a similar $\beta$-elimination. Most of the time[172,173] the respective enamine is released from the imine (Figure 4–28), protonated on the $\beta$-carbon by the enzyme, and released into the solution. Occasionally, the enamine, as a nucleophile, inserts into the pyridoximine of the resting enzyme to inactivate it.[173,174] These $\beta$-substituted reagents are mechanism-based inhibitors, but unfortunately, they do not modify and identify a general acid–base in the active site; rather, they attack the only available electrophile in the active site, the pyridoximine.

The strategy of active-site labeling, in which a reagent is designed to identify a particular general base–acid in the active site responsible for the removal or addition of a particular proton, has been expanded to encompass reagents that are intended to identify amino acids within or near the active site, even though those amino acids may not be catalytic functional groups. An **affinity label** is a compound that is designed to resemble the substrate of an enzyme or the ligand for a protein so closely that it will bind specifically and that also contains an electrophile capable of labeling the conjugate base of an amino acid within or in the vicinity of the site. If the reagent is designed so as to place the electrophile immediately adjacent to an acid–base in the normal mechanism of the enzymatic reaction, the affinity label is an active-site label. In many experiments, however, information about functional groups in the enzymatic mechanism is irrelevant, and all that is of interest is the identification of any amino acid in the vicinity of the active site or the binding site. It is in these instances that an affinity label is not performing the role of an active-site label. A portion of the affinity label is designed to bind to the site, and a different portion, often quite distinct from the portion that binds, provides the electrophile. If the designer is lucky, the binding to the site of the portion responsible for specificity will position the portion providing the electrophile close enough to a suitable nucleophile. Ribulose-bisphosphate carboxylase catalyzes the carboxylation of the enediol of ribulose bisphosphate[175,176]

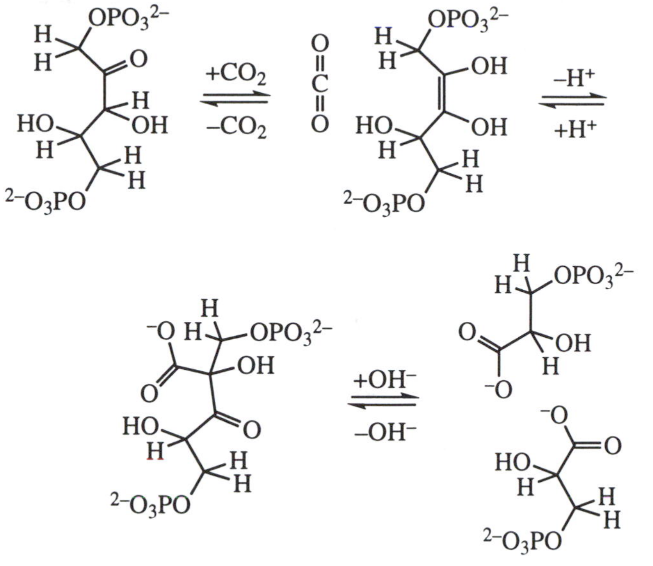

(4–206)

Although there should be two or more general base–acids involved in this reaction, none should be in the vicinity of the electrophile of 2-[*N*-(bromoacetyl-2-)amino]pentitol 1,5-bisphosphate

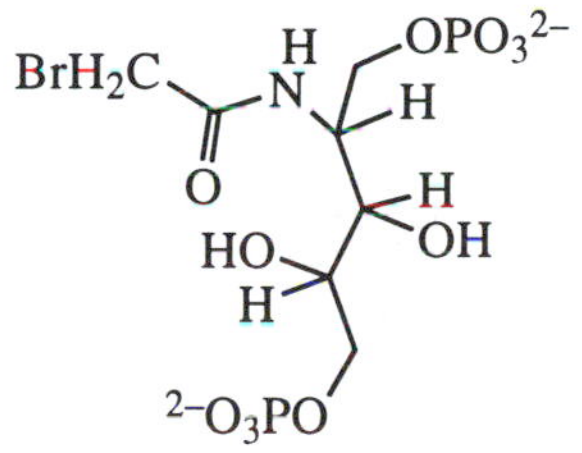

**4–38**

when it binds to the active site in place of the substrate. Nevertheless, 2-[*N*-(bromoacetyl-2-)amino]pentitol irreversibly inactivates ribulose bisphosphate in a reaction the properties of which meet all of the criteria for specific labeling within the active site.[177] It was shown that the covalently modified amino acid was a methionine

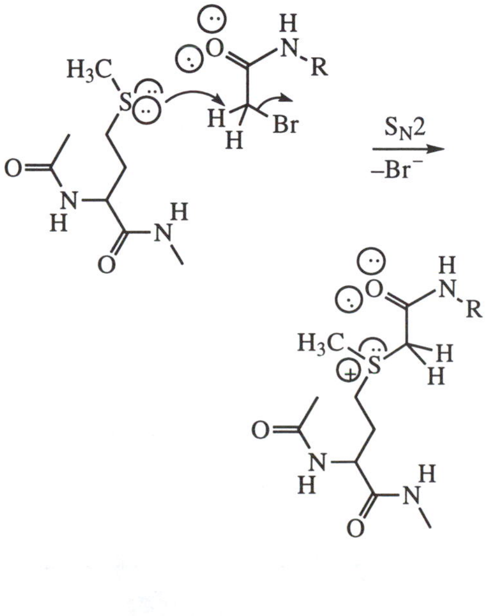

(4–207)

Since methionine can act as neither a general acid nor a general base nor a hydrogen-bond acceptor nor a hydrogen-bond donor, the conclusion reached from these observations is that this methionine is, by chance, one of the amino acids within or immediately adjacent to the active site in this enzyme.

It is also possible to perform affinity labeling with a reagent attached covalently within the active site. A diazoacetyl function was transesterified onto Serine 195 in chymotrypsin[178] by taking advantage of the first step in the normal enzymatic reaction (Figure 3–37)

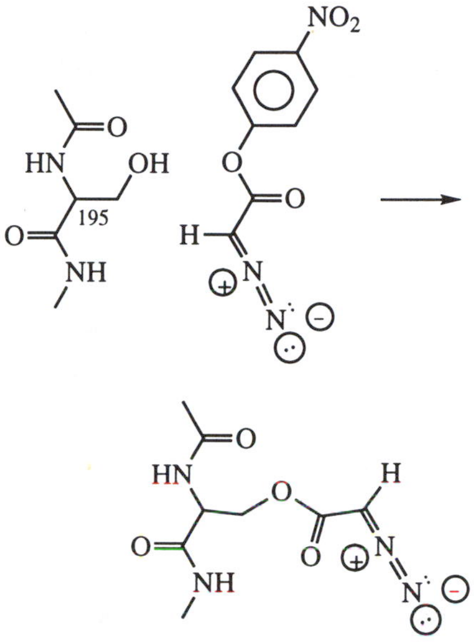

(4–208)

Upon photolysis a carbene was formed that inserted at random into any one of several oxygen–hydrogen bonds of amino acids surrounding the active site.

There are many examples of the use of affinity labels (Table 4–1). Because affinity labels are not directed against specific catalytic residues, they have most frequently been used to identify a particular protein or a particular region within the amino acid sequence of a protein as composing the active site or binding site of interest rather than as a means to identify a specific amino acid. Although functional groups capable of alkylation, acylation, or sulfonylation are often used as electrophiles, the most common choices when the affinity label is not being used as an active-site label have been aryl azides, which form very reactive electrophiles upon photolysis. This increases the chance that a sufficiently reactive nucleophile will be encountered by the electrophile. Nevertheless, it is still necessary that a nucleophile, however unreactive, be accessible by the electrophilic atom of the photoactivated aryl azide. To increase the probability of this happening, the aryl azide can be attached flexibly to the portion of the reagent conferring specificity so that it can reach several amino acids in the vicinity of the active site.

The conjugate bases of the general acid–bases performing catalytic roles within the active site of an enzyme often display **peculiar nucleophilicity**. Presumably, this results from the fact that they are found in special environments that are tailored to their roles in the enzymatic reaction. It is often the case that one of these conjugate bases within the active site will be exceptionally reactive toward an unexceptional electrophile that would otherwise react with every other amino acid of the same type in the protein.

**Table 4–1: Affinity Labels**

| enzyme or receptor | substrate or ligand mimicked | affinity label | reference |
|---|---|---|---|
| pyruvate kinase | ADP | 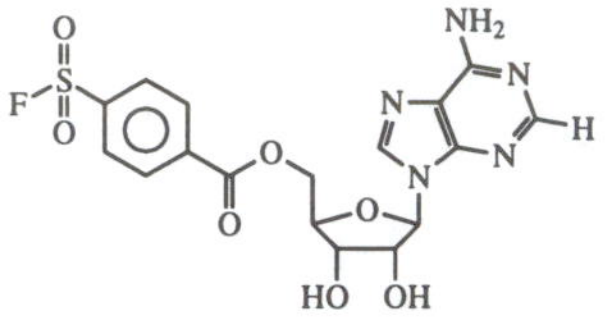 | 179 |
| homoserine dehydrogenase | aspartate semialdehyde | 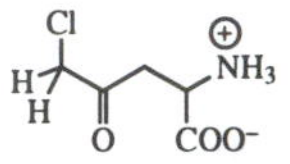 | 180 |
| tryptophan synthase | tryptophan | 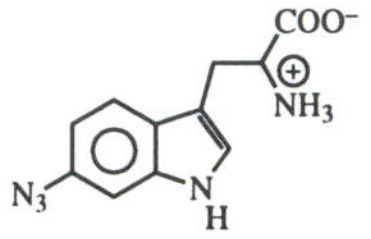 | 181 |
| alcohol dehydrogenase | $NAD^+$ | 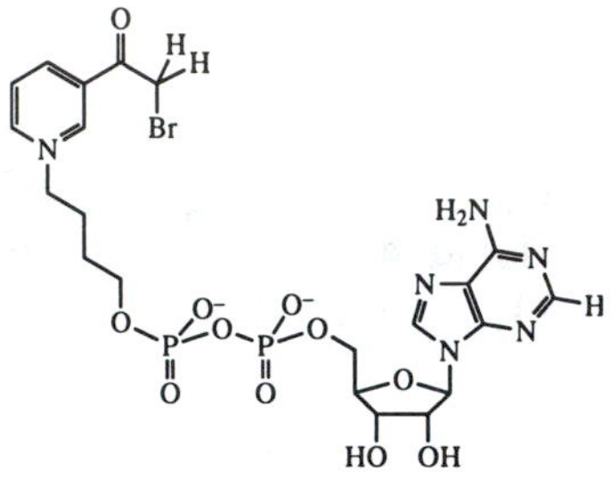 | 182 |
| carnitine acetyltransferase | *O*-acetylcarnitine | 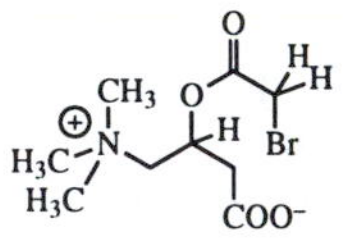 | 183 |
| *β*-adrenergic receptor | pindolol | 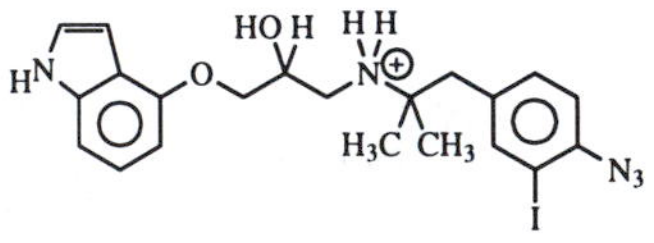 | 184 |
| cyclic 3′,5′-AMP-dependent protein kinase | cyclic 3′,5′-AMP | 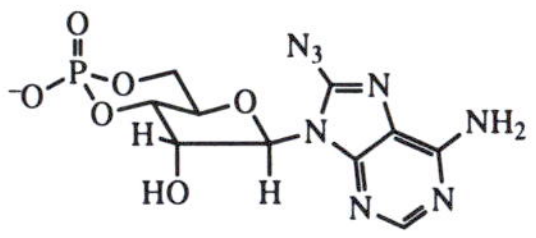 | 185 |
| myosin ATPase | ATP | 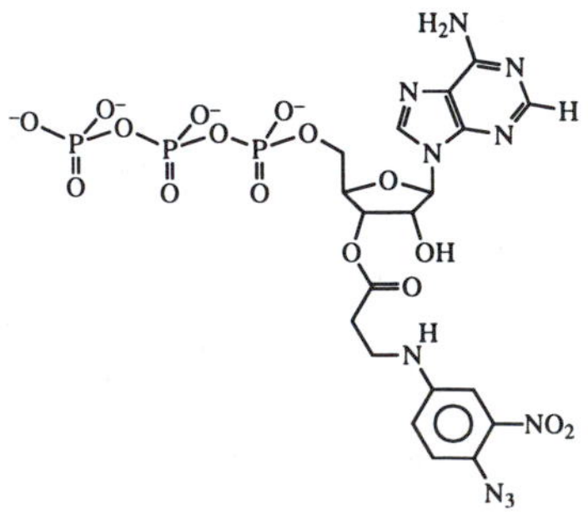 | 186 |
| procollagen-proline, 2-oxoglutarate 3-dioxygenase | procollagen | 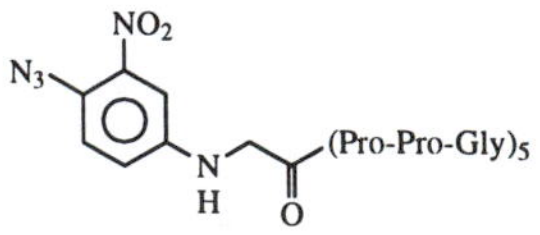 | 187 |

Glyceraldehyde-3-phosphate dehydrogenase from rabbit skeletal muscle is an $\alpha_4$ tetramer composed of folded polypeptides 333 amino acids in length. Each polypeptide contains four cysteines that can be readily titrated when it is unfolded. When the native enzyme was mixed with a concentration of [$^{14}$C]iodoacetic acid equivalent to the concentration of cysteine, the enzymatic activity was rapidly lost. After the inactivation was complete and the reaction was terminated, the protein was digested with trypsin. Only one radioactive tryptic peptide was produced by the digestion, and it could be shown that only Cysteine 149, of the four cysteines present in the protein, had been alkylated by the iodoacetic acid.[188] Cysteine 149 is the cysteine that forms the hemithioacetal with glyceraldehyde 3-phosphate during the enzymatic reaction,[189] and its exceptional nucleophilicity, as demonstrated by its exclusive reaction with iodoacetic acid, may be essential for this role in the mechanism.

Fructose-bisphosphate aldolase from rabbit skeletal muscle is an $\alpha_4$ tetramer composed of polypeptides 361 amino acids in length. Each polypeptide contains 15 arginines. When the native enzyme was mixed with 1,2-cyclohexanedione in sodium borate buffer, the enzymatic activity was rapidly lost.[190] When dihydroxyacetone phosphate or fructose 1,6-bisphosphate was present, however, no activity was lost. During this reaction less than 2 mol of 1,2-cyclohexanedione was incorporated for each mole of polypeptide. Only one of the tryptic peptides in a digest of the enzyme modified with [$^{14}$C]-1,2-cyclohexanedione incorporated radioactivity to a significant level (Figure 4–31),[190] and the incorporation into this peptide was prevented by the addition of fructose 1,6-bisphosphate. The remainder of the incorporation was spread over a number of other peptides indiscriminately. The selectively modified peptide was isolated and sequenced, and from these results it could be concluded that Arginine 148 was the amino acid the modification of which had inactivated the enzyme. Only Arginine 148, of the 15 arginines present in the enzyme, reacted appreciably with 1,2-cyclohexanedione under these circumstances even though 1,2-cyclohexanedione is a nonspecific reagent for arginines in general. It was later found that Arginine 148 is located within the active site in the crystallographic molecular model of rabbit muscle aldolase.[191] In the crystallographic molecular model of the unoccupied enzyme, a sulfate from the solution is found forming hydrogen bonds to this arginine, consistent with the role proposed for it as a hydrogen-bond donor to the phosphate of dihydroxyacetone phosphate while it is bound at the active site.[190]

When ribonuclease was mixed with a molar excess of iodoacetate (8 mol mol$^{-1}$), the enzymatic activity was lost rapidly. When most of the inactivation had occurred, the products were separated chromatographically (Figure 4–32).[192] Aside from some remaining, unalkylated enzyme,

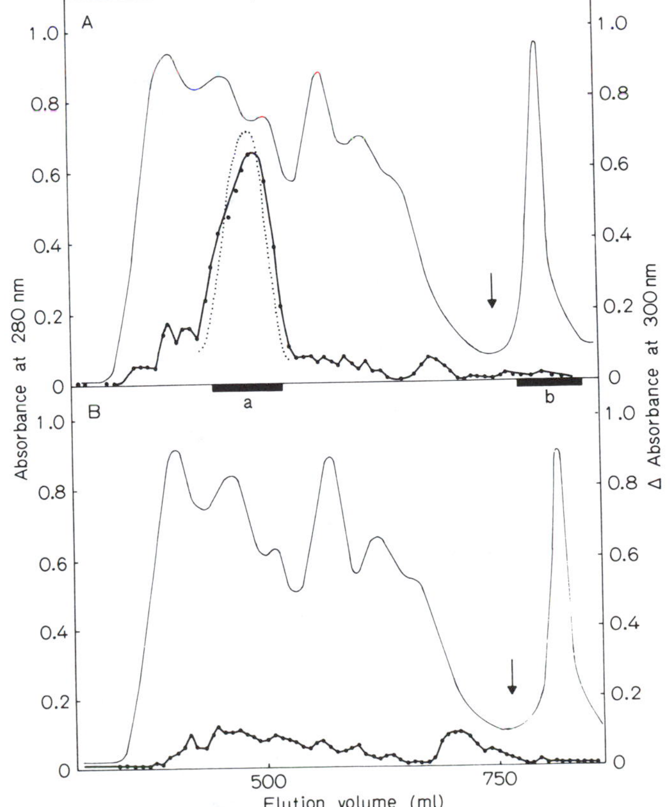


**Figure 4–31:** Distribution of $N^{\omega}$,$N^{\omega}$-(1,2-dihydroxycyclohexyl)arginine residues among the tryptic peptides of rabbit muscle fructose-bisphosphate aldolase inactivated with 1,2-cyclohexanedione.[190] Two solutions (60 mL) containing 50 mM cyclohexanedione and 0.1 M sodium borate, pH 8.0, at 25 °C were prepared. One of the two solutions contained 5 mM fructose 1,6-bisphosphate to protect the active site. The inactivation was initiated by adding fructose-bisphosphate aldolase (0.2 mM final concentration in protomers) and was allowed to proceed until the residual enzymatic activity in the unprotected sample was 10%, and in the protected sample, 90%. The pH of the two solutions was dropped to 4 and both were dialyzed into deionized water and submitted to performic acid oxidation. The protein was then submitted to digestion by trypsin. The resulting tryptic peptides were submitted to molecular exclusion chromatography on Sephadex G-25 in 30% acetic acid. The absorbance of the effluent at 280 nm was monitored continuously. Samples were removed from the various fractions to assay for 1,2-cyclohexanedione released from modified arginines by treatment at alkaline pH. The released cyclohexanedione in samples exposed to alkaline pH was assayed by its reaction with (trimethylamino)acetohydrazide, which forms an adduct with vicinal diones that absorbs at 300 nm. The absorbance at 280 nm (thin line) and the absorbance of the adduct at 300 nm (thick line or dashed line) are plotted as a function of the elution volume (milliliters) from the chromatographic column (3 × 140 cm). (A) Unprotected product. (B) Product from enzyme protected with 5 mM fructose 1,6-bisphosphate. Reprinted with permission from ref 190. Copyright 1979 Springer-Verlag.

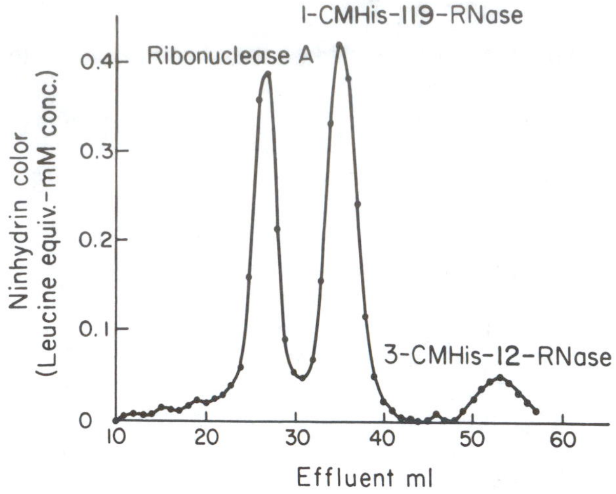


**Figure 4-32:** Separation of the two monoalkylated derivatives of ribonuclease A produced by the alkylation of the protein with iodoacetic acid.[192] A solution containing 16 mM iodoacetic acid and 2 mM ribonuclease was prepared at pH 5.5 and 25 °C. After 6 h, the solution was brought to 0.2 M in sodium phosphate to stop the alkylation, and it was submitted to cation-exchange chromatography on Amberlite IRC 50 (0.9 × 30 cm) eluted with 0.2 M sodium phosphate, pH 6.47. The several products were located by alkaline ninhydrin assay (millimolar, based on the extinction coefficient for leucine in the ninhydrin assay). Ninhydrin color is presented as a function of the volume of the effluent from the column (milliliters). The products were identified as unmodified ribonuclease A, ribonuclease alkylated at nitrogen 1 of Histidine 119, and ribonuclease modified at nitrogen 3 of Histidine 12 by tryptic digestion and isolation of modified peptides. Reprinted with permission from ref 192. Copyright 1963 *Journal of Biological Chemistry*.

only two products were isolated. One was shown to be alkylated only at nitrogen 1 of Histidine 119; the other was shown to be alkylated only at nitrogen 3 of Histidine 12. Neither of the other two histidines in the enzyme, Histidine 48 and Histidine 105, had been alkylated significantly under these circumstances. These results demonstrate that both Histidine 119 and Histidine 12 in native ribonuclease are exceptionally nucleophilic. Furthermore, no bisalkylated product was isolated, and this result suggests that once one of the histidines had reacted, the other became much less nucleophilic. This was later explained by the observation that Histidine 12 and Histidine 119 are adjacent to each other in the active site of the enzyme,[193] and steric hindrance would prevent alkylation of one of them once the other had become alkylated. The product alkylated at Histidine 119 was completely inactive (<1% activity remaining), but the product of alkylation at Histidine 12, although significantly inhibited, retained 7% of the enzymatic activity of native ribonuclease. At the time, this result suggested that Histidine 12 might not be directly involved in the mechanism of the reaction as a general acid–base, and this was later confirmed crystallographically.[194] In the crystallographic molecular model of the complex between ribonuclease and an analogue of the pentavalent intermediate (Figure 3–39), Histidine 12 is providing a hydrogen bond to one of the equatorial oxygens in the pentavalent intermediate rather than acting as a general acid at either of the apical oxygens, which represent the entering and leaving positions to and from the phosphorus.

Various **reagents for the general modification of particular types of amino acids** have been used to identify exceptionally reactive amino acids, the modification of which inactivates the respective enzyme. When a single tryptophan in deoxyribonuclease was modified with *N*-bromosuccinimide, the enzyme was inactivated.[195] When $H^+$-transporting ATP synthase was modified with dicyclohexylcarbodiimide, the enzyme was inactivated concomitantly with the incorporation of the dicyclohexylcarbodiimide into a particular glutamic acid in the amino acid sequence of the $\beta$ polypeptide as the *N*-acyl-*N*,*N*′-dicyclohexylurea.[196] 1,3-Dibromoacetone, a bifunctional alkylating agent, cross-linked a particular cysteine and a particular histidine in papain to inactivate the enzyme.[197,198] The modification of a single arginine in a subunit of ornithine carbamoyltransferase by butanedione led to the inactivation of the enzyme.[199] The exclusive arylation of one exceptionally reactive lysine, the apparent $pK_a$ of which is 8.0, with trinitrobenzenesulfonate led to the inactivation of isocitrate dehydrogenase ($NAD^+$).[200] The bifunctional reagent *p*-[(bromoacetyl)amino]phenyl arsenoxide cross-linked a thiol of lipoamide to a nearby histidine within or adjacent to the active site to inactivate lipoamide dehydrogenase.[201] When phospholipase $A_2$ was methylated with methyl *p*-toluenesulfonate, a product could be isolated chromatographically that had been *N*-methylated exclusively at Histidine 48 and that retained only 1% of the original activity of the original enzyme.[202]

It is unclear why particular amino acids in a protein are extraordinarily reactive relative to other amino acids of the same type in the same protein. Even though the reagent may bear no resemblance to a substrate or a ligand, it nevertheless may be guided by a constellation of hydrogen-bond donors and acceptors to bind, preferentially but adventitiously, adjacent to the one amino acid. Other amino acids in the vicinity may act as general acid–bases to catalyze the modification reaction itself. In any case, amino acids are often found in proteins that are exceptionally nucleophilic in a particular modification reaction.

Whether or not such exceptionally nucleophilic amino acids are always located within an active site or within a ligand-binding site is a separate issue. Usually, in experiments of this type, only two observations are available. First, the enzymatic activity is irreversibly and completely, or almost completely, inhibited when only 1 or 2 mol of reagent have become incorporated for every mole of the enzyme. Second, the inactivation and incorporation are blocked when a substrate, an inhibitor, or a ligand is present. There are, however, ways an enzyme can be inactivated other than through modification at the active site. When a tyrosine in arginine kinase was nitrated with tetranitromethane, the enzyme was irreversibly inhibited;[203] but, at the same time, a significant structural change in the protein was detected spectrally and immunologically. Perhaps this structural change alone, which could have been induced by nitration of a tyrosine anywhere in the enzyme, was responsible for the observed inhibition.

It is usually difficult to decide whether the residual enzymatic activity present after the modification of an enzyme represents as-yet-unmodified enzyme or the modified enzyme itself. This is an important distinction because, if the modified enzyme still possesses significant enzymatic activity, then the amino acid that has been modified must not be essential for catalysis. In the case of the nitration of arginine kinase, a residual enzymatic activity of 10% persisted even at higher than stoichiometric levels of nitration, and this would suggest that the tyrosine nitrated is not essential for the enzymatic reaction. When carbonate dehydratase was alkylated with iodoacetate, Histidine 200 was modified to produce an enzyme that, depending on the pH, had 3–15% the turnover number of the native enzyme.[204] From these results, it can be concluded that Histidine 200 is not essential for enzymatic activity, even though its alkylation does inhibit the enzyme extensively. When 2-nitro-5-(thiocyanato)benzoate was used to inactivate 3-oxoacid-CoA transferase, it produced two products, one of which was enzymatically inactive and the other of which was still enzymatically active. It was concluded that in each product the same cysteine had been modified but that in the enzymatically active product the cysteine was cyanylated while in the enzymatically inactive product it had been oxidized by formation of a mixed disulfide with 2-nitro-5-thiobenzoate. If only the enzymatically inactive product had been obtained, it would have been concluded, incorrectly, that the cysteine was essential for enzymatic activity.[205]

The protection of an enzyme by a substrate against inactivation by an electrophile does not prove that the amino acid the modification of which has caused the inactivation is located within the active site. When tryptophan synthase was irreversibly inhibited with phenylglyoxal, 1 mol of arginine was modified for each mole of enzyme inhibited, and the modification responsible for the inhibition occurred at Arginine 148 in the $\beta$ subunit.[206] Because the enzyme could be protected from irreversible inhibition by pyridoxal phosphate and serine but not by pyridoxal phosphate alone, and because phenylglyoxal was a competitive inhibitor with respect to serine of the normal enzymatic reaction, it was concluded that serine binds to Arginine 148 within the active site during catalysis. In the crystallographic molecular model of tryptophan synthase, however, Arginine 148 is not one of the amino acids that has been identified as an active-site residue.[207] The disadvantage of using a general reagent for protein modification to identify amino acids within the active site is that the electrophile is not carried into the active site and juxtaposed to a particular amino acid but has unbiased access to all of the amino acids on the surface of the protein.

Three of the ultimate goals in chemically modifying an active site are the **determination of the identity of the modified amino acid** or coenzyme, the **identification of the position in the amino acid sequence** of the amino acid that has been modified, and the **deduction of the mechanistic significance** of that amino acid. The structure of the modified amino acid or coenzyme provides information about the specific reaction that occurred and how the reactants were arranged in the active site, which in turn forms the basis of the deduction. When acyl-CoA dehydrogenase with its flavin oxidized was mixed with 3,4-pentadienoyl-coenzyme A, the enzyme was rapidly inactivated.[208] It was concluded from spectral studies of the product that alkylation of nitrogen 5 of the flavin had occurred. This information suggested that the reaction involved an attack at nitrogen 5 by an enol of the allene

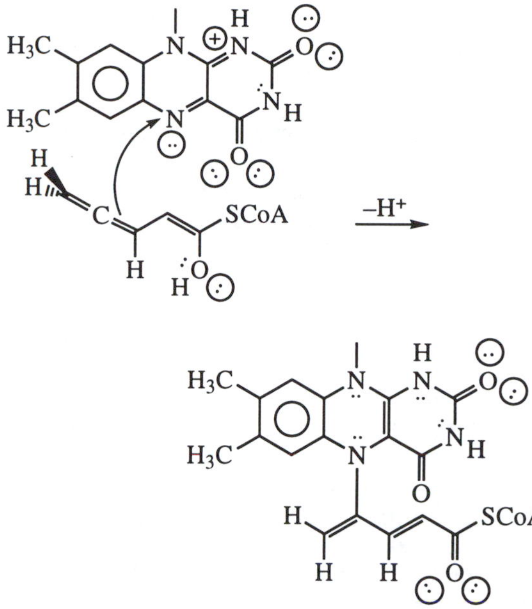

(4–209)

When fructose-bisphosphate aldolase was inactivated with *N*-(bromo[$^{14}$C]acetyl)ethanolamine phosphate and the inactivated enzyme was hydrolyzed in acid, several radioactive peaks were obtained upon amino acid analysis.[209] The major peak was identified as 3-(carboxymethyl)histidine, and it was the only product the yield of which decreased significantly when the modification was performed in the presence of the competitive inhibitor butanediol diphosphate, which protects the enzyme against inactivation. It was concluded that the reaction leading to inactivation was at a histidine in the active site

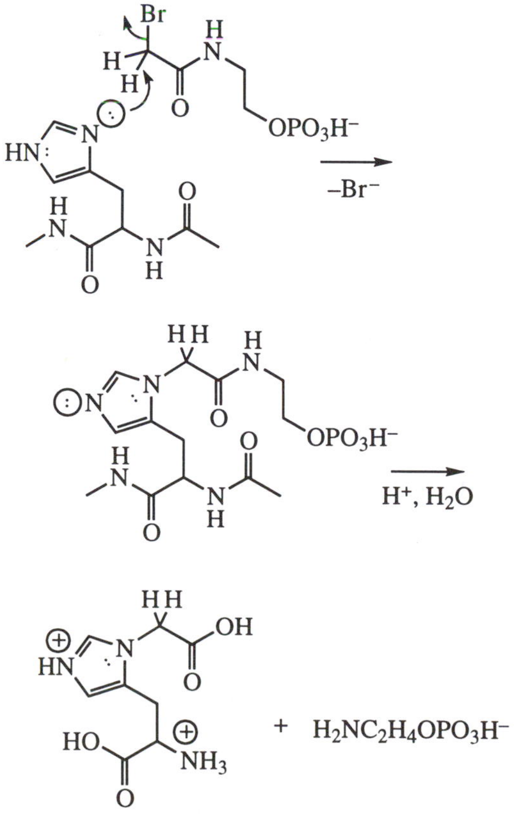

(4–210)

The amino acid in isocitrate dehydrogenase ($NADP^+$) modified by 3-bromo-2-oxoglutarate (Reaction 4–191) has also been identified. First, the modified peptide, DLAGXI-HGLSNVK, was isolated.[210] In this amino acid sequence, X designates the amino acid modified by the 3-bromo-2-oxoglutarate. This amino acid was then identified as a cysteine by isolating the same peptide from enzyme modified by *N*-[2-$^3$H]ethylmaleimide.[211] Finally, it could be shown that modification of this cysteine with *N*-ethylmaleimide produced a form of the enzyme that could no longer be inactivated by 3-bromo-2-ketoglutarate but that was fully active even though the cysteine was now an *S*-(succinimido)cysteine residue.[212] It follows that this cysteine is not a catalytic amino acid within the active site nor even involved in binding the substrates. The most reasonable explanation for these observations is that this cysteine is adjacent to the active site and reacts with the 3-bromo-2-oxoglutarate while it is bound there, and this reaction locks the 2-oxoglutaryl group in the vicinity of the active site so that it effects the 70% irreversible inactivation of the enzyme.

The position of the modified amino acid in the amino acid sequence of an enzyme relates the results of active-site labeling to the structure of the protein. When pepsin was completely inactivated with the active-site label *N*-(diazo[$^{14}$C]acetyl)phenylalanine methyl ester, 1 mol of the reagent was incorporated for every mole of enzyme.[213] The modified protein was digested with pepsin, and a radioactive peptide was partially purified by chromatography. The peptide was then purified to homogeneity by diagonal electrophoresis, taking advantage of the ability of triethylamine to liberate the incorporated, radioactive active-site label. The peptides were submitted to electrophoresis in one dimension, treated with triethylamine, and submitted to the same electrophoresis in the orthogonal dimension. One peptide changed its mobility and moved off the diagonal. The isolated peptide, which no longer was labeled, was found to be IVDT-GTS. It was concluded that the (diazoacetyl)phenylalanine had formed an ester with an aspartic acid in the active site

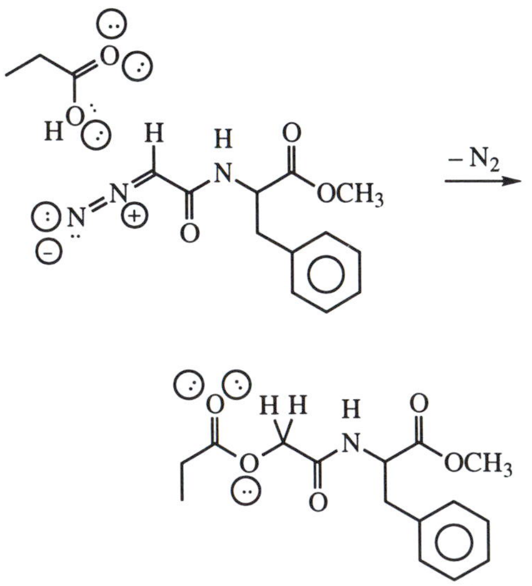

$$\xrightarrow{-N_2}$$

(4–211)

which was readily cleaved in the solution of ethylamine. The sequence of the isolated peptide identified Aspartate 215 as the site of modification.[214] In the crystallographic molecular model of the enzyme,[215] constructed at a later date, Aspartate 215 was found to be hydrogen-bonded to Aspartate 32, an amino acid the alkylation of which with 1,2-epoxy-3-(*p*-nitrophenoxy)propane also leads to inactivation of the enzyme.[216] These chemical modifications were used to define the location of the active site in the crystallographic molecular model.[215] It is presently assumed that these aspartic acid residues are general acid–bases participating in the mechanism,[78, 217] and they have been tentatively identified as the general acid–bases responsible for the pH–rate behavior of the enzyme.

The determination of the structure of the product of a modification reaction and the identification of the amino acid that has been modified are often two separate problems. Complications can arise because the product of a modification reaction is often too unstable to survive the digestion, chromatography, and sequencing required to identify the amino acid modified or because the structure of the phenylthiohydantoin of the modified amino acid produced during the sequencing cannot be established. Phospho-2-dehydro-3-deoxygluconate aldolase catalyzes the reaction

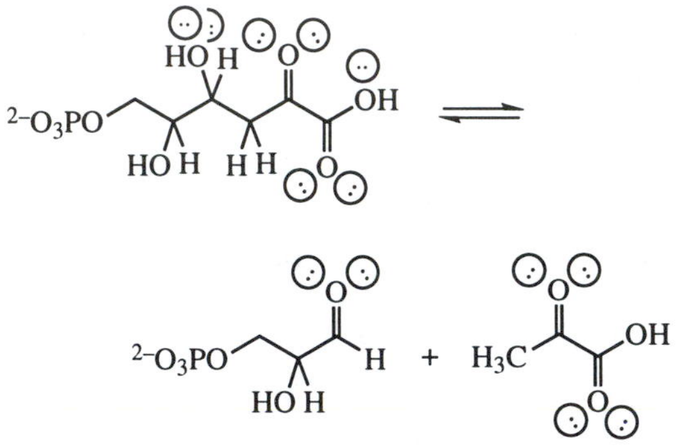

(4–212)

An imine between a lysine in the active site and the 2-keto group on the 2-dehydro-3-deoxy-6-phosphogluconate is formed during the enzymatic reaction to facilitate the retro-aldol cleavage.[218] Therefore, an imine between the same lysine and pyruvate forms as a step in the reverse reaction, and this pyruvyl imine can be reduced to the secondary amine with sodium borohydride[219]

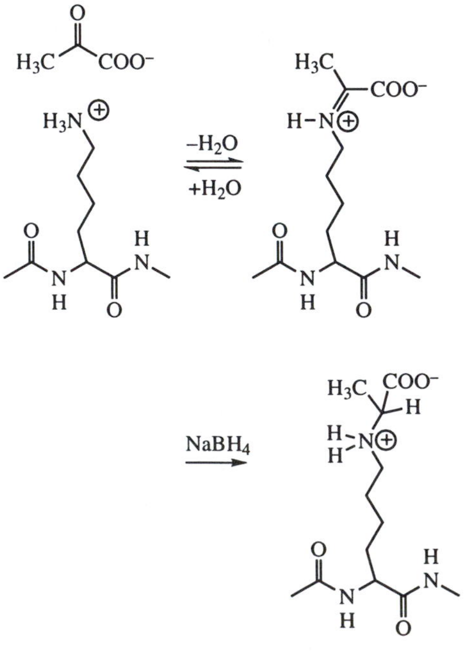

$$\underset{+H_2O}{\overset{-H_2O}{\rightleftharpoons}} \qquad \xrightarrow{NaBH_4}$$

(4–213)

When bromopyruvate is mixed with the enzyme, not only does the imine form but the primary bromide alkylates another amino acid in the vicinity. The other amino acid was tentatively identified as a carboxylate, either aspartate or glutamate, because upon acid hydrolysis of enzyme that had been inactivated with bromopyruvate and reduced with sodium borohydride, $N^{\varepsilon}$-[(1-carboxy-2-hydroxy)ethyl]lysine was produced[220]

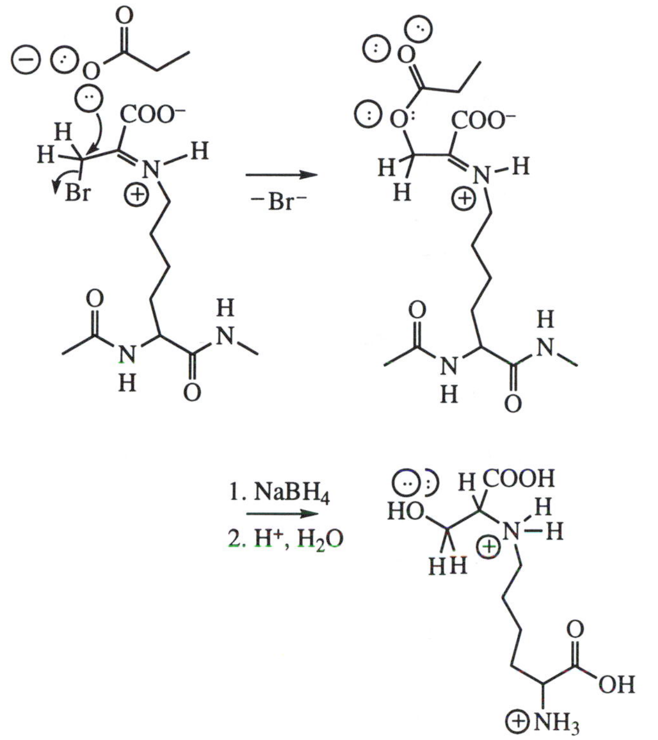

(4–214)

This experiment established the structure of the product of the reaction but did not identify the amino acid, which is presumably the general base removing the proton from pyruvate in the normal condensation. When phospho-2-dehydro-3-deoxygluconate aldolase that had been inactivated with [$^{14}$C]bromopyruvate but not reduced with sodium borohydride was digested with trypsin, a radioactive tryptic peptide could be purified to homogeneity.[221] The sequence of this peptide was TLEVTLR, which identified Glutamate 56 in the complete sequence of the protein[222] as the alkylated carboxylate, but all of the attached glycolate was lost at the first step of the sequencing. To confirm that the glycolate was esterified to the glutamate, the isolated, radioactive peptide was submitted to a Lossen rearrangement, which requires that an ester be present

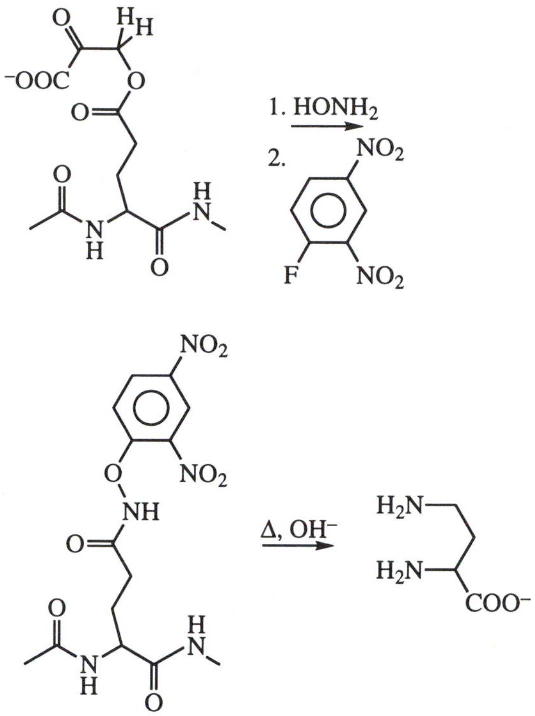

(4–215)

The recovery of 2,4-diaminobutyrate (0.8 mol mol$^{-1}$) was consistent with esterification of the glutamate by glycolate in the radioactive peptide.[221] After the Lossen rearrangement, the peptide was no longer intact so it could not be sequenced to show that the 2,4-diaminobutyrate had replaced the glutamate, but there is no other amino acid in the peptide that could have been converted into 2,4-diaminobutyrate.

Another approach to modification of amino acids within the active site of an enzyme is through **site-directed mutation**. Unlike the use of an active-site label or an affinity label in which the reagent is designed to locate the active site on its own, the use of site-directed mutation requires that the investigator already know, or can make an educated guess, which amino acids occupy the active site. Site-directed mutation permits any one of the amino acids in a protein to be replaced by another amino acid, but the investigator must choose the amino acid to be changed. Therefore, the most definitive uses of site-directed mutation to study active sites of enzymes have been confined to enzymes for which a crystallographic molecular model is available.

A crystallographic molecular model of the active site of tyrosine-tRNA ligase occupied by tyrosinyl adenylate

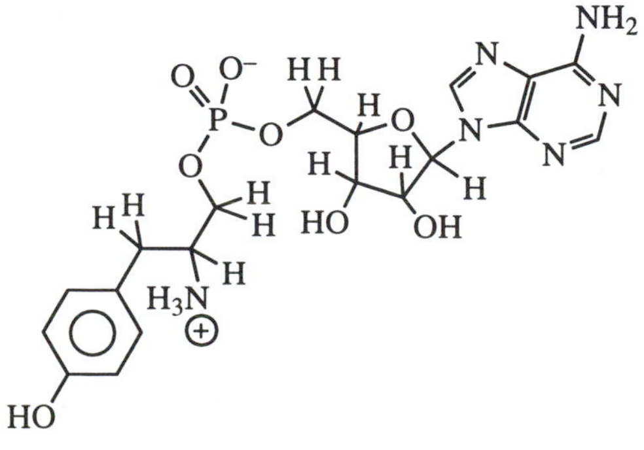

**4–39**

a bisubstrate analogue of tyrosyl adenylate (Structure 3–7), has been constructed (Figure 4–33A)[223] as well as a crystallographic molecular model of the active site of the enzyme occupied with tyrosine alone (Figure 4–33B).[224] Both of these models were constructed from data sets gathered from crystals of the enzyme in a complex with the respective ligands. In each model, the various hydrogen bond donors and acceptors and general acid–bases surrounding the ligands can be readily identified.

The kinetic mechanism for the production of tyrosyl adenylate (Tyr-AMP) bound to the active site of the enzyme (Reaction 3–188) has been shown to be[226]

$$\mathrm{E+Tyr} \underset{K_d^{\mathrm{Tyr}}}{\rightleftharpoons} \mathrm{E\cdot Tyr + ATP} \underset{K_d^{\mathrm{ATP}}}{\rightleftharpoons}$$

$$\mathrm{E\cdot Tyr\cdot ATP} \underset{k_{-3}}{\overset{k_3}{\rightleftharpoons}} \mathrm{E\cdot Tyr-AMP\cdot H_3P_2O_7^-}$$

$$\overset{K_d^{\mathrm{PP}}}{\rightleftharpoons} \mathrm{E\cdot Tyr-AMP + H_3P_2O_7^-} \qquad (4\text{–}216)$$

Methods have been developed to measure each of the dissociation constants ($K_d^{\mathrm{Tyr}}$, $K_d^{\mathrm{ATP}}$, and $K_d^{\mathrm{PP}}$) and each of the rate constants ($k_3$ and $k_{-3}$) in this mechanism.

Many of the amino acids surrounding the active site in tyrosine-tRNA ligase have been modified by site-directed mutation, and the effect of these mutations on the kinetic and equilibrium constants have been measured (Table 4–2). Enzymes in which amino acids have been substituted that associate with the tyrosyl portion of the bisubstrate analogue or with bound tyrosine (Y34, Y169, Q173, D78, and D38) show increased dissociation constants for tyrosine. Two of these (Q173 and D78), however, also show increased dissociation constants for ATP, even though the amino acids mutated are located on the other side of the tyrosine from the adenylate. One particularly interesting region is a cluster of hydrogen bonds surrounding and including the three protons on the primary amino group of the tyrosine (Figure 4–33B). When any of the amino acids in this cluster (Y169, Q173, D78, or D38) are mutated, large increases in the dissociation constant for tyrosine are observed. Enzymes in which amino acids associating with the adenylate portion of the bisubstrate analogue had been mutated (C35, H48, and T51) sometimes displayed alterations in the dissociation constant for ATP but also often displayed alterations of the same magnitude in the dissociation constant for tyrosine.

Originally these changes in the kinetic and equilibrium constants were used to calculate free energies of formation of particular hydrogen bonds between these amino acids and the respective donors or acceptors on the substrates or the transition state in the enzymatic reaction.[232] It became clear, however, that the identity of the amino acid that had been used for the substitution had a significant and unpredictable effect on the kinetic and equilibrium constants and that the situation was clearly not so simple as only the elimination of the donor or the acceptor of the hydrogen bond. For example, when Threonine 51, which forms an apparently simple hydrogen bond with the oxygen of the ribose (Figure 4–33A), is replaced with alanine, proline, or cysteine, remarkably different changes are seen. The proline substitution, which has no detectable effect on the structure of the enzyme,[233] nevertheless decreases the dissociation constant of the tyrosyl adenylate relative to the bound reactants 25-fold while having almost no effect on the dissociation constants of the reactants themselves, even though a hydrogen bond has been eliminated. When Histidine 48, however, which also forms a hydrogen bond with the same oxygen of the ribose, is replaced with glycine, the dissociation constant for ATP increases 2-fold. These observations emphasize the difficulty of distinguishing direct and indirect effects in spite of the absolute specificity of the mutations.

With the exception of the substitutions at Threonine 51, all of the mutations decreased the catalytic capacity of the enzyme either by increasing dissociation constants or by decreasing rate constants. This is not surprising because the wild-type enzyme is a product of natural selection.

The various changes in the kinetic parameters that were observed in these experiments could not have been predicted ahead of time and required detailed knowledge from the crystallographic molecular models to provide explanations for them, after the fact. More useful information was obtained, however, when experiments were designed to test a specific hypothesis.

The intermediate in a phosphoryl exchange reaction such as the formation of tyrosyl adenylate is a pentavalent phosphorus in a trigonal bipyramid. When a model of such a trigonal bipyramid at the $\alpha$-phosphate of the ATP was placed into the vacant active site in the crystallographic molecular model of the enzyme, the phosphate group that was the $\gamma$-phosphate in the ATP could be positioned comfortably so that two of its oxygens formed hydrogen bonds with Threonine 40 and Histidine 45 (Figure 4–33A,C). On the basis of this model-building, it was proposed that these two amino acids, although unable to form hydrogen bonds with any of the phosphates of bound ATP, would nevertheless form strong hydrogen bonds with the $\gamma$-phosphate of the pentavalent intermediate and that this would increase the stability of the intermediate relative to the bound reactants and thereby accelerate the reaction.[225] When Histidine 45 was mutated to alanine, the rate of formation of tyrosyl adenylate ($k_3$) decreased 240-fold (Table 4–2); when Threonine 40 was changed to alanine, the rate decreased 8000-fold; and when both mutations were performed simultaneously, the rate decreased 400,000-fold, even though the dissociation constants for the complexes between these enzymes and ATP were almost unaffected by any of these mutations. These results were taken as evidence for the involvement of these amino acids in binding and stabilizing the pentavalent intermediate.

Both Lysine 82 and Arginine 86 are near the phosphate of the bisubstrate analogue in the crystallographic molecular model (Figure 4–33A) and can be readily rotated to form hydrogen bonds to an oxygen of the former $\alpha$-phosphate and an oxygen of the former $\gamma$-phosphate, respectively, in the model of the pentavalent intermediate when it is placed in the empty active site.[226] When either was replaced by site-directed mutagenesis, large decreases were also seen in the rate of formation of tyrosyl adenylate (Table 4–2). Unfortunately, however, the experiments were carried further and both Lysine 230 and Lysine 233, the amino groups of which can approach within only 0.8 nm of the phosphates in the molecular model of the pentavalent intermediate, were also mutated. When this was done, similar decreases in the rate constant (Table 4–2) for the formation of the tyrosyl adenylate were

**Figure 4–33:** Crystallographic molecular models of the active site of tyrosine-tRNA ligase occupied by tyrosinyl adenylate (A),[223] tyrosine (B),[224] or the pentavalent intermediate in the phosphotransfer reaction (C).[225] (A) Crystals of tyrosine-tRNA ligase were grown in 2.6 M ammonium sulfate, pH 7.0. These crystals were soaked for 2 days in a $10^{-5}$ M solution of tyrosinyl adenylate (**4–39**) in 2.6 M ammonium sulfate. A data set to Bragg spacing of 0.27 nm was gathered from these crystals. A map of difference electron density was calculated with amplitudes from this data set and amplitudes and phases from a crystallographic study of the isomorphous, unliganded enzyme. The positive difference electron density of the tyrosinyl adenylate in the active site was located and a model of the bisubstrate analogue was built into this electron density. Panel A reprinted with permission from ref 223. Copyright 1985 Wiley-Interscience. (B) Crystals of a derivative of tyrosine-tRNA ligase that had been truncated at residue 317 were grown from 6% poly(ethylene glycol) and 1 mM tyrosine, pH 7.5. A data set was gathered to spacing of 0.25 nm. The refined crystallographic molecular model was obtained by molecular replacement. A previous crystallographic molecular model was truncated at amino acid 317 and then rotated and translated in the present unit cell until its calculated amplitudes matched most closely the measured amplitudes. This preliminary crystallographic molecular model was then submitted to refinement. During the course of the refinement, a region of positive electron density established itself in the active site. A model of tyrosine could readily be built within this feature and this tyrosine, surrounded by the amino acids forming the active site, is present in the figure. Panel B reprinted with permission from ref 234. Copyright 1987 Academic Press. (C) A model of tyrosyl adenylate was built into the active site (panel A) in place of tyrosinyl adenylate. A model of pyrophosphate was attached to phosphorus so that its oxygen was coaxial with the oxygen of the tyrosine carboxyl group. The adenylate oxygen and the other two oxygens on phosphorus were moved into a trigonal planar array normal to this axis. Panel C reprinted with permission from ref 225. Copyright 1985 National Academy of Sciences.

A

B

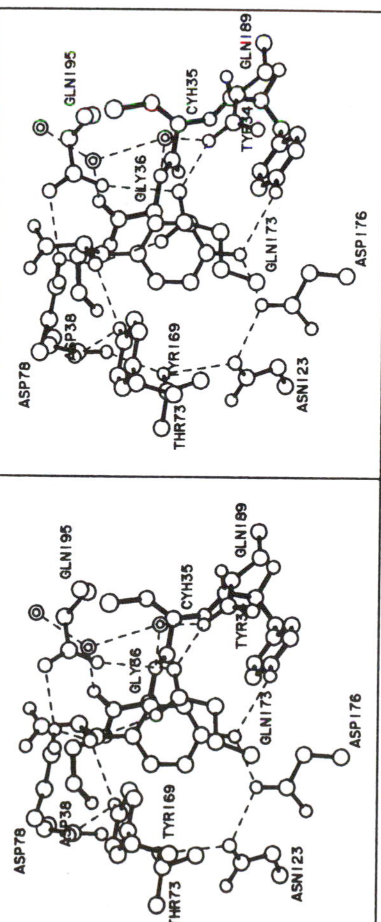


C

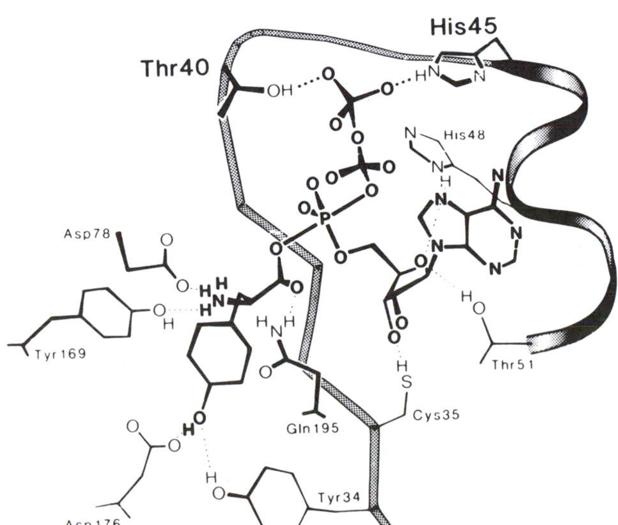

observed, even though these amino acids should not be able to form hydrogen bonds with the oxygens on the phosphates.

One of the steps in the enzymatic transamination of an amino acid catalyzed by pyridoxal phosphate is the tautomerization of the pyridoxamine to the pyridoximine

(4–217)

Because the two carbons are only two atoms apart, a general base must move the proton from one to the other. It is believed that the loss of tritium from [4′-$^3$H]pyridoxamine phosphate when it is reconstituted with the apoenzyme of aspartate aminotransferase

(4–218)

is catalyzed by the same general base that is responsible for the tautomerization in the normal reaction.[234] Lysine 258 is the lysine with which the pyridoxal phosphate forms an imine in the native pyridoximine form of aspartate aminotransferase in the absence of substrates.[235] By the time the tautomerization occurs, however, Lysine 258 is a free, general base that has been shown in model-building studies to be capable of approaching carbon 4′ of the pyridoxal phosphate closely enough to remove the proton. As such, this lysine is a candidate for the general base catalyzing the actual tautomerization. When it was mutated to an alanine, not only the ability of the enzyme to catalyze the overall reaction but even its ability to catalyze hydrogen exchange from [4′-$^3$H]pyridoxamine phosphate was lost.[234] It is possible, however, to restore catalytic activity of this mutant for transamination by adding

**Table 4–2: Kinetic and Equilibrium Constants for Formation of Tyrosyl Adenylate by Mutants of Tyrosine-tRNA Ligase[225–231]**

| mutant | $K_d^{TYR}$ ($\mu$M) | $K_d^{ATP}$ (mM) | $k_3$ ($s^{-1}$) | $k_{-3}$ ($s^{-1}$) | $K_d^{PP}$ (mM) | $K_{eq}$ ($k_3/k_{-3}$) |
|---|---|---|---|---|---|---|
| wild type | 12 | 4.7 | 38 | 17 | 0.6 | 2.3 |
| Y34 → F | 29 | 4.4 | 35 | 22 | 0.4 | 1.6 |
| C35 → G | 11 | 4.5 | 4 | 33 | 0.6 | 0.1 |
| H48 → N | 23 | 3.8 | 27 | 24 | 0.6 | 1.1 |
| H48 → G | 23 | 10 | 10 | 16 | 0.4 | 0.6 |
| T51 → A | 12 | 4.7 | 75 | 11 | 1.0 | 6.8 |
| Y169 → F | 1300 | 4.6 | 35 | 14 | 0.7 | 2.5 |
| Q173 → A | 700 | 100 | 20 | 40 | 1.7 | 0.5 |
| D78 → A | 900 | 5 | 0.04 | 0.12 | 10 | 0.3 |
| D38 → A | 620 | | | | | |
| H45 → G | 43 | 1.3 | 0.16 | | | |
| T40 → A | 11 | 3.8 | 0.005 | | | |
| H45 → G and T40 → A | 4.5 | 1.1 | 0.001 | | | |
| K82 → A | 13 | 40 | 2 | | | |
| K82 → N | 13 | 5.2 | 0.3 | 0.3 | 1.5 | 1.0 |
| R86 → A | 7.5 | 10 | 0.005 | | | |
| R86 → Q | 4 | 3.5 | 0.004 | 0.0035 | 2.5 | 1.1 |
| K230 → N | 19 | 3.4 | 0.03 | | | |
| K233 → A | 11 | | 0.02 | | | |

amines to the solution.[236] The kinetics of the reaction are first-order in the concentration of the complex of mutant enzyme, pyridoximine, and amino acid and first-order in the concentration of the added amine. The resulting second-order rate constant for methylamine (40 $M^{-1}$ $s^{-1}$) is greater than that for ammonia (21 $M^{-1}$ $s^{-1}$), which is greater than that for ethylamine (1.3 $M^{-1}$ $s^{-1}$). It has been concluded that the added amines slide into the active site and take the place of the missing primary amino group of the missing Lysine 258 and act as the catalytic base.

Site-directed mutation can also be used to rule out a mechanism for an enzyme that is based on an examination of the crystallographic molecular model. Within the active site of aspartate aminotransferase in the crystallographic molecular model, Tyrosine 70 is positioned adjacent to Lysine 258. In model-building studies with the crystallographic molecular model, it was possible to form a hydrogen bond between Tyrosine 70 and Lysine 258 that would position Lysine 258 over the respective carbons from which it removes and adds the proton during the tautomerization (Reaction 4–217), and it was proposed that this tyrosine might be essential for enzymatic activity.[235] When Tyrosine 70 was mutated to a phenylalanine, however, the enzyme was still able to catalyze the reaction, albeit with a reduced turnover number (170 $s^{-1}$ for wild type against 30 $s^{-1}$ for the mutant).[237]

In the reaction catalyzed by cytochrome *c* peroxidase, the compound I that forms at the heme (Reaction 2–155) has a radical cation located within the protein rather than within the porphyrin of the heme. Tryptophan 51 is immediately adjacent to the heme in the crystallographic molecular model; and, on the basis of this proximity alone, it was proposed as a candidate for bearing the radical cation. When Tryptophan 51 was mutated to a phenylalanine, however, the enzymatic activity was not diminished, and the same radical cation was formed when the mutant enzyme was converted to compound I.[238] This result ruled out Tryptophan 51 as the location of the radical cation.

## *Suggested Reading*

Holland, P.C., Clark, M.G., & Bloxham, D.P. (1973) Inactivation of Pig Heart Thiolase by 3-Butynoyl-Coenzyme A and 4-Bromocrotonyl-Coenzyme A, *Biochemistry 12*, 3309–3315.

Toney, M.D., & Kirsch, J.F. (1993) Lysine 258 in Aspartate Aminotransferase: Enforcer of the Circe Effect for Amino Acid Substrates and General-Base Catalysis for the 1,3-Prototropic Shift, *Biochemistry 32*, 1471–1479.

## PROBLEM 4–19

Glycine hydroxymethyltransferase is a pyridoxal phosphate-dependent enzyme that catalyzes the reversible interconversion of glycine and serine

L-serine + tetrahydrofolate ⇌
glycine + $N^{5,10}$-methylenetetrahydrofolate (1)

This reaction is the major source of $N^{5,10}$-methylenetetrahydrofolate in mammals. The enzyme also carries out the following transformations

L-allothreonine ⇌ glycine + acetaldehyde (2)

L-phenylserine ⇌ glycine + benzaldehyde (3)

aminomalonate ⇌ glycine + $CO_2$ (4)

$\alpha$-methylserine + tetrahydrofolic acid ⇌
D-alanine + $N^{5,10}$methylenetetrahydrofolate (5)

D-alanine + *E*·pyridoxal-P →
pyruvate + pyridoxamine-P + *E* (6)

D-3-fluoro-2-aminopropionate (D-fluoroalanine) ⇌
pyruvate + HF + $NH_3$ (7)

(A) Write the mechanism for Reaction 7. Start with the pyridoxal phosphate bound to the enzyme in imine linkage to a lysine and the D-fluoroalanine in the active site. In every step, write out the complete structures of the pyridoxal and the amino acid or the adducts formed from the two of them. Indicate acids and bases as HA and B to save time.

(B) Explain the following results from[238a] a study of Reaction 7.

| reactant | $V_{max}/K_m$ ($min^{-1}$ $M^{-1}$) |
|---|---|
| D-[2-$^1$H]fluoroalanine | 22 |
| D-[2-$^2$H]fluoroalanine | 4 |

When tetrahydrofolate is present, D-fluoroalanine is turned into pyruvate 100 times faster than L-fluoroalanine is turned into pyruvate. The normal substrate for the enzyme, however, is L-serine.

(C) Draw two structures, side by side, of the adducts formed at the active site between pyridoxal phosphate and D-fluoroalanine and between pyridoxal phosphate and L-serine, respectively, that explain why D-fluoroalanine is the preferred substrate over L-fluoroalanine but L-serine is the preferred substrate over D-serine.

(D) Explain, by drawing a chemical mechanism involving a nucleophilic amino acid side chain (–X⊙) in the active site, why the enzyme becomes irreversibly inactivated while it is performing Reaction 7.

## PROBLEM 4–20

The following compound was synthesized as an affinity label for the active site of lysozyme

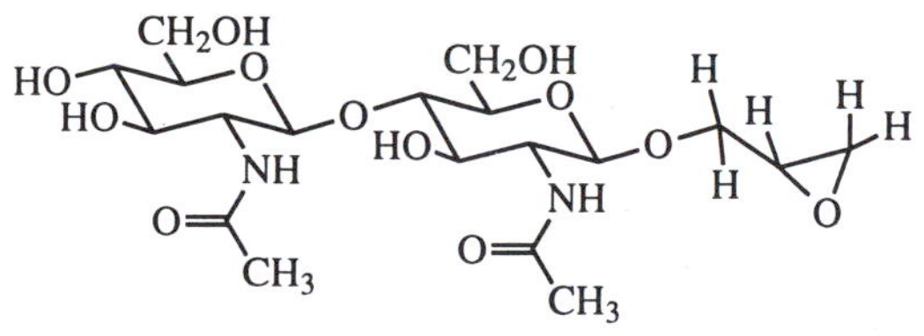

2′,3′-epoxypropyl $\beta$-di(*N*-acetyl-D-glucosamine)glycoside [EDAG]

After binding to the active site, EDAG reacts with lysozyme to form a covalent derivative of the enzyme that contains 1 mol of the affinity reagent attached to Aspartate 52 for each mole of the enzyme and that is inactive.[238b]

(A) Write the chemical mechanism for the reaction between Aspartate 52 and EDAG. Which carbon of the electrophilic center is the most likely electrophile?

Lysozyme, to which EDAG had been attached and which was inactive, was crystallized, and its structure was determined by X-ray diffraction.[238b]

(B) Which subsites (Figure 3–25) on the inactive lysozyme molecule were occupied by the two sugar residues of the affinity label?

When the covalently modified enzyme was reduced with $NaBH_4$, the ester was reduced to an alcohol and a derivative of lysozyme with homoserine at position 52 resulted.[238c]

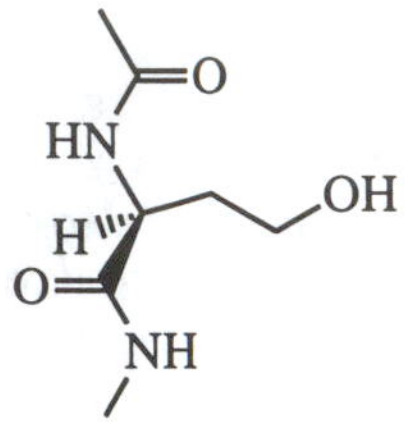

homoserine [Hser]

(C) Lysozyme($Hser_{52}$) is catalytically inactive but binds hexasaccharide substrates such as $(GlcNAc)_6$ with the same affinity as the native enzyme. Explain this observation.

The turnover number of lysozyme, $k_{cat}$, varies considerably with pH (Figure 4–11), but the Michaelis constant is invariant with changes in pH.

(D) Explain in explicit terms, referring to particular amino acid residues, the pH dependence of $k_{cat}$ below pH 6.

**PROBLEM 4–21**

Aldolases are enzymes that catalyze the following chemical reaction

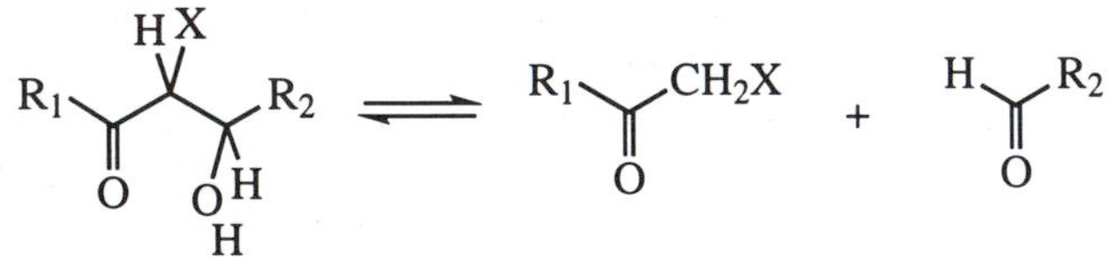

where X = OH or H. It has been shown that when the ketone product alone is mixed with an aldolase, allowed to equilibrate, and then treated with $NaBH_4$ (a mild reducing reagent), the following amino acid usually can be isolated from a hydrolysate of the protein

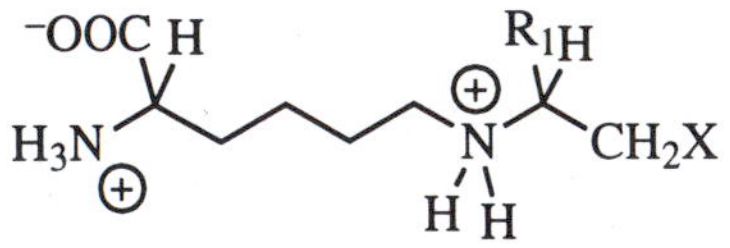

This result is evidence for the existence of the following reactive intermediate

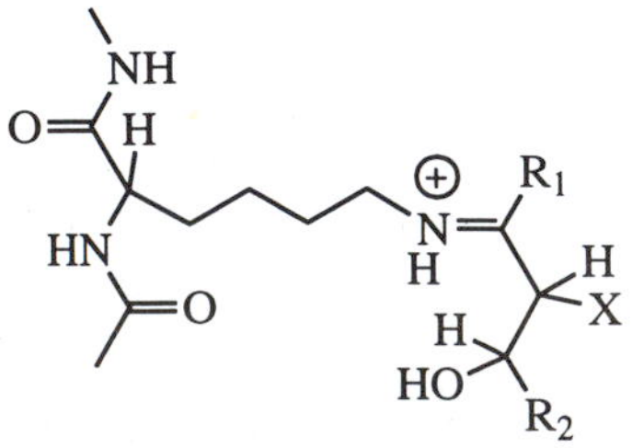

(A) Write out the steps involved in the aldolase mechanism. Where are protons rearranged?

The proposal we shall examine is that whenever a proton is added or removed from a substrate at the active site of an enzyme, an amino acid on that enzyme, in the active site, accepts or donates the proton. Phospho-2-dehydro-3-deoxygluconate aldolase catalyzes the following reaction

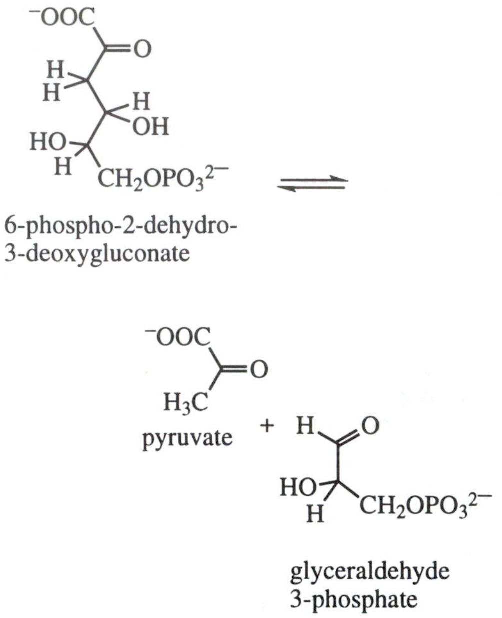

6-phospho-2-dehydro-3-deoxygluconate

pyruvate

glyceraldehyde 3-phosphate

Phospho-2-dehydro-3-deoxygluconate aldolase was mixed with pyruvate in the presence of $^3H_2O$ and incubated for the noted times.[238d] Following the reaction, pyruvate was purified on an ion-exchange column and examined for $^3H$ incorporation.

| enzyme | time (h) | $\mu$mol of $^3H$ ($\mu$mol of pyruvate)$^{-1}$ |
|---|---|---|
| none | 1 | <0.005 |
| boiled | 1 | <0.005 |
| active | 1 | 0.585 |
| none | 16 | 0.073 |
| active | 16 | 2.310 |

(B) To which carbon in pyruvate is the tritium added? How does this process fit into the overall mechanism? Draw out what you think is occurring at the active site. Include proton donors and acceptors.

Phospho-2-dehydro-3-deoxygluconate aldolase was mixed with 1 mM bromopyruvate at pH 6. Samples were removed at various intervals and immediately diluted 200-fold into a pH 8 solution to inactivate the bromopyruvate. The enzyme, following this dilution procedure, was then assayed for aldolase activity.[238e]

| time of initial incubation with bromopyruvate (min) | enzymatic activity remaining (%) |
|---|---|
| 0 | 100 |
| 1 | 78 |
| 2 | 65 |
| 3 | 50 |
| 4 | 41 |
| 5 | 28 |

The bromopyruvate is irreversibly inactivating the enzyme.

(C) Why was bromopyruvate chosen? Examine your diagram for the enzyme mechanism. What group is reacting with the bromopyruvate and what is the type of chemical reaction occurring?

(D) Assume that Mechanism 4–170 describes this inactivation and that

$$\frac{d[E-ASL]}{dt} \equiv v_{inact} \equiv v_i = k_i[E\cdot ASL]$$

and show that

$$v_i = \frac{k_i([E]_0 - [E-ASL])[ASL]}{K_m^{ASL} + [ASL]}$$

where

$$K_m^{ASL} = \frac{k_{-1} + k_i}{k_1}$$

(E) If

$$v_i = \frac{d[E-ASL]}{dt}$$

show by integration that Equation 4–171 describes the inactivation. Show that the reaction corresponds to this expectation by plotting the data from the table in this form. Remember that

$$\%\text{ activity remaining} = \frac{[E]_0 - [E-ASL]}{[E]_0} \times 100$$

[ASL] = 1 mM at all $t$, and $[E]_0 \ll$ [ASL]. If $t_{1/2}$ = time required to inactivate half the enzyme, show that

$$t_{1/2} = \frac{\ln 2}{k_i}\left(\frac{K_m^{ASL}}{[ASL]} + 1\right)$$

(F) The values of $t_{1/2}$ for several bromopyruvate concentrations were calculated from curves such as the one you drew earlier.[238e]

| [bromopyruvate] (mM) | $t_{1/2}$ (min) |
|---|---|
| 0.21 | 5.5 |
| 0.30 | 4.1 |
| 0.46 | 3.1 |
| 0.89 | 2.3 |
| 1.5 | 1.6 |
| 3.5 | 1.3 |

Do these data conform to expectations? What are the numerical values of $K_m^{ASL}$ and $k_i$?

If bromopyruvate is binding to the active site of phospho-2-dehydro-3-deoxygluconate aldolase at the same position as pyruvate, then pyruvate should protect the enzyme. The following values for $t_{1/2}$were measured when 6 mM pyruvate was present in addition to the bromopyruvate.[238e]

| [bromopyruvate] (mM) | $t_{1/2}$ (min) |
|---|---|
| 5.0 | 8.3 |
| 10.0 | 4.7 |
| 20.0 | 2.9 |

(G) What kind of an inhibitor is pyruvate with respect to the inactivation caused by bromopyruvate? Is this expected? Why?

Radioactive [$^{14}$C]bromopyruvate was synthesized by the direct bromination of [$^{14}$C]pyruvate. Phospho-2-dehydro-3-deoxygluconate aldolase was incubated at pH 6 for 20 min in the presence of 1 mM unradioactive bromopyruvate and 40 mM pyruvate. The enzyme was dialyzed exhaustively and then incubated with 1 mM [$^{14}$C]bromopyruvate at pH 6 in the absence of pyruvate for 20 min. The enzyme was precipitated 3–4 times with $(NH_4)_2SO_4$ until constant counts per minute of $^{14}$C (milligram of protein)$^{-1}$ were obtained. It was found that 90% of the activity had disappeared and 1.1 mol of $^{14}$C (mol of enzyme protomer)$^{-1}$ had been covalently incorporated. When the radioactive protein was incubated with 0.5 M NaOH or unfolded in 10 M urea and reacted with hydroxylamine, 90–95% of the radioactivity was removed from the protein. Ethers, phenol ethers, and secondary amines are all stable under these conditions. When the [$^{14}$C]-labeled protein was hydrolyzed in 6 M HCl at 100 °C for 24 h, 99% of the $^{14}$C was released as hydroxypyruvic acid.[238f]

(H) What amino acid side chain has reacted covalently with bromopyruvate to inactivate the enzyme?

The [$^{14}$C]-labeled enzyme inactivated with [$^{14}$C]bromopyruvate was prepared and divided into two fractions. One fraction was unfolded with 1% sodium dodecyl sulfate. Both native and unfolded [$^{14}$C]aldolase were then reacted with $NaBH_4$ for 30 min at 50 °C. When these two fractions were hydrolyzed in 6 M HCl, the unfolded sample yielded only [$^{14}$C]glyceric acid, but the hydrolysate of the native enzyme contained a new unknown [$^{14}$C]amino acid (60% yield). When enzyme, purified from bacteria grown on [$^3$H]lysine, was reacted with [$^{14}$C]bromopyruvate, reduced, and hydrolyzed, the unique unknown amino acid contained both $^{14}$C and $^3$H in a 1:1 molar ratio.[238f]

(I) What is this novel amino acid and what occurred during the inactivation and then the reduction?

(J) Look again at your diagram of the proton movements during the aldolase reaction. Notice that during the cleavage a proton is removed from the hydroxyl on carbon 4 but that during the reverse reaction a proton is removed from the methyl group of carbon 3. Consider the amino acid that performs this function. What is there about its structure that makes it particularly suitable?

## PROBLEM 4–22

2,3-Epoxypropanol 1-phosphate, glycidol phosphate, is an irreversible inhibitor of triosephosphate isomerase. When glycidol phosphate is added to a solution of triosephosphate isomerase, in large excess over the concentration of enzyme, the enzymatic activity disappears in a pseudo-first-order fashion such that a plot of log ([active enzyme]) against time is linear, and a rate constant, $k_i$, could be obtained from the slope. It was found the $k_i$ varied with [glycidol 1-P] and was affected by the addition of the competitive inhibitor, glycerol 1-phosphate.[238g]

| [glycidol 1-P] (mM) | $k_i$ ($min^{-1} \times 10^{-3}$) enzyme alone | enzyme + 0.3 mM glycerol 1-phosphate |
|---|---|---|
| 0.25 | 0.6 | |
| 0.50 | 1.1 | |
| 0.75 | 1.5 | |
| 1.0 | 1.8 | 0.7 |
| 2.0 | 2.2 | 1.3 |
| 3.0 | | 1.7 |
| 4.0 | | 2.0 |
| 5.0 | | 2.2 |

(A) Plot these data in an informative format and determine $K_m^{ASL}$ and $K_d^{GAP}$, where $K_d^{GAP}$ is the dissociation constant of glyceraldehyde 3-phosphate from the active site.
(B) What two features of the behavior of the kinetics of the inactivation reaction lead to the conclusion that glycidol phosphate is an active-site label?
(C) What participant in Mechanism 4–136 is reacting with the glycidol phosphate?

When the enzyme was fully inactivated, 0.9 ± 0.1 mol of [$^{32}$P]glycidol 1-phosphate (mol of active sites)$^{-1}$ was incorporated into the protein.[238h] The radioactive protein was digested with pepsin and a radioactive peptide was isolated. Upon hydrolysis, the amino acid composition of the peptide was

| amino acid | mol (100 mol of aa)$^{-1}$ |
|---|---|
| Glu | 20 |
| Pro | 22 |
| Ala | 18 |
| Val | 21 |
| Tyr | 19 |

The optical density of the peptide at 280 nm was 5800 $M^{-1}$ $cm^{-1}$.

(D) What is the minimum molecular formula for this peptide?

This peptide was radioactive ($^{32}$P) and lost its radioactivity upon either alkali treatment, which will not cleave phosphate esters, or phosphatase treatment, which will. The same peptide was isolated from unmodified enzyme and enzyme modified with glycidol, 2,3-epoxypropan–1-ol, and $HOPO_3^{2-}$. Upon electrophoresis the following behavior was observed.

| active-site label | product | electrophoretic mobility (relative to aspartate) pH 6.5 | (relative to serine) pH 1.9 |
|---|---|---|---|
| none | unlabeled peptide | −0.21 | 0.43 |
| glycidol phosphate | phosphoglyceryl peptide | −0.27 | 0 |
| glycidol + $HOPO_3^{2-}$ | glyceryl peptide | 0 | 0.42 |

(E) Which amino acid in the peptide had reacted with the glycidol 1-phosphate?

## PROBLEM 4–23

Isocitrate dehydrogenase catalyzes the following oxidative decarboxylation

[isocitrate] + $NAD^+$ ⇌ [2-ketoglutarate] + $CO_2$ + NADH

isocitrate    2-ketoglutarate

(A) Write a mechanism for this reaction that involves an intermediate. Include amino acids of your own choice to catalyze all proton removals and additions.
(B) 3-Bromo-2-ketoglutarate is an irreversible inhibitor of the enzyme. Draw a mechanism incorporating one of the amino acids you used in part A to explain how this compound might inactivate the enzyme covalently.

The enzymatic activity remaining in a solution that contained the enzyme and 8.4 mM 3-bromo-2-ketoglutarate was determined as a function of time. The ratios between the enzymatic activities remaining at particular times, referred to as E, and the initial enzymatic acitvity, referred to as $E_c$, were plotted[239] (reprinted with permission from ref 239; copyright 1982 American Chemical Society).

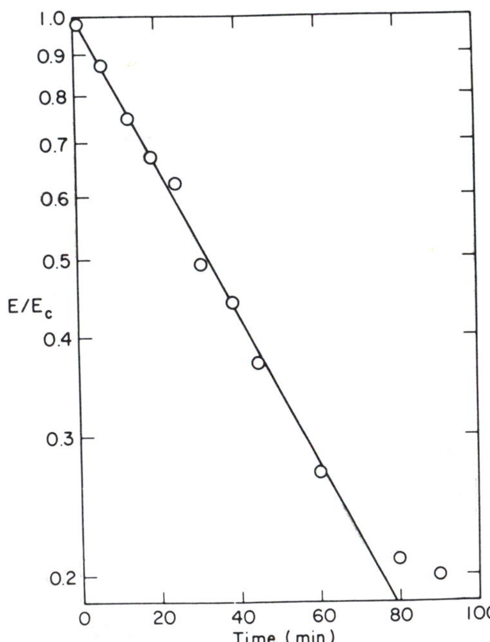


(C) Write a kinetic mechanism and its associated rate equation that are consistent with these observations.

The observed first-order inactivation rate constants, $k_{inact}$, were obtained for various concentrations of 3-bromo-2-ketoglutarate[239] (reprinted with permission from ref 239. copyright 1982 American Chemical Society).

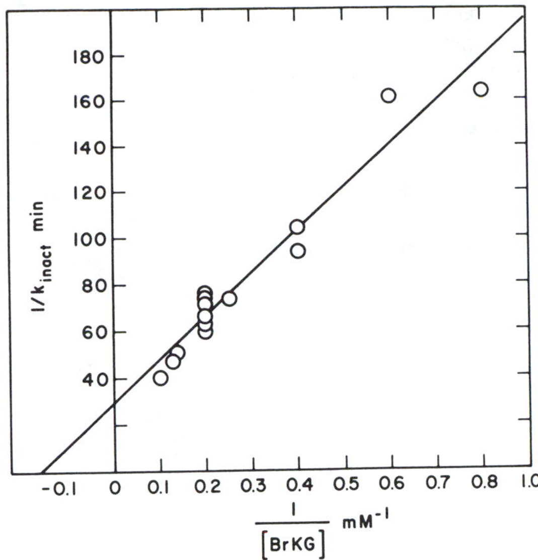


(D) What does the plot in this figure state about the reaction between isocitrate dehydrogenase and 3-bromo-2-keto-glutarate?

When the half-times of inactivation of isocitrate dehydrogenase at 5.0 mM 3-bromo-2-ketoglutarate were determined as a function of the concentration of isocitrate added to the solution, the following results were obtained (reprinted with permission from ref 239; copyright 1982 American Chemical Society).

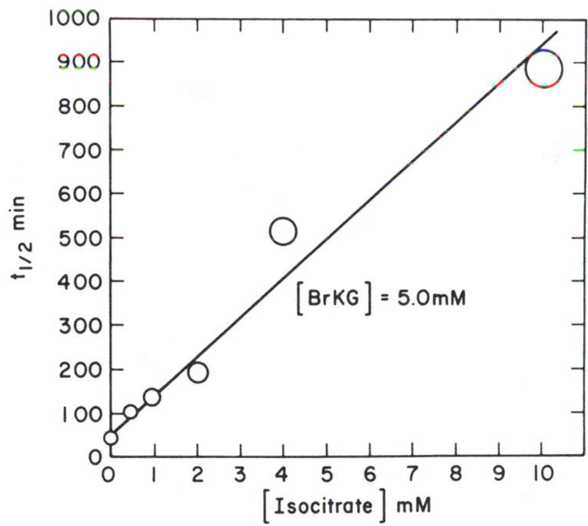


(E) What conclusion can be drawn from this figure?

## PROBLEM 4–24

Within the bacterium *Escherichia coli*, there is an enzyme that can catalyze each of the following three reactions

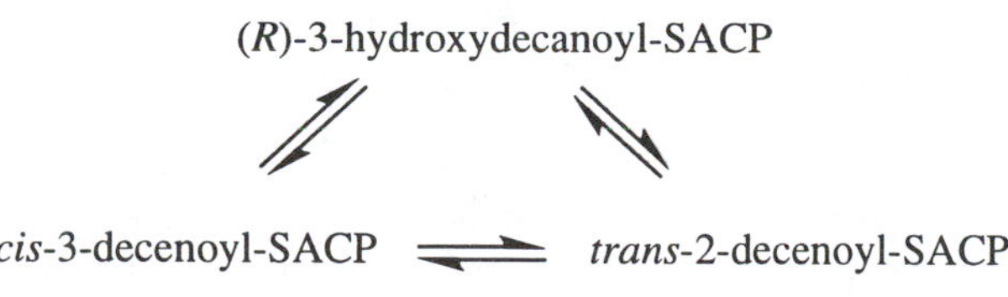


The abbreviation SACP is for acyl carrier protein, a small protein that forms a thioester with the fatty acids. Substrates that are thioesters of *N*-acetylcysteamine (NAC) are as effective as the thioesters of ACP.

(A) If any one of the substrates were mixed with a high enough concentration of the enzyme in $D_2O$, so that the three reactions catalyzed by the enzyme came to equilibrium faster than nonenzymatic exchange could occur, deuterium would be incorporated by the enzyme into some or all of the three substrates. If it is assumed that all dehydrations are *trans*, how many deuteriums are incorporated into each substrate and what are the absolute stereochemistries of each deuterated substrate?

The effects of pH on the kinetic parameters of the enzyme were determined. The following figure describes the details of such an experiment and presents the results.[240]

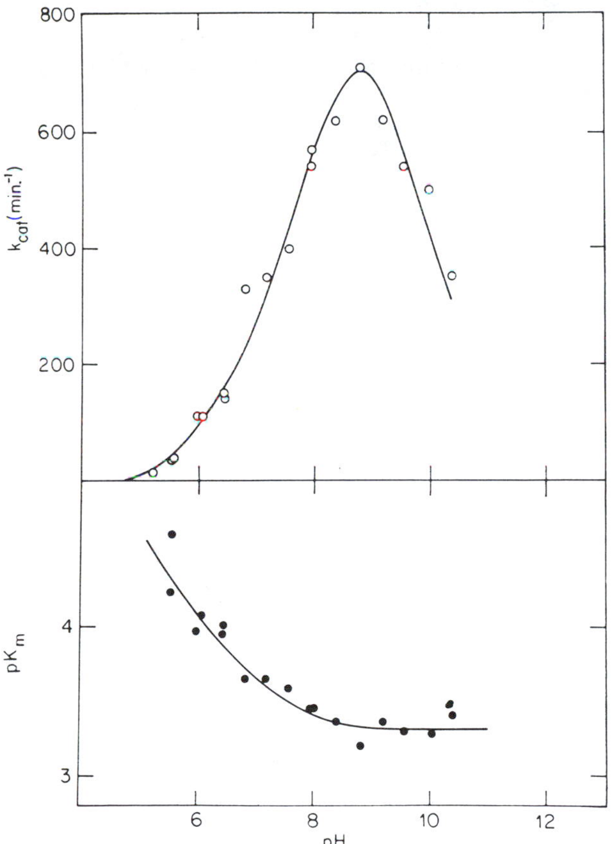


Effect of pH on $k_{cat}$ and $K_m$ of dehydrase.[240] The isomerization of *cis*-3-decenoyl-NAC to *trans*-2-decenoyl-NAC was measured at various pH values and at constant temperature (30 °C) and ionic strength (0.05 M). Reactions were initiated by addition of 4 μg of dehydrase to cuvettes containing 1.0 mL of substrate, the concentration of which had been determined by measuring absorbance at 232 nm. Product formation was monitored at 263 nm. Values of $k_{cat}$ and $K_m$ were calculated from double-reciprocal plots of the data at each pH. At least 10 different substrate concentrations were used for each point. The buffer systems were sodium acetate, imidazolium chloride, potassium phosphate, tris(hydroxymethyl)aminomethane hydrochloride, and sodium glycinate. Reprinted with permission from ref 240. Copyright 1969 *Journal of Biological Chemistry*.

(B) Write a mechanism for this specific reaction as it occurs within the active site of the enzyme. Your mechanism must contain the explanation for the behavior observed in the figure. Use amino acid side chains of your choice to act as catalytic residues. Choose a different one for each role to distinguish them in later answers.

(C) Explain the effects of pH on $k_{cat}$ and $K_m$ observed in the preceding figure in terms of acid dissociations of the general acids and general bases that you chose in part B. Consider both the free enzyme and the enzyme–substrate complex. Remember that in this particular experiment the *cis*-3-decenoyl thioester of *N*-acetylcysteamine (NAC) is the substrate.

(D) Which $pK_a$ apparently shifts into the range covered in the experiment when substrate binds to the enzyme? Identify this $pK_a$ by naming the amino acid you chose to perform this role. How does the binding of substrate affect the other $pK_a$?

Both 3-decynoyl-NAC and 2,3-decadienoyl-NAC are irreversible inhibitors of the enzyme. The following experiment was performed with these affinity reagents.[166]

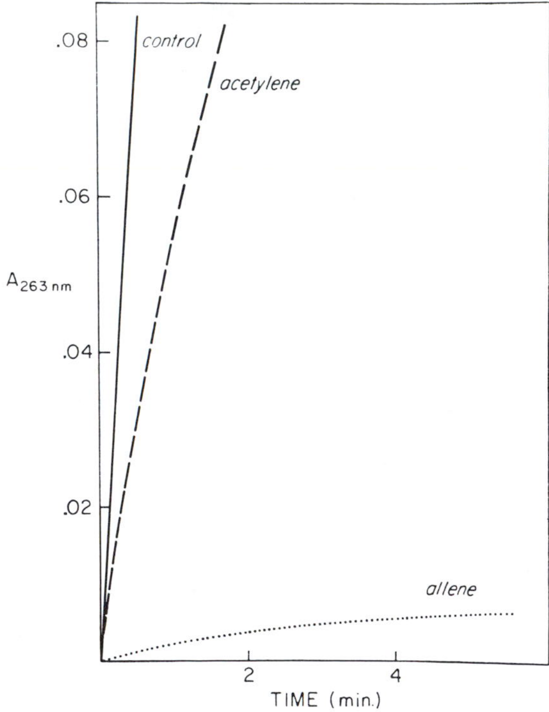


Comparison of 3-decynoyl-NAC and 2,3-decadienoyl-NAC as dehydrase inhibitors.[166] To cuvettes containing $10^{-5}$ M thioester inhibitors and $4 \times 10^{-4}$ M *cis*-3-decenoyl-NAC in 0.01 M potassium phosphate, pH 7.0, were added 6 μg of purified dehydrase at zero time. The rate of *trans*-2-decenoyl-NAC production was measured spectrophotometrically. Reprinted with permission from ref 166. Copyright 1970 *Journal of Biological Chemistry.*

(E) Considering only nucleophilic addition to an olefin, draw the structure of the intermediate that explains why one of these compounds is so much more reactive than the other. Use one of the nucleophiles in your active site for the addition.

(F) Draw at least two possible products of the reaction of 2,3-decadienoyl-NAC with amino acid residues in the active site that you constructed in the answer to part B.

(G) Which enantiomer of 2,3-decadienoyl-NAC should be the reactive one?

2,2-Dideutero-3-decynoyl-NAC was synthesized as well as 3-decynoyl-NAC, and these two compounds were tested together as inactivators of the enzyme.[166]

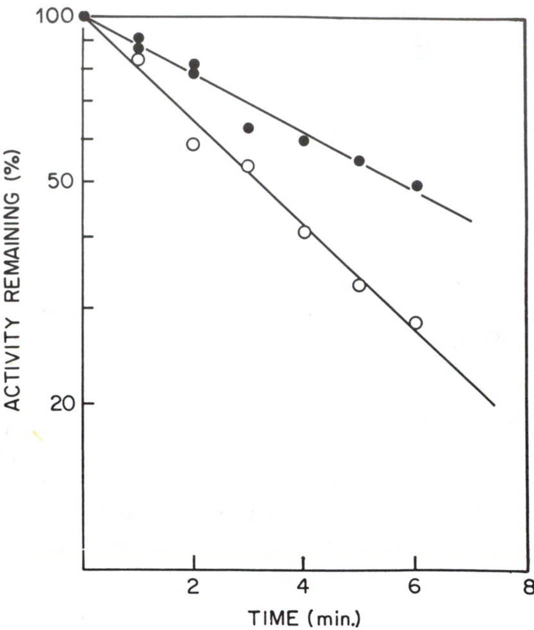


Kinetic comparison of inhibition by deuterated and non-deuterated acetylenic thioesters.[166] To cuvettes containing 0.5 mL of potassium phosphate, pH 6.0 at an ionic strength of 0.05 M, were added, to a final concentration of $8.0 \times 10^{-8}$ M, either 3-decynoyl-NAC (○) or 2,2-dideutero-3-decynoyl-NAC (●) and, at zero time, 33 μg of dehydrase. At various times 0.5 mL of 2–$3.1 \times 10^{-4}$ M *cis*-3-decenoyl-NAC in the same buffer was added and the initial velocity of product formation was recorded. The degree of inactivation was based on comparison with a similar reaction to which no acetylenic thioester had been added. Since the amount of acetylenic thioester was present at a 60-fold molar excess over dehydrase, the inactivation data were treated in a first-order manner. Reprinted with permission from ref 166. Copyright 1970 *Journal of Biological Chemistry.*

The deuterium isotope effect observed (${}^{H}k/{}^{D}k = 2.60$) is similar in magnitude to that seen for 2,2-dideutero-3-hydroxydecanoyl-NAC during its dehydration to *trans*-2-decanoyl-NAC (${}^{H}k/{}^{D}k = 2.25$). 2-Deutero-2,3-decadienoyl-NAC, however, shows no isotope effect during its inactivation of the enzyme.

(H) What does the deuterium isotope effect suggest that the enzyme must be doing to 3-decynoyl-NAC before this active-site label can inactivate the enzyme?

(I) What must the 3-decynoyl-NAC be turned into before it can inactivate the enzyme?

(J) Incorporate the mechanism for this transformation into your active site from part B. Write an overall mechanism that incorporates the three reactions naturally catalyzed by the enzyme, the inactivation by the allene, and the suicidal inactivation by the alkyne.
(K) Is the enzyme catalyzing the reaction through an enol intermediate, through a carbonium ion intermediate, or by a concerted mechanism?

## PROBLEM 4–25

L-Lactate dehydrogenase catalyzes Reaction 3–217.

(A) Write a chemical mechanism for this reaction as it would occur in the active site using A and B to indicate any necessary acids or bases. Abbreviate $NAD^+$ as

When the bisubstrate analogue of the bound reactants, (*S*)-Lac-NAD (Structure **3–36**), was mixed at a final concentration of 2.1 mM with a final concentration of 80 $\mu$M active sites of L-lactate dehydrogenase, the following result was obtained.[241]

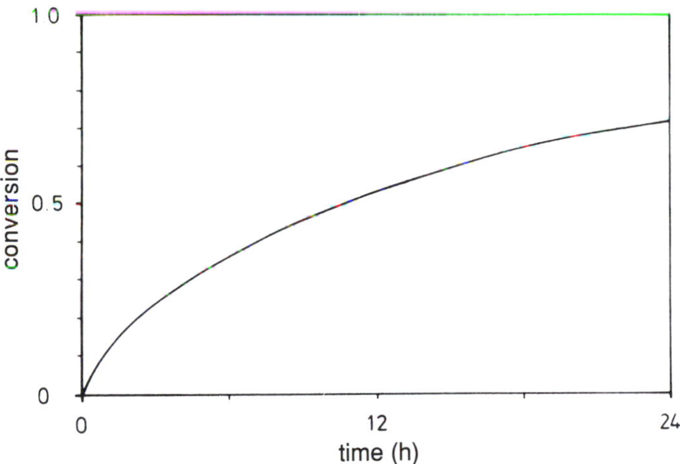


Time course of the conversion of 2.1 mM (*S*)-Lac-$NAD^+$ in the presence of 20 $\mu$M pig heart lactate dehydrogenase in 1 M glycine buffer containing 0.4 M hydrazine (pH 9.5).[241] The relative increase of the 366-nm absorption (1.0 corresponding to 100% conversion) is presented as a function of time in hours. Adapted with permission from ref 241. Copyright 1978 American Chemical Society.

(B) Write a chemical equation for the reaction being catalyzed by the enzyme in this experiment.
(C) Why was 0.4 M hydrazine added to the solution?
(D) Estimate the turnover number of the enzyme in this reaction.

The complex between L-lactate dehydrogenase and (*S*)-Lac-$NAD^+$ has been crystallized, and a map of electron density has been calculated from the X-ray diffraction of the crystals (Figure 3–36).

(E) What are the identities of the acids or bases that you used in the mechanism you wrote in part A?
(F) Draw proper hydrogen bonds, with appropriate bond angles, for the hydrogen bonds between the bisubstrate analogue and Serine 163, Histidine 195, Arginine 109, and Arginine 171.

The steady-state kinetics of the enzymatic reaction catalyzed by L-lactate dehydrogenase, in the direction from (*S*)-lactate to pyruvate, have been shown to conform to the empirical equation

$$\frac{V_{max}}{v_0} = 1 + \frac{K_m^{NAD}}{[NAD]} + \frac{K_m^{Lac}}{[Lac]} + \frac{K_m^{NI}}{[NAD][Lac]}$$

where $v_0$ is the initial velocity, $V_{max}$ is the maximum velocity, and $K_m^{NAD}$, $K_m^{Lac}$, and $K_m^{NI}$ are the respective Michaelis constants.

(G) When $NAD^+$ is held at saturating concentrations, what is the empirical equation for $V_{max}/v_0$ and what is the equation for $v_0$?

The simplest kinetic mechanism that would give this equation for $v_0$ when $NAD^+$ is present at saturation is

$$NAD^+ + E\cdot NAD^+ + Lac \underset{k_{-1}}{\overset{k_1}{\rightleftharpoons}}$$

$$NAD^+ + E\cdot NAD^+ \cdot Lac \xrightarrow{k_2}$$

$$E\cdot NAD^+ + NADH + Pyr$$

Notice that, at saturating $NAD^+$, the enzyme behaves as if there were only one reactant, namely lactate.

The effects of pH on the kinetic constants of lactate dehydrogenase have been determined[241a] at 28.5 °C and an ionic strength of 0.2 M.

| pH | $k_{cat}$ ($min^{-1} \times 10^{-3}$) | $K_m^{Lac}$ (mM) | pH | $k_{cat}$ ($min^{-1} \times 10^{-3}$) | $K_m^{Lac}$ (mM) |
|---|---|---|---|---|---|
| 5.6 | 0.9 | 80 | 8.0 | 4.8 | 9.2 |
| 5.8 | 1.0 | 38 | 9.0 | 8.5 | 11 |
| 7.1 | 1.8 | 17 | 9.5 | 8.5 | 11 |
| 7.5 | 3.2 | 4 | 10.2 | 23.5 | 53 |

(H) Determine graphically an apparent $pK_a$ for the unoccupied binary complex between enzyme and $NAD^+$, $E\cdot NAD^+$, unoccupied by lactate, that is reflected in these data. This is analogous to an apparent macroscopic $pK_a$ for the free enzyme in an enzymatic reaction with only one reactant. What would you plot in that case?
(I) Assume that this apparent macroscopic $pK_a$ is actually the microscopic $pK_a$ of an amino acid on the enzyme in the enzyme·$NAD^+$ complex. What form of that amino acid, the conjugate acid or the conjugate base, must be present for the enzyme·$NAD^+$ complex to be active in the oxidation of lactate? What occurs when the other form of the amino acid is present?

L-Lactate dehydrogenase is inactivated irreversibly by diethyl pyrocarbonate. The following experiment was performed.[45]

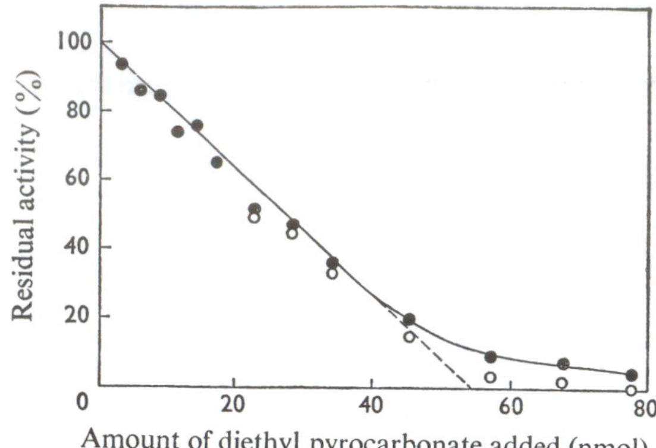


Covalent inhibition of lactate dehydrogenase with diethyl pyrocarbonate.[45] Freshly diluted samples of inhibitor were added to solutions of enzyme (56 nmol of subunits) in 0.08 mL of 0.1 M sodium phosphate buffer, pH 6.0, at 20 °C. After 1 min, the enzyme was diluted 200-fold into ice-cold buffer, and a sample was used to measure the enzyme activity. The activity was measured about 5 min after the addition of inhibitor and was expressed as a percentage of that of an untreated control (●). The open circles (○) have been corrected for the reactivation (about 5%) that occurs during the 5 min necessary to assay the enzyme. Reprinted with permission from ref 45. Copyright 1973 Biochemical Society.

(J) Explain why this experiment was done, what the results demonstrate, and why there must be one particularly reactive amino acid in the protein.

Diethyl pyrocarbonate reacts so rapidly with lactate dehydrogenase that the reaction must be monitored in a stopped-flow apparatus. For example, when the solution in one syringe was 30 $\mu$M lactate dehydrogenase in pyrophosphate buffer (ionic strength = 0.15) at pH 8.4 and the solution in the other syringe was 10.6 mM diethyl pyrocarbonate in the same buffer, the absorbance of the mixture at 250 nm, which monitors the product of the reaction between the enzyme and diethyl pyrocarbonate, showed a first-order relaxation with a rate constant of 0.87 $s^{-1}$.

(K) Write a kinetic mechanism for the reaction between the nucleophilic form of the enzyme and diethyl pyrocarbonate. Do not solve the rate equation.

(L) Why is the observed reaction kinetically first-order, and what is the observed rate constant equal to in terms of your kinetic mechanism?

The concentrations of both lactate dehydrogenase and diethyl pyrocarbonate and the ionic strength were kept constant in a series of stopped-flow experiments in which the pH was varied systematically. The observed first-order rate constant for the reaction was plotted as a function of the pH.[45]

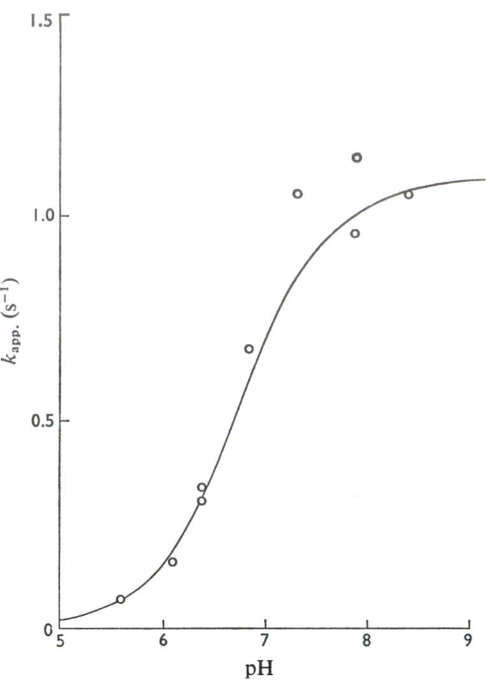


Dependence of the rate of reaction of L-lactate dehydrogenase with diethyl pyrocarbonate on pH.[45] The apparent first-order rate constant for the change in absorbance at 251 nm was determined in the stopped-flow apparatus. The concentration of enzyme after mixing was 15 $\mu$M, and that of diethyl pyrocarbonate was 5.3 mM. Buffers were sodium pyrophosphate ($\Gamma/2$ = 150 mM). The line drawn is for a p$K_a$ of 6.8 and a rate constant for the unprotonated enzyme of 214 $M^{-1}\,s^{-1}$. Reprinted with permission from ref 45. Copyright 1973 Biochemical Society.

(M) Write a kinetic mechanism for this inactivation describing it as the acylation of the conjugate base of an amino acid on the protein. Incorporate the acid dissociation of the conjugate acid of that amino acid in your mechanism. Derive an equation for the observed first-order rate constant, $k_{app}$, in terms of the intrinsic rate constant, $k_{DP}$, between diethyl pyrocarbonate and the conjugate base of the amino acid, the acid dissociation constant, $K_a^{AH}$, of the amino acid, and the total concentration of the enzyme, $[E]_{TOT}$.

(N) Compare the p$K_a$ observed in these experiments with that observed in the graph you prepared in part H.

When NADH is bound to the enzyme, diethyl pyrocarbonate is still able to inactivate it. The apparent p$K_a$ for the rate of the reaction of diethyl pyrocarbonate with the enzyme·NADH complex is 6.7, and the rate constant between diethyl pyrocarbonate and the unprotonated enzyme is 66 $M^{-1}\,s^{-1}$.

(O) What amino acid in lactate dehydrogenase is assumed to react with diethyl pyrocarbonate? Identify it by the type of amino acid it is and its number in the sequence of L-lactate dehydrogenase. What is its apparent p$K_a$ in the free enzyme, in the enzyme·NADH complex, and in the enzyme·$NAD^+$ complex?

# Acid–Bases in Crystallographic Molecular Models

It would seem that, when a crystallographic molecular model of the active site of a particular enzyme became available, the general acid–bases involved in the enzymatic mechanism could be identified immediately by inspection. Unfortunately, this has not been the case. The reasons for the recurring ambiguity that accompanies crystallographic molecular models of active sites are that there are usually too many general acid–bases within an active site and that they are too flexible.

The reason that there are too many general acid–bases in an active site is that substrates, in addition to presenting to the active site the protons that must be removed and the lone pairs of electrons that must be protonated for the reaction to occur, also have significant numbers of donors and acceptors of hydrogen bonds, and these must be occupied by acceptors and donors from the active site. These acceptors and donors within the active site are often general acid–bases in their own right, and they are indistinguishable chemically from the catalytic general acid–bases.

A refined crystallographic molecular model of deoxyribonuclease I has been constructed.[242] The active site has been located by binding the calcium complex of thymidine 3′,5′-bisphosphate to the enzyme in the crystals and calculating a difference map of electron density. Surrounding the bound ligand in the vicinity of the bond cleaved during catalysis are Aspartate 248, Histidine 249, Tyrosine 208, Aspartate 165, Histidine 31, and Aspartate 39. Any of these amino acids is a likely candidate for a general acid–base involved in the mechanism.

The only structures in which the catalytic general acid–bases are revealed unambiguously are the transition states themselves, and these are inaccessible to crystallography. Often an attempt is made to overcome this difficulty by a **model-building** experiment (Figure 4–33C) in which molecular models of transition states or intermediates of high energy are inserted into the active site.[225,235] It should always be recalled that these model-building experiments are based not on direct observations but on personal intuition, and they should always be clearly distinguished, both by the author and by the reader, from direct observations. When such a model-building experiment is conducted, it is usually found that several acid–bases are attached flexibly enough to the polypeptides providing the superstructure of the active site that two or more of them can be positioned convincingly in the model to act as a necessary general acid–base.

It is, however, sometimes the case that the assignment of a catalytic general acid–base is less ambiguous. Lysozyme catalyzes the hydrolysis of an acetal (Figure 3–25). The inescapable first step in the hydrolysis of an acetal is the protonation of the oxygen of the first leaving group

$$R_1{-}O{-}C(R_3)(R_4){-}O{-}R_2 \underset{-H^+}{\overset{+H^+}{\rightleftharpoons}} R_1{-}\overset{+}{O}H{-}C(R_3)(R_4){-}O{-}R_2 \rightleftharpoons R_1{-}O{-}H + R_3R_4C{=}\overset{+}{O}{-}R_2 \qquad (4\text{–}219)$$

and in the case of the reaction catalyzed by lysozyme it must be the bridging glycosidic oxygen that is protonated by a general acid in the active site. A crystallographic molecular model of the strained substrate within the active site (Figure 3–26) was constructed from the results of model building and a crystallographic molecular model of the active site occupied by an analogue of a high-energy intermediate.[243] In this model, the glycosidic oxygen is surrounded by hydrophobic amino acids with the exception of Glutamate 35. It is assumed that this glutamic acid is the general acid responsible for protonating the oxygen. Most of the time, however, there is more than one potential acid–base surrounding the atoms involved in the chemical reaction, and other information is required to make a decision among the different possibilities.

One step in the reaction catalyzed by citrate (*si*) synthase is the formation of citryl-SCoA from acetyl-SCoA and oxaloacetate

$$^{-}OOCH_2C(=O)COO^- + H_3C{-}C(=O){-}SCoA \rightleftharpoons {}^{-}OOCH_2C(OH)(COO^-){-}CH_2{-}C(=O){-}SCoA \qquad (4\text{–}220)$$

For this condensation to occur, a proton must be removed from the acetyl-SCoA in the forward direction and added to the enol of acetyl-SCoA in the reverse direction.

Crystals of citrate (*si*) synthase in which the active sites were occupied by both citrate and coenzyme A could be obtained, and they were submitted to crystallography.[244] In the resulting map of electron density, both the citrate and the coenzyme A were readily identified as unattached masses, and a refined crystallographic molecular model was constructed. In this model three arginines and three histidines surround the citrate (Figure 4–34). The portion of the citrate that arises from acetyl-SCoA was readily identified either from the fact that the synthase proceeds with addition of the enol to the *si* face of the oxaloacetate or from the fact that the sulfur of the coenzyme A was adjacent to Aspartate 375 in the crystallographic molecular model. The other experimental observation required to identify the general base removing the proton from carbon during the condensation is that the reaction proceeds with inversion of configuration at the methyl group of acetyl-SCoA.[112,245] Therefore, the general base responsible must be on the opposite side of the methyl carbon from the *si* face of the oxaloacetate (Reaction 4–220). Only Histidine 274 satisfies this requirement. Because Histidine 320 forms a hydrogen bond with the hydroxyl on carbon 3 of citrate, it should be the general acid that protonates this oxygen during the condensation.

**Figure 4–34:** Crystallographic molecular model of the active site of citrate (*si*) synthase occupied by citrate.[244] Citrate (*si*) synthase from chicken heart was crystallized from a solution of 1.1 M sodium citrate, pH 6.0, containing 1 mM coenzyme A. The crystals were of space group *C*2 and a data set was collected to Bragg spacings of 0.20 nm. The phases of the reflections were estimated by the method of isomorphous replacement. A molecular model of the polypeptide of 437 amino acids was inserted into the map of electron density for a protomer of the dimeric protein and the resulting crystallographic molecular model was refined. At an early step in the refinement, features of electron density suitable for a bound coenzyme A and a bound citrate were located in the map of electron density of each protomer. Models of these two substrates were inserted respectively into these features, and their structures were refined as the whole protein was refined. The figure presents the model of a citrate molecule surrounded by the adjacent amino acids forming this portion of the active site. Reprinted with permission from ref 244. Copyright 1982 Academic Press.

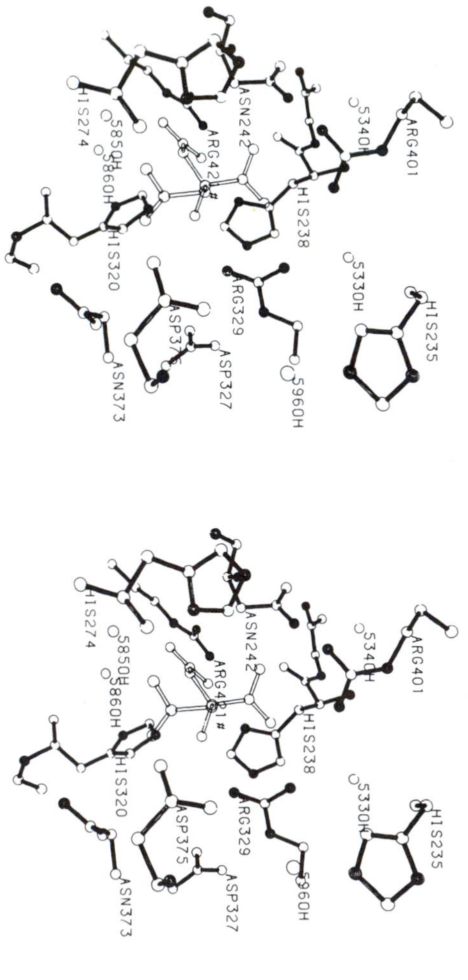

The precise description of the mechanism for the condensation reaction catalyzed by citrate (*si*) synthase, which was possible because its **stereochemistry** had already been determined, can be contrasted with the ambiguity surrounding the participants in the hydrolysis reaction. The other step in the reaction catalyzed by citrate (*si*) synthase is the hydrolysis of citryl-SCoA

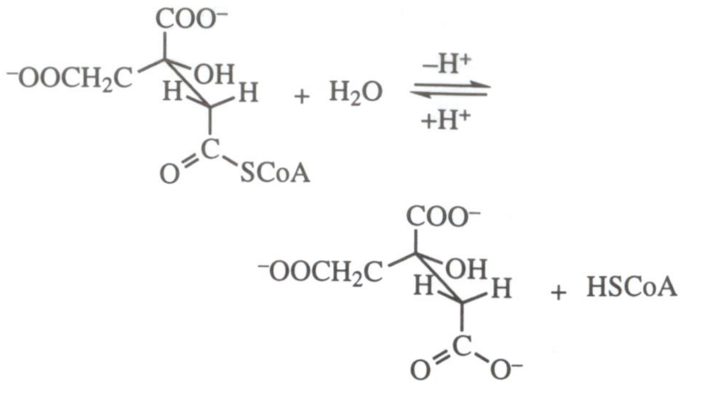

(4–221)

The difficulty in this instance was that nothing was known about this mechanism before the crystallography was performed. Histidine 274, Aspartate 375, Water 585, and Water 586 all surround the acyl carbon that is the victim of the hydrolysis, and there are too many possibilities. Therefore, with the same crystallographic molecular model at the same resolution, an exact description can be given for one step in the reaction but no description can be given for the other step. The disparity arises from the difference in the chemical information available about the two steps.

When substrates or bisubstrate analogues are occupying the active site in crystals of the enzyme, the crystallographic molecular model is more informative. For example, the crystallographic molecular model of lactate dehydrogenase occupied by (3*S*)-5-(3-carboxy-3-hydroxypropyl)NAD$^+$ has a hydrogen bond between Histidine 195 and the 3′-hydroxyl of the analogue (Figure 3–36), and it is assumed that this histidine removes the proton from the hydroxyl of lactate during its conversion to pyruvate. The most reliable and intellectually satisfactory way, however, to identify the general acids and general bases responsible for the proton transfers in an enzymatic mechanism is to obtain a crystallographic molecular model of the active site occupied by an **analogue of a high-energy intermediate**.

Ribonuclease, as one step of its usual mechanism for cleaving RNA, catalyzes the hydrolysis of nucleoside 2′,3′-cyclic phosphates such as cytidine 2′,3′-cyclic phosphate (Reaction 3–221). The high-energy intermediate in this reaction should be a pentavalent trigonal bipyramid

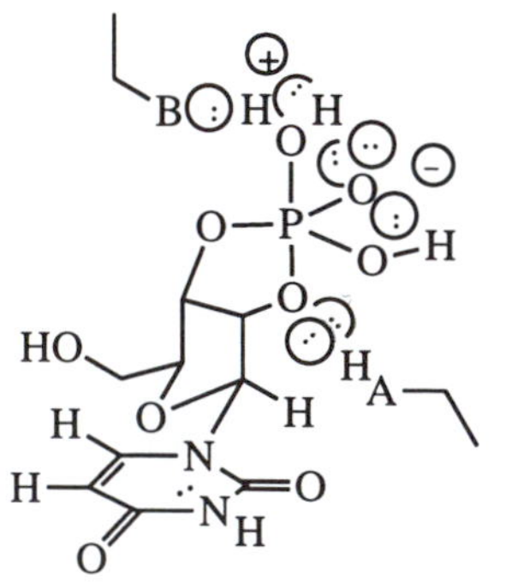

**4–40**

in which a general base is removing a proton from the attacking water at one apex and a general acid is adding a proton to the leaving oxygen at the other apex. Mainly as a result of the active-site modifications performed with iodoacetic acid (Figure 4–32), it was believed that these two amino acids would

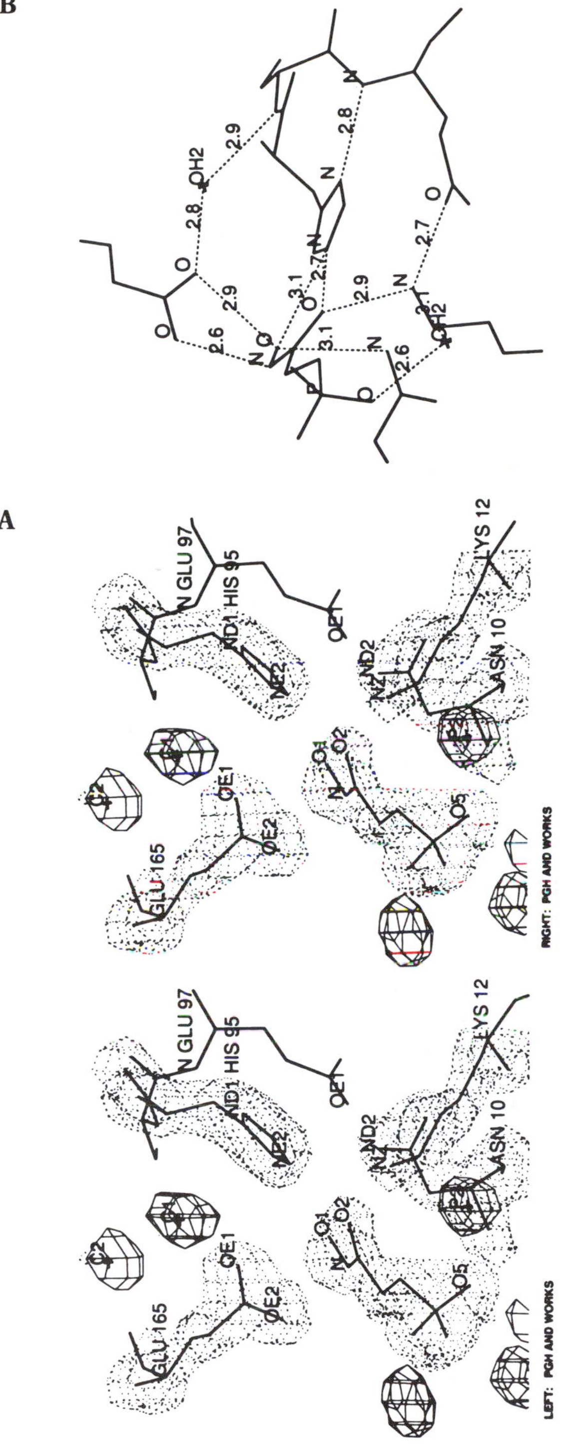


**Figure 4–35:** Crystallographic molecular model of phosphoglycolohydroxamate bound to the active site of triosephosphate isomerase.[12] Triosephosphate isomerase was crystallized from a solution of concentrated poly(ethylene glycol) in the presence of phosphoglycolohydroxamate. Reflections were collected to Bragg spacings of 0.19 nm from the resulting crystals (space group $P2_1$). A data set from an isomorphous replacement with ethyl mercury phosphate provided a map of electron density detailed enough to position an earlier crystallographic molecular model of the unliganded enzyme in the new unit cell. This preliminary molecular model was then submitted to refinement against the new data set. As the refinement progressed, a feature of electron density adjacent to Glutamate 164 increased in detail until it became possible to position a molecular model of phosphoglycolohydroxamate unambiguously within its electron density. Sharp isolated features of difference electron density could be assigned as molecules of water. (A) The figure includes the electron density of the phosphoglycolohydroxamate, and the four catalytic acid–bases, Glutamate 165, Histidine 95, Lysine 12, and Asparagine 10, as well as the skeleton of the refined crystallographic model. The difference electron density of neighboring molecules of water is also included. (B) The distances between selected atoms (in angstroms) are indicated on a skeletal drawing of the crystallographic molecular model in this region. Reprinted with permission from ref 12. Copyright 1991 American Chemical Society.

be Histidine 119 and Histidine 12. It seems to have gone unnoticed that the enzyme modified at Histidine 12 was still active. When a crystallographic molecular model of the active site occupied by an analogue of the pentavalent intermediate (Figure 3–39) became available, the general acid–base operating on oxygen 2′ was identified as Lysine 41 and the general base–acid operating on the oxygen at the opposite apex was identified as Histidine 119. Because this was a neutron diffraction study, the hydrogens on the general acid–bases could be observed directly in the maps of neutron scattering density, and they are included explicitly in the figure.

One of the neutral tautomers of phosphoglycolohydroxamate (**4–41**)[246] is an analogue of the *cis*-enediol intermediate (**4–42**) in the reaction catalyzed by triosephosphate isomerase (Reaction 4–165)

**4–41** **4–42**

and binds tightly to the active site of the enzyme. A crystallographic molecular model of this analogue bound to the active site identifies the general acid–bases involved in the mechanism of the reaction (Figure 4–35).[12]

The *cis*-enediol is formed by the removal of the *pro-R* proton from carbon 1 of dihydroxyacetone phosphate or the lone proton from carbon 2 of D-glyceraldehyde 3-phosphate (Reaction 4–133). In the crystallographic molecular model, Glutamate 165, which had already been identified by chemical modification as the base performing these proton removals,[142,247,248] is positioned on the proper side of the analogue for this role (Figure 4–35). One or the other of the oxygens of Glutamate 165 is 0.28 nm from the nitrogen of the hydroxamate and 0.34 nm from the acyl carbon of the hydroxamate.[249] As these two atoms occupy the positions of carbons 1 and 2 of the *cis*-enediol, respectively, the glutamic acid is positioned to be a competent acid in the addition of a proton to either of these carbons, as expected

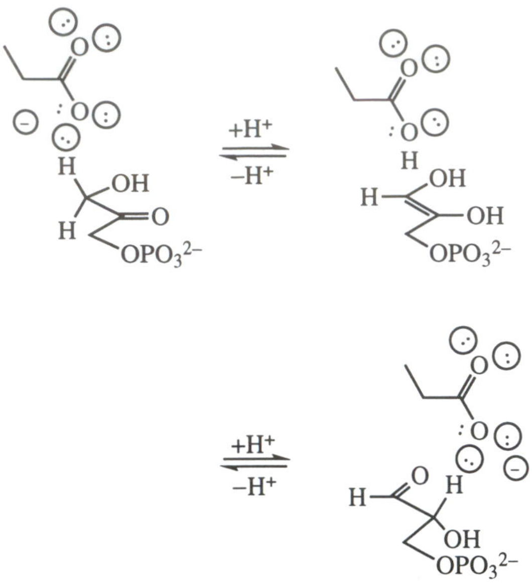

(4–222)

**4–43**

During the reaction, a proton must also be transferred between the oxygen on carbon 1 and the oxygen on carbon 2 of the substrates to complete the isomerization of dihydroxyacetone phosphate to D-glyceraldehyde 3-phosphate. It can be inferred that the sequence of steps at this side of the substrate must be the protonation of the carbonyl oxygen coincident with removal of a proton from carbon to produce the neutral enediol, followed by removal of the proton from one of the oxygens of the enediol coincident with protonation at carbon to produce the appropriate product. This conclusion follows from studies of isotopic exchange of tritium during the enzymatic reaction catalyzed by a mutant lacking Histidine 95.[250] This also makes sense chemically because it is only in the neutral enediol that the other carbon becomes as basic as the carbon from which the proton was originally removed.

It has been concluded from the results of site-directed mutation[249] that this shuttling of a proton between the two oxygens is accomplished by Histidine 95, which first protonates the incipient enediolate coincident with the removal of the proton from carbon to produce enediol **4–42** and then removes the other proton from enediol **4–42** to produce, coincident with protonation at carbon, the other transiently formed enediolate. In the crystallographic molecular model, one of the nitrogens of this histidine is 0.31 nm from one of the oxygens of the hydroxamate and 0.27 nm from the other (Figure 4–35B). This positions the histidine ideally to act as the acid–base mediating the transfer of the protons between these two oxygens.

The peculiar feature of the crystallographic molecular model is that the other nitrogen of the imidazole is the acceptor to a hydrogen bond with the amide nitrogen–hydrogen of Glutamate 97 (Figure 4–35B). This means that before the histidine protonates either incipient enediolate anion, it is the neutral imidazole

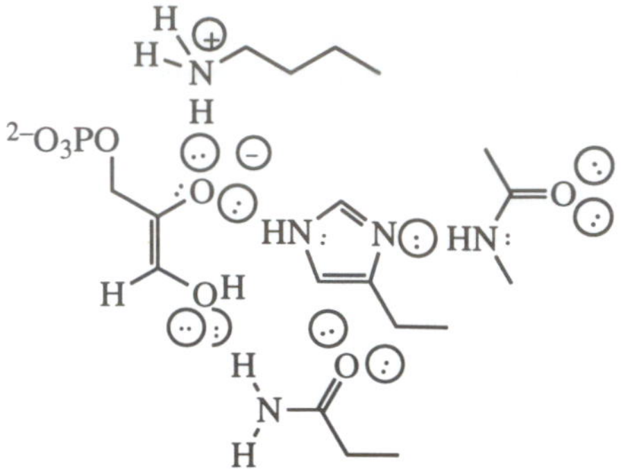

This peculiar fact has been verified by nuclear magnetic resonance spectroscopy.[13] The $pK_a$ of a neutral histidine in a polypeptide is 14[43a] and that of the conjugate acid of the enolate should be 11.[251] Therefore, the required proton transfer in Structure **4–43** from Histidine 95 should be endergonic. This is not too much of a problem in terms of rates of proton transfer, but if this is an accurate appraisal of the situation, then Histidine 95 cannot assist Glutamate 165 in forming the enediol in the first place. If Histidine 95 cannot protonate efficiently the fully formed enediolate, then it will necessarily be incapable of protonating efficiently the transition state leading to the enediolate. If it cannot protonate the transition state, it cannot assist in the removal of the proton from carbon, which is the most difficult step in the entire reaction.

Here again, it will be difficult to interpret the crystallographic structure without further chemical information. At first glance, however, it seems reasonable to conclude that Lysine 12 ($pK_a$ = 10) must protonate the enolate shown in Structure **4–43**. In the reverse reaction, the hydroxyl group on carbon 2 must shuttle a proton from Lysine 12 ($pK_a$ = 10) to the anionic conjugate base of Histidine 95 ($pK_a$ = 14) after it has protonated the enediolate on oxygen of carbon 1 to lock the proton on the enediol. Both of these proton transfers would proceed with overall neutralization of charge within the low dielectric of a closed active site, and in both instances the source of the proton would be the strongest acid in the vicinity.

Proteolytic enzymes catalyze the hydrolysis of amides (Figure 1–7). One inescapable step in the hydrolysis of an amide is the protonation of the nitrogen in the tetravalent intermediate

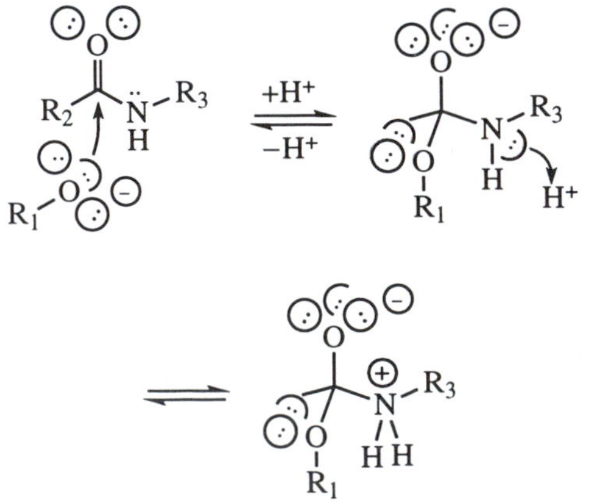

(4–223)

This requirement arises from the two facts that the lone pair of electrons on the nitrogen cannot be protonated before the tetravalent intermediate is formed because they are $\pi$ electrons and that the lone pair of electrons on the nitrogen must be protonated after the tetravalent intermediate is formed because the amide anion cannot be a leaving group. Therefore, the identification of the general acid responsible for this protonation is fundamental to understanding the mechanism of any proteolytic enzyme.

Thermolysin is a proteolytic enzyme that uses zinc as a Lewis acid to activate water

$$Zn^{2+} + OH_2 \rightleftharpoons Zn^{2+}OH_2 \rightleftharpoons Zn^{2+}OH^- + H^+ \qquad (4\text{–}224)$$

The hydroxide ion bound to $Zn^{2+}$ is the nucleophile ($R_1O^-$) that forms the tetravalent intermediate. In the crystallographic molecular model of an analogue of the tetravalent intermediate occupying the active site (Figure 3–34), one of the oxygens of Glutamate 143 is immediately adjacent to the nitrogen of the phosphoamidate[252]

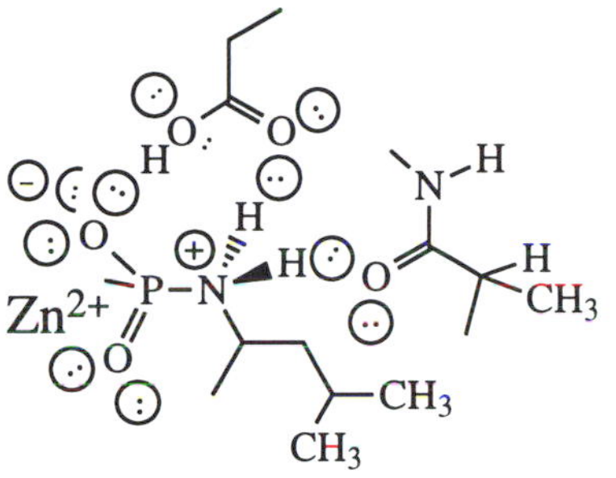

**4–44**

and this complex identifies the conjugate acid of Glutamate 143 as the general acid responsible for protonating the nitrogen of the leaving group

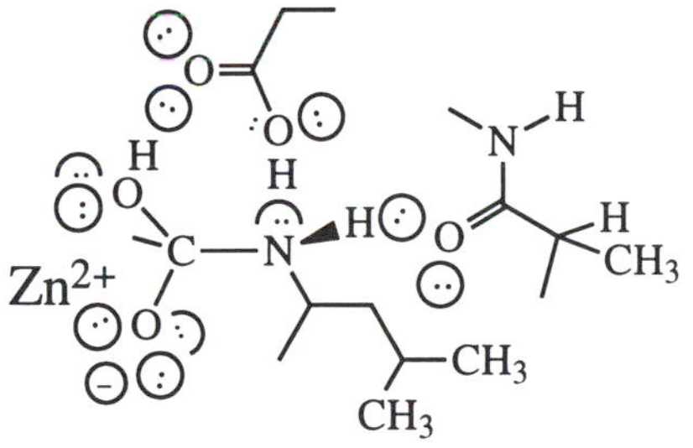

**4–45**

in the tetravalent intermediate. In this complex, the removal of the proton from the hydroxyl of the tetravalent intermediate, just delivered as a hydroxide from the zinc, could occur in concert with the donation of the proton to the nitrogen.

The proteolytic enzymes that use a serine as the alkoxide anion ($R_1O^-$) forming the acyl enzyme intermediate (Figure 3–37) also must provide a general acid to protonate the nitrogen of the leaving group. In the complex of a boronic acid with the active site (Figure 3–38), the complex of an aldehyde with the active site (Figure 3–32), or the ester of isopropyl phosphate with the serine in the active site of one or the other of these enzymes,[43] the position that would be occupied by the nitrogen of the leaving group is readily identified as the free oxygen that is not in the oxyanion hole. This oxygen is invariably forming a hydrogen bond with a histidine and this histidine in turn forms a hydrogen bond at its other nitrogen with an aspartate (Figure 3–38). In these complexes the histidine is present as the protonated cationic acid and the aspartate is present as the unprotonated anionic base.[43]

In this set of crystallographic molecular models, each of these histidines is the histidine homologous to Histidine 57 in chymotrypsin. This is the histidine whose free base also forms a hydrogen bond with the nucleophilic serine in each of the unoccupied enzymes (Figure 4–13), and this histidine must be unprotonated for the enzyme to function. It follows that this histidine removes the proton from the serine during the formation of the tetravalent intermediate, pivots, and then donates a proton to the nitrogen of the leaving group. It is the strong hydrogen bond to the anionic aspartate that acts as the pin during this pivot[43] and aligns the histidine in turn with the hydroxyl of the serine (Figure 4–13) and then with the nitrogen of the leaving group (Figure 3–38).

A very similar hydrogen bond between an aspartate and a histidine in the crystallographic molecular model of malate dehydrogenase has been noted.[253] The histidine in this couple is the homologue of Histidine 195 in the crystallographic molecular model of the active site of lactate dehydrogenase occupied by the bisubstrate analogue (Figure 3–36). Histidine 195 in this latter crystallographic molecular model also forms a hydrogen bond with an aspartate. As these histidines are the general acid–bases in the respective enzymes responsible for proton donation to or removal from the oxygen of the respective substrates, this hydrogen bond with the aspartate also may serve to orient the histidine properly to fulfill this role in the reaction. No pivoting of these histidines is required, however.

One of the more unusual constellations of general acid–bases that has been observed in a crystallographic molecular model is the two tryptophans that orient and activate the leaving group in the reaction catalyzed by haloalkane dehalogenase[254]

$$H_2O + R_2(H)(R_1)C\text{–}Cl \rightleftharpoons R_2(H)(R_1)C\text{–}OH + Cl^- + H^+ \qquad (4\text{–}225)$$

This reaction involves no coenzymes or metallic cations but proceeds by a simple $S_N2$ attack of the carboxylate anion of Aspartate 124 on the electrophilic carbon of the alkyl halide to produce the alkyl ester and chloride ion. The resulting ester is then hydrolyzed directly. In crystallographic molecular models[254] of the active site occupied respectively by reactant, intermediate, and product, the chlorine atom is always found to be pinned between the pyrrole nitrogen–hydrogen bonds of Tryptophan 125 and Tryptophan 175

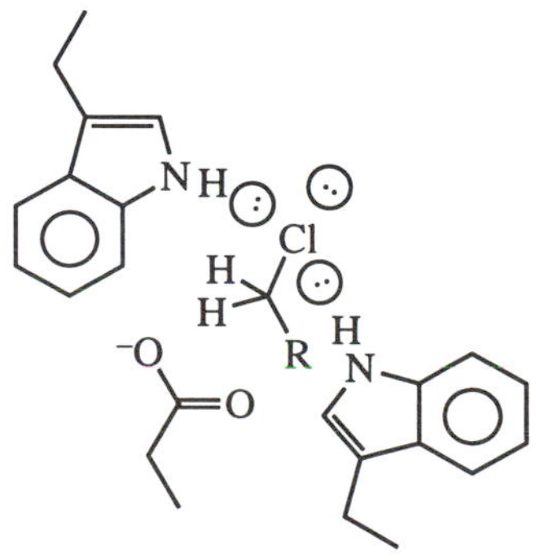

**4–46**

Although the chloride is a good leaving group, the carboxylate is a weak nucleophile, so general acid catalysis of this reaction at chlorine is appropriate.

A remarkable aspect of the general acid–bases in crystallographic molecular models of active sites is that they are often participants in **networks of hydrogen bonds**. A sufficient explanation for this observation is that such a network orients the acidic proton or lone pair on the general acid–base properly for catalysis and eliminates rotational motion that would increase the entropy of approximation and decrease the rate of the reaction. It has also been proposed, however, that such a network, because it is a constellation of general acids and general bases, could also alter the properties of the catalytic group as an acid–base. It was originally proposed, when the catalytic constellation in chymotrypsin was observed (Figure 4–13), that the buried aspartate would displace protons in the network in its direction

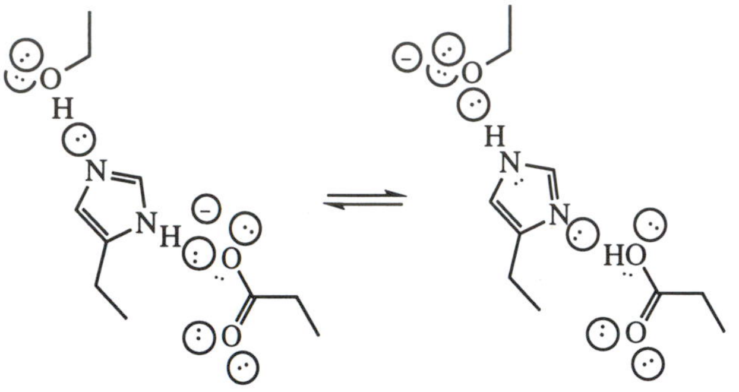

(4–226)

and render the serine more nucleophilic.[41] This proposal was weakened when it was shown by neutron diffraction that when the proximal nitrogen of the histidine is protonated, the proton on the distal nitrogen is not transferred to the carboxylate[43]

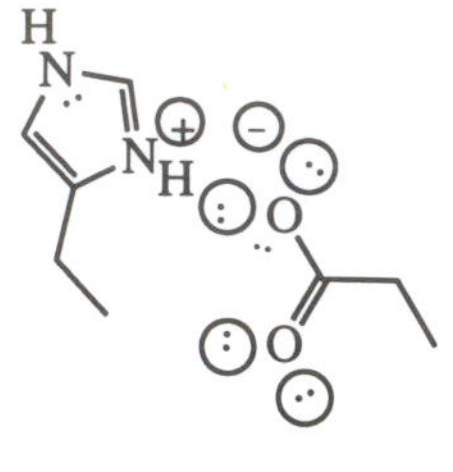

**4–47**

as is required by Reaction 4–226. Nevertheless, the possibility exists that **concerted proton transfers** within such networks of hydrogen bonds may be involved in general acid–base catalysis in the active sites of enzymes.[255]

### *Suggested Reading*

Kossiakoff, A.A., & Spencer, S.A. (1981) Direct Determination of the Protonation States of Asparitic Acid-102 and Histidine-57 in the Tetrahedral Intermediate of the Serine Proteases: Neutron Structure of Trypsin, *Biochemistry 20*, 6462–6474.

# Metalloenzymes

Aside from sodium, lithium, and rubidium ions, which bind adventitiously over the surfaces of proteins, there is a limited set of **metallic cations** that are incorporated into proteins and affect their function advantageously. These are the cations of potassium, calcium, magnesium, manganese, zinc, copper, iron, cobalt, molybdenum, vanadium, nickel, and tungsten. The nontransition metals, potassium, calcium, and magnesium, and the transition metals, manganese and zinc, occur almost exclusively in their most common oxidation states: $K^+$, $Ca^{2+}$, $Mg^{2+}$, $Mn^{2+}$, and $Zn^{2+}$, respectively. Because of the availability of two or more readily accessible oxidation levels, transition metals such as iron or copper are often used as one-electron carriers and in this role alternate between oxidation levels such as $Fe^{2+}$ and $Fe^{3+}$ or $Cu^+$ and $Cu^{2+}$, respectively. Transition metals, such as the $Ni^{2+}$ in urease or the $Fe^{2+}$ in hemoglobin, also fill roles in the active sites of enzymes in which no changes in oxidation level are required and in fact are to be avoided. The participation of all of these metallic cations in enzymatic reactions can be divided into three categories.

First, there are enzymes that **contain coenzymes**, such as hemes, chlorophylls, and coenzyme $B_{12}$, in which a metallic cation is enclosed within a cage of ligands provided by the coenzyme. In this situation, the metallic cation is tightly bound to the coenzyme, which is tightly bound to the active site. Each coenzyme is constructed with a unique and characteristic metallic cation: iron in a heme, magnesium in a chlorophyll, and cobalt in coenzyme $B_{12}$. The catalytic capacity of the metallic cation in enzymes containing one of these coenzymes is characteristic of the coenzyme as a whole.

Second, there are enzymes that **retain a metallic cation** or several metallic cations as a part of their structure even after purification from a biological source. For example, when fructose-bisphosphate aldolase is purified from yeast, the enzyme is found to contain one tightly bound $Zn^{2+}$ for each active site. If the $Zn^{2+}$ is removed from the enzyme, the enzyme loses its enzymatic activity. Fructose-bisphosphate aldolase in which the $Zn^{2+}$ has been replaced by $Ni^{2+}$, $Cu^{2+}$, or $Mg^{2+}$ is inactive (< 0.5% activity).[256] When, however, the $Zn^{2+}$ is replaced[257] by $Co^{2+}$, $Fe^{2+}$, or $Mn^{2+}$, the enzyme regains activity, but its turnover number has decreased 2-, 2.7-, or 7-fold, respectively,[256] relative to the enzyme in which $Zn^{2+}$ has been reincorporated. Even though a limited number of replacements are permitted, there can be little doubt that the metallic cation normally incorporated into fructose-bisphosphate aldolase is $Zn^{2+}$, because that cation is still present following the purification.

Third, there are many enzymes that **require a metallic cation** to be present in solution before they display any catalytic activity but are themselves devoid of any metallic cation upon purification. In these situations, the necessary metallic cation is either bound to the enzyme less tightly and dissociates as the enzyme is purified or the procedures for purifying the enzyme are so harsh that they promote the dissociation of an otherwise tightly bound metallic cation. In the cytoplasm under normal circumstances, however, the site at which the metallic cation binds is occupied because the concentration of metallic cations in the cytoplasm is high enough to saturate it. When phosphoglucomutase is purified from muscle, it contains no metallic cations but displays an absolute requirement for one (activity in the absence of metallic cation is <0.001% that in the presence of $Mg^{2+}$).[258] The requirement for a metallic cation can be fulfilled by $Mg^{2+}$, $Ni^{2+}$, $Co^{2+}$, $Mn^{2+}$, or $Zn^{2+}$. Although the relative capacity of each metallic cation to reactivate the enzyme varies with temperature and pH,[259] $Mg^{2+}$, $Ni^{2+}$, and $Co^{2+}$ are almost equally effective at restoring a high turnover number, while the turnover numbers of the enzyme reconstituted with $Mn^{2+}$ and $Zn^{2+}$ are much less.

In the cytoplasm, $Mg^{2+}$ is probably the metallic cation bound to the active site of phosphoglucomutase because it is more abundant than the others, but this is not certain because competition among several cations for the active site may occur. The issue is further complicated by the fact that it is only the free concentration of the various divalent cations that determines the outcome of such competition. Because divalent metallic cations are usually complexed tightly to proteins, nucleic acids, carbohydrates, and metabolites dissolved in the cytoplasm, the free concentration of any given divalent metallic cation is only a small and unpredictable fraction of its total concentration, which is usually the only known value, and its equilibrium distribution among the numerous binding site in the cytoplasm is also unknown.

When phospho*enol*pyruvate carboxykinase (GTP) is purified from rat liver, it displays a requirement for two different metallic cations for full activity.[260] In the absence of any metallic cation, the enzyme is inactive (<0.4% of maximum activity) in the carboxylation of phospho*enol*pyruvate. Addition of $Mn^{2+}$, $Co^{2+}$, or $Zn^{2+}$ restores some enzymatic activity (3% of maximum activity), but when any one of these metallic cations is added along with $Mg^{2+}$, enzymatic activity is far greater (15–100% of the maximum activity). The addition of $Mg^{2+}$ alone, however, cannot support any enzymatic activity. It can be concluded that there are two sites in the active site that must be occupied by a divalent cation before the enzyme can function properly. When both sites are occupied by $Mn^{2+}$, $Zn^{2+}$, or $Co^{2+}$ the enzyme catalyzes the reaction, but poorly. When one of the sites is occupied by $Mg^{2+}$ and the other by $Mn^{2+}$, $Zn^{2+}$, or $Co^{2+}$, but not by $Mg^{2+}$, the enzyme catalyzes the reaction efficiently. Again, it can be presumed that in the cytoplasm, the first site is always occupied by $Mg^{2+}$, because it is abundant and produces the highest enzymatic activity. It is less clear, however, which metallic cation is normally situated at the other site when the enzyme is in the cell, even though $Co^{2+}$ and $Mn^{2+}$ at saturation produce greater enzymatic activity than does $Zn^{2+}$.

In some instances it is possible to identify the metallic cation involved in the normal enzymatic reaction as it occurs in the cell. Treatment of a crude homogenate of yeast with chelating agents renders the arginase within it inactive, but the addition of $Mn^{2+}$, $Co^{2+}$, $Fe^{2+}$, or $Ni^{2+}$ can restore the enzymatic activity. Only the enzymatic activity that is restored in the presence of $Fe^{2+}$, however, has the same pH–rate behavior, inhibition by $HOPO_3^{2-}$, and Michaelis constant for ornithine as the enzymatic activity in fresh homogenates,[261] and it can be concluded that in the cytoplasm of the yeast, arginase normally has $Fe^{2+}$ within its active site.

Eventually, the role of a metallic cation in **maintaining the structure** of an enzyme must be distinguished from its role as a **catalytic functional group**. In the crystallographic molecular model of aspartate transcarbamylase, there is an atom of zinc tetrahedrally coordinated to the four sulfurs of Cysteines 109, 114, 137, and 140 in each regulatory subunit.[261a,262] This atom of zinc forms a tetrahedral, covalent complex with the structure

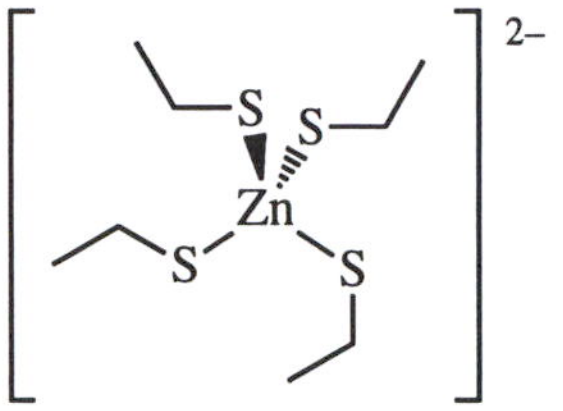

**4–48**

that resembles structures observed in polynuclear clusters that form between $Zn^{2+}$ and small mercaptans. When the zinc is displaced from the thiols by organic mercurials, the enzyme separates into catalytic trimers and regulatory dimers that can be reassociated[263,264] by reincorporating $Zn^{2+}$. Other naturally occurring divalent metal cations such as $Co^{2+}$, $Cu^{2+}$, and $Ni^{2+}$ also promote reassociation and become incorporated in the reconstituted enzyme, but only $Zn^{2+}$ is found in the enzyme when it is purified from the bacteria.[264] In the crystallographic molecular model, the $Zn^{2+}$ is located adjacent to the boundary between the regulatory subunit and its neighboring catalytic subunit but distant from both the active site of the enzyme and the site for binding nucleotides. Therefore, the role of the $Zn^{2+}$ is entirely structural. Its complexation with the thiols creates and stabilizes the proper structure at the surface of a regulatory subunit. This stable structure associates with a complementary structure on the surface of a catalytic subunit. A metallic cation can fulfill such a structural role because the complexes it forms either covalently or ionically with lone pairs of electrons on bases within the protein are quite stable, especially if the protein itself orients the bases advantageously.

In the case of aspartate transcarbamylase, removal of the $Zn^{2+}$ from the enzyme produces catalytic subunits with full enzymatic activity. In most instances, however, the removal of a metallic cation from a protein leads to loss of enzymatic activity, and separating the effect on the structure of the protein, which itself is responsible for the enzymatic activity, from a direct effect at the active site, in which the metal itself can be a catalytic group, is difficult. For example, when mammalian liver arginase, an $\alpha_4$ tetramer, is treated with the chelating agent ethylenediamine tetraacetic acid (EDTA), it loses all of its enzymatic activity.[265] At the same time, however, it dissociates into $\alpha$ monomers. When $Mn^{2+}$ is added to the inactive $\alpha$ monomers, enzymatic activity is regained, and the $\alpha_4$ tetramers re-form. It is possible that the dissociation of the $\alpha_4$ tetramer is responsible for the inactivation and that the $Mn^{2+}$ required for activity is necessary to retain the proper structure of the protein rather than as a catalytic group in the active site. In many instances, however, it has been demonstrated unequivocally that a metallic cation is one of the catalytic groups in the active site of a particular enzyme. If it is acting as a bound cation rather than as a component of a larger coenzyme, the metallic cation usually functions in the enzymatic reaction as a **Lewis acid**. Because a proton is also a Lewis acid, a metallic cation often fulfills a role in an enzymatic mechanism that would otherwise be fulfilled by a proton donated by a general acid.

The metallic cations incorporated directly into proteins in aqueous solution, because they are themselves Lewis acids, are at all times surrounded by Lewis bases, and the strongest Lewis bases present in biological fluids are the lone pairs of electrons on oxygens, nitrogens, and sulfurs. The proton is also a Lewis acid, and in biological fluids every acidic proton is always surrounded by lone pairs of electrons on oxygens, nitrogens, or sulfurs. The proton is so small, however, that it can accommodate directly only two Lewis bases at a time in one hydrogen bond. Because metallic cations have core electrons, they are larger than a proton and can accommodate more Lewis bases simultaneously. The metallic cations incorporated into proteins are always surrounded by 4–8 lone pairs of electrons from nitrogens, oxygens, sulfurs, or halides.

The similarity between a proton and a metallic cation, each acting as a Lewis acid, can be quantified. It has already been noted that $H_3O^+$ is a shorthand for the cation $[H(OH_2)_m]^+$, where $m$ is variously depicted as having a value of 4 or a value of 21. The values for $m$ in the equivalent complexes of water and a metallic cation, $[Me(OH_2)_m]^{n+}$, are between 4 and 8. As a result, $H_3O^+$ and $[Me(OH_2)_m]^{n+}$ are analogous Brønsted acids, and their $pK_a$ values can be compared. The hydronium cation, $[H(OH_2)_m]^+$, as a Brønsted acid displays a $pK_a$ for the reaction

$$[H(OH_2)_m]^+ \rightleftharpoons H^+ + [H(OH_2)_{m-1}(OH)] \qquad (4\text{–}227)$$

of −1.75. Metallic cations surrounded by 4–8 molecules of water in dilute aqueous solution display Brønsted $pK_a$ values for the reaction

$$[Me(OH_2)_m]^{n+} \rightleftharpoons H^+ + [Me(OH_2)_{m-1}(OH)]^{(n-1)+} \qquad (4\text{–}228)$$

These $pK_a$ values have been determined for the various metallic cations incorporated into metalloenzymes (Table 4–3). From these $pK_a$ values it can be concluded that none of these metallic cations, even a trivalent one, is so effective as the proton at lowering the $pK_a$ of the cluster of water to which it is bound. It follows that a proton is a stronger Lewis acid than a metallic cation. The difference is that at pH 7 the molar concentration of hydronium ion is very small, while the molar concentration of the complex between water and a metallic cation will be equal to the free molar concentration of that metallic cation.

**Table 4–3: Acid Dissociation Constants of Aquo Complexes of Metallic Cations[266]**

| cation | $pK_a$ | cation | $pK_a$ |
|---|---|---|---|
| $Ca^{2+}$ | 12.7 | $Zn^{2+}$ | 9.6 |
| $Mg^{2+}$ | 11.4 | $Ni^{2+}$ | 9.4 |
| $Mn^{2+}$ | 10.7 | $Cu^{2+}$ | 7.5 |
| $Fe^{2+}$ | 10.1 | $V^{3+}$ | 2.9 |
| $Co^{2+}$ | 9.6 | $Fe^{3+}$ | 2.2 |
| | | $Co^{3+}$ | 0.7 |

The preferences of a metal ion for a particular type of lone pair of electrons is usually discussed in terms of the **hardness** or **softness** of the Lewis acid and the Lewis base.[267] The rule is that hard acids prefer hard bases and soft acids prefer soft bases. Of the divalent metal ions of importance in enzymology, the series of hardness is $Mg^{2+} > Ca^{2+} > Mn^{2+} > Fe^{2+} > Co^{2+} > Ni^{2+} > Cu^{2+}$, $Zn^{2+}$. For the commonly encountered bases, lone pairs on oxygen are harder than lone pairs on chlorine, which are harder than lone pairs on nitrogen, which are harder than lone pairs on sulfur. These rankings, for example, are consistent with the fact that calcium ion has a strong preference for oxygen ligands while zinc cation has a preference for thiol ligands. It also explains why the bases on metallothionein, a protein responsible for chelating soft, toxic heavy metal cations, are entirely the thiols of cysteine residues in the protein.

The Lewis bases surrounding a metallic cation in solution are held by interactions the characteristics of which span the spectrum between ionic and covalent. **Ionic interactions** are created by the electrostatic forces between the metallic cation and an anion or a dipole on a ligand. If the forces are entirely ionic, the number and orientation of the Lewis bases around the cation are determined solely by steric considerations. The larger the cation or the smaller the bases, the more bases will be gathered. The interactions between hard cationic acids and hard bases are usually ionic. The calcium dication is an example of a hard, purely ionic metallic cation. In biological solutions, $Ca^{2+}$ is surrounded by 6–9 lone pairs of electrons, invariably from oxygen atoms,[268] the hardest of bases. The number and orientation of these lone pairs around the calcium depends entirely on the size and shape of the bases that provide them.[269]

The zinc dication in the crystallographic molecular model of aspartate transcarbamylase (**4–48**) is a good example of a metallic cation participating in **covalent** bonds. In this arrangement, a soft metallic cation is bonding covalently to four soft bases. The interactions between such soft cationic Lewis acids and soft bases are usually covalent. Because zinc is a $d_{10}$ transition metal, its $2d$ shell is filled. As a result, the $Zn^{2+}$–sulfur covalent bonds are formed from $sp^3$ hybrid orbitals, and this produces the usual tetrahedral arrangement.

The major structural difference between ionic interactions and covalent bonds is the directional properties of the arrangements of ligands. Covalent bonds usually position the participating atoms in strict geometric orientations, such as the tetrahedron in the case of the $Zn^{2+}$–sulfur bonding, while ionic interactions are completely malleable, resembling pigs at a trough. The fact that covalent bonds involving metals are so reliable is reflected in the practice during crystallographic refinement of considering them as fixed geometrically as the bonds involving carbon, nitrogen, and oxygen. For example, in the initial crystallographic molecular model of aspartate transcarbamylase built directly from the unrefined map of electron density, the arrangement of the sulfurs around the $Zn^{2+}$ and the four sulfurs was restrained to a tetrahedral geometry, as an $sp^3$ carbon would have been in all subsequent refinements.[270] This can be dangerous, however, particularly if the ligands to the $Zn^{2+}$ are harder, less covalent bases.[271] In such intermediate cases, various mixtures of ionic and covalent behavior are observed. The main indication of such deviations from covalent behavior is the loss of directional ligation.

An important feature of the ability of a metallic cation to act as a Lewis acid in an enzymatic reaction is the rate at which the bases surrounding the cation can be exchanged (Table 4–4). Most rates of exchange are rapid (>1 $\mu s^{-1}$) and could be easily accommodated during an enzymatic reaction, but several are so slow that a complex between substrate and metallic cation would be inexchangeable during the lifetime of the intermediates in an enzymatic reaction. This is particularly true of complexes centered on trivalent cations.

**Table 4–4: Rates of Exchange of Water Bound in Inner-Sphere Complex at Metallic Cations at 25 °C**[272]

| cation | rate constant[a] ($\mu s^{-1}$) | cation | rate constant ($\mu s^{-1}$) |
|---|---|---|---|
| $K^+$ | 4000 | $Mg^{2+}$ | 0.5 |
| $Cu^{2+}$ | 5000 | $Ni^{2+}$ | 0.03 |
| $Ca^{2+}$ | 200 | $V^{2+}$ | 0.0001 |
| $Mn^{2+}$ | 25 | $V^{3+}$ | 0.001 |
| $Zn^{2+}$ | 20 | $Fe^{3+}$ | 0.0002 |
| $Fe^{2+}$ | 3 | $Co^{3+}$ | $1 \times 10^{-7}$ |
| $Co^{2+}$ | 2 | $Cr^{3+}$ | $7 \times 10^{-13}$ |

[a]For the reaction $[Me(OH_2)_m]^{n+} + H_2O^* \rightleftharpoons [Me(OH_2)_{m-1}(^*OH_2)]^{n+} + H_2O$.

In complexes that exchange their ligands slowly, the oxidation–reduction in which a metallic cation participates can be distinguished as **inner-sphere reactions**, in which only directly coordinated ligands can participate so that ligand exchange is required for the reaction to occur, or as **outer-sphere reactions**, in which the directly coordinated ligands remain unchanged and the reactants exchange electrons at a distance. For example, the oxidation–reduction of the iron in cytochrome *c* is an outer-sphere reaction because the heme and the amino acids chelating the iron, none of which is the ultimate oxidant or reductant, remain associated during the reaction.

In reactions in which a metallic cation participates as a Lewis acid, however, the base on the reactant is usually bound in direct coordination to the metallic cation. For example, it has been shown[273] that divalent metallic cations such as $Zn^{2+}$ or $Cu^{2+}$ can increase the rate constant for the decarboxylation of 3-keto acids by factors of at least 100 if the 3-keto acid is designed to promote the formation of a complex with the cation and if the complex coordinates the cation directly to the oxygen of the developing enolate

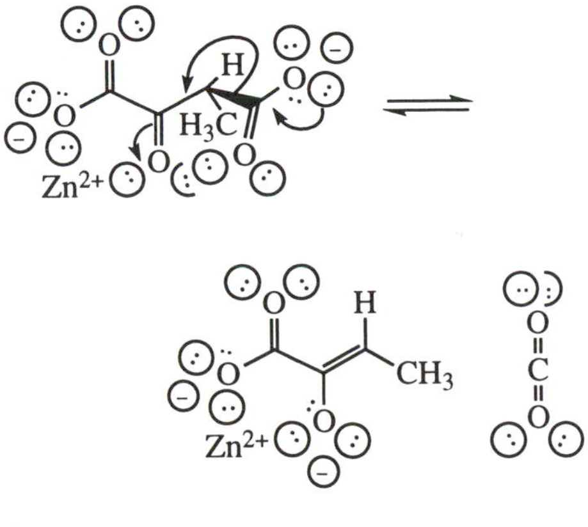

(4–229)

In this example the $Zn^{2+}$ serves the role of a proton by forming the equivalent of an enol. An example of an outer-sphere reaction in which a metallic cation participates as an acid would be a reaction in which a water molecule bound to the metallic cation would act as a general acid by virtue of its lowered $pK_a$ (Table 4–3).

**Potassium** cation is the most abundant cation in the cytoplasm of all normal cells. Because its presence is as reliable as that of water, it is not surprising that many enzymatic reactions are inhibited upon its removal from the solution.[274] Potassium has a large enough ionic radius (0.13 nm) to accommodate 4–8 lone pairs from oxygens or nitrogens as Lewis bases; and, because $K^+$ is a hard cation, the association is entirely ionic. The enzyme in which its role has been examined in most detail is pyruvate kinase. This enzyme has an absolute requirement for a monovalent cation to catalyze its reaction.[275] Although ammonium and rubidium cations can fulfill this role as well, under normal circumstances potassium is the required cation. In steady-state kinetic measurements of the reaction catalyzed by pyruvate kinase, potassium cation behaves as if it were one of the reactants in a multireactant mechanism.[276]

There are calculations from nuclear magnetic resonance studies[277] of the distance between the potassium ion and a tightly bound divalent cation in the active site of the enzyme, and there are kinetic studies suggesting that the potassium cation associates with one of the lone pairs on the carboxylate of phospho*enol*pyruvate.[278] On the basis of these observations, a feature in the map of electron density of the complex between phospho*enol*pyruvate and the enzyme has been assigned to the potassium cation.[279] Although the details are unclear, the potassium is surrounded by one of the oxygens of the carboxylate of phospho*enol*pyruvate and oxygens from the side chains of Threonine 327, Glutamine 328, Serine 361, and Glutamate 363. It is too distant from the location of the bonds being made and broken to act as a general Lewis acid in the reaction, and presumably it participates only as a point of attachment for the carboxylate of the reactant. This type of an association would be analogous to a hydrogen bond; and, as water would occupy the vacant cation in the stead of the oxygen of the reactant, all of the considerations relevant to the strength of ionic interactions and hydrogen bonds in aqueous solution would also apply to this association.

**Calcium** cation, another hard Lewis acid, also participates in purely ionic interactions with no particular geometric requirements. Its small ionic radius (0.11 nm) is more than compensated by its increased charge, and it associates as a Lewis acid with as many as 9 lone pairs on Lewis bases. Calcium has a low affinity for nitrogen,[268] and it can be distinguished from $Mg^{2+}$ by this characteristic. When bound to proteins, $Ca^{2+}$ often serves in a structural role by gathering around itself oxygens from the polypeptide, its side chains, and molecules of water (Figure 4–36).[280] These associations are highly specific, but the specificity is provided by the distribution of the oxygens within the protein and the donors and acceptors of hydrogen bonds between the protein and associated molecules of water, not by the calcium.

Calcium ion occasionally[281,282] plays a catalytic role as a general Lewis acid in enzymatic reactions. Its rate of ligand exchange is fast (Table 4–4), but its acidity is weak (Table 4–3). Its inability to form covalent interactions and thereby distort substrates may also be disadvantageous. In two instances in

**Figure 4–36:** Crystallographic molecular model of the binding site for $Ca^{2+}$ in bovine trypsin.[280] Crystals of bovine trypsin were grown in concentrated ammonium sulfate containing 10 mg $mL^{-1}$ benzamidine, a strong competitive inhibitor of the enzyme, and 10 mM $CaCl_2$. Reflections were collected to Bragg spacings of 0.18 nm. By the use of a Patterson search procedure, an earlier crystallographic molecular model of trypsin was positioned properly in the new unit cell. This preliminary crystallographic molecular model was refined against the new data set to a final *R*-factor of 0.23. During the refinement procedure, a sharp feature of difference electron density equivalent in magnitude to a $Ca^{2+}$ cation became apparent, and adding an atom of calcium to the molecular model at this location eliminated this feature. Water molecules were also added at various stages of the refinement. The six ligands surrounding the $Ca^{2+}$ cation in the final refined crystallographic molecular model are presented in the figure as well as some of the neighboring protein. Two of the ligands are fixed molecules of water. Reprinted with permission from ref 280. Copyright 1975 Academic Press.

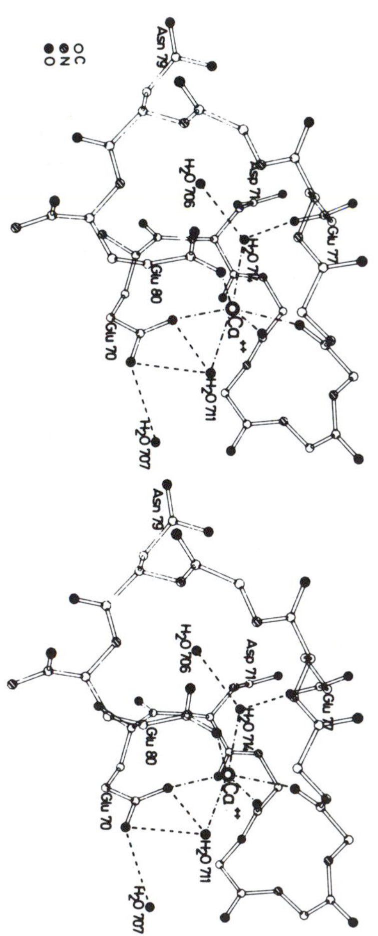


which it acts catalytically as a general Lewis acid, the base on the substrate bound to the $Ca^{2+}$ is the oxygen of a phosphate, the $Ca^{2+}$ forms only this one contact with the substrate, and its role appears to be the stabilization of negative charge that develops on the oxygen as the reaction progresses.

**Magnesium** cation, although also a hard, alkaline earth metallic cation, displays directional bonding reminiscent of a covalent-coordinate metallic cation not because of covalence but because of steric effects associated with its smaller ionic radius. Its preferred coordination is with 6 basic lone pairs in an octahedral arrangement, and the interatomic distances between the metal and the adjacent atoms are shorter and much less variable (0.20–0.21 nm) than those for calcium (0.23–0.26 nm). Its acidity, however, is almost as weak as that of calcium (Table 4–3). The predominant role of magnesium in enzymatic reactions is to form complexes with phosphates. An example of such a complex is that between $Mg^{2+}$ and two molecules of $HATP^{3-}$ (Figure 4–37).[283] In this particular complex, $Mg^{2+}$ is coordinated octahedrally to three oxygens from each of the three phosphates on two molecules of ATP. In the predominant form[284] of magnesium adenosine triphosphate present in solution in the cytoplasm, $(MgATP)^{2-}$, three of the coordination sites on $Mg^{2+}$ would be occupied by the oxygens of the phosphates and the other three by molecules of water. When the complex associates with the active site of an enzyme, the three sites occupied by water can become occupied by Lewis bases from the active site. One advantage of using magnesium rather than a proton as the Lewis acid in reactions involving ATP is that more of the oxygens are neutralized. At pH 7 the predominant forms of free ATP are the unprotonated form, $ATP^{4-}$, and the monoprotonated form, $HATP^{3-}$, in about equal proportions.[284] In the predominant form of the magnesium complex, $(MgATP)^{2-}$, however, two of the oxygens are fully neutralized.

Although $Mg^{2+}$ is a much weaker Lewis acid than the proton (Table 4–3), the effect of its complexation is similar to the enhancement of the ability of a phosphate to act as a leaving group that would result from protonation.[285] For example, at high concentration (0.1 M), $Mg^{2+}$ enhances the solvolysis of geranyl diphosphate, which proceeds through an $S_N1$ mechanism involving formation of an allyl carbocation

$$\mathrm{R(H_3C)C{=}CH{-}CH_2{-}O{-}P(=O)(O^-){-}O{-}P(=O)(O^-){-}OH} \rightleftharpoons \mathrm{R(CH_3)C^{\oplus}{-}CH{=}CH_2} + \mathrm{{}^-O{-}P(=O)(O^-){-}O{-}P(=O)(O^-){-}OH} \qquad (4\text{–}230)$$

From a consideration of the dissociation constant for the complex between geranyl diphosphate and $Mg^{2+}$, it could be concluded that the form of geranyl diphosphate showing enhanced reactivity had two $Mg^{2+}$ bound as Lewis acids to the departing pyrophosphate. Dimethylallyl *trans*-transferase is an enzyme that catalyzes, as the first step in its overall reaction, the formation of the same allyl cation. The enzyme also requires $Mg^{2+}$ as an essential reactant. After geranyl diphosphate has bound as a reactant at the active site of the enzyme, two $Mg^{2+}$ are then bound to produce the productive complex.[286] It has been concluded that they act as Lewis acids, directly coordinated to the oxygens as in the nonenzymatic reaction, to enhance the capacity of pyrophosphate to act as a leaving group.

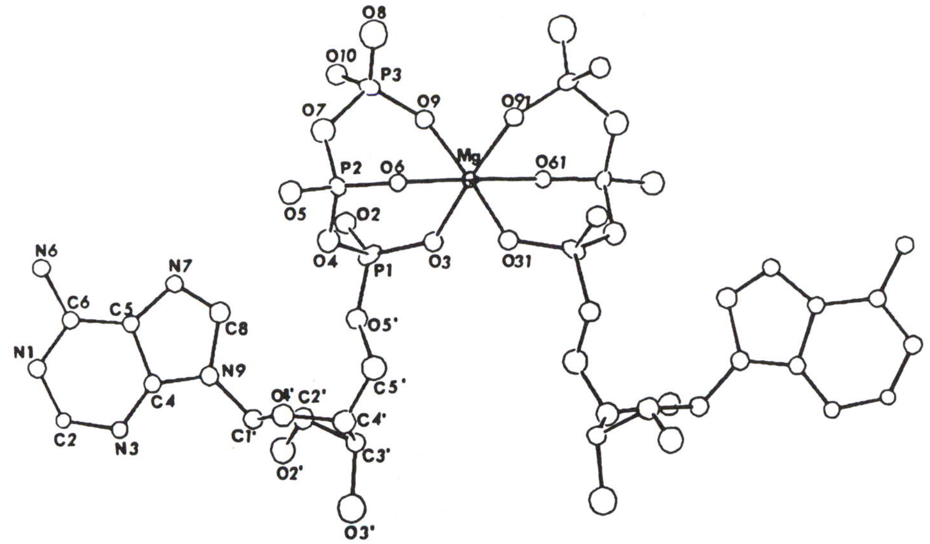

**Figure 4–37:** Crystallographic molecular model of the complex between the triphosphates of two molecules of $HATP^{3-}$ and a $Mg^{2+}$ cation.[283] Crystals of (2,2′-dipyridylammonium)$_2Mg_2(HATP)_2$ were obtained by mixing ATP, 2,2′-dipyridylamine, and $Mg(NO_3)_2$ in 30% ethanol. One of the two magnesiums in the unit cell was chelated octahedrally by six oxygens, three from each of the triphosphates of two molecules of $HATP^{3-}$. The other counterions in the unit cell have been omitted to emphasize the structure around the central magnesium cation. Reprinted with permission from ref 283. Copyright 1983 Adenine Press.

Pyruvate kinase also has two magnesium cations bound to the active site when it is occupied by the reactants in the productive complex. The unliganded enzyme binds one $Mg^{2+}$ at the active site, and the second $Mg^{2+}$ arrives complexed with ATP or ADP.[287] This was shown by using, in place of $(MgATP)^{2-}$, a $Cr^{3+}$ complex of ATP, which is a stable covalent complex almost completely inert to exchange (Table 4–4). The enzyme transferred the $\gamma$-phosphate from $CrATP^-$ to an analogue of *enol*pyruvate but only when $Mg^{2+}$ was also present to occupy the site on the free enzyme for the second, uncomplexed divalent cation. Both magnesium cations involved in the normal reaction, the one bound to the free enzyme and the one brought in by the ATP, have been located in the maps of electron density derived from crystals of the enzyme in which the active site is occupied by substrates.[279] The $Mg^{2+}$ that binds to the free enzyme forms a complex with the *enol* oxygen and one of the other oxygens on the phosphate monoester of phospho*enol*pyruvate and acts as a Lewis acid to activate it for transfer to the oxygen of ADP

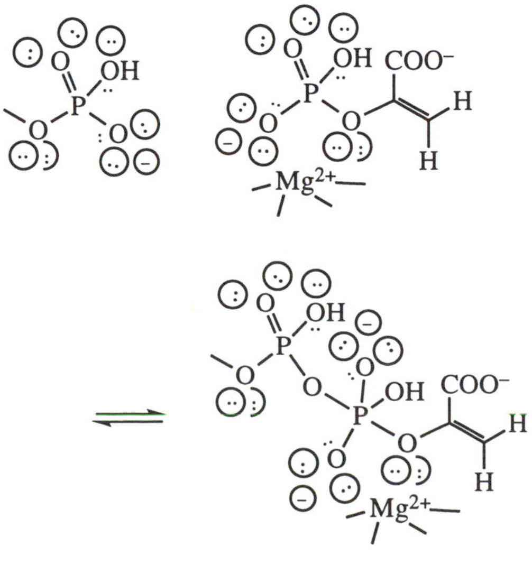

(4–231)

The $Mg^{2+}$ entering the active site bound to ADP or ATP is chelated similarly to the $Mg^{2+}$ in $MgHATP^{2-}$ (Figure 4–37) and probably also acts as a Lewis acid to catalyze the phosphate transfer.

The two oxygens on the phosphate of phospho*enol*pyruvate are directly coordinated to the magnesium ion (Equation 4–231) to form the complex at the active site (P to Mg distance of 0.3 nm).[279] Therefore, as the phospho*enol*pyruvate binds to the enzyme, it must displace two molecules of water from the coordination shell of the already bound magnesium cation. This must not be a very rapid process (Table 4–4) and may become rate-limiting under certain circumstances. An apparent $pK_a$ of 9.1 in the kinetic behavior of pyruvate kinase has been assigned to the dissociation of a proton from a molecule of water bound to the $Mg^{2+}$ in the free enzyme.[288] It has been proposed that when this proton is lost to produce a hydroxide at the magnesium, the exchange necessary to attach one or two of the oxygens on the phospho*enol*pyruvate directly to the bound magnesium is prevented. These considerations emphasize that gathering substrates around a metal ion in an inner sphere complex may not be so rapid a process as gathering them by forming hydrogen bonds to amino acids. Magnesium, however, is one of the slowest divalent cations to exchange its ligands, and this may be why it is seldom used as a Lewis acid other than with phosphates. One of the rare exceptions, however, is isocitrate dehydrogenase,[62] in which $Mg^{2+}$ may perform a role as a Lewis acid that is not coordinated to phosphate.

It appears from crystallographic and spectral[289] observations that $Mg^{2+}$, when bound to the active site of an enzyme, prefers oxygen ligands exclusively, in particular the oxygens of carboxylates. The softer metallic dication of **manganese**, $Mn^{2+}$, however, forms complexes with both nitrogen and oxygen bases such as ammonia, imidazole, 1,2-diaminoethane, water, alcohols, carboxylates, the carbonyl oxygens of ketones and aldehydes, and the acyl oxygens of amides. Consistent with its degree of hardness, oxygen bases and nitrogen bases are roughly equivalent in their affinities for $Mn^{2+}$. In aqueous solution, the hexaammonium complex is observed only at concentrations of ammonia greater than 2 M; hexaimidazole salts can be crystallized from anhydrous ethanol. All of the complexes between such unhindered hard bases and $Mn^{2+}$ are hexacoordinate and octahedral, reminiscent of directional covalent bonds; and, in these complexes, mixtures of various ligands around manganese can occur. For example, each of the species $[Mn(OH_2)_n(NH_3)_{6-n}]^{2+}$ with $0 < n < 6$ is observed in mixtures of ammonia and water. When $Mn^{2+}$ is engaged catalytically in an enzymatic reaction, it is complexed octahedrally by Lewis bases both from amino acids in the active site and from substrates.

Aside from its role as an oxidant–reductant in a few processes such as the oxidation of water during photosynthesis, manganese often substitutes for magnesium as a Lewis acid in enzymatic reactions, and it is sometimes difficult to tell which is the metallic cation used under natural circumstances. Manganese usually activates at lower concentrations than magnesium, but in the cytoplasm its total concentration is lower than that of magnesium. As can magnesium, manganese can form a complex with one of the nominal reactants to produce the actual reactant entering the active site. For example, the requirement for $Mn^{2+}$ in the reaction catalyzed by isocitrate dehydrogenase ($NAD^+$) results from the fact that the actual substrate is the neutral complex between $Mn^{2+}$ and the isocitrate dianion,[290] and the $Mn^{2+}$ enters the active site within this complex just as $Mg^{2+}$ often enters active sites in a complex with $ATP^{4-}$. If one of the ligands to $Mn^{2+}$ when the complex enters the active site is the alcoholic oxygen on isocitrate, the $Mn^{2+}$ as a general Lewis acid (Table 4–3) would promote the removal of the proton during the oxidation of the isocitrate and promote the formation of the enol during the subsequent decarboxylation

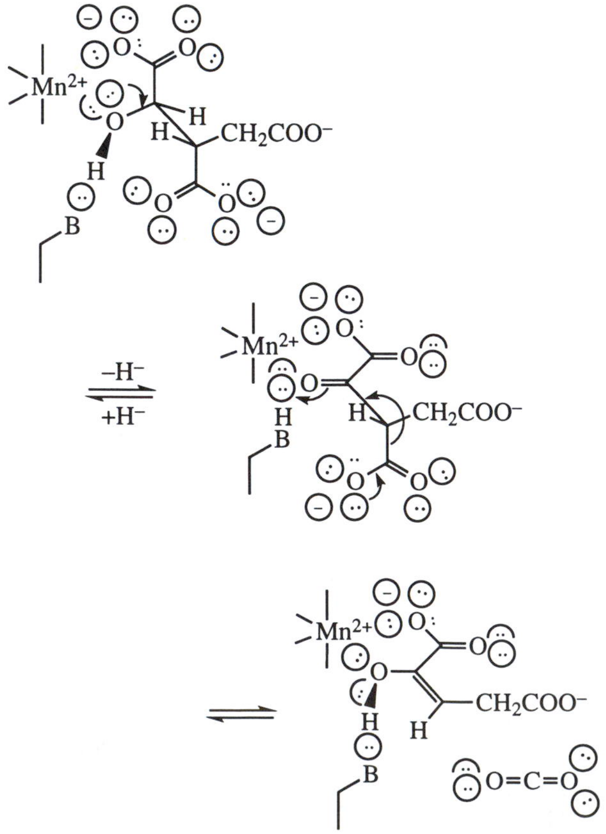

(4–232)

The second step in this reaction would resemble the nonenzymatic catalysis of the decarboxylation of 3-keto acids by divalent metallic cations (Reaction 4–229).[273]

The most versatile metallic cation functioning as a Lewis acid in the active sites of metalloenzymes is **zinc**. Its versatility in this role arises from its ability to form both tetracoordinate and pentacoordinate complexes with Lewis bases and its ability, even though it is a softer metallic cation, to form complexes with lone pairs from oxygen and nitrogen, as well as sulfur. Often a mixture of two or three of these rather different bases forms the binding site on the enzyme for the $Zn^{2+}$. When it is bound by four sulfurs, which are soft bases complementary to the soft $Zn^{2+}$, the interaction is covalent and tetrahedral. The complex between $Zn^{2+}$ and the harder base ammonia is tetracoordinate $[Zn(NH_3)_4]^{2+}$ and tetrahedral, probably because it is also covalent. This tetrahedral covalent form of $Zn^{2+}$ is the most common and is observed when $Zn^{2+}$ forms complexes with 1,2-diaminoethane, cyclic lactams, and imidazole. As the ligands become harder, however, geometries become more variable. For example, the complex $[Zn(OH_2)_6]^{2+}$ between $Zn^{2+}$ and water, a hard base, is hexacoordinate and octahedral; but, as protons are removed, it eventually decreases its coordination to four, as $[Zn(OH)_3(OH_2)]^-$. An ionic, octahedral complex forms with carboxylates

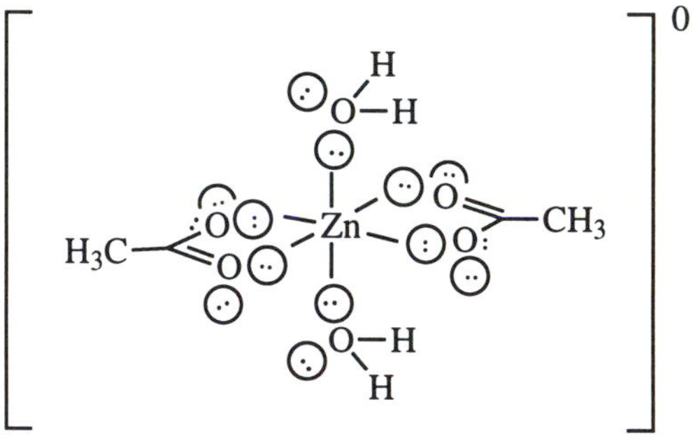

**4–49**

The zinc cation forms a pentacoordinate complex with, among other ligands, 8-aminoquinazoline (Figure 4–38),[291] in which the four nitrogens from two aminoquinazolines and a molecule of water are the five Lewis bases that generate the complex $[Zn(N_2C_9H_8)_2(OH_2)]^{2+}$.

The ability of $Zn^{2+}$ to alternate between tetravalent and pentavalent coordination while bound as a Lewis acid in the active site of an enzyme has been demonstrated in crystallographic studies of thermolysin. In the crystallographic molecular model of the unliganded enzyme, the $Zn^{2+}$ in the active site is tetracoordinate. Histidine 142, Histidine 146, and Glutamate 166 provide the three lone pairs holding the $Zn^{2+}$ permanently in the active site, and the fourth position is occupied by a molecule of water.[292] In the crystallographic molecular model[293] of the complex between the competitive inhibitor β-phenylpropionyl-L-phenylalanine

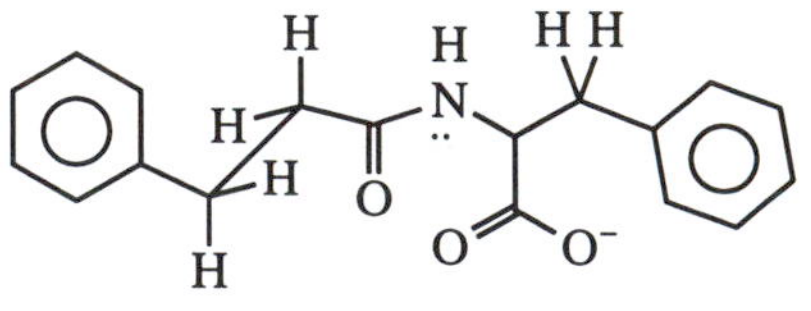

**4–50**

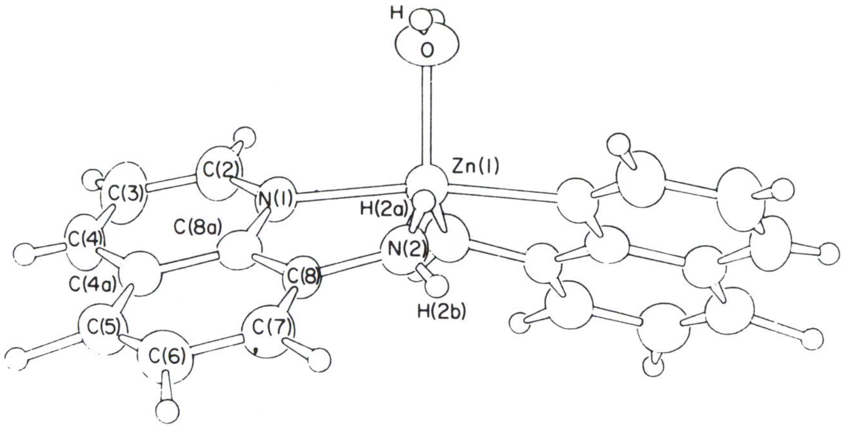

**Figure 4–38:** Crystallographic molecular model of *trans*-bis(8-aminoquinoline)aquozinc (II) tetrachlorozincate.[291] The complex was prepared by mixing 8-aminoquinoline and $ZnCl_2$ in methanol. Crystals were formed by slowly evaporating the methanol. The solid had the empirical formula $(C_9H_8N_2)_2Zn_2Cl_4{\cdot}H_2O$. A map of electron density was prepared crystallographically. Each unit cell contained four of the bis(8-aminoquinolate) complexes, one of which is displayed in the figure. The accompanying $ZnCl_4^{2-}$ ion is not included in the figure. Reprinted with permission from ref 291. Copyright 1981 Gordon & Breach Science Publishers.

and the active site, the acyl oxygen of its amide replaces the oxygen of the water molecule and the $Zn^{2+}$ remains tetracoordinate (Figure 4–39B).[271] The hydroxamate of L-leucine

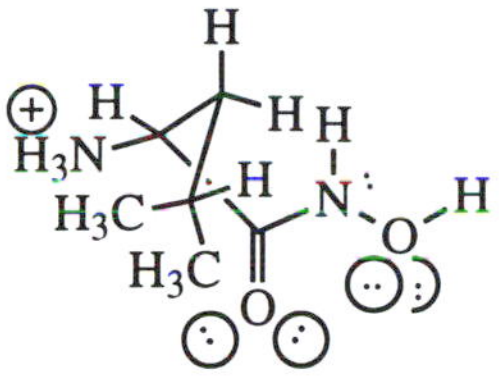

**4–51**

is also a competitive inhibitor for the enzyme ($K_i$ = 0.2 mM); but, in the crystallographic molecular model of the complex between it and the active site, the hydroxamate provides two ligands to the $Zn^{2+}$, a lone pair from the acyl oxygen and a lone pair from the hydroxyl oxygen (Figure 4–39A).[293] In Figure 4–39B the hydroxamate portion of L-leucine hydroxamate and the entire molecule of $\beta$-phenylpropionyl-L-phenylalanine are superposed as they sit in the active sites of the respective crystallographic molecular models. The lone oxygen of the latter bisects the positions of the two oxygens in the former.

These observations have suggested that the acyl oxygen of the peptide bond in a normal reactant for the enzyme binds to the conjugate base (**4–52**) of the zinc–water complex

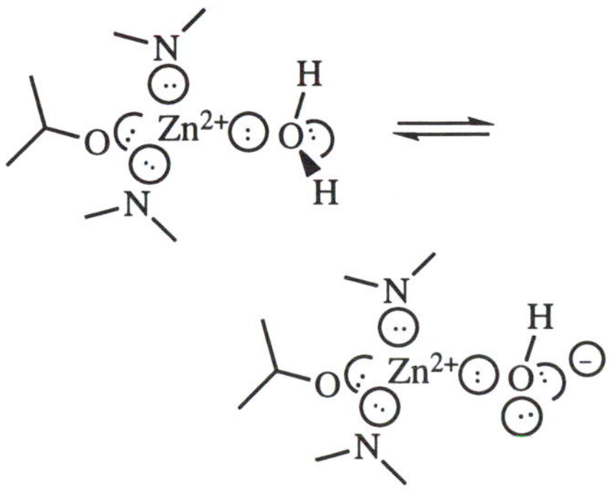

$$\rightleftharpoons \quad + H^+$$

**4–52** (4–233)

pushing to one side the oxygen of the hydroxyl to form the pentacoordinate complex. The hydroxide anion bound on the $Zn^{2+}$ could then attack the acyl carbon nucleophilically

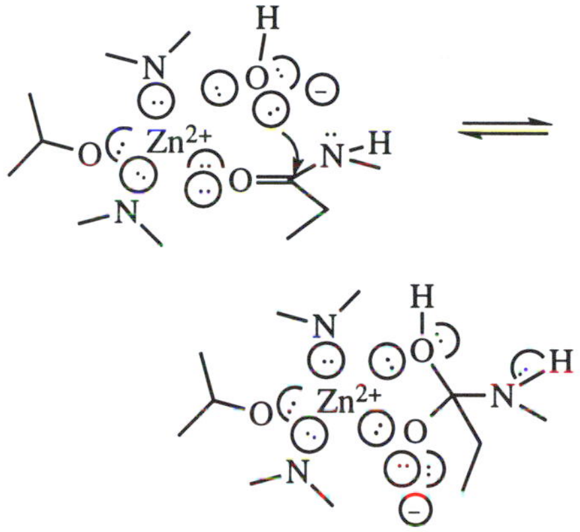

(4–234)

This would produce a four-membered metallocycle, which, unlike the five-membered metallocycle in the case of the hydroxamate, would be strained. In the crystallographic molecular model of the complex between thermolysin and phosphonamidate **4–53**

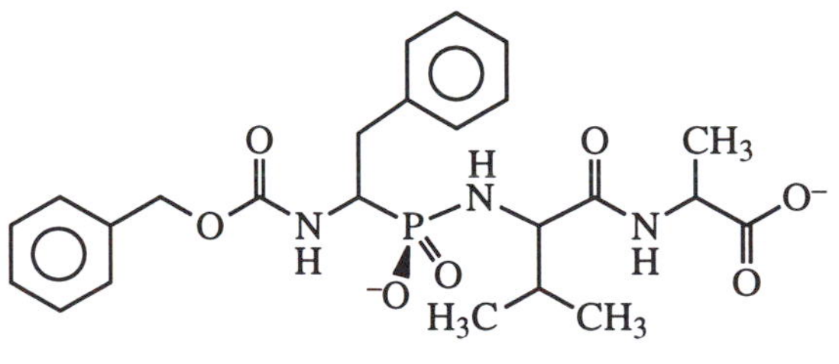

**4–53**

such a four-membered metallocycle is formed at the phosphonamidate[294]

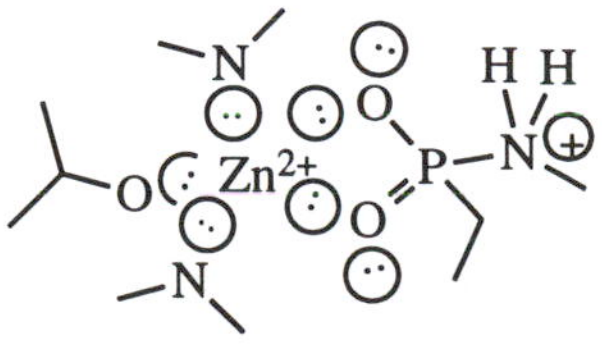

**4–54**

A

B

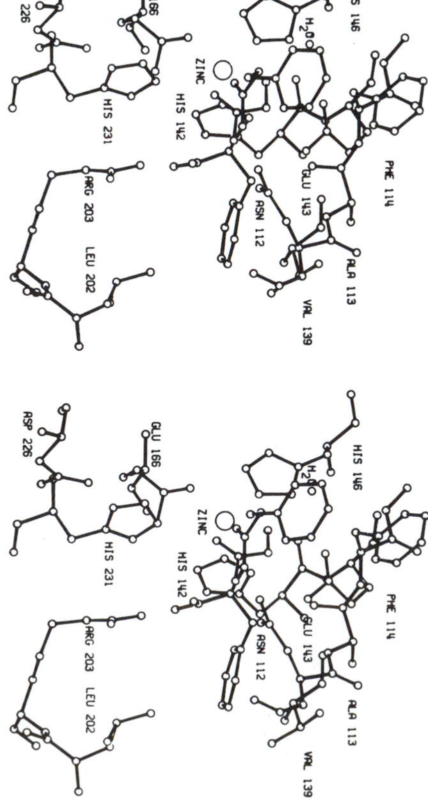


**Figure 4–39:** Comparison of pentacoordinate and tetracoordinate complexes of the $Zn^{2+}$ in the active site of thermolysin.[271,293] Thermolysin was crystallized from 0.01 M calcium acetate and 5% dimethyl sulfoxide, pH 7.2. Complexes were formed between the enzyme and either β-phenylpropionyl-L-phenylalanine (4–50) or L-leucine hydroxamate (**4–51**) by soaking the crystals in respective solutions of these ligands. Data sets were collected from crystals of each of these complexes. For each complex, a map of difference electron density ($F_{complex}$–$F_{uncomplexed}$) was calculated. The data set $F_{complex}$ was the observed data set in each case. The data set $F_{uncomplexed}$ was a data set calculated from a refined crystallographic molecular model of native unliganded thermolysin from which all water molecules in the active site had been removed. The phases used in the calculation were the calculated phases, $\alpha_{calc}$, from this refined crystallographic molecular model of the native unliganded enzyme. A molecular model of each inhibitor was then placed in the positive feature of electron density in the active site of the respective map of difference electron density, and these resulting preliminary crystallographic molecular models of the respective complexes were submitted to refinement against the respective data sets. (A) The active site of the refined crystallographic molecular model between L-leucine hydroxamate and thermolysin. The hydroxamate replaces the water bound to the $Zn^{2+}$ in the crystallographic model of the unliganded enzyme and forms two contacts through its two oxygens with the $Zn^{2+}$. In the process, the coordination of the $Zn^{2+}$ is expanded to five lone pairs. (B) Crystallographic molecular model for the active site of thermolysin in which the hydroxamate portion [ONCO(C)] of L-leucine hydroxamate and the entire molecule of β-phenylpropionyl-L-phenylalanine are superposed at their respective positions in their respective crystallographic molecular models. The single oxygen coordinated to the $Zn^{2+}$ in the latter tetravalent complex is located between the two oxygens coordinated in the former pentavalent complex. Reprinted with permission from ref 271. Copyright 1981 American Chemical Society.

In the normal reaction, however, this metallocycle should break at the weakest bond, that between the hydroxyl of the tetravalent intermediate and the $Zn^{2+}$, to generate the tetracoordinate complex represented in Figure 3–34. In the tetracoordinate complex the hydroxyl that has just left the $Zn^{2+}$ is hydrogen-bonded to Glutamate 143, which would be responsible for removing the proton from the oxygen and adding the proton to the departing nitrogen.

Contrarily, it is also possible that the hydroxyl bound to the $Zn^{2+}$, instead of performing the role of a nucleophile, acts as a general base, with a $pK_a$ around 9, to remove a proton from a molecule of water while it acts as the nucleophile

(4–235)

The advantage of this mechanism is that a five-membered metallocycle is formed, which should be more stable. If this were the mechanism, however, a molecule of water should have been found upon a pentacoordinate $Zn^{2+}$ in the crystallographic molecular model of Figure 3–34, but it was not.

A water molecule bound to $Zn^{2+}$ also acts as a catalytic functional group in the active site of carbonate dehydratase, the enzyme responsible for the hydration of the *gem*-dialdehyde carbon dioxide[295]

(4–236)

In the crystallographic molecular model of the active site of unoccupied carbonate dehydratase,[296] the $Zn^{2+}$ is fixed in position by coordinate covalent interactions with Histidine 94, Histidine 96, and Histidine 119 and the fourth coordinate position is occupied by a molecule of water (Figure 4–40),[297] whose other three hydrogen-bonding positions are in turn occupied by Threonine 199 and two other molecules of water. The maximum velocity of the enzymatically catalyzed hydration reaction increases with pH[298,299] and, in the case of the isoenzyme II that is pictured, reaches a constant value after passing through an apparent p$K_a$ of 6.9. The acid dissociation constant responsible for this behavior has been assigned[295] to the water bound to the $Zn^{2+}$ (Table 4–2). It should be recalled that the apparent p$K_a$ from the pH–rate profile of an enzyme is always less than or equal to the actual p$K_a$ of the acid governing the behavior. In any case, the active form of the enzyme would be the form in which the fourth position on $Zn^{2+}$ is occupied by a hydroxide anion.[300]

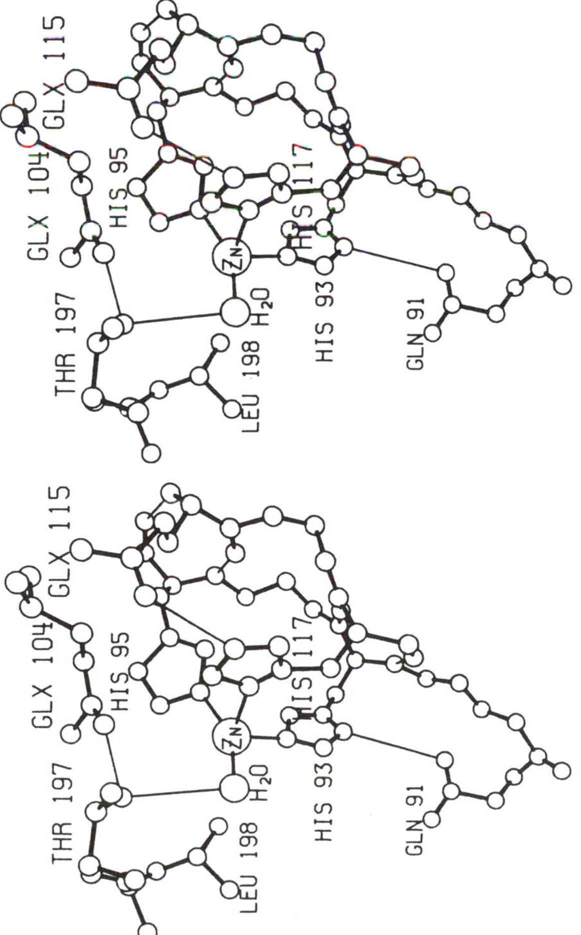


**Figure 4–40:** Crystallographic molecular model of the active site of human carbonate dehydratase II.[297] Data sets were collected from crystals of human carbonate dehydratase II and four heavy-atom derivatives. Phases were determined by the method of isomorphous replacement, and these phases and the data set to spacings of 0.20 nm were used to prepare a map of electron density for the $P2_1$ unit cell. A provisional model of the polypeptide, based on an incomplete amino acid sequence, was inserted into the map of electron density. The $Zn^{2+}$, chelated by three histidines, could be recognized readily as a large sphere of electron density in the original map. The crystallographic molecular model presented is that of the unrefined structure because, at the time, the complete sequence was not known. Nevertheless, the most recent, highly refined crystallographic molecular model[296] differs insignificantly from that presented. The sequence numbers of the various amino acids, however, have changed. Glutamine 91 is now Glutamine 92; Histidine 93 is now Histidine 94; Histidine 95 is now Histidine 96; Glutamate 104 is now Glutamate 106; Glutamate 115 is now Glutamate 117; Histidine 117 is now Histidine 119; Threonine 197 is now Threonine 199; and Leucine 198 is now Threonine 200. Nevertheless, all hydrogen bonds indicated in the figure remain as they are in the latest refined molecular model, except that Leucine 198 is now Threonine 200. Threonine 200 forms a hydrogen bond to a water that in turn is hydrogen-bonded to the water bound to the $Zn^{2+}$. Reprinted with permission from *Nature*, ref 297. Copyright 1972 Macmillan Magazines Limited.

There are two issues that pertain to the role of the $Zn^{2+}$ as a Lewis acid in the active site of a carbonate dehydratase. First, the carbon dioxide could either become an inner-sphere ligand to $Zn^{2+}$ during its hydration or it could remain bound in the outer sphere adjacent to the $Zn^{2+}$ and hydrogen-bonded to surrounding donors. Second, because the water bound to the $Zn^{2+}$ in the crystallographic molecular model is itself hydrogen-bonded to other molecules of water, the hydroxide bound to the $Zn^{2+}$ in the active form of the enzyme could act as a general base to remove a proton from an adjacent molecule of water in the outer sphere, which would be the actual nucleophile (Reaction 4–235), or the hydroxide bound to the $Zn^{2+}$ could itself be the nucleophile.

Imidazole is a competitive inhibitor of isoenzyme I of carbonate dehydratase.[298] In a crystallographic molecular model of the complex between isoenzyme I and imidazole, the imidazole occupies the fifth position on a pentacoordinate $Zn^{2+}$ in addition to three histidines and water.[301] Thiocyanate, S=C=N⁻, resembles carbon dioxide structurally and is an inhibitor of carbonate dehydratase. It is more basic ($pK_a$ = 0.85) than carbon dioxide and would be a stronger ligand for $Zn^{2+}$. In the crystallographic molecular model of the complex between thiocyanate and isoenzyme II of carbonate dehydratase at high pH, the nitrogen of the thiocyanate is a tight fifth ligand to the $Zn^{2+}$ in addition to the three histidines and a molecule of hydroxide anion[302]

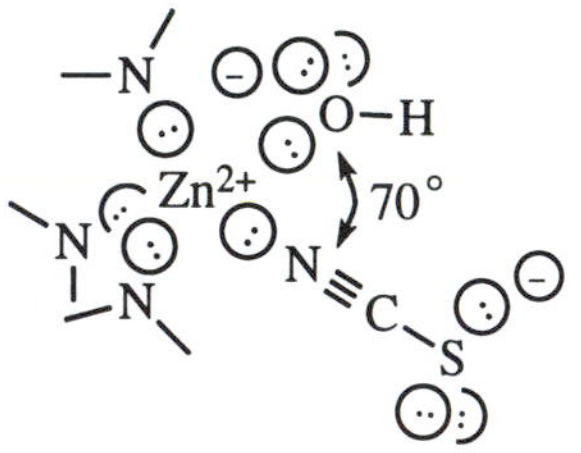

**4–55**

All of these results suggest that carbon dioxide, at some point during the reaction catalyzed by carbonate dehydratase, could become a fifth ligand to the $Zn^{2+}$.

Sulfonamides, in the anionic form[303]

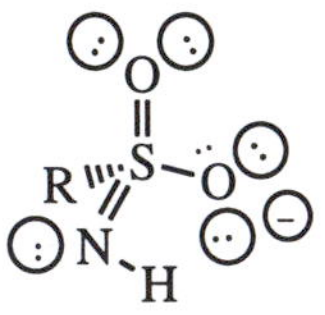

**4–56**

bind tightly to carbonate dehydratase and inhibit the enzymatic activity. If they serve as models for the binding of the substrate carbonate to the enzyme, they might identify catalytic functional groups. In the crystallographic molecular models of sulfonamides bound to the active site of isoenzyme II,[302] the nitrogen is one of the four ligands to a tetracoordinate $Zn^{2+}$ and one of the oxygens of the sulfonamide forms a hydrogen bond to Threonine 199. If the hydroxide anion were only performing the role of a general base and removing a proton from an adjacent molecule of water, which was the actual nucleophile, and if the sulfonamides are adequate models of the bound product, carbonate (**4–57**), then the $Zn^{2+}$ in the sulfonamide complex (**4–58**) should have been pentacoordinate instead of tetracoordinate and a molecule of water bound to the $Zn^{2+}$ should have been the fifth ligand

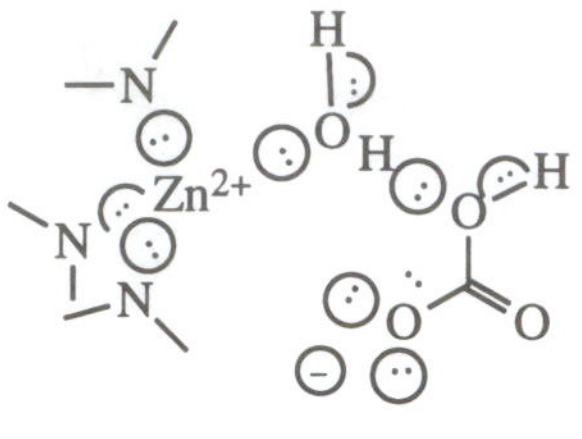

**4–57** 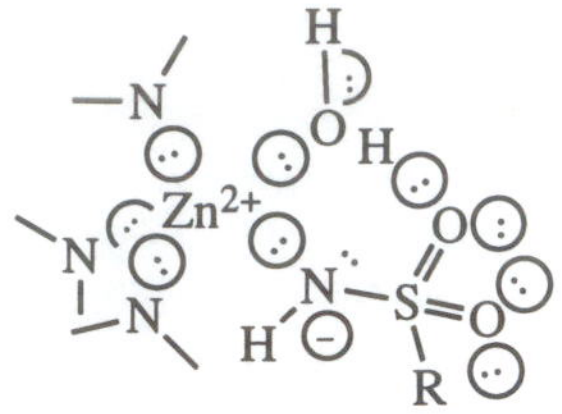 **4–58**

As it was not, the observation is more consistent with the hydroxide bound to the $Zn^{2+}$ being the actual nucleophile[295]

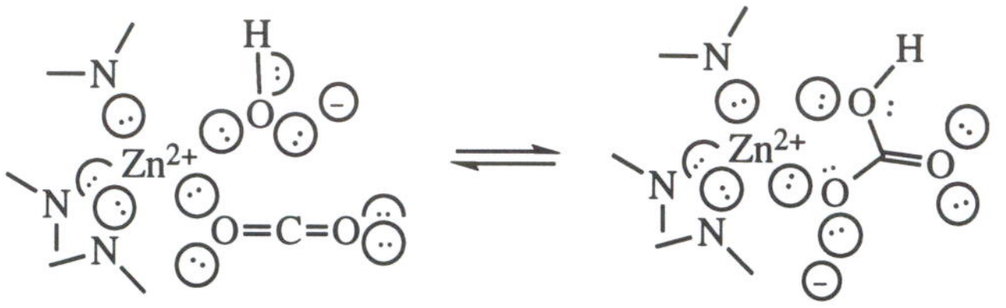

(4–237)

There are enzymes in which $Zn^{2+}$ serves as a simple Lewis acid rather than as a participant in a complex with a water molecule. In alcohol dehydrogenase the $Zn^{2+}$ in the active site is tetrahedral and tetracoordinate. It is attached covalently to the protein by coordination to the soft bases Cysteine 174 and Cysteine 46 and to Histidine 67. In the unliganded enzyme, the fourth coordination site is occupied by a molecule of water, but in the crystallographic molecular model of the active site occupied by NAD⁺ and *p*-bromobenzyl alcohol (Figure 4–41),[304] the oxygen of the alcohol has replaced the water molecule. This crystallographic molecular model is consistent with the $Zn^{2+}$ performing the role of a Lewis acid to lower the $pK_a$ of the oxygen of the alcohol (Table 4–3) and promote the acid–base reaction that occurs coincident with hydride transfer

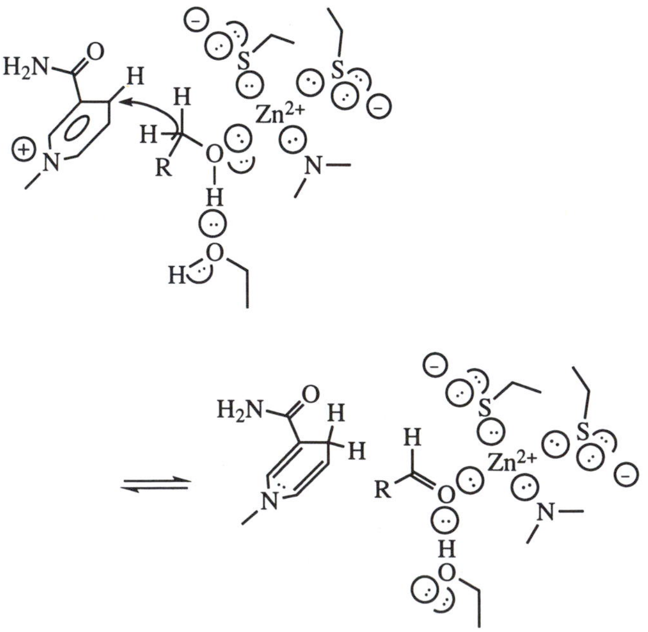

(4–238)

On the basis of the pH–rate profiles of the enzymatic activity, there is an acid in the binary complex between NAD⁺ and alcohol dehydrogenase with an apparent $pK_a$ of 7.6, and this acid has been assigned[305] to the water bound to the $Zn^{2+}$.

Presumably, after the water has been replaced by the alcohol, the p$K_a$ of the hydroxyl of the alcohol is also shifted into this range. The proton removed from the hydroxyl of the alcohol is passed formally but not in fact along a network of hydrogen-bonded hydroxyl groups out of the active site (Figure 4–41).

(4–239)

Alkaline phosphatase hydrolyzes phosphate esters (Reaction 4–160) by transferring the phosphate to Serine 102 in the active site and then to water. Within the crystallographic molecular model of its active site[128] there are two $Zn^{2+}$ that are held only 0.4 nm apart. One of these $Zn^{2+}$ is coordinated to three histidines, Histidines 331, 372, and 412, and has a molecule of water as its fourth ligand. The other $Zn^{2+}$ is coordinated to Aspartate 51, Aspartate 369, and Histidine 370 and has a molecule of water as its fourth ligand. The two molecules of water are on the same side of the adjacent $Zn^{2+}$ cations

(4–240)

and are replaced by two of the oxygens on the phosphate of the reactant upon occupation of the active site because arsenate, a substrate analogue, binds equidistant from the two $Zn^{2+}$ and adjacent to both of them on the side on which the molecules of water are located in the unoccupied enzyme. If

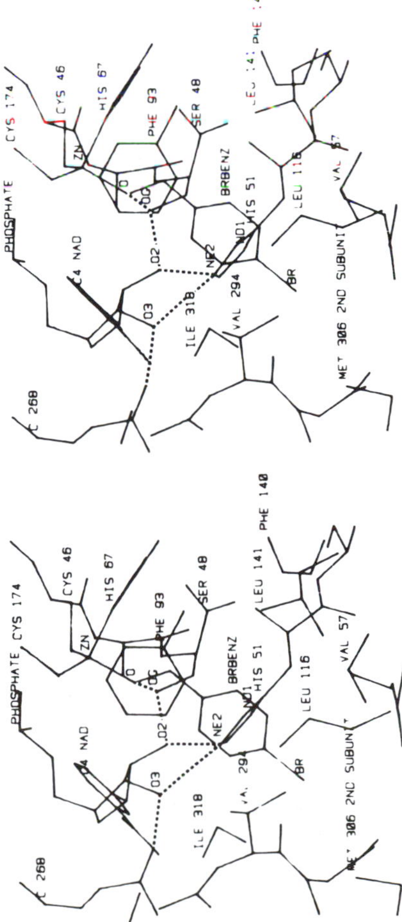


**Figure 4–41:** Crystallographic molecular model of the active site of horse liver alcohol dehydrogenase occupied by bromobenzyl alcohol (BRBENZ).[304] Crystals of alcohol dehydrogenase were formed in a solution of 25% 2-methyl-2,4-pentanediol containing $NAD^+$ and bromobenzyl alcohol. A data set was collected from these crystals to Bragg spacings of 0.29 nm. A map of difference electron density ($F_{complex}$–$F_{uncomplexed}$) was calculated. The data set $F_{complex}$ was the observed data set for the liganded enzyme. The data set $F_{uncomplexed}$ was a data set calculated from a previous refined crystallographic molecular model of the enzyme in complex with NADH and dimethyl sulfoxide. This data set $F_{uncomplexed}$ was calculated by leaving out the molecule of dimethyl sulfoxide in the refined molecular model. The phases used for the calculation of ($F_{complex}$–$F_{uncomplexed}$) were the phases calculated from the previous refined crystallographic molecular model. In the map of difference electron density ($F_{complex}$–$F_{uncomplexed}$) only one feature of positive electron density appeared. It had the unmistakable shape of bromobenzyl alcohol and was located immediately adjacent to the $Zn^{2+}$ in the active site. A molecular model of bromobenzyl alcohol was placed into this feature, and this placement positioned the molecule of bromobenzyl alcohol within the refined crystallographic molecular model of alcohol dehydrogenase. The region surrounding the active site in the final model is presented. Reprinted with permission from ref 304. Copyright 1982 *Journal of Biological Chemistry.*

this is the way that the phosphate of a reactant is bound, the two $Zn^{2+}$ act as Lewis acids to make the phosphorus more electrophilic and more susceptible to nucleophilic attack by Serine 102.

**Copper** exists in biochemical situations as the kinetically stable cations $Cu^{+}$ and $Cu^{2+}$. The cupric form of the metal behaves as a Lewis acid similarly to most of the other divalent metallic cations. It is one of the strongest divalent Lewis acids (Table 4–3) and its rates of ligand exchange are also the most rapid (Table 4–4), and one would expect that it would be ideal for most of the purposes to which $Mn^{2+}$ and $Zn^{2+}$ are put. It is rarely found in these roles, however, in metalloenzymes. Rather, it is usually used as a one-electron carrier, often in reactions involving oxygen activation such as those catalyzed by monooxygenases.

The soft metallic cation $Cu^{2+}$ can form a number of coordination complexes with lone pairs from oxygen, nitrogen, and sulfur. The variety of these complexes defies categorization. They range from dicoordinate to octacoordinate complexes. Even in the more common tetracoordinate and hexacoordinate stereochemistries, the terms used to describe the variations such as square planar, tetrahedral, compressed tetrahedral, elongated tetragonal octahedral, and trigonal octahedral indicate that the arrangement of many Lewis bases around copper, as with calcium, is governed mainly by steric effects among the ligands, rather than by covalent bonding. Examples of complexes between $Cu^{2+}$ and simple biochemical ligands are $[Cu(NH_3)_5]^{2+}$

$NH_3$
$H_3N$ — $Cu^{2+}$ — $NH_3$
$H_3N$ $NH_3$

**4–59**

in which four of the nitrogens are equivalent and the fifth forms a longer bond to $Cu^{2+}$, and $[Cu(imidazole)_4(OH_2)_2]^{2+}$ and $[Cu(formate)_2(OH_2)_2]^{2+}$, which are both elongated tetragonal octahedral structures. Simple thiols such as mercaptans reduce $Cu^{2+}$ to $Cu^{+}$ and form complex polymeric structures with $Cu^{+}$ in which the copper is multidentate and the thiol is bidentate.

The azurins and the plastocyanins are related metalloproteins involved in one-electron transfers in which the single copper passes reversibly from Cu(II) to Cu(I) to carry the electron. In crystallographic molecular models of these proteins,[306,307] the copper is coordinated to two histidines, a methionine, and a cysteine (Figure 4–42)[306] with no particular geometric regularity. In the apoprotein, the four atoms, two nitrogens, and two sulfurs that surround the copper in the holoprotein assume the same orientations even though the copper is not present.[308] This observation is consistent with the conclusion that the stereochemistry around copper is not effected by covalent bonds to the cation but by the steric demands of the protein. This distorted geometry may lower the barrier to the electron transfer reactions in which the protein participates.

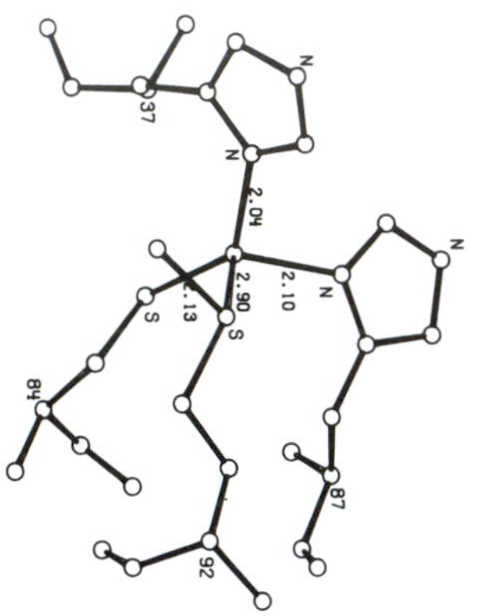


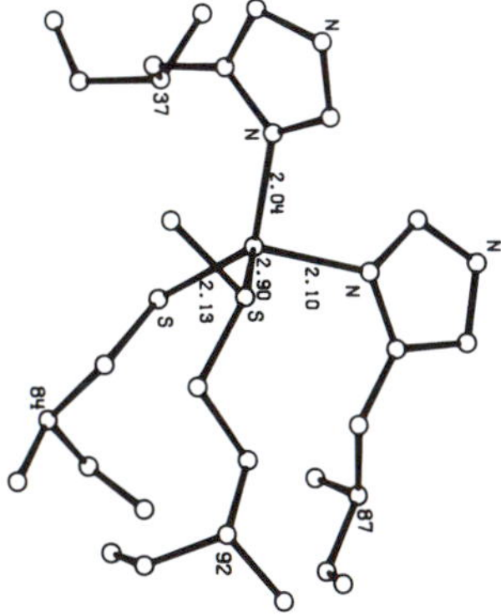


**Figure 4–42:** Crystallographic molecular model of the four ligands, Histidine 37, Cysteine 84, Histidine 87, and Methionine 84, surrounding the copper ion in poplar plastocyanin.[306] Crystals of poplar plastocyanin were grown from 2.6 M $(NH_4)_2SO_4$, pH 6.0. A data set was collected from these crystals to Bragg spacing of 0.16 nm. The phases were obtained by isomorphous replacement using three heavy-atom derivatives. A model of the polypeptide and a copper ion was inserted into the initial map of electron density, and the resulting crystallographic molecular model was refined. The four ligands to the copper in the final refined molecular model are presented. Bond lengths are in Ångstrom units. Reprinted with permission from ref 306. Copyright 1983 Academic Press.

Superoxide dismutase is a metalloprotein containing both a copper cation and a zinc cation. It is also a one-electron carrier shuttling electrons between superoxide anions

$$O_2^{\cdot -} + E\cdot Cu(II) \rightleftharpoons O_2 + E\cdot Cu(I) \qquad (4\text{–}241)$$

$$O_2^{\cdot -} + E\cdot Cu(I) + 2H^{+} \rightleftharpoons H_2O_2 + E\cdot Cu(II) \qquad (4\text{–}242)$$

In the crystallographic molecular model of the enzyme, the $Zn^{2+}$ and the copper are both ligands to the same histidine, Histidine 61 (Figure 4–43).[309,310] In addition to this histidine, the copper is surrounded by three other imidazoles in a tetracoordinate complex. The imidazole bridging the two metals in the oxidized enzyme would be unprotonated at both nitrogens

**4–60**

Based on spectral changes that occur during the enzymatic reaction[311] and the dependence of potentiometric titrations on pH,[312] it has been proposed that in the reduced enzyme the bridging imidazole is displaced from the copper(I) and becomes protonated

$$+e^- \ +H^+ \rightleftharpoons -e^- \ -H^+$$

(4–243)

**4–61**

The fact that the $Zn^{2+}$ lies below the plane of the imidazole in the oxidized form of the active site (Figure 4–43) would provide the torque necessary to tilt the ring away from the $Cu^+$ as the copper–nitrogen bond is broken. The p$K_a$ of an isolated neutral imidazole acting as an acid is high,[250a] but the bound $Zn^{2+}$ in Structure **4–61** should lower the p$K_a$ of the imidazole considerably and make it an effective general acid. It has been proposed[311] that this is the general acid inescapably required in the reduction of superoxide anion

(4–244)

and the structure of the active site in the crystallographic molecular model is consistent with this proposal.[309]

**Iron**, when it is a catalytic functional group in the active site of an enzyme, is almost always in a coenzyme such as a heme or an iron–sulfur cluster. In most of these instances, the iron acts either as an electron carrier because of its ability to convert between Fe(II) and Fe(III) or as a catalytic group that

**Figure 4–43:** Crystallographic molecular model of the active site of superoxide dismutase.[310] Crystals of bovine erythrocyte superoxide dismutase were grown from 58% 2-methyl-2,4-pentanediol, pH 7.5. A data set was collected from these crystals to Bragg spacings of 0.20 nm. Five heavy-atom derivatives were used to estimate phases from the reflections in this data set. A molecular model of the polypeptide and atoms of copper and zinc were inserted into the initial map of electron density, and the initial crystallographic molecular model was refined against the observed data set. The amino acid side chains that are displayed are those of the seven ligands to the copper and the zinc: Histidines 44, 46, 61, 69, 78, and 118 and Aspartate 81. Reprinted with permission from ref 310. Copyright 1982 Academic Press.

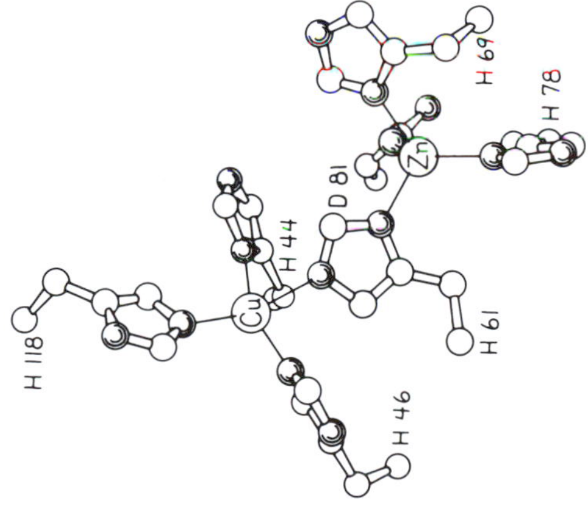

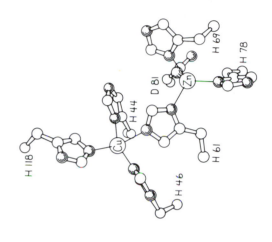

can bind and activate oxygen. An interesting exception to these roles is encountered with aconitate hydratase[313] and maleate hydratase.[314] These two enzymes catalyze the hydration of 3-(carboxymethyl)maleate and maleate, respectively

(4–245)

$R_1 = H; R_2 = -CH_2COO^-$

$R_1 = -CH_2COO^-; R_2 = H$

$R_1 = R_2 = H$

Each of these enzymes has an iron–sulfur cluster in its active site that participates in catalysis of this hydration. The cluster is of the 4Fe–4S type (Figure 2–15B), but one of the four irons in the cluster is held in the active site only by the three inorganic sulfides. The requirement of these enzymes for $Fe^{2+}$ results from the fact that this loosely held $Fe^{2+}$ readily dissociates from the iron–sulfur cluster, but its site in the cluster remains occupied when a sufficient concentration of $Fe^{2+}$ is added to the solution.[315] The enzyme is inactive if the cluster has lost this fourth iron. In a complex between enzyme and an alkenyl reactant, before hydration occurs, two of the six coordination sites on this labile iron are occupied by a molecule of water and an oxygen from the carboxylate on the carbon to which the hydroxyl will be added. After hydration, the coordination sites are occupied by the hydroxyl that has been added and one oxygen from the carboxylate on the same carbon[316]

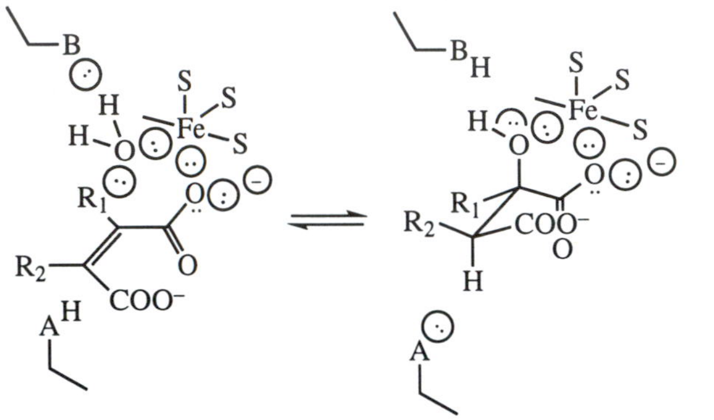

(4–246)

It is assumed that the iron acts as a Lewis acid, lowering the p$K_a$ of the water molecule (Table 4–2) so that a proton can be removed more readily to generate the nucleophilic hydroxide ion while it simultaneously makes the olefin more electrophilic by inductive electron withdrawal.

There are metalloenzymes in which the catalytic iron is not present in either a heme or an iron–sulfur cluster. Examples in which such an iron is present in the active site of an enzyme that is involved in either oxidation–reduction or the activation of oxygen are ribonucleoside-diphosphate reductase[317]

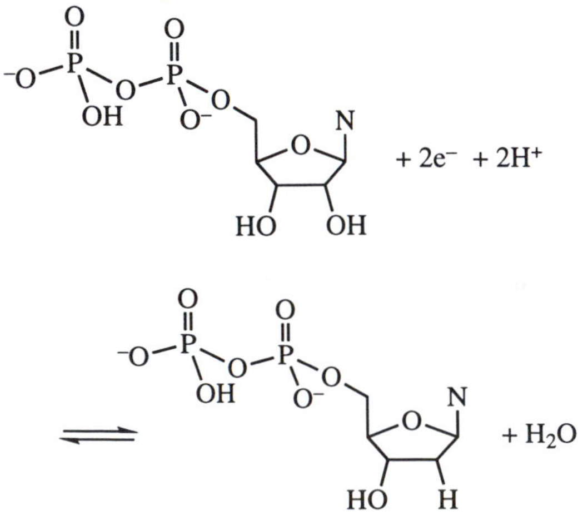

(4–247)

where N is a nucleoside base, and protocatechuate 3,4-dioxygenase[318]

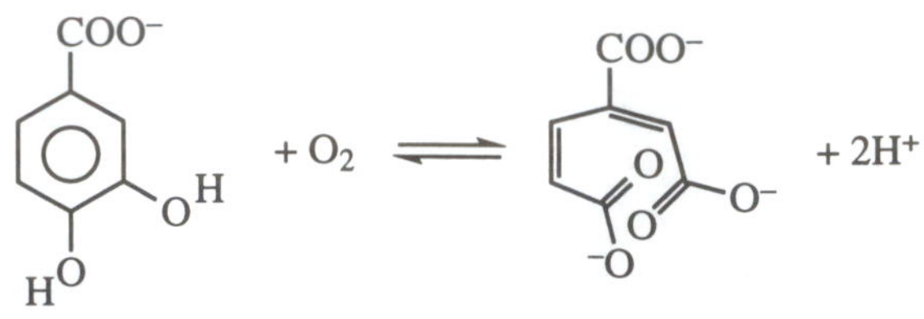

(4–248)

Enzymes in which $Fe^{2+}$ may assume the uncomplicated role of a Lewis acid are the hydrolytic enzymes yeast arginase[261]

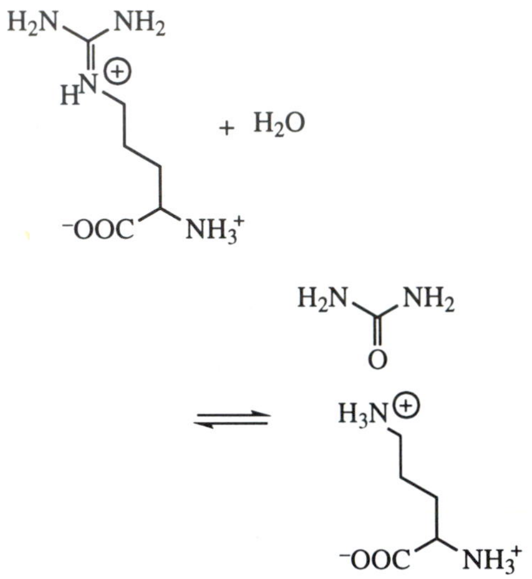

(4–249)

and uteroferrin[319]

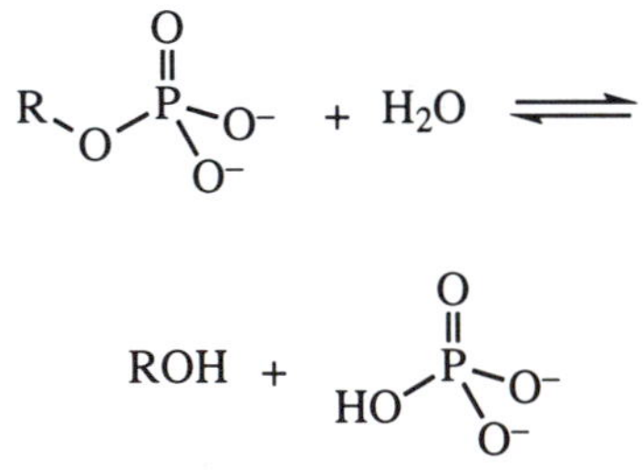

(4–250)

The metalloenzyme lysine 2,3-aminomutase catalyzes an isomerization stereochemically indistinguishable[320] from that catalyzed by an enzyme utilizing coenzyme $B_{12}$

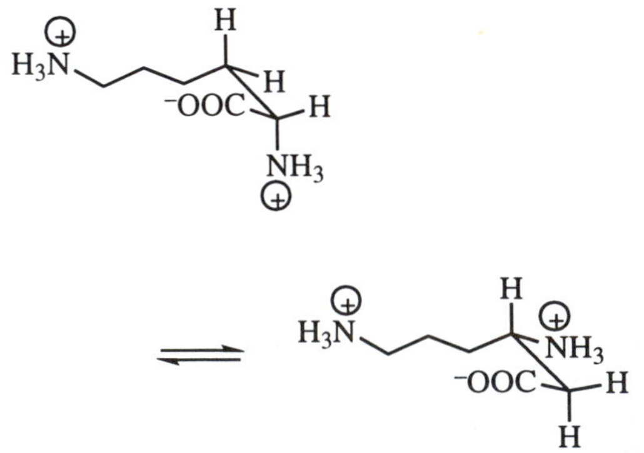

(4–251)

with a puzzling exception. The amino group is transferred from carbon 2 to carbon 3 of the same molecule of substrate while the deuterium is removed from one molecule of substrate and added back to the next molecule of substrate to occupy the active site. The enzyme requires that both $Fe^{2+}$ and pyridoxal phosphate be bound at the active site to display any enzymatic activity but is devoid of coenzyme $B_{12}$.

Aside from its possible involvement as a weakly bound metallic cation in enzymatic reactions requiring a Lewis acid, **cobalt** is engaged in enzymatic reactions only within coenzyme $B_{12}$. **Molybdenum** is a component of metalloenzymes that catalyze oxidation–reductions, such as nitrogenase

$$N_2 + 6e^- + 6H^+ \rightleftharpoons 2NH_3 \qquad (4\text{–}252)$$

nitrate reductase (NADP)

$$H^+ + NADPH + NO_3^- \rightleftharpoons NADP^+ + NO_2^- + H_2O \qquad (4\text{–}253)$$

and xanthine oxidase

$$O_2 + \text{(HN, N, N, NH, O, O; C–H)} + H_2O \rightleftharpoons \text{(HN, NH, NH, NH; =O, O, O)} + H_2O_2 \qquad (4\text{–}254)$$

In each of these enzymatic reactions, the molybdenum is enclosed within a larger coenzyme. **Vanadium** is used in place of molybdenum in the nitrogenases from certain species.[321] Aside from its possible role as a Lewis acid in metalloenzymes that can use any one of an array of metallic dications and its exclusive role, presumably also as a Lewis acid, in urease[322]

$$H^+ + H_2N{-}C(=O){-}NH_2 + 2H_2O \rightleftharpoons HO{-}C(=O){-}O^- + 2NH_4^+ \qquad (4\text{–}255)$$

**nickel** is also a component of the active site in enzymes involved in oxidation–reduction, such as hydrogenase[323]

$$H_2 \rightleftharpoons 2e^- + 2H^+ \qquad (4\text{–}256)$$

carbon monoxide dehydrogenase[324]

$$CO + H_2O \rightleftharpoons CO_2 + 2e^- + 2H^+ \qquad (4\text{–}257)$$

and methyl-coenzyme M methylreductase[325]

$$^-O{-}S(=O)_2{-}CH_2CH_2{-}S{-}CH_3 + 2H^+ + 2e^- \rightleftharpoons {}^-O{-}S(=O)_2{-}CH_2CH_2{-}SH + CH_4 \qquad (4\text{–}258)$$

In this latter enzyme, the nickel is found coordinated within a novel tetrapyrrole[326] that resembles closely the tetrapyrroles of porphyrins, pheophytins, and corrins. The complex between nickel and this tetrapyrrole is a yellow compound analogous to a heme. **Tungsten** is a component of the metalloenzyme formate dehydrogenase ($NADP^+$)[327]

$$^-O{-}C(=O){-}H + NADP^+ \rightleftharpoons O{=}C{=}O + NADPH \qquad (4\text{–}259)$$

also an enzyme catalyzing an oxidation–reduction. In each of these simple oxidation–reduction reactions, the particular metal involved is presumably effective because its oxidation–reduction potential is appropriate for the electron transfers that must occur and because it is particularly efficient at forming and unforming the necessary inner-sphere complexes with reactants or intermediates that are involved in the reactions. None of these latter metals, molybdenum, vanadium, nickel, or tungsten, is widely used, and the impression left is that they are peculiarly suited for the particular roles in which they are found and not so easily adapted to other roles as are some other metallic cations.

## *Suggested Reading*

Tainer, J.A., Getzoff, E.D., Richardson, J.S., & Richardson, D.C. (1983) Structure and mechanism of copper, zinc superoxide dismutase, *Nature 306*, 284–287.

### PROBLEM 4–26

The following is a direct quotation from ref 282.

Staphylococcal nuclease catalyzes the hydrolysis of both DNA and RNA at the 5′ position of the phosphodiester bond yielding a free 5′-hydroxyl group and a 3′-phosphate monoester. The pH optimum is between 8.6 and 10.3 and varies inversely with the $Ca^{2+}$ concentration, but at any pH rather high levels of $Ca^{2+}$, typically 0.01 M, are required for optimal activity. With minor exceptions, which will be discussed later, $Ca^{2+}$ is required for activity with all substrates and cannot be replaced with other ions, although $Ca^{2+}$ and a number of other ions do promote the binding of various inhibitors. The 5′-*p*-nitrophenyl esters of pdT and pdTp are the simplest known good substrates for the nuclease. Both these esters are hydrolized at essentially the same rate via P–O bond cleavage of the 5′–C–O–P bond, but the pdTp-based ester has a $K_m$ showing almost two orders of magnitude tighter binding to enzyme. Both the 5′-methyl ester of pdT and thymidine 5′-fluorophosphate are poor substrates. With these simple substrates, the products of enzyme-catalyzed reaction are exclusively *p*-nitrophenyl phosphate and dT or dTp for the first two and methylphosphate or fluorophosphate plus dT for the second two, a striking and significant contrast to the products of a nonenzymatic hydrolysis of these compounds where *p*-nitrophenol, fluoride ion, or at least some methanol along with pdT or pdTp would result. Simple diesters of phosphate are not substrates, so this enzyme is indeed a nuclease and not a general phosphodiesterase. For a series of dinucleotides, $dN^{\alpha}pdN^{\beta}$, as substrates, there is a distinct order of preference (dT ≥ dA >> dC >> dG) for the base in the $\beta$-nucleotide position, but little base specificity in the $\alpha$-position. The presence of a terminal 3′ phosphate results in a better substrate than the plain dinucleotide; a terminal 5′ phosphate results in a poorer substrate. In the presence of $Ca^{2+}$, or a wide variety of other metal ions, pA, pdA, and pdT are good inhibitors ($K_i \simeq 10^{-5}$); pdTp is the best known inhibitor ($K_i \simeq 10^{-7}$). The binding of the nucleoside 5′-monophosphates involves a single $Ca^{2+}$ ion; the binding of pdTp and of DNA and RNA appears to involve two. Nucleosides themselves and nucleoside 2′- or 3′-monophosphates are not inhibitory. Mapping studies of the active site region with oligonucleotides having a terminal 5′ phosphate $(pdT)_n$, show maximum binding when $n = 3$, suggesting the presence of a third, probably ionic, binding site in addition to those for the 5′- and 3′-phosphates of pdTp. Studies of the interactions among the inhibitors, the simple substrates, DNA, and RNA strongly imply a single, common binding and hydrolytic site on the nuclease, although, as noted elsewhere, there are differences between DNA and RNA as substrates that may indicate different hydrolytic mechanisms.

The point of the preceding brief, but fairly comprehensive, review of the enzymological properties of staphylococcal nuclease is to establish that the crystal structure of the nuclease–pdTp–$Ca^{2+}$ complex should resemble fairly closely that of the actual nuclease-substrate-$Ca^{2+}$ complex. Because a diester substrate will have a single negative charge on the phosphate at the hydrolytic site and pdTp has two negative charges on its 5′ phosphate, some difference is, of course, to be expected.

In the abbreviation used for mononucleotides, a p in front of the capital letter, for example, pdT, stands for a phosphate attached to the 5′-hydroxyl of the ribose; a p behind, for example, dTp, stands for a phosphate attached to the 3′-hydroxyl. The d stands for 2′-deoxy. The capital letters stand for the various nucleosides.

(A) Write the mechanism for the hydrolysis of the 5′-*p*-nitrophenyl ester of pdTp, as it would occur during a nonenzymatic reaction that produces *p*-nitrophenol and pdTp. Draw your intermediates, for example, the trigonal bypyramid, with proper stereochemistry.

(B) Why is *p*-nitrophenol rather than *p*-nitrophenyl phosphate the product of the nonenzymatic reaction?

(C) Write the mechanism for the hydrolysis of the 5′-*p*-nitrophenyl ester of pdTp as it would occur under catalysis by micrococcal nuclease, which produces *p*-nitrophenyl phosphate and dTp. Draw your intermediates with proper stereochemistry, and include general acids (HA) or general bases (B) where appropriate.

Crystals of the complex between micrococcal nuclease, $Ca^{2+}$, and pdTp were submitted to diffraction, and a map of electron density from a data set with minimum Bragg spacings of 0.17 nm was obtained. A molecular model built from the sequence of the protein, the $Ca^{2+}$ ion, and the known structure of the inhibitor were placed into the map. The following picture of the active site has been derived from the final refined molecular structure[328] (reprinted with permission from ref 328; copyright 1989 Alan R. Liss, Inc.).

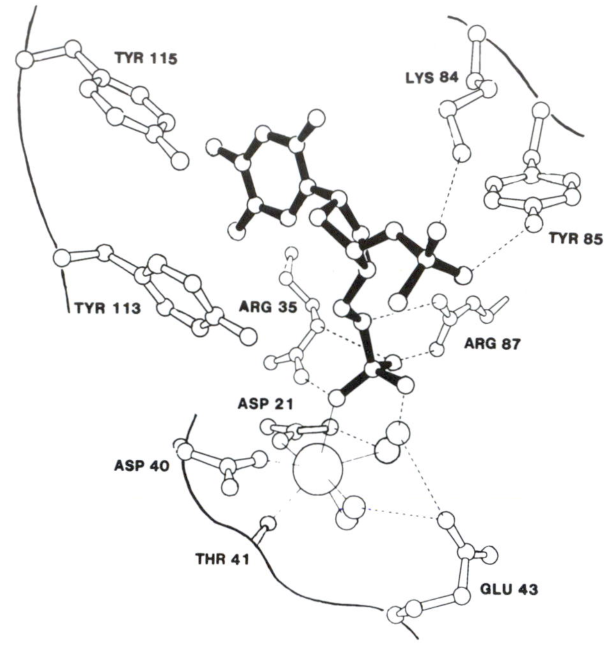


The $Ca^{2+}$ ion is the large sphere.

(D) The diagram contains no hydrogens and no distinction is made among carbon, oxygen, nitrogen, and phosphorus. Draw, in the same orientation and geometry as in the figure, a complete picture of the active site with the pdTp bound to it, indicating clearly the $Ca^{2+}$ and all oxygens, nitrogens, phosphoruses, and acidic hydrogens. Clearly show hydrogen-bond donors and acceptors. Expand the scale so that you have plenty of room. Attach a *p*-nitrophenyl group to the appropriate oxygen.

(E) What are the smaller detached open circles in the crystallographic molecular model?

(F) There are two possible molecules of water for the role of nucleophile in the hydrolysis of 5′-*p*-nitrophenyl pdTp. Write two mechanisms for the hydrolysis, using respectively one and then the other of these molecules of water. In each case, which is the general acid or general base activating the respective molecule of water?

(G) Why is *p*-nitrophenyl phosphate rather than *p*-nitrophenol the product of the enzymatic reaction? Invoke two rules of phosphate chemistry in your explanation. What must the active site be preventing the pentavalent intermediate from undergoing?

(H) What is the identity of the general acid protonating the leaving group in the mechanism that you presented in part C? Why is this a peculiar choice?

(I) When Glutamate 43 is mutated to a serine residue, the $V_{max}$ of the enzymatic reaction decreases 2700-fold at pH 7.4 but the reaction becomes first-order in [$OH^-$]. It was concluded that $OH^-$ from the solution can overcome the fact that the glutamate is missing in the mutant. If this glutamate is acting as a general base, what is its role in the enzymatic reaction?

(J) What is the nucleophilic oxygen that attacks the phosphorus during the enzymatic reaction? Identify it explicitly in the crystallographic molecular model of the active site.

## *References*


1. Loudon, G.M., & Ryono, D.E. (1976) *J. Am. Chem. Soc. 98*, 1900–1907.
2. Jencks, W.P., & Carriulo, J. (1959) *J. Biol. Chem. 234*, 1272–1279.
3. Capon, B., & Zucco, C. (1982) *J. Am. Chem. Soc. 104*, 7567–7572.
4. Funderburk, L.H., Aldwin, L., & Jencks, W.P. (1978) *J. Am. Chem. Soc. 100*, 5444–5459.
5. Jencks, W.P., & Carriulo, J. (1959) *J. Biol. Chem. 234*, 1280–1285.
6. Dunn, B.M., & Bruice, T.C. (1970) *J. Am. Chem. Soc. 92*, 2410–2416.
7. Cordes, E.H. (1967) *Prog. Phys. Org. Chem. 4*, 1–44.
8. Jencks, W.P., & Carriulo, J. (1961) *J. Am. Chem. Soc. 83*, 1743–1750.
9. Jencks, W.P. (1969) *Catalysis in Chemistry and Enzymology*, pp 173–175, MacGraw-Hill, New York.
10. Ewing, S.P., Lockshon, D., & Jencks, W.P. (1980) *J. Am. Chem. Soc. 102*, 3072–3084.
11. Young, P.R., & Jencks, W.P. (1977) *J. Am. Chem. Soc. 99*, 1206–1214.
12. Davenport, R.C., Bash, P.A., Seaton, B.A., Karplus, M., Petsko, G.A., & Ringe, D. (1991) *Biochemistry 30*, 5821–5826.
13. Lodi, P.J., & Knowles, J.R. (1991) *Biochemistry 30*, 6948–6956.
14. Page, M.I., & Jencks, W.P. (1972) *J. Am. Chem. Soc. 94*, 8818–8827.
15. Coward, J.K., & Bruice, T.C. (1969) *J. Am. Chem. Soc. 91*, 5339–5345.
16. Komiyama, M., Roesel, T.R., & Bender, M.L. (1977) *Proc. Natl. Acad. Sci. U.S.A. 74*, 23–25.
17. Mallick, I.M., D'Souza, V.T., Yamaguchi, M., Lee, J., Chalabi, P., Gadwood, R.C., & Bender, M.L. (1984) *J. Am. Chem. Soc. 106*, 7252–7254.
18. Swain, C.G., & Brown, J.F. (1952) *J. Am. Chem. Soc. 74*, 2534–2537.
19. Rony, P.R., McCormack, W.E., & Wunderly, S.W. (1969) *J. Am. Chem. Soc. 91*, 4244–4251.
20. Swain, C.G., & Brown, J.F. (1952) *J. Am. Chem. Soc. 74*, 2538–2543.
21. Hine, J., Cholod, M.S., & Chess, W.K. (1973) *J. Am. Chem. Soc. 95*, 4270–4276.
22. Lienhard, G.E., & Wang, T. (1969) *J. Am. Chem. Soc. 91*, 1146–1153.
23. Knowles, J.R. (1976) *CRC Crit. Rev. Biochem. 4*, 165–173.
24. Herries, D.G., Mathis, A.P., & Rabin, B.R. (1962) *Biochem. J. 85*, 127–134.
25. Cornish-Bowden, A.J., & Knowles, J.R. (1969) *Biochem. J. 113*, 353–362.
26. Plaut, B., & Knowles, J.R. (1972) *Biochem. J. 129*, 311–320.
27. Kiick, D.M., & Cook, P.F. (1983) *Biochemistry 22*, 375–382.
28. Dixon, M. (1953) *Biochem. J. 55*, 161–170.
29. Waley, S.G. (1953) *Biochim. Biophys. Acta 10*, 27–34.
30. Alberty, R.A., & Massey, V. (1954) *Biochim. Biophys. Acta 13*, 347–353.
31. Banerjee, S.K., Kregar, I., Turk, V., & Rupley, J.A. (1973) *J. Biol. Chem. 248*, 4786–4792.
32. Schmidt, D.E., & Westheimer, F.H. (1971) *Biochemistry 10*, 1249–1253.
33. Jencks, W.P. (1959) *J. Am. Chem. Soc. 81*, 475–481.
34. Parsons, S.M., & Raftery, M.A. (1972) *Biochemistry 11*, 1630–1633.
35. Edsall, J.T., & Wyman, J. (1958) *Biophysical Chemistry: Volume I*, pp 477–550, Academic Press, New York.
36. Parsons, S.M., & Raftery, M.A. (1972) *Biochemistry 11*, 1623–1630.
37. Brauenstein, A.E. (1973) in *The Enzymes, Third Edition, Volume IX, Group Transfer* (Boyer, P., Ed.) pp 379–481, Academic Press, New York.
38. Belasco, J.G., Herlihy, J.M., & Knowles, J.R. (1978) *Biochemistry 17*, 2971–2978.
39. Kuo, L.C., Herzberg, W., & Lipscomb, W.N. (1985) *Biochemistry 24*, 4754–4761.
40. Tsukada, H., & Blow, D.M. (1985) *J. Mol. Biol. 184*, 703–711.
41. Blow, D.M. Birktoft, J.J., & Hartley, B.S. (1969) *Nature 221*, 337–340.
42. Cruickshank, W.H., & Kaplan, H. (1972) *Biochem. J. 130*, 1125–1131.

43a. Kyte, J. (1995) *Structure in Protein Chemistry*, p 59, Garland Publishing, New York.

43. Kossiakoff, A.A., & Spencer, S.A. (1981) *Biochemistry 20*, 6462–6474.
44. Hardman, M.J., Valenzuela, P., & Bender, M.L. (1971) *J. Biol. Chem. 246*, 5907–5913.
45. Holbrook, J.J., & Ingram, V.A. (1973) *Biochem. J. 131*, 729–738.
46. Brooks, R.L., Shore, J.D., & Gutfreund, H. (1972) *J. Biol. Chem. 247*, 2382–2383.
47. Stone, S.R., & Morrison, J.F. (1984) *Biochemistry 23*, 2753–2758.
48. Blanchard, J.S., & Cleland, W.W. (1980) *Biochemistry 19*, 3543–3550.
49. Edsall, J.T., & Wyman, J. (1958) *Biophysical Chemistry: Volume I*, pp 450–476, Academic Press, New York.
50. Cleland, W.W. (1977) *Adv. Enzymol. Relat. Areas Mol. Biol. 45*, 273–387.
51. Zannis, V.I., & Kirsch, J.F. (1978) *Biochemistry 17*, 2669–2674.
52. Cohn, E.J., & Edsall, J.T. (1943) *Proteins, Amino Acids, and Peptides as Ions and Dipolar Ions*, pp 105–111, Hafner, New York.
53. Grace, S., & Dunaway-Mariano, D. (1983) *Biochemistry 22*, 4238–4247.
54. Rife, J.E., & Cleland, W.W. (1980) *Biochemistry 19*, 2321–2328.
55. Hupe, D.J., & Pohl, E.R. (1984) *J. Am. Chem. Soc. 106*, 5634–5640.
56. Loewus, M.W. (1977) *J. Biol. Chem. 252*, 7221–7223.
57. Westheimer, F.H. (1961) *Chem. Rev. 61*, 265–273.
58. Bell, R.P., Fendley, J.A., & Hulett, J.R. (1956) *Proc. R. Soc. London, A 235*, 452–468.

59. Barnes, D.J., & Bell, R.P. (1970) *Proc. R. Soc. London, A 318,* 421–440.
60. Swain, C.G., Stivers, E.C., Reuwer, J.F., & Schaad, L.J. (1958) *J. Am. Chem. Soc. 80,* 5885–5893.
61. Northrup, D.B. (1975) *Biochemistry 14,* 2644–2651.
62. O'Leary, M.H., & Limburg, J.A. (1977) *Biochemistry 16,* 1129–1135.
63. Goitein, R.K., Chelsky, D., & Parsons, S.M. (1978) *J. Biol. Chem. 253,* 2963–2971.
64. Rosenberg, S., & Kirsch, J.F. (1981) *Biochemistry 20,* 3196–3204.
65. Cook, P.F., Blanchard, J.S., & Cleland, W.W. (1980) *Biochemistry 19,* 4853–4858.
66. Ngai, K., & Kallen, R.G. (1983) *Biochemistry 22,* 5231–5236.
67. Cleland, W.W. (1975) *Biochemistry 14,* 3220–3224.
68. Schimerlik, M.I., Grimshaw, C.E., & Cleland, W.W. (1977) *Biochemistry 16,* 571–576.
69. Plapp, B.V., Brooks, R.L., & Shore, J.D. (1973) *J. Biol. Chem. 248,* 3470–3475.
70. Miller, S.M., & Klinman, J.P. (1983) *Biochemistry 22,* 3091–3096.
71. Grissom, C.B., & Cleland, W.W. (1985) *Biochemistry 24,* 944–948.
72. Scharschmidt, M., Fisher, M.A., & Cleland, W.W. (1984) *Biochemistry 23,* 5471–5478.
73. Janes, S.M., & Klinman, J.P. (1991) *Biochemistry 30,* 4599–4605.
74. Palcic, M.M., & Klinman, J.P. (1983) *Biochemistry 22,* 5957–5966.
75. Farnum, M., Palcic, M., & Klinman, J.P. (1986) *Biochemistry 25,* 1898–1904.
76. Horenstein, B.A., Parkin, D.W., Estupiñán, B., & Schramm, V.L. (1991) *Biochemistry 30,* 10788–10795.
77. Venkatasubban, K.S., & Schowen, R.L. (1984) *CRC Crit. Rev. Biochem. 17,* 1–44.
78. Hyland, L.J., Tomaszek, T.A., & Meek, T.D. (1991) *Biochemistry 30,* 8454–8463.
79. Shen, T.Y.S., & Westhead, E.W. (1973) *Biochemistry 12,* 3333–3337.
79a. Malhotra, S.K., & Ringold, H.J. (1965) *J. Am. Chem. Soc. 87,* 3228–3236.
80. Gadd, R.E., & Henderson, J.F. (1970) *Biochem. Biophys. Res. Commun. 38,* 363–368.
81. Hanson, K.R. (1966) *J. Am. Chem. Soc. 88,* 2731–2742.
82. Gawron, O., & Fondy, T.P. (1959) *J. Am. Chem. Soc. 81,* 6333–6334.
83. Anet, F.A.L. (1960) *J. Am. Chem. Soc. 82,* 994–995.
84. Asada, Y., Tanizawa, K., Sawada, S., Suzuki, T., Misono, H., & Soda, K. (1981) *Biochemistry 20,* 6881–6886.
85. Keys, L.D., & Johnston, M. (1985) *J. Am. Chem. Soc. 107,* 486–492.
86. Prescott, D.J., & Rabinowitz, J.L. (1968) *J. Biol. Chem. 243,* 1551–1557.
87. Rétey, J., & Lynen, F. (1965) *Biochem. Z. 342,* 256–271.
88. Cohn, M., Pearson, J.E., O'Connell, E.L., & Rose, I.A. (1970) *J. Am. Chem. Soc. 92,* 4095–4098.
89. Bartsch, R.A., & Zavada, J. (1980) *Chem. Rev. 80,* 453–494.
90. Rétey, J., Fierz, H., & Zeylemaker, W.P. (1970) *FEBS Lett. 6,* 203–205.
91. Sawada, S., Tanaka, A., Inouye, Y., Hirasawa, T., & Soda, K. (1974) *Biochim. Biophys. Acta 350,* 354–357.
92. Schroepfer, G.J. (1966) *J. Biol. Chem. 241,* 5441–5447.
93. Mohrig, J.R., Schultz, S.C., & Morin, G. (1983) *J. Am. Chem. Soc. 105,* 5150–5151.
94. Wilton, D.C., & Akhtar, M. (1970) *Biochem. J. 116,* 337–339.
95. Clayton, R.B., & Edwards, A.M. (1963) *J. Biol. Chem. 238,* 1966–1972.
96. Biellman, J.F., & Hirth, C.G. (1970) *FEBS Lett. 8,* 55–56.
97. Biellman, J.F., & Hirth, C.G. (1970) *FEBS Lett. 9,* 335–336.
98. Rose, I.A., O'Connell, E.L., & Mortlock, R.P. (1969) *Biochim. Biophys. Acta 178,* 376–379.
99. Rose, I.A., & O'Connell, E.L. (1961) *J. Biol. Chem. 236,* 3086–3092.
100. Simon, H., & Medina, R. (1966) *Z. Naturforsch. 21B,* 496–497.
101. Chang, M.N.T., & Walsh, C.T. (1981) *J. Am. Chem. Soc. 103,* 4921–4927.
102. Chang, M.N.T., & Walsh, C.T. (1981) *J. Am. Chem. Soc. 103,* 7797.
103. Smissman, E.E., Suh, J.T., Oxman, M., & Daniels, R. (1962) *J. Am. Chem. Soc. 84,* 1040–1041.
104. Hill, R.K., & Newkome, G.R. (1969) *J. Am. Chem. Soc. 91,* 5893–5894.
105. Fisher, H.F., Frieden, C., McKee, J.S.M., & Alberty, R.A. (1955) *J. Am. Chem. Soc. 77,* 4436–4438.
106. Farrar, T.C., Gutowsky, H.S., Alberty, R.A., & Miller, W.G. (1957) *J. Am. Chem. Soc. 79,* 3978–3980.
107. Cornforth, J.W., Redmond, J.W., Eggerer, H., Buckel, W., & Gutshow, C. (1969) *Nature 221,* 1212–1213.
108. Lüthy, J., Rétey, J., & Arigoni, D. (1969) *Nature 221,* 1213–1215.
109. Townsend, C.A., Scholl, T., & Arigoni, D. (1975) *J. Chem. Soc., Chem. Commun. 22,* 921–922.
110. Coates, R.M., Koch, S.C., & Hegde, S. (1986) *J. Am. Chem. Soc. 108,* 2762–2764.
111. Lenz, H., & Eggerer, H. (1976) *Eur. J. Biochem 65,* 237–246.
112. Eggerer, H., Buckel, W., Lenz, H., Wunderwald, P., Gottschalk, G., Cornforth, J.W., Donninger, C., Mallaby, R., & Redmond, J.W. (1970) *Nature 226,* 517–519.
113. Cornforth, J.W., Clifford, K., Mallaby, R., & Phillips, G.T. (1972) *Proc. R. Soc. London, B 182,* 277–295.
114. Popjak, G., & Cornforth, J.W. (1966) *Biochem. J. 101,* 553–568.
115. Rose, I.A. (1970) *J. Biol. Chem. 245,* 6052–6056.
116. Meloche, H.P., & Mehler, L. (1973) *J. Biol. Chem. 248,* 6333–6338.
117. Rétey, J., Suckling, C.J., Arigoni, D., & Babior, B.M. (1974) *J. Biol. Chem. 249,* 6359–6360.
118. Diziol, P., Haas, H., Rétey, J., Graves, S.W., & Babior, B.M (1980) *Eur. J. Biochem. 116,* 211–224.
119. Abbot, S.J., Jones, S.R., Weinman, S.A., Bockhoff, F.M., McLafferty, F.W., & Knowles, J.R. (1979) *J. Am. Chem. Soc. 101,* 4323–4332.
120. Cullis, P.M., & Lowe, G. (1981) *J. Chem. Soc., Perkin Trans. 1,* 2317–2321.
121. Blättler, W.A., & Knowles, J.R. (1979) *Bioochemistry 18,* 3927–3933.
122. Buchwald, S.L., & Knowles, J.R. (1980) *J. Am. Chem. Soc. 102,* 6601–6602.
123. Jarvest, R.L., Lowe, G., & Potter, B.V.L. (1981) *J. Chem. Soc., Perkin Trans. 1,* 3186–3195.
124. Eckstein, F., Romaniuk, P.J., & Connolly, B.A. (1982) *Methods Enzymol. 87,* 197–212.
125. Richard, J.P., & Frey, P.A. (1978) *J. Am. Chem. Soc. 100,* 7757–7758.
126. Richard, J.P., Ho, H., & Frey, P.A. (1978) *J. Am. Chem. Soc. 100,* 7756–7757.
127. Jones, S.R., Kindman, L.A., & Knowles, J.R. (1978) *Nature 275,* 564–565.

128. Sowadski, J.M., Handschumacher, M.D., Krishna Murthy, H.M., Foster, B.A., & Wyckoff, H.W. (1985) *J. Mol. Biol. 186*, 417–433.
129. Sheu, K.R., Richard, J.P., & Frey, P.A. (1979) *Biochemistry 18*, 5548–5556.
130. Garces, E., & Cleland, W.W. (1969) *Biochemistry 8*, 633–640.
131, Pai, E.F., Sachsenheimer, W., Schirmer, R.H., & Schultz, G.E. (1977) *J. Mol. Biol. 114*, 37–45.
132, Webb, M.R., & Trentham, D.R. (1980) *J. Biol. Chem. 255*, 1775–1779.
133. Webb, M., Grubmeyer, C., Penefsky, H.S., & Trentham, D.R. (1980) *J. Biol. Chem. 255*, 11637–11639.
134. Webb, M.R., & Trentham, D.R. (1981) *J. Biol. Chem. 256*, 4884–4887.
135. Bastide, F., Meissner, G., Fleischer, S., & Post, R.L. (1973) *J. Biol. Chem. 248*, 8385–8391.
136. Webb, M.R., & Trentham, D.R. (1980) *J. Biol. Chem. 255*, 8629–8632.
136a. Cheung, Y., & Walsh, C. (1976) *J. Am. Chem. Soc. 98*, 3397–3398.
136b. Ikeda, Y., & Tanaka, K. (1983) *J. Biol. Chem. 258*, 9477–9487.
137. Anderson, V.E., & Hammes, G.G. (1984) *Biochemistry 23*, 2088–2094.
138. Mohrig, J.R., Vreede, P.J., Schultz, S.C., & Fierke, C.A. (1981) *J. Org. Chem. 46*, 4655–4658.
139. Sedgwick, B., & Cornforth, J.W. (1977) *Eur. J. Biochem. 75*, 465–479.
139a. Rieder, S.V., & Rose, I.A. (1959) *J. Biol. Chem. 234*, 1007–1010.
139b. Rose, I.A. (1958) *J. Am. Chem. Soc. 80*, 5835–5836.
139c. Grimshaw, C.E., Sogo, S.G., Copley, S.D., & Knowles, J.R. (1984) *J. Am. Chem. Soc. 106*, 2699–2700.
140. Rose, I.A., & O'Connell, E.L. (1969) *J. Biol. Chem. 244*, 6548–6557.
141. Hartman, F.C. (1968) *Biochem. Biophys. Res. Commun. 33*, 888–894.
142. Coulson, A.F.W., Knowles, J.R., Priddle, J.D., & Offord, R.E. (1970) *Nature 227*, 180–181.
143. Hartman, F.C. (1971) *Biochemistry 10*, 146–154.
144. Corran, P.H., & Waley, S.G. (1975) *Biochem. J. 145*, 335–344.
145. Lolis, E., & Petsko, G.A. (1990) *Biochemistry 29*, 6619–6625.
146. Straus, D., Raines, R., Kawashima, E., Knowles, J.R., & Gilbert, W. (1985) *Proc. Natl. Acad. Sci. U.S.A. 82*, 2272–2276.
147. Kalman, T.J., & Goldman, D. (1981) *Biochem. Biophys. Res. Commun. 102*, 682–689.
148. Gold, A.M., & Fahrney, D. (1964) *Biochemistry 3*, 783–791.
149. Wilde, J., Hunt, W., & Hupe, D.J. (1977) *J. Am. Chem. Soc. 99*, 8319–8321.
150. Motiu-DeGrood, R., Hunt, W., Wilde, J., & Hupe, D.J. (1979) *J. Am. Chem. Soc. 101*, 2182–2190.
151. Kyte, J. (1981) *J. Biol. Chem. 256*, 3231–3232.
152. Meloche, H.P. (1967) *Biochemistry 6*, 2273–2280.
153. Roche, T.E., & McFadden, B.A. (1969) *Biochem. Biophys. Res. Commun. 37*, 239–246.
154. Hass, G.M., & Neurath, H. (1971) *Biochemistry 10*, 3535–3540.
155. Hartman, F.C. (1963) *J. Biol. Chem. 238*, 3036–3047.
156. O'Leary, M., & Diaz, E. (1982) *J. Biol. Chem. 257*, 14603–14605.
157. Hartman, F.C. (1981) *Biochemistry 20*, 894–898.
158. Lienhard, G.E., & Rose, I.A. (1964) *Biochemistry 3*, 185–191.
159. Okamoto, M., & Morino, Y. (1973) *J. Biol. Chem. 248*, 82–90.
160. Vollmer, S.J., Switzer, R.L., Hermodson, M.A., Bower, S.G., & Zalkin, H. (1983) *J. Biol. Chem. 258*, 10582–10585.
161. Hass, G.M., & Neurath, H. (1971) *Biochemistry 10*, 3541–3546.
162. Lipscomb, W.N., Hartsuck, J.A., Reeke, G.N., Quicho, F.A., Bethge, P.H., Ludwig, M.L., Steitz, T.A., Muirhead, H., & Coppola, J.C. (1968) *Brookhaven Symp. Biol. 21*, 24–90.
163. O'Connell, E.L., & Rose, I.A. (1973) *J. Biol. Chem. 248*, 2225–2231.
164. Bloch, K. (1987) *Annu. Rev. Biochem. 56*, 1–19.
165. Rando, R.R. (1974) *Biochem. Pharmacol. 23*, 2328–2331.
166. Endo, K., Helmkamp, G.M., & Bloch, K. (1970) *J. Biol. Chem. 245*, 4293–4296.
167. Johnston, M., Jankowski, D., Marcotte, P., Tanaka, H., Esaki, N., Soda, K., & Walsh, C. (1979) *Biochemistry 18*, 4690–4701.
168. Abeles, R.H., & Walsh, C.T. (1973) *J. Am. Chem. Soc. 95*, 6124–6125.
169. Johnston, M., Jankowski, D., Marcotte, P., Tanaka, H., Esaki, N., Soda, K., & Walsh, C. (1979) *Biochemistry 18*, 4690–4701.
170. Fendrich, G., & Abeles, R.H. (1982) *Biochemistry 21*, 6685–6695.
171. Metcalf, B.W., Bey, P., Danzin, C., Jung, M.J., Casara, P., & Vevert, J.P. (1978) *J. Am. Chem. Soc. 100*, 2551–2553.
172. Wang, E.A., Kallen, R., & Walsh, C. (1981) *J. Biol. Chem. 256*, 6917–6926.
173. Roise, D., Soda, K., Yagi, T., & Walsh, C. (1984) *Biochemistry 23*, 5195–5201.
174. Ueno, H., Likos, J.J., & Metzler, D.E. (1982) *Biochemistry 21*, 4387–4393.
175. Lorimer, G.H. (1978) *Eur. J. Biochem. 89*, 43–50.
176. Pierce, J., Tolbert, N.E., & Barker, R. (1980) *J. Biol. Chem. 255*, 509–511.
177. Fraij, B., & Hartman, F.C. (1982) *J. Biol. Chem. 257*, 3501–3505.
178. Hexter, C.S., & Westheimer, F.H. (1971) *J. Biol. Chem. 246*, 3928–3933.
179. Wyatt, J.L., & Colman, R.F. (1977) *Biochemistry 16*, 1333–1342.
180. Hirth, C.G., Veron, M., Villar-Palasi, C., Hurion, N., & Cohen, G.N. (1975) *Eur. J. Biochem. 50*, 425–430.
181. Miles, E.W., & Phillips, R.S. (1985) *Biochemistry 24*, 4694–4703.
182. Jeck, R., Woenckhaus, C., Harris, J.I., & Runswick, M.J. (1979) *Eur. J. Biochem. 93*, 57–64.
183. Chase, J.A.F., & Tubbs, P.K. (1970) *Biochem. J. 116*, 713–720.
184. Rashidbaigi, A., & Ruoho, A.E. (1981) *Proc. Natl. Acad. Sci. U.S.A. 78*, 1609–1613.
185. Pomerantz, A.H., Rudolph, S.A., Haley, B.E., & Greengard, P. (1975) *Biochemistry 14*, 3858–3862.
186. Jeng, S.J., & Guillory, R.J. (1975) *J. Supramol. Struct. 3*, 448–468.
187. de Waal, A., de Jong, L., Hartog, A.F., & Kemp, A. (1985) *Biochemistry 24*, 6493–6499.
188. Harris, I., Meriwether, B.P., & Park, J.H. (1963) *Nature 198*, 154–157.
189. Buehner, M., Ford, G.C., Moras, D., Olsen, K.W., & Rossmann, M.G. (1974) *J. Mol. Biol. 90*, 25–49.
190. Patthy, L., Varadi, A., Thész, J., & Kovacs, K. (1979) *Eur. J. Biochem. 99*, 309–313.
191. Sygush, J., Beaudry, D., & Allaire, M. (1987) *Proc. Natl. Acad. Sci. U.S.A. 84*, 7846–7850.
192. Crestfield, A.M., Stein, W.H., & Moore, S. (1963) *J. Biol. Chem. 238*, 2413–2419.
193. Wyckoff, H.W., Tsernoglou, D., Hanson, A.W., Knox, J.R., Lee, B., & Richards, F.M. (1970) *J. Biol. Chem. 245*, 305–328.
194. Wlodawer, A., Miller, M., & Sjolin, L. (1983) *Proc. Natl. Acad. Sci. U.S.A. 80*, 3628–3631.
195. Poulos, T.L., & Price, P.A. (1971) *J. Biol. Chem. 246*, 4041–4045.
196. Esch, F.S., Böhlen, P., Otsuka, A.S., Yoshida, M., & Allison, W.S. (1981) *J. Biol. Chem. 256*, 9084–9089.
197. Husain, S.S., & Lowe, G. (1968) *Biochem. J. 108*, 861–866.

198. Husain, S.S., & Lowe, G. (1970) *Biochem. J. 117*, 341–346.
199. Marshall, M., & Cohen, P.P. (1980) *J. Biol. Chem. 255*, 7301–7305.
200. Fan, C.C., & Plaut, G.W.E. (1974) *Biochemistry, 13*, 45–51.
201. Adamson, S.R., Robinson, J.A., & Stevenson, K.J. (1984) *Biochemistry 23*, 1269–1274.
202. Verheij, H.M., Volwerk, J.J., Jansen, E.H.J.M., Puyk, W.C., Dijkstra, B.W., Drenth, J., & de Haas, G.H. (1980) *Biochemistry 19*, 743–750.
203. Kassab, R., Fattoum, A., & Pradel, L.A. (1970) *Eur. J. Biochem. 12*, 264–269.
204. Khalifah, R.G., & Edsall, J.T. (1972) *Proc. Natl. Acad. Sci. U.S.A. 69*, 172–176.
205. Kindman, L.A., & Jencks, W.P. (1981) *Biochemistry 20*, 5183–5187.
206. Tanizawa, K., & Miles, E.W. (1983) *Biochemistry 22*, 3594–3603.
207. Hyde, C.C., Ahmed, S.A., Padlan, E.A., Miles, E.W., & Davies, D.R. (1988) *J. Biol. Chem. 263*, 17857–17871.
208. Wenz, A., Ghisla, S., & Thorpe, C. (1985) *Eur. J. Biochem. 147*, 553–560.
209. Hartman, F.C., Suh, B., Welch, M.H., & Barker, R. (1973) *J. Biol. Chem. 248*, 8233–8239.
210. Ehrlich, R.S., & Colman, R.F. (1987) *J. Biol. Chem. 262*, 12614–12619.
211. Smyth, G.E., & Colman, R.F. (1991) *J. Biol. Chem. 266*, 14918–14925.
212. Smyth, G.E., & Colman, R.F. (1992) *Arch. Biochem. Biophys. 293*, 356–361.
213. Bayliss, R.S., Knowles, J.R., & Wybrandt, G.B. (1969) *Biochem. J. 113*, 377–386.
214. Tang, J., Sepulveda, P., Marciniszyn, J., Chen, K.C.S., Huang, W.Y., Tao, N., Liu, D., & Lanier, J.P. (1973) *Proc. Natl. Acad. Sci. U.S.A. 70*, 3437–3439.
215. Gustchina, A.E., & Andreeva, N.S. (1985) *Mol. Biol. (USSR) 19*, 225–229.
216. Hartsuck, J.A., & Tang, J. (1972) *J. Biol. Chem. 247*, 2575–2580.
217. James, M.N.G., & Sielecki, A.R. (1983) *J. Mol. Biol. 163*, 299–361.
218. Meloche, H.P., & Wood, W.A. (1964) *J. Biol. Chem. 239*, 3511–3514.
219. Grazi, E., Meloche, H., Martinez, G., Wood, W.A., & Horecker, B.L. (1963) *Biochem. Biophys. Res. Commun. 10*, 4–10.
220. Meloche, H.P. (1973) *J. Biol. Chem. 248*, 6945–6951.
221. Meloche, H.P., Monti, C.T., & Hogue-Angeletti, R.A. (1978) *Biochem. Biophys. Res. Commun. 84*, 589–594.
222. Suzuki, N., & Wood, W.A. (1980) *J. Biol. Chem. 255*, 3427–3435.
223. Blow, D.M., & Brick, P. (1985) in *Biological Macromolecules and Assemblies, Volume 2: Nucleic Acids and Interactive Proteins* (Jurnak, F.A., & McPherson, A., Eds.) pp 442–469, Wiley-Interscience, New York.
224. Brick, P., & Blow, D.M. (1987) *J. Mol. Biol. 194*, 287–297.
225. Leatherbarrow, R.J., Fersht, A.R., & Winter, G. (1985) *Proc. Natl. Acad. Sci. U.S.A. 82*, 7840–7844.
226. Fersht, A.R., Knill-Jones, J.W., Bedouelle, H., & Winter, G. (1988) *Biochemistry 27*, 1581–1587.
227. Leatherbarrow, R.J., & Fersht, A.R. (1987) *Biochemistry 26*, 8524–8528.
228. Wells, T.N.C., & Fersht, A.R. (1986) *Biochemistry 25*, 1881–1886.
229. Ho, C.K., & Fersht, A.R. (1986) *Biochemistry 25*, 1891–1897.
230. Fersht, A.R., Leatherbarrow, R.J., & Wells, T.N.C. (1987) *Biochemistry 26*, 6030–6038.
231. Lowe, D.M., Winter, G., & Fersht, A.R. (1987) *Biochemistry 26*, 6038–6043.
232. Fersht, A.R., Shi, J., Knill-Jones, J., Lowe, D.M., Wilkinson, A.J., Blow, D.M., Brick, P., Carter, P., Waye, M.M.Y., & Winter, G. (1985) *Nature 314*, 235–238.
233. Brown, K.A., Brick, P., & Blow, D.M. (1987) *Nature 326*, 416–418.
234. Kochhar, S., Finlayson, W.L., Kirsch, J.F., & Christen, P. (1987) *J. Biol. Chem. 262*, 11446–11448.
235. Kirsch, J.F., Eichele, G., Ford, G., Vincent, M.G., Jansonius, J.N., Gehring, H., & Christen, P. (1984) *J. Mol. Biol. 174*, 497–525.
236. Toney, M.D., & Kirsch, J.F. (1989) *Science 243*, 1485–1488.
237. Toney, M.D., & Kirsch, J.F. (1987) *J. Biol. Chem. 262*, 12403–12405.
238. Fishel, L.A., Villafranca, J.E., Mauro, J.M., & Kraut, J. (1987) *Biochemistry 26*, 351–360.
238a. Wang, E.A., Kallen, R., & Walsh, C. (1981) *J. Biol. Chem. 256*, 6917–6926.
238b. Moult, J., Eshdat, Y., & Sharon, N. (1973) *J. Biol. Chem. 75*, 1–4.
238c. Eshdat, Y., Dunn, A., & Sharon, N. (1974) *Proc. Natl. Acad. Sci. U.S.A. 71*, 1658–1662
238d. Meloche, H.P., & Wood, W.A. (1964) *J. Biol. Chem. 239*, 3511–3514.
238e. Meloche, H.P. (1967) *Biochemistry 6*, 2273–2280.
238f. Meloche, H.P. (1973) *J. Biol. Chem. 248*, 6945–6951.
238g. Rose, I.A., & O'Connell, E.L. (1969) *J. Biol. Chem. 244*, 6548–6557.
238h. Miller, J.C., & Waley, S.G. (1971) *Biochem. J. 123*, 163–170.
239. Bednar, R.A., Hartman, F.C., & Colman, R.F. (1982) *Biochemistry 21*, 3681–3689.
240. Helmkamp, G., & Bloch, K. (1969) *J. Biol. Chem. 244*, 6014–6022.
241. Grau, U., Kapmeyer, H., & Trommer, W.E. (1978) *Biochemistry 17*, 4621–4626.
241a. Winer, A.D., & Schwert, G.W. (1958) *J. Biol. Chem. 231*, 1065–1083.
242. Suck, D., Oefner, C., & Kabsch, W. (1984) *EMBO J. 3*, 2423–2430.
243. Ford, L.O., Johnson, L.N., Machin, P.A., Phillips, D.C., & Tjian, R. (1974) *J. Mol. Biol. 88*, 349–371.
244. Remington, S.R., Wiegand, G., & Huber, R. (1982) *J. Mol. Biol. 158*, 111–152.
245. Rétey, J., Lüthy, J., & Arigoni, D. (1970) *Nature 226*, 519–521.
246. Collins, K.D. (1974) *J. Biol. Chem. 249*, 136–142.
247. Hartman, F.C. (1970) *Biochem. Biophys. Res. Commun. 39*, 384–388.
248. Waley, S.G., Mill, J.C., Rose, I.A., & O'Connell, E.L. (1970) *Nature 227*, 181.
249. Komives, E.A., Chang, L.C., Lolis, E., Tilton, R.F., Petsko, G.A., & Knowles, J.R. (1991) *Biochemistry 30*, 3011–3019.
250. Nickbarg, E.B., Davenport, R.C., Petsko, G.A., & Knowles, J.R. (1988) *Biochemistry 27*, 5948–5960.
251. Bash, P.A., Field, M.J., Davenport, R.C., Petsko, G.A., Ringe, D., & Karplus, M. (1991) *Biochemistry 30*, 5826–5832.
252. Tronrud, D.E., Monzingo, A.F., & Matthews, B.W. (1986) *Eur. J. Biochem. 157*, 261–268.
253. Birktoft, J.J., & Banaszak, L.J. (1983) *J. Biol. Chem. 258*, 472–482.
254. Verschueren, K.H.G., Seljee, F., Rozeboom, H.J., Kalk, K.H., & Dijkstra, B.W. (1993) *Nature 363*, 693–698.
255. Wang, J.H. (1968) *Science 161*, 328–334.
256. Kadonaga, J.T., & Knowles, J.R. (1983) *Biochemistry 22*, 130–136.
257. Kobes, R.D., Simpson, R.T., Vallee, B.L., & Rutter, W.J. (1969) *Biochemistry 8*, 585–588.
258. Ray, W.J. (1969) *J. Biol. Chem. 244*, 3740–3747.
259. Ray, W.J., Goodin, D.S., & Ng, L. (1972) *Biochemistry 11*, 2800–2804.
260. Colombo, G., Carlson, G.M., & Lardy, H.A. (1981) *Biochemistry 20*, 2749–2757.
261. Middlehoven, W.J. (1969) *Biochim. Biophys. Acta 191*, 110–121.
261a. Kyte, J. (1995) *Structure in Protein Chemistry*, p 332, Garland Publishing, New York.

262. Honzatko, R.B., Crawford, J.L., Monaco, H.L., Ladner, J.E., Edwards, B.F., Evans, D.R., Warren, S.G., Wiley, D.C., Ladner, R.C., & Lipscomb, W.N. (1982) *J. Mol. Biol. 160*, 219–263.
263. Rosenbusch, J.P., & Weber, K. (1971) *Proc. Natl. Acad. Sci. U.S.A. 68*, 1019–1023.
264. Nelbach, M.E., Pigiet, V.P., Gerhart, J.C., & Schachman, H.K. (1972) *Biochemistry 11*, 315–327.
265. Carvajal, N., Venegas, A., Oestreicher, G., & Plaza, M. (1971) *Biochim. Biophys. Acta 250*, 437–442.
266. Huheey, J.E. (1972) *Inorganic Chemistry, Principles of Structure and Reactivity*, p 214, Harper & Row, New York.
267. Pearson, R.G. (1966) *Science 151*, 172–177.
268. Martin, R.B. (1984) in *Metal Ions in Biological Systems: Volume 17, Calcium and Its Role in Biology* (Sigel, H., Ed.) pp 1–50, Marcel Dekker, New York.
269. Einspahr, H., & Bugg, C.E. (1984) in *Metal Ions in Biological Systems: Volume 17, Calcium and Its Role in Biology* (Sigel, H., Ed.) pp 51–97, Marcel Dekker, New York.
270. Kim, K.H., Pan, Z., Honzatko, R.B., Ke, H., & Lipscomb, W.N. (1987) *J. Mol. Biol. 196*, 853–875.
271. Holmes, M.A., & Matthews, B.W. (1981) *Biochemistry 20*, 6912–6920.
272. Margerum, D.W., Cayley, G.R., Weatherburn, D.C., & Pagenkopf, G.K. (1978) in *Coordination Chemistry: Volume 2* (Martell, A.E., Ed.) ACS Monograph 174, pp 1–220, American Chemical Society, Washington, DC.
273. Kubala, G., & Martell, A.E. (1982) *J. Am. Chem. Soc. 104*, 6602–6609.
274. Suelter, C.H. (1970) *Science 168*, 789–795.
275. Kachmar, J.F., & Boyer, P.D. (1953) *J. Biol. Chem. 200*, 669–682.
276. Suelter, C.H., Singleton, R., Kayne, F.J., Arrington, S., Glass, J., & Mildvan, A.S. (1966) *Biochemistry 5*, 131–139.
277. Reuben, J., & Kayne, F.J. (1971) *J. Biol. Chem. 246*, 6227–6234.
278. Nowak, T., & Mildvan, A.S. (1972) *Biochemistry 11*, 2819–2828.
279. Muirhead, H., Clayden, D.A., Cuffe, S.P., & Davies, C. (1987) *Biochem. Soc. Trans. 15*, 996–999.
280. Bode, W., & Schwager, P. (1975) *J. Mol. Biol. 98*, 693–717.
281. Suck, D., & Oefner, C. (1986) *Nature 321*, 620–625.
282. Cotton, F.A., Hazen, E.E., & Legg, M.J. (1979) *Proc. Natl. Acad. Sci. U.S.A. 76*, 2551–2555.
283. Cini, R., Sabat, M., Sundaralingam, M., Burla, M.C., Nunzi, A., Polidori, G., & Zanazzi, P.F. (1983) *J. Biomol. Struct. Dyn. 1*, 633–637.
284. Alberty, R.A. (1969) *J. Biol. Chem. 244*, 3290–3302.
285. Brems, D.N., & Rilling, H.C. (1977) *J. Am. Chem. Soc. 99*, 8351–8352.
286. King, H.J., & Rilling, H.C. (1977) *Biochemistry 16*, 3815–3819.
287. Dunaway-Mariano, D., Benovic, J.L., Cleland, W.W., Gupta, R.K., & Mildvan, A.S. (1979) *Biochemistry 18*, 4347–4354.
288. Dougherty, T.M., & Cleland, W.W. (1985) *Biochemistry 24*, 5870–5875.
289. Ray, W.J., & Multani, J.S. (1972) *Biochemistry 11*, 2805–2812.
290. Cohen, P.F., & Colman, R.F. (1974) *Eur. J. Biochem. 47*, 35–45.
291. Kerr, M.C., Preston, H.S., Ammon, H.L., Huheey, J.E., & Stewart, J.M. (1981) *J. Coord. Chem. 11*, 111–115.
292. Matthews, B.W., Weaver, L.H., & Kester, W.R. (1974) *J. Biol. Chem. 249*, 8030–8044.
293. Kester, W.R., & Matthews, B.W. (1977) *Biochemistry 16*, 2506–2516.
294. Holden, H.M., Tronrud, D.E., Monzingo, A.F., Weaver, L.H., & Matthews, B.W. (1987) *Biochemistry 26*, 8542–8553.
295. Silverman, D.N., & Vincent, S.H. (1983) *CRC Crit. Rev. Biochem. 14*, 207–255.
296. Eriksson, A.E., Jones, T.A., & Liljas, A. (1988) *Proteins: Struct., Funct., Genet. 4*, 274–282.
297. Liljas, A., Kannan, K.K., Bergsten, P.C., Waara, I., Fridborg, K., Strandberg, B., Carlbom, U., Jarup, L., Lövgren, S., & Petef, M. (1972) *Nature, New Biol. 235*, 131–137.
298. Khalifah, R.G. (1971) *J. Biol. Chem. 246*, 2561–2573.
299. Pocker, Y., & Bjorkquist, D.W. (1977) *Biochemistry 16*, 5698–5707.
300. Woolley, P. (1975) *Nature 258*, 677–682.
301. Kannan, K.K., Petef, M., Fridborg, K., Cid-Dresdner, H., & Lövgren, S. (1977) *FEBS Lett. 73*, 115–119.
302. Eriksson, A.E., Kylsten, P.M., Jones, T.A., & Liljas, A. (1988) *Proteins: Struct., Funct., Genet. 4*, 283–293.
303. Kanamori, K., & Roberts, J.D. (1983) *Biochemistry 22*, 2658–2664.
304. Eklund, H., Plapp, B.V., Samana, J.P., & Brändén, C.I. (1982) *J. Biol. Chem. 257*, 14349–14358.
305. Evans, S.A., & Shore, J.D. (1980) *J. Biol. Chem. 255*, 1509–1514.
306. Guss, J.M., & Freeman, H.C. (1983) *J. Mol. Biol. 169*, 521–563.
307. Norris, G.E., Anderson, B.F., & Baker, E.N. (1983) *J. Mol. Biol. 165*, 501–521.
308. Garrett, T.P., Clingeleffer, D.J., Guss, J.M., Rogers, S.J., & Freeman, H.C. (1984) *J. Biol. Chem. 259*, 2822–2825.
309. Tainer, J.A., Getzoff, E.D., Richardson, J.S., & Richardson, D.C. (1983) *Nature 306*, 284–287.
310. Tainer, J.A., Getzoff, E.D., Beem, K.M., Richardson, J.S., & Richardson, D.C. (1982) *J. Mol. Biol. 160*, 181–217.
311. McAdam, M.E., Fielden, E.M., LaVelle, F., Calabrese, L., Cocco, D., & Rotilio, G. (1977) *Biochem. J. 167*, 271–274.
312. Lawrence, G.D., & Sawyer, D.T. (1979) *Biochemistry 18*, 3045–3050.
313. Kennedy, S.C., Rauner, R., & Gawron, O. (1972) *Biochem. Biophys. Res. Commun. 47*, 740–745.
314. Dreyer, J.L. (1985) *Eur. J. Biochem. 150*, 145–154.
315. Kennedy, M.C., Emptage, M.H., Dreyer, J., & Beinert, H. (1983) *J. Biol. Chem. 258*, 11098–11105.
316. Kennedy, M.C., Werst, M., Telser, J., Emptage, M.H., Beinert, H., & Hoffman, B.M. (1987) *Proc. Natl. Acad. Sci. U.S.A. 84*, 8854–8858.
317. Brown, N.C., Eliasson, R., Reichard, P., & Thelander, L. (1969) *Eur. J. Biochem. 9*, 512–518.
318. Whittaker, J.W., & Lipscomb, J.D. (1984) *J. Biol. Chem. 259*, 4487–4495.
319. Laufter, R.B., Antanaitis, B.C., Aisen, P., & Que, L. (1983) *J. Biol. Chem. 258*, 14212–14218.
320. Aberhart, D.J., Gould, S.J., Lin, H., Thiruvengadam, T.K., & Weiller, B.H. (1983) *J. Am. Chem. Soc. 105*, 5461–5470.
321. Robson, R.L., Eady, R.R., Richardson, T.H., Miller, R.W., Hawkins, M., & Postgate, J.R. (1986) *Nature 322*, 388–390.
322. Dixon, N.E., Gazzola, C., Blakeley, R.L., & Zerner, B. (1975) *J. Am. Chem. Soc. 97*, 4131–4133.
323. Teixeira, M., Moura, I., Xavier, A.V., Devartanian, D.V., LeGall, J., Peck, H.D., Huynh, B.H., & Moura, J.J.G. (1983) *Eur. J. Biochem. 130*, 481–484.
324. Ragsdale, S.W., Clark, J.E., Ljungdahl, L.G., Lundie, L.L., & Drake, H.L. (1983) *J. Biol. Chem. 258*, 2364–2369.
325. Ellefson, W.L., Whitman, W.B., & Wolfe, R.S. (1982) *Proc. Natl. Acad. Sci. U.S.A. 79*, 3707–3710.
326. Hamilton, C.L., Scott, R.A., & Johnson, M.K. (1989) *J. Biol. Chem. 264*, 11605–11613.
327. Yamamoto, I., Saiki, T., Liu, S.,& Ljungdahl, L.G. (1983) *J. Biol. Chem. 258*, 1826–1832.
328. Loll, P.J., & Lattman, E.E. (1989) *Proteins: Struct., Funct., Genet. 5*, 183–201.

## Chapter FIVE

# Intermediates in Enzymatic Reactions

Almost all of the chemical reactions catalyzed by enzymes pass through two or more transition states and one or more intermediates of high energy. An intermediate in the mechanism of any chemical reaction can be distinguished from a transition state in that mechanism by the length of time it exists. A **transition state** lasts only as long as is necessary to pass over a maximum in the reaction coordinate (Figure 3–23). Because the movement over such a maximum resembles only half the stroke of an intramolecular vibration, the time required to pass over the top is less than the time required for a complete intramolecular vibration ($<10^{-12}$ s). Because the movement through the transition state is over a maximum of potential energy, a transition state does not have a finite lifetime. An **intermediate** in a reaction is an arrangement of atoms that persists longer than the time required for the intramolecular and intermolecular vibrations among those atoms to become established. For any arrangement of atoms to persist for this length of time, the connections holding them together must be full covalent bonds or fully established noncovalent interactions, not partial bonds in the process of forming and unforming as is the case in a transition state. In this sense, an intermediate in any reaction must itself be either one molecule or the noncovalent complex of two or more molecules, even though these molecules may be of higher energy than either the reactants or the products. Because intermediates have finite lifetimes, an intermediate in any reaction must have a potential energy that is less than the potential energy of the transition state immediately preceding it on the reaction coordinate leading from reactants and less than the potential energy of the transition state immediately following it on the reaction coordinate proceeding toward products. In other words, the constellation of atoms defining an intermediate is always an arrangement that occupies a potential well on the reaction coordinate (Figure 5–1). If the potential energy of an intermediate were not such a local minimum on the reaction coordinate, the intermediate would not have a finite lifetime.

Two examples of reactions occurring in solution will serve to illustrate the distinction between transition states and intermediates. An example of a reaction that proceeds without an intermediate is the isomerization of chorismate to prephenate (Figure 1–16). The reactant passes through a single transition state that is cyclic and aromatic* to produce the product directly. An example of a reaction in which there are

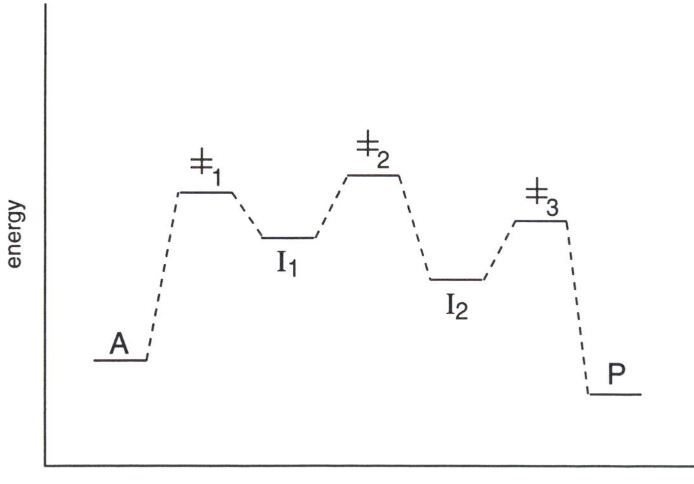


**Figure 5–1:** Diagrammatic representation of a reaction coordinate between reactant (A) and product (P) that passes through three transition states ($\ddagger_1$, $\ddagger_2$, and $\ddagger_3$) and two intermediates ($I_1$ and $I_2$). Potential energies are presented as a function of distance along the reaction coordinate.

*An aromatic transition state, as does an aromatic molecule, contains an unbroken ring of atoms, each possessing a *p* orbital parallel to and overlapping with those of its immediate neighbors and together composing a $\pi$ molecular orbital system containing $4n + 2$ electrons.

intermediates is the hydrolysis of an amide (Figure 1–7). In this reaction, the tetravalent intermediate formed must exist for a period of time sufficient for the protonation of the nitrogen that is on the leaving group to occur. Otherwise, the leaving group would be an amide anion, which is 17 orders of magnitude more basic than the hydroxide anion or 35 orders of magnitude more basic than the molecule of water that attacked the acyl carbon in the first place. Furthermore, if water was the initial nucleophile rather than hydroxide anion, the tetravalent intermediate must also exist long enough for the proton to be removed from the cationic oxygen that was derived from the oxygen of the water. Each of these protonations and deprotonations produces a new tetravalent intermediate of unique structure, each a molecule in its own right, and each protonation or deprotonation itself passes through a transition state, even though the respective free energies of activation for each of these transition states may be only slightly greater than the chemical potentials of the respective tetravalent intermediates.

An alternative to the existence of the tetravalent intermediate in the hydrolysis of an amide in which an intermediate would not be involved would be an $S_N2$ reaction at the acyl carbon of the amide

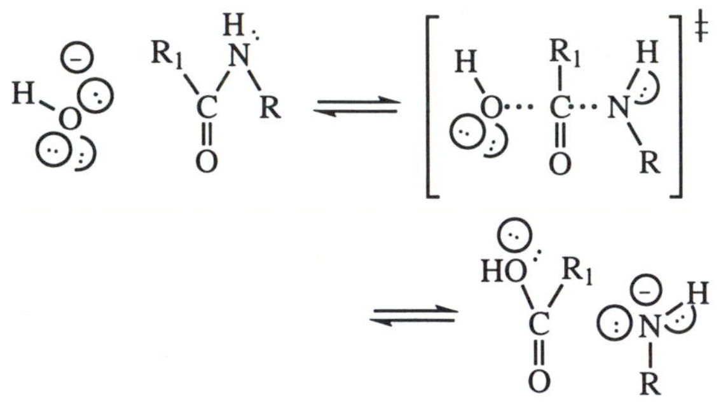

(5–1)

For various reasons such a reaction, although not immediately unreasonable, would have a significantly higher free energy of activation than the free energies of activation for formation of a tetravalent intermediate, its tautomerizations, and its decomposition. Therefore, the chemical reaction in solution proceeds through the tetravalent intermediates and their accompanying transition states rather than through only the one transition state.

An enzymatic reaction involves intermediates for the same reason that a nonenzymatic reaction does. The reaction coordinate passing through a series of successive intermediates encounters a series of free energies of activation that are all smaller than the free energy of activation for even the most exergonic of the reaction coordinates that have only one transition state. It should not be thought that in this way the enzyme divides up the larger free energy of activation associated with the one step; rather it happens that there is an alternative, but independent, reaction coordinate passing through the intermediates, and this reaction coordinate has transition states that are all lower in free energy of activation than that for any reaction coordinate with only a single transition state. This conclusion follows from the fact that it is the mechanism with intermediates that is the one actually observed.

In reactions that occur in solution, an intermediate is usually described as if it were a single molecule, for example, one of the tetravalent intermediates in the hydrolysis of an amide. This may be reasonable in other solvents, but in water the intermediates in most reaction mechanisms have molecules of water or general acid–bases or both noncovalently bound to them through hydrogen bonds, and these bound molecules are usually essential participants in the intermediate both structurally and energetically. For this reason, an intermediate in an enzymatic reaction can be viewed as a simpler structure than the analogous intermediate in the nonenzymatic, aqueous reaction because the necessary molecules of water and molecules of general acid–bases have been replaced in most part by well-oriented general acid–bases from within the folded polypeptide and its associated coenzymes. The intermediate in an enzymatic reaction is the complex between these catalytic functional groups in the active site and a particular covalent arrangement of the atoms that had composed together the reactants and will compose together the products of the enzymatic reaction. As is the case in aqueous solution, an intermediate in an enzymatic reaction is the entire complex. To simplify matters, however, the intermediate in an enzymatic reaction is often discussed as if it were only the molecule or molecules formed from the atoms of the reactants or products, and the active site is usually not explicitly mentioned.

The intermediates encountered in enzymatic reactions can be divided into two classes. On the one hand, an enzymatic reaction can pass through the **same series of intermediates** that the same chemical reaction would pass through in the same aqueous solution in the absence of the enzyme. This situation is the one upon which is based the paradigm of the enzyme binding the transition state or the intermediate of high energy in the reaction more tightly than the reactants (Mechanism 3–211). An example of such a mechanism is that catalyzed by thermolysin, in which a hydroxide anion stabilized by $Zn^{2+}$, acting as a Lewis acid, attacks the acyl carbon of the amide to form the tetravalent intermediate (Reaction 4–234), and Glutamic Acid 143 protonates the nitrogen to produce the leaving group. On the other hand, because the active site of an enzyme resembles an aqueous solution only remotely, **unique intermediates**, which do not participate in the uncatalyzed reaction, may participate in the enzymatically catalyzed reaction. For example, although the proteinases in the family represented by trypsin and chymotrypsin catalyze a reaction identical to that catalyzed by thermolysin, these enzymes use a serine in the active site as the initial nucleophile to produce an ester between the enzyme and the acyl carbon of the original amide and then the internal ester is hydrolyzed by water (Figure 3–37).[1,2] Both the formation of the acyl enzyme and its hydrolysis must pass through the tetravalent intermediates. It is not immediately obvious, however, that the reaction catalyzed by a serine proteinase can be described by the algorithm of the enzyme being able to bind a transition state that would exist in solution more tightly than it binds the reactants.

The serine proteinases, represented by trypsin and chymotrypsin, use an amino acid in the active site to form an intermediate unique to their enzymatic reactions relative to the mechanism of the reaction in solution, but adenosylhomocysteinase uses a coenzyme in the active site to form an intermediate unique to its enzymatic reaction. The reaction catalyzed by the enzyme

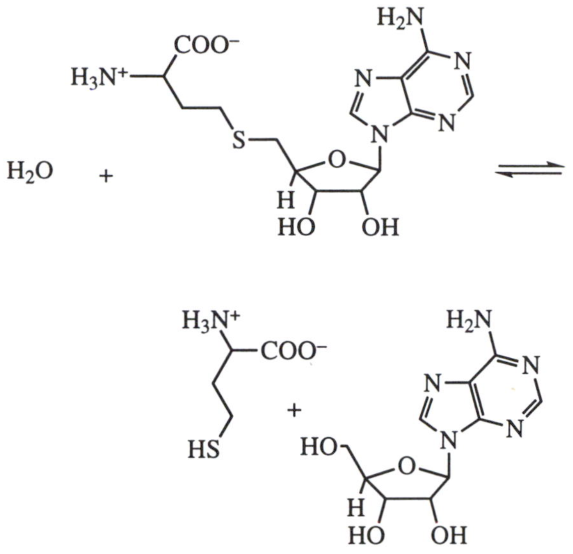

(5–2)

is formally a nucleophilic substitution at a primary carbon. In solution, it would probably proceed by an $S_N2$ mechanism without an intermediate, but it would be a difficult reaction because the thiolate anion is a poor leaving group. The enzyme contains $NAD^+$ bound tightly to its active site as a coenzyme.[3] When *S*-adenosylhomocysteine is added to the enzyme, NADH is transiently formed. When the reaction is run in $^2H_2O$, deuterium appears at carbon 4′ in the unreacted substrate and the hydrolysis of [4′-$^2$H]-*S*-adenosylhomocysteine by the enzyme displays a deuterium kinetic isotope effect of 1.5. 4′,5′-Dehydroadenosine is converted by the enzyme to adenosine. All of these results are consistent[3] with a mechanism (Figure 5–2) in which 3′-oxo-4′,5′-dehydroadenosine is the central intermediate. If this is the case, the substitution occurs indirectly rather than directly, by an elimination–addition *β* to a carbonyl, a reaction in which thiols readily participate. This mechanism is reminiscent of the mechanism of *O*-succinylhomoserine (thiol)lyase, in which a substitution was accomplished by an even more complicated elimination–addition (Reaction 2–17). In the case of adenosylhomocysteinase, the fact that the enzyme can catalyze hydrogen exchange at carbon 4′ of 5′-deoxyadenosine is consistent with the enzymatic reaction proceeding through the two respective enolates as intermediates in the elimination and addition, respectively, as shown, and this enzymatic reaction has five unique intermediates.

Intermediates in enzymatic reactions can also be divided into two other classes: covalent or noncovalent. A **covalent intermediate,** such as the acyl enzyme of the serine proteinases or the pyridoximine in enzymes using pyridoxal phosphate as a coenzyme, is an intermediate in which a covalent bond is formed between a reactant and a permanent functional group in the active site, either an amino acid or a coenzyme. A **noncovalent intermediate**, such as 3′-oxo-4′,5′-dehydroadenosine, is an intermediate formed from the reactant during an enzymatic reaction and is held so tightly in the active site by noncovalent forces that it cannot escape into the solution and short-circuit the enzymatic reaction. For example, if the 3′-oxo-4′,5′-dehydroadenosine were to escape from the active site, the enzyme would be left in its reduced form incapable of reaction with homocysteinyladenosine, and the 3′-oxo-4′,5′-dehydroadenosine would probably form Michael adducts with proteins in the cytoplasm; both of these consequences would be unfortunate.

# Chemical Identification of Intermediates

If a covalent intermediate between an amino acid on the protein and a portion of a reactant is formed during the overall enzymatic reaction, the enzyme is transiently and covalently modified during the reaction. The chemical identification of this modification can be used as evidence for the involvement of the covalent intermediate in the enzymatic reaction. For example, when chymotrypsin catalyzes the hydrolysis of *p*-nitrophenyl acetate, the fast step in the overall reaction is the formation of the **acetyl enzyme,** followed by its slow hydrolysis. When *p*-nitrophenyl [$^{14}$C]acetate was used as a substrate, the [$^{14}$C]acetyl enzyme was formed, and the acetyl group was shown to be incorporated as an ester with Serine 195 by isolation of the acetylated peptide, GDS(acetyl)GGPL,[1] where (acetyl) indicates modification of the serine.

Purified citrate (*pro*-3*S*)-lyase catalyzes the Claisen condensation

$$\text{citrate} \rightleftharpoons \text{acetate} + \text{oxaloacetate} \qquad (5\text{–}3)$$

Following purification, the enzyme slowly loses activity in a process hastened by the addition of citrate and $Mg^{2+}$ or of hydroxylamine.[4] The inactive enzyme can be reactivated by exposure to acetic anhydride[4] or acetyl-SCoA.[5] During the reactivation, the enzyme becomes acetylated on the thiol of a molecule of coenzyme A covalently bound to the protein. The coenzyme A is attached through a (*β*1,2) glycosidic linkage to a ribose that in turn is bound in a phosphodiester to a serine of the polypeptide.[6] Only the acetylated enzyme is active as citrate lyase.

The enzyme lacking the covalently bound acetyl group, however, will carry out reactions in which it uses extraneously added acetyl-SCoA or citryl-SCoA as if they were covalently bound to the protein.[7] Presumably this occurs because the extraneous acetyl-SCoA or citryl-SCoA displaces the covalently bound coenzyme A from the active site. From the products of these reactions, it could be inferred that the normal enzymatic reaction proceeds through a covalent citryl intermediate in the sequence

$$\text{E-CoAS-acetyl} + \text{citrate} \rightleftharpoons \text{E-CoAS-citryl} + \text{acetate} \qquad (5\text{–}4)$$

$$\text{E-CoAS-citryl} \rightleftharpoons \text{E-CoAS-acetyl} + \text{oxaloacetate} \qquad (5\text{–}5)$$

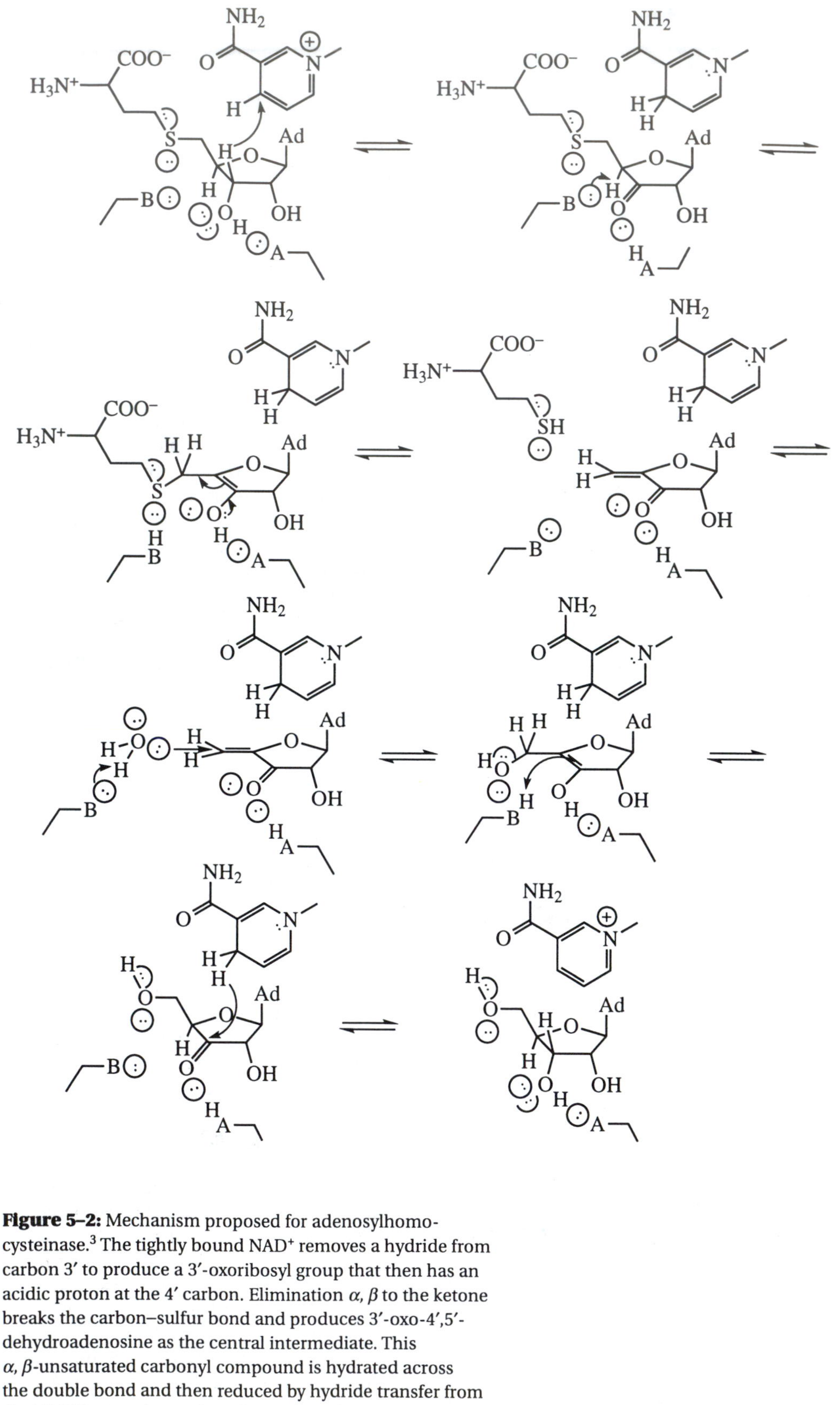

**Figure 5–2:** Mechanism proposed for adenosylhomocysteinase.[3] The tightly bound $NAD^+$ removes a hydride from carbon 3′ to produce a 3′-oxoribosyl group that then has an acidic proton at the 4′ carbon. Elimination $\alpha, \beta$ to the ketone breaks the carbon–sulfur bond and produces 3′-oxo-4′,5′-dehydroadenosine as the central intermediate. This $\alpha, \beta$-unsaturated carbonyl compound is hydrated across the double bond and then reduced by hydride transfer from the NADH to produce adenosine as the other product of the enzymatic reaction.

where E-CoASH is the covalently bound coenzyme A. The advantage of forming the citryl thioester in Reaction 5–4 is obvious because the carbon–carbon bond cleaved or formed during the retrocondensation or the condensation, respectively (Reaction 5–5), is weakened considerably (Table 1–4).

It is more difficult, however, to understand the mechanism of the exchange of the citryl thioester for the acetyl thioester (Reaction 5–4), even though there is isotopic evidence that this reaction definitely occurs.[8] The most likely intermediate in this partial reaction is a mixed anhydride formed between citrate and acetate[9]

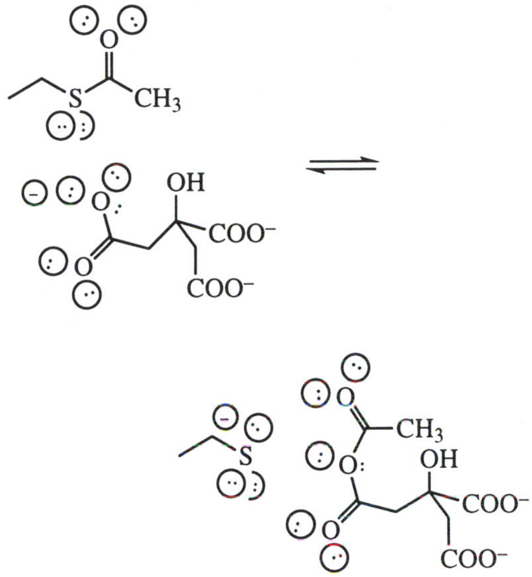

(5–6)

This anhydride is then trapped at the acyl carbon of the citryl group either by the central thiolate of the coenzyme A or by the thiolate of a cysteine known to be located nearby in the active site[10] and then transferred to the central thiolate of the coenzyme A. The involvement of an anhydride similar to the one in Reaction 5–6 has also been proposed for the mechanism of 3-oxo acid CoA-transferase[11]

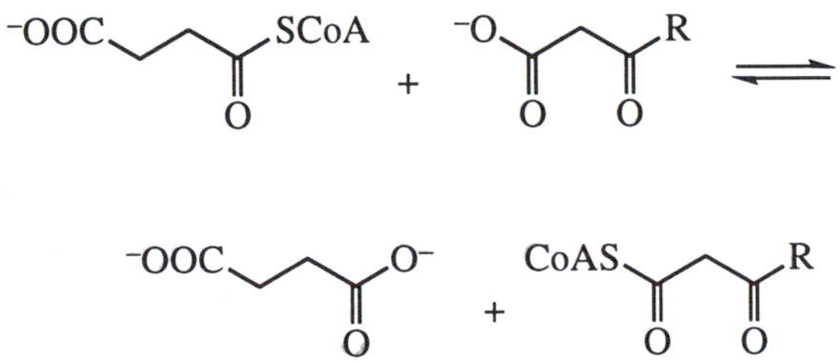

(5–7)

In this example, the formation of a mixed anhydride between the acyl group of the acyl-SCoA and a glutamate on the enzyme followed by the formation of the documented[12] coenzyme A thioester at that glutamate by two successive acyl exchanges

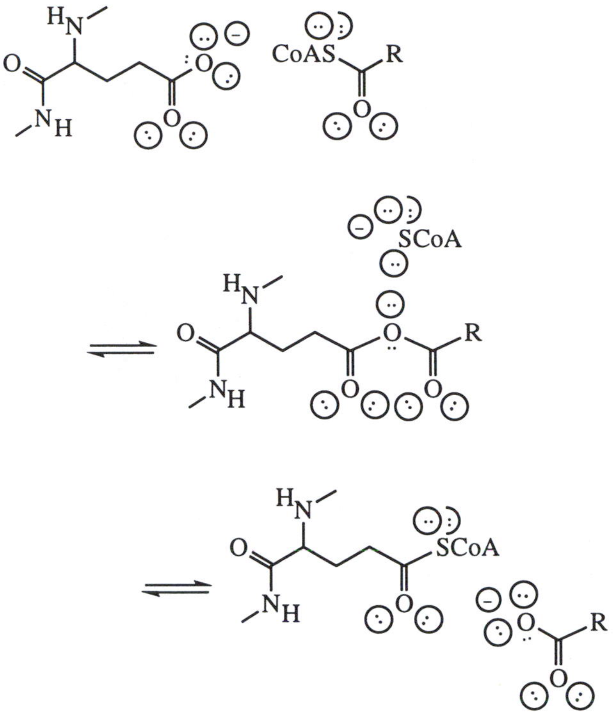

(5–8)

is sufficient to explain the exchange of the coenzyme A between the two carboxylates. This mechanism is consistent with the effects of substitutions in the acyl-SCoA chosen as the reactant on the rate of the enzymatic reaction.

Citramalate lyase

$$(S)\text{-citramalate} \rightleftharpoons \text{acetate} + \text{pyruvate} \tag{5–9}$$

also proceeds through a mechanism in which a citramalyl thioester is formed at a covalently bound coenzyme A in the active site at the expense of an acetyl thioester.[13] In both citrate (*pro*-3*S*)-lyase and citramalate lyase, large hetero-oligomeric enzymes with covalently bound coenzymes A must be used to provide the intermediate thioester. That this might be only marginally advantageous is suggested by the fact that isocitrate lyase, which also catalyzes a similar Claisen condensation

$$\text{isocitrate} \rightleftharpoons \text{succinate} + \text{glyoxylate} \tag{5–10}$$

proceeds without forming any thioesters,[14] presumably by firmly protonating the carboxylate adjacent to the carbon–carbon bond to be cleaved so that the 4,4-dihydroxy-3-butenoate can be stabilized as the enol

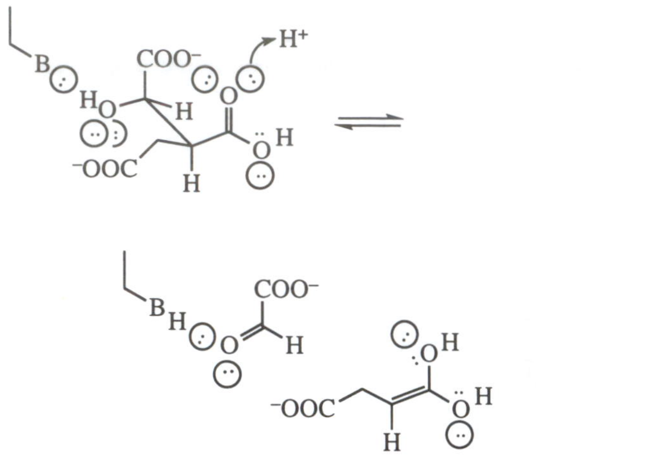

(5–11)

As with the proteolytic enzymes, these examples of Claisen condensations illustrate the fact that different strategies involving different intermediates are often used to catalyze the same reaction.

Fatty acid synthase also uses the thioester of a pantetheine covalently bound to the enzyme to promote a Claisen condensation[15]

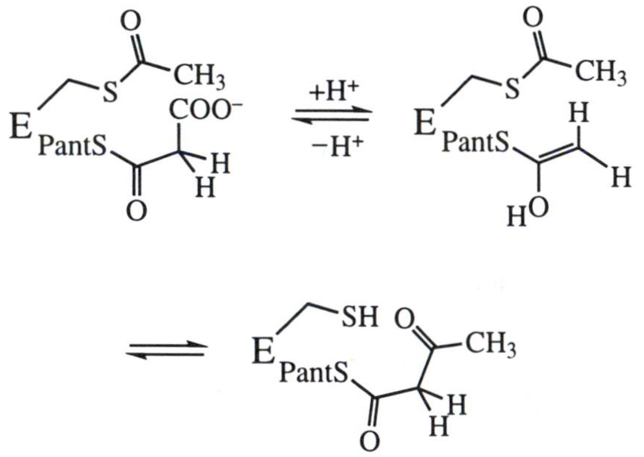

(5–12)

The acetyl group with which the enol is condensed is bound also as a thioester but to a cysteine in the active site of the enzyme. Unlike the situation that exists with citrate (*pro*-3*S*)-lyase, in which the initial reactant is an unactivated carboxylic acid, both the malonyl group and the acetyl group arrive at the respective active sites in fatty acid synthase as thioesters of coenzyme A. Therefore, they are already activated. Nevertheless, the transfer of the two groups to their respective locations for the condensation (Reaction 5–12) is not a direct one but passes through a sequence of other covalent intermediates, each formed by an acyl transfer reaction. The acetyl group of acetyl-SCoA is transferred first to a serine to form an ester in the active site of the acetyltransferase of the multienzyme complex,[16] then to the thiol of the pantetheine to form a thioester again, and then to the thiol of the cysteine in the active site to situate it for the condensation. The malonyl group of malonyl-SCoA is transferred first to a serine to form an ester in the active site of the malonyltransferase of the multienzyme complex[17] and then to the thiol of the pantetheine to form the necessary thioester. The two respective serines that pass on the acetyl group and the malonyl group to the thiol of the pantetheine were identified by alkylating the thiols to prevent the transfers, forming the esters with [$^{14}$C]acetyl-SCoA or [$^{14}$C]malonyl-SCoA, respectively, and isolating the unique peptides containing the esters.[16,17]

The mechanism just described, however, has one drawback. If a malonyl group arrived first, its occupation of the thiol of the pantetheine would block the transfer of the acetyl to the thiol of the cysteine. For this reason the presence of free coenzyme A is an absolute requirement for fatty acid synthetase. In its absence the enzyme rapidly becomes occupied by malonyl groups, which cannot be removed because the back reaction is blocked, and the acetyl groups cannot be transferred onto the active site.[18]

In addition to the covalent acyl enzyme intermediates of the serine proteinases and the covalent thioester intermediates involved in the Claisen condensations, covalent intermediates formed between an amino acid in the active site of an enzyme and inorganic phosphate are commonly encountered. These are usually formed by a phosphonyl transfer from a phosphate ester or phosphoric anhydride to a nucleophile in the active site

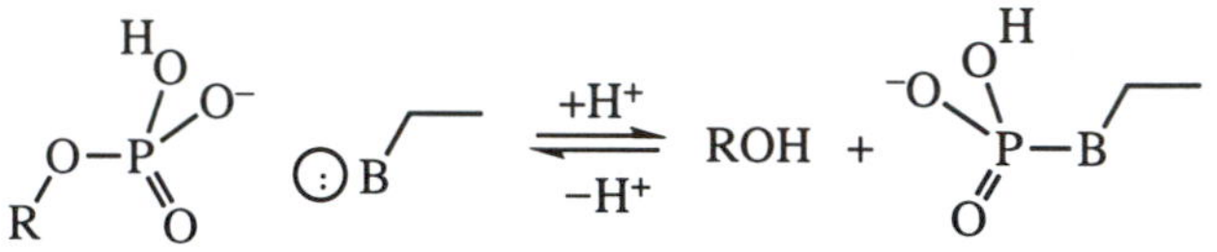

(5–13)

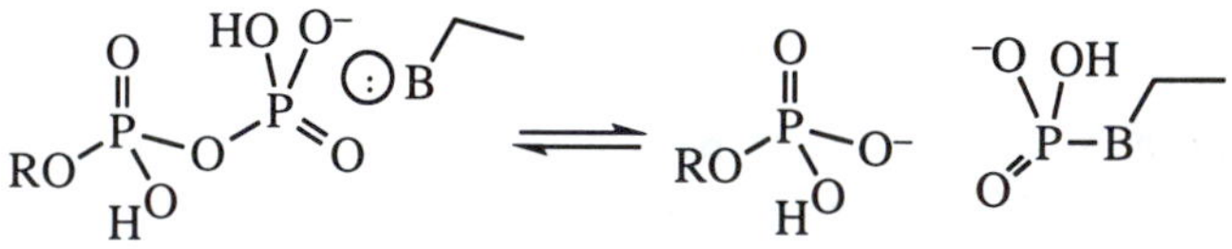

(5–14)

to form a phosphorylated intermediate. A **phosphorylated intermediate** is an ester, an anhydride, or an amide formed between an amino acid in the active site and phosphoric acid. The phosphorylated intermediate is the phosphonoamino acid, for example, *O*-phosphonoserine, $O^{\delta}$-phosphonoglutamate, or *N*-phosphonohistidine.

One advantage of such a phosphorylated intermediate is that the same functional groups in the active site used to activate the phosphate donor by general acid–general base catalysis can serve to activate the acceptor in an identical manner. The chemical reaction catalyzed by nucleoside-diphosphate kinase is

$$\text{MgXTP} + \text{MgYDP} \rightleftharpoons \text{MgXDP} + \text{MgYTP} \qquad (5\text{–}15)$$

where XTP and YTP are any of a number of nucleoside 5′-triphosphates and XDP and YDP are the corresponding nucleoside 5′-diphosphates. In this reaction, the oxygens acting as phosphate donor and phosphate acceptor are chemically identical. As a result, the enzymatic reaction proceeds through a phosphorylated intermediate[19] formed by transfer of a phosphate from the one nucleoside triphosphate to an amino acid in the active site and discharged by transfer of the phosphate to the other nucleoside diphosphate. Because the enzyme is only marginally selective as to the nucleotide, it must recognize mainly the complex between magnesium and the triphosphate and carbon 3′ through carbon 5′ of the ribose or deoxyribose. These structures are common to both donor and acceptor. Therefore, the same constellation of amino acids in the active site can recognize and activate the donor as well as the acceptor. To do this, the donor must leave the active site as MgXDP before the acceptor is bound as MgYDP, and the phosphate transferred must stay in the active site. By remaining in the active site as the phosphonoamino acid, the free energy of hydrolysis of the phosphoric anhydride can be retained to facilitate transfer to the acceptor.

This mechanism can be contrasted with that of adenylate kinase

$$\text{MgATP} + \text{AMP} \rightleftharpoons \text{MgADP} + \text{ADP} \qquad (5\text{–}16)$$

This enzyme is specific for adenine nucleotides and consequently the active site must form hydrogen bonds over the complete structure of both donor and acceptor. The donor and acceptor are quite different structurally when viewed from the oxygens donating and accepting phosphate. For both of these reasons, the active site could not be designed so that the same constellation of amino acids is able to recognize both donor and acceptor, and adenylate kinase catalyzes the direct transfer of the phosphate from donor to acceptor while both are bound simultaneously to different regions of the active site.[20,21]

In the reaction catalyzed by alkaline phosphatase (Reaction 4–160), the various donors of the phosphate and the water, the ultimate acceptor, may not resemble each other at first glance, but they are all alcohols. The crystallographic molecular model suggests that only the phosphate itself forms contacts with the active site and is activated,[22] and for this reason, the enzyme has a broad specificity for phosphate monoesters. It is not surprising that there is a phosphorylated intermediate in this enzymatic reaction

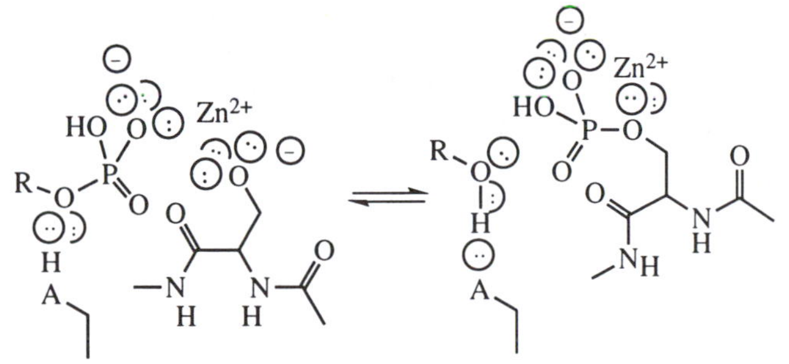

(5–17)

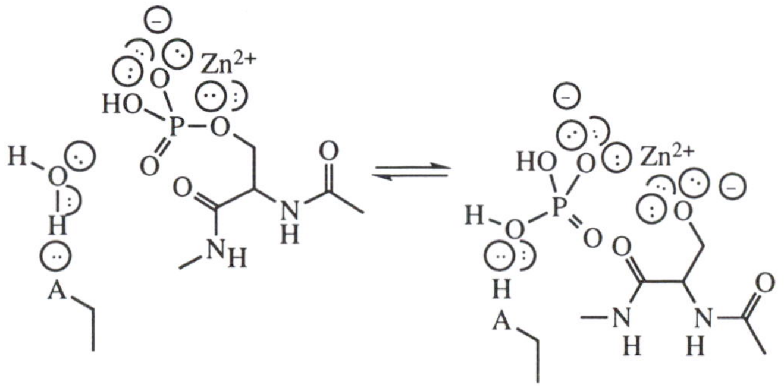

(5–18)

This is most clearly demonstrated by the fact that the reaction proceeds with retention of configuration at phosphorus.[23]

The demonstration of the existence of phosphorylated intermediates in enzymatic reactions begins with the observation that [$^{32}$P]phosphate is transferred from a donor in the reaction to the enzyme itself to form a covalently phosphorylated protein. The phosphonoamino acid formed in this transfer can sometimes be identified directly by hydrolyzing the modified protein and isolating the covalently modified amino acid. For example, succinate-CoA ligase (ADP-forming) proceeds through a phosphorylated intermediate,[24] and the [$^{32}$P]phosphate incorporated during this reaction could be isolated as *N*-phosphonohistidine after alkaline hydrolysis or enzymatic digestion of the protein.[25] The phosphorylated intermediate in the reaction catalyzed by nucleoside diphosphate kinase has been identified in the same way as $N^1$-phosphonohistidine.[26] The phosphorylated protein formed as an intermediate in the reaction catalyzed by alkaline phosphatase has been digested with pepsin at low pH, to prevent alkaline hydrolysis, and an analysis of the peptides produced showed that the phosphate is covalently attached to the first serine in the sequence TGKPDYVTDSAASA.[27] The phosphorylated intermediate in the enzymatic reaction catalyzed by $Ca^{2+}$-transporting ATPase[28] has been identified indirectly as a $\gamma$-phosphonoaspartate by reducing it to [$^3$H]homoserine with sodium [$^3$H]borohydride

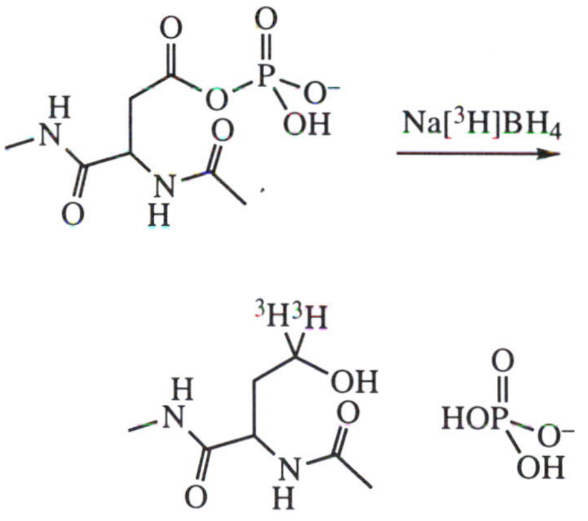

$\text{Na}[^3\text{H}]\text{BH}_4$

(5–19)

and isolating the [$^3$H]homoserine produced upon hydrolysis of the polypeptide in acid.[29] The phosphorylated intermediate in the enzymatic reaction catalyzed by enzyme II of the phosphotransferase system has been identified as an *S*-phosphonocysteine both by isolation of the phosphonopeptide and by nuclear magnetic resonance analysis.[30]

To describe a phosphonoamino acid in a protein as a phosphorylated intermediate, its kinetic competence must be demonstrated. The **kinetic competence** of any intermediate in an enzymatic reaction is its capacity to be formed and discharged at rates equal to or greater than the overall rate of the normally occurring reaction. All of the actual intermediates in the mechanism of an enzyme must be kinetically competent. If a particular compound, either a noncovalently bound molecule or a covalent adduct with the enzyme, is not kinetically competent, that compound cannot be an intermediate on the main pathway of the enzymatic mechanism.

In the enzymatic reaction catalyzed by phosphoglucomutase

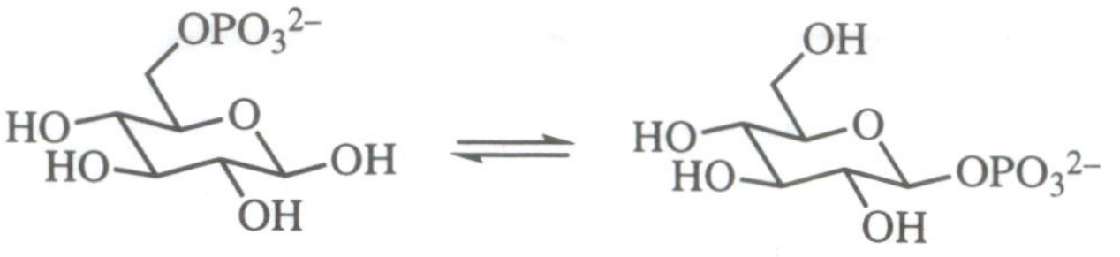

(5–20)

only enzyme already covalently phosphorylated at its active site is enzymatically active. If enzyme that has [$^{32}$P]phosphate covalently attached at its active site is mixed with glucose 1-phosphate, the [$^{32}$P]phosphate is replaced by nonradioactive phosphate with a rate constant[31] of 5900 min$^{-1}$. Under the same conditions the turnover of the enzyme with glucose 1-phosphate as reactant is 6200 min$^{-1}$. Because these two numbers are identical within experimental error, the phosphoenzyme intermediate is kinetically competent. When the $Mg^{2+}$ used as an obligatory reactant is replaced with $Zn^{2+}$, the rate constant for the loss of the [$^{32}$P]phosphoenzyme intermediate upon addition of glucose 1-phosphate decreases to 25 s$^{-1}$ but the turnover of the enzyme decreased to 10 s$^{-1}$. Again, the phosphoenzyme intermediate is kinetically competent. The mechanism of this enzymatic reaction has been proposed to be[32,33]

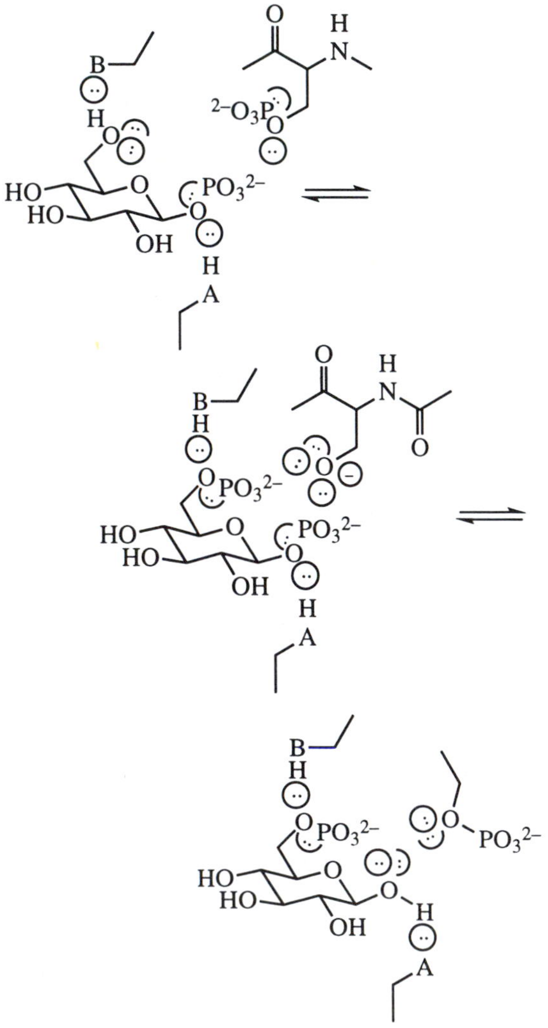

(5–21)

which requires that the phosphate on the enzyme be replaced by a phosphate from one of the substrates during each turnover. The empty enzyme that binds the reactant is phosphorylated and the enzyme from which the product dissociates is also phosphorylated.

The stereochemistry of an enzymatic reaction can also be consistent with the existence of a covalent phosphorylated intermediate.[34] If the overall enzymatic reaction is observed to proceed with inversion of configuration at the phosphorus transferred, this observation is consistent with direct in-line transfer between donor and acceptor without the involvement of a phosphorylated intermediate. If the overall reaction is observed to proceed with retention of configuration at phosphorus, this is consistent with two successive phosphotransfers both taking place with inversion, one from donor to the enzyme and the other from the enzyme to the acceptor.

Covalent acyl, thioester, and phosphorylated intermediates are sometimes stable enough to be identified directly. There are other covalent intermediates formed between amino acids in active sites and portions of the substrates that must be **trapped** by chemical conversion to a stable species. An example of this is the previously described reduction of the aspartyl phosphate to [$^{3}$H]homoserine by Na[$^{3}$H]$BH_4$ (Reaction 5–19). Acetoacetate decarboxylase catalyzes the Lewis acid–base reaction

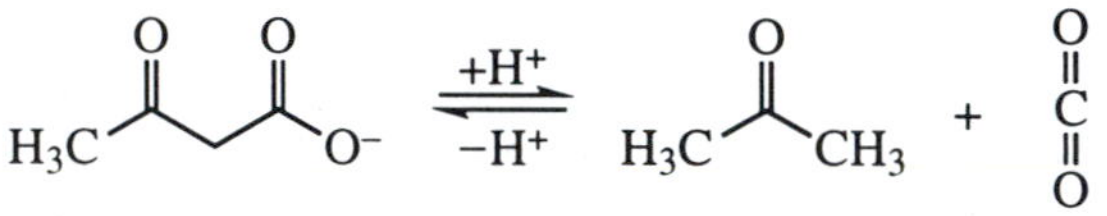

(5–22)

When [3-$^{14}$C]acetoacetate was mixed with the enzyme in the presence of sodium borohydride, radioactivity was incorporated into the protein.[35] Digestion of this modified protein with trypsin produced the radioactive peptide ELSAYPK([$^{14}$C]isopropyl)K in which the radioactive modification was $\varepsilon$-[$^{14}$C]isopropyllysine. From this observation, it was concluded that the covalent intermediate trapped by the sodium borohydride was the imine between the modified lysine and acetone

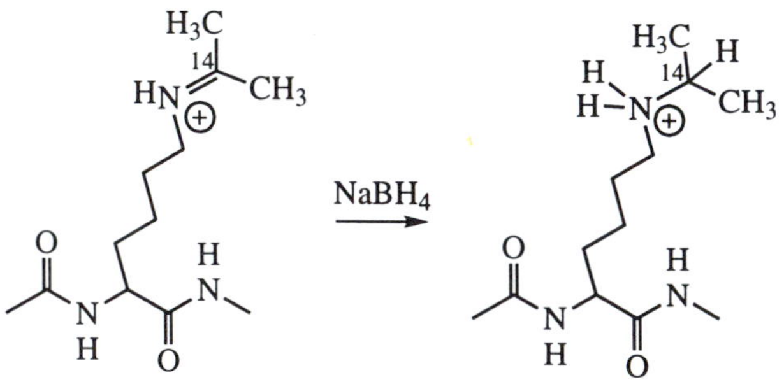

(5–23)

This is consistent with a mechanism in which the iminium cation of the neutral ketimine of acetoacetate is formed in the active site to expedite the decarboxylation

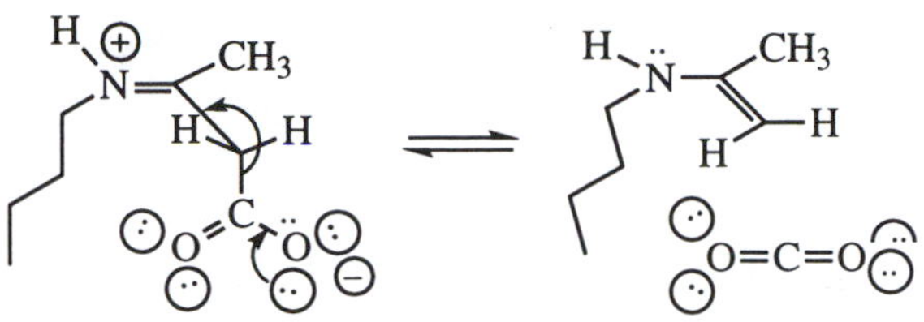

(5–24)

by positioning the carbon–carbon bond to be broken parallel to the $\pi$ bond of the iminium cation, so that the $\pi$ system of the iminium cation can withdraw electron density from the carbon–carbon $\sigma$ bond (Reaction 1–78).

These experiments were based on earlier observations with fructose-bisphosphate aldolase in which a very similar imine between a lysine in the active site and dihydroxyacetone phosphate had been trapped with sodium borohydride.[36] In this latter case, the carbon–carbon bond weakened by the iminium cation breaks to produce an enamine (as in Reaction 5–24) and an aldehyde rather than the *gem*-dicarbonyl. When fructose-bisphosphate aldolase is mixed with dihydroxyacetone phosphate, an equilibrium is established on the active site between the imine and the enamine

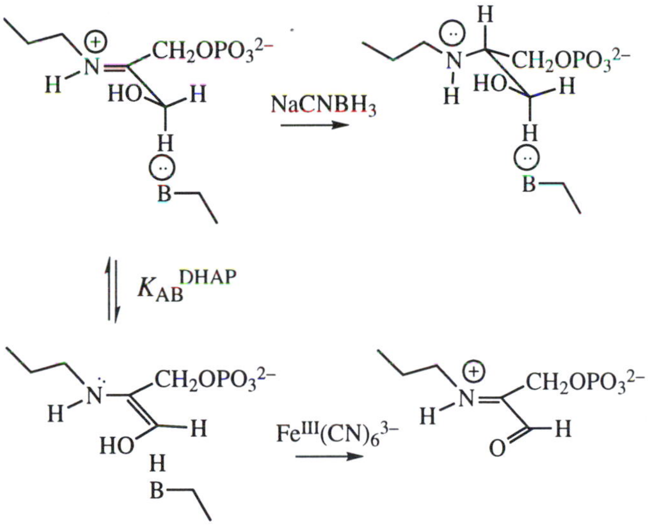

(5–25)

The enzyme could be denatured rapidly in acid and the 1,2-enaminol could be trapped by oxidation, either with tetranitromethane[37] or with ferricyanide, or the imine could be trapped with sodium cyanoborohydride.[38] In this way the equilibrium constant for the acid–base reaction at the active site ($K_{AB}^{DHAP} = 3$) could be ascertained.

Covalent intermediates between coenzymes incorporated into an active site and portions of a substrate are also involved in enzymatic mechanisms. Urocanate hydratase catalyzes an apparent 1,4-hydration of a diene in urocanate

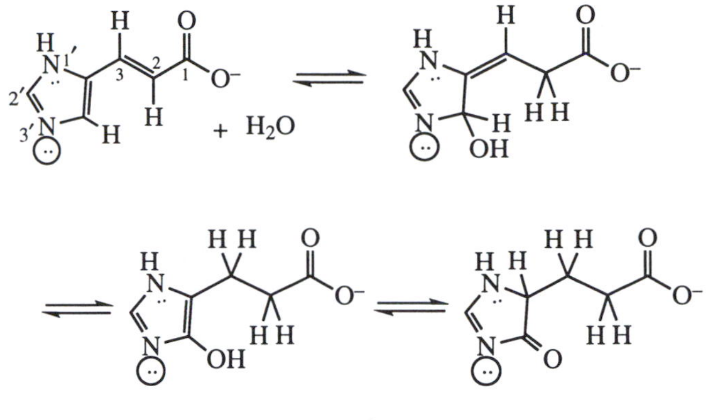

(5–26)

A covalent adduct between 3-imidazolylpropionate, a competitive inhibitor of the enzyme, and the tightly bound $NAD^+$ present in the active site as a coenzyme can be trapped by oxidation with phenazine methosulfate.[39] On the basis of the structure of the product of this oxidation, the structure of the original covalent adduct between inhibitor and coenzyme was deduced to be

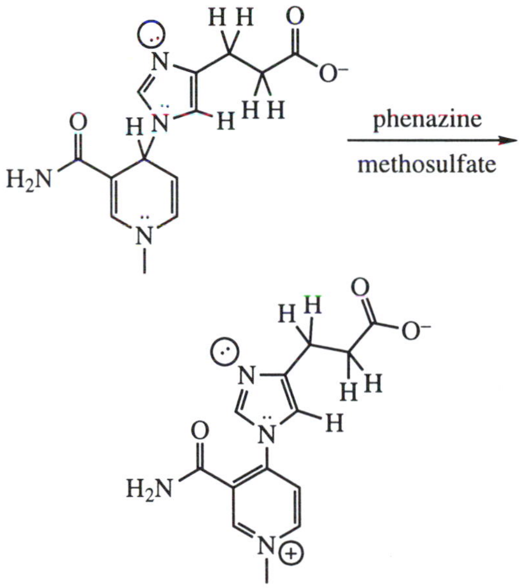

(5–27)

A very similar covalent adduct is formed transiently during the normal reaction,[40] and its participation in the reaction could be to alkylkate nitrogen 3′ of the imidazole and perhaps cause carbon 4′ to be more electrophilic[41] during the addition of water

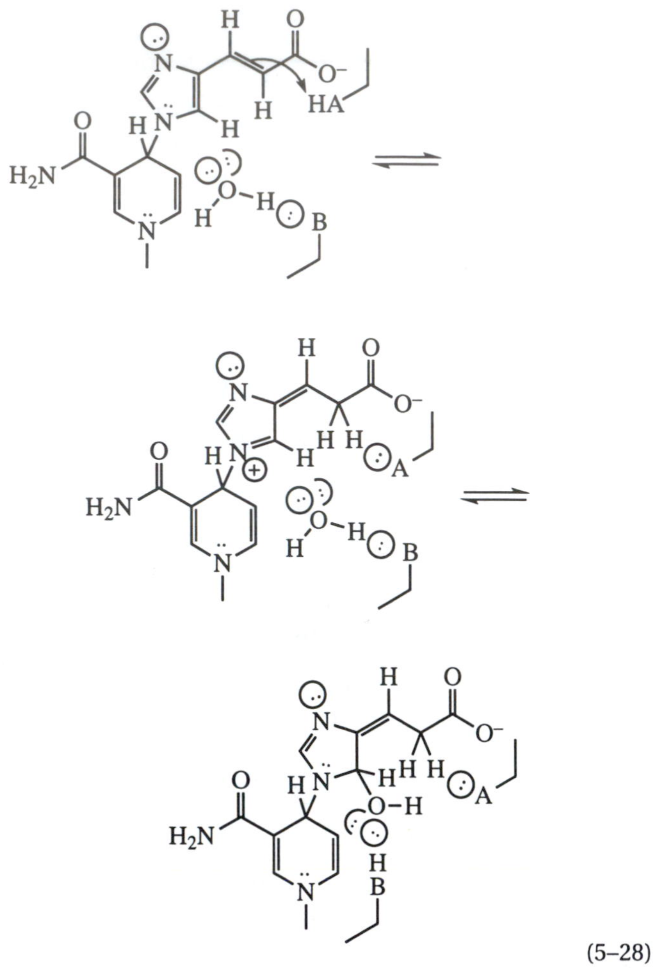

(5–28)

Covalent intermediates between enzymes and reactants have also been observed crystallographically. Thymidylate synthase is responsible for providing the methyl group on thymidine

2′-deoxyuridine monophosphate +

5,10-methylenetetrahydrofolate ⇌

dihydrofolate + thymidine monophosphate

(5–29)

During this reaction the methylene carbon of 5,10-methylenetetrahydrofolate becomes the methyl group of thymidine monophosphate, replacing the hydrogen at carbon 5 of 2′-deoxyuridine monophosphate.

The enzyme is covalently inactivated by 2′-deoxy-5-fluorouridine monophosphate in a reaction involving the addition of Cysteine 146 (in the active site of the enzyme from *Escherichia coli*) to the $\alpha,\beta$-unsaturated amide of the 2′-deoxy-5-fluorouridine monophosphate[42]

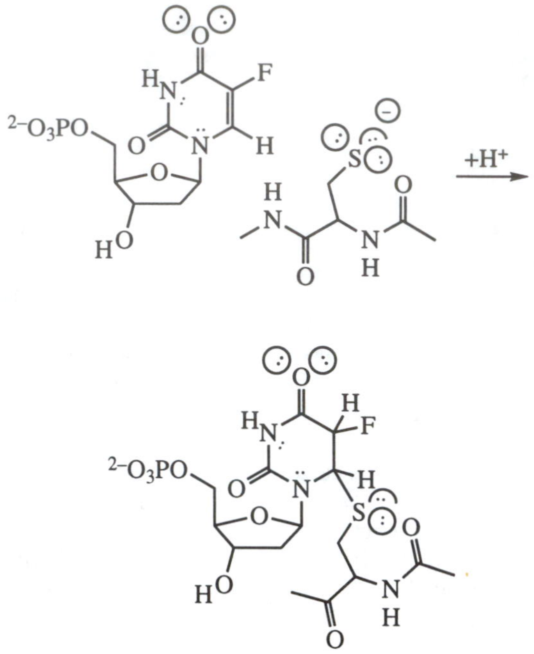

(5–30)

**5–1**

It is believed that this addition also occurs with the natural reactant 2′-deoxyuridine monophosphate during the normal reaction, but the normal reaction can proceed to completion because there is not a fluorine at carbon 5. The advantage of this addition in the normal reaction is to turn carbon 5 into the saturated $\alpha$-carbon of an amide unconjugated to carbon 6. This transformation leads to an increase in the acidity of the proton on carbon 5, a fact which could be demonstrated in a model system by noting the appearance of hydrogen exchange at carbon 5 of uridine upon addition of thiols[43]

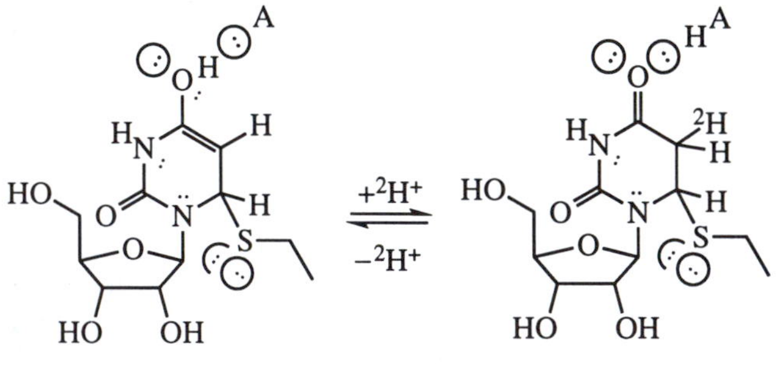

**5–2**

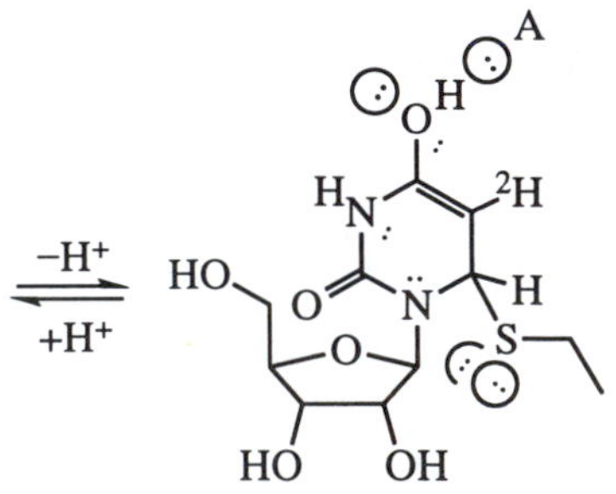

(5–31)

as would be expected from an $\alpha$-proton on an amide. On the basis of these observations, it can be concluded that the addition of the cysteine to carbon 6 of 2′-deoxyuridine monophosphate produces enol **5–2** as a covalent intermediate at the active site in the normal mechanism.

In the crystallographic molecular model of the complex formed from thymidylate synthase of *E. coli*, 2′-deoxy-5-fluorouridine monophosphate, and 5,10-methylenetetrahydrofolate,[44] the covalent bond formed between Cysteine 146 from the active site and the 5-fluorouracil is clearly observed, as well as a covalent bond between carbon 5 of the 5-fluorouracil and the methylene carbon from the 5-methyleneiminium cation of tetrahydrofolate. Such an intermediate is formed by nucleophilic addition of the enol of adduct **5–1** to the 5-methyleneiminium cation formed from the 5,10-methylenetetrahydrofolate

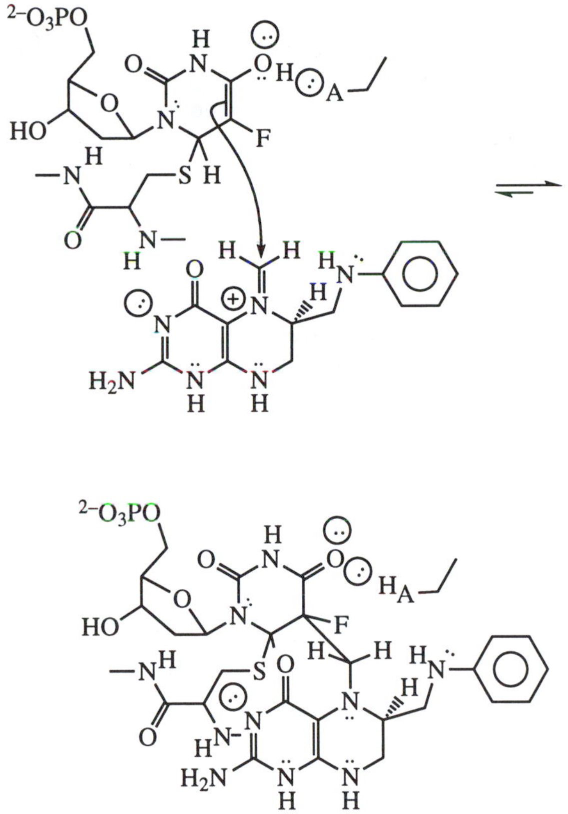

(5–32)

**5–3**

Therefore, the crystallographic molecular model verifies the capacity of the covalent intermediate between the cysteine and the uracil to function as an enol as well as the ability of that enol to add to the 5-methyleneiminium cation of tetrahydrofolate during the normal enzymatic reaction.

The next step in the normal enzymatic reaction would be the formation of an exocyclic olefin at carbon 5 of the uracil by an $E1_{CB}$ elimination (Figure 5–3), which is forbidden to the 5-fluorouracil. The crystallographic ternary complex **5–3** formed from Cysteine 146, the 2′-deoxy-5-fluorouridine 5′-phosphate, and the 5-methyleneiminium cation has the stereochemistry **5–4** expected for the immediate product of the addition of the enolate (Reaction 5–32)

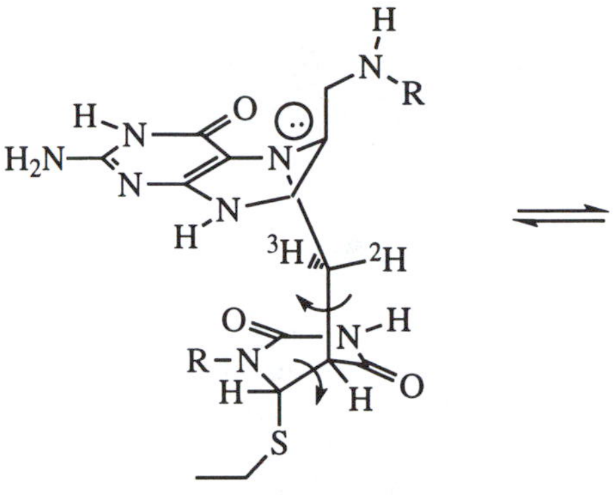

**5–4**

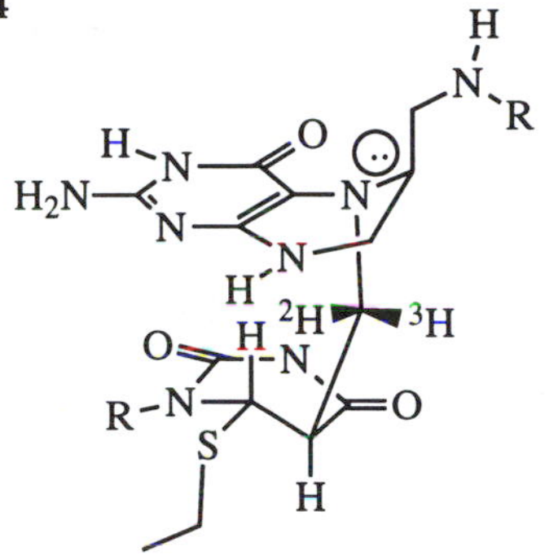

**5–5**

(5–33)

Both the carbon–sulfur bond and the bond just formed between the methylene carbon and carbon 5 of the uracil are parallel to the $\pi$ system of the acyl group at carbon 4 of the uracil, and the latter is antiperiplanar to the lone pair of electrons on nitrogen 5 of the tetrahydrofolate. For the $E1_{CB}$ elimination to occur (Figure 5–3) in the next step, however, the bond to the hydrogen on carbon 5 of the uracil must be parallel to the $\pi$ system of the acyl group at carbon 4 of the uracil and the bond between nitrogen 5 of the tetrahydrofolate and the methylene carbon must be antiperiplanar to the bond to the hydrogen on carbon 5 of the uracil. This stereochemistry (**5–5**) can be acceptably achieved in the crystallographic molecular model by rotating around several of the bonds in the intermediate[45] (Reaction 5–33). Aside from these electronic considerations, the intermediacy of the new rotational isomer **5–5** is also consistent with the known stereochemistry of the reaction. (6*R*,11*S*)-[11-$^3$H,11-$^2$H]-5,10-methylenetetrahydrofolate produces (5*S*)-[5-$^3$H,5-$^2$H]thymidine monophosphate.[46,47] Because tetrahydrofolate has the 6*R* conformation, when the 5-methyleneiminium cation is formed in this reaction it will be the *Z* isomer[47]

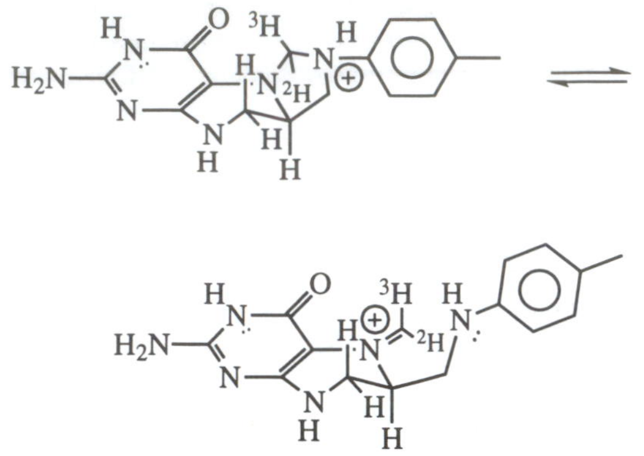

(5–34)

This isomer would produce adduct **5–4** in Reaction 5–33 with the noted stereochemistry. Upon rotational isomerization, elimination, and direct intramolecular hydride transfer[48,49] from carbon 6 of tetrahydrofolate (Figure 5–3), which would occur from above in the orientation of Reaction 5–33, the (*S*)-[$^3$H,$^2$H]methyl would be generated. If the rotational isomerization of Reaction 5–33 did not occur, the (*R*)-methyl would be generated by a *syn* elimination, an outcome other than the one observed.

In the latter steps of the reaction, the necessity for the covalent intermediate between the uracil and Cysteine 146 is also evident. This adduct continues to permit carbons 4 and 5 to act as an acyl carbon and an $\alpha$-carbon to an acyl carbon, respectively, so that the necessary enols and $\alpha,\beta$-unsaturated acyl compounds can be formed. Such chemistry at the $\alpha$ position to an acyl carbon would not be possible if the olefin of uracil were intact. Although model compounds of carbonyl adduct **5–6** in which the olefin of the uracil is intact will produce thymine and the respective analogue of dihydrofolate,[50] these oxidation–reductions have the properties of free radical reactions and proceed very poorly in polar environments. The addition of the thiol to the uracil provides an alternative based on the familiar reactions of acyl compounds and acyl enols.

Intermediates in enzymatic reactions that are not covalently bound to the protein are more commonly encountered than those that are covalently bound. Usually, these noncovalent intermediates are unstable molecules that are confined tightly to the active site to prevent their dissociation into the solution and subsequent decomposition. On the basis of this premise, it is often assumed that all intermediates in enzymatic reactions are too unstable to be studied directly. There are, however, instances in which a proposed intermediate has

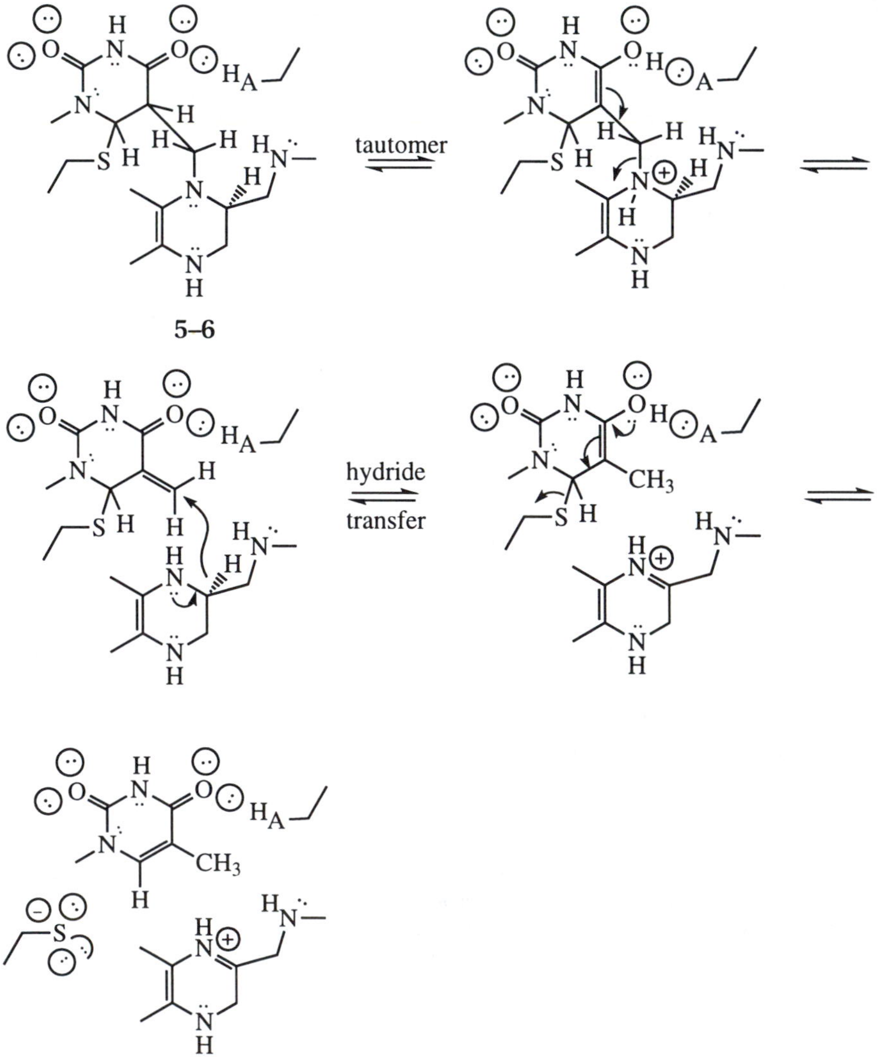

**Figure 5–3:** Final steps in the mechanism of the reaction catalyzed by thymidylate synthase. The ternary complex **5–6** formed from Cysteine 146, 2′-deoxyuridine 5′-phosphate, and 5,10-methylenetetrahydrofolate is protonated at the uracil oxygen to promote the tautomerization between carbon 5 of the uracil and nitrogen 5 of the tetrahydrofolate. This sets up an $E1_{CB}$ elimination to produce the exocyclic olefin of the uracil. Hydride is transferred from tetrahydrofolate to this exocyclic olefin to produce the enol, which then undergoes a $\beta$-elimination of cysteine in a step producing the 5-methyluracil (i.e., thymine) as the product.

been chemically synthesized and its competence in the enzymatic reaction has been demonstrated.

Amino enol ether **5–7** is an intermediate in the reaction catalyzed by anthranilate synthase

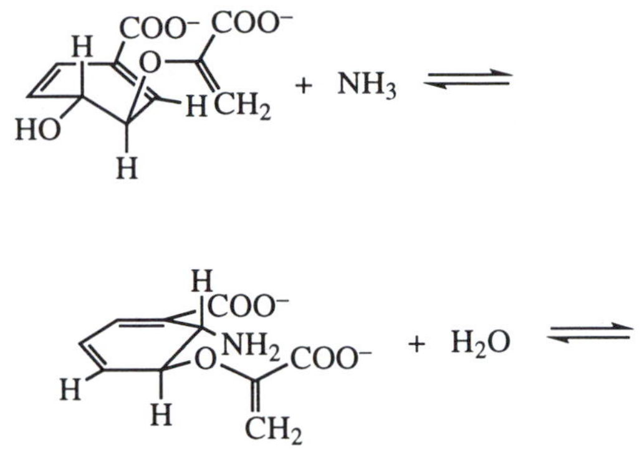

**5–7**

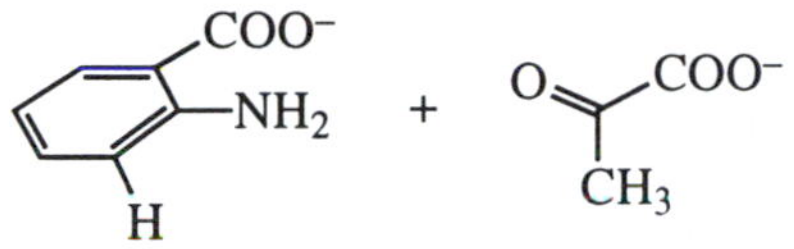

(5–35)

It is stable enough to be synthesized chemically.[51] When it was mixed with anthranilate synthase, the enzyme could convert it to anthranilate in a reaction that displayed a $V_{max}$ of 500 nmol $min^{-1}$ $mg^{-1}$ and a $K_m$ of 0.1 mM. These observations demonstrated that the proposed intermediate is kinetically competent.

The enol of pyruvate is unstable in solution relative to the ketone

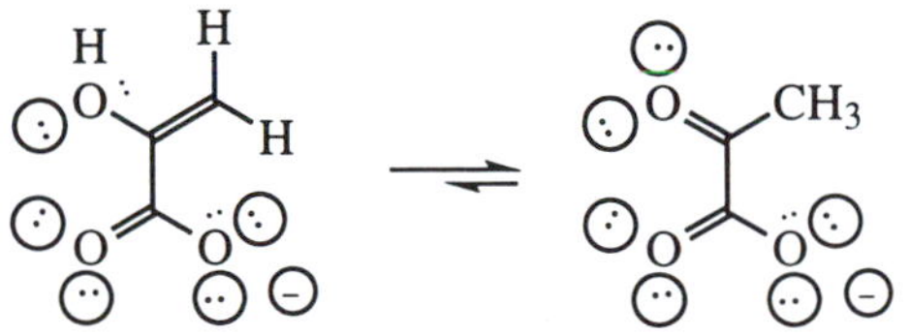

(5–36)

but if the enol is generated rapidly by treatment of phospho*enol*pyruvate with alkaline phosphatase, its lifetime (4 min at 20 °C and pH 6.4) is long enough that it can be examined as an intermediate in an enzymatic reaction.[52] When the biotinyl enzyme methylmalonyl-CoA carboxyltransferase

(*S*)-2-carboxypropanoyl-SCoA + pyruvate

⇌ propanoyl-SCoA + oxaloacetate (5–37)

is present in a solution, made from $^2H_2O$, in which either (*Z*)-[3-$^3$H]*enol*pyruvate or (*E*)-[3-$^3$H]*enol*pyruvate is generated, the enzyme catalyzes the stereospecific conversion of the *enol*pyruvate to [3-$^3$H,3-$^2$H]pyruvate by protonating its 2-*si* face in a reaction requiring the presence of (*S*)-2-carboxypropanoyl-SCoA.[53] This ability of the enzyme to accept *enol*pyruvate as a reactant is consistent with the participation of *enol*pyruvate as an intermediate in the enzymatic reaction

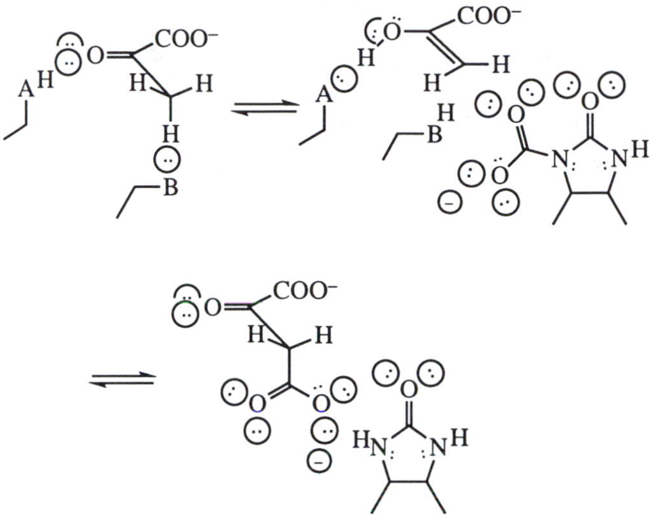

(5–38)

If a noncovalent intermediate in an enzymatic reaction is too unstable to obtain as a synthetic product, another approach to its identification is to trap it by a chemical reaction that produces a stable derivative of it. The phosphonocarbonate **5–8** formed as an intermediate at the active site in the reaction catalyzed by carbamyl-phosphate synthase (glutamine-hydrolyzing)

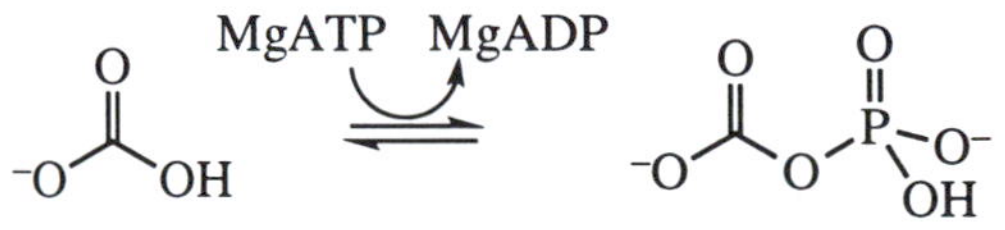

**5–8**

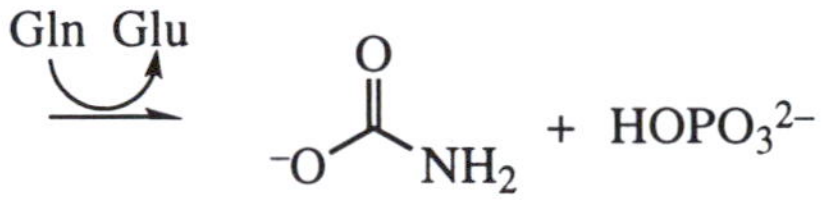

(5–39)

can be trapped by mixing a solution of the enzyme, MgATP, and bicarbonate with diazomethane[54]

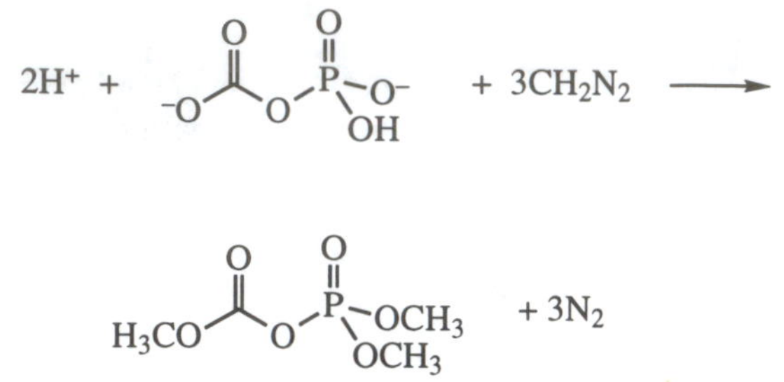

$$2H^+ + {}^-O-C(=O)-O-P(=O)(OH)-O^- + 3CH_2N_2 \longrightarrow H_3CO-C(=O)-O-P(=O)(OCH_3)-OCH_3 + 3N_2 \quad (5\text{–}40)$$

The trimethyl phosphonocarbonate is a stable compound, unlike the phosphonocarbonate itself.

Kynureninase catalyzes the reaction

$$\text{L-kynurenine} + H_2O \rightleftharpoons \text{anthranilate} + \text{L-alanine} \quad (5\text{–}41)$$

Enamine **5–9**, formed during the reaction

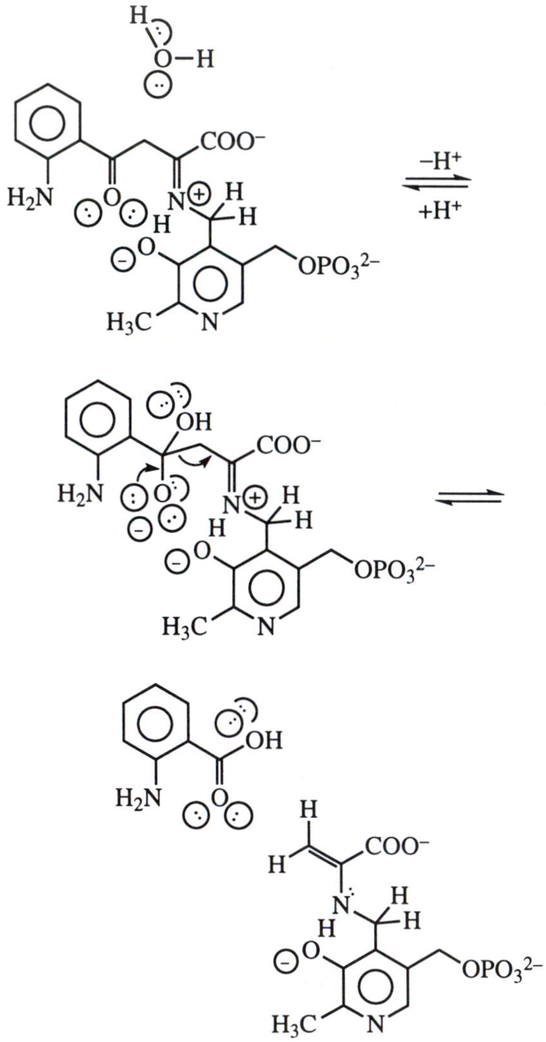

$-H^+$ / $+H^+$

(5–42)

**5–9**

can be trapped by benzaldehyde[55]

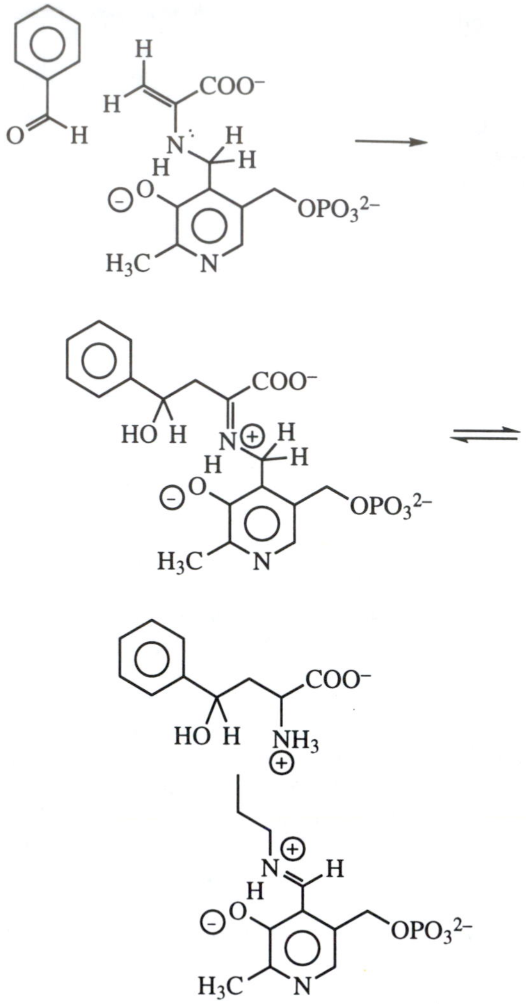

(5–43)

before it is converted to alanine, the normal product.

2-Carboxy-3-ketoribitol 1,5-bisphosphate, 5–10

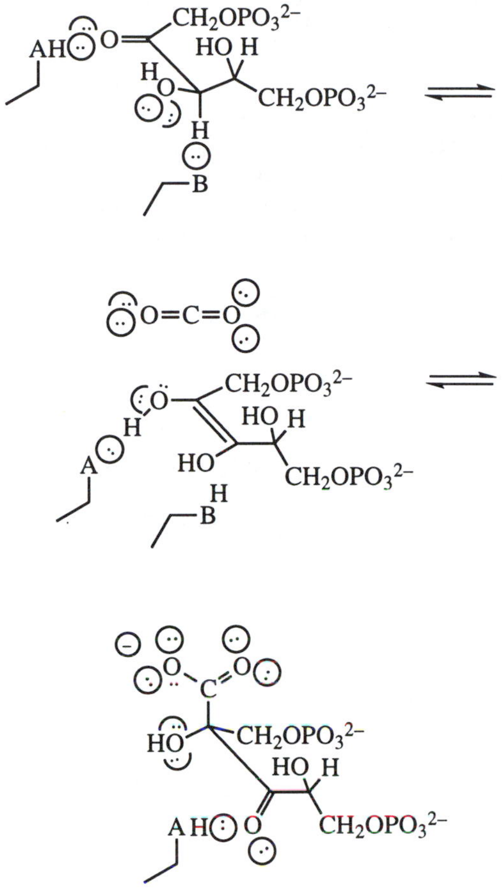

(5–44)

**5–10**

is formed as an intermediate[56] at the active site during the production of two molecules of 3-phosphoglycerate catalyzed by ribulose-bisphosphate carboxylase

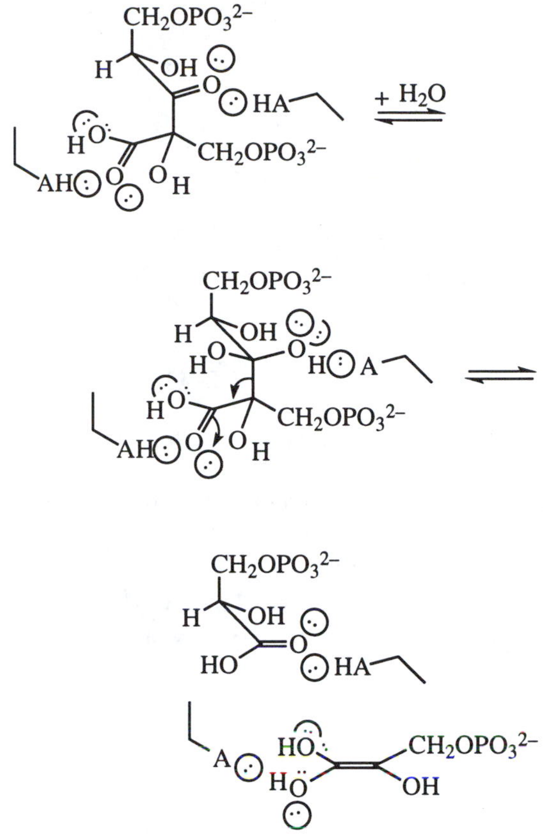

(5–45)

It can be trapped by reduction with $Na[^3H]BH_4$

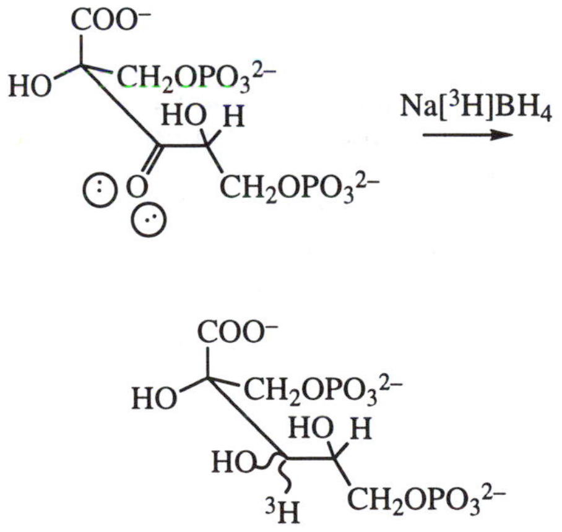

$Na[^3H]BH_4$

(5–46)

by denaturing the enzyme in the presence of the reductant during the course of its reaction.[57] The reduction produces a racemic mixture of products because the intermediate is reduced after the active site has been peeled away.

The 2-dehydro-*myo*-inositol 1-phosphate (**5–11**) produced as an intermediate in the reaction catalyzed by *myo*-inositol-1-phosphate synthase

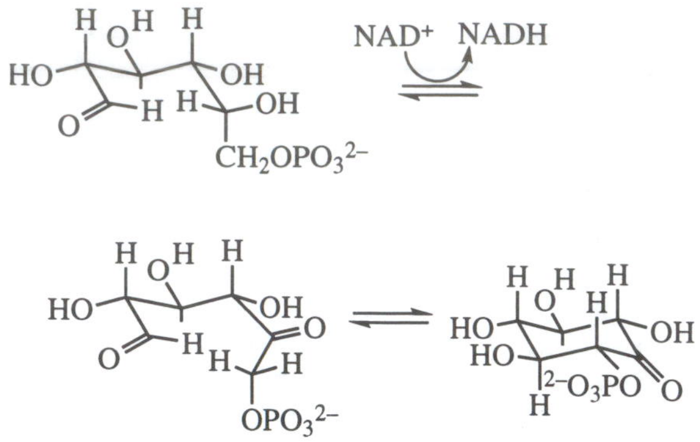

NAD+ NADH

$CH_2OPO_3^{2-}$

$OPO_3^{2-}$

(5–47)

**5–11**

is the product of carbon–carbon bond formation by an aldol condensation. It can also be trapped nonstereospecifically[58] with $Na[^3H]BH_4$ before it is reduced in the normal reaction to *myo*-inositol 1-phosphate by the coenzymatic NADH

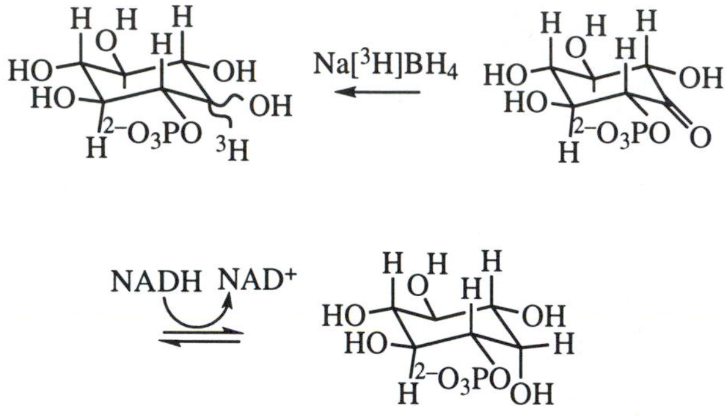

$Na[^3H]BH_4$

NADH NAD+

(5–48)

There are many instances in which the outcomes of chemical reactions other than the complete reaction catalyzed by the enzyme on its natural substrates can be used to provide evidence for a specific noncovalent intermediate in an enzymatic reaction. For example, adenosylhomocysteinase, in addition to catalyzing its normal reaction (Reaction 5–2), also catalyzes the unnatural oxidative elimination of adenine from 2′-deoxyadenosine

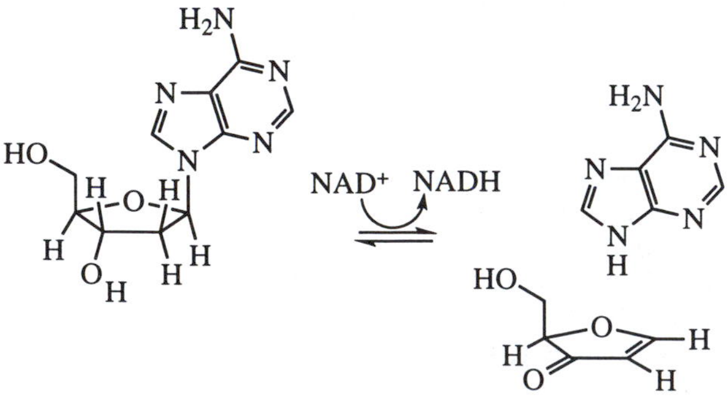

NAD+ NADH

(5–49)

concomitant with the reduction of the bound coenzyme to NADH.[59] This reaction is consistent with the initial formation of 2′-deoxy-3′-ketoadenosine during a reaction analogous to the first step in the normal enzymatic reaction (Figure 5–2). Glycine hydroxymethyltransferase, in addition to its normal reaction of transferring formaldehyde from the $\alpha$-carbon of glycine to nitrogens at positions 5 and 10 of tetrahydrofolate (Reaction 2–6), catalyzes both the decarboxylation of aminomalonate and the aldol cleavage of allothreonine.[60] The former reaction is a reflection of the ability of the enzyme to weaken a $\sigma$ bond on the $\alpha$-carbon of an amino acid in either the normal reaction (Reaction 5–50) or the unnatural reaction (Reaction 5–51) to produce the same intermediate

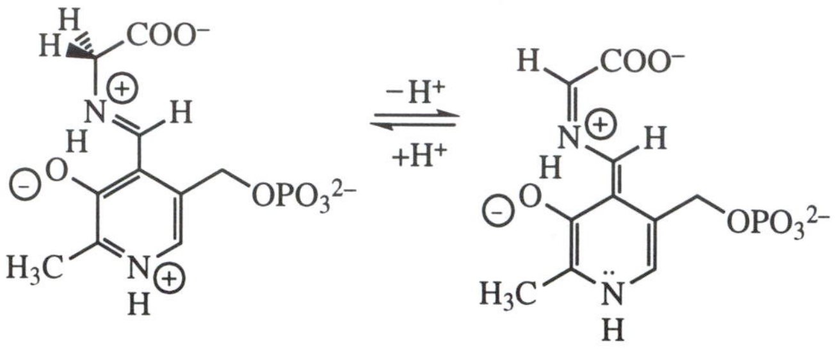

$-H^+$

$+H^+$

(5–50)

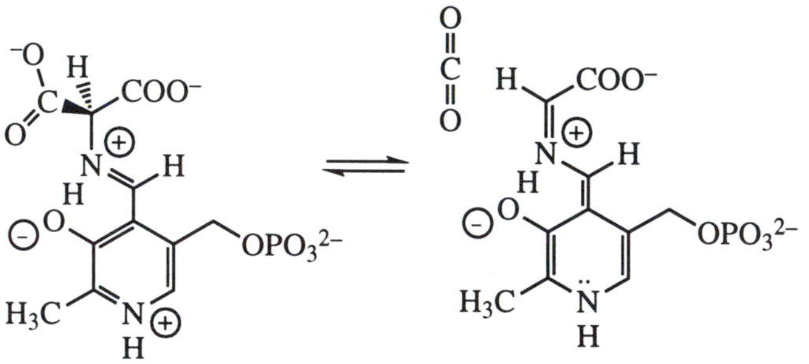

(5–51)

The latter reaction demonstrates that the intermediate is an enamine capable of reacting with electrophiles other than the naturally involved 5,10-methylenetetrahydrofolate

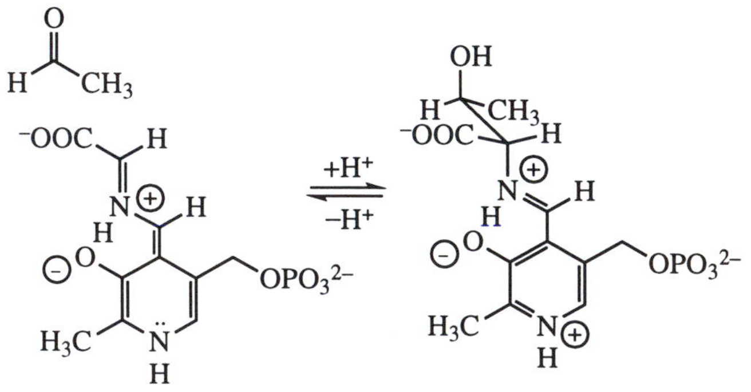

$+H^+$

$-H^+$

(5–52)

In the enzymatic reaction catalyzed by phosphogluconate dehydrogenase (decarboxylating), 3-keto-6-phosphogluconate **5–12** is the central, noncovalent intermediate

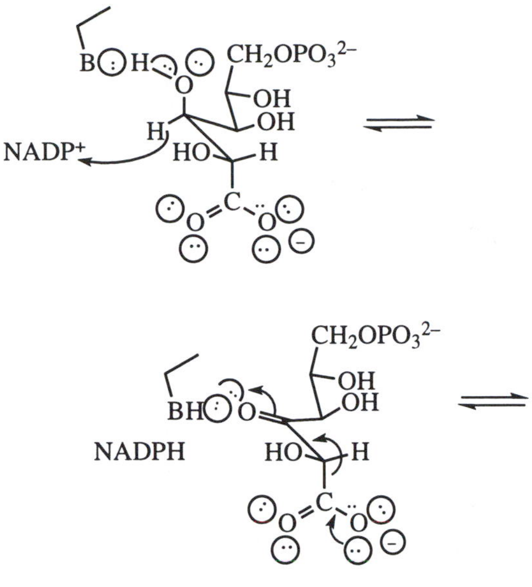

**5–12**

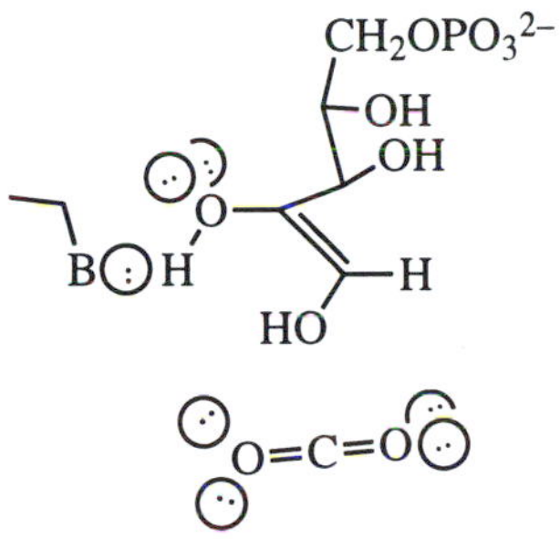

(5–53)

It had not been trapped chemically, in part because the rate of the dehydrogenation is slower than the rate of the decarboxylation so little of the intermediate can accumulate on the active site during the normal reaction.[61] Fortuitously, it was discovered that 2-deoxy-6-phosphogluconate is converted into 2-deoxy-3-keto-6-phosphogluconate

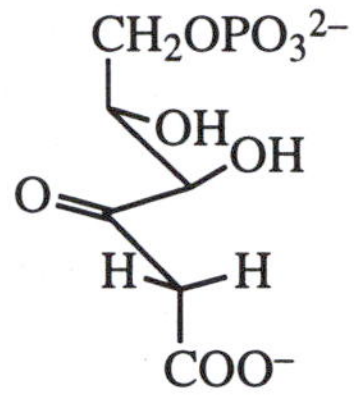

**5–13**

by the enzyme. Unlike the natural intermediate, which remains tightly bound, the rate at which 3-oxo-6-phospho-2-deoxygluconate dissociates from the active site is almost the same as the rate at which it is decarboxylated while on the active site. Consequently, 3-oxo-6-phospho-2-deoxygluconate appears in the solution as the reaction progresses (Figure 5–4).[61] This 3-oxo-6-phospho-2-deoxygluconate is eventually rebound at the active site and decarboxylated to 2-deoxyribulose 5-phosphate. Simply by replacing the hydroxyl at carbon 2 with a hydrogen, an otherwise undetectable intermediate was produced and released into the solution in significant yield by the enzyme.

A somewhat different strategy was used to force the noncovalent intermediate in the reaction catalyzed by farnesyl-diphosphate farnesyltransferase[62] to dissociate from the active site. The intermediate, presqualene diphosphate **5–14**,[63–65] is formed in an initial step by an acid–base reaction and then reduced by NADPH to the final product squalene

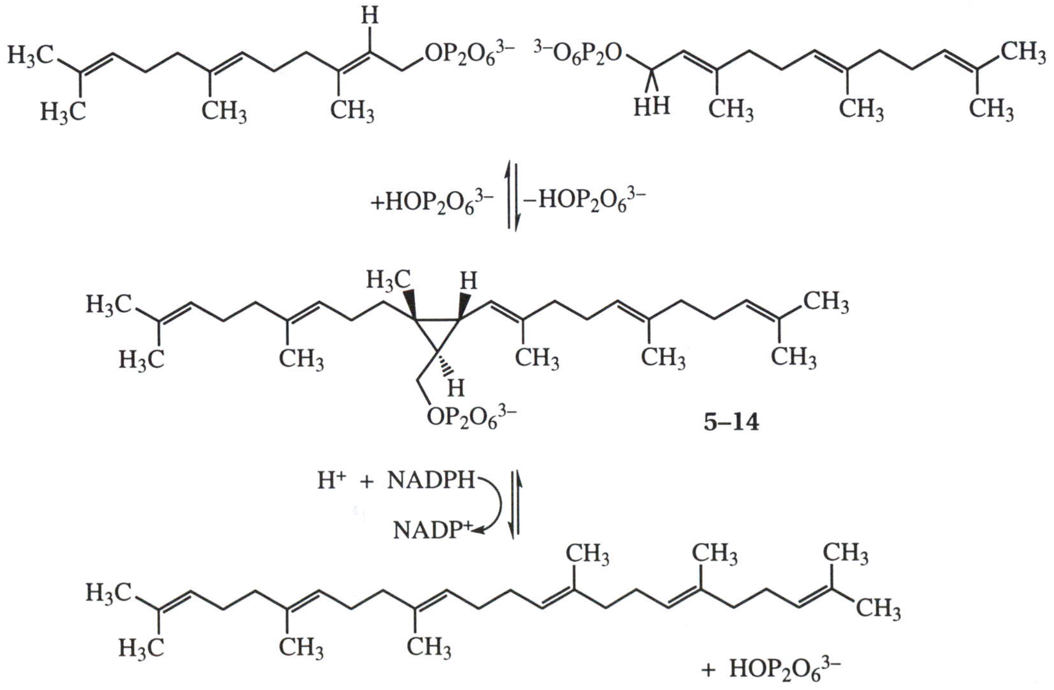

**5–14**

(5–54)

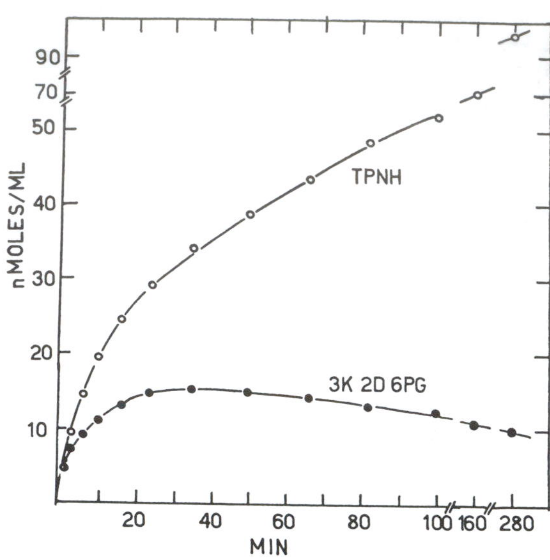


**Figure 5–4:** Kinetics of the formation of 3-oxo-6-phospho-2-deoxygluconate (3K2D6PG) and NADPH (TPNH) in the unnatural reaction catalyzed by phosphogluconate dehydrogenase (decarboxylating).[61] To the reaction mixture at 20 °C (final volume of 1 mL buffered at pH 8.0) was added 1.0 $\mu$mol of 2-deoxy-6-phosphogluconate, 1.2 $\mu$mol of $NADP^+$, and 0.68 unit of phosphogluconate dehydrogenase. The increase in the concentration of NADPH (nanomoles milliliter$^{-1}$) was followed as a function of time (minutes) by the absorbance of the solution at 340 nm (○). The concentration of 3-keto-2-deoxy-6-phosphogluconate (nanomoles milliliter$^{-1}$) was monitored by withdrawing samples and forming formazan derivatives from the $\beta$-ketocarboxylic acid by reaction with diazotized *p*-nitroaniline (●). Reprinted with permission from ref 61. Copyright 1973 *Journal of Biological Chemistry.*

In the absence of NADPH, the second step in the enzymatic reaction is blocked, and the intermediate dissociates as a stable compound from the active site.[66] Under normal circumstances sufficient NADPH is available to prevent the release of the intermediate, but the enzyme is membrane-bound; and, after it has been purified in solutions of detergent, the major product of the enzymatic reaction becomes the presqualene diphosphate even in the presence of saturating NADPH.[62] In this instance, the intermediate in the enzymatic reaction was sufficiently complex that total organic synthesis was required to define its structure unambiguously.[63–64]

As with phosphogluconate dehydrogenase (decarboxylating), an alternative reactant has been used with glutamate–ammonia ligase

$$\text{glutamate} + NH_4^+ + \text{MgATP} \rightleftharpoons \text{glutamine} + \text{MgADP} + HOPO_3^{2-} \qquad (5\text{–}55)$$

to identify an intermediate in the enzymatic reaction. When the enzyme is mixed with glutamate and MgATP in the absence of ammonia, a large change in the fluorescence of the protein occurs that is consistent with the formation of an intermediate.[67] If this intermediate sits in the active site in the absence of ammonia, pyrrolidone carboxylate is formed by cyclization of an activated glutamate

+ HX

(5–56)

where X is a good leaving group, presumably inorganic phosphate. It has been shown that $^{18}O$ is transferred stoichiometrically from L-[$\gamma$-$^{18}O$]glutamate to inorganic phosphate during enzymatic reaction,[68] but mechanisms not involving the presumed intermediate 5-phosphoglutamate, yet incorporating this transfer of $^{18}O$, can be formulated.[69] Attempts to isolate directly the expected intermediate, 5-phosphoglutamate, usually have been unsuccessful owing to the spontaneous formation of pyrrolidone carboxylate. *cis*-1-Amino-1,3-dicarboxycyclohexane, however, is a reactant for glutamate–ammonia ligase that resembles glutamate closely[70] but cannot cyclize as glutamate can. When *cis*-1-amino-1,3-dicarboxycyclohexane is incubated with the enzyme, in the absence of ammonia, the analogous acyl phosphate is formed, and it can be isolated and identified[71]

+ MgATP ⇌

$OPO_3^{2-}$ + MgADP

(5–57)

These results demonstrate the ability of the enzyme to form an acyl phosphate at the proper position in a reactant. It is also possible to identify 5-phosphoglutamate indirectly as an intermediate in the reaction catalyzed by glutamate–ammonia ligase by adding sodium borohydride to a solution containing the complex between the enzyme and the intermediate formed in the absence of ammonia. The borohydride reduces the 5-phosphoglutamate to 2-amino-5-hydroxyvalerate[72] (Reaction 5–19). The identification of this product, however, demonstrates only that the 5-carboxy group is activated sufficiently to react with sodium borohydride, not that it is necessarily phosphorylated.

Formate–tetrahydrofolate ligase

MgATP + formate + tetrahydrofolate ⇌

10-formyltetrahydrofolate + MgADP + $HOPO_3^{2-}$ (5–58)

has formyl phosphate, formed from MgATP and formate

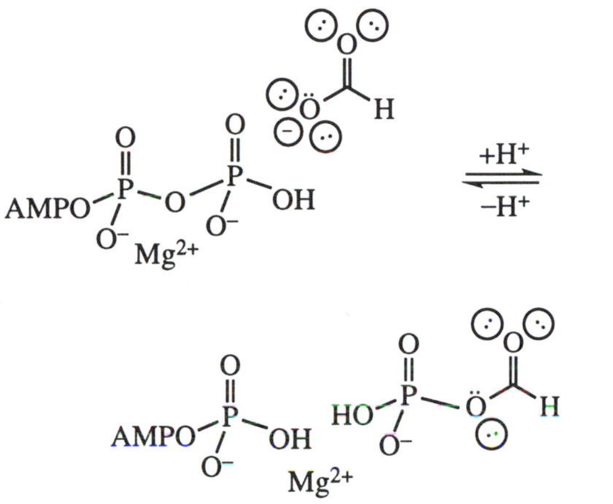

(5–59)

as an unstable intermediate in its reaction. Carbamyl phosphate

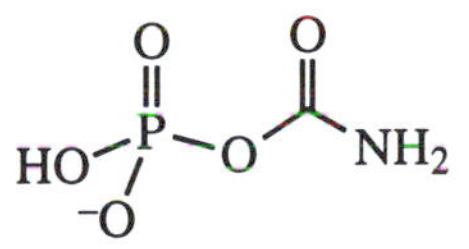

**5–15**

resembles formyl phosphate and is mistaken by the enzyme for formyl phosphate. The enzyme catalyzes phosphotransfer from carbamyl phosphate to MgADP to form MgATP (the reverse of Reaction 5–59), consistent with the role of formyl phosphate as an intermediate in the enzymatic reaction,[73] but the rate of the unnatural reaction is only about 1% that of the reaction of $HOPO_3^{2-}$ and 10-formyltetrahydrofolate with MgADP to form MgATP. The purpose for producing formyl phosphate would be to activate the carboxyl group by providing an excellent leaving group. This permits the nitrogen of tetrahydrofolate to form the anilide in an acyl exchange

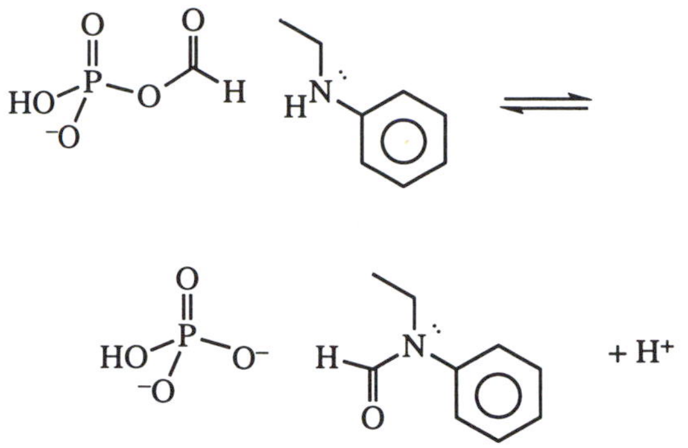

(5–60)

It has already been noted that analogues of intermediates of high energy are often competitive inhibitors with high affinity for the active sites of enzymes, and several examples of such analogues have already been presented. The earlier discussion focused on these inhibitors as evidence for the proposal that the active sites of enzymes are constructed to bind intermediates of high energy tightly and thereby lower the free energy required to produce these intermediates at the active site. Another view of this observation is that if a particular compound has been designed to mimic a postulated intermediate in a proposed mechanism for the enzymatic reaction, the fact that it binds tightly to the active site is evidence for the existence of that intermediate. For example, the fact that the difluoromethyl ketal phosphate **5–16** is a competitive inhibitor of 3-phosphoshikimate 1-carboxyvinyltransferase with high affinity for the active site ($K_i$ = 4 nM)[74] has been presented as evidence that methyl ketal phosphate **5–17** is an intermediate in the enzymatic reaction

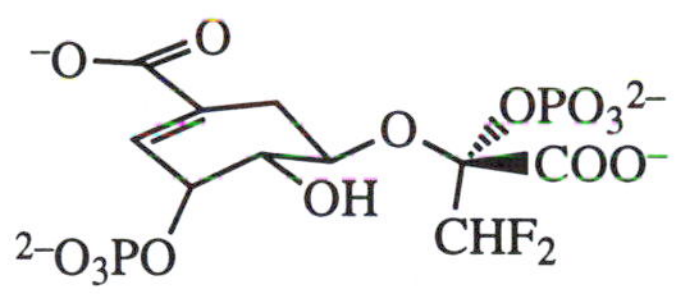

**5–16**

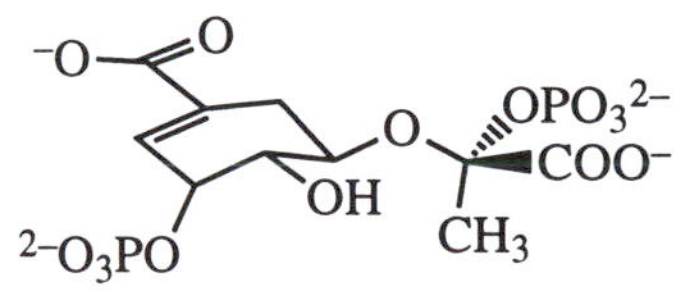

**5–17**

In this instance, the introduction of the two fluorines at the methyl carbon provided enough electron withdrawal to stabilize a ketal phosphate that would otherwise be too unstable to synthesize or isolate from the enzyme. The methyl ketal phosphate would be formed in the first step of the enzymatic reaction

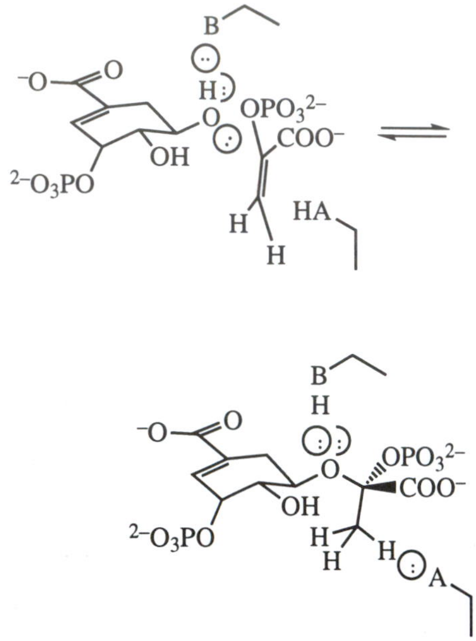

(5–61)

from shikimate 3-phosphate and phospho*enol*pyruvate.

When a $\sigma$ bond between carbon and another atom is broken, the two electrons in the bond can remain on carbon to produce a carbanionic intermediate, almost always stabilized in an enzymatic reaction as an **enamine** or an **enol**,[75] for example

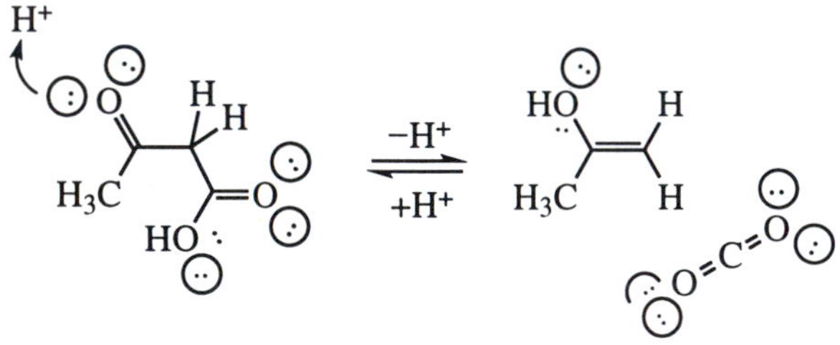

(5–62)

or the two electrons can depart with the leaving group to produce a **carbocationic intermediate**, for example

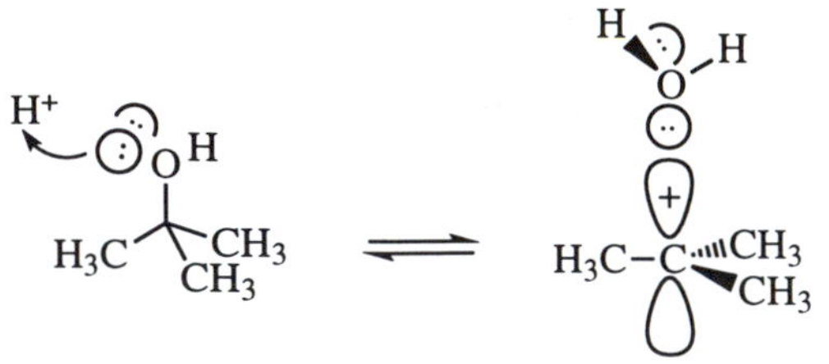

(5–63)

or the two electrons can be split between the carbon and the departing atom to produce a **radical intermediate** (Reaction 2–214). There are enzymatic examples of each of these possibilities, and it is often difficult to distinguish among them. Addition–elimination reactions in solution provide the clearest examples of reactions in which either carbocationic (Reaction 1–88) or carbanionic (Reaction 1–94) intermediates can participate. Carbocationic or carbanionic mechanisms dominate when addition–eliminations in solution occur under acidic or basic conditions, respectively, but at neutral pH either is a possibility. Carbocationic intermediates in enzymatic reactions, as in reactions in solution, are generally stabilized by hyperconjugation from immediately adjacent alkyl groups or by conjugation with either neighboring $\pi$ systems among carbons or neighboring lone pairs on oxygen or nitrogen. Carbanionic intermediates are almost always formed on the $\alpha$-carbon of either a protonated carbonyl group or a protonated acyl group and so are actually enols or enamines.

The addition–elimination catalyzed by the enzyme aconitate $\Delta$-isomerase could have proceeded either way[76]

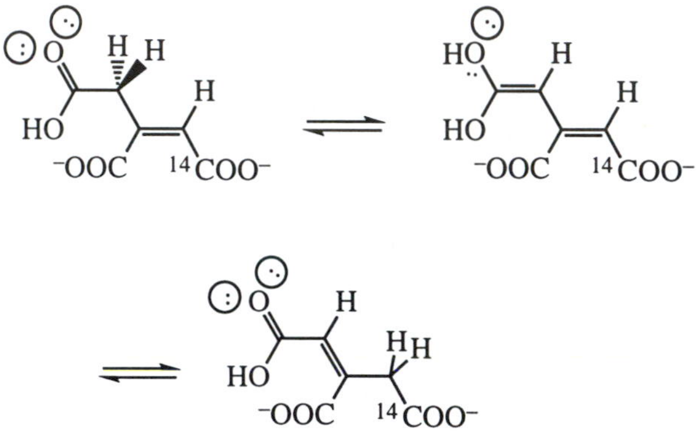

(5–64)

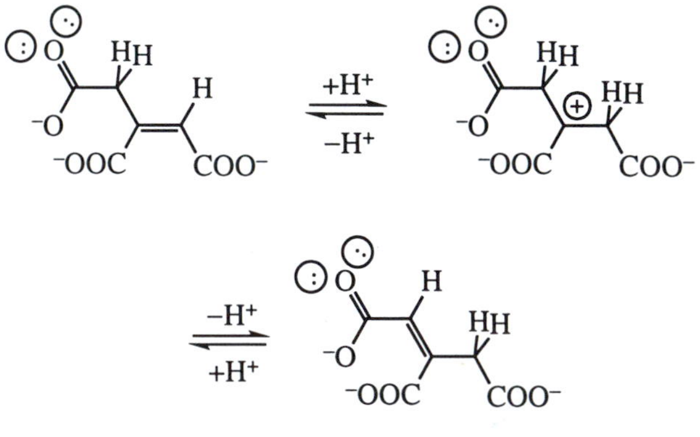

(5–65)

although the carbocation is not so appealing as the enol because it is forced to reside next to the electropositive carbon of a carboxylate. It was observed that during the isomerization of *cis*-[4-$^3$H]aconitate in $^2H_2O$, the *pro-S* tritium was removed from the reactant and a small fraction of that tritium (5%) was transferred to carbon 2. The tritium or deuterium added to carbon 2 occupied the *S* position in the product. These observations are consistent with the removal of a proton from carbon 4 of the reactant by a base on the enzyme and its transfer across that same face of the intermediate by the base and its subsequent addition to carbon 2 from that same face. If this is the case, the intermediate must be the enol (Reaction 5–64).

Enzymatic addition–eliminations that occur adjacent to carboxylates as in the case of aconitate Δ-isomerase have been observed uniformly to proceed through intermediate enols[77]

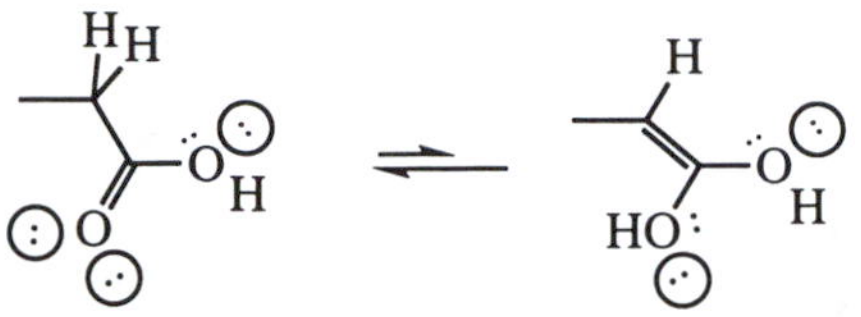

(5–66)

rather than carbocations. A simple demonstration of this generalization is that these enzymes are strongly and competitively inhibited by the conjugate bases of nitrate analogues of the respective enolates

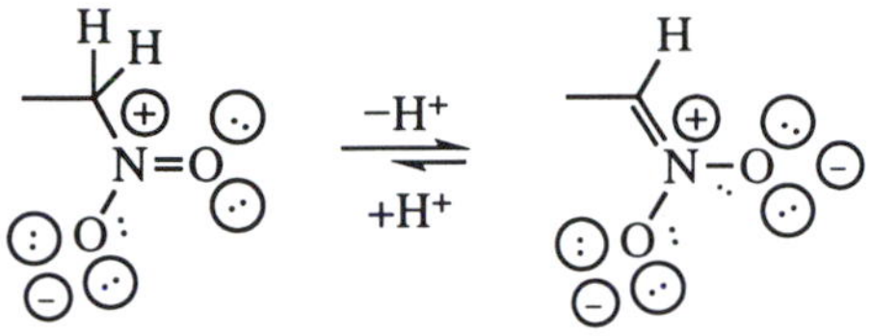

(5–67)

For example, the intermediate enol formed during the dehydration–hydration of fumarate by fumarate hydratase

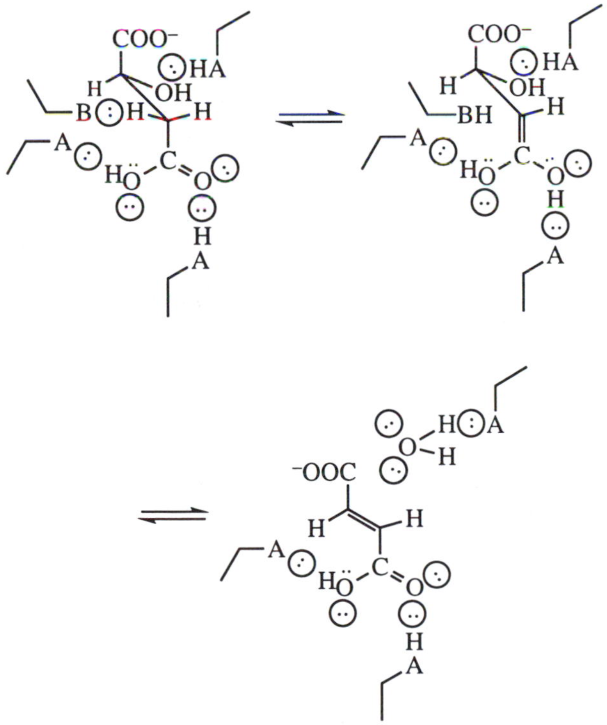

(5–68)

can be mimicked by the conjugate base of 3-nitro-2-hydroxypropionate[78]

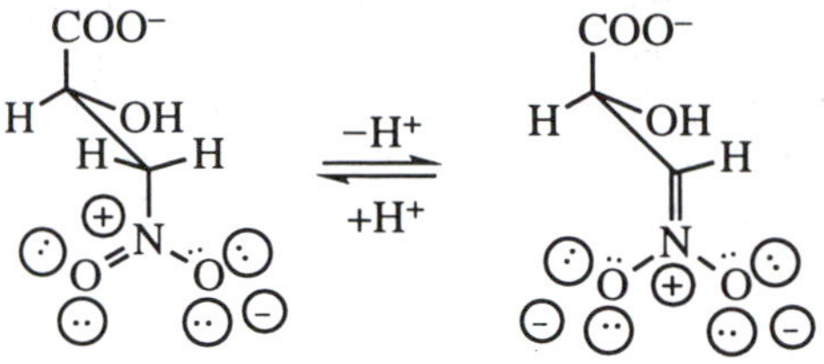

(5–69)

The conjugate base of 3-nitro-2-hydroxypropionate is a competitive inhibitor of fumarate hydratase with a dissociation constant $10^4$ times smaller than the dissociation constant for succinate, a simple competitive inhibitor of the enzyme, and $10^3$ times smaller than the Michaelis constant for malate as a reactant. Similarly, the conjugate base of 2-amino-3-nitropropionate is a potent competitive inhibitor ($K_m^{Asp}K_i^{-1} = 200$) of aspartate ammonia-lyase,[78] the conjugate base of (3-hydroxy-2-nitropropyl)phosphonate is a potent competitive inhibitor ($K_m^{2PGA}K_i^{-1} = 10^3$) of enolase,[77] and the conjugate base of 2-hydroxy-3-nitro-1,2-propanedicarboxylate is a potent competitive inhibitor ($K_m^{citrate}K_i^{-1} = 3000$) for aconitate hydratase.[79]

Intermediate enols and enamines in enzymatic reactions have also been identified by their susceptibility to oxidation. It has already been noted that the electron-rich 1,2-enaminol produced between aldolase and dihydroxyacetone phosphate can be oxidized to the iminaldehyde (Reaction 5–25) with $Fe^{III}(CN)_6^{3-}$ as well as other oxidants.[80] The production of the vicinal dione, 3-hydroxypyruvaldehyde phosphate, by the enzyme from dihydroxyacetone phosphate and an oxidant provides evidence for the existence of this 1,2-enaminol. The rapid reduction of $Fe^{III}(CN)_6^{3-}$ in the presence of aspartate aminotransferase and 3-hydroxyaspartate is thought to represent oxidation of the electron-rich vinylogous enamine at the active site

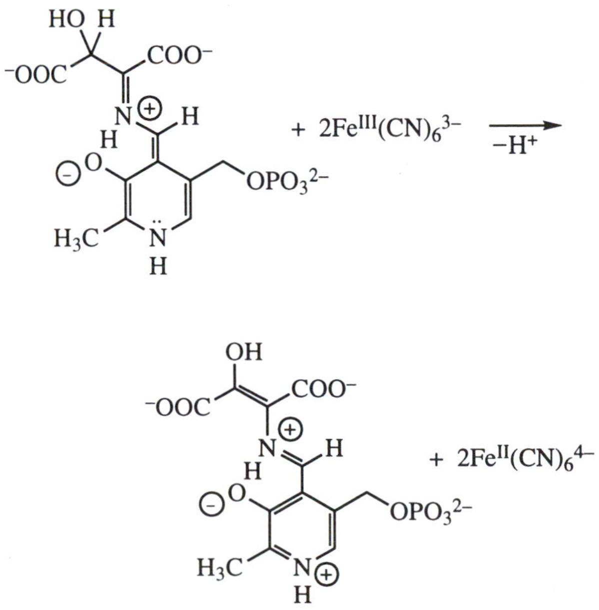

$$+ 2Fe^{III}(CN)_6^{3-} \xrightarrow{-H^+} + 2Fe^{II}(CN)_6^{4-}$$

(5–70)

and the reduction of tetranitromethane in the presence of phosphogluconate dehydrogenase (decarboxylating), ribulose 5-phosphate, and NADH is thought to represent the oxidation of the electron-rich *cis*-enediol intermediate in the reaction catalyzed by the enzyme[80] (Reaction 5–53)

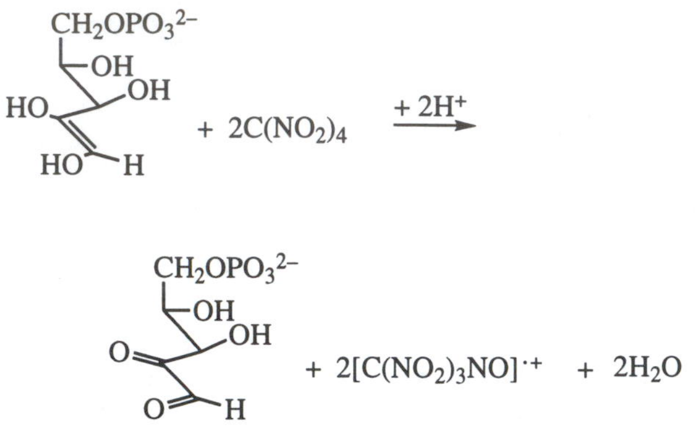

$CH_2OPO_3^{2-}$ ... $+ 2C(NO_2)_4 \xrightarrow{+2H^+}$ ... $+ 2[C(NO_2)_3NO]^{\cdot +} + 2H_2O$

(5–71)

In these oxidation reactions, the oxidants used are one-electron oxidants that must remove electrons one at a time from the electron-rich enols or enamines, with concomitant loss of one or two protons.

If an analogue of the reactant in an enzymatic reaction is synthesized to incorporate a good leaving group, such as a halogen on a carbon adjacent to the carbanionic carbon in the enol formed as an intermediate in the reaction, and if the active site will recognize the analogue as a reactant, then when the enol forms, the leaving group next to the carbanionic carbon will eliminate

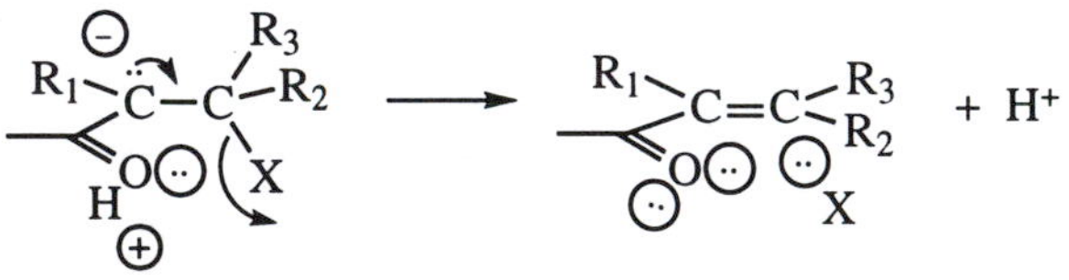

(5–72)

The product of the elimination will identify the carbanionic carbon of the enol. D-Amino acid oxidase converts 2-amino-3-chlorobutyrate to 2-ketobutyrate,[81] a product that arises from the elimination of chloride from an intermediate enol

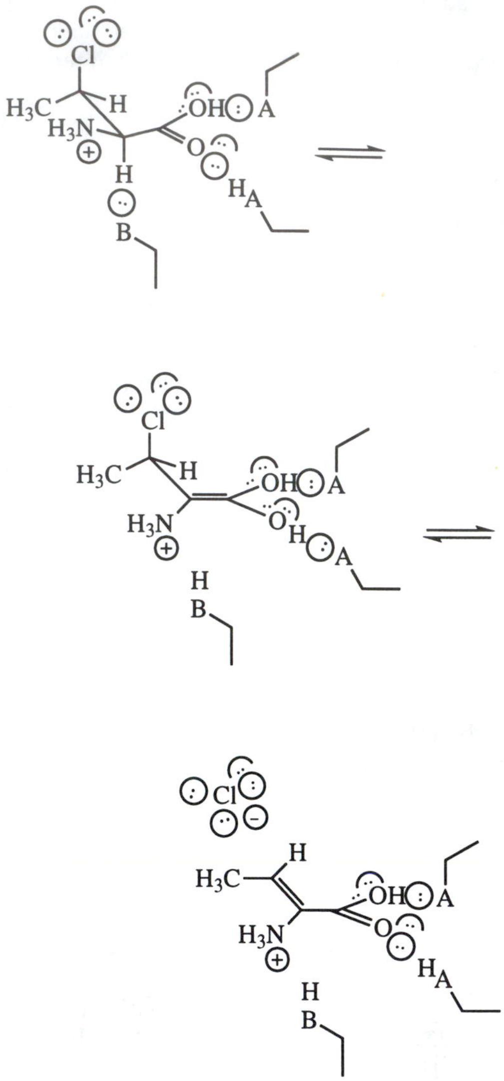

(5–73)

In the absence of a good leaving group on the adjacent carbon, the intermediate, electron-rich enol formed in the normal enzymatic reaction is oxidized by flavin

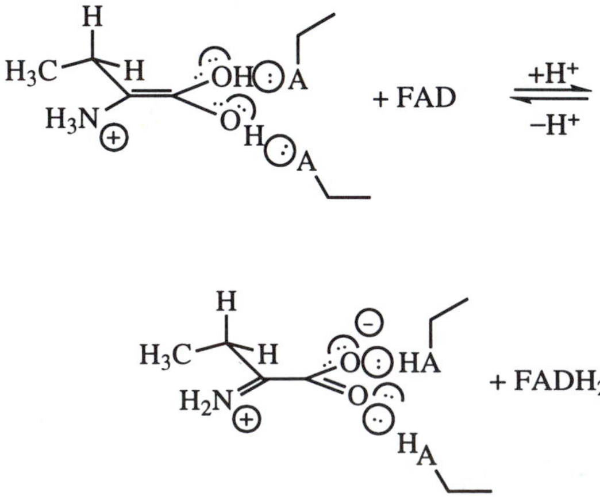

$+ FAD \underset{-H^+}{\overset{+H^+}{\rightleftharpoons}}$ ... $+ FADH_2$

(5–74)

Propionyl-CoA carboxylase converts 3-fluoropropionyl-SCoA to acrylyl-SCoA,[82] a product that arises from elimination from an intermediate enol

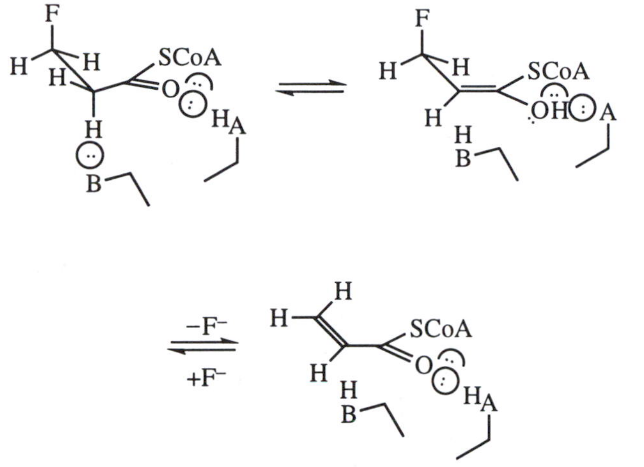

(5–75)

In the absence of a good leaving group, the intermediate enol in the normal reaction catalyzed by the enzyme is carboxylated by *N*-carboxybiotin bound at the active site as a coenzyme

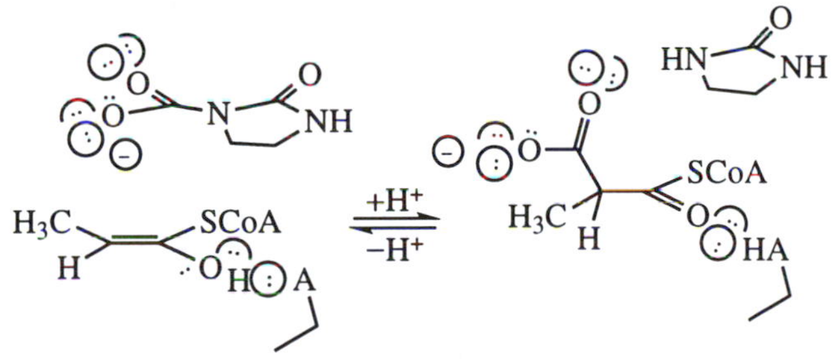

(5–76)

Although **carbocationic intermediates** are common in organic reactions under acidic conditions, they are commonly encountered in enzymatic reactions only in glycosyl transfer and polyisoprenoid biosynthesis. In a glycosyl transfer reaction such as that catalyzed by lysozyme (Figure 3-25), the carbocationic intermediates are stabilized as oxonium cations

**5–18**

The carbocationic intermediates involved in polyisoprenoid biosynthesis are all allylic carbocations

**5–19**

*β*-Galactosidase is a galactosyltransferase that displays little specificity for the alcohols between which the galactose is transferred

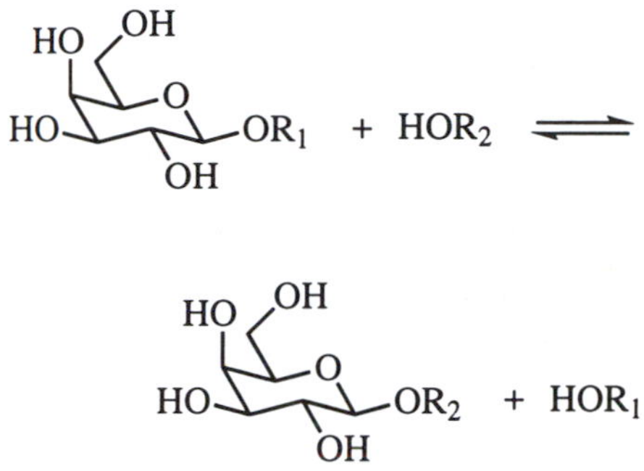

(5–77)

When the enzyme processed any one of a series of *β*-galactosides in a solution of 0.25 M methanol in water, the products were galactose, arising from transfer to water ($R_2$ = H), and methyl *β*-galactoside, arising from transfer to methanol ($R_2$ = $CH_3$), in a molar ratio of 1:2.[83] The molar ratio between the two products remained 1:2 within the experimental error of the measurements when the $V_{max}$ of the enzyme was varied over a 50-fold range by choosing a set of reactants that differed in the properties of the leaving group, $HOR_1$ (Table 5–1).[83] The most reasonable explanation for this observation is that either acceptor, the water or the methanol, reacts with the galactosyl group only after the leaving group has departed from the active site. The production of an oxonium cation as an intermediate in the reaction

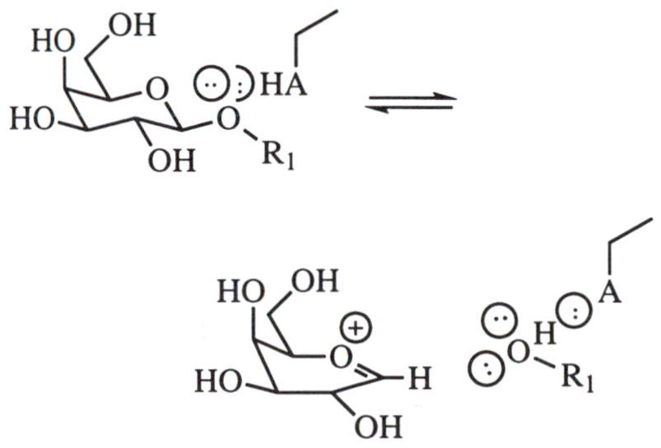

(5–78)

would explain these observations, as would a covalent galactosyl-enzyme intermediate. In either case, either acceptor, methanol or water, would be reacting with the same intermediate, regardless of the particular reactant used to produce that intermediate.

**Table 5–1: Ratio of Products for the Solvolysis of β-Galactosides by β-Galactosidase[a]**

| β-galactoside | $V_{max}$[b] (μmol $min^{-1}$ $mg^{-1}$) | ratio of methyl β-galactosides[c] |
|---|---|---|
| *o*-nitrophenyl | 360 | 1.97 ± 0.05 |
| *m*-nitrophenyl | 327 | 1.96 ± 0.13 |
| *m*-chlorophenyl | 180 | 2.08 ± 0.24 |
| *p*-nitrophenyl | 69 | 1.99 ± 0.05 |
| phenyl | 40 | 1.94 ± 0.20 |
| *p*-methoxyphenyl | 33 | 2.14 ± 0.13 |
| *p*-chlorophenyl | 7.5 | 2.02 ± 0.16 |
| *p*-bromophenyl | 7.0 | 2.03 ± 0.08 |

[a]Reaction run at 25 °C in 0.25 M methanol at pH 7.5. [b]Maximum velocity in the absence of methanol. [c]Ratio of products formed in the presence of 0.25 M methanol.[83]

The same type of oxonium cation is an intermediate in the reaction catalyzed by phosphorylase

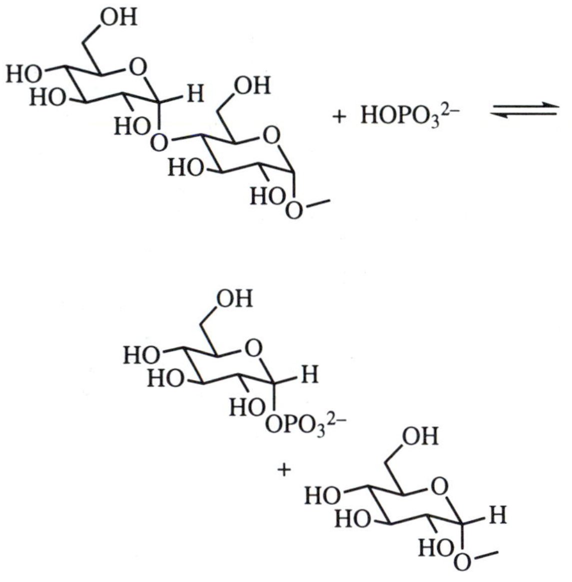

(5–79)

In this reaction, the phosphate acts as the enforced acceptor for the oxonium cation

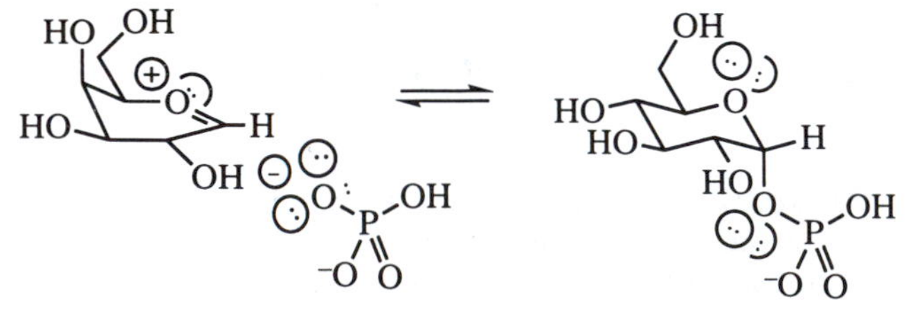

(5–80)

2,6-Anhydro-1-deoxy-D-*gluco*-hept-1-enitol and inorganic phosphate are converted to 1-deoxy-α-D-*gluco*-heptulose 2-phosphate by phosphorylase in a reaction in which the planar olefin is mistaken for the planar oxonium cation[84]

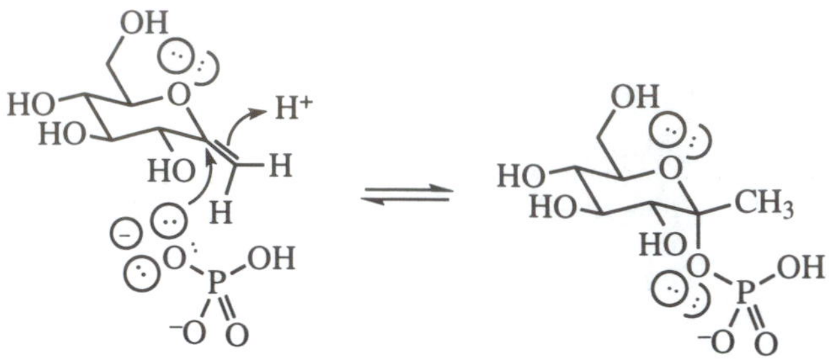

(5–81)

The existence of this reaction is consistent with the existence of the oxonium cation in the normal reaction.

The initial enzyme in polyisoprenoid biosynthesis is dimethylallyl *trans*-transferase

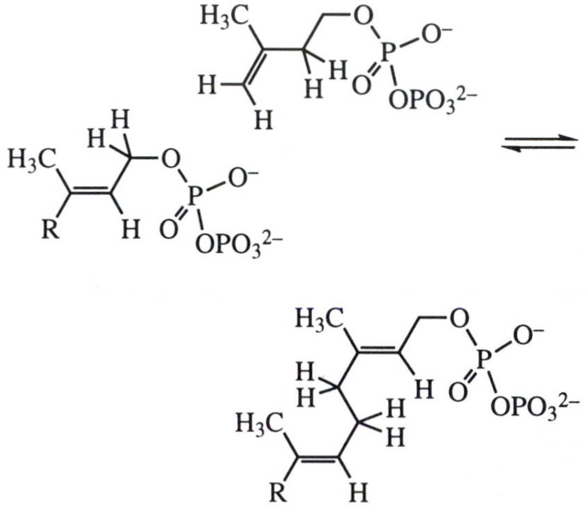

(5–82)

where R is a methyl group in dimethylallyl diphosphate or a 5-methylpent-4-enyl group in geranyl diphosphate. The reaction can be formulated with an $S_N1$ mechanism (Reaction 1–16) resembling that for the cationic polymerization of alkenes at low pH. In this mechanism an allylic carbocation is formed by loss of pyrophosphate from the allylic diphosphate

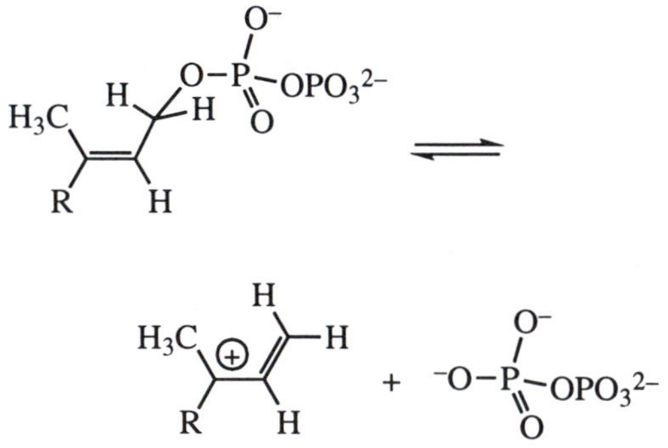

(5–83)

Because the diphosphate is on a primary carbon and because the addition of the alkene proceeds with inversion of configuration at the allylic carbon bearing the diphosphate, it was also possible that an $S_N2$ reaction could occur at the allylic carbon rather than an $S_N1$ reaction.[85]

A decision has been made between these two possibilities by replacing one of the methyl groups of the dimethylallyl diphosphate with a trifluoromethyl group.[86] In model studies, it was shown that such a replacement would enhance an $S_N2$ reaction by a factor of 10 but would inhibit an $S_N1$ reaction by a factor of $10^{-6}$ because of the difficulty of forming the carbonium ion next to an electron-withdrawing substituent

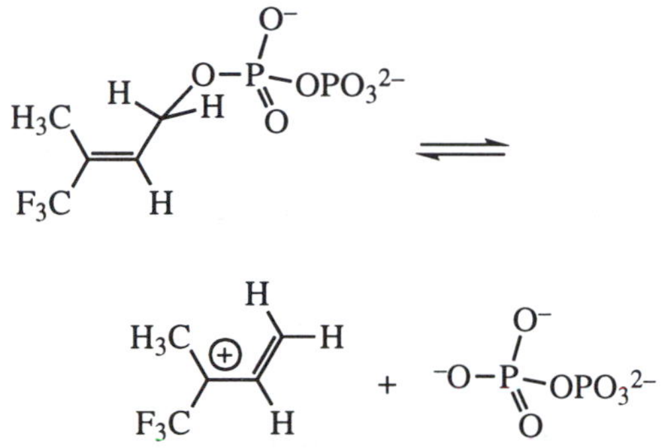

(5–84)

In the enzymatic reaction, (trifluoromethyl)methylallyl diphosphate, although it is a competitive inhibitor, shows no reaction ($V_{max}^{CF_3} < 10^{-6} V_{max}^{CH_3}$). These results rule out an $S_N2$ mechanism and are consistent with the existence of a carbocationic intermediate in the normal reaction. A series of fluorinated derivatives of geranyl diphosphate were also synthesized, and it was shown that the rate constants for their nonenzymatic solvolysis at low pH, a reaction definitely proceeding by an $S_N1$ mechanism, correlated closely with the turnover rates of the same compounds as reactants in the reaction catalyzed by dimethylallyl *trans*-transferase (Reaction 5–82).[87] These results also support the existence of an allyl carbocationic intermediate in this enzymatic reaction.

The other carbocationic intermediate in this same enzymatic reaction is the tertiary carbocation formed upon addition of the initial carbocation to the olefin. This other carbocation eliminates a proton to yield the other double bond in the product. Evidence for the existence of this other carbocationic intermediate was obtained similarly by showing that 2,2-difluoroisopentenyl diphosphate was not utilized as an acceptor in the enzymatic reaction even when it was incubated for a period of 3 days with the enzyme and geranyl diphosphate serving as the donor of the initially formed carbocation.[88] Again, the fluorination prevented the reaction from occurring by hindering the formation of the tertiary carbocation on the acceptor.

The linear polyisoprenoids produced in such reactions end up with an allylic diphosphate at one end (Reaction 5–82). These diphosphates in turn are susceptible to an array of cyclization reactions. Geranyldiphosphate cyclase I produces (+)-$\alpha$-pinene, (+)-limonene, (+)-camphene, and (+)-borneol diphosphate. The same active site produces all of these products[89,90] and all four can be explained as arising from the same carbocationic intermediate

**5–20**

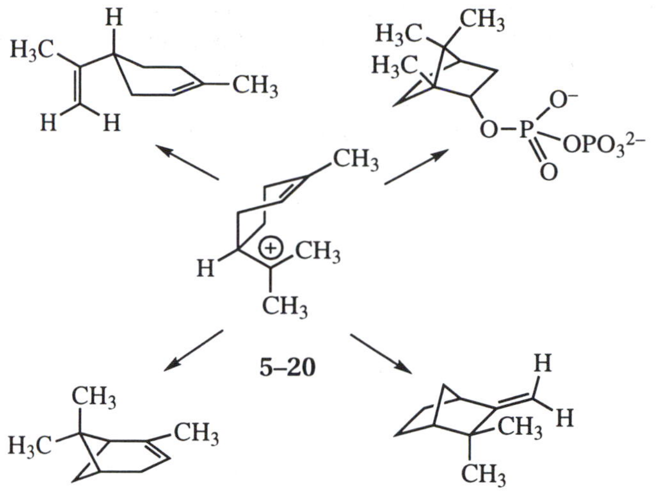

(5–85)

which in turn can be derived from the carbocationic intermediate produced during the dissociation of geranyl diphosphate

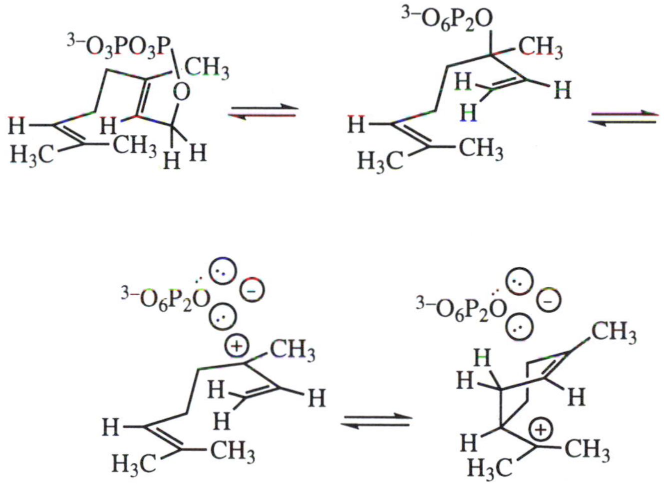

(5–86)

Therefore, the identity of the products is consistent with a series of carbocationic intermediates in the enzymatic reaction. The (+)-borneol diphosphate formed during the reaction should arise by the quenching of rearranged carbocation **5–20** by pyrophosphate[91]

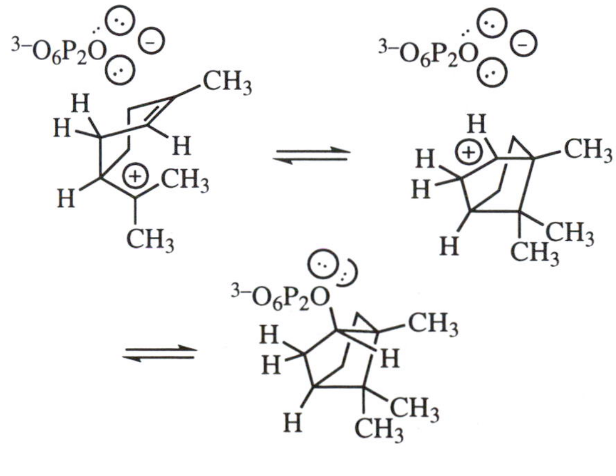

(5–87)

In fact, the same oxygen, out of the seven oxygens in pyrophosphate, that left to produce the initial carbocation (Reaction 5–86) reassociates with the rearranged carbocation to produce the (+)-borneol diphosphate.[91,92]

An analogue of the intermediate, sulfonium cation **5–21**

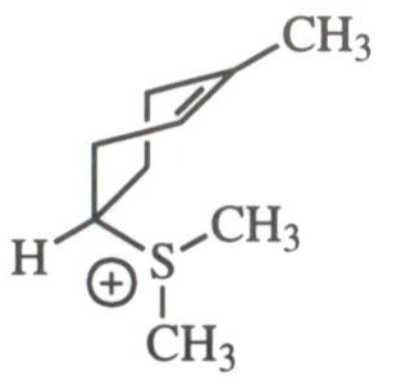

**5–21**

was synthesized[90] as a mimic of the common carbocationic intermediate (Reaction 5–85). Sulfonium cation **5–21** in the presence of pyrophosphate is a good inhibitor ($K_i$ = 1 $\mu$M) of geranyldiphosphate cyclase II, consistent with the involvement of carbocationic intermediate **5–20** in the enzymatic reaction.

The chemistry of almost all enzymatic reactions is heterolytic because it is based on acid–base reactions, which are by definition heterolytic. There are certain coenzymes, in particular flavins, hemes, and coenzyme $B_{12}$, and a few enzymes lacking such coenzymes that may engage in homolytic chemistry. The difficulty of forming **radical intermediates** in an enzymatic reaction is their ability to participate in hydrogen abstraction.

Most organic radicals can be formally defined as the product of a hydrogen abstraction. For example, the tyrosine radical formed upon X-irradiation of a crystal of tyrosine hydrochloride has the structure

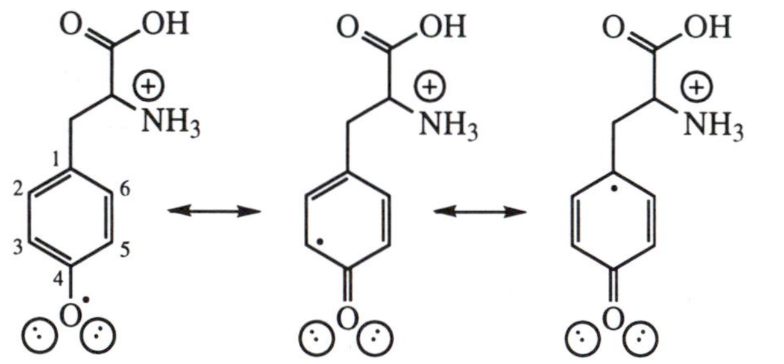

**5–22**

with spin densities in the solid state of 0.24–0.32 on the oxygen and carbons 1, 3, and 5, respectively.[93] It can be thought of as the product of electron transfer from the $\pi$ system followed by proton dissociation from the radical cation

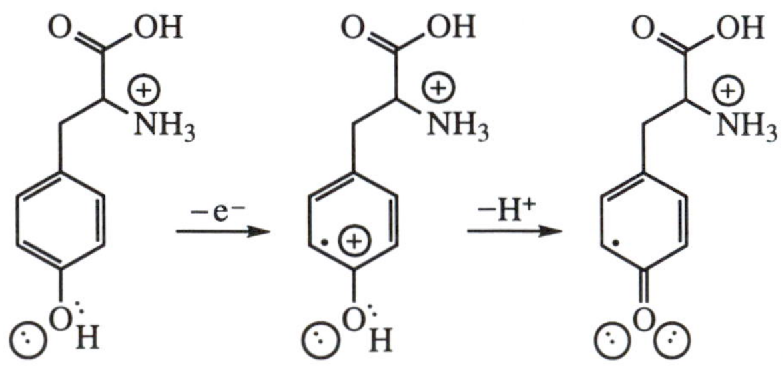

(5–88)

or the product of hydrogen abstraction from the oxygen

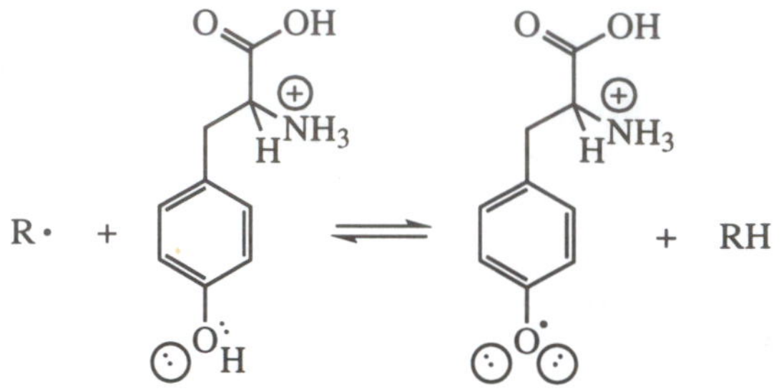

(5–89)

Regardless of the mechanism of its formation, it is capable of hydrogen abstraction at oxygen from a suitable donor (RH in Reaction 5–89).

In marked contrast to acid–base chemistry, any hydrogen in a protein is susceptible to hydrogen abstraction to form a radical. The hydrogenated form of any free radical is the functional group produced when that free radical abstracts a hydrogen from another location. The products of this hydrogen abstraction are a new radical on the atom from which the hydrogen has been abstracted and a new $\sigma$ covalent bond between the hydrogen and the atom on which the unpaired electron was previously located. If the hydrogenated form of a radical formed as an intermediate in an enzymatic reaction has a bond dissociation energy (BDE) greater than the bond dissociation energies of the hydrogens that necessarily surround it in the active site, the radical would uncontrollably abstract one of these hydrogens and produce a radical that would spread over the structure of the protein by successive hydrogen abstractions or electron transfers. This consideration, for example, is the reason that the existence of a radical as reactive as the 5′-deoxyadenosyl radical in the mechanism of coenzyme $B_{12}$ seems so unlikely. It is possible, however, that enzymatic active sites could be constructed so as to surround free radicals with hydrogens that are difficult to abstract, such as the hydrogens of water (BDE = 498 kJ mol$^{-1}$), phenyl hydrogens (BDE = 460 kJ mol$^{-1}$), or alcoholic hydrogens (BDE = 435 kJ mol$^{-1}$). It is the unavoidable nature of hydrogen abstraction that may explain why most radical intermediates that have been proposed for enzymatic reactions, with the notable exception of the 5′-deoxyadenosyl radical and the alkyl radicals thought to be involved in the reactions catalyzed by cytochrome P-450 (Reaction 2–167),[94] have low values for the bond dissociation energies of their hydrogenated forms. It may also explain why radical intermediates seem so rare.

Ribonucleoside-diphosphate reductase (R = $-PO_3^{2-}$) and ribonucleoside-triphosphate reductase (R = $-P_2O_6^{3-}$) are two enzymes that catalyze the same oxidation–reduction

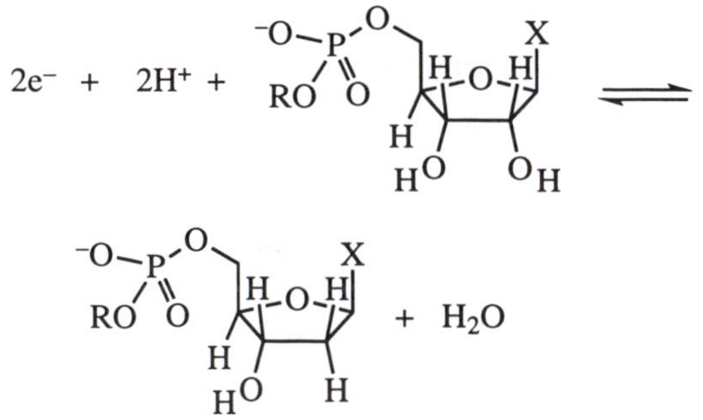

(5–90)

where X is uracil, cytosine, adenine, or guanine and R is either a phosphate or a diphosphate, respectively. In both cases, the two electrons are provided by two adjacent cysteines in the active site

$$\text{–SH} \quad \text{HS–} \rightleftharpoons \text{–S–S–} + 2H^+ + 2e^- \qquad (5\text{–}91)$$

The cystine formed at the active site is reduced back to two cysteines by two other adjacent cysteines on the enzyme, which in turn are reduced back again to cysteines by two adjacent cysteines on the ultimate reductant, thioredoxin, through a sequence of disulfide exchanges.[95] The two cysteines on thioredoxin, a small protein acting as an electron carrier, end up as a cystine at the end of the reaction.

Ribonucleoside-triphosphate reductase has a coenzyme $B_{12}$ within its active site. Ribonucleoside-diphosphate reductase, although it lacks coenzyme $B_{12}$, has a unique and perhaps more informative feature. Within native, enzymatically active ribonucleoside-diphosphate reductase, in the absence of substrates, Tyrosine 122 of the folded B2 polypeptide from *E. coli*[96] is a stable neutral radical (**5–22**) whose unpaired spin can be detected by electron spin resonance (Figure 5–5).[97] That the unpaired spin is on a tyrosine can be demonstrated by replacing specific hydrogens on the tyrosines in the protein with deuteriums. From these experiments, it can be calculated that carbons 3 and 5 of the phenol of the tyrosine each bear 0.26 unpaired electron and carbon 1 bears 0.51 unpaired electron. The difference in spin distribution between tyrosyl radical in ribonucleoside-triphosphate reductase and the tyrosyl radical in crystalline tyrosine hydrochloride may be due to magnetic coupling to one or both of two non-heme ferric ions in the protein. These two non-heme ferric irons form a bidentate cluster[98] similar to that in hemerythrin (Figure 2–29), and there is one bidentate cluster and one tyrosyl radical for each active site on the enzyme. Although the distance between the closest ferric ion and the oxygen of the tyrosyl radical is 0.53 nm, and the tyrosyl radical is not a ligand to either iron,[98] it is thought that this magnetically coupled system is responsible for the peculiar absorption spectrum[99] of the enzyme between wavelengths of 300 and 700 nm. It is noteworthy that a tyrosyl radical is not very reactive (BDE = 360 kJ $mol^{-1}$). This and its coupling to the bidentate cluster may explain its docility.

The presence of this preexisting free radical in ribonucleoside-diphosphate reductase and the presence of coenzyme $B_{12}$, which is thought to be capable of dissociating to produce a 5′-deoxyadenosyl radical, in the active site of ribonucleoside-triphosphate reductase have led to the proposal that rad-

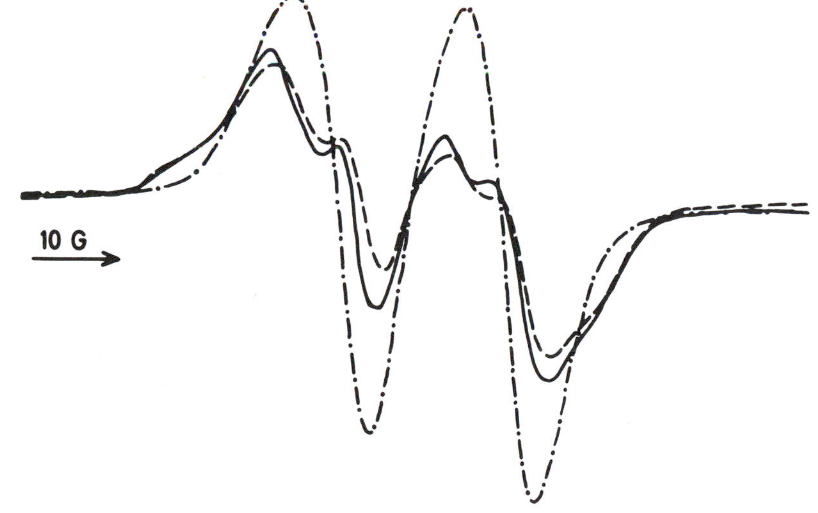


**Figure 5–5:** Electron spin resonance spectra of the tyrosyl radical in ribonucleoside-diphosphate reductase.[97] A strain of the bacterium *Escherichia coli* that overproduces ribonucleoside-diphosphate reductase was used for these experiments. In this strain about 3% of the cytoplasmic protein is the enzyme. Suspensions of intact bacteria were used for the measurement of the electron spin resonance spectrum at 77 K. The spectrum observed from unsupplemented cells (– – –) results from the overlap of several resonances from unpaired spins at carbons 1, 3, and 5 of a tyrosine. This was demonstrated by the fact that when the bacteria were grown on [3,5-$^2H_2$]tyrosine, the complexity of the spectrum disappeared (– · – · –) but not when they were grown on [2,6-$^2H_2$]tyrosine (—). When Tyrosine 122 in the enzyme was mutated to phenylalanine, or when the plasmid responsible for overproduction was omitted, all of these resonances disappeared.[96] Reprinted with permission from ref 97. Copyright 1978 *Journal of Biological Chemistry.*

**Figure 5–6:** Possible mechanism for ribonucleoside-diphosphate reductase or ribonucleotide-triphosphate reductase involving free radical intermediates. In the first step, a hydrogen is abstracted from the 3′-carbon of the ribose to initiate the production of the neutral oxallylic radical 5–23 by β-elimination of water and deprotonation of the 3′-oxygen. The oxallylic radical abstracts hydrogen from cysteine (BDE = 371 kJ $mol^{-1}$), and the resulting cysteinyl radical forms a cystinyl radical with an adjacent cysteine. The cystinyl radical transfers its unpaired electron density to the ketone at carbon 3′ of the ribose to permit it to abstract a hydrogen to regenerate the original radical at the active site and the 2′-deoxyribose.

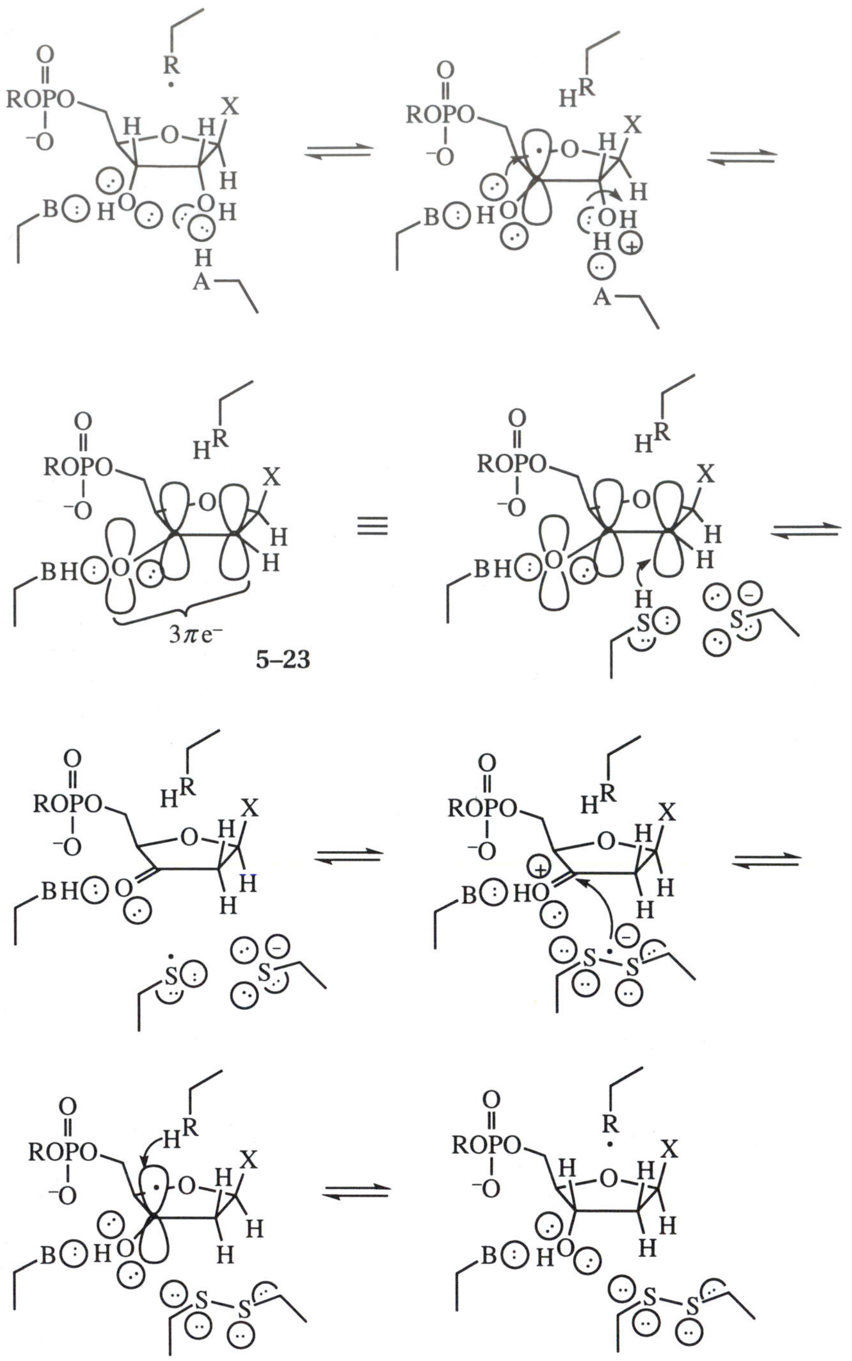

ical intermediates are involved in the common enzymatic mechanism of these two enzymes. In particular, a mechanism can be formulated (Figure 5–6) that involves oxallylic radical 5–23 as the central intermediate.[100] The mechanism is drawn to incorporate the retention of configuration[101] that occurs at carbon 2′.

From crystallographic studies it has been learned that in ribonucleoside-diphosphate reductase, the tyrosyl radical is greater than 1.0 nm from the active site,[98] and it follows that it cannot directly abstract a hydrogen from the substrate. For this reason it is believed that an electron is transferred at long range, through the protein from an oxidizable amino acid that is within the active site to the tyrosyl radical during catalysis to produce a radical (-R·) immediately adjacent to the hydrogen on the 3′-carbon. In ribonucleoside-triphosphate reductase the immediately adjacent radical (-R·) is either the 5-deoxyadenosyl radical itself or another radical produced by the coenzyme $B_{12}$ at an oxidizable amino acid in the active site.

In either enzyme, the immediately adjacent radical would abstract a hydrogen from carbon 3′ of the reactant to form an alkyl radical on the substrate. Although this alkyl radical should be quite reactive in its own right, it is immediately susceptible to dehydration at carbon 2′ because the product of this dehydration is the stable oxallylic radical **5–23**. This radical is then reduced by the adjacent cysteines, perhaps through an intermediate disulfide radical anion (Figure 2–6).

There are several experimental observations consistent with this mechanism. When [3′-$^3$H]UDP is used as a reactant in the reaction catalyzed by ribonucleoside-diphosphate reductase, a kinetic isotope effect ($^{1H}k/^{3H}k = 3$)[100] is observed, consistent with the breaking of the bond between carbon 3′ and its hydrogen. Most of the tritium remains on carbon 3′ in the product, [3′-$^3$H]-2′-deoxyuridine diphosphate, but about 1% is lost to solvent, consistent with loss from RH in Figure 5–6 if, for example, R were yet another cysteine.[102]

2′-Chloro-2′-deoxycytidine diphosphate[103] and 2′-chloro-2′-deoxyuridine diphosphate[103,104] are converted by ribonucleoside-diphosphate reductase to 2′,3′-dideoxy-3′-oxocytidine diphosphate and 2′,3′-dideoxy-3′-oxouridine diphosphate, respectively.[105] If [3′-$^3$H]-2′-chloro-2′-deoxyuridine diphosphate is used as a reactant, the tritium is transferred[105] intramolecularly to carbon 2′

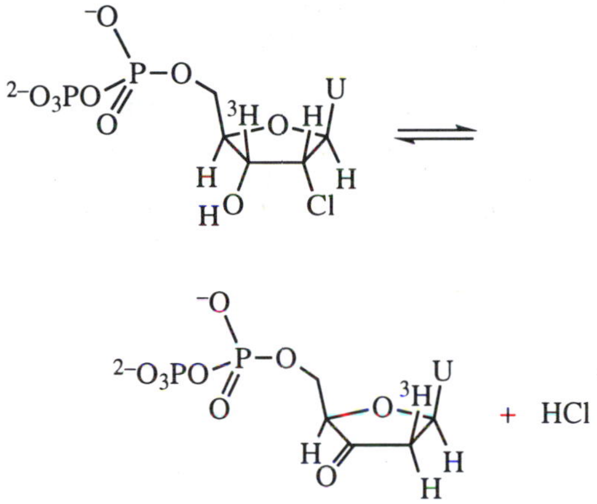

(5–92)

For this reaction to occur, the enzyme must be in its reduced, dithiol form even though the reaction is formally a dehydrohalogenation and not a reduction. The product formed in this reaction is consistent with a radical-assisted dehalogenation[105]

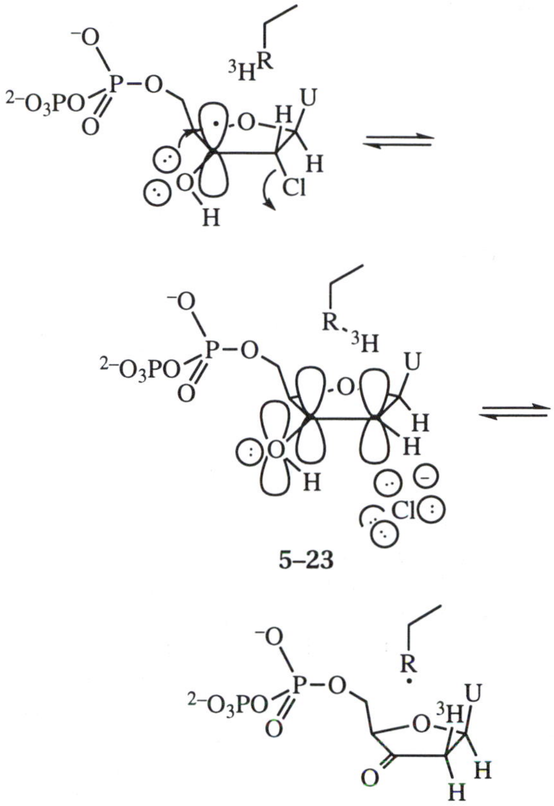

**5–23**

(5–93)

The difficulty with this proposal, however, is that the same oxallylic radical **5–23** is formed as is formed during the normal reaction (Figure 5–6), and there is no satisfactory explanation of why it would be quenched and released as the ketone in this instance while it is reduced to the alcohol in the normal reaction. The same dehydrohalogenation (Reaction 5–92) is catalyzed also by ribonucleoside-triphosphate reductase,[106] an additional observation that demonstrates that the fundamental mechanisms of the reactions catalyzed by ribonucleoside-diphosphate reductase and by ribonucleoside-triphosphate reductase are the same.

Dopamine $\beta$-monooxygenase, in addition to its normal reaction of hydroxylating 2-(3,4-dihydroxyphenyl)ethylamine

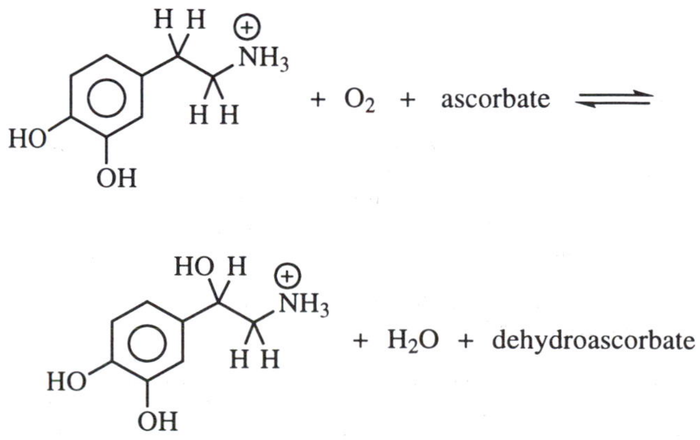

$+ O_2 +$ ascorbate $\rightleftharpoons$

$+ H_2O +$ dehydroascorbate

(5–94)

also catalyzes the reaction[107]

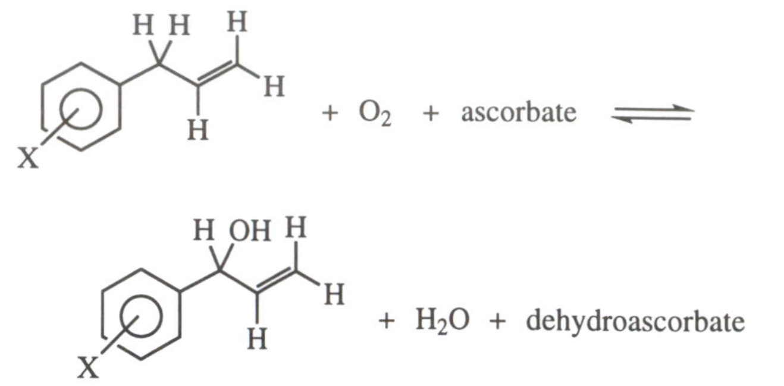

(5–95)

where X is any one of a series of *meta* or *para* substituents. The deuterium kinetic isotope effect on $V_{max}/K_m^{O_2}$ for 3-(*p*-hydroxyphenyl)propene as a reactant in this enzymatic reaction is 2.0, an observation demonstrating that carbon–hydrogen bond cleavage is a rate-determining step in the reaction. When a series of substituted 3-phenylpropenes was examined, $V_{max}/K_m^{O_2}$ varied with the Hammett coefficient $\sigma^+$ with a $\rho$ value of –1.2.* This demonstrates that the reaction occurs more rapidly as the phenyl ring becomes more capable of donating electron density to the adjacent carbon. Therefore, the intermediate formed upon removal of the hydrogen cannot be carbanionic. The value of $\rho$ is similar to that observed for abstraction of hydrogen from one of the benzylic carbons of a toluene to produce a benzyl radical. In such a homolytic cleavage, the electron density at the benzylic carbon decreases and thus the reaction is favored by electron donation. A $\rho$ value of –1.2 would also be consistent with a rate-determining formation of a carbocation by hydride removal. A similar correlation ($\rho = -1.5$) is also observed between the logarithm of the rate constant for hydrogen removal from the reactant and the Hammett coefficient $\sigma_p$ for a series of *para*-substituted 2-phenylethylamines as reactants (Figure 5–7)[109]

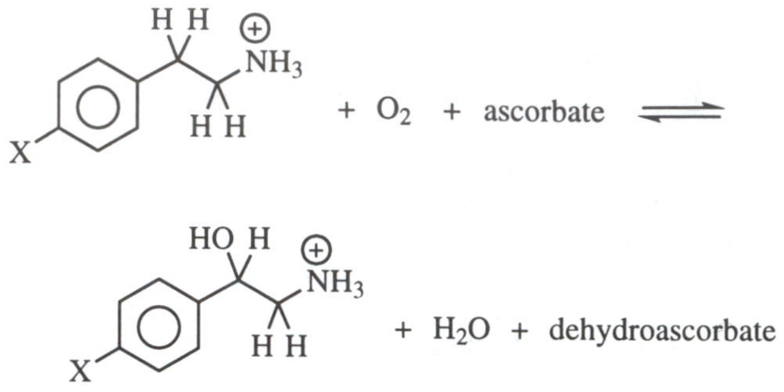

(5–96)

In this case a better correlation was observed with $\sigma_p$ than $\sigma^+$. This difference was presented as further support for a radical intermediate in the reaction.

Dopamine $\beta$-monoxygenase contains a binuclear copper center in the active site, and it has been proposed that a hydroxyl radical generated from a hydrogen peroxide anion bound at this center abstracts the hydrogen from the reactant

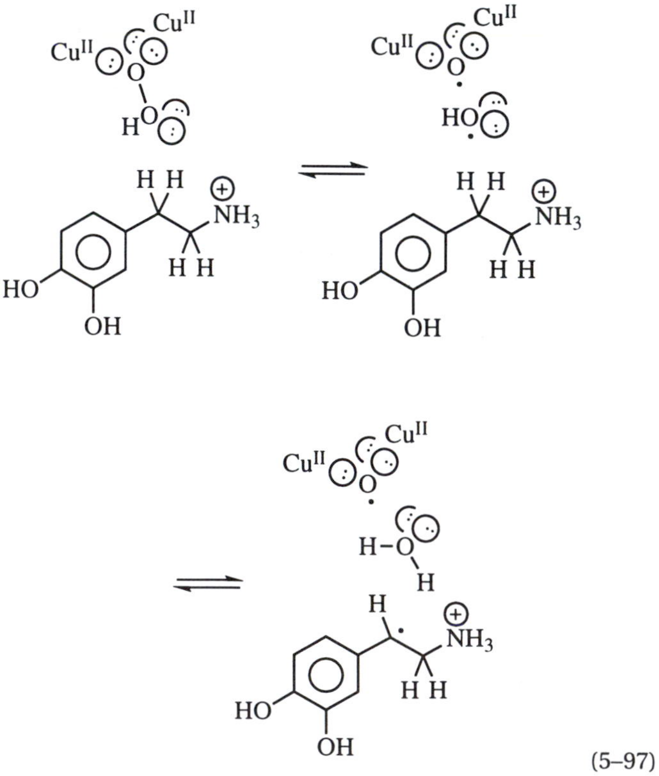

(5–97)

The hydrogen peroxide complexed to the two $Cu^{II}$ would be generated from $O_2$ and two $Cu^{I}$ of the reduced enzyme. The oxygen radical chelated by the two $Cu^{II}$ and the benzyl radical (the products of Reaction 5–97) could collapse to form the cupric alkoxide of the product

*The Hammett relationship is a correlation between the rate or equilibrium constant for a particular reaction and the ability of an adjacent functional group, usually a substituted phenyl ring, to donate or withdraw electron density from an atom participating in the reaction. Each of a series of adjacent functional groups can be assigned a Hammett coefficient, $\sigma$, the numerical value of which is based on the extent to which that same functional group can donate (negative numbers) or withdraw (positive numbers) electron density from the atom that participates in the reaction and to which the functional group is attached. These numbers in turn are based on the effects of the various functional groups on the rates or equilibrium constants of a particular reference reaction. Different sets of values of $\sigma$ can be used depending on the reference reaction chosen. For example, the set $\sigma_p$ is based on the effects of *para* substituents on the acid dissociation constants of benzoic acids, and the set $\sigma^+$ is based on the effects of substituents on the $S_N1$ solvolysis of (dimethylphenyl)methyl chlorides.[108] A series of functional groups for which values of $\sigma$ are available is chosen for the experiment. The magnitude of the effect of electron donation or electron withdrawal on the rate or equilibrium of the reaction being examined and the direction of its influence are expressed in the value of $\rho$, the slope of the line relating $\sigma$ to the logarithm of the rate constant or the equilibrium constant for that reaction. If $\rho$ is a negative number, the reaction is assisted by electron donation to the participating atom; if $\rho$ is a positive number, the reaction is assisted by electron withdrawal from the participating atom. The magnitude of $\rho$ expresses the extent of the effect. Values of $\rho$ greater than +1 or less than –1 indicate strong effects of electron withdrawal or electron donation, respectively.

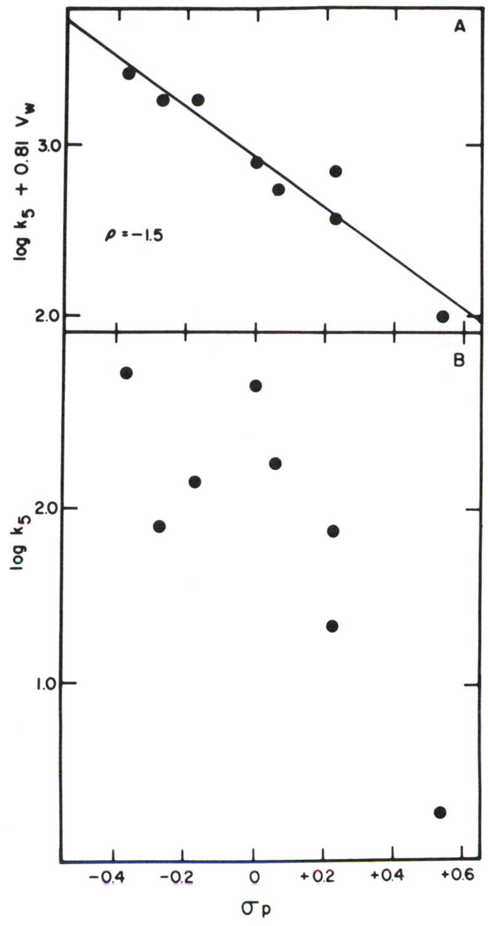

**Figure 5–7:** Correlation of the rate constant $k_5$ for the step involving carbon–hydrogen bond dissociation in the reaction catalyzed by dopamine β-monooxygenase with the Hammett coefficient $\sigma_p$ for a series of *para*-substituted 2-phenylethylamines.[109] Initial velocities of the enzymatic reaction were measured as the initial rates of oxygen consumption at 35 °C and pH 6.2 at ionic strength of 0.15 M. For each *para*-substituted 2-phenylethylamine, the initial velocities at a series of concentrations were measured to determine the Michaelis constant for each, $K_m^{phnethNH_2}$, and the maximum initial velocity, $V_{max}^{phnethNH_2}$. 2,2-Dideutero derivatives of each 2-phenylethylamine were also prepared and deuterium isotope effects on $V_{max}/K_m$ were measured for each of them. From all of these measurements together, values for $k_5$, the rate constant for the step converting the 2-phenylethylamine to the 2-hydroxy-2-phenylethylamine at the active site could be calculated for each 2-phenylethylamine. (A) By performing a two-parameter analysis, using $\sigma_p$ and the van der Waals volumes, $V_w$, as the two parameters, a good correlation was obtained with the parameter $\sigma_p$. This linear correlation has a slope, ρ, of –1.5. The coefficient for the parameter $V_w$, was 0.81. (B) The logarithms of these rate constants were also plotted alone as a function of $\sigma_p$, but a poor correlation was observed. Reprinted with permission from ref 109. Copyright 1985 American Chemical Society.

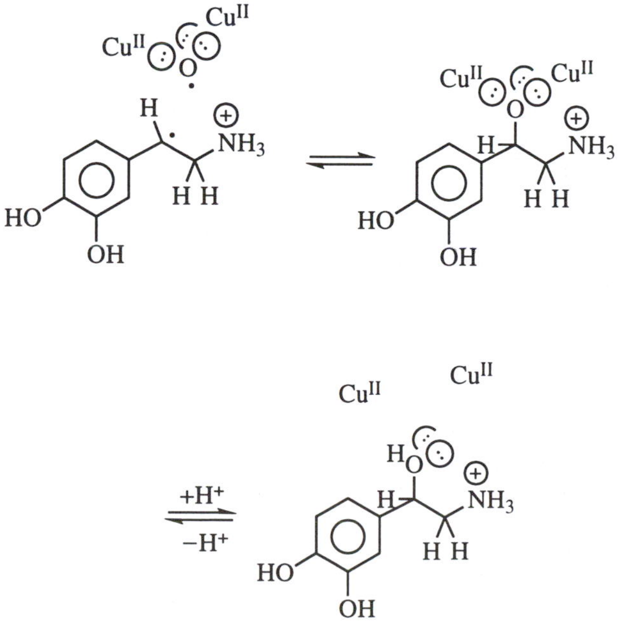

(5–98)

Ketones or **ketimines** transiently formed from alcohols or amines, respectively, are intermediates in a number of enzymatic reactions. In these instances, $NAD^+$, tightly bound at the active site, acts as a coenzyme oxidizing and then reducing the substrates. Adenosylhomocysteinase employs such a strategy as a complicated alternative to a nucleophilic substitution (Figure 5–2).

A simpler example of this mechanism[110] is that of UDP-glucose 4-epimerase

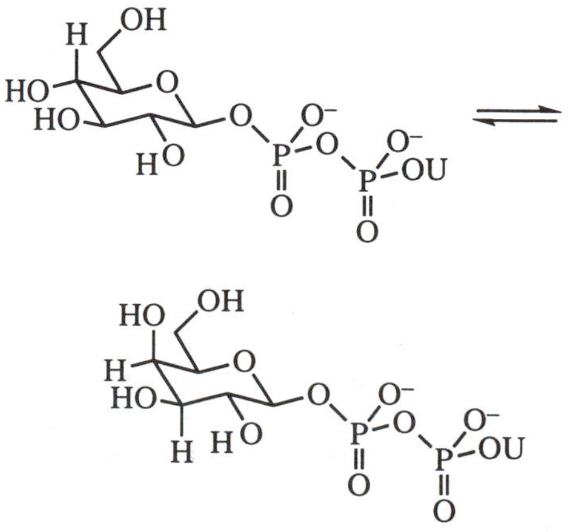

(5–99)

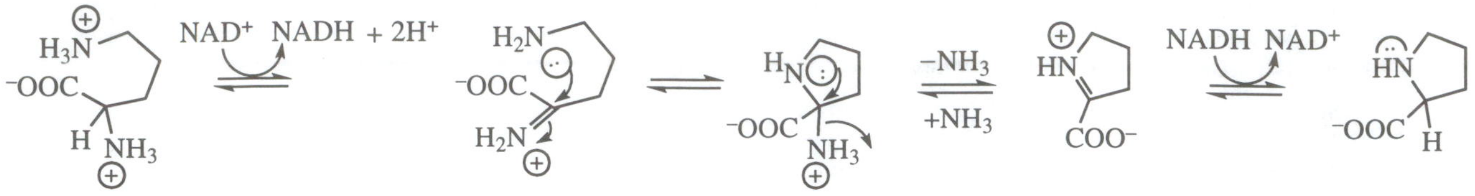

**Figure 5–8:** Mechanism for the reaction catalyzed by ornithine cyclodeaminase involving a ketimine intermediate. L-Ornithine is oxidized to the $\alpha$-ketimine by removal of the $\alpha$-hydrogen as a hydride. The removal of this hydride is by NAD$^+$ bound at the active site as a coenzyme. Intramolecular transimination produces the cyclic imine that is reduced by the NADH from the *re* face, the same face from which it was removed in L-ornithine, to produce L-proline.

The tightly bound NAD$^+$ at the active site oxidizes UDP-glucose to the UDP-4-ketohexose

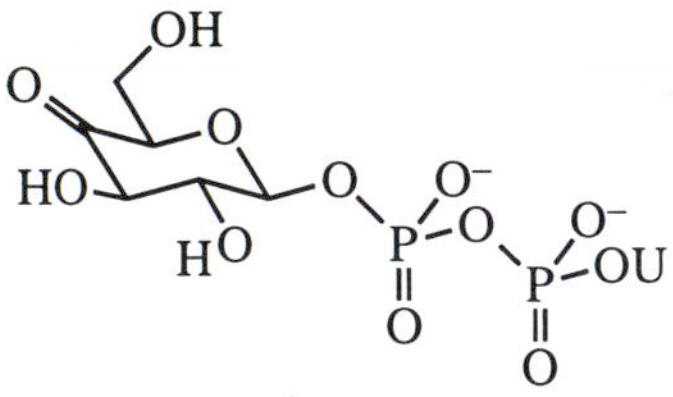

**5–24**

and then the NADH reduces the UDP-4-ketohexose to UDP-galactose. There are several observations consistent with the existence of such an intermediate in the enzymatic reaction. When either UDP-glucose or UDP-galactose is mixed with the enzyme, a portion of the bound NAD$^+$ is reduced to NADH.[111] When TDP-4-keto-6-deoxyglucose is mixed with enzyme the coenzyme of which has been reduced to NADH, the TDP-4-keto-6-deoxyglucose is reduced to a mixture of TDP-6-deoxyglucose and TDP-6-deoxygalactose.[112] When high concentrations of the enzyme are mixed with UDP-galactose and Na[$^3$H]BH$_4$ is subsequently added, the ketonic intermediate can be trapped as a mixture of UDP-[4-$^3$H]glucose and UDP-[4-$^3$H]galactose.[113,114]

Ornithine cyclodeaminase, which catalyzes the reaction

$$\text{ornithine} \rightleftharpoons \text{proline} + NH_4^+ \qquad (5\text{–}100)$$

also contains tightly bound NAD$^+$ as a coenzyme; and, upon addition of ornithine, a portion of the NAD$^+$ is reduced to NADH.[115] In the reaction catalyzed by the enzyme, the $\delta$ nitrogen of the ornithine is retained in the proline. These results are consistent with the formation of pyrrol-1-ine-2-carboxylate as an intermediate in the enzymatic reaction (Figure 5–8). In this reaction the hydride is removed from and returned to the same carbon with the same stereochemistry.

The examples discussed so far are intermediates in enzymatic reactions that have been identified chemically. Intermediates can also be defined by double kinetic isotope effects, by isotopic exchange, and by pre-steady-state kinetics.

## *Suggested Reading*

Buckel, W., & Bobi, A. (1976) The Enzyme Complex Citramalate Lyase from *Clostridium tetanomorphum, Eur. J. Biochem. 64,* 255–262.

Buckel, W. (1976) Acetic Anhydride: An Intermediate Analogue in the Acyl-Exchange Reaction of Citramalate Lyase, *Eur. J. Biochem. 64,* 263–267.

## PROBLEM 5–1

The following reaction is catalyzed by ATP-citrate (*pro*-3*S*)-lyase:

$$\text{citrate} + \text{CoASH} + \text{MgATP} \rightleftharpoons$$
$$\text{acetyl-SCoA} + \text{oxaloacetate} + \text{MgADP} + \text{HOPO}_3^{2-}$$

On the basis of the following data,[115a] write a detailed mechanism for the reaction. Give explanations for all the experiments.

(A) When the enzyme is incubated with [1,5-$^{14}C_2$]citrate, ATP, and MgCl$_2$, radioactivity is found to elute with the protein peak on molecular exclusion chromatography.

(B) When the protein peak containing the radioactivity, which is obtained from step A, is incubated with CoASH in the absence of MgCl$_2$ and ATP, [$^{14}$C]acetyl-SCoA is obtained. In the absence of CoASH, no acetyl-SCoA is found.

(C) When [8-$^{14}$C]ATP is incubated with the enzyme in the presence of MgCl$_2$, no radioactivity is associated with the protein that is isolated by molecular exclusion chromatography.

(D) When [$\gamma$-$^{32}$P]ATP is incubated with the enzyme in the presence of MgCl$_2$, radioactivity is associated with the isolated protein. When this radioactive protein is incubated with [1,5-$^{14}C_2$]citrate in the presence of CoASH but in the absence of MgCl$_2$, [$^{14}$C]oxaloacetate is formed and all of the radioactivity is released from the enzyme.

(E) When the enzyme is incubated with ATP and HO$^{32}$PO$_3^{-2}$, in the presence of MgCl$_2$, no radioactivity is found in the enzyme or in the ATP. When the enzyme is incubated with ATP and [8-$^{14}$C]ADP, in the presence of MgCl$_2$, no radioactivity is found in the enzyme, but the ATP becomes radioactive.

(F) When cityl 1-phosphate is incubated with the enzyme in the presence of CoASH, oxaloacetate and acetyl-SCoA are formed.

(G) When citryl 1-phosphate and CoASH are incubated without the enzyme, nothing happens.
(H) When [$^{14}$C]citryl 1-phosphate is incubated with the enzyme in the absence of CoASH and the enzyme is purified by molecular exclusion chromatography, radioactivity is associated with the protein peak.
(I) When the radioactive protein isolated in step H is incubated with CoASH, oxaloacetate and acetyl-SCoA are formed.

## PROBLEM 5–2

*β*-Galactosidase catalyzes Reaction 5–77.

(A) The enzyme acts only on the *β* anomer of a galactoside and releases only the *β* anomer of the product. It cares little, however, what $R_1$ and $R_2$ are. Write a mechanism for this reaction, as it would occur in the active site, which is consistent with these observations. Position catalytic residues to explain both the stringent stereochemistry of the reaction at carbon 1 and the carelessness at R. Neglect the portion of galactose from carbon 2 to carbon 6. Use specific amino acid side chains of your choice for catalysis.
(B) The mechanism you have written in part A can be symbolically abbreviated:

$$E + A \underset{k_1}{\overset{k_{-1}}{\rightleftharpoons}} E \cdot A \xrightarrow{k_2} E \cdot C + P$$

$$E \cdot C + B \underset{k_{-3}}{\overset{k_3}{\rightleftharpoons}} E \cdot C \cdot B \xrightarrow{k_4} E + Q$$

Write out the complete initial velocity equation for this mechanism.
(C) Take the following limit and write an expression for $V_{max}$ containing only kinetic constants ($k_i$) and $[E]_{TOT}$.

$$V_{max} = \lim_{[A],[B] \to \infty} v_0$$

(D) Refer to your molecular mechanism for *β*-galactosidase and determine which rate constant in the kinetic mechanism would be most affected by changes in R.
(E) Consider the following table of relative maximum velocities. The various R groups are listed in order of increasing ability to leave.[115b–d]

| R | $V_{max}/V_{max}^{2NO_2Phe}$ |
|---|---|
| methyl | 0.06 |
| 4-bromophenyl | 0.02 |
| 4-chlorophenyl | 0.02 |
| 4-methoxyphenyl | 0.1 |
| phenyl | 0.1 |
| 4-nitrophenyl | 0.2 |
| 3-chlorophenyl | 0.5 |
| 3-nitrophenyl | 0.9 |
| 2-nitrophenyl | 1.0 |
| 2,5-dinitrophenyl | 1.1 |
| 3,5-dinitrophenyl | 1.1 |
| 2,4-dinitrophenyl | 1.3 |

Explain, in terms of the expression you derived in part C, why $V_{max}/V_{max}^{2NO_2Phe}$ increases as R becomes a better leaving group and then levels off, showing no further increase.

## PROBLEM 5–3

Methanol in water reacts about 100-fold more rapidly with the acyl intermediate in the reaction catalyzed by *α*-chymotrypsin than does water itself. If this is the case, explain why the catalytic constant, $k_{cat}$, for the combined hydrolysis and methanolysis of *p*-nitrophenyl acetate by chymotrypsin is markedly increased in methanol–water solutions but that for *N*-acetyl-L-tryptophan amide is unchanged from the value in water alone. Estimate the value of $k_{cat}$ for the combined hydrolysis and methanolysis of *p*-nitrophenyl acetate in 1 M methanol in water relative to $k_{cat}$ in water alone and the fraction of the product that would be methyl acetate.

## PROBLEM 5–4

Adenosine deaminase catalyzes the hydrolysis of a number of 6-substituted purine ribosides to inosine

[6-X-purine riboside] + $H_2O$ ⇌ [inosine] + HX

The enzyme has no coenzyme or metal ion requirements. The following values have been obtained for $V_{max}$ in a pH range where $V_{max}$ is independent of pH.

| X | $V_{max}$[a] ($\mu$mol min$^{-1}$ mg$^{-1}$) |
|---|---|
| $NH_2$ (adenosine) | 16 |
| $NH_2NH$ | 15 |
| Cl | 19 |
| $CH_3NH$ | 15 |
| $CH_3O$ | 18 |

[a]Accurate ±20% of given value. Data from ref 115e.

What conclusion can be drawn from these kinetic data? Propose a mechanism for adenosine hydrolysis involving an intermediate that is consistent with the kinetic data and that includes general acid or general base catalysis or both at whatever points it would be useful for catalysis. Write out only that portion of the substrate structure that is reacting.

## PROBLEM 5–5

UDP-Glucose–hexose-1-phosphate uridylyltransferase catalyzes the reaction

$$\text{UDP-glucose} + \text{galactose 1-phosphate} \rightleftharpoons \text{UDP-galactose} + \text{glucose 1-phosphate}$$

(A) Suppose that the reaction proceeded through the two ternary complexes

$$\text{E·UDP-glucose·galactose 1-phosphate} \rightleftharpoons \text{E·glucose 1-phosphate·UDP-galactose}$$

Draw the structure of the pentavalent intermediate at phosphorus that would have to be bound tightly by the active site. Indicate the entering oxygen and the leaving oxygen.

The following results have been observed[116]

**A**

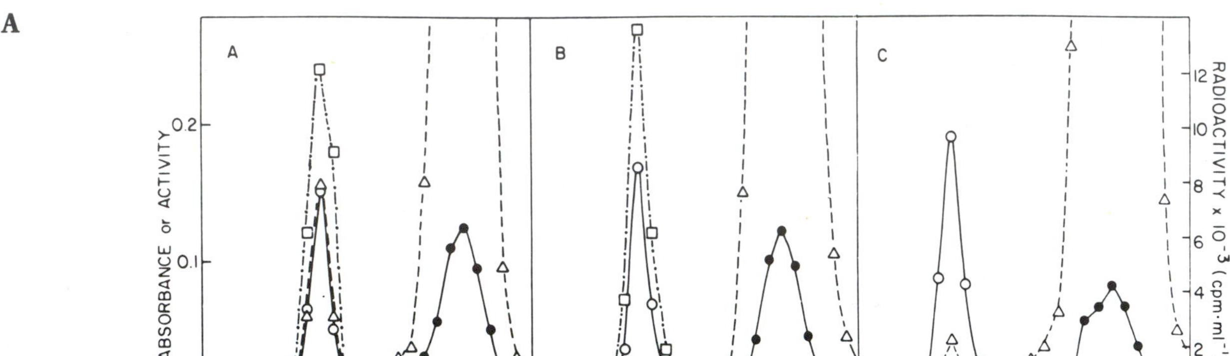

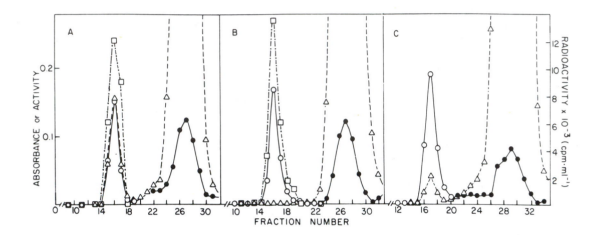

**Figure A:** (A) The complete reaction mixture contained initially 0.33 mg of UDP-glucose–hexose-1-phosphate uridylyltransferase and 32 nmol of [*uracil*-5,6-$^3H_2$]UDP-glucose (specific radioactivity $2.0 \times 10^7$ cpm $\mu mol^{-1}$) in 0.2 mL of a buffer consisting of 5 mM potassium phosphate and 50 $\mu$M cysteine at pH 7.5. After 5 min at room temperature the solution was passed thorough a column (1 × 41 cm) of Sephadex G-25 equilibrated and eluted with the same buffer at 4 °C. Fractions of approximately 1 mL were collected at 5-min intervals. (B) Conditions were identical with those of panel A except that tritiated UDP-glucose was replaced with 34 nmol of UDP-[U-$^{14}$C]glucose (specific radioactivity $5.8 \times 10^6$ cpm $\mu mol^{-1}$). (C) Conditions were identical with those of panel A except that the reaction mixture also contained 32 nmol of glucose-1-P. The symbols are (○), $A_{280}$ (protein), (●) $A_{260}$ (nucleotides), (△) radioactivity, and (□) catalytic activity (units milliliter$^{-1}$, left-hand ordinate). Reprinted with permission from ref 116. Copyright 1974 *Journal of Biological Chemistry.*

**B**

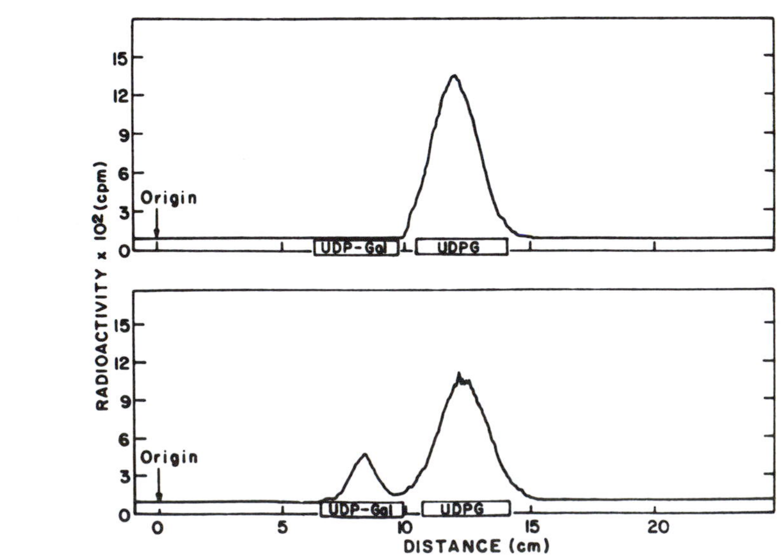


**Figure B:** The complete reaction mixture contained initially 32 nmol of [*uracil*-5,6-$^3H_2$]UDP-glucose ($6.32 \times 10^5$ cpm), 36 nmol of UDP-galactose, and 0.56 unit of uridylyltransferase in 25 $\mu$L of a buffer consisting of 5 mM potassium phosphate and 0.5 mM cysteine at pH 7.5. After 25 min at 25 °C, half of the solution was subjected to descending paper chromatography for 92 h using ethanol–methyl ethyl ketone–0.5 M morpholinium tetraborate, pH 8.6 (7:2:3), as the mobile phase. The upper panel is the control without enzyme and the lower is the complete reaction mixture. Reprinted with permission from ref 116. Copyright 1974 *Journal of Biological Chemistry.*

(B) On the basis of these results, write a mechanism for the enzymatic reaction involving a covalent uridylyl-enzyme intermediate.
(C) On the basis of the mechanism you wrote in part B, explain the uridylyl exchange observed in Figure B.
(D) The covalent uridylyl intermediate is formed between a portion of the substrate and a histidine in the active site of the enzyme.[117] Draw the complete structure of the covalent uridylyl intermediate.
(E) The overall enzymatic reaction proceeds with retention of configuration of phosphorus.[34] To which of the two phosphorus atoms in a uridylyl substrate must this statement refer? Draw a structure of UDP-glucose and indicate the phosphorus with an arrow. Show how this observation rules out a mechanism in which this phosphorus is transferred directly in a ternary complex as in part A and is consistent with the formation of the covalent intermediate you drew in part C. Show this by drawing structures of the intermediates in the key steps.
(F) Explain why the mechanism of part B that passes through a covalent intermediate is more economical than the mechanism of part A that passes through a ternary complex. Consider the amino acids in the active site that must form hydrogen bonds to the sugars of the substrates and those that activate the phosphate for transfer. Why can they be called upon to perform twice the work?

## PROBLEM 5–6

The enzyme nucleoside-diphosphate kinase (NDP kinase) catalyzes Reaction 5–15. When ATP is used as the phosphate donor (MgXTP) and dGDP as the phosphate acceptor (MgYDP) in the enzymatic reaction, the following kinetic behavior is observed.

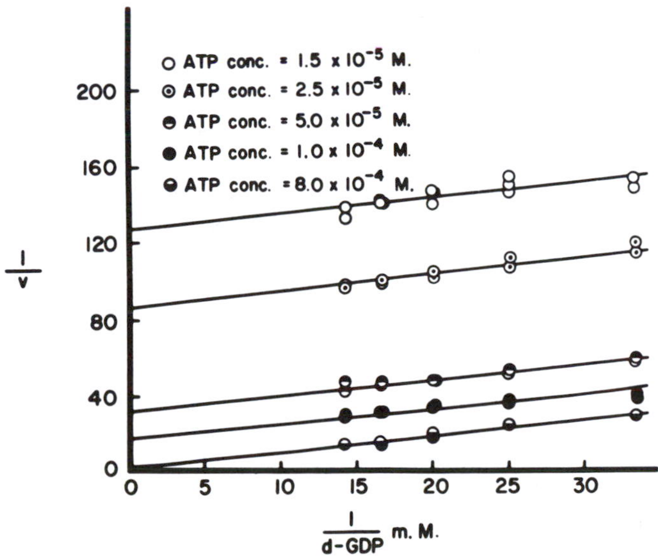


**Figure A:** Plot of the reciprocal of the initial velocity ($\Delta A_{340}$ minute$^{-1}$) against the reciprocal of concentration of dGDP (millimolar).[118] Reprinted with permission from ref 118. Copyright 1965 Academic Press.

(A) Write a kinetic mechanism that is consistent with these observations. In your mechanism explicitly name substrates and products as ATP, dGDP, ADP, and dGTP. Abbreviate the enzyme as E.

When NDP kinase was mixed with [$^{32}$P]ATP for 1 min at 30 °C and then added to a Sephadex G-100 column, the following elution profile was obtained.

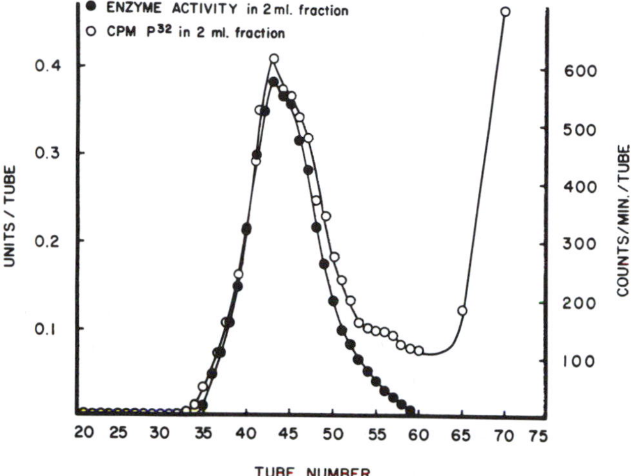


**Figure B:** Separation of [$^{32}$P]NDP kinase from [$^{32}$P]ATP on Sephadex G-100 columns.[119] Reprinted with permission from ref 119. Copyright 1966 *Journal of Biological Chemistry.*

The authors point out in the publication that NDP kinase elutes from Sephadex G-100 at a position where the stationary phase for the molecular exclusion is able to separate proteins on the basis of their size.

(B) How do these results rule out the possibility that the $^{32}$P is associated with proteins other than NDP kinase?
(C) Which phosphate of the [$^{32}$P]ATP was certainly radioactive?
(D) Write a chemical mechanism for the NDP kinase reaction that incorporates both your kinetic mechanism and an explicit explanation of the results in Figure B.
(E) When enzyme from fractions 40–45 from the Sephadex G-100 column was mixed with GDP, what radioactive nucleotide was subsequently isolated from the mixture? Indicate precisely where the radioactive phosphorus resides on that nucleotide.

**PROBLEM 5–7**

Riboflavin synthase catalyzes the reaction

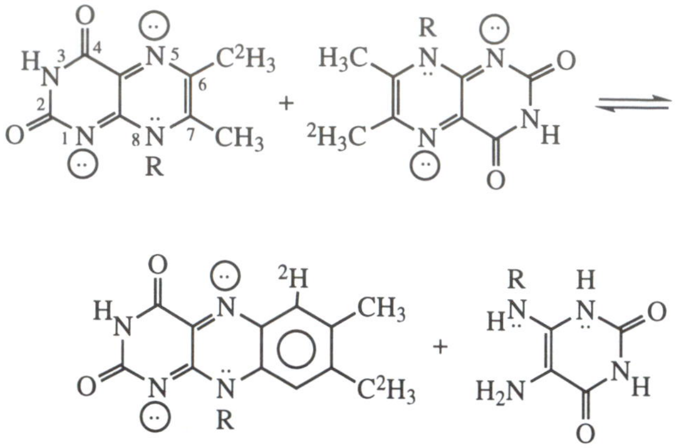

where R is ribityl. Note that the two reactants, 6,7-dimethyl-8-(1-ribityl)lumazine, are identical. That the two molecules of 6,7-dimethyl-8-(1-ribityl)lumazine are joined in the noted opposite orientations has been demonstrated by determining the stereochemistry of the product formed from 6-deuteriomethyl-7-methyl-8-(1-ribityl)lumazine.[120]

It has been shown that the hydrogens on the 7-methyl group of the lumazine are acidic and susceptible to exchange under basic conditions in the absence of the enzyme

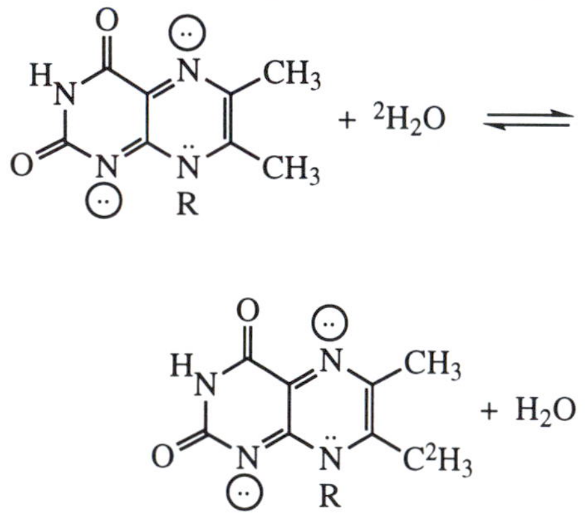

(A) Write a mechanism for this exchange with an enol as an intermediate.

(B) This enol is an intermediate in the enzymatic reaction. Use this enol and a tautomer of the other lumazine that is electrophilic at carbon 6 to form the intermediate

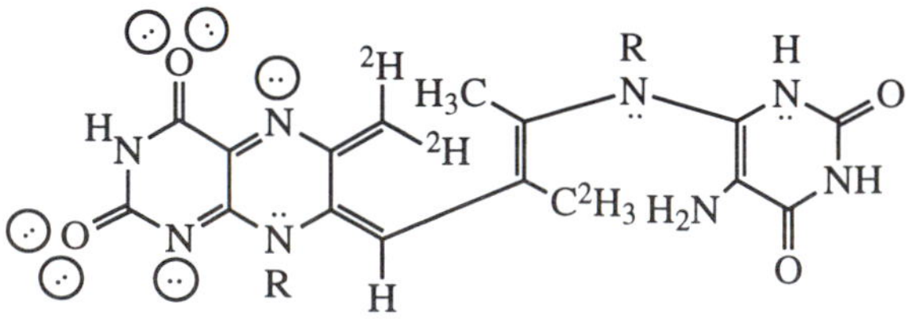

in a series of steps involving a condensation, a tautomerization, and a retro Michael addition.

(C) What type of reaction proceeding through what type of transition state would convert the preceding intermediate into

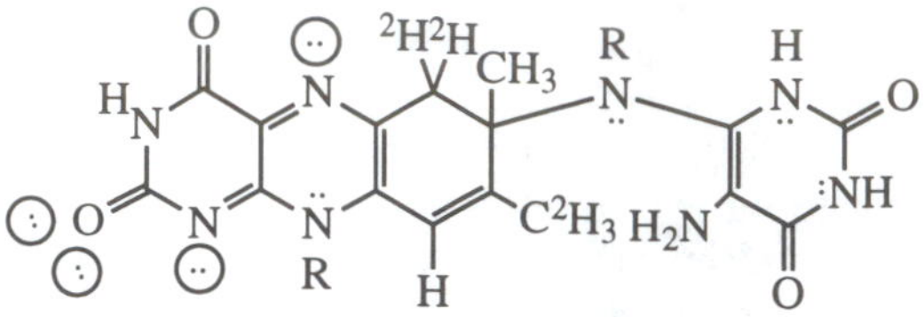

(D) Convert this intermediate into the products observed in the enzymatic reaction by a tautomerization followed by a retro Michael addition.

**PROBLEM 5–8**

Read the following direct quotation adapted with permission from reference 121.

The catalytic mechanism of glycogen phosphorylase has remained a tantalizing mystery in spite of the intensive investigations which are summarized in a recent review. The transfer of glucose from glucose-1-P to the terminal glucose of glycogen or other glucose polymers occurs with retention of configuration, with the bond breakage occurring between the anomeric carbon and the oxygen, and with no detectable partial exchange reactions. Strong inhibition by 1,5-gluconolactone suggests an intermediate glucosyl cation in the half-chair conformation of an oxonium ion in the transition state. As predicted, "scrambling" of the anomeric oxygen of glucose-1-P with the other nonbridging phosphate oxygens has been observed, in the absence of the total reaction, in the tertiary complex with potato phosphorylase and cyclodextrin.

Extensive studies with coenzyme analogues suggest that only the phosphate of PLP [pyridoxal phosphate] is likely to be involved in the catalytic mechanism. Shimomura and Fukui extended this approach with PLP derivatives modified at the phosphate moiety and showed that the rate of reconstitution increased with the increments in the number of anions at this position. Significantly, the rate of reconstitution with PLPP was more than 6 times that with PLP. High resolution x-ray crystallographic maps from two laboratories confirmed that only the phosphate of the coenzyme was near the substrate but that it was 7 Å from the phosphate of the substrate and further from other parts. Nevertheless, it was demonstrated that phosphite or fluorophosphate confer activity on pyridoxal-reconstituted enzyme and that pyrophosphate, a potent inhibitor of pyridoxal phosphorylase, binding with a stoichiometry of 1/monomer, was competitive with both phosphite and glucose-1-P. The phosphates of the coenzyme and substrate may thus be closer together in the active enzyme than is seen in the crystal structures of the allosterically inactive conformers.

Two attractive hypotheses have been proposed for the involvement of the coenzyme phosphate in the catalytic mechanism. On the basis of model building experiments, Johnson *et al.* have suggested that glucose-1-P binds very differently in the productive mode of the enzyme, so that the coenzyme phosphate may function as a dianion to carry out a nucleophilic attack on the C-1 carbon while His 376 acts as a general acid to protonate the glycosidic oxygen. A formal covalent intermediate is inconsistent, however, with the subsequent finding of Takagi *et al.* that enzyme reconstituted with PLP-β-Glc shows no enzymatic activity even after prolonged incubation with substrates and activator, although this experiment does not necessarily rule out a mere stabilization of the carbonium ion by the phosphate. An alternative proposal by Helmreich and Klein suggests that the coenzyme dianionic phosphate acts as a proton acceptor during the glucosyl transfer reaction. The major evidence against this role for the phosphate is that fluorophosphate, with a pK of 4.8, is as effective as phosphite ($pK_a$ = 6.6) in activating the pyridoxal-reconstituted enzyme at pH 6.8.

In a recent investigation, Withers *et al.* studied the effect of the substrate analogue, α-D-glucopyranosyl cyclic 1,2-phosphate on the $^{31}P$ NMR spectrum of the coenzyme. The mobile dianionic form of the coenzyme phosphate is replaced in the presence of this substrate analogue by either a monoprotonated form with some exchange broadening or by a tightly bound and constrained dianion. The latter possibility suggests that the phosphorus of the constrained dianion could be an electrophile which, by direct interaction with the phosphate of the substrate, would facilitate the bond breakage at the anomeric carbon of glucose. This hypothesis has been presented at various conferences in 1980.

This hypothesis would predict that a compound which combines the phosphates of the substrate and coenzyme in a pyrophosphate linkage would prove to have interesting properties when used to reconstitute the enzyme. The synthesis of PLPP-α-Glc and its behavior when bound to the protein of phosphorylase *b* are described in this communication. The results suggest that the coenzyme phosphate does indeed act as a catalyst by direct interaction with the phosphate of the substrate.

(A) In the chemical reaction catalyzed by glycogen phosphorylase (Reaction 5–80) there are two reactants and two products. Draw in an unambiguous, three-dimensional representation the structures of the reactant and the product that confirm the statement that the enzymatic reaction "occurs with retention of configuration." Only two structures, one reactant and one product, should be drawn. Enclose within a rectangle the atom whose configuration is retained as well as the bond broken and formed during the reaction.

(B) Draw the unambiguous, three-dimensional structures of 1,5-gluconolactone and the intermediate glucosyl cation it is supposed to mimic.

(C) What kind of an inhibitor against what reactant or product must 1,5-gluconolactone be?

(D) Start with the following substrate, and write a series of chemical reactions that involve the oxonium ion and also explain the " 'scrambling' of the anomeric oxygen of glucose-1-P."

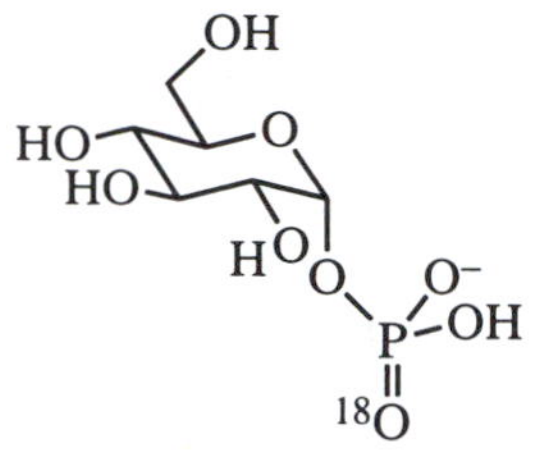

(E) Cyclodextrin is a ring of six glucose molecules, attached head to tail (α1→4) around the ring so that no 4-OH of glucose is available. Using the concept of induced fit, explain why the "'scrambling' of the anomeric oxygen of glucose-1-P" is enhanced by at least 5-fold when cyclodextrin is added to an incubation containing only glycogen phosphorylase and glucose 1-[$^{18}O$]phosphate.

When glycogen phosphorylase is incubated in 0.2 M imidazolium citrate, pH 7.0, the pyridoxal phosphate present in the native enzyme is released and can then be removed from the protein by gel filtration to produce apoenzyme. The holoenzyme can then be "reconstituted" with a series of pyridoxal phosphate analogues, including pyridoxal itself. This table presents the properties of a holoenzyme reconstituted with pyridoxal, the unphosphorylated form of the native coenzyme and the effect of anions on that activity.

| anion added | $pK_{a2}$ | specific activity[a] pyridoxal[a] | native[b] |
|---|---|---|---|
| none | | 2.4 | 60 |
| phosphite | 6.6 | 9.1 | 56 |
| fluorophosphate | 4.8 | 7.6 | 60 |
| nitrate | | 1.9 | 62 |
| sulfate | 1.9 | 2.0 | 60 |
| bicarbonate | 6.4 | 0.4 | 57 |
| thiophosphate | 6.2 | 0.2 | 62 |
| pyrophosphate | 1.5 | 0.0 | 61 |

[a]Micromoles of phosphate minute$^{-1}$ (milligram protein)$^{-1}$; the reaction mixture contained pyidoxal-reconstituted phosphorylase, 0.075 M glucose-1-P, 1% glycogen, and the indicated phosphate analogues at a concentration of 0.01 M. After incubation for 5 min at 30 °C, the phosphate liberated was measured. Data from ref 122. [b]Conditions were identical with those for footnote *a* except that the reaction mixture contained native phosphorylase (4.9 μg).

Either native enzyme containing pyridoxal phosphate or enzyme reconstituted with pyridoxal was assayed in the glycogen synthesis reaction, and release of $HOPO_3^{2-}$ was followed. The pyridoxal form of the enzyme requires added anions to display catalytic activity because it was proven that the activity in the absence of any anion is due to activation by contaminating $HOPO_3^{2-}$. The behavior of phosphite as an activator and pyrophosphate as an inhibitor were examined more closely.[122]

| parameters | pyridoxal-reconstituted | native |
|---|---|---|
| $K_m$ phosphite | $9 \times 10^{-4}$ M | none |
| $K_m$ AMP | $2.5 \times 10^{-5}$ M | $(3–8) \times 10^{-5}$ M |
| $K_m$ glucose-1-P | $2.8 \times 10^{-2}$ M | $7 \times 10^{-3}$ M |
| $K_m$ glycogen | 0.01% | 0.02% |
| cooperative binding of AMP | yes | yes |
| sp act. | 3.4 (8.7%) | 39 |
| max velocity | 12 (19%) | 65 |

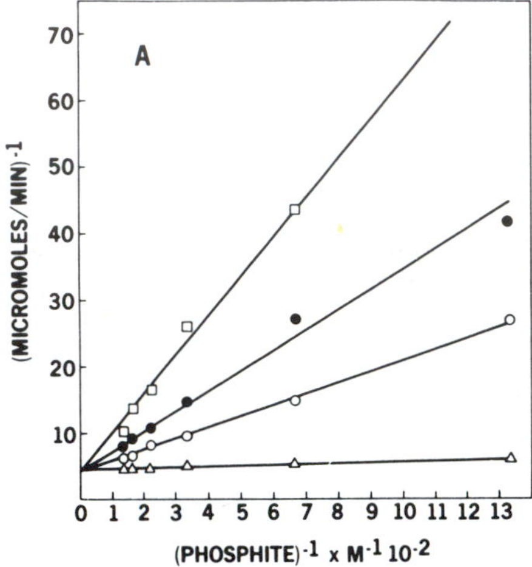


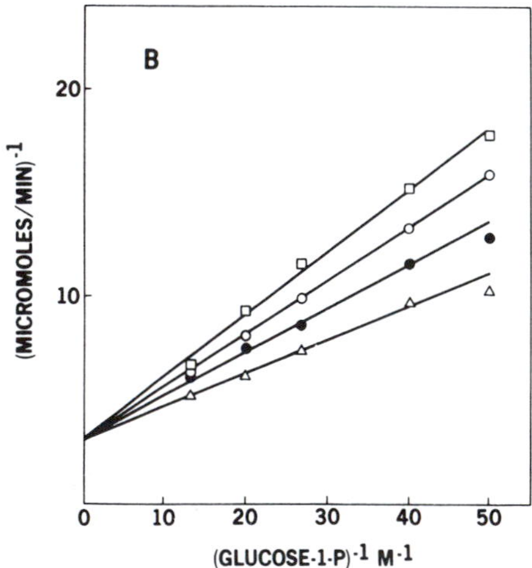


Double-reciprocal plots of pyrophosphate inhibition of pyridoxal-reconstituted phosphorylase.[122] The reaction mixtures at pH 6.8 consisted of pyridoxal-reconstituted phosphorylase, 20 mM $\beta$-glycerophosphate, 15 mM 2-mercaptoethanol, 1 mM AMP, and 1% glycogen. (A) Enzyme, 64 $\mu$g, 16 mM glucose-1-P, phosphite at the indicated concentrations, and pyrophosphate at 0 (△), 0.3 mM (○), 0.6 mM (●), and 1.2 mM (□).
(B) Enzyme, 22 $\mu$g, 7.5 mM phosphite, glucose-1-P at the indicated concentrations, and pyrophosphate at 0 (△), 0.1 mM (●), 0.2 mM (○), and 0.3 mM (□).


(F) Draw the two structures of the active site occupied by pyridoxal and either phosphite and glucose-1-P or pyrophosphate.

(G) Explain how pyrophosphate, as an inhibitor, can compete for the active site with both phosphite and glucose-1-P while phosphite functions as a required activator.

(H) The two crystallographic studies mentioned above were examinations of the native holoenzyme complexed with glucose-1-P in the absence of "glucose polymer." The distance between the two phosphorus atoms in pyrophosphate is about 0.23 nm. Using the arguments you formulated in parts E–G, explain how the phosphate of pyridoxal phosphate could get close enough to act as a direct catalytic group in the enzymatic mechanism. Start by considering explicitly the form of the enzyme actually observed in the crystallographic studies. Finally, give what you believe to be a reasonable explanation for this feature of the enzymatic reaction.

(I) On the basis of the short description presented here, draw the steps in the mechanism proposed for the enzymatic reaction by Johnson et al. Include both the coenzyme and His 376.

(J) Draw the complete structure of "PLP-$\beta$-Glc," the "formal covalent intermediate" proposed by Johnson et al. and synthesized by Takagi et al.

(K) Present a drawing of the "mere stabilization of the carbonium ion by the phosphate."

(L) The results presented in the first table were the only published observations of the activation of pyridoxal enzyme by fluorophosphate at the time. As such, the statement that the "major evidence against this role for phosphate is that fluorophosphate, with a pK of 4.8, is as effective as phosphite (p$K_a$ = 6.6) in activating pyridoxal-reconstituted enzyme at pH 6.8" is unjustified. What missing piece of essential information, about the activation of pyridoxal enzyme by fluorophosphate and phosphite, would be required to make such an argument against the phosphate of pyridoxal phosphate acting as a general base?

(M) Draw the mechanism described by the statement "that the phosphorus [atom] of the constrained dianion could be an electrophile which, by direct interaction with the phosphate of the substrate, would facilitate the bond breakage at the anomeric carbon of glucose." Use one of the oxygens on the phosphate of glucose-1-P as the nucleophile and write the reaction in the direction of glycogen synthesis from glucose-1-P.

The following is also a direct quotation adapted with permission from reference 121.

> PLPP-$\alpha$-Glc was bound to apophosphorylase *b* from rabbit muscle, in the same mode as other PLP analogues modified at the 5′-phosphate group, as determined in the CD titration experiments. The CD spectrum of the PLPP-$\alpha$-Glc-reconstituted enzyme is similar to that of the PLP-reconstituted enzyme as well as the intact holoenzyme. The second order rate constant of reconstitution of PLPP-$\alpha$-Glc determined from the CD change at 335 nm was 140 $M^{-1}$ $s^{-1}$, which is smaller than that of PLP (420 $M^{-1}$ $s^{-1}$) but much larger than that of PLP-$\beta$-Glc (3.6 $M^{-1}$ $s^{-1}$). After reconstitution with PLPP-$\alpha$-Glc, enzyme

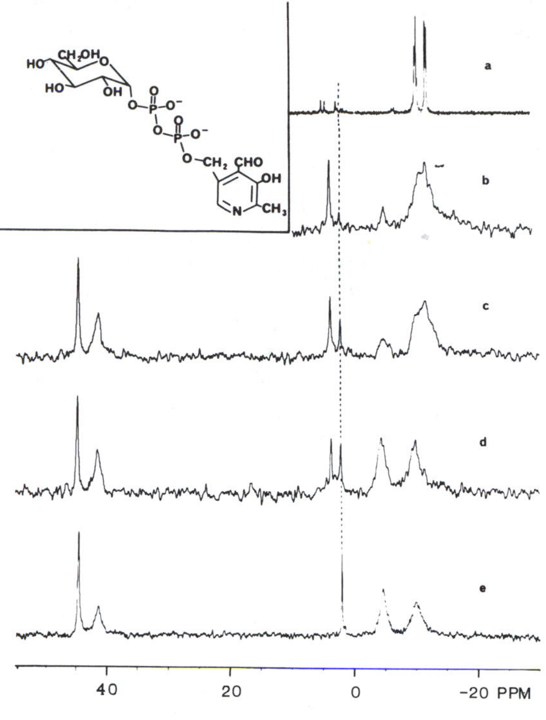


**Figure 1:** $^{31}P$ NMR spectra of various PLP derivatives, alone or bound to the phosphorylase *b* apoenzyme in place of the natural coenzyme. *a.* PLPP-$\alpha$-Glc, 500 acquisitions. *b.* PLPP-$\alpha$-Glc-reconstituted phosphorylase *b* (0.78 mM monomers), 23,000 acquisitions. *c.* PLPP-$\alpha$-Glc-reconstituted phosphorylase *b* (0.76 mM monomers) plus 1.4 mM AMPS, 24,000 acquisitions. *d.* PLPP-$\alpha$-Glc-reconstituted phosphorylase *b* (0.76 mM monomers) plus 1.4 AMPS plus 60 mM maltopentaose, 20,000 acquisitions. *e.* PLPP-reconsatituted phosphorylase *b* (1.1 mM monomers) plus 2.2 mM AMPS, 24,000 acquistions. Conditions were as described under "Materials and Methods." The specific activity of the apoenzyme was 0.1 $\mu$mol/min/mg. The initial specific actiity of the PLPP-$\alpha$-Glc-reconstituted enzyme was 0.2 $\mu$mol/min/mg. After 40 h of dialysis and concentration, at the beginning of the collection of spectrum *b,* the specific activity was 1.6; after a further 26 h (when collection of spectrum *c* was completed), the specific activity was 3.4; maltopentoase was then added and the specific activity remained constant (3.2 after a further 24 h and collection of spectrum *d*). *Inset,* structure of PLPP-$\alpha$-Glc. Reprinted with permission from ref 121. Copyright 1981 *Journal of Biological Chemistry.*

activity slowly appeared upon prolonged incubation. The appearance of activity is accelerated by raising the pH and markedly repressed by the presence of maltopentaose or glycogen, as shown in Table I. No retardation was observed with 0.1 M glucose-1-P, 1 mM AMP, 60 mM glucose, or 3 mM caffeine (data not shown). The reaction observed with the reconstiuted enzyme could be caused by the slow decomposition of the enzyme-bound PLPP-$\alpha$-Glc to PLPP followed by loss of $P_i$. On the other hand, the PLPP-reconstituted enzyme restored its activity much more slowly than the PLPP-$\alpha$-Glc-reconstituted enzyme; specific activities after 4 days, 2.8 *versus* 6.0 at pH 6.6 and 2.2 *versus* 45 at pH 7.9. PLPP-$\alpha$-Glc and PLPP free in solution are stable at these pH values. Although the mechanism of the activation of the PLPP-$\alpha$-Glc-reconstituted enzyme is not yet clear, no glucose-1-P was seen in the NMR experiments described for Fig. 1, while a steady increase in $P_i$ was observed.

TABLE I

*Reactivation of rabbit muscle apophosphorylase reconstituted with PLPP-$\alpha$-Glc*

Rabbit muscle apophosphorylase *b* (9.3 $\mu$M) was mixed with 8.4 $\mu$M PLPP-$\alpha$-Glc (or PLP) in a solution containing 0.1 M NaCl, 10 mM 2-(*N*-morpholino)ethanesulfonic acid, 10 mM 4-(2-hydroxyethyl)-1-piperazineethanesulfonic acid, 10 mM triethanolamine, 20 mM 2-mercaptoethanol at the indicated pH and at 20 °C. The apoenzyme used has a specific activity of 3.6 $\mu$mol/min/mg. Enzyme activities were measured in the direction of glycogen synthesis. Time course of the reactivation is shown at the top. The numbers in parentheses show the enzyme activities of the PLP-reconstituted enzyme for reference. Effect of substrate glucan on the reactivation is shown at bottom. Either maltopentaose or glycogen was added to the above incubation mixture at zero time. The enzyme activities were measured after incubation for 72 h.

| | enzyme activity | |
|---|---|---|
| | pH 6.8 | pH 7.6 |
| *h* | *µmol/min/mg* | |
| Time | | |
| 3 | 2.2 (47) | 2.2 (49) |
| 24 | 4.6 (63) | 13 (64) |
| 48 | 5.5 (72) | 19 (70) |
| 72 | 7.9 (66) | 25 (62) |
| Addition | | |
| No addition | 7.9 (57) | 25 (58) |
| + maltopentaose, 0.1 M | 4.6 (62) | 3.3 (68) |
| + glycogen, 1 mg/ml | 3.5 (57) | 3.8 (58) |

The $^{31}P$ NMR spectrum of PLPP-$\alpha$-Glc is shown in Fig. 1*a.* Two sets of peaks are observed, centered at −10.7 and −12.3 ppm. The upfield peak is apparently a doublet of doublets, identifying it as the glucose-proximal phosphate. The down-field peak appears to be a doublet of triplets, as expected for the pyridoxal-proximal phosphate.

Fig. 1*b* shows the spectrum of PLPP-$\alpha$-Glc reconstituted phosphorylase *b.* The two individual peaks observed in free PLPP-$\alpha$-Glc have merged as a single asymmetric, broad peak centered at ~ −11.5 ppm. The upfield edge of the peak corresponds well to the resonance assigned to the glucose-proximal phosphate. This may imply very little change in the environment of this phosphate upon binding to the protein. The resonance assigned to the pyridoxal-proximal phosphate, however, appears to have shifted upfield slightly, resulting in a single, broadened peak. This implies a change in the environment of this phosphate. The sharp resonance at 3.5 ppm is probably due to contaminating AMP. These spectra were accumulated in blocks of 2000 acquisitions and this resonance was present in the first block and remained at constant intensity throughout the experiment. The very small peak at 1.9 ppm is due to free $P_i$ and this peak was observed to increase slowly with time. The small broad peak at $\delta = -4.9$ ppm will be refered to shortly.

On addition of nucleotide activator (AMPS) to this sample, the spectrum shown in Fig. 1*c* was obtained. The two downfield peaks are due to free and bound AMPS. The observation of separate signals of relatively narrow line width for the free and bound species indicates tight nucleotide binding. The spectrum was analyzed by computer-assisted line shape analysis, as described earlier and an off-rate constant $k_{off,}$ for AMPS of 36 $s^{-1}$ and a dissociation constant of ≈30 $\mu$M was obtained. The upfield region of the spectrum is changed very little upon activation. The peak due to $P_i$ has increased (and was observed to do so in a time-dependent manner in the individual blocks). The small broad peak at $\delta = -4.9$ ppm has also increased slightly, but the rest of the spectrum is essentially identical with that of the nonactivated enzyme.

Subsequent addition of maltopentaose to this activated enzyme, however, produces drastic changes in the observed spectrum (Fig. 1*d*). The nucleotide is still observed to be tightly bound ($k_{off} = 32\ s^{-1}$, $K_D \approx 30 \mu M$), but the upfield region has changed markedly, with the production of two major resonances at –4.9 and –10.3 ppm. These two resonances are assigned to the two phosphate residues of PLPP bound to phosphorylase. The species has arisen via cleavage of the gylosidic linkage producing PLPP covalently bound to the active site. This assignment of the spectrum is based upon a comparison with the $^{31}P$ NMR spectrum of phosphorylase *b* reconstituted with PLPP, Fig. 1*e*. The two spectra are essentially identical in this region. Binding constants and off-rate constants for AMPS to the PLPP enzyme in Fig. 1*e* ($k_{off} = 38\ s^{-1}$, $K_D \approx 40$ $\mu M$) are equivalent to those observed in Fig. 1*d*, providing further evidence for the identity of this new enzyme form. The resonances at 1.9 and 3.5 ppm assigned to $P_i$ and possibly AMP, respectively, are superimposed upon a small broad peak. This broad resonance is tentatively assigned to the phosphate of PLP present in the small amount of native enzyme arising, presumably, from decomposition of the PLPP bound to the enzyme. Such a time-dependent release of phosphate and production of active enzyme were also observed with PLPP-phosphorylase *b*, both in the recovery of activity, as mentioned previously, and appearance of the two $^{31}P$ NMR resonances after a longer period (spectrum not shown). The small broad peak observed ($\delta = -4.9$ ppm) in Fig. 1, *b* and *c*, is similarly assigned to the downfield resonance of PLPP bound to phosphorylase. This resonance is assigned to the terminal ($\beta$) phosphate of the compound.

The data for Fig. 1*d* were accumulated in blocks of 2000, each representing approximately 1 h of data accumulation time. Conversion of PLPP-$\alpha$-Glc phosphorylase *b* to PLPP-phosphorylase *b* was essentially complete within the first hour. The insensitivity of this NMR technique does not allow us to estimate any times shorter than this.

The fate of the glucose freed from PLPP-$\alpha$-Glc-reconstituted enzyme was investigated by incubating the latter with a 50% excess of maltopentaose and chromatographing the deproteinized reaction mixture (data not shown). No free glucose could be detected, while the maltopentaose had been disproportionated to the expected series of oligomers.

(N) What step or steps in the mechanism you drew in part M are similar to the reaction that actually occurs upon addition of maltopentaose to the enzyme reconstituted with PLPP-$\alpha$-Glc?

(O) What is the difference between "PLPP-phosphorylase *b*" and the intermediate that it mimics in the reaction you drew in part M that prevents the PLPP enzyme from releasing $HOPO_3^{2-}$ while the intermediate in part M would do so readily? Hint: there is a difference of one atom.

(P) Present your own discussion of the evidence for and against both the electrophilic phosphorus mechanism and the general base mechanism presented by Helmreich and Klein.

# Double Kinetic Isotope Effects

In almost every enzymatically catalyzed reaction more than one bond is formed or broken. If each of these bonds is formed or broken in a different kinetic step, then intermediates are necessarily produced during the reaction. If two or more of the bonds are formed or broken in the same step, then these formations or cleavages are concerted and fewer or perhaps no intermediates are produced. Therefore, the order in which or simultaneity with which specific bonds are made or broken can define the chemical structures and the number of the intermediates in a particular reaction.

If a particular bond is broken during a particular step in an enzymatic reaction and one of the two atoms participating in that bond is changed to a heavier isotope, that step will experience a kinetic isotope effect and the microscopic rate constants for that step, both forward and reverse, will decrease. The influence of the intrinsic kinetic isotope effect for that specific step in the kinetic mechanism on the rate of the overall enzymatic reaction is determined by the extent to which that step is a rate-determining step in the overall kinetic mechanism (Equation 4–115). If each of two separate bond cleavages or bond formations independently exerts a kinetic isotope effect on the overall enzymatic reaction, in other words, if isotopic substitution at either bond decreases the overall rate of the enzymatic reaction, it is possible to determine from the influence of the one kinetic isotope effect on the other kinetic isotope effect whether the two bond cleavages or bond formations occur in two kinetically discrete steps or in the same step.[123]

Malate dehydrogenase (oxaloacetate-decarboxylating) ($NADP^+$)

$$\text{malate} + NADP^+ \rightleftharpoons \text{pyruvate} + CO_2 + NADPH \qquad (5\text{–}101)$$

catalyzes a reaction in which, in the direction written, the bond between carbon 2 of malate and its hydrogen is necessarily broken and the bond between carbons 4 and 3 of malate is necessarily broken. There are both a deuterium kinetic isotope effect and a $^{13}$carbon kinetic isotope effect on the overall enzymatic reaction resulting from substitutions at these two positions, respectively. Based on the fact that $\beta$-keto carboxylic acids readily decarboxylate, a possible mechanism for the reaction catalyzed by the enzyme is one involving two steps

(5–102)

It is also possible, however, that the two bond cleavages, carbon–hydrogen and carbon–carbon, occur in the same step. In this latter instance the reaction would be concerted, while in the two-step process oxaloacetate would be formed on the active site as an intermediate.

A **double kinetic isotope effect** is the change in one of the kinetic isotope effects associated with the overall enzymatic reaction brought about by substituting an isotope at another position in the molecule. In the case of malate dehydrogenase (oxaloacetate-decarboxylating) ($NADP^+$), a double kinetic isotope effect would be either the change in the overall $^{13}$carbon kinetic isotope effect at carbon 4 of malate brought about by substituting a deuterium at carbon 2 or the change in the overall deuterium kinetic isotope effect at carbon 2 of malate brought about by substituting a $^{13}$carbon at carbon 4. Because, however, heavy atom kinetic isotope effects, such as the $^{13}$carbon kinetic isotope effect in malate dehydrogenase (oxaloacetate-decarboxylating) ($NADP^+$) are so small, it is technically impossible to detect the change in a deuterium kinetic isotope effect produced by the substitution of an isotope of a heavy atom, such as $^{13}$carbon for $^{12}$carbon. Consequently, only the effect of substituting a deuterium for a hydrogen on a $^{13}$carbon, $^{15}$nitrogen, or $^{18}$oxygen kinetic isotope effect can be measured accurately.

If the reaction catalyzed by malate dehydrogenase (oxaloacetate-decarboxylating) ($NADP^+$) were concerted and both bonds were broken in the same step, the substitution of hydrogen for deuterium at carbon 2 would decrease the rate of this single step, causing it to become more rate-determining in the overall reaction, and the $^{13}$carbon kinetic isotope effect on the overall reaction would have to increase in magnitude. If each of the two bonds is broken in a separate step, however, the substitution of hydrogen for deuterium at carbon 2 will decrease the rate only of the step involving hydride transfer, causing it to become more rate-determining at the expense of the other steps in the reaction. Consequently, the $^{13}$carbon kinetic isotope effect on the overall reaction would have to decrease in magnitude because the step in which this bond is broken has become less rate-determining. It should be obvious that such an argument can apply to any two bond cleavages or bond formations in any enzymatic reaction.

This intuition can be verified mathematically.[123] The $^{13}$carbon kinetic isotope effect, $^{13}(V/K)$, on the overall reaction of malate dehydrogenase (oxaloacetate-decarboxylating) ($NADP^+$) is determined by

$$^{13}(V/K) = \frac{^{12}V_{max}/K_m^{[4\text{-}^{12}\text{C}]\text{malate}}}{^{13}V_{max}/K_m^{[4\text{-}^{13}\text{C}]\text{malate}}} = \frac{\dfrac{^{12}k_n}{^{13}k_n} + c_f + c_r\dfrac{^{12}K_{eq}}{^{13}K_{eq}}}{1 + c_f + c_r} \qquad (5\text{–}103)$$

where $k_n$ is the forward rate constant for the step in which the bond between carbons 4 and 3 is broken (Equations 4–115 and 4–116). If the step in the reaction in which the bond between hydrogen and carbon 2 is broken precedes or follows the step in which the bond between carbons 3 and 4 is broken, then either $c_r$ or $c_f$, respectively, will be increased by the substitution of deuterium at carbon 2 (Equation 4–114), while $^{12}k_i/^{13}k_i$ and $^{12}K_{eq}/^{13}K_{eq}$ will be unchanged. Therefore, the observed quantity $^{13}(V/K)_H$, the $^{13}$carbon kinetic isotope effect measured when carbon 2 has a hydrogen, would have to be greater than the observed quantity $^{13}(V/K)_D$, the $^{13}$carbon kinetic isotope effect measured when carbon 2 has a deuterium. If the bond between carbon 2 and the hydrogen is broken in the same step as that in which the bond between carbons 4 and 3 is broken, then $c_f$ or $c_r$ would both be decreased by the substitution of deuterium at carbon 2 because $k_n$ and $k_{-n}$, the two forward and reverse rate constants for the step involving the cleavage of the carbon–carbon bond, would both decrease (Equations 4–113 and 4–114). Therefore, the observed quantity $^{13}(V/K)_H$ would have to be less than the observed quantity $^{13}(V/K)_D$.

This mathematical development is not unique to $^{13}$carbon–deuterium double kinetic isotope effects and can be applied to the double kinetic isotope effect between deuterium and any other isotope as long as both the deuterium and the other isotope produce separately a kinetic isotope effect on the overall enzymatic reaction. If the two bonds broken or formed both involve hydrogen, but the hydrogens are at different locations, a deuterium–deuterium double kinetic isotope effect can also be assessed.[124]

For malate dehydrogenase (oxaloacetate-decarboxylating) ($NADP^+$) at pH 8.0 and 25 °C, the kinetic quantity $V_{max}/K_m^{malate}$, when carbon 4 is $^{12}$carbon, has a numerical value 1.0302 ± 0.0005 times greater than it has when carbon 4 is $^{13}$carbon.[123] If the hydrogen at carbon 2 is changed to a deuterium, this $^{13}$carbon kinetic isotope effect decreases to 1.0250 ± 0.0007. Therefore, the bond between carbon 2 and hydrogen is broken in a different step in the kinetic mechanism than that in which the bond between carbons 4 and 3 is broken.

It is also possible, in advantageous cases, to determine which step precedes the other if the cleavages or formations of the two bonds, respectively, are not concerted. If the equilibrium constant for the reaction is significantly shifted by the substitution of one isotope for another, $c_f$ and $c_r$ are not symmetric in Equation 5–103. Because the order in which the steps occur determines whether $c_f$ or $c_r$ is changed by the other isotopic substitution, if at least one of the equilibrium isotope effects is significantly different from 1.0, different relationships will exist among the various observed kinetic isotope effects depending on which step comes first.[123]

Malate dehydrogenase (oxaloacetate-decarboxylating) ($NADP^+$) has a deuterium equilibrium isotope effect[125] of 1.18. In other words, the equilibrium constant for the reaction when a hydride is transferred between oxaloacetate and $NADP^+$ is 1.18 times greater, in favor of oxaloacetate and NADPH, than the equilibrium constant when a deuteride is transferred between oxaloacetate and $NADP^+$ (Reaction 5–102). Taken together, the magnitudes of the deuterium equilibrium isotope effect, the $^{13}$carbon equilibrium isotope effect, the deuterium kinetic isotope effect, the $^{13}$carbon kinetic isotope effect, and the $^{13}$carbon kinetic isotope effect after substitution of deuterium at carbon 2 are consistent only with a kinetic mechanism in which hydride transfer precedes decarboxylation.[123] Therefore, oxaloacetate must be an intermediate in the enzymatic reaction.

It should be reiterated that double kinetic isotope effects are observed only when each of the two isotopic substitutions alone displays a significant kinetic isotope effect on the overall enzymatic reaction. Unless the step or each of the steps involving the cleavage or the formation of the bonds is rate-determining, the shift in the relative importance of the rate-determining steps will not be observed. Therefore, to be reassured of the interpretations of the results of these experiments, both of the single kinetic isotope effects must be demonstrated. In the case of malate dehydrogenase (oxaloacetate-decarboxylating) ($NADP^+$) the deuterium kinetic isotope effect $K_m^{[2\text{-}^2H]malate} V_{max}^{^1H} / K_m^{[2\text{-}^1H]malate} V_{max}^{^2H}$ is 1.47, and as noted, the $^{13}$carbon kinetic isotope effect $K_m^{[4\text{-}^{13}C]malate} V_{max}^{^{12}C} / K_m^{[4\text{-}^{12}C]malate} V_{max}^{^{13}C}$ is 1.302.[123]

Glucose-6-phosphate dehydrogenase

$$\text{glucose 6-phosphate} + NADP^+ \rightleftharpoons \text{glucono-1,5-lactone 6-phosphate} + NADPH \qquad (5\text{–}104)$$

also displays a double kinetic isotope effect.[123] The hydrogen on carbon 1 of glucose 6-phosphate can be substituted with a deuterium ($K_m^{[1\text{-}^2H]glucose\ 6\text{-}P} V_{max}^{^1H} / K_m^{[1\text{-}^1H]glucose\ 6\text{-}P} V_{max}^{^2H} = 3.0 \pm 0.1$) and the $^{12}$carbon of carbon 1 of glucose 6-phosphate can be substituted with $^{13}$carbon ($K_m^{[1\text{-}^{13}C]glucose\ 6\text{-}P} V_{max}^{^{12}C} / K_m^{[1\text{-}^{12}C]glucose\ 6\text{-}P} V_{max}^{^{13}C} = 1.017 \pm 0.001$). When the hydrogen at carbon 1 is substituted with deuterium the $^{13}$carbon kinetic isotope effect increases to 1.032 ± 0.001. Because the two isotopes were substituted at either side of the same bond, the cleavage of the same bond produces each of the two kinetic isotope effects, the reaction must be concerted, and the expected increase in the $^{13}$carbon kinetic isotope effect was observed.

Phosphogluconate dehydrogenase (decarboxylating) ($NADP^+$)

$$\text{6-phosphogluconate} + NADP^+ \rightleftharpoons \text{ribulose 5-phosphate} + CO_2 + NADPH \qquad (5\text{–}105)$$

catalyzes formally the same reaction as does malate dehydrogenase (oxaloacetate-decarboxylating) ($NADP^+$). The $^{13}$carbon kinetic isotope effect for $^{13}$carbon substitution at carbon 1 of 6-phosphogluconate is 1.0096 ± 0.0006. When the hydrogen on carbon 3 of the 6-phosphogluconate is replaced with a deuterium, the $^{13}$carbon kinetic isotope effect decreases to 1.0081 ± 0.0002. Therefore, hydride transfer from carbon 3 and decarboxylation of carbon 1 are not concerted.[126] The magnitudes of the various kinetic isotope effects are together consistent with a kinetic mechanism in which hydride transfer precedes decarboxylation and in which 3-keto-6-phosphogluconate is the intermediate in the enzymatic reaction (Reaction 5–53).

Prephenate dehydrogenase also catalyzes a reaction in which a hydride transfer and a decarboxylation occur. The reaction can be written with a reasonable intermediate, **5–25**

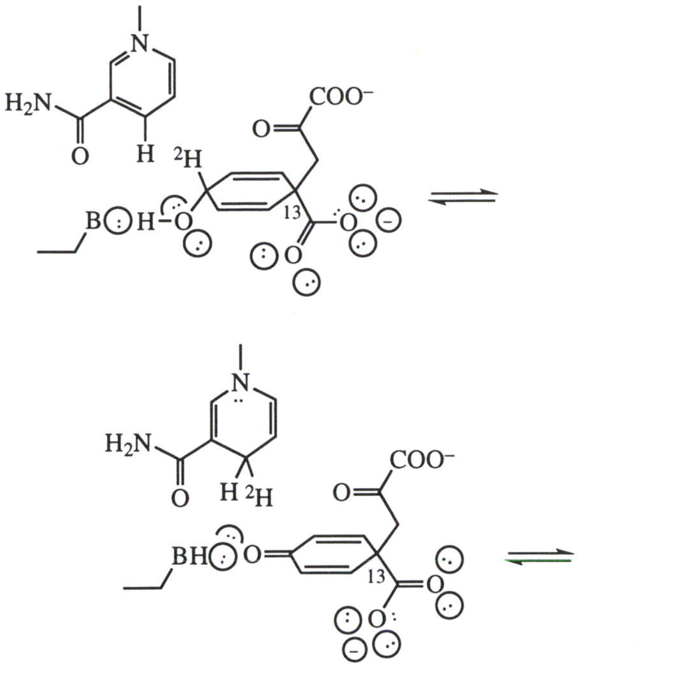

5–25

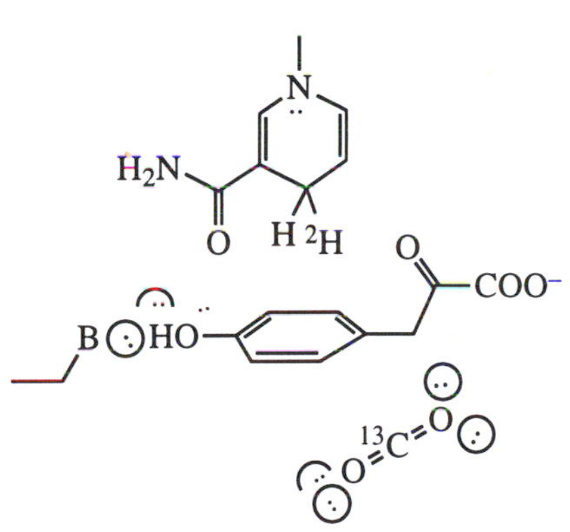

(5–106)

that resembles the intermediate in the electrophilic aromatic substitution of a phenol

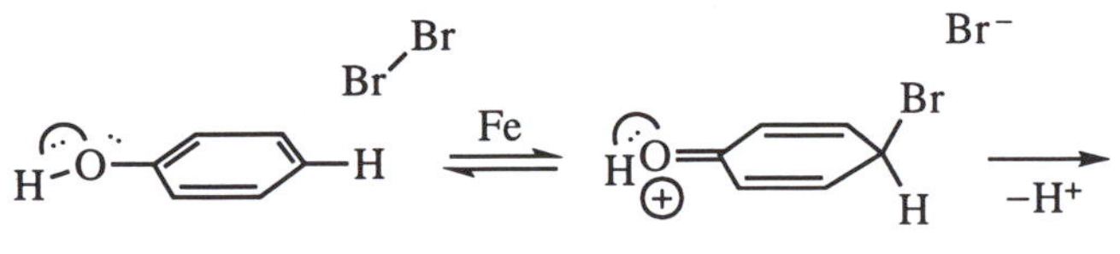

(5–107)

It has also been shown[127] that the related acid-catalyzed 1,4-elimination of water and carbon dioxide from prephenate passes through a carbocationic intermediate even though it is even less stable than the intermediate written in Reaction 5–106

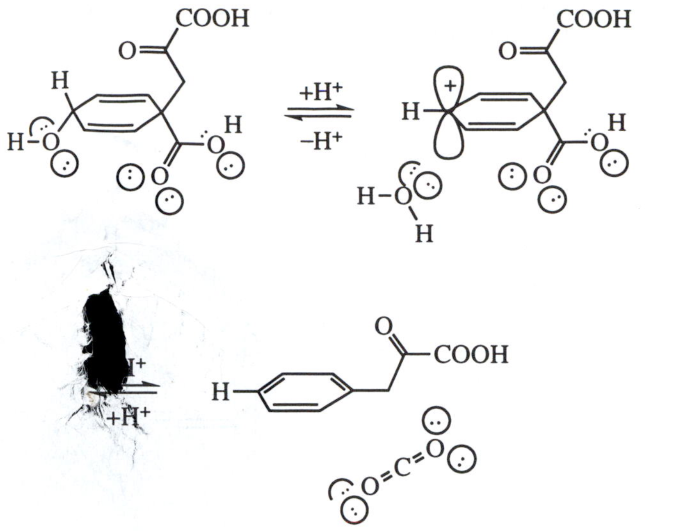

(5–108)

Furthermore, the 1,4-elimination catalyzed by prephenate dehydrogenase is antarafacial because prephenate has the hydrogen on carbon 4 in a *trans* orientation to the carboxylate at carbon 1.[128] This also speaks against a concerted mechanism for this reaction. Because the two electrons in the $\sigma$ bond between carbon 1 and the carboxylate would enter the $\pi^*$ molecular orbitals of the olefins from below the plane of the ring and the pairs of electrons in the $\pi$ molecular orbitals of the olefins would push into the $\sigma^*$ orbital of the $\sigma$ bond between carbon and oxygen also from below the plane of the ring, a concerted 1,4-antarafacial elimination is supposed to be less favored than a concerted 1,4-suprafacial elimination.

Nevertheless, when the hydrogen on carbon 4 of the alternate reactant for prephenate dehydrogenase, 1-carboxy-4-hydroxy-2,5-cyclohexadiene-1-propanoic acid

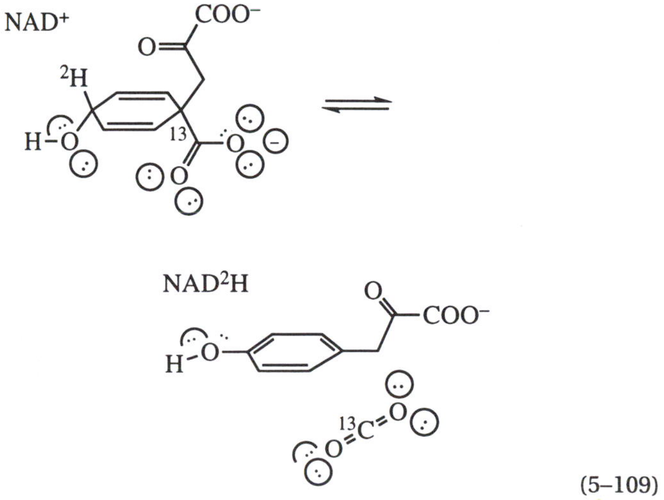

(5–109)

is replaced with a deuterium, the $^{13}$carbon kinetic isotope effect on the overall enzymatic reaction for [$^{13}$C]carboxy substitution at carbon 1 increases from 1.0033 ± 0.0008 to 1.0103 ± 0.0013.[127] These results are consistent only with a concerted mechanism for prephenate dehydrogenase in which hydride transfer and decarboxylation occur in the same transition state. This mechanism is different from the one expected, and the intermediate in Reaction 5–106 must not be formed. Therefore, the carboxylate anion must provide push to the hydride transfer antarafacially

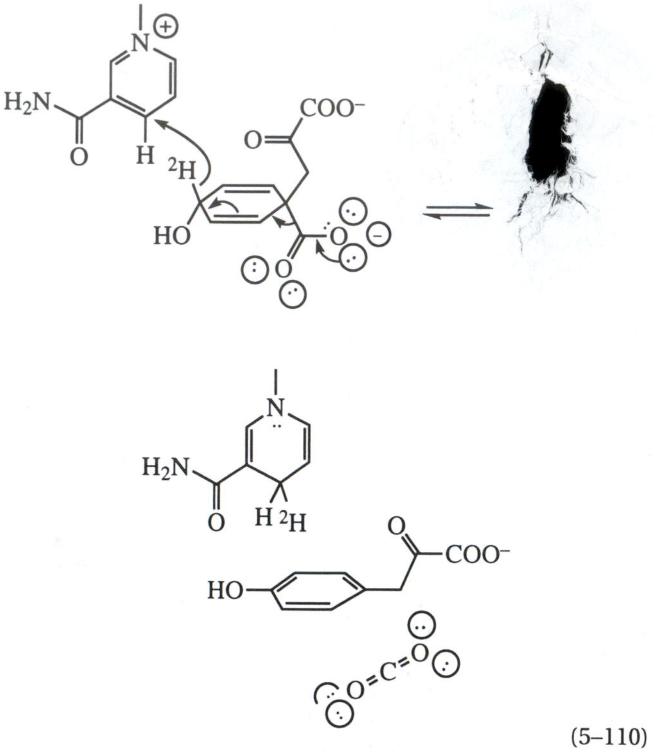

(5–110)

Phenylalanine-ammonia lyase catalyzes an elimination–addition of ammonia to an olefin

$$\text{phenylalanine} \rightleftharpoons \textit{trans}\text{-cinnamate} + NH_4^+ \qquad (5\text{–}111)$$

During this elimination–addition, the bond between carbon 2 of phenylalanine and the nitrogen is broken or formed, respectively, and the bond between carbon 3 and the *pro-S* hydrogen is broken or formed, respectively. At several values of pH (pH 8.4, 9.1, and 9.5) and with either phenylalanine or 1,4-dihydrophenylalanine as reactants, a $^{15}$nitrogen kinetic isotope effect is observed ($K_m^{[2\text{-}^{15}N]Phe} V_{max}^{14N} / K_m^{[2\text{-}^{14}N]Phe} V_{max}^{15N}$ = 1.007–1.020).[129] At all three pH values, when the hydrogen on carbon 3 of the reactant was replaced with deuterium the $^{15}$nitrogen kinetic isotope effect of 1.007–1.020 decreased by 0.0005–0.014 units. These observations are inconsistent with the cleavage of both the carbon–nitrogen bond and the carbon–hydrogen bond occurring in a concerted fashion but are consistent with the existence of an intermediate. Other results suggest that the intermediate is the carbanion

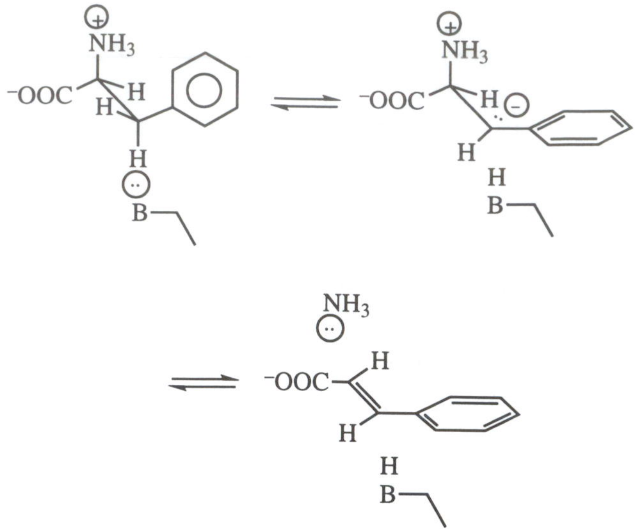

(5–112)

and that the elimination proceeds by an E1$_{CB}$ mechanism,[129] even though the carbanion is not an obviously advantageous intermediate.

The examples just discussed were for situations in which one kinetic isotope effect increases or decreases when another isotope is changed. It is also possible for no change to occur in the kinetic isotope effect upon isotopic substitution at the other position even though individual kinetic isotope effects are observed for isotopic substitution at each position independently. Formate dehydrogenase

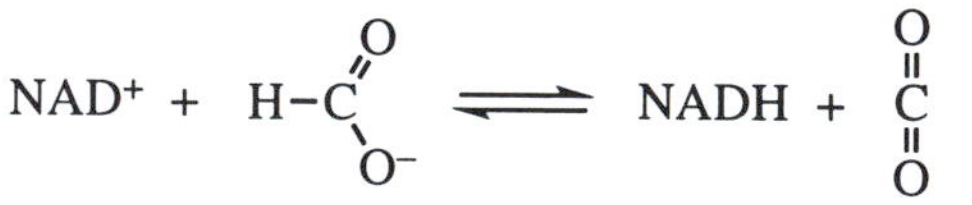

$$NAD^+ + H\text{–}C(=O)O^- \rightleftharpoons NADH + O{=}C{=}O \qquad (5\text{–}113)$$

displays both a $^{13}$carbon kinetic isotope effect (1.0413 ± 0.0006) and a deuterium kinetic isotope effect (2.85 ± 0.04) for [$^{13}$C]formate and [$^{2}$H]formate, respectively, when the pyridine aldehyde derivative of NAD$^+$ is used as the other reactant.[130] When the two deuterated formates, [$^{12}$C,$^{2}$H]formate and [$^{13}$C, $^{2}$H]formate, were used as reactants, the $^{13}$carbon kinetic isotope effect was identical (1.414 ± 0.0001). In this instance the bond involving the deuterium is the same bond as that involving $^{13}$carbon, namely, the carbon–hydrogen bond of formate, so these results can say nothing about intermediates in the reaction. They do, however, speak to the issue of the rate-limiting step in the enzymatic reaction. If any steps other than hydride transfer contributed to the rate constant $V_m/K_m^{formate}$, in other words, if the commitments $c_f$ and $c_r$ in Equation 5–103 were other than nearly zero, the $^{13}$carbon kinetic isotope effect would have to increase when deuterium was substituted for hydrogen. Only if hydride transfer is the rate-limiting step in the enzymatic reaction and all steps preceding it are so much faster that they do not contribute to the value of $V_m/K_m^{formate}$ will the slowing of this step by substituting deuterium for hydrogen have no effect on the observed $^{13}$carbon kinetic isotope effect. Because no effect was seen, it could be concluded[131] that both the deuterium kinetic isotope effect and the $^{13}$carbon kinetic isotope effect were the intrinsic isotope effects for the step in which the carbon–hydrogen bond was broken heterolytically. From these intrinsic kinetic isotope effects, arguments could be made concerning the structure of the transition state for hydride transfer at the active site.[130]

When steps other than the one involving bond cleavage are slow enough to be rate-determining in the overall rate of the enzymatic reaction, in other words, when either the commitment forward or the commitment in reverse or both are significantly greater than zero, the results of double kinetic isotope studies, because they provide additional independent equations, can be used in concert with other results to estimate intrinsic kinetic isotope effects. For example, intrinsic kinetic isotope effects for malate dehydrogenase (oxaloacetate-decarboxylating)(NADP$^+$) could be calculated from the $^{13}$carbon–deuterium double kinetic isotope effects and the tritium, deuterium, and $^{13}$carbon kinetic isotope effects.[123] For carbon–hydrogen bond cleavage during hydride transfer, $^1k/^2k$ was estimated to be 5.6 ± 0.5, and for carbon–carbon bond cleavage during decarboxylation, $^{12}k/^{13}k$ was estimated to be 1.05 ± 0.01. These values were confirmed by later intermediate partitioning experiments.[132]

### *Suggested Reading*

Hermes, J.D., Roeske, C.A., O'Leary, M.H., & Cleland, W.W. (1982) The Use of Multiple Isotope Effects to Determine Enzyme Mechanisms and Intrinsic Isotope Effects: Malic Enzyme and Glucose-6-phosphate Dehydrogenase, *Biochemistry 21*, 5106–5114.

## Isotopic Exchange

A simple example of isotopic exchange is that observed when dihydroxyacetone phosphate was mixed with fructose-bisphosphate aldolase in tritiated water.[133] When the specific radioactivity of the water was 9800 cpm ($\mu$mol of hydrogen)$^{-1}$, the specific radioactivity of the dihydroxyacetone phosphate in the solution monotonically increased from 0 to 9800 cpm ($\mu$mol of dihydroxyacetone phosphate)$^{-1}$, after which only a slow nonenzymatic increase in its specific radioactivity was seen. Because no glyceraldehyde 3-phosphate was added, no fructose 1,6-bisphosphate was produced. The enzyme, however, was able to transform the one reactant by exchanging a tritium for a protium. This serves as an example of **isotopic exchange in the absence of one or more reactants.** If an intermediate in the enzymatic reaction can be formed in the absence of the missing reactant or reactants and the intermediate has lost one or more of the atoms present in the added reactant, isotopic exchange with the solution of those atoms in that reactant may be observed.

All of the observed properties of the production of the tritiated dihydroxyacetone phosphate are consistent with the mechanism (Figure 5–9) in which an enamine is formed at the active site (Reaction 5–25), and the reaction can proceed no further because the normal electrophile, glyceraldehyde 3-phosphate, has been omitted. The *pro-S* hydrogen of the unlabeled reactant is removed by a base on the active site, the proton on the base exchanges for a triton in the water by an acid–base reaction, and the triton is added to the enamine by the conjugate acid to produce tritiated dihydroxyacetone phosphate. Even though there are four acidic hydrogens on the iminium adduct, the final specific radioactivity of the dihydroxyacetone phosphate at the completion of the enzymatic reaction is not 49,000 cpm $\mu$mol$^{-1}$ because only one of the four protons is removed and replaced with a triton by the general base–acid in the active site. When the enzymatic reaction comes to equilibrium, the specific radioactivity of that one hydrogen, and hence of the dihydroxyacetone phosphate, must be equivalent to the specific radioactivity of a hydrogen in the water.

This simple example illustrates the fact that the rate of the observed isotopic exchange is determined by the rate at which the bound reactant is transformed into the intermediate, the rate at which the detached isotope is transferred out of the active site, the rate at which the other isotope is transferred into the active site, the rate at which the intermediate is returned to the isotopically substituted reactant, and the rate at which that reactant dissociates from the active site. Usually, however, the only step in this process that is of kinetic interest is the one in which the intermediate is formed from the bound reactant. It follows that the rate of the isotopic exchange establishes only a lower limit to the rate at which the intermediate itself is formed. The importance of this consideration is illustrated by the equilibrium isotope exchanges catalyzed by enolase.[134]

**Equilibrium isotopic exchanges** are the isotopic exchanges within a reactant observed when all of the reactants and all of the products of the enzymatic reaction are present in the solution and the respective concentrations of reactants and products are purposely set so that they satisfy the equation for the equilibrium constant. If the absolute molar concentrations of reactants and products present in solution are such that their ratio is equal to the equilibrium constant, the reaction does not cease; rather, the rate at which reactants are converted to products will necessarily be equal to the rate at which products are converted to reactants. This is an important feature of the state of equilibrium that is never stressed sufficiently, so the mistaken impression is often left that a reaction at equilibrium is quiescent. Actually, it is at equilibrium that the most is happening. It is when appropri-

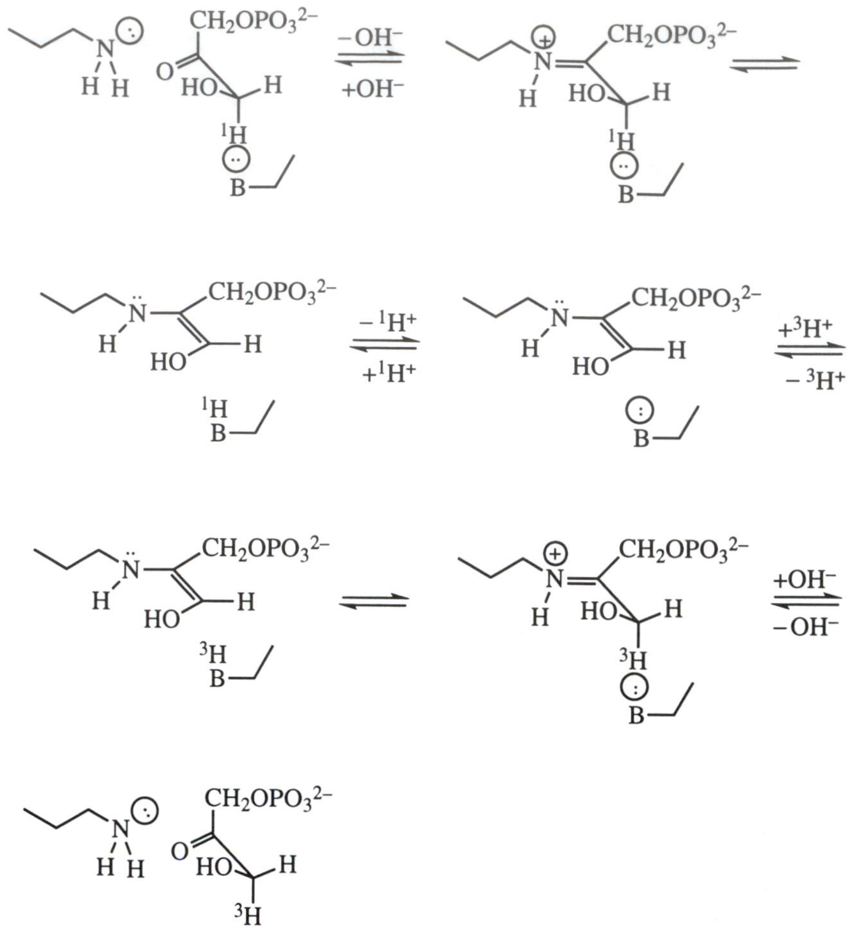

**Figure 5–9:** Full mechanism for the exchange of a tritium for a protium at carbon 3 of dihydroxyacetone phosphate by fructose-bisphosphate aldolase. Dihydroxyacetone phosphate forms the iminium intermediate at the active site, and the proton is removed by the base. The proton dissociates from the base into the solution and a triton associates in two acid–base reactions. The base protonates the enamine to produce the tritiated iminium cation. The tritiated dihydroxyacetone phosphate dissociates from the active site into the solution. As the reaction reaches equilibrium, the specific radioactivity of the *pro-S* hydrogen of the dihydroxyacetone phosphate becomes equivalent to the specific radioactivity of the hydrogens in the water.

ate isotopes are used that the rate of these otherwise hidden reactions contributing to this active interconversion of reactants and products can be determined.

For the reaction catalyzed by enolase

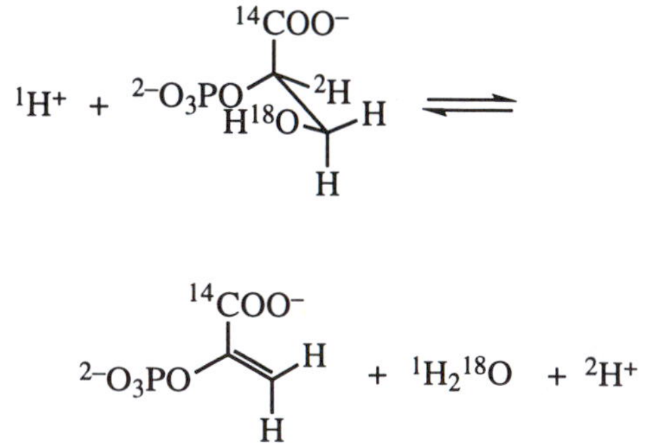

$$^{1}H^{+} + 2\text{-phospho}[1\text{-}^{14}C,2\text{-}^{2}H,3\text{-}^{18}O]\text{glycerate} \rightleftharpoons \text{phospho}[1\text{-}^{14}C]\textit{enol}\text{pyruvate} + {}^{1}H_2{}^{18}O + {}^{2}H^{+} \tag{5–114}$$

the equilibrium constant between pH 6 and pH 8 is

$$K_{eq} = 6.3 = \frac{[\text{phospho}\textit{enol}\text{pyruvate}]}{[2\text{-phosphoglycerate}]} \tag{5–115}$$

Equilibrium concentrations of phospho*enol*pyruvate (0.11 M) and 2-phospho[2-$^{2}$H]glycerate (0.018 M) were mixed in $H_2{}^{18}O$ and trace amounts of 2-phospho[1-$^{14}$C]glycerate and 2-phospho[2-$^{3}$H]glycerate were added (isotopic labels in Reaction 5–114). Because the concentrations of phospho*enol*pyruvate and 2-phosphoglycerate remained constant, the rate at which phospho[1-$^{14}$C]*enol*pyruvate was formed is a direct measure of the isotopic exchange

$$\text{phospho}[1\text{-}^{12}C]\textit{enol}\text{pyruvate} + 2\text{-phospho}[1\text{-}^{14}C]\text{glycerate} \rightleftharpoons 2\text{-phospho}[1\text{-}^{12}C]\text{glycerate} + \text{phospho}[1\text{-}^{14}C]\textit{enol}\text{pyruvate} \tag{5–116}$$

The rate of this isotopic exchange at equilibrium is the overall rate for the conversion of 2-phosphoglycerate into phospho*enol*pyruvate coincident with the conversion of phospho*enol*pyruvate into 2-phosphoglycerate. The rate at which 2-phospho[3-$^{18}$O]glycerate was formed is a direct measure of the rate of the isotopic exchange

$$2\text{-phospho}[3\text{-}^{16}O]\text{glycerate} + H_2{}^{18}O \rightleftharpoons H_2{}^{16}O + 2\text{-phospho}[3\text{-}^{18}O]\text{glycerate} \tag{5–117}$$

The rate of this isotopic exchange is the overall rate for the cleavage of the carbon–oxygen bond in 2-phosphoglycerate to produce water, the release of that water, the binding of another water to the active site, and the insertion of that water back into 2-phosphoglycerate. The rate at which $^{3}H_2O$ was formed is a direct measure of the isotopic exchange

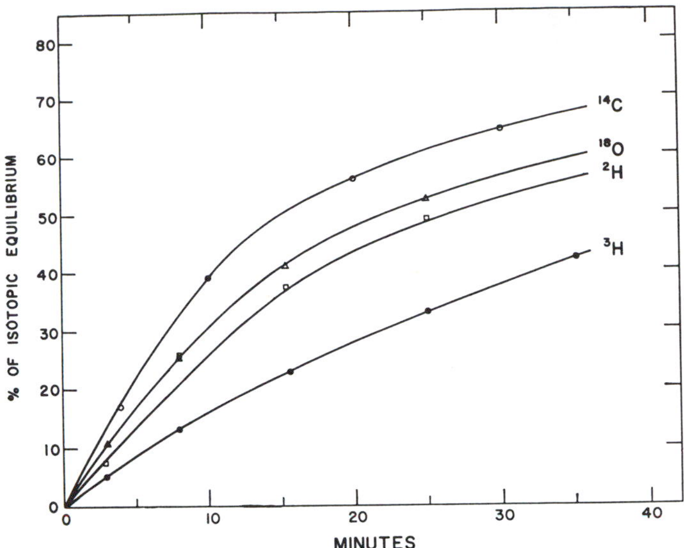


**Figure 5–10:** Time course for the equilibrium isotopic exchange of $^{14}C$, $^{18}O$, $^{2}H$, and $^{3}H$ into and out of phosphoglycerate catalyzed by enolase.[134] The reaction mixture at pH 6.5 contained 0.11 M phospho*enol*pyruvate, 0.018 M phos-pho[2-$^{2}$H]glycerate (97% $^{2}H$), 25 mM $MgCl_2$, and traces of phospho[1-$^{14}$C]glycerate and phospho[2-$^{3}$H]glycerate in a total volume of 1.5 mL of 1.33 atom % $H_2{}^{18}O$. The reaction was initiated by adding 50 $\mu$g of enolase. At various times, samples (50 $\mu$L) were removed and quenched at 115 °C for 40 s. Values for the isotopic distributions at equilibrium were measured at 18 h. The $^{2}H$ and $^{3}H$ exchanged into the $H_2O$ in each sample was determined by distilling the water. The $^{18}O$ exchanged into the phosphoglycerate was determined from the residue following distillation of the water. The $^{14}$carbon exchanged into phospho*enol*pyruvate was determined by separating the two substrates chromatographically. For each isotope, the values at a particular time (minutes) are expressed as a percentage of their values at equilibrium (18 h). The data were fit to the general equation for isotopic exchange at equilibrium,[135] rate = $\{[A][P]/([A] + [P])\}\{[\ln(1-f)]/t\}$, where rate is the rate of the exchange, [A] and [P] are the total molar concentrations of each of the two species exchanging the isotope (see Table 5–2), and $f$ is the fraction of isotopic exchange achieved at time $t$. From these fits, the rates (millimolar minute$^{-1}$) can be calculated for each isotopic exchange (Table 5–2). When one of the species exchanging is $H_2O$, the factor to the left of the expression becomes unity. Reprinted with permission from ref 134. Copyright 1971 *Journal of Biological Chemistry.*

2-phospho[2-$^{3}$H]glycerate + $^{1}H_2O$ ⇌
$^{3}H_2O$ + 2-phospho[2-$^{1}$H]glycerate (5–118)

The rate for this isotopic exchange is the overall rate for the cleavage of the carbon–tritium bond in 2-phosphoglycerate to produce a triton, the dissociation of that triton, the association of a proton, and the insertion of that proton back into 2-phosphoglycerate. The rate at which $^{2}H_2O$ was formed is a direct measure of the homologous isotopic exchange

2-phospho[2-$^{2}$H]glycerate + $^{1}H_2O$ ⇌
$^{2}H_2O$ + 2-phospho[2-$^{1}$H]glycerate (5–119)

Each of the increases in the levels of these four products behaved as approaches to equilibrium (Figure 5–10),[134] from which rates for each of the isotopic exchanges could be calculated (Table 5–2).[134] The rates of the observed tritium and deuterium exchanges were each slower than the rate of hydrogen exchange would have been because of the kinetic isotope effect, but from the rates of the tritium and deuterium exchanges, the value for the hydrogen exchange

2-phospho[2-$^{1}$H]glycerate + $^{1}H_2O$ ⇌
$^{1}H_2O$ + 2-phospho[2-$^{1}$H]glycerate (5–120)

could be calculated (Equations 4–106 and 4–115).

**Table 5–2: Rates of Equilibrium Isotopic Exchange for Enolase at pH 6.5[a]**

| isotopic equilibrium | | | observed rate (mM min$^{-1}$) | relative rates |
|---|---|---|---|---|
| [$^{14}$C]PGA | ⇌ | [$^{14}$C]PEP | 0.68 | 1.0 |
| [$^{18}$O]$H_2O$ | ⇌ | [$^{18}$O]PGA | 0.71 | 1.1 |
| [$^{3}$H]PGA | ⇌ | [$^{3}$H]$H_2O$ | 0.34 | 0.5 |
| [$^{2}$H]PGA | ⇌ | [$^{2}$H]$H_2O$ | 0.62 | 0.9 |
| [$^{1}$H]PGA | ⇌ | [$^{1}$H]$H_2O$ | | 6.2[b] |

[a]Reaction conditions and measurements as in Figure 5–10. [b]Estimated from $^{2}H$ and $^{3}H$ exchange rates.

The rate of the isotopic exchange for a hydrogen at carbon 2 of 2-phosphoglycerate was much faster than the rate of the isotopic exchange for the oxygen at carbon 3 of 2-phosphoglycerate, which in turn was faster than the rate of isotopic exchange accomplished by the forward and then the backward operation of the entire reaction. One interpretation[134] of these observations is that the carbon–hydrogen bond is broken before the carbon–oxygen bond in the normal enzymatic reaction and that an intermediate enol is formed

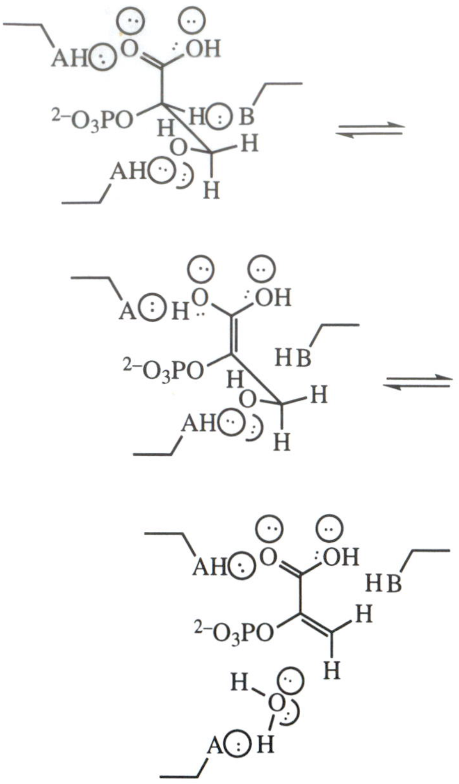

(5–121)

Although this interpretation is chemically reasonable, it requires that the relative rates of isotopic exchange be equated with the relative rates of primary bond cleavage, but this assumption is not necessarily so. The proton removed from carbon 2 of the 2-phosphoglycerate and the water removed from carbon 3 of the 2-phosphoglycerate (Reaction 5–121) are both ultimate products of the enzymatic reaction. For a particular isotopic exchange to occur, the product carrying the initial isotope must dissociate from the active site and a molecule of the same product carrying the other isotope must associate with the active site. Even though most of the water bound by a protein is in very rapid exchange with the bulk solvent, it is possible that the molecule of water produced during the enzymatic reaction is trapped within the active site. In this case, it would be reasonable to assume that the rate at which the proton departs from the base in exchange for a proton in the solvent is fast while the rate at which the water departs from the active site in exchange for water in the solvent is slow. This assumption is reasonable because proton transfer should be even more rapid than exchange of water bound to a protein. If this is so, the difference in rate observed between hydrogen exchange and oxygen exchange (Table 5–2) may reflect only the differences in the rates of these subsequent processes, and the carbon–oxygen bond could break in concert with the carbon–hydrogen bond or even before it during the actual enzymatic reaction. The important point is that the rate of hydrogen exchange is the lower limit of the rate of cleavage of the carbon–hydrogen bond and the rate of oxygen exchange is the lower limit of the rate of cleavage of the carbon–oxygen bond.

Isotopic exchange in the absence of one of the reactants in an enzymatic reaction that has more than one reactant, as with the isotopic exchange catalyzed by fructose-bisphosphate aldolase, can provide unequivocal evidence for a particular intermediate. Many enzymes, however, fail to catalyze any reaction, let alone isotopic exchange, in the absence of all of their reactants. In such situations equilibrium isotopic exchange can provide useful information. Glutamate–ammonia ligase (Reaction 5–55) catalyzes a reaction with three reactants and three products. The central intermediate in the reaction is glutamyl phosphate (Reaction 5–57). The formation of the glutamyl phosphate is known to require only glutamate and MgATP; yet, in the presence of glutamate but in the absence of ammonia, no isotopic exchange (< 0.01% that seen in the presence of ammonia) between MgATP and MgADP

$$\text{Mg}[^{12}\text{C}]\text{ATP} + \text{Mg}[^{14}\text{C}]\text{ADP} \rightleftharpoons \text{Mg}[^{14}\text{C}]\text{ATP} + \text{Mg}[^{12}\text{C}]\text{ADP} \quad (5\text{–}122)$$

was observed.[136] This exchange would result from formation of glutamyl phosphate ($\text{GluOPO}_3\text{H}^-$) from Mg[$^{12}$C]ATP and glutamate (Glu)

$$\text{E} + \text{Glu} + \text{Mg}[^{12}\text{C}]\text{ATP} \rightleftharpoons \text{E}\cdot\text{GluOPO}_3\text{H}^-\cdot\text{Mg}[^{12}\text{C}]\text{ADP} \quad (5\text{–}123)$$

dissociation of the Mg[$^{12}$C]ADP formed

$$\text{E}\cdot\text{GluOPO}_3\text{H}^-\cdot\text{Mg}[^{12}\text{C}]\text{ADP} \rightleftharpoons \text{E}\cdot\text{GluOPO}_3\text{H}^- + \text{Mg}[^{12}\text{C}]\text{ADP} \quad (5\text{–}124)$$

association of Mg[$^{14}$C]ADP, and formation of Mg[$^{14}$C]ATP from the Mg[$^{14}$C]ADP and the glutamyl phosphate

$$\text{E}\cdot\text{GluOPO}_3\text{H}^- + \text{Mg}[^{14}\text{C}]\text{ADP} \rightleftharpoons \text{E} + \text{Glu} + \text{Mg}[^{14}\text{C}]\text{ATP} \quad (5\text{–}125)$$

The most reasonable explanation for why no exchange occurs in the absence of ammonia is that the active site does not close around the reactants and become properly structured until all three of them are present. When equilibrium concentrations of all six substrates were present, however, the MgATP–MgADP exchange described in Equation 5–122 was observed as a simple result of the overall reaction.

The magnitudes of the equilibrium isotopic exchanges observed with glutamate–ammonia ligase are also contrary to expectation. They present an even better example of the danger of correlating relative isotopic exchange rates with the sequence in which bonds are formed or broken at the active site. Both the rate of the isotopic exchange

$$[^{14}\text{C}]\text{glutamate} + [^{12}\text{C}]\text{glutamine} \rightleftharpoons [^{12}\text{C}]\text{glutamate} + [^{14}\text{C}]\text{glutamine} \quad (5\text{–}126)$$

at equilibrium and the rate of the isotopic exchange

$$[^{15}\text{N}]\text{NH}_4^+ + [^{14}\text{N}]\text{glutamine} \rightleftharpoons [^{15}\text{N}]\text{glutamine} + [^{14}\text{N}]\text{NH}_4^+ \quad (5\text{–}127)$$

at equilibrium were greater than the rate of the MgATP–MgADP exchange (Reaction 5–122) at equilibrium.[136] Because glutamyl phosphate is the intermediate in the enzymatic reaction, MgADP must be formed in a step in the overall enzymatic reaction that precedes the step in which glutamine is formed, yet the magnitudes of the rates of the isotopic exchanges were in the opposite relationship. The most reasonable explanation is that the isotopic exchanges at equilib-

**Figure 5–11:** Hydrogen exchange at carbon 4′ of urocanate during the reaction catalyzed by the enzyme urocanate hydratase.[41] The reaction mixture (3 mL) contained 4 mM urocanate in $^2H_2O$ (99.9% deuterium) at p$^2$H 7.5 and 25 °C. The reaction was initiated with 0.4 mg of urocanate hydratase. The total concentration of urocanate (millimolar) was followed by its absorbance at 317 nm (▲). Samples were removed at various times, quickly frozen, and lyophilized. The fraction of the remaining urocanate that had incorporated one deuterium at carbon 4′ was assessed by gas chromatography followed by mass spectrometry. That the deuterium substitution was at carbon 4′ was confirmed by nuclear magnetic resonance. From the fraction of the urocanate that was deuterated and the total concentration of urocanate, the millimolar concentrations of [4′-$^2$H]urocanate (●) and [4′-$^1$H]urocanate (○) could be calculated and presented as a function of time (minutes). Reprinted with permission from ref 41. Copyright 1981 American Chemical Society.

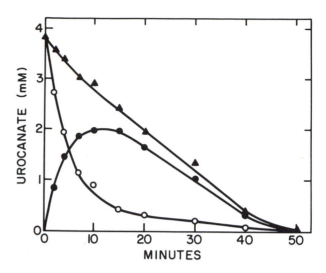

rium are reflecting only the relative order in which reactants associate with the active site and products dissociate from the active site, either because of differences in the relative rates of the random association and dissociation for each of the individual substrates to and from the active site or because there is a compulsory order in which substrates must associate with or dissociate from the active site.

**Isotopic exchange at initial velocity** is the isotopic exchange observed at an atom in one of the reactants when all of the reactants and none of the products are present and the reaction is proceeding only in one direction rather than remaining at equilibrium. Urocanate hydratase (Reaction 5–26) has a tightly bound NAD$^+$ at its active site, and the ultimate product of the reaction has been formally oxidized at carbon 4′. These facts raised the possibility that the hydrogen at carbon 4′ was removed by NAD$^+$ as a hydride ion. [4′-$^1$H]Urocanate was mixed with enzyme in $^2H_2O$. At various times, the reaction was stopped, and the concentrations of [4′-$^1$H]urocanate and [4′-$^2$H]urocanate were determined (Figure 5–11).[41] Although urocanate was continuously disappearing, [4′-$^2$H]urocanate was being produced by hydrogen exchange at a rate significantly in excess of the rate of the overall reaction so that it could accumulate before it was in turn converted to product. Because the hydrogens at carbon 4 of the nicotinamide in NADH are not acidic, the existence of this rapid hydrogen exchange suggests that the hydrogen at carbon 4′ of urocanate is removed as a proton by a general base in the active site

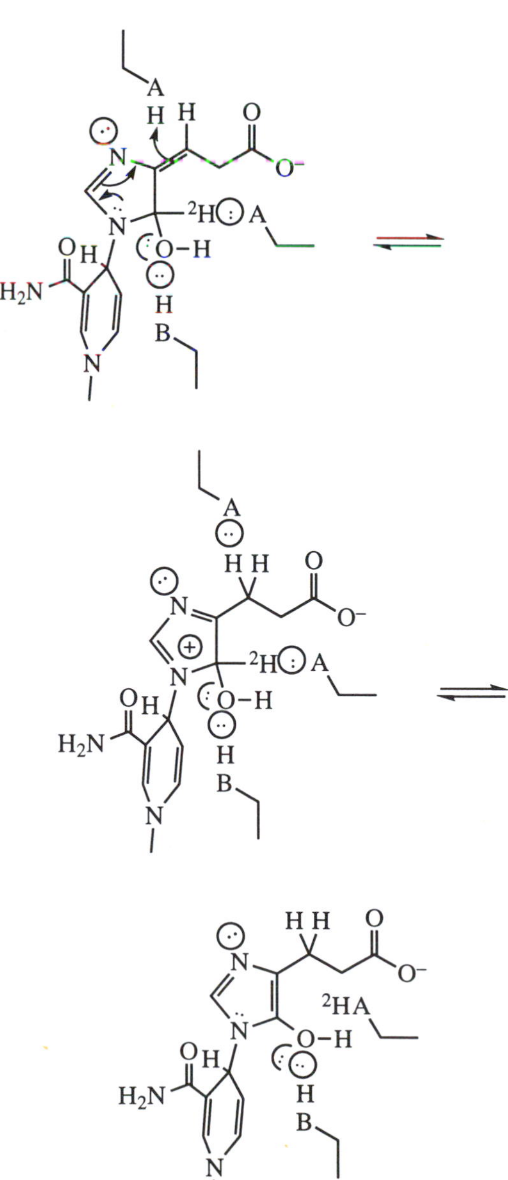

(5–128)

Glutamate dehydrogenase (NADP$^+$) catalyzes a reaction with three reactants and three products

$$\text{2-oxobutyrate} + NH_4^+ + \text{NADPH} \rightleftharpoons \text{glutamate} + H_2O + \text{NADP}^+ \quad (5\text{–}129)$$

The intermediate in the reaction is the iminium cation that in turn is reduced by the NADPH

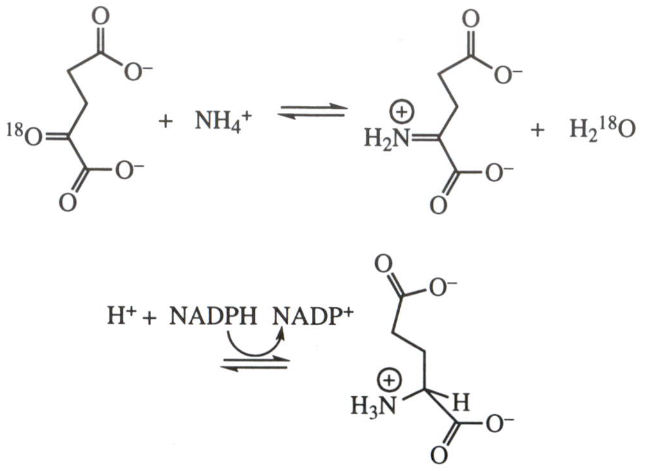

$$^{18}O{=}\text{2-oxoglutarate} + NH_4^+ \rightleftharpoons H_2N^+{=}\text{iminium} + H_2{}^{18}O$$

$$\xrightleftharpoons[]{H^+ + \text{NADPH} \rightarrow \text{NADP}^+} H_3N^+\text{–CH–glutamate}$$

(5–130)

No enzymatically catalyzed isotopic exchange, however, of the type

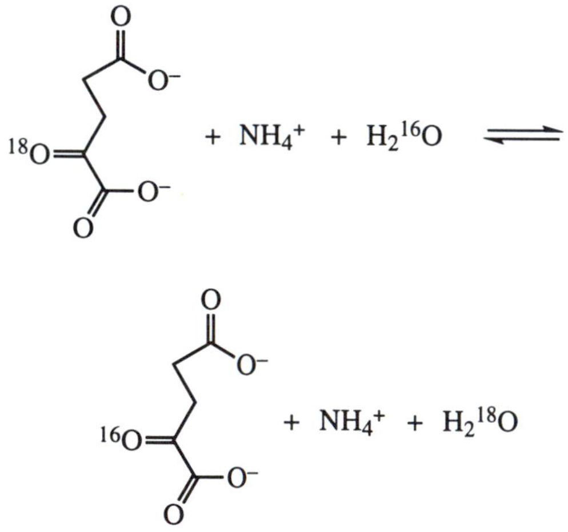

$$^{18}O{=}\text{2-oxoglutarate} + NH_4^+ + H_2{}^{16}O \rightleftharpoons {}^{16}O{=}\text{2-oxoglutarate} + NH_4^+ + H_2{}^{18}O$$

(5–131)

could be observed in the absence of NADPH.[137] The following strategy was designed to reveal this isotopic exchange. The enzyme (0.25 mM in active sites) was mixed with [2-$^{18}$O]-2-oxoglutarate (0.3 mM), NADPH (1.1 mM), and limiting amounts of ammonia (< 0.2 mM), so that when the reaction had reached equilibrium, most of the 2-oxoglutarate would still remain in solution. Under these conditions, a rapid increase of [2-$^{16}$O]-2-oxobutyrate at the expense of [2-$^{18}$O]-2-oxobutyrate (Reaction 5–131) was observed (Figure 5–12) as a result of the oxygen exchange resulting from the first step of Reaction 5–130.[137] The isotopic exchange occurred only during the approach to equilibrium of the enzymatic reaction. After equilibrium had been established, almost all of the ammonia had been converted to glutamate. Because the rate of the exchange decreased dramatically as the ammonia was depleted, it could be concluded that the exchange observed in less than 10 s was not due to the reversal of the complete reaction, because if this were the case it would have proceeded at the same rate after equilibrium had been reached. At equilibrium only the normal uncatalyzed isotopic exchange, resulting from hydration of the ketone

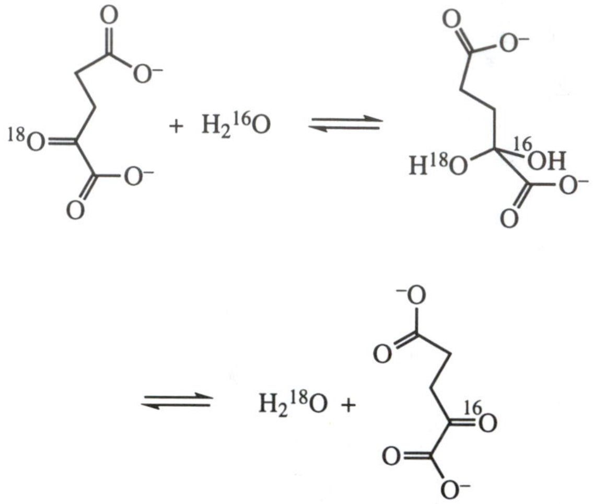

$$^{18}O{=}\text{2-oxoglutarate} + H_2{}^{16}O \rightleftharpoons H^{18}O\text{–C(}^{16}OH\text{) hydrate} \rightleftharpoons H_2{}^{18}O + {}^{16}O{=}\text{2-oxoglutarate}$$

(5–132)

persisted. Because this uncatalyzed reaction itself was so rapid ($t_{1/2}$ = 5 min), large concentrations of the enzyme (0.25 mM) were used to complete the approach to equilibrium immediately (<10 s) so that the almost instantaneous enzymatically catalyzed exchange could be clearly distinguished from the uncatalyzed exchange.

Methionine $\gamma$-lyase uses pyridoxal phosphate as a coenzyme to catalyze an elimination

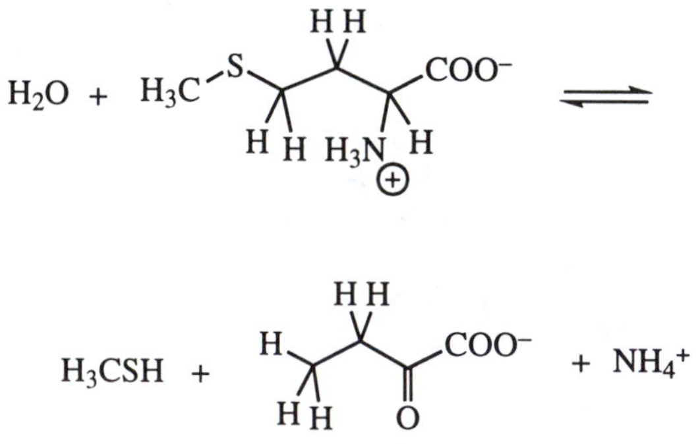

$$H_2O + H_3C\text{–S–CH}_2\text{–CH}_2\text{–CH(NH}_3^+\text{)–COO}^- \rightleftharpoons H_3CSH + H\text{–CH}_2\text{–CH}_2\text{–C(=O)–COO}^- + NH_4^+$$

(5–133)

The reaction is based on the ability of the pyridoxal phosphate to project a $\pi$ molecular orbital system over the carbon skeleton of the methionine (Reaction 2–17). The individual steps in the expansion of this $\pi$ molecular orbital system could be followed by isotopic exchange at initial velocity. When methionine was mixed with the enzyme in $^2H_2O$ and the $\alpha$-hydrogen and the $\beta$-hydrogens were monitored by nuclear magnetic resonance spectroscopy, the initial rate at which the $\alpha$-hydrogen was replaced by deuterium was more rapid than the initial rate at which one of the two $\beta$-hydrogens was replaced by deuterium, and the formation of 2-oxobutyrate was much slower than both isotopic exchanges (Figure 5–13).[138] This provides evidence for the sequential formation of the two intermediates

**Figure 5–12:** Rate of carbonyl oxygen exchange of 2-oxoglutarate catalyzed by glutamate dehydrogenase ($NADP^+$).[137] Solutions at pH 7.6 and 25 °C contained 0.25 mM glutamate dehydrogenase ($NADP^+$), 1.1 mM NADPH, and 6 $\mu$M (○), 55 $\mu$M (△), 118 $\mu$M (□), and 180 $\mu$M (▲) ammonia. As ammonia was always present at concentrations less than that of the enzyme, each enzyme molecule on the average could complete less than one turnover before equilibrium was reached. The reaction was initiated by adding [2-$^{18}$O]-2-oxoglutarate (58% isotopic enrichment) to a final concentration of 0.31 mM. Samples of the reaction mixture were withdrawn and quenched with an equal volume of 50% $H_2O_2$. The $H_2O_2$ converted the [2-$^{18}$O]-2-oxoglutarate to [$^{18}$O]succinate by oxidative decarboxylation, and the $^{18}$oxygen content of the succinate was determined by mass spectroscopy. This value was used to calculate the percent isotopic enrichment in the carbonyl oxygen of the [2-$^{18}$O]-2-oxoglutarate (percent $^{18}$O content) as a function of time (seconds). For each set of data, the solid line extending from 160 to 10 s represents a single first-order curve fit to the loss of $^{18}$O due to uncatalyzed exchange. These lines are extrapolated to zero time by dotted lines to indicate the level as $^{18}$O in the 2-oxoglutarate if the enzymatically catalyzed reaction had reached equilibrium instantaneously following the addition of reactant. The difference between these values and the initial value of 58% isotopic enrichment is the amount of $^{18}$oxygen exchange that occurred before the ammonia had been depleted by the enzymatic reaction. The uncatalyzed $^{18}$O exchange from [2-$^{18}$O]-2-oxoglutarate in the absence of enzyme, ammonia, and NADPH is also plotted (●). Reprinted with permission from ref 137. Copyright 1984 National Academy of Sciences.

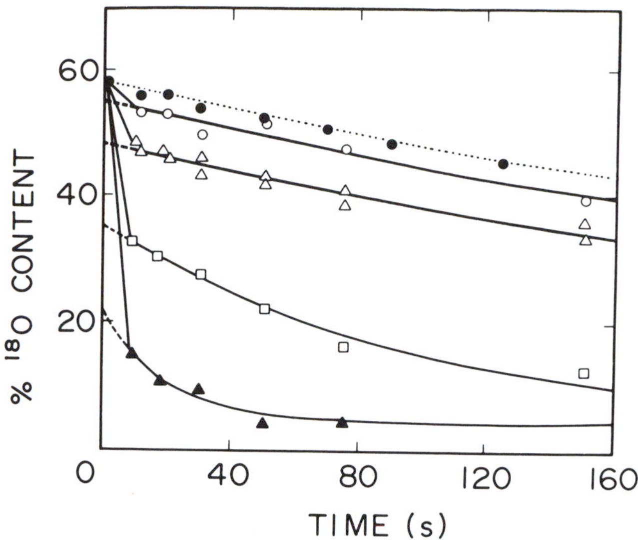


**Figure 5–13:** Isotopic exchange of the hydrogens on the $\alpha$- and $\beta$-carbons of L-methionine, as well as the loss of the methanethiol during the overall reaction, in the enzymatic reaction catalyzed by methionine $\gamma$-lyase.[138] A solution (0.5 mL) was prepared at p$^2$H 8.6 and 28 °C containing 0.10 M L-methionine in $^2H_2O$ (99.8% isotopic enrichment). The reaction was initiated by adding 34 $\mu$g of methionine $\gamma$-lyase, and the solution was placed in a nuclear magnetic resonance spectrometer. At the indicated intervals, spectra were recorded. The intensities ($I$) of the absorptions from the $\alpha$-hydrogen (curve 5), the $\beta$-hydrogens (curve 4), the $\gamma$-hydrogens (curve 3), and the $S$-methyl hydrogens (curve 2) were determined by integration of the respective peaks on each spectrum. The logarithms of the intensities [log $I(t)$] are presented (●) as a function of time (minutes). The two $\beta$-hydrogens exchange at different rates, and this heterogeneity accounts for the nonlinear behavior of their absorption. At various intervals, samples were withdrawn and quenched, and the appearance of 2-oxobutyrate was followed by reacting the ketone with 3-methyl-2-benzothiozolone. The disappearance of L-methionine as a result of the enzymatic reaction was assumed to be equal to the difference between the concentration of 2-oxobutyrate at any time [$C(t)$] and its concentration at infinite time ($C_\infty$). The logarithm of this difference (○) is plotted as a function of time (minutes). Reprinted with permission from ref 138. Copyright 1985 American Chemical Society.

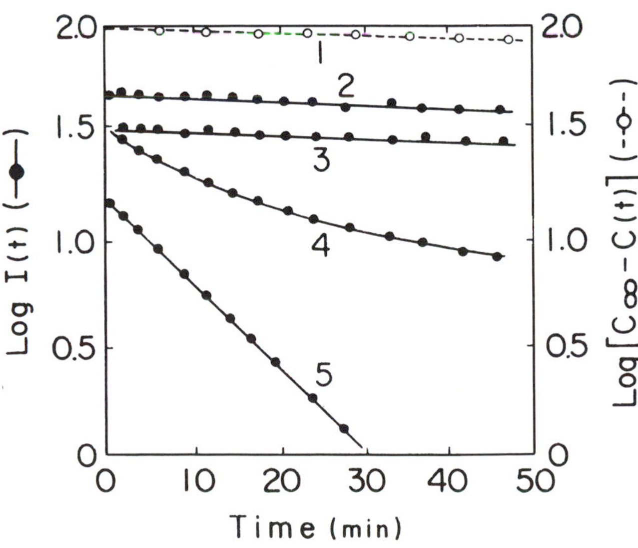

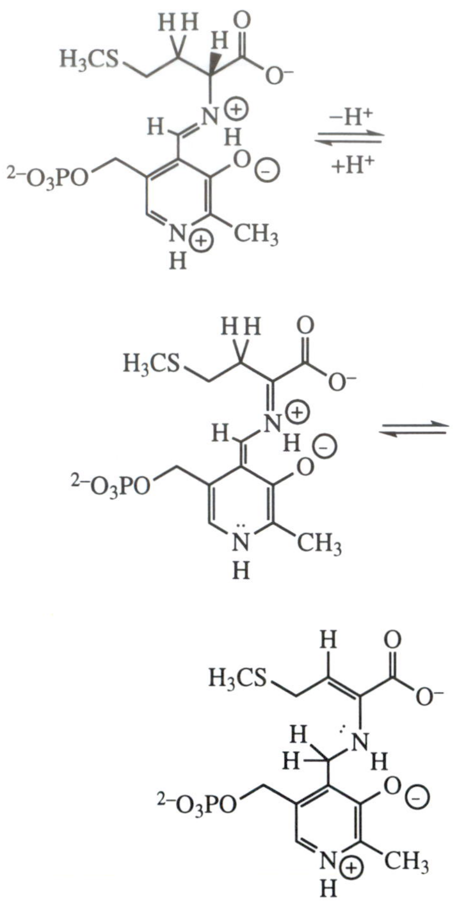

(5–134)

in the enzymatic reaction.

The three types of isotopic exchange discussed so far, isotopic exchange in the absence of a reactant, equilibrium isotopic exchange, and isotopic exchange at initial velocity can be distinguished from each other by examining a general kinetic mechanism

$$E_1 + A \underset{k_{-1}'[P]}{\overset{k_1'[A]}{\rightleftharpoons}} E_i + P \underset{k_{-i}'[Q]}{\overset{k_i'}{\rightleftharpoons}} E_n + Q \qquad (5\text{–}135)$$

The term $k_1'[A]$ is the forward composite rate constant (Equation 4–110) for all of the steps between the form of the enzyme $E_1$ and the form of the enzyme $E_i$. The form $E_1$ is the form that binds the reactant, reactant A, at which the isotopic exchange occurs. The form $E_i$ is the form from which the molecule of product P, carrying the original isotope away from the active site, has just dissociated. The term $k_{-1}'[P]$ is the backward composite rate constant (Equation 4–111) for all of the steps between the form of the enzyme $E_i$ and the form of the enzyme $E_1$. The form $E_i$ is the form to which the molecule of product P carrying the other isotope binds, and the form $E_1$ is the form from which the isotopically exchanged reactant A has just dissociated. In the case of fructose-bisphosphate aldolase, product P is the proton that is removed from carbon 1 of dihydroxyacetone phosphate (Figure 5–9) and that is replaced by a triton, and reactant A is the dihydroxyacetone phosphate itself. In the case of enolase, product P is the $H_2{}^{16}O$, the deuteron, or the triton, depending on the measurement (Figure 5–10), that has been removed from the 2-phosphoglycerate and that is replaced by $H_2{}^{18}O$, the proton, or the proton, respectively, and reactant A is the 2-phosphoglycerate itself. In the case of urocanate hydratase, product P is the proton that is removed from carbon 4′ of the urocanate (Reaction 5–26) and that is replaced by a deuteron, and reactant A is the urocanate itself.

The term $k_i$ in Equation 5–135 is the forward composite rate constant (Equation 4–112) for all steps between the form of the enzyme $E_i$ and the form of the enzyme $E_n$ from which product Q, which defines the last step in the enzymatic reaction, has just dissociated. The term $k_{-i}'[Q]$ is the backward composite rate constant for all of the steps in the mechanism between the form of the enzyme $E_n$ to which the product Q binds and the form of the enzyme $E_i$. In the case of enolase, product Q was the phospho*enol*pyruvate because its dissociation from the enzyme was the ultimate step measured in the experiment. In the case of urocanate hydratase, product Q was 5-carboxyethyl-4,5-dihydro-4-oxoimidazole because its dissociation from the enzyme was kinetically irreversible under conditions of initial velocity and therefore terminated the reaction.

Isotopic exchange in the absence of one or more of the reactants provides evidence for the formation of an intermediate bound to the active site in the form of the enzyme $E_i$. For this exchange to occur, this intermediate must be produced at the active site before the omitted reactant or reactants are required to associate with the active site. It is the omission of the reactant or reactants that prevents the reaction from proceeding beyond $E_i$ because $k_i$ is 0 in the absence of the omitted reactant. In the experiments with fructose-bisphosphate aldolase, for example, the form of the enzyme $E_i$ is the enamine covalently bound to the active site. The experiment was designed to render $k_i$ equal to 0 because no glyceraldehyde 3-phosphate, the electrophile whose presence is necessary to proceed with the next step in the reaction, was added. Therefore the isotopic exchange did not have to compete for enamine $E_i$ (Mechanism 5–135) with the later steps in the normal enzymatic reaction.

Isotopic exchange at initial velocity is observed if the composite rate from $E_i$ backward to free reactant A, governed by $k_{-1}'[P]$, is comparable in magnitude to the composite rate from $E_i$ forward to free product Q, governed by $k_i'$. In the case of urocanate hydratase, the form of the enzyme $E_i$ is the form from which the proton, which had been removed from carbon 4 of the imidazole, has just been released into solution. The composite rate backward from $E_i$, involving the deuteronation of carbon 4 followed by the release of the [4-$^2$H]urocanate, is by chance faster than the composite rate forward from $E_i$ to $E_n$, which is responsible for release of the 5-carboxyethyl-4,5-dihydro-4-oxoimidazole. As a result, deuterium appears in the urocanate at a rate in excess of the rate at which urocanate disappears (Figure 5–11). The partition ratio $k_{-1}'[P](k_i')^{-1}$ determines the ratio between the initial rate of isotopic exchange and the initial rate of the overall enzymatic reaction because all steps preceding the formation of $E_i$, which is governed only by $k_{-1}[A]$, are identical in the two processes. Because the reaction is proceeding at initial velocity and no product Q has been added, the rate determined by $k_{-i}'[Q]$, at least initially, is equal to 0.

Equilibrium isotopic exchange is designed so that all steps in the overall reaction are operating simultaneously. Each of the one or more equilibrium isotopic exchanges measured results from a segment of the complete pathway. The segment responsible for a particular isotopic exchange covers the steps from the point at which the reactant undergoing exchange, reactant A, enters the active site to the step at which the atom undergoing isotopic substitution leaves the active site on the respective product, product P. If the segment responsible for one isotopic exchange overlaps all of the steps in the segment responsible for another isotopic exchange but has additional steps as well, the first isotopic exchange must have a rate that is less than or equal to the rate of the second. For example, if all steps in Mechanism 5–135 are operating continuously from the first instant, the rate of any isotopic exchange arising from the dissociation and reassociation of product P must be greater than or equal to the rate of any isotopic exchange arising from the dissociation and reassociation of product Q because there are additional steps between the release of product P and the release of product Q. If the rate constant $k_i'$ is much greater than the rate constant $k_{-1}'[\mathrm{P}]$, both products will effectively dissociate simultaneously, and no difference in the rates of isotopic exchange will be observed. Because both hydrogen exchange and oxygen exchange in the reaction catalyzed by enolase start with the association of 2-phosphoglycerate, the fact that the rate of hydrogen exchange is greater than the rate of oxygen exchange (Table 5–2) requires that the proton leave the active site in a step preceding the step at which the water leaves the active site.

From these considerations, it is clear that equilibrium isotopic exchange can be used to measure the sequence in which reactants bind and products are released. Even though MgADP must be formed at the active site of glutamate–ammonia ligase before glutamine is formed, the fact that glutamate–glutamine exchange (Reaction 5–126) and ammonia–glutamine exchange (Reaction 5–127) are more rapid than MgATP–MgADP exchange (Reaction 5–122) states that either MgATP is slower to dissociate from the enzyme than glutamate and ammonia in the reverse direction or MgADP is slower to dissociate than glutamine in the forward direction or both of these relationships hold.

A distinction, however, can be made by isotopic exchange at equilibrium between a difference in the relative rates of dissociation from and association with the active site and an absolute obligation that one reactant associate or one product dissociate before another can.[139] Suppose that reactant A must associate with the active site before the active site is capable of associating with reactant B and that product P must dissociate from the active site before product Q can gain access to the solvent and dissociate

$$\mathrm{E} \underset{k_{-1}}{\overset{k_1[\mathrm{A}]}{\rightleftharpoons}} \mathrm{E\cdot A} \underset{k_{-2}}{\overset{k_2[\mathrm{B}]}{\rightleftharpoons}} \mathrm{E\cdot A\cdot B} \underset{k_{-3}}{\overset{k_3}{\rightleftharpoons}} \mathrm{E\cdot P\cdot Q} \underset{k_{-4}[\mathrm{P}]}{\overset{k_4}{\rightleftharpoons}} \mathrm{E\cdot Q} \underset{k_{-5}[\mathrm{Q}]}{\overset{k_5}{\rightleftharpoons}} \mathrm{E} \qquad (5\text{–}136)$$

and suppose that one equilibrium isotopic exchange can be measured between reactant B and product P, the proximal pair, and another equilibrium isotopic exchange can be measured between reactant A and product Q, the distal pair. When the absolute concentrations of all reactants and products are raised in concert, the distal equilibrium isotopic exchange, A ⇌ Q will increase, reach a maximum, and decrease to zero while the proximal equilibrium isotopic exchange, B ⇌ P, will increase and approach a maximum asymptotically. If, however, there is no obligate order and reactants can associate with the vacant enzyme and products can dissociate from the ternary complex, E·P·Q, at random

$$\mathrm{E} \underset{k_{-1}}{\overset{k_1[\mathrm{A}]}{\rightleftharpoons}} \mathrm{E\cdot A} \underset{k_{-4}}{\overset{k_4[\mathrm{B}]}{\rightleftharpoons}} \mathrm{E\cdot A\cdot B} \qquad \mathrm{E} \underset{k_{-2}}{\overset{k_2[\mathrm{B}]}{\rightleftharpoons}} \mathrm{E\cdot B} \underset{k_{-3}}{\overset{k_3[\mathrm{A}]}{\rightleftharpoons}} \mathrm{E\cdot A\cdot B}$$

$$\mathrm{E\cdot A\cdot B} \underset{k_{-5}}{\overset{k_5}{\rightleftharpoons}} \mathrm{E\cdot P\cdot Q}$$

$$\mathrm{E\cdot P\cdot Q} \underset{k_{-6}[\mathrm{Q}]}{\overset{k_6}{\rightleftharpoons}} \mathrm{E\cdot P} \underset{k_{-9}[\mathrm{P}]}{\overset{k_9}{\rightleftharpoons}} \mathrm{E} \qquad \mathrm{E\cdot P\cdot Q} \underset{k_{-7}[\mathrm{P}]}{\overset{k_7}{\rightleftharpoons}} \mathrm{E\cdot Q} \underset{k_{-8}[\mathrm{Q}]}{\overset{k_8}{\rightleftharpoons}} \mathrm{E} \qquad (5\text{–}137)$$

all of the isotopic exchanges will increase and approach a maximum asymptotically.

Because glutamate–glutamine exchange (Reaction 5–126) is catalyzed by glutamate–ammonia ligase more rapidly than MgATP–MgADP exchange (Reaction 5–122), it was possible that MgATP was required to associate with the active site before glutamate was able to and that glutamine was required to leave before MgADP was able to. Both equilibrium isotopic exchanges, however, increased and approached their respective maxima asymptotically when the concentrations of reactants and products were increased in concert.[136,140] It follows that either reactant may associate or either product may dissociate first but that the microscopic rate constants for the associations or dissociations of the nucleotides are slower than those for the amino acids, as would be expected from their respective sizes.

Similar results were observed with the equilibrium isotopic exchanges catalyzed by fructokinase.[141] Although the equilibrium isotopic exchange

$$\mathrm{Mg[^{14}C]ATP + Mg[^{12}C]ADP \rightleftharpoons Mg[^{12}C]ATP + Mg[^{14}C]ADP} \qquad (5\text{–}138)$$

is 4 times more rapid at saturation than the equilibrium isotopic exchange

$$\mathrm{[^{14}C]fructose + [^{12}C]fructose\ 1\text{-}phosphate \rightleftharpoons [^{12}C]fructose + [^{14}C]fructose\ 1\text{-}phosphate} \qquad (5\text{–}139)$$

the rates of both equilibrium isotopic exchanges approach a maximum asymptotically as concentrations of reactants and products are increased. Therefore, either reactant can associate first or either product can dissociate first, but the rates of association or dissociation for the nucleotides are greater than those for the sugars.

If the mechanism for an isotopic exchange for the particular enzyme being studied can be abbreviated by the composite mechanism of Mechanism 5–135, an expression for the **rate of isotopic exchange at equilibrium** can be derived[139] for the process in which the form $\mathrm{E}_i$ is the intermediate produced immediately by the loss to the solution of the isotope originally in reactant A. The rate of any particular isotopic exchange at equilibrium ($rate_{ie}$) can be defined by the equation

$$rate_{ie} = \frac{d[^iX]}{dt} \qquad (5\text{–}140)$$

where $d[^iX]/dt$ is the initial rate at which one of the two isotopes (i) exchanging between a pair of molecules appears in the molecule X into which it is being exchanged. In Mechanism 5–135, the molecule X could be either A or P, depending on which of the two was being followed in the experiment. For example, the rate of isotopic exchange could be the initial rate at which deuterium appears in the water in the reaction catalyzed by enolase on phospho[2-$^2$H]glycerate (Figure 5–10). By defining the rate of isotopic exchange as an initial rate, the intention is to indicate that the rate in question is the rate of exchange in the absence of any of the products of that exchange, in other words, the rate in the absence of back exchange. For example, this would be the rate at which deuterium appears in the water from phospho[2-$^2$H]glycerate before enough $^2H_2O$ has accumulated that deuterium becomes reincorporated into the phosphoglycerate. Although it is defined as an initial rate, the rate of exchange can be determined experimentally (Figure 5–10) from the observed concentrations of the products of isotopic exchange by the integrated rate equation[135]

$$rate_{ie} = -\frac{[A][P]}{[A]+[P]}\left[\frac{\ln(1-f)}{t}\right] \qquad (5\text{–}141)$$

where [A] and [P] are the total molar concentrations of each of the two species exchanging the isotope and $f$ is the fraction of the isotopic exchange achieved at time $t$ relative to the isotopic exchange achieved at isotopic equilibrium. If Mechanism 5–135 is consistent with the actual mechanism of the enzymatic reaction in question, then the equation defining the rate of isotopic exchange at equilibrium in terms of the composite rate constants is[139]

$$rate_{ie} = \frac{k_i'[E]_{TOT}}{\left(1+\dfrac{k_i'}{k_{-1}'[P]}\right)\left(1+\dfrac{k_{-1}'[P]}{k_1'[A]}\right)} \qquad (5\text{–}142)$$

It should be emphasized that Equation 5–141 is used to determine the numerical value of a rate from a set of experimental observations, while Equation 5–142 explains the value of the rate constant in terms of a particular mechanism.

Because the rates of equilibrium isotopic exchanges are significantly influenced by the rates at which reactants and products associate with and dissociate from the active site, isotopic exchange in the absence of one or more reactant, if it occurs, is usually more informative about the existence of intermediates in the enzymatic reaction. In the absence of the appropriate transfer RNA but the presence of the appropriate amino acid, amino acid-tRNA ligases catalyze the isotopic exchange[142]

$$[^{32}P]HOP_2O_6^{4-} + Mg[^{31}P]ATP \rightleftharpoons [^{31}P]HOPO_6^{4-} + Mg[^{32}P]ATP \qquad (5\text{–}143)$$

and it was the existence of this isotopic exchange that led to the identification of the amino acid adenylates[143] that are the central intermediates in these enzymatic reactions (Reaction 3–188).

Pyruvate carboxylase catalyzes the carboxylation of pyruvate

$$MgATP + HCO_3^- + pyruvate \rightleftharpoons MgADP + HOPO_3^{2-} + oxaloacetate \qquad (5\text{–}144)$$

The central intermediate is *N*-carboxybiotin formed in a reaction involving activation of bicarbonate by phosphorylation (Figure 2–8). Consistent with this fact, the enzyme catalyzes the isotopic exchange

$$Mg[^{32}P]ADP + Mg[^{31}P]ATP \rightleftharpoons Mg[^{31}P]ADP + Mg[^{32}P]ATP \qquad (5\text{–}145)$$

that requires the addition of bicarbonate[144,145] but proceeds in the absence of pyruvate. This exchange is, however, eliminated by removal of biotin from the active site,[146] so it could not be determined whether it results from the reversal of the phosphorylation of bicarbonate, the more informative result, or if the reaction must proceed to *N*-carboxybiotin before the exchange that requires the dissociation of Mg[$^{31}$P]ADP from the active site can occur

$$E{\cdot}MgATP{\cdot}HCO_2^-{\cdot}biotin \rightleftharpoons E{\cdot}MgADP{\cdot}HOPO_3CO_2^{2-}{\cdot}biotin \rightleftharpoons E{\cdot}MgADP{\cdot}HOPO_3^{2-}{\cdot}N\text{-carboxybiotin} \qquad (5\text{–}146)$$

In the absence of $HCO_3^-$, MgATP, MgADP, and $HPO_3^{2-}$, the enzyme catalyzes the isotopic exchange

$$[3\text{-}^{14}C]pyruvate + [3\text{-}^{12}C]oxaloacetate \rightleftharpoons [3\text{-}^{12}C]pyruvate + [3\text{-}^{14}C]oxaloacetate \qquad (5\text{–}147)$$

that results from reversible formation of the *N*-carboxybiotin

$$E{\cdot}oxaloacetate{\cdot}biotin \rightleftharpoons E{\cdot}pyruvate{\cdot}N\text{-carboxybiotin} \qquad (5\text{–}148)$$

The design of these experiments was based on very similar isotopic exchanges that had been observed with propionyl-CoA carboxylase[147] and methylcrotonyl-CoA carboxylase.[148]

Pyruvate, orthophosphate dikinase catalyzes the double transphosphorylation

$$pyruvate + HOPO_3^{2-} + MgATP \rightleftharpoons phospho\textit{enol}pyruvate + MgO_3POPO_3^{2-} + AMP \qquad (5\text{–}149)$$

The enzyme catalyzes the isotopic exchange[149]

$$[^{14}C]AMP + Mg[^{12}C]ATP \rightleftharpoons [^{12}C]AMP + Mg[^{14}C]ATP \qquad (5\text{–}150)$$

in the absence of all four of the other substrates. It catalyzes the isotopic exchange

$$[^{32}P]HOPO_3^{2-} + Mg[^{31}P]O_3POPO_3^{2-} \rightleftharpoons [^{31}P]HOPO_3^{2-} + Mg[^{32}P]O_3POPO_3^{2-} \qquad (5\text{–}151)$$

in the presence of MgATP but the absence of the three other substrates. It catalyzes the isotopic exchange

$$[^{14}C]pyruvate + phospho[^{12}C]\textit{enol}pyruvate \rightleftharpoons [^{12}C]pyruvate + phospho[^{14}C]\textit{enol}pyruvate \qquad (5\text{–}152)$$

in the absence of all four other substrates. A mechanism[150] involving a pyrophosphoryl-enzyme intermediate[151]

$$E + MgATP \rightleftharpoons E\text{-}OPO_3PO_3^{3-} \cdot Mg^{2+} + AMP \qquad (5\text{–}153)$$

which first phosphorylates inorganic phosphate

$$E\text{-}OPO_3PO_3^{3-} \cdot Mg^{2+} + HOPO_3^{2-} \rightleftharpoons E\text{-}OPO_3^{2-} + HO_3POPO_3^{3-} \cdot Mg^{2+} \qquad (5\text{–}154)$$

and then *enol*pyruvate

$$E\text{-}OPO_3^{2-} + \text{pyruvate} \rightleftharpoons E + \text{phospho}\textit{enol}\text{pyruvate} \qquad (5\text{–}155)$$

is consistent with the observation of each of these isotopic exchanges in the absence of one or more of the reactants.

In the isotopic exchanges discussed so far, a molecule of the reactant is required to enter the active site, a molecule of the product that bears the atom to be isotopically exchanged—for example, a molecule of water or a proton—is required to leave the active site, another molecule of that product bearing the other isotope is required to enter the active site, and the isotopically exchanged molecule of the reactant is required to leave the active site. Often, the requirement that a product leave and reenter the active site cannot be fulfilled because the release of that product is slow or purposely prevented. For example, because glutamyl phosphate is an intermediate in the reaction catalyzed by glutamate–ammonia ligase (Reaction 5–57), and because it is formed in the absence of ammonia, it should have been possible to measure an isotopic exchange between MgATP and MgADP (Reaction 5–122) in the absence of ammonia. It has already been noted that this exchange cannot be detected.[136] One reasonable explanation for the absence of this exchange is that the product, MgADP, cannot dissociate from the active site until glutamine is formed from the ammonia and the glutamyl phosphate even though glutamyl phosphate can be produced in the absence of ammonia. Because, however, the bond between the $\beta$- and the $\gamma$-phosphate of the MgATP has to break during the formation of the glutamyl phosphate, positional isotopic exchange could be used to detect this step in the reaction.

**Positional isotopic exchange** is the exchange of one atom within a molecule with another atom of the same kind within the same molecule, but of a different isotope, in such a way that the product is chemically identical to the reactant but the two isotopes have been interchanged. The interchange of the isotopes within the same molecule shows that a reaction has occurred even though no chemical change has taken place. When MgATP containing $^{18}$oxygen between the $\beta$- and $\gamma$-phosphorus atoms was mixed with glutamate–ammonia ligase and glutamate, in the absence of ammonia, the $^{18}$O was found to exchange to the peripheral positions on the $\beta$-phosphate of the MgATP[152]

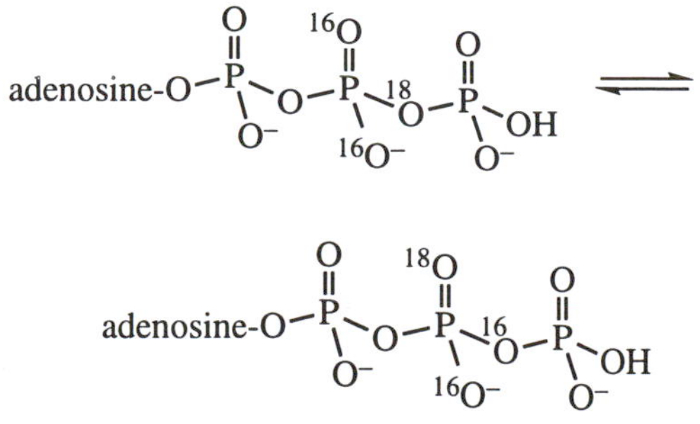

(5–156)

The explanation for this positional isotopic exchange is that the $\gamma$-phosphate is transferred to the oxygen of the glutamate to form glutamyl phosphate, the $\beta$-phosphate of the tightly bound MgADP rotates about the oxygen–phosphorus bond to interchange the $^{18}$oxygen and one of the two $^{16}$oxygens

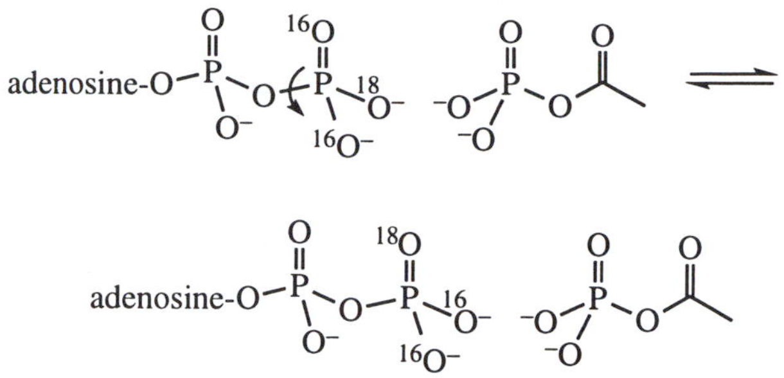

(5–157)

and the phosphate is transferred back from the glutamate to this $^{16}$oxygen of the rotationally isomerized MgADP. This isotopic exchange in the absence of ammonia can occur because the reaction cannot proceed beyond the complex of the active site with MgADP and glutamyl phosphate; when ammonia was added to permit the reaction to proceed forward, the rate of the positional isotopic exchange was diminished. In the absence of ammonia, the initial rate of the positional isotopic exchange was equal to the initial velocity of the complete enzymatic reaction in the presence of ammonia, and this demonstrated its kinetic competence. All of these results are consistent with the existence of glutamyl phosphate as an intermediate in the enzymatic reaction.

Guanosine-monophosphate synthase catalyzes the reaction

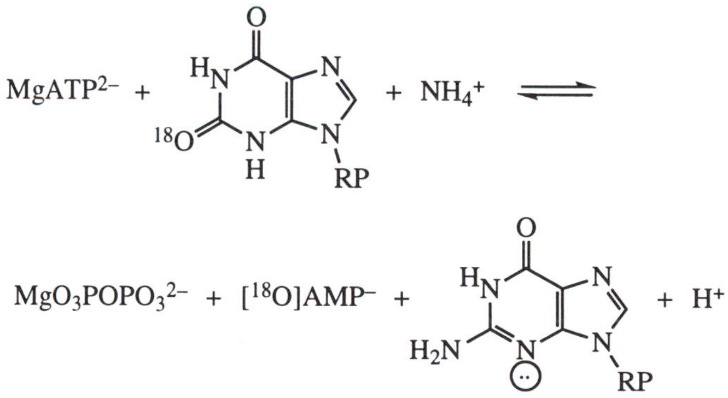

$$MgATP^{2-} + [2\text{-}^{18}O]\text{xanthosine 5'-monophosphate} + NH_4^+ \rightleftharpoons MgO_3POPO_3^{2-} + [^{18}O]AMP^- + \text{GMP} + H^+ \qquad (5\text{–}158)$$

where RP is ribosyl 5′-monophosphate. Formally, the reaction is an acyl exchange in which the nitrogen of the ammonia substitutes for oxygen 2 of the xanthosine 5′-monophosphate. When [2-$^{18}$O]xanthosine 5′-monophosphate was used as a reactant (Reaction 5–158), $^{18}$oxygen was found in the phosphate of the AMP formed during the reaction.[153] This led to the suggestion that xanthosine adenylate[154] is formed as an intermediate

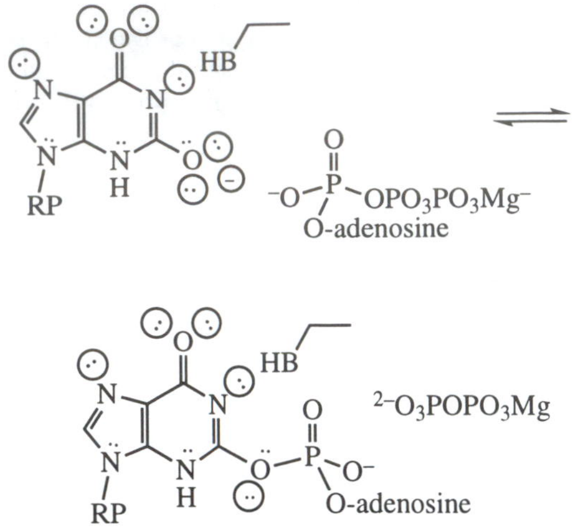

(5–159)

This would create an activated acyl derivative and in the process couple directly the hydrolysis of MgATP to the acyl exchange reaction. While the direct MgATP–pyrophosphate exchange

$$\mathrm{Mg}[\gamma\text{-}^{31}\mathrm{P}]\mathrm{ATP} + [^{32}\mathrm{P}]\mathrm{HOPO_3PO_3}^{3-} \rightleftharpoons \mathrm{Mg}[\gamma\text{-}^{32}\mathrm{P}]\mathrm{ATP} + [^{31}\mathrm{P}]\mathrm{HOPO_3PO_3}^{3-} \quad (5\text{–}160)$$

that might have accompanied this mechanism could not be detected in the absence of ammonia, it was possible to detect a positional isotopic exchange requiring xanthosine but not ammonia[155] in which pyrophosphate is not required to depart from the enzyme

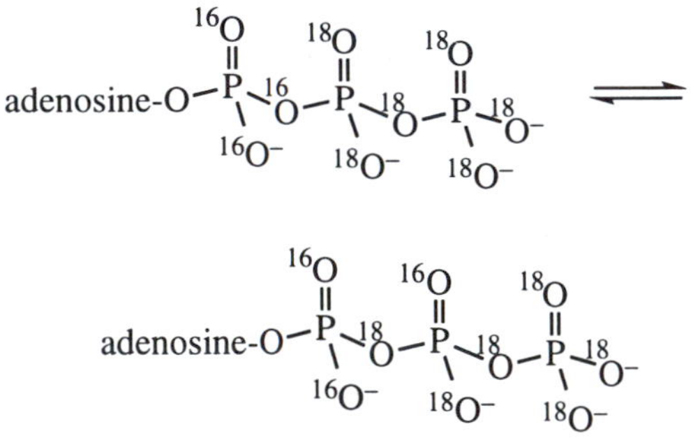

(5–161)

This positional isotopic exchange is consistent with the existence of xanthosine adenylate as an intermediate.

The exchange that occurs between substrates and isotopes of hydrogen or oxygen in the solvent water has two possible outcomes. **Medium isotopic exchange** is isotopic exchange in which a reactant binds, undergoes exchange with the solvent, and returns otherwise unchanged to the medium.[156] Medium exchange in this sense resembles all of the isotopic exchanges that have been discussed so far. Medium isotopic exchange can be distinguished from intermediate isotopic exchange.[157] **Intermediate isotopic exchange** is isotopic exchange in which a reactant enters the active site, isotopic exchange occurs with the solvent water, and the product leaves the active site bearing the exchanged isotope.

Intermediate isotopic exchange could be observed with citrate (*si*) synthase

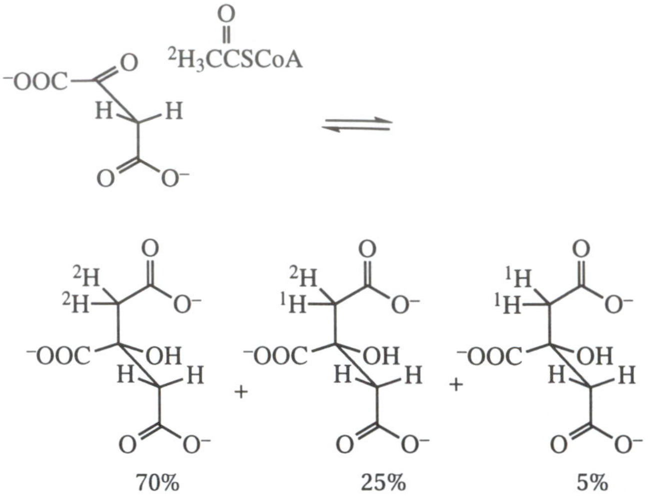

70% 25% 5%

(5–162)

When trideuteroacetyl-SCoA was used as a reactant in $^{1}H_2O$, 0.35 atom of protium was incorporated at carbon 2 of the [2-$^{2}$H]citrate produced.[158] When [$^{18}$O]acetyl-SCoA was used as a reactant in $H_2{}^{16}O$, 0.26 atom of $^{16}$O, beyond the one $^{16}$oxygen necessarily incorporated, was found in the carboxylate of citrate derived from the thioester of the intermediate citryl-SCoA

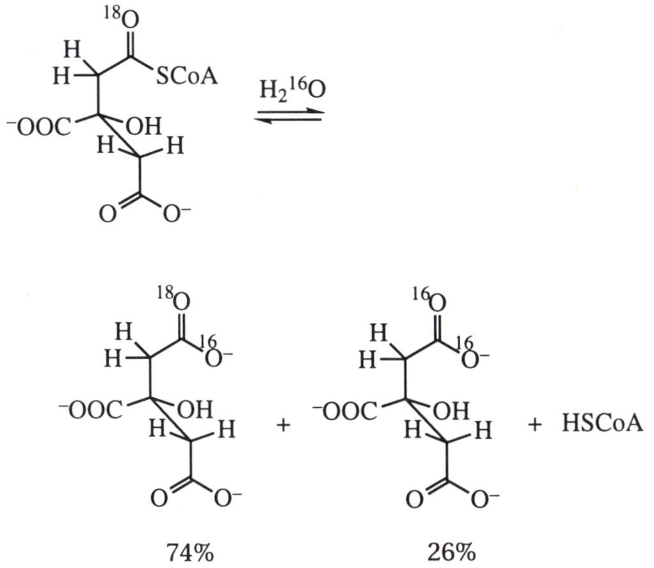

74% 26%

(5–163)

These two intermediate isotopic exchanges occur because the step in which the carbon–carbon bond is formed reverses after the deuteron on Histidine 274 in the active site (Figure 4–34) is exchanged for a proton but before citrate can be released as a product, and the step at which the citryl-SCoA is hydrolyzed reverses after the carboxylate flips over in the active site but before citrate can be released as a product, respectively. The magnitude of the isotopic exchanges can be used to calculate that about 0.7 reversal of each of these steps occurs prior to the release of citrate.

Myosin hydrolyzes MgATP to MgADP and $HOPO_3{}^{2-}$ to obtain the free energy necessary for contraction of muscle. On the active site of myosin the hydrolysis occurs without any intermediates by a direct in-line attack of water

$$\text{AMP-O-P(=O)(O^-)-}\ddot{\text{O}}\text{-P(=}^{16}\text{O)(}^{16}\text{O}^-\text{)-}^{16}\text{OH} + \text{H}^{18}\text{OH} \underset{+\text{H}^+}{\overset{-\text{H}^+}{\rightleftharpoons}} \text{AMP-O-P(=O)(O^-)-O}^- + \text{H}^{16}\text{O-P(=}^{16}\text{O)(}^{16}\text{O}^-\text{)-}^{18}\text{OH} \qquad (5\text{–}164)$$

with inversion of configuration at phosphorus.[159] When myosin, in the absence of actin, hydrolyzes MgATP to MgADP and $HOPO_3^{2-}$ in $[^{18}O]H_2O$, instead of containing only one atom of $^{18}$oxygen, as expected (Reaction 5–164), the $HOPO_3^{2-}$ contains almost four atoms of $^{18}$oxygen.[157,160] The reason for this observation is that the release of products from the active site of myosin is very slow in the absence of actin[161] and the hydrolysis (Reaction 5–164) reverses many times as the inorganic phosphate tumbles at the active site before the products depart. In this instance, it is one of the reactants, $H_2O$, rather than one of the products that enters and leaves the active site during these reversals carrying away, one at a time and at random,[162] the $^{16}$oxygens on phosphorus and replacing them with $^{18}$oxygens. The existence of this rapid reversal of a reaction that is significantly exergonic in solution is one demonstration that the standard free energy of the reaction has been dramatically altered by the environment of the active site.

As actin is added to the solution, the rate of the overall hydrolysis reaction increases dramatically and the magnitude of the intermediate isotopic exchange decreases[160] by a relationship predicted for random exchange of all three oxygens on the $\gamma$-phosphate of ATP, until at high concentrations of actin only one atom of $^{18}$oxygen is incorporated into the phosphate during the hydrolysis, and no intermediate isotopic exchange is observed (Figure 5–14).[163] This result is one illustration of the fact that the main effect of actin on the rate at which myosin hydrolyzes MgATP is to accelerate dramatically the rate at which the two products dissociate from the active site. At high enough concentrations of actin, the prod-

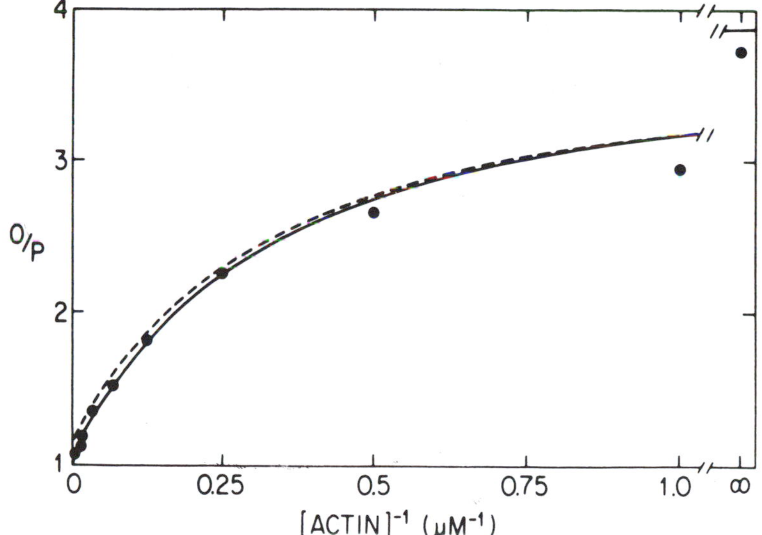


**Figure 5–14:** Effect of actin concentration on the intermediate exchange of $^{18}$oxygen for $^{16}$oxygen into the inorganic phosphate produced from $Mg[^{16}O]ATP$ in $[^{18}O]H_2O$ by subfragment 1 of myosin.[163] Myosin was digested with chymotrypsin, and the large globular fragment (subfragment 1) missing the coiled coil of the intact protein,[163a] but still containing the active site at which MgATP is hydrolyzed, was purified by anion-exchange chromatography. Solutions (3 mL) containing 1 mM $MgCl_2$, 0.5 mM MgATP, and 1.75 mM phospho*enol*pyruvate and 200 $\mu$g of pyruvate kinase (to regenerate the MgATP as it was hydrolyzed) were prepared at pH 7 and 23 °C. Various amounts of actin were added, and the hydrolysis of the MgATP was initiated in each case by adding enough subfragment 1 to produce 80% hydrolysis of the MgATP and phospho*enol*pyruvate over 2 h. Reaction mixtures were quenched at 2 h with perchloric acid, and the remaining nucleotide and protein were removed by adding activated charcoal and submitting the samples to centrifugation. The inorganic phosphate was isolated and the oxygens in the inorganic phosphate were converted to gaseous carbon dioxide by reaction with guanidinium chloride. The isotopic composition of the $[^{18}O]CO_2$ was determined by mass spectroscopy. From these measurements, the ratio of $^{18}$oxygen and phosphorus (O/P) in the inorganic phosphate produced during the enzymatic reaction could be calculated. This ratio is presented as a function of the inverse of the concentration of actin (micromolar$^{-1}$). Adapted with permission from ref 163. Copyright 1978 American Chemical Society.

ucts are released from the active site so rapidly that no reversal can occur before their dissociation. As these examples illustrate, intermediate isotopic exchange detects the reversibility of those intermediate steps in the enzymatic reaction that occur after the association of the reactant with the active site.

Such measurements of intermediate isotopic exchange can also rule out the existence of two pathways as an explanation for the decrease in incorporation of $^{18}$oxygen into phosphate upon acceleration of an ATPase reaction. $H^+$-Transporting ATP synthase also catalyzes the hydrolysis of MgATP. As the rate of the reaction decreases when the concentration of MgATP is lowered, the number of $^{18}$oxygens incorporated into the product $HOPO_3^{2-}$ increases, just as when the concentration of actin is decreased with actomyosin ATPase. One explanation for such an observation is that at high rates of hydrolysis the reaction proceeds through a pathway in which reversal of hydrolysis cannot occur, while at low rates of hydrolysis the reaction proceeds through a different pathway in which complete reversal and consequently complete exchange occurs at the hydrolytic step. By determining the relative concentrations of the isomeric forms of $HOPO_3^{2-}$ with one $^{18}$oxygen, two $^{18}$oxygens, three $^{18}$oxygens, and four $^{18}$oxygens at several different concentrations of MgATP, the possibility that two pathways were operating could be ruled out.[164]

Isotopic exchange between a hydrogen on a substrate and hydrogen on the water in the solvent is usually taken as evidence that the hydrogen is removed from the substrate as a proton. For example, when mandelate racemase catalyzes the epimerization of D-mandelate in $^2H_2O$

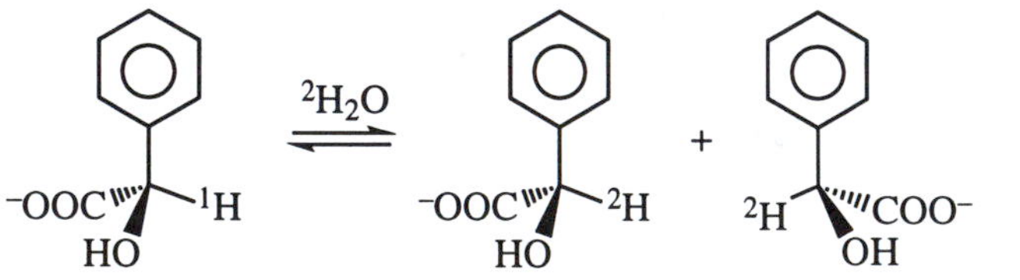

(5–165)

deuterium appears on the α-carbon of the mandelate.[165] This observation has been presented as evidence that the enzyme proceeds through an intermediate enol formed by direct proton removal from the α-carbon. The observation that the enzyme catalyzes the $E_{1CB}$ elimination of HBr from *p*-(bromomethyl)-mandelate is also consistent with the participation of the enol

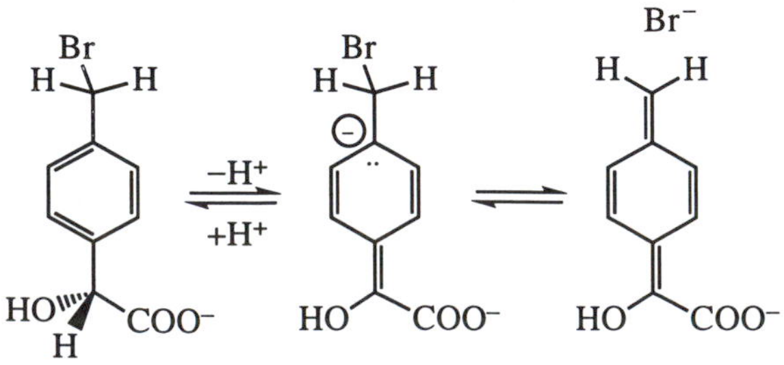

(5–166)

An enzyme in the cytoplasm of *Chlorella pyrenoidosa* is able to interconvert GDP-D-mannose and GDP-L-galactose, a reaction requiring a double epimerization

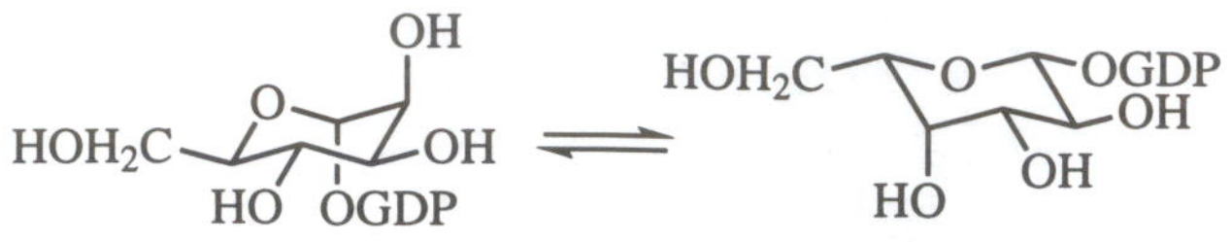

(5–167)

one at carbon 3 and one at carbon 5. When the reaction is carried out in $^3H_2O$, tritium is incorporated into both GDP-D-mannose and GDP-L-galactose at both carbon 3 and carbon 5.[166] The existence of this isotopic exchange suggests a mechanism in which a 4-keto intermediate is formed and the two adjacent epimerizations occur through the two respective enols

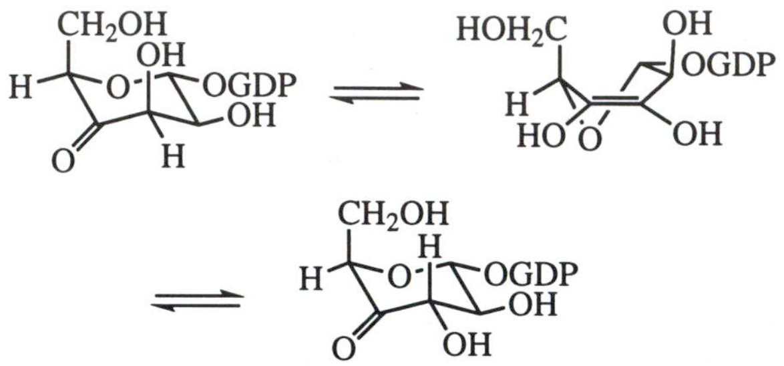

(5–168)

formed by removals of protons from the respective carbons α to the carbonyl.

The existence of hydrogen exchange with the solvent, however, does not necessarily demonstrate that the hydrogen was initially removed as a proton. During the reaction catalyzed by the flavoenzyme glutaryl-CoA dehydrogenase

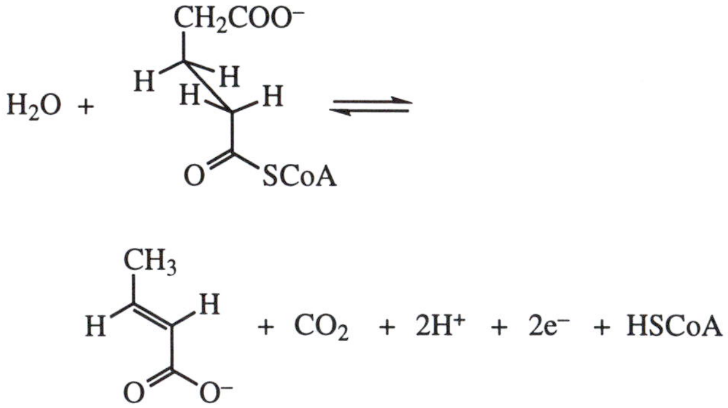

$$H_2O + {}^{-}OOCCH_2CH_2CH_2COSCoA \rightleftharpoons CH_3CH{=}CHCOO^- + CO_2 + 2H^+ + 2e^- + HSCoA$$

(5–169)

isotopic exchange of hydrogen at carbon 3 of the reactant is observed.[167] The first step in this reaction is the removal of a proton from carbon 2 of the reactant to produce the enol, which is then oxidized to the olefin by flavin

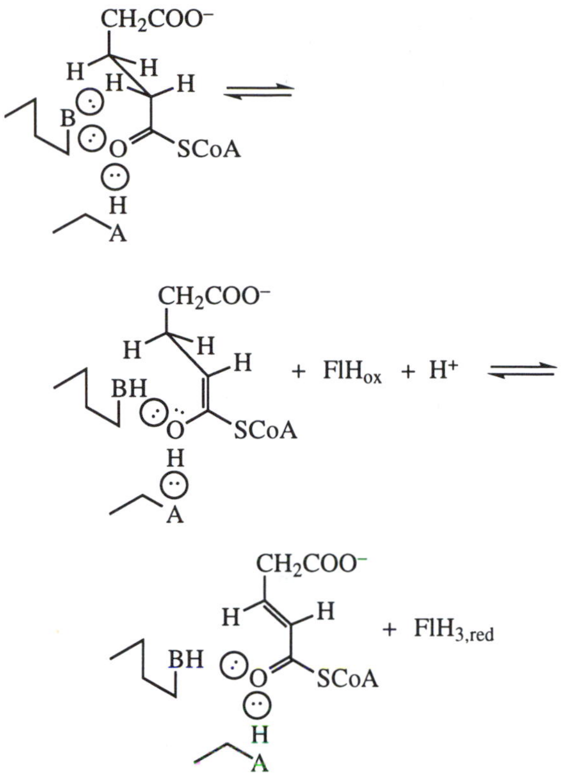

(5–170)

It is possible that, during this step, the hydrogen at carbon 3 could be transferred to the nitrogen of the oxidized flavin as a hydride ion and only then become acidic and susceptible to exchange.

**Intramolecular isotopic exchange** is sometimes observed in enzymatic reactions. In the reaction catalyzed by glutaryl-CoA dehydrogenase, the proton removed to generate the initial enol is the same proton added to carbon 4 after the decarboxylation[167]

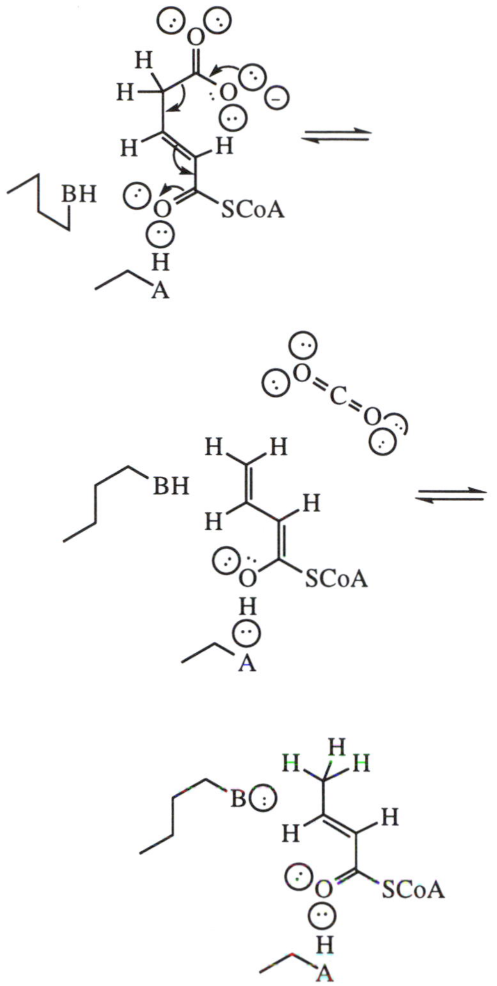

(5–171)

This suggests that the conjugate acid of the base that removes the proton in the first step is the acid providing the proton in the last step of the reaction.

Aspartate 4-decarboxylase catalyzes the reaction

$$\text{aspartate} \rightleftharpoons \text{alanine} + CO_2 \qquad (5\text{–}172)$$

Pyridoxal phosphate is the coenzyme catalyzing the reaction. Occasionally (0.05% of the time) the enzyme fails to complete the normal reaction and the following side reaction occurs

$$\text{aspartate} + \text{pyridoxal phosphate} \rightleftharpoons$$
$$CO_2 + \text{pyridoxamine phosphate} + \text{pyruvate} \qquad (5\text{–}173)$$

The pyridoxamine phosphate formed at the active site is replaced by pyridoxal phosphate from the solution so that in the abortive reaction the coenzyme becomes a substrate. When [2-$^3$H]aspartate is used as a reactant, a significant amount of tritium (equivalent to 17% intramolecular transfer)[168] is found at carbon 4′ of the pyridoxamine phosphate formed as a result of Reaction 5–173. The existence of this intramolecular isotopic exchange suggests that the same base that removes the $\alpha$-proton from the aspartate adduct protonates the 4′-carbon of the pyridoxal phosphate to form the external imine necessary to promote the decarboxylation

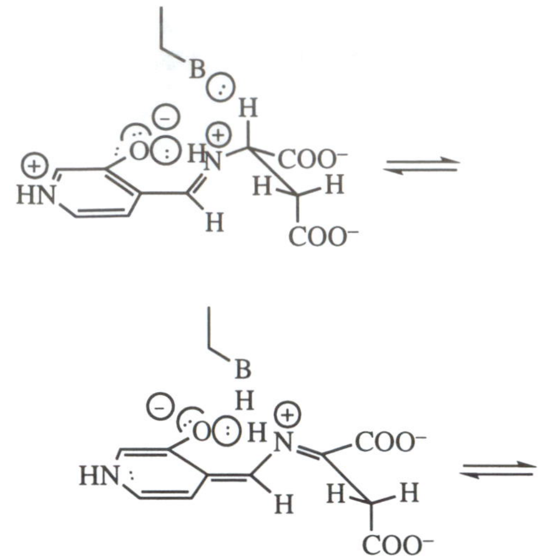

5–26

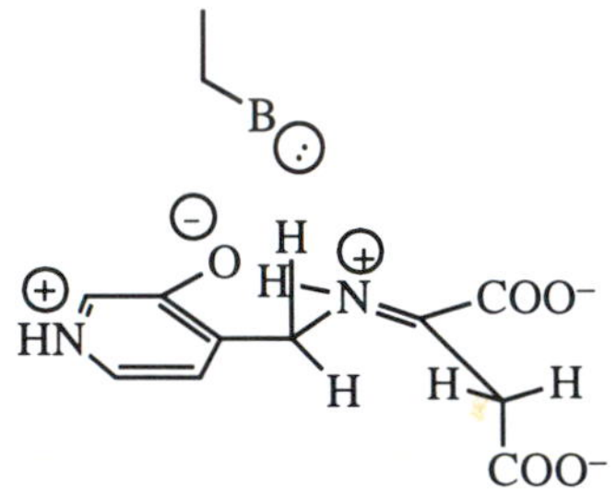

(5–174)

Following decarboxylation, the imine hydrolyzes occasionally before the proton can be transferred back to the $\alpha$-carbon of the amino acid

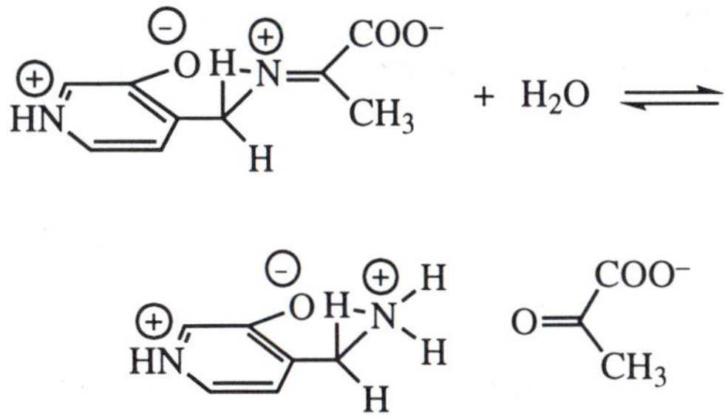

$$+ H_2O \rightleftharpoons$$

(5–175)

This could occur either on the active site or after the adduct has dissociated intact from the active site. The existence of the intramolecular isotopic exchange and the abortive reaction itself provide evidence that the external imine is an obligatory intermediate in the enzymatic reaction. This is reasonable because quininoid **5–26** should be too electron-rich to promote decarboxylation, a reaction that requires $\pi$ electron withdrawal. In other words, the carbanion that would be formed if quininoid **5–26** decarboxylated directly would have 10 $\pi$ molecular orbitals and 12 $\pi$ electrons, a circumstance that would require two $\pi$ electrons to occupy antibonding orbitals.

Isotopic exchange between a substrate of the enzymatic reaction and the enzyme itself has also been observed. The mechanism of 3-oxo acid CoA-transferase (Reaction 5–8) involves the formation of an anhydride between a glutamate at the active site and the carboxylic acids that are substrates for the enzyme (Reaction 5–8). When [$^{18}$O]succinate and acetoacetyl-SCoA were used as reactants in the enzymatic reaction, the isotopic exchange

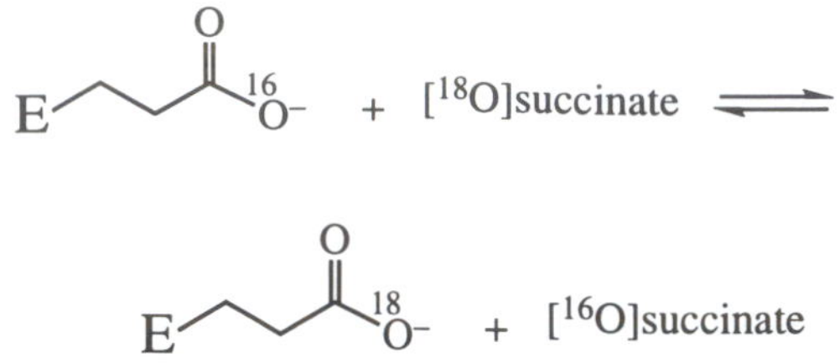

$$+ [^{18}O]\text{succinate} \rightleftharpoons$$

$$+ [^{16}O]\text{succinate} \quad (5\text{–}176)$$

could be detected.[169] This was consistent with the existence of the anhydride as an intermediate in the enzymatic reaction.

## *Suggested Reading*

Fisher, H.F., & Viswanathan, T.S. (1984) Carbonyl oxygen exchange evidence of imine formation in the glutamate dehydrogenase reaction and identification of the "occult role" of NADPH, *Proc. Natl. Acad. Sci. U.S.A. 81*, 2747–2751.

### PROBLEM 5–9

2-Amino-$\varepsilon$-caprolactam racemase is an enzyme that uses pyridoxal phosphate as a coenzyme to catalyze the racemization of the lactam of lysine

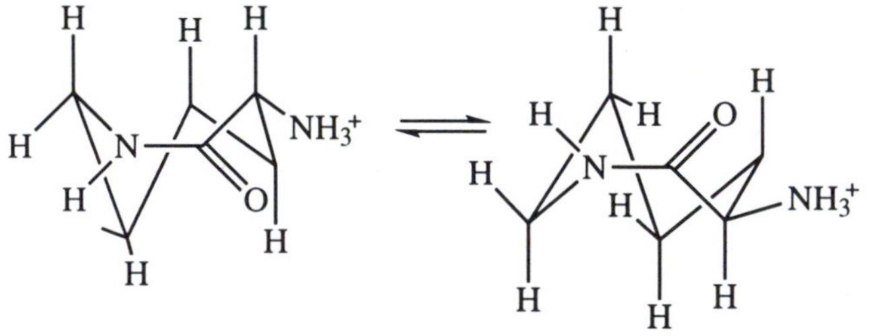

When D-[2-$^1$H]-2-amino-$\varepsilon$-caprolactam was mixed with the enzyme in $^2H_2O$, D-[2-$^2$H]-2-amino-$\varepsilon$-caprolactam was formed at an initial rate that was 80% the initial rate at which L-2-amino-$\varepsilon$-caprolactam was formed.[169a]

(A) Write a mechanism for the enzymatic reaction, incorporating the pyridoxal phosphate, that explains this isotopic exchange at initial velocity. Explicitly show the release of a product that is necessary to observe the exchange.

D-[2-$^1$H]-2-amino-$\varepsilon$-caprolactam was converted with 2-amino-$\varepsilon$-caprolactam racemase to L-2-amino-$\varepsilon$-caprolactam in $^2H_2O$ in the presence of excess L-2-amino-$\varepsilon$-caprolactam hydrolase so that the product of the first reaction was immediately hydrolyzed. A significant fraction (17%) of the lysine retained the $^1$H at carbon 2.

(B) What does this observation suggest about the general acid and general base catalyzing the racemization?

(C) There are two reasons why one particular amino acid is used as a general acid–base in this reaction. What is the amino acid and what are the two reasons? Use this amino acid in your answer to part A.

## PROBLEM 5–10

Ribulose-bisphosphate carboxylase catalyzes the sum of Reactions 5–44 and 5–45

ribulose 1,5-bisphosphate + $CO_2 \rightleftharpoons$
3-phosphoglycerate + 3-phosphoglycerate

The reaction was initiated by mixing enzyme with ribulose 1,5-bisphosphate and $CO_2$ in $^3H_2O$. As the reaction progressed, tritium appeared at carbon 3 of the ribulose 1,5-bisphosphate remaining in the solution.[169b]

| extent of reaction (%) | specific radioactivity of the remaining ribulose bisphosphate (cpm mmol$^{-1}$ × $10^{-5}$) |
|---|---|
| 21 | 0.9 |
| 38 | 1.9 |
| 56 | 3.3 |
| 70 | 5.2 |
| 82 | 7.4 |
| 86 | 8.4 |
| 98 | 15.4 |

(A) Write a mechanism demonstrating why this hydrogen exchange at initial velocity is evidence for an enediol intermediate in the enzymatic reaction.

(B) The specific radioactivity of the water was $7.0 \times 10^{-5}$ cpm ($\mu$mol of hydrogen)$^{-1}$. Why does the specific radioactivity of the remaining ribulose bisphosphate exceed this value at 98% reaction?

## PROBLEM 5–11

$Na^+/K^+$-Transporting ATPase catalyzes the following reaction

$$3Na_i^+ + 2K_o^+ + MgATP \rightleftharpoons 3Na_o^+ + 2K_i^+ + MgADP + HOPO_3^{2-}$$

in which the subscripts i and o refer to the inside and the outside of a cell, respectively. This reaction will be referred to as overall turnover. Overall turnover is always measured in, and requires, the presence of all substrates: MgATP, $Na^+$, and $K^+$. In the cell, the cations are moved against their concentration gradients at the expense of the hydrolysis of ATP. When the cell is fragmented, the enzyme continues to move the cations but accomplishes nothing even though it still hydrolyzes the MgATP. Thus, it is named an adenosinetriphosphatase.

The enzyme catalyzes several isotopic exchanges in the absence of overall turnover. In the absence of $Na^+$ and MgATP, the enzyme catalyzes the following exchange reaction

$$H_2{}^{18}O + H^{16}OPO_3^{2-} \rightleftharpoons H^{18}OPO_3^{2-} + H_2{}^{16}O$$

This reaction occurs at a rate greater than that of the ATPase activity seen when $Na^+$ and MgATP are added. In the absence of $K^+$, the enzyme very slowly catalyzes the following exchange reaction

$$Mg[^{14}C]ADP + Mg[^{12}C]ATP \rightleftharpoons Mg[^{14}C]ATP + Mg[^{12}C]ADP$$

When the enzyme is treated with *N*-ethylmaleimide this ATP $\rightleftharpoons$ ADP exchange reaction increases greater than 100-fold and becomes 3 times faster than the overall turnover was before the treatment. The overall turnover, however, is eliminated by the treatment with the *N*-ethylmaleimide. *N*-Ethylmaleimide has no effect on the $P_i \rightleftharpoons H_2O$ exchange reaction observed in the absence of $Na^+$ and MgATP. When $Na^+$ and $K^+$ are both omitted, none of the exchange reactions occur.

(A) Write a kinetic mechanism for the overall reaction, involving an intermediate, that explains both the $P_i \rightleftharpoons H_2O$ exchange and the ATP $\rightleftharpoons$ ADP exchange.

(B) Write the steps in the overall mechanism that lead to $Na^+$-dependent ATP $\rightleftharpoons$ ADP exchange.

(C) Write the steps in the overall mechanism that lead to $K^+$-dependent $P_i \rightleftharpoons H_2O$ exchange.

(D) Considering the rates of the exchanges before the enzyme is treated with *N*-ethylmaleimide, decide whether the intermediate is a high-energy or a low-energy intermediate.

(E) Show that the effects of *N*-ethylmaleimide can be explained by adding to your mechanism a step which involves a conformational change between two forms of the intermediate and which is blocked by reaction with *N*-ethylmaleimide.

(F) Considering your mechanism and assuming that all biochemical phosphostransfers are in line, draw the two products, with proper stereochemistry, that result when the following substrate is hydrolyzed by the enzyme in $[^{17}O]H_2O$

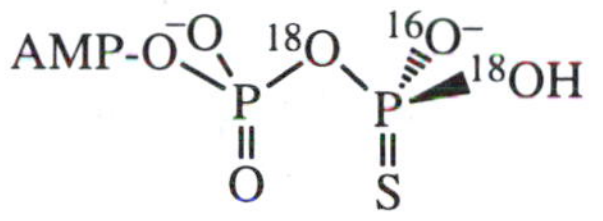

AMP-O–P(=O)(O⁻)–$^{18}$O–P(=S)($^{16}$O⁻)($^{18}$OH)

The $^{17}$oxygen ends up in one of the products.

## PROBLEM 5–12

Carbamyl-phosphate synthase (ammonia) catalyzes the reaction

$2MgATP + NH_4^+ + HCO_3^- \rightleftharpoons$
$2MgADP + 2HOPO_3^{2-}$ + carbamyl phosphate

In the presence of $HCO_3^-$ but the absence of ammonia, the enzyme catalyzes the isotopic exchange[169c]

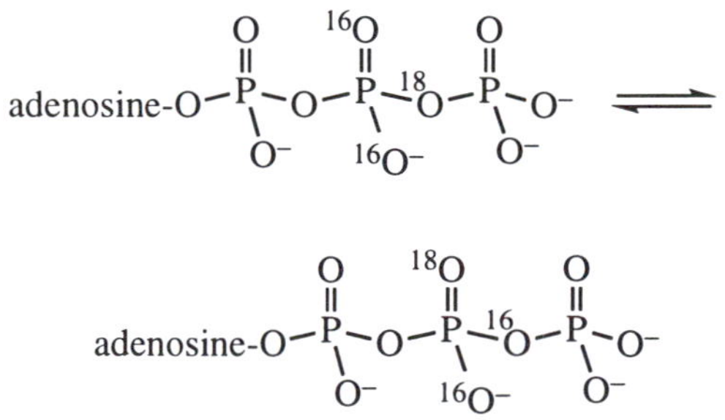

adenosine-O–P(=O)(O⁻)–O–P(=$^{16}$O)($^{16}$O⁻)–$^{18}$O–P(=O)(O⁻)–O⁻ $\rightleftharpoons$

adenosine-O–P(=O)(O⁻)–O–P(=$^{18}$O)($^{16}$O⁻)–$^{16}$O–P(=O)(O⁻)–O⁻

When the enzyme is mixed with MgATP and $HC^{18}O_3^-$ in the absence of ammonia, it catalyzes a slow ATPase activity in which 1 mol of $^{18}$oxygen appears in $HOPO_3^{2-}$ for every MgATP hydrolyzed.

(A) Write a mechanism for the overall enzymatic reaction containing an intermediate that can explain these observations.

(B) Why is the isotopic exchange inhibited by ammonia?

## PROBLEM 5–13

The enzyme 3-phosphoshikimate 1-carboxyvinyltransferase catalyzes the reaction

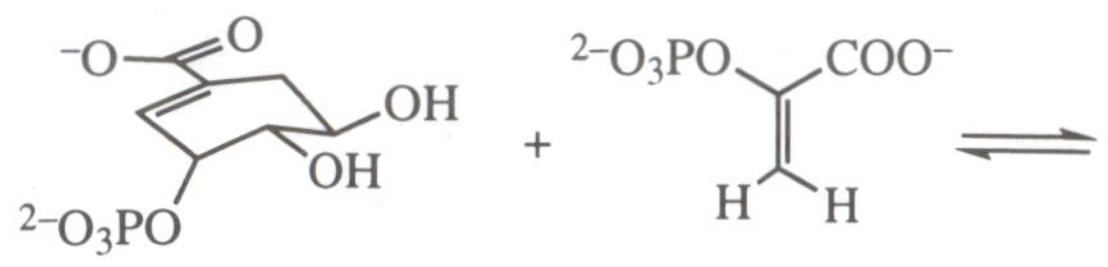

shikimate 3-phosphate (S3P) + phospho*enol*pyruvate (PEP) ⇌

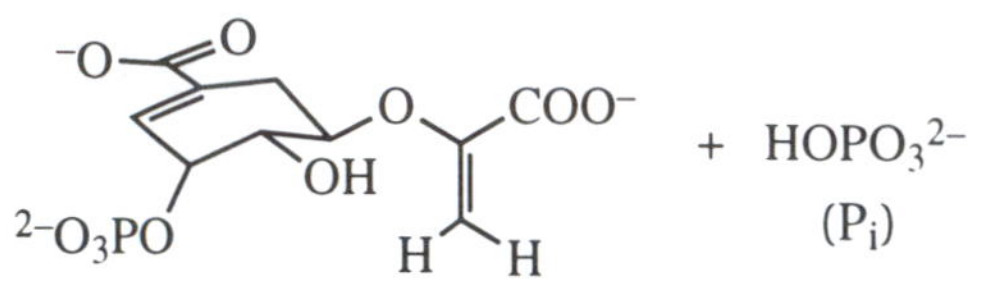

*enol*pyruvylshikimate 3-phosphate (EPSP) + $HOPO_3^{2-}$ ($P_i$)

From here on you should abbreviate shikimate 3-phosphate as ROH. The enzyme is present in plants and bacteria and has been purified from *Klebsiella pneumoniae*. Phospho*enol*pyruvate labeled with $^{18}O$ in the oxygen between the phosphorus and carbon 2 produces $[^{18}O]HOPO_3^{2-}$ and unlabeled *enol*pyruvylshikimate 3-phosphate.

(A) When phospho*enol*pyruvate labeled with deuterium at carbon 3 was used as a substrate in the enzymatic reaction and the reaction was allowed to proceed in $[^1H]H_2O$ only a short way (7–18%) toward equilibrium, the isotopic composition of the product *enol*pyruvylshikimate 3-phosphate and of the remaining phospho*enol*pyruvate were as follows[169d]

| pH | extent of reaction (%) | | initial PEP (%) | 5-*enol*pyruvyl shikimate 3-phosphate (%) | final PEP (%) |
|---|---|---|---|---|---|
| | | $d_0$ | <1 | 8.8 | <1 |
| 6.25 | 14 | $d_1$ | 7.6 | 46.8 | 6.8 |
| | | $d_2$ | 92.4 | 44.4 | 93.2 |
| | | $d_0$ | <1 | 6.4 | <1 |
| 6.25 | 13 | $d_1$ | 5.2 | 47.2 | 6.4 |
| | | $d_2$ | 94.8 | 46.3 | 93.2 |
| | | $d_0$ | <1 | 6.0 | <1 |
| 9.35 | 18 | $d_1$ | 7.6 | 41.6 | 9.6 |
| | | $d_2$ | 92.4 | 52.4 | 89.3 |
| | | $d_0$ | <1 | 5.8 | <1 |
| 10.4 | 7 | $d_1$ | 5.2 | 40.4 | 6.6 |
| | | $d_2$ | 94.8 | 53.7 | 93.4 |

In this table, $d_0$, $d_1$, and $d_2$ signify the percent of the molecules of phospho*enol*pyruvate, either as intially present or after the reaction was quenched, or *enol*pyruvylshikimate 3-phosphate produced during the reaction that contain no deuterium, one deuterium, or two deuteriums, respectively, as determined by mass spectroscopy. By nuclear magnetic resonance spectroscopy, it was shown that the $^1H$ introduced into the product was uniformly distributed between the *E* and *Z* positions. Write a mechanism to explain the appearance of $^1H$ in the product.

(B) Your mechanism should contain a step in which a hydrogen or deuterium is removed from carbon. For the sake of argument, assume every step in your mechanism is irreversible, and from the percentages of deuterium in the products when the reaction is run at pH 10.4, calculate the deuterium isotope effect for the step in the reaction during which hydrogen or deuterium is removed from carbon. Do not forget that intermediates with two deuteriums and one hydrogen are twice as likely to lose a deuterium as a hydrogen and intermediates with two hydrogens and one deuterium are twice as likely to lose a hydrogen as a deuterium even when no isotope effect is operating.

(C) 3-Phosphoshikimate 1-carboxyvinyltransferase was mixed with phospho*enol*pyruvate in $[^3H]H_2O$ and incubated under the noted conditions. The phospho*enol*pyruvate was isolated, and its radioactivity was determined.

| additions[a] | reaction time (h) | total cpm of $^3H$ in PEP ($\times 10^{-3}$) |
|---|---|---|
| PEP, 1 mM | 3 | 0 |
| PEP, 5 mM | 3 | <1.5 |
| PEP, 1 mM, and ddS3P, 20 mM | 3 | 76.4 |
| PEP, 5 mM, and ddS3P, 20 mM | 3 | 185 |
| PEP, 1 mM, and ddS3P, 20 mM | 1 | 28.7 |
| PEP, 1 mM, ddS3P, 20 mM, and glyphosate, 5 mM | 3 | 54.5 |

[a]The reaction contained in addition to the noted compounds the following: NaF, 10 mM, Tris-HCl, pH 7.6, 100 mM, in a total volume of 0.25 mL, $[^3H]H_2O$, $3 \times 10^{11}$ cpm $mL^{-1}$, and enzyme, 0.29 unit, all at 37 °C. Data from ref 169e.

In this experiment, ddS3P is 4,5-dideoxyshikimate 3-phosphate

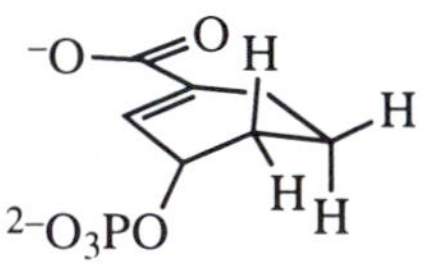

4,5-dideoxyshikimate 3-phosphate

Write a mechanism to explain the tritium incorporation into the phospho*enol*pyruvate in the presence of dideoxyshikimate 3-phosphate.

(D) The dideoxyshikimate 3-phosphate is acting by an induced-fit mechanism. Explain what this means in this situation.

(E) What damaging side reaction is the induced-fit mechanism of this enzyme designed to prevent?

Glyphosate, *N*-phosphonomethylglycine, $^{-}OOCCH_2{}^{+}NH_2CH_2PO_3{}^{2-}$, inhibits the transferase. When the kinetics of its inhibition were examined, the following results were obtained.[170]

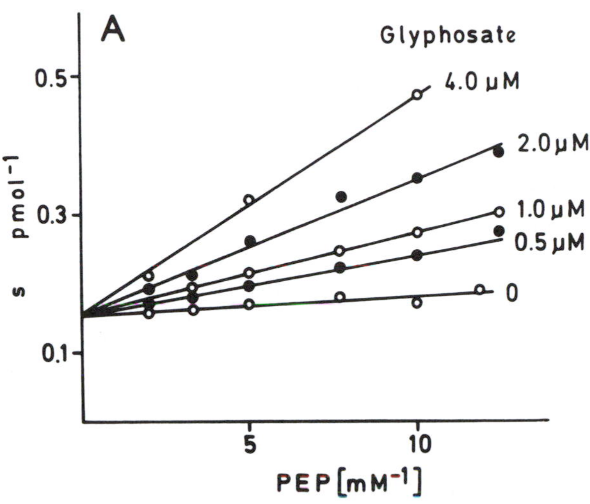


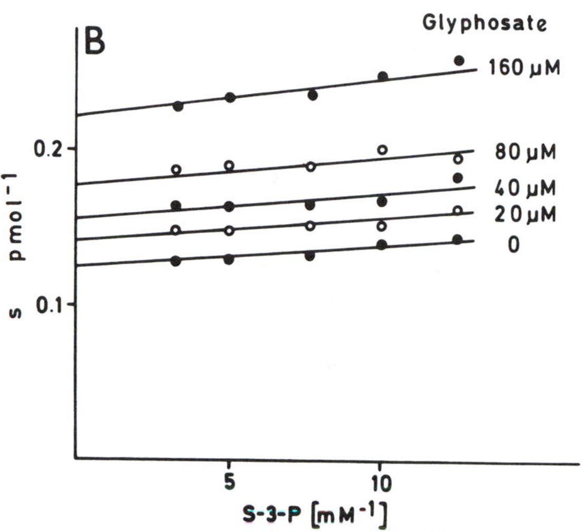


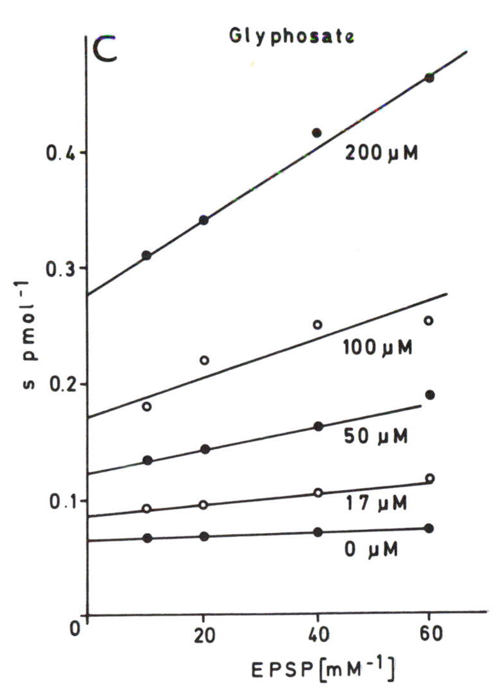


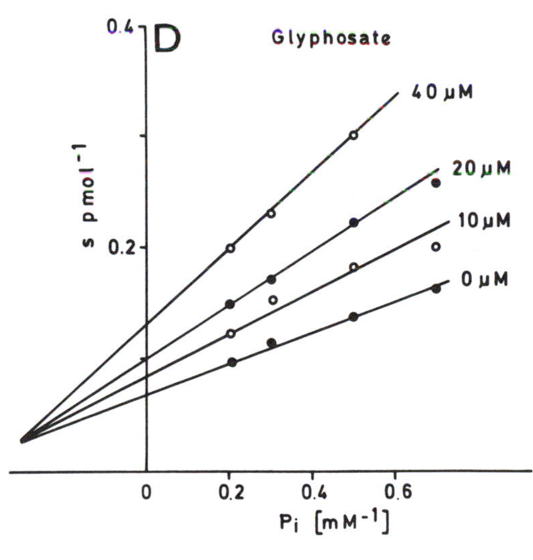


**Figure A:** Kinetic analysis of inhibition of 3-phosphoshikimate 1-carboxyvinyltransferase by glyphosate.[170] Double-reciprocal plots of initial velocities against variable reactant concentration (millimolar $^{-1}$) are given for (A) PEP ([S3P], 5 mM); (B) S3P ([PEP], 5 mM); (C) EPSP ([$P_i$], 20 mM); (D) $P_i$ ([EPSP], 0.2 mM). The unit s·pmol$^{-1}$ is the reciprocal of the initial velocity in picomoles second$^{-1}$. Reprinted with permission from ref 170. Copyright 1984 Springer-Verlag.

(F) Against which substrate is glyphosate a competitive inhibitor?

The inhibition caused by glyphosate was examined as a function of pH. Both substrates, phospho*enol*pyruvate and shikimate 3-phosphate, were held at constant concentrations of 5 mM throughout.[170]

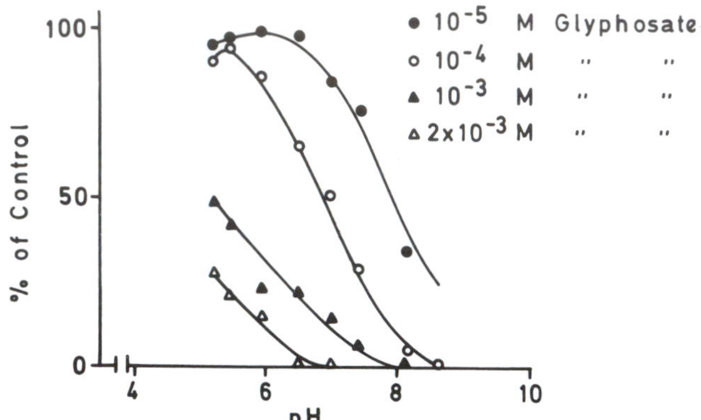


**Figure B:** Dependence inhibition of 3-phosphoshikimate 1-carboxyvinyltransferase by glyphosate on pH.[170] Initial velocities of the forward reaction were measured in 100 mM tris(hydroxymethyl)methylammonium maleate at different pH values in the absence and presence of glyphosate at the indicated concentrations. Activities are given as a percentage of the control (100% = activity at a given pH in the absence of glyphosate) as a function of pH. Reprinted with permission from ref 170. Copyright 1984 Springer-Verlag.

The parameter $I_{50}$ is defined as the concentration of inhibitor necessary to inhibit an enzyme by 50% under a set of fixed conditions. Complete this table for the experiments described in Figure B.

| pH | $I_{50}$ |
|---|---|
| | $2 \times 10^{-3}$ M glyphosate |
| | $1 \times 10^{-3}$ M glyphosate |
| | $1 \times 10^{-4}$ M glyphosate |
| | $1 \times 10^{-5}$ M glyphosate |

(H) To explain this behavior, assume the kinetic mechanism

$$\mathrm{I \cdot E} \xrightleftharpoons{K_i} \mathrm{I + E} \xrightleftharpoons{K_a} \mathrm{EH + A} \underset{k_{-1}}{\overset{k_1}{\rightleftharpoons}} \mathrm{EHA} \xrightarrow{k_2} \mathrm{E + P}$$

assume that A is the reactant with which the glyphosate, I, competes, and assume that the other reactant is saturating and does not enter the mechanism. Determine an equation for $v_0$ from this mechanism.

(I) When [I] = 0, $v_0 = v_{\text{uninhibited}}$, and when [I] = $I_{50}$, $v_0 = 0.5v_{\text{uninhibited}}$. Using the equation you derived in part H, solve for $I_{50}$ as a function of [H$^+$] and [A].

(J) Show that the behavior of $I_{50}$ with pH seen in part G can be explained by the equation you derived in part I.

(K) What intermediate in your mechanism for part C is glyphosate supposed to resemble?

(L) Why does glyphosate bind to a form of the enzyme with one less proton than the form to which phospho*enol*pyruvate binds? Answer this question by drawing two structures of the active site, one occupied by the intermediate and one by glyphosate, and include a general acid–base from the enzyme in your drawing.

## PROBLEM 5–14

Phospho*enol*pyruvate carboxylase catalyzes the reaction

$$\text{phospho}\textit{enol}\text{pyruvate} + HCO_3^- \rightleftharpoons \text{oxaloacetate} + HOPO_3^{2-}$$

When the reaction is carried out with [$^{18}$O]$HCO_3^-$, one atom of $^{18}$oxygen is incorporated into the $HOPO_3^{2-}$ formed. When the reaction is performed with chiral [$^{16}$O,$^{17}$O] thiol phospho*enol*pyruvate, inversion of configuration at phosphorus occurs. Phospho*enol*-$\alpha$-ketobutyrate does not substitute for phospho*enol*pyruvate in the normal reaction but is slowly hydrolyzed in a reaction catalyzed by the enzyme and requiring the presence of $HCO_3^{2-}$. When this bicarbonate-dependent hydrolysis is performed in the presence of [$^{18}$O]$HCO_3^{2-}$, $^{18}$oxygen is incorporated into the $HOPO_3^{2-}$ produced during the hydrolysis. The molar proportions of phosphate having one, two, or three atoms of $^{18}$oxygen were 70%, 25%, and 5%.[170a] Write a mechanism for the hydrolysis of phospho*enol*-$\alpha$-ketobutyrate that explains the observed intermediate isotopic exchange and incorporates the inversion of configuration at phosphorus observed in the normal reaction.

# Flow Methods

The turnover numbers of most enzymes between 20 and 37 °C lie in the range between 1 and $10^6$ $s^{-1}$. This means that once the reactants are assembled on the active site of the enzyme, each of the individual steps between the successive intermediates in the enzymatic reaction will usually have rate constants in excess of 1 $s^{-1}$ and each step will take less than a second to occur. If sufficiently high molar concentrations of active sites of the enzyme are used in an experiment, measurable quantities of the intermediates in the enzymatic reaction can be generated. Because, however, the steps of the reaction will be so fast, the device used to observe the formation, interconversion, or decay of these intermediates must be able to instantaneously initiate a particular step in an enzymatic reaction and immediately (<10 ms) monitor its progress.

For kinetic studies, reactions are usually initiated by mixing two solutions, each of which contains one or more of the reactants. If the two solutions to be mixed are introduced through separate ports of a mixing chamber at rapid, continuous, and uniform flows, the fluid emerging immediately adjacent to the mixing chamber will have been mixed for only a very short time.[171] If the flow through the mixing chamber is stopped abruptly, the reaction occurring in the fluid immediately adjacent to the mixing chamber will have progressed during the short time it was passing through the mixing chamber but will continue to proceed in the fluid now at rest.[172] If an observation chamber is located as close to the mixing chamber as possible and the flow is stopped as abruptly as possible,[173,174] the fluid within the observation chamber will have been mixed only a short time before, and the progress of the reaction can then be followed within the still fluid. The interval between the time at which the fluid in the observation chamber was mixed and the time of the complete cessation of the flow is referred to as the **dead time** because no observation can be made of changes occurring during this interval.

A device that accomplishes this strategy is a stopped-flow apparatus (Figure 5–15).[174,175] The two solutions are delivered to the mixers by two syringes that are driven by a hydraulic

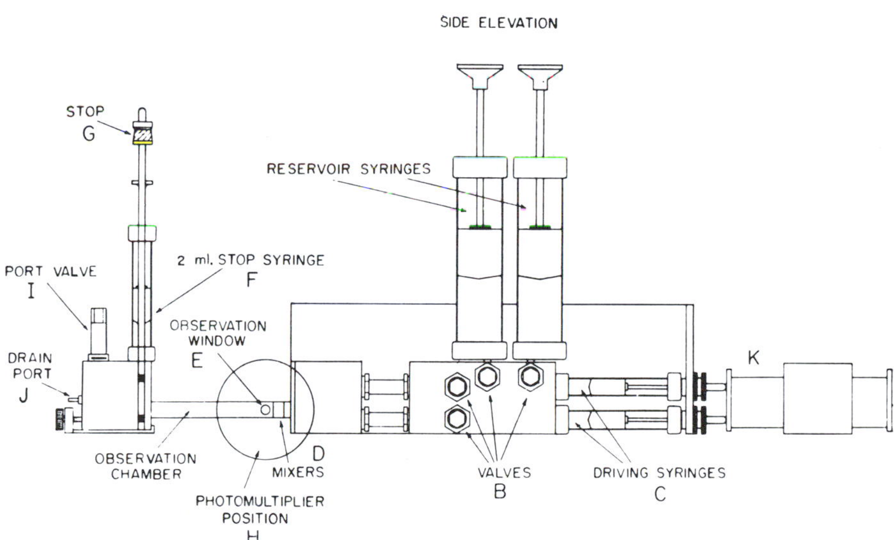


**Figure 5–15:** Stopped-flow spectrophotometer.[175] The two reservoir syringes store large quantities of the two solutions to be mixed. Small amounts of each of these two solutions are introduced through valves (B) into the two respective drive syringes (C). After the drive syringes are filled, the valves are turned so that the two drive syringes become connected directly to the two ports of the mixing chamber (D) that efficiently mixes the two fluids passing through it at high velocity. Just to the downstream side of the mixing chamber is a window (E) through which observations can be made. The observation chamber that contains this window is connected directly to a syringe (F) that fills as the fluid flows through the chamber until the end of its shaft hits a solid barrier or stop (G). After the stop has been struck by the shaft, all flow through the apparatus ceases. A photomultiplier (H) monitors the change in absorbance in the still fluid inside the observation window. Light of the desired wavelength passing through the observation window is produced by a monochromator. The stop syringe is drained through a port (J) by turning a valve (I). The hydraulic plunger (K) develops significant hydrostatic pressure in the fluids to promote rapid flow, so all valves and tubes must be designed to withstand this high pressure. Reprinted with permission from ref 175. Copyright 1963 Wiley-Interscience.

plunger. The window to the observation chamber is immediately adjacent to the mixers. During the flowing phase, the continuously emerging fluid fills and expands a stop syringe until its distending plunger strikes a solid barrier. This collision stops the flow abruptly within the completely sealed system. The progress of the reaction in the observation chamber is followed spectroscopically, and the output of the photomultiplier is displayed on an oscilloscope. Stopped-flow apparatuses of this type can have dead times of less than 1 ms.[176]

From an examination of the stopped-flow apparatus, it is clear that the contents of the two syringes define the reaction observed. For example, one of the syringes in a stopped-flow apparatus was filled with a solution 30 $\mu$M in active sites of lactate dehydrogenase and 20 $\mu$M in NADH. At these concentrations, all of the NADH added was already bound to active sites of the enzyme, and the solution only contained the complex, E·NADH, and free enzyme. The other syringe was filled with 2.0 mM pyruvate. After cessation of flow, the fluid in the observation chamber was monitored by its absorbance at 340 nm, a wavelength at which NADH absorbs light strongly but NAD$^+$ does not. Therefore, the decrease in absorbance at 340 nm (Figure 5–16A)[177] directly monitored the transfer of the hydride from NADH to pyruvate

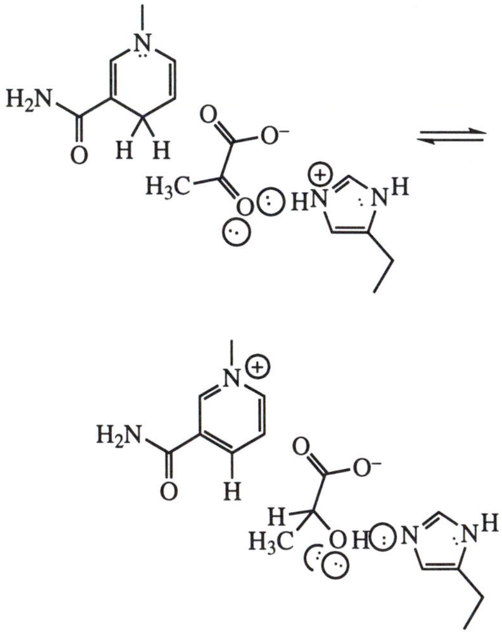

(5–177)

as it occurred on the active site. Because the dead time of the apparatus was 1.3 ms, this reaction could be monitored from 1.3 ms after mixing to its completion. The loss of absorbance was a first-order process (Figure 5–16B)

$$-\ln\left(\frac{A_t - A_\infty}{A_0 - A_\infty}\right) = k_{obs}t \qquad (5\text{–}178)$$

where $A_t$ is the absorbance at any time $t$, $A_0$ is the absorbance at time zero, and $A_\infty$ is the absorbance at equilibrium. The first-order rate constants ($k_{obs}$) measured at various concentrations of pyruvate displayed saturation (Figure 5–16C), an observation consistent with the kinetic mechanism

$$\mathrm{E{\cdot}NADH} \underset{k_{-1}}{\overset{k_1[\mathrm{Pyr}]}{\rightleftharpoons}} \mathrm{E{\cdot}NADH{\cdot}Pyr} \xrightarrow{k_2} \mathrm{E{\cdot}NAD^+{\cdot}lactate} \qquad (5\text{–}179)$$

If $k_1$ and $k_{-1}$ are so large that the complex E·NADH·Pyr is in equilibrium at all times with E·NADH and free pyruvate, then the observed first-order rate constant, $k_{obs}$, will be

$$k_{obs} = \frac{k_2[\mathrm{Pyr}]}{\dfrac{k_{-1}}{k_1} + [\mathrm{Pyr}]} \qquad (5\text{–}180)$$

Therefore, at saturation with pyruvate, the observed first-order rate of the reaction should be $k_2$, the rate constant for the transfer of the hydride from NADH to pyruvate after both the NADH and the pyruvate are bound to the active site. It is this process that seems to have been measured directly by the stopped-flow observation. At pH 9.0 and 22 °C, $k_2$ had a value of 260 s$^{-1}$. The rate of hydride transfer at any concentration of pyruvate measured directly in these stopped-flow experiments equaled the steady-state turnover rate for the overall reaction at the same concentration of pyruvate (Figure 5–16C).

If the outlet from a mixing chamber fed by two syringes that are compressed at a constant rate by a hydraulic plunger is attached to a length of tubing that is connected to a second mixing chamber, the time that the fluid spends between the first mixing chamber and the second mixing chamber is determined only by the flow rate and the length of the tubing. If the reaction is initiated in the first mixing chamber, in analogy with the stopped-flow apparatus, and abruptly halted at the second mixing chamber, for example, by quenching with strong acid, the final solution emerging from the device will provide a measure of the progress of the reaction over the time spent between the two mixing chambers. This strategy is referred to as **quenched-flow**. Its advantage is that the products of the reaction are collected and can be assayed chemically rather than only spectroscopically. Its disadvantage is that only one time point in the development of the product is provided. To obtain the whole course of the reaction a sequence of tubes that vary in length or a plunger of variable speed is used to provide a sequence of reaction times.

One of the syringes of a quenched-flow apparatus was filled with a solution of active, phosphorylated phosphoglucomutase (Reaction 5–20), and the other was filled with glucose 1-[$^{32}$P]phosphate. After these solutions were mixed in a rapid mixing chamber at a constant flow rate, the mixture passed through a tube of a particular length and was then mixed with trichloroacetic acid. Using a series of tubes of different length, the amount of $^{32}$phosphate attached covalently to the protein (Reaction 5–21) as a function of the time spent between the mixing chamber and the quenching solution could be assessed (Figure 5–17A).[31] As the time increased, the yield of [$^{32}$P]phosphoenzyme increased in a first-order process (Figure 5–17B)

$$\ln\left(\frac{{}^{32}\mathrm{P}_\infty - {}^{32}\mathrm{P}_t}{{}^{32}\mathrm{P}_\infty - {}^{32}\mathrm{P}_0}\right) = -k_{obs}t \qquad (5\text{–}181)$$

where the terms $^{32}$P are the yields of [$^{32}$P]phosphoenzyme at completion of the reaction ($^{32}\mathrm{P}_\infty$), at a particular time

A

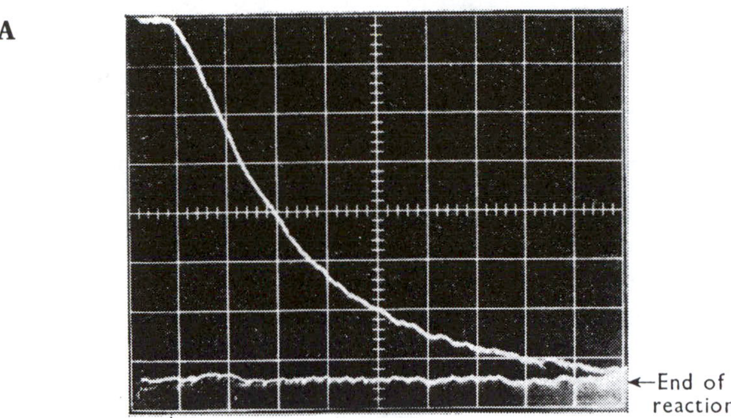


B

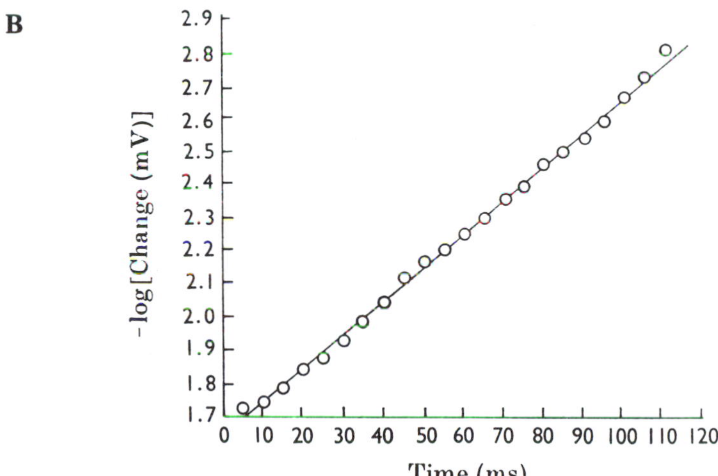


C

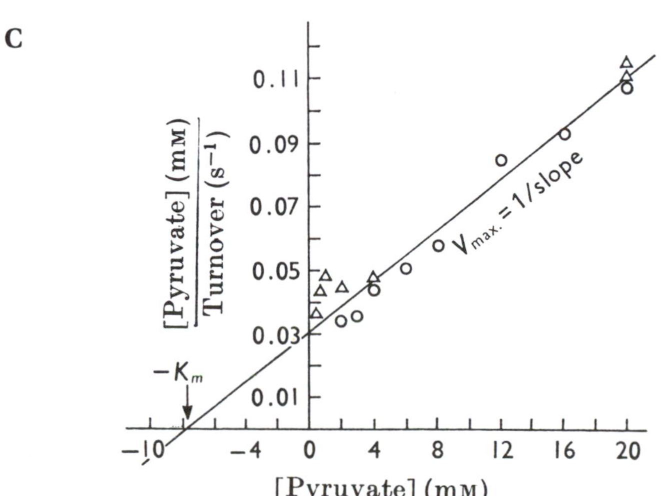


**Figure 5–16:** Monitoring the hydride transfer from NADH to pyruvate at the active site of lactate dehydrogenase with a stopped-flow spectrophotometer.[177] A solution of porcine lactate dehydrogenase (28 $\mu$M in active sites) and NADH (20 $\mu$M) in one syringe of the stopped-flow apparatus was mixed with an equal volume of a solution of pyruvate (2.0 mM) in the other syringe. Both solutions were at pH 9.0 and 22 °C. The dead time of the apparatus was 1.3 ms. (A) The flow of fluid through the apparatus ceased at the point indicated by the arrow (flow stops) and the absorbance of the solution in the observation chamber was monitored at 340 nm, the wavelength at which NADH absorbs. The upper oscilloscope trace displays the voltage from the photomultiplier tube as a function of time (milliseconds) immediately following cessation of flow. The lower, flat trace is a second sweep after all change in voltage had ceased (end of reaction). (B) The values of the negative logarithms of the differences [change (mV)] between the millivolts at any time on the upper trace in panel A and the millivolts at the end of the reaction are plotted as a function of time (milliseconds) to show that the relaxation observed in panel A is a first-order relaxation and to obtain a first-order rate constant (22 $s^{-1}$) for concentration of pyruvate of 1.0 mM. (C) The values for the first-order rate constant were measured directly by stopped-flow spectroscopy at several different concentrations of pyruvate. Each of these directly measured values of the first-order rate constant for hydride transfer (seconds$^{-1}$) is the turnover rate (turnover) for the active site in catalyzing hydride transfer at that concentration of pyruvate. The initial velocities of the enzymatic reaction, $v_0$, were also measured at several different concentrations of pyruvate at saturating concentrations of NADH and at pH 9.0 and 22 °C. These values of initial velocity (micromolar second$^{-1}$) were each divided by the respective molar concentration of active sites (micromolar) in the solution to obtain turnover rates for the active site in catalyzing the overall reaction at that concentration of pyruvate. The quantity [pyruvate] turnover$^{-1}$ (millimolar second$^{-1}$) could be calculated for each of the turnover rates of hydride transfer measured directly by stopped-flow spectroscopy (△) or for each of the turnover rates of the overall reaction measured by initial velocities at steady state (○). These values are plotted as a function of [pyruvate] (millimolar). The slope of the line on such a plot is $V_{max}^{-1}$(260 $s^{-1}$). Adapted with permission from ref 177. Copyright 1971 Biochemical Society.

A

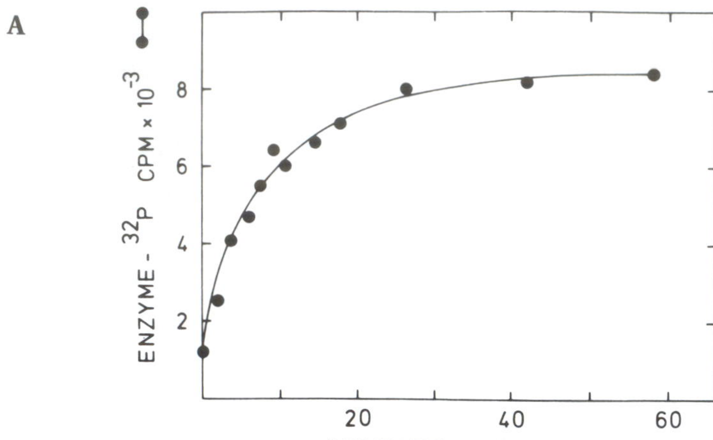


B

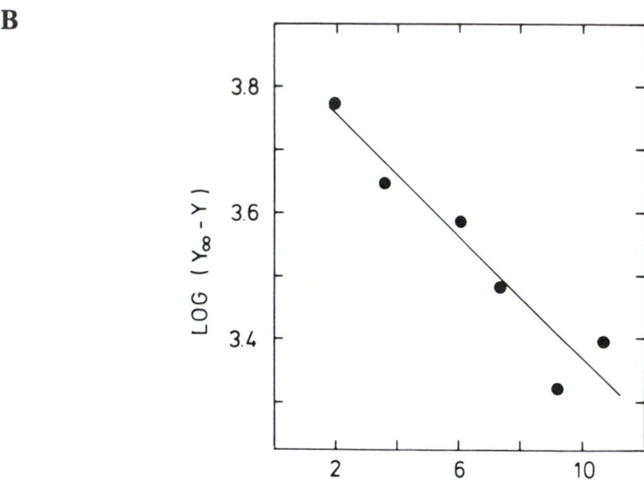


**Figure 5–17:** Increase of [$^{32}$P]phosphoenzyme after active, fully phosphorylated phosphoglucomutase was mixed with glucose 1-[$^{32}$P]phosphate.[31] Solutions of [$^{32}$P]phosphorylated phosphoglucomutase [$8.1 \times 10^3$ cpm (nmol of active sites)$^{-1}$] at a concentration of 0.01 mg mL$^{-1}$ were mixed with equal volumes of a solution of 2 $\mu$M glucose 1-[$^{32}$P]phosphate ($75 \times 10^3$ cpm nmol$^{-1}$). All solutions contained 1 mM $MgCl_2$, were buffered at pH 7.5, and were at room temperature. The two solutions were mixed in each instance in a mixing chamber, the mixture flowed through a tube of a particular length, and the solution was then introduced through the end of the tube into a rapidly swirling solution of 5% trichloracetic acid. The protein that precipitated was collected by filtration and the counts per minute attached to the protein was determined by liquid scintillation counting. The time each mixture spent between the mixing chamber and the quench solution was varied by choosing different lengths of tubing. (A) The counts per minute of $^{32}$phosphorus attached to protein (enzyme-$^{32}$P in counts per minute $\times 10^{-3}$, ●) is presented as a function of the time the mixture spent in the tubing between the mixing chamber and the quenching solution (incubation time in milliseconds). Because the enzyme had some $^{32}$P attached at the beginning of the reaction, the cpm at time zero is not 0. (B) The values of the logarithm of the difference between the cpm attached to protein at long times ($Y_\infty$) and the counts per minute at any particular time ($Y$) are plotted as a function of time to show that the relaxation in panel A was a first-order process and to obtain a rate constant (110 s$^{-1}$). Adapted with permission from ref 31. Copyright 1974 *Journal of Biological Chemistry.*

$t$ ($^{32}P_t$), and at the initial time point ($^{32}P_0$). The first-order rate constant for formation of the $^{32}$P-phosphorylated enzyme was 110 s$^{-1}$.

Each of the processes being followed in the two experiments just described is both a **first-order relaxation** and an **approach to equilibrium**. Suppose that by designing the experiment properly the kinetic mechanism for the reaction being followed is

$$E_1 \underset{k_{-1}}{\overset{k_1}{\rightleftharpoons}} E_2 \qquad (5\text{–}182)$$

where $E_1$ and $E_2$ are two forms of the enzyme, for example, E·NADH·Pyr and E·NADH·Lac or E-$^{31}$P·Glc-1-$^{32}$P and E-$^{32}$P·Glc-6-$^{31}$P. Suppose that, over the duration of the reaction, the sum

$$[E_1] + [E_2] = [E]_{TOT} \qquad (5\text{–}183)$$

remains constant. Suppose also, for the sake of argument, that the physical property being monitored is absorbance, $A$, but any other property, such as counts per minute incorporated into protein, that is proportional to the molar concentration of either form of the enzyme would be equivalent. The absorbance due to the two forms of the enzyme is then

$$A = \varepsilon_1[E_1] + \varepsilon_2[E_2] \qquad (5\text{–}184)$$

where $\varepsilon_1$ and $\varepsilon_2$ are the respective extinction coefficients. From Equations 5–182 and 5–183

$$-\frac{d[E_1]}{dt} = k_1[E_1] - k_{-1}[E_2] \qquad (5\text{–}185)$$

$$-\frac{d[E_1]}{(k_1+k_{-1})[E_1]-k_{-1}[E]_{TOT}}=dt \quad (5\text{–}186)$$

$$-\ln\frac{(k_1+k_{-1})[E_1]-k_{-1}[E]_{TOT}}{(k_1+k_{-1})[E_1]_0-k_{-1}[E]_{TOT}}=(k_1+k_{-1})t \quad (5\text{–}187)$$

where $[E_1]_0$ is the concentration of form $E_1$ at the beginning of the reaction. If $[E_1]_{eq}$ and $[E_2]_{eq}$ are the concentrations of the two forms at the end of the reaction

$$k_1[E_1]_{eq}=k_{-1}[E_2]_{eq} \quad (5\text{–}188)$$

$$k_{-1}[E]_{TOT}=(k_1+k_{-1})[E_1]_{eq} \quad (5\text{–}189)$$

$$-\ln\frac{[E_1]-[E_1]_{eq}}{[E_1]_0-[E_1]_{eq}}=(k_1+k_{-1})t \quad (5\text{–}190)$$

$$-\ln\frac{(\varepsilon_1-\varepsilon_2)[E_1]-(\varepsilon_1-\varepsilon_2)[E_1]_{eq}}{(\varepsilon_1-\varepsilon_2)[E_1]_0-(\varepsilon_1-\varepsilon_2)[E_1]_{eq}}=(k_1+k_{-1})t \quad (5\text{–}191)$$

$$-\ln\frac{\varepsilon_1[E_1]+\varepsilon_2[E_2]-\varepsilon_2[E]_{TOT}-\varepsilon_1[E_1]_{eq}-\varepsilon_2[E_2]_{eq}+\varepsilon_2[E]_{TOT}}{\varepsilon_1[E_1]_0+\varepsilon_2[E_2]_0-\varepsilon_2[E]_{TOT}-\varepsilon_1[E_1]_{eq}-\varepsilon_2[E_2]_{eq}+\varepsilon_2[E]_{TOT}}=(k_1+k_{-1})t \quad (5\text{–}192)$$

$$\ln\left(\frac{A-A_{eq}}{A_0-A_{eq}}\right)=(k_1+k_{-1})t \quad (5\text{–}193)$$

Equation 5–193 states that if the reaction observed is described by Mechanism 5–182, the behavior of any physical property the magnitude of which is proportional both to the concentration of product and to the concentration of reactant will display first-order behavior regardless of whether the measurements were initiated when all the enzyme was $E_1$ or in midreaction ($[E_2]_0 > 0$), regardless of whether the reaction goes to completion ($k_{-1} = 0$) or reaches an equilibrium, regardless of whether both product and reactant display the physical property ($\varepsilon_1 > 0$ and $\varepsilon_2 > 0$) or only one displays the physical property, and regardless of whether the rate constants $k_{-1}$ and $k_1$ are unimolecular rate constants, pseudo-first-order rate constants ($k_1 = k_1'[A]$ or $k_{-1} = k_{-1}'[P]$), or composite rate constants. The only requirement for observing the reaction is that the two extinction coefficients (constants of proportionality) for the two forms of the enzyme must be different enough that a significant change occurs in the magnitude of the physical property during the measurement. Because of this generality, such first-order processes are commonly observed in kinetic measurements. They are referred to as **relaxations** and the reciprocal of the first-order rate constant observed for a given relaxation is referred to as a **relaxation constant** or **relaxation time**, $\tau$. If a first-order relaxation is not observed, the system is usually altered by varying the initial conditions until one is. If two relaxations occur sequentially and the relaxation times differ from each other sufficiently, the observed behavior can be analyzed numerically, and the two relaxation times can be extracted from the biphasic behavior. By using stopped-flow or quenched-flow measurements, it is possible in ideal situations to step systematically through the relaxations in an enzymatic mechanism.

In a naturally occurring posttranslational modification, glutamate–ammonia ligase (Reaction 5–55) is adenylated at a tyrosine in each of its 12 subunits[178]

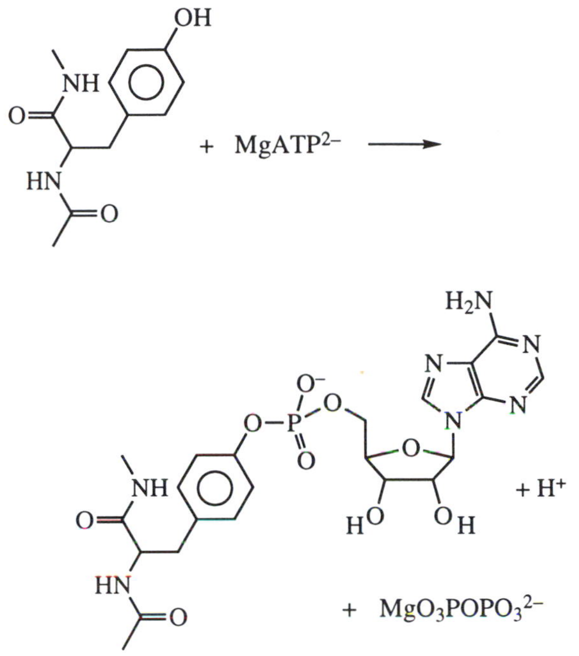

[Chemical structures: tyrosine residue + $MgATP^{2-}$ → adenylated tyrosine residue + $H^+$ + $MgO_3POPO_3^{2-}$]

(5–194)

If the $Mg^{2+}$ complex of 2-aza-1-$N^6$-ethenoadenosine triphosphate

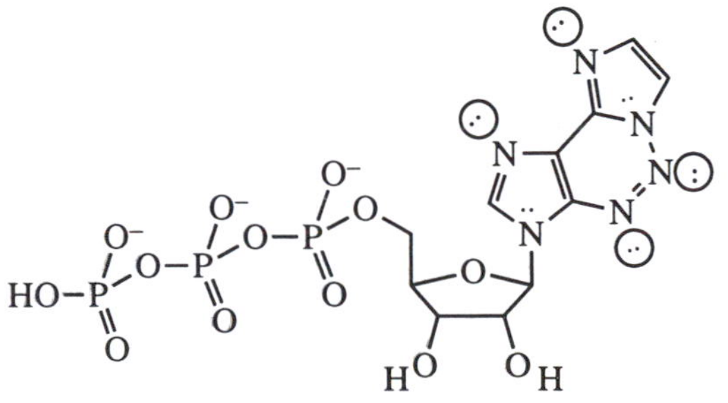

[Chemical structure]

**5–27**

is used to adenylate the enzyme rather than MgATP, the resulting protein is highly fluorescent,[179] and its fluorescence changes significantly as substrates occupy the active site. These changes in fluorescence can be used to monitor the processes occurring at the active site by stopped-flow spectroscopy.[176]

When a solution of the $Mn^{2+}$ complex of the modified glutamate–ammonia ligase (aza-ε-GS) was added to one syringe and a solution of glutamate to the other, a first-order relaxation was observed upon mixing (Figure 5–18).[176] The first-order rate constant, $k_{obs}$, for the observed relaxation was a function of the concentration of glutamate added to the syringe. At low concentrations of glutamate, the behavior of the rate constant, as expected from Equation 5–193, satisfied the equation

$$k_{obs} = k_{-1} + k_1[\text{Glu}] \tag{5–195}$$

where $k_1 = 1.3 \times 10^5\ \text{M}^{-1}\ \text{s}^{-1}$ and $k_{-1} = 100\ \text{s}^{-1}$, but as the concentration of glutamate was raised, the rate constant $k_{obs}$ approached a maximum value asymptotically. This behavior is consistent with the mechanism

$$\text{E}\cdot\text{Mn}^{2+} + \text{Glu} \underset{k_{-1}}{\overset{k_1}{\rightleftharpoons}} \text{E}\cdot\text{Mn}\cdot\text{Glu} \underset{k_{-2}}{\overset{k_2}{\rightleftharpoons}} \text{E}'\cdot\text{Mn}\cdot\text{Glu} \tag{5–196}$$

in which the bimolecular association of glutamate with the active site is followed by an isomerization of the initial complex before equilibrium is reached. At high concentrations of glutamate, equilibrium was reached in times shorter than the dead time (0.5 ms) of the apparatus so that $(k_2 + k_{-2})$ could not be estimated. At concentrations of ATP in excess of 0.1 mM, the association of ATP with the active site also reached equilibrium in times shorter than the dead time of the apparatus.

When a solution of the complex formed from the enzyme, manganese, and ATP was added to one syringe and a solution of glutamate at a concentration where its association would be complete within the dead time of the apparatus was added to the other syringe, the fluorescence was found to change in a process that displayed three successive relaxations (Figure 5–19)[176]

$$\text{E}\cdot\text{Mn}\cdot\text{ATP}\cdot\text{Glu} \underset{k_{-1}}{\overset{k_1}{\rightleftharpoons}} \text{I}_1 \underset{k_{-2}}{\overset{k_2}{\rightleftharpoons}} \text{I}_2 \underset{k_{-3}}{\overset{k_3}{\rightleftharpoons}} \text{I}_3 \tag{5–197}$$

with rate constants of $14\ \text{s}^{-1}$, $1.4\ \text{s}^{-1}$, and $0.4\ \text{s}^{-1}$. The intermediates $I_1$, $I_2$, and $I_3$ were formed during the successive relaxations. One of these intermediates is thought to be the complex formed from the active site, manganese, ADP, and glutamyl phosphate.

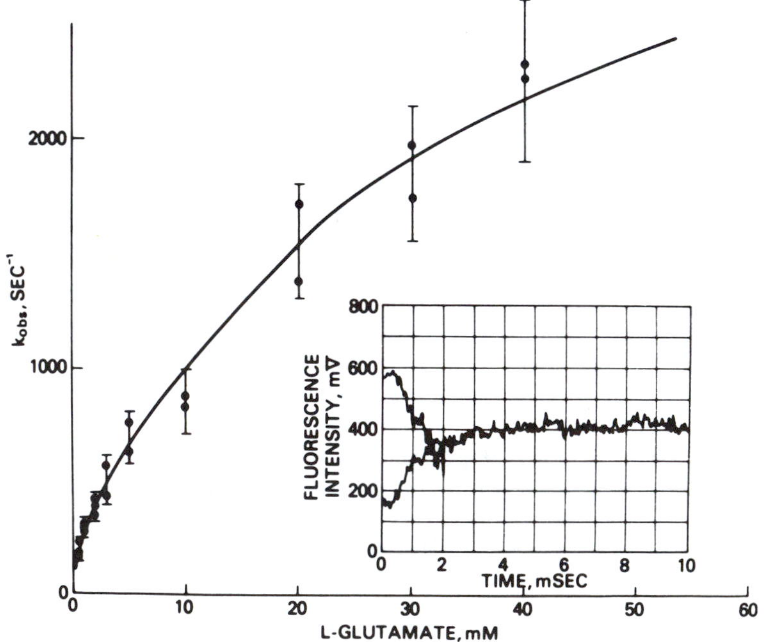


**Figure 5–18:** Rate constants for the first-order relaxation observed when the $Mn^{2+}$ complex of glutamate–ammonia ligase, adenylated with 2-aza-1-$N^6$-ethenoadenosine triphosphate, was mixed with various concentrations of glutamate.[176] (Inset) A solution of the $Mn^{2+}$ complex of the fluorescent enzyme (12 μM) was mixed with an equal volume of a solution of 2 mM L-glutamate in a stopped-flow fluorometer. Both solutions were at 10 °C, contained 5 mM $MnCl_2$ and 0.1 M KCl, and were buffered at pH 7.2. The emission of fluorescence at 470 nm ($\lambda_{excitation}$ = 300 nm) from the modified enzyme was monitored as a function of time following cessation of flow. The dead time of the apparatus was 0.5 ms. The lower trace is the increase in fluorescence intensity measured by the photomultiplier (millivolts) as a function of time (milliseconds). The upper trace is the logarithm of the difference between the voltage at any time and the fluorescence at long times as a function of time. From the slope of the upper curve, a first-order rate constant, $k_{obs}$, can be calculated (seconds$^{-1}$). In the main body of the figure, the individual first-order rate constants $k_{obs}$ are plotted as a function of the concentration of L-glutamate (millimolar). Reprinted with permission from ref 176. Copyright 1982 *Journal of Biological Chemistry.*

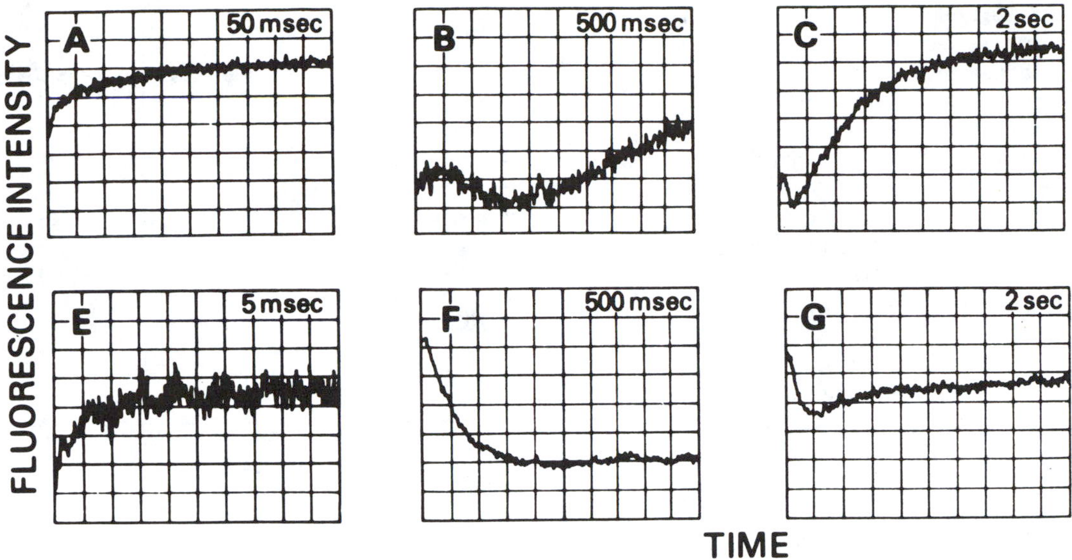


**Figure 5–19:** The three successive relaxations in glutamate–ammonia ligase that occur when the complex between enzyme, $Mn^{2+}$, and ATP is mixed with glutamate at saturating concentrations.[176] A solution containing 5 mM ATP and 15 $\mu$M $Mn^{2+}$ complex of glutamate–ammonia ligase adenylated with 2-aza-1-$N^6$-ethenoadenosine triphosphate was mixed with an equal volume of a solution containing 50 mM glutamate. Both solutions were at 10 °C, contained 0.1 M KCl and 5 mM $MnCl_2$, and were buffered at pH 7.2. The fluorescence of the mixed solutions, arising from the ethenoadenosine, was monitored at 470 nm, and the excitation wavelengths were 300 nm for panels A–C and 360 nm for panels E–G. The oscilloscope registered the voltage (millivolts) from the photomultiplier as a function of time (milliseconds). The traces in panels C and G are the full traces at the noted wavelength of excitation over 20 s (2 s division$^{-1}$). Panels B and F are expanded tracings of those in panels C and G, respectively, spanning 5 s (500 ms division$^{-1}$), panel A is an expanded tracing of that in panel C spanning 500 ms (50 ms division$^{-1}$), and panel E is an expanded tracing of that in panel G spanning 50 msec (5 ms division$^{-1}$). Reprinted with permission from ref 176. Copyright 1982 *Journal of Biological Chemistry.*

A solution of the complex formed from the fluorescent enzyme, manganese, ATP, and ammonia was added to one syringe and a solution of glutamate was added to the second syringe so that upon the formation of the glutamyl phosphate at the active site, ammonia would be present to react immediately with it. The slowest relaxation observed had a first-order rate constant of 2.5 s$^{-1}$, and its magnitude was not a function of the concentration of glutamate. This result suggests that the even slower relaxation of 0.4 s$^{-1}$ observed in the absence of ammonia does not occur under normal circumstances when ammonia is present to scavenge the glutamyl phosphate. Therefore, glutamyl phosphate is probably present in either $I_1$ or $I_2$ or in both (Reaction 5–197).

When a solution of intermediate $I_3$, formed by mixing enzyme, manganese, ATP, and glutamate, was added to one syringe and a solution of ammonia was added to the other syringes, a rapid first-order relaxation ($k_{obs}$ > 20 s$^{-1}$) was observed upon mixing. The rate of this relaxation was dependent on the concentration of ammonia. This result suggests that it was due to the association of the ammonia with the enzyme. This concentration-dependent relaxation was followed by a slower relaxation ($k_{obs}$ = 3.7 s$^{-1}$), the rate of which was independent of the ammonia concentration (Figure 5–20).[176] This latter relaxation and the similar relaxation of 2.5 s$^{-1}$ observed when glutamate was added last could both be monitoring the formation of glutamine at the active site from glutamyl phosphate and ammonia. If this is the case, however, some step following the formation of the products, phosphate, ADP, and glutamine, at the active site must limit the rate of the overall enzymatic reaction because under the conditions chosen for the experiment the turnover number of the enzyme was 0.34 s$^{-1}$. In this example it has been possible to step through the mechanism of the reaction catalyzed by glutamate–ammonia ligase.

Protocatechuate 3,4-dioxygenase catalyzes the oxygenation of protocatechuate

$$\text{(COO}^- \text{, HO, OH-substituted ring)} + O_2 \rightleftharpoons \text{(COO}^- \text{, }^-\text{O–C(=O), C(=O)–O}^- \text{ ring-opened product)} + 2H^+ \qquad (5\text{–}198)$$

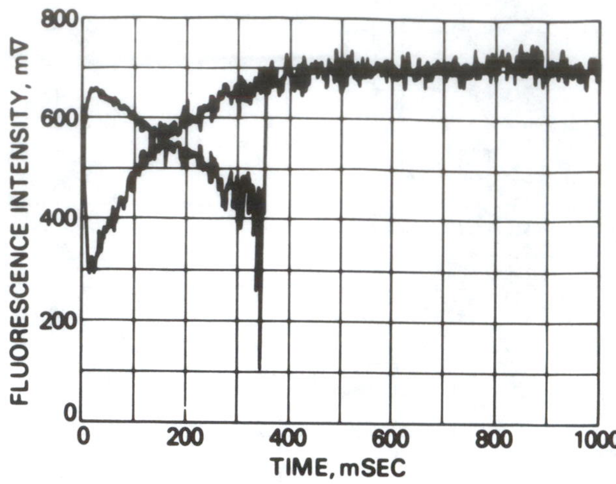


**Figure 5–20:** Concentration-independent relaxation observed when the complex of $Mn^{2+}$, ADP, glutamyl phosphate, and glutamate–ammonia ligase, adenylated with 2-aza-1-$N^6$-ethenoadenosine triphosphate ($E{\cdot}Mn{\cdot}ADP{\cdot}GluPO_3^{2-}$), was mixed with ammonia.[176] A solution containing 13 $\mu$M fluorescent glutamate–ammonia ligase, 5 mM $MnCl_2$, 100 mM glutamate, and 5 mM ATP was mixed with an equal volume of a solution containing 5 mM $NH_4Cl$. Both solutions were at 10 °C, contained 0.1 M KCl, and were buffered at pH 7.2. The intensity of the fluorescence at 470 nm ($\lambda_{excitation}$ = 300 nm) was monitored by a photomultiplier. The oscilloscope registered the voltage from the photomultiplier (millivolts) as a function of time (milliseconds). The rising trace is the direct output of the photomultiplier. The falling linear trace is the logarithm of the difference in the voltage at any time and the voltage at long times. That this latter trace is linear shows that the relaxation is first-order and provides a first-order rate constant (3.7 $s^{-1}$). Reprinted with permission from ref 176. Copyright 1982 *Journal of Biological Chemistry.*

The enzyme contains tightly bound iron and forms a complex with protocatechuate in the absence of oxygen that absorbs visible light in a broad range from 500 to 1000 nm. Changes in the structure of this complex could be followed by monitoring absorbance in this range. When enzyme and protocatechuate were mixed at equivalent concentrations of 60 $\mu$M, greater than 85% of the protocatechuate formed a complex with the active site. When a solution of this complex between the enzyme and protocatechuate in one syringe of a stopped-flow spectrophotometer was mixed with a solution of oxygen in the other, a three-step process was observed (Figure 5–21).[180]

The steps of this process could be dissected by performing measurements at two different wavelengths (403 and 460 nm). The initial absorbance at either wavelength at the dead time (1.3 ms) following mixing was substantially different from the absorbance of the complex of enzyme and protocatechuate that would have been observed in the absence of oxygen. Therefore an initial intermediate must have formed within the dead time of the apparatus ($k_{obs} > 1000\ s^{-1}$). Upon cessation of flow, the first change observed was the decay of this intermediate to form a second intermediate ($k_{obs} = 450\ s^{-1}$). Finally, this second intermediate decayed at a rate ($k_{obs} = 55\ s^{-1}$) that was identical to the turnover number for the enzyme under these conditions. From this identity it could be concluded that coincident with the decay of the second intermediate, the products are released from the active site. From the traces taken of the same reaction at several different wavelengths, the absorption spectra of the first intermediate ($ESO_2$) and the second intermediate ($ESO_2^{\bullet}$) could be determined (Figure 5–22).

At 750 nm, the intermediates and the free enzyme show much less absorbance than the enzyme–reactant complex (ES), and both the intermediates and the free enzyme have very similar extinction coefficients. Therefore, the decrease in absorbance at 750 nm upon mixing the complex of enzyme and protocatechuate with oxygen should monitor only the disappearance of this complex. At this wavelength the rate of decay of the enzyme–reactant complex (ES) as a function of the concentration of oxygen could be followed (Figure 5–23). The rate constant for this relaxation should be (in analogy with Equation 5–195)

$$k_{obs} = k_2[O_2] + k_{-2} \tag{5–199}$$

Because the line intersects the origin, no significant rate of dissociation was evident ($k_{-2} < 5\ s^{-1}$), and the second-order rate constant for the association of oxygen, $k_1$, was $5 \times 10^5\ M^{-1}s^{-1}$.

From all of these measurements the kinetic mechanism for the enzymatic reaction must be at least

$$E + PC \underset{k_{-1}}{\overset{k_1}{\rightleftharpoons}} E{\cdot}PC \xrightarrow{k_2[O_2]} E{\cdot}SO_2 \underset{k_{-3}}{\overset{k_3}{\rightleftharpoons}} E{\cdot}SO_2^{\bullet} \xrightarrow{k_4} E + M \tag{5–200}$$

where PC is protocatechuate, M is 3-carboxy-*cis,cis*-muconate, $k_2$ equals $5 \times 10^5\ M^{-1}\ s^{-1}$, ($k_3 + k_{-3}$) equals 450 $s^{-1}$, and $k_4$ equals 55 $s^{-1}$. Because the visible spectra of the successive intermediates are so different, they probably differ among themselves covalently.

In the experiments just described with protocatechuate 3,4-dioxygenase, the absorption spectra of the various intermediates in the enzymatic reaction were reconstructed from

**Figure 5–21:** The three successive relaxations that occur when the complex between protocatechuate and protocatechuate 3,4-dioxygenase from *Pseudomonas putida* is mixed with oxygen.[180] A solution containing 0.24 mM protocatechuate and 0.24 mM active sites of protocatechuate 3,4-dioxygenase was mixed with an equal volume of a solution containing 12 mM $O_2$. The solutions were at 10 °C and were buffered at pH 8.3. The absorbance of the mixture ($\Delta A$) was followed at either 403 nm (top trace) or 460 nm (bottom trace). The absorbance (0.02 absorbance unit division$^{-1}$ at 403 nm and 0.04 unit division$^{-1}$ at 460 nm) was monitored as a function of time (milliseconds) in the still fluid just beyond the mixing chamber after flow stopped (at 0 ms). The absorbance that would have been observed had an oxygen-free solution been mixed is indicated on the ordinate (460 and 403 for $A_{460}$ and $A_{403}$, respectively). That these levels differ from those observed at cessation of flow shows that a relaxation faster than the dead time of the apparatus (1.3 ms) had occurred immediately upon mixing. The second relaxation is observed at both 403 and 460 nm and is complete after 5 ms. The third relaxation is only observed at 460 nm and is complete at 100 ms. Reprinted with permission from ref 180. Copyright 1981 *Journal of Biological Chemistry.*

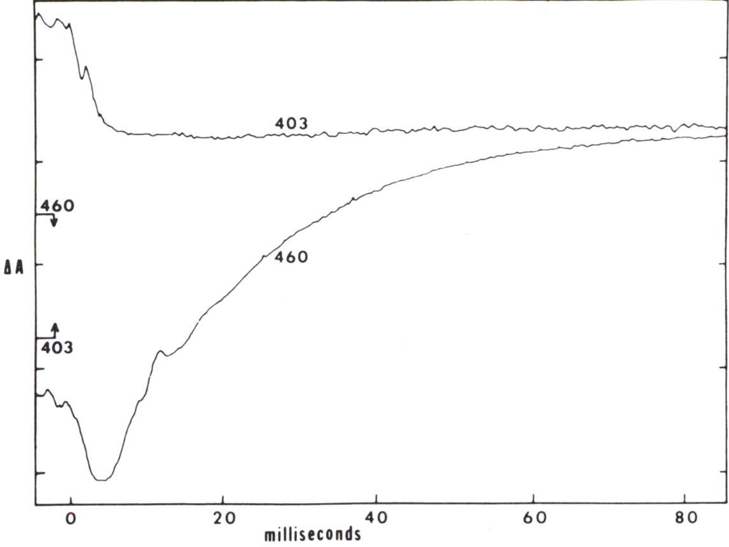


**Figure 5–22:** Spectra of the various forms of protocatechuate 3,4-dioxygenase formed during catalysis of the oxygenation of protocatechuate.[180] The spectra of the unoccupied enzyme (E) and the complex between the enzyme and protocatechuate (ES) could be measured directly at pH 8.3 and 10 °C. The spectra for the two successive intermediates, $ESO_2$ and $ESO_2^*$, formed in the first relaxation ($k_{obs} > 1000$ s$^{-1}$) and in the second relaxation ($k_{obs} > 450$ s$^{-1}$) were reconstructed from stopped-flow traces like those in Figure 5–21 gathered at successive wavelengths. The spectrum of $ESO_2^*$ was assembled from the absorbances of the mixtures at 5 ms. The absorbances of the mixtures as a function of time were then extrapolated to zero time to obtain the absorbance of $ESO_2$ at each wavelength from which its spectrum could be assembled. From the molar concentrations of active sites, the extinction coefficients could be calculated for each absorbance at each wavelength and the extinction coefficient (millimolar$^{-1}$ centimeter$^{-1}$) is presented as a function of wavelength (nanometers) for each complex. Reprinted with permission from ref 180. Copyright 1981 *Journal of Biological Chemistry.*

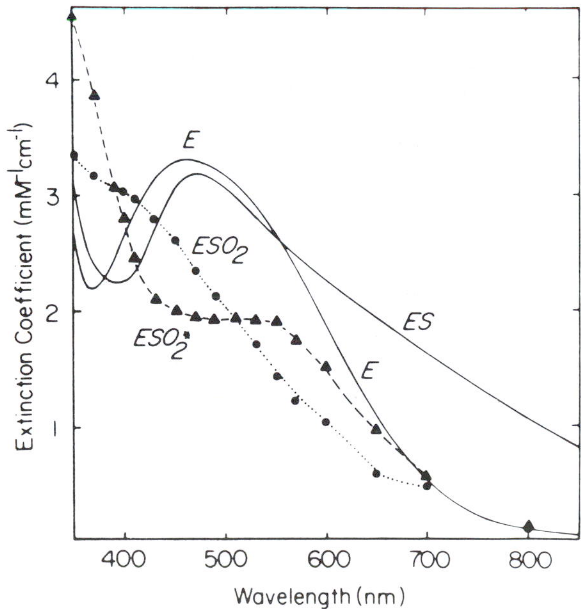

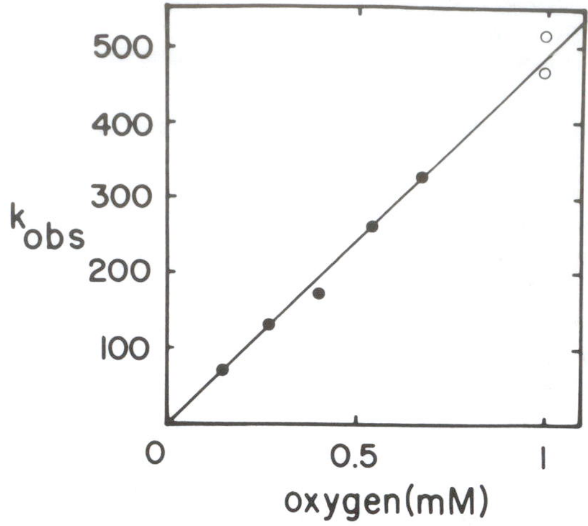


**Figure 5–23:** Dependence of the decay of the absorbance of the complex between protocatechuate and protocatechuate 3,4-dioxygenase on the concentration of oxygen.[180] A solution of 50 $\mu$M protocatechuate and 100 $\mu$M active sites of protocatechuate 3,4-dioxygenase was mixed with equal volumes of solutions containing various concentrations of oxygen (millimolar). The solutions were at 4 °C and were buffered at pH 8.3. Upon mixing, a single first-order relaxation was observed for each concentration of oxygen in the stopped-flow spectrophotometer at a wavelength of 750 nm. The first-order rate constants for these relaxations (second$^{-1}$) are presented as a function of the concentrations of oxygen (millimolar). Reprinted with permission from ref 180. Copyright 1981 *Journal of Biological Chemistry.*

individual stopped-flow traces at different wavelengths (Figure 5–22). It is also possible to gather absorption spectra so rapidly that a series of spectra can be accumulated over the duration of a rapid relaxation. For example, full absorption spectra from 325 to 550 nm could be gathered at intervals of 4.7 ms, beginning 2 ms after cessation of flow, to follow two successive relaxations in the reaction catalyzed by tryptophan synthase, a pyridoxal phosphate enzyme (Reaction 3–244). The first half of the reaction catalyzed by the enzyme is the dehydration of serine to form the aldimine between 1-amino-1-carboxyethene and the pyridoxal phosphate at the active site. The two relaxations that could be observed following the rapid mixing of the enzyme with serine were associated with this process (Figure 5–24).[181] The first relaxation ($k_{obs1}$ = 200 s$^{-1}$) is associated with a shift in maximum absorption from 412 to 425 nm, and the second relaxation ($k_{obs2}$ = 70 s$^{-1}$) is associated with a decrease in absorption at 425 nm and an increase in absorption from 325 to 370 nm coincident with an increase in absorption between 470 and 550 nm. It is the second transition that displays the clean isosbestic points. These two relaxations have been assigned to the formation of the serine aldimine ($\lambda_{max}$ = 425) on the pyridoxal phosphate in the active site, followed by the establishment of an equilibrium between the aldimine of 1-amino-1-carboxyethene ($\lambda_{max}$ = 350) and the quinoid ($\lambda_{max}$ = 460 nm)

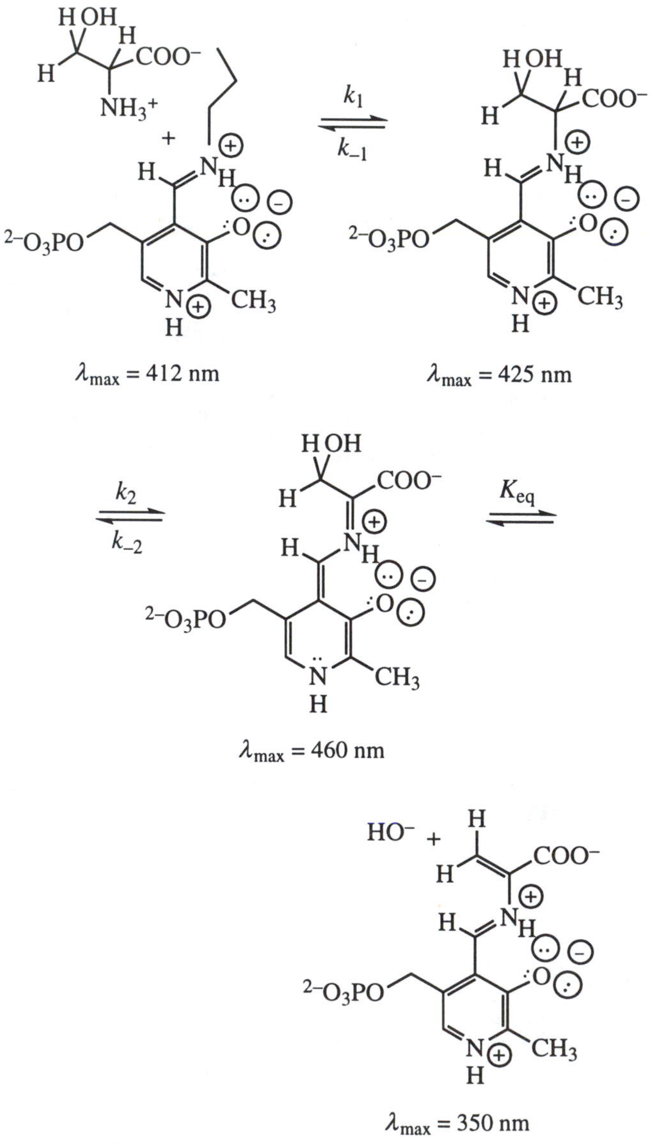

$\lambda_{max}$ = 412 nm $\underset{k_{-1}}{\overset{k_1}{\rightleftharpoons}}$ $\lambda_{max}$ = 425 nm

$\underset{k_{-2}}{\overset{k_2}{\rightleftharpoons}}$ $\lambda_{max}$ = 460 nm $\overset{K_{eq}}{\rightleftharpoons}$

$HO^-$ + $\lambda_{max}$ = 350 nm

(5–201)

A

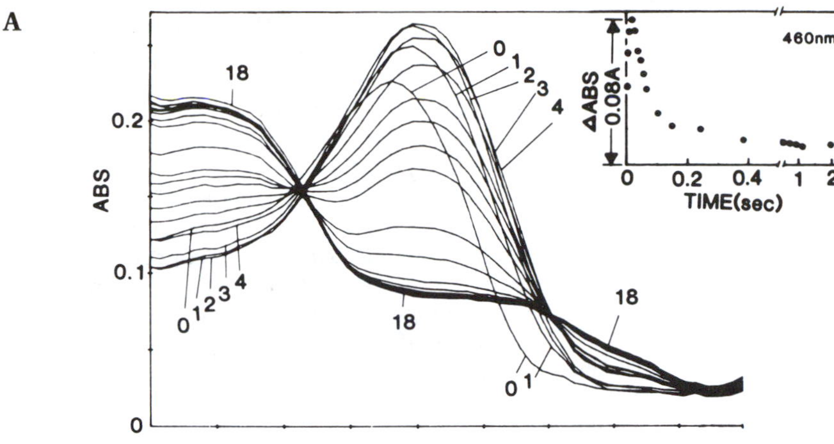


B

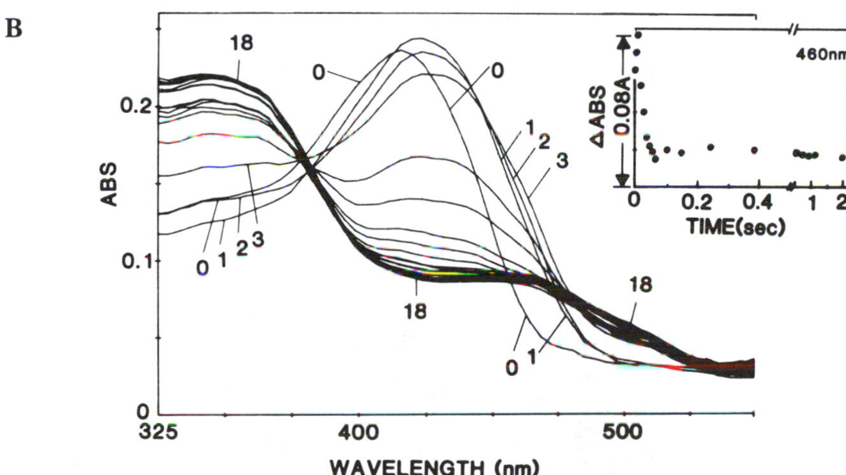


**Figure 5–24:** Spectral analysis of two relaxations that occur when the enzyme tryptophan synthase from *E. coli* is mixed with serine, one of its two reactants.[181] A solution of tryptophan synthase (26 $\mu$M) was mixed with an equal volume of a solution of either (A) DL-[$\alpha$-$^1$H]serine (16 mM) or (B) DL-[$\alpha$-$^2$H]serine (16 mM). The solutions were at 25 °C and buffered at pH 7.8 with 0.1 M sodium phosphate. Spectral scans of absorbance (ABS) against wavelength (nanometers) between 325 and 550 nm were taken every 4.7 ms, commencing 2 ms after flow stopped until scan 8 at 33 ms. Thereafter scans were taken at (9) 42, (10) 51, (11) 61, (12) 70, (13) 107, (14) 154, (15) 247, (16) 387, (17) 761, (18) 1135, and (19) 1980 ms. The insets show the change in absorbance (ΔABS) as a function of time (seconds) at one wavelength (460 nm) for each run to illustrate the decrease in the rate of the second relaxation when the $\alpha$-protium on the serine (A) was replaced by a deuterium (B). Reprinted with permission from ref 181. Copyright 1985 American Chemical Society.

In support of this assignment was the fact that the rate constant of the second relaxation decreased more than 4-fold ($k_{obs2}{}' = 15\ s^{-1}$) when the protium on the serine was replaced with a deuterium (Figure 5–24B). This last observation illustrates the ability of stopped-flow experiments to measure intrinsic kinetic isotope effects directly.

Intermediates in the reaction catalyzed by glutathione reductase (Reaction 2–95) have also been observed by stopped-flow spectroscopy.[182] Because the enzyme has a flavin bound at the active site as a coenzyme, it exhibits rapid changes in absorbance throughout the visible range from 400 to 750 nm. In the oxidized, resting form of the enzyme, the two cysteines in the active site are oxidized to cystine and the flavin is in the oxidized form. When a solution of the oxidized enzyme, $E{<}^{S}_{S}{\cdot}FlH_{ox}$, in one syringe of the apparatus was mixed with a solution of NADPH in the other syringe at a 10-fold molar excess over active sites, the changes in absorbance over the visible range were consistent with the kinetic mechanism

$$E{<}^{S}_{S}{\cdot}FlH_{ox} \underset{k_{-1}}{\overset{k_1[NADPH]}{\rightleftharpoons}} E{<}^{S}_{S}{\cdot}FlH^{\cdot-}{\cdot}NADPH^{\cdot+} \underset{k_{-2}}{\overset{k_2}{\rightleftharpoons}}$$

$$E{<}^{S}_{S}{\cdot}FlH\ NADP \underset{k_{-3}}{\overset{k_3}{\rightleftharpoons}} E{<}^{SH}_{S\cdot}{\cdot}FlH^{\cdot-} \qquad (5\text{–}202)$$

In the first step ($k_1$[NADPH] + $k_{-1}$), which takes place within the dead time of the apparatus, an intermediate, $E{<}^{S}_{S}{\cdot}FlH^{\cdot-}{\cdot}NADPH^{\cdot+}$, was formed that had an absorbance spectrum indicative of a charge transfer complex in which an electron equilibrates between the NADPH and the oxidized flavin. In the next step, a kinetic deuterium isotope effect was observed with [$^2$H]NADPH, indicating that the hydrogen is removed from the NADPH at this point. The spectral changes observed upon forming this intermediate, $E\langle^S_S \cdot Fl \cdot NADP$, however, were not consistent with the formation of either reduced flavin, $FlH_{3red}$, or an adduct between one of the cysteines and the flavin at its 4a carbon. This is consistent with the conclusion drawn from the crystallographic molecular model that both of these intermediates are sterically prohibited.[183] The fact that the intermediate formed in this step, $E\langle^S_S \cdot Fl \cdot NADP$, had excess absorbance at 420 nm that decreased in the next step ($k_3 + k_{-3}$) is consistent with the presence of a disulfide radical anion (Figure 2–6).[184] In the final step, the stable reduced form of the enzyme was produced. This form has a visible spectrum expected for a charge transfer complex between a thiol radical and flavin semiquinone, $E\langle^{SH}_{S\cdot} \cdot FlH\cdot^-$, in which an electron equilibrates between the sulfur and the flavin. When a solution of this reduced form of the enzyme was placed in one syringe of the stopped-flow apparatus and a solution of the missing reactant, oxidized glutathione, was placed in the other, the reoxidation of the enzyme occurred with an observed first-order rate constant of 1200 $s^{-1}$. Because this rate constant is greater than the rate constants, ($k_2 + k_{-2}$) and ($k_3 + k_{-3}$), observed for the formation of each intermediate, the rate-limiting step of the enzymatic reaction must be the formation of the reduced form followed by its rapid reoxidation.

If the appearance of a certain product in an enzymatic reaction is monitored in a stopped-flow or quenched-flow experiment and that product is formed in a step followed by significantly slower steps, then a burst of that product will be observed. A **burst** of product is the rapid appearance of an amount of product that is equal in molar concentration to the molar concentration of active sites in the solution and that can be distinguished kinetically from the much slower accumulation of product that follows the burst and proceeds at the normal steady-state rate. For example, when a solution of the complex between L-lactate dehydrogenase and $NAD^+$, with $NAD^+$ in excess over active sites, was mixed with a solution of lactate, an increase in the molar concentration of NADH proceeded rapidly in a burst (Figure 5–25)[185] the size of which was equivalent to the molar concentration of active sites, and then NADH was slowly produced at a rate equal to the steady-state rate of the enzyme. The explanation for this behavior is that there is a step in the enzymatic reaction following hydride transfer that is slower than all of the steps preceding hydride transfer, and for this reason, the very first hydride transfer at each active site is more rapid than the steady-state rate of subsequent hydride transfers.

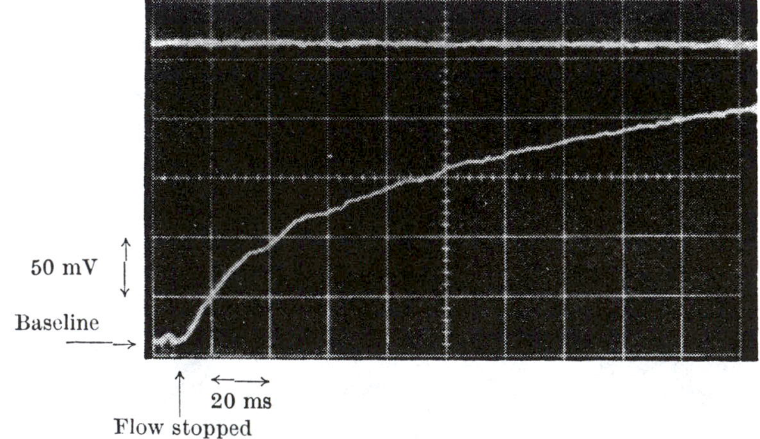


**Figure 5–25:** Burst of NADH formation observed in a stopped-flow spectrophotometer immediately after the complex between L-lactate dehydrogenase and $NAD^+$ was mixed with a solution of lactate.[185] A solution containing 40 mM $NAD^+$ and 70 $\mu$M active sites of porcine cardiac L-lactate dehydrogenase was mixed with an equal volume of a solution of 80 mM lactate in a stopped-flow spectrophotometer. The solutions were at 20 °C, were buffered at pH 6.7, and contained 0.1 M NaCl. The absorbance of the solution adjacent to the mixing chamber was monitored at 340 nm, the wavelength at which NADH absorbs. The voltage of the photomultiplier (millivolts) is presented as a function of time (milliseconds). A first-order burst of NADH production ($t_{1/2}$ = 10 ms) was followed by a zero-order steady-state production of NADH (from about 80 ms onward). The amplitude of the burst (after subtracting steady-state production) was equivalent to the absorbance of a molar concentration of NADH equal to the molar concentration of active sites. Adapted with permission from *Nature*, ref 185. Copyright 1968 Macmillan Magazines Limited.

Argininosuccinate synthase catalyzes the reaction

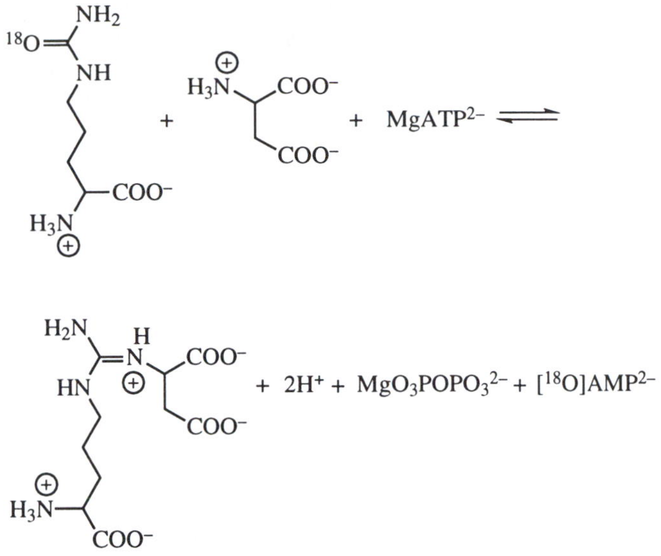

$$^{18}\text{O}{=}\text{C(NH}_2\text{)NH(CH}_2)_3\text{CH(NH}_3^+)\text{COO}^- + \text{H}_3\text{N}^+\text{CH(COO}^-)\text{CH}_2\text{COO}^- + \text{MgATP}^{2-} \rightleftharpoons$$

$$\text{H}_2\text{N(HN=)C{-}N}^+\text{H{-}CH(COO}^-)\text{CH}_2\text{COO}^- \ldots + 2\text{H}^+ + \text{MgO}_3\text{POPO}_3^{2-} + [^{18}\text{O}]\text{AMP}^{2-}$$

(5–203)

During the reaction, $^{18}$oxygen in [*ureido*-$^{18}$O]citrulline is transferred to the phosphate of the product AMP.[186] This observation suggests that citrulline adenylate is formed as an intermediate in the enzymatic reaction

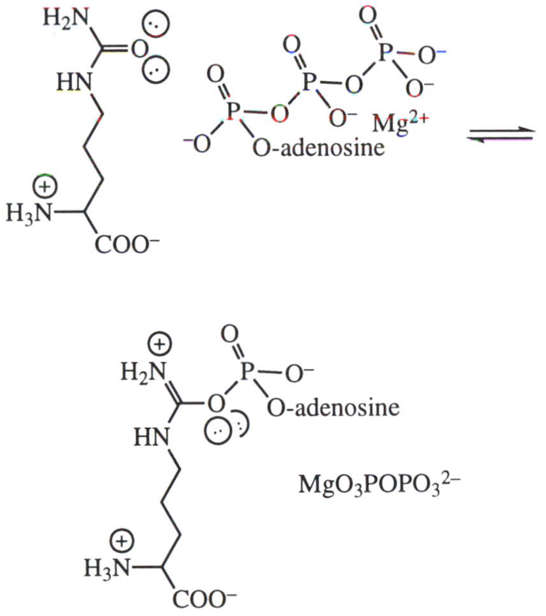

(5–204)

This would activate the citrulline for acyl exchange and explain the capture of the free energy of hydrolysis of MgATP. No isotopic exchanges in the absence of aspartate, which would be expected from Reaction 5–204, have been observed.[187] Solutions of argininosuccinate synthase were mixed in a quenched-flow apparatus with solutions containing citrulline, MgATP, and aspartate; and, after each had reacted for a specific interval, the mixtures were quenched with 0.25 M $H_2SO_4$. The molar concentrations of AMP and argininosuccinate in the acid solutions emerging from the second mixing chamber increased as the time spent between the two mixing chambers was increased (Figure 5–26).[188] The concentration of argininosuccinate increased at a rate equal to the steady-state rate of the enzymatic reaction from the first instant, but a burst in the concentration of the AMP occurred before the rate of its production became equal to that of the argininosuccinate. An explanation for these observations is that the burst of AMP in the acid solution resulted from a burst of citrulline adenylate in the initial mixture that decomposed rapidly to AMP in the 0.25 M $H_2SO_4$ after being released from the active site by the denaturation of the protein. If this is so, citrulline adenylate is formed rapidly ($k_{obs} = 9.7\ s^{-1}$) early in the enzymatic reaction and argininosuccinate is formed subsequently in a slow step ($k_{obs} = 0.6\ s^{-1}$) that is rate-limiting for the turnover of the enzyme.

A burst is also observed in the hydrolysis of MgATP catalyzed by myosin.[189] When myosin was mixed with MgATP, at pH 8 and 20 °C, in a quenched-flow apparatus and the mixture was quenched with 1 M hydrochloric acid after various lengths of time, the yield of $HOPO_3^{2-}$ increased rapidly ($k_{obs} = 14\ s^{-1}$) as the time between the mixing chambers was systematically increased, in a burst equal to 1.2 mol of $HOPO_3^{2-}$ (mol of myosin)$^{-1}$. Following the burst, phosphate was produced at the steady-state rate of the enzymatic reaction ($k_{cat} = 0.02\ s^{-1}$). The reason for the burst is that MgATP associates rapidly with the active site and is hydrolyzed rapidly to form a complex of the enzyme with both MgADP and $HOPO_3^{2-}$, but both MgADP and $HOPO_3^{2-}$ dissociate from the active site only slowly at rates equal to the steady-state rate of turnover of the enzyme.[190]

While bound at the active site of myosin, the MgATP is reversibly hydrolyzed and re-formed, and the details of this process could be studied by combining isotopic exchange and quenched-flow spectroscopy. In one syringe of the quenched-flow apparatus was a solution of Mg[$\gamma,\gamma,\gamma$-$^{18}O_3$]ATP and in the other syringe was a 2-fold molar excess of myosin. The ATP contained $^{18}$oxygen in the three peripheral oxygens of the $\gamma$-phosphate. All of the solutions were made with $H_2{}^{16}O$. At various times the reaction was quenched, and the distribution of [$^{18}O_3$]$O_3POH^{2-}$, [$^{18}O_2$]$O_3POH^{2-}$, and [$^{18}$O]$O_3POH^{2-}$ was determined (Figure 5–27).[191] From the rate of replacement of $^{18}$oxygen with $^{16}$oxygen in the inorganic phosphate bound at the active site during this intermediate isotopic exchange, the kinetic constants for the mechanism

$$\text{E} \xrightarrow{k_1[\text{MgATP}]} \text{E}\cdot\text{MgATP} \underset{k_{-2}}{\overset{k_2}{\rightleftharpoons}}$$

$$\text{E}\cdot\text{MgADP}\cdot\text{HOPO}_3^{2-} \xrightarrow{k_3} \text{E}+\text{MgADP}+\text{HOPO}_3^{2-} \qquad (5\text{–}205)$$

could be calculated to be $k_1 = 2 \times 10^6\ M^{-1}\ s^{-1}$, $k_2 = 80\ s^{-1}$, $k_{-2} = 15\ s^{-1}$, and $k_3 = 0.06\ s^{-1}$. Therefore, the hydrolytic reaction rapidly reverses, and does so many times ($k_2 = 1000k_3$), while the substrates are bound at the active site.

It is also possible in stopped-flow or quenched-flow experiments to observe a **lag** in a reaction. If the process being directly monitored is rapid but is preceded by a slow step, the change will begin slowly and accelerate as the slow step is completed. When a solution of the complex between fatty acid

**Figure 5–26:** Burst of AMP formation catalyzed by bovine liver argininosuccinate synthase upon mixing the enzyme with citrulline, aspartate, and MgATP as measured in a quenched-flow apparatus.[188] Solutions of argininosuccinate synthase (34 $\mu$M active sites) were mixed with equal volumes of a solution of 5.0 mM citrulline, 2.0 mM ATP, 20 mM $MgCl_2$, and 5.0 mM aspartate in a quenched-flow apparatus. Both solutions were at 22 °C and were buffered at pH 8.0. The time between the first mixing chamber, in which the two solutions were combined, and the second mixing chamber, in which the initial mixtures were quenched with two volumes of 0.25 M $H_2SO_4$, was varied by using tubing of varying length. The quenched solutions emerging from the second mixing chamber were assayed for AMP by analytical high-pressure anion-exchange chromatography. The quenched solutions were also assayed for argininosuccinate by converting the argininosuccinate to fumarate with argininosuccinate lyase and quantifying the fumarate by analytical high-pressure anion-exchange chromatography. The number of moles of each product, either AMP (curve A) or argininosuccinate (curve B), in units of moles of product for every mole of active sites in the initial mixtures (product/enzyme), is presented as a function of time (seconds). Reprinted with permission from ref 188. Copyright 1985 American Chemical Society.

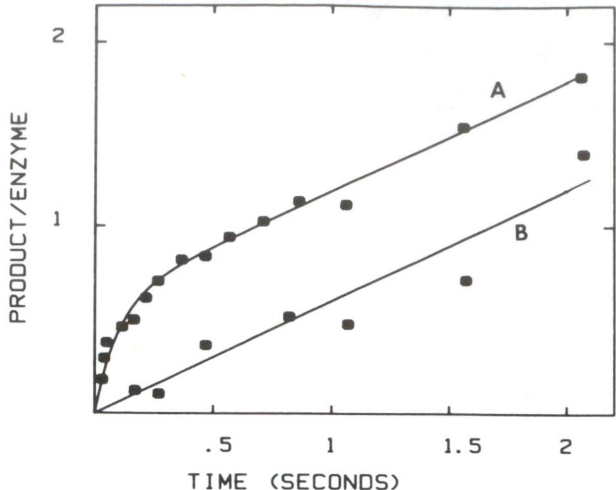


**Figure 5–27:** Exchange of $^{16}$oxygen into the $HOPO_3^{2-}$ bound at the active site of rabbit skeletal myosin in the complex between enzyme, MgADP, and $HOPO_3^{2-}$.[191] Solutions containing subfragment 1 of myosin (80 $\mu$M active sites) were mixed with equal volumes of solutions containing 2 mM $MgCl_2$ and 40 $\mu$M [$\gamma,\gamma,\gamma$-$^{18}O_3$]ATP in a quenched-flow apparatus. The ATP was labeled with $^{18}$oxygen at the three peripheral oxygens of the $\gamma$-phosphate. The solutions were at 20 °C and were buffered at pH 8.0. The time the mixtures spent between the mixing chamber and the outlet in a rapidly stirring solution of 7% $HClO_4$ was varied by choosing different lengths of tubing. The $HOPO_3^{2-}$ in each of the final acidic solutions was purified by extraction of the molybdate complex into organic solvent followed by anion-exchange chromatography. The purified $HOPO_3^{2-}$ was converted to trimethyl phosphate and the relative amounts of [$^{18}O_3$]$OP(OCH_3)_3$, [$^{18}O_2$]$OP(OCH_3)_3$, [$^{18}O$]$OP(OCH_3)_3$, and [$^{16}O_4$]$OP(OCH_3)_3$ were determined by mass spectroscopy. The amount of [$^{16}O_4$]$OPO_3^{2-}$ contaminating the preparation of enzyme (91% of the total $HOPO_3^{2-}$) was subtracted from these amounts. The values were then corrected for the $HOPO_3^{2-}$ released as a product from the active site to obtain the isotopic distribution in only the $HOPO_3^{2-}$ bound at the active site of the enzyme. The relative amounts (percent of total $HOPO_3^{2-}$) of [$^{18}O_3$]$HOPO_3^{2-}$ (3, ●), [$^{18}O_2$]$HOPO_3^{2-}$ (2, ○), [$^{18}O$]$HOPO_3^{2-}$ (1, ■), and [$^{16}O_4$]$HOPO_3^{2-}$ (0, ▲) determined in this analysis are presented as a function of time (milliseconds). The curves drawn are based on Mechanism 5–205 with $k_1 = 2 \times 10^6\ M^{-1}\ s^{-1}$, $k_2 = 80\ s^{-1}$, $k_{-2} = 15\ s^{-1}$, and $k_3 = 0.06\ s^{-1}$. Reprinted with permission from ref 191. Copyright 1981 *Journal of Biological Chemistry.*

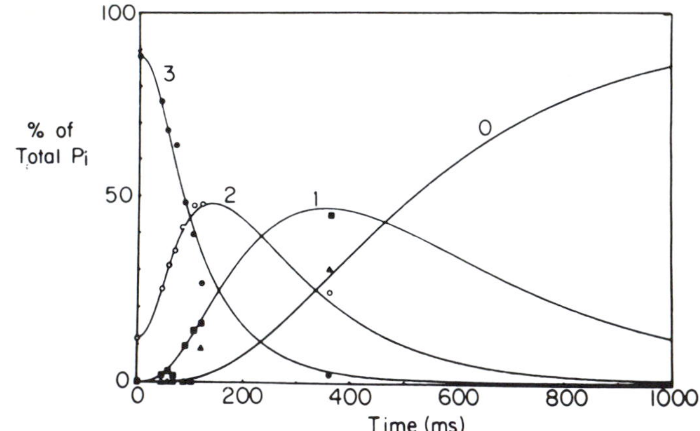

**Figure 5–28:** Lag in the oxidation of NADPH when the complex between NADPH and chicken liver fatty acid synthase is mixed with acetyl-SCoA and malonyl-SCoA.[192] A solution of fatty acid synthase (4.3 $\mu$M active sites) and 1.5 $\mu$M NADPH was mixed with a solution of 110 $\mu$M acetyl-SCoA and 200 $\mu$M malonyl-SCoA in a stopped-flow fluorometer (dead time 2 ms). The two solutions were at 25 °C and were buffered at pH 7.0. The fluorescence (arbitrary units) of the solution in the observation chamber at wavelengths greater than 420 nm ($\lambda_{excitation}$ = 367 nm), which was due to the NADPH, was monitored as a function of time (milliseconds). The solid trace is the direct observation of the fluorescence, and the open circles are values calculated by Equation 5–207 with $k_1 = 4.3\ s^{-1}$, $k_2 = 9.5\ s^{-1}$, and $[NADPH]_0$ = 237 arbitrary units. It was necessary to add a background of 7 arbitrary units to all of the calculated values. Reprinted with permission from ref 192. Copyright 1983 American Chemical Society.

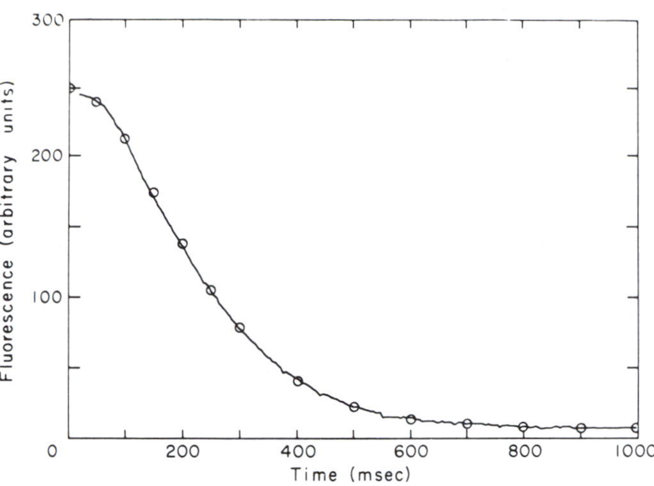


synthase and NADPH at a 3-fold molar excess of enzyme over NADPH was placed in one syringe of a stopped-flow apparatus and a solution of acetyl-SCoA and malonyl-SCoA was placed in the other, the disappearance of the NADPH, as followed by its fluorescence, displayed a lag after mixing (Figure 5–28).[192] The behavior is consistent with the kinetic mechanism

$$\mathrm{E{\cdot}NADPH + Ac\text{-}SCoA + Mal\text{-}SCoA} \xrightarrow{k_1} \mathrm{E}{\cdot}\begin{matrix}\mathrm{SAcAc}\\ \mathrm{NADPH}\end{matrix} \xrightarrow{k_2} \mathrm{E}{\cdot}\begin{matrix}\mathrm{SButOH}\\ \mathrm{NADP^+}\end{matrix} \qquad (5\text{–}206)$$

in which an acetoacetyl (AcAc) bound as a thioester to the pantetheine must be formed first from acetyl-SCoA (Ac-SCoA) and malonyl-SCoA (Mal-SCoA) before it can be reduced, with loss of NADPH, to (hydroxybutyryl)pantetheine (SButOH). The kinetic equation derived from this mechanism is[193]

$$[\mathrm{NADPH}] = [\mathrm{NADPH}]_0 \exp\{-k_2[t + \frac{1}{k_1}(e^{-k_1 t} - 1)]\} \qquad (5\text{–}207)$$

which fits the observations if $k_1 = 4.3\ s^{-1}$ and $k_2 = 9.5\ s^{-1}$ (open circles in Figure 5–28). The rate constant $k_1$ is a complicated function of the concentrations of acetyl-SCoA and malonyl-SCoA. If, however, acetoacetyl-SCoA is used to form the enzyme-bound acetoacetylpantetheine, the rate constant $k_1$ is a simple function as expected for rapid reversible association of the acetoacetyl-SCoA followed by transesterification.

One of the more remarkable results of studying enzymatic reactions by flow methods is the discovery that steps other than those at which the actual chemical transformations occur are often kinetically rate-limiting. When experiments are performed at saturating concentrations of reactants, the associations of the reactants cannot be rate-limiting, but dissociations of products often are. Myosin is an extreme case of this situation, in which hydrolysis of reactant is 1000 times more rapid than dissociation of products (Mechanism 5–205). The rate-determining step in the oxidative phosphorylation catalyzed by glyceraldehyde-3-phosphate dehydrogenase is the release of NADH from the active site.[194] The rate-determining step in the reaction catalyzed by hexokinase is the release of MgADP as a product from the active site.[195]

In many instances, however, relaxations have been observed that cannot be assigned to either the chemical transformation of the substrates bound at the active site or to the dissociation of products from the active site. For example, such unexplained relaxations have been observed in the reductive dephosphorylation catalyzed by glyceraldehyde-3-phosphate dehydrogenase,[194] in at least one of the steps during the formation of glutamyl phosphate at the active site of glutamate–ammonia ligase in the absence of ammonia (Figure 5–19), in a step preceding hydrogen transfer in the reaction catalyzed by dihydropteridine reductase,[196] and in a step immediately preceding the association of the reactant with alkaline phosphatase.[197] Other similar intramolecular relaxations occur following the association of competitive inhibitors such as proflavin, which is a competitive inhibitor of chymotrypsin,[198] or 2-hydroxy-5-nitrobenzylphosphonate, which is a competitive inhibitor of alkaline phosphatase,[199] as well as following the association of glutamate with glutamate–ammonia ligase (Figure 5–18). Relaxations for which there are no other conceivable explanations are usually assigned by exclusion to conformational changes in the protein.

The power of flow methods is their ability to monitor individual steps in an enzymatic reaction. One of the drawbacks is that because everything happens so rapidly, it is necessary to design the experiments imaginatively to identify chemically what process is being observed, and even then, identification may not be possible. All of the experiments are performed in such a way that first-order relaxations occur, so it is important to try to distinguish relaxations arising from association of reactants from relaxations arising from the making and breaking of bonds from relaxations arising from the dissociation of products. Ideally, the actual bonds made or broken in a particular relaxation should be identified. Usually, however, one or more of the relaxations remain unidentified, and they are usually ascribed, for want of a better ascription, to conformational changes of the protein.

## Suggested Reading

Rhee, S.G., Ubom, G.A., Hunt, J.B., & Chock, P.B. (1982) Catalytic Cycle of the Biosynthetic Reaction Catalyzed by Adenylylated Glutamine Synthetase from *Escherichia coli*, *J. Biol. Chem. 257*, 289–297.

## PROBLEM 5–15

Lactate dehydrogenase catalyzes the reaction

$$H^+ + \text{pyruvate} + \text{NADH} \rightleftharpoons \text{NAD}^+ + \text{lactate}$$

The substrate, NADH, absorbs light with a $\lambda_{max}$ = 340 nm, while $NAD^+$ has negligible absorbance at this wavelength.

(A) Write the simplest kinetic mechanism that could describe this enzymatic reaction. Do not derive an equation for $v_0$. Include in your kinetic mechanism only steps involving binding of substrates and the interconversion of substrates and products once they are bound.

This enzymatic reaction has been studied by stopped-flow spectrophotometry. A solution containing a higher concentration of active sites (28 $\mu$M) than NADH (20 $\mu$M) was added to one syringe of a stopped-flow apparatus, and a solution of pyruvate at a saturating concentration (2 mM) was added to the other. The decrease in $A_{340}$ was followed (Figure 5–16A). At the original concentrations of enzyme (28 $\mu$M) and NADH (20 $\mu$M) in the experiment described in Figure 5–16, almost all (>90%) of the NADH was bound to active sites of the enzyme.

(B) Write the step or steps in your kinetic mechanism of part A that are being directly monitored in Figure 5–16.

This stopped-flow experiment was repeated at several different concentrations of pyruvate, and first-order rate constants ($k_{obs}$) were calculated for each of them. Initial velocity measurements were also made at saturating NADH, and turnover numbers ($k_{cat} = v_0/[E]_{TOT}$) at various pyruvate concentrations were calculated from these measurements. Both the first-order rate constants ($k_{obs}$) from the stopped-flow measurements and the turnover numbers from initial velocity measurements were plotted on the same graph as $[S]_0 k_{obs}^{-1}$ against $[S]_0$ or $[S]_0 k_{cat}^{-1}$ against $[S]_0$, respectively (Figure 5–16C). The value of $V_{max}/[E]_{TOT}$ determined from Figure 5–16C is 260 $s^{-1}$.

(C) The bimolecular rate constant for the association of pyruvate with the enzyme under the conditions of Figure 5–16 is greater than $10^6$ $M^{-1}$ $s^{-1}$. Calculate the lower limit of the pseudo-first-order rate constant for the binding of pyruvate to the enzyme at the smallest concentration (1 mM) used in these experiments.

(D) According to the results in Figure 5–16 and the calculation of part C, which step in the mechanism that you wrote in part A is rate-limiting for lactate dehydrogenase in the direction of pyruvate reduction?

(E) When the experiment shown in Figure 5–16 was repeated with $NAD^2H$, $V_{max}/[E]_{TOT}$ was found to be 200 $s^{-1}$. Why does this result suggest that the actual kinetic mechanism is more complicated than what you wrote in part A?

Enzyme and a saturating concentration of $NAD^+$ were added to one syringe of a stopped-flow apparatus, and a saturating concentration of lactate was added to the other. The solutions were rapidly mixed in a stopped-flow spectrophotometer that had a dead time of 1.3 ms. An increase in $A_{340}$, whose rate was so fast that it was complete within the dead time ($k > 1000$ $s^{-1}$), occurred upon mixing. The size of this burst of NADH production could be calculated from the extinction coefficient of enzyme-bound NADH. Above pH 8 the nanomoles of NADH formed in the burst were equal to the nanomoles of active sites present.[177]

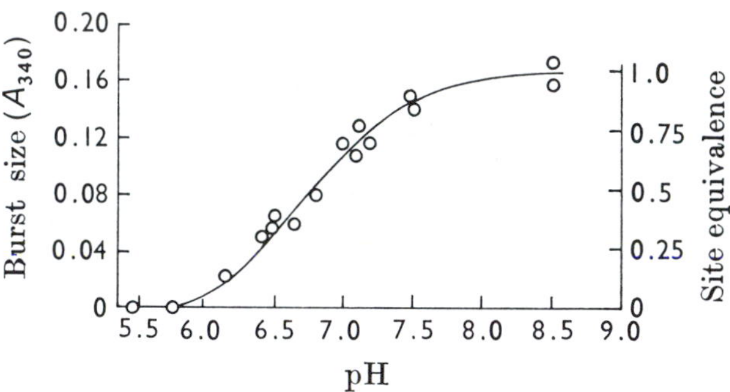


Effect of pH on the magnitude of the initial burst of NADH production.[177] A solution of 40 mM $NAD^+$ and 55 $\mu$M lactate dehydrogenase was mixed with an equal volume of 0.50 M lithium DL-lactate. The buffer was 67 mM $KH_2PO_4$ and the temperature was 22 °C. The pH value was recorded after the reaction. Reprinted with permission from ref 177. Copyright 1971 Biochemical Society.

(F) Write the step or steps in the kinetic mechanism in part A that are being monitored in the experiments described in the figure.

The turnover number at saturation, $V_{max}/[E]_{TOT}$, of lactate dehydrogenase for the oxidation of lactate by $NAD^+$ was determined to be 80 $s^{-1}$ at pH 8 when initial velocity studies were performed under the conditions of those for the figure, but the rate of the burst described in the figure is greater than the dead time of the instrument ($k > 1000$ $s^{-1}$). It was calculated that pyruvate release from the enzyme is greater than 10,000 $s^{-1}$ under the conditions of these experiments. The rate constant of the reaction E·NADH → E + NADH was measured in a stopped-flow experiment under the conditions of the figure at pH 8 and found to be 460 $s^{-1}$.

(G) The results just described and those presented in part E are inconsistent with the mechanism of part A but can be explained if a conformational change of the enzyme is added as a step of the kinetic mechanism. Write an improved kinetic mechanism that incorporates the conformational change in the proper place and makes the mechanism consistent with the results.

## PROBLEM 5–16

A kinetic mechanism for glyceraldehyde-3-phosphate dehydrogenase is presented in the following scheme[194]

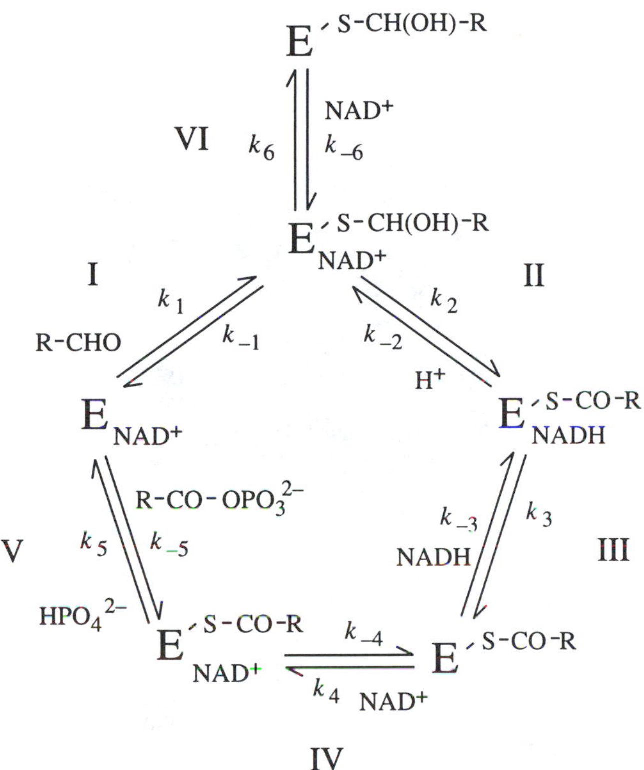


where R-CHO refers to D-glyceraldehyde 3-phosphate and Roman numerals refer to individual processes. This scheme was reprinted with permission from ref 194 (copyright 1971 Biochemical Society). The enzyme also catalyzes the simpler but analogous reaction

$$CH_3CHO + HOPO_3^{2-} + NAD^+ \rightleftharpoons CH_3COOPO_3^{2-} + NADH + H^+$$

The same acetyl enzyme intermediate as that involved in this reaction can be produced by mixing the enzyme with *p*-nitrophenyl acetate. The enzyme was mixed with *p*-nitrophenyl [$^{14}$C]acetate at pH 7.0 in the absence of $NAD^+$ and incubated at 0 °C for 15 min. The protein was then precipitated with cold acetone and washed. For every mole of active sites, 0.6 mol of $^{14}$C was present in this precipitate. The [$^{14}$C]enzyme was digested with pepsin, and a radioactive peptic peptide was isolated whose sequence was

Lys-Ile-Val-Ser-Asn-Ala-Ser-X-Thr-Thr-Asp

The amino acid X contained all of the $^{14}$C in the peptide.[199a] In the total amino acid sequence of the enzyme this region had the sequence KIVSNASCTTD.

(A) Draw the structure of amino acid X in the modified peptide. In your drawing place it within a peptide backbone (–CONHCHXCONH–). Indicate the location of $^{14}$carbon.

The lobster enzyme, with $NAD^+$ bound to each active site, can be crystallized from a concentrated solution of $(NH_4)_2SO_4$. These crystals were submitted to X-ray crystallography. The known sequence of the protein could be placed into the map of electron density that resulted. A large continuous volume of electron density remained that could be satisfactorily filled with a molecule of $NAD^+$. After this was done, two other significant volumes of electron density remained unaccounted for. From an examination of difference Fourier maps made between the original crystals and crystals that had been soaked in 1.4 M sodium citrate, it was concluded that these two volumes of electron density in the original crystals were sulfate anions.

In the final crystallographic molecular model, one of the sulfate anions was located very close to both the $NAD^+$ and the amino acid from which the X in the peptide discussed above had been formed. It was assumed that this was the location of the active site. A "model-building" experiment was then performed in which a model of the substrate glyceraldehyde 3-phosphate was placed into this region of the three-dimensional model. A diagram of this region of the hypothetical molecular model is shown in the following figure[200] (reprinted with permission from ref 200; copyright 1975 *Journal of Biological Chemistry*).

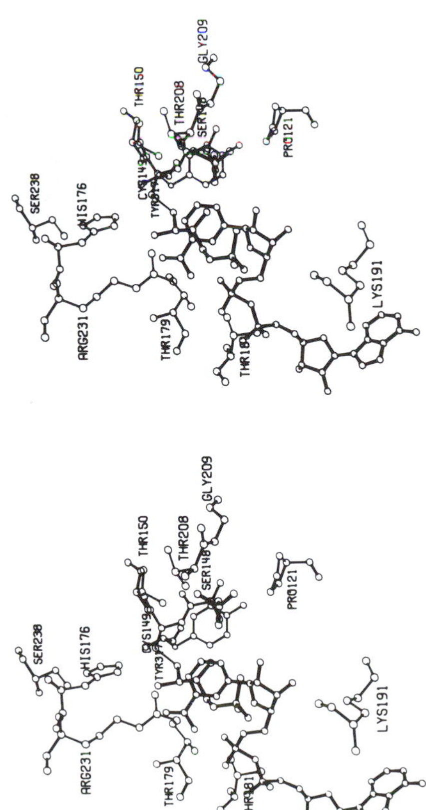

(B) Label the following, by drawing arrows (on a xerographic copy) to the appropriate atoms:
(1) the amide carbon on the nicotinamide ring,
(2) carbon 1 of the glyceraldehyde 3-phosphate,
(3) the heteroatom in the side chain of amino acid X,
(4) the phosphorus atom of the 3-phosphate on the substrate,
(5) the adenine ring of the NAD.
Identify each arrow by the appropriate number (1–5).

(C) In the model-building experiment displayed in the figure, two pieces of real information were used to position carbon 1 of the substrate and rotate it into a reasonable orientation, respectively, and a third piece of real information was used to position the 3-phosphate. What were these three pieces of information?

Several stopped-flow kinetic experiments have been performed with glyceraldehyde-3-phosphate dehydrogenase. When enzyme and glyceraldehyde 3-phosphate were mixed in one syringe, allowed to come to equilibrium, and then rapidly mixed with a solution of $NAD^+$ in the second syringe, the results displayed in the following figure were obtained.[194]

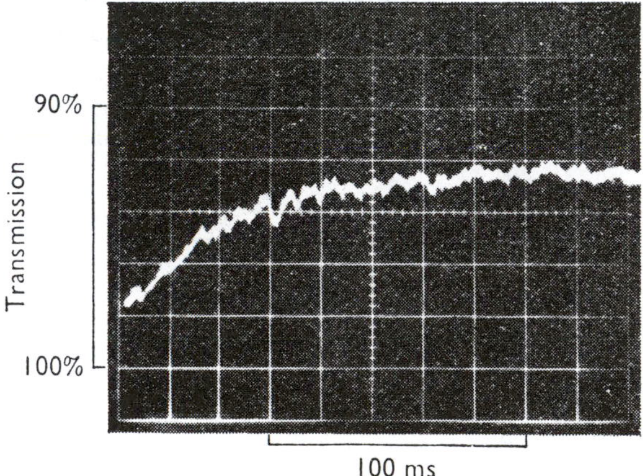


Spectrophotometric record of the reaction of $NAD^+$ and sturgeon glyceraldehyde-3-phosphate dehydrogenase preincubated with glyceraldehyde 3-phosphate.[194] The stopped-flow trace was recorded at 340 nm. The final reaction mixture contained enzyme (1.73 mg $mL^{-1}$), glyceraldehyde 3-phosphate (0.1 mM), $NAD^+$ (2.8 mM), and EDTA (1 mM) at pH 8.4. One syringe contained enzyme that had been preincubated with reactant to ensure formation of the aldehyde–enzyme complex, and the other syringe contained $NAD^+$. The percentage transmission of the solutions before mixing was greater than 100% and is off scale. Reprinted with permission from ref 194. Copyright 1971 Biochemical Society.

The substrate, $NAD^+$, absorbs weakly at 340 nm, while NADH absorbs strongly at this wavelength. Recall the relationship between transmission and absorption.

(D) What two steps in the kinetic mechanism presented above could contribute to the kinetic trace in the figure?

(E) Assume that only one of these two steps is rate-determining under the conditions of the figure. How would you experimentally change the rate of each step independently to show which is rate-determining? Two different experiments should be described.

When lobster enzyme, in which every active site already contains an $NAD^+$, was mixed with 1,3-diphosphoglycerate in one syringe and allowed to come to equilibrium and then rapidly mixed with a solution containing NADH and glyceraldehyde 3-phosphate in the second syringe, the results displayed in the next figure were obtained.

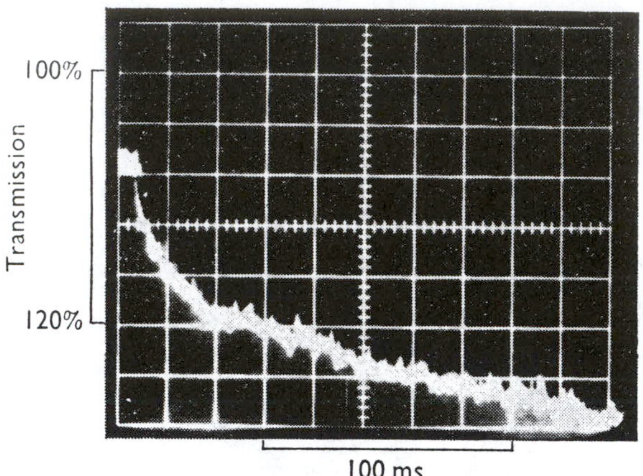


Spectrophotometric record of the reaction of 1,3-diphosphoglycerate and lobster enzyme with NADH in the presence of glyceraldehyde 3-phosphate.[194] The stopped-flow trace was recorded at 340 nm. The final reaction mixture contained enzyme (0.34 mg $mL^{-1}$), 1,3-diphosphoglycerate (0.25 mM), NADH (0.12 mM), glyceraldehyde 3-phosphate (3.0 mM), 2-mercaptoethanol (0.01 mM), and EDTA (150 mM) at pH 5.2. One syringe contained enzyme, 1,3-diphosphoglycerate, and 2-mercaptoethanol; and the other syringe contained NADH and glyceraldehyde 3-phosphate. Reprinted with permission from ref 194. Copyright 1971 Biochemical Society.

(F) Which two steps in the kinetic mechanism presented above are contributing to the fast relaxation seen in the second figure?

(G) Why does the trace in the first figure flatten after the fast relaxation, while the trace in the second figure does not but continues to descend?

## PROBLEM 5–17

Succinate thiokinase catalyzes the following reaction

$$\text{succinyl-SCoA} + HOPO_3^{2-} + \text{MgADP} \rightleftharpoons \text{succinate} + \text{MgATP} + \text{HSCoA}$$

The reaction proceeds through a phosphorylated intermediate, E~P, which is a phosphonohistidine residue on the enzyme. The following three steps have been proposed for the overall reaction

$$\text{succinyl-SCoA} + \text{HOPO}_3^{2-} + \text{E} \rightleftharpoons \text{E·succinyl phosphate} + \text{HSCoA}$$

$$\text{E·succinyl phosphate} \rightleftharpoons \text{E\textasciitilde P} + \text{succinate}$$

$$\text{E\textasciitilde P} + \text{MgADP} \rightleftharpoons \text{E} + \text{MgATP}$$

Two quenched-flow experiments were performed with succinate thiokinase. In both experiments, the two syringes of the apparatus were connected to a mixing chamber, the outlet of which was attached to a tube that terminated in a solution of 8% phenol, which inactivated the enzyme instantaneously as it entered. The time between the mixing of the two solutions and the entry into the 8% phenol was varied between 20 and 100 ms by changing the flow rates of the fluids that entered the mixing chamber. The results are displayed in the following two figures.

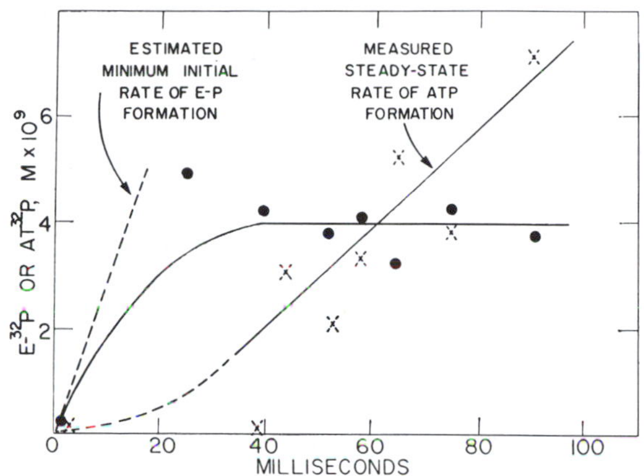


Rate of E~[$^{32}$P]P and [$\gamma$-$^{32}$P]ATP formation from [$^{32}$P]$HOPO_3^{2-}$. Measurements were made at 24 °C. One syringe contained 0.05 mM [$^{32}$P]$HOPO_3^{2-}$ (1.2 × 10$^7$ cpm $\mu$mol$^{-1}$), 0.1 mM ADP, and 0. 2 mM succinyl-SCoA at pH 6.6. The other syringe contained enzyme (0.5 mg mL$^{-1}$) in 0.1 M KCl, 8 mM $MgCl_2$, and 0.1 mM EDTA at pH 7.4. The steady-state velocity of ATP formation as indicated in the figure[201] was measured separately in reactions of 2–15 s duration. Observed [$\gamma$-$^{32}$P]ATP (×); observed E~[$^{32}$P]P (●). Reprinted with permission from ref 201. Copyright 1968 American Chemical Society.

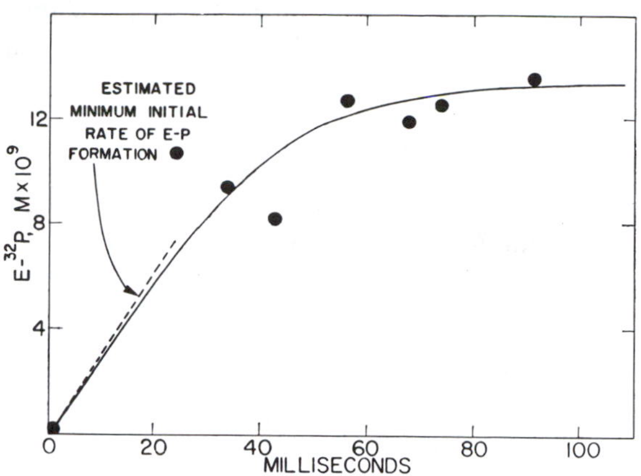


Rate of E~[$^{32}$P]P formation from [$\gamma$-$^{32}$P]ATP.[201] One syringe contained enzyme (0.25 mg mL$^{-1}$) in 0.1 M KCl, 8 mM $MgCl_2$, and 0.1 mM EDTA at pH 7.4. The other syringe contained 0.02 mM HSCoA, 0.08 mM [$\gamma$-$^{32}$P]ATP (9 × 10$^5$ cpm $\mu$mol$^{-1}$), and 0.05 M sodium succinate buffered at the same pH as the enzyme solution. The rate of [$^{32}$P]$HOPO_3^{2-}$ formation was measured separately in reactions of 2–15 s duration. Observed E~[$^{32}$P]P (●). Reprinted with permission from ref 201. Copyright 1968 American Chemical Society.

(A) Write the step-by-step mechanism for the enzymatic reaction that produces the E~[$^{32}$P]P and [$\gamma$-$^{32}$P]ATP observed in the first figure. Draw the structures of all of the substrates explicitly, abbreviating only the coenzyme A as CoAS-, the MgATP as MgADP-O- plus the final phosphate, and the MgADP as MgADP-O. Use the explicit structure of the imidazole of the histidine.

(B) Write the step-by-step mechanism for the enzymatic reaction that produces the E~[$^{32}$P]P observed in the second figure as well as [$^{32}$P]$HOPO_3^{2-}$ at completion of the overall reaction.

The two mechanisms you write in parts A and B, respectively, must be identical to each other, except for the isotopes, because of the principle of microscopic reversibility.

In the reaction displayed in the first figure, the rate of E~P formation was ≥300 nM s$^{-1}$ because equilibrium had been reached at the earliest times measured. The steady-state rate of [$\gamma$-$^{32}$P]ATP production was measured in a separate experiment, which did not use the apparatus and which was run over 2–15 s, and this rate was found to be 90 nM s$^{-1}$. The solid line in the first figure has this slope but has been shifted to coincide with the actual measurements of [$\gamma$-$^{32}$P]ATP in the quenched-flow apparatus.

In the case of the experiment displayed in the second figure, the rate of E~[$^{32}$P]P formation was ≥300 nM s$^{-1}$ and the steady-state rate of [$^{32}$P]$HOPO_3^{2-}$ formation, determined with the same preparations in a separate experiment, was 290 nM s$^{-1}$. In both experiments, the steady-state rates are the usual initial velocities of enzymatic kinetics.

(C) What important fact concerning the formation of E~P do these two experiments establish? Why was quenched flow needed to establish this fact?

Succinate thiokinase was mixed with several substrates, the phosphate bearing the noted isotopic labels, as described in the following paragraph and table, which are quoted with adaptation from the original report.[202]

> The incubation of the purified enzyme with $^{18}O$-labeled phosphate in the presence of succinate, ATP, CoA led to a rapid exchange of oxygen between succinate and phosphate (table). The phosphate was additionally labeled with $^{32}$phosphorus to indicate the extent of over-all reversal of reaction 1. The use of relatively large amounts of enzyme in this experiment to achieve 50% equilibration between inorganic phosphate and ATP revealed a slow P-ATP exchange in the absence of succinate (line 2, table). This probably is due to the presence of an unknown contaminating enzyme since the succinic-thiokinase enzyme isolated from other sources has an absolute succinate requirement for the phosphate-ATP exchange. Within experimental error, the $^{18}O$ exchange reaction between inorganic phosphate and succinate is completely dependent upon ATP, CoA and enzyme. A comparison of the rate of $^{18}O$ exchange into succinate and the $^{32}P$ exchange into ATP indicates that the oxygen exchange reaction can occur independently of the over-all reaction 1. Assuming that the one succinate oxygen would be replaced during complete reversal of reaction 1, the theoretical per cent. $^{18}O$ Exchange into succinate would be 1/4 the value of the $^{32}P$ exchange into ATP (100% exchange representing replacement of all four succinate oxygen atoms). In contrast to this prediction, the $^{18}O$ exchange into succinate is some fivefold greater than the theoretical value (table). This finding implies the reversible formation and cleavage of an intermediate compound, independent of the over-all reversal of reaction 1.

| | $[^{32}P]HOPO_3$–ATP exchange | | $[^{18}O]HOPO_3^{2-}$–succinate exchange | | |
|---|---|---|---|---|---|
| additions[a] | (cpm $\mu mol^{-1}$) | % exchange | mass ratio of $CO_2$ (44/46) | $^{18}$oxygen atom % excess | % exchange[b] |
| (1) complete system | 20,650 | 56 | 0.00762<br>0.00744 | 0.338 | 66.9 |
| (2) complete system – enzyme | <100 | <1 | 0.00422<br>0.00421 | <0.001 | <0.1 |
| (3) complete system – CoA | 2000 | 5 | 0.00439<br>0.00441 | 0.024 | 4.7 |
| (4) complete system – ATP | <100 | <1 | 0.00441<br>0.00447 | 0.024 | 4.7 |
| (5) complete system – succinate | 5900 | 16 | 0.00433<br>0.00425[c] | 0.010 | 2.0 |

[a]The complete system contained 10 mM $MgCl_2$, 10 mM cysteine, 5 mM ATP, 50 mM succinate, 1 mM HSCoA, 1.6 mg $mL^{-1}$ enzyme, and either 5 mM $K_2[^{32}P]HOPO_3$ (73,400 cpm $\mu mol^{-1}$) or 5 mM $K_2[^{18}O]HOPO_3$ (5.05% excess $^{18}O$). The reaction was run at 30 °C for 1 h. Table is from ref 202. [b]Calculated on the basis of 100% exchange being equal to replacement of four $^{16}$oxygens on succinate by $^{18}$oxygen from phosphate and corrected for the 10-fold excess of succinate over phosphate. [c]Succinate added after acidification and extraction.

(D) Write a step-by-step mechanism to explain the transfer of $^{18}$oxygen from $[^{18}O]HOPO_3^{2-}$ to succinate. Expand your mechanism to show why this exchange is 5 times faster than the overall reaction. What is the "intermediate compound"?

A solution containing succinate thiokinase and MgADP, the standard incubation mixture, was mixed with other substrates, as described in the following figure, and the synthesis of MgATP was followed.[203]

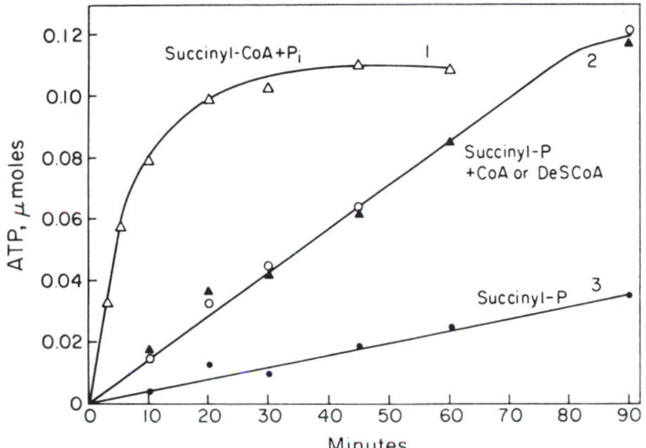


Synthesis of ATP from succinyl phosphate and from succinyl-SCoA plus inorganic phosphate.[203] To the standard incubation mixture at 17 °C of the radiochemical assay were added the following (in micromoles milliliter$^{-1}$): curve 1, succinyl-SCoA (0.20) and potassium phosphate (0.20); curve 2, succinyl phosphate (1.0) and HSCoA or desulfoCoA (0.025); curve 3, succinyl phosphate (1.0). Reprinted with permission from ref 203. Copyright 1969 *Journal of Biological Chemistry.*

(E) Write a step-by-step mechanism to explain the synthesis of MgATP from succinyl phosphate (curve 3). Expand your mechanism to explain synthesis of MgATP from succinyl-SCoA and $HOPO_3^{2-}$ (curve 1). How could a reaction with fewer steps (curve 3) be slower than one with more steps (curve 1)?

(F) Desulfocoenzyme A is coenzyme A lacking the sulfhydryl. How can it activate the formation of ATP from succinyl phosphate?

(G) What other isotopic exchange reactions, such as the $[^{18}O]HOPO_3^{2-} \rightleftharpoons$ succinate reaction described above, should succinate thiokinase catalyze if the three-step mechanism presented in the first paragraph of the problem is correct?

## *References*


1. Oosterbahn, R.A., & van Adrichem, M.E. (1958) *Biochim. Biophys. Acta 27,* 423–425.
2. Bender, M.L., & Kezdy, F.J. (1964) *J. Am. Chem. Soc. 86,* 3704–3714.
3. Palmer, J.L., & Abeles, R.H. (1979) *J. Biol. Chem. 254,* 1217–1226.
4. Buckel, W., Buschmeier, V., & Eggerer, H. (1971) *Hoppe-Seyler Z. Physiol. Chem. 352,* 1195–1205.
5. Schmellenkamp, H., & Eggerer, H. (1972) *Hoppe-Seyler Z. Physiol. Chem. 353,* 1563–1564.
6. Oppenheimer, N.J., Singh, M., Sweeley, C.C., Sung, S., & Srere, P.A. (1979) *J. Biol. Chem. 254,* 1000–1002.
7. Buckel, W., Ziegert, K., & Eggerer, H. (1973) *Eur. J. Biochem. 37,* 295–304.
8. Wood, T.G., Weisz, O.A., & Kozarich, J.W. (1984) *J. Am. Chem. Soc. 106,* 2223–2224.
9. Buckel, W. (1976) *Eur. J. Biochem. 64,* 263–267.
10. Basu, A., Subramanian, S., Hiremath, L.S., & SivaRaman, C. (1983) *Biochem. Biophys. Res. Commun. 114,* 310–317.
11. White, H., & Jencks, W.P. (1976) *J. Biol. Chem. 251,* 1688–1699.
12. Solomon, F., & Jencks, W.P. (1969) *J. Biol. Chem. 244,* 1079–1081.
13. Buckel, W., & Bobi, A. (1976) *Eur. J. Biochem. 64,* 255–262.
14. Dimroth, P., Mayer, K., & Eggerer, H. (1975) *Eur. J. Biochem. 51,* 267–273.
15. Lynen, F. (1967) *Biochem. J. 102,* 381–400.
16. Ziegenhorn, J., Niedermeier, R., Nüssler, C., & Lynen, F. (1972) *Eur. J. Biochem. 30,* 285–300.
17. Schweizer, E., Piccinini, F., Duba, C., Gunther, S., Ritter, E., & Lynen, F. (1970) *Eur. J. Biochem. 15,* 483–499.
18. Stern, A., Sedgwick, B., & Smith, S. (1982) *J. Biol. Chem. 257,* 799–803.
19. Colomb, M.G., Chéruy, A., & Vignais, P.V. (1972) *Biochemistry 11,* 3378–3386.
20. Lienhard, G.E., & Secemski, I. (1973) *J. Biol. Chem. 248,* 1121–1123.
21. Richard, J.P., & Frey, P.A. (1978) *J. Am. Chem. Soc. 100,* 7757–7758.
22. Sowadski, J.M., Handschumacher, M.D., Krishna Murthy, H.M., Foster, B.A., & Wyckoff, H.W. (1985) *J. Mol. Biol. 186,* 417–433.

23. Jones, R.S., Kindman, A.L., & Knowles, J.R. (1978) *Nature 275*, 564–565.
24. Ramaley, R.F., Bridger, W.A., Moyer, R.W., & Boyer, P.D. (1967) *J. Biol. Chem. 242*, 4287–4298.
25. DeLuca, M., Ebner, K.E., Hultquist, D.E., Kreil, G., Peter, J.B., Moyer, R.W., & Boyer, P.D. (1963) *Biochem. Z. 338*, 512–525.
26. Edlund, B., Rask, L., Olsson, P., Walinder, O., Zetterqvist, O., & Engstrom, L. (1969) *Eur. J. Biochem. 9*, 451–455.
27. Schwartz, J.H., Crestfield, A.M., & Lipmann, F. (1963) *Proc. Natl. Acad. Sci. U.S.A. 49*, 722–729.
28. Yamamoto, T., & Tonomura, Y. (1967) *J. Biochem. (Tokyo) 62*, 558–575.
29. Degani, C., & Boyer, P.D. (1973) *J. Biol. Chem. 248*, 8222–8226.
30. Pas, H., Meyer, G.H., Kruizinga, W.H., Tamminga, K.S., van Weeghel, R.P., & Robillard, G.T. (1991) *J. Biol. Chem. 266*, 6690–6692.
31. Walinder, O., & Joshi, J.G. (1974) *J. Biol. Chem. 249*, 3166–3169.
32. Britton, H.G., & Clarke, J.B. (1968) *Biochem. J. 110*, 161–175.
33. Ray, W.J., Hermodson, M.A., Puvathingal, J.M., & Mahony, W.C. (1983) *J. Biol. Chem. 258*, 9166–9174.
34. Sheu, K.R., Richard, J.P., & Frey, P.A. (1979) *Biochemistry 18*, 5548–5556.
35. Laursen, R.A., & Westheimer, F.H. (1966) *J. Am. Chem. Soc. 88*, 3426–3430.
36. Grazi, E., Rowley, P.T., Cheng, T., Tchola, O., & Horecker, B.L. (1962) *Biochem. Biophys. Res Commun. 9*, 38–43.
37. Healy, M.J., & Christen, P. (1972) *J. Am. Chem. Soc. 94*, 7911–7916.
38. Kuo, D.J., & Rose, I.A. (1985) *Biochemistry 24*, 3947–3952.
39. Matherly, L.H., DeBrosse, C.W., & Phillips, A.T. (1982) *Biochemistry 21*, 2789–2794.
40. Matherly, L.H., Johnson, K.A., & Phillips, A.T. (1982) *Biochemistry 21*, 2795–2798.
41. Egan, R.M., Matherly, L.H., & Phillips, A.T. (1981) *Biochemistry 20*, 132–137.
42. Bellisario, R.L., Maley, G.F., Galivan, J.H., & Maley, F. (1976) *Proc. Natl. Acad. Sci. U.S.A. 73*, 1848–1852.
43. Kalman, T.I. (1971) *Biochemistry 10*, 2567–2573.
44. Matthews, D.A., Appelt, K., Oatley, S.J., & Xuong, N.H. (1990) *J. Mol. Biol. 214*, 923–936.
45. Matthews, D.A., Villafranca, J.E., Janson, C.A., Smith, W.W., Welsh, K., & Freer, S. (1990) *J. Mol. Biol. 214*, 937–948.
46. Tatum, C.M., Vederas, J., Schleicher, E., Benkovic, S.J., & Floss, H.G. (1977) *J. Chem. Soc., Chem. Commun.*, 218–220.
47. Slieker, L.J., & Benkovic, S.J. (1984) *J. Am. Chem. Soc. 106*, 1833–1838.
48. Pastore, E.J., & Freidkin, M. (1962) *J. Biol. Chem. 237*, 3802–3810.
49. Lorenson, M.Y., Maley, G.F., & Maley, F. (1967) *J. Biol. Chem. 242*, 3332–3344.
50. Wilson, R.S., & Mertes, M.P. (1973) *Biochemistry 12*, 2879–2886.
51. Teng, C.P., & Ganem, B. (1984) *J. Am. Chem. Soc. 106*, 2463–2464.
52. Kuo, D.J., O'Connell, E.L., & Rose, I.A. (1979) *J. Am. Chem. Soc. 101*, 5025–5030.
53. Kuo, D.J., & Rose, I.A. (1982) *J. Am. Chem. Soc. 104*, 3235–3236.
54. Powers, S.G., & Meister, A. (1976) *Proc. Natl. Acad. Sci. U.S.A. 73*, 3020–3024.
55. Bild, G.S., & Morris, J.C. (1984) *Arch. Biochem. Biophys. 235*, 41–47.
56. Siegel, M.I., & Lane, M.D. (1973) *J. Biol. Chem. 248*, 5486–5498.
57. Schloss, J.V., & Lorimer, G.H. (1982) *J. Biol. Chem. 257*, 4691–4694.
58. Chen, C.H., & Eisenberg, F. (1975) *J. Biol. Chem. 250*, 2963–2967.
59. Abeles, R.H., Fish, S., & Lapinskas, B. (1982) *Biochemistry 21*, 5557–5562.
60. Palekar, A.G., Tate, S.S., & Meister, A. (1973) *J. Biol. Chem. 248*, 1158–1167.
61. Rippa, M., Signori, M., & Dallochio, F. (1973) *J. Biol. Chem. 248*, 4920–4925.
62. Sasiak, K., & Rilling, H.C. (1988) *Arch. Biochem. Biophys. 260*, 622–627.
63. Altman, L.J., Kowerski, R.C., & Rilling, H.C. (1971) *J. Am. Chem. Soc. 93*, 1782–1783.
64. Coates, R.M., & Robinson, W.H. (1971) *J. Am. Chem. Soc. 93*, 1785–1786.
65. Popjak, G., Edmond, J., & Wong, S. (1973) *J. Am. Chem. Soc. 95*, 2713–2714.
66. Rilling, H.C. (1966) *J. Biol. Chem. 241*, 3233–3246.
67. Timmons, R.B., Rhee, S.G., Luterman, D.L., & Chock, P.B. (1974) *Biochemistry 13*, 4479–4485.
68. Kowalsky, A., Wyttenbach, C., Langer, L., & Koshland, D.E. (1956) *J. Biol. Chem. 219*, 719–725.
69. Villafranca, J.J., & Wedler, F.C. (1974) *Biochemistry 13*, 3286–3291.
70. Gass, J.D., & Meister, A. (1970) *Biochemistry 9*, 842–846.
71. Tsuda, Y., Stephani, R.A., & Meister, A. (1971) *Biochemistry 10*, 3186–3189.
72. Todhunter, J.A., & Purich, D.L. (1975) *J. Biol. Chem. 250*, 3505–3509.
73. Buttlaire, D.H., Himes, R.H., & Reed, G.H. (1976) *J. Biol. Chem. 251*, 4159–4161.
74. Alberg, D.G., Lauhon, C.T., Nyfeler, R., Fässler, A., & Bartlett, P.A. (1992) *J. Am. Chem. Soc. 114*, 3535–3546.
75. Gerlt, J.A., Kozarich, J.W., Kenyon, G.L., & Gassman, P.G. (1991) *J. Am. Chem. Soc. 113*, 9667–9669.
76. Klinman, J.P., & Rose, I.A. (1971) *Biochemistry 10*, 2259–2266.
77. Anderson, V.E., Weiss, P.M., & Cleland, W.W. (1984) *Biochemistry 23*, 2779–2786.
78. Porter, D.J.T., & Bright, H.J. (1980) *J. Biol. Chem. 255*, 4772–4780.
79. Schloss, J.V., Porter, D.J.T., Bright, H.J., & Cleland, W.W. (1980) *Biochemistry 19*, 2358–2362.
80. Healy, M.J., & Christen, P. (1973) *Biochemistry 12*, 35–41.
81. Walsh, C.T., Krodel, E., Massey, V., & Abeles, R.H. (1973) *J. Biol. Chem. 248*, 1946–1955.
82. Stubbe, J., Fish, S., & Abeles, R.H. (1980) *J. Biol. Chem. 255*, 236–242.
83. Stokes, T.M., & Wilson, I.B. (1972) *Biochemistry 11*, 1061–1064.
84. Klein, H.W., Im, M.J., & Palm, D. (1986) *Eur. J. Biochem. 157*, 107–114.
85. Cornforth, J.W., Cornforth, R.H., Popjak, G., & Yengoyan, L. (1966) *J. Biol. Chem. 241*, 3970–3987.
86. Poulter, C.D., Satterwhite, D.M., & Rilling, H.C. (1976) *J. Am. Chem. Soc. 98*, 3376–3377.
87. Poulter, C.D., Wiggins, P.L., & Le, A.T. (1981) *J. Am. Chem. Soc. 103*, 3926–3927.
88. Poulter, C.D., Mash, E.A., Argyle, J.C., Muscio, O.J., & Rilling, H.C. (1979) *J. Am. Chem. Soc. 101*, 6761–6763.
89. Gambiel, H., & Croteau, R. (1984) *J. Biol. Chem. 259*, 740–748.
90. Croteau, R., Wheeler, C.J., Askela, R., & Oehlschlager, A.C. (1986) *J. Biol. Chem. 261*, 7257–7263.
91. Croteau, R.B., Shaskus, J.J., Renstrom, B., & Felton, N.M. (1985) *Biochemistry 24*, 7077–7085.
92. Cane, D.E., Saito, A., Croteau, R., Shaskus, J., & Felton, M. (1982) *J. Am. Chem. Soc. 104*, 5831–5833.
93. Fasanella, E.L., & Gordy, W. (1969) *Proc. Natl. Acad. Sci. U.S.A. 62*, 299–304.

94. Groves, J.T., McClusky, G.A., White, R.E., & Coon, M.J. (1978) *Biochem. Biophys. Res. Commun. 81*, 154–160.
95. Aberg, A., Hahne, S., Karlsson, M., Larsson, A., Ormo, M., Ahgren, A., & Sjöberg, B. (1989) *J. Biol. Chem. 264*, 12249–12252.
96. Larsson, A., & Sjöberg, B. (1986) *EMBO J. 5*, 2037–2040.
97. Sjöberg, B., Reichard, P., Gräslund, A., & Ehrenberg, A. (1978) *J. Biol. Chem. 253*, 6863–6865.
98. Nordlund, P., Sjöberg, B., & Eklund, H. (1990) *Nature 345*, 593–598.
99. Petersson, L., Gräslund, A., Ehrenberg, A., Sjöberg, B., & Reichard, P. (1980) *J. Biol. Chem. 255*, 6706–6712.
100. Stubbe, J., Ator, M., & Krenitsky, T. (1983) *J. Biol. Chem. 258*, 1625–1630.
101. Batterham, T.J., Ghambeer, R.K., Blakley, R.L., & Brownson, C. (1967) *Biochemistry 6*, 1203–1208.
102. Stubbe, J. (1990) *J. Biol. Chem. 265*, 5329–5332.
103. Thelander, L., Larsson, B., Hobbs, J., & Eckstein, F. (1976) *J. Biol. Chem. 251*, 1398–1405.
104. Stubbe, J.A., & Kozarich, J.W. (1980) *J. Am. Chem. Soc. 102*, 2505–2507.
105. Ator, M.A., & Stubbe, J. (1985) *Biochemistry 24*, 7214–7221.
106. Ashley, G.W., Harris, G., & Stubbe, J. (1988) *Biochemistry 27*, 7841–7845.
107. Fitzpatrick, P.F., Flory, D.R., & Villafranca, J.J. (1985) *Biochemistry 24*, 2108–2114.
108. Ritchie, C.D., & Sager, W.F. (1964) *Prog. Phys. Org. Chem. 2*, 323–400.
109. Miller, S.M., & Klinman, J.P. (1985) *Biochemistry 24*, 2114–2127.
110. Maxwell, E.S. (1957) *J. Biol. Chem. 229*, 139–151.
111. Wilson, D.B., & Hogness, D.S. (1964) *J. Biol. Chem. 239*, 2469–2481.
112. Nelsestuen, G., & Kirkwood, S. (1970) *Fed. Proc. 29*, 337.
113. Maitra, U.S., & Ankel, H. (1971) *Proc. Natl. Acad. Sci. U.S.A. 68*, 2660–2663.
114. Maitra, U.S., & Ankel, H. (1973) *J. Biol. Chem. 248*, 1477–1479.
115. Muth, W.L., & Costilow, R.N. (1974) *J. Biol. Chem. 249*, 7463–7467.
115a. Inoue, H., Suzuki, F., Tanioka, H., & Takeda, Y. (1968) *J. Biochem. (Tokyo) 63*, 89–100.
115b. Stokes, T.M., & Wilson, I.B. (1972) *Biochemistry 11*, 1061–1064.
115c. Viratelle, O.M., & Yon, J.M. (1973) *Eur. J. Biochem. 33*, 110–116.
115d. Sinnott, M.L., & Viratelle, O.M. (1973) *Biochem. J. 133*, 81–87.
115e. Wolfenden, R. (1966) *J. Am. Chem. Soc. 88*, 3157–3158.
116. Wong, L., & Frey, P.A. (1974) *J. Biol. Chem. 249*, 2322–2324.
117. Yang, S.L., & Frey, P.A. (1979) *Biochemistry 18*, 2980–2984.
118. Mourad, N., & Parks, R.E. (1965) *Biochem. Biophys. Res. Commun. 19*, 312–316.
119. Mourad, N., & Parks, J.R. (1966) *J. Biol. Chem. 241*, 3838–3844.
120. Beach, R.L., & Plaut, G.W.E. (1970) *J. Am. Chem. Soc. 92*, 2913–2916.
121. Withers, S.G., Madsen, N.B., Sykes, B.D., Takagi, M., Shimomura, S., & Fukui, T. (1981) *J. Biol. Chem. 256*, 10759–10762.
122. Parrish, R.F., Uhing, R.J., & Graves, D.J. (1977) *Biochemistry 16*, 4824–4831.
123. Hermes, J.D., Roeske, C.A., O'Leary, M.H., & Cleland, W.W. (1982) *Biochemistry 21*, 5106–5114.
124. Belasco, J.G., Albery, W.J., & Knowles, J.R. (1983) *J. Am. Chem. Soc. 105*, 2475–2477.
125. Cook, P.F., Blanchard, J.S., & Cleland, W.W. (1980) *Biochemistry 19*, 4853–4858.
126. Rendina, A.R., Hermes, J.D., & Cleland, W.W. (1984) *Biochemistry 23*, 6257–6262.
127. Hermes, J.D., Tipton, P.A., Fisher, M.A., O'Leary, M.H., Morrison, J.F., & Cleland, W.W. (1984) *Biochemistry 23*, 6263–6275.
128. Danishefsky, S., Hirama, M., Fritsch, N., & Clardy, J. (1979) *J. Am. Chem. Soc. 101*, 7013–7018.
129. Hermes, J.D., Weiss, P.M., & Cleland, W.W. (1985) *Biochemistry 24*, 2959–2967.
130. Hermes, J.D., Morrical, S.W., O'Leary, M., & Cleland, W.W. (1984) *Biochemistry 23*, 5479–5488.
131. Blanchard, J.S., & Cleland, W.W. (1980) *Biochemistry 19*, 3543–3550.
132. Grissom, C.B., & Cleland, W.W. (1985) *Biochemistry 24*, 944–948.
133. Rose, I.A., & Rieder, S.V. (1958) *J. Biol. Chem. 231*, 315–329.
134. Dinovo, E.C., & Boyer, P.D. (1971) *J. Biol. Chem. 246*, 4586–4593.
135. Myers, O.E., & Prestwood, R.J. (1951) in *Radioactivity Applied to Chemistry* (Wahl, A.C., & Bonner, N.A., Eds.) pp 6–43, John Wiley, New York.
136. Wedler, F.C., & Boyer, P.D. (1972) *J. Biol. Chem. 247*, 984–992.
137. Fisher, H.F., & Viswanathan, T.S. (1984) *Proc. Natl. Acad. Sci. U.S.A. 81*, 2747–2751.
138. Esaki, N., Nakayama, T., Sawada, S., Tanaka, H., & Soda, K. (1985) *Biochemistry 24*, 3857–3862.
139. Boyer, P.D. (1959) *Arch. Biochem. Biophys. 82*, 387–410.
140. Allison, R.D., Todhunter, J.A., & Purich, D.L. (1977) *J. Biol. Chem. 252*, 6046–6051.
141. Raushel, F.M., & Cleland, W.W. (1977) *Biochemistry 16*, 2176–2181.
142. Hoagland, M.B. (1955) *Biochim. Biophys. Acta 16*, 288–289.
143. DeMoss, J.A., Genuth, S.M., & Novelli, G.D. (1956) *Proc. Natl. Acad. Sci. U.S.A. 42*, 325–332.
144. McClure, W.R., Lardy, H.A., & Cleland, W.W. (1971) *J. Biol. Chem. 246*, 3584–3590.
145. Scrutton, M.C., Keech, D.B., & Utter, M.F. (1965) *J. Biol. Chem. 240*, 574–581.
146. Scrutton, M.C., & Utter, M.F. (1965) *J. Biol. Chem. 240*, 3714–3723.
147. Kaziro, Y., Hass, L.F., Boyer, P.D., & Ochoa, S. (1962) *J. Biol. Chem. 237*, 1460–1468.
148. Lynen, F., Knappe, J., Lorch, E., Jütting, G., Ringelmann, E., & Lachance, J.P. (1961) *Biochem. Z. 335*, 123–167.
149. Evans, H.J., & Wood, H.G. (1968) *Proc. Natl. Acad. Sci. U.S.A. 61*, 1448–1453.
150. Milner, Y., & Wood, H.G. (1976) *J. Biol. Chem. 251*, 7920–7928.
151. Milner, Y., & Wood, H.G. (1972) *Proc. Natl. Acad. Sci. U.S.A. 69*, 2463–2468.
152. Midelfort, C.F., & Rose, I.A. (1976) *J. Biol. Chem. 251*, 5881–5887.
153. Lagerkvist, U. (1958) *J. Biol. Chem. 233*, 143–149.
154. Fukuyama, T.T. (1966) *J. Biol. Chem. 241*, 4745–4749.
155. von der Saal, W., Crysler, C.S., & Villafranca, J.J. (1985) *Biochemistry 24*, 5343–5350.
156. Rosing, J., Kayalar, C., & Boyer, P.D. (1977) *J. Biol. Chem. 252*, 2478–2485.
157. Levy, H.M., & Koshland, D.E. (1959) *J. Biol. Chem. 234*, 1102–1107.
158. Myers, J.A., & Boyer, P.D. (1984) *Biochemistry 23*, 1264–1269.
159. Webb, M.R., & Trentham, D.R. (1980) *J. Biol. Chem. 255*, 8629–8632.
160. Shukla, K.K., & Levy, H.M. (1977) *Biochemistry 16*, 132–136.
161. Lymn, R.W., & Taylor, E.W. (1971) *Biochemistry 10*, 4617–4624.
162. Sleep, J.A., Hackney, D.D., & Boyer, P.D. (1980) *J. Biol. Chem. 255*, 4094–4099.
163. Sleep, J.A., & Boyer, P.D. (1978) *Biochemistry 17*, 5417–5422.

163a. Kyte, J. (1995) *Structure in Protein Chemistry*, pp 496–497, Garland Publishing, New York.
164. Hutton, R.L., & Boyer, P.D. (1979) *J. Biol. Chem. 254*, 9990–9993.
165. Kenyon, G.L., & Hegeman, G.D. (1970) *Biochemistry 21*, 4036–4043.
166. Barber, G.A. (1979) *J. Biol. Chem. 254*, 7600–7603.
167. Gomes, B., Fendrich, G., & Abeles, R.H. (1981) *Biochemistry 20*, 1481–1490.
168. Chang, C., Laghai, A., O'Leary, M., & Floss, H.G. (1982) *J. Biol. Chem. 257*, 3564–3569.
169. Benson, R.W., & Boyer, P.D. (1969) *J. Biol. Chem. 244*, 2366–2371.
169a. Ahmed, S.A., Esaki, N., Tanaka, H., & Soda, K. (1986) *Biochemistry 25*, 385–388.
169b. Saver, B.G., & Knowles, J.R. (1982) *Biochemistry 21*, 5398–5403.
169c. Rubio, V., Britton, H.G., Grisolia, S., Sproat, B.S., & Lowe, G. (1981) *Biochemistry 20*, 1969–1974.
169d. Grimshaw, C., Sogo, S.G., & Knowles, J.R. (1982) *J. Biol. Chem. 257*, 596–598.
169e. Anton, D.L., Hedstrom, L., Fish, S.M., & Abeles, R.H. (1983) *Biochemistry 22*, 5903–5908.
170. Steinrücken, H.C., & Amrhein, N. (1984) *Eur. J. Biochem. 143*, 351–357.
170a. Fujita, N., Izui, K., Nishino, T., & Katsuki, H. (1984) *Biochemistry 23*, 1774–1779.
171. Hartridge, H., & Roughton, F.J.W. (1923) *Proc. R. Soc. (London) A104*, 376–394.
172. Roughton, F.J.W. (1934) *Proc. R. Soc. (London) B115*, 495–503.
173. Chance, B. (1940) *J. Franklin Instit. 229*, 455–476.
174. Gibson, Q.H., & Roughton, F.J.W. (1955) *Proc. R. Soc. (London) B143*, 310–334.
175. Roughton, F.J.W., & Chance, B. (1963) in *Techniques of Organic Chemistry: Volume VIII, Part II, Investigation of Rates and Mechanisms of Reactions* (Friess, S.L., Lewis, E.S., & Weissberger, A., Eds.) pp 703–792, Wiley-Interscience, New York.
176. Rhee, S.G., Ubom, G.A., Hunt, J.B., & Chock, P.B. (1982) *J. Biol. Chem. 257*, 289–297.
177. Stinson, R.A., & Gutfreund, H. (1971) *Biochem. J. 121*, 235–240.
178. Shapiro, B.M., & Stadtman, E.R. (1968) *J. Biol. Chem. 243*, 3769–3771.
179. Rhee, S.G., Ubom, G.A., Hunt, J.B., & Chock, P.B. (1981) *J. Biol. Chem. 256*, 6010–6016.
180. Bull, C., Ballou, D.P., & Otsuka, S. (1981) *J. Biol. Chem. 256*, 12681–12686.
181. Drewe, W.F., & Dunn, M.F. (1985) *Biochemistry 24*, 3977–3987.
182. Huber, P.W., & Brandt, K.G. (1980) *Biochemistry 19*, 4568–4575.
183. Pai, E.F., & Schulz, G.E. (1983) *J. Biol. Chem. 258*, 1752–1757.
184. Asmus, K.D. (1983) in *Radioprotectors and Anticarcinogens* (Nygaard, O.F., & Simic, M.G., Eds.) pp 23–42, Academic Press, New York.
185. Criddle, R.S., McMurray, C.H., & Gutfreund, H. (1968) *Nature 220*, 1091–1095.
186. Rochovansky, O., & Ratner, S. (1961) *J. Biol. Chem. 236*, 2254–2260.
187. Hilscher, L.W., Hanson, C.D., Russell, D.H., & Rauschel, F.M. (1985) *Biochemistry 24*, 5888–5893.
188. Ghose, C., & Rauschel, F.M. (1985) *Biochemistry 24*, 5894–5898.
189. Lymn, R.W., & Taylor, E.W. (1970) *Biochemistry 9*, 2975–2983.
190. Taylor, E.W., Lymn, R.W., & Moll, G. (1970) *Biochemistry 15*, 2984–2991.
191. Webb, M.R., & Trentham, D.R. (1981) *J. Biol. Chem. 256*, 10910–10916.
192. Cognet, J.A.H., Cox, B.G., & Hammes, G.G. (1983) *Biochemistry 22*, 6281–6287.
193. Cognet, J.A.H., & Hammes, G.G. (1985) *Biochemistry 24*, 290–297.
194. Trentham, D.R. (1971) *Biochem. J. 122*, 71–77.
195. Wilkinson, K.D., & Rose, I.A. (1979) *J. Biol. Chem. 254*, 12567–12572.
196. Poddar, S., & Henkin, J. (1984) *Biochemistry 23*, 3143–3148.
197. Halford, S.E. (1971) *Biochem. J. 125*, 319–327.
198. Havsteen, B.N. (1967) *J. Biol. Chem. 242*, 769–776.
199. Halford, S.E., Bennett, N.G., Trentham, D.R., & Gutfreund, H. (1969) *Biochem. J. 114*, 243–251.
199a. Harris, I., Meriwether, B.P., & Park, J.H. (1963) *Nature 198*, 154–157.
200. Moras, D., Olsen, K.W., Sabesan, M.N., Buehner, M., Ford, G.C., & Rossmann, M.G. (1975) *J. Biol. Chem. 250*, 9137–9162.
201. Bridger, W.A., Millen, W.A., & Boyer, P.D. (1968) *Biochemistry 7*, 3608–3616.
202. Hager, L.P. (1957) *J. Am. Chem. Soc. 79*, 4864–4866.
203. Hildebrand, J.G., & Spector, L.B. (1969) *J. Biol. Chem. 244*, 2606–2613.

## Chapter SIX

# Conformational Changes

Because many instances have been documented crystallographically, there is no doubt that conformational changes in the molecular structure of a protein can be promoted by the binding of a substrate, an inhibitor, or a ligand. There are several types of conformational changes that have been observed. They differ in the extent to which the elements of the protein participate in the change.

A conformational change can alter the spatial relationships among the subunits in an oligomer. The conformational change between deoxyhemoglobin, in which all the hemes are unoccupied, and oxyhemoglobin, in which all four hemes have oxygen bound, involves the rotation of one $\alpha\beta$ dimer with respect to the other by 15° and an increase in the distance between the two dimers of 0.1 nm along a screw axis (Figure 6–1).[1] The conformational change between unliganded aspartate transcarbamylase and aspartate transcarbamylase in which all six active sites have *N*-(phosphonacetyl)-L-aspartate bound involves a rotation of one catalytic trimer[1a] with respect to the other by 10° and a widening of the distance between the two by 1.2 nm along the 3-fold rotational axis of symmetry (Figure 6–2).[2] During this change, the three regulatory dimers[1a] turn about their respective 2-fold rotational axes of symmetry by 15°. The points of contact between the regulatory dimers and the catalytic trimers act as pivots during this conformational change. It is the rotation of the pivoted regulatory dimers that spreads apart the two catalytic trimers. In such conformational changes, the subunits can, to a first approximation, be treated as rigid bodies that realign with respect to each other by definable geometric movements.

Conformational changes can also occur between domains in the same folded polypeptide. The induced fit performed by a subunit of hexokinase (Figure 3–43) upon binding reactants is one of the clearest examples of such a conformational change.

Local conformational changes in the immediate vicinity of the active site are also observed frequently. After dihydroxyacetone phosphate associates with the active site of triosephosphate isomerase, a short loop of polypeptide comprising amino acids 168–177 closes over the substrate to lock it in the active site (Figure 6–3).[3] The closing of this lid embeds the phosphate of the substrate within a mold of amino acids that prevents it from moving during the reaction. Because the phosphate is held in the plane of the *cis*-enediol (Figure 6–3) and because it binds to the enzyme as the dianion,[4] the $\beta$-elimina-

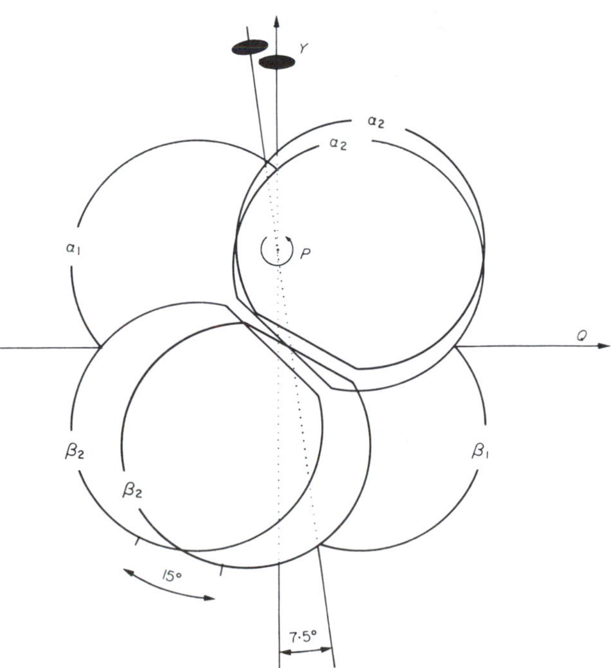


**Figure 6–1:** Diagrammatic representation of the global structural differences between the crystallographic molecular models of the deoxy conformation of human hemoglobin and the oxy conformation.[1] Human deoxyhemoglobin was crystallized in 2.4 M phosphate buffer at pH 6.4 under nitrogen. The space group of the crystals was $P2_1$, and reflections were gathered from crystals of native protein and two isomorphous derivatives to Bragg spacings of 0.25 nm. The phases were estimated from these multiple isomorphous replacements. Crystals of hemoglobin liganded with carbon monoxide were grown under an atmosphere of CO in a similar manner. The space group of these crystals was $P4_12_12$, and a data set was collected to Bragg spacings of 0.27 nm. Phases were estimated for an initial map of electron density by using a refined crystallographic molecular model of methemoglobin inserted into the $P4_12_12$ unit cell by molecular replacement. The initial crystallographic molecular models of deoxyhemoglobin and carbonmonoxyhemoglobin were then refined against the respective data sets. The map of electron density for the $\alpha_1\beta_1$ dimer of deoxyhemoglobin was superposed on the map of electron density for the $\alpha_1\beta_1$ dimer of carbonmonoxyhemoglobin. When these dimers were superposed, it was found that rotation of the $\alpha_2\beta_2$ dimer of deoxyhemoglobin 7.5° counterclockwise about an axis *P* normal to the plane of the diagram and a movement of 0.1 nm out of the plane of the page was necessary to superpose it onto the $\alpha_2\beta_2$ dimer of the carbonmonoxyhemoglobin. Reprinted with permission from ref 1. Copyright 1979 Academic Press.

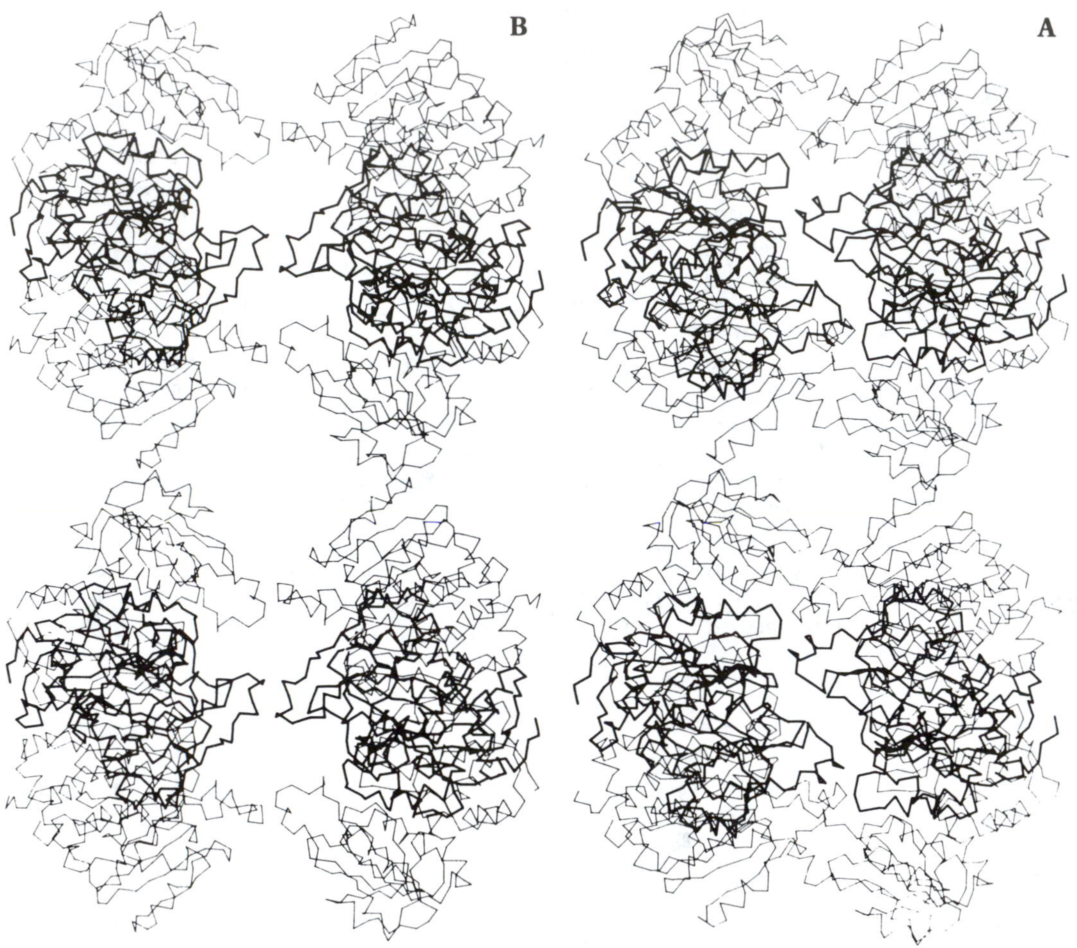


**Figure 6–2:** Crystallographic molecular models of the two catalytic trimers[1a] in the crystallographic molecular models of the unliganded conformation (A) and the conformation promoted by the binding of *N*-(phosphonacetyl)-L-aspartate (B) to aspartate transcarbamylase from *Escherichia coli*.[2] Crystals of unliganded enzyme were formed by dialysis against 50 mM tris(hydroxymethyl)aminomethane hydrochloride and 50 mM imidazolium chloride at pH 6.35. Crystals of space group *R*32 were obtained. Reflections from crystals of native enzyme and four isomorphous heavy-atom derivatives were collected to Bragg spacings of 0.3 nm. Crystals of liganded enzyme were grown from 50 mM *N*-ethylmorpholinium maleate at pH 5.9 in the presence of 1 mM *N*-(phosphonacetyl)-L-aspartate. Crystals of the space group *P*321 were obtained. Reflections were collected from crystals of the liganded enzyme and two isomorphous heavy-atom derivatives to Bragg spacings of 0.25 nm. The molecular models presented are $\alpha$-carbon tracings of the two refined, complete crystallographic molecular models. The regulatory dimers were left out for the sake of clarity. The two views are down one of the 2-fold rotational axes of symmetry, and the orthogonal 3-fold rotational axis of symmetry of the catalytic trimers is vertical in the plane of the page. Reprinted with permission from ref 2. Copyright 1985 National Academy of Sciences.

tion from the *cis*-enediol that would produce methyl glyoxal

(6–1)

is prevented. When the loop was shortened by site-directed mutation, the mutant enzyme, in addition to being a very poor catalyst, produced significant quantities of this undesired side product.[5]

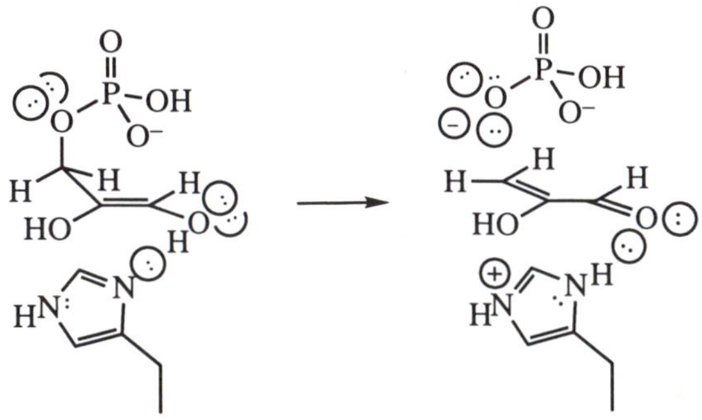

It is usually the case that local conformational changes accompany the larger ones involved in rearrangements of subunits or reorientations of domains. For example, the local conformational changes on the distal side of the heme in hemoglobin (Figure 2–28) occur during the global conformational change promoted by the binding of ligand. The repositioning of various amino acids to produce a functional active site occurs coincident with the closing of the hinged domains in citrate (*si*) synthase.[6]

It is seldom possible to perform the crystallographic studies necessary to define particular conformational changes in atomic detail, and other less precise methods must be used to detect conformational changes. These are the chemical and physical methods normally used to study the structure of native proteins. When isopropyl β-D-thiogalactoside, an analogue of a natural ligand, is bound by *lac* repressor, a reproducible change in the circular dichroic spectrum of the protein is observed, as well as a change in its sedimentation coefficient[7] and a change in its optical spectrum indicative of alterations in the environments around its tyrosines.[8] In 6-phosphofructokinase there is a cysteine residue in each subunit that reacts preferentially with fluorodinitrobenzene. When activators of the enzymatic reaction such as AMP are bound to the enzyme, the rate of this arylation increases; however, when inhibitors of the enzyme such as citrate are bound, its rate decreases.[9] It is assumed that the rate of this reaction responds to shifts in the orientation of amino acids surrounding this cysteine that result from a conformational change in the protein. When a histidine in the active site of succinate-CoA ligase (ADP-forming) is phosphorylated to form the covalent intermediate in the enzymatic reaction, the rate at which trypsin inactivates the enzyme decreases 100-fold.[10] This decrease in rate is assumed to result from a change in the conformation of the protein, although it may be the result of a local dampening of the vibrational modes of some of the accessible loops in the folded polypeptide. Phosphorylase can be modified at a cysteine in each of its subunits by *N*-(1-oxyl-2,2,6,6-tetramethyl-4-piperidinyl)iodoacetamide to introduce a stable free radical[10a] into the protein. The electron spin resonance spectrum of this free radical shifts, presumably in response to a conformational change in the protein, when AMP, an activator of the enzymatic activity, is bound by the enzyme.[11] The changes in the conformation of hemoglobin documented crystallographically can be followed by observing changes in the nuclear magnetic resonance spectrum of the protein.[12] Dramatic changes in the rate of hydrogen exchange of peptide bonds in the first 12 amino acids of the α subunit also occur during the conformational change produced by the binding of oxygen to hemoglobin,[13] even though this region does not differ in its structure when the crystallographic molecular models are compared.[1]

An understanding of the established ability of a ligand to promote a change in the conformation of a protein relies upon casting the situation as a **linkage relationship**. If it is the case that all the information necessary to produce one folded native structure of lowest free energy is encoded in the amino acid sequence of a polypeptide, it must also be the case that the information necessary to produce two or more different folded native structures is encoded in the same amino acid sequence. The native structure of a protein is simply the most thermodynamically stable conformation accessible to the folding polypeptide or the assembling protomers. Because a folding polypeptide or a set of assembling protomers has so many degrees of freedom, it seems likely that for every protein there are several other accessible conformations almost as low in standard free energy of formation as the native structure. This situation would produce a population of rapidly equilibrating conformations. The other accessible conformations would be minor contributors to this population under a given set of circumstances because their standard free energies of formation would be greater than that of the native structure. In the simplest situation, one could imagine two conformations of the protein, the native structure and another conformation, in equilibrium with each other. The two conformations would be, for example, the two that have been observed crystallographically for hemoglobin, aspartate transcarbamylase, hexokinase, or triosephosphate isomerase. If the two conformations E and E′ are accessible to each other, the difference in free energy between them defines an equilibrium constant*

$$E \xrightarrow{K_{eq}^{E}} E' \tag{6–2}$$

$$K_{eq}^{E} = \frac{[E']}{[E]} \tag{6–3}$$

In what follows, as well as in the crystallographic observations just described, the two conformations of the protein being discussed will be, and have been, the unliganded and the liganded conformations. The **unliganded conformation** of a protein is that conformation of the folded polypeptide and assembled protomers that is most stable in the absence of all of the specific ligands of the protein. The **liganded conformation** of a protein is the conformation that is most stable when one or more of the specific ligands of that protein are bound at their respective binding sites on the protein, if that conformation is structurally distinct from the unliganded conformation.

In the instances of large global conformational changes, such as those observed with hemoglobin and aspartate transcarbamylase, in which the ligand is a minor player in the structure, it must be the case that these two conformations, the unliganded and the liganded, have finite lifetimes at local minima of free energy and exist as definable stable states in equilibrium with each other, both when the ligand is bound and in its absence. For example, when phosphofructokinase from *E. coli* was crystallized from a solution in which no ligands were present, so that in this solution the unliganded conformation was the predominant species; nevertheless, it was the small amount of the protein present in the global conformation that would normally have been promoted by the binding of MgADP that crystallized.[14] Even in the case of local conformational changes such as the one observed with triosephosphate isomerase (Figure 6–3), the closed conformation in the absence of ligand should have a finite lifetime because, in this case, the lid makes contacts with the amino acids that form the rim of the box into which the substrate fits.

It is possible, however, that a local conformational change between the unliganded conformation and a liganded conformation could be so limited that the amino acids participating in it make contact only with portions of the ligand itself in the liganded conformation and not with any of the other amino acids in the protein. In this case, the liganded conformation in the absence of the ligand might not have a finite lifetime at a local minimum of free energy because in the absence of ligand, this conformation would be passed through only during the sweeps of certain vibrational modes of the folded

*From here on a unidirectional arrow, rather than two opposing arrows, will be used to designate an equilibrium so that products and reactants are clearly defined.

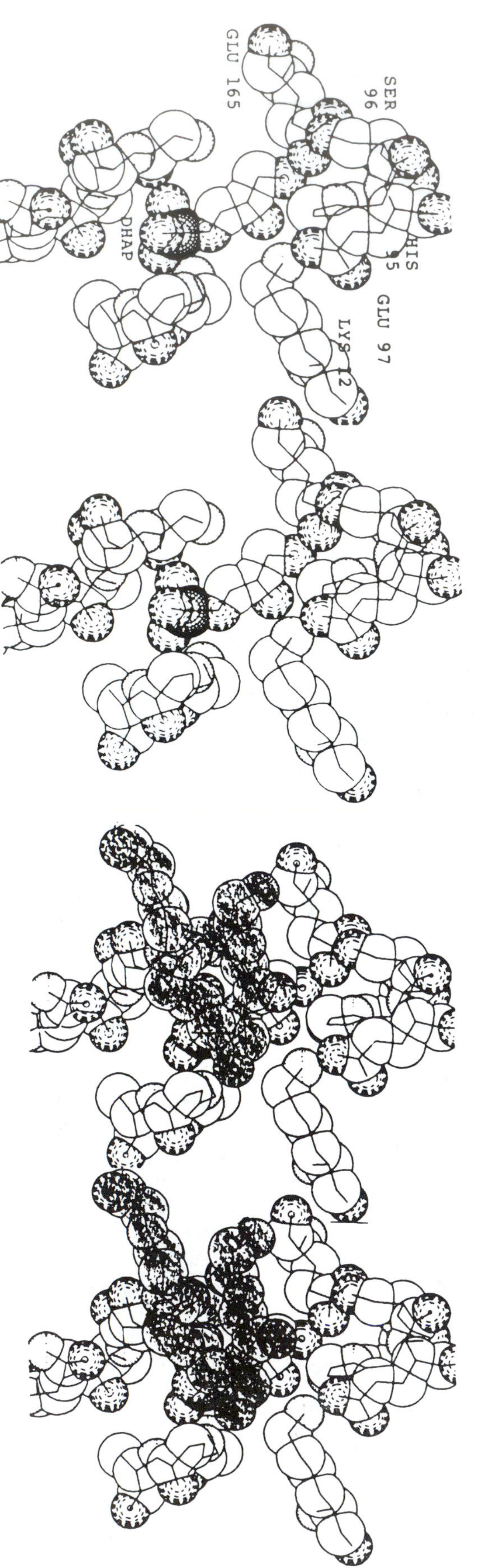


A

B

**Figure 6–3:** Loop of 10 amino acids from Tryptophan 168 to Threonine 177 that closes over the substrates after they occupy the active site of chicken muscle triosephosphate isomerase.[3] The unliganded enzyme was crystallized from 60% ammonium sulfate at pH 7.4. The space group of the crystals was $P2_12_12_1$. Reflections were gathered from crystals of the native protein and five isomorphous heavy-atom derivatives to Bragg spacings of 0.25 nm. From these data sets, a map of electron density and a crystallographic molecular model were prepared. Crystals of triosephosphate isomerase were transferred to a solution containing 10 mM dihydroxyacetone phosphate, the substrate was allowed to diffuse into the crystal at –15 °C for 10 min, and a new data set was collected. From this new data set a map of difference electron density was calculated. In addition to a feature of positive electron density corresponding to bound substrate, presumably an equilibrium mixture of dihydroxyacetone phosphate and glyceraldehyde 3-phosphate, there was a prominent feature of negative electron density where the loop from Tryptophan 168 to Threonine 177 had been in the unliganded molecular model and a feature of positive electron density of equivalent magnitude lying like a flap or a lid over the phosphate of the substrate. The loop had pivoted over 0.6 nm at its most peripheral edge to shut the active site from access to the solvent. (A) Active site of the liganded crystallographic molecular model occupied by a molecular model of dihydroxyacetone phosphate with the lid removed. (B) Same crystallographic molecular model of the active site with the lid (dark atoms) included. Reprinted with permission from ref 3. Copyright 1987 Cold Spring Harbor Laboratory.

polypeptide. In this case, the polypeptide would spend less than $10^{-12}$ s in this conformation on each sweep. Only the probability of its attainment would define a difference between the chemical potential of the unoccupied liganded conformation and the chemical potential of the native structure. Nevertheless, this difference in chemical potential would define a difference in standard free energy.

If two conformations of a protein, the liganded conformation E′ and the unliganded conformation E, exist as definable states in equilibrium with each other so that each is accessible from the other and both are populated in either the presence or absence of the ligand, then the following linkage relationship can be written (see also Equation 3–222)

$$\begin{array}{ccc} \mathrm{E} + \mathrm{L} & \xleftarrow[\Delta G_{\mathrm{L}}^{\circ}]{K_{\mathrm{d}}^{\mathrm{EL}}} & \mathrm{E}\cdot\mathrm{L} \\ K_{\mathrm{eq}}^{\mathrm{E}} \Big\downarrow \Delta G_{\mathrm{E}}^{\circ} & & \Delta G_{\mathrm{EL}}^{\circ} \Big\downarrow K_{\mathrm{eq}}^{\mathrm{EL}} \\ \mathrm{E}' + \mathrm{L} & \xleftarrow[K_{\mathrm{d}}^{\mathrm{E'L}}]{\Delta G_{\mathrm{L}}^{\circ\prime}} & \mathrm{E}'\cdot\mathrm{L} \end{array} \tag{6–4}$$

If the unliganded conformation E is the most stable in the absence of the ligand ($K_{\mathrm{eq}}^{\mathrm{E}} < 1$) and the liganded conformation E′·L is the most stable when the ligand is bound ($K_{\mathrm{eq}}^{\mathrm{EL}} < 1$), it necessarily follows that the ligand binds more tightly to the liganded form E′ than it does to unliganded form E ($K_{\mathrm{d}}^{\mathrm{E'L}} < K_{\mathrm{d}}^{\mathrm{EL}}$). It is the difference in the standard free energy of binding that provides the free energy necessary to shift the equilibrium between the two conformations. This coupling is referred to as **linkage**.

These differences in the standard free energies, reflected in the differences in the equilibrium constants, are not dependent on the paths available between the various states. In the case of hemoglobin, there is no reason to doubt that the deoxy conformation and the oxy conformation can be directly interconverted either when the hemes are unliganded or when they are liganded or that oxygen can associate directly with and dissociate directly from either conformation. In the case of triosephosphate isomerase, however, the closed, unliganded conformation E′ cannot associate directly with dihydroxyacetone phosphate because the loop covers the active site (Figure 6–3). Therefore, the reaction defined by $K_d^{E'L}$ (Equation 6–4) must have as its steps the opening of the lid, the association of the substrate with the open active site, and the closing of the lid.[3] Nevertheless, because the initial and final states exist, the equilibrium constant exists regardless of the path.

If the conformational equilibrium of a particular protein shifts in response to the binding of a ligand, the observed macroscopic dissociation constant for the ligand, $K_{d,obs}$, will be a function of the microscopic equilibrium constants[15]

$$K_{d,obs} = \frac{([E]+[E'])[L]}{[E\cdot L]+[E'\cdot L]} \quad (6\text{–}5)$$

$$K_{d,obs} = \frac{[E](1+K_{eq}{}^{E})[L]}{[E\cdot L](1+K_{eq}{}^{EL})} = \frac{K_d{}^{EL}(1+K_{eq}{}^{E})}{(1+K_{eq}{}^{EL})} \quad (6\text{–}6)$$

Because $K_{d,obs}$ is not a function of any concentration, it follows that normal linear behavior with no positive or negative deviations will be observed when the binding of ligand is followed as a function of the molar concentration of the ligand. If both $K_{eq}{}^{E}$ and $K_{eq}{}^{EL}$ are significantly less than 1, no significant shift in the distribution of conformations takes place upon binding of the ligand, only the conformations E and E·L are present in significant concentration, and $K_{d,obs} = K_d{}^{EL}$. If both $K_{eq}{}^{E}$ and $K_{eq}{}^{EL}$ are significantly larger than 1, no shift in the distribution takes place upon binding of the ligand, only the conformations E′ and E′·L are present in significant concentration, and $K_{d,obs} = K_d{}^{E'L}$. It is only when $K_{eq}{}^{E} < 1$ and $K_{eq}{}^{EL} > 1$ that the numerical value of the observed dissociation constant lies between the numerical values of the two microscopic dissociation constants and that the identity of the dominant conformation of the protein changes upon the binding of ligand.

The advantage of such a ligand-promoted conformational change for increasing the specificity of an enzymatic reaction has already been noted. Ligand-promoted conformational changes are also used to control the rate of enzymatic reactions.

### *Suggested Reading*

Baldwin, J., & Chothia, C. (1979) Hemoglobin: The Structural Changes Related to Ligand Binding and Its Allosteric Mechanism, *J. Mol. Biol. 129*, 175–220.

# Allosteric Control of Enzymatic Reactions

Aspartate transcarbamylase catalyzes the first committed reaction in the metabolic pathway leading to the pyrimidine nucleotides: orotidine 5′-phosphate, uridine 5′-triphosphate, cytidine 5′-triphosphate, thymidine 5′-triphosphate, and deoxycytidine 5′-triphosphate. The **first committed reaction** in any metabolic pathway is the earliest step in the pathway in which a metabolite is produced that can only be used in that pathway. For example, *N*-carbamylaspartic acid, one of the products of the reaction catalyzed by aspartate transcarbamylase, can be used to synthesize only pyrimidine nucleotides. Metabolic regulation can be exerted at the branch points of major metabolic pathways; for example, citrate activates acetyl-CoA carboxylase to divert acetyl-SCoA from the Krebs cycle to fatty acid biosynthesis. The simplest form of metabolic regulation, however, is when the end products of a metabolic pathway inhibit the first committed reaction in that pathway. For example, when the cytoplasmic concentrations of pyrimidine nucleotides are too high, the flux through the pathway for their biosynthesis is diminished by decreasing the rate of the reaction catalyzed by aspartate transcarbamylase. This occurs because several of the pyrimidine nucleotides, with the interesting exception of uridine 5′-triphosphate, are inhibitors of the catalysis performed by the enzyme (Figure 6–4A).[16] If, however, the concentrations of certain purine nucleotides, in particular ATP, are in significant excess over those of the pyrimidines, the biosynthesis of pyrimidines is increased by accelerating the rate of the reaction catalyzed aspartate transcarbamylase. This occurs because ATP is an activator of the catalysis performed by the enzyme (Figure 6–4B). There are two remarkable features of the behavior of aspartate transcarbamylase: the effects of the nucleotides observed when they are acting as inhibitors or activators and the positive deviations from ideal behavior observed when the initial velocity is followed as a function of the concentration of aspartate in the absence of nucleotides.

It is known from studies in which the trimeric catalytic subunits were separated[17] from the dimeric regulatory subunits[1a] that the active sites are on the catalytic subunits[17] of aspartate transcarbamylase and the binding sites for nucleotides are on the regulatory subunits.[18,19] In the native enzyme, the six active sites are symmetrically displayed on the two catalytic trimers[2] and the six binding sites for nucleotides are symmetrically displayed on the three regulatory dimers. In the crystallographic molecular model for the intact enzyme,[20] each of the sites on a folded regulatory polypeptide at which nucleotides, either CTP or ATP, are bound[21,22] is 6.0 nm distant from and makes no direct contacts with the closest active site.[2] As a result, the active sites are distinct and separate from the regulatory sites, and the inhibition of catalysis is an allosteric effect.[23] An **allosteric effect** is an indirect interaction between distinct and separate sites on a protein. When the two interacting sites bind different ligands, for example, aspartate and carbamyl phosphate on the one hand and CTP or ATP on the other, the interaction between those sites, be it inhibition or activation, is referred to as a **heterotropic allosteric effect.**

It is important, as far as the mechanism is concerned, to distinguish between indirect allosteric interactions between distant, distinct sites and the simpler interactions among inhibitors and activators binding to the same site. For example, the enzymatic activity of glutamate dehydrogenase is affected by both GTP and ADP, neither of which resembles the reactants, yet spectroscopic studies suggest that they bind at subsites within the active site or immediately adjacent to it to exert their effects.[24] This would be simple enzymatic inhibition. Phosphofructokinase is activated by MgADP so that glycolysis will accelerate as levels of MgADP rise. It has been demonstrated crystallographically that the site at which the effector ADP is bound by the enzyme is separate and distinct from the active site (Figure 6–5).[14] This is allosteric activation.

The positive deviation from ideal behavior observed when the initial velocity of aspartate transcarbamylase is determined as a function of the concentration of aspartate in the absence of nucleotides (Figure 6–4A) is a homotropic cooperative effect. A **homotropic effect** is an allosteric interaction between two or more sites on the same protein, each of which binds identical ligands.[23] One way to explain the observed positive deviation from linearity is to assume that the binding of the first one or more aspartates to a molecule of the enzyme increases the affinity of that molecule of the enzyme for the last or the last few aspartates that bind to it. In other words, the aspartates cooperate in occupying the active sites on the enzyme. For this reason, the positive deviation from ideal behavior is referred to as a **homotropic cooperative effect.**

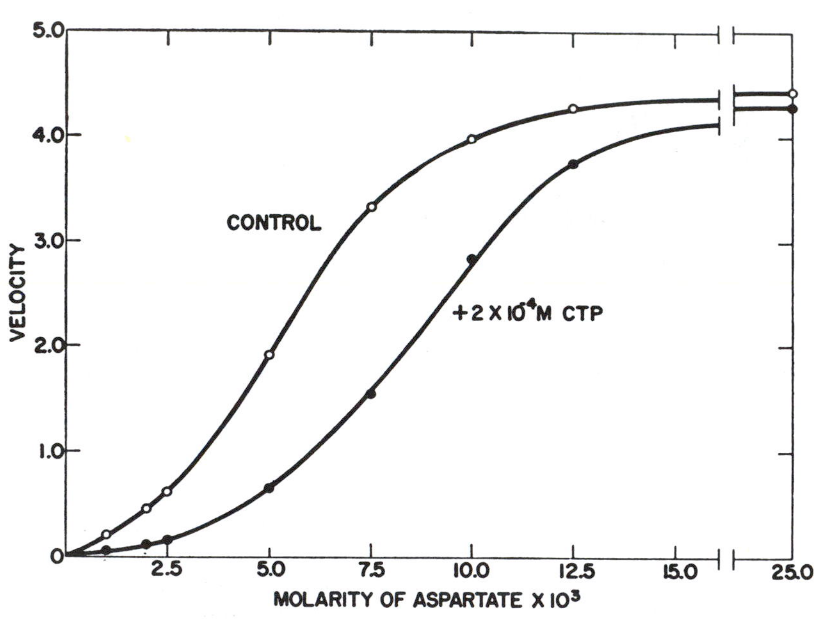


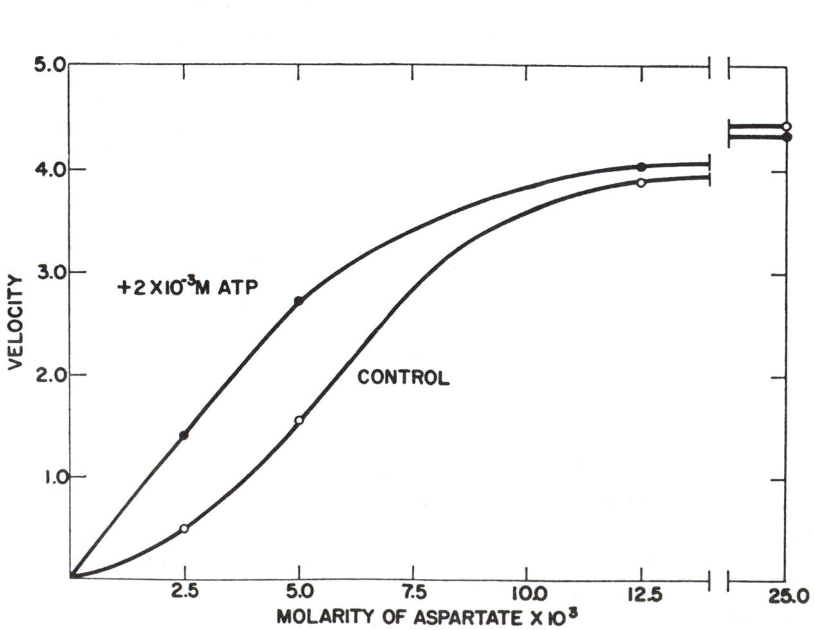


**Figure 6–4:** Inhibition (A) of aspartate transcarbamylase from *E. coli* by the pyrimidine nucleotide cytidine 5′-triphosphate (CTP) and activation (B) of the enzyme by the purine nucleotide ATP.[16] Individual reaction mixtures (0.50 mL) were prepared containing 3.6 mM carbamyl phosphate, variable concentrations of aspartate as indicated, and either no other additions (open circles in both panels), 0.2 mM CTP (closed circles in panel A), or 2.0 M ATP (closed circles in panel B). The pH of the mixtures was 7.0, and the temperature was 28 °C. The enzymatic reaction was initiated by adding aspartate transcarbamylase to 90 ng $mL^{-1}$. After 30 min the reaction was quenched and the yield of carbamylaspartate was determined colorimetrically. The velocity of the enzymatically catalyzed reaction (in micromoles $hour^{-1}$ $microgram^{-1}$) is presented as a function of the concentration (molarity × $10^3$) of aspartate. Reprinted with permission from ref 16. Copyright 1962 *Journal of Biological Chemistry.*

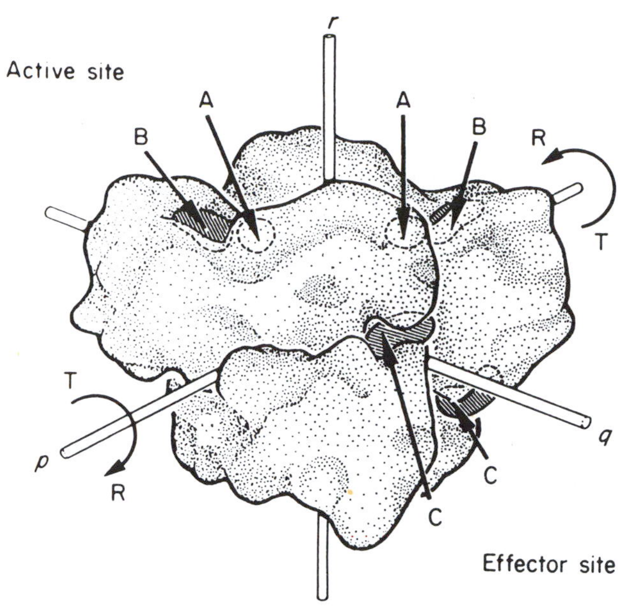


**Figure 6–5:** Schematic view of the crystallographic molecular model of the tetramer of phosphofructokinase from *E. coli* showing the elongated subunits arranged around the three orthogonal axes of symmetry.[14] Phosphofructokinase was crystallized from 14% poly(ethylene glycol) and 1.0 M NaCl, pH 7.9. The crystals belonged to space group *C*2. Reflections were gathered to Bragg spacings of 0.24 nm. An available refined crystallographic molecular model of the liganded enzyme was used to phase the reflections by molecular replacement, and the initial molecular model of the unliganded enzyme was refined against the data set. A drawing was made of the van der Waals surface of the final molecular model. The molecular 2-fold rotational axes of symmetry are labeled *p*, *q*, and *r*. Sites A bind fructose 6-phosphate and sites B bind MgATP, and together one site A and one site B form an active site. Sites C are the allosteric regulatory sites that bind either the activators ADP and GDP or the inhibitor phospho*enol*pyruvate. Reprinted with permission from ref 14. Copyright 1989 Academic Press.

Obviously, for this explanation to be valid, each molecule of the enzyme must have more than one active site. Aspartate transcarbamylase has six symmetrically displayed active sites,[2] one for each folded catalytic polypeptide. Each folded catalytic polypeptide with its associated, folded regulatory polypeptide can be designated a **protomer** of the enzyme. Enzymes that display homotropic cooperative effects are oligomers of identical protomers.[23]

Homotropic cooperative behavior is also of regulatory significance. Because the increase in enzymatic activity as a function of the concentration of reactant occurs over a narrower range of concentration (Figure 6–4B) in the cooperative situation relative to the noncooperative situation, changes in the concentration of reactant in the cytoplasm exert greater control over the rate of enzymatic reaction. In the specific case of aspartate transcarbamylase, the activity of the enzyme is shut down much more effectively when aspartate, which is used widely in other metabolic reactions, is present at a low concentration in the cytoplasm.

It has been pointed out that most enzymes that display heterotropic allosteric interactions also display homotropic cooperative interactions.[23] There is, however, no requirement that this be the case. A mutant of aspartate transcarbamylase has been described in which the enzyme remains as a hexamer of protomeric heterodimers and in which the enzymatic activity is still inhibited by CTP but which no longer displays homotropic cooperative behavior.[25] Ribonucleoside-triphosphate reductase (Reaction 5–90) catalyzes the reduction of several ribonucleoside triphosphates. Many of the deoxyribonucleoside triphosphates that are products of the reaction are in turn heterotropic allosteric activators of the reduction of ribonucleotide triphosphates with different heterocyclic bases from the respective bases on those activators.[26] These interactions serve to balance the outputs of the various products. The enzyme, however, cannot display homotropic cooperativity because it is monomeric.[27] Phosphorylase *b* catalyzes the reaction

$$\text{glycogen}_n + \text{glucose 1-phosphate} \rightleftharpoons \text{glycogen}_{n+1} + \text{HOPO}_3^{2-} \qquad (6\text{–}7)$$

The reaction is activated by AMP acting as a heterotropic allosteric effector[28] that binds to the enzyme at a site distant from the active site.[29] This is another mechanism by which glycolysis is activated as levels of MgATP drop. The effect of AMP on the initial velocity of the reaction shows homotropic cooperative behavior, manifested in a positive deviation from ideal behavior (Figure 6–6).[30] When the enzyme is treated with glutaraldehyde, a cross-linking agent, but not when it is treated with the monofunctional homologue butyraldehyde, the homotropic cooperativity disappears in both this instance (Figure 6–6) and in the other homotropic cooperative effects observed with this enzyme.[31] Yet the enzymatic activity increases slightly, and the heterotropic allosteric activation observed with AMP remains. If phosphorylase *b* is treated with sodium borohydride and then transferred to 7% formamide, it dissociates into monomers.[32] These monomers no longer display homotropic cooperativity, but AMP remains a heterotropic allosteric activator.

These examples demonstrate that heterotropic allosteric effects can be dissociated from homotropic cooperative effects. The reason that the two properties are usually present together is that both require that the enzyme be able to adopt at least two conformations whose relative stabilities are linked to the binding of ligands. Once the ability to adopt two such ligand-linked conformations arises, both heterotropic allosteric behavior and homotropic cooperative behavior can exploit the same conformational change because most proteins are oligomeric. Many, if not the majority, of the enzymes that are responsible for the first committed reactions in metabolic pathways display both heterotropic allosteric behavior and homotropic cooperativity, so it seems that there is no particular difficulty to the evolution of ligand-linked conformational changes able to affect the functional properties of an

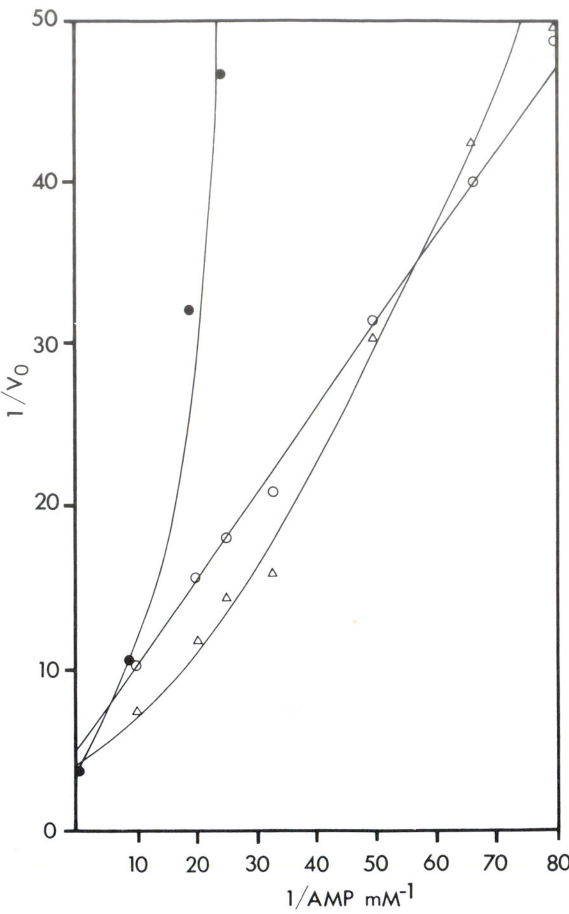


**Figure 6–6:** Elimination of the homotropic cooperativity in the regulation of rabbit muscle phosphorylase *b* by AMP as the result of treatment of the enzyme with glutaraldehyde.[30] Assay mixtures were prepared containing 1% glycogen, 16 mM glucose 1-phosphate, and various concentrations of AMP. The enzymatic reaction was initiated by adding either native phosphorylase *b* (△) or phosphorylase *b* that had been treated with either 0.05% glutaraldehyde (○) or 1.0% butyraldehyde (●) for 10 min at 0 °C in 40 mM glycerophosphate, pH 7.5. After 5 min the enzymatic reaction was quenched, and the concentration of $HOPO_3^{2-}$ produced during the synthesis of glycogen by the enzyme from glucose 1-phosphate was assayed colorimetrically. The initial velocity, $v$, is expressed in units of absorbance at 660 nm $(5\ \text{minutes})^{-1}$ $(\text{milligram of enzyme})^{-1}$. The reciprocal of the initial velocity ($v_0$) is expressed as the reciprocal of the concentration of AMP ($\text{millimolar}^{-1}$). Note that the initial velocity increases as the concentration of the obligatory activator, AMP, increases. Reprinted with permission from ref 30. Copyright 1969 American Chemical Society.

enzyme. The interesting fact, however, is that the majority of the enzymes, because they are in the middle or at the peripheries of metabolic pathways, display neither heterotropic allosteric behavior nor homotropic cooperative behavior even though they are invariably oligomeric. This suggests that selective pressure against such properties can be quite strong.

The inhibition of pyruvate kinase

$$\text{phospho}\textit{enol}\text{pyruvate} + \text{MgADP} \rightleftharpoons \text{pyruvate} + \text{MgATP} \qquad (6\text{–}8)$$

by phenylalanine (Figure 6–7A)[33] regulates the concentration of phospho*enol*pyruvate, a precursor of the aromatic amino acids. This inhibition offers a simple example of a heterotropic allosteric effect. Formally, phenylalanine is a competitive inhibitor for phospho*enol*pyruvate because it changes only $K_m$ and not $V_{max}$ regardless of its concentration.[34] Furthermore, it has been demonstrated, by directly measuring the binding of phospho*enol*pyruvate to pyruvate kinase (Figure 6–7B),[33] that the observed Michaelis constant for this reactant is equal to the observed dissociation constant either in the absence of phenylalanine or at the same concentration of phenylalanine.[34] Therefore, because the binding of phenylalanine to the enzyme increases the Michaelis constant, it must also increase the dissociation constant for phospho*enol*-pyruvate.

If one disregards for the moment the cooperativity, this behavior can be explained by a linkage cube[33]

$$
\begin{array}{ccccccc}
 & \text{E} & \xleftarrow{\;K_d^{P}\;} & \text{E}\cdot\text{P} & \xrightarrow{\;k_{cat}^{P}\;} & & \\
 & {\scriptstyle K_d^{F}}\nearrow\big| & & {\scriptstyle K_d^{FP}}\nearrow\big| & & & \\
\text{F}\cdot\text{E} & \xleftarrow{\;K_d^{PF}\;} & & \text{F}\cdot\text{E}\cdot\text{P} & \xrightarrow{\;k_{cat}^{FP}\;} & & \\
\big\downarrow K^{FE} & & \big\downarrow K^{E} & \big\downarrow K^{FEP} & \big\downarrow K^{EP} & & \\
 & \text{E}' & \xleftarrow{\;K_d'^{P}\;} & \text{E}'\cdot\text{P} & \xrightarrow{\;k_{cat}'^{P}\;} & & \\
 & {\scriptstyle K_d'^{F}}\nearrow & & {\scriptstyle K_d'^{FP}}\nearrow & & & \\
\text{F}\cdot\text{E}' & \xleftarrow{\;K_d'^{PF}\;} & & \text{F}\cdot\text{E}'\cdot\text{P} & \xrightarrow{\;k_{cat}'^{FP}\;} & &
\end{array}
\qquad (6\text{–}9)
$$

each of whose edges is governed either by a dissociation constant

$$K_d^{IJ} + \frac{[\text{I}][\text{E}\cdot\text{J}]}{[\text{I}\cdot\text{E}\cdot\text{J}]} \qquad (6\text{–}10)$$

or the equilibrium constant for an isomerization

$$K^{IEJ} = \frac{[\text{I}\cdot\text{E}'\cdot\text{J}]}{[\text{I}\cdot\text{E}\cdot\text{J}]} \qquad (6\text{–}11)$$

It has been shown that each protomer of the $\alpha_4$ tetramer has one active site to which phospho*enol*pyruvate binds (Figure 6–7B) and one allosteric site to which phenylalanine binds. The conformation E′ binds phenylalanine more strongly than does the conformation E, the conformation E binds phospho-

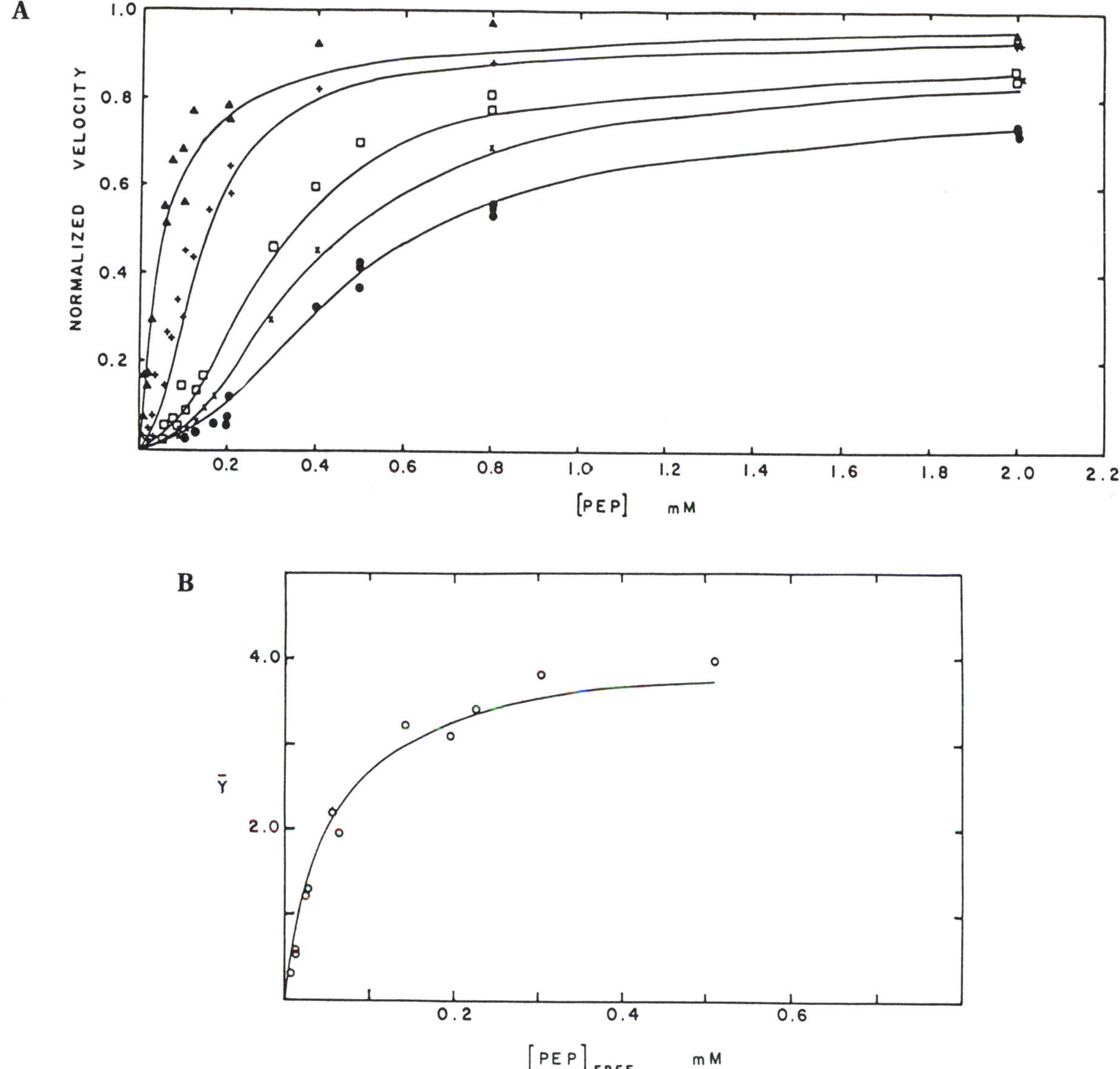


**Figure 6–7:** Heterotropic allosteric inhibition of rabbit muscle pyruvate kinase by phenylalanine and demonstration that the Michaelis constant and the dissociation constant of phospho*enol*pyruvate from the enzyme are the same.[33] (A) Individual reaction mixtures were prepared with 72 mM KCl, 7.2 mM $MgSO_4$, 2.0 mM ADP, variable amounts of phospho*enol*pyruvate as indicated, and no phenylalanine (▲) or 3 mM (+), 7 mM (□), 9 mM (×), or 12 mM (●) phenylalanine. The mixtures were at pH 7.5 and 23 °C. The enzymatic reaction was initiated by adding pyruvate kinase, and the production of pyruvate was monitored by coupling it to the oxidation of NADH by lactate dehydrogenase. Initial velocities were calculated from the decrease in absorbance at 340 nm. The initial velocities were all normalized so that a value of 1.0 was the initial velocity of uninhibited enzyme saturated with phospho*enol*pyruvate (1000 $s^{-1}$). The normalized initial velocities are presented as a function of the concentration of phospho*enol*pyruvate (millimolar). (B) Binding of phospho*enol*pyruvate to pyruvate kinase was measured by a molecular exclusion method at pH 7.5 and 23 °C. Solutions (0.55 mL) containing pyruvate kinase (10 to 20 mg $mL^{-1}$) and various concentrations of [$^{14}$C]phospho*enol*pyruvate were mixed with individual aliquots of 50 mg of dehydrated Sephadex G-50. The Sephadex imbibed a certain amount of the fluid. It was assumed that the free phospho*enol*pyruvate equilibrated between the fluid in the interior of the beads of the molecular exclusion medium and the solution surrounding the beads while the protein remained entirely on the outside of the beads. From the volume of fluid added, the total amount of protein added, the total amount of phospho*enol*pyruvate added, the molar concentration of phospho*enol*pyruvate in the fluid outside the beads, as determined by scintillation counting, and the molar concentration of protein in the fluid outside the beads, as determined by the absorbance at 280 nm, the moles of phospho*enol*pyruvate bound to each mole of tetrameric enzyme ($\bar{Y}$) could be calculated as a function of the free concentration of phospho*enol*pyruvate ($[PEP]_{free}$ in millimolar). The points are the experimental determinations. The line is drawn on the basis of the assumption that the dissociation constant for phospho*enol*pyruvate is the same as the numerical value (0.44 mM) of the Michaelis constant observed kinetically (▲ in panel A). Reprinted with permission from ref 33. Copyright 1984 American Chemical Society.

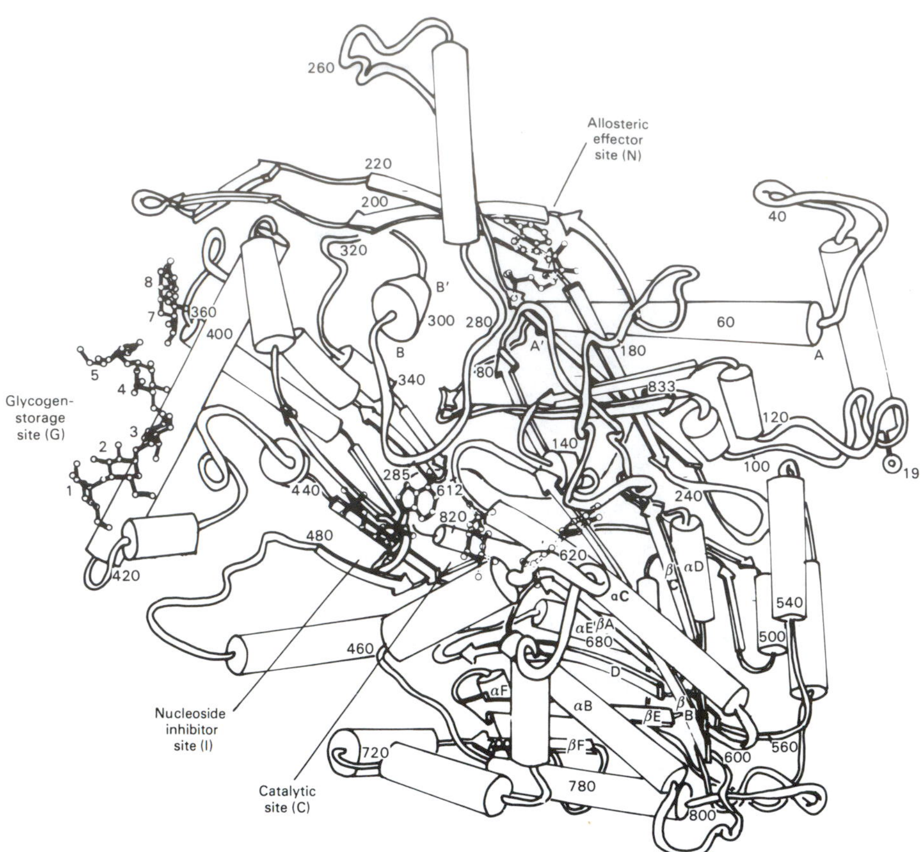


**Figure 6–8:** Schematic drawing of the crystallographic molecular model of one subunit of rabbit muscle phosphorylase *b* indicating all of the sites for the various allosteric effectors.[39] The drawing summarizes a large number of crystallographic studies in which data sets from various liganded forms of the crystalline enzyme were used to provide maps of electron density or difference electron density. In these studies the crystals were often stabilized by cross-linking to prevent them from shattering as a result of the conformational changes promoted by the binding of certain of the ligands. Although this shortcut placed in question whether the observed changes in conformation were indicative, it did allow the various binding sites for ligands to be identified. In the drawing, AMP is shown bound at the allosteric effector site N, maltoheptaose at the glycogen storage site G, glucose 1-phosphate and pyridoxal phosphate at the catalytic site C, and AMP at the nucleoside inhibitor site I. Reprinted with permission from ref 39. Copyright 1984 Biochemical Society.

*enol*pyruvate more strongly than does the conformation E′, and most of the enzyme is in the conformation E in the absence of phenylalanine. By linkage, as phenylalanine is added to the solution, the equilibrium between the two conformations shifts in favor of the conformation E′; and, as phospho*enol*pyruvate is added to the solution, it shifts in favor of the conformation E. Because phenylalanine shifts the apparent equilibrium constant between conformations E and E′ ($K_{eq}{}^{E}$ in Equation 6–6) in favor of the latter, the apparent dissociation constant for phospho*enol*pyruvate increases as the concentration of phenylalanine is increased (Equation 6–6).

This description of the situation predicts that phenylalanine alone should be able to shift the equilibrium distribution between the two conformations in one direction while phospho*enol*pyruvate should be able to shift it in the other. Phenylalanine decreases the sedimentation velocity of unliganded pyruvate kinase ($\Delta s_{20,w}$ = −0.24 S ), and phospho*enol*pyruvate increases its sedimentation velocity ($\Delta s_{20,w}$ = +0.11 S); phenylalanine increases by 10-fold the rate of reaction of cysteines in the protein with dithiobis(2-nitrobenzoate), and phospho*enol*pyruvate decreases the rate by 5-fold.[34] These observations are consistent with the two respective ligands pulling by linkage in opposite directions on the conformational equilibrium, which in the absence of any ligand favors the conformation E.

The fact that $V_{max}$ is the same regardless of the concentration of phenylalanine could be most simply explained by assuming that once phospho*enol*pyruvate is bound, all forms of the enzyme would have the same turnover number. In other words, the effect of phenylalanine is only on $K_m$.[23] To fit all the data with a kinetic model,[33] however, it was necessary in the case of pyruvate kinase to set $k_{cat}^{P'}$ and $k_{cat}^{FP'}$ at the same value but at one-tenth the value of $k_{cat}^{P}$ and $k_{cat}^{FP}$. The $V_{max}$ in this instance is invariant not because every one of the four turnover numbers is the same but because at saturating phospho*enol*pyruvate the enzyme is almost completely in the conformation E under all circumstances.

Heterotropic allosteric effectors can also exert their influence only on $V_{max}$ without affecting $K_m$.[23] For example, fructose 1,6-bisphosphate is a heterotropic allosteric inhibitor of glycerol kinase. It is a noncompetitive inhibitor with respect to both glycerol and MgATP.[35] An explanation for this behavior also follows from a linkage cube (Mechanism 6–9). Fructose 1,6-bisphosphate binds more tightly to an enzymatically inactive conformation E′ than an active conformation E, but in its absence the enzyme is mostly in the active conformation E. As fructose 1,6-bisphosphate is added, the concentration of the conformation E′ increases at the expense of conformation E, and the activity of the enzyme decreases. If each of the normal reactants, either glycerol or MgATP, has the same affinity for each of the two conformations of the enzyme, the binding of neither will be able to shift the equilibrium back in favor of conformation E. In this case, $V_{max}$ will decrease while the values of $K_m$, to the extent that they reflect only the dissociation constants for the reactants, which are necessarily the same for each conformation, will remain unchanged.

One of the more remarkable features of heterotropic allosteric effects is the fact that several, often very different, ligands can control the activity of the same enzyme. Phospho*enol*pyruvate carboxylase is allosterically activated by CDP and acetyl-SCoA and allosterically inhibited by aspartate.[36] Malate dehydrogenase (oxaloacetate-decarboxylating) ($NADP^+$) is allosterically inhibited by acetyl-SCoA, NADH, and adenosine 3′,5′-cyclic monophosphate.[37]

It is also possible for several very similar ligands to exert different effects. Cytidine 5′-triphosphate and ATP bind to the same site on the regulatory dimers of aspartate transcarbamylase[21] but exert opposite effects on the enzymatic activity (Figure 6–4). In ribonucleoside-diphosphate reductase, dATP, ATP, dGTP, and dTTP bind to two distinct sites on the protein. Depending upon what nucleotides are in which sites, the enzymatic activity is stimulated or inhibited and the relative specificity for the reactants, ADP, GDP, UDP, and CDP, is determined.[38] Presumably, it is the linkage of the binding of the effectors to one or more conformational equilibria that can explain these diverse capabilities.

Phosphorylase (Reaction 6–7) is controlled by a large collection of different heterotropic allosteric effectors (Table 6–1) that associate with one or more of five distinct sites (Figure 6–8).[39] The protein can exist in at least two conformations, one of high and one of low enzymatic activity. The allosteric effector site (N) binds AMP, an allosteric activator of the enzyme,[28,40] but the enzyme is inhibited when this site is occupied by ATP.[41] It is believed that the opposite effects from the same site can occur because AMP binds more tightly to the active conformation of the enzyme than to the inactive conformation, while ATP prevents the binding of AMP but has no effect on the conformational equilibrium. A similar situation occurs in the active (catalytic) site. Glucose 1-phosphate, acting as a substrate, shifts the conformational equilibrium in favor of the active conformation,[41] while glucose, also binding at the active site as a simple competitive inhibitor,[29] shifts it in favor of the inactive conformation to shut down glycogen utilization when levels of glucose are high.[42] Sandwiched between Phenylalanine 285 and Tyrosine 612 is another site, the nucleoside inhibitor site (I), at which planar heterocycles such as those in caffeine, adenine, inosine, adenosine, ATP, or flavin mononucleotide can bind.[43] These ligands are all heterotropic allosteric inhibitors of the enzyme. Because either glucose, binding to the active site, or any one of these inhibitors, binding to the nucleoside inhibitor site, shifts the conformational equilibrium in favor of the inactive form, when both glucose and one of these other inhibitors are present, they inhibit the enzyme synergistically.[43,44] Glucose decreases the apparent dissociation constant (Equation 6–6) for any one of these inhibitory heterocyclic compounds by independently shifting the equilibrium constant for the isomerization between the active and inactive conformations. Glucose 6-phosphate binds at a site that is adjacent to the allosteric effector site. When it is bound at this site, one of the phosphate oxygens of the glucose 6-phosphate overlaps the location at which one of the phosphate oxygens of AMP is bound.[39] When glucose 6-phosphate is bound at this site, AMP cannot bind to the allosteric effector site and activate the enzyme because of steric exclusion. This explains the inhibition of AMP activation manifested by glucose 6-phosphate. This makes sense because glycogenolysis should not be activated when the concentration of glycolytic intermediates is high. Finally, there is a site at which the enzyme normally binds to glycogen, the glycogen storage site (G). When a polysaccharide occupies this site, in the absence of AMP, the conformational equilibrium is shifted in favor of the active conformation and the enzymatic reaction is accelerated.[45]

**Table 6–1: Effects of Various Ligands on the Conformational Equilibrium between Enzymatically Active and Inactive Conformations of Phosphorylase**

| ligand | site[a] | conformer stabilized |
|---|---|---|
| AMP | N | active |
| ATP | N | neither[b] |
| glucose 1-phosphate | C | active |
| glucose | C | inactive |
| caffeine | I | inactive[c] |
| glucose 6-phosphate[d] | | neither |
| glycogen | G | active |

[a]Named as in Figure 6–8. [b]Competes with AMP for nucleotide site. [c]Synergistic with glucose. [d]Binds at a site adjacent to nucleotide site and competes with AMP.

Just as a global conformational change links the binding of various heterotropic allosteric effectors at the various sites to the binding of each other and to the enzymatic activity of phosphorylase, a conformational change can link the two unique active sites in an enzyme that carries out two different reactions. Aspartate kinase–homoserine dehydrogenase is an $\alpha_4$ tetramer, each of whose identical protomers possesses separate active sites for the following two reactions[46]

$$\text{MgATP} + \text{aspartate} \rightleftharpoons \text{MgADP} + \text{aspartyl 4-phosphate} \qquad (6\text{–}12)$$

$$\text{homoserine} + \text{NAD}^+ \rightleftharpoons \text{aspartate 4-semialdehyde} + \text{NADH} \qquad (6\text{–}13)$$

Reaction 6–12 is the first committed step in the biosynthesis of a set of amino acids, including threonine. Therefore, it makes sense that it is inhibited by threonine acting as an allosteric inhibitor. Both enzymatic activities, however, each occurring at different active sites, are inhibited by threonine, and this fact suggests that the enzyme undergoes a global conformational change linked to the binding of threonine. Threonine is a competitive inhibitor with respect to aspartate[47] in Reaction 6–12 and a noncompetitive inhibitor with respect to homoserine[48] in Reaction 6–13. This result suggests that aspartate has a greater affinity for the active conformation of the enzyme than for the inactive conformation while homoserine has no preference and that the inactive conformation has a lower turnover number, at least in Reaction 6–13, than the active conformation.

The conformational change also couples events at the two respective active sites on a protomer. When the competitive inhibitor *p*-aminosalicylate was bound to the location in the active site of homoserine dehydrogenase to which NADPH normally binds, the rate of alkylation by iodoacetamide of a cysteine in the active site for aspartate kinase increased 10-

fold.[46] When either MgATP or aspartate was bound to the active site for aspartate kinase, the Michaelis constant for potassium cation, an essential activator of homoserine dehydrogenase, decreased as much as 50-fold.[49]

## Suggested Reading

Oberfelder, R.W., Lee, L.L., & Lee, J.C. (1984) Thermodynamic Linkages in Rabbit Muscle Pyruvate Kinase: Kinetic, Equilibrium, and Structural Studies, *Biochemistry 23*, 3813–3821.

Oberfelder, R.W., Barisas, B.G., & Lee, J.C. (1984) Thermodynamic Linkages in Rabbit Muscle Pyruvate Kinase: Analysis of Experimental Data by a Two-State Model, *Biochemistry 23*, 3822–3826.

## PROBLEM 6–1

The reaction catalyzed by phospho-2-dehydro-3-deoxyheptonate aldolase is the first committed reaction in phenylalanine and tyrosine biosynthesis:

phospho*enol*pyruvate + D-erythrose 4-phosphate + $H_2O \rightleftharpoons$ 7-phospho-2-dehydro-3-deoxyheptonate + $HOPO_3^{2-}$

Its two reactants are phospho*enol*pyruvate and erythrose 4-phosphate. The enzymatic reaction is inhibited by both phenylalanine and tyrosine. Below are some of the kinetic results which describe this heterotropic allosteric inhibition.

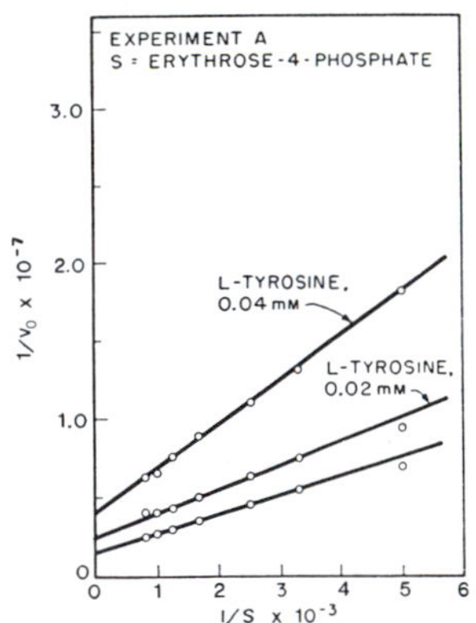


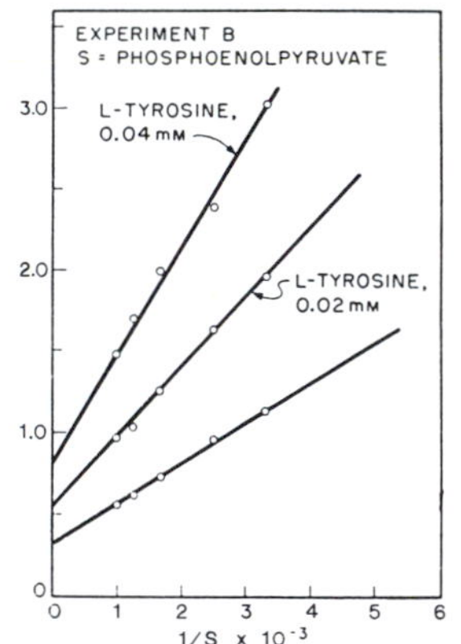


Inhibition of phospho-2-dehydro-3-deoxyheptonate aldolase by L-tyrosine.[50] In experiment A, the reaction mixture contained phospho*enol*pyruvate, 1 μmol; erythrose-4-P, as indicated; and 64 μg of protein in a total volume of 0.5 mL. In experiment B, the reaction mixture contained erythrose-4-P, 1 μmol; phospho*enol*pyruvate, as indicated; and 50 μg of protein. Both reaction mixtures were incubated for 10 min at 37 °C, and the amount of 7-phospho-2-dehydro-3-deoxyheptonate formed was determined. The concentration of substrate (S) is expressed as moles of substrate liter$^{-1}$ and initial velocity ($v_0$) as moles of 7-phospho-2-dehydro-3-deoxyheptonate produced liter$^{-1}$ second$^{-1}$. Reprinted with permission from ref 50. Copyright 1962 *Journal of Biological Chemistry.*

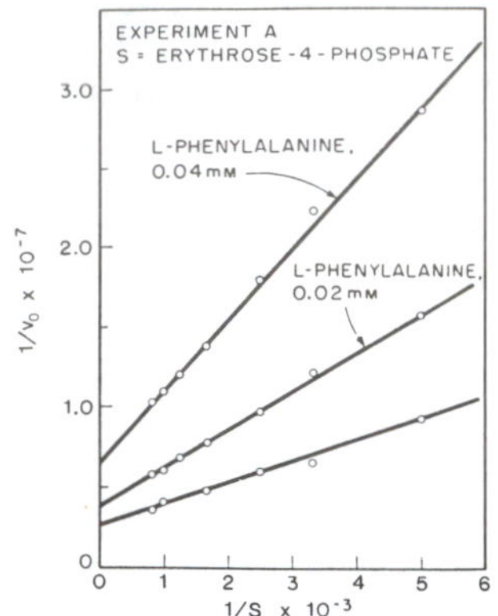


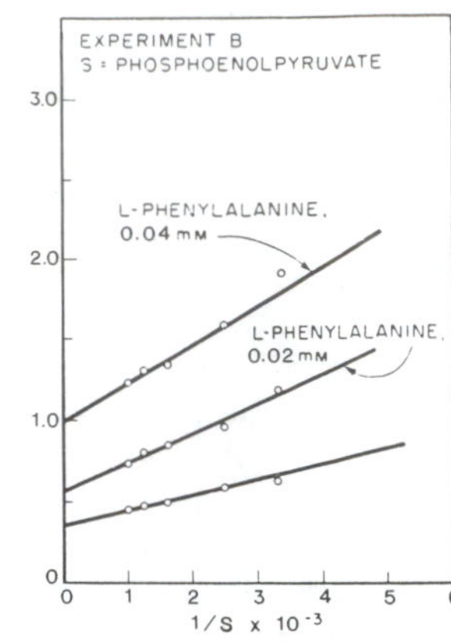


Inhibition of phospho-2-dehydro-3-deoxyheptonate aldolase by L-phenylalanine.[50] The procedure was the same as described above with the exceptions that the reaction mixture contained 63 μg of protein in experiment A and 47 μg of protein in experiment B. Reprinted with permission from ref 50. Copyright 1962 *Journal of Biological Chemistry.*

(A) Do the heterotropic allosteric inhibitors, phenylalanine and tyrosine, compete with any of the reactants?
(B) Extend the lines in each of the figures and decide whether the heterotropic allosteric inhibitors might affect the binding of substrates to the enzyme.
(C) What kinetic parameter of the enzymatic reaction is affected by the binding of the inhibitors to the enzyme?
(D) Assume that there are two forms of the enzyme, E and E′, in equilibrium with each other and that phenylalanine affects this equilibrium. Draw the linkage relationship between the binding of phenylalanine and the conformational equilibrium of the enzyme.
(E) If $K^{E}$ = [E′]/[E], $K^{E\cdot Phe}$ = [E′·Phe]/[E·Phe], $K_d^{Phe}$ = [E][Phe]/[E·Phe], and $K_d'^{Phe}$ = [E′][Phe]/[E′·Phe], what is the relationship between these four equilibrium constants? If $K_d^{E4P}$ = [E4P][E]/[E·E4P] and $K_d'^{E4P}$ = [E4P][E′]/[E′·E4P], where E4P is erythrose 4-phosphate, how are $K_d^{E4P}$ and $K_d'^{E4P}$ probably related to each other to explain the experimental results?
(F) In order to explain the effect of phenylalanine on the properties of the enzyme, what probably are the values of $K^{E}$ and $K^{E\cdot Phe}$ with respect to 1.00 if $K_d^{Phe} << K_d'^{Phe}$?
(G) How do E and E′ probably differ in their functional properties?

## PROBLEM 6–2

The Bohr effect of hemoglobin is the change in oxygen affinity of hemoglobin (Hb) with a change in pH. It is due to heterotropic allosteric interaction between $H^+$ and $O_2$. The main experimental facts can be approximated in the following way. An approximation is made to simplify the situation and elucidate underlying principles.

(a) The reaction between Hb and $O_2$ can be described by the following equilibrium

$$HbO_2 \rightleftharpoons Hb + O_2$$

(b) At pH 7 the dissociation constant for this reaction is

$$K^{O_2}_{d,app} = 1.8\times10^{-5}\,M$$

(c) The dissociation constant changes with pH, reaching a constant minimum value above pH 9.5 and a constant maximum value below pH 6.0. These values are $2.5 \times 10^{-6}$ M and $5 \times 10^{-5}$ M, respectively.
(d) If a 0.1 mM solution of oxygenated hemoglobin in 0.2 M NaCl at pH 7 is deoxygenated by bubbling with $N_2$, the pH of the resulting solution increases by 0.26 pH unit.
(e) Assume that the reaction of interest between hemoglobin and hydrogen ion can be considered to be linked to the ionization of a single amino acid residue in each $\alpha\beta$ dimer.
(f) The oxygen binding curves for Hb have been determined at several different pH values. According to the scheme above the reaction measured experimentally is

$$HbO_{2,TOT} \overset{K^{O_2}_{d,app}}{\rightleftharpoons} Hb_{TOT} + O_2$$

where $[HbO_2]_{TOT}$ = $[HbO_2]$ + $[HbH^+O_2]$, $[Hb]_{TOT}$ = $[Hb]$ + $[HbH^+]$, and $K^{O_2}_{d,app} = [Hb]_{TOT}[O_2]/[HbO_2]_{TOT}$. The values are

| $K^{O_2}_{d,app}$ (M) | pH | $K^{O_2}_{d,app}$ (M) | pH |
|---|---|---|---|
| $2.5 \times 10^{-6}$ | 10.0 | $2.5 \times 10^{-5}$ | 7.0 |
| $3.0 \times 10^{-6}$ | 9.0 | $3.7 \times 10^{-5}$ | 6.5 |
| $4.8 \times 10^{-6}$ | 8.25 | $4.5 \times 10^{-5}$ | 6.0 |
| $1.3 \times 10^{-6}$ | 7.6 | $5.0 \times 10^{-5}$ | 5.0 |

(A) Define the four microscopic equilibrium constants: $K_d^{O_2}$, $K_d^{O_2H^+}$, $K_a^{Hb}$, and $K_a^{HbO_2}$. Write the reactions to which they refer. Label each microscopic equilibrium with appropriate changes in standard free energy ($\Delta G°$). Indicate the direction of each reaction to which these quantities apply. Show that the four values for $\Delta G°$ must be related by a simple summation.
(B) Show that $K_d^{O_2}K_a^{HbO_2} = K_d^{O_2H^+}K_a^{Hb}$
(C) How many equilibrium constants are independent ones?
(D) By using the relationships between the equilibrium constants derived above, show that $K^{O_2}_{d,app}$ is a function of only $[H^+]$ and give the mathematical relationship.
(E) Plot $\log K^{O_2}_{d,app}$ against pH. Explain the shape of the curve in terms of the equation in part D.
(F) If the value of the $pK_a$ for the oxygen-linked group in deoxyhemoglobin ($pK_a^{Hb}$) were 8.5, what would be the value of the $pK_a$ of this group in oxyhemoglobin, $pK_a^{HbO_2}$?
(G) Explain observation d.

# Cooperativity

Before elaborate explanations for the observation of positive deviations from ideal behavior are discussed, it should be noted that there can also be simple explanations not requiring homotropic cooperative interactions among the subunits in an oligomeric protein. Tryptophan is both a substrate and a heterotropic allosteric activator of tryptophan 2,3-dioxygenase because the reaction is the first step in the catabolism of this amino acid. The two roles of tryptophan were distinguished by using analogues. The competitive inhibitor 5-fluorotryptophan binds only at the active site,[51] while $\alpha$-methyltryptophan cannot bind at the active site but acts as a heterotropic allosteric activator at a separate site.[52] When the heterotropic allosteric site is occupied by $\alpha$-methyltryptophan, the initial velocity of the enzymatic reaction is a hyperbolic function of the concentration of tryptophan; and, even in the absence of $\alpha$-methyltryptophan, the binding of the 5-fluorotryptophan to the active site is a hyperbolic function of its concentration. In the absence of either 5-fluorotryptophan or $\alpha$-methyltryptophan, however, the initial velocity of the enzyme as a function of the concentration of tryptophan displays positive deviation from hyperbolic behavior. An explanation for this behavior consistent with all of the facts is that the positive deviation arises because, as the active sites become occupied, the allosteric sites are also becoming occupied. As a result, the initial velocities at higher concentrations of tryptophan are greater than those predicted from the initial slope of the curve at concentrations of tryptophan where the activator site is vacant. This explanation does not invoke homotropic cooperativity among active sites. Originally, it was observed that the initial velocity of glycine hydroxymethyltransferase as a function of the concentration of the reactant, tetrahydrofolate, displayed positive deviations from hyperbolic behavior. It was later demonstrated that this observation was artifactual and arose from instability of the tetrahydrofolate at low concentrations.[53] As a result, the actual concentration of reactant in the lower ranges was less than the nominal concentrations, causing the initial velocities to be lower than they should have been. At higher concentrations, the tetrahydrofolate was stable and the initial velocities were accurate, but higher relative to those at the lower concentrations. The majority of the instances of positive deviations from ideal behavior, however, probably do involve allosteric processes.

The positive deviation from ideal behavior observed for the initial velocity of many enzymatic reactions (Figures 6–4A and 6–6A) or the binding of ligands to many proteins can be described experimentally in different ways. Because hemoglobin is a protein that displays this type of behavior and because it has been studied for a longer period of time than any of the others, it serves as a useful example. When the fraction of the hemes in hemoglobin that are occupied by oxygen is measured as a function of the concentration of oxygen, positive deviation from ideal behavior is observed (Figure 6–9).[54] Such positive deviation is often referred to as **sigmoid** behavior.

Consider a protein, P, with $n$ sites, ST, for ligand L. Suppose each molecule of that protein, for some unknown reason, were required to bind simultaneously all $n$ molecules of L to those $n$ sites and that each molecule of that protein, consequently, was always either fully occupied or completely unoccupied

$$P + nL \overset{K_d^L}{\longleftarrow} P\cdot L_n \tag{6–14}$$

The dissociation constant for this reaction would be

$$K_d^L = \frac{[P][L]^n}{[P\cdot L_n]} \tag{6–15}$$

and if $Y$ were defined (Equation 3–144) as the fraction of the sites, ST, occupied, then

$$\frac{[ST\cdot L]}{[ST]_{TOT}} \equiv Y = \frac{[P\cdot L_n]}{[P]_{TOT}} = \frac{[L]^n}{[L]^n + K_d^L} \tag{6–16}$$

Equation 6–16 displays sigmoid behavior. If it is rearranged

$$\frac{Y}{1-Y} = \frac{[L]^n}{K_d^L} \tag{6–17}$$

$$\log\left(\frac{Y}{1-Y}\right) = n\log[L] + pK_d^L \tag{6–18}$$

This is the **Hill equation.**[55] The parameter $n$ would be the slope of a plot of log $[Y/(1-Y)]$ against log [L]. Only a very peculiar protein, however, unlike any that currently exists, would bind all of its ligands simultaneously.

The preceding derivation is based on a particular mechanism, the mechanism in which each molecule of the protein was required to bind all of its ligands simultaneously (Equation 6–14). The uselessness of this as an explanation for the sigmoid behavior of the binding of ligands to proteins or the behavior of the initial velocity of an enzymatic reaction as a function of the concentration of reactant is apparent because no molecule of protein could be constructed so that all of its ligands were required to bind simultaneously. Nevertheless, it is possible to use the Hill equation empirically by plotting experimental data for the binding of a ligand to a protein in the format log $[Y/(1-Y)]$ against log [L]; and, over a limited range, the plot is usually linear. The value of the slope of such a plot is known as the **Hill coefficient**. Even though a Hill coefficient provides no insight into the reason for the sigmoid behavior, it is still useful to tabulate values of the Hill coefficient derived from such plots as semiquantitative measures of the degree of sigmoidicity. If the number of sites for the ligand on a molecule of the protein is known, the closer the value of $n$ is to that number, the greater is the cooperativity. In theory, $n$ can never exceed the number of cooperating sites because Equation 6–14 describes, by definition, a completely cooperative situation.

There are several reasons that plotting experimental data by the Hill equation (Equation 6–16) is unsatisfactory. The value of $n$ rarely equals the known number of sites, as it should. The value of $n$ is almost never an integer, as it should be if the mechanism on which the equation was derived were actually in operation. The value of $n$ varies depending on the position

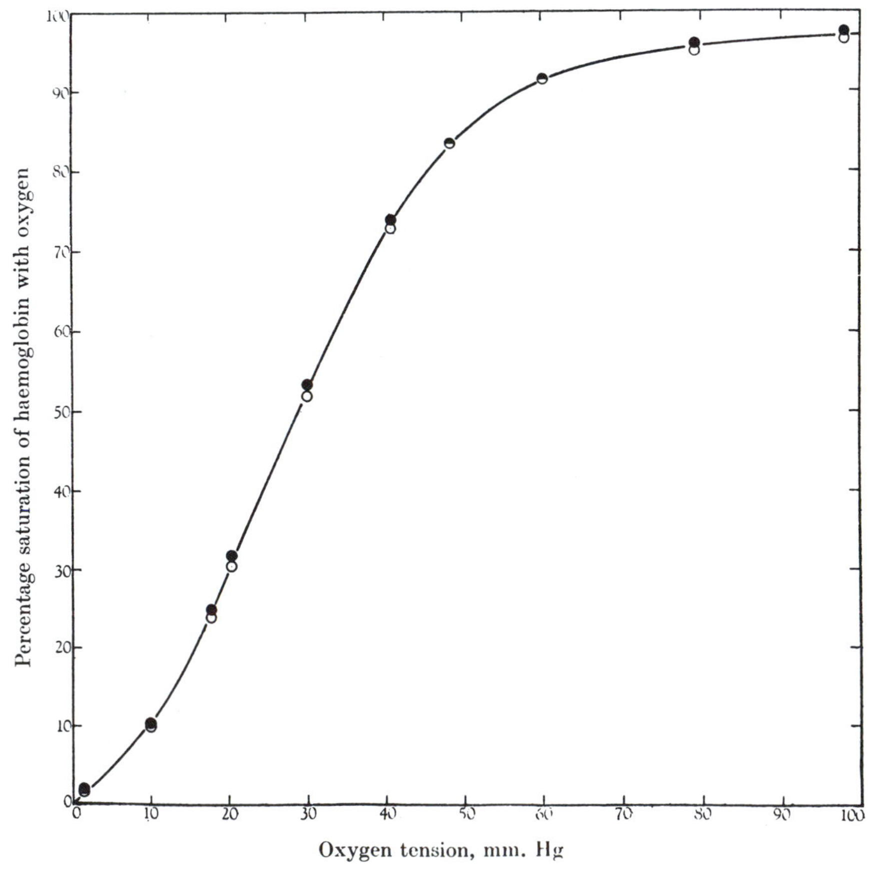


**Figure 6–9:** Dissociation curve for oxygen from human hemoglobin.[54] Samples of blood from C.G.D. were equilibrated with various mixtures of nitrogen and oxygen and 40 mmHg of $CO_2$ at 38 °C. After equilibrium was reached, the oxygen bound to the hemoglobin at the noted partial pressure of $O_2$ (oxygen tension in millimeters of mercury) was driven off the hemes by adding potassium ferricyanide. The $O_2$ gas liberated into the vapor phase above the blood was then measured manometrically in either a Haldane apparatus (●) or a van Slyke apparatus (○). The same amount of blood was used for each reading and the concentration of bound oxygen was tabulated relative to the concentration of bound oxygen at saturation (720 mmHg). The percentage of saturation of the hemoglobin with oxygen is presented as a function of the partial pressure of $O_2$ (oxygen tension in millimeters of mercury). Reprinted with permission from ref 54. Copyright 1947 Cambridge University Press.

on the curve at which the slope is determined. According to Equation 6–18, a plot of log [$Y/(1-Y)$] against log [L] should be a straight line. Most observations are made around the midregion (Figure 6–9) of the saturation curve ($0.10 < Y < 0.90$), and within this range the behavior, within the errors of the experiment, is usually linear. When measurements are taken over a broader range, however, it is always found that log [$Y/(1-Y)$] is not a linear function of log [L] (Figure 6–10).[56]

It is possible to describe the complete set of observations for hemoglobin, as well as for other proteins and enzymes displaying sigmoid behavior, with a somewhat more realistic model.[57] Hemoglobin contains four hemes and binds four oxygens at saturation. If this process occurs in four simple steps

$$\text{Hb} \overset{K_{d1}}{\longleftarrow} \text{Hb}\cdot\text{O}_2 \overset{K_{d2}}{\longleftarrow} \text{Hb}\cdot(\text{O}_2)_2 \overset{K_{d3}}{\longleftarrow} \text{Hb}\cdot(\text{O}_2)_3 \overset{K_{d4}}{\longleftarrow} \text{Hb}\cdot(\text{O}_2)_4 \tag{6–19}$$

where the $K_{di}$ are all dissociation constants for oxygen, then

$$Y = \frac{K_{d2}K_{d3}K_{d4}[O_2] + 2K_{d3}K_{d4}[O_2]^2 + 3K_{d4}[O_2]^3 + 4[O_2]^4}{4(K_{d1}K_{d2}K_{d3}K_{d4} + K_{d2}K_{d3}K_{d4}[O_2] + K_{d3}K_{d4}[O_2]^2 + K_{d4}[O_2]^3 + [O_2]^4)} \tag{6–20}$$

This is an **Adair equation.**

If every site on a molecule of hemoglobin were binding oxygen with the same intrinsic dissociation constant, $K_{d0}$, and every site were unlinked to any other site, then by statistical correction

$$K_{d0} = 4K_{d1} = \tfrac{3}{2}K_{d2} = \tfrac{2}{3}K_{d3} = \tfrac{1}{4}K_{d4} \tag{6–21}$$

and by simplification of Equation 6–20

$$Y = \frac{[O_2]}{K_{d0} + [O_2]} \tag{6–22}$$

All of the sites would be binding oxygen independently and no sigmoid behavior would be observed, only the hyperbolic behavior of binding to unlinked sites. The macroscopic dissociation constant for the first step in Mechanism 6–19 always has one-fourth the value of its intrinsic dissociation constant because the reactant has four sites at which association can occur while the product has only one site from which dissociation can occur. The macroscopic dissociation constant for the second step in Mechanism 6–19 always has two-thirds of the value of its intrinsic dissociation constant because the reactant has three sites for association but the product has only two sites for dissociation, and so forth. If the dissociation constants in Equation 6–20 are allowed to differ from the relationship in Equation 6–21 (in other words, the sites are not independent) Equation 6–20 can be fit to the observation of sigmoid behavior.

From Equation 6–20, it can be shown[58] that

$$\lim_{[O_2]\to 0} \log\left(\frac{Y}{1-Y}\right) = \log\frac{[O_2]}{4K_{d1}} \tag{6–23}$$

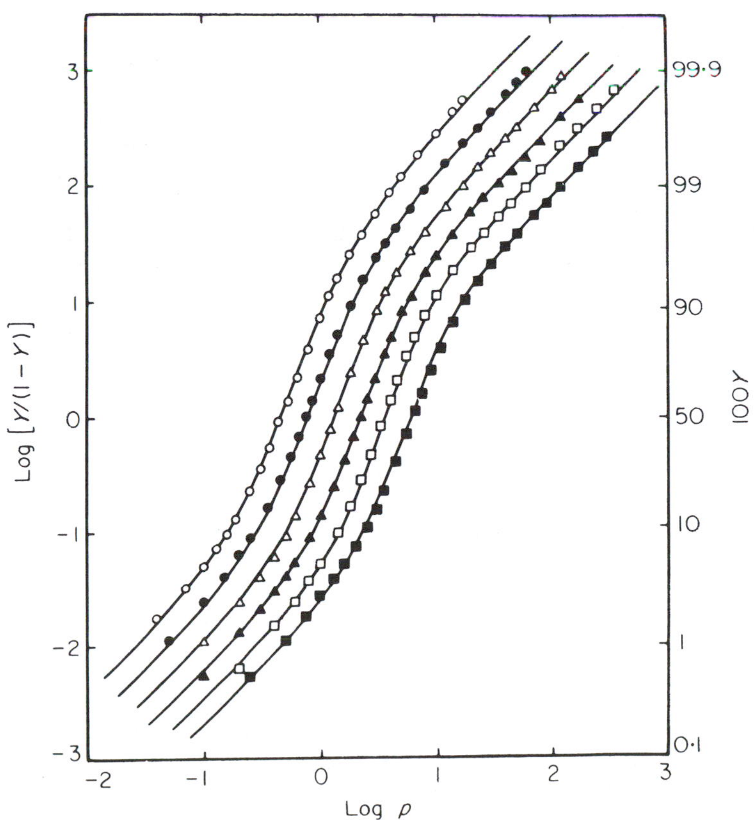


**Figure 6–10:** Dissociation of oxygen from human hemoglobin[56] measured over a range of greater than 3 orders of magnitude in concentration of $O_2$. Hemoglobin A was prepared from human blood and was stripped of all organic phosphates. Solutions of this hemoglobin at a concentration of 0.6 mM in heme were prepared at pH 9.1 and 0.1 M NaCl. They were equilibrated respectively at the noted temperatures, and the fractional saturation of the hemoglobin, $Y$, was determined as a function of the partial pressure of oxygen ($p$) in the gas phase above the solution. For each measurement the concentration of oxygen in the solution of hemoglobin was measured directly with an oxygen electrode that had been calibrated with standard gas mixtures at the same time that the transmittance of the solution was measured at a wavelength of 620 nm. The absorbance of the solution purged with $N_2$ was assumed to be the absorbance of unliganded hemoglobin, and the absorbance extrapolated to infinite oxygen concentration was assumed to be that of fully liganded hemoglobin. The fractional saturation, $Y$, was assumed to be directly proportional to the observed absorbance relative to these limits. After a measurement was taken at one concentration, the mixture of nitrogen and oxygen was changed. All consecutive readings were taken on the same solution at a particular temperature. The reversibility of the process was demonstrated by returning frequently to concentrations of earlier measurements. From the individual measurements of $Y$, the function log [$Y/(1-Y)$] was calculated and is presented as a function of the logarithm of the partial pressure of oxygen (log $p$ in millimeters of mercury). The percent of the hemes occupied (100$Y$) is also presented. The temperatures were 10 °C (○), 15 °C (●), 20 °C (△), 25 °C (▲), 30 °C (□), and 35 °C (■). Reprinted with permission from ref 56. Copyright 1979 Academic Press.

and

$$\lim_{[O_2]\to\infty} \log\left(\frac{Y}{1-Y}\right) = \log\frac{4[O_2]}{K_{d4}} \tag{6–24}$$

It follows that the asymptotes at either end of the curves (Figure 6–10), which have slopes of 1 as required by Equations 6–23 and 6–24, fix the values for $K_{d1}$ and $K_{d4}$. The values for $K_{d2}$ and $K_{d3}$ can then be obtained[58] by fitting the data to Equation 6–20. For example, the data describing the dissociation of oxygen from hemoglobin at 30 °C (Figure 6–10) could be fit with Equation 6–20, if $K_{d1} = 8.4 \pm 0.2$ mmHg, $K_{d2} = 13 \pm 1$ mmHg, $K_{d3} = 16 \pm 1$ mmHg, and $K_{d4} = 0.90 \pm 0.04$ mmHg. When the statistical factors (Equation 6–21) are applied, these correspond to intrinsic dissociation constants, $K_{d0}$, for each step of 34 ± 1 mmHg, 20 ± 2 mmHg, 11 ± 1 mmHg, and 0.22 ± 0.01 mmHg, respectively. On the basis of this analysis, the largest increase in intrinsic affinity would occur between the finding of the third and fourth oxygens. Because the data are of such high accuracy and extend over almost 4 orders of magnitude (Figure 6–1), the errors of estimate are quite small (<10%) even though the data are fit with four parameters. Part of this is due to the fact that the asymptotes, at each end of the curve, assess either $K_{d1}$ alone or $K_{d4}$ alone. If data were available for only the midrange, as is usually the case, the errors of estimate would be much greater.

From the curves, the difference in intrinsic free energy of dissociation between the first and last step can be directly calculated.[59] The vertical distance between the two asymptotes is $\log(4K_{d1}) - \log(K_{d4}/4)$ and

$$\Delta G_4^\circ - \Delta G_1^\circ = 2.303RT\,[\log(4K_{d1}) - \log(K_{d4}/4)] \tag{6–25}$$

where $\Delta G_i^\circ$ is the intrinsic free energy of dissociation for step $i$ in Mechanism 6–19.

The difficulty with the Adair equation is that it provides no explanation of the phenomenon because it is only analytic. An explanation of homotropic cooperativity has, however, been provided that is based on an understanding of the symmetry of oligomeric proteins and the established crystallographic structures of proteins displaying homotropic cooperativity.

The crystals of aspartate transcarbamylase that form in solutions of the enzyme in the absence of ligands are of the space group $R32$ and the crystallographic asymmetric unit is one folded catalytic polypeptide and one folded regulatory polypeptide. All molecular axes of symmetry[1a] coincide with crystallographic axes of symmetry and the protein is necessarily perfectly symmetric.[20] Crystals grown of the protein in which all six of its catalytic sites are occupied by *N*-(phosphonacetyl)-L-aspartate have the space group $P321$ and the crystallographic asymmetric unit is two folded catalytic polypeptides and two regulatory polypeptides. The 3-fold molecular axis of symmetry coincides with the crystallographic axis of symmetry and is precise. Although the 2-fold molecular axes of symmetry do not coincide with crystallographic axes of symmetry, only small deviations from exact symmetry, thought to result from crystal packing forces, are observed.[2] The conclusion that can be drawn is that both unliganded and fully liganded aspartate transcarbamylase are fully symmetric oligomers even though they differ dramatically from each other in conformation (Figure 6–2).

Phosphorylase *a*, with its active site occupied by the allosteric inhibitor glucose, crystallizes from solution in the space group $P4_32_12$. One folded polypeptide of the $\alpha_2$ dimer is the asymmetric unit, and all molecular axes of symmetry coincide with crystallographic axes of symmetry and are precise.[60] Although the conformation of the enzyme induced by occupying the active sites with substrates has not been crystallized directly, cross-linked crystals of the form of the enzyme occupied by glucose can be perfused with a virtual substrate, 5-thio-α-D-pyranosyl 1-phosphate, which causes the crystals to undergo an easily observed macroscopic change in structure, representing a significant conformational change in the protein.[61] The space group and asymmetric unit, however, remain unchanged.[62] Although the protein is constrained by the crystal during the transition, these observations suggest that the inactive and the active conformations of phosphorylase *a* are both fully symmetric oligomers even though they differ from each other in conformation.

Phosphofructokinase, occupied by the heterotropic allosteric inhibitor 2-phosphoglycolate, crystallizes in the space group $P2_12_12_1$ and the asymmetric unit is necessarily the $\alpha_4$ tetramer. The three molecular 2-fold rotational axes of symmetry in the tetramer, however, are exact within the resolution of the measurements.[63] When the enzyme is fully occupied by the substrate fructose 6-phosphate, it crystallizes in the space group $I222$ with a single folded polypeptide of the $\alpha_4$ tetramer as the asymmetric unit, and all molecular axes of symmetry coincide with crystallographic axes of symmetry and are precise.[64] Even though the two conformations crystallized under these two conditions differ significantly in conformation, each of the two tetramers is fully symmetric.

Deoxyhemoglobin crystallizes from solution in the space group $P2_1$ with the tetramer necessarily the asymmetric unit.[65] Within the resolution of the map of electron density the one molecular 2-fold rotational axis of symmetry in the $(\alpha\beta)_2$ tetramer is exact.[66] When these crystals are exposed to oxygen, they shatter; and, as had been predicted by Haurowitz,[67] this easily observed change of state represents the significant change in conformation of the molecule itself that occurs upon ligation (Figure 6–1). Fully liganded hemoglobin, in which all four irons are occupied by oxygen, crystallizes in the space group $P4_12_12$ with the molecular 2-fold rotational axis of symmetry coinciding with the crystallographic 2-fold rotational axis of symmetry.[68] Therefore the unliganded and the fully liganded conformations of hemoglobin, although they differ dramatically from each other,[1] are both fully symmetric oligomers.

These are experimental observations of the symmetry of the conformations of oligomeric proteins that display homotropic cooperativity. They demonstrate that the conformation of each of these proteins in the absence of ligands differs significantly from the conformations of the proteins when all of the sites have been occupied. They also demonstrate that both the unliganded protein and the fully liganded protein are fully symmetric oligomers. No crystallographic molecular structures of proteins displaying homotropic cooperativity in which the oligomer can be shown to be only partially liganded rather than fully liganded are available, so intermediate states of ligation have not been directly observed.

It was pointed out by Monod, Wyman, and Changeux that if ligand-linked conformational changes in oligomeric proteins were to occur with conservation of molecular symmetry, the homotropic cooperativity often observed in the association of ligands to homooligomeric proteins could be readily explained.[23] Consider the simplest case, that of an $\alpha_2$ dimer that has one site for ligand L on each subunit and that displays a ligand-linked conformational change with conservation of symmetry. The dominant conformation in the population of unoccupied, fully symmetric dimers is designated as T·⊥, where T stands for the tight conformation, and the dominant conformation in the population of fully occupied, fully symmetric dimers is designated as L·R·ᴚ·⅂ where R stands for the relaxed conformation.[23] Because the conformational change occurs with conservation of symmetry, only two conformations, or two states, of the protein are possible, T·⊥ and R·ᴚ. The accessible states are

$$\begin{array}{ccccc} \text{T·⊥} & \xleftarrow{\ \frac{1}{2}K_d^{TL}\ } & \text{L·T·⊥} & \xleftarrow{\ 2K_d^{TL}\ } & \text{L·T·⊥·⅂} \\ K_{RT}\ \Big\uparrow & & K_{RT}\dfrac{K_d^{RL}}{K_d^{TL}}\ \Big\uparrow & & K_{RT}\left(\dfrac{K_d^{RL}}{K_d^{TL}}\right)^2\ \Big\uparrow \\ \text{R·ᴚ} & \xleftarrow[\ \frac{1}{2}K_d^{RL}\ ]{} & \text{L·R·ᴚ} & \xleftarrow[\ 2K_d^{RL}\ ]{} & \text{L·R·ᴚ·⅂} \end{array} \tag{6–26}$$

where

$$K_{RT} = \frac{[\text{T·⊥}]}{[\text{R·ᴚ}]} \tag{6–27}$$

and $K_d^{TL}$ and $K_d^{RL}$ are the intrinsic dissociation constants for ligand L from a site in the T and the R conformation, respectively. The fact that the association with the ligand L is cooperative must mean that a molecule of the unoccupied protein, on average, has a lower affinity for the ligand than a molecule of the protein with, on average, one site occupied and the other unoccupied. If the model of Equation 6–26 is to explain this behavior, the equilibrium between the conformations of the enzyme must shift after the first ligand has been bound but before the second is bound. Specifically, the T conformation must dominate in the population in the absence of ligand ($K_{RT} > 1$), and the R conformation must dominate after, on average, one ligand has bound [$K_{RT}K_d^{RL}/K_d^{TL} < 1$]. For this to occur, $K_d^{RL}$ must be less than $K_d^{TL}$ by a sufficient factor to accomplish the necessary shift in the equilibrium between the two conformations. If $K_d^{RL}$ must be less than $K_d^{TL}$ to achieve the shift in the conformational equilibrium, the second ligand on average will necessarily bind more tightly than the first, and this produces the homotropic cooperativity. The appeal of this hypothesis is that the same property that tips the conformational equilibrium, the fact that $K_d^{RL} < K_d^{TL}$, also causes the second ligand to bind more tightly than the first once the equilibrium between the conformations has shifted. It is the requirement that symmetry be conserved during the conformational change responsible for the homotropic cooperativity that couples these two forces together because it necessitates that the dissociation constant of the unoccupied active site decrease in concert with the dissociation constant for the occupied active site.

The major, global conformational change of an oligomeric protein that involves reorientation of the protomers (Figure 6–1 or 6–2) and that permits it in this way to display homotropic cooperative behavior among its widely separated sites should be distinguished from the local conformational changes that occur around the sites themselves upon their occupation by the respective ligands. In crystallographic molecular models of aspartate transcarbamylase, the global conformational change between the T conformation present in the absence of ligands and the R conformation present when all of the active sites are occupied involves a significant reorientation of the protomers (Figure 6–2). When the unliganded enzyme, however, already in the T conformation becomes occupied by CTP, the relative orientation of the protomers does not change, but significant shifts (as much as 1.5 nm) in the location of segments of polypeptide surrounding the binding site for CTP occur as these segments close around the ligand.[20] Presumably, these local changes in conformation are involved in the increase in stability of the T conformation brought about by the binding of CTP, although they are not propagated beyond the surroundings of the binding site once the enzyme is already in the T conformation.

The modification of the isolated, enzymatically active catalytic trimer of aspartate transcarbamylase with tetranitromethane converts tyrosines in the protein to nitrotyrosines, which have an absorbance at 430 nm. When the active site in such a modified catalytic trimer is first occupied with carbamyl phosphate, the binding of succinate, a competitive inhibitor of the enzymatic activity, causes a significant decrease in the absorbance at 430 nm. This decrease can be used to monitor the proportion of active sites occupied by the succinate.[69] When these nitrated catalytic subunits are reconstituted with regulatory dimers to produce the intact $(\alpha\beta)_6$ hexamer of aspartate transcarbamylase, the same change in absorbance upon occupation of the active site persists.[70] It still monitors directly the binding of the succinate rather than the global conformational change involved in homotropic cooperativity. For this reason, it can be assumed that whatever conformational change leads to the change of absorbance of the nitrotyrosine is confined to the immediate vicinity of the active site and is not propagated globally.

Phosphofructokinase has been crystallized in two active conformations, each with the same global conformation but differing in the occupation of the site for the heterotropic allosteric activator, MgADP. Extensive local differences in conformation around the effector site to which MgADP binds (Figure 6–5) are observed when the two molecular models are compared, but these conformational changes were not widespread enough to extend to the active site.[14] Presumably these are the local changes in conformation that alter the relative stabilities of the global conformations.

The arguments presented to support the hypothesis that symmetry is conserved during a global conformational change of an oligomeric enzyme,[23] as distinguished from a local unpropagated conformational change, are the same arguments that can be used to explain why oligomeric proteins are built around rotational axes of symmetry in the first place. If a symmetrical arrangement is advantageous in one conformation it is also advantageous in another.

A general equation, derived from the proposal that symmetry is conserved, has been presented.[23] If the system has the structure

$$
\begin{array}{ccccccccc}
\mathrm{T}_n & \xleftarrow{\frac{1}{n}K_d^{\mathrm{TL}}} & \mathrm{T}_n\cdot\mathrm{L} & \xleftarrow{\frac{2}{n-1}K_d^{\mathrm{TL}}} & \mathrm{T}_n\cdot\mathrm{L}_2 & \cdots\cdots & \mathrm{T}_n\cdot\mathrm{L}_{n-1} & \xleftarrow{nK_d^{\mathrm{RL}}} & \mathrm{T}_n\cdot\mathrm{L}_n \\
K_{\mathrm{RT}}\uparrow & & \uparrow K_{\mathrm{RT}}\dfrac{K_d^{\mathrm{RL}}}{K_d^{\mathrm{TL}}} & & \uparrow K_{\mathrm{RT}}\left(\dfrac{K_d^{\mathrm{RL}}}{K_d^{\mathrm{TL}}}\right)^2 & & \uparrow & & K_{\mathrm{RT}}\left(\dfrac{K_d^{\mathrm{RL}}}{K_d^{\mathrm{TL}}}\right)^n \uparrow \\
\mathrm{R}_n & \xleftarrow[\frac{1}{n}K_d^{\mathrm{RL}}]{} & \mathrm{R}_n\cdot\mathrm{L} & \xleftarrow[\frac{2}{n-1}K_d^{\mathrm{RL}}]{} & \mathrm{R}_n\cdot\mathrm{L}_2 & \cdots\cdots & \mathrm{R}_n\cdot\mathrm{L}_{n-1} & \xleftarrow[nK_d^{\mathrm{RL}}]{} & \mathrm{R}_n\cdot\mathrm{L}_n
\end{array}
\qquad (6\text{–}28)
$$

and only conformational changes that conserve symmetry $[\mathrm{T}_n\cdot\mathrm{L}_i \rightleftharpoons \mathrm{R}_n\cdot\mathrm{L}_i]$ are permitted, then the fraction of the sites occupied

$$
Y \equiv \frac{([\mathrm{R}_n\cdot\mathrm{L}]+2[\mathrm{R}_n\cdot\mathrm{L}_2]+\cdots+n[\mathrm{R}_n\cdot\mathrm{L}_n])+([\mathrm{T}_n\cdot\mathrm{L}]+2[\mathrm{T}_n\cdot\mathrm{L}_2]+\cdots+n[\mathrm{T}_n\cdot\mathrm{L}_n])}{n([\mathrm{R}_n]+[\mathrm{R}_n\cdot\mathrm{L}]+\cdots+[\mathrm{R}_n\cdot\mathrm{L}_n])+n([\mathrm{T}_n]+[\mathrm{T}_n\cdot\mathrm{L}]+\cdots+[\mathrm{T}_n\cdot\mathrm{L}_n])} \qquad (6\text{–}29)
$$

is

$$
Y = \frac{K_{\mathrm{RT}}\dfrac{[\mathrm{L}]}{K_d^{\mathrm{TL}}}\left(1+\dfrac{[\mathrm{L}]}{K_d^{\mathrm{TL}}}\right)^{n-1}+\dfrac{[\mathrm{L}]}{K_d^{\mathrm{RL}}}\left(1+\dfrac{[\mathrm{L}]}{K_d^{\mathrm{RL}}}\right)^{n-1}}{K_{\mathrm{RT}}\left(1+\dfrac{[\mathrm{L}]}{K_d^{\mathrm{TL}}}\right)^{n}+\left(1+\dfrac{[\mathrm{L}]}{K_d^{\mathrm{RL}}}\right)^{n}} \qquad (6\text{–}30)
$$

If the initial velocity of an enzymatic reaction as a function of reactant A is being followed and it can be assumed that the turnover number of the R conformation is $k_{cat}^{\mathrm{R}}$ and that of the T conformation is $k_{cat}^{\mathrm{T}}$, then

$$
\frac{v_0}{[\mathrm{E}]_{\mathrm{TOT}}} = \frac{k_{cat}^{\mathrm{T}}K_{\mathrm{RT}}\dfrac{[\mathrm{A}]}{K_d^{\mathrm{TA}}}\left(1+\dfrac{[\mathrm{A}]}{K_d^{\mathrm{TA}}}\right)^{n-1}+k_{cat}^{\mathrm{R}}\dfrac{[\mathrm{A}]}{K_d^{\mathrm{RA}}}\left(1+\dfrac{[\mathrm{A}]}{K_d^{\mathrm{RA}}}\right)^{n-1}}{K_{\mathrm{RT}}\left(1+\dfrac{[\mathrm{A}]}{K_d^{\mathrm{TA}}}\right)^{n}+\left(1+\dfrac{[\mathrm{A}]}{K_d^{\mathrm{RA}}}\right)^{n}} \qquad (6\text{–}31)
$$

if the reactant associates and dissociates from the active site much more rapidly than it is turned over.

Most if not all instances of homotropic cooperativity examined so far can be quantitatively described by Equation 6–30 or 6–31. This perhaps is not surprising because these equations have three or five adjustable parameters, respectively. For example, the continuous curves in Figure 6–10 that connect the observed values for the binding of $O_2$ to hemoglobin are all solutions to Equation 6–30, and they fit the data as well as do curves generated with the Adair equation. For the binding of $O_2$ at 30 °C and pH 7.4, the values for the parameters are $K_{\mathrm{RT}} = (1.3 + 0.2) \times 10^6$, $K_d^{\mathrm{TO_2}} = 31 \pm 2$ mmHg, and $K_d^{\mathrm{RO_2}} = 0.19 \pm 0.01$ mmHg. As expected, the dissociation constant for $O_2$ from the R conformation is significantly smaller than for $O_2$ from the T conformation, and $K_{\mathrm{RT}}$ is greater than 1 so that in the absence of $O_2$, the T conformation dominates.

Both the homotropic cooperativity of the initial velocity of pyruvate kinase as a function of the concentration of phospho*enol*pyruvate in the presence of phenylalanine and the heterotropic inhibition of the activity of the enzyme by phenylalanine (Figure 6–7) can be explained simultaneously by Mechanism 6–28, if it is amended to include the heterotropic allosteric inhibition by phenylalanine (Mechanism 6–9). If it is assumed that phenylalanine inhibits pyruvate kinase because it has a greater affinity for the T conformation (E′ in Mechanism 6–9), which has a lower affinity for phospho*enol*pyruvate and a lower turnover number, than it does for the R conformation (E in Mechanism 6–9), which has a higher affinity for phospho*enol*pyruvate and a higher turnover number, then an equation analogous to Equation 6–31 but including the binding of phenylalanine can be derived. A set of parameters can be found that permits this equation to reproduce the observed behavior of the initial velocity (smooth curves in Figure 6–7A)[33] at pH 7.5 and 23 °C and the binding of phospho*enol*pyruvate (smooth curve in Figure 6–7B) as well as the binding of phenylalanine (smooth curve in Figure 6–11).[33] The dissociation constants for phenylalanine from the two conformations ($K_d^{\mathrm{TF}} = 0.6$ mM and $K_d^{\mathrm{RF}} = 14$ mM) and phospho*enol*pyruvate from the two conformations ($K_d^{\mathrm{TP}} = 440\ \mu$M and $K_d^{\mathrm{RP}} = 44\ \mu$M) and the equilibrium constant for isomerization of the free enzyme ($K_{\mathrm{RT}} = 0.09$) differ in the expected directions from each other or with respect to 1, respectively. Because $K_{\mathrm{RT}} = 0.09$, the enzyme in the absence of any ligands is already mostly in the R conformation, and the initial velocity is a hyperbolic function of the concentration of phospho*enol*pyruvate (Figure 6–7A) because the binding of phospho*enol*pyruvate cannot affect significantly the absolute amounts of the two conformations. Because phenylalanine, however, does shift the conformational equilibrium significantly in favor of the T conformation, its binding displays sigmoid behavior (Figure 6–11). The difference in the two values for $k_{cat}$ ($k_{cat}^{\mathrm{R}} = 10k_{cat}^{\mathrm{T}}$) has already been discussed.

Following the publication of the explanation of Monod, Wyman, and Changeux, it was pointed out that if a simple model involving only two states of the protein is satisfactory to explain homotropic cooperativity, a model involving more than two states will be as satisfactory.[71] Therefore, to establish the central hypothesis that because of the conservation of symmetry, there are only two conformations, results other than measurements of the binding of ligands or the behavior

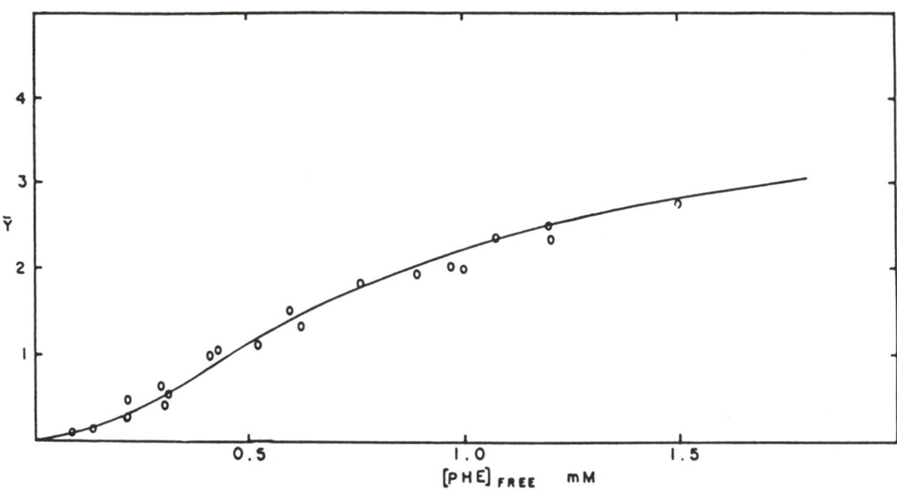


**Figure 6–11:** Binding of phenylalanine to rabbit muscle pyruvate kinase.[33] The binding of [$^3$H]phenylalanine to the enzyme was measured by equilibrium dialysis at pH 7.5 and 23 °C. A solution containing 12 mg mL$^{-1}$ pyruvate kinase was placed in one compartment, and a solution of [$^3$H]phenylalanine of an appropriate molar concentration was placed in the other compartment. After equilibrium had been reached (20 h), the final concentration of the protein in its compartment, as determined by absorbance at 280 nm, and the final concentrations of phenylalanine, as determined by scintillation counting, in each compartment were assayed. From these concentrations the free concentration of phenylalanine as well as the moles of phenylalanine bound for each mole of the tetrameric enzyme ($\bar{Y}$) could be calculated for each measurement. The fractional saturation ($\bar{Y}$) is presented as a function of the free concentration of phenylalanine ($[Phe]_{free}$ in millimolar). The smooth curve was calculated from an equation derived for a combination of Mechanisms 6–9 and 6–28. Reprinted with permission from ref 33. Copyright 1984 American Chemical Society.

of initial velocities as a function of the concentrations of reactants must be gathered. Experiments addressing this point seek to demonstrate that only two conformations of the protein exist over the complete range of concentrations of a ligand that displays cooperativity. Then, because two symmetric conformations have been observed crystallographically, they must be the only two.

The change in the distribution of conformations that occurs in a solution of aspartate transcarbamylase upon the addition of a ligand can be monitored by sedimentation velocity.[72] The sedimentation coefficient of native aspartate transcarbamylase decreases by 3.5% as the unliganded enzyme becomes fully occupied either with carbamyl phosphate and succinate[7,72] or with *N*-(phosphonacetyl)-L-aspartate.[73] These measurements have been corrected for the small increase in mass due solely to the occupation of the six active sites. This decrease in the sedimentation coefficient is consistent with a change in the frictional coefficient resulting from the expansion of the molecule observed crystallographically (Figure 6–2) and monitors directly the shift in the equilibrium between the global conformational states of the enzyme.

The binding of succinate, a competitive inhibitor with respect to aspartate, to aspartate transcarbamylase, when the active sites are saturated with carbamyl phosphate, displays homotropic cooperativity. The binding of succinate can be monitored directly by the change in absorbance at 430 nm in intact aspartate transcarbamylase that has been reconstituted from nitrated catalytic subunits. The absorbance of this modified enzyme changes in direct proportion to the fraction of the active sites occupied by succinate.[70] When the change in sedimentation coefficient as a function of the concentration of succinate is compared to the occupation of the active sites as a function of the concentration of succinate, both at saturating carbamyl phosphate (Figure 6–12A),[70] it can be seen that the majority of the shift in the distribution of conformations has occurred significantly before the majority of the active sites are occupied.[72] This is as expected because the last succinates bind to the R conformation with high affinity in order to explain the homotropic cooperativity; and, therefore, the conformation must have changed before the last succinates are bound.

This observation, that the completion of the conformational change in the protein precedes the full occupation of the active sites, is a commonly observed property of proteins that display homotropic cooperativity. For example, the allosteric activator AMP binds to phosphorylase *a* with homotropic cooperativity. Its occupation of the binding sites can be followed directly by equilibrium dialysis, and the change in the distribution of conformations it promotes can be followed separately by monitoring change in the electron spin resonance spectrum of a covalently bound, stable free radical, 1-oxyl-2,2,6,6-tetramethylpiperidine. When the distribution of conformations is displayed as a function of the fraction of the sites occupied (Figure 6–13),[11] the conformational change is also seen to precede the occupation of the sites.

This observation is the expected result regardless of the theory used to explain homotropic cooperativity. The last ligands to bind must bind to the conformation of high affinity stabilized by the binding of the first ligands. For this to occur, the population of the conformation of high affinity must become dominant before all of the ligands have been bound by the protein. The more important issue is whether there are only two conformations populated to any significant degree in these ligand-promoted transitions or more than two conformations.

Catalytic subunits of aspartate transcarbamylase, nitrated

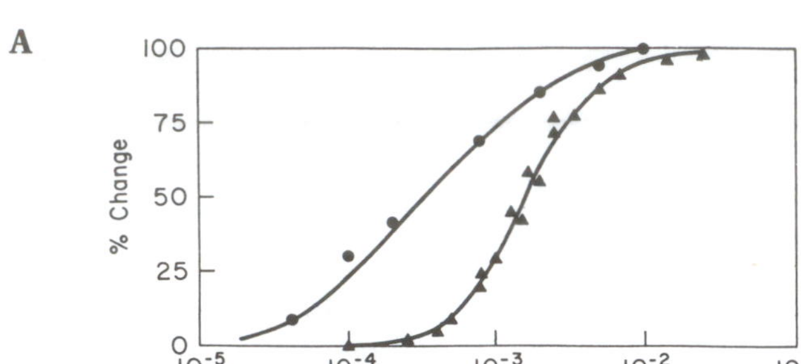


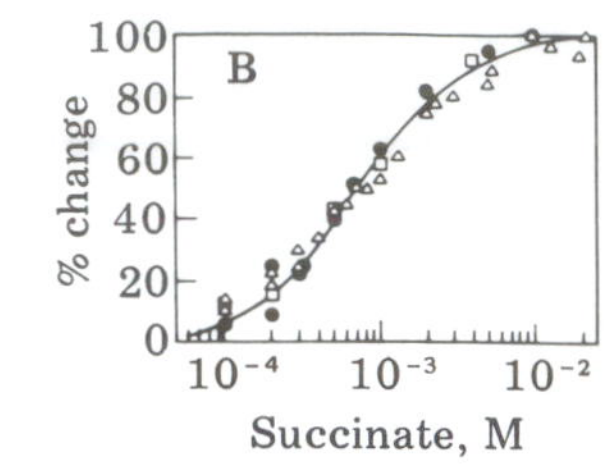


**Figure 6–12:** Relationship between the change in global conformation of *E. coli* aspartate transcarbamylase, as monitored by sedimentation velocity, changes in absorbance of nitrotyrosines, or the rate of mercuration of cysteines, and the occupation of the active sites by succinate, as monitored spectrophotometrically. (A) Catalytic trimers, made from purified $(\alpha\beta)_6$ hexamers of aspartate transcarbamylase, were nitrated with tetranitromethane in the presence of carbamyl phosphate and succinate so that 0.7 nitrotyrosine was formed for every catalytic polypeptide and 85% of the enzymatic activity remained. These nitrated catalytic trimers were reconstituted with unmodified regulatory dimers to form $(\alpha^{Nit}\beta)_6$ hexamers that were enzymatically active and displayed homotropic cooperativity. A change in the spectrum of the nitrotyrosine at 430 nm was shown to be directly proportional to the concentration of bound succinate when carbamyl phosphate was present at saturating concentration. The occupation of the active site (▲) by succinate in percent, where 0% is unoccupied and 100% is fully occupied with six succinates for each hexamer, is presented as a function of succinate concentration (molar). The change in the sedimentation coefficient ($\Delta s_{max} = -2.9\%$) of these reconstituted $(\alpha^{Nit}\beta)_6$ hexamers (●) was also followed as a function of the molar concentrations of succinate at saturating carbamyl phosphate under the same conditions and is presented in units of percent change.[70] Panel A reprinted with permission from ref 70. Copyright 1973 American Chemical Society. (B) Nitrated catalytic trimers $\alpha_3^{Nit}$ were covalently inactivated by modifying their active sites with pyridoxal phosphate and $NaBH_4$. These inactive catalytic trimers $\alpha_3^{Nit,Pyr}$ were then modified with 3,4,5,6-tetrahydrophthalic anhydride to increase their negative charge. They were reconstituted with native, unmodified catalytic trimers and regulatory dimers. The $(\alpha_3^{Nit,Pyr}\alpha_3^{native}\beta_6)$ heterohexamers resulting from this reconstitution were isolated by anion-exchange chromatography followed by removal of the phthalyl groups. The nitrated catalytic trimer in this hybrid could not bind succinate but registered the global conformational change of the hexamer through a difference spectrum with a maximum at 450 nm. The global conformation change was monitored by this change in absorbance at 450 nm (△), by the change in sedimentation coefficient (●), and by the change in the rate at which the enzyme was modified by *p*-hydroxymercuribenzoate (□). All assays were performed at saturating carbamyl phosphate and at 20 °C. The percent change in each of these three properties is presented as a function of the concentration of succinate (molar).[74] Panel B reprinted with permission from ref 74. Copyright 1980 National Academy of Sciences.

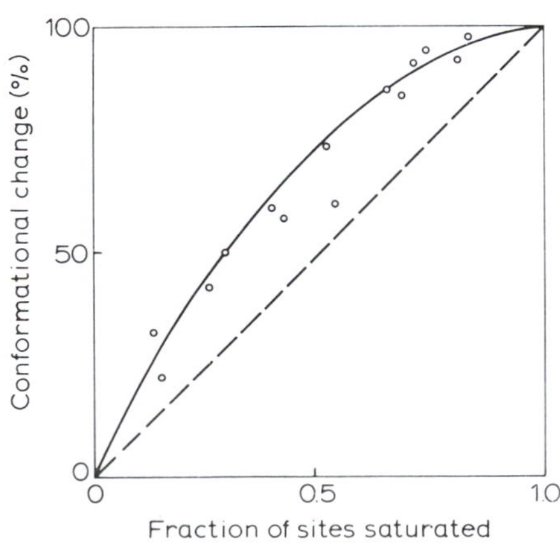


**Figure 6–13:** Plot of the extent of the conformational change of spin-labeled rabbit muscle phosphorylase *a* as a function of the fraction of the sites occupied by AMP.[11] Phosphorylase *a* was modified with *N*-(1-oxyl-2,2,6,6-tetramethyl-4-piperidinyl)iodoacetamide to introduce one stable nitroxyl free radical into each subunit of the protein. Binding of AMP was followed by equilibrium dialysis by measuring at equilibrium the molar concentrations of [U-$^{14}$C]adenosine 5′-monophosphate in the two chambers, one containing 4 $\mu$M phosphorylase *a* and one lacking phosphorylase *a*, for a series of concentrations of the ligand. The binding displayed positive deviation from ideal behavior and could be fit with Equation 6–30 with $K_{RT} = 3$, $K_d^{RL} = 2\ \mu M$, $K_d^{TL} = 10\ \mu M$, and $n = 4$. As the occupation of the enzyme by AMP was increased by increasing the free concentrations of the ligand, the electron spin resonance spectrum changed as the result of a broadening of the outer two absorbances of the triplet. The binding of AMP was measured and the ESR spectra were recorded for a series of samples with the same free concentrations of AMP. For each free concentration of AMP, the percentage of the total broadening of the spectrum observed [conformational change (%)], where the total broadening of the spectrum observed was taken to be that between the limits of no ligand and saturation with AMP, is presented as a function of the fraction of the sites for AMP that were occupied. Reprinted with permission from ref 11. Copyright 1974 Elsevier Science Publishers.

with tetranitromethane[73a] to give them a visible absorption spectrum, were then fully and permanently inactivated by reaction with pyridoxal phosphate and sodium borohydride

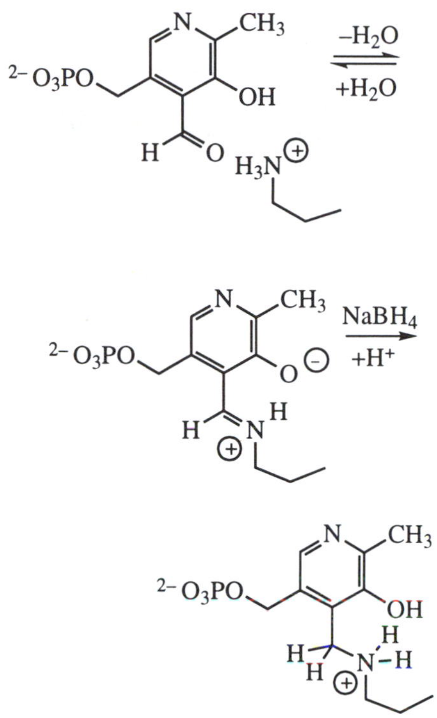

(6–32)

They were then rendered more negatively charged by modification of their lysines with 3,4,5,6-tetrahydrophthalic anhydride[74]

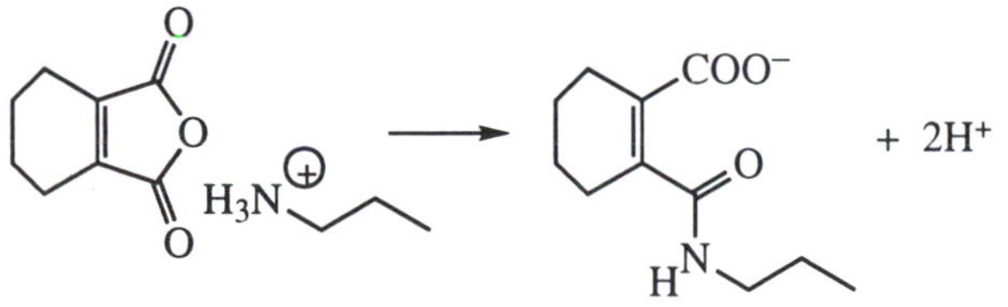

(6–33)

These yellow, inactive, negatively charged subunits were mixed with an equal amount of native unmodified catalytic subunits, and regulatory subunits were added to reconstitute $(\alpha\beta)_6$ hexamers. $(\alpha\beta)_6$ Hexamers with one native catalytic subunit and one yellow, inactive catalytic subunit could be purified by chromatography on (diethylaminoethyl)-Sephadex, and the tetrahydrophthalyl groups necessary for this separation could then be removed by hydrolysis at pH 6 under intramolecular general base catalysis. When succinate was added to these hybrid molecules saturated with carbamyl phosphate, an increase in their absorbance at 450 nm was observed. Because neither succinate nor carbamyl phosphate could bind to the active sites of the nitrated catalytic subunit, the change in absorbance must have been monitoring a shift in the distribution of the conformations of the nitrated catalytic subunit brought about by the binding of succinate to the other, unnitrated catalytic subunit in the hybrid hexamer. The change in absorbance of this hybrid hexamer at 450 nm coincided precisely with the change in sedimentation coefficient. It was also found that the change in the rate constant for the reaction of cysteines in the protein with *p*-hydroxymercuribenzoate coincided with the other two changes (Figure 6–12B) when each of these three independent monitors of the distribution of conformations of the protein was followed as a function of the concentration of succinate.[72,74]

The bisubstrate analogue *N*-(phosphonacetyl)-L-aspartate (3–4) binds with homotropic cooperativity to aspartate transcarbamylase but binds so tightly to the active site that it can be used as a stoichiometric ligand. When occupation of the active site is followed by ultraviolet difference spectroscopy as a function of the moles of *N*-(phosphonacetyl)-L-aspartate added for every mole of hexameric enzyme, a clean titration that inflects at 6 mol $mol^{-1}$ is observed (Figure 6–14A).[73] When the sedimentation coefficient was followed over the same range of stoichiometries, it was again observed that the change in the distribution of conformations monitored by sedimentation preceded the occupation of the active sites, as expected for a conformational change producing homotropic cooperativity.

When catalytic subunits of aspartate transcarbamylase were reconstituted with nitrated regulatory subunits, an increase in absorbance at 445 nm and a corresponding decrease in absorbance at 380 nm was observed when *N*-(phosphonacetyl)-L-aspartate was bound to the active sites of the reconstituted enzyme.[75] As the nitrotyrosines were on the regulatory subunits, well removed from the active sites,[2] the nitrated tyrosines must have been responding to a global conformational change in the intact enzyme. The changes in absorbance at either 380 or 445 nm coincided with the change in sedimentation coefficient when presented as a function of the stoichiometry between *N*-(phosphonacetyl)-L-aspartate and enzyme (Figure 6–14B).[75]

A zinc cation is found at the interface between each folded regulatory and catalytic polypeptide in crystallographic molecular models of aspartate transcarbamylase.[20] The zinc can be replaced by nickel. The bound nickel displays absorption maxima[76] at 360, 440, 660, and 720 nm as well as maxima and minima in the difference circular dichroic spectrum[77] at 360, 405, and 460 nm. Both the absorption spectrum[76] and the difference circular dichroic spectrum[77] of a solution of aspartate transcarbamylase display significant changes as the active sites of the modified enzyme are occupied by *N*-(phosphonacetyl)-L-aspartate. The progress of these changes as a function of the stoichiometry of the binding of the bisubstrate analogue coincides with that of the change in the distribution of conformations detected by sedimentation (Figure 6–14C).[77]

It follows from all of the results just described (Figures 6–12 and 6–14) that the change in the distribution of conformations detected by sedimentation, the change in the distribution of conformations detected by a cysteine in the protein, the change in the distribution of conformations resulting from binding of ligand to one catalytic subunit but detected by nitrotyrosines in the other, the change in the distribution of conformations detected by nitrotyrosines in the regulatory subunits, and the change in the distribution of conformations detected at metal ions distant from the active sites all occur in concert as the active sites of the enzyme become occupied. One explanation for these coincidences is that all of these physical measurements are monitoring the same single conformational change between only two states. In other words,

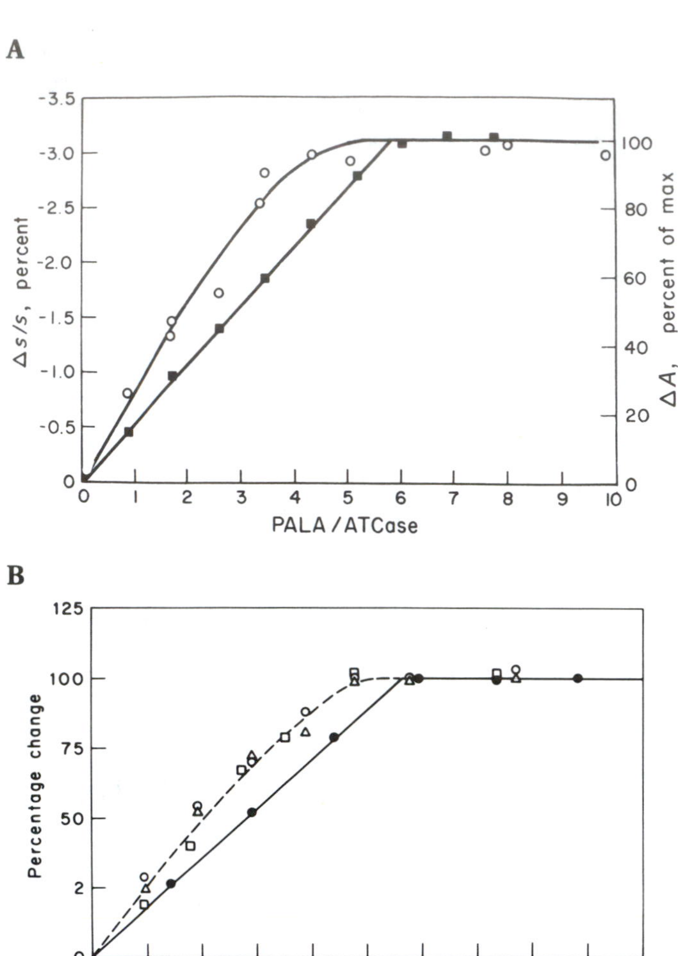


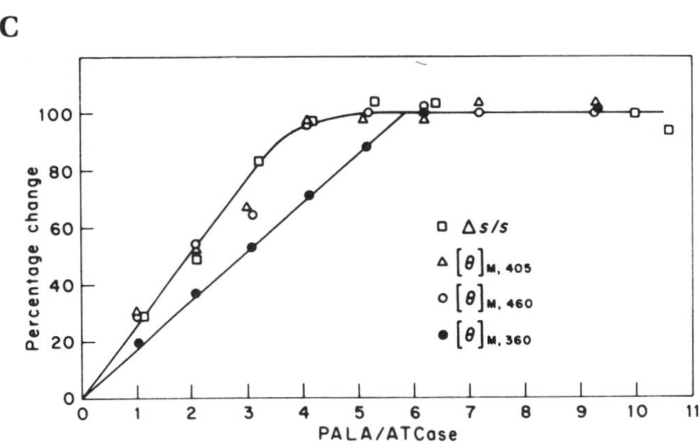


**Figure 6–14:** Coincidence of several physical measurements monitoring the global conformational change in aspartate transcarbamylase from *E. coli* produced by *N*-(phosphonacetyl)-L-aspartate. In all panels, *N*-(phosphonacetyl)-L-aspartate was used to occupy the active sites to varying degrees. This ligand displays homotropic cooperative behavior. In all panels, the change in the sedimentation coefficient (○, panel A; □, panel B; □, panel C) at various levels of occupation of the active site was used as the common monitor of the change in global conformation of the enzyme. The association of *N*-(phosphonacetyl)-L-aspartate was measured directly by following either the difference spectrum of the enzyme at 290 and 286 nm (panel A, $\Delta A$, ■, and panel B, ●) or the circular dichroism at 360 nm (panel C, ●), each of which registers only occupation of the active site with ligand. These measurements are presented as the percentage change, relative to that between unliganded enzyme and fully saturated enzyme, as a function of the molar ratio between *N*-(phosphonacetyl)-L-aspartate (PALA) and enzyme (ATCase; 2.4 mg $mL^{-1}$ in panel A, 2.1 mg $mL^{-1}$ in panel B, and 5 mg $mL^{-1}$ in panel C). The differences in sedimentation coefficient ($\Delta s/s$ in percent in panel A, ○; and as percent change relative to the total change at saturation in panel B, □, and panel C, □) are also presented as a function of the molar ratio between *N*-(phosphonacetyl)-L-aspartate and enzyme (3–4 mg $mL^{-1}$). (A) Solutions of native protein were monitored at pH 7.0 and 20 °C.[73] Panel A reprinted with permission from ref 73. Copyright 1977 American Chemical Society. (B) The protein used for these experiments was reconstituted from native catalytic subunits and regulatory subunits modified with tetranitromethane [0.9 nitro group (polypeptide)$^{-1}$]. The change in absorbance (at 17 mg $mL^{-1}$ protein) at 380 nm (△) or at 445 nm (○), both of which are due to the absorbance of the nitrotyrosines, as a percentage change relative to the total change between unliganded and fully saturated enzyme, is presented. The changes were monitored at pH 7.0 and 25 °C.[75] Panel B reprinted with permission from ref 75. Copyright 1981 *Journal of Biological Chemistry*. (C) The protein used for these experiments was reconstituted from native catalytic subunits and stripped regulatory subunits in the presence of $Ni(NO_3)_2$ (10-fold molar excess over sites) to replace the $Zn^{2+}$ of the native enzyme with $Ni^{2+}$. The circular dichroism (at 5 mg protein $mL^{-1}$) at 405 nm (△) or 460 nm (○), arising from absorption bands of the complexed $Ni^{2+}$, is presented as a percentage change as a function of the molar ratio between *N*-(phosphonacetyl)-L-aspartate and enzyme. Measurements were made at pH 7.0.[77] Panel C reprinted with permission from ref 77. Copyright 1983 *Journal of Biological Chemistry*.

no combination of these observations was able to detect more than one conformational change involving only two conformations of the enzyme. If there were a third conformation present at significant concentration, the only way that all of the observations could coincide is if the changes in each of these observed quantities in the third conformation were in the same proportion relative to the respective changes in these observed quantities between the T and the R conformations. This seems unlikely. Formally, the coincidence of all of these physical measurements is equivalent to observing an isosbestic point in a series of spectra, an observation that has always been accepted as evidence for only two states. If there is only one conformational change between unoccupied and fully occupied aspartate transcarbamylase, then the conformational change must occur with conservation of symmetry because both unoccupied and fully occupied aspartate transcarbamylase, are symmetric oligomers.

Consistent with these conclusions is the fact that both the positive cooperativity and the changes in the distribution of conformations of aspartate transcarbamylase can be explained by Mechanism 6–28 with the same set of parameters.[78] By Mechanism 6–28, the fraction of the enzyme in the R conformation[23]

$$\bar{R} \equiv \frac{[R_n]+[R_n \cdot L]+\cdots+[R_n \cdot L_n]}{([R_n]+[R_n \cdot L]+\cdots+[R_n \cdot L_n])+([T_n]+[T_n \cdot L]+\cdots+[T_n \cdot L])} \tag{6–34}$$

is

$$\bar{R} \equiv \frac{\left(1+\frac{[L]}{K_d^{RL}}\right)^n}{K_{RT}\left(1+\frac{[L]}{K_d^{TL}}\right)^n+\left(1+\frac{[L]}{K_d^{RL}}\right)^n} \tag{6–35}$$

If it is assumed that the fractional change in sedimentation coefficient is equal to $\bar{R}$, a unique set of parameters can be found that can fit not only the data for the initial velocity as a function of aspartate concentration (Equation 6–31) but also the change in sedimentation coefficient as a function of *N*-(phosphonacetyl)-L-aspartate concentration (Equation 6–35). Furthermore, the changes caused by adding either CTP or ATP to the solution, both in initial velocity and in the conformations detected by all of these physical measurements, can be fit by the same set of parameters, appropriately adjusted for the presence of the inhibitor or activator, respectively. The observed behavior of the distribution of conformations as a function of the occupation of the heterotropic allosteric site in phosphorylase by AMP could also be described by Equation 6–35 (smooth curve in Figure 6–13), using the same parameters used to fit the binding data.

There is another result suggesting that only one conformational change between only two states occurs during the homotropically cooperative occupation of the active site of aspartate transcarbamylase by ligands. When the initial velocity of the reaction was measured in the direction from carbamyl aspartate and phosphate[79] to carbamyl phosphate and aspartate, *N*-(phosphonacetyl)-L-aspartate was a potent activator (Figure 6–15).[80] One explanation for this observation is that the T conformation has a very low enzymatic activity; and neither reactant, unlike the reactants for the reaction in the opposite direction, can shift the conformational equilibrium in favor of the more active R conformation. As has already been noted, however, *N*-(phosphonacetyl)-L-aspartate can shift the equilibrium in favor of the R conformation. When it occupies one or more of the active sites on a molecule of the enzyme, the distribution of conformations changes and the enzymatic activity at the other, empty, active sites increases dramatically. When the concentration of enzyme is in sufficient excess over the concentration of *N*-(phosphonacetyl)-L-aspartate, each molecule of enzyme to which inhibitor is bound will have only one molecule of inhibitor bound. The increase in activity in this range of low concentration of *N*-(phosphonacetyl)-L-aspartate (<0.05 mol $mol^{-1}$) will then be a function only of the number of active sites converted to the R conformation by one molecule of *N*-(phosphonacetyl)-L-aspartate.[80] Several conditions were chosen in which the equilibrium constant between the T and the R conformations was shifted by other ligands sufficiently toward the R conformation that the binding of only one molecule of *N*-(phosphonacetyl)-L-aspartate would convert the T conformation to the R conformation almost quantitatively. Under these conditions the experimental value for the number of active sites converted was between 4 and 5. Therefore, under these conditions, the binding of one molecule of *N*-(phosphonacetyl)-L-aspartate to one active site was able to convert 4–5 of the remaining 5 active sites to the R conformation. This suggests that the

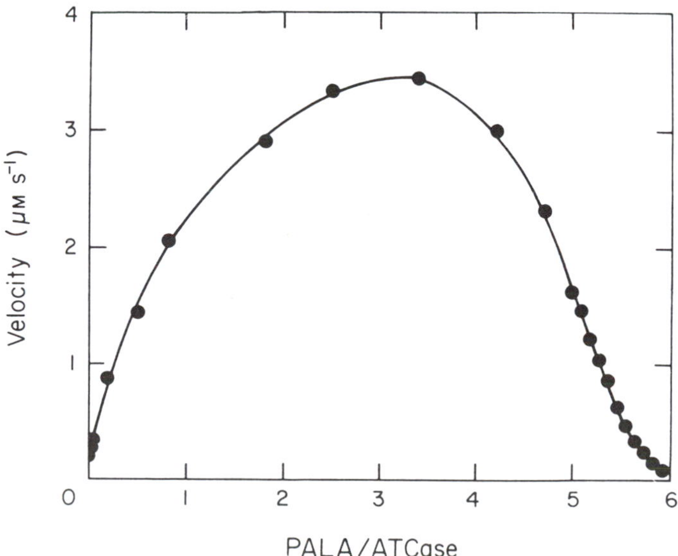


**Figure 6–15:** Activation of the phosphorolysis of *N*-carbamyl-L-aspartate catalyzed by aspartate transcarbamylase from *E. coli* by *N*-(phosphonacetyl)-L-aspartate binding stoichiometrically to the active site.[80] The enzymatic reaction was assayed in the direction from *N*-carbamyl-L-aspartate to aspartate by using arsenate in place of phosphate so that the carbamyl arsenate formed in the reaction would hydrolyze immediately upon formation and drive the reaction in its usually unfavorable direction. The enzyme (12 μM final concentration) was added to final concentrations of 3 mM arsenate and 5 mM *N*-carbamyl-L-aspartate at pH 7.0 and 30 °C to initiate the reaction. The production of aspartate was monitored continuously by coupling it to the oxidation of NADH with glutamate–oxaloacetate transaminase and malate dehydrogenase. As a result, the reaction mixture contained these two enzymes as well as 0.5 mM 2-oxoglutarate and 0.16 mM NADH. The rate of decrease in the absorbance at 340 nm was used to calculate directly the initial velocity of the enzymatic reaction (micromolar $second^{-1}$), which is presented as a function of the molar ratio of *N*-(phosphonacetyl)-L-aspartate (PALA) and aspartate transcarbamylase (ATCase). At these concentrations of enzyme and *N*-(phosphonacetyl)-L-aspartate (>200 nM), all (>90%) of the *N*-(phosphonacetyl)-L-aspartate was bound by the enzyme ($K_d^{PALA}$ = 25 nM). Reprinted with permission from ref 80. Copyright 1985 Academic Press.

conformational change is highly concerted and that only two conformations can exist in significant concentrations.

One reason that many oligomeric enzymes display both heterotropic allosteric interactions and homotropic cooperativity is that the conformational change used for communicating the heterotropic effects can be the same conformational change used for communicating the homotropic effects. When this is the case, the conformational equilibrium is shifted in favor of the less active T conformation by a heterotropic allosteric inhibitor, in favor of the more active R conformation by a heterotropic allosteric activator, and in favor of the more active R conformation by reactants. An intact $(\alpha\beta)_6$ hexamer of aspartate transcarbamylase was reconstituted from regulatory subunits and nitrated catalytic subunits, and the active sites were saturated with carbamyl phosphate. Both the homotropic cooperativity observed with aspartate as a reactant and the change in sedimentation coefficient upon adding *N*-(phosphonacetyl)-L-aspartate were those expected if the protein was 60% in the T conformation and 40% in the R conformation ($K_{RT} = 1.5$).[81] This significant shift in the equilibrium constant in favor of the R conformation resulted fortuitously from the nitration. When the heterotropic allosteric inhibitor CTP was added to the solution, the absorbance of the nitrated enzyme at 450 nm decreased; and, when the heterotropic allosteric activator ATP was added, the absorbance increased. These were in the directions expected for an increase in the concentration of the less active T conformation or the more active R conformation, respectively. In the presence of CTP, the percentage of T conformation calculated from both the homotropic cooperativity displayed by aspartate and the change in sedimentation coefficient produced by *N*-(phosphonacetyl)-L-aspartate increased to 80% ($K_{RT} = 4$), and in the presence of ATP it decreased to 40% ($K_{RT} = 0.7$). These are the changes expected if the heterotropic allosteric effects are coupled to the same global conformational change as the homotropic cooperativity.

A similar experiment was performed on a site-directed mutant of aspartate transcarbamylase in which Lysine 143 in the regulatory subunit had been mutated to alanine.[82] In this mutant, the majority (70%) of the protein was in the T conformation in the absence of ligands ($K_{RT} = 2.6$), as judged by the differential sedimentation coefficient and the cooperativity in the binding of *N*-(phosphonacetyl)-L-aspartate. This shifted to 97% ($K_{RT} = 35$) upon addition of saturating concentrations of CTP but to 12% ($K_{RT} = 0.17$) upon addition of saturating concentrations of ATP. In this mutant, therefore, the shifts in equilibrium constant were somewhat more dramatic than in the nitrated protein, but the directions were the same. In addition, the values for the equilibrium constant ($K_{RT}$) for the conformational change between the R and the T conformations in this mutant aspartate transcarbamylase could be calculated independently from either the binding of *N*-(phosphonacetyl)-L-aspartate (Equation 6–30) or the observed changes in sedimentation coefficient (Equation 6–35), either in the absence of nucleotides or in the presence of CTP or ATP. In all cases the values for $K_{RT}$ calculated by these two independent methods agree. This provides further support for the two-state model.

If the effects manifested by a heterotropic allosteric effector are exerted through the same conformational change as those controlled by the homotropic cooperative ligands, the effects observed will depend on which conformation of the enzyme is favored in the absence of ligands. In the usual situation, it is the T conformation that dominates in the absence of all ligands and the binding of a ligand or the transformation of a reactant displays homotropic cooperativity. Because an allosteric inhibitor usually increases the stability of the T conformation (increases the value of $K_{RT}$), it will increase the degree of the homotropic cooperativity and increase the sigmoidicity. For example, addition of the heterotropic allosteric inhibitor acetyl-SCoA causes the initial velocity of malate dehydrogenase (decarboxylating) as a function of the concentration of malate to become more sigmoid than it is in its absence.[37] Another example of this behavior is that of aspartate transcarbamylase (Figure 6–4A). Because an allosteric activator, however, usually increases the stability of the R conformation (decreases the value of $K_{RT}$), it will decrease the degree of homotropic cooperativity. In the extreme, where the difference in the affinity of the two conformations for the allosteric activator is sufficient to invert the value of the equilibrium constant for the conformational change, the protein will be mostly in the R conformation in the presence of the activator even in the absence of other ligands. When ligands normally displaying homotropic cooperativity are then added, they will display no homotropic cooperativity because the protein is now almost entirely in the R conformation (Figure 6–4B). For example, acetyl-SCoA is a heterotropic allosteric activator of phospho*enol*pyruvate carboxylase. In this way, flow through the Krebs cycle is increased when levels of acetyl-SCoA are high. Acetyl-SCoA in its role as allosteric activator is able to eliminate the sigmoid behavior of the initial velocity of phospho*enol*pyruvate carboxylase as a function of the concentration of phospho*enol*pyruvate (Figure 6–16).[36]

It has also been observed, as one would expect, that a heterotropic allosteric inhibitor can turn the binding of a ligand or the transformation of a reactant that does not display homotropic cooperativity into a homotropically cooperative process. In this case, the capacity to assume a T conformation must be a property of the protein; but, in the absence of ligands, the R conformation is dominant ($K_{RT} < 1$). Because the ligand or the reactant binds only to the dominant R conformation and no significant shift in the conformational equilibrium occurs, the binding of the ligand or the transformation of the reactant is a hyperbolic function of its concentration. When the heterotropic allosteric inhibitor is added, however, it shifts the conformational equilibrium in favor of the T conformation (increasing $K_{RT}$ until it is greater than 1), and homotropic cooperativity can now be displayed. One example of this effect is the appearance of sigmoid kinetics when pyruvate kinase is inhibited by phenylalanine (Figure 6–7). Another example is the behavior of threonine dehydratase. Isoleucine is a heterotropic allosteric inhibitor of threonine dehydratase, the enzyme that catalyzes the first committed reaction in isoleucine biosynthesis. Isoleucine, in its role as an allosteric inhibitor, induces homotropic cooperative behavior for the reactant threonine (Figure 6–17).[83]

In yet another example, the heterotropic allosteric inhibitor GTP converts the hyperbolic behavior of the initial velocity of AMP deaminase as a function of the concentration of AMP into sigmoid behavior.[84]

Although the simplest strategy for a protein that displays both homotropic cooperativity and heterotropic allosteric behavior is to link these two processes to the same global conformational change between two states, it appears that evolu-

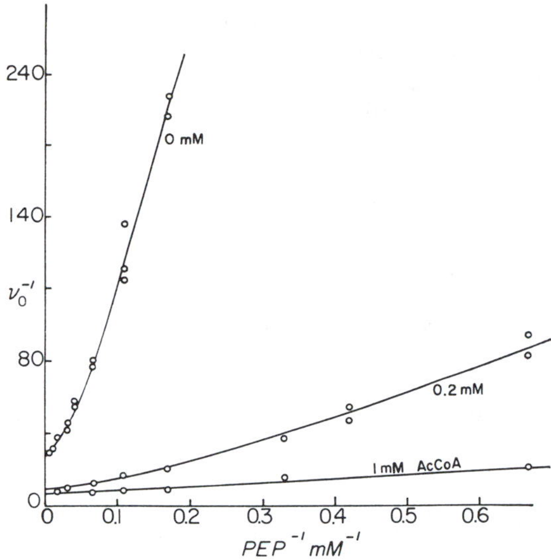


**Figure 6–16:** Effect of the heterotropic allosteric activator acetyl-SCoA on the homotropic cooperativity of phospho*enol*pyruvate carboxylase from *Salmonella typhimurium*.[36] Initial velocities of the enzymatic production of oxaloacetate from 10 mM $HCO_3^-$ and varying concentrations of phospho*enol*pyruvate were determined. The reaction was initiated by adding the enzyme (15 $\mu$g), and the production of oxaloacetate was monitored by coupling it to the oxidation of NADH (0.15 mM) by malate dehydrogenase. The loss of absorbance at 340 nm was followed as a function of time. Assay mixtures at 25 °C also contained 10 mM $MgCl_2$ and were buffered at pH 9.0. The reciprocal of the initial velocity of the enzyme ($v_0^{-1}$) is presented as a function of the reciprocal of the concentration (millimolar$^{-1}$) of phospho*enol*pyruvate (PEP) at 1 and 0.2 mM acetyl-SCoA and in the absence of acetyl-SCoA (AcCoA). Reprinted with permission from ref 36. Copyright 1969 *Journal of Biological Chemistry.*

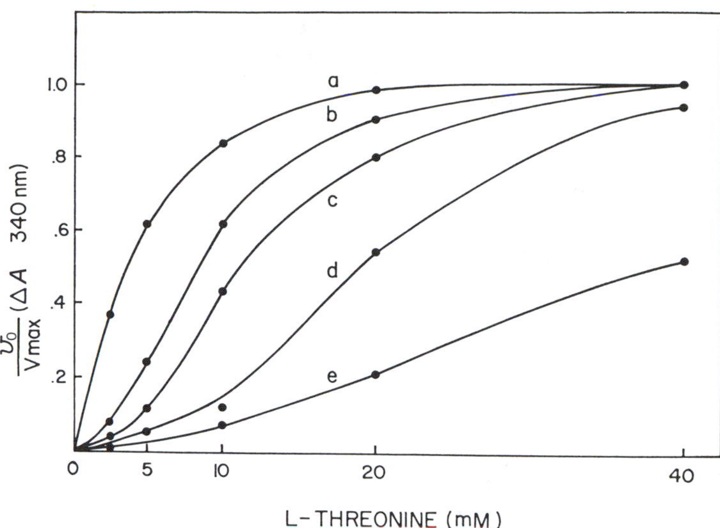


**Figure 6–17:** Appearance of sigmoid behavior in the steady-state kinetics of threonine dehydratase from *E. coli* upon addition of the allosteric inhibitor isoleucine.[83] Solutions were prepared at 37 °C and pH 8.2 containing various concentrations of threonine and no isoleucine (a), 0.02 mM isoleucine (b), 0.05 mM isoleucine (c), 0.1 mM isoleucine (d), or 0.2 mM isoleucine (e), where all concentrations are those of the final mixture. The reactions were initiated by adding threonine dehydratase, and the production of 2-oxobutyrate was monitored by coupling it to the oxidation of NADH (0.2 mM) catalyzed by L-lactate dehydrogenase. The decrease in absorbance ($\Delta A$) at 340 nm was used to calculate initial velocities ($v_0$). The relative initial velocities $v_0/V_{max}$ are presented as a function of the concentration of L-threonine (millimolar). Reprinted with permission from ref 83. Copyright 1973 *Journal of Biological Chemistry.*

tion by natural selection has produced some proteins that have the ability to assume more than two conformations, the conformational changes among which are linked arbitrarily to the binding of the various ligands. The stable free radical 1-oxyl-2,2,6,6-tetramethylpiperidine, when it is covalently bound to phosphorylase *a* (Figure 6–8), detects changes in the distribution of conformations in solutions of the protein (Figure 6–13).[11] The changes detected by the spin label when the heterotropic allosteric activator AMP is bound by the enzyme differ significantly from those promoted by the substrate glucose 1-phosphate.[85] The results suggest that the enzyme can assume four distinct conformations variously linked to the binding of ligands.

Hemoglobin is the protein displaying homotropic cooperativity, as well as heterotropic allosteric behavior, that has been studied most extensively. It has already been noted that a global conformational change is promoted by the complete oxygenation of deoxyhemoglobin (Figure 6–1). The conformation dominant in the absence of oxygen is referred to as the deoxy conformation, and the conformation dominant when all four hemes are saturated with oxygen is referred to as the oxy conformation. The conformation present when all four hemes are occupied by carbon monoxide is indistinguishable from the oxy conformation and is assumed also to be the oxy conformation.[68] This identity further illustrates the fact that these two conformations are two structural designations that are independent of the state of ligation. The oxy conformation can exist in solution in the absence of any ligands and the deoxy conformation can exist even when the protein is fully occupied by oxygen. It is just that the ligands shift the equilibrium constant for the conformational change.

The binding of either oxygen (Figure 6–9) or carbon monoxide by hemoglobin is homotropically cooperative, and the binding of either is affected by protons acting as either heterotropic allosteric inhibitors or activators and several organic phosphates acting as heterotropic allosteric inhibitors. The effect of pH on the binding of oxygen is known as the **Bohr effect**. As the pH of the solution is decreased and the concentration of protons increases, the average number of protons bound to a molecule of hemoglobin, each at a specific base on its surface, increases. In this instance, it is these bound protons that are the heterotropic allosteric ligands to which the conformational changes of the protein are linked. The affinity of hemoglobin for oxygen, as quantified by the pressure of oxygen necessary to occupy half of the hemes, decreases between pH 11 and 6.5 and then increases again below pH 6 (Figure 6–18).[86] These changes arise from the fact that the oxy conformation of hemoglobin differs in structure from the deoxy conformation and from the fact that the acid dissociation constants of several amino acids change when the conformational change takes place. The changes in affinity for oxygen with pH can be quantitatively explained (triangles in Figure 6–18) by considering only the differences between the acid–base titration curve of deoxyhemoglobin and the acid–base titration curve of fully oxygenated hemoglobin brought about by these shifts in the acid dissociation constants. If conformations other than these two exist, they do not contribute significantly to the Bohr effect.

At 20 °C and 0.3 M ionic strength, the titration curves of fully oxygenated hemoglobin and deoxyhemoglobin intersect at pH 6.3. This follows from the fact that at this pH protons are neither taken up nor released when oxygen is added to deoxyhemoglobin. Above pH 6.3, the deoxy conformation has a higher affinity for protons than the oxy conformation because it binds more protons at a given pH. Therefore, increasing the pH (lowering the concentration of protons) favors the oxy conformation and causes the affinity of hemoglobin for oxygen to increase. Below pH 6.3, the oxy conformation has a higher affinity for protons than the deoxy conformation because it binds more protons at a given pH. Therefore, lowering the pH (raising the concentration of protons) also causes the affinity of hemoglobin for oxygen to increase (Figure 6–18).

Various amino acids have been implicated as those whose acid dissociation constants change upon the change in the global conformation of hemoglobin. When the amino-terminal amino groups of the $\alpha$ polypeptides, but not those of the $\beta$ polypeptides, were carbamylated with cyanate, the difference in the acid–base titration curves between deoxyhemoglobin and oxygenated hemoglobin[87] decreased by 25%, and this suggests that the $pK_a$ of these amines shifts upon oxygenation. When the $pK_a$ of the amino-terminal amino group of the $\alpha$ subunit in native hemoglobin was directly determined, it was found to be 7.8 in deoxyhemoglobin and 7.0 in hemoglobin saturated with carbon monoxide.[88]

The binding of oxygen and carbon monoxide, with the exception of absolute affinity, show identical Bohr effects and cooperativity, and the effect of ligation with carbon monoxide on the $pK_a$ should be the same as that of ligation with oxygen. The individual acid–base titration curves of each of the 11 unique histidines in each $\alpha\beta$ asymmetric unit of hemoglobin have been measured by nuclear magnetic resonance in both deoxyhemoglobin and hemoglobin saturated with carbon monoxide.[89] The shifts in the titration curves observed upon ligation were in sum sufficient to explain the magnitude of the entire Bohr effect. This last result suggests that small changes in $pK_a$ in a large array of acids and bases over the surface of the protein add together to create the overall difference in the two titration curves.

Naturally occurring organic phosphates such as 2,3-bisphosphoglycerate and inositol hexaphosphate act as heterotropic allosteric inhibitors of the binding of oxygen to hemoglobin.[90] This inhibition results from the fact that any one of these compounds binds at least 1000-fold more tightly to deoxyhemoglobin than to oxygenated hemoglobin. There is only one site to which these organic phosphates bind for every tetramer of hemoglobin. This site lies upon the true 2-fold rotational axis of symmetry of the deoxyhemoglobin tetramer.[91] Donors of hydrogen bonds to 2,3-bisphosphoglycerate when it is bound at this site are contributed by the same amino acids in the two $\beta$ subunits displayed symmetrically about this axis. Because significant displacements of the $\beta$ subunits occur around this 2-fold rotational axis of symmetry during the global conformational change that occurs during oxygenation (Figure 6–1), it is not surprising that if the deoxy conformation binds 2,3-bisphosphoglycerate with high affinity, the oxy conformation should bind it poorly if at all.

The kinetics of the binding of oxygen and carbon monoxide by hemoglobin have been exhaustively determined. In a sense, the choice of hemoglobin for these studies was unfortunate for several reasons. First, many of the measurements were made before the role of 2,3-bisphosphoglycerate was appreciated. Second, hemoglobin is not an $\alpha_4$ homotetramer but an $(\alpha\beta)_2$ heterotetramer, and the hemes in the folded $\alpha$ polypeptides associate with and dissociate from oxygen and carbon monoxide with significantly different rates,[92,93] a fact that further complicates quantitative analysis. Finally, the $(\alpha\beta)_2$ tetramer of hemoglobin dissociates into $\alpha\beta$ dimers, and these $\alpha\beta$ dimers have significantly different kinetic behavior than $(\alpha\beta)_4$ tetramers.[94] This also increases the heterogeneity of the system. Because much of the kinetic work with hemoglobin has been focused on a search for evidence of conformations other than the oxy conformation and the deoxy conformation,[95] the unavoidable heterogeneity due to these other factors defeats the purpose of these experiments. Any claim that an additional conformation or conformations of hemoglobin have been observed[96] should be regarded with suspicion.

There are, however, several educational features of the kinetics of binding of oxygen or carbon monoxide to hemoglobin. When deoxyhemoglobin is mixed with oxygen in a stopped-flow apparatus, the rate at which the first molecule of oxygen combines with the deoxy conformation of hemoglobin can be obtained from the initial rate of the reaction.[97] When hemoglobin fully saturated with oxygen is exposed to a bright flash of light, a portion of the oxygen is driven off the hemes instantaneously.[98] Depending on the magnitude of the flash, as much as 99% of the $O_2$ can be removed.[99] When only a small fraction of the oxygen is removed, however, the rate at which oxygen recombines with the hemoglobin to reestablish saturation provides a value for the rate constant for combination of oxygen with the oxy conformation of hemoglobin.[100] When these rate constants are corrected for the fact that a deoxyhemoglobin molecule has four hemes open while the partially unoccupied hemoglobin in the oxy conformation has only one heme open, the intrinsic rate constant at which a heme in deoxyhemoglobin associates with oxygen is about one-tenth the rate at which a heme in the oxy conformation combines with oxygen.[97,101] The rate constant at which oxygen dissociates from a heme in the oxy conformation of hemoglobin can be calculated from the initial rate at which carbon monoxide, at high concentration, displaces $O_2$ from fully oxygenated hemoglobin, and the rate constant at which oxygen dissociates from a heme in the deoxy conformation can be determined from the rate constant at which oxygen associates and the value for the dissociation constant for the first oxygen

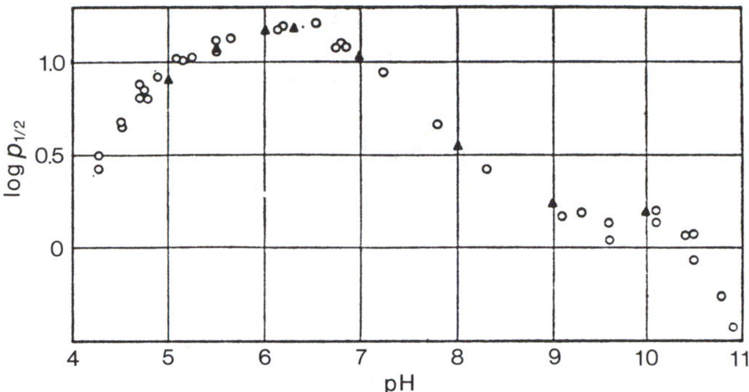


**Figure 6–18:** Bohr effect of pH on the affinity of hemoglobin for oxygen.[86] A solution of human hemoglobin (3–5 mg mL$^{-1}$) at 20 °C and at a particular value of pH was equilibrated with a series of partial pressures of oxygen from no oxygen to saturating levels of oxygen. The pH was assessed both before and after the measurements. The absorbance of the solution at several wavelengths (460–600 nm) was followed. From these spectrophotometric measurements, the partial pressure of oxygen at which half of the complete change in absorbance had occurred ($p_{1/2}$) could be determined, where the complete change is the change between fully deoxygenated and fully oxygenated hemoglobin. The various values of these partial pressures at half-saturation are presented as a function of the observed values of pH of the solutions at half-saturation. The change in pH produced by saturating deoxygenated hemoglobin with oxygen was also measured. Unbuffered solutions of hemoglobin (8–16 mg mL$^{-1}$) at various values of pH were flushed with argon until they were deoxygenated. The change in pH resulting from the addition of saturating levels of oxygen was then recorded. From the change in pH, the concentration of the hemoglobin, and the buffering capacity of the solution, the uptake or release of protons [in moles (mole of heme)$^{-1}$] by the hemoglobin could be calculated for each pH. From these observed values of proton uptake, the value of $p_{1/2}$ at a particular pH could be calculated. The directly observed values of the logarithm of $p_{1/2}$ (log $p_{1/2}$, ○) and the values of the logarithm of $p_{1/2}$ (log $p_{1/2}$) calculated from proton uptake (▲) are presented as a function of the pH. Adapted with permission from ref 86. Copyright 1963 *Journal of Biological Chemistry.*

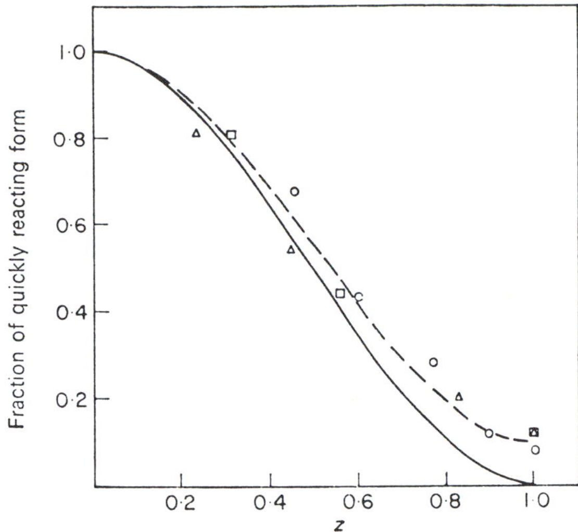


**Figure 6–19:** Fraction of the population of hemoglobin molecules that converts to the deoxy conformation as a function of the fraction of the carbon monoxides that are removed from carbonmonoxyhemoglobin by flash photolysis.[101] Solutions of human hemoglobin (50 $\mu$M in heme) were mixed with a 2-fold molar excess of carbon monoxide at pH 7 and 20 °C. At these concentrations of hemoglobin and carbon monoxide the hemoglobin was fully saturated. Flashes of increasing intensity were applied to the solutions to dissociate increasing fractions of the bound carbon monoxide. This instantaneously created a population of hemoglobin molecules initially all in the oxy conformation with varying amounts of carbon monoxide bound. As carbon monoxide remained in molar excess, it then reoccupied the vacant hemes. The reoccupation took place, however, in two phases, a fast phase ($t_{1/2} \cong 5$ ms) and a slow phase ($t_{1/2} \cong 50$ ms). As the fraction of the hemes unoccupied by carbon monoxide after the flash increased, the fraction of the unoccupied hemes binding CO also slowly increased. The fraction of the unoccupied hemes reacting quickly is presented as a function of the fraction ($z$) of the total hemes vacated by the flash. The solid line is calculated from Mechanism 6–28 and other measurements of rate and equilibrium constants. The dashed line is also calculated from Mechanism 6–28 but is corrected for the finite molar concentrations of carbon monoxide and hemoglobin. Reprinted with permission from ref 101. Copyright 1971 Academic Press.

on the hemoglobin (Figure 6–10). When these rate constants for dissociation are corrected for the fact that fully oxygenated hemoglobin has four hemes occupied while the singly occupied deoxy conformation has only one heme occupied, the intrinsic rate constant at which oxygen dissociates from a heme in the deoxy conformation of hemoglobin is about 10 times faster than the rate constant at which it dissociates from a heme in the oxy conformation.[97,101] These two differences in the respective rate constants account for the 160-fold difference in the intrinsic dissociation constants for the first and the last oxygens on liganded hemoglobin (Figure 6–10). The interesting point is that both association and dissociation are affected by the conformational change.

If the intensity of the flash that is applied to a solution of hemoglobin saturated with carbon monoxide is systematically increased, the proportion of the bound carbon monoxide that is ejected from the hemes increases. As the amount of unoccupied heme is systematically increased, a new component appears at the expense of the oxy conformation (Figure 6–19).[101] This new component associates more slowly with the reassociating carbon monoxide (by a factor of about 10) than does the oxy conformation. This new component is hemoglobin in the deoxy conformation, the concentration of which increases as the degree of saturation decreases. When solutions of deoxyhemoglobin are mixed with carbon monoxide in a stopped-flow apparatus and then the several solutions are

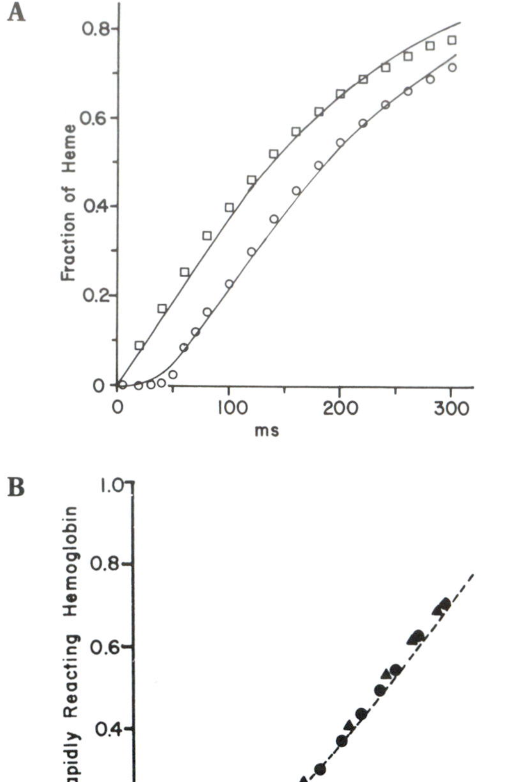

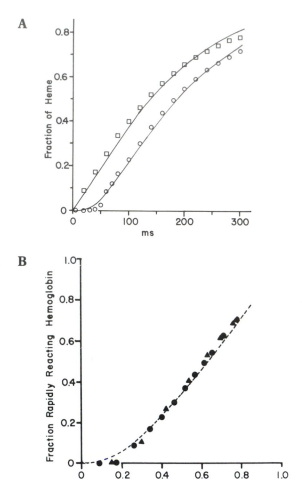


**Figure 6–20:** Appearance of the oxy conformation of hemoglobin after human deoxyhemoglobin is mixed with carbon monoxide.[102] (A) A solution of hemoglobin at pH 7 and 20 °C was rapidly mixed in a stopped-flow apparatus (dead time 1 ms) so that the final concentrations in the observation chamber were 18.5 $\mu$M in heme and 45 $\mu$M in carbon monoxide. The fractional saturation of the hemoglobin with CO was followed at 436 nm, which is the isosbestic point for the spectra of deoxy and oxy conformations of hemoglobin. The change in absorbance at this wavelength only registers occupation of the iron with carbon monoxide, regardless of any conformational changes taking place in the protein. The formation of the oxy conformation of hemoglobin was followed at a wavelength of 425 nm, which is an isosbestic point for the spectra of deoxyhemoglobin and carbonmonoxyhemoglobin. In other experiments, two forms of hemoglobin were defined kinetically (Figure 6–19) as rapidly reacting hemoglobin, assumed to be hemoglobin in the oxy conformation, and slowly reacting hemoglobin, assumed to be hemoglobin in the deoxy conformation. It had also been shown kinetically that these two forms of hemoglobin, when they are unoccupied, differ in their absorbance at 425 nm. Therefore, to measure the proportions of the deoxy and the oxy conformations at any time, the initial absorbance excursion at this wavelength after full photolysis with a laser at the indicated times was taken to be directly proportional to the concentration of the oxy conformation. Both the fraction of the heme occupied by CO (□) and the fraction of the heme in the oxy conformation (○) are presented as a function of time (milliseconds). (B) A series of experiments were performed at two different pairs of concentrations of heme and carbon monoxide (18.5 and 45 $\mu$M, ●, and 20 and 90 $\mu$M, ▲, respectively) as in panel A. The fraction of the rapidly reacting hemoglobin (open circles in panel A) is plotted as a function of the fractional ligation (open squares in panel A) for each pair of concentrations. Adapted with permission from ref 102. Copyright 1978 Rockefeller University Press.

exposed to bright flashes at various time intervals after mixing, the proportion of the hemoglobin that, after the flash, rapidly rebinds the carbon monoxide increases more slowly with time than does the degree of saturation (Figure 6–20A).[99,102] This proportion of rapidly reacting hemoglobin should be the proportion of the hemoglobin that has been converted into the oxy conformation as the ligation with CO progresses. The fraction of the hemoglobin in the oxy conformation is a function only of the degree of occupation of the hemes with carbon monoxide (Figure 6–20B),[102] and a certain proportion of the hemes must be occupied before the conformational change from the deoxy to the oxy conformation can occur. All of these results are consistent with the existence of two conformations for the hemoglobin molecule, the oxy and the deoxy conformations. When a sufficient number of oxygen or carbon monoxide molecules are removed from a molecule of saturated hemoglobin, the oxy conformation converts to the deoxy conformation (Figure 6–19). As soon as a sufficient number of oxygen or carbon monoxide molecules are bound to a molecule of hemoglobin, the deoxy conformation converts to the oxy conformation (Figure 6–20).

When fully oxygenated hemoglobin in the oxy conformation is exposed to a flash bright enough to remove all of the oxygen (99%) and the solution is monitored at the isosbestic wavelength (436 nm) for deoxygenated and oxygenated iron in the hemes, a rapid (10,000 $s^{-1}$ at 25 °C) first-order relaxation is observed that corresponds to the conformational change from the unliganded oxy conformation immediately present to the deoxy conformation.[98] The yield of deoxy conformation decreases as the concentration of oxygen is increased and more and more oxy conformation is trapped by oxygen before it can convert. This behavior,[98] as well as all of the other kinetic observations, can be explained quantitatively[101,102] (curves in Figures 6–19 and 6–20) by Mechanism 6–28 if the various equilibrium constants are decomposed into forward and reverse rate constants. Only two conformations, the oxy conformation, corresponding to the R conformation of Mechanism 6–28, and the deoxy conformation, corresponding to the T conformation, are required if it is assumed, as can be demonstrated independently, that the residual deviations from predicted behavior result from differences between the hemes in $\alpha$ and $\beta$ subunits.[103]

Many naturally occurring, mutant forms of hemoglobin have provided altered molecules with properties different from those of the native protein. For example, in hemoglobins Bethesda, Nancy, and Rainier, Tyrosine 145 in the $\beta$ subunits of the human protein is replaced with a histidine, an aspartate, or a cysteine, respectively. In these mutants, in the absence of organic phosphates, the oxygen affinity has increased, the cooperativity is entirely lost, and the kinetics of ligation resemble those of isolated $\alpha$ chains or the oxy conformation of normal hemoglobin.[104–106] The only symptom in heterozygous patients with one of these three abnormal hemoglobins, however, is an increase in total red cell mass in response to the inability of half of their hemoglobin to release oxygen in the tissues because of this high affinity. Presumably, homozygous individuals could not occur. In these three mutant proteins, the equilibrium between the deoxy and the oxy conformations no longer favors the deoxy conformation in the absence of oxygen but favors the oxy conformation. If the allosteric effector inositol hexaphosphate is added, however, it can shift the conformational equilibrium of the unliganded protein sufficiently in favor of the deoxy conformation to cause some of the properties of native hemoglobin to be regained partially.[105,106] When Histidine 146 in the $\beta$ subunits is changed to an arginine as in hemoglobin Cochin-Port Royal, however, the only effect is an alteration in the Bohr effect.[106] Because the oxygen affinity of hemoglobin Cochin-Port Royal is almost normal, heterozygous patients carrying this mutation are asymptomatic. If Histidine 146 in the $\beta$ subunits is changed instead to an aspartate as in hemoglobin Hiroshima, behavior similar to that of the three mutants at Tyrosine 145 is observed but the decrease in affinity for oxygen is not so large.[107] Even so, it is large enough to cause similar increases in total red cell mass. These mutant forms of hemoglobin illustrate not only that similar structural changes at the same location lead to similar functional changes but also that the ability to display a ligand-linked conformational change can be destroyed by a single change in the amino acid sequence of a protein.

## *Suggested Reading*

Wang, C., Yang, Y.R., Hu, C.Y., & Schachman, H.K. (1981) Communication between Subunits in Aspartate Transcarbamoylase. Effect of Active Site Ligands on the Tertiary Structure of the Regulatory Chains, *J. Biol. Chem. 256*, 7028–7034.

### PROBLEM 6–3

Derive Equation 6–16 from Equation 6–15 and conservation of protein

$$[P] + [P \cdot L_n] = [P]_{TOT}$$

### PROBLEM 6–4

Adenosine-phosphate deaminase (AMP deaminase) catalyzes the following reaction:

$$AMP + H_2O \rightleftharpoons IMP + NH_3$$

The enzyme is a tetramer. It is activated by potassium ion ($K^+$). The kinetics of this activation, with respect to AMP concentration, are displayed in panel A of the following figure. The open circles are the data gathered in the presence of a nonactivating cation (tetramethylammonium); the closed circles, the data gathered in the presence of $K^+$. Panel B of the figure is simply a replot of the data by the Hill equation.

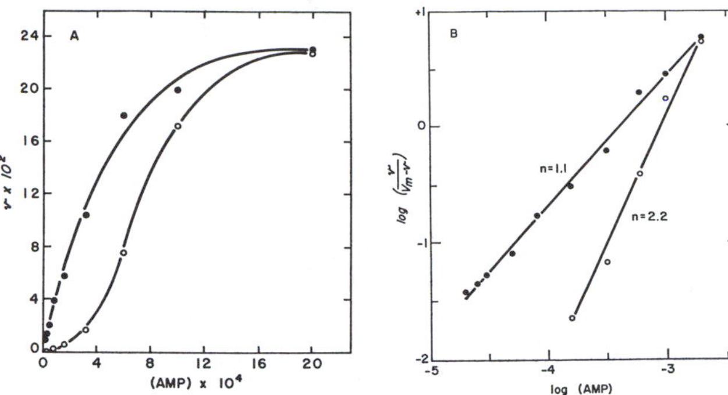


Effect of AMP concentration on the initial velocity of AMP deaminase[108] in the presence and absence of $K^+$. Assay conditions were 0.05 M imidazolium chloride, pH 6.5, and 0.15 M tetramethylammonium chloride (○) or KCl (●). (A) Initial velocity ($v$) with respect to AMP concentration (molarity × $10^4$); (B) Hill plots of data from panel A. Reprinted with permission from ref 108. Copyright 1967 *Journal of Biological Chemistry*.

Two conclusions can be drawn from the data. Potassium ion activates the enzyme by increasing its apparent affinity for AMP without altering its maximum velocity, and potassium ion eliminates the cooperativity of AMP binding.

(A) Assume that there are two forms of the enzyme, E and E′, in equilibrium with each other and that $K^+$ and AMP affect that equilibrium. Draw the two linkage relationships that connect the binding of either $K^+$ or the first AMP to the conformational equilibrium of the enzyme.

(B) If $K_A = [E']/[E]$, $K_A^K = [E' \cdot K]/[E \cdot K]$, $K_A^{AMP} = [E' \cdot AMP]/[E \cdot AMP]$, $K_d^{AMP} = [AMP][E]/[E \cdot AMP]$, $K_d'^{AMP} = [E'][AMP]/[E' \cdot AMP]$, $K_d^K = [E][K]/[E \cdot K]$, and $K_d'^K = [E'][K]/[E' \cdot K]$, write two fundamental equalities relating some of these equilibrium constants with others.

(C) Assume that E has a higher affinity for AMP than E′; in other words, $K_d^{AMP} << K_d'^{AMP}$. What must be true for the value of $K_A$ to explain the cooperativity of the enzyme in the absence of $K^+$?

(D) What must be true for the values of $K_A$ and $K_A^K$ to explain $K^+$ activation?

(E) Why does the activation by $K^+$ result in the loss of cooperativity?

(F) Why does $K^+$ only affect $K_m$ for AMP rather than $V_{max}$?

### PROBLEM 6–5

The initial velocity of the reaction catalyzed by aspartate transcarbamylase is a sigmoid function of the aspartate concentration at saturating carbamyl phosphate (Figure 6–4).

(A) Write a series of linkage boxes that describes the binding of each of the six aspartates to the enzyme and that can explain the observed cooperativity.

Maleate ($^-$OOCCHCHCOO$^-$) is an inhibitor of aspartate transcarbamylase that is competitive with aspartate. The following behavior was observed when maleate was added to the enzyme along with the aspartate.

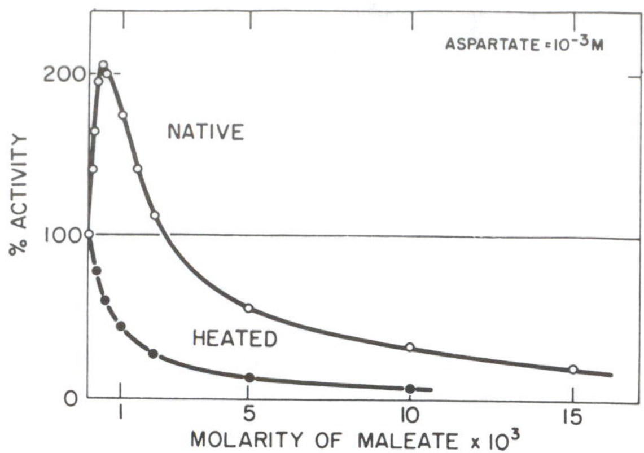


Activation and inhibition of aspartate transcarbamylase by maleate, a substrate analogue.[109] The reaction mixtures contained 3.6 mM carbamyl phosphate; 1.0 mM aspartate; 0.04 M potassium phosphate, pH 7.0; maleate at the concentrations indicated; and 0.4 $\mu$g mL$^{-1}$ native aspartate transcarbamylase (○) or 0.2 $\mu$g mL$^{-1}$ heated aspartate transcarbamylase (●). Heated aspartate transcarbamylase was prepared by heating native aspartate transcarbamylase at a concentration of 20 $\mu$g mL$^{-1}$ in 4 mM potassium phosphate buffer, pH 7.0, for 4 min at 60 °C, followed by rapid cooling in an ice–water bath. The enzymatic activity (percent of control) is presented as a function of the concentration of maleate (molarity × 10$^3$). Reprinted with permission from ref 109. Copyright 1963 Cold Spring Harbor Laboratory.

Note the concentration of aspartate and where this falls on the curve in Figure 6–4.

(B) Where is the enzyme in the linkage boxes before maleate is added?
(C) Why does the maleate first stimulate the enzymatic activity?
(D) Why does it eventually inhibit enzymatic activity?
(E) What has happened during the heat treatment described in the legend to the figure?

# Conformational Changes and Oligomeric Associations

There is a relationship between the binding of a ligand by any oligomeric protein and the strength of the interactions that hold its subunits together.[110] Consider first the description of the binding of the ligand, L, to an oligomeric protein, $S_n$, by the Adair mechanism (Mechanism 6–19) combined with the dissociation of the oligomeric protein

$$\begin{array}{ccccccc} S_n & \xleftarrow{K_{d1}{}^{L}} & S_n{\cdot}L & \xleftarrow{K_{d2}{}^{L}} & S_n{\cdot}L_2 \cdots S_n{\cdot}L_{n-1} & \xleftarrow{K_{dn}{}^{L}} & S_n{\cdot}L_n \\ {}^{0}K_{oligo}\downarrow & & {}^{1}K_{oligo}\downarrow & & {}^{(n-1)}K_{oligo}\downarrow & & {}^{n}K_{oligo}\downarrow \\ nS & \xleftarrow[\frac{1}{n}K^{L}_{d,int}]{} & S{\cdot}L+(n-1)S & \cdots & (n-1)S{\cdot}L+S & \xleftarrow[nK^{L}_{d,int}]{} & nS{\cdot}L \end{array} \tag{6–36}$$

The oligomeric protein is composed from $n$ subunits of S, and $K^{L}_{d,int}$ is the intrinsic dissociation constant for ligand L from the free subunit. By linkage

$$ {}^{0}K_{oligo}\prod_{i=1}^{n}K_{di}{}^{L} = {}^{n}K_{oligo}(K^{L}_{d,int})^{n} \tag{6–37}$$

If the average dissociation constant for a ligand from the intact oligomer is greater than the intrinsic dissociation constant for that ligand from the free subunit, or in other words, if the average affinity of oligomer for the ligand is less than that of the monomer, then

$$\prod_{i=1}^{n}K_{di}{}^{L} > (K^{L}_{d,int})^{n} \tag{6–38}$$

and ${}^{0}K_{oligo} < {}^{n}K_{oligo}$. As the ligands occupy the sites on the subunits, the interactions among the subunits become weaker. If, however, the average dissociation constant for the ligand from the oligomer is less than the intrinsic dissociation constant of the free subunit, then ${}^{0}K_{oligo} > {}^{n}K_{oligo}$, and as the ligands occupy the sites, the interactions among the subunits become stronger.

If the subunits in an oligomer bind the ligand independently and the intrinsic dissociation constant for the ligand from a site (Equation 6–21) within the intact oligomer is the same as the intrinsic dissociation constant for the ligand from a site when it is present in a disassembled subunit, then the interactions among the subunits are unchanged as the occupation of the site is increased. In this case, the binding of ligands and the interactions among the subunits are unlinked. If, however, the protein displays homotropic cooperativity so that $K_{d1}$ is greater than the value assigned by Equation 6–21 and $K_{dn}$ is less than the value assigned by Equation 6–21 and if the intrinsic dissociation constant for the ligand from the free subunit is equal to the average dissociation constant for the ligand from the oligomer, then the successive occupations of the oligomer by the first ligands will weaken the interactions among the subunits while the successive occupations of the oligomer by the last ligands will return the strength of the interactions to that of the unliganded oligomer.

These considerations also can be superimposed upon Mechanism 6–28

$$
\begin{array}{ccccccccc}
T_n & \xleftarrow{\frac{1}{n}K_d^{TL}} & T_n{\cdot}L & \xleftarrow{\frac{2}{n-1}K_d^{TL}} & T_n{\cdot}L_2 & \cdots & T_n{\cdot}L_{n-1} & \xleftarrow{nK_d^{TL}} & T_n{\cdot}L_n \\
K_{oligo}^{T} & K_{RT} & K_{oligo}^{T}\left(\frac{K_d^{TL}}{K_d^{ML}}\right) & K_{RT}\left(\frac{K_d^{RL}}{K_d^{TL}}\right) & & & K_{oligo}^{T}\left(\frac{K_d^{TL}}{K_d^{ML}}\right)^n & & K_{RT}\left(\frac{K_d^{RL}}{K_d^{TL}}\right)^n \\
nM & \xleftarrow{\frac{1}{n}K_d^{ML}} & (n-1)\,M + M{\cdot}L & & \cdots & & (n-1)M{\cdot}L + M & \xleftarrow{nK_d^{ML}} & nM{\cdot}L \\
K_{RT}K_{oligo}^{T} & & K_{RT}K_{oligo}^{T}\left(\frac{K_d^{RL}}{K_d^{ML}}\right) & & & & K_{RT}K_{oligo}^{T}\left(\frac{K_d^{RL}}{K_d^{ML}}\right)^n & & \\
R_n & \xleftarrow{\frac{1}{n}K_d^{RL}} & R_n{\cdot}L & \xleftarrow{\frac{2}{n-1}K_d^{RL}} & R_n{\cdot}L_2 & \cdots & R_n{\cdot}L_{n-1} & \xleftarrow{nK_d^{RL}} & R_n{\cdot}L_n
\end{array}
\qquad (6\text{–}39)
$$

If the protein displays homotropic cooperativity, then in the absence of ligand, the operating equilibrium constant for disassembly of the oligomer will be $K_{oligo}^{T}$; and, at saturation with ligand the operating equilibrium constant for the disassembly of the oligomer will be $K_{RT}K_{oligo}^{T}(K_d^{RL}/K_d^{ML})^n$. Again it is the relative magnitudes of the dissociation constant for the ligand from the site in a free subunit, $K_d^{ML}$, and the dissociation constants for the ligand from the two conformations of the oligomer, $K_d^{RL}$ and $K_d^{TL}$, that determine the outcome. If a free disassembled subunit has a dissociation constant for ligand, $K_d^{ML}$, equal to the intrinsic dissociation constant for the ligand from the R conformation, $K_d^{RL}$, then the interactions among the subunits are weaker when the oligomer is fully saturated than when it is unoccupied because $K_{RT}$ is greater than 1. If, however, the free disassembled subunits have a dissociation constant for ligand, $K_d^{ML}$, equal to the intrinsic dissociation constant of the ligand for the T conformation, $K_d^{TL}$, then the interactions among the subunits are stronger when the oligomer is fully saturated than when it is unoccupied because, under these circumstances, $K_{RT}(K_d^{RL}/K_d^{ML})^n$ is equal to $K_{RT}(K_d^{RL}/K_d^{TL})^n$, which is necessarily less than 1, if the protein displays homotropic cooperativity.

From the arguments just presented, it is clear that there is no necessity for a ligand that displays homotropic cooperative behavior either to increase or to decrease the strength of the interactions among the subunits of an oligomer. In most instances that have been studied, however, in which a ligand does affect the dissociation constant of the oligomer, the unliganded T conformation turns out to be a more stable oligomer than the fully liganded R conformation.

An evolutionary explanation can be provided for this coincidence. For an oligomer to display homotropic cooperativity and heterotropic allosteric behavior, there must be at least two global conformations, one of low affinity and one of high affinity. An extant oligomeric protein existed first as a monomer before it evolved the ability to oligomerize. It can be assumed that, as a monomer, it performed the same function that it presently performs and that immediately after the oligomer was fixed by natural selection, it had the same affinities for substrates as the parental monomer. It is much easier to produce a conformation of lower affinity from an already functioning protein than to produce a conformation of higher affinity. This is borne out by recent experience with site-directed mutation. For this reason, the new global conformation arising after the oligomer was fixed by natural selection would usually have been one of lower affinity than either the separated monomer or the original oligomer. This new conformation, therefore, would have to be the T conformation and the conformation of the original oligomer would be the R conformation. If the affinity of the R conformation does not drift too far from that of the original oligomer, the unliganded T conformation would be more stable than the fully liganded R conformation because $K_{RT}(K_d^{RL}/K_d^{ML})^n$ would remain greater than 1.

The impression created by the observed preference is that oligomers displaying homotropic cooperativity are oligomers in which a relatively unstable conformation of the monomer, with low affinity for ligand, is produced by contorting the subunit. The forces that contort each subunit into this conformation of low affinity are the strong interactions among the subunits in the oligomeric T conformation. As ligand is bound and the global conformational change occurs, it is to the conformation of high affinity displayed by the free subunit that the protein relaxes coincident with the decrease in these forces among the subunits.[110]

Both hemoglobin and aspartate transcarbamylase offer examples of oligomers that display homotropic cooperative behavior in which the affinity of the disassembled subunit for ligand is decreased when it is assembled into the unliganded, T conformation. The binding of oxygen to the isolated $\alpha$ and $\beta$ subunits of hemoglobin displays no cooperativity (Figure 6–21),[111] and the affinities of the free subunits for oxygen are much greater (60-fold) than the affinity of the unliganded $(\alpha\beta)_2$ tetramer of hemoglobin for the first molecule of oxygen. When aspartate transcarbamylase is locked in the T conformation by treatment with tartryl diazide, it no longer displays homotropic cooperativity. The Michaelis constant for aspartate ($K_m^{Asp}$ = 120 mM) is significantly larger than the Michaelis constant for $\alpha_3$ trimeric catalytic subunits made from the same intact, modified enzyme ($K_m^{Asp}$ = 20 mM).[112] In both hemoglobin and aspartate transcarbamylase, incorporating the subunit into the T conformation decreases the affinity of the site for its ligand significantly, while the affinity of the site for the ligand when the oligomer is in the R conformation is similar to that of the site in the free subunit. It necessarily follows that these oligomers must be less stable in the liganded form than the unliganded form.

There are, however, examples in which the oligomer is more stable in the liganded form. Ribulose-phosphate 3-epimerase in its active form is a dimer, but in the absence of

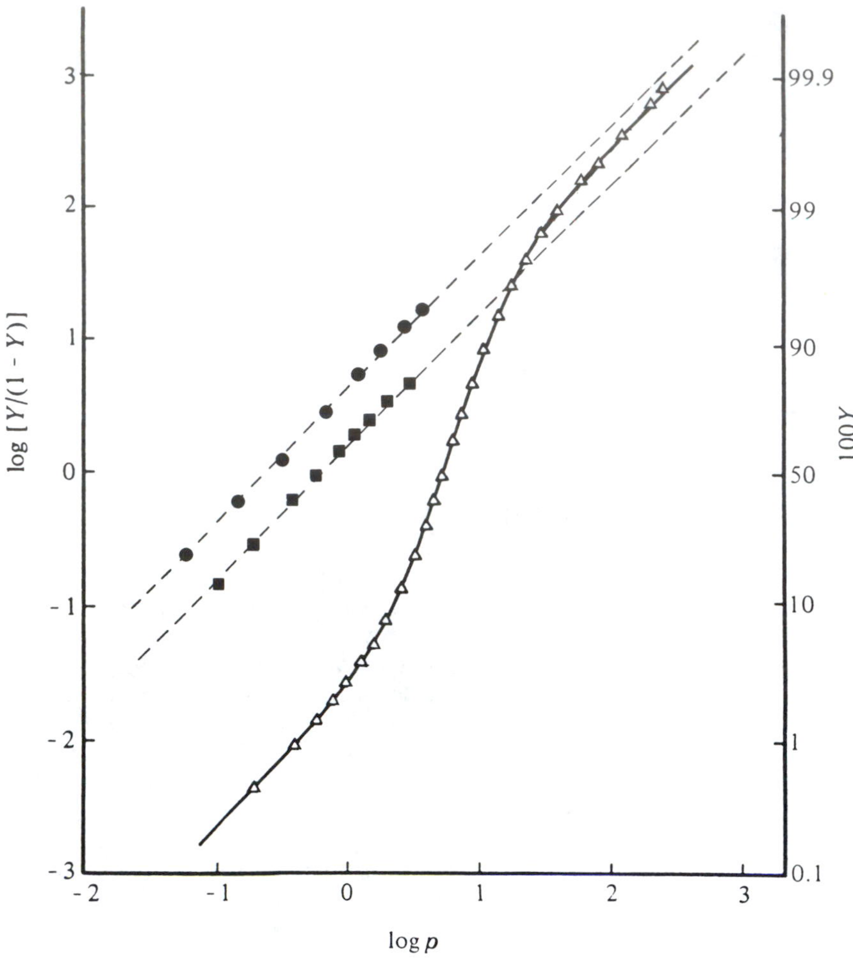


**Figure 6–21:** Comparison of the affinity of isolated $\beta$ subunits (●) and isolated $\alpha$ subunits (■) of human hemoglobin for oxygen with the affinity of the intact $(\alpha\beta)_2$ tetramer (△) for oxygen.[111,111a] Hemoglobin was dissociated into its constituent subunits by treatment with *p*-mercuribenzoate, separation of the mercurated subunits on (carboxymethyl)cellulose, and removal of the mercury coincident with rereduction of the cysteines in the protein. The binding of oxygen to either the isolated subunits or the intact $(\alpha\beta)_2$ tetramer was followed spectrophotometrically as described in the legend to Figure 6–10. The values of log $[Y/(1-Y)]$ are presented, as well as the percent occupancy ($100Y$), as functions of the logarithms of the partial pressure ($p$) of oxygen. Neither separated $\alpha$ subunits nor separated $\beta$ subunits displayed homotropic cooperativity ($n_{\text{Hill}} = 1.0$). The dissociation constant for oxygen from $\alpha$ subunits (0.6 mmHg) is somewhat larger than that for oxygen from $\beta$ subunits (0.2 mmHg). The first intrinsic dissociation constant in the fit of the Adair equation (Equation 6–40) to these data is 46 mmHg, and the last is 0.29 mmHg. If the affinity of the $\beta$ subunit remains greater than that of the $\alpha$ subunit in the $(\alpha\beta)_2$ tetramer, then the first oxygen binding to the deoxy conformation should preferentially bind to the $\beta$ subunits and the last preferentially to the $\alpha$ subunits, and the first dissociation constant should be for dissociation of oxygen from $\beta$ subunits in the deoxy conformation and the last for dissociation from $\alpha$ subunits in the oxy conformation. Adapted with permission from ref 111, copyright 1982 Cambridge University Press, and from ref 111a, copyright 1971 Academic Press.

substrates the enzyme dissociates into monomers of low enzymatic activity.[113] Because the substrate ribulose 5-phosphate has a higher affinity for the active dimeric species, it induces the inactive monomer to dimerize. If initial velocities are monitored at low concentrations of enzyme, ribulose 5-phosphate acting as a reactant displays homotropic cooperative behavior because the extent of dimerization increases as its concentration is increased. In this case, the ligand displaying homotropic cooperativity strengthens the interactions between the subunits rather than weakening them because it has a higher affinity for the oligomeric form of the enzyme than for the isolated subunits.

One of the most exhaustively studied examples of the relationship between the strength of subunit interactions and the ligation of a protein displaying homotropic cooperativity is hemoglobin. Hemoglobin saturated with either oxygen or carbon monoxide dissociates readily into $\alpha\beta$ heterodimers.[114] A particularly striking demonstration of this fact is that when human and canine carbonmonoxyhemoglobins were mixed and immediately submitted to isoelectric focusing, a high yield of the hybrid, $(\alpha\beta)_{\text{human}}(\alpha\beta)_{\text{canine}}$, was observed at short times, before it rapidly disappeared as it disproportionated.[115] When human deoxyhemoglobin was mixed with canine deoxyhemoglobin, however, the mixture had to sit for several hours before being submitted to isoelectric focusing before any of the hybrid was observed. The hybrid that was then separated by isolectric focusing under anaerobic conditions disproportionated only slightly over several hours. This experiment shows that in liganded hemoglobin the interactions between $\alpha\beta$ heterodimers are weak, while in unliganded hemoglobin the interactions are strong.

The $\alpha\beta$ heterodimers formed upon dissociation of tetrameric hemoglobin bind oxygen noncooperatively and with an affinity similar to the affinities of the isolated subunits.[94,116] Therefore, the system can be simplified to consider only the dissociation of the homotropically cooperative tetramer to the noncooperative dimers because this is the only dissociation linked to the homotropic cooperativity. The changes in standard free energy for the various equilibria have been evaluated[117] at pH 7.4 and 21 °C. They can be presented either in the Adair formulation

$$\begin{array}{ccccccc}
(\alpha\beta)_2 & \xleftarrow{+23} & (\alpha\beta)_2\cdot(\mathrm{O}_2) & \xleftarrow{+52} & (\alpha\beta)_2\cdot(\mathrm{O}_2)_3 & \xleftarrow{+39} & (\alpha\beta)_2\cdot(\mathrm{O}_2)_4 \\
& \searrow^{+60} & & & & \swarrow^{+33} & \\
& 2\alpha\beta & & \xleftarrow{+140} & & 2[\alpha\beta\cdot(\mathrm{O}_2)_2] &
\end{array} \qquad (6\text{–}40)$$

or in the formulation of Mechanism 6–39

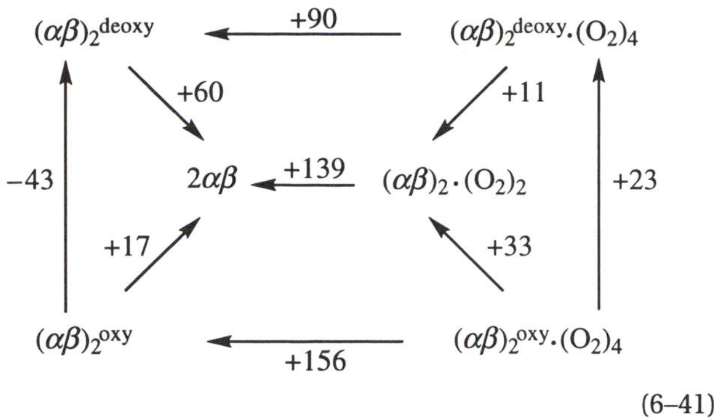


(6–41)

where the numerical values are for standard free energies in kilojoules mole$^{-1}$ with concentrations of 1 M as the standard states. The intrinsic dissociation constant of the $\alpha\beta$ dimer for oxygen ($\Delta G°$ = +35 kJ mol$^{-1}$) is intermediate[118] between the intrinsic dissociation constant of the tetramer for the first molecule of oxygen ($\Delta G°$ = +23 kJ mol$^{-1}$) and for the last molecule of oxygen ($\Delta G°$ = +39 kJ mol$^{-1}$); and, because of this, the binding of oxygen decreases the stability of the tetramer when it is in the deoxy conformation ($\Delta\Delta G°$ = 60–11 kJ mol$^{-1}$) but increases the stability of the tetramer when it is in the oxy conformation ($\Delta\Delta G°$ = 17–33 kJ mol$^{-1}$). Therefore, when deoxyhemoglobin tetramers are assembled from dimers, the dissociation constant of the hemes is increased ($\Delta\Delta G°$ = 23–35 kJ mol$^{-1}$), while when oxyhemoglobin tetramers are assembled from dimers, the dissociation constant of the hemes is decreased ($\Delta\Delta G°$ = 39–35 kJ mol$^{-1}$). It is the fact that the affinity of the $\alpha\beta$ dimers for oxygen is closer to the affinity of the oxy conformation than to that of the deoxy conformation that causes the unliganded tetramer in the deoxy conformation to be more stable than the fully liganded tetramer in the oxy conformation ($\Delta\Delta G°$ = 60–33 kJ mol$^{-1}$).

As with hemoglobin, the free catalytic $\alpha_3$ subunits of aspartate transcarbamylase display a much higher affinity for aspartate than when they are confined in the unliganded $(\alpha_3)_2(\beta_2)_3$ hexamer. Unlike hemoglobin, however, the $(\alpha_3)_2(\beta_2)_3$ hexamer of aspartate transcarbamylase remains intact as ligands bind and can be dissociated only under extreme conditions. It is possible, however, to produce an $(\alpha_3)_2(\beta_2)_2$ oligomer. This complex is the hexamer missing one of its regulatory dimers.[1a,119] This incomplete oligomer still displays both homotropic cooperativity and allosteric inhibition by CTP. It is, however, unstable enough that its rate of dissociation can be measured.[120] The fact that occupation of the active site in this oligomer increases its rate of dissociation by more than 250-fold is consistent with the conclusion that the affinities of isolated catalytic subunits for substrates are much closer to the affinities of the R conformation than to the affinities of the T conformation.

In the case of hemoglobin, because the affinity of the hemes in the isolated $\alpha\beta$ dimers is intermediate between those in the oxy conformation of the tetramer and those in the deoxy conformation of the tetramer, the conformation of the folded polypeptides in the dimer must be different from the conformations of the folded polypeptides in either conformation of the tetramer. This would mean that the effect of a heterotropic allosteric effector on the dissociation of the oligomer might be unrelated to the effect of the ligand displaying homotropic cooperativity on the dissociation of the oligomer. If, however, the conformation of the dissociated subunits of a protein resembles closely one of the two conformations of the oligomer, usually the R conformation, then the affinities of the isolated subunits for heterotropic allosteric effectors and for ligands displaying homotropic cooperativity should correspond to the affinities displayed by that conformation of the oligomer. Adenosine monophosphate is a heterotropic allosteric inhibitor of amidophosphoribosyltransferase

$$\text{glutamine} + \text{5-phosphoribose 1-diphosphate} \rightleftharpoons \text{5-phosphoribosyl-1-amine} + HOP_2O_6^{3-} + \text{glutamate} \tag{6–42}$$

because this enzyme catalyzes the first committed reaction in purine biosynthesis. The reactant 5-phosphoribose 1-diphosphate displays homotropic cooperativity.[121] These two observations could be explained if AMP has a higher affinity for the unliganded T conformation than it does for the fully liganded R conformation and the R conformation has a higher affinity for 5-phosphoribose 1-diphosphate than does the T conformation. If it is also the case that the dissociated subunits are similar in conformation to the R conformation of the oligomer, rather than the T conformation, and share the higher affinity of the R conformation for 5-phosphoribose 1-diphosphate and its lower affinity for AMP, then this would explain the observation that AMP increases the stability of the oligomer while 5-phosphoribose 1-diphosphate decreases its stability.[122]

In the example just cited, the heterotropic allosteric inhibitor increases the stability of the oligomer. This is consistent with the prejudice that the T conformation, by definition the conformation of low enzymatic activity or low affinity for reactants and consequently the form stabilized by the inhibitor, is the tight conformation. This, however, is only a prejudice and not a law. Leucine is a heterotropic allosteric inhibitor of 2-isopropylmalate synthase.[123] The enzyme in the absence of any ligands participates in an equilibrium between a tetramer and a monomer that occurs in accessible ranges of concentration.[124] Leucine, because it has a higher affinity for the monomer than the tetramer, shifts this equilibrium in the direction of dissociation.[125] From all of these considerations it is clear that heterotropic allosteric effectors and homotropic cooperative ligands will not necessarily have influences on the stability of an oligomer, relative to its separated protomers, that can be predicted.

There is a further connection between homotropic cooperative behavior and the equilibrium between oligomers and protomers. Consider the following mechanism

$$\begin{array}{ccccc}
2S + 2L & \xleftarrow{\frac{1}{2}K_{d,int}^{SL}} & S\cdot L + S + L & \xleftarrow{2K_{d,int}^{SL}} & 2S\cdot L \\
\uparrow K_S & & & & \uparrow K_{SL} \\
S_2 + 2L & \xleftarrow[\frac{1}{2}K_{d,int}^{S_2L}]{} & S_2\cdot L + L & \xleftarrow[2K_{d,int}^{S_2L}]{} & S_2\cdot L_2
\end{array} \tag{6–43}$$

where the dimerization of subunits, S, is linked to the binding of ligand, L. Neither the monomer, by necessity, nor the dimer, by design, alone displays homotropic cooperativity.

$$[S]_{TOT} = [S] + [S{\cdot}L] + 2[S_2] + 2\,[S_2{\cdot}L] + 2[S_2{\cdot}L_2] \tag{6–44}$$

$$[S]_{TOT} = [S]\left(1+\frac{[L]}{K_{d,int}^{SL}}\right)+2\frac{[S]^2}{K_s}\left(1+\frac{[L]}{K_{d,int}^{S_2L}}\right)^2 \tag{6–45}$$

$$[S]=\frac{\sqrt{\left(1+\frac{[L]}{K_{d,int}^{SL}}\right)^2+8\frac{[S]_{TOT}}{K_s}\left(1+\frac{[L]}{K_{d,int}^{S_2L}}\right)^2}-\left(1+\frac{[L]}{K_{d,int}^{SL}}\right)}{\frac{4}{K_s}\left(1+\frac{[L]}{K_{d,int}^{S_2L}}\right)^2} \tag{6–46}$$

If the fraction of the sites occupied is $Y$, then

$$Y=\frac{[S{\cdot}L]+[S_2{\cdot}L]+2[S_2{\cdot}L_2]}{[S]_{TOT}} \tag{6–47}$$

$$Y=\frac{\frac{2[S][L]}{K_sK_{d,int}^{S_2L}}\left(1+\frac{[L]}{K_{d,int}^{S_2L}}\right)+\frac{[L]}{K_{d,int}^{SL}}}{1+\frac{[L]}{K_{d,int}^{SL}}+\frac{2[S]}{K_s}\left(1+\frac{[L]}{K_{d,int}^{S_2L}}\right)^2} \tag{6–48}$$

If it is assumed that the ligand L binds more tightly to the dimer than the monomer ($K_{d,int}^{S_2L} < K_{d,int}^{SL}$), then in the range of total concentration of protein $K_{SL} < [S]_{TOT} < K_S$, addition of ligand will turn a solution of unliganded monomers into a solution of liganded dimers. Because of the linkage relationship

$$\left(K_{d,int}^{SL}\right)^2 K_{SL} = \left(K_{d,int}^{S_2L}\right)^2 K_S \tag{6–49}$$

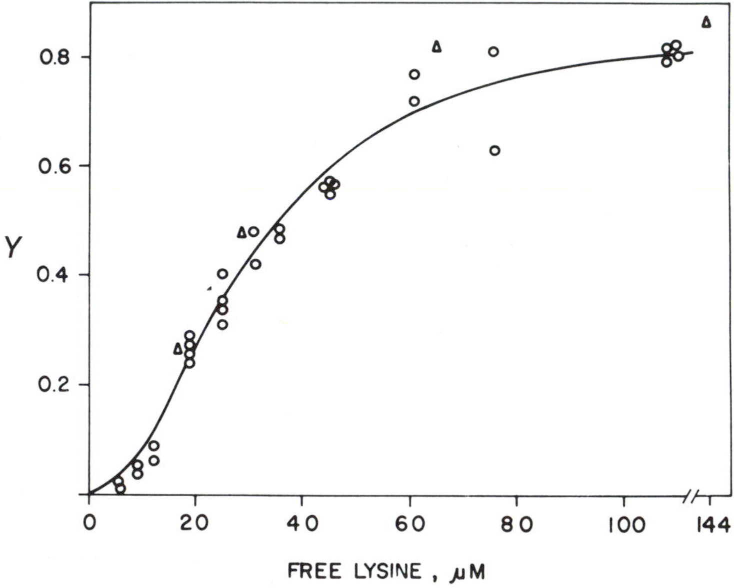


**Figure 6–22:** Binding of lysine to aspartate kinase from *E. coli.*[126] Binding of lysine to the enzyme was measured either by ultrafiltration (○) or by equilibrium dialysis (△). For ultrafiltration, solutions (0.3 mL) containing 20–40 μg of enzyme at pH 7.0, 0.21 M KCl, and 25 °C were prepared. Each solution also contained a different concentration of [$^{14}$C]lysine. After equilibrium was reached, each solution was submitted to ultrafiltration through a filter that retained the protein. The excess counts per minute of $^{14}$carbon trapped by the filter above control levels in the absence of enzyme was assumed to be the amount of lysine bound to the protein. From the initial moles of protein added to each solution and the moles of lysine bound, the fractional saturation of the protein could be calculated. For equilibrium dialysis, a series of chambers was assembled in each of which a solution (0.5 mL) containing the enzyme (1.25 mg mL$^{-1}$) was separated from an identical solution lacking enzyme. All solutions were at pH 7, 0.24 M KCl, and 25 °C. [$^{14}$C]Lysine at different concentrations was added to the successive chambers, and after 16 h, the moles of lysine in the two solutions on either side of each membrane were determined. The difference in concentration was assumed to represent lysine bound to the enzyme. Fractional saturation of the enzyme ($Y$), based on molarity of sites, is presented as a function of the free concentration of lysine (micromolar). Adapted with permission from ref 126. Copyright 1974 *Journal of Biological Chemistry.*

however, the range in which this shift from monomer to dimer will be observed is actually determined by the difference in affinity of monomer and dimer for ligand L

$$\left(\frac{K_{\text{d,int}}^{\text{S}_2\text{L}}}{K_{\text{d,int}}^{\text{SL}}}\right)^2 K_\text{S} < [\text{S}]_\text{TOT} < K_\text{S} \qquad (6\text{–}50)$$

When the concentration of the protein is too small, no dimer will form upon addition of ligand and the apparent dissociation constant for the ligand will be $K_{\text{d,int}}^{\text{SL}}$. When the concentration of the protein is too high, the protein is always a dimer, both in the absence and in the presence of saturating ligand, and the apparent dissociation constant will be $K_{\text{d,int}}^{\text{S}_2\text{L}}$. When the concentration of the protein is within the proper range, the binding of the ligand L will display homotropic cooperativity according to Equations 6–46 and 6–48.

Lysine is another member of the set of amino acids the biosynthesis of which has aspartate kinase (Reaction 6–12) as its first committed step. One of the aspartate kinases responds to threonine as a heterotropic allosteric inhibitor and another responds to lysine as a heterotropic allosteric inhibitor. The enzyme responding to lysine exists in an equilibrium between $\alpha$ monomers and an $\alpha_2$ dimer within a particular range of protein concentrations.[126] Addition of lysine shifts this equilibrium in favor of the dimer. Consistent with Equation 6–48, lysine binds to the enzyme in an association that displays homotropic cooperativity (Figure 6–22).[126] When the concentration of the enzyme is decreased to a level at which the binding of lysine is no longer able to convert the monomers to the dimer, the homotropic cooperativity disappears because the monomer cannot display homotropic cooperativity.

The binding of ligands to a protein promotes either oligomerization or disassembly because the ligands shift a conformational equilibrium in favor of a conformation of the protein in which the interactions among the subunits across the interfaces are either stronger or weaker, respectively, than they were in the previously dominant conformation. The oligomerization or disassembly of the protein that occurs subsequent to the conformational change may or may not be necessary for the functional changes that are observed in the protein. In many instances, the oligomerization or disassembly simply monitors indifferently the fact that the conformational change has occurred. Citrate is a heterotropic allosteric activator of acetyl-CoA carboxylase

$$\text{acetyl-SCoA} + \text{MgATP} + \text{HCO}_3^- \rightleftharpoons \text{malonyl-SCoA} + \text{MgADP} + \text{HPO}_4^{2-} \qquad (6\text{–}51)$$

because acetyl-SCoA is in this way diverted to fatty acid when levels of citrate are high. Citrate causes the enzyme to polymerize into long filaments.[127] It can be demonstrated by stopped-flow experiments that the rate constant with which the enzymatic activity is increased by citrate (1.0 $s^{-1}$) can significantly exceed the rate constant with which the polymerization takes place at low concentrations of the enzyme (0.02 $s^{-1}$). Bovine glutamate dehydrogenase

$$\text{glutamate} + \text{H}_2\text{O} + \text{NAD}^+ \rightleftharpoons \text{2-oxoglutarate} + \text{NH}_4^+ + \text{NADH} \qquad (6\text{–}52)$$

is either activated or inhibited by various nucleotides,[128] and these same nucleotides also produce complicated oligomerizations and dissociations of the protein.[129] The same enzyme from the rat, however, although its enzymatic activity is if anything more sensitive to the same nucleotides, displays none of the changes in quaternary structure seen with the bovine enzyme.[128,130] In both of these examples, the subsequent change in quaternary structure of the protein seems irrelevant to the change in function produced by the conformational change.

### *Suggested Reading*

Funkhouser, J.D., Abraham, A., Smith, V.A., & Smith, W.G. (1974) Kinetic and Molecular Properties of Lysine-Sensitive Aspartokinase. Factors Influencing the Lysine-Mediated Association Reaction and Their Relationship to the Cooperativity of Lysine Inhibition, *J. Biol. Chem. 249*, 5478–5484.

### PROBLEM 6–6

Because 2,3-bisphosphoglycerate binds to a site created by the two $\beta$ subunits in tetrameric hemoglobin, it cannot bind to the dissociated $\alpha\beta$ dimers.

(A) Draw a relationship between the binding of 2,3-bisphosphoglycerate and the dissociation of deoxyhemoglobin into dimers.

(B) How would the presence of 2,3-bisphosphoglycerate affect the apparent dissociation constant between dimers and tetramers of deoxyhemoglobin? Derive formulas relating the dissociation constant of the oligomer and the molar concentration of 2,3-bisphosphoglycerate.

# Rates of Conformational Changes

The conformational changes that produce changes in the affinity of a protein for its ligands, in its enzymatic activity, or in its state of oligomerization are all unimolecular isomerizations of the folded polypeptides and alterations in the orientations of the subunits. The first-order rate constants for these processes have a wide range of values. The rate constant at which the unliganded oxy conformation of hemoglobin, produced during a bright flash of light, relaxes to the deoxy conformation[98] is 10,000 $s^{-1}$ at 25 °C and pH 9.0. The rate constant for the conformational relaxation that occurs after aspartate transcarbamylase in the R conformation binds CTP[131] is 10,000 $s^{-1}$ at 25 °C and pH 7.0. These processes are quite rapid even when compared to turnover rates of enzymes. The opening and closing of the active sites of enzymes such as hexokinase (Figure 3–43) and citrate synthase must also proceed at rates greater than the turnover numbers for these enzymes.

There are, however, many conformational changes that have very slow rates. When 1-phosphofructokinase

$$\text{fructose 6-phosphate} + \text{MgATP} \rightleftharpoons \text{fructose 1,6-bisphosphate} + \text{MgADP} \quad (6\text{–}53)$$

is preincubated with heterotropic allosteric activators and then diluted for enzymatic assay, its enzymatic activity is quite high immediately but decreases slowly to a steady-state rate.[132] When the enzyme is preincubated with heterotropic allosteric inhibitors and then diluted for assay, its enzymatic activity is low immediately but then increases slowly to the steady-state rate. These decreases and increases in enzymatic activity reflect the slow rate constant for the conformational changes responsible for the heterotropic allosteric effects, and the first-order rate constants for these relaxations are between 0.003 and 0.03 $s^{-1}$ at 25 °C and pH 7.5. Very similar behavior is also observed[133] with amidophosphoribosyltransferase (Reaction 6–42). The slow conformational change in this instance has a rate constant of about 0.005 $s^{-1}$ at 25 °C. Formally, there is no distinction except in magnitude between a rapid conformational relaxation and a slow conformational relaxation. It may, however, be the case that the rate of the conformational change has evolved to match the rates of the particular fluctuations in concentration of effectors to which the enzyme is adapted to respond.

## Crystallographic Studies of Conformational Changes

The molecular details of the conformational changes that produce homotropic cooperativity or heterotropic allosteric behavior can be gathered from examinations of crystallographic molecular models of the two or more conformations involved. Usually the greatest changes that occur in the transitions between the two molecular models are at the interfaces among the subunits. There is no necessity for this, however. The interfaces among the subunits could remain exactly the same and all the changes could occur within the subunits. All that is necessary is for the two conformations to retain symmetry and differ in affinity for the ligand. Therefore, the attention paid to the changes at the interfaces may be exaggerated.

The changes that occur at the active site or at the site for the binding of a ligand and that explain the differences in enzymatic activity or differences in affinity are also of importance. There is, however, an ambiguity involved in these evaluations as well. Because the active conformation of high affinity studied crystallographically is usually produced by occupying the sites on the protein with ligand, changes around the active site that result from the global conformational change are difficult to distinguish from local conformational changes that result only from the binding of the ligand itself. For example, Histidine E7, which lies on the side of the heme to which the ligand binds, has a different position in the crystallographic molecular model of carbonmonoxyhemoglobin than in that of oxyhemoglobin,[68] resulting only from a difference in the local conformation brought about by binding the particular ligand. In aspartate transcarbamylase, a short loop of polypeptide, comprising amino acids 80–87 of an adjacent folded catalytic polypeptide, lies over the closed, occupied active site in the crystallographic molecular model of the R conformation but is shifted away to open the active site in the model of the unoccupied, T conformation.[134] Because the active site is shut off from the solution when this loop covers it, it must be the case that the loop closes over the active site after the substrates have been bound. Therefore, this rearrangement must be a local conformational change that occurs coincident with the binding of the substrates and not during the global conformational change because the conformation change occurs at concentrations of ligands well below saturation. If the closing of the lid were strongly coupled to the global conformational change, then the last ligands, rather than binding with high affinity, could not bind at all.

Crystallographic molecular models are available for hemoglobin in both its unliganded deoxy conformation[66] and its oxy conformation either in a complex with carbon monoxide[135] or in a complex with oxygen.[68] The interfaces between subunits $\alpha_1$ and $\beta_1$ and between subunits $\alpha_2$ and $\beta_2$ (Figure 6–1) are the same in each conformation. The two interfaces that change, namely, the interfaces between subunits $\alpha_1$ and $\beta_2$ and between subunits $\alpha_2$ and $\beta_1$, are identical to each other both before and after the conformational change because of the 2-fold rotational axis of symmetry. These interfaces are formed by the third $\alpha$ helix (helix C) of one of the two subunits interdigitating with a loop of polypeptide (the FG corner) between the sixth and seventh $\alpha$ helices of the other subunit (Figure 6–23).[1] Passing through the center of each of these two identical interfaces is a 2-fold rotational axis of pseudosymmetry (normal to the page in Figure 6–23), so each of these two interfaces has two of these sets of contacts. Because of the pseudosymmetry, however, the two sets adjust differently to the conformational change. At the contacts between the FG corner of subunit $\alpha_1$ and helix C of subunit $\beta_2$, the arrangement of the interlocking contacts among the three amino acids from subunit $\alpha_1$ and the two amino acids from subunit $\beta_2$ remain the same during the conformational change, but the shift in the orientations of the subunits is accommodated by shifting plastically the conformations and atomic contacts among these five amino acids.[1] At the contacts between helix C of subunit $\alpha_1$ and the FG corner of subunit $\beta_2$, however, the order of the amino acids in the interdigitations shifts to accommodate the new orientation as the two faces slide across each other. For example, Histidine 97 from the FG corner of the $\beta_2$ subunit packs between Threonine 41 and Proline 44 from the C helix of the $\alpha_1$ subunit in the deoxy conforma-

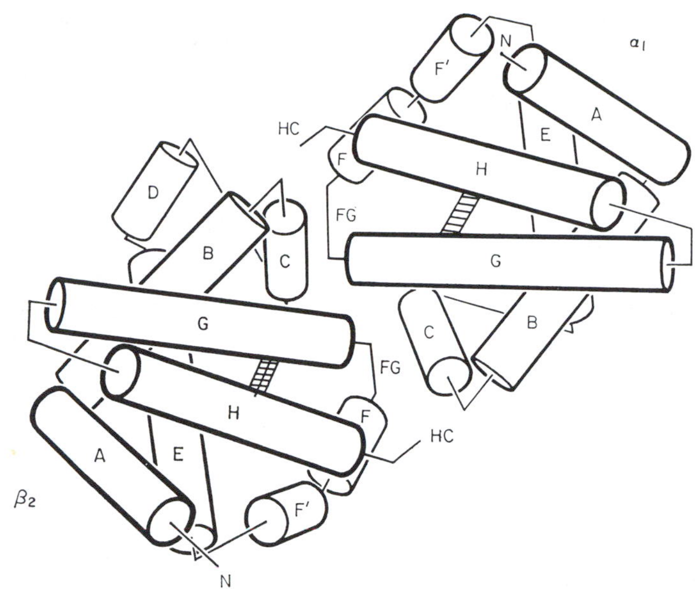


**Figure 6–23:** Schematic diagram of the arrangement of an $\alpha_1$ subunit and a $\beta_2$ subunit that form the $\alpha_1\beta_2$ interface in the crystallographic molecular model of hemoglobin.[1] The crystallographic molecular models of deoxyhemoglobin and carbonmonoxyhemoglobin described in Figure 6–1 were used to draw the diagram. Because the $\alpha_1\beta_2$ interface shown here rearranges significantly during the conformational change, the diagram represents neither model but is an average structure that is intended simply to display the disposition of secondary structures at the interface in both crystallographic molecular models. The hemes are seen side-on in this view as hatched trapezoids. The $\alpha$ helices are labeled consecutively A–H from the amino terminus of each subunit. The $\alpha_1\beta_2$ interface is formed from the FG corners and the C helices. The 2-fold rotational axis of pseudosymmetry is normal to the page and passes through the center of the structure. Reprinted with permission from ref 1. Copyright 1979 Academic Press.

tion, but it shifts over to pack between Threonine 38 and Threonine 41 in the oxy conformation. It has been proposed that this discrete alternation between only two stable arrangements is the reason that there are only two conformations of the oligomer.[1]

In the deoxy conformation of the $(\alpha\beta)_2$ tetramer, the $\alpha_1\beta_1$ dimer and the $\alpha_2\beta_2$ dimer are distorted from their normal conformations when they are free in solution more than they are contorted in the oxy conformation of the tetramer (Scheme 6–41). The free energy for this extra contortion results from the fact that the interfaces of the deoxy conformation are stronger than those of the oxy conformation. This difference in the strength of the interfaces arises from the fact that the contacts between the $\alpha_1\beta_1$ dimer and the $\alpha_2\beta_2$ dimer are more extensive in the deoxy conformation than in the oxy conformation.[136]

At the hemes, the difference in affinity for oxygen between the deoxy conformation and the oxy conformation arises from the shift in the position of the sixth helix (helix F) that causes the iron to be pulled out of the plane of the heme in the deoxy conformation while residing within the plane of the heme in the oxy conformation (Figure 2–28). The connection between this shift and the global conformational change can be thought of in two equivalent ways. A change in global conformation causes the interfaces involving the FG corners to change their structure, pushing or pulling helix F across the heme and changing the affinity for oxygen. The binding of oxygen pulls the iron into the plane of the heme, which in turn pulls on helix F and lowers the stability of the deoxy conformation because the shifted helix can be accommodated only in the oxy conformation.

Crystallographic molecular models are available for aspartate transcarbamylase in the T conformation to which CTP is bound[137] and in the R conformation to which *N*-(phosphonacetyl)-L-aspartate is bound.[22,134] In this case, however, the differences between the two conformations are much more extensive than with hemoglobin and cannot be so economically described. During the global conformational change (Figure 6–2), involving the rotation of the catalytic subunits relative to each other and the opening up of the space between them, several changes[22] occur in the interfaces among the subunits.[1a] An extensive, symmetrically displayed interface between folded catalytic polypeptide C1 and folded regulatory polypeptide R4 is present in the T conformation but not in the R conformation and presumably would act to stabilize the T conformation. The interface between adjacent folded catalytic polypeptides in the C subunit becomes more extensive in the R conformation than in the T conformation, which would operate in the opposite direction. These interfaces, however, are not supposed to be involved in the global conformational change. The symmetrically displayed interface between folded catalytic polypeptide C1 and folded regulatory polypeptide R1, which acts as the pivot during the global conformational change, sustains a number of changes in atomic contacts; but, overall, the number of contacts increases between the T conformation and the R conformation. Finally, an extensive symmetrically displayed interface between folded catalytic polypeptide C1 and folded catalytic polypeptide C4 in the T conformation is drastically altered and almost eliminated during the conformational change.

When the atomic contacts across the interfaces are examined in detail, 31 are lost and 25 different ones are gained upon the transition from the T conformation to the R conformation. This, however, is a deceptive tally because large changes in the number of contacts (34 lost and 48 gained) also occur within the folded polypeptides themselves, and there is no reason to focus on only one category of changes to explain the overall differences in stability between the two global conformations.

The most remarkable change in the active site between the two global conformations of aspartate transcarbamylase (Figure 6–24),[134] aside from the closing of the lid formed by amino acids 80–87 of an adjacent folded catalytic polypeptide, is a narrowing caused by the movement of one of the two hinged domains forming the cleft in which the active site lies (Figure 6–25).[134] This narrowing of the active site brings the

A

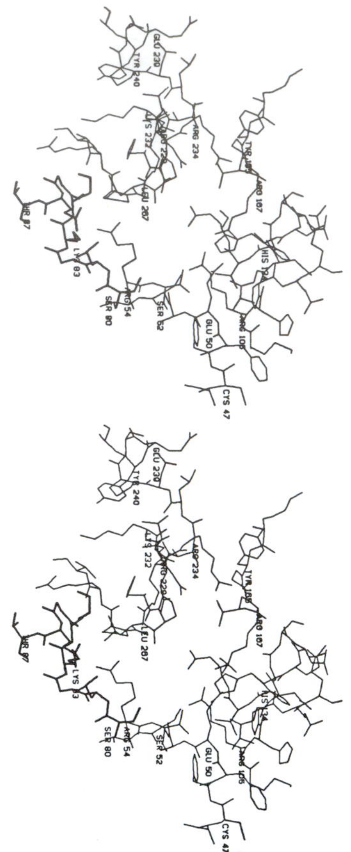

**Figure 6–24:** Comparison of the active sites in the crystallographic molecular models of aspartate transcarbamylase from *E. coli* crystallized in the presence of CTP (A) and in the presence of saturating concentrations of *N*-(phosphonacetyl)-L-aspartate (B).[134] The two refined crystallographic molecular models have been described in the legend to Figure 6–2. In panel B, the *N*-(phosphonacetyl)-L-aspartate (PALA, dark lines) is seen within the active site surrounded by the catalytic residues. It has been trapped within the active site by the closing of a lid formed from Serine 80–Threonine 87 from the adjacent catalytic polypeptide. In panel A, the active site in the unliganded crystallographic molecular model is open to the solvent because this lid has swung away, and the active site is much wider (compare the distances between the side chains of Glutamate 50 and Arginine 234) because the two hinged domains forming its sides have swung open (Figure 6–25). This figure offers an additional opportunity to observe the details of an active site surrounding an analogue of a substrate. It is unusual in that it is formed from portions of two separate polypeptides. Reprinted with permission from ref 134. Copyright 1987 Academic Press.

B

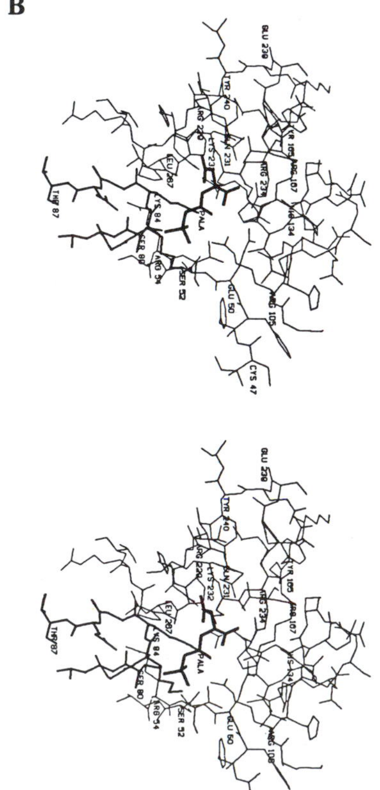

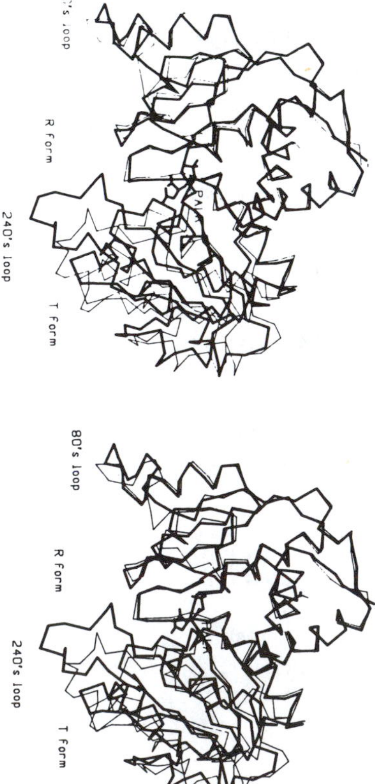


**Figure 6–25:** Superposition of the $\alpha$-carbon tracings of the folded catalytic polypeptides from the crystallographic molecular models of aspartate transcarbamylase from *E. coli* crystallized in the presence of CTP (T form, thin lines) or in the presence of saturating concentrations of *N*-(phosphonacetyl)-L-aspartate (R form, thick lines), respectively.[134] The two refined crystallographic molecular models have been described in the legend to Figure 6–2. The domains to the right of each structure in the drawing were superposed. The resulting displacement of the domains to the left in each structure with respect to each other demonstrates quite clearly that the two domains are hinged. The location of the active site in the jaws of the hinge is identified by the molecule of *N*-(phosphonacetyl)-L-aspartate (PALA). The 240's loop is a large loop at the farthest edge of the left domain that makes the largest sweep as the hinge closes. The 80's loop is the small loop that closes over the neighboring active site upon its occupation (Figure 6–24). Reprinted with permission from ref 134. Copyright 1987 Academic Press.

various amino acids involved in it into proper register to bind the reactants and should explain the increase in affinity and enzymatic activity observed in the transition from the T conformation to the R conformation. At the tip of the domain that shifts is a loop (240's loop) that sustains the greatest translation in position. This is the prominent loop symmetrically displayed in the space between the two catalytic subunits in Figure 6–2. In the T conformation the 240's loop makes few contacts at the interface between the two C subunits; but, in the R conformation, each of these loops makes contacts across the space between the two C subunits with the symmetrically displayed loop on the opposite side (Figure 6–2). It has been proposed[138] that it is these contacts, only possible after the two C subunits have rotated relative to each other about the 3-fold axis of symmetry, that locks the two domains in the proper orientation to form a narrowed active site of high affinity and high activity.

If the closing of the hinge observed crystallographically is representative of the global conformational change of the unliganded enzyme that is involved in the homotropic cooperativity displayed by aspartate transcarbamylase in solution, then predictions based on the two models about the effects of certain site-directed mutations should be validated by experiment.[138] Tyrosine 240 in the catalytic subunit participates in a hydrogen bond with Aspartate 271 in the crystallographic molecular model of the T conformation but does not participate in a hydrogen bond in the R conformation. It is the breaking of this hydrogen bond, among many other changes, that permits the contacts to occur between the 240's loops in the space between the C subunits (Figure 6–2). When Tyrosine 240 is mutated to a phenylalanine so Aspartate 271 has no donor in the T conformation, the apparent affinity of the enzyme for aspartate increases and its cooperativity ($n_{\text{Hill}} = 1.8$) decreases relative to the native enzyme ($n_{\text{Hill}} = 2.2$). This is consistent with a destabilization of the T conformation. Glutamate 50 forms a hydrogen bond with Arginine 105 in the T conformation[22] but forms hydrogen bonds with both Arginine 167 and Arginine 234 in the R conformation, one at each of its oxygens (Figure 6–24). When it is mutated to a glutamine, so that one of these latter arginines necessarily lacks an acceptor, the affinity of the enzyme for aspartate decreases as well as the cooperativity ($n_{\text{Hill}} = 1.0$), consistent with a destabilization of the R conformation relative to the T conformation.[138] When both Tyrosine 240 and Glutamate 50 are mutated, the enzyme displays a higher affinity and greater cooperativity ($n_{\text{Hill}} = 1.6$) than when only Glutamate 50 is mutated, consistent with a destabilization of both the R and the T conformations. These results suggest that the changes in association involving these amino acids observed crystallographically are indeed involved in the global conformational change responsible for cooperativity rather than simply in unpropagated local conformational changes arising from the binding of the ligand.

One of the most remarkable aspects of the structure of the interfaces involved in the conformational changes producing homotropic cooperativity is their relatively small surface area. The major changes in the interfaces of aspartate transcarbamylase are the elimination of the small C1–R4 interface and the creation of the even smaller interface between the 240's loops. The only interface that changes when the deoxy conformation of hemoglobin turns into the oxy conformation is that between an $\alpha_1$ subunit and a $\beta_2$ subunit (Figure 6–23), which involves only 10 amino acids. The interface between

**Figure 6–26:** Tracing of the $\alpha$-carbon positions in the crystallographic molecular model[29] of the $\alpha_2$ dimer of glycogen phosphorylase *a* from rabbit muscle in the inactive conformation stabilized by glucose (Table 6–1). Phosphorylase *a* (20 mg $mL^{-1}$) was crystallized from a solution of 10 mM $Mg^{2+}$ and 50 mM glucose at pH 6.7. The crystals were of space group $P4_32_12_1$, and a single subunit ($n_{aa} = 841$) occupied the asymmetric unit. Reflections from native crystals and five heavy-atom derivatives were collected to Bragg spacings of 0.25 nm. Phases from these isomorphous replacements were used to calculate a map of electron density into which a molecular model of the polypeptide was inserted, and the initial molecular model was submitted to refinement. In the figure, the two subunits (heavy and light lines, respectively) of the $\alpha_2$ dimer are displayed around one of the crystallographic 2-fold rotational axes of symmetry, which in this view is normal to the page. The interface between the subunits is peculiar in that it is formed by interdigitating strands that create a neck between the two globular structures. The amino terminus of each subunit is the dead-end strand of polypeptide protruding from each neck. The molecules of glucose in the crystallographic molecular models of each subunit are included, and molecules of AMP, not in the models, are included to indicate the locations of the sites for AMP. Reprinted with permission from ref 29. Copyright 1979 Academic Press.

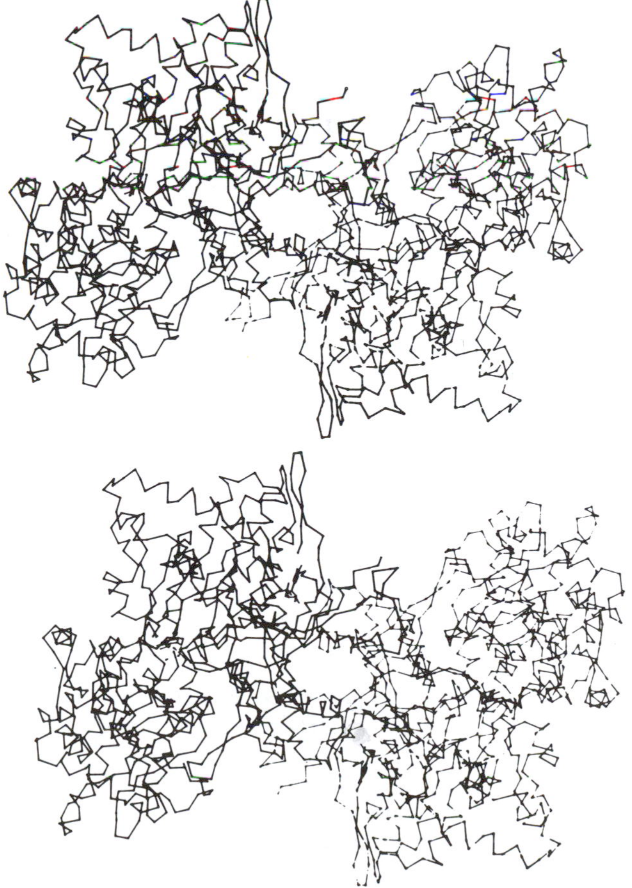

the two subunits in phosphorylase is at a neck connecting the two much wider globular subunits (Figure 6–26).[29] The 2-fold rotational axis of symmetry passes through the center of the neck, which is formed from several loops of polypeptide that interdigitate symmetrically around the axis. On the 2-fold axis itself, however, is a large vacant cavity in the middle of the neck (Figure 6–26). This cavity gives the neck the appearance of flexibility rather than rigidity. Both the narrow cross section of the neck relative to the centers of the two subunits and the cavity in the middle of the neck cause this interface to be less extensive than those usually encountered in noncooperative oligomers.

It is at these interfaces that the changes in structure occur, ensuring that symmetry is maintained. Of necessity, 2-fold rotational axes of symmetry run through the centers of each of these interfaces and changes in structure at the interfaces come in identical pairs related by these axes. That the pairs of structural changes are required to occur simultaneously is the reason the symmetry is maintained. It is possible that if the interfaces were too extensive and too strong, this would interfere with the coordination of these changes.

It is not the case, however, that the rearrangements of the interactions at the interfaces during the conformational change provide a mechanical force sufficient to distort the subunits into the new conformations. These interactions are never extensive enough to provide forces of this magnitude. It is rather that the subunits themselves have within their amino acid sequences the information necessary to dictate the two or more global conformations linked to the binding of the ligands. Within the oligomer in the absence of ligands, one of these two or more conformations of each subunit is favored, and the overall conformation of the oligomer is of necessity symmetric. The binding of ligand to one subunit causes local conformational changes that destabilize the unliganded global conformation of that subunit. The interfaces, however, remain symmetric and in the unliganded conformation until the number of ligands bound is sufficient to shift the equilibrium constant for the global conformational change of the oligomer to the other side of unity. Then, as the global conformational change of the oligomer commences, it cannot stop at some intermediate asymmetric state because the interfaces among the subunits do not mesh properly until a symmetrical oligomer of the other conformation is formed.[23] It is the difference of a few kilojoules mole$^{-1}$ of standard free energy between the imprecise fit at the interfaces in an asymmetrical oligomer and the precise fit in a symmetrical oligomer that dictates that the ligand-linked global conformational changes built into the amino acid sequences of each subunit occur in concert during the global conformational change of the oligomer.

### *Suggested Reading*

Ladjimi, M.M., & Kantrowitz, E.R. (1988) A Possible Model for the Concerted Allosteric Transition in *Escherichia coli* Aspartate Transcarbamylase as Deduced from Site-Directed Mutagenesis Studies, *Biochemistry 27*, 276–283.

# Systems for Pleiotropic Regulation

The metabolic activity of the cytoplasm of a cell is often affected dramatically by the increase or decrease in the concentration of a particular molecule in the extracellular medium. Such extracellular molecules that regulate the metabolism of a cell can be referred to as agonists. Circulating hormones are the most familiar of such regulatory agonists. For example, when the concentration of glucagon increases in the extracellular fluid around a liver cell, the rate of gluconeogenesis in the independent cell increases. As with most instances of the effect of an extracellular agonist, however, glucagon produces several independent effects. In addition to the acceleration of gluconeogenesis, the uptake of amino acids is enhanced, the synthesis of several enzymes is increased, and the enzymatic activities of several of the enzymes in the pathway of fatty acid synthesis are decreased. In analogy to the effects of a pleiotropic mutation in genetics, the multiple effects produced by an extracellular agonist represent pleiotropic regulation of cellular function. **Pleiotropic regulation** is the exertion of control over several distinct cellular processes by the same stimulus. This is usually accomplished by systems that employ second messengers such as calcium or cyclic AMP.

Calmodulin is a small monomeric protein able to control the catalytic activity of an extensive set of enzymes. It was originally described as a protein capable of stimulating the enzymatic activity of 3′,5′-cyclic-nucleotide phosphodiesterase.[140] As its name implies, it couples the levels of enzymatic activity of the many enzymes that it controls to the molar concentration of free calcium ion in the cytoplasm. This results from the fact that when the protein is unoccupied it has a different conformation than when it is occupied with calcium ion. In almost every instance,[141] the unliganded conformation cannot affect any of the enzymes controlled by calmodulin, but the liganded conformation can control each of them. The monomeric protein binds four calcium ions, each with almost equal intrinsic affinity.[140,142,143] As the intrinsic dissociation constant for these sites is 11 $\mu$M, the fluctuations of cytoplasmic calcium concentration that can be detected by calmodulin fall in the concentration range between 100 nM (unliganded) and 100 $\mu$M (fully liganded).

A crystallographic molecular model is available for calmodulin in the fully liganded conformation (Figure 6–27A).[144] The four calcium ions are bound by four internally repeated domains, two at each end of the molecule (Figure 6–27B),[145] whose superposable structures and homologous sequences clearly indicate that they were derived from a common ancestor. Each of these four domains is built on a helix-loop-helix motif originally identified as the motif of two of the domains in parvalbumin, a protein that also binds calcium ions.[146,147] Other members of this superfamily include troponin C and certain light chains of myosin, some of which contain domains that have lost their abilities to bind calcium. In the crystallographic molecular models of domains from

**Figure 6–27:** Crystallographic molecular model of calmodulin from bovine brain. Calmodulin (5 mg $mL^{-1}$) was recrystallized from 20–30% 2-methyl-2,4-pentanediol and 2 mM $CaCl_2$ between pH 4 and 6. The space group of the crystals was *P*1, with one molecule of calmodulin in each asymmetric unit. Two isomorphous derivatives as well as anomolous dispersion were used to phase a data set to Bragg spacings of 0.22 nm. A model of the polypeptide was inserted into the initial map of electron density and the initial crystallographic molecular model was submitted to refinement. (A) Complete crystallographic molecular model[144] showing all second- and third-row atoms. The lysines in the model are highlighted. Lysine 75 and Lysine 77 are within the portion of the central $\alpha$ helix that unfolds to allow the molecule to wrap around the $\alpha$-helical binding sites in its targets. Panel A reprinted with permission from ref 144. Copyright 1988 Academic Press. (B) $\alpha$-Carbon tracing of the complete crystallographic molecular model.[145] This presentation demonstrates that the two globular domains at each end of the central $\alpha$ helix are superposable on each other and that within each of these globular domains is a 2-fold rotational axis of pseudosymmetry superposing two internally repeating domains. Therefore, all four domains are superposable and each binds a $Ca^{2+}$ (black and white circles). Panel B reprinted with permission from *Nature,* ref 145. Copyright 1985 Macmillan Magazines Limited.

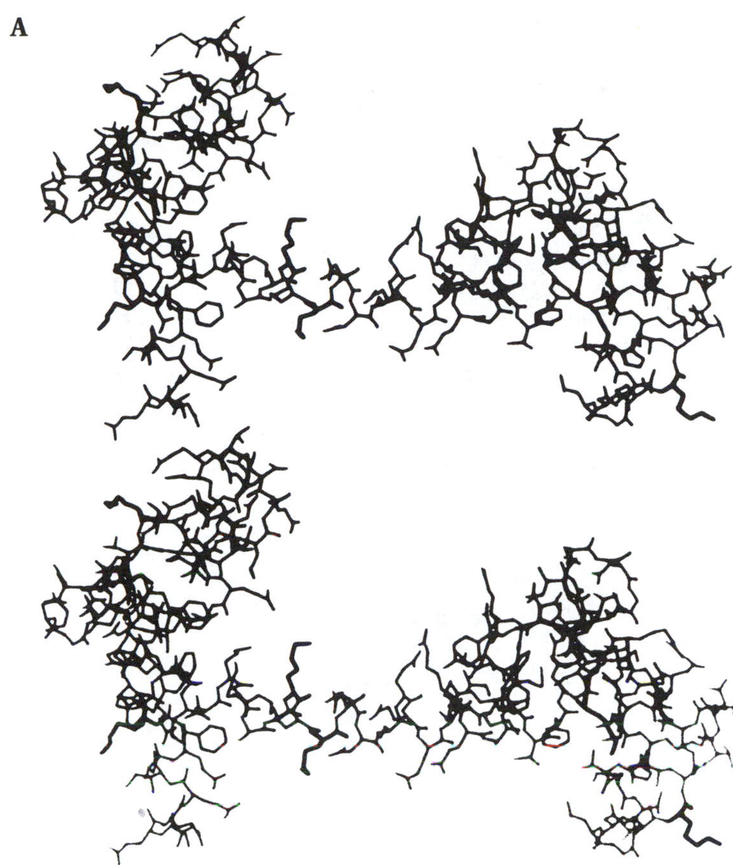

members of this superfamily that do bind calcium, the calcium ions are enfolded by the helix-loop-helix structures (Figure 6–28).[148] Within each site, the calcium ion is chelated directly by an oxygen atom from the side chain of either an aspartate or an asparagine at position *n* in the amino acid sequence, an oxygen atom from the aspartate at position ($n$ + 2), the acyl oxygen of the peptide bond from the amino acid at position ($n$ + 4), both oxygen atoms from the side chain of the glutamate at position ($n$ + 9), and the oxygen atom of a molecule of water.[144] This surrounds the calcium ion with an irregular shell of seven oxygens with no particular geometric arrangement other than the one enforced arbitrarily by the polypeptide. The strong electrostatic interactions between the calcium ion and this shell of lone pairs of electrons from oxygen atoms probably provides significant free energy of formation to stabilize the domains that seem too small to assume a defined structure otherwise.

The most peculiar feature of the structure of calmodulin is a bare $\alpha$ helix connecting the two ends of the molecule (Figure 6–27A). Because unsupported $\alpha$ helices are unstable in aqueous solution, this was an unexpected observation. It must be the case that, at its two ends, the globular domains provide stable segments of $\alpha$ helix that nucleate and brace an otherwise unfavorable structure, if it does remain helical in the free protein in solution.

The atomic details of the conformational change that occurs when the calcium dissociates from calmodulin are unknown. Circular dichroic measurements[149–151] suggest that the content of $\alpha$ helix decreases by about 20%. Measurements of scattering of X-radiation by solutions of calmodulin demonstrate that the radius of gyration decreases by about 6%

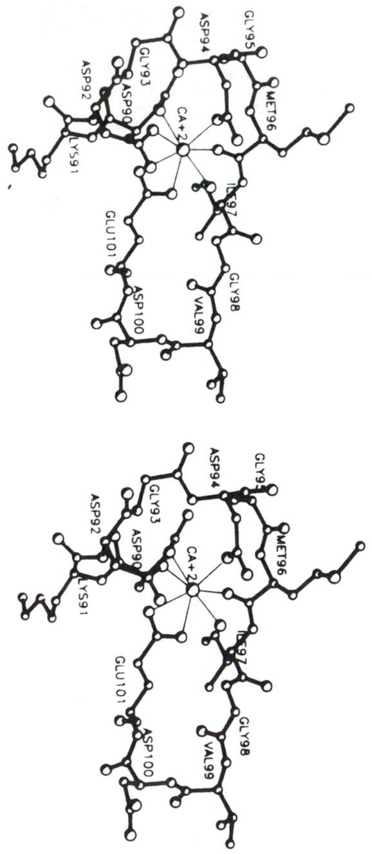

**Figure 6–28:** Atomic details of one of the binding sites for $Ca^{2+}$ in the crystallographic molecular model of carp parvalbumin, a member of the calmodulin superfamily.[148] Parvalbumin was crystallized from 2 M $(NH_4)_2SO_4$ at pH 7. The crystals were of space group *P*1, and a data set was collected to Bragg spacings of 0.19 nm. A crystallographic molecular model of a closely related parvalbumin was placed in the unit cell and rotated and translated until its calculated data set matched the observed data set most closely. After the side chains were switched to those of the parvalbumin used in the present study, the crystallographic molecular model was submitted to refinement against the observed data set. The structure of the carboxy-terminal calcium-binding site of the model is presented, formed from residues Aspartate 90–Glutamate 101. The $Ca^{2+}$ is surrounded by seven oxygens in an irregular arrangement. Reprinted with permission from ref 148. Copyright 1988 Academic Press.

upon dissociation of the calcium,[152,153] but this seems inconsistent with the decrease in the sedimentation coefficient observed upon dissociation of the calcium.[149]

Various correlations have been made between occupation of the sites on calmodulin by calcium and its conformation and ability to be recognized by the enzymes it activates. The change in conformation detected by far-ultraviolet circular dichroism is directly proportional to the occupation of sites for calcium ion on the protein.[143] That the intrinsic affinities of all the sites for calcium are very similar suggests that this change represents similar local conformational changes occurring independently around each of the individual sites. The activation by calmodulin of myosin light-chain kinase,[154] 3′,5′-cyclic-nucleotide phosphodiesterase,[155] adenylate cyclase,[156] and the calmodulin-sensitive $Ca^{2+}$-transporting ATPase[157] have been studied by varying systematically both the concentration of free calcium and the concentration of calmodulin. The results, when analyzed with the separately determined dissociation constants for the binding of calcium to calmodulin,[143] are consistent with the conclusion that the only active species of calmodulin is the one fully occupied by four calcium ions. Therefore, the conformation in solution that is recognized by the enzymes only becomes the dominant one after all of the sites have been occupied. That conformation presumably is the one observed crystallographically (Figure 6–27), but not necessarily. It has been shown, however, also crystallographically, that the liganded conformation of calmodulin (Figure 6–27), as it binds to the enzymes it activates, wraps around a particular $\alpha$ helix found in each of these proteins.[158] In the process the central $\alpha$ helix of the calmodulin unfolds from Alanine 73 to Lysine 77 to permit the structure to enfold the $\alpha$ helix on the target with a right-handed twist.

In analogy with the mechanism by which other enzymes are controlled, it can be assumed that each of the enzymes activated by calmodulin exists in at least two conformations, an inactive one dominant in the absence of calmodulin and an active one dominant when the complex between calcium and calmodulin is bound. In most instances, it is the complex between calcium and the regulatory protein of the calmodulin superfamily that is able to bind as the allosteric activator for the target enzyme. In this way, the target enzyme is activated only when the molar concentration of free calcium is high enough to produce the complex and is inactive when the concentration of calcium falls significantly below its dissociation constant from calmodulin. This ligand-linked conformational change of calmodulin couples the catalytic activity of the target enzyme to the levels of free calcium, which in this case is the second messenger. At least one member of the calmodulin superfamily, however, activates its target enzyme only when it is unliganded. Recoverin ($n_{aa}$ = 202) is a regulatory protein of the calmodulin superfamily that has two helix-loop-helix motifs. It controls the enzyme guanylate cyclase.[141] At low concentrations of calcium, recoverin activates guanylate cyclase maximally, but at saturating concentrations of free calcium (>400 nM), it is unable to activate guanylate cyclase.[141,159]

Among the group of proteins controlled by calmodulin are several protein kinases. The **protein kinases** are a large family of enzymes that regulate other enzymes by covalently phosphorylating particular threonines, serines,[160] or tyrosines[161] on their surfaces. The reaction catalyzed by these protein kinases is the transfer of the $\gamma$-phosphate from the respective oxygen of MgATP to the oxygen on one of the serines, threonines, or tyrosines on the particular enzyme controlled by the particular protein kinase. The enzymatic activity of a protein kinase is regulated in turn by the concentration of a heterotropic allosteric effector specific for that protein kinase. For example, there are protein kinases activated by adenosine 3′,5′-cyclic monophosphate,[162] by calmodulin saturated with calcium,[163] or by circulating hormones.[164] The site phosphorylated on the enzyme or enzymes modified by a protein kinase appears to be recognized by certain features of the sequence of amino acids surrounding it. This follows from the fact that short synthetic peptides that have the same sequence of amino acids as those surrounding the normal site of covalent phosphorylation on a targeted enzyme are often recognized and phosphorylated by the protein kinase.[165,166]

The mechanisms by which the enzymatic activities of the protein kinases themselves are controlled by their particular heterotropic allosteric effectors, such as calmodulin or cyclic AMP, have turned out to be quite unpredictable. They often are different from the mechanisms by which heterotropic allosteric effectors usually control proteins. Part of this difference is explained by the fact that the active site of a protein kinase is designed necessarily to recognize specific amino acid sequences on its protein reactants. For this reason, one strategy that has evolved, referred to as **intrasteric regulation**,[167] is to include within the amino acid sequences of the protein kinase itself an amino acid sequence that resembles the sequence on the reactant recognized by the active site of the protein kinase.[168] This internal pseudosubstrate is found in a region of the folded polypeptide of the protein kinase that is in a suitable location and accessible enough to bind intramolecularly to its own active site. When the internal pseudosubstrate is occupying the active site, the protein kinase is inactive. To accomplish this self-contained intrasteric inhibition, the segment containing the pseudosubstrate can be located on the same folded polypeptide that forms the active site of the protein kinase, as in the case of myosin light-chain kinase,[168,169] or on a folded polypeptide of completely different sequence within a heterooligomer, as in the cyclic AMP-dependent protein kinase.[170]

The internal amino acid sequence recognized as the pseudosubstrate usually is difficult to identify conclusively within the amino acid sequence of the enzyme because protein kinases do not recognize a very specific amino acid sequence in the first place. For example, the canonical motif[171] recognized by the cyclic AMP-dependent protein kinase in other proteins is –RRXS–, but this occurs only in about 30% of the instances in which a serine on a target enzyme is phosphorylated by this kinase.[172] The pseudosubstrate site within the cyclic AMP-dependent protein kinase,[170] however, is –FDRRVSVCA–.

The mechanism by which the intrasteric inhibition of the protein kinase by the internal pseudosubstrate is relieved by the heterotropic allosteric activator varies significantly among the different protein kinases. Smooth muscle myosin light-chain kinase[169] is the enzyme responsible for transferring a phosphate from the $\gamma$ position on MgATP to Serine 19 of the regulatory light chain ($n_{aa}$ = 171) of smooth muscle myosin.[173] The sequence surrounding the serine phosphorylated by the enzyme is AcSSKRAKTKTTKKRPQRATSNVFAMFDQSQIQ–,[174] where Ac is the acetylated amino terminus of the myosin light chain. The enzyme has an absolute requirement for fully liganded calmodulin to display activity.[154,169] A peptide has been isolated from a cyanogen bromide digest of skeletal muscle myosin light-chain kinase that could compete with the native skeletal muscle enzyme for calmodulin and that by itself bound calmodulin tightly.[175] That cyanogen bromide fragment was derived from the segment –RRLKSQILLKKYLMKRRWKKNFIAVSAANRFKKISSSGALMALGV, where the last valine is the carboxy terminus of the intact polypeptide ($n_{aa}$ = 608) that folds to produce the monomeric myosin light-chain kinase from skeletal muscle.[176,177] The homologous region of monomeric smooth muscle myosin light-chain kinase ($n_{aa}$ = 972) has the sequence[178] –KKLSKDRMKKYMARRKWQKTGHAVRAIGRLSSMAMISGMSGRKA– from Lysine 784 to Alanine 827. It has been shown[178,179] that the amino acid sequence recognized by calmodulin within this region is –ARRKWQKTGHAVRAIGRLSS–. In the complex between calmodulin and this segment ($K_d$ = 0.94 $\mu$M),[180] this amino acid sequence assumes the structure of an $\alpha$ helix that is completely enfolded by the calmodulin.[158]

From comparing this amino acid sequence in myosin light-chain kinase with that of the myosin light chain itself, it became clear that the binding site for calmodulin on smooth muscle myosin light-chain kinase overlaps a region of its sequence that has a modest degree of similarity to the sequence surrounding the serine in its substrate,[178] but in the position homologous to Serine 19 in the smooth muscle myosin light chain there is a histidine in the amino acid sequence of the myosin light-chain kinase (Figure 6–29).[168] It was proposed that in the absence of calmodulin, the sequence within the polypeptide of smooth muscle myosin light-chain kinase that is homologous to the sequence around Serine 19 in the smooth muscle myosin light chain is bound intramolecularly in the active site as an intrasteric inhibitor and that this causes the enzyme to be inactive. When calmodulin enfolds tightly the overlapping segment next to this internal pseudosubstrate, the internal pseudosubstrate is prevented from binding as an inhibitor to the active site, and the enzyme becomes active. This proposal is consistent with the observations that the synthetic peptide AKKLSKDRMKKYMARRKWQKTG is a competitive inhibitor of smooth muscle myosin light-chain kinase.[178] It is also consistent with the observation that when smooth muscle myosin light-chain kinase was cleaved with trypsin at the arginine in the sequence –HAVRAIG–, a shortened, inactive enzyme, incapable of being activated by calmodulin, was produced; but, when the enzyme was then cleaved with trypsin at a site to the amino-terminal side of this arginine, an even shorter form of the enzyme was produced that had high activity regardless of whether or not calmodulin was present.[168] The longer form still possessed the intrasterically inhibitory sequence but not the sequence to which calmodulin binds, while the shorter did not possess the inhibitory sequence and was constitutively active.

Cyclic AMP-dependent protein kinase is an $\alpha_2\beta_2$ heterodimer in its inactive form. The smaller ($n_{aa}$ = 350)[182] of the two polypeptides folds to form a subunit containing the active site for phosphorylating proteins and is referred to as the catalytic subunit. The larger of the two polypeptides ($n_{aa}$ = 390)[183] folds and oligomerizes to form an $\alpha_2$ dimer that contains four binding sites for cyclic AMP, two on each of these regulatory subunits.[184] Each regulatory subunit is composed of three domains: two internally repeating domains,[183] each of which forms one of the binding sites for the heterotropic allosteric activator, cyclic AMP; and an amino-terminal domain of about 100 amino acids that folds to produce the faces that form the interface between the two regulatory subunits.[185] Between the domain responsible for dimerization and the first domain responsible for the binding of cyclic AMP, the regulatory subunit also contains, between Glycine 88 and Threonine 100, a pseudosubstrate site, –GRFDRRVSVCAET–.[170] Thermolysin cleaves the regulatory subunit at the peptide bond between Arginine 93 and Valine 94 to produce a

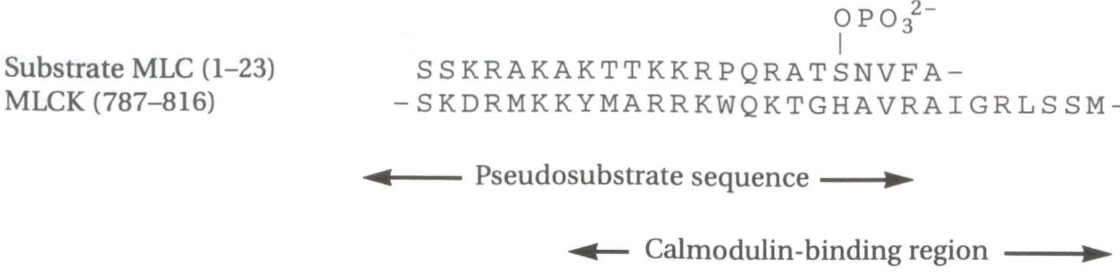


**Figure 6–29:** Overlap between the intramolecular pseudosubstrate sequence and the sequence providing the calmodulin-binding site in smooth muscle myosin light-chain kinase (MLCK).[168] Myosin light-chain kinase ($n_{aa}$ = 972) is responsible for the phosphorylation of Serine 19 on the light chain of myosin (MLC). The site of phosphorylation was determined by digesting phosphorylated myosin light chain and isolating the phosphopeptide. The sequence of the first 23 amino acids of the light chain of myosin is presented on the top line and the serine phosphorylated is identified. Within the amino acid sequence of myosin light-chain kinase,[181] between Serine 787 and Alanine 809, there is a sequence that is homologous (6 identities out of 23 amino acids as well as several conservative substitutions) to the first 23 amino acids of the light chain of myosin. This is the pseudosubstrate site that binds intramolecularly to the active site of the enzyme and inhibits its activity.[168] The cyanogen bromide fragment of myosin light-chain kinase beginning at Alanine 796 and continuing to Methionine 816 was found to bind strongly to calmodulin occupied by calcium ($K_d$ < 1 nM),[175,179] and this observation identified this region of the protein as the calmodulin-binding site.

protein that can no longer form a stable heterooligomer with a catalytic subunit, but chymotrypsin cleaves the peptide bond between Phenylalanine 90 and Aspartate 91 to produce a protein that still forms a stable heterooligomer with a catalytic subunit.[170] Therefore, the intact, internal pseudosubstrate is necessary for the formation of the $(\alpha\beta)_2$ heterooligomer, and it is assumed that the heterooligomer is formed by the binding of the pseudosubstrate sequence on the regulatory subunit to the active site on the catalytic subunit to intrasterically inhibit the enzymatic activity.

When cyclic AMP binds to the binding sites on the regulatory subunits, each regulatory subunit in the dimer releases its associated catalytic subunit, which becomes a separate, monomeric protein in solution.[186,187] Once released from the complex, each catalytic subunit, is fully active as a protein kinase. Presumably it is a conformational change induced by the binding of cyclic AMP to the two sites on a regulatory subunit that alters the structure of the protein around the internal pseudosubstrate and dramatically lowers its affinity for the active site.

An enzyme controlled by covalent phosphorylation mediated by a protein kinase or by several different protein kinases may be phosphorylated by that kinase or by those kinases at only one particular location[160] or at multiple locations throughout its sequence. For example, glycogen synthase is controlled by phosphorylation. It has seven serines scattered over its amino acid sequence that are capable of being phosphorylated,[188] and there are five distinct protein kinases that can each phosphorylate one or more of these serines.[189] Some of them can be phosphorylated by more than one of these protein kinases. It is clear that in such cases the effects of phosphorylation on the enzymatic activity are gratifyingly complex.

The phosphorylation of an enzyme by a protein kinase, as with the control of an enzyme by a noncovalently bound heterotropic allosteric effector, can either increase or decrease its enzymatic activity. Acetyl-CoA carboxylase is inactivated upon its phosphorylation[190] by cyclic AMP-dependent protein kinase.[191] When phenylalanine 4-monooxygenase, however, is phosphorylated by a calmodulin-dependent protein kinase,[192] its enzymatic activity is increased.[193]

Because the enzymes controlled by covalent phosphorylation are often controlled by other heterotropic allosteric effectors as well, the particular effect of a given phosphorylation on the enzymatic activity depends upon the concentration of these other effectors in a complicated fashion. When glycogen synthase is unphosphorylated, its Michaelis constant for the reactant UDP-glucose is decreased 5-fold by the heterotropic allosteric activator glucose 6-phosphate, but its maximum initial velocity is unaffected by the allosteric activator.[194] When, however, the enzyme is phosphorylated by protein kinases,[195,196] its Michaelis constant for UDP-glucose no longer decreases when glucose 6-phosphate is added, but its maximum initial velocity increases 10-fold or more when the activator is added. The maximum initial velocity of the phosphorylated enzyme in the presence of glucose 6-phosphate is equal to the maximum initial velocity of the unphosphorylated enzyme in its absence. Because of such complications, comparisons of the enzymatic activities of the phosphorylated and unphosphorylated forms of the enzymes must be carefully described.

An allosteric effector can also alter the rate at which the enzyme is phosphorylated by a protein kinase. For example, the rate at which acetyl-CoA carboxylase is phosphorylated by cyclic AMP-dependent protein kinase is decreased almost completely by saturation of the enzyme with citrate, a heterotropic allosteric activator.[190]

When an enzyme is regulated by being phosphorylated by a protein kinase, the phosphorylation is formally equivalent to the binding of a heterotropic allosteric effector except that the effector is covalently rather than noncovalently bound. The presence of the phosphate increases the stability of a particular global conformation so that it becomes dominant relative to the global conformation dominant in the unphosphorylated enzyme. When the $\alpha_2$ dimer of phosphorylase is phosphorylated at Serine 14 in the amino acid sequence of each of its identical constituent polypeptides, its enzymatic properties change significantly. The enzymatic activity of the unphosphorylated enzyme has an absolute requirement[28] for the heterotropic allosteric activator adenosine monophosphate, but the enzymatic activity of the phosphorylated enzyme is only stimulated by 50% when adenosine monophosphate is added. The maximum initial velocities of both forms are indistinguishable when saturating concentrations of adenosine monophosphate are present, but the dissociation constant for adenosine monophosphate to the phosphorylated form of the enzyme is 30-fold smaller than that for the unphosphorylated form.[197] That the enzyme has different conformations in these two states, phosphorylated and unphosphorylated, is demonstrated by the fact that at high concentration (>0.1 mg $mL^{-1}$) phosphorylated phosphorylase is an $(\alpha_2)_2$ tetramer while unphosphorylated phosphorylase remains an unaggregated $\alpha_2$ dimer,[198] even though both forms crystallize as $\alpha_2$ dimers.[29]

Crystallographic molecular models of the $\alpha_2$ dimers of phosphorylated and unphosphorylated phosphorylase have been obtained, and their structures have been compared.[199] The phosphorylated serine is only 14 amino acids from the amino terminus, and, in the crystallographic molecular model of unphosphorylated phosphorylase, the majority of the first 16 amino acids are disordered. The fact that Serine 14 is splayed randomly away from the main globular structure is probably essential for its access to the active site of phosphorylase kinase, the relevant protein kinase. The location of the seventeenth amino acid in the sequence is at the junction between the globular regions of the dimer and the neck that forms the interface (Figure 6–26). The first 16 amino acids in the crystallographic molecular model of phosphorylated phosphorylase have an ordered structure with a short stretch of $3_{10}$ helix. The two identical strands, related by the 2-fold rotational axis of symmetry, lie across the back of the neck in Figure 6–26. In the change from the disordered to the ordered and tightly pinned strand, the phosphate on Serine 14 has formed a cluster of hydrogen bonds with two arginines, an aspartate, and a glutamine from two other interdigitated segments of polypeptide, each from a different one of the two subunits (Figure 6–30).[199] In the process of forming this entirely new arrangement, the helices that entwine to form the neck shift slightly and two or three hydrogen bonds in the interface between the two subunits break as the phosphorylated conformation is attained. It is these changes at the interface that appear to be the major contributors to the change in stability of the global phospho conformation relative to the global dephospho conformation.

When crystals of phosphorylated phosphorylase in the conformation stabilized by the heterotropic allosteric inhibitor, glucose, are perfused with the reactants, inorganic phosphate and maltopentaose, a conformational change occurs that may represent the conformational change involved in the heterotropic allosteric behavior and the homotropic cooperativity displayed by this enzyme.[61] These changes that are observed in the crystallographic molecular model are significantly different from those that are observed between the models of phosphorylated and unphosphorylated phosphorylase, particularly within the first 350 amino acids, which provide all of the contacts at the interface between the two subunits. It seems, therefore, that the global conformational change producing the heterotropic allosteric behavior and the homotropic cooperativity is a different global conformational change from the one responsible for the functional differences between the phosphorylated and unphosphorylated forms of the enzyme.

One of the main differences between the conformational changes observed upon the phosphorylation of a protein and those observed during the binding of heterotropic allosteric ligands are in the character of the local conformational changes. When a ligand binds noncovalently, there is a pocket on the surface of the protein into which it enters, and the pocket usually then tightens around the ligand as the noncovalent interactions are formed between ligand and protein. When a protein is phosphorylated, however, the site or sites of phosphorylation must be accessible to the active site of the protein kinase, which usually means they are on flexible loops. Upon phosphorylation, if phosphorylase may act as a paradigm, the loops then adhere to the surface of the protein by forming, among others, interactions among amino acids and the phosphate. In either the binding of the ligand or the phosphorylation of a threonine, serine, or tyrosine, however, it is these local changes that stabilize one global conformation relative to the other and promote the shift in the equilibrium constant between them. The global conformational changes themselves, involving rearrangements of subunits and movements of domains and shifts of secondary structures, do not differ qualitatively whether the heterotropic effector is noncovalently or covalently bound.

There is another set of proteins regulating cellular processes that relies upon conformational changes to perform its functions. The proteins with the central role in these regulatory systems are a closely related family of guanosine triphosphatases known as G-proteins.[200] G-Proteins are $\alpha\beta\gamma$ heterotrimers in which the $\alpha$ subunit is variable and determines the specific role of the G-protein and the $\beta$ and $\gamma$ subunits are always the same. The enzymatic reaction catalyzed by a G-protein[201] is

$$\mathrm{MgGTP} + \mathrm{H_2O} \rightleftharpoons \mathrm{MgGDP} + \mathrm{HOPO_3}^{2-} \qquad (6\text{–}54)$$

No metabolic purpose is served by this reaction; rather, it is used as a timing device.

The steps in an abbreviated kinetic mechanism of this reaction are

$$\mathrm{G + MgGTP} \underset{k_{-1}}{\overset{k_1}{\rightleftharpoons}} \mathrm{G \cdot MgGTP} \underset{k_{-2}}{\overset{k_2}{\rightleftharpoons}}$$

$$\mathrm{G \cdot MgGDP \cdot HOPO_3}^{2-} \underset{k_{-3}}{\overset{k_3}{\rightleftharpoons}}$$

$$\mathrm{G \cdot MgGDP + HOPO_3}^{2-} \underset{k_{-4}}{\overset{k_4}{\rightleftharpoons}}$$

$$\mathrm{G + MgGDP + HOPO_3}^{2-} \qquad (6\text{–}55)$$

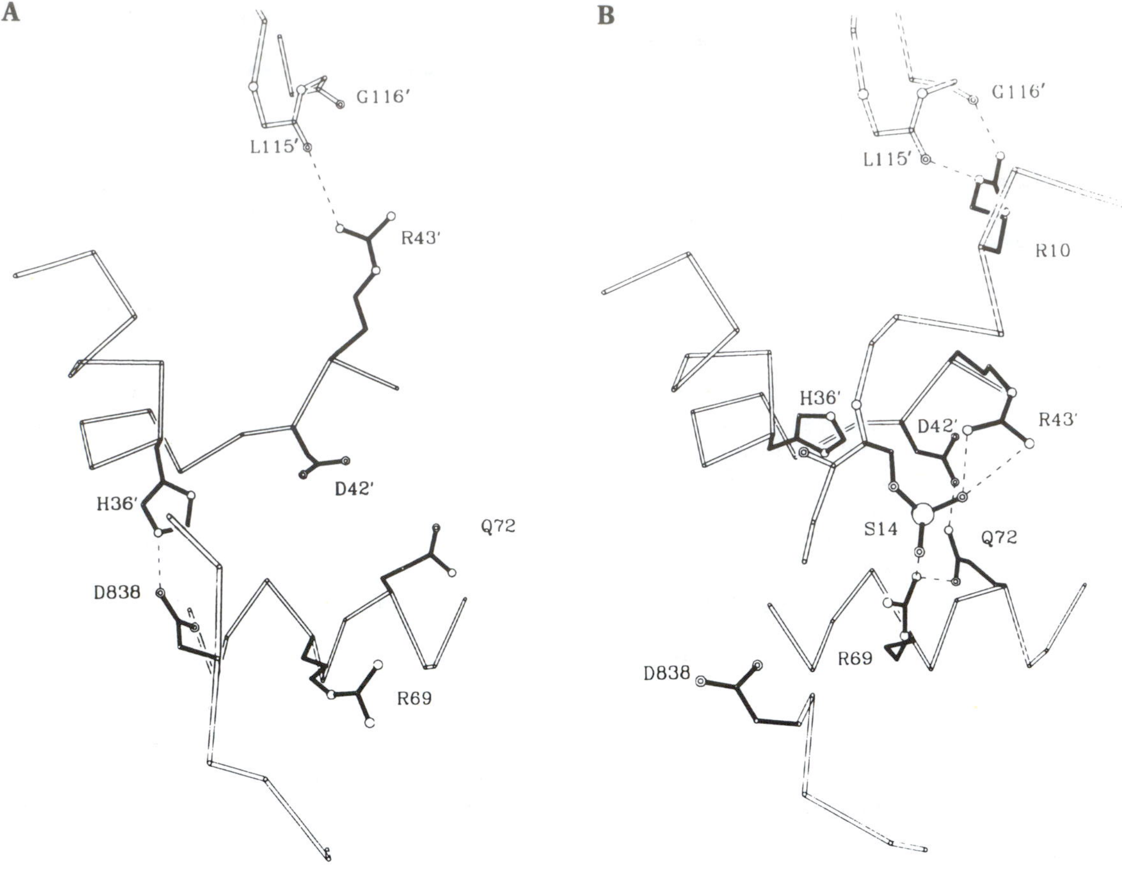


**Figure 6–30:** Diagrammatic representations[199] of the segments of polypeptide surrounding the phosphorylated serine (Serine 14) in the crystallographic molecular model of phosphorylase *a* from rabbit muscle (B) and the same segments in the crystallographic molecular model of the unphosphorylated protein, phosphorylase *b* (A). The two conformations were crystallized as described in the legend for Figure 6–26. They crystallize in the same space group, $P4_32_12$, with almost identical unit cells. The maps of electron density for the two forms were calculated independently in separate laboratories with separate isomorphous replacements, and the two crystallographic molecular models are almost identical except in regions where presumably the conformational changes have occurred. In the figure the $\alpha$-carbon tracings both for amino acids 30′–44′ from one polypeptide of the $\alpha_2$ dimer and for the adjacent interdigitating segments of amino acids 65–75 and 836–840 of the other polypeptide are displayed. The primes are used to distinguish amino acids from one subunit from those from the other subunit. Several side chains emerging from the $\alpha$-carbons are included to indicate interactions. A short segment of polypeptide (the backbone for Leucine 115′ and Glycine 116′) is also shown. In the unphosphorylated protein (A) the first 16 amino acids of each polypeptide are disordered and do not appear in the map of electron density. In the protein phosphorylated at Serine 14, the binding of the phosphate in the pocket formed from the segments of polypeptide pins the amino terminus, and it is ordered and present in the map of electron density. The segment amino acid 8 through amino acid 16 in this ordered conformation is displayed in panel B. Reprinted with permision from *Nature*, ref 199. Copyright 1988 Macmillan Magazines Limited.

where G is the G-protein. The turnover number of the G-protein in this enzymatic reaction is very slow, and depending on the circumstances, it is limited by the rate of the hydrolysis itself ($k_2$) or by the dissociation of the MgGDP ($k_4$). When the G-protein is present alone, the turnover number of the enzyme[202,203] at initial velocity is about 1 min$^{-1}$ at 25 °C. The rate-limiting step under these circumstances is the dissociation of the product MgGDP, but the rate constant for the hydrolysis on the active site is also quite slow ($k_2$ = 10 min$^{-1}$).

G-Proteins couple the reception of a signal at a receptor to the change in the enzymatic activity of another protein. The receptors, the occupation of which initiate the process, are also a closely related family[204,205] of integral membrane proteins inserted into the plasma membrane of the cell. Each of these proteins has seven very hydrophobic sequences long enough to form membrane-spanning $\alpha$ helices. Usually, the receptor has an extracellular site for a hormone or other agonist that appears in the solution surrounding the cell. The binding of the agonist promotes a conformational change in the receptor in the usual manner. This conformational change then permits the receptor to act as a heterotropic allosteric activator of the G-protein or G-proteins that it controls. When the complex between agonist and receptor is bound by the G-protein, the G-protein then experiences a conformational change that causes the rate of the GTPase reaction (Equation 6–54) to increase.[200] This is due to an increase in both the rate of dissociation of MgGDP ($k_4$ in Scheme 6–55) and the rate of association of MgGTP ($k_1$ in Scheme 6–55). These changes are sufficient to make hydrolysis at the active site ($k_2$ in Scheme 6–55), which remains unchanged at 10 min$^{-1}$, rate-limiting.[206] This causes the dom-

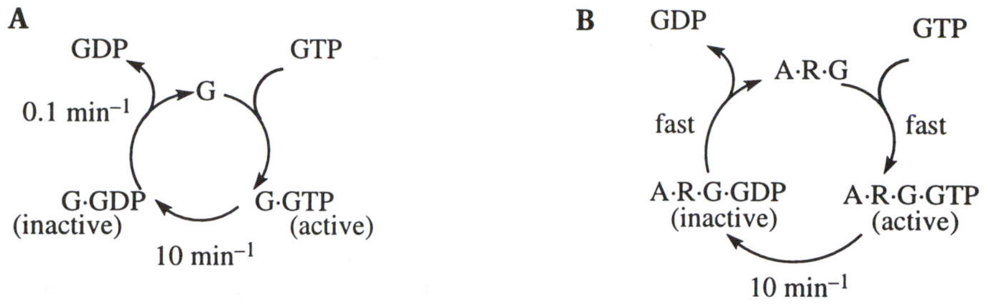

**Figure 6–31:** Effect of forming the complex between agonist (A), receptor (R), and G-protein (G) on the rates of the individual steps in the GTPase reaction catalyzed by the G-protein. (A) In the absence of the complex between agonist and receptor (A·R), the GTPase reaction is limited by the rate of dissociation of MgGDP (GDP). This causes the dominant form of the G-protein in solution to be the complex with MgGDP, which is unable to activate or inactivate the enzymes controlled by the G-protein. (B) When the ternary complex of agonist, receptor, and G-protein (A·R·G) is formed, the rates of dissociation of MgGDP and association of MgGTP (GTP) are accelerated sufficiently that the hydrolysis of MgGTP at the active site becomes the rate-limiting step in the reaction. This causes the dominant form of the G-protein in solution to be the complex of the G-protein with unhydrolyzed MgGTP, which has a conformation that acts as a heterotropic allosteric effector of the enzymes controlled by the G-protein.

inant form of the G-protein at steady state to become the complex G·MgGTP rather than a dominant mixture of G·MgGDP and unliganded G-protein (Figure 6–31).

The complex between agonist and receptor participates in rapidly associating and dissociating interactions in which many G-proteins, one after the other, can be converted from G·MgGDP to G·MgGTP.[207,208] As long as the receptor remains occupied with its agonist, the dominant form of the accessible G-protein in the cell is G·MgGTP because as soon as a molecule of G·MgGDP·$HOPO_3^{2-}$ is formed, as the result of the unchanged rate of hydrolysis, the complex between agonist and receptor returns it to the form G·MgGTP by catalyzing the dissociation of MgGDP and association of MgGTP. As soon as the receptor becomes unoccupied, however, it reverts to the conformation dominant in the absence of agonist, and the concentration of G·MgGTP decreases at a rate of 10 min$^{-1}$ ($k_2$ in Scheme 6–55) to produce the initial mixture of G·MgGDP and unliganded G-protein. This is the timing device that turns off the response when the agonist disappears.

The conformation of the G-protein when MgGTP is bound is different from the conformation when MgGDP is bound or when the G-protein is unoccupied.[209] It is the conformation of the G-protein when MgGTP is bound that is in turn the allosteric effector for the enzymes responsible for the ultimate response to the binding of the agonist to the receptor. This explains why analogues of GTP such as 5′-guanylyl imidodiphosphate[210] and guanosine 5′-*O*-(3-thiotriphosphate),[211] which cannot be hydrolyzed by G-proteins, produce much higher and more persistent activation of the enzymatic activities controlled by G-proteins even in the absence of a complex between agonist and receptor. Among the enzymes controlled allosterically by G-proteins are adenylate cyclase,[212] 3′,5′-cyclic GMP phosphodiesterase,[213,214] and the phospholipase responsible for the production of inositol 1,4,5-triphosphate.[215]

### Suggested Reading

Pearson, R.B., Wettenhall, R.E.H., Means, A.R., Hartshorne, D.J., & Kemp, B.E. (1988) Autoregulation of Enzymes by Pseudosubstrate Prototypes: Myosin Light-Chain Kinase, *Science 241*, 970–973.

# Conformational Conversion of Energy

The majority of the MgATP produced in an animal cell from MgADP and $HOPO_3^{2-}$ is produced by $H^+$-transporting ATPase, and the majority of the MgATP that is consumed in an animal cell by hydrolysis to form MgADP and $HOPO_3^{2-}$ is consumed by actomyosin ATPase and $Na^+/K^+$-transporting ATPase. All three of these enzymes[215a] interconvert the chemical free energy available from the hydrolysis of MgATP into conformational changes that perform work.

Actomyosin is the complex of proteins responsible for all of the rapid movements displayed by the cells from which animals are formed. The proteins contained in actomyosin can be divided into those that produce the movement, actin and myosin, and those that regulate the movement, troponin, tropomyosin, and the myosin light chains. The two discrete components of actomyosin are thick filaments of myosin[215b] and thin filaments. Thin filaments are helical polymers of actin[215c] to which tropomyosin and troponin are often added.

The original description of the mechanism by which actomyosin generates the force that results in movement was based on studies of striated skeletal muscle[216,217] because in this tissue thick filaments and thin filaments are aligned next to each other in repeating patterns that permit their movements relative to each other to be observed in a light microscope. These observations led to a description of muscular contraction in which thin filaments of actin slide past thick filaments of myosin. This mechanism, however, is not confined to skeletal muscle but can explain all movement performed by actomyosin in all cells.[218,219]

The actin filament is a helical polymeric protein that is produced by the assembly of actin monomers from a point of initiation and that reaches a length determined by the association of a protein that caps the elongating filament. Because the actin filament is a helical polymeric protein defined by a screw operation, it is polar; one end is different from the other and all monomers are oriented in the same recognizable direction. The point of initiation of each filament of actin is affixed to the cellular structure against which the force is to be exerted. This structure can be a complex of proteins that connect the filament of actin to a cellular membrane[220] or a large polymeric assembly of proteins such as the Z-bands in skeletal muscle.[217] Actin filaments can be of any length and the suitable length for a particular purpose is determined by rates of initiation, elongation, and capping.

While actin filaments can be of any length, thick filaments of myosin are confined to a limited range (1.5–2.5 $\mu$m) of lengths. A thick filament of myosin is bipolar with a 2-fold rotational axis of pseudosymmetry in the center of the bare zone.[215b] All of the globular heads of the myosin molecules in one end of a thick filament are pointed in the same direction, and that end will slide along a filament of actin when MgATP is present. The active movement of a thick filament of myosin along a filament of actin is always in one direction, from the capped end of the filament of actin toward the attached point of nucleation,[221] and the active movement of the actin filament across the one end of the thick filament is always from the tip of that end of the thick filament toward the bare zone. As one end of a thick filament slides up one filament of actin, or several filaments of the same alignment, the other end slides up another filament of actin, or set of filaments. Because the two ends of the thick filament have the opposite polarity, the filaments of actin sliding across one end of the thick filament are drawn across that end in the opposite direction from the filaments drawn across the other end. As a result, the respective ropes of actin radiating from each end of the thick filament pull in opposite directions on the points to which they are attached and contraction takes place. The thick filament is a simple mechanical device that can pull objects attached to ropes running out from its two ends toward itself and in the process pull the objects attached to the two ends closer together. It is the bipolarity with which the molecules of myosin are assembled into the thick filament that creates this ability. Each individual molecule of myosin,[215b] however, is unipolar with a globular head and a helical tail. By itself, a molecule of myosin in a thick filament can pull on a filament of actin only in one direction.

The capacity of myosin to slide up a filament of actin can be demonstrated by attaching purified myosin to beads that can be directly observed in a microscope and watching these beads slide up oriented filaments of purified actin.[221] In this way, it could be demonstrated that the only requirements for movement at a rate (4 $\mu$m $s^{-1}$) comparable to the rate at which skeletal muscle can contract (6 $\mu$m $s^{-1}$) are filaments of pure actin, the pure myosin itself, and MgATP.

The MgATP is necessary because its hydrolysis provides the free energy for the directional movement of myosin along the filament of actin and for the performance of work during contraction under tension. Myosin is able to catalyze the hydrolysis of MgATP, and this myosin ATPase activity is enhanced by actin. In an intact muscle fiber pulling against a load, the tension developed is directly proportional to the rate of the ATPase of the myosin in that fiber.[222]

There is a sequence of mechanical steps that explain the ability of one end of a thick filament to pull a thin filament of actin across itself.[223] In the absence of MgATP, actomyosin assumes a state referred to as rigor. In this state cross-bridges can be observed between the thick filament and adjacent thin filaments.[224] These cross-bridges are the globular myosin heads arrayed around the thick filament.[215b] At one end, each of these cross-bridges is affixed permanently by its tail to the thick filament; at the other, it is bound very tightly through a specific interface to a molecule of actin in the thin filament. In rigor,[225,226] the angle made between the axis of the thick filament and the axis of the cross-bridge connecting the thick and the thin filaments is 50°. Upon binding MgATP, the cross-bridge releases from the filament of actin and swings, while free, until the angle between thick filament and cross-bridge increases to 90°. During this swing, the tip of the cross-bridge, which bears the face that associates with actin, moves up the actin filament. This movement up the actin filament can occur only if, at this instant, other cross-bridges in the thick filament are in different steps in the cycle and remain attached tightly to the actin filament while the cross-bridge that has released swings upward. For this reason only aggregates of myosin can sustain sliding movements. Following the hydrolysis of the MgATP, the loose cross-bridge then reassociates with the actin molecule one step closer to the attached end of the filament of actin. Upon dissociation of the products, MgADP and $HOPO_3^{2-}$, the cross-bridge swings back to an angle of 50°. Because this angle between thick filament and cross-bridge is acute relative to the bare zone of the thick filament, this stroke pulls the actin in the correct direction across the thick filament[221,223] and generates the force of the movement. Ironically, the force-generating step is the relaxation of the vacant cross-bridge from the conformation (90°) produced originally by binding MgATP to the conformation (50°) found in rigor. Everyone who moves her muscles has first-hand experience of the mechanical properties of a conformational change in a protein.

The enzymatic details of this cycle performed by myosin have been studied by following the relationships between the interactions of myosin with actin and the steps in the hydrolysis of MgATP. Because coordinated and synchronous interactions between proteins as large as an actin filament and a thick filament cannot be established experimentally, myosin is represented in these investigations by the globular head alone. This fragment, referred to as myosin subfragment 1 or myosin S1, is produced by controlled proteolytic digestion of myosin and is a monodisperse, soluble protein.[227] The kinetics of its ATPase reaction and the details of its interactions with filamentous actin suggest that these two properties are unaltered by the digestion necessary to remove it from intact myosin.[228,229]

The initial insight into the interactions between MgATP, myosin, and actin came with the demonstration[230] that when MgATP was bound by myosin, the affinity of myosin for actin decreases by a factor of more than 100 and the demonstration[231] that when actin was present in the solution of myosin, the maximum velocity of its ATPase reaction increased by a factor of at least 200.

In the absence of MgATP, the dissociation constant for the complex between filamentous actin and myosin S1 is $2 \times 10^{-7}$ M (pH 7.0, 22 °C);[232,233] but, in the presence of saturating MgATP, it is greater than $5 \times 10^{-4}$ M (pH 7.0, 15 °C).[234] This change in affinity between myosin and actin is linked to a difference in the affinity of myosin for MgATP. The dissociation constant for the complex between MgATP and myosin S1 bound to filamentous actin is greater than $10^{-6}$ M (pH 7.5, 22 °C)[235,236] while that for the complex between MgATP and myosin S1 alone is $6 \times 10^{-11}$ M (pH 7.0, 21 °C).[237] The difference in these two dissociation constants is entirely in the rates of dissociation of MgATP. The rates of association are very similar.[236] Therefore, the change in affinity between actin and myosin S1 cannot be due to a direct competition between MgATP and actin for the face of myosin S1 that forms the interface between myosin S1 and actin; rather, the binding of MgATP and the binding of actin must be linked by an allosteric conformational change in the myosin S1 itself. This may be the same conformational change that changes the angle of the cross-bridge once the myosin head releases from the actin filament.

Because of this conformational change that links the binding of MgATP to the weakening of the interaction between myosin and actin, the binding of MgATP to the rigor complex between a myosin head in a thick filament and a protomer of actin in a thin filament promotes dissociation of the interface between the myosin and the actin.[226] Once the interface is broken, however, the MgATP is locked into the active site of the dissociated head, for all practical purposes irreversibly ($k_{dissociation} = 1 \times 10^{-4}\ s^{-1}$).[237]

It has been demonstrated that this release of the myosin head from the thin filament is not a necessary precondition for the next steps in the hydrolysis of MgATP,[238] but for contraction to occur it must be. Therefore, the issue is whether or not the substantial decrease in dissociation constant between myosin and actin is sufficient to ensure dissociation of the myosin from actin at the normal physiological orientations of thick and thin filaments in an actomyosin complex. This can be determined from measurements of the rate of dissociation of MgATP from a small number of the myosin heads in myofibrils in which the majority of the myosin heads have no MgATP bound to them and are in rigor.[229] These measurements demonstrate that the equilibrium constant for the reaction

$$\mathrm{A{\cdot}M + MgATP \rightleftharpoons A + M{\cdot}MgATP} \quad (6\text{–}56)$$

where A is filamentous actin and M is a myosin head protruding from a thick filament, is $1 \times 10^4$. Therefore, the binding of MgATP does ensure the dissociation of actin and myosin in a functioning myofibril.

Once the myosin head to which MgATP has been bound releases from actin, it becomes equivalent to myosin S1 in the absence of actin. It has been shown by a number of quenched-flow experiments that once MgATP is bound to isolated myosin S1, the hydrolysis of the phosphate anhydride to yield tightly bound MgADP and $HOPO_3^{2-}$ is rapid ($k_1 = 50\ s^{-1}$; pH 8, 20 °C).[239,240] Remarkably, an equilibrium between tightly bound MgATP and tightly bound MgADP and $HOPO_3^{2-}$

$$\mathrm{M{\cdot}MgATP} \underset{k_{-1}}{\overset{k_1}{\rightleftharpoons}} \mathrm{M{\cdot}MgADP{\cdot}HOPO_3^{\,2-}} \quad (6\text{–}57)$$

is rapidly established,[241] and the equilibrium constant ($K_{eq}$ = 10; pH 8, 22 °C)[240] is much closer to unity than for hydrolysis of MgATP in solution. This is another example of the difference between an equilibrium constant in solution and at an active site.

In the absence of actin, the reactant, MgATP, is not released from the active site of myosin S1, and the release of the products, MgADP[242] and inorganic phosphate,[243] is very slow. This is due to a rate-limiting conformational change ($k = 0.06\ s^{-1}$; pH 8, 21 °C) in the M·MgADP·$HOPO_3^{2-}$ complex that is necessary to produce a conformation that then releases both products rapidly, phosphate faster than MgADP.[244] When filamentous actin is present, however, the release of products is very fast.[236] The acceleration of this release is due to the rapid association of actin with the M·MgADP·$HOPO_3^{2-}$ complex ($k = 2 \times 10^7\ M^{-1}\ s^{-1}$; pH 7, 15 °C)[235] and, presumably, a consequent acceleration of the same rate-limiting conformational change. Because a myosin head has about the same dissociation constant from filamentous actin when MgADP and $HOPO_3^{2-}$ are bound to it[234] as when MgATP is bound to it, the firm association between myosin and actin found in the rigor complex is only reestablished after the MgADP and $HOPO_3^{2-}$ dissociate. Because actin must be bound, however, for the dissociation of products to occur, as the rigor complex is re-formed, subsequent to the dissociation of products, actin will necessarily be bound to the myosin head. It is during the reestablishment of the rigor complex that the power stroke occurs and the angle of the cross-bridge shifts from 90° to 50°. If there were not some mechanism to ensure that an actin in a thin filament was bound during the power stroke, it would be wasted. It is the ability of the myosin head to wait until actin is bound before releasing the products that ensures this coupling.

The ability of myosin in skeletal muscle to alternately associate and dissociate with the actin filament during the sliding of the thick filament along the thin filament is controlled by cytoplasmic levels of cationic calcium. At low concentrations of $Ca^{2+}$ (<200 nM), actin and myosin cannot form a complex in either the presence or the absence of MgATP, and there are no points of attachment between thick filaments and thin filaments. In this state the thin filaments can be pulled readily in the direction opposite to the one in which they would actively slide. At high levels of calcium (>1 $\mu$M), active sliding progresses, coincident with the hydrolysis of MgATP. In both the activated state (high $Ca^{2+}$) and the relaxed state (low $Ca^{2+}$),[245,246] the predominant form of the myosin is M·MgADP·$HOPO_3^{2-}$, and it is only the inability of actin to associate with myosin and promote release of products and complete the cycle that differs between these two states. It is the tropomyosin bound to the filament of actin within the intact thin filament that prevents this association.

Tropomyosin is a segment of rope formed from two identical $\alpha$-helical strands, each 284 amino acids in length, that twist around each other in a coiled coil. It is, in fact, the paradigm of this structure. From image reconstruction of paracrystals of filaments containing actin and tropomyosin in one of their natural arrangements, a low-resolution model of the structure of these filaments could be calculated (Figure 6–32).[247–249] The basic actin helix[215c] could be readily distinguished in the center of the structure, but two narrow tubular structures could be observed winding up the two grooves, one on either side of the actin helix. These are two continuous ropes of tropomyosin formed from individual segments each 42 nm in length that overlap with each other at their ends over a short segment of about 9 amino acids.[250] The length of each segment of tropomyosin, the fact that it lies in a helical groove, and the rise for each actin in the actin helical polymer are such that each tropomyosin segment associates with seven actins arrayed along the one side of the actin helix. This means that each segment of tropomyosin spans a length of 14 actins overall. In the specimen used for the image reconstruction in Figure 6–32, the thin filament is in its active form, able to associate with myosin. Both X-ray diffraction patterns from intact muscle[251,252] and image reconstruction of tropomyosin–actin filaments[247,248] suggest that, upon inactivation of the thin filament, the tropomyosin ropes shift their position along the entire length of the helical polymer of actin by moving farther out of each groove and in the process flattening themselves against the actin monomers.[248] This movement creates steric hindrance preventing myosin from associating with actin.

The length of the segment of the thin filament controlled on one of the two sides by one molecule of tropomyosin is 38 nm. Troponin, the protein that coordinates the movement of tropomyosin into and out of the grooves of actin with the changes in the cytoplasmic concentration of $Ca^{2+}$, is distributed along the thin filament at intervals of 38 nm.[253] From this, it can be concluded that each protomer of tropomyosin in each continuous rope is controlled by one molecule of troponin.[254] That molecule of troponin is located in the central region of a protomer of tropomyosin about one-third of the way from one end.[255]

Troponin has been purified from muscle and is identified by its ability to restore sensitivity to calcium of mixtures of actin, myosin, and tropomyosin.[256] The protein is formed from three distinct polypeptides[257] designated C ($n_{aa}$ = 160),[258] I ($n_{aa}$ = 180),[259] and T ($n_{aa}$ = 260).[260] As these three folded polypeptides can be separated from each other and individually display functional properties,[261] they can be considered to be separate subunits in the overall complex. Troponin C is homologous in sequence to calmodulin, shares a common structure with calmodulin,[262] and is the component responsible[263] for binding $Ca^{2+}$. Troponin T binds tightly to tropomyosin[261] and is thought to form the major link between troponin and tropomyosin. Troponin I by itself binds to neither tropomyosin nor actin separately but binds tightly[264] to actin–tropomyosin filaments (Figure 6–32). When troponin I is added to actin–tropomyosin filaments, which support normal levels of actomyosin ATPase in its absence, it produces complete inhibition of this actomyosin ATPase activity[265] and promotes the reorientation of the tropomyosin ropes characteristic of normal relaxation.[247,248]

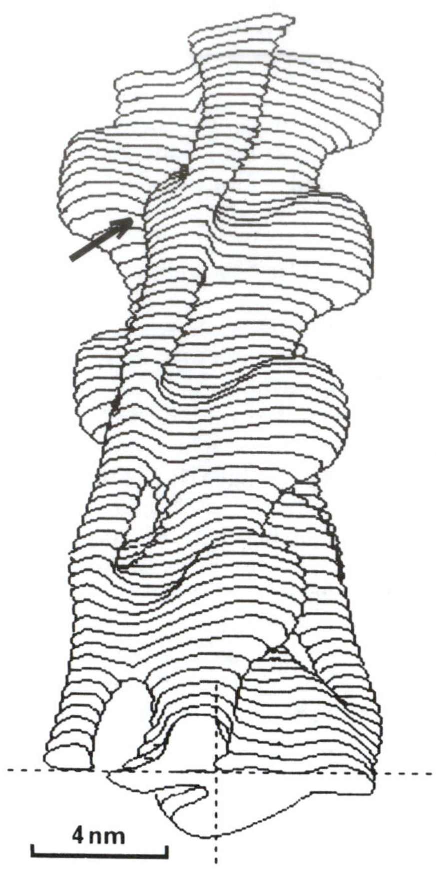


**Figure 6–32:** Reconstructed image of a filament of actin and tropomyosin obtained by Fourier analysis of an electron micrograph of a paracrystalline array.[247] Purified rabbit actin and tropomyosin were mixed at a molar ratio of 4:1 at pH 7.3 and 4 °C. The concentration of $Ca^{2+}$ was raised to about 10 $\mu$M, and $Mg^{2+}$ was added to 50 mM. These cations induced side-to-side aggregation of filaments of actin and tropomyosin to produce, among other structures, ribbons of filaments in crystalline array referred to as **paracrystals**. These paracrystals were transferred to a grid for electron microscopy that had been coated with a meshwork of collodion and a continuous layer of amorphous carbon. The adsorbed paracrystals were negatively stained with a glass of uranyl acetate. In one of the holes of the meshwork of collodion, a ribbon of 10 actin–tropomyosin filaments in width and only one filament in depth was located. A micrograph of this ribbon was taken, and the optical density of the negative was digitized. The Fourier transform of this image had the distribution of amplitudes expected from a crystalline array of structures with 28 subunits in 13 turns of a left-handed helix. Reflections on this reciprocal lattice in the Fourier transform were gathered, and the transform of these reflections contained the map of electron scattering density shown in the figure. Two ropes of tropomyosin (front and back) occupy grooves on the surface of the basic actin helix. They are associated with the smaller lobe (arrow) of each actin monomer. Reprinted with permission from ref 247. Copyright 1983 Academic Press.

On the basis of these observations it has been concluded that, in the absence of calcium, troponin I or troponins I and T in the troponin complex move the tropomyosin ropes into the location that blocks sterically the binding of myosin heads to actin. Upon the binding of calcium to troponin C, the force exerted by troponin I is relieved and the tropomyosin ropes return to their location in the complex between tropomyosin and actin alone (Figure 6–32). This relieves the steric effect and permits normal contraction to occur. In the absence of troponin, thin filaments of actin only or of actin and tropomyosin display about the same functional properties.

In smooth muscle and other cells, the sliding of myosin along actin is also regulated by changes in the concentration of $Ca^{2+}$ but not through the mediation of troponin. Thin filaments from smooth muscle lack this regulatory protein, and concentrations of calcium regulate myosin directly.[266] In these cells, the myosin has a small polypeptide associated with it known as the light chain.[267] This polypeptide is phosphorylated[267,268] by a protein kinase that is controlled by the concentration of $(Ca^{2+})_4$-calmodulin,[169] and this myosin light-chain kinase has an absolute requirement for $(Ca^{2+})_4$-calmodulin to display its enzymatic activity. When the myosin light chain is in its phosphorylated state, high levels of actomyosin ATPase[269,270] and the consequent contraction[271] are displayed. When it is dephosphorylated, actomyosin ATPase is inhibited and relaxation occurs.

The molecular mechanism of actomyosin ATPase coordinates specific conformational changes in the proteins involved—release of myosin from actin, the increase in the angle of the cross-bridge, the reassociation of myosin with actin, and the power stroke that decreases the angle of the cross-bridge—with specific changes that occur at the active site—binding of MgATP, hydrolysis of the bound MgATP, and release of the products. In this way, the free energy of hydrolysis of MgATP is coupled irrevocably to the creation of mechanical energy. Another enzyme, $Na^+/K^+$-transporting ATPase, also couples conformational changes in the protein to steps in the hydrolysis of MgATP to convert the free energy of hydrolysis of MgATP into the movement of cations against their concentration gradients across the plasma membrane.

$Na^+/K^+$-Transporting ATPase[272] catalyzes the reaction

$$3Na_c^+ + 2K_e^+ + MgATP_c \rightleftharpoons 3Na_e^+ + 2K_c^+ + MgADP_c + HOPO_3^{2-} \quad (6\text{–}58)$$

where the subscripts c and e refer to the cytoplasm and the extracytoplasmic space, respectively.[273,274] It is a protein spanning the plasma membranes of all animal cells and catalyzing the active transport of the alkali cations, $Na^+$ and $K^+$, across itself.

$Na^+/K^+$-Transporting ATPase is a member of a family of enzymes, each of which spans a bilayer and catalyzes the primary active transport of particular inorganic cations across the plasma membrane or the endoplasmic reticular membranes of a eukaryotic cell between the cytoplasm and an extracytoplasmic space. The particular reactions catalyzed by these enzymes serve to introduce the other members of the family. Each couples the hydrolysis of MgATP to the movement of cations across a membrane, but the identity and stoichiometry of the cations moving out of and into the cytoplasm differs. $H^+/K^+$-Transporting ATPase from the plasma membranes of the gastric mucosa[275] transports two $H^+$ out of the cytoplasm, rather than three $Na^+$, but still in exchange for two $K^+$. Endoplasmic reticular[276] $Ca^{2+}$-transporting ATPase catalyzes a reaction in which two $Ca^{2+}$ are transported out of the cytoplasm[277] and one or more $H^+$ are transported into the cytoplasm.[278] Calmodulin-dependent $Ca^{2+}$-transporting ATPase from animal plasma membranes[279,280] catalyzes a reaction in which only one $Ca^{2+}$ is transported out of the cytoplasm[281,282] in exchange for two $H^+$. $H^+$-Transporting ATPases in the plasma membranes of fungi[283,284] and plants[285,286] are responsible for transporting protons out of the cytoplasm.

$Na^+/K^+$-Transporting ATPase is composed of two subunits,[287] $\alpha$ ($n_{aa}$ = 1020)[288,289] and $\beta$ ($n_{aa}$ = 305),[290] each present in the active complex in one copy.[291,292] The $\beta$ subunit is a cell surface glycoprotein uninvolved in catalysis, and the $\alpha$ subunit performs the active transport. The $\alpha$ subunit is a large protein[293] that dwarfs the bilayer of the membrane. Each of the other five enzymes of the family contains a polypeptide homologous in amino acid sequence to the $\alpha$ polypeptide of $Na^+/K^+$-transporting ATPase.[286,294,295] The fact that all of the amino acid sequences of all of these catalytic polypeptides are homologous to each other requires that they have descended from a common ancestor, that they all fold respectively to produce superposable structures, and that they all catalyze active transport by the same general mechanism.

In each of these enzymes, this large protein is fixed in the bilayer, as are all membrane-spanning proteins, and catalyzes transport not by moving back and forth across the membrane but by passing the cations back and forth through itself.[296] In this regard, active transport is not transport through a membrane but transport through a protein.

During the catalysis of active transport, each of the members of this family passes through a homologous sequence of steps coupling the hydrolysis of MgATP to the movement of the cations. A fundamental aspect of this scheme is the existence of at least four distinct conformations of the enzyme that have been given the designations $E_2$, $E_1$, $E_1{\sim}P$, and $E_2$-P. These four conformations differ in structure from each other both at the active site where the MgATP is hydrolyzed and at a compartment that has access alternately to each side of the membrane and through which the transported cations pass.

$Na^+/K^+$-Transporting ATPase is the one enzyme in which all of these conformational changes have been observed and correlated both with changes in the accessibility of the compartment for the cations and with changes at the active site for hydrolysis of MgATP. The easiest place to start (Figure 6–33) is with the enzyme in the conformation designated $E_2{\cdot}K_2$, which is the conformation of the protein in which two $K^+$ are enclosed in the compartment for the cations and the compartment is occluded, or inaccessible, to the solution on either side of the membrane.[297,298] These occluded cations are the two potassiums that would have entered the compartment from the extracytoplasmic solution during the normal sequence of events and would be in transit to the cytoplasmic solution. During a conformational change, which can be detected by following either intrinsic properties of the protein[299] or the response of fluorescent, covalently attached groups that do not otherwise affect the enzyme,[300] the compartment then opens to the cytoplasmic surface. This $E_1$ conformation, which is open to the cytoplasmic surface, is the one from which the two previously occluded $K^+$ depart[301] and to which the three $Na^+$ bind during the normal cycle. This

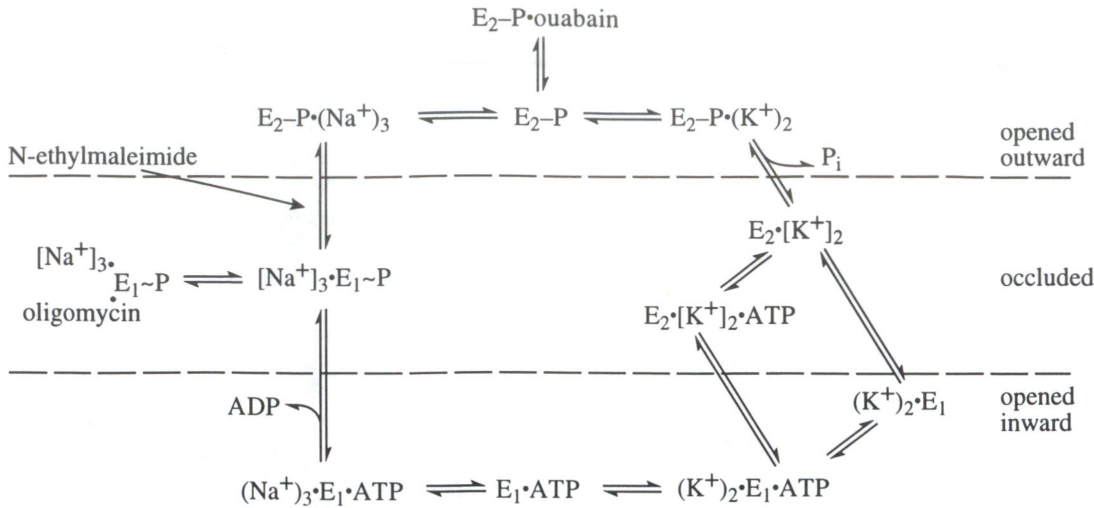


**Figure 6–33:** Kinetic mechanism for the active transport of $Na^+$ and $K^+$ catalyzed by $Na^+/K^+$-transporting ATPase. The four major conformations of the enzyme are $E_2$, $E_1$, $E_1$~P, and $E_2$-P. Changes at the active site are indicated by the binding of ATP, release of ADP, formation of the two phosphorylated forms $E_1$~P and $E_2$-P, and dephosphorylation of $E_2$-P. Changes in occupation and accessibility of the compartment for cations are indicated by the binding and release of $K^+$ and $Na^+$, respectively, and the major shifts between the conformation opened inward ($E_1$), the two occluded conformations ($E_1$~P and $E_2$), and the conformation opened outward ($E_2$-P) are designated by their locations with respect to the dashed lines representing the bilayer. The inhibitors oligomycin and ouabain are noncovalently bound by specific conformations, $E_1$~P and $E_2$-P, respectively, and modification of the enzyme with *N*-ethylmaleimide prevents the conformational change from $E_1$~P to $E_2$-P.

exchange of three $Na^+$ for two $K^+$ produces the competition displayed by these cations at the cytoplasmic surface[302] in the kinetics of the enzymatic reaction; $K^+$ is a competitive product inhibitor of $Na^+$ in its role as a reactant.[272,303] When the three $Na^+$ have entered the compartment through the opening to the cytoplasm, the compartment again closes and the three $Na^+$ become locked within the compartment in a different occluded state[304] that defines in part the $Na_3{\cdot}E_1$~P conformation. The conformational change that then occurs between the $Na_3{\cdot}E_1$~P conformation and the $E_2$-P conformation can also be followed by covalently attached fluorescent reporters of the structure of the enzyme.[300] In the $E_2$-P conformation the compartment is open to the extracytoplasmic solution and the three $Na^+$ depart to be replaced by two $K^+$. It is during this exchange that $Na^+$ at the extracytoplasmic surface acts as a competitive product inhibitor of $K^+$ in its role as a reactant.[272,303] In the final step, the protein closes around the two $K^+$ to regenerate the $E_2{\cdot}K_2$ conformation.[298]

The existence of the four discrete global conformations assumed by the enzyme, $E_2{\cdot}K_2$, $E_1$, $Na_3{\cdot}E_1$~P, and $E_2$-P, has been established in several ways other than their ability to occlude particular cations. There are several short segments of the folded polypeptide that are susceptible to proteolytic cleavage, presumably because they form exposed loops on the surface of the native enzyme, and the accessibilities of these segments to the proteolytic enzymes change dramatically as the conformations change.[305–307] The affinity of the different conformations of the enzyme for ligands such as MgATP[299] and the cardiac glycosides,[308] a set of natural products that inhibit the enzyme by binding at its extracytoplasmic surface,[309] are dramatically different. Finally, a number of fluorescent molecules have been covalently attached to various locations in the protein, and their fluorescence emission changes as changes are made in the conformation of the enzyme.[300,310] Because these various reporters of the structure of the protein are scattered over its surface, it is assumed that the changes in structure being detected are global conformational changes.

As in actomyosin, steps in the mechanism for the hydrolysis of MgATP at the active site of the enzyme are coordinated to each of these mechanical movements at the compartment through which the cations pass. When the enzyme is in the $E_2{\cdot}K_2$ conformation, two $K^+$ are occluded within the compartment and the active site is empty. There are two parallel paths through which the mechanism can proceed from this point (Figure 6–33). In a slow step, the enzyme can open to the cytoplasm, and then MgATP at low concentration can bind with high affinity to the active site; or MgATP at high concentration can bind first, and then the enzyme can open to the cytoplasm in a fast step.[299,311] Physiological concentrations of MgATP are sufficient to saturate the active site when the enzyme is in the $E_2{\cdot}K_2$ form, even though it displays low affinity for MgATP; and, under normal circumstances, the binding of MgATP accelerates the opening of the compartment to the cytoplasm.[300] Because MgATP can bind to the active site of the enzyme by either of these two paths, the kinetics of the enzyme display negative deviations from linear behavior.[312] These steps resemble the release of myosin from actin that results from the binding of MgATP to the active site of the myosin. Once the compartment is open to the cytoplasm, the two $K^+$ are eventually replaced by three $Na^+$. At this point, while the compartment closes around the cations, the γ-phosphate of MgATP is transferred to an aspartate in the enzyme[313] during the formation of the $Na_3{\cdot}E_1{\sim}P$ conformation. In the $Na_3{\cdot}E_1{\sim}P$ conformation, the aspartyl phosphate is in a form of high intrinsic free energy of hydrolysis because it is in rapid equilibrium with MgADP

$$E_1{\sim}PO_3^{2-}{\cdot}MgADP \rightleftharpoons E_1{\cdot}MgATP \qquad (6\text{–}59)$$

as would be expected from the free energy of hydrolysis for acyl phosphates in solution.[314,315] If the next step in the mechanism is blocked by the inhibitor oligomycin or by covalent modification of the enzyme with *N*-ethylmaleimide, this rapid equilibrium is manifested in a rapid exchange of [$^{14}$C]MgADP with [$^{12}$C]MgATP.[316]

If, however, the next step proceeds spontaneously, as it normally does, the $Na_3{\cdot}E_1{\sim}P$ conformation converts rapidly to the $E_2$-P conformation, in which the compartment is open to the extracytoplasmic surface. The acyl phosphate, although still at the same aspartate,[317] has changed from one of high intrinsic free energy of hydrolysis to one of low intrinsic free energy of hydrolysis because in this conformation it is in rapid equilibrium with inorganic phosphate[318]

$$E_2\text{-}PO_3^{2-} + H_2O \rightleftharpoons E_2 + HOPO_3^{2-} + H^+ \qquad (6\text{–}60)$$

This is analogous to what occurs on the active site of myosin where MgADP and $HOPO_3^{2-}$ are in equilibrium with MgATP, which was an unexpected observation. Here an acyl phosphate is in equilibrium with $H_2O$ and $HOPO_3^{2-}$, also contrary to expectation. Again the structure of the active site in this conformation has dramatically shifted the value of an equilibrium constant relative to what it would be in free solution.

In the $E_2$-P conformation, the compartment for cations is open to the extracytoplasmic surface and the three $Na^+$ eventually are exchanged for two $K^+$. Coincident with the closing of the compartment to occlude the two $K^+$, the acyl phosphate is hydrolyzed to regenerate the empty active site.[315,319] This step can be followed by observing the effect of the concentration of $K^+$ on the rate of hydrolysis of the aspartyl phosphate[320] formed from [γ-$^{32}$P]MgATP in the presence of $Na^+$. Because the aspartyl phosphate in the $E_2$-P conformation hydrolyzes rapidly upon addition of $K^+$ while the aspartyl phosphate in the $Na_3{\cdot}E_1{\sim}P$ conformation rapidly transfers its phosphate to MgADP, the relative amounts of these two conformations can be determined under any set of circumstances.[315]

The steps in the process at which the standard free energy changes that represent the standard free energy of hydrolysis of MgATP are manifested, as in the case with actomyosin, are not the steps in which the covalent bonds are made and broken, which are in the transitions between $E_1{\cdot}MgATP$ and $(Na^+)_3{\cdot}E_1{\sim}PO_3^{2-}{\cdot}MgADP$ and $E_2$-P and $E_2(K^+)_2$, but steps in which only conformational changes of the protein occur, which are in the transitions between $(Na^+)_3{\cdot}E_1{\sim}PO_3^{2-}$ and $E_2\text{-}PO_3^{2-}$ and between $E_2(K^+)_2$ and $E_1{\cdot}MgADP$.

The Michaelis constants, and presumably the dissociation constants, for $Na^+$ or $K^+$ entering the enzyme as reactants from the cytoplasmic or the extracytoplasmic surface, respectively, are significantly smaller than the dissociation constants for the $Na^+$ and $K^+$ leaving the enzyme as products from the extracytoplasmic or the cytoplasmic surface, respectively.[272,303] Although this is not necessary for active transport to occur by this mechanism (Figure 6–33), in this way the rates of the exchange of two $K^+$ for three $Na^+$ in the $E_1$ conformation and three $Na^+$ for two $K^+$ in the $E_2$-P conformation are accelerated. The time between the opening to one side of the membrane of the compartment loaded with the proper number of one type of cation, the cationic products, and the closing of the compartment around the proper number of the other type of cation, the cationic reactants, can be viewed as the time it takes for the cationic products to vacate and the cationic reactants to occupy the compartment. As these exchanges proceed by random collisions of cations with the open compartment, the probability that the cationic reactants will be bound in the correct number and in the absence of bound cationic products increases as the affinity of the site for the cationic products decreases and the affinity of the site for the cationic reactants increases. This is because of the fact that, under normal circumstances, the concentrations of the cationic products are always greater than the concentrations of cationic reactants on the respective sites of the membrane because the cations are being transported against their gradients of concentration. For example, at the cytoplasmic side, $K^+$ is the cationic product and $Na^+$ is the cationic reactant but the concentration of $K^+$ is significantly greater than the concentration of $Na^+$.

The entire process of active transport summarized in Equation 6–58 is reversible, as are all enzymatic reactions. If sufficient concentration gradients of $Na^+$ and $K^+$ are created across a cell and the concentrations of MgADP and $HOPO_3^{2-}$ in the cytoplasm are high and the concentration of MgATP is low, $Na^+/K^+$-transporting ATPase will synthesize MgATP[321] from MgADP and $HOPO_3^{2-}$. The rapid reversibility of the two halves of the mechanism, that in which $Na^+$ is transported and that in which $K^+$ is transported, can be demonstrated even more easily. The enzyme in an intact cell in the absence of $K^+$ catalyzes a rapid three-for-three exchange of cytoplasmic [$^{24}$Na]$Na^+$ for extracytoplasmic [$^{23}$Na]$Na^+$ that requires the presence of both MgATP[322] and MgADP[323] as expected if the mechanism is as described (Figure 6–33). It also catalyzes, in the absence of $Na^+$, a rapid two-for-two exchange of cytoplasmic [$^{42}$K]$K^+$ for extracytoplasmic [$^{39}$K]$K^+$ that requires the presence of $HOPO_3^{2-}$ as expected (Figure 6–33) but also the presence of MgATP.[324] The latter requirement results from the fact that the binding of MgATP at the active site accelerates the rate-limiting step in the $K^+/K^+$ exchange, that of opening the compartment to the cytoplasm to release the occluded $K^+$.

This description of the mechanism of $Na^+/K^+$-transporting ATPase can be extended to the other members of the family by simply changing the identity of the actions. The mechanism explains the process of active transport by coupling specific events that take place at the active site where the MgATP is hydrolyzed to distinct, but also specific, events that take place in the compartment through which the cations pass into and out of the cytoplasm. This compartment is identified by its ability to occlude the cations. The distance between this compartment and the active site is 3.5 nm or more,[325,326] and the only reasonable explanation for this rigid coupling of events is through extensive global conformational changes in the enzyme.

Because occluded forms exist, the actual transport of the cations involves the closing of the compartment through which they pass at one side of the membrane, after they are bound, before it opens to the other side of the membrane. Because the barrier at the closed end of an open compartment is at that instant the barrier between the cytoplasmic and extracytoplasmic spaces, it follows that, as they are transported, it is not the cations that move but the barrier that moves. One of the most striking observations consistent with this picture of the events is that, in endoplasmic reticular $Ca^{2+}$-transporting ATPase, the first $Ca^{2+}$ to enter the compartment from the cytoplasm is the first $Ca^{2+}$ to leave the compartment to the extracellular space[327] as if the two $Ca^{2+}$ remained in fixed positions as the cytoplasmic barrier closed and then the extracytoplasmic barrier opened. If this is an accurate picture of the mechanism by which the cations are transported through one of these enzymes, then it follows that any amino acids that are within the compartment will also formally move across the electrical barrier when the compartment closes at one end and then opens at the other. By following the effects of voltage on the transition between the $E_1$~P and the $E_2$-P conformations[328] and between the $E_2$ and the $E_1$ conformations[329] in $Na^+/K^+$-transporting ATPase, it has been concluded that no movement of charge occurs in the latter. If this is the case, the compartment must bear no net charge when it is occupied by two $K^+$ in the $E_2$ conformation. It necessarily follows that there are carboxylates within the compartment for these cations. These results and the fact that binding sites for $Ca^{2+}$ in proteins usually contain carboxylates have led to the assumption that the compartments for cations in these enzymes will contain glutamates or aspartates or both.[330]

The distance between the compartment for cations and the active site[325,326] is at least 3.5 nm and the diameter of the cytoplasmic globular portion of one of these enzymes is about 6.0 nm.[292,331–333] These dimensions and one's intuition suggest that, in any one of these enzymes, the active site is within the cytoplasmic globular region and the compartment for the cations is in the middle of the membrane-spanning region of the folded catalytic polypeptide and that it is the global conformational changes that couple the events at these distant locations.

## *Suggested Reading*

Steinberg, M., & Karlish, S.J.D. (1988) Studies on the Conformational Changes in Na, K-ATPase Labeled with 5-Iodoacetamidofluorescein, *J. Biol. Chem. 264*, 2726–2734.

## *References*


1. Baldwin, J., & Chothia, C. (1979) *J. Mol. Biol. 129*, 175–220.

1a. Kyte, J. (1995) *Structure in Protein Chemistry*, p 332, Garland Publishing, New York.

2. Krause, K.L., Volz, K.W., & Lipscomb, W.N. (1985) *Proc. Natl. Acad. Sci. U.S.A. 82*, 1643–1647.

3. Alber, T.C., Davenport, R.C., Giammona, D.A., Lolis, E., Petsko, G.A., & Ringe, D. (1987) *Cold Spring Harbor Symp. Quant. Biol. 52*, 603–613.

4. Belasco, J.G., Herlihy, J.M., & Knowles, J.R. (1978) *Biochemistry 17*, 2971–2978.

5. Pompliano, D.L., Peyman, A., & Knowles, J.R. (1990) *Biochemistry 29*, 3186–3194.

6. Remington, S.R., Wiegand, G., & Huber, R. (1982) *J. Mol. Biol. 158*, 111–152.

7. Kirschner, M.W., & Schachman, H.K. (1971) *Biochemistry 10*, 1900–1919.

8. Oshima, Y., Matsuura, M., & Horiuchi, T. (1972) *Biochem. Biophys. Res. Commun. 47*, 1444–1450.

9. Mathias, M.M., & Kemp, R.G. (1972) *Biochemistry 11*, 578–584.

10. Moffet, F.J., Wang. T., & Bridger, W.A. (1972) *J. Biol. Chem. 247*, 8139–8144.

10a. Kyte, J. (1995) *Structure in Protein Chemistry*, p 513, Garland Publishing, New York.

11. Griffiths, J.R., Price, N.C., & Radda, G.K. (1974) *Biochim. Biophys. Acta 358*, 275–280.

12. Perutz, M.F., Ladner, J.E., Simon, S.R., & Ho, C. (1974) *Biochemistry 13*, 2163–2173.

13. Ray, J., & Englander, W. (1986) *Biochemistry 25*, 3000–3007.

14. Rypniewski, W.R., & Evans, P.R. (1989) *J. Mol. Biol. 207*, 805–821.

15. Weber, G. (1972) *Biochemistry 11*, 864–878.

16. Gerhart, J.C., & Pardee, A.B. (1962) *J. Biol. Chem. 237*, 891–896.

17. Gerhart, J.C., & Holoubek, H. (1967) *J. Biol. Chem. 242*, 2886–2892.

18. Changeux, J., Gerhart, J.C., & Schachman, H.K. (1968) *Biochemistry 7*, 531–538.

19. Suter, P., & Rosenbusch, J.P. (1977) *J. Biol. Chem. 252*, 8136–8141.

20. Honzatko, R.B., Crawford, J.L., Monaco, H.L., Ladner, J.E., Edwards, B.F.P., Evans, D.R., Warren, S.G., Wiley, D.C., Ladner, R.C., & Lipscomb, W.N. (1982) *J. Mol. Biol. 160*, 219–263.

21. Honzatko, R.B., Monaco, H.L., & Lipscomb, W.N. (1979) *Proc. Natl. Acad. Sci. U.S.A. 76*, 5105–5109.

22. Ke, H., Lipscomb, W.N., Cho, Y., & Honzatko, R.B, (1988) *J. Mol. Biol. 204*, 725–747.

23. Monod, J., Wyman, J., & Changeux, J. (1965) *J. Mol. Biol. 12*, 88–118.

24. Cross, D.G., & Fisher, H.F. (1970) *J. Biol. Chem. 245*, 2612–2621.

25. Kerbiriou, D., & Hervé, G. (1972) *J. Mol. Biol. 64*, 379–392.

26. Beck, W.S. (1967) *J. Biol. Chem. 242*, 3148–3158.

27. Panagou, D., Orr, M.D., & Blakely, R.L. (1972) *Biochemistry 11*, 2378–2388.

28. Helmreich, E., & Cori, C.F. (1964) *Proc. Natl. Acad. Sci. U.S.A. 51*, 131–138.

29. Sprang, S., & Fletterick, R.J. (1979) *J. Mol. Biol. 131*, 523–551.

30. Wang, J.H., & Tu, J. (1969) *Biochemistry 8*, 4403–4410.

31. Wang, J.H., & Tu, J.I. (1970) *J. Biol. Chem. 245*, 176–182.

32. Tu, J., & Graves, D.J. (1973) *J. Biol. Chem. 248*, 4617–4622.

33. Oberfelder, R.W., Barisas, B.G., & Lee, J.C. (1984) *Biochemistry 23*, 3822–3826.

34. Oberfelder, R.W., Lee, L.L., & Lee, J.C. (1984) *Biochemistry 23*, 3813–3821.

35. Thorner, J., & Paulus, H. (1973) *J. Biol. Chem. 248*, 3922–3932.

36. Maeba, P., & Sanwal, B.D. (1969) *J. Biol. Chem. 244*, 2549–2557.

37. Spina, J., Bright, H.J., & Rosenbloom, J. (1970) *Biochemistry 9*, 3794–3801.

38. Brown, N.C., & Reichard, P. (1969) *J. Mol. Biol. 46*, 39–55.

39. Lorek, A., Wilson, K.S., Sansom, M.S.P., Stuart, D.I., Stura, E.A., Jenkins, J.A., Zanotti, G., Hajdu, J., & Johnson, L.N. (1984) *Biochem. J. 218*, 45–60.

40. Cori, C.F., & Cori, G.T. (1936) *Proc. Soc. Exp. Biol. Med. 34*, 702–705.
41. Fletterick, R.J., & Madsen, N.B. (1980) *Annu. Rev. Biochem. 49*, 31–61.
42. Stalmans, W., Laloux, M., & Hers, H. (1974) *Eur. J. Biochem. 49*, 415–427.
43. Sprang, S., Fletterick, R., Stern, M., Yang, D., Madsen, N., & Sturtevant, J. (1982) *Biochemistry 21*, 2036–2048.
44. Kasvinsky, P.J., Schechosky, S., & Fletterick, R.J. (1978) *J. Biol. Chem. 253*, 9102–9106.
45. Kasvinsky, P.J., Madsen, N.B., Fletterick, R.J., & Sygush, J. (1978) *J. Biol. Chem. 253*, 1290–1296.
46. Véron, M., Saari, J.C., Villar-Palasi, C., & Cohen, G.N. (1973) *Eur. J. Biochem. 38*, 325–335.
47. Stadtman, E.R., Cohen, G.N., LeBras, G., & deRobichon-Szulmajster, H. (1961) *J. Biol. Chem. 236*, 2033–2038.
48. Patte, J., leBras, G., Loviny, T., & Cohen, G.N. (1963) *Biochim. Biophys. Acta 67*, 16–30.
49. Broussard, L., Cunningham, G.N., Starnes, W.L., & Shive, W. (1972) *Biochem. Biophys. Res. Commun. 46*, 1181–1186.
50. Smith, L.C., Ravel, J.M., Lax, S.R., & Shive, W. (1962) *J. Biol. Chem. 237*, 3566–3570.
51. Koike, K., & Feigelson, P. (1971) *Biochemistry 10*, 3385–3390.
52. Koike, K., Poillon, W.N., & Feigelson, P. (1969) *J. Biol. Chem. 244*, 3457–3462.
53. Schirch, L., & Quashnock, J. (1981) *J. Biol. Chem. 256*, 6245–6249.
54. Courtice, F.C., & Douglas, C.G. (1947) *J. Physiol. 105*, 345–356.
55. Hill, A.V. (1910) *J. Physiol. 40*, iv-vii.
56. Imai, K. (1979) *J. Mol. Biol. 133*, 233–247.
57. Adair, G.S. (1925) *J. Biol. Chem. 63*, 529.
58. Edsall, J.T., Felsenfeld, G., Goodman, D.S., & Gurd, F.R.N. (1954) *J. Am. Chem. Soc. 76*, 3054–3061.
59. Saroff, H.A., & Minton, A.P. (1972) *Science 175*, 1253–1255.
60. Fletterick, R.J., Sygusch, J., Murray, N., Madsen, N.B., & Johnson, L.N. (1976) *J. Mol. Biol. 103*, 1–13.
61. Goldsmith, E.J., Sprang, S.R., Hamlin, R., Xuong, N., & Fletterick, R.J. (1989) *Science 245*, 528–532.
62. Madsen, N.B., Kasvinsky, P.J., & Fletterick, R.J. (1978) *J. Biol. Chem. 253*, 9097–9101.
63. Evans, P.R., Farrants, G.W., & Lawrence, M.C. (1986) *J. Mol. Biol. 191*, 713–720.
64. Evans, P.R., & Hudson, P.J. (1979) *Nature 279*, 500–504.
65. TenEyck, L.F., & Arnone, A. (1976) *J. Mol. Biol. 100*, 3–11.
66. Fermi, G. (1975) *J. Mol. Biol. 97*, 237–256.
67. Haurowitz, F. (1938) *Hoppe-Seyler's Z. Physiol. Chem. 254*, 266–274.
68. Shaanan, B. (1983) *J. Mol. Biol. 171*, 31–59.
69. Kirschner, M.W., & Schachman, H.K. (1973) *Biochemistry 12*, 2987–2997.
70. Kirschner, M.W., & Schachman, H.K. (1973) *Biochemistry 12*, 2997–3004.
71. Koshland, D.E., Némethy, G., & Filmer, D. (1966) *Biochemistry 5*, 365–385.
72. Gerhardt, J.C., & Schachman, H. (1968) *Biochemistry 7*, 538–552.
73. Howlett, G.J., & Schachman, H.K. (1977) *Biochemistry 16*, 5077–5083.
73a. Kyte, J. (1995) *Structure in Protein Chemistry*, p 365, Garland Publishing, New York.
74. Yang, Y., & Schachman, H.K. (1980) *Proc. Natl. Acad. Sci. U.S.A. 77*, 5187–5191.
75. Wang, C., Yang, Y.R., Hu, C., & Schachman, H.K. (1981) *J. Biol. Chem. 256*, 7028–7034.
76. Johnson, R.S., & Schachman, H.K. (1980) *Proc. Natl. Acad. Sci. U.S.A. 77*, 1995–1999.
77. Johnson, R.S., & Schachman, H.K. (1983) *J. Biol. Chem. 258*, 3528–3538.
78. Howlett, G.J., Blackburn, M.N., Compton, J.G., & Schachman, H.K. (1977) *Biochemistry 16*, 5091–5099.
79. Foote, J., & Lipscomb, W.N. (1981) *J. Biol. Chem. 256*, 11428–11433.
80. Foote, J., & Schachman, H.K. (1985) *J. Mol. Biol. 186*, 175–184.
81. Hensley, P., & Schachman, H.K. (1979) *Proc. Natl. Acad. Sci. U.S.A. 76*, 3732–3736.
82. Eisenstein, E., Markby, D.M., & Schachman, H.K. (1990) *Biochemistry 29*, 3724–3731.
83. Calhoun, D.H., Rimerman, R.A., & Hatfield, G.W. (1973) *J. Biol. Chem. 248*, 3511–3516.
84. Tomozawa, Y., & Wolfenden, R. (1970) *Biochemistry 9*, 3400–3404.
85. Campbell, I.D., Dwek, R.A., Price, N.C., & Radda, G.K. (1972) *Eur. J. Biochem. 30*, 339–347.
86. Antonini, E., Wyman, J., Brunori, M., Bucci, E., Fronticelli, C., & Rossi-Fanelli, A. (1963) *J. Biol. Chem. 238*, 2950–2957.
87. Kilmartin, J.V., & Rossi-Bernardi, L. (1969) *Nature 222*, 1243–1246.
88. Garner, M.H., Bogardt, R.A., & Gurd, F.R.N. (1975) *J. Biol. Chem. 250*, 4398–4404.
89. Russu, I.M., Ho, N.T., & Ho, C. (1982) *Biochemistry 21*, 3031–3043.
90. Benesch, R., Benesch, R.E., & Yu, C.I. (1968) *Proc. Natl. Acad. Sci. U.S.A. 59*, 526–532.
91. Arnone, A. (1972) *Nature 237*, 146–149.
92. Hoffman, B.M., Gibson, Q.H., Bull, C., Crepeau, R.H., Edelstein, S.J., Fisher, R.G., & McDonald, M.J. (1975) *Ann. N.Y. Acad. Sci. 244*, 174–186.
93. Blough, N.V., Zemel, H., Hoffman, B.M., Lee, T.C.K., & Gibson, Q.H. (1980) *J. Am. Chem. Soc. 102*, 5685–5686.
94. Andersen, M.E., Moffat, J.K., & Gibson, Q.H. (1971) *J. Biol. Chem. 246*, 2796–2807.
95. Gibson, Q.H. (1982) in *Hemoglobin and Oxygen Binding* (Ho, C., Ed.) pp 321–327, Elsevier, New York.
96. Marden, M.C., Hazard, E.S., & Gibson, Q.H. (1986) *Biochemistry 25*, 7591–7596.
97. Gibson, Q.H. (1970) *J. Biol. Chem. 245*, 3285–3288.
98. Sawicki, C.A., & Gibson, Q.H. (1977) *J. Biol. Chem. 252*, 5783–5788.
99. Gibson, Q.H., & Parkhurst, L.J. (1968) *J. Biol. Chem. 243*, 5521–5524.
100. Gibson, Q.H. (1956) *J. Physiol. 134*, 123–134.
101. Hopfield, J.J., Shulman, R.G., & Ogawa, S. (1971) *J. Mol. Biol. 61*, 425–443.
102. Sawicki, C.A., & Gibson, Q.H. (1978) *Biophys. J. 24*, 21–33.
103. Sawicki, C.A., & Gibson, Q.H. (1977) *J. Biol. Chem. 252*, 7538–7547.
104. Hayashi, A., & Stamatoyannopoulos, G. (1972) *Nature New Biol. 235*, 70–72.
105. Olson, J.S., & Gibson, Q.H. (1972) *J. Biol. Chem. 247*, 3662–3670.
106. Arnone, A., Gacon, G., & Wajcman, H. (1976) *J. Biol. Chem. 251*, 5875–5880.
107. Imai, K., Hamilton, H.B., Miyaji, T., & Shibata, S. (1972) *Biochemistry 11*, 114–121.
108. Smiley, K.L., & Suelter, C.H. (1967) *J. Biol. Chem. 242*, 1980–1981.
109. Gerhart, J.C., & Pardee, A.B. (1963) *Cold Spring Harbor Symp. Quant. Biol. 28*, 491–496.

110. Noble, R.W. (1969) *J. Mol. Biol. 39*, 479–491.
111. Imai, K. (1982) *Allosteric Effects in Haemoglobin*, p 113, Cambridge University Press, Cambridge.
111a. Tyuma, I., Shimizu, K., & Imai, K. (1971) *Biochem. Biophys. Res. Commun. 43*, 423–428.
112. Enns, C.A., & Chan, W.W. (1979) *J. Biol. Chem. 254*, 6180–6186.
113. Karmali, A., Drake, A.F., & Spencer, N. (1983) *Biochem. J. 211*, 617–623.
114. Guidotti, G. (1967) *J. Biol. Chem. 242*, 3685–3693.
115. Bunn, H.F., & McDonough, M. (1974) *Biochemistry 13*, 988–993.
116. Mills, F.C., Johnson, M.L., & Ackers, G.K. (1976) *Biochemistry 15*, 5350–5362.
117. Ackers, G.K. (1980) *Biophys. J. 32*, 331–346.
118. Mills, F.C., & Ackers, G.K. (1979) *Proc. Natl. Acad. Sci. U.S.A. 76*, 273–277.
119. Yang, Y.R., Syvanen, J.M., Nagel, G.M., & Schachman, H.K. (1974) *Proc. Natl. Acad. Sci. U.S.A. 71*, 918–922.
120. Subramani, S., Bothwell, M.A., Gibbons, I., Yang, Y.R., & Schachman, H.K. (1977) *Proc. Natl. Acad. Sci. U.S.A. 74*, 3777–3781.
121. Wood, A.W., & Seegmiller, J.E. (1973) *J. Biol. Chem. 248*, 138–143.
122. Holmes, E.W., Wyngaarden, J.B., & Kelley, W.N. (1973) *J. Biol. Chem. 248*, 6035–6040.
123. Kohlhaw, G., Leary, T.R., & Umbarger, H.E. (1969) *J. Biol. Chem. 244*, 2218–2225.
124. Leary, T.R., & Kohlhaw, G.B. (1972) *J. Biol. Chem. 247*, 1089–1095.
125. Leary, T.R., & Kohlhaw, G. (1970) *Biochem. Biophys. Res. Commun. 39*, 494–501.
126. Funkhouser, J.D., Abraham, A., Smith, V.A., & Smith, W.G. (1974) *J. Biol. Chem. 249*, 5478–5484.
127. Beatty, N.B., & Lane, M.D. (1983) *J. Biol. Chem. 258*, 13043–13050.
128. Sedgwick, K.A., & Frieden, C. (1968) *Biochem. Biophys. Res. Commun. 32*, 392–397.
129. Frieden, C., & Colman, R.F. (1967) *J. Biol. Chem. 242*, 1705–1715.
130. Arnold, H., & Maier, K.P. (1971) *Biochim. Biophys. Acta 251*, 133–140.
131. Eckfeldt, J., Hammes, G.G., Mohr, S.C., & Wu, C. (1970) *Biochemistry 9*, 3353–3362.
132. Ramaiah, A., & Tejwani, G.A. (1973) *Eur. J. Biochem. 39*, 183–192.
133. Singer, S.C., & Holmes, E.W. (1977) *J. Biol. Chem. 252*, 7959–7963.
134. Krause, K.L., Volz, K.W., & Lipscomb, W.N. (1987) *J. Mol. Biol. 193*, 527–553.
135. Baldwin, J.M. (1980) *J. Mol. Biol. 136*, 103–128.
136. Lesk, A.M., Janin, J., Wodak, S., & Chothia, C. (1985) *J. Mol. Biol. 183*, 267–270.
137. Kim, K.H., Pan, Z., Honzatko, R.B., Ke, H., & Lipscomb, W.N. (1987) *J. Mol. Biol. 196*, 853–875.
138. Ladjimi, M.M., & Kantrowitz, E.R. (1988) *Biochemistry 27*, 276–283.
139. Stura, E.A., Zanotti, G., Babu, Y.S., Sansom, M.S.P., Stuart, D.I., Wilson, K.S., Johnson, L.N., & Van de Werve, G. (1983) *J. Mol. Biol. 170*, 529–565.
140. Lin, Y.M., Liu, Y.P., & Cheung, W.Y. (1974) *J. Biol. Chem. 249*, 4943–4954.
141. Dizhoor, A.M., Ray, S., Kumar, S., Niemi, G., Spencer, M., Brolley, D., Walsh, K.A., Philipov, P.P., Hurley, J.B., & Stryer, L. (1991) *Science 251*, 915–918.
142. Jarrett, H.W., & Kyte, J. (1979) *J. Biol. Chem. 254*, 8237–8244.
143. Burger, D., Cox, J.A., Comte, M., & Stein, E.A. (1984) *Biochemistry 23*, 1966–1971.
144. Babu, Y.S., Bugg, C.E., & Cook, W.J. (1988) *J. Mol. Biol. 204*, 191–204.
145. Babu, Y.S., Sack, J.S., Greenhough, T.J., Bugg, C.E., Means, A.R., & Cook, W.J. (1985) *Nature 315*, 37–40.
146. Kretsinger, R.H., & Nockolds, C.E. (1973) *J. Biol. Chem. 248*, 3313–3326.
147. Moews, P.C., & Kretsinger, R.H. (1975) *J. Mol. Biol. 91*, 201–228.
148. Declercq, J., Tinant, B., Parello, J., Etienne, G., & Huber, R. (1988) *J. Mol. Biol. 202*, 349–353.
149. Kuo, I.C., & Coffee, C.J. (1976) *J. Biol. Chem. 251*, 6315–6319.
150. Dedman, J.R., Potter, J.D., Jackson, R.L., Johnson, J.D., & Means, A.R. (1977) *J. Biol. Chem. 252*, 8415–8422.
151. Hennessey, J.P., Manavalan, P., Johnson, W.C., Malencik, D.A., Anderson, S.R., Schimerlik, M.I., & Shaltin, Y. (1987) *Biopolymers 26*, 561–571.
152. Heidorn, D.B., & Trewhella, J. (1988) *Biochemistry 27*, 909–915.
153. Seaton, B.A., Head, J.F., Engelman, D.M., & Richards, F.M. (1985) *Biochemistry 24*, 6470–6743.
154. Blumenthal, D.K., & Stull, J.T. (1980) *Biochemistry 19*, 5608–5614.
155. Cox, J.A., Malnoe, A., & Stein, E.A. (1981) *J. Biol. Chem. 256*, 3218–3222.
156. Malnoe, A., Cox, J.A., & Stein, E.A. (1982) *Biochim. Biophys. Acta 714*, 84–92.
157. Cox, J.A., Comte, M., & Stein, E.A. (1982) *Proc. Natl. Acad. Sci. U.S.A. 79*, 4265–4269.
158. Meador, W.E., Means, A.R., & Quiocho, F.A. (1992) *Science 257*, 1251–1255.
159. Lambrecht, H., & Koch, K. (1991) *EMBO J. 10*, 793–798.
160. Fischer, E.H., Graves, D.J., Crittenden, E.R.S., & Krebs, E.G. (1959) *J. Biol. Chem. 234*, 1698–1704.
161. Sefton, B.M., Hunter, T., Beemon, K., & Eckhart, W. (1980) *Cell 20*, 807–816.
162. Walsh, D.A., Perkins, J.P., & Krebs, E.G. (1968) *J. Biol. Chem. 243*, 3763–3774.
163. Dabrowska, R., Sherry, J.M.F., Aromatorio, D.K., & Hartshorne, D.J. (1978) *Biochemistry 17*, 253–258.
164. Carpenter, G., King, L., & Cohen, S. (1978) *Nature 276*, 409–410.
165. Kemp, B.E., Benjamini, E., & Krebs, E.G. (1976) *Proc. Natl. Acad. Sci. U.S.A. 73*, 1038–1042.
166. Kemp, B.E., Pearson, R.B., & House, C. (1984) *J. Biol. Chem. 257*, 13349–13353.
167. Kemp, B.E., & Pearson, R.B. (1991) *Biochim. Biophys. Acta 1094*, 67–76.
168. Pearson, R.B., Wettenhall, R.E.H., Means, A.R., Hartshorne, D.J., & Kemp, B.E. (1988) *Science 241*, 970–973.
169. Adelstein, R.S., & Klee, C.B. (1981) *J. Biol. Chem. 256*, 7501–7509.
170. Weldon, S.L., & Taylor, S.S. (1985) *J. Biol. Chem. 260*, 4203–4209.
171. Kemp, B.E., Graves, D.J., Benjamini, E., & Krebs, E.G. (1977) *J. Biol. Chem. 252*, 4888–4894.
172. Pearson, R.B., & Kemp, B.E. (1991) *Methods Enzymol. 200*, 62–81.
173. Pearson, R.B., Jakes, R., John, M., Kendrick-Jones, J., & Kemp, B.E. (1984) *FEBS Lett. 168*, 108–112.
174. Maita, T., Chen, J., & Matsuda, G. (1981) *Eur. J. Biochem. 117*, 417–424.
175. Blumenthal, D.K., Takio, K., Edelman, A.M., Charbonneau, H., Titani, K., Walsh, K.A., & Krebs, E.G. (1985) *Proc. Natl. Acad. Sci. U.S.A. 82*, 3187–3191.

176. Takio, K., Blumenthal, D.K., Walsh, K.A., Titani, K., & Krebs, E.G. (1986) *Biochemistry 25*, 8049–8057.
177. Herring, B.P., Stull, J.T., & Gallagher, P.J. (1990) *J. Biol. Chem. 265*, 1724–1730.
178. Kemp, B.E., Pearson, R.B., Guerriero, V., Bagchi, I.C., & Means, A.R. (1987) *J. Biol. Chem. 262*, 2542–2548.
179. Lukas, T.J., Burgess, W.H., Prendergast, F.G., Lau, W., & Watterson, D.M. (1986) *Biochemistry 25*, 1458–1464.
180. van Berkum, M.F.A., & Means, A.R. (1991) *J. Biol. Chem. 266*, 21488–21495.
181. Olson, N.J., Pearson, R.B., Needleman, D.S., Hurwitz, M.Y., Kemp, B.E., & Means, A.R. (1990) *Proc. Natl. Acad. Sci. U.S.A. 87*, 2284–2288.
182. Shoji, S., Ericsson, L.H., Walsh, K.A., Fischer, E.H., & Titani, K. (1983) *Biochemistry 22*, 3702–3709.
183. Takio, K., Smith, S.B., Krebs, E.G., Walsh, K.A., & Titani, K. (1984) *Biochemistry 23*, 4200–4206.
184. Rannels, S.R., & Corbin, J.D. (1981) *J. Biol. Chem. 256*, 7871–7876.
185. Potter, R.L., & Taylor, S.S. (1979) *J. Biol. Chem. 254*, 9000–9005.
186. Tao, M., Salas, M.L., & Lipmann, F. (1970) *Proc. Natl. Acad. Sci. U.S.A. 67*, 408–414.
187. Kumon, A., Yamamura, H., & Nishizuka, Y. (1970) *Biochem. Biophys. Res. Commun. 41*, 1290–1297.
188. Picton, C., Aitken, A., Bilham, T., & Cohen, P. (1982) *Eur. J. Biochem. 124*, 37–45.
189. Cohen, P., Yellowlees, D., Aitken, A., Donella-Deana, A., Hemmings, B.A., & Parker, P.J. (1982) *Eur. J. Biochem. 124*, 21–35.
190. Lent, B.A., Lee, K., & Kim, K. (1978) *J. Biol. Chem. 253*, 8149–8156.
191. Hardie, D.G., & Guy, P.S. (1980) *Eur. J. Biochem. 110*, 167–177.
192. Doskeland, A.P., Schworer, C.M., Doskeland, S.O., Chrisman, T.D., Soderling, T.R., Corbin, J.D., & Flatmark, T. (1984) *Eur. J. Biochem. 145*, 31–37.
193. Donlon, J., & Kaufman, S. (1978) *J. Biol. Chem. 253*, 6657–6659.
194. Rosell-Perez, M., Villar-Palasi, C., & Larner, J. (1962) *Biochemistry 1*, 763–768.
195. Friedman, D.L., & Larner, J. (1963) *Biochemistry 2*, 669–675.
196. Itarte, E., Robinson, J.C., & Huang, K. (1977) *J. Biol. Chem. 252*, 1231–1234.
197. Cori, G.T., & Green, A.A. (1943) *J. Biol. Chem. 151*, 31–38.
198. Seery, V.L., Fischer, E.H., & Teller, D.C. (1967) *Biochemistry 6*, 3315–3327.
199. Sprang, S.R., Acharya, K.R., Goldsmith, E.J., Stuart, D.I., Varvill, K., Fletterick, R.J., Madsen, N.B., & Johnson, L.N. (1988) *Nature 336*, 215–221.
200. Gilman, A.G. (1987) *Annu. Rev. Biochem. 56*, 615–649.
201. Cassel, D., & Selinger, Z. (1977) *Proc. Natl. Acad. Sci. U.S.A. 74*, 3307–3311.
202. Brandt, D.R., & Ross, E.M. (1985) *J. Biol. Chem. 260*, 266–272.
203. Higashijima, T., Ferguson, K.M., Smigel, M.D., & Gilman, A.G. (1987) *J. Biol. Chem. 262*, 757–761.
204. Dixon, R.A.F., Kobilka, B.K., Strader, D.J., Benovic, J.L., Dohlman, H.G., Frielle, T., Bolanowski, M.A., Bennett, C.D., Rands, E., Diehl, R.E., Mumford, R.A., Slater, E.E., Sigal, I.S., Caron, M.G., Lefkowitz, R.J., & Strader, C.D. (1986) *Nature 321*, 75–79.
205. Nathans, J., & Hogness, D.S. (1984) *Proc. Natl. Acad. Sci. U.S.A. 81*, 4851–4855.
206. Brandt, D.R., & Ross, E.M. (1986) *J. Biol. Chem. 261*, 1656–1664.
207. Tolkovsky, A.M., & Levitski, A. (1978) *Biochemistry 17*, 3795–3810.
208. Pedersen, S.E., & Ross, E.M. (1982) *Proc. Natl. Acad. Sci. U.S.A. 79*, 7228–7232.
209. Higashijima, T., Ferguson, K.M., Sternweis, P.C., Ross, E.M., Smigel, M.D., & Gilman, A.G. (1987) *J. Biol. Chem. 262*, 752–756.
210. Lefkowitz, R.J. (1974) *J. Biol. Chem. 249*, 6119–6124.
211. Pfeuffer, T., & Helmreich, E.J.M. (1975) *J. Biol. Chem. 250*, 867–876.
212. Harwood, J.P., Low, H., & Rodbell, M. (1973) *J. Biol. Chem. 248*, 6239–6245.
213. Uchida, S., Wheeler, G.L., Yamazaki, A., & Bitensky, M.W. (1981) *J. Cyclic Nucleotide Res. 7*, 95–104.
214. Fung, B.K., Hurley, J.B., & Stryer, L. (1981) *Proc. Natl. Acad. Sci. U.S.A. 78*, 152–156.
215. Litosch, I., Wallis, C., & Fain, J.N. (1985) *J. Biol. Chem. 260*, 5464–5471.
215a. Abrahams, J.P., Leslie, A.G.W., Lutter, R., & Walker, J.E. (1994) *Nature 370*, 621–628.
215b. Kyte, J. (1995) *Structure in Protein Chemistry*, p 496, Garland Publishing, New York.
215c. Kyte, J. (1995) *Structure in Protein Chemistry*, p 318, Garland Publishing, New York.
216. Huxley, A.F., & Niedergerke, R. (1954) *Nature 173*, 971–973.
217. Huxley, H., & Hanson, J. (1954) *Nature 173*, 973–977.
218. Ashton, F.T., Somlyo, A.V., & Somlyo, A.P. (1975) *J. Mol. Biol. 98*, 17–29.
219. Weber, K., & Groeschel-Stewart, U. (1974) *Proc. Natl. Acad. Sci. U.S.A. 71*, 4561–4564.
220. Tokuyasu, K.T., Dutton, A.H., Geiger, B., & Singer, S.J. (1981) *Proc. Natl. Acad. Sci. U.S.A. 78*, 7619–7623.
221. Spudich, J.A., Kron, S.J., & Sheetz, M.P. (1985) *Nature 315*, 584–587.
222. Levy, R.M., Umazume, Y., & Kushmerick, M.J. (1976) *Biochim. Biophys. Acta 430*, 352–365.
223. Huxley, H.E. (1969) *Science 164*, 1356–1366.
224. Reedy, M.K. (1968) *J. Mol. Biol. 31*, 155–176.
225. Marston, S.B., Tregear, R.T., Rodger, C.D., & Clarke, M.L. (1979) *J. Mol. Biol. 128*, 111–126.
226. Reedy, M.K. (1967) *Am. Zool. 7*, 465–481.
227. Lowey, S., Slayter, H.S., Weeds, A.G., & Baker, H. (1969) *J. Mol. Biol. 42*, 1–29.
228. Margossian, S., & Lowey, S. (1973) *J. Mol. Biol. 74*, 313–340.
229. Sleep, J.A. (1981) *Biochemistry 20*, 5043–5051.
230. Eisenberg, E., & Moos, C. (1968) *Biochemistry 7*, 1486–1489.
231. Eisenberg, E., & Moos, C. (1970) *J. Biol. Chem. 245*, 2451–2456.
232. Margossian, S.S., & Lowey, S. (1978) *Biochemistry 17*, 5431–5439.
233. Marston, S., & Weber, A. (1975) *Biochemistry 14*, 3868–3873.
234. Stein, L.A., Schwarz, R.P., Chock, P.B., & Eisenberg, E. (1979) *Biochemistry 18*, 3895–3909.
235. Sleep, J.A., & Hutton, R.L. (1978) *Biochemistry 17*, 5423–5430.
236. Lymn, R.W., & Taylor, E.W. (1971) *Biochemistry 10*, 4617–4624.
237. Cardon, J.W., & Boyer, P.D. (1978) *Eur. J. Biochem. 92*, 443–448.
238. Stein, L.A., Chock, P.B., & Eisenberg, E. (1981) *Proc. Natl. Acad. Sci. U.S.A. 78*, 1346–1350.
239. Lymn, R.W., & Taylor, E.W. (1970) *Biochemistry 9*, 2975–2983.
240. Bagshaw, C.R., & Trentham, D.R. (1973) *Biochem. J. 133*, 323–328.
241. Bagshaw, C.R., Trentham, D.R., Wolcott, R.G., & Boyer, P.D. (1975) *Proc. Natl. Acad. Sci. U.S.A. 72*, 2592–2596.
242. Taylor, E.W., Lymn, R.W., & Moll, G. (1970) *Biochemistry 9*, 2984–2991.
243. Trentham, D.R., Bardsley, R.G., Eccleston, J.F., & Weeds, A.G. (1972) *Biochem. J. 126*, 635–644.
244. Bagshaw, C.R., & Trentham, D.R. (1974) *Biochem. J. 141*, 331–349.

245. Maruyama, K., & Weber, A. (1972) *Biochemistry 11*, 2990–2998.
246. Marston, S.B., & Tregear, R.T. (1972) *Nature New Biol. 235*, 23–24.
247. O'Brien, E.J., Couch, J., Johnson, G.R.P., & Morris, E.P. (1983) in *Actin: Structure and Function in Muscle and Non-muscle Cells* (dos Remedios, C.G., & Barden, Eds.) pp 3–16, Academic Press, Sydney.
248. Wakabayashi, T., Huxley, H.E., Amos, L.A., & Klug, A. (1975) *J. Mol. Biol. 93*, 477–497.
249. Toyoshima, C., & Wakabayashi, T. (1985) *J. Biochem. (Tokyo) 97*, 245–263.
250. McLaghlin, A.D., & Stewart, M. (1975) *J. Mol. Biol. 98*, 293–304.
251. Haselgrove, J.C. (1972) *Cold Spring Harbor Symp. Quant. Biol. 37*, 341–352.
252. Huxley, H.E. (1972) *Cold Spring Harbor Symp. Quant. Biol. 37*, 361–376.
253. Ohtsuki, I., Masaki, T., Nonomura, Y., & Ebashi, S. (1967) *J. Biochem. (Tokyo) 61*, 817–819.
254. Spudich, J.A., Huxley, H.E., & Finch, J.T. (1972) *J. Mol. Biol. 72*, 619–632.
255. Ohtsuki, I. (1974) *J. Biochem. (Tokyo) 75*, 753–765.
256. Ebashi, S., Kodama, A., & Ebashi, F. (1968) *J. Biochem. (Tokyo) 64*, 465–477.
257. Greaser, M.L., & Gergely, J. (1971) *J. Biol. Chem. 246*, 4226–4233.
258. Collins, J.H., Potter, J.D., Horn, M.J., Wilshire, G., & Jackman, N. (1973) *FEBS Lett. 36*, 268–272.
259. Wilkinson, J.M., & Grand, R.J.A. (1975) *Biochem. J. 149*, 493–496.
260. Pearlstone, J.R., Carpenter, M.R., Johnson, P., & Smillie, L.B. (1976) *Proc. Natl. Acad. Sci. U.S.A. 73*, 1902–1906.
261. Greaser, M.L., & Gergely, J. (1973) *J. Biol. Chem. 248*, 2125–2133.
262. Herzberg, O., & James, M.N.G. (1985) *Nature 313*, 653–659.
263. Potter, J.D., & Gergely, J. (1975) *J. Biol. Chem. 250*, 4628–4633.
264. Potter, J.D., & Gergely, J. (1974) *Biochemistry 13*, 2697–2703.
265. Eisenberg, E., & Kielley, W.W. (1974) *J. Biol. Chem. 249*, 4742–4748.
266. Sobieszek, A., & Bremel, R.D. (1975) *Eur. J. Biochem. 55*, 49–60.
267. Adelstein, R.S., Conti, M.A., & Anderson, W. (1973) *Proc. Natl. Acad. Sci. U.S.A. 70*, 3115–3119.
268. Frearson, N., Focant, B.W.W., & Perry, S.V. (1976) *FEBS Lett. 63*, 27–32.
269. Sobieszek, A., & Small, J.V. (1977) *J. Mol. Biol. 112*, 559–576.
270. Chacko, S., Conti, M.A., & Adelstein, R.S. (1977) *Proc. Natl. Acad. Sci. U.S.A. 74*, 129–133.
271. deLanerolle, P., & Stull, J.T. (1980) *J. Biol. Chem. 255*, 9993–10000.
272. Skou, J.C. (1960) *Biochim. Biophys. Acta 42*, 6–23.
273. Whittam, R. (1962) *Biochem. J. 84*, 110–118.
274. Whittam, R., & Ager, M.E. (1964) *Biochem. J. 93*, 337–348.
275. Rabon, E.C., McFall, T.L., & Sachs, G. (1982) *J. Biol. Chem. 257*, 6296–6299.
276. Lytton, J., & MacLennan, D.H. (1988) *J. Biol. Chem. 263*, 15024–15031.
277. Hasselbach, W., & Makinose, M. (1963) *Biochem. Z. 339*, 94–111.
278. Chiesi, M., & Inesi, G. (1980) *Biochemistry 19*, 2912–2918.
279. Niggli, V., Adunyah, E.S., Penniston, J.T., & Carafoli, E. (1981) *J. Biol. Chem. 256*, 395–401.
280. Greeb, J., & Shull, G.E. (1989) *J. Biol. Chem. 264*, 18569–18576.
281. Schatzmann, H.J., & Vincenzi, F.F. (1969) *J. Physiol. 201*, 369–395.
282. Niggli, V., Sigel, E., & Carofoli, E. (1982) *J. Biol. Chem. 257*, 2350–2356.
283. Slayman, C.L., Long, W.S., & Lu, C.Y. (1973) *J. Membr. Biol. 14*, 305–338.
284. Scarborough, G.A. (1976) *Proc. Natl. Acad. Sci. U.S.A. 73*, 1485–1488.
285. Hodges, T.K., Leonard, R.T., Bracker, C.E., & Keenan, T.W. (1972) *Proc. Natl. Acad. Sci. U.S.A. 69*, 3307–3311.
286. Pardo, J.M., & Serrano, R. (1989) *J. Biol. Chem. 264*, 8557–8562.
287. Kyte, J. (1972) *J. Biol. Chem. 247*, 7642–7649.
288. Shull, G.E., Swartz, A., & Lingrel, J.B. (1985) *Nature 316*, 691–695.
289. Kawakami, K., Noguchi, S., Noda, M., Takahashi, H., Ohta, T., Kawamura, M., Nojima, H., Nagano, K., Hirose, T., Inayama, S., Hayashida, H., Miyata, T., & Numa, S. (1985) *Nature 316*, 733–736.
290. Shull, G.E., Lane, L.K., & Lingrel, J.B. (1986) *Nature 321*, 429–431.
291. Craig, W.S. (1982) *Biochemistry 21*, 5707–5717.
292. Zampighi, G., Kyte, J., & Freytag, W. (1984) *J. Cell Biol. 98*, 1851–1864.
293. Nicholas, R.A. (1984) *Biochemistry 23*, 888–898.
294. Shull, G.E., & Lingrel, J.B. (1986) *J. Biol. Chem. 261*, 16788–16791.
295. Shull, G.E., & Greeb, J. (1988) *J. Biol. Chem. 263*, 8646–8657.
296. Kyte, J. (1974) *J. Biol. Chem. 249*, 3652–3660.
297. Beaugé, L.A., & Glynn, I.M. (1979) *Nature 280*, 510–512.
298. Glynn, I.M., & Richards, D.E. (1982) *J. Physiol. 330*, 17–43.
299. Karlish, S.J.D., & Yates, D.W. (1978) *Biochim. Biophys. Acta 527*, 115–130.
300. Steinberg, M., & Karlish, S.J.D. (1989) *J. Biol. Chem. 264*, 2726–2734.
301. Forbush, B. (1987) *J. Biol. Chem. 262*, 11104–11115.
302. Hoffman, J.F. (1962) *J. Gen. Physiol. 45*, 837–859.
303. Post, R.L., Merritt, C.R., Kinsolving, C.R., & Albright, C.D. (1960) *J. Biol. Chem. 235*, 1796–1802.
304. Glynn, I.M., Hara, Y., & Richards, D.E. (1984) *J. Physiol. 351*, 531–547.
305. Jorgensen, P.L., & Andersen, J.P. (1988) *J. Membr. Biol. 103*, 95–120.
306. Helmich-deJong, M.L., vanEmst-deVries, S.E., & dePont, J.J.H.H.M. (1987) *Biochim. Biophys. Acta 905*, 358–370.
307. Jorgensen, P.L. (1977) *Biochim. Biophys. Acta 466*, 97–108.
308. Kyte, J. (1972) *J. Biol. Chem. 247*, 7634–7641.
309. Carilli, C.T., Farley, R.A., Perlman, D.M., & Cantley, L.C. (1982) *J. Biol. Chem. 257*, 5601–5606.
310. Taniguchi, K., Suzuki, K., Kai, D., Matsuoka, I., Tomita, K., & Iida, S. (1984) *J. Biol. Chem. 259*, 15228–15233.
311. Moczydlowski, E.G., & Fortes, P.A.G. (1981) *J. Biol. Chem. 256*, 2357–2366.
312. Neet, K.E., & Green, N.M. (1977) *Arch. Biochem. Biophys. 178*, 588–597.
313. Bastide, F., Meissner, G., Fleischer, S., & Post, R.L. (1973) *J. Biol. Chem. 248*, 8385–8391.
314. Post, R.L., Kume, S., Tobin, T., Orcutt, B., & Sen, A.K. (1969) *J. Gen. Physiol. 54*, 306s-326s.
315. Hara, Y., & Nakao, M. (1981) *J. Biochem. (Tokyo) 90*, 923–931.
316. Fahn, S., Koval, G.J., & Albers, R.W. (1968) *J. Biol. Chem. 243*, 1993–2002.
317. Siegel, G.J., Koval, G.J., & Albers, R.W. (1969) *J. Biol. Chem. 244*, 3264–3269.
318. Dahms, A.S., & Boyer, P.D. (1973) *J. Biol. Chem. 248*, 3155–3162.
319. Albers, R.W., Fahn, S., & Koval, G.J. (1963) *Proc. Natl. Acad. Sci. U.S.A. 50*, 474–481.

320. Kanazawa, T., Saito, M., & Tonomura, Y. (1970) *J. Biochem. (Tokyo) 67*, 693–711.
321. Lew, V.L., Glynn, I.M., & Ellory, J.C. (1970) *Nature 225*, 866–867.
322. Cavieres, J.D., & Glynn, I.M. (1976) *J. Physiol. 263*, 214P-215P.
323. Glynn, I.M., & Hoffman, J.F. (1971) *J. Physiol. 218*, 239–256.
324. Simons, T.J.B. (1974) *J. Physiol. 237*, 123–155.
325. Scott, T. (1985) *J. Biol. Chem. 260*, 14421–14423.
326. Squier, T.C., Bigelow, D.J., deAncos, J.G., & Inesi, G. (1987) *J. Biol. Chem. 262*, 4748–4754.
327. Inesi, G. (1987) *J. Biol. Chem. 262*, 16338–16342.
328. Rephaeli, A., Richards, D.E., & Karlish, S.J.D. (1986) *J. Biol. Chem. 261*, 12437–12440.
329. Rephaeli, A., Richards, D., & Karlish, S.J.D. (1986) *J. Biol. Chem. 261*, 6248–6254.
330. Clarke, D.M., Loo, T.W., & MacLennan, D.H. (1990) *J. Biol. Chem. 265*, 6262–6267.
331. Taylor, K.A., Dux, L., & Martonosi, A. (1986) *J. Mol. Biol. 187*, 417–427.
332. Mohraz, M., Simpson, M.V., & Smith, P.R. (1987) *J. Cell Biol. 105*, 1–8.
333. Herbert, H., Skriver, E., Söderblom, M., & Maunsbach, A.B. (1988) *J. Ultrastruct. Mol. Struct. Res. 100*, 86–93.

# Index

## B

## C

## D

## E

## F



## G

## H

## I

## K

## L



## M

## N

## O

## P

## Q

## R



## S

## T

## U



## V



## W



## X,Y,Z